AEROSOL MEASUREMENT

Principles, Techniques, and Applications

AEROSOL MEASUREMENT

Principles, Techniques, and Applications

Edited by
Klaus Willeke
Paul A. Baron

VNR VAN NOSTRAND REINHOLD
New York

1993

Library of Congress Catalog Card Number 92-11371

ISBN 0-442-00486-9

Printed in the United States of America

Van Nostrand Reinhold
115 Fifth Avenue
New York, New York 10003

Chapman and Hall
2-6 Boundary Row
London, SE1 8HN, England

Thomas Nelson Australia
102 Dodds Street
South Melbourne 3205
Victoria, Australia

Nelson Canada
1120 Birchmount Road
Scarborough, Ontario MIK 5G4, Canada

16 15 14 13 12 11 10 9 8 7 6 5 4 3 2 1

Library of Congress Cataloging-in-Publication Data

Aerosol measurement: principles, techniques, and applications/
edited by Klaus Willeke, Paul A. Baron.
p. cm.
Includes bibliographical references and index.
ISBN 0-442-00486-9
1. Aerosols—Measurement. 2. Air—Pollution—Measurement.
I. Willeke, Klaus. II. Baron, Paul A., 1944–
TD884.5.A33 1992 92-11371
628.5′3′0287—dc20 CIP

TABLE OF CONTENTS

Preface

The measurement of aerosols has been practiced widely for several decades. Until recently, the development of new measurement methods was primarily motivated by the need to evaluate particulate pollution control devices, and to find better means of monitoring indoor and outdoor aerosols. During the past several years, industry has become increasingly interested in modern aerosol measurement methods, not only to protect the health of their workers, as required by law, but to increase productivity and, thereby, gain competitive advantage. For instance, in the production of semiconductor circuit boards a single submicrometer-sized particle may spoil the circuit if it adheres to the board where a circuit of submicrometer dimensions is being deposited. As a consequence, the number of undergraduate and graduate students taking courses in aerosol science and measurement has increased dramatically in recent years. The increased importance of this field is also evidenced by the creation and rapid growth of aerosol research associations, such as the American Association for Aerosol Research, the European Association for Aerosol Research (Gesellschaft für Aerosolforschung, Association pour la Recherche des Aérosols), Japan Association of Aerosol Science and Technology, and several other national associations.

In Part I of this book we present the fundamentals relevant for novices to this field, utilizing approaches developed in over twenty years of teaching university courses on aerosol science and measurement. Since we expect many readers to be air pollution regulators, industrial hygienists and environmental scientists or engineers, we have applied our experience in teaching short courses to practitioners: the chapters of Part I stress the physical understanding and give useful equations but avoid lengthy scientific derivations. We have authored or co-authored several of the first chapters to provide models for the remaining chapter contributions in order to achieve a uniform style and a consistent structure in the book. Readers familiar with the principles of aerosol measurement can find details on specific instrumental techniques in Part II. Many of the chapters in Parts I and II offer sample calculations, thus making the book suitable for use as a teaching text. The practitioner concerned with the special requirements of his or her field, such as industrial hygiene or industrial aerosol processing, can find aerosol measurement applications in Part III. The bringing together of many application fields by experts enable the reader to look into the practices of related fields, so that technology transfer and adaptations may result.

Bruce Appel has dedicated his chapter to the memory of Thomas Dzubay, whose untimely death has saddened the aerosol research community. We join in offering our respects to his family and friends.

We also thank our wives Audrone (KW) and Diane (PB) for their support during the many evening and weekend hours needed to assemble this book.

List of Principal Symbols

Roman Symbols

B	particle mobility (cm/s dyn), Eq. 3-14
c_m	mass concentration (g/cm^3, mg/m^3, μg/m^3)
c_n	number concentration (particles/cm^3)
c_q	particle charge concentration (statC/cm^3)
C	concentration of solute in solvent (cm^3/cm^3)
C_c	slip correction factor, Eqs. 3-8, 3-9
C_d	drag coefficient, Eqs. 3-19, 3-23, 3-24
CV	coefficient of variation, Eq. 5-19
d	diameter of an object, such as a particle (cm, μm), characteristic dimension of an object (cm)
d_a	aerodynamic diameter (cm, μm), Eqs. 2-2, 2-3, 3-30
d_p	particle diameter (cm, μm)
d_e	envelope-equivalent diameter (cm, μm)
d_m	mass-equivalent diameter (cm, μm), Eq. 3-21
$\bar{d}_p$	mean particle diameter
d_s	Stokes diameter (cm, μm), Eq. 3-30
d_{50}	median particle diameter (cm, μm)
D	diffusion coefficient of particle (cm^2/s), Eq. 3-13, fractal dimension of particle
D_v	diffusion coefficient of vapor molecule (cm^2/s), Eq. 4-4
e	charge on an electron (4.8×10^{-10} statC, 1.6×10^{-19} C)
f	frequency (Hz, s^{-1}), friction factor, Eq. 6-57
E	electric field (statV/cm), total efficiency of a filter, Eq. 10-1
F	force on a particle (dyn)
g	gravitational constant (cm/s^2)
H	height of a chamber or duct (cm), molecular accomodation coefficient, Eq. 3-51
I	intensity of light or radiation (erg/s cm^2), electric current (statamp)
J	flux of gas molecules (number/s cm^2)
k	Boltzmann's constant (1.38×10^{-16}dyn cm/K), thermal conductivity (cal/cm s K)
K	coagulation coefficient (cm^3/s), Eq. 4-9
L	length, light path length (cm, m)
m	refractive index
m_p	particle mass (g, mg, μg, ng)
M	gram molecular weight (g/cm^3)
N	number concentration (number/cm^3)
N_p	particle concentration (number/cm^3)
n	molecular concentration (number/cm^3), number of unit charges, number of particles, number of measurements
p	partial pressure (dyn/cm^2, Pa, atm)
p_s	saturation vapor pressure (dyn/cm^2, Pa, atm)
P	pressure (dyn/cm^2, Pa, atm), penetration fraction
Pe	Peclet number, Eq. 3-16
q	charge on a particle (statC, C)
Q	flow rate (cm^3/s)
Q_e	particle extinction efficiency
r	distance between two particles (cm), radial distance (cm)

R	specific gas constant (dyn cm/K g), Eq. 3-3, ratio
R_u	universal gas constant (8.31×10^7 dyn cm/K mol), Eq. 3-3
Re_f	flow Reynolds number, Eqs. 3-1, 3-2
Re_p	particle Reynolds number, Eqs. 3-1, 3-2
Re_0	particle Reynolds number under initial conditions, Eq. 3-37
S	stopping distance (cm, μm), Eqs. 3-36, 3-37, Sutherland constant (K), Eq. 3-10
Sc	Schmidt number, Eq. 3-17
Sh	Sherwood number, Eqs. 6-43, 6-45
S_R	saturation ratio, Eq. 4-2
Stk	Stokes number, Eq. 3-39
$t_{1/2}$	half-life or half-time, Eq. 5-12
T	temperature (K, °C)
U	gas velocity (cm/s), sampling velocity (cm/s), output signal of a photometer (V)
U_0	ambient gas velocity (cm/s)
v_a	sampled air volume (cm^3)
v_p	particle volume (cm^3)
V	velocity of particle relative to gas (cm/s), potential (statV)
$\bar{V}$	average molecular velocity (cm/s), Eq. 3-5
V_p	particle velocity (cm/s)
V_0	initial velocity of a particle (cm/s)
V_{ts}	terminal settling velocity (cm/s), Eq. 3-28
x	distance in the x direction (cm)
x_{rms}	root mean square Brownian motion in the x direction (cm), Eq. 3-15
Z	electrical mobility (cm^2/statV s), Eq. 3-45

Greek Symbols

α	coefficient in slip correction equation, Eq. 3-8, thermal diffusivity (cm^2/s)
β	coefficient in slip correction equation, Eq. 3-8, length to width ratio, aspect ratio
γ	coefficient in slip correction equation, Eq. 3-8, surface tension (dyn/cm), specific heat ratio
λ	mean free path (cm, μm), wavelength (cm, μm)
τ	relaxation time of a particle (s), Eq. 3-34
η	dynamic viscosity (P, poise = dyn cm/s), efficiency
θ	angle (rad, °)
ν	kinematic viscosity (cm^2/s)
ρ_e	effective density that includes voids (g/cm^3)
ρ_f	fluid density (g/cm^3)
ρ_g	gas density (g/cm^3)
ρ_p	particle density (g/cm^3)
σ	standard deviation, Eq. 8-2
σ_g	geometric standard deviation
ϕ	angle (rad, °)
χ	dynamic shape factor
ω	angular velocity (rad/s)

Subscripts

a	air or gas, aspiration
ac	alternating current
B	mobility equivalent
c	cylinder
d	droplet, droplet surface, drag
dc	direct current
dep	deposition
diff	in a diffusiophoretic field
e	effective
elec	in an electric field
ev	equivalent volume
f	flow, fluid, fiber
g	geometric, gas
grav	in a gravity field
i	initial, individual
j	jet
m	mass
n	number
p	particle
r	reference to NTP
s	saturation condition
th	in a thermal gradient field
ts	terminal settling under influence of gravity
trans	transmission
0	initial condition
∞	far from particle surface

I

Principles

1

Bridging Science and Application in Aerosol Measurement

Klaus Willeke

Department of Environmental Health
University of Cincinnati
Cincinnati, OH, U.S.A.

and

Paul A. Baron

National Institute for Occupational Safety and Health
Centers for Disease Control, Public Health Service
Department of Health and Human Services, Cincinnati, OH, U.S.A.

INTRODUCTION

An aerosol is an assembly of liquid or solid particles suspended in a gaseous medium (e.g., air) long enough to enable observation and measurement. The need to measure aerosols has increased dramatically in recent years. This need has arisen in several disciplines. While a lot of information is available in the published literature, the scientists and practitioners making aerosol measurements in one application may have little or no awareness of the knowledge and experience gained in other applications.

For instance, environmental engineers and industrial hygienists make aerosol measurements in order to ensure that the public and the industrial work force are not exposed to hazardous aerosols at undesirable concentration levels. Faced by increasingly complex and demanding regulations, aerosol measurements are becoming more costly in terms of equipment and time, and may require more than an elementary knowledge of the subject for an interpretation of the data. The results are critical as expensive control measures may have to be taken as a result of such measurements.

In contrast, the scientists and engineers concerned with industrial materials are developing an ever-increasing number of manufacturing processes in which the material passes through an aerosol phase that needs to be monitored and controlled. For instance, powders and pigments may be produced by passing the feed materials into a flame, plasma, laser, or flow furnace, where they evaporate. Upon cooling, a very high concentration of very small particles are formed. With time, the aerosols may agglomerate to a lower concentration of larger particles. At any time, the trajectories of these particles may be directed by external forces, e.g., gravity or an applied electric field, to deposit the particles in a predetermined manner, thus forming products such as ceramics or optical fibers. While producing desirable materials in this manner, the aerosol may have to be measured, not only to ensure a uniform product but also to avoid exposure of humans to processing materials which may be quite hazardous.

As a consequence, the novice and the experienced scientist or practitioner alike may have to become familiar with new principles, aerosol measurement techniques, and applications. This book attempts to address all of these aspects by dealing with aerosol measurement in three parts. Part I is devoted to basic concepts such as aerosol generation, transformation, transport, and data presentation. Also, the basic principles of aerosol measurement are introduced. Part II expands the latter by devoting a chapter to each principal instrumental technique or group of techniques. Finally, a wide range of applications is discussed in Part III. Each application requires a specific set of aerosol properties to be measured, thus dictating the type of measurement technique or group of associated techniques that can be used.

The book attempts to give the fundamental principles in sufficient detail so that scientists and practitioners may use them in deciding which aerosol properties to measure and how to interpret the results. The technique and application chapters attempt to guide them in performing the actual measurements. As such, the book bridges science and application in aerosol measurement.

Supplementary references are cited at the end of each chapter. A summary list of books and journals is given below. (Adapted from: Education Committee of the American Association for Aerosol Research (1990), a bibliography of aerosol science and technology, *Aerosol Science and Technology*, Vol. 14, pp 1–4.)

BOOK REFERENCES

General References

Abraham, F. F. 1974. *Homogeneous Nucleation Theory: The Pretransition Theory of Vapor Condensation, Supplement I: Advances in Theoretical Chemistry*. New York: Academic Press.

Bailey, A. G. 1988. *Electrostatic Spraying of Liquids*. New York: Wiley.

Beddow, J. K. 1980. *Particulate Science and Technology*. New York: Chemical Publishing Co.

Bohren, C. F. and D. R. Huffman. 1983. *Absorption and Scattering of Light by Small Particles*. New York: Wiley.

Davies, C. N. (ed.). 1966. *Aerosol Science*. New York: Wiley.

Dennis, R. 1976. *Handbook on Aerosols. Publication TID-26608*. Springfield, VA: National Technical Information Service, U.S. Dept. of Commerce.

Einstein, A. 1956. *Investigations on the Theory of Brownian Motion*. New York: Dover.

Friedlander, S. K. 1977. *Smoke, Dust and Haze*. New York: Wiley.

Fuchs, N. A. 1989. *The Mechanics of Aerosols*. New York: Wiley.

Fuchs, N. A. and A. G. Sutugin. 1970. *Highly Dispersed Aerosols*. Ann Arbor, MI: Ann Arbor Science Publishers.

Green, H. L. and W. R. Lane. 1964. *Particulate Clouds, Dust, Smokes and Mists*, 2nd edn. Princeton, NJ: Van Nostrand Reinhold.

Happel, J. and H. Brenner. 1973. *Low Reynolds Number Hydrodynamics with Special Applications to Particulate Media*, 2nd revised edn. Leiden: Noordhoff.

Hesketh, H. E. 1977. *Fine Particles in Gaseous Media*. Ann Arbor, MI: Ann Arbor Science Publishers.

Hidy, G. M. 1972. *Aerosols and Atmospheric Chemistry*. New York: Academic Press.

Hidy, G. M. and J. R. Brock. 1970. *The Dynamics of Aerocolloidal Systems*. New York: Pergamon.

Hidy, G. M. and J. R. Brock (eds.). 1971. *Topics in Recent Aerosol Research*. New York: Pergamon.

Hidy, G. M. and J. R. Brock (eds.). 1972. *Topics in Current Aerosol Research, Part 2*. New York: Pergamon.

Hinds, W. C. 1982. *Aerosol Technology*. New York: Wiley.

Kerker, M. 1969. *The Scattering of Light and Other Electromagnetic Radiation*. New York: Academic Press.

Lefebvre, A. H. 1989. *Atomization and Sprays*. New York: Hemisphere.

Liu, B. Y. H. (ed.). 1976. *Fine Particles*. New York: Academic Press.

Marlow, W. H. (ed.). 1982*a*. *Aerosol Microphysics I*. Chemical Physics of Microparticles. Berlin: Springer.

Marlow, W. H. (ed.). 1982*b*. *Aerosol Microphysics II. Chemical Physics of Microparticles*. Berlin: Springer.

Mason, B. J. 1971. *The Physics of Clouds*. Oxford: Clarendon Press.

McCrone, W. C., et al. 1980. *The Particle Atlas, Vols. I–VII*. Ann Arbor, MI: Ann Arbor Science Publishers.

Mednikov, E. P. 1980. *Turbulent Transport of Aerosols* (in Russian). Science Publishers.

Orr, C., Jr. 1966. *Particulate Technology*. New York: Macmillan.

Reist, P. C. 1984. *Introduction to Aerosol Science*. New York: Macmillan.

Sanders, P. A. 1979. *Handbook of Aerosol Technology*. Melbourne, FL: Krieger Publishing.

Sedunov, Y. S. 1974. *Physics of Drop Formation in the Atmosphere* (translated from Russian). New York: Wiley.

Twomey, S. 1977. *Atmospheric Aerosols*. Amsterdam: Elsevier.

Van de Hulst, H. C. 1957. *Light Scattering by Small Particles*. New York: Wiley. (Republished in 1981, unabridged and corrected. New York: Dover.)

Vohnsen, M. A. 1982. *Aerosol Handbook*, 2nd edn. Mendham, NJ: Dorland.

Whytlaw-Grey, R. W. and H. S. Patterson. 1932. *Smoke: A Study of Aerial Disperse Systems*. London: E. Arnold.

Willeke, K. (ed.). 1980. *Generation of Aerosols and Facilities for Exposure Experiments*. Ann Arbor, MI: Ann Arbor Science Publishers.

Williams, M. M. R. and S. K. Loyalka. 1991. *Aerosol Science Theory and Practice: With Special Application to the Nuclear Industry*. Oxford: Pergamon.

Withers, R. S. 1979. *Transport of Charged Aerosols*. New York: Garland.

Yoshida, T., Y. Kousaka, and K. Okuyama. 1979. *Aerosol Science for Engineers*. Tokyo: Power Co.

Zimon, A. D. 1976. *Adhesion of Dust and Powders*, 2nd edn. (in Russian). Moscow: Khimia. (First edition, 1969 (in English). New York: Plenum.)

Measurement Techniques

Allen, T. 1968. *Particle Size Measurement*. London: Chapman & Hall.

Allen, T. 1981. *Particle Size Measurement*, 3rd edn. New York: Methuen.

Barth, H. G. (ed.). 1984. *Modern Methods of Particle Size Analysis*. New York: Wiley.

Beddow, J. K. 1980. *Testing and Characterization of Powders and Fine Particles*. New York: Wiley.

Beddow, J. K. 1984. *Particle Characterization in Technology*. Boca Raton, FL: CRC Press.

Cadle, R. D. 1965. *Particle Size: Theory and Industrial Applications*. New York: Van Nostrand Reinhold.

Cadle, R. D. 1975. *The Measurement of Airborne Particles*. New York: Wiley.

Cheremisinoff, P. N. (ed.). 1981. *Air Particulate Instrumentation and Analysis*. Ann Arbor, MI: Ann Arbor Science Publishers.

Dallavalle, J. M. 1948. *Micromeritics*, 2nd edn. New York: Pitman.

Dzubay, T. G. 1977. *X-ray Fluorescence Analysis of Environmental Samples*. Ann Arbor, MI: Ann Arbor Science Publishers.

Herdan, G. 1953. *Small Particle Statistics*. New York: Elsevier.

Jelinek, Z. K. (translated by W. A. Bryce). 1974. *Particle Size Analysis*. New York: Halstead Press.

Lodge, J. P., Jr. and T. L. Chan (ed.). 1986. *Cascade Impactor, Sampling and Data Analysis*. Akron, OH: American Industrial Hygiene Association.

Malissa, H. (ed.). 1978. *Analysis of Airborne Particles by Physical Methods*. Boca Raton, FL: CRC Press.

Orr, C. and J. M. Dallavalle. 1959. *Fine Particle Measurement*. New York: Macmillan.

Rahjans, G. S. and J. Sullivan. 1981. *Asbestos Sampling and Analysis*. Ann Arbor, MI: Ann Arbor Science Publishers.

Silverman, L., C. Billings, and M. First. 1971. *Particle Size Analysis in Industrial Hygiene*. New York: Academic Press.

Stockham, J. D. and E. G. Fochtman. 1977. *Particle Size Analysis*. Ann Arbor, MI: Ann Arbor Science Publishers.

Vincent, J. H. 1989. *Aerosol Sampling: Science and Practice*. New York: Wiley.

Gas Cleaning

Clayton, P. 1981. *The Filtration Efficiency of a Range of Filter Media for Submicrometer Aerosols*. New York: State Mutual Book and Periodical Service.

Davies, C. N. 1973. *Air Filtration*. London: Academic Press.

Dorman, R. G. 1974. *Dust Control and Air Cleaning*. New York: Pergamon Press.

Mednikov, E. P. 1965. *Acoustic Coagulation and Precipitation of Aerosols*. New York: Consultants Bureau.

Ogawa, A. 1984. *Separation of Particles from Air and Gases, Vols. I and II*. Boca Raton, FL: CRC Press.

White, H. J. 1963. *Industrial Electrostatic Precipitation*. Reading, MA: Addison–Wesley.

Environmental Aerosols/Health Aspects

Air Sampling Instruments. 7th edn. 1990. Cincinnati, OH: American Conference of Governmental Industrial Hygienists.

Brenchly, D. L., C. D. Turley, and R. F. Yarmae. 1973. *Industrial Source Sampling*. Ann Arbor, MI: Ann Arbor Science Publishers.

Cadle, R. D. 1966. *Particles in the Atmosphere and Space*. New York: Van Nostrand Reinhold.

Drinker, P. and T. Hatch. 1954. *Industrial Dust*. New York: McGraw-Hill.

Flagan, R. C. and J. H. Seinfeld. 1988. *Fundamentals of Air Pollution Engineering*. New York: Prentice-Hall.

Hidy, G. M. 1972. *Aerosols and Atmospheric Chemistry*. New York: Academic Press.

Junge, C. 1963. *Air Chemistry and Radioactivity*. New York: Academic Press.

McCartney, E. J. 1976. *Optics of the Atmosphere*. New York: Wiley.

Mercer, T. T. 1973. *Aerosol Technology in Hazard Evaluation*. New York: Academic Press.

Middleton, W. E. K. 1952. *Vision through the Atmosphere.* Toronto: University of Toronto.

Muir, D. C. F. (ed.). 1972. *Clinical Aspects of Inhaled Particles.* London: Heinemann.

National Research Council, Subcommittee on Airborne Particles. 1979. *Airborne Particles.* Baltimore, MD: University Park Press.

Perera, F. and A. K. Ahmen. 1979. *Respirable Particles: Impact of Airborne Fine Particles on Health and Environment.* Cambridge, MA: Ballinger.

Seinfeld, J. H. 1986. *Atmospheric Chemistry and Physics of Air Pollution.* New York: Wiley.

Whitten, R. C. (ed.). 1982. *The Stratospheric Aerosol Layer.* Berlin: Springer.

Industrial Applications and Processes

Andonyev, S. and O. Filipyev. 1977. *Dust and Fume Generation in the Iron and Steel Industry.* Chicago, IL: Imported Publications.

Austin, P. R. and S. W. Timmerman. 1965. *Design and Operation of Clean Rooms.* Detroit, MI: Business News Publishing Co.

Boothroyd, R. G. 1971. *Flowing Gas–Solids Suspensions.* London: Chapman & Hall.

Donnet, J. B. and A. Voet. 1976. *Carbon Black.* New York: Marcel Dekker.

Marshall, W. R., Jr. 1954. *Atomization and Spray Drying,* Chemical Engineering Progress Monograph Series, Vol. 50, No. 23. New York: AIChE.

Proceedings of Meetings

Advances in Air Sampling. 1988. Papers from the American Conference of Governmental Industrial Hygienists Symposium. Ann Arbor, MI: Lewis Publishers.

Air and Waste Management Association. 1989. *Visibility and Fine Particles.* Proceedings of the 1989 EPA/A&WMA International Specialty Conference, Pittsburgh, PA.

ASTM Symposium on Particle Size Measurement. 1959. ASTM Special Technical Publication No. 234.

Barber, D. W. and R. K. Chang. 1988. *Optical Effects Associated with Small Particles.* Singapore: World Scientific Publishing Co.

Beddow, J. K. and T. P. Meloy (eds.). 1980. *Advanced Particulate Morphology.* Boca Raton, FL: CRC Press.

Davies, C. N. 1964. *Recent Advances in Aerosol Research.* New York: Macmillan.

Dodgson, J., R. I. McCallum, M. R. Bailey, and D. R. Fisher (eds.). 1989. *Inhaled Particles VI.* Oxford: Pergamon.

Fedoseev, V. A. 1971. *Advances in Aerosol Physics* (*translation of Fizika Aerodispersnykh Sistem*). New York: Halsted Press.

Gerber, H. E. and E. E. Hindman (eds.). 1982. *Light Absorption by Aerosol Particles.* Hampton, VA: Spectrum Press.

Israel, G. 1986. *Aerosol Formation and Reactivity.* Proceedings of the Second International Aerosol Conference, 22–26 September 1986, Berlin (West). Oxford: Pergamon.

Kuhn, W. E., H. Lamprey and C. Sheer (eds.). 1963. *Ultrafine Particles.* New York: Wiley.

Lee, S. D., T. Schneider, L. D. Grant, and P. J. Verkerk (eds.). 1986. *Aerosols: Research, Risk Assessment and Control Strategies.* Proceedings of the Second U.S.–Dutch International Symposium, 19–25 May 1985, Williamsburg, VA. Chelsea, MI: Lewis Publishers.

Liu, B. Y. H., D. Y. H. Pui, and H. J. Fissan. 1984. *Aerosols: Science, Technology and Industrial Applications of Airborne Particles.* 300 Extended Abstracts from the First International Aerosol Conference, 17–21 September 1984, Minneapolis, MN. New York: Elsevier.

Lundgren, D. A., et al. (eds.). 1979. *Aerosol Measurement.* Gainesville, FL: University Presses of Florida.

Marple, V. A. and B. H. Y. Liu (eds.). 1983. *Aerosols in the Mining and Industrial Work Environments.* Ann Arbor, MI: Ann Arbor Science Publishers.

Mercer, T. T., P. E. Morrow, and W. Stober (eds.). 1972. *Assessment of Airborne Particles.* Springfield, IL: C.C. Thomas Publishers.

Mittal, K. L. (ed.). 1988. *Particles on Surfaces 1: Detection, Adhesion, and Removal.* Proceedings of a Symposium held at the Seventeenth Annual Meeting of the Fine Particle Society, 28 July–2 August 1986. New York: Plenum.

Mittal, K. L. (ed.). 1990. *Particles on Surfaces 2: Detection, Adhesion, and Removal.* New York: Plenum.

Richardson, E. G. (ed.). 1960. *Aerodynamic Capture of Particles.* New York: Pergamon.

Shaw, D. T. (ed.). 1978*a*. *Fundamentals of Aerosol Science.* New York: Wiley.

Shaw, D. T. (ed.). 1978*b*. *Recent Developments in Aerosol Technology.* New York: Wiley.

Siegla, P. C. and G. W. Smith (eds.). 1981. *Particle Carbon: Formation during Combustion.* New York: Plenum.

Spurny, K. 1965. *Aerosols: Physical Chemistry and Applications.* Proceedings of the First National Conference on Aerosols. Prague: Publishing House of the Czechoslovak Academy of Sciences.

Walton, W. H. (ed.). 1971. *Inhaled Particles III.* Surrey: Unwin Brothers.

Walton, W. H. (ed.). 1977. *Inhaled Particles IV.* Oxford: Pergamon.

Walton, W. H. (ed.). 1982. *Inhaled Particles V.* Oxford: Pergamon.

SELECTED JOURNALS ON AEROSOL SCIENCE AND APPLICATIONS

Aerosol Science and Technology

American Industrial Hygiene Association Journal

Annals of Occupational Hygiene
Atmospheric Environment
Environmental Science and Technology
International Journal of Multiphase Flow
Journal of Aerosols in Medicine
Journal of Aerosol Research, Japan
Journal of Aerosol Science
Journal of the Air and Waste Management Association (formerly Journal of the Air Pollution Control Association)
Journal of Colloid and Interface Science
Langmuir
Particle Characterization
Particulate Science and Technology
Powder Technology
Staub Reinhaltung der Luft

CITATION LISTING

Current Awareness in Particle Technology (Loughborough, UK: University of Technology).

2

Aerosol Fundamentals

Paul A. Baron

Department of Health and Human Services, Public Health Service
Centers for Disease Control
National Institute for Occupational Safety and Health
Cincinnati, OH, U.S.A.

and

Klaus Willeke

Department of Environmental Health
University of Cincinnati
Cincinnati, OH, U.S.A.

INTRODUCTION

The term aerosol refers to an assembly of liquid or solid particles suspended in a gaseous medium long enough to enable observation or measurement. The term originated as the gas phase equivalent of the term hydrosol, which refers to a suspension of particles in a liquid (from the Greek word combination "water particle"). Manufactured and naturally produced particles, found in ambient and industrial air environments or in industrial process gas streams, may have a great diversity in size, shape, density, and chemical composition. The diversity of microscopic particles includes what we might think are ideal shapes, such as spheres (droplets of water or oil) or cylinders (glass fibers). However, it also includes more complex shapes such as crystalline particles, which have some regular and some fractured surfaces; asbestos fibers, which are often bundles of finer fibrils, or may be matted clumps of fibrils; and carbon black particles, which often consist of an extended framework of very small spheroids. All these have a different chemical constitution. Even if all these particles have the same microscopically observed diameter, the mass, surface area and other properties of each particle are likely to be quite different.

A variety of techniques are available for obtaining useful information about these particles. One way to detect the particles is by scattering light from them. The scattering pattern produced by spheres is quite predictable, while that from glass fibers is predictable only if their orientation relative to the light beam is known. The light pattern from asbestos fibers might be similar to that from glass fibers in some cases, but may be much more complex in cases where the fibers are splayed bundles or matted clumps. Much less light might be scattered by carbon black due to its light-absorbing properties.

To characterize the size of spherical droplets, a single parameter, such as the diameter, may suffice. With fibers, two dimensions may be sufficient, although often two may not be enough for less regular shapes. Clearly, for more complex particles we have to choose the particle properties which are important for our purposes in order to make a meaningful measurement. We can characterize particles by: (i) shape parameters, which can be observed under a microscope; (ii) light scattering properties, measured using an optical particle counter or a nephelometer; (iii) elemental or chemical properties, measured by X-ray fluorescence or infrared spectroscopy; (iv) surface properties, determined using a pycnometer or from adsorption measurements; or (v) dynamic behavior, determined from measurements of settling velocity or diffusion. Note that one characteristic, e.g., physical size, may not correlate well with another characteristic, e.g., chemical composition. However, any or all of these types of characterization may be important for the scientist. For example, in order to estimate the toxicity of an aerosol entering the lung, one needs to know the size-dependent diffusion, gravitational settling, impaction, and interception properties of the particles to determine the deposition rate within the lung. In addition, the chemistry, surface area, and fibrosity of the particles may indicate their interactions with the lung tissues once they are deposited.

Sand found on a beach also comes in many different sizes, but here the shapes tend to be closer to those of spheres because the material has been subjected to erosive forces of wind and water motion. When measuring these macroscopic materials as to their physical, chemical, and biological composition, we must use different methods and techniques. Similarly, when measuring microscopic materials, a diversity of methods and techniques is used. In this chapter, common aerosol characteristics are introduced in preparation for more detailed discussions of specific measurement techniques, which are discussed in later chapters.

DESIRABLE VERSUS UNDESIRABLE AEROSOLS

The development of many aerosol sampling and analysis techniques has been stimulated by a variety of applications. Since around the 1950s, advances in aerosol measurement have been motivated by investigations into the health effects of radioactive aerosols and industrial aerosols in the work place and the environment. More recently, a great deal of effort has gone into trying to understand the effect of various natural and man-made aerosols on global warming. The production of high-speed integrated circuits has required increasingly cleaner environments to reduce contamination by aerosol particles. These efforts have resulted not only in more refined and sensitive instruments but also have increased the understanding of particle generation and transport mechanisms. All these efforts have largely been aimed at reducing contaminants. In contrast, a great deal of knowledge is gained today by researchers working with "desirable" aerosols used to produce high technology materials such as ceramic powders, superconducting materials, and optical fibers.

UNITS AND USE OF EQUATIONS

The metric system of units (Weast, 1989) will be used throughout the book. Since aerosol particles range in diameter from about 10^{-9} to about 10^{-4} m, the unit of micrometer ($1\ \mu m = 10^{-6}$ m) is generally used when discussing fiber dimensions. For instance, a particle most hazardous to the human respiratory system is of the order of 10^{-6} m in diameter and is conveniently described as a 1 μm particle. The term "micron" has been used in the past as a colloquial version of micrometer, but is no longer accepted in technical writing. Researchers manufacturing aerosols through evaporation and subsequent condensation processes may deal with particles in the 0.01–0.1 μm range, or even smaller, and, therefore, prefer to express particle sizes in nanometers ($1\ nm = 10^{-9}$ m).

The aerosol mass concentration, i.e., the mass of particulate matter in a unit volume of gas, is expressed in g/m^3 or, since the amount of aerosol mass is generally very low, in mg/m^3 or $\mu g/m^3$. Particle velocity, e.g., under the influence of gravity or an electric field, is generally expressed in cm/s. When using these cgs (centimeter, gram, second) units, the particle size given is multiplied by 10^{-4} to convert μm to cm or by 10^{-7} to convert nm to cm.

Pressure is generally expressed in pascal (1 Pa = 1 N/m^2). Atmospheric pressure (101 kPa = 1.01×10^6 dyn/cm^2) may also be referred to as 1 atm (= 14.7 psig = 760 mm Hg = 1040 cm H_2O = 408 inch H_2O). Gas and particle properties are listed at normal temperature and pressure (NTP), which refers to 1 atm of pressure and 20°C (= 293 K = 68°F). Many handbooks list values at 1 atm and 0°C (standard temperature and pressure = STP), which are less useful since most aerosol measurements in the environment are at temperatures close to 20°C.

When dealing with engineering systems and applications regulated by the U.S. Environmental Protection Agency, the alternate commonly used units, such as millions of particles per cubic foot (mppcf), are given in parentheses. The tables in Appendix B give the conversion factors for the major units used by the practitioner or researcher dealing with aerosols.

Example 2-1

A miner drilling into a rock face during his work shift hits a seam of quartz (a form of crystalline silica extremely hazardous to the lung). The X-ray diffraction analysis on the personal sample taken over a period of 240 min at 1.7 l/min indicates 240 μg of respirable crystalline silica. Assume that this represents pure silica particles and that they are 2.8 μm diameter spherical particles with a density of 2.66 g/cm^3. What was the airborne exposure concentration in particle/cm^3, mppcf (million particles per cubic foot), g/m^3, mg/m^3, and μg/m^3?

Answer. The volume, v_p, of a single silica particle of physical diameter d_p is:

$$v_p = \frac{\pi}{6} d_p^3 = \frac{\pi}{6}(2.80\ \mu m)^3 = 11.5\ \mu m$$

$$= 1.15 \times 10^{-11}\ cm^3$$

The mass, m_p, of a particle with density ρ_p is

$$m_p = \rho_p v_p = (2.66\ g/cm^3)(1.15 \times 10^{-11}\ cm^3)$$

$$= 3.06 \times 10^{-11}\ g$$

$$= 3.06 \times 10^{-8}\ mg$$

$$= 3.06 \times 10^{-5}\ \mu g$$

The number of particles, n_p, in 240 μg of silica mass is

$$n_p = \frac{\text{silica mass}}{\text{single particle mass}}$$

$$= \frac{240\ \mu g}{3.06 \times 10^{-5}\ \mu g}$$

$$= 7.84 \times 10^6\ \text{particles}$$

At flow rate Q and sampling time t the volume of sampled air, v_a, is

$$v_a = Q \times t = (1.7\ l/min)(240\ min) = 408\ l$$

$$= 4.08 \times 10^5\ cm^3 = 0.408\ m^3$$

The number concentration, c_n, of silica particles is, therefore,

$$c_n = \frac{n_p}{v_a} = \frac{7.84 \times 10^6\ \text{particles}}{4.08 \times 10^5\ cm^3}$$

$$= 19.2\ \text{particles/cm}^3$$

$$= 1.92 \times 10^7\ \text{particles/m}^3$$

$$= 0.545\ \text{mppcf}$$

The particle mass concentration, c_m, is

$$c_m = m_p c_n$$

$$= (3.06 \times 10^{-11}\ g)(1.92 \times 10^7\ \text{particles/m}^3)$$

$$= 5.88 \times 10^{-4}\ g/m^3$$

$$= 0.588\ mg/m^3 = 588\ \mu g/m^3$$

Silica is one of the most toxic dusts encountered in the work place. The exposure standard in the U.S. is 50 μg/m^3. Thus, this measurement indicates excessive exposure. In comparison, the work place standard for the least toxic materials is 10 mg/m^3. The environmental air quality standard in the U.S. is 80 μg/m^3. The units of mppcf are no longer in common use for air measurements. They were popular for the measurement of dust concentrations by light microscopy.

Electrostatic calculations will be given in cgs or in Système Internationale (SI) units. Calculations, occasionally, will also be performed in both these systems to facilitate conversion, since each system has its advantages. Electrostatic calculations are convenient in the cgs units because the proportionality constant in Coulomb's law is unity. In this system, all electrical units are defined with the prefix "stat." The elementary unit of charge, *e*, is equal to 4.8×10^{-10} statcoulomb. The electric field is expressed in statvolt/cm. One statvolt equals 300 V in SI units. Also, particle motions expressed in cm/s reflect convenient magnitudes of particle velocity in an electric field. The advantage of the SI system is that it uses the familiar units of volts and amperes. The elementary unit of charge, *e*, is equal to 1.6×10^{-19} C in SI units.

It is assumed that the reader has available a computer to perform calculations. The reader is encouraged to program the equations given in this book and to calculate the results with a variety of input parameters to gain a feeling for the resulting values and how they behave. Spreadsheet programs are particularly useful for this purpose and many of the equations given in this text have been set up in a spreadsheet template (Baron 1984). The equations often represent idealized situations and it is, therefore, important to understand the approximations and assumptions, as well as the accuracy, that accompany the use of these equations.

COMMON TECHNICAL AND DESCRIPTIVE TERMS

Various names are used to describe airborne particulate matter. The name "particle" refers to a single unit of matter, having, generally, a density approaching the intrinsic density of the bulk material. Individual particles may be chemically homogeneous or may contain a variety of chemical species as well as consist of solid or liquid materials or both. Particle shapes may be simple, as in spherical liquid droplets, or complex, as in fiber bundles or agglomerated smoke. Many of the following terms do not have strict scientific definitions but rather are in common use as merely descriptive terms, often indicating the appearance or source of the particles.

Aerosol An assembly of liquid or solid particles suspended in a gaseous medium long enough to enable observation or measurement. Generally, the sizes of aerosol particles are in the range 0.001–100 μm. If the particle concentration is high enough that the density of the aerosol is more than about 1% greater than the gas alone, the assembly is considered a cloud and has bulk properties that differ from a more dilute aerosol.

Dust Solid particles formed by crushing or other mechanical breakage of a parent material. These particles generally have irregular shapes and are larger than about 0.5 μm.

Fog or mist Liquid-particle aerosol. These can be formed by condensation of supersaturated vapors or by physical shearing of liquids, such as in nebulization, spraying, or bubbling.

Fume Particles that are usually the result of condensed vapor with subsequent agglomeration. Solid fume particles typically consist of complex chains of submicrometer-sized particles (usually < 0.05 μm) of similar dimension. Fumes are often the result of combustion and other high temperature processes.

Particle A small, discrete object.

Particulate A particle; this term is also used as an adjective, indicating that the material in question has particle-like properties.

Smog An aerosol consisting of solid and liquid particles, created, at least in part, by the action of sunlight on vapors. The term smog is a combination of the words smoke and fog and often refers to the entire range of such pollutants, including the gaseous constituents.

Smoke A solid or liquid aerosol, the result of incomplete combustion or condensation of supersaturated vapor. Most smoke particles are submicrometer in size.

A number of terms describe the shape and origin of particles in an aerosol. These include:

Aggregate A heterogeneous particle in which the various components are not easily broken up. The term heterogeneous indicates that the individual components may differ from each other in size, shape, and chemical composition.

Agglomerate A group of particles held together by van der Waals forces or surface tension.

Flocculate A group of particles very loosely held together, usually by electrostatic forces. Flocculates can easily be broken up by shear forces within the air.

Primary particle A particle introduced into the air in solid or liquid form. A primary particle is often contrasted with a secondary particle.

Secondary particle Usually, a particle formed in the air by gas to particle conversion. This term is sometimes used to describe agglomerated or redispersed particles.

PARTICLE SIZE AND SHAPE

Particle size is important because it largely determines the behavior of the particle in gas suspension. Particles behave differently in different size ranges and are even governed by different physical laws. For example, on the earth's surface, particles only slightly larger than gas molecules are governed primarily by Brownian motion, while large, visible particles are affected primarily by gravitational and inertial forces.

Particle size and shape can be quite complex and are often defined only to the extent that one can measure or calculate them. Therefore, there are numerous definitions of particle size and shape that depend on the measurement technique or on the use to which the parameter is put. For instance, an electron microscope is a common means of measuring the size and shape of a particle. To accomplish this type of measurement, a particle is collected on a substrate, a process which may place the particle on the surface in some preferred orientation. The analyst measures the particle by comparison with standard-sized objects within the observation area. Except for ideally shaped spherical particles, the analyst usually reduces a complex shape to one or two measured parameters, e.g., width, or diameter and length. With an image analysis system, one may be able to extract more features of a particle's shape. The usual aim in collecting this type of information is to reduce the data collected from each particle to the fewest numbers that can adequately characterize the particle.

Size Parameters

A commonly used term in aerosol science and technology is that of equivalent diameter. This refers to a diameter that is a measurable index of a particle. When a particle is reported by a technique, the measurement usually corresponds to a specific physical property. Thus, an equivalent diameter is reported as the diameter of a sphere having the same value of a specific physical property as the irregularly shaped particle being measured (Fig. 2-1). When the motion of a particle is of concern, the mobility-equivalent diameter, d_B, is the diameter of a sphere with the same mobility as the particle in question.

For instance, aerodynamic (equivalent) diameter (equivalent is sometimes left out or implied) is the diameter of a unit-density sphere having the same gravitational settling velocity as the particle being measured. This definition is often used for characterizing particles that move primarily by settling, as opposed to diffusion in still air (i.e., diameters larger than about 0.3 μm at normal atmospheric temperature and pressure). Reference to the aerodynamic diameter of a particle is useful for describing particle settling and inertial behavior in the respiratory tract, one of

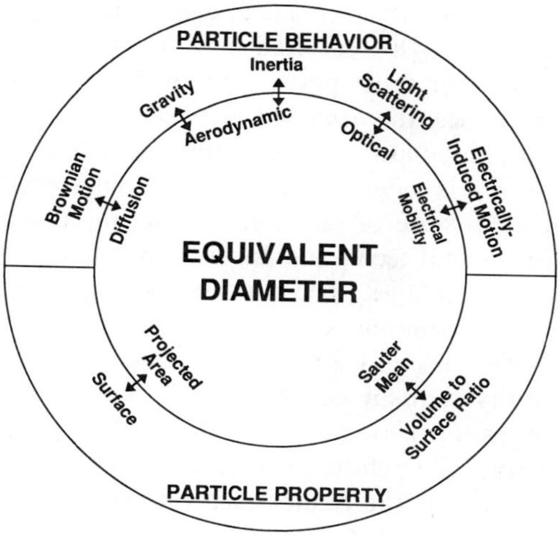

FIGURE 2-1. Particle Size Definitions that Depend on Observations of Particle Properties or Behavior.

the body organs most at risk upon exposure to toxic aerosols. The behavior of particles in other devices such as filters, cyclones, and impactors is also often governed by the aerodynamic flows around the particles and the sizes are, therefore, reported in terms of the particles' aerodynamic diameter. As will be seen in the next chapter, a solid, spherical particle's gravitational settling velocity is proportional to the particle density, ρ_p, the square of the physical particle diameter, d_p and the Cunningham slip correction factor, C_c. The latter is introduced because the suspending gas is not a continuous fluid, but consists of discrete molecules. The slip correction factor, described in more detail in the next chapter, is a function of particle diameter, i.e., $C_c = C_c(d_p)$. Thus,

$$\rho_p C_c(d_p) d_p^2 = \sqrt{1}\, C_c(d_a) d_a^2 \qquad (2\text{-}1)$$

One of the conditions chosen here is that of a unit-density sphere ($\rho_p = 1$ g/cm^3), which defines the particle diameter as the aerodynamic diameter, d_a:

$$d_a = \sqrt{\frac{C_c(d_p)}{C_c(d_a)}}\sqrt{\rho_p}\, d_p \qquad (2\text{-}2)$$

When the particle density is close to unity, $C_c(d_p)$ differs little from $C_c(d_a)$ and the ratio of the two can be approximated by one. For particles above about 1 μm, the slip correction factor is close to unity, so this ratio can be approximated by one even for particle densities very different from unity. Therefore, for many applications, Eq. 2-2 reduces to

$$d_a = \sqrt{\rho_p}\, d_p \qquad (2\text{-}3)$$

Example 2-2

What is the aerodynamic (equivalent) diameter of a spherical particle that is 3 μm in diameter and has a particle density of 4 g/cm^3? Ignore the slip correction factors.

Answer. From Eq. 2-3,

$$d_a = \sqrt{\rho_p}\, d_p = \sqrt{4}(3\ \mu\text{m}) = 6\ \mu\text{m}$$

This indicates that a 6 μm unit-density particle gravitationally settles at the same velocity as the 3 μm particle with the higher density.

A particle may be extremely complex in shape, such as an agglomerate. In this case, a significant part of the internal volume of the particle is made up of voids. When describing the properties or behavior of such a particle, two additional definitions are available: the mass-equivalent diameter, for which the particle is compressed into a spherical particle without voids, and the envelope-equivalent diameter, for which the particle voids are included in the sphere. The mass-equivalent diameter is convenient because it uses the bulk density of the material, a parameter often available in the literature (e.g., Weast 1989) or easily measured.

Microscopes and other particle-imaging systems are often used to measure particles. For instance, observing a particle's silhouette and calculating the diameter of a circle that has the same area gives the projected area (equivalent) diameter. Collecting an aerosol particle for measurement in a microscope can cause a number of biases to occur in the assessment of the original aerosol. For instance, the collected particle may be oriented by the surface. A fiber usually settles onto a surface with its long axis parallel to the surface. An agglomerate may collapse onto the surface from gravity or from surface tension of adsorbed water and appear more spread out than in its original form. The continued air flow over the particle may desiccate it, thus reducing it in size and mass. The collected particle may also react with the collection substrate, which may change the particle's size and chemical composition. The analyst needs to consider these possibilities in using data from methods involving sample collection.

We note here a dichotomy in measurement technique: namely, that of collection of an aerosol particle for laboratory measurement versus direct, *in situ* measurement of the particle. Traditionally, collection followed by measurement was often the most readily available. This approach still has its advantages, since it brings to bear the many powerful analytical techniques available in the laboratory. However, this approach has the disadvantages that the particles may be modified by the transport and collection processes and that the analytical result is not immediately available. *In situ* techniques, on the other hand, provide only a limited degree of particle characterization.

In situ techniques can be subdivided further into extractive and external-sensing techniques. Extractive techniques require the aerosol to be brought to the instrument sensor, while external-sensing techniques measure the aerosol in its undisturbed state.

A common *in situ* technique is the measurement of light scattered from the particles. The amount of light scattered from individual particles is a complex function of particle parameters characterizing the size, shape, and refractive index as well as instrumental parameters such as the wavelength of light and the scattering angle. The usual approach is to define an optical equivalent diameter, which is the diameter of a calibration particle that scatters as much light in a specific instrument as the particle being measured. For simple particle shapes, such as spheres, ellipsoids, and rods of known chemical composition, the amount of light scattered may be calculated exactly. For most particles with more complex shapes, the association between optical equivalent diameter and a physically useful property is often difficult to establish precisely. In spite of this, light scattering as an instrumental technique has a number of distinct advantages. These advantages include rapid, continuous and sensitive detection of particles, often at a relatively low cost.

Spray aerosol droplets used as fuels in combustion processes burn or react at their surfaces. Therefore, a useful measurement parameter is the Sauter mean diameter, i.e., the diameter of a droplet whose surface-to-volume ratio is equal to that of all the droplets in the spray distribution.

Since submicrometer particles move primarily by Brownian diffusion, it is natural to define their size by a diffusion-equivalent diameter, i.e., the diameter of a unit-density spherical particle with the same rate of diffusion as the particle being measured. For compact particles, the diffusion-equivalent diameter is very close to the physical diameter, as might be measured with an electron microscope.

The measurement of small particles by diffusion-based techniques is often relatively slow and has poor resolution. In an electric field, a particle of known charge moves along a predictable trajectory. Therefore, the electrical mobility of a charged particle in an electric field is the basis for defining the electrical mobility-equivalent-diameter. Particle motion in an electric field can yield high-resolution measurements as well as separation of desired particle sizes.

In addition to the various equivalent diameters mentioned above, any other physical property, such as mobility in a magnetic field, external surface area, radioactivity, and chemical or elemental concentration, can be used to determine an equivalent diameter.

Size Ranges

Although it is customary to discuss particulate clouds in terms of particle size, rarely is such a cloud composed of single-diameter particles—only in the laboratory, and then only with great care, can single-sized aerosols be produced. Such single-sized particulate aerosols are referred to as monodisperse. These aerosols are useful for studying their size-dependent properties or for calibrating instruments. Whether dust, mist, or fume, virtually all naturally occurring aerosols are a mixture of a wide variety of particle sizes, i.e., they are polydisperse.

A large airborne molecule can be considered as a very small aerosol particle. Although air consists of nitrogen, oxygen, and other gases, air molecules can be considered for most calculations as having an average diameter of 0.37 nm (0.00037 μm). In comparison, aerosol particles are generally 1 nm (0.001 μm) in diameter, or larger. Fume particles of this size can be seen only immediately upon condensation from the vapor state. A short time later, the high concentration of these very small particles causes coagulation into larger entities, ultimately reaching sizes near 1 μm.

Conversely, dust particles result from a size reduction of larger materials. Generally, one considers particles less than 100 μm (0.1 mm) in diameter to stay airborne long enough to be observed and measured as aerosols. For example, human hairs range from about 50 μm to about 100 μm in diameter. If they were cut into small pieces and released into the air, they would be near the upper limit of the aerosol size range. Size reduction of bulk material by mechanical forces, be they natural or induced by human action, can occur only for sizes where the externally applied forces are greater than the internal cohesion forces. Particles smaller than about 0.5 μm are rare in dust distributions for this reason.

PARTICLE SUSPENSIONS

Since an aerosol is a system of airborne particles suspended in a gas medium, one generally considers the gas properties and flow dynamics first and then evaluates how individual particles follow or deviate from the gas motion. The difference in trajectories between particles and gas molecules is the basis for many aerosol particle size measurement techniques. It is also the basis for many devices controlling aerosol contaminants and for techniques manipulating aerosol particles for manufacturing purposes. Changes in gas properties generally affect the particle trajectories.

As an example, one may appreciate the need for dealing with air flow characteristics first by asking how much aerosol deposition will occur 50 km from an aerosol-emitting power plant. The wind velocity determines the speed with which the aerosol is transported away from the power plant. Large particles gravitationally settling in a shorter time than is available for transport to the 50 km distant site will not be found at the receptor site. The mechanism of settling and dispersion is determined by the degree and mode of turbulence. Returning to aerosol measurement principles, a "horizontal elutriator" size selectively removes particles in a horizontal flow channel. Here, the gas flow is generally "well behaved" by the careful avoidance of air turbulence.

INSTRUMENT CONSIDERATIONS

In general, one cannot obtain particle size information on the entire five-decade size range of 0.001–100 μm with a single instrument. On a macroscopic scale, this would be equivalent to measuring a 1 cm distance with a small scale rule and then using the same scale for measuring a 1 km distance (which is 5 orders of magnitude larger than 1 cm). When sensing with optical techniques utilizing white light, the wavelength of visible light from about 0.4 to 0.7 μm limits the observation of particles to about this size range and larger. Inertial techniques become inefficient below about 0.5 μm at normal temperature and pressure. In an electron microscope, the observational tool is electromagnetic radiation (electrons) with a much smaller wavelength that can "see" much smaller particles. Therefore, one expects to apply different instrumental techniques, measuring different size parameters, for submicrometer versus supermicrometer-sized aerosols.

Most aerosol-sizing instruments effectively measure over a size range no larger than $1\frac{1}{2}$ orders of magnitude. Thus, the largest measurable size may be about 50 times the smallest measurable size for a given instrument. Since most of the size parameters measured relate to the particle surface, volume, or mass, this size range corresponds to a surface range of 2500 and a volume or mass range of 125,000. Instruments measuring a cumulative value, e.g., total mass or number, can cover a wider size range. Preferably, each aerosol-sizing instrument should give a monotonically increasing response to increases in particle size. Unfortunately, some optical devices may detect the same amount of scattered light for more than one particle size, resulting in a significant loss of size resolution.

When a single-source aerosol is measured by any of the above-mentioned size parameters, the representative particle size is usually quoted as the mean size (average of all sizes), median size (equal number of particles above and below this size), or the mode size (size with the maximum number of particles). The spread of the particle size distribution is characterized by an arithmetic or geometric (logarithmic) standard deviation. Typically, the particle size distribution is "lognormal," i.e., the particle concentration versus particle size curve looks "normal" (also referred to as bell-shaped or Gaussian) when the particle size is plotted on a logarithmic scale.

The reason for the use of this logarithmic or geometric size scale can be conceptualized by breaking a piece of blackboard chalk. For example, a 64 mm long piece of chalk would break up into two pieces, each of 32 mm length. Subsequent breakup yields pieces of 16, 8, 4, 2, 1 mm, etc. length until the internal forces resist further breakage. The ratio of adjacent sizes is always two, thus appearing at the same linear distance on a logarithmic or geometric size scale. Since with each breakage step, more and more particles are produced, the distribution is skewed, so that there are many more small particles than large ones. This exercise of breaking up a piece of chalk mimics the way, in which natural and man-made forces generate aerosols. Generally, aerosol particle sizes are, therefore, plotted on a logarithmic size scale.

Many aerosols measured in ambient or industrial air environments or in industrial process streams are a mixture of aerosols, resulting in more than one particle mode and covering a wide size range. This may make the measurement and analysis of the aerosols considerably more complex. In general, one should attempt to first identify all aerosol sources and decide what information is needed and for what purpose. This decision will then point the way to the best available instrument to reach the desired objective.

Aerosol instruments not only differ in the size parameter that they measure, but each size parameter may "weigh" the particle size differently. A grocery store analogy can help elucidate this concept. If ten large apples and 100 small raisins are purchased, the median size "by number count" is somewhat larger than the size of the raisins. The median is close to that of the raisins because the median size divides the "population" into two, but in this case, most of the "population" consists of small raisins. If each piece of fruit is weighed on a scale, the weight of the apples dominates and the median size "by mass" is considerably larger. Thus, aerosol measurement "by mass" results in a larger median size than aerosol measurement "by count," although the same particle size distribution is measured. Therefore, any size result should by accompanied by a description of the "weighing" factor, or the "weighting" as it is commonly called.

If many particles in an aerosol are measured and the particles are grouped into discrete, contiguous, size bins, the size distribution can be represented by plotting particle number versus size. The lower and upper particle diameter limits, d_l and d_u, of each size bin need to be chosen with some care in order to get a useful description of the overall size distribution. The number of particles in each bin will depend on the size of the bin, i.e., $d_u - d_l$. To remove this bin width dependence, the number of particles in each bin is usually normalized by dividing the number of particles in the bin by the bin width.

PARTICLE SHAPE MEASUREMENT

Traditionally, particle shape has been acknowledged by including a "shape factor" in the particle motion equations. For a nonspherical particle, inclusion of this factor in the equation allows one to calculate the desired parameter while characterizing complex particle shapes by a single dimension. While this provides an indication of the particle's behavior under certain conditions, it does not provide sufficient details to characterize fully the particle. For instance, the particle's reactivity is a sensitive function of the particle's surface, and often the shape and texture provide clues as to the particle's formation and history. Since powerful computers are now available to researchers and practitioners, methods will be described that characterize the shape more directly. Since it is generally difficult to measure the shape of all particles in an aerosol, careful measurements on a few particles are often assumed representative of the entire aerosol.

A variety of schemes have been developed to measure the outline shape and detailed texture of particles. Simple shapes, such as spheres (droplets) and rods (simple fibers), can be completely described by one or two dimensions, respectively. More complex shapes are difficult to characterize. For instance, measuring the distance of the particle perimeter from the particle centroid and analyzing this distance as a function of angle, using Fourier analysis, has been proposed as a classification scheme for particle shapes (Beddow, Philip, and Vetter 1977). A number of similar shape description techniques have also been described (Kaye 1981). However, such techniques are generally limited to the outline profile of a particle and cannot characterize all the surface complexities and convolutions of many particles.

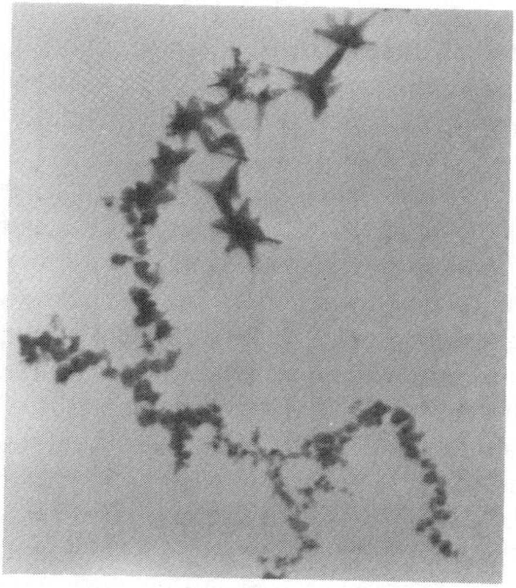

FIGURE 2-2. A Fume Particle Has Different Degrees of Complexity, Depending on Magnification. This Complexity Can be Characterized by the Measured Fractal Dimension.

Fractal Dimension

A sphere is a three-dimensional object and has certain mathematical properties that we are familiar with. Liquid droplets, for instance, assume a spherical shape under many circumstances and are, thus, relatively easy to measure and characterize. It is often convenient to assume that a particle has a simple shape, but many particles have relatively complex shapes. A fume particle, for example, may consist of thousands of smaller, nearly spherical particles, or spherules, arranged in a complex branched-chain configuration and is therefore more difficult to characterize (Fig. 2-2). When such a particle is collected on a surface, only a two-dimensional representation of the particle, i.e., the outline of the particle, is usually observed. If we try to measure the perimeter of such a particle, we find that the perimeter increases as the magnification is increased.

This changing perimeter with magnification occurs in much the same way as when we try to estimate the coastline of an island. If we observe the island from a satellite in space, the island may look almost circular. As we fly above the island in an airplane, we see more details of the island's complex shape and the measurable length of the coastline increases. If we land on the island and try to measure the coastline on foot, the length may be still longer as we follow the contours of rocks and crevices. On the other hand, for a smooth, spherical particle that we observe under a microscope, the measured perimeter is essentially constant, regardless of the magnification.

If the particle were a long, straight chain, it might be characterized as a one-dimensional object having, primarily, length. Complex branching, however, causes the particle to take up more space than a linear object, while not meeting the criteria for a sheet-like (two-dimensional) or a spheroidal (three-dimensional) object. It can, however, be characterized by assigning it a fractional, or fractal, dimension. The term fractal was coined by Mandelbrot (1983) and has found use in a wide range of applications, including particle shape, turbulence, lung structure, and fibrous filter structure (Kaye 1989). The principal characteristic of a fractal object is that a measure of complexity is similar on several measurement scales. For instance, if one looks at a fume particle at several different magnifications, the variation, or complexity, may appear very similar. This property of similar complexity at several scales, or scale invariance, is described by the mathematical concept of self-similarity.

A particle's perimeter fractal dimension can be measured in a number of ways: either by hand, using pictures taken at a series of magnifications, or with an image analysis system connected to a microscope. The particle dimension, D, is $D = 1 + |m|$, where m is the slope of a log–log plot of the perimeter versus a "measurement length." The latter term refers to the size of the ruler or the distance between the caliper points used to trace the outline of the particle.

Returning to the island example (which looks like a large particle under a microscope), we can measure the length of its coastline using maps of several scales. Starting at the largest scale, a fixed-length ruler is "walked" around the coastline. The number of steps around the island is a measure of the perimeter. This measurement is repeated a number of times, each time increasing the map scale or decreasing the ruler length. The perimeter of the island is plotted against the caliper step size on a log–log scale, producing a line with slope m. The perimeter increases with decreasing step size because a smaller ruler is able to follow the details of the outline more closely.

The resulting dimension, D, which is for this example between one and two, is indicative of the complexity of the perimeter. A dimension near one indicates a smooth outline with a clearly defined length. A dimension near two indicates a particle with a very complex shape that nearly fills the area within the outer envelope of the object. If we use, instead of a ruler that measures just length, an instrument that measures area, e.g., gas molecules of different sizes adsorbing on the particle surface, the surface complexity of the particle can be estimated from the change in apparent surface area with gas molecule size. In this case, the fractal dimension will be between two and three. We are measuring a surface (two dimensions) that is complex enough to partially fill a volume (three dimensions) that is dependent on the size of the measurement tool.

The observed shape of a particle is the result of its history. The fume particle, for instance, begins as a vapor condensing into spherules. The spherules, being very small, diffuse rapidly and coagulate into branched chains. As the chains increase in size and the number of individual spherules decreases, the chains may intercept one another and form larger agglomerates. Such an agglomerate might be observed at several magnifications, ranging from the structure of the entire particle down to the chemical structure of the surface. Thus, at lowest magnifications, the complex structure can be represented by a fractal dimension; at intermediate magnifications, the spherules have nearly integral dimension; and at still higher magnifications, the spherule surface may be rough and characterized by another fractal dimension (Fig. 2-3).

Fractals have been used to characterize the mechanism of particle formation. Fume particles, for example, are usually formed by the diffusion-limited aggregation of single particles and clusters of particles and have been

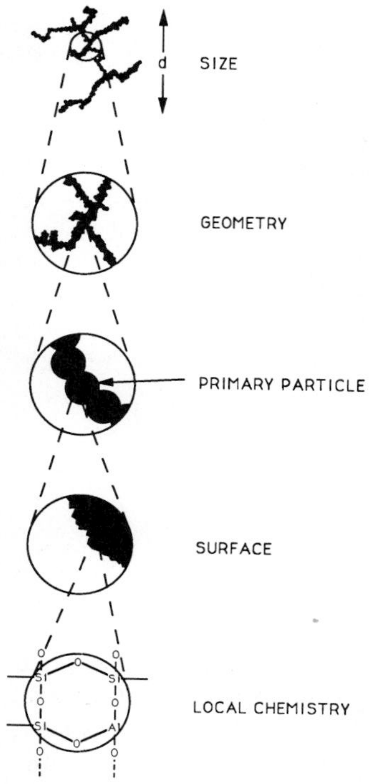

FIGURE 2-3. Schematic Structure of an Aggregate Embedded in a Two-Dimensional Space. Scanning Down the Figure Corresponds to Viewing the Object at Ever-Higher Magnification (Adapted from Schaefer and Hurd 1990.)

microscopically measured to have fractal dimensions in the range 1.3–1.5, although, theoretically, their dimension can be as high as 1.7. Other physical measurements of particles have also been related to their fractal dimension. For instance, scattered radiation ranging in wavelength from visible light to X-rays has been used to characterize the structure of fumed silica and to explain the structure in terms of growth processes (Schaefer and Hurd 1990).

PARTICLE FORCES

The intraparticle and interparticle forces that hold particles together or to a surface, and the forces that detach particles from each other or from a surface are difficult to quantitate for use by the practitioner. These forces may depend on particle bulk and surface parameters (size, shape, roughness, chemistry), the properties of the surrounding gas (temperature, humidity) and the mechanics of the contacting particles (relative particle velocity, contact time). These forces will, therefore, be described qualitatively.

When particles are subjected to an external force, such as gravity or an electrical force, the particles will move in the force field. The migration velocity in the force field is particle-size-dependent, a fact that is exploited by most aerosol size spectrometers for particle size discrimination.

Adhesion Forces

In contrast to gas molecules, aerosol particles that contact one another generally adhere to each other and form agglomerates. If they contact a surface, such as a filter or any other particle collection device, they are assumed to adhere to the surface, i.e., particle adhesion is the working hypothesis of these devices.

The London–van der Waals forces, which are attractive in nature, act over very short distances relative to particle dimensions (Friedlander 1977, 44). According to the theory of their origin, random motion of electrons in an electrically neutral material creates instantaneous dipoles which may induce complementary dipoles in neighboring material and, thus attract the surfaces to each other.

Most particles 0.1 μm or larger carry some small net charge which exerts an attractive force in the presence of a particle with an opposite charge (Hinds 1982, 129). For two charged particles (point charges), this force is inversely proportional to the square of the separation distance. After two surfaces have made contact with each other by either or both of the above forces, the surfaces may deform with time, thereby increasing the contact area and decrease the separation distance, and thus increasing the force of adhesion.

Figure 2-4a exemplifies how air humidity may affect particle adhesion. At high humidity, liquid molecules are adsorbed on the particle surface and fill the capillary spaces at

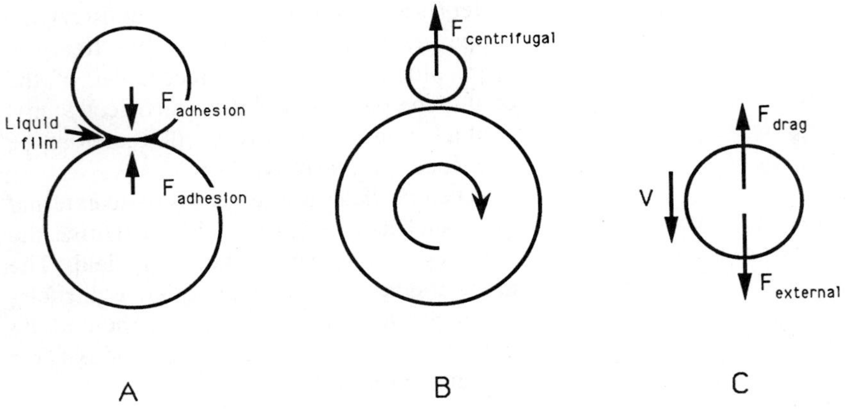

FIGURE 2-4. Examples of Particle Forces: (a) Adhesion due to Liquid Film, (b) Detachment due to Centrifugal Force, (c) Particle Motion at Velocity V due to a Balance between Drag Force and an External Force.

and near the point of contact. The surface tension of this liquid layer increases the adhesion between the two surfaces.

Detachment Forces and Particle Bounce

Figure 2-4b exemplifies the detachment of a particle from a rotating body. The centrifugal force is proportional to the particle's mass or volume, i.e., particle diameter cubed (d^3). Detachment by other types of motion, such as vibration, is similarly proportional to d^3, while detachment by air currents is proportional to the exposed surface area, i.e., d^2. In contrast, most adhesion forces are linearly dependent on particle diameter. Thus, large particles are more readily detached than small ones. While individual particles less than 10 μm are not likely to be easily removed, e.g., by vibration, a thick layer of such particles may be easily dislodged in large (0.1–10 mm) chunks (Hinds 1982, 130). Reentrainment of particles from a surface into an aerosol flow may, therefore, create measurement problems after a significant number of particles has been deposited from the aerosol.

If an aerosol flow is directed towards a surface, e.g., in filters and impactors, particles with sufficient inertia will deviate from the air streamlines and move towards the surface. Liquid and sticky small particles will deposit on the surface. Upon contact, a solid particle and the surface may deform. If the rebound energy is greater than the adhesion energy, a condition that may occur for sufficiently high impact velocity, a solid particle will "bounce," i.e., move away after contact with the surface. On contact with the surface, some or all of the particle's kinetic energy is converted to thermal energy, resulting in reduced kinetic energy on rebound or heating of the particle/surface interface on sticking, respectively. Grease or oil on the surface will generally increase the likelihood of adhesion, but after a layer of particles has been deposited, the incoming particles may bounce from the top surface of the previously deposited particles.

Externally Applied Forces

When an airborne particle is subjected to an externally applied force, e.g., gravity, it will be moved by that force. Opposing this external force is the aerodynamic drag force, as shown in Fig. 2-4c. When the two forces are in equilibrium, which happens almost instantaneously (there is a very short relaxation period, having consequences to be discussed later), the particle moves in the force field with migration velocity V. A knowledge of the two opposing forces allows the determination of this velocity. Particle velocity is important for estimating collection on surfaces as well as for separating particles by size. Quite often, aerosol measurements are designed to simulate some natural process, such as particle deposition in the respiratory system. Thus, it is important to understand the aerosol behavior in the original system as well as in the instrument in order to make accurate measurements. Utilization of the forces involved in the original process to make the measurement is one way of achieving this accuracy.

In space, astronauts must pay special attention to the dust generated by their clothing and the activities they engage in. Otherwise, their living space quickly becomes polluted with aerosols. On earth, gravity has a major cleaning effect on ambient and industrial aerosols. Larger particles tend to settle out more rapidly. Since the gravitational force is readily accessible for measurement applications, it is the basis for the definition of aerodynamic diameter.

We are familiar with the attraction of lint particles to clothing. This is due to charge differences between the lint and the clothing. Similarly, charged aerosol particles can be attracted to or repelled from charged surfaces or other particles. Few particles carry no charge, although the magnitude of charge can vary greatly. Particles that are freshly aerosolized tend to carry greater charge levels than particles that have been airborne for hours or longer. This aging effect is due to the attraction of oppositely charged airborne ions produced by natural radiation. For aerosol particles that are highly charged, the electric force may exceed the gravitational force by several orders of magnitude. This readily generated force can be used for air cleaning as well as particle separation and measurement, e.g., with electrical mobility analyzers.

If there is a gradient in the number of particles present in the air, a diffusion force can be defined that moves the particles from the high-concentration to the low-concentration environment. It is often the dominant motive force for particles smaller than about 0.2 μm diameter. The diffusion battery, for example, is commonly used for measuring submicrometer particles. Diffusion is also important for understanding the particle and gas deposition properties of the human lung. If the suspending gas is a nonuniform mixture of gases, the particles may be moved also by diffusiophoretic forces caused by the concentration gradient of the gas components.

Inertial forces can be applied to particles by forcing the suspending air to change direction. Size-dependent inertial effects are used for particle separation, collection and measurement in such devices as impactors, cyclones, and acceleration nozzles. Impaction is an important mechanism for particle deposition in the respiratory system.

If there is a temperature gradient in the aerosol-containing space between two surfaces, the higher activity of the air molecules near the hot surface pushes the particles towards the colder surface (thermophoretic force). This property is exploited in the thermal precipitator, which is used to collect particles onto a desired surface. A special case of thermophoresis, but generally not very useful, is produced by light. Illumination of a particle heats up one side of the particle as well as gas molecules nearby which push the particle towards the colder side. Illumination can also produce radiation pressure, whereby the stream of photons exerts a force on the particle (photophoresis).

References

Cited

Baron, P. A. 1984. *Aerosol Calculator*. Computer program, TSI, Inc. St. Paul, MN.

Beddow, J. K., G. C. Philip, and A. F. Vetter. 1977. On relating some particle profile characteristics to the profile Fourier coefficients. *Powder Technol.* 18:19–25.

Hinds, W. 1982. *Aerosol Technology*. New York: Wiley.

Kaye, B. H. 1981. *Direct Characterization of Fine Particles*. New York: Wiley.

Kaye, B. H. 1989. *A Random Walk through Fractal Dimensions*. Weinheim: VCH Verlagsgesellschaft mbH.

Mandelbrot, B. B. 1983. *The Fractal Geometry of Nature.* New York: Freeman.

Friedlander, S. K. 1977. *Smoke, Dust and Haze.* New York: Wiley.

Schaefer, D. W. and A. J. Hurd. 1990. Growth and structure of combustion aerosols. *Aerosol Sci. Technol.* 12:876–90.

Weast, R. C. (ed.). 1989. *Handbook of Physics and Chemistry.* Cleveland: The Chemical Rubber Company.

Additional

Bird, R. B., W. E. Stewart, and E. N. Lightfoot. 1960. *Transport Phenomena.* New York: Wiley.

Cross, J. 1987. *Electrostatics: Principles, Problems and Applications.* Bristol: Adam Hilger.

Moore, A. D. 1973. *Electrostatics and its Applications.* New York: Wiley.

3

Gas and Particle Motion

Paul A. Baron

National Institute for Occupational Safety and Health
Centers for Disease Control, Public Health Service
Department of Health and Human Services, Cincinnati, OH, U.S.A.

and

Klaus Willeke

Department of Environmental Health
University of Cincinnati
Cincinnati, OH, U.S.A.

INTRODUCTION

Aerosols consist of two components: a gas or gas mixture, most commonly air, and the particles suspended in it. The behavior of the particles within the aerosol depends to a large extent on the motion and intrinsic properties of the suspending gas. Submicrometer-sized particles, especially those < 0.1 μm diameter, are affected by the motion of individual gas molecules (the free molecular regime). Thus, the kinetic theory of gases is useful in understanding the behavior of these particles. Larger particles can be treated as being submersed in a continuous gaseous medium or, more broadly, a fluid (the continuum regime). The tools of gas or fluid dynamics are more useful for this size range. Intermediate-sized particles can usually be treated by an adjustment of equations from the continuum regime. This intermediate range is termed the transition or slip regime. Whether considering a molecular ensemble or a continuous fluid, the motion of the gas will largely dictate the behavior of the suspended particles. In this chapter, concepts and parameters that affect gas and particle motion will be discussed and quantified as migration and deposition parameters in specific force fields.

BULK GAS MOTION

Reynolds Number

When measuring an aerosol, elucidating the gas flow patterns is critical to understanding what happens to the aerosol in the environment, on its way into the sensor of the measuring instrument, or in the actual sensor. While the aerosol particles follow the overall gas flow, their trajectories can deviate from the gas flow due to various external forces as well as changes in gas direction and velocity. Gas motion can be visualized by observing the streamlines, i.e., tracing the motion of miniscule volumes of gas. Gas flowing around an aerosol particle can have the same flow streamline pattern as gas passing around a large object such as a basketball.

The equivalence of fluid motion for various-sized objects can be described in terms of the forces involved. The flow pattern, whether it is smooth or turbulent, is governed by the ratio of the inertial force of the gas to the friction force of the gas moving over the surface. This ratio is expressed by the Reynolds number, *Re*, an extremely useful parameter when dealing with aerosols:

$$Re = \frac{\rho_g V d}{\eta} = \frac{Vd}{\nu} \qquad (3\text{-}1)$$

where V is the velocity of the gas, η the dynamic gas viscosity, ν the kinematic viscosity ($= \eta/\rho_g$), and d a characteristic dimension of the object, such as the diameter of a sphere. Since this dimensionless number characterizes the flow, it depends on gas density, ρ_g, and not on the particle density. At normal temperature and pressure (NTP), i.e., 20°C (293 K) and 101 kPa (1 atm), $\rho_g = 1.192 \times 10^{-3}$ g/cm^3 and $\eta = 1.833 \times 10^{-4}$ dyn s/cm^2, which reduces Eq. 3-1 to

$$Re = 6.5Vd \qquad (3\text{-}2)$$

for V in cm/s and d in cm.

A distinction must be made between the flow Reynolds number, Re_f, and the particle Reynolds number Re_p. The flow Reynolds number defines the gas flow in a tube or channel of cross-sectional dimension d. The particle Reynolds number defines the gas flow around a particle that may be found in this tube or channel flow. The characteristic dimension in the latter is particle diameter, d_p, and V expresses the relative velocity between the particle and the gas flow. Since the difference between these velocities is generally small, and the particle's dimension is very small, the particle Reynolds number usually has a very small numerical value.

Common Gas Flows

When friction forces dominate the flow, i.e., at low Reynolds numbers, the flow is smooth, or laminar. Under laminar flow, no streamlines loop back on themselves. At higher Reynolds numbers, the inertial forces dominate and loops appear in the streamlines until at still higher Reynolds numbers the flow becomes chaotic, or turbulent. The actual values of the Reynolds number depend on how the gas flow is bounded. For instance, laminar flow occurs in a circular duct when the flow Reynolds number is less than about 2000, while turbulent flow occurs for Reynolds numbers above 4000. In the intermediate range, the gas flow is sensitive to the previous history of the gas motion. For instance, if the gas velocity is increased into this intermediate range slowly, the flow may remain laminar. When a gas passes around a suspended object, such as a sphere, flow is laminar for particle Reynolds numbers below about 0.1.

Since, often, it is expensive and difficult to test collection and measurement systems at full scale and *in situ*, small-scale water (or other liquid) models operating at the same Reynolds number as the system being studied are a useful alternative. Dye injection into the flow stream allows visualization of the streamlines. Such models can operate on a smaller physical scale with a slower time response, so that it is easy to observe the time evolution of flow patterns. The same technique can be used to model the behavior of particles.

Example 3-1

Silica dust of 10 μm diameter is removed by a 30 cm diameter ventilation duct at 2000 cm/s (about 4000 fpm). An old rule of thumb in industrial hygiene is that silica dust of this size gravitationally settles at 1 cm/s. Calculate the flow and particle Reynolds numbers at 20°C.

Answer. The relevant parameters for the flow Reynolds number are the duct diameter and the gas flow velocity in the duct. From Eq. (3-2),

$$Re_f = 6.5\,Vd = 6.5(2000 \text{ cm/s})(30 \text{ cm}) = 3.9 \times 10^5$$

The relevant parameters for the particle Reynolds number are the particle diameter and the gravitational-settling velocity perpendicular to the gas flow:

$$Re_p = 6.5\,Vd = 6.5(1.0 \text{ cm/s})(10 \times 10^{-4} \text{ cm})$$
$$= 6.5 \times 10^{-3}$$

The flow Reynolds number exceeds 4000, indicating turbulent flow in the ventilation duct. The particle Reynolds number is less than one, indicating that the flow around the particle can be laminar. However, it is not so in this case because the gas flow is turbulent.

Many gas-handling systems for instruments use cylindrical tubing to carry the aerosol from one place to another. Understanding the flow patterns within the tubing is important for predicting the losses that occur within the tubing as well as predicting the distribution of particles within the tubing. If a gas begins to flow in a cylindrical tube, the friction at the wall slows the gas velocity relative to the motion at the center of the tube. At low Reynolds numbers, the dominating friction force produces a characteristic laminar parabolic velocity profile. The gas velocity at the center of the tube for this "Poiseuille flow" is twice that of the average velocity in the tube. Poiseuille flow does not get established immediately. A common rule of thumb is to assume that it takes ten tube diameters for this equilibrium flow to be effectively established.

The region near a surface where the flow is dominated by frictional forces is termed the boundary layer. When the flow starts along a surface, either in time or space, the boundary layer consists only of the gas at the surface, where the relative velocity is zero. At low Reynolds numbers, the boundary layer grows until steady-state conditions are reached. For the cylinder flow example above, the boundary layer grows into a parabolic flow profile that fills the cylinder, as indicated in the lower part of Fig. 3-1. At higher Reynolds numbers (in the turbulent regime) or during abrupt changes in flow conditions, the boundary layer can become separated from the surface. The development of the boundary layer and its relationship to the overall flow depends upon the object immersed in the fluid. This type of behavior has been described in many fluid mechanics texts (e.g., White 1986) and especially in *Boundary-Layer Theory* by Schlichting (1979).

There are a wide variety of flow situations for which empirical or experimentally verified theoretical solutions exist. For instance, when a gas passing through a cylindrical tube under laminar flow conditions negotiates a 90° bend, the cylindrical symmetry of the flow pattern in the tube is reduced to a plane of symmetry. Thus, the flow symmetry must also be reduced. Two parallel circulation patterns are set up on either side of the plane of the bend, as shown in Fig. 3-1. This secondary pattern causes mixing of the gas as well as increased inertial forces on particles suspended in the gas (Tsai and Pui 1990). In tubing used to transport aerosols, bends are generally undesirable because of increased particle loss.

For various reasons, there are often constrictions or expansions in a tube carrying a gas. A constriction will force the gas to increase in velocity and be focused at the center of the tubing, even more than the constriction in size. After this contraction region, or "vena contracta," the gas flow eventually expands again to fill the tubing and reestablishes an equilibrium pattern. These disturbances will also cause increased particle deposition.

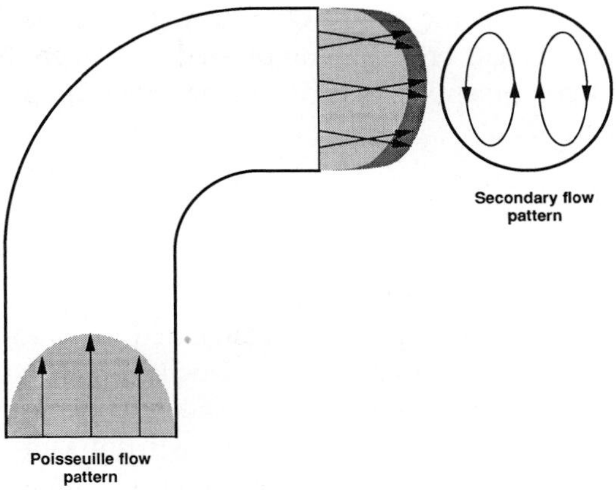

FIGURE 3-1. Flow Circulation After a 90° Bend in Tubing.

When a gas flows from an initial tube diameter into a suddenly expanded section or into free space, the flow pattern may persist for many initial tube diameters downstream. If the expansion of the tube is very slight, the flow does not separate from the walls and the flow pattern can expand smoothly to fill the increased diameter of the tube. In general, the angle between the wall and the tube axis needs to be less than 7° to avoid flow separation from the tube wall.

Gas Density and Mach Number

The density of a gas, ρ_g, is related to its temperature, T, and pressure, P, through the equation of state

$$P = \rho_g \frac{R_u}{M} T = \rho_g R T \qquad (3\text{-}3)$$

where ρ_g is the gas density ($= 1.192 \times 10^{-3}$ g/cm^3 for air at NTP), T is the absolute gas temperature in K, M is the molecular weight in g/mol and R_u is the universal gas constant ($= 8.31 \times 10^7$ dyn cm/mol K). In air, the effective molecular weight is 28.9 g/mol. Thus, the specific gas constant for air is $R = 2.88 \times 10^6$ dyn cm/K g or 2.84 atm cm^3/K g. One atmosphere equals 101 kPa, where 1 Pa = 1 N/m^2 = 10 dyn/cm^2.

When this gas moves at a high velocity relative to the acoustic velocity, U_g, in that gas, the gas becomes compressed. The degree of compression depends on the Mach number, Ma:

$$Ma = \frac{U}{U_{\text{sonic}}} \qquad (3\text{-}4)$$

Here the gas velocity is designated as U to distinguish it from particle velocity V. When $Ma \ll 1$, the gas flow is considered incompressible. This is true in most aerosol sampling situations. In air, the sonic or sound velocity at ambient temperature is about 340 m/s (1100 ft/s).

TRANSITION AND GAS MOLECULAR FLOW

Knudsen Number

Large aerosol particles are constantly bombarded from all directions by a large number of gas molecules. When a particle is small, less than 1 μm in size, its location in space may be affected by the bombardment of individual gas molecules. Its motion is then no longer determined by continuum flow considerations, but by gas kinetics.

The average velocity of a molecule, $\bar{V}$, is a function of its molecular weight, M, and the gas temperature, T. In air ($M_{\text{air}} = 28.9$ g/mol) at normal temperature and pressure (NTP, 20°C, 1 atm), this molecular velocity is 463 m/s. Using these air reference values, the average velocity can be estimated for other gases and temperatures:

$$\bar{V} = \bar{V}_r \left(\frac{T}{T_r}\right)^{1/2} \left(\frac{M_r}{M}\right)^{1/2} \qquad (3\text{-}5)$$

The mean free path, λ, is the mean distance a molecule travels before colliding with another molecule. In air at 20°C and atmospheric pressure, the mean free path, λ_r, is 0.0665 μm. The mean free path is an abstraction that is determined from a kinetic theory model which relates it to the coefficient of viscosity. Using these reference values, λ is determined for other pressures and temperatures (Willeke 1976):

$$\lambda = \lambda_r \left(\frac{101.3}{P}\right)\left(\frac{T}{293.15}\right)\left(\frac{1 + 110/293.15}{1 + 110/T}\right) \qquad (3\text{-}6)$$

where P is in kPa, T is in K, and S (in K) is the Sutherland constant (Table 3-1). If the unit of atmosphere is used for pressure, the factor of 101 used in Eq. (3-6) is substituted by one. The mean free path and the average molecular velocity are parameters that are frequently used to predict bulk properties of a gas, such as thermal conductivity, diffusion, and viscosity. Mean free paths for other gases are presented in Table 3-1.

TABLE 3-1 Gas Properties for Several Gases at NTP (293.15 K and 101.3 kPa)

Gas	η (μP)	S (K)	ρ_g (10^{-3} g/cm³)	λ (μm)
Air	182.03	110.4	1.205	0.0665
Ar	222.92	141.4	1.662	0.0694
He	195.71	73.8	0.167	0.192
H_2	87.99	66.7	0.835	0.123
CH_4	109.77	173.7	0.668	0.0537
C_2H_6	92.49	223.2	1.264	0.0328
iso-C_4H_{10}	74.33	255.0	2.431	0.0190
N_2O	146.46	241.0	1.837	0.0433
CO_2	146.73	220.5	1.842	0.0432

Source: Rader (1990).

The Knudsen number, Kn, relates the gas molecular mean free path to the physical dimension of the particle, usually the particle radius:

$$Kn = \frac{2\lambda}{d_p} \tag{3-7}$$

where d_p is the physical diameter of the particle. The Knudsen number is somewhat counterintuitive as an indicator of particle size since it has an inverse size dependence. $Kn \ll 1$ indicates continuum flow and $Kn \gg 1$ indicates free molecular flow. The intermediate range, approximately $Kn = 0.4$–20, is usually referred to as the transition or slip flow regime.

Slip Flow Regime and Correction Factor

If a particle is much smaller than the gas molecular mean free path ($Kn \gg 1$), it can travel past an obstacle at a very small distance from the object since no gas molecule may impede it. If the particle is very large ($Kn \ll 1$), many gas molecular collisions occur near the surface and the particle is decelerated. When the Knudsen number is of the order of unity, the particle may slip by the obstacle. When the particle size is in this "slip flow regime," it is convenient to assume that the particle is still moving in a continuum gas flow. To accommodate the difference, a slip correction factor, C_c, also referred to as the "Cunningham slip correction factor," is introduced into the equations. An empirical fit to air data for particles gives (Allen and Raabe 1985)

$$C_c = 1 + Kn[\alpha + \beta \exp(-\gamma/Kn)] \tag{3-8}$$

Various values for α, β, and γ have been reported. However, it is important to use the mean free path with which these constants were determined. The value of λ_r used in Eq. 3-6 should also be consistent with the derivation of the slip coefficient constants. For solid particles, $\alpha = 1.142$, $\beta = 0.558$, and $\gamma = 0.999$ (Allen and Raabe 1985). For oil droplets, $\alpha = 1.207$, $\beta = 0.440$, and $\gamma = 0.596$ (Rader 1989). C_c for other gases such as CO_2 and He are similar within a few percent. The slip correction and viscosity values are better determined than most other aerosol-related parameters and are, therefore, reported with a higher degree of precision.

For pressures other than atmospheric, the slip correction changes because of the pressure dependence of mean free path in Kn and the following may be used for solid particles:

$$C_c = 1 + \frac{1}{Pd_p}[15.39 + 7.518 \exp(-0.0741 Pd_p)] \tag{3-9}$$

where P is pressure in kPa and d_p is the particle diameter in μm (Hinds 1982).

C_c equals one in the continuum regime and and becomes greater than one for decreasing particle diameter in the transition regime. For instance, $C_c = 1.02$ for 10 μm particles, 1.15 for 1 μm particles and 2.9 for 0.1 μm particles. Note that the shape factor and the slip correction factor must be consistent with the type of equivalent diameter used in the same equation (Brockmann and Rader 1990).

Gas Viscosity

Gas viscosity is primarily due to the momentum transfer that occurs during molecular collisions. These frequent and rapid collisions tend to damp out differences in bulk gas motion as well as impede the net motion of particles relative to the gas. Thus, the mobility of a particle in a force field depends on the aerodynamic drag exerted on the particle through the gas viscosity. Fluid dynamic similitude, as expressed by the Reynolds number, depends on gas viscosity, η. Therefore, a knowledge of the gas viscosity is important when dealing with aerosol particle mechanics. The viscosity can be related to a reference viscosity, η_r, and a reference temperature, T_r, as:

$$\eta = \eta_r \left(\frac{T_r + S}{T + S} \right) \left(\frac{T}{T_r} \right)^{3/2} \qquad (3\text{-}10)$$

where S is the Sutherland interpolation constant (Schlichting 1979). Note that viscosity is independent of pressure.

In cgs units, viscosity is expressed in dyn s/cm^2, also referred to as poise or P. For air at 293.15 K, the viscosity is 182.03 μP and $S = 110.4$ K. The interpolation formula is fitted to the data over the range 180–2000 K (Schlichting 1979). Reference values of viscosity and Sutherland constants for other gases are presented in Table 3-1.

GAS AND PARTICLE DIFFUSION

The random movement of gas molecules causes gas and particle diffusion if there is a concentration gradient. For instance, in a diffusion denuder, SO_2 gas molecules may diffuse to an absorbing surface due to their high diffusivity. Sulfate particles, which are larger and, therefore, have lower diffusivity, will mostly be transported through the device. Thus, the SO_2 gas molecules are separated from the sulfate particles.

Gas Diffusion

Diffusion always causes net movement from a higher concentration to a lower one. The net flux of gas molecules, J, is in the direction of lower concentration. Thus, in simple one-dimensional diffusion,

$$J = -D\frac{\partial N}{\partial x} \qquad (3\text{-}11)$$

where x is the direction of diffusion, N is the concentration, and D is a proportionality constant, referred to as the diffusion coefficient. The diffusion coefficient for a gas with molecular weight M is (Hinds 1982, 24)

$$D = -\left(\frac{3\sqrt{2\pi}}{64Nd_{\mathrm{molec}}^2} \right) \sqrt{\frac{RT}{M}} \qquad (3\text{-}12)$$

where N is the number of gas molecules/cm^3 and d_{molec} is the molecular collision diameter (3.7×10^{-8} cm for air). The diffusion coefficient of air molecules at 293 K is 0.18 cm^2/s.

Particle Diffusion

Small particles can achieve significant diffusive motion in much the same fashion as described for gas molecules. The difference is only in the particle size and shape. Because of their increase in inertia with particle mass and the larger surface area over which the bombardment by the gas molecules is averaged, large particles will diffuse more slowly than small particles. For particles in a gas, the diffusion coefficient or diffusivity, D, can be computed by

$$D = \frac{kTC_c}{3\pi\eta d_p} = kTB \qquad (3\text{-}13)$$

where k, the Boltzmann constant, is 1.38×10^{-16} dyn cm/K and the mechanical mobility, B (cm/s dyn), is a convenient aerosol property that combines particle size with some of the properties of the suspending gas:

$$B = \frac{C_c}{3\pi\eta d_p} \qquad (3\text{-}14)$$

Particle diffusion, also referred to as Brownian motion, occurs because of the relatively high velocity of small particles and it is sometimes useful to estimate how far, on an average, these particles move in a given time. The root mean square (rms) distance, x_{rms}, that the particles can travel in time t is

$$x_{rms} = \sqrt{2Dt} \tag{3-15}$$

Example 3-2

Fume aerosols of 0.01 μm diameter are drawn into the deep lung regions of a worker whose alveoli can be approximated by 0.2 mm diameter spheres. Estimate if these particles are likely to deposit in this area of the lung during a breath-holding period of 4 s. Assume the body temperature to be 37°C.

Answer. We note that calculation of x_{rms} (Eq. 3-15) requires a knowledge of the diffusion coefficient, which in turn requires the slip correction factor and viscosity. To simplify the calculation, let us assume for the moment that diffusion is taking place at room temperature. Thus, the air viscosity is 183 μP and the mean free path is 0.0665 μm. (For a more exact estimate of these parameters at body temperature, use Eqs. 3-6 and 3-10, respectively.)

The slip correction factor can be determined from Eq. 3-8 using constants for solid particles:

$$C_c = 1 + Kn[1.142 + 0.558 \exp(-0.999/Kn)]$$

$$C_c = 1 + \frac{2 \times 0.0665\ \mu\text{m}}{0.01\ \mu\text{m}}\left[1.142 + 0.558 \exp\left(-0.999\frac{0.01\ \mu\text{m}}{2 \times 0.0665\ \mu\text{m}}\right)\right]$$

$$= 23.1$$

We then estimate the diffusion coefficient, using Eq. 3-13.

$$D = \frac{kTC_c}{3\pi\eta d_p}$$

$$D = \frac{(1.36 \times 10^{-16}\ \text{dyn cm/K})(293\ \text{K})(23.1)}{3 \times 3.14(1.83 \times 10^{-4}\ \text{P})(1 \times 10^{-6}\ \text{cm})}$$

$$= 5.33 \times 10^{-4}\ \text{cm}^2/\text{s}$$

Finally, using Eq. 3-15,

$$x_{rms} = \sqrt{2Dt} = \sqrt{2(5.33 \times 10^{-4}\ \text{cm}^2/\text{s})(4\ \text{s})}$$

$$= 0.0653\ \text{cm} = 0.653\ \text{mm}$$

We find that at room temperature, the rms displacement by diffusion is much larger than the alveolar size. At the elevated temperature in the lung (37°C), the particles are expected to move faster and diffuse further. If this air temperature is used in the calculation of the diffusion coefficient, x_{rms} is 0.697 mm. Thus, we know that most of these particles are likely to be collected in the alveolar space of the lung. A more exact analysis can be made by considering such factors as the spherical geometry of the alveoli, the location of the particles within the alveoli, and the air temperature.

By including the air temperature dependence also in the calculation of viscosity and mean free path, the rms displacement calculation results in 0.681 mm.

Table 3-2 includes examples of x_{rms} for various sized particles after a period of 10 s.

Peclet Number

The amount of convective transport of particles towards an object may be related to the diffusive transport through the dimensionless Peclet number, Pe:

$$Pe = \frac{Ud_c}{D} \tag{3-16}$$

where d_c is the significant dimension of the particle-collecting surface and U is the upstream gas velocity towards the surface. The larger the value of Pe, the less important is the diffusional process (Licht 1988, 226). Pe is often used in the description of diffusional deposition on filters.

Schmidt Number

The ratio of the Peclet number (Eq. 3-16) to the Reynolds number (Eq. 3-1) is referred to as the Schmidt number, *Sc*. It expresses the ratio of kinematic viscosity to diffusion coefficient:

$$Sc = \frac{\eta_g}{\rho_g D} = \frac{\nu}{D} \tag{3-17}$$

As the Schmidt number increases, convective mass transfer increases relative to Brownian diffusion of particles. It has been used for describing diffusive transport in flowing fluids (convective diffusion), especially in the development of filtration theory (Friedlander 1977). *Sc* is relatively independent of temperature and pressure near standard conditions.

AERODYNAMIC DRAG ON PARTICLES

Externally applied forces on an aerosol particle are opposed and rapidly balanced by the aerodynamic drag force. An example of this is a sky diver: the gravitational force pulling the sky diver towards the earth is eventually balanced by the air resistance and the diver reaches a final falling speed of about 63 m/s (140 miles per hour).

A particle's drag force, F_{drag}, relates the resistive pressure of the gas to the velocity pressure and is determined by the relative motion between the particle and the surrounding gas. When the particle dimensions are much larger than the distance between the gas molecules, the surrounding gas can be considered as a continuous fluid (continuum regime). Under this condition, the drag force is given by

$$F_{drag} = \frac{\pi}{8} C_d \rho_g V^2 d_p^2 \tag{3-18}$$

Note that the aerodynamic drag is related to the gas density, ρ_g, and not the particle density. The drag coefficient, C_d, relates the drag force to the velocity pressure. When the inertial force pushing the gas aside, due to the velocity difference between the gas and the particle, is much smaller than the viscous resistance force, the drag coefficient, C_d, is expressed in terms of gas flow parameters:

$$C_d = \frac{24}{Re_p}, \quad Re_p < 0.1 \tag{3-19}$$

where Re_p is the particle Reynolds number. This relationship is accurate within 1% in the Re_p range indicated. If 10% accuracy is acceptable, Eq. 3-19 can be used up to $Re_p < 1.0$. Combining Eqs. 3-1, 3-18, and 3-19 results in

$$F_{drag} = 3\pi\eta V d_p \tag{3-20}$$

This equation is also known as the Stokes law. For the Stokes law flow of gas around the particle, the drag on the particle depends only on gas viscosity, η, particle velocity, V, and particle diameter, d_p. This assumes that the particle is spherical. The particle drag for shapes other than spheres is usually difficult to predict theoretically. Therefore, for particles of other shapes, a dynamic shape factor, χ, is introduced that relates the motion of the particle under consideration to that of a spherical particle

$$F_{drag} = 3\pi\eta\chi V d_m \tag{3-21}$$

where d_m is now the mass-equivalent diameter, defined as the diameter of a sphere composed of the particle bulk material with no voids that has the same mass as the particle in question. The shape factor is sometimes related to the equivalent volume or volume-equivalent diameter, d_{ev}, defined as the diameter of a sphere of equivalent volume. This term may be ambiguous. When the equivalent volume is composed of particle bulk material with no void, $d_{ev} = d_m$. However, if the material includes voids, $d_{ev} > d_m$. If we determine the shape factors and equivalent diameters for particles that we wish to measure, the behavior of the particles can be predicted when they are influenced by various force fields, e.g., gravity or electrostatic.

We know that gases are not continuous fluids as indicated above, but consist of discrete molecules. Therefore, when the particle size approaches the mean free path of the gas molecular motion (transition or slip regime), we can apply a correction that takes the "slip" between the particle and the gas into account. Thus, the Cunningham slip correction factor,

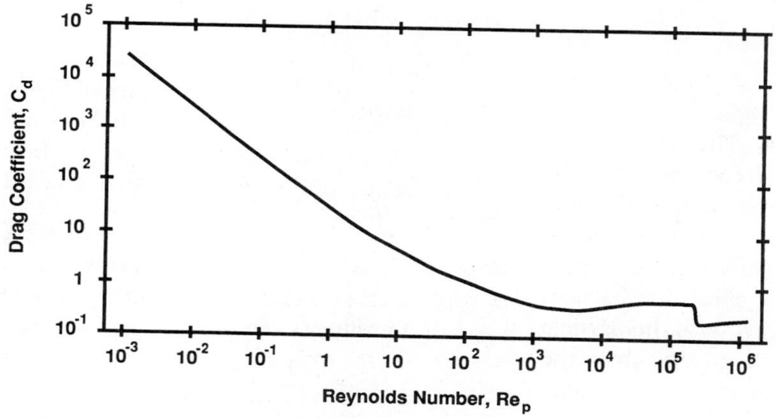

FIGURE 3-2. Drag Coefficient as a Function of Particle Reynolds Number for Spherical Particles.

C_c, is introduced into Eq. 3-21:

$$F_{drag} = \frac{3\pi\eta\chi V d_e}{C_c} \tag{3-22}$$

Equation 3-22 assumes that the flow around the particle is laminar. As particles move faster (i.e., have a larger Re_p), the above relationships must be modified further. As indicated above, the range over which Eq. 3-19 is accurate defines the Stokes regime. For larger Re_p, empirical relationships for C_d have been developed to extend Stokes law. The drag coefficient, C_d, for a spherical particle is the ratio of the resistance pressure due to aerodynamic drag (drag force/cross-sectional area) to the velocity pressure of the flow towards the sphere, based on the relative velocity between the particle and the suspending gas. Figure 3-2 shows the relationship of the drag coefficient to the particle Reynolds number over a wide range of Reynolds numbers.

For Re_p above 0.1, Sartor and Abbott (1975) developed the following empirical relationship:

$$C_d = \frac{24}{Re_p}(1 + 0.0916 Re_p), \quad 0.1 \le Re_p < 5 \tag{3-23}$$

and Serafini (Friedlander 1977, 105), the following:

$$C_d = \frac{24}{Re_p}(1 + 0.158 Re_p^{2/3}), \quad 5 \le Re_p < 1000 \tag{3-24}$$

Note that these relationships have been derived from data taken with smooth spheres. Similar relationships have been derived and reviewed for particles such as droplets, solid spheroids, disks, and cylinders (Clift, Grace, and Weber 1978, 142). Typically, Re_p is based on the equatorial diameter for disks and spheroids and on the cylinder diameter for cylinders, although other definitions can be used. Particles with extreme shapes may have a significantly different drag coefficient. For instance, C_d for fibers is up to 4 times lower than for spheres with $Re_p < 100$ when the fiber diameter is used as the significant dimension in the Reynolds number expression (Eq. 3-1).

Thus, by using the appropriate form of the drag coefficient (Eq. 3-19, 3-23, or 3-24) and including the shape factor and slip coefficient, the drag force can be calculated over a wide range of particles and conditions:

$$F_{drag} = \frac{\pi C_d \rho_g \chi V^2 d_p^2}{8 C_c} \tag{3-25}$$

PARTICLE MOTION DUE TO GRAVITY

The gravitational force, F_{grav}, is proportional to particle mass, m_p, and gravitational acceleration, g:

$$F_{\text{grav}} = m_{\text{p}}g = (\rho_{\text{p}} - \rho_{\text{g}})v_{\text{g}}p \approx \rho_{\text{p}}v_{\text{p}}g \tag{3-26}$$

where ρ_{g} is the gas density. The gravitational pull depends on the difference between the density of the particle and that of the surrounding medium. For a particle in water, this buoyancy effect is significant. For a particle in air, the buoyancy effect can be neglected for compact particles since the particle density is generally much greater than the density of the gas. If the particle is spherical, particle volume, v_{p}, can be replaced with $\pi d_{\text{p}}{}^3/6$:

$$F_{\text{grav}} = \frac{\pi}{6} d_{\text{p}}^3 \rho_{\text{p}} g \tag{3-27}$$

The gravitational field of the earth was mentioned in Chapter 2. This field exerts a force pulling a particle down. As the particle begins to move, the gas surrounding the particle exerts an opposing drag force, which, after a short period of acceleration, equals the gravitational force and the particle reaches its terminal settling velocity. By equating the two forces, the following relationship is obtained for the gravitational-settling velocity, V_{grav}, in the Stokes regime (Eq. 3-19):

$$V_{\text{grav}} = V_{\text{ts}} = \frac{\rho_{\text{p}} d^2 g C_{\text{c}}}{18\eta}, \quad Re_{\text{p}} < 0.1 \tag{3-28}$$

In order to reflect the equilibrium between the two opposing forces, this velocity is also referred to as terminal settling velocity, V_{ts}. Assuming negligible slip ($C_{\text{c}} = 1$), this equation reduces to the following at NTP:

$$V_{\text{grav}} = 0.003 \rho_{\text{p}} d_{\text{p}}^2 \tag{3-29}$$

where ρ_{p} is in g/cm^3 and d_{p} in μm. Spherical particles (e.g., droplets) are common in nature and their motion can be described mathematically. Therefore, the behavior of nonspherical particles is often referenced to such particles through a comparison of their behavior in a gravitational field.

Example 3-3

An open-faced filter cassette samples at 2 l/min (33.3 cm^3/s) over its inlet face of about 35 mm diameter. If the cassette is held facing downward, can a 25 μm diameter particle with a density of 3 g/cm^3 be drawn upward onto the filter in calm air?

Answer. The cassette samples at a flow rate Q over a cross-sectional filter area A. The upward air velocity, U, is

$$U = \frac{Q}{A} = \frac{33.3 \text{ cm}^3/\text{s}}{\pi\left(\dfrac{3.5 \text{ cm}}{2}\right)^2} = 3.46 \text{ cm/s}$$

The gravitational settling velocity of the 25 μm particle is, from Eq. 3-29,

$$\begin{aligned} V_{\text{grav}} &= 0.003\ \rho_{\text{p}} d_{\text{p}}^2 \\ &= 0.003(3 \text{ g/cm}^3)(0.0025 \text{ cm})^2 \\ &= 5.63 \text{ cm/s} > 3.46 \text{ cm/s} \end{aligned}$$

The particle cannot be drawn upward into the sampler.

This is also the principle of a vertical elutriator, which prevents particles above a certain size from passing through the device. However, in some implementations of this device, e.g., the cotton dust elutriator, inlet effects complicate the penetration efficiency.

The aerodynamic diameter, d_{a}, of a particle is the diameter of a unit-density sphere that has the same settling velocity as the particle in question, as shown in Eqs. 2-1–2-3 of Chapter 2. Another definition that is also commonly used is the Stokes diameter, d_{s}, which is the diameter of a spherical particle with the same density and settling velocity as the particle in question. The aerodynamic diameter can be related to the Stokes diameter through the settling velocity equation

$$\rho_{\text{p}} d_{\text{s}}^2 = 1\, d_{\text{a}}^2 = d_{\text{a}}^2 \tag{3-30}$$

A number of instruments, including the horizontal and the vertical elutriators, use settling velocity to separate particles according to size. For instance, aerosol particles of a certain size (d_p), initially spread throughout a quiescent rectangular chamber or room of height H, will settle at a constant velocity, V_{grav}. After some time, t, the particle concentration in the chamber, $N(t)$, will be

$$N(t) = N_0\left(1 - \frac{V_{grav}t}{H}\right) \quad (3\text{-}31)$$

where n_0 is the initial particle concentration in the chamber. The same relationship determines the concentration of particles in a rectangular channel with air flowing through it (a horizontal elutriator). At some distance downstream of the entrance to the channel (where the aerosol concentration is n_0), the concentration will be $n(t)$, where t is the time needed to reach that distance.

The above discussion of particle settling describes the behavior of particles in still air, a condition that is not often achieved in the environment or even in the laboratory. When the gas in a container undergoes continual and random motion, such as in a room with several randomly directed fans, the particles undergo "stirred settling." The time-dependent concentration, $n(t)$, under these conditions is also expressed in terms of an initial particle concentration, n_0, the gravitational-settling velocity in still air, V_{grav}, and the height of the container, H:

$$N(t) = N_0 \exp\left(\frac{-V_{grav}t}{H}\right) \quad (3\text{-}32)$$

This equation applies to any container shape with vertical walls and a horizontal bottom. This indicates that even under stirred or turbulent conditions, larger particles (higher settling velocities) will settle down more rapidly than smaller particles, even though some of the large particles may persist in the air for a long time because of the exponential decay. Note that the form of Eqs. 3-31 and 3-32 are similar except for the exponential decay when stirring takes place during the settling. This similarity in form occurs for all such comparisons of uniform and stirred settling.

Gravitational Settling at Higher Reynolds Numbers

Particle-settling velocity can be calculated accurately for $Re_p < 0.1$ using Eq. 3-28. At higher Reynolds numbers, the observed settling velocity is lower than predicted by this equation because the drag coefficient is higher than predicted by Eq. 3-19. For spherical particles ($\chi = 1$), the gravitational-settling velocity can be expressed as a function of C_d by equating the drag force (Eq. 3-25) to the gravitational force (Eq. 3-27):

$$V_{grav} = \sqrt{\frac{4\,\rho_p C_c d_p g}{3\,\rho_g C_d \chi}} \quad (3\text{-}33)$$

The drag coefficient has a complex dependence on the settling velocity and, therefore, Eq. 3-33 cannot be solved in closed form. Graphical (Licht 1988, 160) and tabular (Hinds 1982, 51) determinations of the settling velocity at high Reynolds numbers have been used. Using Eq. 3-33 and the drag coefficient equation for the appropriate Reynolds number, e.g., Eq. 3-23 or 3-24, an iterative solution for the settling velocity can readily be obtained using a computer or calculator. A guess for C_d allows the calculation of an initial value for V_{grav}, which is then used to calculate a new value of C_d. The new value of C_d is then used in Eq. 3-33 and the iteration is continued until the values converge.

PARTICLE PARAMETERS

The gravitational force effectively removes large particles from the suspending gas. Particles of 1 µm or smaller take a long time to settle (see Table 3-2). To settle these, the removal force is increased, e.g., by rotating the gas volume, as in a centrifuge. Other devices channel the gas flow in a circular fashion (e.g., cyclones) or through bends (e.g., impactors) to create an increased force field. The following parameters are useful for describing the inertial and settling behavior of particles.

Relaxation Time and Stopping Distance

Using the Stokes settling velocity relationship (Eq. 3-28), several useful particle parameters can be defined. The first is the particle relaxation time,

$$\tau = \frac{\rho_p d_p^2 C_c}{18\eta} \tag{3-34}$$

This is the time a particle takes to reach $1/e$ of its final velocity when subjected to a gravitational field. The relaxation period is typically quite short, as indicated in Table 3-2, and can, therefore, be neglected for most practical applications. Use of this parameter simplifies the expression for gravitational settling to

$$V_{\text{grav}} = \tau g \tag{3-35}$$

Quite often, a particle, rather than starting from rest in a gravitational field, is injected into the air with an initial velocity, V_0. For instance, such a particle might be released from a rotating grinding wheel. The product of the relaxation time and the initial particle velocity is referred to as the stopping distance, S:

$$S = V_0 \tau \tag{3-36}$$

where S is in cm. Values of S for an initial velocity of 1000 cm/s are given in Table 3-2. The concept of stopping distance is useful, e.g., in impactors when evaluating how far a particle moves across the air streamlines when the flow makes a right-angle bend.

Example 3-4

A grinding wheel dislodges many wheel and workpiece particles and projects them from the contact point towards the receiving hood of the ventilation system. A particle of a certain size and density is projected 1 cm away. How far will a particle twice this size be projected? Estimate the projected distance when the speed of the grinding wheel is doubled.

Answer. The projected distance is proportional to the stopping distance. From Eqs. 3-34 and 3-36,

$$S = V_0 \tau \propto V_0 \rho_p d_p^2$$

The stopping distance depends on the square of the particle diameter, so a two-times-larger particle will project the distance four times, to 4 cm. At twice the grinding wheel speed, the particle will come off at approximately twice the initial velocity, resulting in a doubling of the distance to 2 cm.

The above stopping distance equation assumes that the particle is in the Stokes regime. If the particle diameter and velocity are such that Re_p is larger than 0.1, the stopping distance will be somewhat less than quadrupled for the larger particle. This is because outside

TABLE 3-2 Particle Parameters for Unit-Density Particles Under Standard Conditions

Particle Diameter, d_p (μm)	Slip Correction Factor, c_c	Settling Velocity, V_{grav} (cm/s)	Relaxation Time, τ (s)	Stopping Distance, S $V_0 = 1000$ cm/s (cm)	Mobility, B (cm/dyn s)	Diffusion Coefficient, D (cm^2/s)	rms Brownian Displacement in 10s (cm)
0.00037[a]			2.6×10^{-10}	2.5×10^{-7}	9.7×10^{12}	0.18[b]	2.80
0.01	23.04	6.95×10^{-6}	7.1×10^{-9}	7.1×10^{-6}	1.4×10^{10}	5.5×10^{-4}	0.10
0.1	2.866	8.65×10^{-5}	8.8×10^{-8}	8.8×10^{-5}	1.7×10^{8}	6.8×10^{-6}	1.2×10^{-2}
1	1.152	3.48×10^{-3}	3.5×10^{-6}	3.5×10^{-3}	6.8×10^{6}	2.7×10^{-7}	2.3×10^{-3}
10	1.015	3.06×10^{-1}	2.3×10^{-4}	0.23	6.0×10^{5}	2.4×10^{-8}	7.0×10^{-4}
100	1.002	2.61×10^{1}	1.3×10^{-2}	13	5.9×10^{4}	2.4×10^{-9}	2.2×10^{-4}

a. Average diameter of a molecule in air
b. Calculated using Eq. 3-12

the Stokes regime the drag increases faster with diameter (see Fig. 3-2). Similarly, increasing the initial velocity also increases Re_p and results in somewhat less than doubling of the distance.

Since Eq. 3-28 is accurate only in the Stokes regime, the following empirical relationship can be used at higher Re_p (Mercer 1973, 41):

$$S = \frac{\rho_p d_p}{\rho_g}\left[Re_0^{1/3} - \sqrt{6}\arctan\left(\frac{Re_0^{2/3}}{\sqrt{6}}\right)\right],$$
$$1 < Re_p < 400 \qquad (3\text{-}37)$$

where Re_0 is the Reynolds number of the particle at the initial velocity.

Stokes Number

When gas flow conditions change suddenly, as at the particle collection surface of an impactor, the ratio of the stopping distance to a characteristic dimension, d, is defined as the Stokes number, Stk:

$$Stk = \frac{S}{d} \qquad (3\text{-}38)$$

The characteristic dimension depends on the application, e.g., in fibrous filtration it is the diameter of the fiber and in impaction flows it is the radius of the impactor nozzle. For a given percentage of particle removal, the Stokes number value is, therefore, application-specific. For example, the Stokes number of an impactor with one or several identical circular nozzles is

$$Stk = \frac{\rho_p d_p^2 V C_c}{9\eta d_j} \qquad (3\text{-}39)$$

where d_j is the impactor jet diameter in cm and V is the particle velocity in the jet. V is assumed to be equal to the gas velocity in the jet.

Shape Factor

As described above, particle aerodynamic diameter and the Stokes diameter have been defined using ideal spherical particles. Apart from liquid droplets or particles produced from liquid droplets, few particles in nature are spheres. It is convenient to describe more complex shapes by a single diameter and have the additional flow resistance or drag represented by a factor. This dynamic shape factor, χ, is the ratio of the drag force of the particle in question (particle diameter d_p) to that of a sphere of equivalent volume (volume-equivalent diameter d_{ev}). The expression for gravitational settling (Eq. 3-28) thus becomes

$$V_{grav} = \frac{\rho_p d_p^2 g C_c(d_p)}{18\eta} = \frac{d_a^2 g C_c(d_a)}{18\eta}$$
$$= \frac{\rho_p d_{ev}^2 g C_c(d_{ev})}{18\eta\chi}, \quad Re_p < 0.1 \qquad (3\text{-}40)$$

The value of the Cunningham slip factor, C_c, depends on the chosen diameter, d_p, d_a or d_{ev}. The shape factor is always equal to or greater than one. Compact shapes typically have values between one and two, while more extreme shapes, such as fibers and high-volume aggregates, may have larger values. Shape factors are useful for converting a readily measurable equivalent diameter to one that depends on particle behavior, such as aerodynamic diameter or diffusion-equivalent diameter. Thus, shape factors have been defined in a variety of ways that have to do with the available means of measuring the physical and equivalent particle diameter as well as the means of measuring particle drag. Therefore, when applying published shape factors, it may be important to understand the experimental basis for their development.

Some particles have relatively regular shapes with volumes that can be calculated or compact shapes that can be measured with a microscope to determine a volume-equivalent diameter. For such particles, the shape factor is, from Eq. 3-40,

$$\chi = \frac{\rho_p d_{ev}^2 C_c(d_{ev})}{d_a^2 C_c(d_a)}, \quad Re_p < 0.1 \qquad (3\text{-}41)$$

Three variables need to be measured: ρ_p, d_{ev} and d_a. The volume-equivalent diameter may be measured microscopically or determined from the mass (measured chemically or using radioactive tracers) and the number of particles (Barbe-le Borgne et al. 1986). The aerodynamic diameter can be measured in a settling chamber or centrifuge and, if the particle contains no voids, the density is the bulk density of the particle material.

Shape factors have also been measured by settling macroscopic models of regularly shaped particles in liquids. For instance, this technique has been used to measure shape factors for cylinders and chains of spheres (Kasper, Niida, and Yang 1985), for rectangular prisms (Johnson, Leith, and Reist 1986), and for modified rectangular prisms (Sheaffer 1987). These particles have two or three distinct symmetry axes and, therefore, may have two or three shape factors, depending on their orientation. Shape factors have also been derived for oblate and prolate spheroids (Fuchs 1964, 37). Table 3-3 exemplifies a few.

Porous particles, aggregates, and fume particles may have an effective density (ρ_e, including internal voids) that is quite different from the bulk material density, ρ_p. In this case, a shape factor defined as a function of the mass-equivalent diameter (d_m) may be more appropriate (Brockmann and Rader 1990), thus replacing d_{ev} with d_m in Eq. 3-41. The shape factor χ may be further broken down as the product of envelope shape factor, κ, and a second component, δ [defined as $(\rho_p/\rho_e)^{1/3}$], which is due to the porosity of the particle. Theoretically or empirically derived shape factors as described above can be used to match the approximate envelopes of the observed particles. For relatively compact particles, the porosity component of χ dominates, while for more sparse, branched-chain aggregates the envelope factor dominates.

TABLE 3-3 Dynamic Shape Factors for Various Types of Compact Particles (No Internal Voids)

Shape	Dynamic Shape Factor, χ
Sphere	1.00
Cluster of spheres	
2-sphere chain	1.12
3-sphere chain	1.27
4-sphere chain	1.32
Prolate spheroid ($L/D = 5$)[a]	
Axis: horizontal	1.05
Axis: vertical	1.39
Glass fiber ($L/D = 5$)	1.71
Dusts	
Bituminous coal	1.05–1.11
High-ash soft coal	1.95
Quartz	1.36–1.82
Sand	1.57
Talc	2.04
UO_2	1.28
ThO_2	0.99

Source: Davies (1979)

a. Calculated values; all others are experimental

PARTICLE MOTION IN AN ELECTRIC FIELD

The application of electrostatic forces is particularly effective for submicrometer-sized particles, for which gravity forces are weak because of the d_p^3 dependence (Eq. 3-27). On a large scale, removal of aerosols by electrostatic forces is practiced in electrostatic precipitators (also called electrofilters). In aerosol sampling and measuring instruments, electrostatic forces are applied to precipitate or redirect either all aerosol particles or those in a specific size range. Such particle motion caused by an electric field is called *electrophoresis*.

For a particle with a total charge equal to n times the elementary unit of charge, e, the electrostatic force, F_{elec}, in an electric field of intensity E is

$$F_{elec} = neE \tag{3-42}$$

Example 3-5

A 0.5 μm diameter unit density particle has been diffusion-charged with 18 elementary units of charge. Calculate the electrical force on the particle when it passes between two flat parallel plates (e.g., an electrostatic precipitator) which have 5 kV applied across a 2 cm gap. Compare the electrical to the gravitational force.

Answer. Expressed in cgs units, the electric field between the plates is

$$E = \left(\frac{5000\ \text{V}}{2\ \text{cm}}\right)\left(\frac{1\ \text{statV}}{300\ \text{V}}\right) = 8.33\,\frac{\text{statV}}{\text{cm}}$$

Using Eq. 3-42,

$$\begin{aligned} F_{\text{elec}} &= neE \\ &= 18(4.8 \times 10^{-16}\ \text{statC})(8.33\ \text{statV/cm}) \\ &= 7.2 \times 10^{-8}\ \text{dyn} \end{aligned}$$

In SI units, the same calculation is:

$$E = \frac{5000\ \text{V}}{2 \times 10^{-2}\ \text{m}} = 2.5 \times 10^{5}\,\frac{\text{V}}{\text{m}}$$

$$\begin{aligned} F_{\text{elec}} &= neE \\ &= 18(1.6 \times 10^{-16}\ \text{C})(2 \times 10^{5}\ \text{V/m}) \\ &= 7.2 \times 10^{-13}\ \text{N} \end{aligned}$$

One newton (N) in SI units equals 10^5 dyn in the cgs system of units. Using Eq. 3-27,

$$\begin{aligned} F_{\text{grav}} &= \frac{\pi}{6} d_{\text{p}}^{3} \rho_{\text{p}} g \\ &= \frac{\pi}{6}(0.5 \times 10^{-4}\ \text{cm})^{3}\left(\frac{1\ \text{g}}{\text{cm}^{3}}\right)\left(980\,\frac{\text{cm}}{\text{s}^{2}}\right) \\ &= 6.41 \times 10^{-11}\ \text{dyn} \end{aligned}$$

Comparing the two forces,

$$\frac{F_{\text{elec}}}{F_{\text{grav}}} = \frac{7.2 \times 10^{-8}}{6.41 \times 10^{-11}} = 1120$$

The electric force exceeds the gravity force by over a thousand times.

Electrostatic forces can affect particle motion, and to a certain extent, gas motion as well. These forces can be important during particle generation, transport, and measurement. If a particle is placed in an electric field described by Eq. 3-42, it will reach a terminal velocity, V_{elec}, when the field and drag forces are equal:

$$V_{\text{elec}} = \frac{neEC_{\text{c}}}{3\pi\eta d_{\text{p}}} \tag{3-43}$$

The electronic charge, e, is 4.80296×10^{-10} statcoulomb (statC). This terminal, or drift, velocity can also be written in terms of the particle mobility, B, as

$$V_{\text{elec}} = neEB \tag{3-44}$$

or, including the electric charge, in terms of the particle electrical mobility, $Z = neB$, as

$$V_{\text{elec}} = ZE \tag{3-45}$$

where the electrical mobility, Z, is in cm/s per statvolt/cm (or cm^2/statV s), i.e., unit electrical mobility is a drift velocity of 1 cm/s in a 1 statV/cm field. One statV is equal to 300 volts (V).

Example 3-6

Foundry fumes are sampled into an electrostatic precipitator for collection onto an electron microscope grid. A power supply is used to apply a potential of 5000 V across the condenser with a plate spacing, H, of 1 cm. The aerosol flows through the condenser at a uniform velocity of 1 cm/s. The particles of concern have an electrical mobility of 0.01 cm^2/statV s. What is the minimum plate length, L, that will precipitate all of these particles?

Answer. The potential is (5000 V)/(300 V/statV) = 16.7 statV. The precipitation time, t_{e}, in the electric field is

$$\begin{aligned} t_{\text{e}} &= \frac{H}{V_{\text{elec}}} = \frac{H}{ZE} \\ &= \frac{1\ \text{cm}}{\left(0.01\,\frac{\text{cm}^{2}}{\text{statV s}}\right)\left(\frac{16.7\ \text{statV}}{1\ \text{cm}}\right)} = 6\ \text{s} \end{aligned}$$

The transit time, t_{t}, for the air flow at velocity U must equal or exceed this time:

$$t_{\text{t}} = \frac{L}{U} \geq t_{\text{e}}$$

$$L \geq (6\ \text{s})(1\ \text{cm/s}) = 6\ \text{cm}$$

In SI units, the mobility is converted to 3.33 $\times 10^{-9}$ m^2/V s and the spacing is 0.01 m:

$$t_e = \frac{0.01\ \text{m}}{\left(3.33\times10^{-9}\ \dfrac{\text{m}^2}{\text{V s}}\right)\left(\dfrac{5000\ \text{V}}{0.01\ \text{m}}\right)} = 6\ \text{s}$$

The simplest electric field is uniform, e.g., between two large parallel plates

$$E = \frac{\Delta V}{x} \tag{3-46}$$

where x (cm) is the distance between the plates and ΔV is the difference in potential (statV or V). The field between two concentric tubes or between a tube and a concentric wire is also used for electrostatic precipitation. In this case the field is dependent on the distance, r, from the axis:

$$E = \frac{\Delta V}{r \ln(r_o/r_i)} \tag{3-47}$$

where ΔV is the difference in potential between the outer and the inner tube (or wire) of radius r_o and r_i, respectively.

The force on each of two particles with n_1 and n_2 unit charges on them is described by Coulomb's law. In cgs units this is

$$F_{\text{elec}} = \frac{n_1 n_2 e^2}{r^2} \tag{3-48}$$

where r (in cm) is the distance between the particles. This equation applies strictly only to point charges. However, it is a good approximation for the force between two particles or a particle at some distance from a charged object, such as a sampler, and indicates that the force drops off rapidly with distance. Aerosol particles, which typically carry a limited amount of charge because of their small surface area, are, in general, affected electrically only when they are quite close to another charged particle or close to a charged object.

In SI units, Eq. 3-48 is converted to give the force in newtons (N) as

$$F_{\text{elec}} = \frac{n_1 n_2 e^2}{4\pi\varepsilon_0 r^2} \tag{3-49}$$

where r is in meters (m), the electronic charge is 1.602×10^{-19} coulomb (C), and ε_0 is the permittivity constant (8.854×10^{-12} C^2/N m^2 for air or vacuum).

PARTICLE MOTION IN OTHER FORCE FIELDS

Particle motion is governed by a variety of other forces. Very small particles approach the behavior of the molecules of the surrounding gas, i.e., they diffuse readily and have little inertia and they can be affected by light pressure, acoustic pressure, and thermal pressure. In a fashion similar to gravitational and electrical forces, other forces can be used to cause particle motion and, thus, size-selective measurement. These same forces can also cause particles to be lost rapidly in the sampling inlet or on measurement instrument surfaces. Other forces not mentioned may have some effects but are generally much weaker than the ones mentioned here. For instance, magnetic forces are typically several orders of magnitude smaller than electrostatic forces, but have been used for fiber alignment.

Thermophoresis

Particles in a thermal gradient are bombarded more strongly by gas molecules on the hotter side and are, therefore, forced away from a heat source. Thus, heated surfaces tend to remain clean, while relatively cool surfaces tend to collect particles. For particles smaller than the mean free path (λ), the thermophoretic velocity, V_{th}, is independent of particle size and is (Waldmann and Schmitt 1966)

$$V_{\text{th}} = \frac{0.55\eta}{\rho_g} \nabla T, \quad d_p < \lambda \tag{3-50}$$

where ∇T is the thermal gradient in K/cm. There is a slight increase (of the order of 3%) in the velocity of rough-surfaced particles vs. spherical solids or droplets.

For particles larger than λ, the thermophoretic velocity depends on the ratio of the thermal conductivity of the gas to that of the particle and also on the particle size. For large conductive aerosol particles, the thermophoretic velocity may be about 5 times lower than for small, nonconductive ones. To calculate the thermophoretic velocity, the molecular accommodation coefficient (H) is needed:

$$H \cong \left(\frac{1}{1 + 6\lambda/d_p}\right)\left(\frac{k_g/k_p + 4.4\lambda/d_p}{1 + 2k_g/k_p + 8.8\lambda/d_p}\right) \quad (3\text{-}51)$$

where k_g and k_p are the thermal conductivities of gas and particle, respectively. The thermal conductivity of air is 5.6×10^{-5} cal/cm s K, while that for particles ranges from 0.16 for a metal (iron) to 1.9×10^{-5} cal/cm s K for an insulator (asbestos) (Mercer 1973, 166). The thermally induced particle velocity is then (Waldmann and Schmitt 1966)

$$V_{th} = \frac{-3\eta H}{2\rho_g T}\nabla T, \quad d_p > \lambda \quad (3\text{-}52)$$

Thermophoresis is relatively independent of particle size over a wide range and has been used for collecting small samples, such as for electron microscope measurements, in thermal precipitators. The sampling rate of these instruments is low because of the difficulty of maintaining a thermal gradient and, thus, thermal precipitators have not been scaled up for large volume use.

Photophoresis

Photophoresis is similar to thermophoresis in that particle motion is caused by thermal gradients at the particle surface except that in this case the heating is caused by light absorption by the particle rather than by an external source. Light shining on a particle may be preferentially absorbed by the side nearer to the light source or, under certain circumstances of weak absorption and focusing, by the far side of the particle. Thus, in the former case the particle will be repelled from the light source while in the latter, called reverse photophoresis, it will be attracted.

Electromagnetic Radiation Pressure

Electromagnetic radiation can have a direct effect on particle motion by transferring momentum to the particle. Light impinging on a particle can be reflected, refracted or absorbed. The fraction of momentum transfer from the light beam to the particle depends on the geometric cross section of the particle as well as on the average direction of the scattered light. If a significant fraction of the light is absorbed by the particle, photophoresis, as described above, will be more important in deciding particle motion. Radiation pressure has been used to trap particles in focused laser beams and manipulate them for further study.

Acoustic Pressure

Acoustic waves, either stationary, as in a resonant box, or traveling in open space can be reflected, diffused, or absorbed by particles. Particle motion in an acoustic field includes oscillation in response to the gas motion, circulation in the acoustic field, or net drift in some direction. Such waves have been used to increase particle coagulation or agglomeration or, in other cases, to enhance droplet evaporation or condensation (Hesketh 1977, 97). A resonant acoustic system has also been used to measure particle aerodynamic diameter by measuring a particle's ability to oscillate in response to the air motion (Mazumder et al. 1979).

Diffusiophoresis and Stephan Flow

When the suspending gas differs in composition from one location to another, diffusion of the gas takes place. This gas diffusion results in suspended particles acquiring a net velocity as a function of the gas diffusion, i.e., diffusiophoresis. The particles are pushed in the direction of the larger molecule flow. The force is a function of the molecular weight and diffusion coefficients of the diffusing gases and is largely independent of particle size.

A special case of diffusiophoresis occurs near evaporating or condensing surfaces. A net flow of the gas–vapor mixture away from an evaporating surface is set up that creates a drag on particles. The converse situation holds for a condensing surface, i.e., gas and particles will flow towards the surface. This net motion of the gas–vapor mixture is called the Stephan (also spelled Stefan) flow and can cause the motion of particles near these surfaces (Fuchs 1964, 67). The Stephan flow can affect particle collection in industrial scrubbers and scavenging of the environment by growing cloud droplets. In order to increase particle collection by the Stephan flow, the vapor must be supersaturated. Diffusiophoretic velocities are generally only significant for very small particles. For instance, diffusiophoresis of 0.005–0.05 μm diameter particles was found to have the following net deposition velocity, V_{diff}, towards surfaces condensing water vapor (Goldsmith and May 1966):

$$V_{\mathrm{diff}} = 1.9 \times 10^{-3} \frac{\mathrm{d}P}{\mathrm{d}x} \tag{3-53}$$

where the deposition velocity is in cm/s and $\mathrm{d}P/\mathrm{d}x$ is the pressure gradient of the diffusing vapor in kPa/cm. Note that in condensing and evaporating droplets, thermophoretic effects can also be important.

References

Allen, M. D. and O. G. Raabe. 1985. Slip correction measurements of spherical solid aerosol particles in an improved Millikan apparatus. *Aerosol Sci. Technol.* 4:269–86.

Barbe-le Borgne, M., D. Boulaud, G. Madelaine, and A. Renoux. 1986. Experimental determination of the dynamic shape factor of the primary sodium peroxide aerosol. *J. Aerosol Sci.* 17:79–86.

Brockmann, J. E. and D. J. Rader. 1990. APS response to nonspherical particles and experimental determination of dynamic shape factor. *Aerosol Sci. Technol.* 13:162–72.

Clift, R., J. R. Grace, and M. E. Weber. 1978. *Bubbles Drops and Particles.* New York: Academic Press.

Davies, C. N. 1945. Definitive equations for the fluid resistance of spheres. *Proc. Phys. Soc.* 57:259.

Friedlander, S. K. 1977. *Smoke, Dust and Haze.* New York: Wiley.

Fuchs, N. 1964. *The Mechanics of Aerosols.* Oxford: Pergamon. (Reprinted in 1989, Mineola, NY: Dover.)

Goldsmith, P. and F. G. May. 1966. In *Aerosol Science*, ed. C. N. Davies. London: Academic Press.

Hesketh, H. E. 1977. *Fine Particles in Viscous Media.* Ann Arbor: Ann Arbor Science Publishers.

Hinds, W. C. 1982. *Aerosol Technology.* New York: Wiley.

Johnson, D. L., D. Leith, and P. C. Reist. 1987. Drag on non-spherical, orthotropic aerosol particles. *J. Aerosol Sci.* 18:87–97.

Kasper, G., T. Niida, and M. Yang. 1985. Measurements of viscous drag on cylinders and chains of spheres with aspect ratios between 2 and 50. *J. Aerosol Sci.* 16:535–56.

Licht, W. 1988. *Air Pollution Control Engineering: Basic Calculations for Particulate Collection.* New York: Marcel Dekker.

Mazumder, M. K., R. E. Ware, J. D. Wilson, R. G. Renninger, F. C. Hiller, P. C. McLeod, R. W. Raible, and M. K. Testerman. 1979. SPART analyzer: Its application to aerodynamic size measurement. *J. Aerosol Sci.* 10:561–69.

Mercer, T. T. 1973. *Aerosol Technology in Hazard Evaluation.* New York: Academic Press.

Rader, D. J. 1990. Momentum slip correction factor for small particles in nine common gases. *J. Aerosol Sci.* 21:161–68.

Sartor, J. D. and C. E. Abbott. 1975. Prediction and measurement of the accelerated motion of water drops in air. *J. Appl. Meteorol.* 14(2):232–39.

Schlichting, H. 1979. *Boundary-Layer Theory.* New York: McGraw Hill.

Sheaffer, A. W. 1987. Drag on modified rectangular prisms. *J. Aerosol Sci.* 18:11–16.

Tsai, C. J. and D. Y. H. Pui. 1990. Numerical study of particle deposition in bends of a circular cross-section—laminar flow regime. *Aerosol Sci. Technol.* 12:813–31.

Waldmann, L. and K. H. Schmitt. 1966. Thermophoresis and diffusiophoresis of aerosols. In *Aerosol Science*, ed. C. N. Davies. London: Academic Press.

White, F. M. 1986. *Fluid Mechanics.* New York: McGraw-Hill.

Willeke, K. 1976. Temperature dependence of particle slip in a gaseous medium. *J. Aerosol Sci.* 7:381–87.

4

Physical and Chemical Changes in the Particulate Phase

William C. Hinds

Department of Environmental Health Sciences
UCLA School of Public Health
Los Angeles, CA, U.S.A.

INTRODUCTION

Aerosols, by their nature, are somewhat unstable in the sense that concentration and particle properties change with time. These changes can be the result of external forces, such as the loss of larger particles by gravitational settling, or they may be the result of physical and chemical processes that serve to change the size or composition of the particles. This chapter addresses the latter category of processes. All these processes involve mass transfer to or from the particle. This transfer may be the result of molecular transfer between the particle and the surrounding gas, for example, condensation, evaporation, nucleation, adsorption, absorption, and chemical reaction, or it may result from interparticle transfer, such as coagulation.

Processes that cause physical or chemical changes in the particulate phase influence the particle size distribution of nearly all aerosols. These processes contribute in an essential way to the earth's hydrological cycle. They are involved in the formation of photochemical smog and are the key to shaping the atmospheric aerosol size distribution. These processes play a significant role in many occupational aerosol exposures and in the operation of the condensation nuclei counters, described in Chapter 19. They are central to industrial aerosol processing and for the generation of test aerosols.

Condensation, thermal coagulation, and adsorption are related processes that rely on the diffusion of molecules or particles to a particle's surface. Evaporation is the opposite of condensation and is governed by the same laws. Reactions may be nongrowth processes that change the composition or density of an aerosol particle with little or no change in particle size. Because the processes discussed in this chapter are related and may be going on simultaneously, it is necessary to look at each process separately to obtain an accurate picture of the changes that occur. Furthermore, these processes depend on particle size in a complex way, so we adopt a single-particle approach for much of the analysis that follows. We rely on the concepts of mean free path and diffusion coefficient, which were defined in Chapter 3.

Definitions

The partial pressure of a vapor is a way of expressing the concentration of that vapor in a volume of gas. It is the pressure the vapor would exert if it were the only component present. This pressure, expressed as a fraction of the ambient pressure, is the fractional concentration of vapor. Air at 20°C (68°F) and 50% relative humidity has a partial pressure of water vapor of 8.8 mmHg, which means that the air–water mixture is 8.8/760 = 1.2% water vapor on a volume basis.

The vapor pressure or saturation vapor pressure is a unique property for any liquid at a given temperature. It represents the minimum partial pressure of that liquid's vapor that must be maintained at the gas–liquid interface to prevent evaporation. This is a condition required for mass equilibrium: no net transfer of molecules at the liquid surface, that is, no net condensation or evaporation. Vapor pressure as defined here is for a flat liquid surface but, as will be explained below, a slightly greater partial pressure is required to maintain mass equilibrium around an aerosol particle. The partial pressure of vapor in a sealed chamber containing a liquid will eventually reach the vapor pressure of the liquid at the temperature of the container.

The vapor pressure of water in mmHg at a temperature T in K is given by

$$p_s = 10^{8.11 - 1750/(T-38)} \quad \text{for } T = 273\text{–}330 \text{ K} \tag{4-1}$$

For aerosol condensation and evaporation processes, it is the ratio of the partial pressure of vapor to the saturation vapor pressure that is important. This ratio is called the saturation ratio S_R. When the saturation ratio is equal to one the mixture is described as saturated; when it is greater than one the mixture is supersaturated; and when less than one it is unsaturated.

Nucleation or nucleated condensation refers to the process of initial formation of a particle from vapor. This process is usually facilitated by the presence of small particles, called condensation nuclei, that serve as sites for condensation.

Adsorption is the process whereby vapor molecules attach to solid surfaces. It is most important for porous solids, such as activated charcoal, that have large surface areas. Absorption refers to the process of vapor molecules transferring from the gas phase to the liquid phase.

For aerosol particles, condensation occurs when more vapor molecules arrive at a particle's surface than leave it. It results in a net growth of the particle. Evaporation is the reverse of condensation, and results in a net loss of molecules and a shrinkage of the particle.

Example 4-1

Saturated air coming from the ocean at 20°C is carried by air currents up the side of a mountain to an altitude of 1 km. Assuming this represents adiabatic expansion to a pressure of 670 mm Hg, what would be the saturation ratio of this air mass if no condensation occurred?

Answer. The absolute temperature change as a result of an adiabatic expansion of saturated air from p_1 to p_2 is given by

$$T_2 = T_1\left(\frac{p_2}{p_1}\right)^{0.28} = 293\left(\frac{670}{760}\right)^{0.28} = 283 \text{ K}$$

At 283 K the saturation vapor pressure for water is given by Eq. 4-1:

$$p_s = 10^{8.11 - 1750/(283-38)} = 9.3 \text{ mm Hg}$$

The saturation ratio is the ratio of the actual partial pressure of vapor, 17.6 mm Hg, to the saturation vapor pressure for the ambient temperature, 9.3 mm Hg.

$$S_R = \left(\frac{17.6}{9.3}\right) = 1.89$$

The Kelvin Effect

Vapor pressure has been defined as the partial pressure required for mass equilibrium (no net evaporation or condensation) for a flat liquid surface. Because liquid aerosol particles have a sharply curved surface, a greater partial pressure is required to maintain mass equilibrium at a given temperature than for a flat liquid surface. This increase in the partial pressure of vapor required increases with decreasing particle size. This effect is called the Kelvin effect. The saturation ratio required for mass equilibrium (no net condensation or evaporation) for a droplet of diameter d_p is given by the Kelvin equation:

$$S_R = e^{4\gamma M/\rho_p R T d_p} \tag{4-2}$$

where γ, M, and ρ_p are the surface tension, molecular weight, density of the liquid, and R is the gas constant. Thus, 0.1 and 0.01 μm diameter water droplets require an environment with a saturation ratio of at least 1.022 and 1.24, respectively, to prevent evaporation. Evaporation will occur if the saturation ratio is less than that given by Eq. 4-2, even if the saturation ratio is greater than one. Likewise, if the saturation ratio is greater than that required by the Kelvin equation, then condensation and growth will occur. For a given supersaturation, the minimum droplet size required to prevent evaporation is given by Eq. 4-2 and is referred to as the Kelvin diameter for that condition.

CONDENSATION

Growth Rate

When a droplet of pure liquid is in a supersaturated environment that exceeds the requirement given by the Kelvin equation, the droplet grows by condensation of vapor on its surface. The rate of growth depends on the saturation ratio and the particle size. It is controlled by the rate of arrival of vapor molecules at the droplet surface. Initially, the droplet diameter will usually be less than the mean free path, λ, of the surrounding gas, (0.066 μm under standard conditions, see Chapter 3) and the rate of arrival of vapor molecules is governed by the kinetic theory of gases. The growth rate, i.e., the rate of increase in droplet diameter, is given by Hinds (1982) as

$$\frac{d(d_p)}{dt} = \frac{2(p - p_d)}{\rho_p \sqrt{2\pi R T/M}} \quad \text{for } d_p < \lambda \tag{4-3}$$

where p is the partial pressure of vapor in the neighborhood of the droplet and p_d is the partial pressure of vapor at the droplet surface, as given by the Kelvin equation. The application of Eq. 4-3 to obtain the growth rate in cm/s requires that pressure be expressed in dyn/cm^2 (note that mmHg × 1330 = dyn/cm^2), density of the liquid in g/cm^3, temperature in K, molecular weight in g/mol, and the gas constant R as 8.31×10^7 dyn cm/K mol.

Once a droplet's size exceeds the dimensions of the mean free path, the rate of arrival of vapor molecules is governed by the rate of molecular diffusion to the droplet surface. Under these conditions, the rate of growth is given by

$$\frac{d(d_p)}{dt} = \frac{4 D_v M (p - p_s)}{\rho_p d_p R T} \quad \text{for } d_p > \lambda \tag{4-4}$$

where D_v is the diffusion coefficient of the vapor molecules (0.24 cm^2/s for water vapor at 20°C) and p_s is given by Eq. 4-1.

Equations 4-3 and 4-4 apply only to pure materials, i.e., single-component liquids without any dissolved salts or impurities. Note that the growth rate for droplets less than the mean free path is independent of droplet size, but is inversely proportional to droplet size for droplets larger than the mean free path.

Time Required for Growth

The time required for a droplet to grow from d_1 to d_2 can be obtained by integrating Eq. 4-4 over the size limits:

$$t = \frac{\rho_p R T (d_2^2 - d_1^2)}{8 D_v M (p - p_s)} \quad \text{for } d_1 > \lambda \tag{4-5}$$

Example 4-2

What is the rate of growth by condensation for 0.05, 0.5, and 5 μm water droplets at a saturation ratio of 1.08 and 20°C?

Answer. Since 0.05 μm is less than the mean free path (0.066 μm), we can use Eq. 4-3:

$$\frac{d(d_t)}{dt} = \frac{2(p - p_d)}{\rho_p \sqrt{2\pi R T/M}}$$

with $p = 1.08 \times p_s$, where p_s is given by

Eq. 4-1 for $T = 273 + 20 = 293$ K, i.e., $p_s = 10^{8.11-1750/(293-38)} = 17.6$ mm Hg $= 1330 \times 17.6 = 23{,}400$ dyn/cm^2. Therefore, $p = 1.08 \times 23{,}400 = 25{,}300$ dyn/cm^2. p_d is given by the Kelvin equation, Eq. 4-2,

$$p_d = p_s e^{4\gamma M/\rho_p RTd}$$

where

$$\begin{aligned} \gamma &= 72.7 \text{ dyn/cm} \\ M &= 18 \text{ g/mol} \\ \rho_p &= 1.0 \text{ g/cm}^3 \\ R &= 8.31 \times 10^7 \text{ dyn cm/K mol} \\ T &= 293 \text{ K} \\ d &= 5 \times 10^{-6} \text{ cm} \end{aligned}$$

Substituting these values in the expression for p_d gives

$$p_d = 23{,}400 \exp\left(\frac{4(72.7)\,18}{1(8.31 \times 10^7)293(5 \times 10^{-6})}\right)$$

$$= 24{,}400 \text{ dyn/cm}^2$$

Finally, substituting into the original equation gives

$$\frac{d(d_t)}{dt} = \frac{2(25{,}300 - 24{,}400)}{1\sqrt{2\pi(8.31 \times 10^7)293/18}}$$

$$= 0.020 \text{ cm/s} = 200\ \mu\text{m/s}$$

Since 0.5 µm droplets are larger than the mean free path, we can use Eq. 4-4:

$$\frac{d(d_p)}{dt} = \frac{4D_v M(p - p_s)}{\rho_p d_p RT}$$

where $D_v = 0.24$ cm^2/s for water vapor. Substituting this along with the values given above for other quantities, we obtain

$$\frac{d(d_p)}{dt} = \frac{4(0.24)\,18(25{,}300 - 23{,}400)}{1(5 \times 10^{-5})(8.31 \times 10^7)(293)}$$

$$= 0.027 \text{ cm/s} = 270\ \mu\text{m/s}$$

Finally, for 5 µm droplets everything is the same as in the preceding calculation, except that the particle size is ten times greater; consequently, the growth rate is one-tenth that given above or

$$\frac{d(d_p)}{dt} = 0.0027 \text{ cm/s} = 27\ \mu\text{m/s}$$

NUCLEATION

Homogeneous

The preceding section describes the growth process for pure materials once the droplets have been formed. The initial formation of the droplet from vapor is a more complicated process. Droplets can be formed in the absence of condensation nuclei. This process, called homogeneous nucleation or self-nucleation, requires large saturation ratios, usually in the range of 2–10, which normally occur only in special laboratory or chemical process situations. Pure water vapor at 20°C and at a saturation ratio of 3.5 or greater forms droplets spontaneously by homogeneous nucleation. This corresponds to a Kelvin diameter of 0.0017 µm and suggests that molecular clusters of about 90 molecules are necessary for this process. A detailed description of homogeneous nucleation is given by Springer (1978).

Heterogeneous

The more common formation mechanism is nucleated condensation or heterogeneous nucleation. This process relies on existing submicrometer particles, called condensation nuclei, to serve as sites for condensation. Our natural atmosphere contains thousands of these nuclei in each cubic centimeter of air. To a first approximation, insoluble nuclei serve as passive sites on which condensation occurs for supersaturated conditions. Under supersaturated conditions, a solid nucleus will have on its surface an adsorbed layer of vapor molecules. If the nucleus has a diameter greater than the Kelvin diameter for a particular condition of supersaturation, the nucleus "looks like" a droplet to the surrounding vapor molecules, and vapor will condense on its surface. Once condensation starts, droplet growth can be described by Eqs. 4-3 and 4-4.

The situation with soluble nuclei is more complex and more important. Our normal atmosphere contains large numbers of soluble nuclei, formed as the solid residue left behind after the water has evaporated from a droplet containing dissolved material. Many are sodium chloride nuclei formed from droplets of seawater created by the action of waves and bubbles in the oceans. Because these soluble nuclei have a strong affinity for water they facilitate the initial formation of droplets and enable their growth to occur at lower saturation ratios than would be the case for insoluble nuclei.

Because of the complex effect the presence of dissolved salt has on the rate of growth of a droplet, Eqs. 4-3 and 4-4 cannot be used to determine growth rates for such droplets. The stabilization time for droplets containing salt is described by Ferron and Soderholm (1990). In general, dissolved salts increase the rate of growth and decrease the rate of evaporation compared to that for pure liquids. As a droplet grows by the addition of water vapor the concentration of salt becomes increasingly dilute. Consequently, it is convenient to characterize the amount of salt in a droplet not by its concentration but by the mass of salt in the droplet a quantity that remains constant during condensation or evaporation processes. The mass of salt is also equal to the mass of the original nucleus upon which the droplet formed.

When a dissolved salt is present in a droplet there are two competing effects at work as the droplet evaporates or grows. As the droplet evaporates, the concentration of salt increases, because only the water leaves. This enhances the affinity of the dissolved salt to hold water in the droplet. The other effect is the Kelvin effect, which results in an increase in the equilibrium vapor pressure required for a droplet as it decreases in size. The relationship between saturation ratio and particle size for droplets containing dissolved salts is illustrated in Fig. 4-1 by the three lines labeled with their indicated mass of dissolved salt.

Equilibrium Conditions

As with pure materials, the region above a given curve in Fig. 4-1 represents a growth region and that below the curve represents an evaporation region. Thus, if the saturation ratio is greater than 1.002, any droplet (or

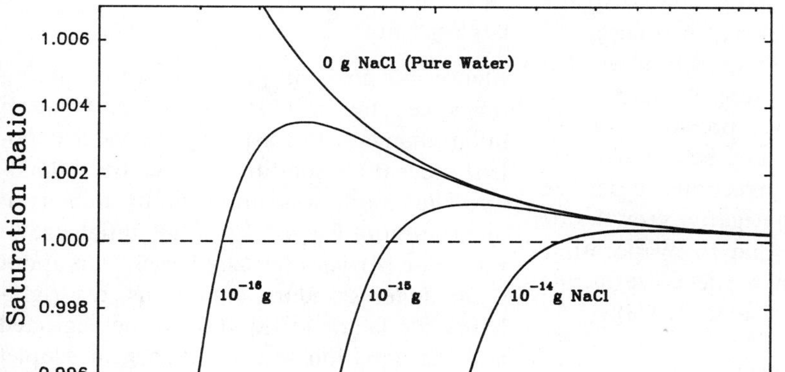

FIGURE 4-1. Saturation Ratio versus Droplet Size for Pure Water and Droplets Containing the Indicated Mass of Sodium Chloride at 20°C. (Adapted from Hinds 1982.)

nuclei) with more than 10^{-15} g of sodium chloride will grow to a large droplet, although its growth rate will slow down as it gets larger, as predicted by Eq. 4-4. When environmental conditions and particle size give a location on Fig. 4-1 that is below and to the left of the peak for a given curve, the droplet will either grow or evaporate until it reaches the curve. This portion of the line represents a true equilibrium region and the droplet will remain at that size as long as environmental conditions stay constant. This is true even if the saturation ratio is less than 1.0. Thus, there are a large number of particles in the atmosphere that will experience an increase in their size with an increase in relative humidity and a decrease with a decrease in relative humidity. The line for pure water does not have this type of equilibrium region. It represents only a demarcation between the growth (above) and evaporation (below) regions. As droplets continue to grow, the concentration of dissolved salts decreases, eventually reaching a point where the droplets behave similar to pure water, and their curves in Fig. 4-1 merge with that for pure water.

EVAPORATION

Rate of Evaporation

The process of evaporation of a pure liquid droplet (no dissolved salts) is similar to the process of growth, except that it proceeds in the opposite direction. Evaporation will occur when the partial pressure of vapor is less than the saturated vapor pressure ($p < p_s$). The rate of evaporation can be predicted by Eq. 4-4 with proper corrections, described below, and noting that a negative growth rate represents a shrinking due to evaporation. The rate of decrease in particle size during evaporation is given by Davies (1978):

$$\frac{d(d_p)}{dt} = \frac{4D_v M}{R\rho_p d_p}\left(\frac{p_\infty}{T_\infty} - \frac{p_d}{T_d}\right) \times \left(\frac{2\lambda + d_p}{d_p + 5.33(\lambda^2/d_p) + 3.42\lambda}\right) \quad (4\text{-}6)$$

where the subscript ∞ refers to conditions removed from the particle and the subscript d refers to conditions right at the particle surface. The last factor in Eq. 4-6 corrects for complications in the calculation of mass transfer by diffusion within one mean free path of the particle surface. This correction is known as the Fuchs correction. This factor can be omitted with little error for evaporating droplets larger than about 2 μm.

For volatile particles such as water or alcohol, a correction must be included for the cooling of the droplet due to the rapid evaporation. This self-cooling effect reduces the droplet temperature and, consequently, the partial pressure at the droplet surface. Figure 4-2 gives the reduction in droplet temperature for a range of ambient temperatures and relative humidities. The quantities p_d and T_d must be evaluated at the cooler conditions prevailing at the droplet surface. The self-cooling effect is independent of particle size.

For particles larger than about 50 μm, an additional correction must be included to account for the disruption in the diffusion of vapor away from the droplet surface caused by the settling of the droplet. This effect increases the rate of evaporation of 50 and 100 μm droplets by 10% and 31%, respectively. More detailed information on this effect is given by Davies (1978) and Fuchs (1959).

Drying Time

Figure 4-3 gives droplet lifetimes or drying times, i.e., the time for evaporation from an initial diameter to zero for pure water droplets under three conditions of relative humidity. The graph was obtained by numerical integration of Eq. 4-6 from the initial size to zero. For particles initially larger than about 2 μm under standard conditions, the right-hand-side factor in Eq. 4-6 can be neglected and the equation integrated to give droplet lifetimes:

$$t = \frac{R\rho_p d_p^2}{8D_v M\left(\frac{p_d}{T_d} - \frac{p_\infty}{T_\infty}\right)} \quad \text{for initial } d_p > 2\ \mu\text{m} \quad (4\text{-}7)$$

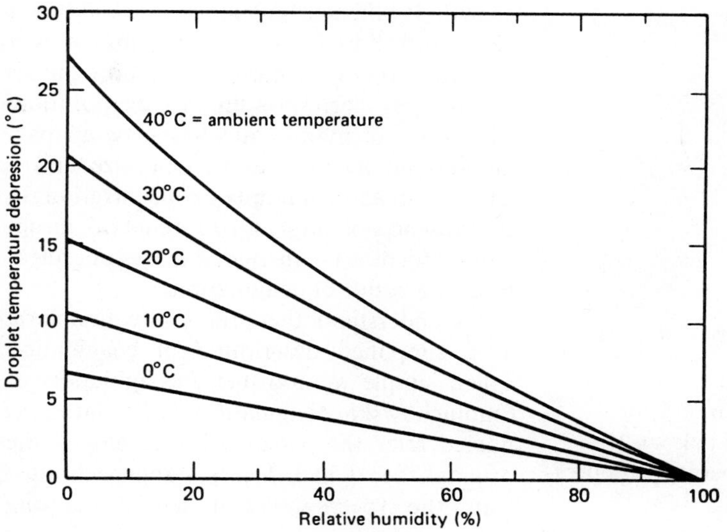

FIGURE 4-2. Water Droplet Temperature Depression versus Relative Humidity. (Source: Hinds 1982. Reprinted with the Permission of John Wiley & Sons, Inc.)

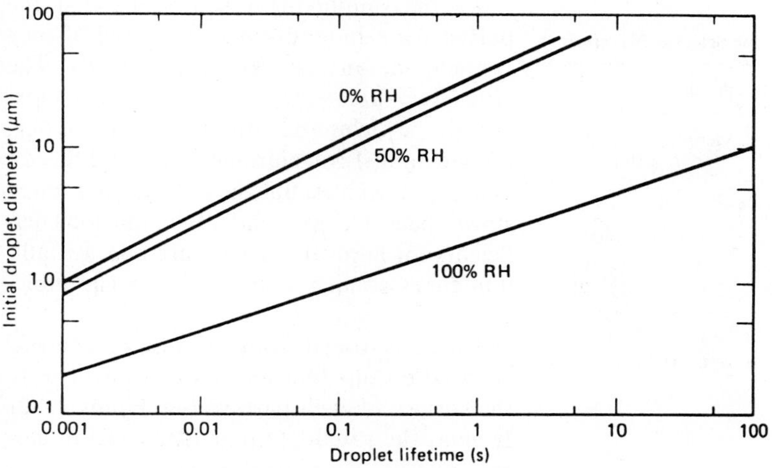

FIGURE 4-3. Drying Times for Pure Water Droplets at 20°C. (Source: Hinds 1982. Reprinted with the Permission of John Wiley & Sons, Inc.)

Table 4-1 gives droplet lifetimes for four materials under standard conditions. It illustrates the wide range of droplet lifetimes for different materials. The effect of material properties on droplet lifetime is, to a first approximation, proportional to $\rho_p/D_v M p_s$.

Example 4-3

Water spray droplets 60 μm in diameter are sprayed into 50% relative humidity air at 20°C. How long before they evaporate completely?

Answer. Use Eq. 4-7:

$$t = \frac{R\rho_p d^2}{8D_v M\left(\frac{p_d}{T_d} - \frac{p_\infty}{T_\infty}\right)}$$

where

$R = 82.1$ atm cm^3/K mol
$\rho = 1.0$ g/cm^3
$d = 6 \times 10^{-3}$ cm
$D_v = 0.24$ cm^2/s
$M = 18$
$p_d = 11.3/760 = 0.149$ atm (Eq. 4-1 for $T = 286$, see below)
$T_d = 293 - 7 = 286$ K (temperature depression from Fig. 4-2)
$p_\infty = 0.5 \times 17.6/760 = 0.0116$ atm
$T_\infty = 293$ K

$$t = \frac{82.1 \times 1(6 \times 10^{-3})^2}{8 \times 0.24 \times 18(0.0149/286 - 0.0116/293)} = 6.8 \text{ s}$$

TABLE 4-1 Droplet Lifetimes for Selected Materials[1]

Initial Droplet Diameter (μm)	Droplet Lifetime (s)			
	Ethyl Alcohol	Water	Mercury	Dioctyl Phthalate
0.01	4×10^{-7}	2×10^{-6}	0.005	1.8
0.1	9×10^{-6}	3×10^{-5}	0.3	740
1	3×10^{-4}	0.001	1.4	3×10^4
10	0.03	0.08	1200	2×10^6

Source: Hinds (1982)

1. Calculated by Eq. 4-6 for vapor-free air at 20°C

COAGULATION

Coagulation is an aerosol growth process that results from the collision of aerosol particles with each other. If the collisions are the result of Brownian motion, the process is called thermal coagulation; if they are the result of motion caused by external forces, the process is termed kinematic coagulation. Thermal coagulation is in some ways analogous to growth by condensation except that it is the other particles diffusing to a particle's surface rather than molecules that cause the growth. It differs from condensation in that a supersaturation is not required and it is a one-way process of growth with no equivalent process corresponding to evaporation. The result of many collisions between particles is an increase in particle size and a decrease in aerosol number concentration. In the absence of any loss or removal mechanisms there is no change in mass concentration as a result of coagulation.

To understand the process we first look at a simplified description of coagulation called simple monodisperse coagulation or Smoluchowski coagulation. The latter is named after the person who developed the original theory in 1917. This approach illustrates the process well, is useful for analyzing many situations, and is the basis for further refinements.

Simple Monodisperse Coagulation

For simple monodisperse coagulation we make the simplifying assumptions that the particles are monodisperse, will stick if they contact one another, and grow slowly. The latter two are valid assumptions for most aerosol particles and situations. Aerosol particles exhibit Brownian motion and diffusion like gas molecules, but these occur at a much slower pace. Consequently, the diffusion coefficients for aerosol particles are about a million times smaller than those for gas molecules.

The derivation developed by Smoluchowski is based on the diffusion of other particles to the surface of each particle (see Hinds 1982). It gives the rate of change (decrease) in aerosol number concentration as

$$\frac{dN}{dt} = -KN^2 \tag{4-8}$$

where N is the particle number concentration and K is the coagulation coefficient. For particles larger than the gas mean free path, K is given by

$$K = 4\pi d_p D = \frac{4kTC_c}{3\eta} \tag{4-9}$$

TABLE 4-2 Coagulation Coefficients for Selected Particle Sizes at 20°C

Particle Diameter (μm)	Coagulation Coefficient (cm^3/s)
0.05	15×10^{-10}
0.1	8.6×10^{-10}
0.5	4.0×10^{-10}
1	3.5×10^{-10}
5	3.1×10^{-10}

Source: Hinds (1982)

where D is the particle diffusion coefficient, η is the gas viscosity in g/cm s, and k is Boltzmann's constant, 1.38×10^{-16} dyn cm/K. The coagulation coefficient has units of cm^3/s when number concentration is expressed in particles/cm^3. The coagulation coefficient is only slightly dependent on particle size, being proportional only to slip correction factor C_c. Table 4-2 gives coagulation coefficients for different size particles under standard conditions. In the usual situation, the extent of particle size increase is sufficiently limited that the coagulation coefficient can be considered a constant, and the rate of coagulation is proportional to number concentration squared. Thus, coagulation is a rapid process at high number concentrations and a slow one at low concentrations.

As a practical matter, the net effect of coagulation over some period of time is a more useful quantity to know than the rate of coagulation. The change in number concentration over a period of time t is obtained by integrating Eq. 4-8 to get

$$N(t) = \frac{N_0}{1 + N_0 K t} \qquad (4\text{-}10)$$

where $N(t)$ is the number concentration at time t and N_0 is the initial number concentration. Number concentration must be expressed in particles/cm^3 and K in cm^3/s.

As number concentration decreases, particle size increases, but, for a contained system with no losses, particle mass will remain constant. If number concentration decreases to one-half of its original value then the same mass (and volume) will be contained in half as many particles, so each particle will have twice its original mass (and volume). For liquid particles, particle size is proportional to the cube root of particle volume, and, consequently, it is also proportional to the inverse cube root of number concentration:

$$d(t) = d_0 \left(\frac{N_0}{N(t)} \right)^{1/3} \qquad (4\text{-}11)$$

Thus, an eightfold reduction in particle number concentration results in a doubling of particle size. Equations 4-10 and 4-11 can be combined to give a more direct expression for the change in particle size due to coagulation over a period of time t:

$$d(t) = d_0 (1 + N_0 K t)^{1/3} \qquad (4\text{-}12)$$

Equations 4-11 and 4-12 are correct for liquid droplets and approximately correct for solid particles that form compact clusters. Table 4-3 gives the time required for various initial concentrations to reach one-half their number concentration and the time for the particle size to double. It is apparent from Table 4-3 that whether or not coagulation can be neglected depends on the time scale under consideration. Thus, over a period of a few minutes, coagulation is important only if

TABLE 4-3 Time Required for Selected Coagulation Processes[1]

Initial Number Concentration (cm^{-3})	Time for Number Concentration to Halve (s)	Time for Particle Size to Double (s)
10^{12}	0.002	0.014
10^{10}	0.2	1.4
10^{8}	20	140
10^{6}	2000 (33 min)	14,000 (4 h)
10^{4}	200,000 (55 h)	1,400,000 (16 d)

Source: Hinds (1982)

1. Assumes simple monodisperse coagulation with $K = 5 \times 10^{-10}\ cm^3/s$

particle number concentration exceeds $10^6/\text{cm}^3$.

Polydisperse Coagulation

The previous description of coagulation is accurate enough for a wide variety of situations, but it requires the assumption of a monodisperse aerosol. In the real case, we usually have a polydisperse aerosol and the situation is more complicated. Because the coagulation process is governed by the rate of diffusion of particles to the surface of each particle, the process is enhanced when small particles with their high diffusion coefficients diffuse to a large particle with its large surface. A tenfold difference in particle size produces a threefold increase in coagulation, and a 100-fold difference results in a more than 25-fold increase in coagulation rate. The use of Eq. 4-10 or Eq. 4-12 for polydisperse aerosols requires the use of numerical methods because the coagulation for every combination of particle sizes has a different value of K and has to be calculated separately (Zebel 1966). For the case of coagulation of an aerosol with a lognormal size distribution having a count median diameter CMD and a geometric standard deviation σ_g, an equation derived by Lee and Chen (1984) can be used to calculate the average coagulation coefficient $\bar{K}$ for a given polydisperse aerosol:

$$\bar{K} = \frac{2kT}{3\eta}\left[1 + e^{\ln^2\sigma_g} + \left(\frac{2.49\lambda}{\text{CMD}}\right)\left(e^{0.5\ln^2\sigma_g} + e^{2.5\ln^2\sigma_g}\right)\right] \quad (4\text{-}13)$$

This value of $\bar{K}$ can be used in place of K in Eq. 4-10 to predict the change in number concentration over a period of time t for which there is only a modest change in CMD. Eq. 4-12 can be used with $\bar{K}$ to predict the increase in CMD over a time period for which $\bar{K}$ is approximately constant. For this type of calculation it is reasonable to assume that σ_g remains constant for modest changes in particle size. For large changes in particle size, calculation can be done as a series of steps with a constant but different $\bar{K}$ for each step.

Example 4-4

(a) An iron oxide fume has an initial number concentration of $10^7/\text{cm}^3$. Assuming the aerosol is monodisperse with a diameter of 0.2 μm, what will be the number concentration and particle size after two minutes? Assume standard conditions. (b) Repeat the above example for a polydisperse aerosol having a CMD of 0.2 μm and a σ_g of 2.0.

Answer. (a) Use Eq. 4-10:

$$N(t) = \frac{N_0}{1 + N_0Kt}$$

where

$$N_0 = 10^7\ \text{cm}^3$$

$$K = \frac{4kTC_c}{3\eta} = \frac{4(1.38 \times 10^{-16})293 \times 1.88}{3(1.8 \times 10^{-4})}$$

$$= 5.6 \times 10^{-10}/\text{cm}^3$$

$$t = 120\ \text{s}$$

Substituting in Eq. 4-10,

$$N(t) = \frac{10^7}{1 + 10^7(5.6 \times 10^{-10})120}$$

$$= 5.98 \times 10^6/\text{cm}^3$$

Use Eq. 4-11 or Eq. 4-12 to determine the change in diameter:

$$d(t) = d_0\left(\frac{N_0}{N(t)}\right)^{1/3} = 0.2\left(\frac{10^7}{5.98 \times 10^6}\right)^{1/3}$$

$$= 0.24\ \mu\text{m}$$

(b) Use Eq. 4-13 to get $\bar{K}$:

$$\bar{K} = \frac{2kT}{3\eta}\left[1 + e^{\ln^2\sigma_g} + \left(\frac{2.49\lambda}{\text{CMD}}\right)\left(e^{0.5\ln^2\sigma_g} + (e^{2.5\ln^2\sigma_g})\right]$$

$$= \frac{2(1.38\times10^{-16})\times 293}{3(1.8\times10^{-4})}\Bigg[1 + e^{\ln^2 2.0} + \left(\frac{2.49\times 0.066}{0.2}\right)(e^{0.5\ln^2 2.0} + e^{2.5\ln^2 2.0})\Bigg]$$

$$= 9.57\times10^{-10}\ \text{cm}^3/\text{s}$$

Substituting into Eq. 4-10, using $\bar{K}$ instead of K, gives

$$N(t) = \frac{N_0}{1 + N_0\bar{K}t}$$

$$= \frac{10^7}{1 + 10^7(9.57\times10^{-10})120}$$

$$= 4.65\times10^6/\text{cm}^3$$

$$\text{CMD}_2 = 0.2\left(\frac{10^7}{4.65\times10^6}\right)^{1/3} = 0.26\ \mu\text{m}$$

Kinematic Coagulation

Kinematic coagulation is a coagulation process where the relative motion between particles is created by external forces rather than by Brownian motion. Brief descriptions of several such mechanisms are given below. In all cases the greater the particle number concentration the greater the rate of coagulation. In all cases there is a relative motion between particles, created by an external force that results in particles contacting one another. In general, there are no simple equations that describe these processes in a complete way. More detailed information is given by Fuchs (1964) or Hinds (1982).

Because particles of different sizes settle at different rates, there is a relative motion between settling particles of different sizes. The aerodynamics of the collision process is complicated and the collision efficiency is low except for the case of very large particles, such as raindrops, settling through micrometer or larger-sized aerosol particles.

A similar process occurs when particles are projected through an aerosol at high velocities. This is an important mechanism for the capture of particles by spray droplets in certain kinds of wet scrubbers used for gas cleaning.

Gradient or shear coagulation occurs for particles moving in a flow velocity gradient. Particles on slightly different streamlines in a velocity gradient travel at different velocities and faster particles eventually overtake the slower ones. If the particles are large enough, particle contact occurs by interception.

In turbulent flow, particles follow a complex path having strong velocity gradients. Relative motion between particles arises from these gradients and from an inertial projection of the particles. The resulting coagulation is called turbulent coagulation. This mechanism is most effective when the turbulent eddy size is of the same order of magnitude as the particle stopping distance. This mechanism is only important for particles larger than about 1 μm. Generally, the more intense the turbulence, the more extensive is the coagulation that results from this mechanism.

Finally, there is acoustic coagulation, where intense sound waves are used to create relative motion between particles. Depending upon their size, particles respond to high-intensity sound waves differently—large particles are unaffected, whereas small particles oscillate with the sound waves. The relative motion that results leads to collisions and is called acoustic coagulation. Generally, sound pressure levels exceeding 120 dB are required to produce significant coagulation.

REACTIONS

When compared to bulk materials, aerosol particles have very high ratios of surface area to mass. For example, 1 g of unit-density material when divided into 0.1 μm particles has a surface area of 60 m^2. Because of their large specific surface (surface area per gram), aerosols participate actively in many kinds of interactions between gas molecules and liquid or solid particles. Particles can undergo three kinds of reactions: reactions between compounds within a particle, reactions between particles of different chemical composition, and reactions between the particle and one or more chemical species in the surrounding gas phase. In the first case, reactions are governed by the usual chemical kinetics. The second case is most likely controlled by the rate of arrival of other particles, which is described by the coagulation process given above. Once dissimilar particles contact each other, reactions proceed by chemical kinetics. The third case, depending on the circumstances, may be controlled by the rate of arrival of the appropriate gas molecules at the particle surface. The rate of arrival of gas molecules is described by the condensation growth equations given in this chapter. Absorption and adsorption are related processes that also have as one of their necessary steps the arrival of gas molecules at the particle surface.

These processes can be thought of as having three mass transfer steps in series, any one of which may be the rate-controlling step. First there is diffusion of specific gas molecules to the surface of the particle. Next is the transfer across the interface or reaction at the interface and, finally, there is diffusion into the solid or liquid particle.

Reaction

In the case of a chemical reaction between the gas and the particle any of the three steps may control the rate of reaction. For solid particles, the diffusion into the interior can be relatively slow even though the distances involved are small. Diffusion into the interior of liquid particles will be more rapid and may be augmented by internal circulation. If the reaction is controlled by the rate of arrival of gas molecules at the particle surface then the maximum rate of reaction is given by Hinds (1982) as

$$R_R = \frac{2\pi d_p D_v p}{kT} \quad \text{for } d_p > \lambda \tag{4-14}$$

where R_R is the rate of reaction in molecules/s. This situation is called a diffusion-controlled reaction. The process can continue until all the molecules of the particle have reacted.

Absorption

The process whereby gas molecules become dissolved in a liquid droplet is called absorption. In this process the transfer at the interface is usually not controlling, but diffusion in either the gas or liquid phase may be. The process can continue until the limit of solubility of the gas in the liquid is reached. This limit may change with temperature or in the presence of other dissolved components.

Adsorption

Adsorption is the transfer of gas molecules from the surrounding gas to a solid surface. There are two types of adsorption that can occur on the surface of a solid particle: physical adsorption (physisorption) and chemical adsorption (chemisorption). Physisorption is a physical process where gas molecules are held to a particle's surface by van der Waals forces. It occurs for all gases when the ambient temperature is below their critical temperature. It is a rapid and readily reversible process. Because the adsorption process is rapid, the diffusion of gas molecules to the particle surface is usually the rate-limiting step. The relationship between the amount of adsorbed gas and the partial pressure of the gas or vapor at a given temperature is called the adsorption isotherm. Physisorption is usually not significant if the saturation ratio is below 0.05, but can lead to an adsorbed layer several molecules thick when the saturation ratio is 0.8 or greater. For a particle in adsorption equilibrium, a reduction in the partial pressure of the vapor will lead to a transfer of adsorbed vapor molecules from the particle's surface to the gas.

The process of adsorption is very similar to the process of condensation. Highly porous materials, like activated carbon, have enormous surface areas and contain numerous small pores and capillaries that facilitate condensation on their surface and inhibit evaporation. The isotherms for highly porous materials will differ significantly from those for smooth solids.

Chemisorption is similar, except that chemical bonds are formed to hold the gas molecules on the particle's surface. It can occur above or below the critical temperature of the gas. In chemisorption only a monolayer can form and, unlike physisorption, the process is not easily reversible because the chemical bonds are much stronger than van der Waals forces. Either the rate of gas phase diffusion or the rate of reaction can control the rate of this process. The rate of transfer slows down as a complete monolayer is approached. In some cases molecules are first held to the surface by physisorption and then slowly react to attach by chemisorption. In other cases a physisorption layer may form on top of a chemisorption layer.

References

Adamson, A. W. 1982. *Physical Chemistry of Surfaces.* New York: Wiley.

Bird, R. B., W. E. Stewart, and E. N. Lightfoot. 1960. *Transport Phenomena.* New York: Wiley.

Davies, C. N. 1978. Evaporation of airborne droplets. In *Fundamentals of Aerosol Science*, ed. T. Shaw. New York: Wiley.

Davies, C. N. 1979. Coagulation of aerosols by Brownian motion. *J. Aerosol Sci.* 10:151–61.

Ferron, G. A. and S. C. Soderholm. 1990. Estimation of the times for evaporation of pure water droplets and for stabilization of salt solution particles. *J Aerosol Sci.* 21:415–29.

Fuchs, N. A. 1959. *Evaporation and Droplet Growth in Gaseous Media.* Oxford: Pergamon.

Fuchs, N. A. 1964. *The Mechanics of Aerosols.* Oxford: Pergamon.

Hinds, W. C. 1982. *Aerosol Technology.* New York: Wiley.

Lee and Chen. 1984. Coagulation rate of polydisperse particles. *Aerosol Sci. Technol.* 3:327–34.

Moore, W. J. 1962. *Physical Chemistry*, 3rd edn. New Jersey: Prentice-Hall.

Springer, G. S. 1978. Homogeneous nucleation. *Advances in Heat Transfer.* 14:281–346.

Zebel, G. 1966. Coagulation of aerosols. In *Aerosol Science.* ed. C. N. Davies. London: Academic Press.

5

The Characteristics of Environmental and Laboratory-Generated Aerosols

Walter John

Air and Industrial Hygiene Laboratory
California Department of Health Services
Berkeley, CA, U.S.A.

INTRODUCTION

Aerosols are produced by natural processes and by people, both intentionally and unintentionally as a byproduct of their activities. Smoke from forest fires, windblown dust, airborne sea salt, and the haze in the Smoky Mountains are examples of natural aerosols. Particles in automotive exhaust, smoke from power plants or agricultural burning, and test aerosols generated in a laboratory are examples of anthropogenic aerosols. The generation process determines whether the particles are in the liquid or solid phase, the chemical composition, the particle size distribution, particle morphology, and other particle properties.

In this chapter we discuss the characteristics of the aerosols which are encountered in several environments: the ambient atmosphere, indoors in offices and residences, and in industrial settings. We then consider the methods of generating laboratory aerosols having some of the key properties of environmental aerosols, since laboratory aerosols are used to test and calibrate samplers and instruments to be used in specific environments. Laboratory aerosols with selected properties are also used in specialized test facilities such as calibration chambers, wind tunnels, and exposure chambers.

ATMOSPHERIC AEROSOLS

Modes in the Size Distribution

Although there are countless sources of atmospheric aerosols, it is possible to simplify the discussion by first considering the particle size distributions. In an important generalization, Whitby (1978) pointed out that the mass distribution of atmospheric aerosols consists of modes (peaks in the distribution) which can be described by lognormal functions (normal distribution of a logarithmic variable), as illustrated in Fig. 5-1. The three modes are the nuclei mode at a particle diameter of 0.018 μm, the accumulation mode at 0.21 μm, and the coarse mode at 4.9 μm. The modal sizes vary somewhat with location and time; the parameters for some typical distributions are listed in Table 5-1. The nuclei mode is formed by the condensation of atmospheric gases into primary particles which then coagulate into aggregates. Coagulation increases the particle size; however, the nuclei mode does not tend to grow over into the size range of the accumulation mode. Instead,

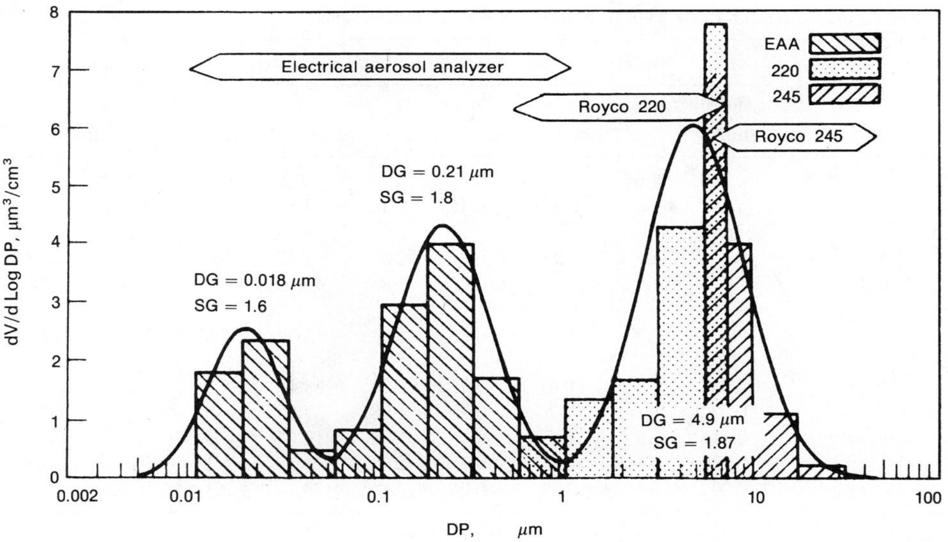

FIGURE 5-1. Trimodal Particle Volume Distribution Measured at the General Motors Milford Proving Grounds, 29 October 1975. The Size Range Measured with each Instrument is Indicated; the Roycos are Optical Particle Counters. DG is the Geometric Mean Particle Diameter and SG is the Geometric Standard Deviation. (Reprinted from Whitby (1978) with the Permission of Pergamon Press, Inc.)

particles from the nuclei mode move into the accumulation mode by coagulating with accumulation mode particles, this being favored over coagulation with other nuclei mode particles because of the greater surface area of the larger particles. The nuclei mode has a relatively short lifetime and is usually not prominent except in the vicinity of combustion sources, such as automobiles.

The accumulation mode is formed by gas-to-particle conversion through chemical reactions and condensation as well as coagulation. The rate of particle growth from condensation slows down with increasing particle diameter. However, the growth of droplets in this mode continues until deposition occurs, usually within a day or two, thus limiting the ultimate particle size. As a result, the accumulation mode does not extend much beyond a few micrometers in diameter, and remains distinct from the larger particles in the coarse mode.

The coarse mode consists primarily of particles generated by mechanical processes. This mode contains windblown dust, sea salt spray, and plant material. The modal size and composition is variable, depending on the nature of the surface cover and atmospheric conditions, especially the wind speed. The origin and composition of the coarse mode is qualitatively different from the nuclei and accumulation modes, giving rise to the characterization of particles in the nuclei and accumulation modes as "fine" particles and those in the coarse mode as "coarse" particles. The dividing line between fine and coarse is usually taken to be at about 2 μm particle diameter, which is the approximate location of the minimum between the accumulation and coarse modes.

Recently, evidence has been obtained for the existence of two modes rather than one in the accumulation mode size range (John et al. 1990). The two modes are a condensation mode at about 0.2 μm and a droplet mode at 0.6 μm. It is believed that the droplet mode grows out of the condensation mode by the addition of sulfate and water. The droplet

TABLE 5-1 Modal Parameters for Eight Typical Atmospheric Size Distributions

	Nuclei			Accumulation			Coarse		
Atmospheric Distribution	d_g (μm)	σ_g	V^1 (μm/cm)^3	d_g (μm)	σ_g	V (μm/cm)^3	d_g (μm)	σ_g	V (μm/cm)^3
Marine, surface	0.019	1.6[c]	0.0005	0.3	2[c]	0.10	12	2.7	12
Clean continental background	0.03	1.6	0.006	0.35	2.1	1.5	6	2.2	5
Average background	0.034	1.7	0.037	0.32	2.0	4.45	6.04	2.16	25.9
Background and aged urban plume	0.028	1.6	0.029	0.36	1.84	44	4.51	2.12	27.4
Background and local sources	0.021	1.7	0.62	0.25	2.11	3.02	5.6	2.09	39.1
Urban average	0.038	1.8	0.63	0.32	2.16	38.4	5.7	2.21	30.8
Urban and freeway	0.032	1.74	9.2	0.25	1.98	37.5	6.0	2.13	42.7
Labadie coal power plant	0.015	1.5	0.1	0.18	1.96	12	5.5	2.5	24
Average	0.029 ± 0.007	1.66 ± 0.1	0.26[a] ± 0.33	0.29 ± 0.06	2.02 ± 0.1	21.5[b] ± 20	6.3 ± 2.3	2.26 ± 0.22	25.9 ± 13

Source: Whitby and Sverdrup (1980) with the permission of John Wiley & Sons, Inc.

1. Volume of particles per volume of air
a. Average omitting marine, urban and freeway, and labadie.
b. Average omitting marine.
c. Assumed

mode is in the optimum particle size range for light scattering and consequent visibility reduction. The sulfate size distribution in Fig. 5-2 has condensation, droplet, and coarse modes.

The justification for the use of lognormal functions to fit the atmospheric size distributions is mainly empirical. Available data are adequately fit by lognormals and the lognormal functions have convenient mathematical properties. For example, if the number distribution is lognormal, the surface and volume distributions will also be lognormal, with the same geometric standard deviation. Table 5-1 lists some typical parameters of the lognormal modes in atmospheric aerosols.

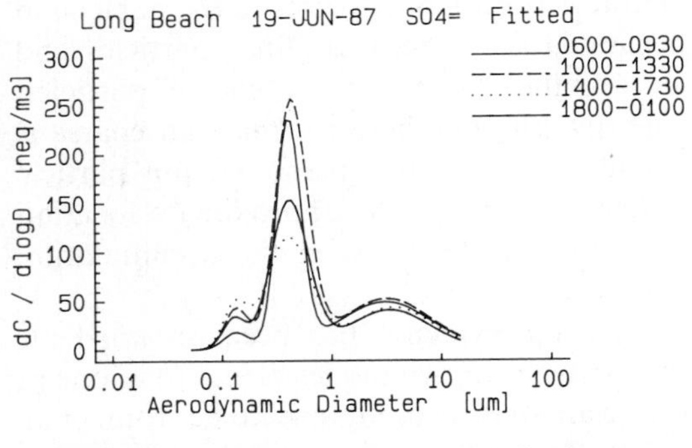

FIGURE 5-2. Sulfate Particle Size Distribution Measured in Ambient Air with a Berner Cascade Impactor. The Sulfate Ion Concentrations are Given in Units of Nanoequivalents Per Cubic Meter of Air. (Reprinted from John et al. (1990) with the Permission of Pergamon Press, Inc.)

Chemical Composition

The chemical compositions of the nuclei and accumulation modes, i.e., the fine particle fraction, reflect their formation from inorganic gases, such as sulfur dioxide and oxides of nitrogen, as well as from organic gases. Acidic sulfur and nitrogen compounds are, typically, at least partially neutralized by ambient ammonia gas to form ammonium salts. Ammonium sulfate and ammonium nitrate are the principal inorganic aerosols in urban areas. Both organic and elemental carbon are found in the fine-particle modes. "Organic carbon" refers to carbon in compounds; "elemental carbon" is also known as soot. Elemental carbon accounts for the typical black appearance of fine particle deposits. It reduces visibility by optical absorbtion, while inorganic particles and organic carbon particles reduce visibility by light scattering. Particles in the accumulation mode (more precisely, the droplet mode) are most efficient at light scattering because the particle diameters are of the same order of magnitude as the wavelength of visible light.

The multimodal size distribution is a universal characteristic of the atmospheric aerosol. The chemical composition can also be described in fairly general terms. Worldwide, three types of aerosols are typical: a continental aerosol, as the term implies, is that over land masses; a marine aerosol, that over oceans; and an urban aerosol, that in the cities. In Fig. 5-3, the chemical compositions of the fine and coarse fractions of a typical urban aerosol are shown.

From the foregoing, it is clear that it is unlikely that any single measurement technique will suffice to characterize the atmospheric aerosol. The particle size distribution spans some four decades or so in particle diameter. The smallest particles (less than about 0.5 μm diameter) diffuse like gases while the largest particles (greater than 0.5 μm diameter) respond to inertial forces. The particles may be liquid or solid and may be semivolatile. Particle shapes may be spherical, crystalline, fibrous, or irregular. Materials may be inorganic, organic, or biological. The applicable technique must be chosen on the basis of the purpose of the measurement. Particle size selection and the type of analysis are important considerations. A detailed discussion of sampling in ambient air can be found in Chapter 28.

Example 5-1

The number distribution of an aerosol is fitted by the function

$$\frac{dN_p}{d(\ln d_p)} = \frac{1}{\sqrt{2\pi}\ln\sigma_g}\exp\left[-\frac{\ln(d_p - \mathrm{CMD})^2}{2(\ln\sigma_g)^2}\right]$$

where $\sigma_g = 1.7$ and CMD = 0.30 μm. Here N_p is the number of particles/cm^3, CMD is the count median diameter, and σ_g is the

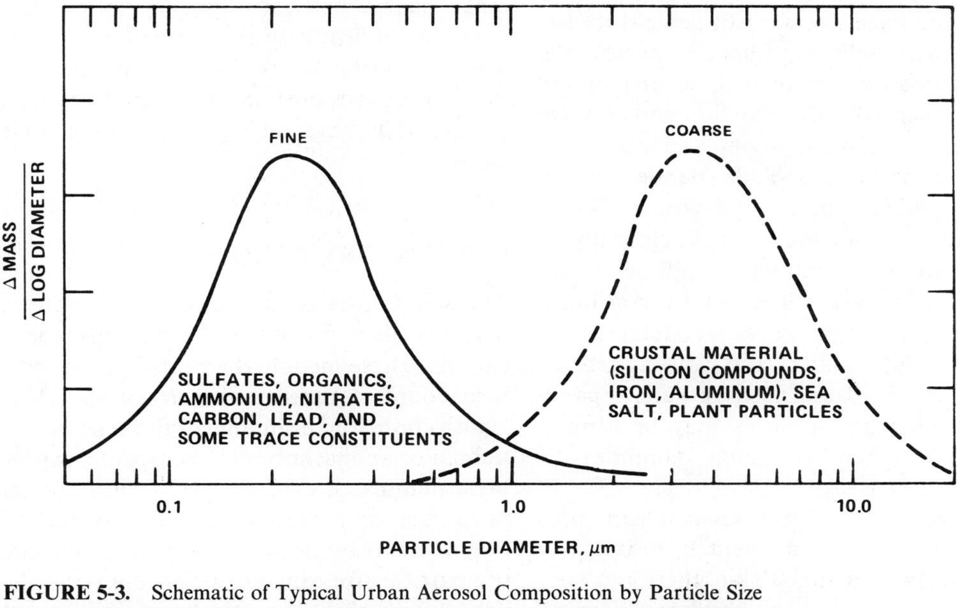

FIGURE 5-3. Schematic of Typical Urban Aerosol Composition by Particle Size Fraction. The Chemical Species are Listed in Approximate Order of Relative Mass Contribution. (From U.S. Environmental Protection Agency 1982)

geometric standard deviation. (a) What is the mass median diameter? (b) If σ_g were 20% larger, how much larger would MMD be?

Answer. (a) The distribution is lognormal, so that the mass median diameter, MMD, can be calculated from CMD and σ_g, using the relation

$$\text{MMD} = \text{CMD} \exp(3 \ln^2 \sigma_g) = 0.70\ \mu\text{m}$$

(b) $\sigma_g = 2.04$ and MMD = 1.38 μm, which is nearly double the original value. This illustrates the sensitivity to errors in the data when conversions are made from number-weighted to mass-weighted parameters.

INDOOR AEROSOLS

In recent years, increasing emphasis has been laid on indoor aerosols, since people average 80–90% of their time indoors. An indoor aerosol usually refers to that in residences and offices as distinguished from that in industrial workplaces. Some of the indoor aerosol derives from infiltration of atmospheric aerosol, although indoor concentrations tend to be lower. Household dust is resuspended by ventilation and by people's activities; vacuum cleaners are excellent aerosol generators because of their inefficient filters. If a smoker is present, tobacco smoke will be one of the dominant aerosols. Combustion aerosols are produced by cooking stoves and fireplaces. Biological particles include dander, fungi, bacteria, pollens, spores, and viruses. Walls, floors, and ceilings may release glass fibers, asbestos fibers, mineral wool, and metal particles. Aerosols are generated by consumer product spray cans. Paper products are a source of cellulose fibers, and clothing articles are sources of natural and synthetic organic fibers. Radioactive aerosols may be formed by the attachment of radon daughters to suspended particles.

Sampling in the indoor environment presents some problems different from outdoor sampling. As indicated above, there is a variety of types of aerosols leading to more variability in the aerosol composition than outdoors. There is a need to avoid disturbing the occupants by sampler noise, etc. The sampling of indoor air is discussed in Chapter 30.

INDUSTRIAL AEROSOLS

The characteristics of aerosols produced in industry are determined by the type of industry, the nature of the product, and the industrial operations. Detailed discussions of emissions from basic industries have been given (Stern 1968). Basic industries include petroleum refineries, nonmetallic mineral product industries, ferrous metallurgical operations, nonferrous metallurgical operations, inorganic chemical industry, pulp and paper industry, and food and feed industries. Power plants and incinerators are examples of stationary combustion sources. The properties of the aerosols emitted to the atmosphere depend on the material burned, combustion conditions, and the type of controls on the stacks.

Within an industry, aerosols are generated by processing activities. Welding produces fumes, chain aggregates of very fine particles. Mechanical operations such as grinding make coarse particles. Spray painting produces liquid droplets in the aerosol size range. The transport and handling of powdered materials produces dust aerosols. Ore piles, coal piles, and tailings piles give rise to fugitive emissions.

LABORATORY AEROSOLS

Aerosols can be generated in the laboratory, with specified properties for various applications. These include the calibration of particle counters, testing of aerosol samplers, health effect studies involving the exposure of human or animal subjects, therapeutic exposures, manufacture of materials with special properties, and many others. The associated requirements of the aerosol generator include the particle size range to be covered and the size distribution, i.e., the mean size and distribution of particle sizes about the mean. Another important consideration is the particle concentration necessary and the constancy of the concentration. The gas flow rate may also be important.

For some applications, such as instrument calibration, a monodisperse aerosol is desired. Any aerosol will have a finite spread of particle sizes about the mean size, depending on the method of generation. There is no precise definition of a monodisperse aerosol, i.e., how small a spread of particle sizes constitutes "monodisperse." In practice, the application will determine what degree of monodispersity is acceptable. The size distributions of the aerosol from most generators can be fairly well fitted by a lognormal function. The geometric standard deviation, σ_g, is a convenient index of the spread of particle sizes. Some authors have considered that a monodisperse aerosol should have a σ_g less than about 1.2. This arbitrary definition has no rigorous justification, but can be used as a rough indicator. The smallest σ_g readily achievable in the laboratory is about 1.01. Most so-called monodisperse aerosol generators can produce aerosols with σ_g's < 1.1. On the other hand, nebulizers produce sprays with σ_g's > 1.7, and the aerosol is considered polydisperse.

The particle properties can be selected for a particular application. In some cases, a specific material is necessary; for example, ammonium sulfate aerosol for a human exposure study, asbestos fiber aerosol for an animal exposure study, and a selected material in the case of manufacturing materials. In other cases, certain material properties suffice; for example, in testing a particle sampler for wall losses, it is customary to use a liquid aerosol, since this represents a worst case. Conversely, solid particles are used to test a sampler for particle bounce and reentrainment. Other particle properties which may be important include particle density, particle morphology, the index of refraction, and surface composition.

Examples of the major types of aerosol generators available are discussed below. However, prior to that a discussion on safety precautions is worthwhile.

Safety Precautions

When generating aerosols, it should always be borne in mind that a respiratory health hazard may be created. A primary consideration is containment of the aerosol. A chemistry hood is a good location for an aerosol generator. Even if the exhaust from the generator is vented, there are usually times when the apparatus is open or there may be leaks. A walk-in hood is especially convenient to accommodate an auxiliary apparatus. If a hood cannot be used, the exhaust should be vented or filtered. Hazardous substances require more stringent containment measures.

Care should be exercised in the choice of aerosol materials. For example, in the past DOP (dioctyl phthalate) was commonly used as a test aerosol since it has nearly unit density and is an oil with low volatility. However, animal tests have implicated DOP as a possible carcinogen. A good substitute is oleic acid, also a nonvolatile oil, which is available in food grade. A side benefit is that uranine, which is frequently added as a fluorescent tracer, is soluble in oleic acid whereas it is insoluble in DOP. This means that the uranine is uniformly dispersed in the oleic acid droplets. Uranine is commonly used to trace waterways and is presumably harmless. Of course, even when the aerosol material is believed to be safe, it is prudent to avoid exposure.

Another hazard is associated with the use of radioactive sources to "neutralize" the electrical charges on aerosols resulting from the generation process. Kr-85, a beta (high-energy electron) emitter, is commonly used in source strengths up to 10 mCi. Unfortunately, Kr-85 also emits gamma rays. Whereas the beta rays are absorbed by the walls of the container, the gamma rays penetrate. It is recommended that a qualified health physicist check the radiation level to evaluate the adequacy of the shielding. Alpha particle sources, such as Po-210, represent a hazard when ingested and must be handled with care.

Example 5-2

The half-life ($t_{1/2}$) of a Po-210 radioactive source is 138 days. If the source is considered ineffective when its activity is reduced by a factor of ten, how long can the source be used?

Answer. The activity or rate of decay of the number of atoms decreases exponentially with time, t, according to the equation

$$dN/dt = (dN/dt)_0 \exp(-t \ln 2/t_{1/2})$$

where N is the number of atoms and the subscript 0 denotes initial time. When the activity has decreased by a factor 0.1,

$$\exp[(-t \ln 2)/t_{1/2}] = 0.1$$

Solving for t,

$$\begin{aligned} t &= -(t_{1/2} \ln 0.1)/\ln 2 \\ &= -(138 \ln 0.1)/0.693 \\ &= 459 \text{ days} \end{aligned}$$

This example is a reminder that the Po-210 static eliminator sources need to be replaced fairly frequently.

Generating Laboratory Aerosols from Spray-Dried Solutions and Suspensions

Vibrating-Orifice Generators

The vibrating-orifice aerosol generator can produce highly monodisperse aerosols in the approximate size range from 0.5–50 μm. The particle diameter can be calculated from the generator's operating conditions so that the aerosol can be considered a primary particle size standard. Also, the aerosol concentration is very stable.

In the vibrating orifice generator (Fig. 5-4), a liquid is forced through an orifice ranging in diameter from 5–30 μm, but, typically, 20 μm diameter. The resulting liquid jet is inherently unstable, tending to break up into nonuniform droplets, but the breakup can be made to occur uniformly by subjecting the jet to a mechanical disturbance of constant frequency. The drop diameter, d_d, is then given by

$$d_d = \left(\frac{6Q}{\pi f}\right)^{1/3} \quad (5.1)$$

where Q is the liquid flow rate and f is the

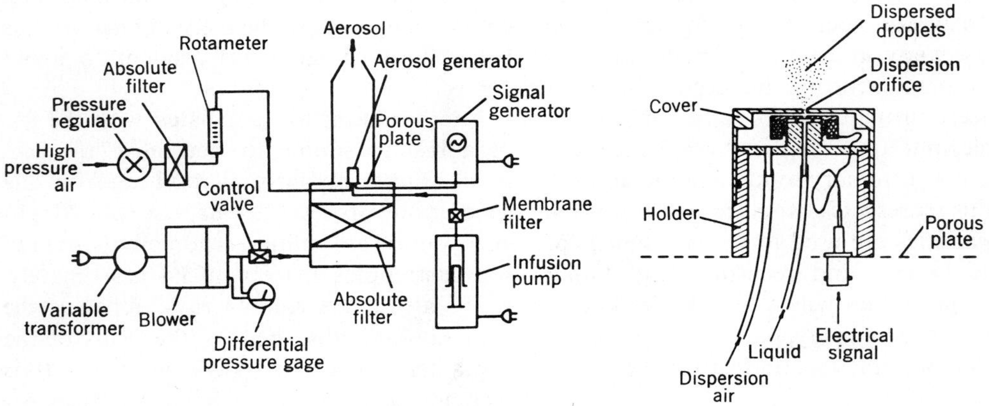

FIGURE 5-4. Diagram of a Vibrating-Orifice Aerosol Generator of the Berglund–Liu Design. (Reprinted from Liu (1974) with the Permission of Air and Waste Management Assoc.)

vibrating frequency. The droplet diameter is typically some tens of micrometers. In order to generate smaller particles, a nonvolatile solute can be dissolved in a volatile solvent. After evaporation of the solvent, the particle diameter is related to the concentration, C, by

$$d_p = C^{1/3} d_d \tag{5.2}$$

Liquid particles can be produced, for example, from a solution of oleic acid in isopropyl alcohol, or solid particles of sodium chloride can be produced from an aqueous solution of sodium chloride. The minimum particle size attainable, in practice, depends on the purity of the solvent. For example, a 20 µm droplet of a solvent containing 1 ppm v/v of total dissolved impurity, a purity difficult to obtain, will dry to a particle of 0.2 µm diameter consisting wholly of impurities. It should be noted that the impurities listed by the supplier of a solvent are generally only those of concern for chemical analysis. Solvents pick up silicates from glass containers and plasticizers from plastic containers. The maximum practicable particle size is not well defined. Particles larger than about 20 µm diameter become progressively more difficult to generate as the diameter is increased. It also becomes more difficult to avoid particle losses in transport. Therefore, the generation of particles larger than about 50 µm requires special effort.

Referring again to Fig. 5-4, the solution is forced through the orifice by a syringe pump. An alternating voltage from a frequency generator is applied to a piezoelectric crystal, which then vibrates the assembly holding the orifice plate. A turbulent jet of air issuing from the hole in the cover above the orifice disperses the droplets before they can coagulate; otherwise, no appreciable output can be obtained. Filtered, dry dilution air is introduced to dry the droplets and transport the aerosol from the generator.

According to Rayleigh, the optimum wavelength, λ, for the disturbance to break up the jet is

$$\lambda = 4.508\, d_j \tag{5.3}$$

where d_j is the jet diameter. Experiment shows that there is a range of about a factor of two in the frequency producing uniform droplets. Within this range, certain frequencies may be encountered which produce satellite droplets, i.e., droplets much smaller than the main drops which are being produced. These satellites are formed in the necking-off region of the jet when a drop is being formed. They can be eliminated by adjustments of the vibrating frequency. Another undesirable characteristic is the production of multiplets because the coagulation of the droplets cannot be completely suppressed. The coagulation of the liquid droplets results in particles with diameters greater than the normal diameter by factors of $2^{1/3}$, $3^{1/3}$, etc. In practice, the operating conditions are adjusted to minimize the multiplets, and corrections are applied to the data taken with the aerosol, if necessary.

Particles from the vibrating orifice typically carry several thousand elementary charges. Since the presence of a high electrical charge may affect subsequent processes involving the aerosol, it is common practice to "neutralize" the charge by exposure of the aerosol to a radioactive source. The source creates charged ions in the gas which are attracted to charges of the opposite sign on the particles, reducing the particle charge distribution to a Boltzmann equilibrium. The radioactive sources commonly used include beta emitters such as Kr-85 or tritium and alpha emitters such as Po-210 or Am-241.

For best results, the output of the aerosol generator should be monitored. This can be readily accomplished with an optical counter. The concentration, particle diameter, and spread of the diameter (monodispersity) can be checked by analyzing the pulse height distribution from the optical counter. Additionally, the amount of multiplets can be determined, and the onset of satellite production can be seen. To check the absolute diameter, a sample of the particles can be deposited on a glass microscope slide. For liquid particles, the slide should be coated with a fluorocarbon compound (such as FC 721, 3M OH & ESD, St. Paul, MN) to prevent the droplets from wetting the slide or, more precisely, to increase the liquid–surface contact angle. A correction for the flattening of the liquid drop on the slide can be made by multiplying the apparent diameter by a correction factor depending on the contact angle. For DOP-(FC 721), the correction factor is 0.74 (John and Wall 1983). The correction factor for oleic acid is nearly the same.

In general, solid particles dry with the formation of voids, resulting in a density less than that of the bulk material. To some extent, the drying process can be controlled by varying the volatility of the solvent, for example, by varying the proportions of water and alcohol and by controlling the amount of dilution air. If the drying is too rapid, the particles tend to have more voids. The density of the particles can be determined by the following method: the mass of the particle, m_p, is calculated from the operating parameters of the vibrating-orifice generator:

$$m_p = c_m \frac{Q}{f} \tag{5.4}$$

where c_m is the mass concentration of the solute. The average density of the particle, including voids, is

$$(\rho_p)_{av} = \frac{6 m_p}{\pi d_g^3} \tag{5.5}$$

where d_g is the particle geometric diameter determined with a microscope.

One example of a solid particle aerosol, namely ammonium fluorescein, deserves special mention because of the useful particle properties and because the generation conditions are somewhat different from the ordinary. Ammonium fluorescein particles are very smooth and have essentially bulk density (1.35 g/cm³). The material has low hygroscopicity and the fluorescence can be used for detection with high sensitivity. The particles are quite bouncy and are, therefore, useful for checking samplers for particle bounce. The solution is prepared by dissolving fluorescein in ammonium hydroxide. A reaction takes place with an ammonium group replacing a hydrogen on the fluorescein molecule. As a result, the molecular weight increases by 5%; thus, 12.8 g fluorescein per liter of solution is a 1% volume concentration. Stoichiometrically, one mole of ammonium hydroxide is required for each mole of fluorescein; however, it is found that an excess of ammonium hydroxide, as much as 50%, facilitates the preparation. When particles of ammonium fluorescein larger than about 10 μm are generated, the droplets tend to dry too fast. To produce smooth particles, it is necessary to humidify the dilution air to slow down the drying. By this method, particles as large as 70 μm have been generated (Vanderpool and Rubow 1988).

The vibrating-orifice aerosol generator can be used to produce uniformly charged, monodisperse aerosols (Reischl, John, and Devor 1977). The technique involves insulating the cap over the orifice assembly and applying a dc voltage to the cap. Charges are induced onto the top of the jet and are trapped on the droplets when they separate from the jet. The monodispersity of the droplets leads to uniform charging. The application of modest voltages, i.e., up to $\pm$ 10 V produces particle charges up to $\pm 10^4$ elementary charges. Reischl, John, and Devor (1977) present the theory and data demonstrating the method, which is useful for experimentation with charged aerosols.

Example 5-3

Oleic acid aerosol is produced by a vibrating-orifice aerosol generator from a solution of oleic acid in isopropyl alcohol, with a volume concentration of 6.30×10^{-4}. The liquid flow rate is 0.2 cm³/min and the vibrating frequency is 118 kHz. The dilution air flow rate is 20 l/min. (a) What is the droplet diameter? (b) What is the oleic acid particle diameter? (c) What is the particle number concentration?

Answer. From Eq. 5-1, the droplet diameter, d_d, is given by:

$$d_d = \left(\frac{6Q}{\pi f}\right)^{1/3} \tag{5-1}$$

where Q is the liquid flow rate and f is the vibrating frequency. Substituting, we find that the droplet diameter is 26.2 μm.

(b) From Eq. 5-2, the particle diameter, d_p, is related to the concentration, C, by

$$d_p = C^{1/3} d_d \tag{5-2}$$

Substituting, we find the particle diameter is 2.25 μm.

(c) The rate of particle production is the same as the vibrating frequency, 118,000 s^{-1}. Dividing this by the flow rate of the dilution air, we find the particle number concentration is 3.5×10^8 m^{-3}.

Spinning-Disk Aerosol Generators

The spinning-disk aerosol generator produces monodisperse aerosols from solutions or suspensions. Liquid is fed to the center of a disk rotating at speeds of the order of 1000 revolutions per second. Centrifugal force causes the liquid to flow to the edge of the disk, where it accumulates until the inertial force overcomes the surface tension. Liquid filaments separate from the disk, breaking up into drops. The drop diameter is given by

$$d_d = K\sqrt{\frac{\gamma}{\rho_f \omega^2 d}} \tag{5.6}$$

where γ is the surface tension, ρ_f the fluid density, ω the angular velocity of rotation and d is the diameter of the disk. The constant K varies in the range 2–7. The formation of the primary drops is accompanied by the production of satellite droplets. Spinning-disk designs include an air flow perpendicular to the edge of the disk to remove the satellites, without appreciable loss of the primary drops. A version of the spinning disk called the spinning top features a free rotor which is suspended and spun by compressed air, allowing the attainment of higher rotational speeds than a rotor on a shaft.

As in the case of the vibrating-orifice generator, particles smaller than the primary droplets can be produced by the use of suitable solutions; the practical particle size range of the spinning disk is roughly the same as that of the vibrating orifice. The spinning disk, however, is capable of producing particle concentrations an order of magnitude higher than the vibrating orifice. The monodispersity is not as high as that of the vibrating orifice, the geometric standard deviation being typically 1.05–1.10.

Latex Aerosols

A convenient and widely used method of generating a monodisperse calibration aerosol consists of atomizing suspensions of polystyrene latex particles. The latex particles are smooth, spherical, and highly monodisperse. Latex aerosols are especially convenient for the calibration of optical particle counters and other particle detectors. Latex aerosols can also be used to calibrate size-selective samplers such as cascade impactors. However, the latex suspensions are relatively expensive, which generally restricts their use to particle size calibrations.

Latex particles are available with diameters ranging from 0.03 μm to 1000 μm. In the range 0.03–3 μm, they are produced by emulsion polymerization, with coefficients of variation (CV, relative standard deviation) of the order of 1%. Large particles, 4–1000 μm diameter, with CVs of 3.5–30%, are produced by suspension polymerization. An intermediate process is used to produce relatively large, uniform particles, 2–20 μm diameter, with CVs of 2–3%. Several materials are used to make the particles, including polystyrene latex, polystyrene DVB (polystyrene cross-linked with divinylbenzene), and polyvinyl toluene. The density of polystyrene latex is 1.05 g/cm³; the range of densities of other available materials is from 0.99 to 1.25 g/cm³.

Latex aerosols are typically produced with compressed-air nebulizers such as the DeVilbiss Model 40, which is made of glass (see the Nebulizers section). A complication arises due to the statistical distribution of the number of particles within the droplets from the nebulizer. The number ranges from zero up to a large number, depending on the suspension concentration. When the droplet dries, an aggregate of the particles in the droplet will be formed. The suspensions are typically supplied with 10% solids, requiring a high dilution to ensure that most of the particles are singlets. The probability that a droplet of a given size will contain *n* particles is given by the Poisson distribution. However, because the nebulizers produce polydisperse droplet distributions, it is necessary to integrate over the droplet size distribution. Raabe (1976) has derived an expression for the singlet ratio, *R*, defined as the ratio of droplets with only one monodisperse particle to all droplets that contain one or more particles:

$$c_n = \frac{6(1 - R)\exp(-4.5 \ln^2\sigma_g)}{\pi(\text{VMD})^3[1 - 0.5\exp(\ln^2\sigma_g)]} \tag{5.7}$$

where c_n is the number of particles per ml of suspension and VMD and σ_g are the volume median diameter and the geometric standard deviation, respectively, of the droplet size distribution. While c_n can be reduced to obtain $R > 0.9$, the concentration of singlets will also be reduced. For $R > 0.95$, more than 99% of the droplets contain no latex spheres. This can lead to another problem. The latex suspensions are typically stabilized against coagulation by the addition of anionic surfactants. The empty droplets from the nebulizer dry to form fine particles of the residual surfactant. Since these greatly outnumber the latex particles, they can interfere with the measurement, depending on the nature of the measurement. In recent years, latex particles having surface functional groups, such as sulfate ions, have become available. The charges on the functional groups stabilize the suspension without the need for a surfactant. Latex particles can also be obtained with fluorescent dyes to facilitate particle detection and analysis.

Example 5-4

A suspension of 1 μm latex particles is nebulized with a droplet volume median diameter, VMD, of 3 μm and a σ_g of 2.0. If a singlet ratio of 0.95 is desired, what should be the number of particles per ml of the suspension? What is the volume percentage of the suspension?

Answer. From Eq. 5-7,

$$c_n = \frac{6(1 - R)\exp(-4.5 \ln^2\sigma_g)}{\pi(\text{VMD})^3[1 - 0.5\exp(\ln^2\sigma_g)]}$$

$$= 2.12 \times 10^9 \text{ ml}^{-1}$$

$$\text{Volume percentage} = (\pi d_p^3/6)\, c_n = 0.11\%$$

Nebulizers

The atomization of solutions and suspensions with compressed air is one of the simplest ways to generate an aerosol. A nebulizer is an atomizer from which only the smallest droplets are allowed to escape. The air enters the nebulizer through an orifice, expanding at high speed past the liquid inlet. The pressure drop from the air flow draws a filament of liquid into the airstream which, subsequently, breaks up into droplets. The droplet spray is directed towards a wall or a baffle, where the larger droplets are removed by impaction. Nebulizers produce aerosols with mass median diameters ranging from about 1–6 μm, with geometric standard deviations of 1.7–2.3, and concentrations of 5–50 g/m^3. Smaller particles can be produced by nebulizing solutions or suspensions and evaporating the solvent.

Since most of the sprayed solution does not leave the nebulizer, the solvent constantly evaporates and the solution becomes more concentrated. This causes the particle size to increase with time. Presaturating the supply air and cooling the reservoir will reduce, but not eliminate, evaporation. One way to avoid the problem is to discard the runoff from the impacted droplets, nebulizing only fresh solution (Liu and Lee 1975).

A drawing of the DeVilbiss Model 40 glass nebulizer (DeVilbiss Health Care Co., Somerset, PA) is shown in Fig. 5-5. When the vent is open, additional air is drawn in at a rate exceeding that of the jet air. Some common types of nebulizers are listed in Table 5-2 with their operating parameters. The Collison nebulizer has a vertical metal feed tube surrounded by a hollow cylinder serving as a

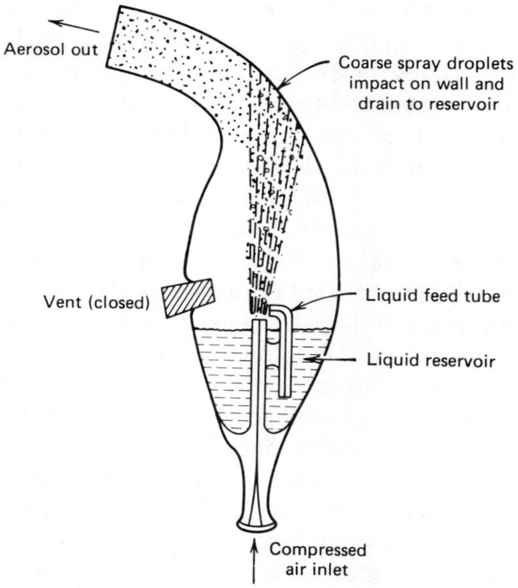

FIGURE 5-5. Drawing of a DeVilbiss Model 40 Glass Nebulizer. (Reprinted from Hinds (1982) with the Permission of John Wiley & Sons, Inc.)

baffle. The Lovelace nebulizer has a baffle consisting of the rounded end of a screw which can be adjusted to optimize the output. Baffles placed close to the jet increase the output by secondary production of droplets when collected liquid is blown off the baffle. The Lovelace nebulizer is highly efficient, producing a relatively high output per unit volume of air flow. The Laskin nebulizer is unique in that the jet impacts on the liquid surface. The agitation produces bubbles, which in turn produce an additional aerosol when they burst. The mass median diameter of the aerosol is the lowest of those listed in the table. Finally, the Babington nebulizer is based on a different operating principle. A hollow sphere having several slit orifices is pressurized from the inside. The liquid is fed to the outside top of the sphere; it then flows down to form liquid films over the orifices. Droplets are formed from the surfaces of the films.

Example 5-5

A solution is nebulized, producing a droplet aerosol with a mass median diameter (MMD) of 6 μm. What should the dilution factor be to reduce the MMD to 1.5 μm?

Answer. Since the volume of a spherical droplet is proportional to the cube of the diameter, the dilution factor $= (6/1.5)^3 = 4^3 = 64$.

TABLE 5-2 Characteristics of Selected Compressed-Air Nebulizers[1]

Type	Operating Pressure (psig)	Flow Rate (l/min)	Output Conc. (g/m^3)	MMD (μm)	σ_g	Reservoir Volume (Ml)
Babington[2]	50	15	~ 5	~ 4	—	200–400
Collison[3]	20	7.1	7.7	2.0	2.0	20–1000
	30	9.4	5.9	—	—	—
DeVilbiss 40[4]	10	11	16	4.2	1.8	10
	20	16	14	3.2	1.8	—
	30	20	12	2.8	1.9	—
Laskin[5]	10	48	3.8	—	—	500–5000
	20	84	4.8	0.7	2.1	—
Lovelace[6]	20	1.5	40	5.8	1.8	4
	30	1.6	31	4.7	1.9	—

Based on Hinds (1982) with the permission of John Wiley & Sons, Inc.

1. Data primarily from Raabe (1976) for water
2. McGraw Respiratory Therapy, Irvine, CA
3. BGI, Inc., Waltham, MA
4. The DeVilbiss Co., Somerset, PA
5. ATI, Baltimore, MD. Hinds data for DOP
6. Sandia Research and Development, Albuquerque, NM

Ultrasonic Nebulizers

In an ultrasonic nebulizer, the energy to aerosolize the liquid is supplied by a piezoelectric crystal. The air flow serves only to transport the aerosol and not to form the droplets. The mechanism of aerosol formation is complicated. The piezoelectric crystal is set into vibration by an electric signal with a frequency of the order of 1 MHz. The vibrational energy causes a fountain to form on the surface of the liquid. The emission of droplets from the fountain is believed to involve capillary waves on the liquid surface. A cup can be placed in the larger reservoir to generate an aerosol from a smaller volume of the solution.

Ultrasonic nebulizers have very high outputs. Commercially available instruments use 1–6 ml/min of solution and produce aerosol concentrations from 70 to 150 g/m^3. The concentration can be varied by controlling the air flow rate since the aerosol production rate is independent of air flow. Evaporative water loss is relatively high because of the high input of energy to the fluid. The mass median diameter of the droplets is in the range 6–9 μm, with geometric standard deviations of 1.4–1.7.

The Sono-Tek (Ponghkeepsie, NY) ultrasonic atomizer allows the control of droplet air concentration by controlling the liquid feed rate. The droplet sizes are relatively large, ranging from a number median diameter of 30–70 μm, depending on the nozzle frequency.

Condensation Aerosol Generators

In this class of aerosol generators, a vapor of the aerosol material is condensed onto nuclei under controlled conditions. A wide range of particle sizes can be produced, ranging from 0.04 to 1.4 μm. The aerosol is fairly monodisperse, with σ_g's typically between 1.1 and 1.4. Output concentrations are high, of the order of 10^6/cm^3.

The condensation nuclei can be produced by a variety of means. In older versions of the generator, such as that developed by Sinclair and LaMer (1949), nuclei were produced by an electric spark or by heating a salt or other substances on a wire or in a combustion boat. The nuclei were then mixed with vapor from a boiler. With this type of generator the particle size is not very reproducible, the equilibration time is long, and there can be thermal degradation of the aerosol material which contacts the boiler walls. These difficulties were circumvented by Rapaport and Weinstock (1955), who introduced the use of a nebulizer as the source of nuclei. The droplets from the nebulizer are passed through a heated tube to evaporate the liquid, leaving a residue particle of impurities or of a low-volatility substance, such as anthracene, which is added to the nebulizer solution. Since each droplet results in a particle, the output of the condensation generator is as stable as that of the nebulizer.

An example of a condensation aerosol generator is shown in Fig. 5-6. A Collison nebulizer produces droplets of, for example, DOP containing a trace of anthracene. The droplet aerosol is vaporized in the top, heated section of the vertical column. The aerosol passes downward to minimize flow instabilities due to heating effects. The vapor condenses on the nuclei in the lower, unheated portion of the vertical tube. For the highest monodispersity, the aerosol is sampled only from the central portion of the tube. Finally, the aerosol is neutralized to remove the electrical charges from the atomization.

This type of generator works well with low-vapor-pressure oils such as DOP or oleic acid. The output particle size can be varied by varying the concentration of the oil in an alcohol solution. Solid particles can be generated from triphenyl phosphate or stearic acid. Other organic materials and some inorganics, such as NH_4Cl, have been used.

Aerosols of Powders or Dusts from Fluidized-Bed Generators

For some applications it is necessary to generate an aerosol from the dry powder of a material. For example, windblown dust may be simulated by generating an aerosol from a soil sample, or fugitive dust from a coal pile may be simulated by using coal dust. Toxic particles such as silica or asbestos can be aerosolized for animal exposure studies.

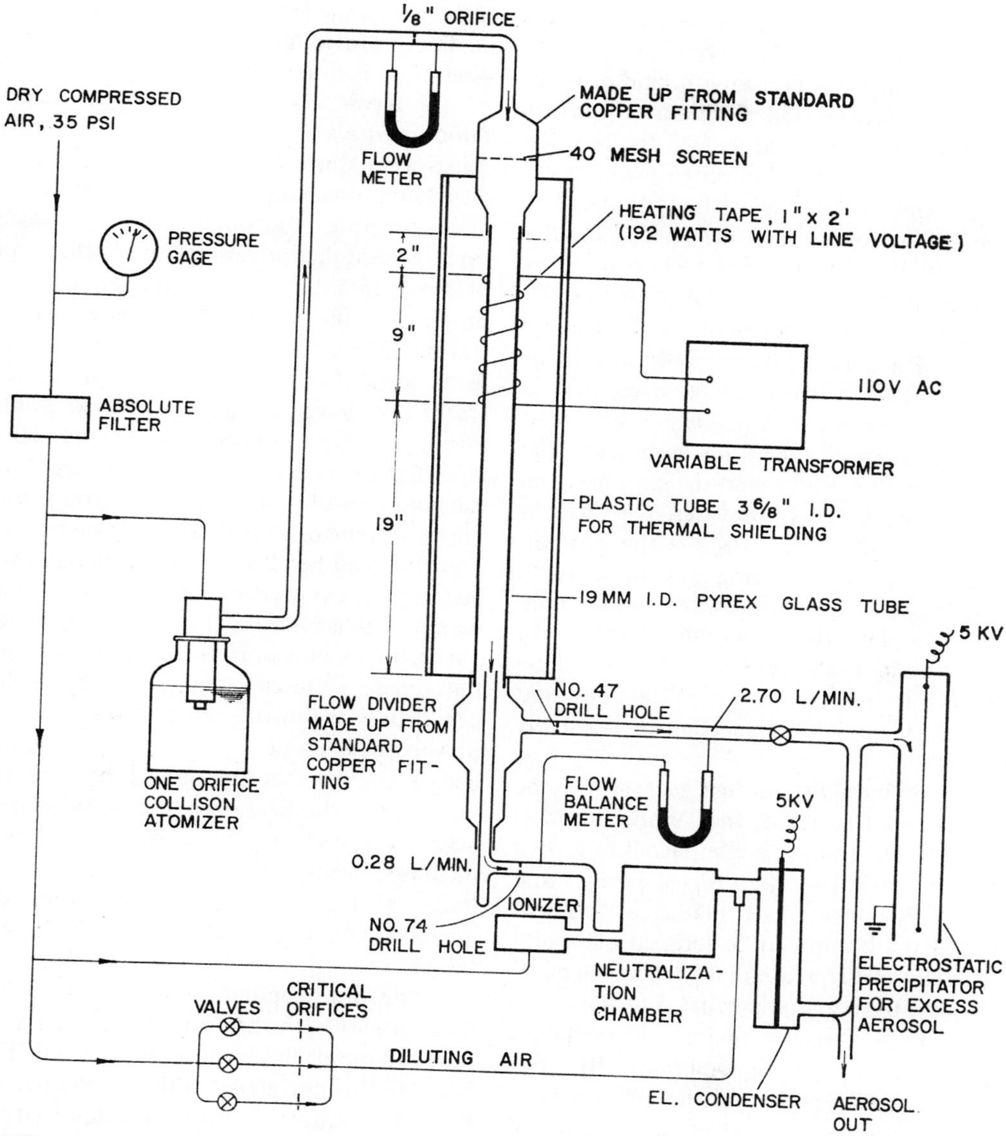

FIGURE 5-6. Condensation-Type Monodisperse Aerosol Generator. (Reprinted from Tomaides, Liu, and Whitby (1971) with the Permission of Pergamon Press, Inc.)

The particles in a powder sample will be agglomerated. To form an aerosol, it is necessary to disperse the powder in air and to break up the agglomerates to obtain the desired particle size distribution. The simplest method to aerosolize a powder is by an air blast. Unfortunately, this method is relatively ineffective in breaking up agglomerates because the forces of cohesion between fine particles are large compared to the shear forces from air flow. A jet of the particles can be directed towards an impaction plate to further break up agglomerates. The remaining large particles can be removed by passing the aerosol through a cyclone. This method may suffice for relatively undemanding applications.

For some applications, the powder can be aerosolized in a batch. The aerosol concentration will then decrease with time as the supply is depleted. For other purposes, it may be necessary to have a relatively constant output, or an output over a long period of time, requiring a powder feed mechanism. Various methods have been used to feed the powder from a reservoir to the generator. Simple gravity feed, accompanied by vibration, is successful in some cases. Screw feed can be used, but the powder tends to jam between the screw and the enclosing tube. This can be alleviated by using a screw made of a flexible material. Pressurizing the reservoir also helps the powder flow. Brushes have been used to transport the powder. A ball chain, similar to a bathtub stopper chain, has been used. The chain fits within a tube, and the gaps between the balls are filled with powder. The chain is moved on pulleys to transport the powder. Another type of metering system involves the filling of grooves in a wheel, which then rotates to place the powder-filled groove in the air stream to the generator. None of these methods is entirely free of difficulties and frequent cleaning of the mechanism is usually required.

One commonly used dust generator is the Wright dust feed (BGI, Inc., Waltham, MA) (Fig. 5-7). The sample is compacted to form a plug in a cylinder. The plug is rotated and advanced against a fixed blade, causing material at the bottom to be scraped off. Compressed air is directed radially along the blade to entrain the released powder. The flow exits axially, to be directed against an impaction plate to break up agglomerates. By controlling the gear ratio and the air flow rate, the output can be varied over a wide range, of the order of 0.01–10 g/m^3. The output is relatively constant and reproducible over a long period of time. The Wright dust feed works well with hard dusts, such as mineral dusts, but is unsuitable for soft, sticky dusts.

The Wright dust feed has been used successfully with standard samples of asbestos. A somewhat similar generator designed specifically for asbestos fibers was developed by Timbrell, Hyett, and Skidmore (1968). A blade rotating at 1500 rpm scrapes fibers from a compacted plug of asbestos. The fibers entrained by an air flow are well dispersed; 48% of the chrysotile fibers and 70% of the amosite fibers are in the respirable mass. Fiber generation is discussed further in Chapter 25.

Fundamentals

Aerosol generators using fluidized beds as the dispersing mechanism have the ability to deagglomerate thoroughly the powdered samples. Another advantage over other types of dust generators is the relatively constant output, i.e., the absence of bursts in concentration. When equipped with a suitable dust feed mechanism, fluidized beds can operate stably over long periods of time as well. Fluidized beds can be scaled over a wide range of sizes, from very small to extremely large, producing aerosol concentrations from milligrams to tens of grams per cubic meter.

A fluidized bed consists of relatively large bed particles, typically of the order of 100 μm diameter, in a cylindrical container. The floor of the bed is of a porous material, such as a fine screen, sintered metal, or a filter, which can retain the bed particles but allow an upward flow of air. In operation, a fluidized bed resembles a boiling liquid, but the best indicator of fluidization is the pressure drop across the bed. As the upward air flow is increased from zero, the pressure drop initially rises linearly with flow rate. Eventually, a condition is reached where the air drag on the particles is equal to the weight of the bed. The pressure drop curve then levels off. The flow velocity at the break in the curve is called the minimum fluidization velocity (MFV). The MFV is correlated with the bed particle Reynolds number. For a typical fluidized-bed aerosol generator, the MFV is of the order of 10 cm/s.

In the two-component fluidized bed initially described by Guichard (1976), a relatively fine powder of the material to be aerosolized is added to a fluidized bed containing larger bed particles. An air flow velocity sufficient to fluidize the bed will exceed the elutriation velocity of the powder particles. When such a two-component bed is turned on, a relatively high aerosol concentration is initially produced, which rapidly decreases to a more steady, but still decreas-

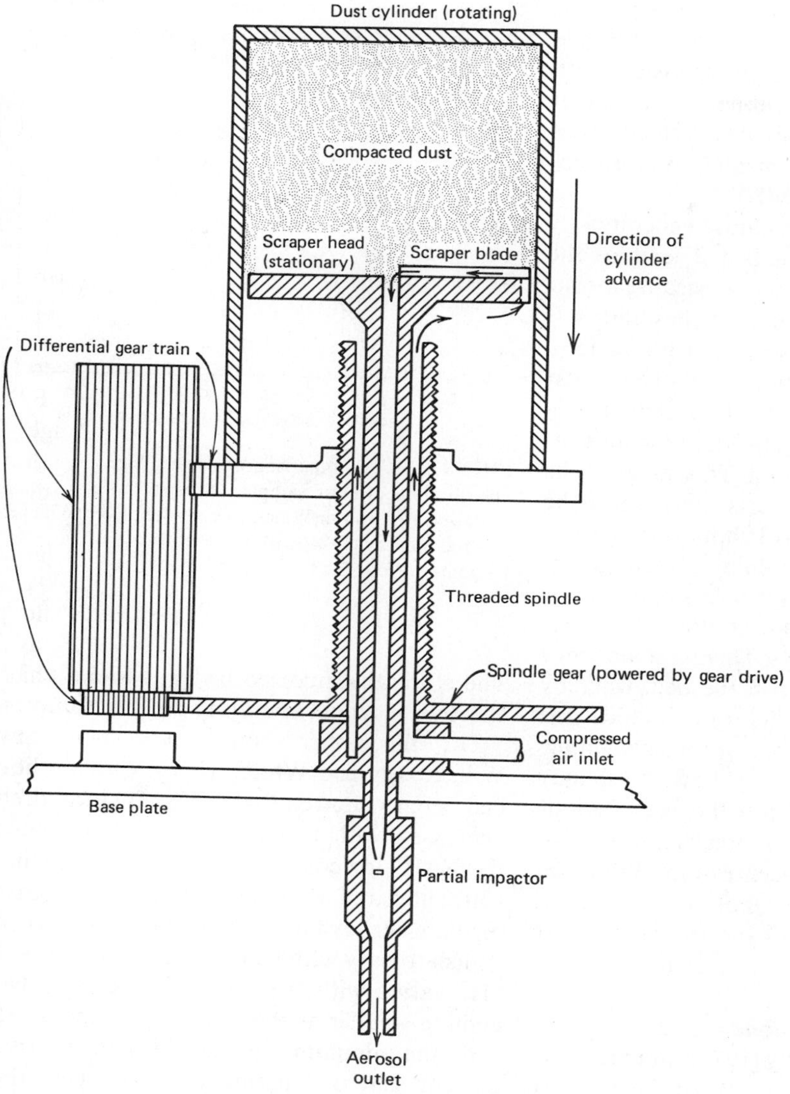

FIGURE 5-7. Wright Dust Feed. (Reprinted from Hinds (1982) with the Permission of John Wiley & Sons, Inc.)

ing, output. It is believed that the initial high rate is due to the elutriation of fine particles which are not attached to bed particles. Subsequent emission is from particles which have adhered to the surfaces of bed particles. The details of the aerosolization are not known, but a plausible scenario is that in the collisions between bed particles, adhering powder or dust particles are knocked off the bed particles. Most of the dust particles reattach to another bed particle, but some are elutriated from the bed. The constant action of the bed promotes the uniform coating of the bed particles with a layer of dust particles, accounting for the deagglomeration which is observed. It can be inferred that the best operation, in terms of the quality and constancy of the aerosol, will occur when the loading of the bed is such as to coat the bed particles with a monolayer of dust particles.

This implies a loading of about one volume percent of dust particles relative to the bed particles. The aerosol output increases with loading and the total volume of the bed. It increases initially with air flow velocity past the MFV, but reaches a maximum at a velocity of about twice the MFV.

Fresh surfaces are constantly generated by the grinding action in the bed. Therefore, the use of dry air is recommended because moisture promotes oxidation of particle surfaces. If oxidation is not a problem, air with > 15% RH may provide a better control of electrical charging (Baron and Deye 1990). Strong electrical charging is usually produced from contact and triboelectrification. This suggests the use of grounded-metal parts. Because of the high charging, the aerosol should be neutralized before use. Some fluidized-bed generators incorporate strong sources of sound or vibration. Vibration can improve the performance in several ways. There is a tendency for channeling to occur in the bed, whereby the air flows at a higher rate in localized areas. Vibration can suppress the channeling, promoting a more uniform flow. Dust may collect on the walls above the bed. Periodically, the accumulation breaks loose, producing a burst in concentration. Vibration inhibits such an accumulation of dust. Vibration also improves the feeding of dust into the bed from a screw or chain.

Particle Size Distributions

Willeke, Lo, and Whitby (1974) compared the particle size distributions of air cleaner test dust, coal dust, and fly ash after dispersion in water containing a wetting agent and after dispersion by a fluidized bed. The particles were sampled onto filters and sized by an optical microscope. The results are shown in Fig. 5-8. The number distributions are lognormal with number median diameters (NMD) and geometrical standard deviations, which agree closely for both methods of dispersion. For fly ash, which had the smallest NMD, dispersion in water was slightly more efficient; however, the difference was considered to be within experimental uncertainties. Also, no particle agglomerates were seen.

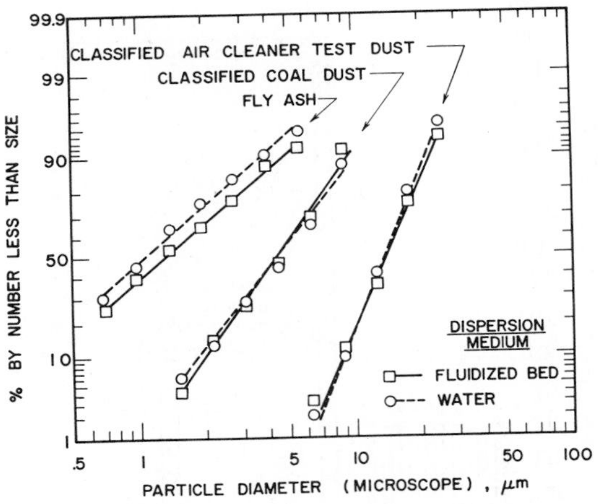

FIGURE 5-8. Comparison of Particle Size Distributions of Dust Samples Dispersed by a Fluidized Bed and in Water. (Reprinted from Willeke, Lo, and Whitby (1974) with the Permission of Pergamon Press, Inc.)

Therefore, the fluidized bed was found to be effective in deagglomerating the dust samples.

On the other hand, experiments by Willeke, Lo, and Whitby (1974) with varying bed heights showed that the NMD for fly ash decreased with increasing bed height while the NMD for coal dust increased somewhat. Carpenter and Yerkes (1980) found the mass median aerodynamic diameter (MMAD) of fly ash to vary with bed loading. The MMAD also varied with the superficial velocity, becoming smaller as the velocity was increased from the minimum fluidization velocity. It is evident that bed parameters can affect the dispersion of the aerosol. The degree of deagglomeration which is desirable depends on the application. In many cases the aim is to simulate the aerosol produced by a source such as a smoke stack or a fugitive dust source. The aerosol from these sources will have a certain degree of agglomeration which can only be determined by measurement. For example, Gillette (1974) has shown that soil aerosols have a size distribution which agrees with that obtained by dispersing soil samples in water. If detergent is added to the water, the resulting size distribution has more small particles than does airborne soil.

Types of Generators

An example of a two-component fluidized-bed aerosol generator is shown in Fig. 5-9. The fluidized bed has a 1.4 cm thick layer of 100 μm brass beads (stainless steel is also frequently used) in a 5.1 cm diameter chamber. This type of generator is commercially available (Model 3400, TSI, Inc., St. Paul, MN). While such a generator is useful for many purposes, the author has some cautions to offer, based on experience. When the generator is turned on with fresh metal bed particles, an aerosol is produced from the bed particles themselves. Initially, the source of particles is the fraction of small particles which are invariably present, i.e., the bed particles are never truly monodisperse. These small particles are rapidly cleared from the bed. Therefore, the bed should be operated first at a higher than normal flow velocity to remove them. Even after the small particles are removed, a fine aerosol persists for a long period of time, of the order of a week of operation. An examination of the bed particles under a microscope reveals that the bed particles become smoother with time. The grinding action in the bed removes asperities, which become aerosolized. This type of background is not observed with glass beads, which are smoother and which have less violent collisions because of their lower mass. After operation with a dust sample, a fluidized bed cannot be cleaned up effectively. If the same type of aerosol is needed again, the bed can be emptied and the material stored for future use.

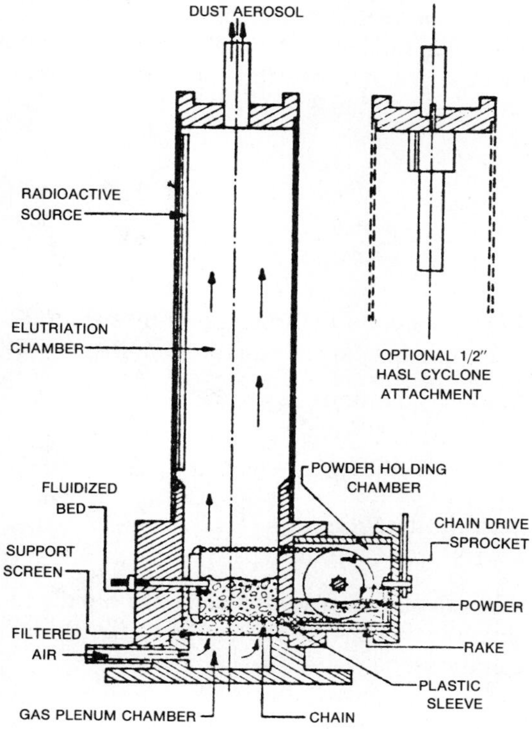

FIGURE 5-9. Two-Component Fluidized-Bed Aerosol Generator. (Reprinted from Marple, Liu, and Rubow (1978) with the Permission of American Industrial Hygiene Assoc.)

In the author's laboratory, the properties of some dusts were found to be altered in a generator having large metal bed particles. Aluminum aerosol from such a bed was found to consist of flattened disks, indicating that the collisions between the bed particles produced pressures exceeding the yield strength of aluminum. Aluminum oxide particles appeared to be broken up in the bed. Such alterations of particle properties could be significant for the subsequent use of the aerosol. The problem can be alleviated by using smaller bed particles obtained by sieving the metal bed particles before use.

John and Wall (1983) developed a sonic fluidized bed, which avoids some of the problems of the large beds and which is useful for some applications. The bed's main feature is its small size, 25 mm diameter at the base, requiring less than one gram of bed particles. The bed is funnel shaped so that the fluidizing velocity is higher than the exit velocity, favoring control of the elutriation velocity. For bed particles, 200 μm glass beads of the type used in gas chromatographic columns can be used. Such beads are highly uniform and clean. Because of the small mass required, they can be discarded when changing dust samples. The sonic fluidized bed (Fig. 5-10) is vibrated by inexpensive piezoelectric crystals which are driven by an electronic oscillator at approximately 9 kHz. Because the bed lacks a feed system, it can only be used in a batch

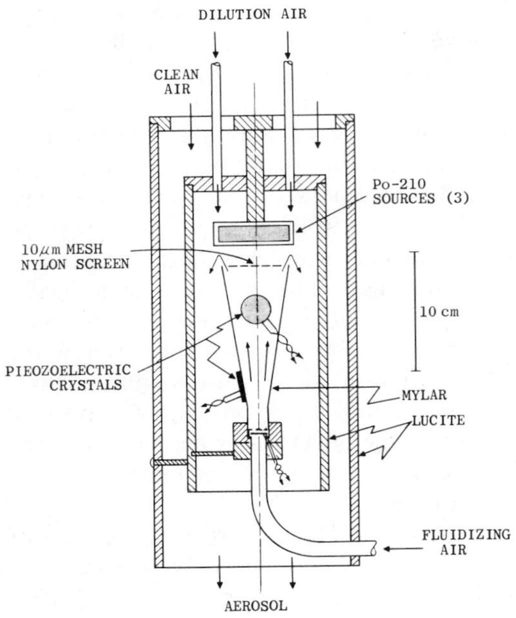

FIGURE 5-10. Sonic Fluidized-Bed Aerosol Generator. (Reprinted from John and Wall (1983) with the Permission of Pergamon Press, Inc.)

mode. It has been used successfully to generate aerosols of glass beads, A/C test dust and Pinole soil for the testing of aerosol samplers. The Pinole soil was simply passed through a coarse screen and placed in the bed without bed particles, the coarse soil particles functioning as bed particles.

Spurny, Boose, and Hochrainer (1975) developed a fluidized bed for the generation of aerosols of asbestos fibers. A special feature of this generator is a mechanical vibrator with adjustable amplitude and frequency. The effect of these vibration parameters on the aerosol concentration, fiber diameter, and fiber length was explored for several varieties of asbestos. It was found that the aerosol characteristics can be controlled to some extent by adjusting the vibration parameters. The generator was found capable of producing useful asbestos aerosols.

Example 5-6

A fluidized bed contains 200 μm glass beads with a density of 2.2 g/cm^3. What is the maximum fluidizing velocity which can be used to avoid elutriation of the bed particles?

Answer. The limiting fluidizing velocity will be equal to the settling velocity of the bed particles. As a first estimate, we calculate the settling velocity, v_s, from the Stokes law:

$$v_s = \frac{\rho_p d_p^2 g}{18\eta} = 266 \text{ cm/s}$$

Then the particle Reynolds number, Re_p, can be calculated from

$$Re_p = \frac{\rho_g v_s d_p}{\eta} = 35$$

Since the particle Reynolds number exceeds five, the motion is out of the Stokes regime, requiring the use of the drag coefficient (see Chapter 3):

$$C_D = \frac{24}{Re_p}(1 + 0.158 Re_p^{2/3}) = 1.8$$

Then the settling velocity at the high Reynolds number can be obtained by equating the drag force to the gravitational force:

$$F_D = C_D \frac{\pi}{8} \rho_g d_p^2 v_s^2 = mg = \frac{\pi}{6} d_p^3 \rho_p g$$

$$v_s = \left(\frac{4\rho_p g d_p}{3 C_D \rho_g}\right)^{1/2} = 163 \text{ cm/s}$$

We now iterate the calculation by substituting the new v_s in the equation for Re_p.

Second iteration: $v_s = 163$, $Re_p = 21$, $C_D = 2.5$, $v_s = 138$ cm/s
Third iteration: $v_s = 138$, $Re_p = 18$, $C_D = 2.8$, $v_s = 131$ cm/s
Fourth iteration: $v_s = 131$, $Re_p = 17.5$, $C_D = 2.83$, $v_s = 130$ cm/s.

The calculation is seen to converge rapidly to give the velocity to better than 1% in four iterations.

AEROSOLS FOR TEST FACILITIES

Some uses of aerosols require specialized, permanent facilities with capabilities beyond those normally available in an aerosol laboratory. For example, it may be necessary to test and calibrate a group of aerosol instruments under controlled and reproducible conditions. A calibration chamber may be used for this purpose. If the performance of an aerosol sampler is to be determined under varying wind speeds, a wind tunnel is required. To investigate the health effects of toxic aerosols on animals, an exposure chamber may be used. The general considerations for the design of such test facilities include the type of aerosols to be used and the means to transport the aerosols within the facility. Several categories of test facilities will be discussed below.

Calibration Chambers

A calibration chamber may be used to measure the efficiency of an aerosol sampler for a particular type of aerosol or for a given particle size distribution. Another application is the calibration of an aerosol instrument such as an optical counter for particles of known size. It may be desired to compare the responses of different instruments sampling the same aerosol side by side. The chamber is basically an enclosure which can be filled with aerosol and which can accommodate the instruments to be tested. The test aerosol is generated and transported to the chamber, mixed within the chamber, and monitored with suitable instrumentation.

The choice of the aerosol generator depends upon the type of aerosol required, e.g., whether it is a calibration aerosol such as monodisperse oleic acid, or a specific kind of aerosol such as coal fly ash. Other considerations are the aerosol concentration, the time duration of the testing, the particle size distribution, etc. The generator is typically located outside the chamber, for convenience of access, and the aerosol transported to the chamber. The mixing of the aerosol within the chamber is critically important so that the response of the instrument to be tested can be calibrated in terms of the reference sampler. A fan may be used to stir the air. Because the air speed near the samplers is very low, it is extremely difficult to avoid local effects due to convection currents. For example, convection caused by heat from an optical particle counter can dominate the air flow within a chamber. The air circulation can be visualized by the use of smoke or by directing a laser beam across the chamber to scatter light from the aerosol. The control of near-zero air speeds is much more difficult than the control of a directed flow at a given air velocity. For this reason, serious consideration should be given to alternative test arrangements, such as sampling from a manifold or using a small, low-wind-speed tunnel. Aerosol instruments are seldom used in a static environment, so that testing in a defined air flow may be an advantage. Nevertheless, for some applications, such as the testing of bulky instruments or many samplers simultaneously, a chamber may be preferable. Such a chamber has been described by Marple and Rubow (1983).

Wind Tunnels

The wind speed dependence of the efficiency of an aerosol sampler may need to be determined if the aerosol is in the inertial size range and if the environment to be sampled presents significant wind speeds. For this type of testing, specialized wind tunnel facilities have been developed. Wind tunnels are large, expensive facilities and, for this reason, few in number. Some of the major existing wind tunnels for aerosol testing were developed or used in connection with the development of the PM-10 (particulate matter, 10 μm) standard by the U.S. Environmental Protection Agency (EPA). Such facilities include the EPA wind tunnel at Research Triangle Park, NC (Ranade et al. 1990) (Fig. 5-11), and the facilities at Colorado State University (now at Wedding & Associates, Fort Collins) (Wedding et al. 1980), Texas A & M University (McFarland and Ortiz 1982), and Warren Spring Laboratory in UK. The EPA standard establishes performance standards for PM-10 samplers, including requirements for wind tunnel testing. The technology was advanced significantly as a consequence of the experimentation and intercomparisons in various facilities during the PM-10 program.

The air is moved through a wind tunnel by fans driven by variable-speed motors. In

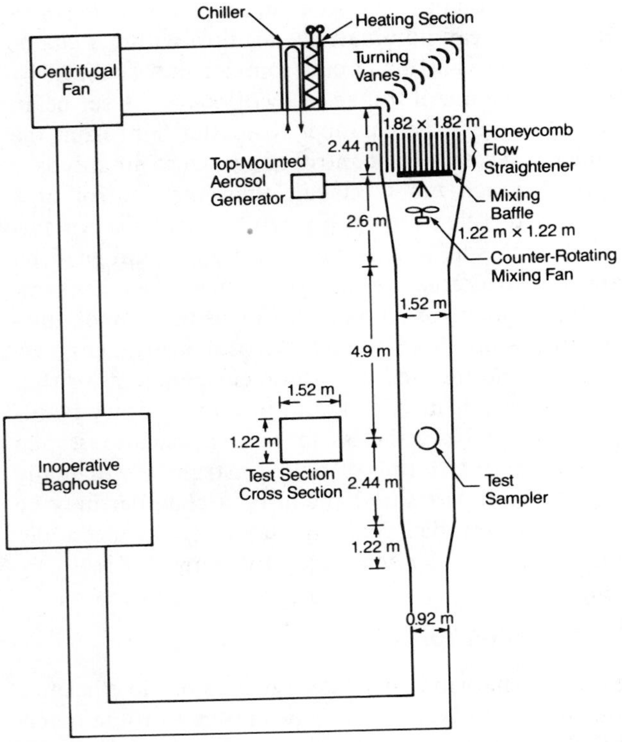

FIGURE 5-11. Schematic of the U.S. EPA Wind Tunnel Test Facility at Research Triangle Park, North Carolina. (Reprinted from Ranade et al. (1990) with the Permission of Elsevier Science Publishing Co., Inc.)

some wind tunnels the air is drawn from the room and exhausted from the other end; other tunnels are constructed in a closed loop so that the air is recirculating. This allows a buildup of the aerosol concentration, removal of ambient particulate matter, and humidity control. Grid structures in the tunnels mix the air and create a level of turbulence. The turbulence intensity is measured by a hot wire anemometer and is expressed in terms of the root-mean-square fluctuation of the wind speed, as a percentage of the wind speed. This is typically less than 10%. The scale and intensity of turbulence in the wind tunnels are generally not representative of those in the ambient air. Limited tests in wind tunnels with varied turbulent intensities do not appear to show any marked effects on measured sampling efficiencies.

For measurements of sampling efficiency, most of the wind tunnel testing is done with monodisperse liquid particles such as oleic acid containing uranine tracer. Solid particles, usually ammonium fluorescein, are used for comparison, in order to test for bounce and reentrainment. These aerosols are generated by a vibrating orifice or spinning disk. Some testing has been done with graded aluminum oxide particles dispensed by a dust generator. Injection of the aerosol into the wind tunnel presents some challenges because of the large cross section, the high settling velocities of the larger particles, and the stringent requirements on the spatial uniformity of the aerosol concentration at the test section. In the Texas A & M facility, the vibrating-orifice aerosol generator is located at the tunnel entrance. Grids downstream provide mixing. In the Colorado State facility, the vibrating-orifice generator is operated upside down. The aerosol is injected from a pipe near the top of the tunnel and mixed by cylindrical baffles downstream. The Warren Spring facility has an unusually wide cross section, 4.3 m, with an array of 14 aerosol injection nozzles, each with an adjustable iris diaphragm.

The test section has a square or rectangular cross section which must be large enough to minimize the effects on the air flow caused by the presence of the sampler under test. A rule of thumb is that the blockage should be less than 5% of the cross sectional area. For PM-10 testing, the maximum variation of the aerosol concentration over the area of the inlet must be less than 10% of the mean. The aerosol concentration is determined by an array of isokinetic filter samplers which are placed in the tunnel at the sampler test location. Sampling is done before and after the sampler test run. Efficiency measurements can be made with a coefficient of variation of a few percent. The cutpoint (particle diameter for 50% efficiency) of a PM-10 sampler can be determined to within less than 1 μm, but the results obtained in different wind tunnel facilities frequently differ by more. It appears that there are still some uncontrolled effects which have not been identified.

PM-10 testing is carried out at wind speeds from 2 to 20 km/h (0.56–5.6 m/s). This spans the normal ambient wind speeds, which in the U.S. average 15 km/h. The maximum wind speed attainable in the wind tunnels is about 40 km/h. Samplers to be used in workplaces have not yet been regulated with respect to windspeed testing, but estimates of appropriate wind speeds are: 0.3–1.5 m/s for indoors and 0.5–4.0 m/s for outdoors.

Exposure Chambers

An important kind of testing for health effects from breathing of aerosols involves animal exposures. Large numbers of animals must be exposed to aerosols for long periods of time, requiring special exposure chambers. An additional consideration is the need for safety precautions because of the toxic nature of the aerosols which include radioactive and carcinogenic substances. The aerosol must be transported to the exposure site for each animal and the exposure must be quantitatively determined.

Exposure chambers are discussed in detail in Chapter 38.

References

Baron, P. A. and G. J. Deye. 1990. Electrostatic effects in asbestos sampling I: Experimental measurements. *Am. Ind. Hyg. J.* 51:51–62.

Berglund, R. N. and B. Y. H. Liu. 1973. Generation of monodisperse aerosol standards. *Environ. Sci. Technol.* 7:147–53.

Carpenter, R. L. and K. Yerkes. 1980. Relationship between fluid bed aerosol generator operation and the aerosol produced. *Am. Ind. Hyg. Assoc. J.* 41:888–94.

Gillette, D. A. 1974. On the production of wind erosion aerosols having the potential for long range transport. *J. de Recherches Atmospheriques* 8:735–44.

Guichard, J. C. 1976. Aerosol generation using fluidized beds. In *Fine Particles*, ed. B. Y. H. Liu. New York: Academic Press.

Hinds, W. C. 1982. *Aerosol Technology*. New York: Wiley.

John, W. and S. M. Wall. 1983. Aerosol testing techniques for size-selective samplers. *J. Aerosol Sci.* 14:713–27.

John, W., S. M. Wall, J. L. Ondo, and W. Winklmayr. 1990. Modes in the size distributions of atmospheric inorganic aerosol. *Atmos. Environ.* 24A:2349–59.

Lippmann, M. 1980. Aerosol exposure methods. In *Generation of Aerosols and Facilities for Exposure Experiments*, Chap. 21, ed. K. Willeke. Ann Arbor: Ann Arbor Science.

Liu, B. Y. H. 1974. Laboratory generation of particulates with emphasis on submicron aerosols. *APCA J.* 24(12).

Liu, B. Y. H. and K. W. Lee. 1975. An aerosol generator of high stability. *Am. Ind. Hyg. Assoc. J.* 36:861–65.

Marple, V. A., B. Y. H. Liu, and K. L. Rubow. 1978. A dust generator for laboratory use. *Am. Ind. Hyg. Assoc. J.* 39:26–32.

Marple, V. A. and K. L. Rubow. 1983. An aerosol chamber for instrument evaluation and calibration. *Am. Ind. Hyg. Assoc. J.* 44:361–67.

May, K. R. 1949. An improved spinning top homogeneous spray apparatus. *J. Appl. Phys.* 20:932–38.

McFarland, A. R. and C. A. Ortiz. 1982. A 10 μm cutpoint ambient aerosol sampling inlet. *Atmos. Environ.* 16:2959–65.

Raabe, O. G. 1976. The generation of aerosols of fine particles. In *Fine Particles*, ed. B. Y. H. Liu. New York: Academic Press.

Ranade, M. B., M. C. Woods, F.-L. Chen, L. J. Purdue, and K. A. Rehme. 1990. Wind tunnel evaluation of PM_{10} samplers. *Aerosol Sci. Technol.* 13:54–71.

Rapaport, E. and S. G. Weinstock. 1955. *Experientia* XI(9):363.

Reischl, G., W. John, and W. Devor. 1977. Uniform electrical charging of monodisperse aerosols. *J. Aerosol Sci.* 8:55–65.

Sinclair, D. and V. K. LaMer. 1949. Light scattering as a measure of particle size in aerosols. *Chem. Rev.* 44:245–67.

Spengler, J. D. and K. Sexton. 1983. Indoor air pollution: A public health perspective. *Science* 221:9–17.

Spurny, K., C. Boose, and D. Hochrainer. 1975. Zur zerstaubung von abesfasern in einem fliessbett-aerosolgenerator. *Staub-Reinhalt. Luft* 35:440–45.

Stern, A. C. (ed.). 1968. *Air Pollution, Vol. III, Sources of Air Pollution and their Control*, 2nd edn. New York: Academic Press.

Timbrell, V., A. W. Hyett, and J. W. Skidmore. 1968. A simple dispenser for generating dust clouds from standard reference samples of asbestos. *Ann. Occup. Hyg.* 11:273–81.

Tomaides, M., B. Y. H. Liu, and K. T. Whitby. 1971. Evaluation of the condensation aerosol generator for producing monodispersed aerosols. *J. Aerosol Sci.* 2:39–46.

U.S. Environmental Protection Agency. 1982. Air quality criteria for particulate matter and sulfur. EPA-600/8-82-029b, December 1982.

Vanderpool, R. W. and K. L. Rubow. 1988. Generation of large, solid monodisperse calibration aerosols. *Aerosol Sci. Technol.* 9:65–69.

Wedding, J. B., A. R. McFarland, and J. E. Cermak. 1977. Large particle collection characteristics of ambient aerosol samplers. *Environ. Sci. Technol.* 11:387–90.

Wedding, J. B., M. Weigand, W. John, and S. Wall. 1980. Sampling effectiveness of the inlet to the dichotomous sampler. *Environ. Sci. Technol.* 14:1367–70.

Whitby, K. T. 1978. The physical characteristics of sulfur aerosols. *Atmos. Environ.* 12:135–59.

Whitby, K. T. and G. M. Sverdrup. 1980. California aerosols: their physical and chemical characteristics. In *The Character and Origins of Smog Aerosols*, eds. G. M. Hidy, et al., p. 495. New York: Wiley.

Willeke, K., C. S. K. Lo, and K. T. Whitby. 1974. Dispersion characteristics of a fluidized bed. *J. Aerosol Sci.* 5:449–55.

Wright, B. M. 1950. A new dust-feed mechanism. *J. Sci. Instr.* 27:12–15.

6

Sampling and Transport of Aerosols*

John E. Brockmann

Sandia National Laboratories
Albuquerque, NM, U.S.A.

INTRODUCTION

Aerosol measurement frequently requires that an aerosol sample be conveyed to a measurement device. This is accomplished by withdrawing a sample from its environment and transporting it through sample lines to the device. It is not uncommon for a sample to be transported to a chamber or bag for storage and subsequent measurement. An aerosol sampling system generally consists of

1. a sample inlet, where the aerosol sample is extracted from its ambient environment (the inlet shape and geometry may vary and although this variety is briefly discussed this chapter will focus on sampling through thin-walled tubes),
2. a sample transport system consisting of the necessary plumbing to convey the aerosol sample to the measuring instrument or to a storage chamber (the components, or flow elements, consist of such items as tubes, elbows, and constrictions), and
3. a sample storage volume (although this item is optional and its presence is determined by necessity rather than by choice) that will have an additional sample inlet and transport system to the measuring instrument (the storage volume is usually an inflatable bag that is filled with the aerosol sample over a time scale that is short compared to the time spent measuring the sample).

It is desirable that the sample is representative of the aerosol in its original environment and is not affected by the sampling process. Such characteristics as particle mass and number concentration and size distribution should remain unchanged between the point at which the aerosol is sampled and the instrument performing the measurement: this is representative sampling. It is, however, difficult to prevent changes from occurring during aerosol sampling and transport. Particles, because of their inertia, do not always enter the sampling inlet representatively. They can be lost from the sample flow by contact with the walls of the sampling system. Inertial, gravitational, and diffusional forces are among the mechanisms that can act to move the particles towards a wall. Any changes should be assessed quantitatively so that the

* This work was supported by the United States Department of Energy under contract DE-AC04-76DP00789.

measurements may be corrected. Sampling practices that introduce uncharacterized changes should be avoided.

Many of the mechanisms that inhibit representative sampling depend on the aerosol particle size, so that a given sampling system may exhibit representative sampling over some range of particle size but not for particles larger or smaller than that range. Generally speaking, larger particles are more strongly influenced by gravitational and inertial forces and are more difficult to sample representatively; smaller particles with higher diffusion coefficients are more easily lost to the walls of the sampling system by diffusion. Employing an aerosol sampling system that samples representatively for the particle size range of interest is of paramount importance.

The potential factors that can cause changes in aerosol characteristics during the sampling process or can otherwise contribute to a nonrepresentative sample are:

1. Aspiration efficiency and deposition in the sampling inlet during sample extraction
2. Deposition during transport through a sampling line or during storage
3. Extremes (high or low) or inhomogeneity in the ambient aerosol concentration
4. Agglomeration of particles during transport through the sampling line
5. Evaporation and/or condensation of aerosol material during transport through the sampling line
6. Reentrainment of deposited aerosol material back into the sample flow
7. High local deposition causing flow restriction or plugging

Each of these factors will be addressed in the following sections of this chapter. The first two items are extensively dealt with in the Aerosol Extraction and Aerosol Transport sections, respectively. The third, fourth, and fifth items are addressed in the Sample Conditioning portion of the Other Sampling Issues section. The sixth and seventh items are addressed in respective sections of the Other Sampling Issues section.

The representative extraction and transport of an aerosol sample is inhibited by loss and deposition mechanisms that are driven by gravitational, inertial, and diffusive forces. The aerodynamic equivalent diameter (Eq. 3-30) is appropriate when deposition is driven by gravitational or inertial forces. Correlations describing this type of deposition are often functions of the particle terminal settling velocity, V_{ts}, the particle Stokes number, Stk, and the flow Reynolds number, Re_f:

$$V_{ts} = \tau g \tag{6-1}$$

$$Stk = \tau U/d \tag{6-2}$$

$$Re_f = \rho_g U d/\eta \tag{6-3}$$

where

τ = particle relaxation time (Eq. 3-34)
g = gravitational acceleration
U = characteristic gas velocity
d = characteristic system dimension
ρ_g = gas density
η = gas absolute viscosity

The mobility-equivalent, or diffusion-equivalent, diameter is appropriate when deposition is driven by diffusive forces. Correlations describing the diffusive deposition of particles are functions of the particle diffusion coefficient, D:

$$D = kTB \tag{6-4}$$

where

k = Boltzmann's constant
T = gas absolute temperature
B = particle dynamic mobility (Eq. 3-14)

The Stokes number is the ratio of the particle stopping distance (a measure of how quickly a particle can accommodate itself to a flowing gas) to the characteristic dimension of the flow geometry. The inertial behavior of particles is characterized by the Stokes number. Particles with large stopping distances have high inertia and large Stokes numbers. Representative sample extraction and transport become more difficult for larger particles because their higher inertia makes them less susceptible to influence by the sample flow.

Calibration

Sampling systems should be calibrated for aerosol sampling and transport efficiency at the gas flows and over the size range of interest. Ideally, a sampling system should be calibrated fully assembled under the conditions in which it will be operating. Often, a calibration of component sections at operational conditions is adequate. Calibration at other than operational conditions may be sufficient if a defensible means, such as models and correlations from the literature, is employed to apply the calibration and predict the performance under operational conditions.

Under some circumstances, a user may not have the means for aerosol calibration. The use of a specified sampling protocol or the use of calibration data found in the literature for some of the commercial samplers or for components in the sampling system may be adequate to ensure that operation will be in a range with acceptable sampling efficiency for the user's application. In this case, at least a flow calibration should be performed.

A system's sampling efficiency can be estimated when the system is composed of components that have well-characterized efficiency data and models available in the literature. A number of these models are reviewed in this chapter. In this case, operation at an estimated sampling efficiency much different from 100% increases the uncertainty in the estimate of the actual sampling efficiency.

Sample Extraction

An aerosol sample is extracted from its environment into an inlet for transport to the measuring instrument. Drawing a representative aerosol sample into an inlet is not trivial. The velocity and direction of the gas from which the sample is being drawn, the orientation of the aerosol sampling probe, the size and geometry of the inlet, the velocity of the sample flow, and the particle size are important factors in the representativeness of an extracted sample. In extracting a sample, a particle must be sufficiently influenced by the sample gas flow to be drawn into the inlet. The particle must also be transported through the inlet without being deposited in the inlet. Particle inertia and gravitational settling are impediments to representative sample extraction, and representative sampling is more difficult with increasing aerodynamic particle size.

The aspiration efficiency, η_{asp}, of a given particle size is defined as the concentration of the particles of that size in the gas entering the inlet divided by their concentration in the ambient environment from which the sample is taken. The transmission efficiency, η_{trans}, of a given particle size is defined as the fraction of aspirated particles of that size that are transmitted through the inlet to the rest of the sampling system. The inlet efficiency, η_{inlet}, is the product of the inlet and transmission efficiencies and is the fraction of the ambient concentration that is delivered to the aerosol transport section of the sampling system by the inlet:

$$\eta_{inlet} = \eta_{asp}\,\eta_{trans} \tag{6-5}$$

Sample Transport

The transport of the aerosol sample through sample lines from the inlet may occur directly to the measurement instrument or into a temporary storage volume for subsequent transport to instruments via sample lines. These sample lines may contain bends, inclines, contractions, and other flow elements; flow may be laminar or turbulent. The deposition of particles during residence in a bag and during transport will alter the characteristics of the aerosol reaching the measurement instrument. Other phenomena that will change the characteristics of the aerosol in the sample flow are particle growth by agglomeration or condensation, particle evaporation, and re-entrainment of previously deposited material into the sample flow. These phenomena are discussed later; sample transport will address aerosol deposition.

A number of deposition mechanisms may be operating and several can be operating in each flow element. Some are not well characterized and conditions where these are encountered should be avoided wherever possible. Deposition mechanisms can be dependent on the flow regime (laminar or turbulent), the flow rate, the tube size and orientation, temperature gradients, vapor condensation onto the walls of the system, and particle size. Various deposition mechanisms depend on different particle equivalent diameters. Settling and inertial deposition depend on the particle's aerodynamic diameter while diffusional deposition depends on the particle's diffusion, or mobility, diameter. The transport efficiency for a given particle size, through a given flow element, under the action of a given deposition mechanism, $\eta_{\text{flow element, mechanism}}$, is defined as the fraction of those particles entering the flow element that are not lost by that deposition mechanism during the transit of that flow element. It is defined as a function of particle size for specific deposition mechanisms that are operating in the flow element. The total transport efficiency, $\eta_{\text{transport}}$, for a given particle size is the product of the transport efficiencies for each mechanism in each flow element of the sample transport system for that particle size:

$$\eta_{\text{transport}} = \prod_{\substack{\text{flow} \\ \text{elements}}} \prod_{\text{mechanisms}} \eta_{\text{flow element, mechanism}} \qquad (6\text{-}6)$$

The sampling efficiency, η_{sample}, is the product of the inlet and total transport efficiencies:

$$\eta_{\text{sample}} = \eta_{\text{inlet}}\,\eta_{\text{transport}} \qquad (6\text{-}7)$$

This chapter presents correlations for the transport efficiencies for various mechanisms operating in various flow elements so that the reader can estimate the total transport efficiency for a sampling system.

Other Sampling Issues

There can be times when the sampled aerosol concentration (either mass or number) is too high for the sampling instrument. Under these circumstances, the sample must be diluted with clean gas to bring the concentration within the measurement range of the instrument. Uncertainty in the dilution and sample flows will produce uncertainty in the calculated concentration, which must be addressed. High number concentrations may drive the aerosol to undergo rapid coagulation, which alters the distribution; the number concentration decreases and the mean particle size increases. Dilution of the sample will arrest the coagulation process so that a representative sample can be measured.

The sampled aerosol may be in a condensing or evaporating environment. Condensation or evaporation of material on or from aerosol particles will change the size of the particles and the total suspended mass of aerosol material. To obtain a representative sample from an environment in which material (such as water vapor) is condensing on the particles, the sample may have to be conditioned by dilution or heating. Obtaining a representative sample from an environment in which particle material is evaporating from the particle is more difficult and can be addressed by minimizing the time between sampling and measurement to keep evaporation to a minimum.

In sampling from the ambient atmosphere, from a room, or from a duct, one must be concerned with the homogeneity of the aerosol throughout the volume of gas. A representative sample requires sampling at a sufficient number of points to give an accurate picture of the aerosol throughout the volume of interest (Fissan and Schwientek 1987). In the case of duct sampling, the American National Standards Institute (ANSI) standard N13.1 (ANSI 1969) provides agreed-upon sampling locations to obtain a representative sample. In room sampling, convection in the room can cause considerable inhomogeneity in the aerosol. This is especially significant in situations corresponding to very low concentrations, such as in clean rooms, where long sampling times are required for meaningful particle counting statistics to be obtained (Fissan and

Schwientek 1987). Sampler placement in this situation may be made on the basis of flow modeling or by the use of tracer smokes or fogs. A further discussion of sampler placement and sample inhomogeneity is given in Chapter 30 on indoor aerosols. Having made the reader aware of the pitfalls of inhomogeneity of the aerosol in attempting to obtain a representative sample, it is assumed that this problem has been addressed and, therefore, attention can now be focused on a single sampling point.

Summary

Correlations describing aspiration efficiency, transmission efficiencies, and transport efficiencies will be given in subsequent sections. These correlations can be used to evaluate the performance of an existing sampling system or to aid in the design of a sampling system. Because these correlations are based on assumptions and experiments that are not always the same as the reader's application, they may not be applicable for calculated efficiencies much different from 1. Because the efficiencies are particle-size-dependent, the range of particle sizes over which sampling is representative (sampling efficiency close to 1) can be estimated with a fair degree of confidence using these correlations. In designing a sampling system, parameters such as flow, line size, orientation, and length can be adjusted using the correlations to estimate the efficiency for the particle size range of interest to achieve representative sampling. Of course, the sampling system should be experimentally evaluated whenever possible. At the end of the presentation for each type of efficiency correlation, a short qualitative discussion is given on how the efficiency changes with the dependent parameters. Sampling situations to be avoided and the methods of avoiding them are also discussed.

While some of the phenomena discussed have been extensively investigated and characterized, others have not. It is the purpose of this chapter to provide the reader with some background information on aerosol sampling and transport, so that sampling systems may be accurately evaluated and appropriately designed and sampling pitfalls may be avoided. For additional information on aerosol sampling, the reader is referred to the review paper on sampling of aerosols by Fuchs (1975), to the more recent review on aerosol sampling and transport by Fissan and Schwientek (1987), and to the recent book by Vincent (1989) on aerosol sampling.

SAMPLE EXTRACTION

Aerosol sampling arises from a number of requirements. Some of them are:

1. Monitoring the ambient air for pollution
2. Monitoring air in the workplace for hazardous materials
3. Monitoring of exhaust stacks and lines to monitor pollution control equipment
4. Monitoring clean rooms for particulate contamination
5. Monitoring from manufacturing or industrial processes
6. Monitoring in experimental research

In all these applications the first step in obtaining a sample is sample extraction.

There are two basic situations in aerosol sampling. They are:

1. Sampling of particles from a quiescent environment
2. Sampling from a gas flow which carries aerosol particles

Ambient air sampling must deal with both quiescent sampling and sampling from flowing gas. It often does so by the use of an inlet coupled with an inertial particle size fractionator. The inlet samples representatively for particles smaller than some specified diameter over a specified range of ambient wind velocities and the internal particle size fractionator passes 50% of the particles of that specified diameter. Ambient air sampling is usually performed with commonly available samplers that incorporate an inlet and size fractionator in conjunction with the measurement device, usually a filter or an impactor, located immediately after the size fractionator so that transport of the sampled aerosol and the attendant losses are minimized. Liu and Pui (1981) and Armbruster and Zebel (1985) discuss inlet design and performance for ambient air samplers. These designs are tested in wind tunnels to determine their sampling efficiency as a function of particle size and wind speed. In 1987, the U.S. Environmental Protection Agency (EPA) set a standard for airborne particulate matter called PM-10 (U.S. Environmental Protection Agency 1987). PM-10 required, among other things, samplers to sample 10 μm aerodynamic diameter particles with 50% efficiency. This can be accomplished by employing an inlet with a high inlet efficiency and a fractionator that allows 50% of the 10 μm aerodynamic diameter particles to pass into the rest of the sampler. The PM-10 regulations require that for a sampler to be officially accepted as a PM-10 sampler, it must pass specified tests in a wind tunnel. The intent was to allow flexibility in sampler design while maintaining a consistency in sampler performance.

While the inlets may perform as required, the internal processes of the samplers, specifically their size fractionators, may cause them to yield results different from those expected from their qualification testing. John, Winklmayr, and Wang (1991) and John and Wang (1991) present a comparison between the Sierra–Andersen model 321A PM-10 sampler and the Wedding high-volume PM-10 sampler. They show that loading of the samplers, and whether or not the fractionator was oiled, produced an effect on the sampling effectiveness. Deagglomeration and reentrainment of collected material, caused by bombardment with sampled aerosols, was found to produce anomalous results in the Sierra–Andersen sampler. These difficulties with PM-10 samplers appear to be more in the area of instrument response but they illustrate some of the pitfalls in sampler design.

PM-10 inlets may be too bulky for applications in which an aerosol sample must be extracted from a duct or in which room air must be sampled at a number of locations. Other inlets are required for these applications.

One type of inlet is the blunt sampler. This term encompasses a number of sampler inlets ranging from what could be called thick-walled nozzles to those in which the inlet is small compared to the overall sampler dimension. Vincent, Hutson, and Mark (1982) describe a blunt sampler as one in which the sampler and inlet configuration present a large physical obstruction to the flow. An example of this type of sampling nozzle is given by Vincent, Emmett, and Mark (1985) as a 40 mm diameter flat disc with a centrally located 4 mm diameter sampling orifice. The sampling orifice need not be in a flat plate. It can be in a spherical body or some intermediate shape (Vincent 1984; Vincent and Gibson 1981). Drawbacks to the blunt sampler and the thick-walled nozzle are particle deposition on the lip or face of the sampler and subsequent reentrainment of material into the inlet, difficult-to-characterize particle bounce, and difficulty in obtaining representative sampling of larger particles.

Another inlet type is the thin-walled nozzle. This nozzle is an idealized sampling nozzle which does not disturb ambient flow and which has no rebound of particles from the leading edge into the nozzle. Sampling with a thin-walled nozzle has received more extensive study than sampling with a blunt sampler or a thick-walled nozzle. For practical usage, a nozzle can be regarded as "thin-walled" when the ratio of its external to internal diameter is less than 1.1 (Belyaev and Levin 1972). This chapter deals with sample extraction employing thin-walled nozzles.

Sampling situations in which the flow velocities are varying present a problem. Generally, the sampling velocity is not a variable quantity. A variation in the sample flow rates introduces variable transmission efficiency through the sampling lines that can effectively result in nonrepresentative sampling. In fact, most instrumentation has a measurement response that is dependent on flow rate and, consequently, operates at a fixed flow rate. Exceptions are the instruments that perform a total integral collection, such as filters. These instruments can be positioned close to the inlet, minimizing the transport distances and losses, making them relatively independent of sample flow rate. This situation still requires some type of intergal flow measurement should a variation of sample flow occur.

Still, sample flow rate may vary because a sampler is turned on and then off. If the sampling velocity is constant during the period of time in which the sampler is on, and the sampler is on for a long period of time compared to the sampled gas residence time in the sampler's inlet and sampling lines (at least a factor of 10 greater), then the dead volume of gas in the inlet and lines is cleared and is small compared to the total volume of gas sampled. The assumption of constant sampling velocity is valid.

Ambient free-stream gas velocity variations may be beyond the control of the user because of flow adjustments or conditions in the duct from which the sample is being drawn. This situation is commonly encountered. Under these conditions, one may sample at a constant sample flow rate over a range of free-stream flows and note the largest particle size for which representative sampling still occurs over this range. The measurements made with this sampling system would need to disregard particles larger than this noted size since their sampling efficiency would be effectively unknown. This is a similar approach to that used in ambient sampling but may not be optimized for large particles. One may develop an inlet along the lines of an ambient air sampler to optimize performance for larger particles over a wide range of free-stream velocities. This has been done by McFarland et al. (1988), in which they present a shrouded aerosol sampling probe that representatively samples 10 μm and smaller particles from duct flow ranging from 2 to 4 m/s.

An alternative approach to the problem of varying free-stream flows is to vary the sample flow so that representative sampling is maintained over the range of free-stream flow variation. This entails variable sample flow and should only be used under conditions in which the particle loss in sampling lines and the instrument response do not depend strongly on sample flow rate. Null-type nozzles—in which pressure measurements responding to the flow inside and outside of the nozzle are balanced at a null condition, so that the sampling and free-stream velocities are matched (Paulus and Thron 1976; Orr and Keng 1976)—can be used to obtain representative sampling. This is an active system in which the sample flow is adjusted to obtain the null condition. The null condition may, because of local fluctuation in the flow, not reflect equality in the sample and free-stream velocities. Kurz and Ramey (1988) suggest an active sampling nozzle that employs a flow sensor and flow controller to maintain representative sampling conditions over a range of duct velocities.

There are specific sampling protocols given by the Environmental Protection Agency (EPA) for source sampling of particles in stacks (U.S. Environmental Protection Agency 1974) when the sampling data are required to verify or test compliance with rules. Sample trains and procedures are specified. The reader is referred to appropriate EPA documentation for this type of sampling.

This chapter concentrates on aerosol sampling through thin-walled nozzles and the transport of the sampled aerosol through sampling lines to the instrument.

Efficiency

Withdrawal of an aerosol sample from its environment into the sampling system requires making a particle enter the sampling inlet and conveying it to the transport portion of the system. The efficiency with which this is accomplished is called the inlet efficiency. There are two components to the inlet efficiency, η_{asp} and η_{trans} of Eq. 6-5. These efficiencies are dependent on the ambient gas velocity, U_0, the inlet geometry, size, and position, the sampling gas velocity, U, and the particle's aerodynamic diameter, d_a.

In efficiently extracting a sample, the sampling gas velocity must be low enough so that the sampled particle can accommodate itself to the sampling gas flow within a distance comparable to the inlet diameter. This is an inertial condition. The sampling gas velocity must also be high enough so that the sampled particle does not settle appreciably in the time that sampling occurs. This is a gravitational-settling condition (Davies 1968).

In sampling from a flowing gas with a nozzle, it is implicitly assumed that the flow velocities are large compared to the settling velocity of the particles being sampled, i.e., that the gravitational-settling condition is met. Grinshpun et al. (1990) point out that in low-velocity sampling, the aspiration efficiency will depend on the ratio of the settling velocity to the ambient gas velocity. It is prudent to determine this ratio for the particle size of interest to ensure that the gravitational-settling condition is met.

A nozzle sampling from still air or from flowing gas may be used in various orientations with respect to gravity and the ambient gas stream flow direction. A nozzle is said to face in the direction opposite to the inlet sample flow direction. Thus, a nozzle facing upwards draws the sample downwards and a nozzle facing the gas flow draws a sample in the same direction as the gas flow. A nozzle facing the gas flow where the direction of the sample flow is aligned with that of the gas flow is said to be sampling isoaxially. Anisoaxial, or nonisoaxial, sampling occurs when the gas flow and sample flow directions are not parallel.

Sampling is said to be isokinetic when it is isoaxial and the mean sample flow velocity through the face of the inlet is equal to the gas flow velocity. Strictly speaking, the term isokinetic applies only to laminar flow in the ambient free stream. The more general term, iso-mean-velocity, is applicable to both laminar and turbulent flow conditions in the free stream. Convention, however, applies the term isokinetic to both flow regimes. This chapter employs conventional terminology but the reader should be aware of the distinction. Sampling with a sampling velocity not equal to the gas velocity is anisokinetic (aniso-mean-velocity) sampling. When the sampling velocity is higher than the gas velocity, the sampling is super-isokinetic (super-iso-mean-velocity) and when the sampling velocity is lower than the gas velocity, the sampling is sub-isokinetic (sub-iso-mean-velocity).

Figure 6-1 is a schematic diagram of isoaxial sampling with a thin-walled nozzle for isokinetic ($U = U_0$), sub-isokinetic ($U < U_0$), and super-isokinetic ($U > U_0$) flow conditions. The limiting streamline represents the boundary between gas which enters the inlet and gas which does not. Gas is always sampled representatively and particles that do not deviate from the gas streamlines will also be sampled representatively. Particles with sufficient inertia to deviate from

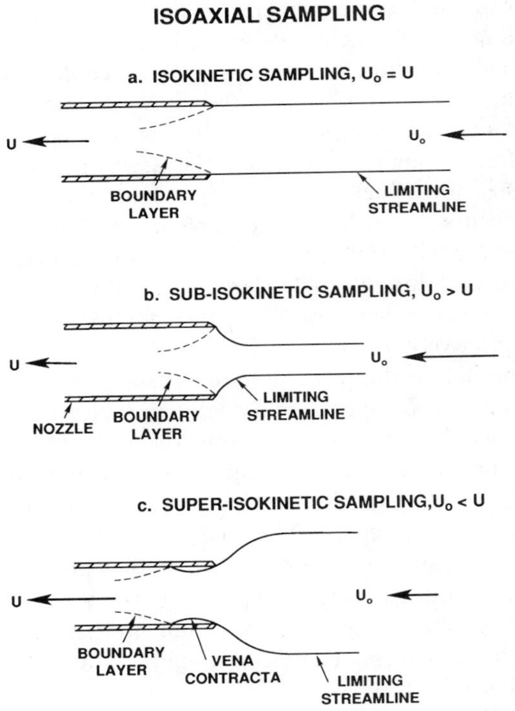

FIGURE 6-1. Schematic Diagram of Isoaxial Sampling with a Thin-Walled Nozzle, with Sample Flow Gas Velocity U and Free-Stream Ambient Gas Velocity U_0 under (a) Isokinetic ($U = U_0$) Sampling Conditions, (b) Sub-isokinetic ($U < U_0$) Sampling Conditions, and (c) Super-isokinetic ($U > U_0$) Sampling Conditions.

the streamlines may not be sampled representatively. The figures are, strictly speaking, for laminar flow in the ambient gas stream. This condition is not always encountered. Turbulent flow in the ambient gas stream introduces a lateral component to the gas velocity that in turn influences the particle motion. However, these figures are qualitatively correct in their depiction of flow and particle transport to and through the inlet for both laminar and turbulent flow conditions.

Figure 6-1a shows isokinetic sampling, in which the limiting streamline flows directly into the nozzle without deviation. In this case, the aspiration efficiency is 1 (100%). Transmission losses arise from gravitational settling inside the nozzle (Okazaki, Wiener, and Willeke 1987*b*). Losses in the inlet can also be caused by free-stream turbulence (Wiener, Okazaki, and Willeke 1988), in which the particles' lateral motion caused by the turbulence causes them to impact the inlet internal wall.

Figure 6-1b shows sub-isokinetic sampling, in which the limiting streamline must diverge from the ambient free-stream flow into the nozzle. Particles with sufficient inertia that lie outside the limiting streamline can cross the limiting streamline to be aspirated by the nozzle. In this case, the aspiration efficiency is 1 or more for all particles, increasing from 1 to a limit of U_0/U for larger particles. Transmission losses arise from gravitational settling in the nozzle (Okazaki, Wiener, and Willeke 1987*b*), from free-stream turbulent effects (Wiener, Okazaki, and Willeke 1988), and from inertial impaction on the inner wall of the nozzle by particles with velocity vectors toward the wall, caused by the expanding streamlines (Liu, Zhang, and Kuehn 1989).

Figure 6-1c shows super-isokinetic sampling, in which the limiting streamline must converge from the ambient free-stream flow into the nozzle. Particles with sufficient inertia that lie within the limiting streamline can cross the limiting streamline and not be aspirated by the nozzle. In this case, the aspiration efficiency is 1 or less for all particles, decreasing from 1 to a limit of U_0/U for larger particles. Transmission losses arise from gravitational settling in the nozzle (Okazaki, Wiener, and Willeke 1987*b*), from free-stream turbulent effects (Wiener, Okazaki, and Willeke 1988), and from turbulent deposition of particles in the vena contracta, formed in super-isokinetic sampling (Hangal and Willeke 1990*b*).

Figure 6-2 is a schematic diagram of anisoaxial sampling for flow conditions where $U_0 = U$, $U_0 > U$, and $U_0 < U$. The angle θ is the angle between the direction of the ambient free-stream gas velocity and the sampling gas velocity. Particles with sufficient inertia to cross the limiting streamlines will be aspirated at efficiencies different from 1. Transmission losses still arise from gravitational

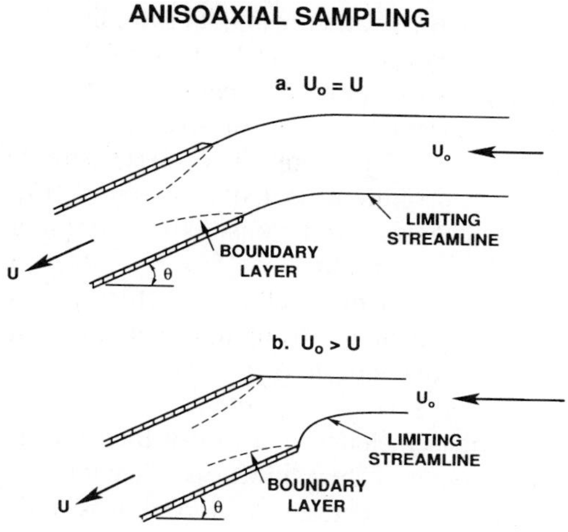

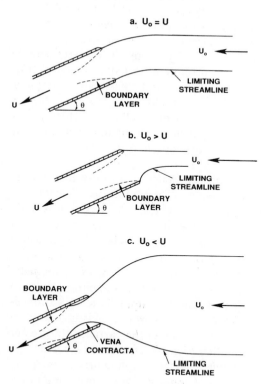

FIGURE 6-2. Schematic Diagram of Anisoaxial Sampling with a Thin-Walled Nozzle, with Sample Flow Gas Velocity U, Inclined at Sampling Angle θ to the Direction of the Free-Stream Ambient Gas Velocity U_0 under Sampling Conditions in which (a) $U = U_0$, (b) $U < U_0$, and (c) $U > U_0$.

settling in the inlet, from free-stream turbulence effects, and from losses in the vena contracta (Hangal and Willeke 1990*b*). An additional transmission loss arises from the impaction of particles on the inside lip of the nozzle facing the free-stream velocity (Hangal and Willeke 1990*b*).

Sampling from Flowing Gas with a Thin-Walled Nozzle

The correlations for aspiration efficiency and transmission efficiency are listed below, along with their equation numbers and the conditions for which they apply:

- η_{asp} — Aspiration efficiency for isoaxial sampling (Eq. 6-8), anisoaxial sampling between 0° and 60° (Eq. 6-20), and anisoaxial sampling between 45° and 90° (Eq. 6-22)
- $\eta_{trans,\,inert}$ — Transport efficiency for inertial deposition in sub-isokinetic isoaxial sampling (Eq. 6-16), super-isokinetic isoaxial sampling (Eq. 6-18), and anisoaxial sampling (Eq. 6-25)
- $\eta_{trans,\,grav}$ — Transport efficiency for gravitational settling in the inlet region of a nozzle (Eq. 6-23)

The inlet efficiency of a thin-walled nozzle is the product of the aspiration efficiency and the transmission efficiency. Isoaxial, isokinetic sampling is the ideal sampling configuration and will aspirate all particle sizes with nearly 100% efficiency. A departure from this ideal configuration into the regions of anisokinetic sampling and anisoaxial sampling results in nonrepresentative sampling; the aspiration efficiency for large particles is different from 100%, and the larger the particles, the greater the difference. Transmission losses in isokinetic, isoaxial sampling arise principally from gravitational settling in horizontal flow and the effects of free-stream turbulence. If the flow is upward or downward with respect to gravity, the transmission losses from settling will be negligible. However, the sampling velocity needs to be large compared to the particle settling velocity. If the flow is neither isokinetic nor isoaxial, then losses in the inlet from inertial effects can occur; the flow can change direction in the course of entering the inlet and the larger particles that do not follow the streamlines can be deposited on the walls.

Several researchers have theoretically and experimentally examined sampling from a flowing gas with thin-walled nozzles. They have examined isokinetic and anisokinetic sampling in isoaxial flow (Belyaev and Levin 1972, 1974; Jayasekera and Davies 1980; Davies and Subari 1982; Lipatov et al. 1986; Stevens 1986; Vincent 1987; Okazaki, Wiener, and Willeke 1987*a*, *b*; Rader and Marple 1988; Liu, Zhang, and Kuehn 1989; Zhang and Liu 1989; Hangal and Willeke 1990*a*, *b*) and in anisoaxial flow (Lundgren, Durham, and Mason 1978; Durham and Lundgren 1980; Okazaki, Wiener, and Willeke 1987*c*; Davies and Subari 1982; Lipatov et al. 1986, 1988; Vincent et al. 1986; Grinshpun, Lipatov, and Sutugin 1990; Hangal and Willeke 1990*a*, *b*). Rader and Marple (1988) give a concise summary of the work on isoaxial sampling. Hangal and Willeke (1990*a*, *b*) present a comprehensive summary of the correlations for thin-walled sampling nozzles under conditions of isoaxial and anisoaxial sampling. They further identify correlations that are applicable under each of the conditions.

It is implicit that the ambient free-stream gas velocity and the sample gas velocity remain constant over the period of sampling for the correlations to apply. The reader is cautioned that the following correlations apply only to conditions of constant gas velocities.

Isoaxial Sampling

For isoaxial sampling where the ambient gas stream velocity is U_0 and the sampling velocity is U, the well-known correlation of Belyaev and Levin (1972) for aspiration efficiency, η_{asp}, has proven satisfactory with an accuracy to within 10%:

$$\eta_{asp} = 1 + (U_0/U - 1)\,[1 - (1 + k\,Stk)^{-1}] \tag{6-8}$$

for $0.18 \leq Stk \leq 2.03$ and $0.17 < U_0/U < 5.6$, where

$$Stk = \tau U_0/d \tag{6-9}$$

$$k = 2 + 0.617(U_0/U)^{-1} \tag{6-10}$$

Stevens (1986) has reviewed the data of Belyaev and Levin (1972, 1974), Jayasekera and Davies (1980), and Davies and Subari (1982) and reported good agreement with the Belyaev and Levin correlation, extending the range of applicability down to a Stokes number of 0.05. At velocity ratios of $U_0/U < 0.2$ (super-isokinetic sampling), considerable discrepancy between the data of Davies and Subari (1982) and the correlation of Belyaev and Levin is seen. Lipatov et al. (1986) have reported experimental data for velocity ratios of U_0/U down to 0.029 (highly super-isokinetic sampling) that are in agreement with the Belyaev and Levin correlation. Lipatov et al. (1986, 1988) have concluded that the differences seen in these data are attributable to particle rebound and entrainment in the course of interaction with the outer surfaces of the sampling nozzle. Their data minimized the effects of bounce while the data of Davies and Subari (1982) did not. These results indicate that for purely aspiration efficiency, the data of Belyaev and Levin (1974) are good for U_0/U values down to 0.029, but for $U_0/U < 0.2$ particle interactions with the walls of the sampling nozzle may begin to occur. The theoretical results of Rader and Marple (1988) support the use of the Belyaev and Levin (1974) correlation over the Stokes number range of $0.005 \leq Stk \leq 10$ and the velocity ratio range of $0.2 \leq U_0/U \leq 5$ with an accuracy to within 10%. Figure 6-3 gives the aspiration efficiency as a function of Stokes number, as calculated by the correlation of Belyaev and Levin (1974). At small Stokes numbers, the efficiency is close to 1 and at large Stokes numbers the efficiency is seen to approach the limiting value of U_0/U.

Rader and Marple (1988) give a correlation for isoaxial aspiration efficiency that includes interception by the nozzle lip:

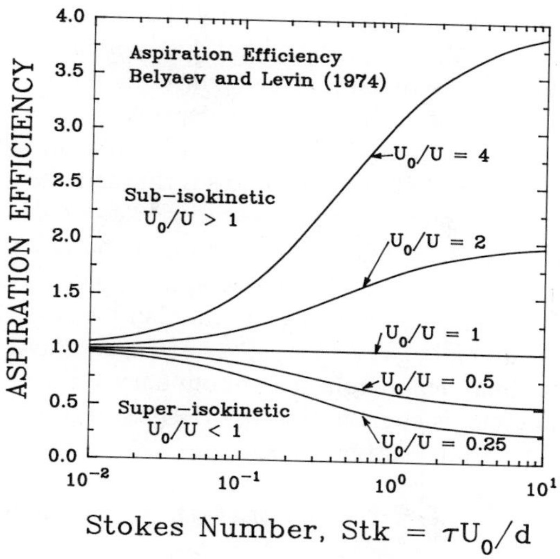

FIGURE 6-3. Plot of the Aspiration Efficiency, η_{asp}, for a Thin-Walled Nozzle as a Function of the Stokes Number (Based on the Free-Stream Ambient Gas Velocity, U_0, and the Nozzle Diameter, d) for Various Values of the Free-Stream to Sampling Gas Velocities, U_0/U, as Given by the Correlation of Belyaev and Levin (1974).

$$\eta_{asp} = 1 + (U_0/U - 1)\left[1 - (1 + 3.77\, Stk^{0.883})^{-1}\right] \tag{6-11}$$

for $0.005 \leq Stk \leq 10$ and $0.2 \leq U_0/U \leq 5$.

Liu, Zhang, and Kuehn (1989) and Zhang and Liu (1989) give a correlation for isoaxial aspiration efficiency based on numerical data:

$$\begin{aligned} \eta_{asp} &= 1 + (U_0/U - 1)/(1 + 0.418/Stk), \quad U_0/U > 1 \\ \eta_{asp} &= 1 + (U_0/U - 1)/[1 + 0.506(U_0/U)^{1/2}/Stk], \quad U_0/U < 1 \end{aligned} \tag{6-12}$$

for $0.01 \leq Stk \leq 100$ and $0.1 \leq U_0/U \leq 10$.

These correlations are very close to the correlation of Belyaev and Levin (1974) and may be used interchangeably.

Particles are deposited in the nozzle inlet by gravitational settling and by inertial effects. The particle losses in the nozzle are accounted for by the transmission efficiency.

Okazaki, Wiener, and Willeke (1987*a*, *b*) assume that a particle that has penetrated into the boundary layer formed in the entrance region of the inlet will deposit by gravitational settling on the inside wall of the inlet. The boundary layer thickness is characterized by the inlet Reynolds number, $Re = Ud/v$ (Schlichting 1968). The fraction of particles which penetrate into the boundary layer is characterized by the Stokes number based on the ambient gas velocity, $Stk = \tau U_0/d$. Gravitational deposition in the boundary layer of the inlet is characterized by the gravitational deposition parameter, Z:

$$Z = (LV_{ts})/(Ud) \qquad (6\text{-}13)$$

where L is the inlet region length.

The gravitational deposition parameter is the ratio of the particle-settling distance during transport in the inlet region, LV_{ts}/U, to the diameter of the inlet, d. Okazaki, Wiener, and Willeke (1987*a*, *b*) have performed experiments to measure the deposition inside an inlet for various particle sizes, velocity ratios, and nozzle diameters. They assume that deposition inside an inlet will be correlated to a combination of the quantities Z, Re, and Stk. The results are correlated as the gravitational-settling transmission efficiency, $\eta_{\text{trans, grav}}$, for horizontal isoaxial sampling by

$$\eta_{\text{trans, grav}} = \exp(-4.7K^{0.75}) \qquad (6\text{-}14)$$

$$K = Z^{1/2}\, Stk^{1/2}\, Re^{-1/4} \qquad (6\text{-}15)$$

The inlet diameters ranged from 0.32 to 1.59 cm but the inlet length was 20 cm for all the test data. Yamano and Brockmann (1989) point out that the analysis is performed for a laminar boundary layer in the inlet tube and that it may not apply once turbulent flow in the tube has developed. Further, if the flow remains laminar, once the laminar boundary layer fills the tube, deposition may no longer be part of the inlet effect but rather should be considered as deposition from tube flow. The single inlet length used in the data may have masked these effects and implicitly included in the correlation gravitational settling for this fixed inlet length. A comparison of the results of Eq. 6-14 to correlations of gravitational deposition in laminar and turbulent pipe flow shows that, generally, Eq. 6-14 gives a lower transmission efficiency through the inlet length. It should, therefore, provide a conservative estimate of the transmission efficiency for gravitational deposition in the inlet.

Inertial losses have been examined by Liu, Zhang, and Kuehn (1989) and by Hangal and Willeke (1990*b*).

In the case where $U_0/U > 1$ (sub-isokinetic sampling), some particles with velocity vectors directed toward the nozzle walls are deposited and the transmission efficiency is less than 1 (Liu, Zhang, and Kuehn 1989). Liu, Zhang, and Kuehn (1989) give an inertial transmission efficiency, $\eta_{\text{trans, inert}}$, for sub-isokinetic isoaxial sampling of

$$\eta_{\text{trans, inert}} = \frac{1 + (U_0/U - 1)/(1 + 2.66/Stk^{2/3})}{1 + (U_0/U - 1)/(1 + 0.418/Stk)} \qquad (6\text{-}16)$$

for $0.01 \le Stk \le 100$ and $1 < U_0/U \le 10$. Hangal and Willeke (1990) assume no inertial transmission losses for sub-isokinetic isoaxial sampling.

Liu, Zhang, and Kuehn (1989) maintain that for $U_0/U < 1$ (super-isokinetic sampling), particle velocities are not directed toward the walls and no particles are deposited. They give an inertial transmission efficiency for super-isokinetic sampling as

$$\eta_{\text{trans, inert}} = 1.0 \qquad (6\text{-}17)$$

for $0.01 \le Stk \le 100$ and $0.01 \le U_0/U < 1$. Hangal and Willeke (1990*b*), however, maintain that in super-isokinetic sampling a vena contracta is formed in the nozzle inlet and that turbulence in the vena contracta will deposit particles contained in it. They give an inertial transmission efficiency for super-isokinetic sampling as

$$\eta_{\text{trans, inert}} = \exp(-75I_v^2) \qquad (6\text{-}18)$$

$$I_v = 0.09[Stk(U - U_0)/U_0]^{0.3} \qquad (6\text{-}19)$$

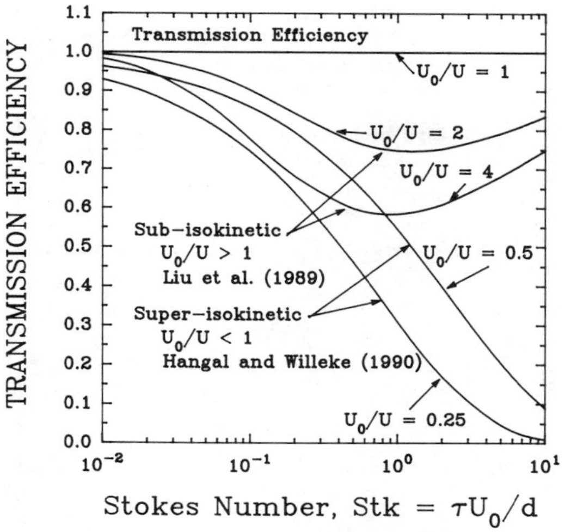

FIGURE 6-4. Plot of the Transmission Efficiency, $\eta_{\text{trans, inert}}$, for a Thin-Walled Nozzle as a Function of the Stokes Number (Based on the Free-Stream Ambient Gas Velocity, U_0, and the Nozzle Diameter, d) for Various Values of the Free-Stream to Sampling Gas Velocities, as Given by the Inertial Deposition Correlation of Liu, Zhang, and Kuehn (1989) for Sub-isokinetic Sampling and the Inertial Deposition Correlation of Hangal and Willeke (1990) for Super-isokinetic Sampling. Gravitational Deposition is Not Included.

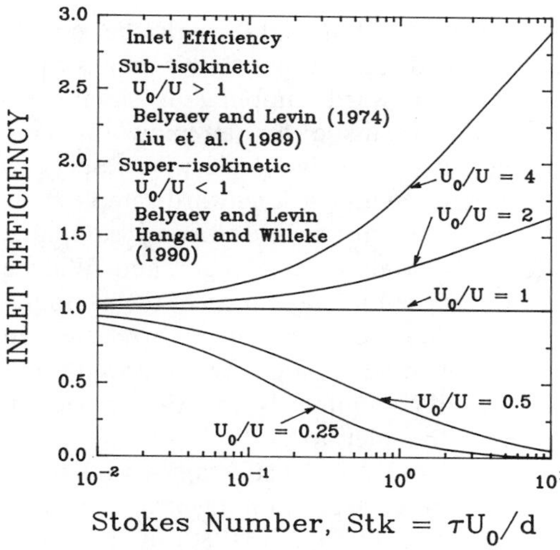

FIGURE 6-5. Plot of the Inlet Efficiency, η_{inlet}, for a Thin-Walled Nozzle as a Function of the Stokes Number (Based on the Free-Stream Ambient Gas Velocity, U_0, and the Nozzle Diameter, d) for Various Values of the Free-Stream to Sampling Gas Velocities, U_0/U, as Given by the Correlation of Belyaev and Levin (1974) Multiplied by the Inertial Deposition Correlation of Liu, Zhang, and Kuehn (1989) for Sub-isokinetic Sampling and the Inertial Deposition Correlation of Hangal and Willeke (1990) for Super-isokinetic Sampling. Gravitational Deposition is Not Included.

for $0.02 \leq Stk \leq 4$ and $0.25 \leq U_0/U < 1.0$, where I_v is the parameter describing inertial losses in the vena contracta.

Figure 6-4 shows the transmission efficiencies as a function of the Stokes number, as calculated by the Liu, Zhang, and Kuehn (1989) correlation for sub-isokinetic sampling and by the Hangal and Willeke (1990) correlation for super-isokinetic sampling. Gravitational-settling loss is not included in this figure.

The inlet efficiency for isoaxial sampling is the product of the aspiration efficiency and all applicable transmission efficiencies. Figure 6-5 plots the inlet efficiency as a function of the Stokes number. These results are calculated from the Belyaev and Levin (1974) correlation for aspiration efficiency and the transmission efficiencies given in Fig. 6-4.

Anisoaxial Sampling

Hangal and Willeke (1990*a*, *b*) have surveyed the literature on anisoaxial sampling and have performed experiments to establish a database on anisoaxial sampling. They have identified correlations with ranges of applicability for anisoaxial sampling. Deposition of particles in the inlet occurs from gravitational settling and from vena contracta deposition, as discussed in the section on Isoaxial Sampling. Anisoaxial sampling has an additional deposition mechanism that Hangal and Willeke (1990*b*) refer to as direct wall impaction. This occurs on the inside nozzle wall facing the ambient free sream; particles with sufficient inertia cross streamlines and impact on the wall. This is similar to the inlet inertial deposition identified by Liu, Zhang, and Kuehn (1989); however, their investigation dealt only with isoaxial sampling.

The anisoaxial data of Hangal and Willeke (1990*a*, *b*) were taken in horizontal free-stream flow, with the nozzle inclined upward or downward with respect to the horizontal. Their conventions indicate that a nozzle facing downward has a negative angle with respect to the horizontal, with the sample flow being directed upward; this case is referred to as upward sampling. A similar explanation is made for an upward-facing nozzle with a positive angle with respect to the horizontal and a downward sample flow; this case is referred to as downward sampling. In their correlations, Hangal and Willeke (1990*a*, *b*) use the magnitude of the sampling angle in degrees. The only correlation in which they differentiate between upward and downward sampling is for the impaction losses in the inlet lip.

Hangal and Willeke (1990*b*) found that the correlation for aspiration efficiency given by Durham and Lundgren (1980) fit their data for sampling angles from 0° to 60°. This expression is

$$\eta_{asp} = 1 + [(U_0/U)\cos\theta - 1] \times \frac{1 - \{1 + [2 + 0.617(U/U_0)]\,Stk'\}^{-1}}{1 - (1 + 2.617\,Stk')^{-1}} \times \{1 - [1 + 0.55 Stk' \exp(0.25\,Stk')]^{-1}\} \quad (6\text{-}20)$$

$$Stk' = Stk \exp(0.022\theta) \quad (6\text{-}21)$$

for $0.02 \le Stk \le 4$ and $0.5 \le U_0/U \le 2.0$ and $0° \le \theta \le 60°$. They extended the correlation of Laktionov (1973) for angles between 45° and 90° and gave the correlation for aspiration efficiency as

$$\eta_{asp} = 1 + [(U_0/U)\cos\theta - 1] \times [3\,Stk^{(U/U_0)^{0.5}}] \quad (6\text{-}22)$$

for $0.02 \le Stk \le 0.2$ and $0.5 \le U_0/U \le 2.0$ and $45° \le \theta \le 90°$.

The aspiration efficiency as a function of the Stokes number for sampling at 0°, 45°, and 90° for a range of U_0/U is shown in Fig. 6-6. Equation 6-20 is used for the 0° and 45° sampling angles and Eq. 6-22 is used for the 90° sampling angle. The 0° curves are essentially those of Belyaev and Levin (1974); one may see the rapid departure from representative sampling for anisoaxial sampling.

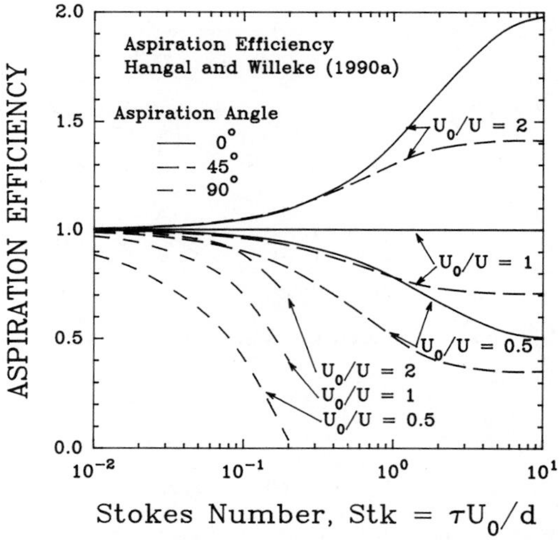

FIGURE 6-6. Plot of the Aspiration Efficiency, η_{asp}, for a Thin-Walled Nozzle at 0°, 45°, and 90° Sampling Angles as a Function of the Stokes Number (Based on the Free-Stream Ambient Gas Velocity, U_0, and the Nozzle Diameter, d) for Various Values of the Free-Stream to Sampling Gas Velocities, U_0/U, as Given by the Correlations Presented in Hangal and Willeke (1990).

Hangal and Willeke (1990*b*) have modified the expression of Okazaki, Wiener, and Willeke (1987*b*) for gravitational settling in the nozzle inlet to account for the inclination of the nozzle. The transmission efficiency for gravitational settling, $\eta_{trans,grav}$, is

$$\eta_{trans,grav} = \exp(-4.7K_\theta^{0.75}) \quad (6\text{-}23)$$

$$K_\theta = K(\cos\theta)^{0.5} \quad (6.24)$$

It is apparent that for horizontal sampling ($\theta = 0°$), K_θ is identical to K (Eq. 6-15) and Eq. 6-23 reduces to Eq. 6-14; for vertical sampling ($\theta = \pm 90°$), $K_\theta = 0$ and there are no gravitational losses. The gravitational-settling transmission efficiency depends only on the orientation of the sampling direction with respect to gravity and not on isoaxial or nonisoaxial sampling.

Hangal and Willeke (1990*b*) give the transmission efficiency for inertia that includes the losses in the vena contracta, accounted for by the parameter I_v, and the losses from direct impaction on the inner wall of the nozzle facing the ambient free-stream gas velocity, accounted for by the parameter I_w. The inertial losses from deposition in the vena contracta and from direct impaction on the nozzle inner wall are combined in the correlation for inertial transmission efficiency:

$$\eta_{\text{trans, inert}} = \exp[-75(I_w + I_v)^2] \quad (6\text{-}25)$$

for $0.02 \le Stk \le 4$ and $0.25 \le U_0/U < 4$. The vena contracta loss parameter is defined as

$$I_v = 0.09[Stk \cos\theta(U - U_0)/U_0]^{0.3} \quad (6\text{-}26)$$

for $0.25 \le U_0/U \le 1.0$ and $I_v = 0$ otherwise. The losses from direct impaction are the only losses that depend strongly on whether or not the nozzle faces upward or downward. In downward sampling, the nozzle faces upward and gravitational settling acts to move the particles away from the wall, thus reducing impaction on the wall. This is accommodated in the correlation by subtracting a quantity, α, from the sampling angle, θ. Similarly, in upward sampling, the nozzle faces downward and gravitational settling acts to move particles toward the wall, increasing the impaction losses. In this case the quantity α is added to the sampling angle, θ, which is always taken as a positive quantity in the calculations. The direct impaction loss parameter is defined as

$$I_w = Stk(U_0/U)^{0.5} \sin(\theta - \alpha) \times \sin[(\theta - \alpha)/2] \quad (6\text{-}27)$$

for downward sampling (nozzle faces upward) and as

$$I_w = Stk(U_0/U)^{0.5} \sin(\theta + \alpha) \times \sin[(\theta + \alpha)/2] \quad (6\text{-}28)$$

for upward sampling (nozzle faces downward), where

$$\alpha = 12[(1 - \theta/90) - \exp(-\theta)] \quad (6\text{-}29)$$

Figure 6-7 shows the transmission efficiency for inertial effects as a function of the Stokes number for sampling angles of 0°, 45°, and 90° at various values of U_0/U. The curve for $U_0/U = 2$ at $\theta = 0°$ is from Liu, Zhang, and Kuehn (1989) and the remaining curves are calculated from Hangal and Willeke (1990*b*). Figure 6-8 shows the inlet efficiency for the same conditions shown in Figs. 6-6 and 6-7. It can be seen that anisoaxial sampling is less representative than isoaxial sampling.

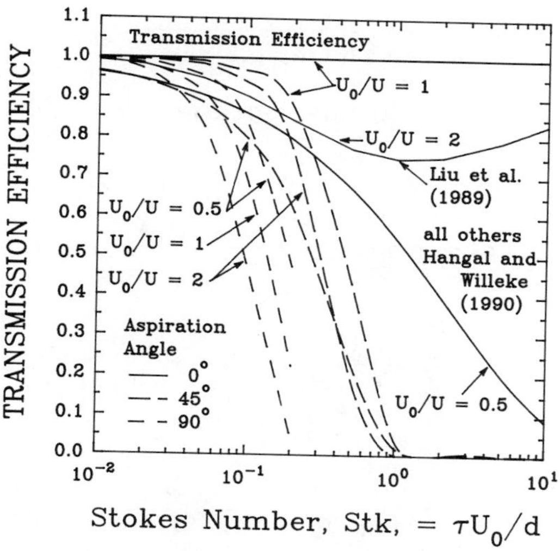

FIGURE 6-7. Plot of the Transmission Efficiency, η_{trans}, for a Thin-Walled Nozzle at 0°, 45°, and 90° Sampling Angles as a Function of the Stokes Number (Based on the Free-Stream Ambient Gas Velocity, U_0, and the Nozzle Diameter, d) for Various Values of the Free-Stream to Sampling Gas Velocities, U_0/U, as Given by the Inertial Deposition Correlation of Liu, Zhang, and Kuehn (1989) (for Sub-isokinetic Sampling at 0°) and the Inertial Deposition Correlations Presented in Hangal and Willeke (1990). Gravitational Deposition is Not Included.

Free-Stream Turbulence Effects

The limited amount of research on the effects of free-stream turbulence in sampling with thin-walled nozzles seems to indicate that there is little effect on the isoaxial aspiration efficiency (Rader and Marple 1988; Vincent, Emmett, and Mark 1985). Wiener, Okazaki, and Willeke (1988) note that although there does appear to be little effect on the aspiration efficiency, there is a measurable effect on the transmission efficiency that can increase

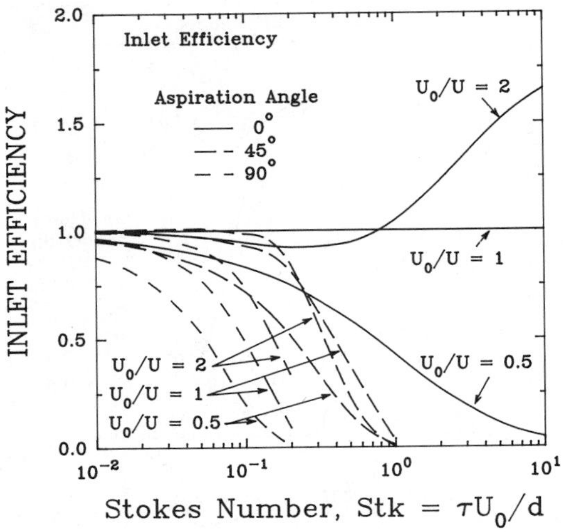

FIGURE 6-8. Plot of the Inlet Efficiency, η_{inlet}, for a Thin-Walled Nozzle at 0°, 45°, and 90° Sampling Angles as a Function of the Stokes Number (Based on the Free-Stream Ambient Gas Velocity, U_0, and the Nozzle Diameter, d) for Various Values of the Free-Stream to Sampling Gas Velocities, U_0/U, as Given by the Aspiration Efficiency Correlations Presented in Hangal and Willeke (1990) Multiplied by the Inertial Deposition Correlation of Liu, Zhang, and Kuehn (1989) (for Sub-isokinetic Sampling at 0°) and the Inertial Deposition Correlations Presented in Hangal and Willeke (1990). Gravitational Deposition is not Included.

or decrease the deposition in the nozzle inlet. Larger nozzle inlets (of the order of a centimeter in diameter) were less susceptible to these effects. They observed that for a Stokes number of less than 1 and a turbulence intensity of less than 7.5%, the spread in sampling efficiency caused by turbulence was less than 15%. This is of the order of the uncertainty in the sampling efficiency correlations.

Summary

The transmission efficiency, η_{trans}, is the product of the gravitational and inertial transmission efficiencies:

$$\eta_{trans} = \eta_{trans,\,grav}\,\eta_{trans,\,inert} \qquad (6\text{-}30)$$

The inlet efficiency, η_{inlet}, is the product of the aspiration efficiency, η_{asp}, and the transmission efficiency, η_{trans}, as given in Eq. 6.5.

The inlet efficiency for sampling with a thin-walled nozzle depends on the Stokes number based on ambient gas velocity and the nozzle inlet diameter, the ratio of ambient gas velocity to sampling gas velocity, and the sampling angle. To obtain a representative sample, the sampling should be isoaxial and isokinetic (iso-mean-velocity) and the Stokes number ($\tau U_0/d$) should be kept small. The ambient free-stream and sampling gas velocities should be large compared to the particle-settling velocity. Larger inlet diameters (of the order of a centimeter) are less susceptible to deposition caused by free-stream turbulence.

Example 6-1

Particles of 15 μm aerodynamic diameter in air at 1 atm and 20°C are sampled sub-isokinetically ($U_0 = 300$ cm/s, $U = 150$ cm/s) from horizontal flow by an isoaxial thin-walled nozzle of diameter 1.27 cm and length 10 cm. What are the aspiration, transmission, and inlet efficiencies for this particle size? If the nozzle is inclined 30° downward from the horizontal (upward sampling) what are the efficiencies?

Answer. For isoaxial sampling, the aspiration efficiency is calculated from the Belyaev and Levin correlation (Eq. 6-8), the transmission efficiency for loss from gravitational settling from Okazaki, Wiener, and Willeke (Eq. 6-14), and the transmission efficiency for loss from inertial deposition from Liu, Zhang, and Kuehn (Eq. 6-16). The inlet efficiency is the product of these three efficiencies. For the above conditions, $\tau = 6.8 \times 10^{-4}$, $Stk = 0.161$, $Re = 1230$, $Z = 0.035$, and $U_0/U = 2$. This gives the following:

Aspiration efficiency = 1.27
Transmission efficiency for gravitational deposition = 0.84
Transmission efficiency for inertial deposition = 0.86
Inlet efficiency = 0.92

For anisoaxial sampling, the efficiencies are calculated from the correlations given in

Hangal and Willeke, the aspiration efficiency is from Eq. 6-20, the transmission efficiency for loss from gravitational settling is from Eq. 6-23, and the transmission efficiency for loss from inertial deposition is from Eq. 6-25. The inlet efficiency is the product of these three efficiencies. For the above conditions, $\tau = 6.8 \times 10^{-4}$, $Stk = 0.161$, $Stk' = 0.31$, $Re = 1230$, $Z \cos\theta = 0.030$, $U_0/U = 2$, $\theta = 30°$, and $\alpha = 8°$. This gives the following:

Aspiration efficiency = 1.11
Transmission efficiency for gravitational deposition = 0.85
Transmission efficiency for inertial deposition = 0.86
Inlet efficiency = 0.81

Sampling in Calm Air

Davies (1968) points out that in sampling from calm air with a small tube at an arbitrary orientation, two conditions must be met for representative sampling. The first is an inertial condition to ensure that particles are drawn into the nozzle. This is expressed as

$$Stk \leq 0.016 \tag{6-31}$$

where the Stokes number is based on the average inlet sampling velocity, U, and the inlet diameter, d. The second is a particle-settling velocity condition to ensure that the orientation of the nozzle has no influence on sampling. This is expressed in terms of the ratio of the settling velocity to the sampling velocity:

$$V_{ts}/U \leq 0.04 \tag{6-32}$$

These two conditions constitute the Davies criterion for representative sampling through a tube in arbitrary orientation. This criterion has proven to be a sufficient condition for representative sampling.

Agarwal and Liu (1980) have established a somewhat more relaxed criterion than Davies. They have developed a theoretical prediction based on the solution of the Navier–Stokes equations for the flow field around an upward-facing inlet and a calculation of the particle trajectories and sampling efficiencies. Their prediction is supported by the experimental results of a number of researchers. The Agarwal and Liu criterion for accurate sampling (a sampling efficiency of 90% or higher) with an upward-facing nozzle is

$$Stk\, V_{ts}/U \leq 0.05 \tag{6-33}$$

or

$$\tau V_{ts}/d \leq 0.05 \tag{6-34}$$

This criterion depends only on particle relaxation time, τ, particle-settling velocity, V_{ts}, and nozzle diameter, d; it does not depend on the sampling flow velocity. Agarwal and Liu (1980) note that the experimental data indicate a dependence on the sampling gas velocity but that at higher sampling efficiencies this dependence is reduced and the criterion is adequate. Grinshpun et al. (1990) have reviewed work on sampling from calm air. They present data for $V_s' \geq 0.005$ and $Stk \geq 2.5$ that show lower efficiencies than the data of Agarwal and Liu (1980) indicate. Grinshpun et al. (1990) point out that although the Agarwal and Liu (1980) analysis is qualitatively correct, it is a first-order approximation. The supporting experimental data of Agarwal and Liu (1980) fall in a region in which V_s' is less than about 10^{-3} and the Stokes number is less than about 1000. These data are outside the Grinshpun et al. (1990) data range. This would suggest that the use of the Agarwal and Liu criterion might not apply for values of V_s' greater than 10^{-3} when the Stokes number is larger than about 1.

Figure 6-9 is a plot of these two criteria showing the regions of representative sampling indicated by each. The relative settling velocity, $V_s' = V_{ts}/U$, is on the horizontal axis and the Stokes number, Stk, is on the vertical axis.

Summary

The Davies criterion is stringent and applies to any nozzle orientation. The Agarwal and Liu criterion is more relaxed and is supported by experimental data for $V_s' \leq 10^{-3}$ and

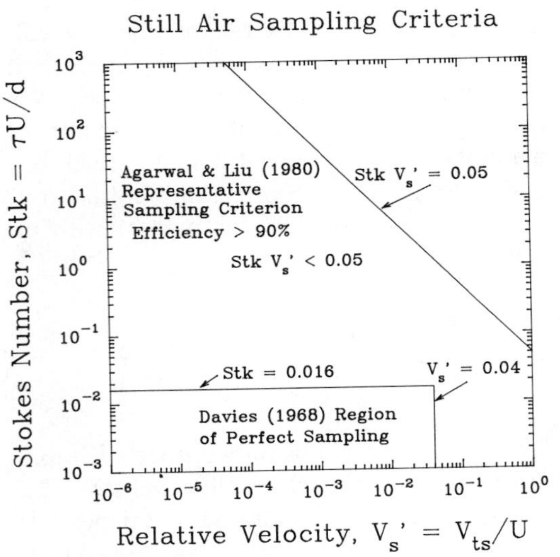

FIGURE 6-9. Plot of the Stokes Number (Based on the Inlet Velocity, U, and the Inlet Diameter, d) with Respect to the Relative Velocity, V'_s (the Ratio of Particle-Settling Velocity, V_{ts}, to Inlet Velocity, U), Showing the Regions of Representative Sampling for a Tube from Still Air as Given by the Sampling Criteria of Davies (1968) and Agarwal and Liu (1980).

$Stk \leq$ ca. 1000. However, it applies to an upward-facing nozzle and its use for other nozzle orientations is not recommended.

Example 6-2

Particles of 15 μm aerodynamic diameter must be sampled representatively from still air at 1 atm and 20°C by an instrument with a sampling flow rate of 5 l/min. What is the inlet diameter required to meet the Davies criterion and the Agarwal and Liu criterion?

Answer. The Davies criterion consists of an inertial condition (Eq. 6-31) and a gravitational-settling condition (Eq. 6-32), which can be expressed in terms of volumetric flow and solved to give the following conditions for the inlet diameter:

$$d \geq [4Q\tau/(0.016\pi)]^{1/3}$$
inertial condition

$$d \leq [0.16Q/(\pi V_{ts})]^{1/2}$$
gravitational-settling condition

For the conditions given in the example, $\tau = 6.8 \times 10^{-4}$ s, $V_{ts} = 0.67$ cm/s, and $Q = 83$ cm^3/s. This gives 1.65 cm $\leq d \leq$ 2.5 cm for the inlet size range that will meet the Davies criterion for representative sampling.

The Agarwal and Liu criterion is a single condition based on the inlet size, particle relaxation time, and settling velocity (Eq. 7-15). The condition for inlet diameter is

$$d \geq \tau V_{ts}/0.05$$

which gives $d \geq 0.009$ cm for the inlet diameter to meet the Agarwal and Liu criterion for representative sampling.

SAMPLE TRANSPORT

A sample transport system consists of the necessary plumbing to convey the aerosol sample to the measuring instrument or to a storage chamber. The components, or flow elements, consist of such items as tubes, elbows, and constrictions. The transport system should be designed to minimize the particle losses; this is often accomplished by minimizing the distance and the time of traversal between the inlet and the destination. A sample storage volume such as a bag or a chamber may be necessary in a number of applications. One example is the case of an airborne sampling system in which a sample must be taken over a short period of time, too short for the instruments to make a measurement. A sample is drawn into a bag over a short period of time and the instrument samples from the bag to make the measurement. Another example is the case of testing high-efficiency filters when the aerosol concentration exiting the filter is low and the flow rate through the filter is high compared to the measurement instrument's sample flow rate. In this case, the low concentration requires a long sampling time, long enough so that the loading of the filter is appropriate for the measurement. A sample from downstream of the filter is drawn into a bag at a high flow rate for a short period of time and the instrument samples from the bag.

During transport through the sampling line or residence time in a storage chamber, particles are lost by various deposition mechanisms. The more common ones that will be discussed in this section are listed below:

1. Gravitational settling
2. Diffusional deposition
3. Turbulent inertial deposition
4. Inertial deposition at a bend
5. Inertial deposition at flow constrictions
6. Electrostatic deposition
7. Thermophoretic deposition
8. Diffusiophoretic deposition

Correlations exist for many of the above mechanisms in laminar or turbulent flow; however, these correlations may not apply to transition flow regimes. It is prudent to avoid these transition flow regimes in the sampling lines. In the following review of particle deposition during transport, the sampling lines are assumed to be tubes of circular cross section. The overall transport efficiency, $\eta_{\text{transport}}$, is the product of the transport efficiencies in each flow element for each mechanism, $\eta_{\text{flow element, mechanism}}$, as given in Eq. 6-6.

The mechanisms and the flow elements where they operate (the correlations are given in the Sample Transport section) are listed below along with the corresponding equation numbers:

$\eta_{\text{tube, grav}}$	Gravitational settling in sample lines for laminar (Eq. 6-37) or turbulent (Eq. 6-41) flow
$\eta_{\text{tube, diff}}$	Diffusional deposition in sample lines for laminar (Eqs. 6-42 and 6-43) or turbulent flow (Eqs. 6-42 and 6-45)
$\eta_{\text{tube, turb inert}}$	Turbulent inertial deposition in sample lines for turbulent flow (Eq. 6-47)
$\eta_{\text{bend, inert}}$	Inertial deposition in sample line bends for laminar (Eq. 6-52) or turbulent flow (Eq. 6-53)
$\eta_{\text{cont, inert}}$	Inertial deposition in an abrupt contraction for laminar flow (Eq. 6-56)
$\eta_{\text{tube, th}}$	Thermophoretic deposition in sample lines for turbulent flow (Eq. 6-57)
$\eta_{\text{tube, dph}}$	Diffusiophoretic deposition in sample lines for turbulent flow (Eq. 6-61)
$\eta_{\text{bag, grav diff}}$	Combined gravitational and diffusional deposition in a storage chamber with a well-mixed volume (Eq. 6-62)

This is by no means an exhaustive list, but it covers conditions commonly encountered in aerosol transport.

Gravitational Settling in Sampling Lines

Particles settle due to gravitational force and deposit on the lower wall of nonvertical lines in a sampling system during transport. Correlations for gravitational settling and deposition under various tube orientations and flow conditions are available.

For laminar flow in a straight horizontal tube with circular cross section, Fuchs (1964) and Thomas (1958) independently solved the problem of gravitational settling by assuming a parabolic flow distribution. Their result for the transport efficiency for gravitational deposition, $\eta_{\text{tube, grav}}$, from laminar flow in a circular horizontal tube is

$$\eta_{\text{tube, grav}} = 1 - \frac{2}{\pi}[2\varepsilon\sqrt{1-\varepsilon^{2/3}} - \varepsilon^{1/3}\sqrt{1-\varepsilon^{2/3}} + \arcsin(\varepsilon^{1/3})] \tag{6-35}$$

$$\varepsilon = (3/4)Z = (3/4)(L/d)(V_{ts}/U) \tag{6-36}$$

where

Z = gravitational deposition parameter
L = length of the tube
d = inside diameter of the tube

Pich (1972) has shown that Eq. 6-35 can be used for laminar flow in elliptical channels by substituting the length of the minor axis of the ellipse for d in the definition of ε in Eq. 6-36. This correlation is applicable for either the major or the minor axis horizontal.

In a circular tube which has an angle of inclination, θ, with respect to the horizontal, the component of the sedimentation velocity of a particle perpendicular to the wall of the pipe is $V_{ts} \cos\theta$. Heyder and Gebhart (1977) modified Eq. 6-37 and obtained the following equation for gravitational settling from laminar flow in a circular inclined pipe. This correlation is in good agreement with their experimental results and can be used as the general correlation for the transport efficiency, $\eta_{\text{tube, grav}}$, for gravitational deposition from laminar flow in a circular tube:

$$\eta_{\text{tube, grav}} = 1 - \frac{2}{\pi}[2\kappa\sqrt{1-\kappa^{2/3}} - \kappa^{1/3}\sqrt{1-\kappa^{2/3}} + \arcsin(\kappa^{1/3})] \tag{6-37}$$

$$\kappa = \varepsilon \cos\theta \tag{6-38}$$

for

$$V_{ts} \sin\theta / U \ll 1 \tag{6-39}$$

The criterion expressed by Eq. 6-39 states that the axial component of a particle's settling velocity should be small compared to the average gas velocity in the tube. For a horizontal flow ($\theta = 0°$) Eq. 6-37 reduces to Eq. 6-35.

Wang (1975) has derived a theoretical formula for gravitational settling from laminar flow in an inclined tube which shows that the deposition rate depends on the flow direction in the tube; deposition from uphill flow is different from deposition from downhill flow. This observation is made when the criterion in Eq. 6-39 is not satisfied. Heyder and Gebhart (1977) show that Wang's formula reduces to Eq. 6-37 when the criterion is satisfied.

The transport efficiency for gravitational settling from laminar flow in a vertical tube is given by Eq. 6-37 as 1 (100%). There is no horizontal area on which particles will deposit by gravitation in a vertical tube. The criterion expressed in Eq. 6-39 must still be kept in mind to avoid sampling situations in which the sample gas flow is not high enough to convey the particles through a vertical sampling line.

Another consideration in laminar flow in a vertical tube is the Saffman force, which can drive particles toward the wall in downward flow or from the wall in upward flow (Saffman 1965, 1968). This force arises from the lift on a spherical particle produced by the velocity gradient in the boundary layer and leads to particle motion across streamlines. Lipatov, Grinshpun, and Semenyuk (1989) and Lipatov, Semenyuk, and Grinshpun (1990) discuss this effect. It appears to be noticeable for particles of the order of 15 μm and increases with an increasing magnitude of velocity gradient and particle size.

In the case of turbulent flow in a horizontal tube, gravitational-settling loss is assumed to occur from a well-mixed volume through the boundary layer. The transport efficiency, $\eta_{\text{tube, grav}}$, is expressed as (Schwendiman, Stegen, and Glissmeyer 1975)

$$\eta_{\text{tube, grav}} = \exp(-4Z/\pi) = \exp(-dLV_{ts}/Q) \tag{6-40}$$

where Q is the volumetric flow rate of gas through the tube.

In the case of an inclined or vertical tube, a similar modification to the one applied to gravitational settling from laminar flow may also be applied to turbulent flow. The transport efficiency for turbulent pipe flow for particles undergoing gravitational deposition is

$$\eta_{\text{tube, grav}} = \exp(-4Z\cos\theta/\pi) = \exp(-dLV_{ts}\cos\theta/Q) \tag{6-41}$$

The criterion expressed by Eq. 6-39 is also applicable in this case as in the case of laminar flow and the same caveats with respect to vertical flow apply.

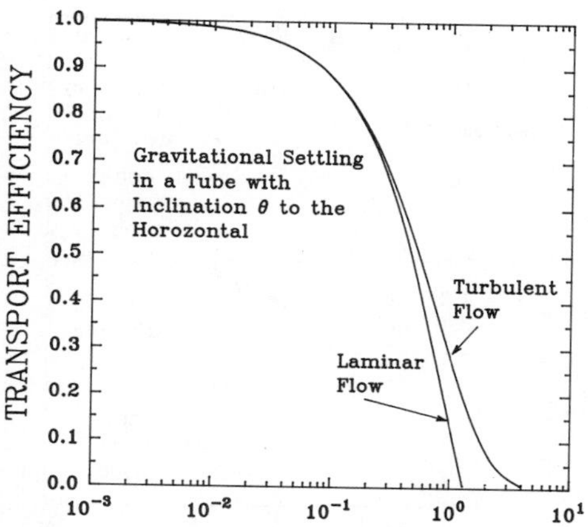

FIGURE 6-10. Plot of the Transport Efficiency, $\eta_{\text{tube,grav}}$, of Particles Undergoing Gravitational Settling Flowing through a Tube Inclined at an Angle θ from the Horizontal, as a Function of the Gravitational-Settling Parameter Z. The Correlations for Laminar and Turbulent Flows are Both Given.

Figure 6-10 shows the transport efficiency through a tube at incline θ with gravitational settling for both laminar and turbulent flow as calculated by Eqs. 6-37 and 6-41, respectively. The two flow conditions give essentially the same results for transport efficiencies greater than 0.5. An implication of this is that for a given tube diameter and length, the higher the flow, the lower the loss from gravitational settling. For both laminar and turbulent flow, gravitational deposition depends on the parameter $Z\cos\theta = (V_{\text{ts}}L/Ud)\cos\theta = (\pi d V_{\text{ts}} L/4Q)\cos\theta$. Decreasing Z yields higher transport efficiency; Z can be decreased by decreasing the transport length, L, by increasing the volumetric flow, Q, by decreasing the tube diameter for a given volumetric flow, and by increasing the inclination angle of the tube, θ.

Yamano and Brockmann (1989) show that for turbulent flow through a bending tube the transport efficiency for gravitational loss is the same as that for a horizontal tube with the same diameter and a length equal to the projected length of the bending tube onto the horizontal plane. If the flow in a bend is laminar, the velocity profile differs from that in a straight pipe, and the transport efficiency is not necessarily the same as that in a horizontal pipe whose length is the projected length of the bend onto the horizontal plane.

Diffusion in Sampling Lines

Small particles undergoing Brownian motion will diffuse from high particle concentrations to low particle concentrations. The wall of a tube acts as a sink for these diffusing particles and the concentration at the wall is taken as zero; particles will diffuse toward a wall and be deposited there. In tube flow, the transport efficiency with diffusive particle loss, $\eta_{\text{tube,diff}}$, may, in general, be expressed as

$$\eta_{\text{tube,diff}} = \exp(-\pi d L V_{\text{diff}}/Q) = \exp(-\xi Sh) \quad (6\text{-}42)$$

where

V_{diff} = deposition velocity for particle diffusion loss to the wall
Sh = Sherwood number

The diffusive deposition velocity, V_{diff}, also called the mass transfer coefficient, can be determined from the available heat and mass transfer correlations. The Sherwood number ($Sh = V_{\text{diff}}d/D$) is a dimensionless mass transfer coefficient that contains the diffusive deposition velocity in the definition. The Sherwood number is correlated with the Reynolds number ($Re_{\text{f}} = \rho_{\text{f}}Ud/\eta$) and the Schmidt number ($Sc = \eta/\rho_{\text{f}}D$) for both laminar and turbulent tube flow. For laminar flow, Holman (1972) gives

$$Sh = 3.66 + \frac{0.0668(d/L)\,Re_{\text{f}}\,Sc}{1 + 0.04[(d/L)\,Re_{\text{f}}\,Sc]^{2/3}} = 3.66 + \frac{0.2672}{\xi + 0.10079\,\xi^{1/3}} \quad (6\text{-}43)$$

$$\xi = \pi D L/Q \quad (6\text{-}44)$$

where

D = particle diffusion coefficient as given in Eq. 6-4
L = tube length
Q = volumetric flow rate through the tube

For turbulent flow, Friedlander (1977) gives

$$Sh = 0.0118\, Re^{7/8}\, Sc^{1/3} \tag{6-45}$$

The diffusional deposition velocity can be determined from the appropriate correlation above and used in Eq. 6-42 to compute the transport efficiency for tube flow in which particles are being deposited by diffusion.

This expression for the laminar flow case gives very good agreement with the analytic solution by Gormley and Kennedy (1949) for diffusional deposition from laminar tube flow. The well-known formulation of Gormley and Kennedy (1949) for the transport efficiency in laminar tube flow for particles undergoing diffusive deposition is

$$\eta_{\text{tube, diff}} = 1 - 2.56\xi^{2/3} + 1.2\xi + 0.77\xi^{4/3} \tag{6-46a}$$

for $\xi < 0.02$ and

$$\eta_{\text{tube, diff}} = 0.819 \exp(-3.657\xi) + 0.097 \exp(-22.3\xi) + 0.032 \exp(-57\xi) \tag{6-46b}$$

for $\xi > 0.02$.

Gormley and Kennedy (1949) also give an expression for the transport efficiency for diffusive deposition from laminar flow between two parallel plates.

Calculated results from Eqs. 6-43 and 6-42 and the Gormley and Kennedy solution in Eq. 6-46 differ by only a few percent over the range of interest.

In diffusive deposition from laminar flow, the transport efficiency is a function of only $\xi = \pi DL/Q$. There is no dependence on tube diameter. To increase the transport efficiency,

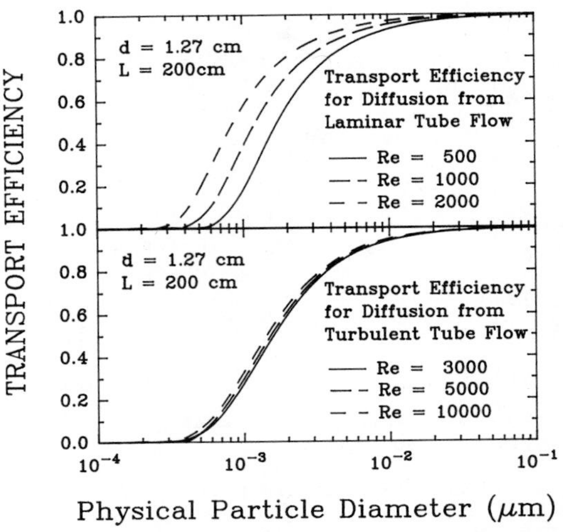

FIGURE 6-11. Plots of the Transport Efficiency, $\eta_{\text{tube, diff}}$, of Particles Undergoing Diffusional Deposition Flowing through a Tube as a Function of Particle Diameter in Laminar Tube Flow (Top) and Turbulent Tube Flow (Bottom) at 1 atm and 20°C in a 1.27 cm Diameter, 200 cm Long Tube.

ξ should be kept small by keeping the transport distance, L, small or by increasing the flow rate, Q. Figure 6-11 shows the transport efficiency for particles deposited by diffusion as a function of particle diameter. The results for laminar flow (top portion) and for turbulent flow (bottom portion) are shown for conditions of air 1 atm and 20°C in a 200 cm long, 1.27 cm diameter tube. In the laminar flow portion of the figure, the increase in the Reynolds number is caused by an increase in Q and one can see the transport efficiency increase at smaller-sized particles for increasing Q.

The experimental work of Brockmann, McMurry, and Liu (1982) supports the use of Eqs. 6-45 and 6-42 to calculate the diffusional deposition of particles from turbulent tube flow. An examination of the ξSh term for turbulent flow in Eq. 6-42 shows that transport efficiency in turbulent flow is not strongly dependent on flow rate, Q. This can be seen in Fig. 6-11. Decreasing the transport distance, L, or increasing the tube diameter, d, will act to increase the transport efficiency.

Example 6-3

A diluter has been designed to measure the sample flow by use of a capillary flow meter through which the aerosol flows. The sample flow is 5 cm^3/s through the 0.181 cm diameter 5 cm long capillary and is drawn from the air stream at 1 atm and 20°C. No particles larger than 0.1 μm are present. What is the transport efficiency for 0.1, 0.01, and 0.001 μm particles through the sampling capillary?

Answer. For the above conditions, the Reynolds number is 510 and the flow is laminar. The gravitational-settling parameter, Z, for a 0.1 μm particle is 3.9×10^{-6} and gravitational deposition is negligible. The only losses come from diffusional deposition from laminar flow (Eqs. 6-42 and 6-43). The transport efficiency for 0.1 μm particles is 1.0, for 0.1 μm particles 0.96, and for 0.001 μm particles 0.45.

Turbulent Inertial Deposition in Sampling Lines

Turbulent inertial deposition occurs when the turbulence in the central region of the pipe flow propels a particle into the laminar sublayer of the turbulent boundary layer. If the particle's inertia is sufficiently high, it will fully penetrate the sublayer and be collected on the wall.

The transport efficiency, $\eta_{\text{tube, turb inert}}$, in a tube with turbulent inertial deposition of particles using the turbulent inertial deposition velocity, V_t, is expressed as

$$\eta_{\text{tube, turb inert}} = \exp(-\pi d L V_t / Q) \quad (6\text{-}47)$$

Liu and Agarwal (1974) found that the dimensionless turbulent deposition velocity, V_+, increases rapidly with increasing dimensionless particle relaxation time, τ_+, reaching a peak of 0.14 at a τ_+ value of approximately 30. Above $\tau_+ = 30$, V_+ shows only a moderate dependence on τ_+, decreasing to 0.085 at $\tau_+ = 1000$. In the region $\tau_+ \leq 12.9$, Liu and Agarwal (1974) give the following correlation between V_+ and τ_+:

$$V_+ = (0.0006)(\tau_+)^2 \quad (6\text{-}48)$$

where

$$V_+ = 5.03(V_t/U)Re_f^{1/8} \quad (6\text{-}49)$$

$$\tau_+ = 0.0395\, Stk\, Re_f^{3/4} \quad (6\text{-}50)$$

where

V_+ = dimensionless deposition velocity
τ_+ = dimensionless particle relaxation time
V_t = deposition velocity for turbulent inertial deposition
Stk = Stokes number formulated with the tube diameter and average gas velocity in the tube

Experimental results by Liu and Agarwal also show that V_+ may be regarded as constant for τ_+ larger than 12.9. That is,

$$V_+ = 0.1 \quad \text{for } \tau_+ > 12.9 \quad (6\text{-}51)$$

Reeks and Skryme (1976) offer an explanation for the decline in deposition velocity with particle size in this region, as reflected by the decline in V_+, based on a particle's declining susceptibility to turbulence transporting it to the boundary layer. Larger particles are less influenced by turbulence than smaller ones. They present a model which accounts for the decline in V_+ at higher values of τ_+. Im and Ahluwalia (1989) also present a model that accounts for this decline and is also applicable to rough as well as smooth surfaces. Both of these models involve more than a few correlational equations and do not differ greatly from the Liu and Agarwal (1974) results.

Figure 6-12 shows the transport efficiency of particles undergoing turbulent inertial deposition as a function of particle aerodynamic diameter for various combinations of volumetric flow and tube diameter. Transport efficiency is increased by increasing the tube

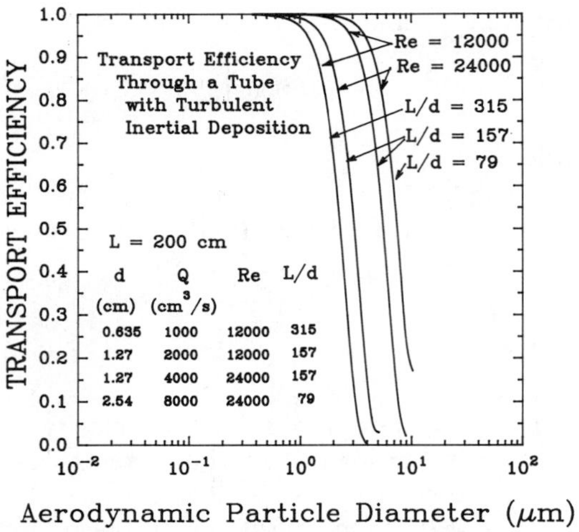

FIGURE 6-12. Plot of the Transport Efficiency, $\eta_{\text{tube, turb inert}}$, of Particles Undergoing Turbulent Inertial Deposition Flowing through a 200 cm Long Tube with Diameters of 0.635 and 1.27 cm at 500 and 1000 cm^3/s as a Function of Particle Aerodynamic Diameter at 1 atm and 20°C.

diameter, d, by decreasing the volumetric flow, Q, or by decreasing the transport length, L.

Example 6-4

Particles of 15 μm aerodynamic diameter in air at 1 atm and 20°C are transported through a 1.27 cm diameter 100 cm long tube inclined at 45° to the horizontal. Determine the total transport efficiency through the tube for flows of 5 l/min and 50 l/min.

Answer. Particle deposition could be caused by gravitational settling, diffusional deposition, and turbulent inertial deposition. For the above conditions, the Reynolds number at 5 l/min is 540 (laminar flow) and at 50 l/min is 5400 (turbulent flow). The diffusion coefficient for a 15 μm diameter particle is 1.6×10^{-8} cm^2/s and diffusional deposition is negligible in both cases.

For the 5 l/min case, the flow is laminar and the only deposition is from gravitational settling (Eq. 6-37). For the above conditions, $\kappa = 0.42$, and the transmission efficiency is 0.42. For the 50 l/min case, the flow is turbulent and deposition is from gravitational settling (Eq. 6-41) and from turbulent inertial deposition (Eq. 6-47). For the above conditions, $Z \cos\theta = 0.056$ and the transport efficiency for gravitational settling is 0.93. The Stokes number is 0.35 and the transport efficiency for turbulent inertial deposition is 0.37. This gives a total transport efficiency of 0.34.

Inertial Deposition in a Bend

When the direction of sampling gas flow is diverted in a bend, an aerosol particle may deviate from the gas flow due to its inertia and deposit on the wall of the bend. In laminar flow, secondary recirculation flow patterns develop that push the axial flow core to the outside of the bend and are responsible for particle deposition on both the inside and outside of the bend. These recirculation flows influence particle deposition and are, in turn, dependent on the flow Reynolds number and the radius of curvature of the bend. Laminar flow is more stable through a bend than through a straight tube; flow can remain laminar up to Reynolds numbers of 5000. The pertinent parameters in characterizing particle deposition from laminar flow through a bend are the curvature ratio, R_0, defined as the radius of the bend divided by the radius of the tube, the flow Reynolds number ($Re_f = \rho_f dU/\eta$), and the Stokes number formulated with the tube diameter and the average gas velocity ($Stk = \tau U/d$).

The simple empirical correlation given by Crane and Evans (1977) for transport efficiency, $\eta_{\text{bend, inert}}$, of particles undergoing inertial deposition in a 90° bend is adequate for estimates of particle deposition from laminar flow in bends and can be extended to bends of other angles:

$$\eta_{\text{bend, inert}} = 1 - Stk\,\phi \tag{6-52}$$

where ϕ is the angle of the bend in radians.

This correlation is an approximation of experimental data (Crane and Evans 1977). It assumes zero deposition only at a Stokes number of zero. The numerically calculated deposition rates of particles from laminar flow in a circular bend (Cheng and Wang 1981) and the experimental work of Pui et al. (1987) indicate that for small Stokes numbers ($Stk < 0.05$–0.1, depending on the Reynolds number) no deposition occurs. If Stokes numbers in bends in laminar flow can be kept below this limit, then no inertial deposition is expected.

Cheng and Wang (1981) present only the results of their numerical calculations for Reynolds numbers of 100 and 1000. Pui, Romay-Novas, and Liu (1987) found that the Cheng and Wang (1981) numerical results for a Reynolds number of 1000 agreed well with their experimental data whereas the calculations for a Reynolds number of 100 did not. This disagreement was attributed to differences in the assumed and actual flow fields at the lower Reynolds number. The experimental results support the use of the Cheng and Wang (1981) model for a Reynolds number of 1000 and curvature ratios, R_0, between 4 and 30. No correlation is given but the transport efficiency for laminar flow through a 90° bend at $Re = 1000$ and $R_0 = 8$ is 1 at $Stk \simeq 0.05$, 0.5 at $Stk \simeq 0.16$, and 0.1 at $Stk \simeq 0.32$. These results may be used as a guide to estimate the transport efficiency.

Pui, Romay-Novas, and Liu (1987) also found that deposition from turbulent flow in a bend can be expressed independently of the Reynolds number. They present an analysis showing that particles are deposited through the boundary layer from a well-mixed core at a constant rate, with total deposition proportional to the angle through which the aerosol flows. This analysis is in good agreement with the data. A data-based correlation for the transport efficiency of particles through a bend in turbulent flow, $\eta_{\text{bend, inert}}$, is given as

$$\eta_{\text{bend, inert}} = \exp(-2.823\, Stk\, \phi) \quad \text{(6-53)}$$

Figure 6-13 shows the transport efficiency of particles undergoing inertial deposition in a 90° bend in both laminar and turbulent flow as a function of the Stokes number. Also shown are the three values from Cheng and Wang (1981) mentioned above for $Re = 1000$ and $R_0 = 8$.

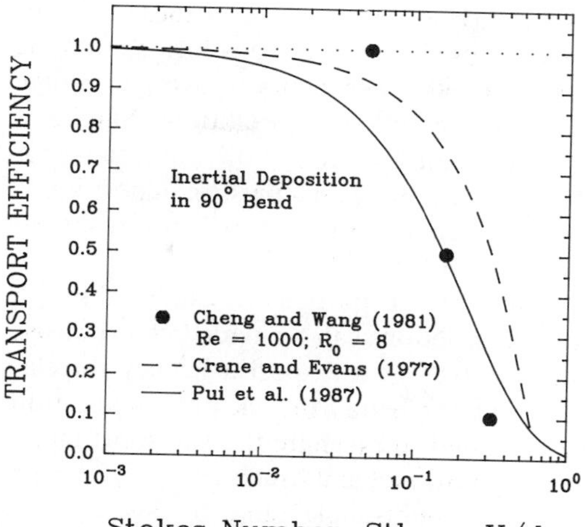

FIGURE 6-13. Plot of the Transport Efficiency, $\eta_{\text{bend, inert}}$, of Particles Undergoing Inertial Deposition in a Bend as a Function of The Stokes Number (Based on the Average Gas Velocity through the Bend, U, and the Tube Diameter, d).

When sampling lines must go through bends, the curvature ratio should be of the order of 4 or higher; abrupt turns should be avoided. To obtain a higher efficiency of transport through bends, the Stokes number should be kept small. If the flow is laminar and the Stokes number can be kept less than 0.05, losses from inertial impaction in the bend will be minimal. If the flow is turbulent, Eq. 6-53 is an adequate correlation for predicting the losses.

Inertial Deposition in Flow Constrictions in Sampling Lines

Flow constrictions in a sampling line should be avoided whenever possible; they produce losses that are difficult to characterize. Examples of constrictions are flow through a valve (although some ball valves present a little more disturbance to flow than the commonly used tube unions and connections), in-line orifices, abrupt changes in flow direction such as in the case of a tee or a cross or a right-angle bend with a curvature ratio close to 1, and changes from a large to small tube diameter either through an abrupt contraction or through a converging tube. If a sampling system having one or more of these features must be used, particle transport should be characterized experimentally over the range of applicable operating conditions. Similarly, sudden expansions in a flow path can produce eddies, from which particle deposition is difficult to characterize. These features should also be avoided.

Estimates for particle transport through flow constrictions can be made in some cases. For example, the transport efficiency correlations for a 90° bend from the previous section may be used to estimate the transport efficiency for flow elements such as a tee, a cross, or a sharp right-angle bend. In this case the highest velocity in the element should be used in the calculations.

Ye and Pui (1990) have developed a correlation for inertial particle deposition from laminar flow in a tube with an abrupt contraction. The correlation is based on numerical calculation and has been compared favorably with experimental data found in the literature. The applicable geometry consists of a large (diameter d_i) and a small (diameter d_o) coaxial circular tube, with a step reduction in cross section from the large to the small tube. This type of geometry is found or approximated in common tube and pipe fittings. Ye and Pui (1990) give a correlation for the transport efficiency, $\eta_{\text{cont, inert}}$, for particles undergoing inertial deposition in an abrupt contraction. It is a function of the tube diameter ratio (d_o/d_i) and the Stokes number based on the small-tube diameter and the average velocity in the large tube, $Stk = \tau U_0/d_o$:

$$\eta_{\text{cont, inert}} = 1 - [1 - (d_o/d_i)^2] \times [1 - \exp[(1.721 - 8.557x + 2.227x^2)] \tag{6-54}$$

$$x = Stk^{0.5}(d_o/d_i)^{0.31} \tag{6-55}$$

for $0.213 \leq x \leq 1.95$. This range of x covers the range of maximum to minimum penetration. The transport efficiency, $\eta_{\text{cont, inert}}$, is 1 for x smaller than this range and $(d_o/d_i)^2$ for x larger than this range. The numerical data of Ye and Pui have been fit to a function that goes continuously to these limits so that any value of x may be used:

$$\eta_{\text{cont, inert}} = 1 - \left[1 - \left(\frac{d_o}{d_i}\right)^2\right] \times \tanh\left[\left(\frac{x}{0.413}\right)^{(0.557 + 0.117/x^2)}\right] \tag{6-56}$$

This formulation does not fit the numerical data at high penetration as well as the function given by Ye and Pui (1990) and calculates a lower penetration than the Ye and Pui's functional form. It does agree to within a few percent.

Figure 6-14 shows the transport efficiency as a function of the Stokes number for particles flowing through an abrupt contraction for three different diameter ratios. It is seen that the large Stokes number limit of transport efficiency is the square of the diameter ratio. Transport efficiency may be kept high by keeping the Stokes number small.

The Ye and Pui (1990) correlation is applicable to the case of deposition on the upstream side of a coaxial orifice and to the case where

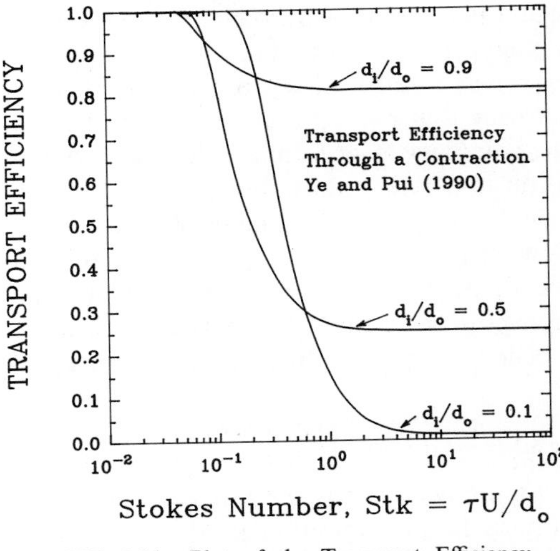

FIGURE 6-14. Plot of the Transport Efficiency, $\eta_{\text{cont, inert}}$, of Particles through a Sudden Contraction in a Tube as a Function of The Stokes Number (Based on the Gas Velocity in the Large Tube, U_0, and the Diameter of the Small Tube, d_o) for contraction ratios of 0.9, 0.5, and 0.1.

the contraction tapers at 60° instead of at 90°. In the case of an orifice, deposition downstream is not considered. In the case of a less than abrupt contraction, some deposition could occur and the Ye and Pui (1990) correlation provides at least a bounding estimate on transmission efficiency.

Electrostatic Deposition in Sampling Lines

The deposition of charged particles by electrostatic forces can occur in sampling lines during transport. Even in an aerosol that is, on an average, neutral, some particles are charged. This charging occurs from diffusion of ions to the particles. Ions are constantly produced in our environment by cosmic rays. For a given ion concentration, there is an equilibrium charge distribution over the aerosol particles. Particles may also be charged by the mechanism that produces them. Static charge in the sampling line or an externally imposed electrical field in the line can produce particle deposition. Because it is not always possible to know the distribution of charge on aerosol particles or the electrical fields in a sampling line that is subject to static charge, electrostatic deposition of particles in sampling lines is most difficult to characterize.

Electrostatic deposition is largely avoidable. Metal or electrically conductive lines set up to avoid electrical fields by having no electrically isolated sections will obviate the problem of electrostatic deposition. If metal lines cannot be used, Tygon™ is an acceptable substitute (Liu et al. 1985). Materials to be avoided for aerosol transport are Teflon™ and Polyflo™ (Liu et al. 1985).

Thermophoretic Deposition in Sampling Lines

A temperature gradient in a gas will cause a suspended particle to move down the gradient from higher toward lower temperatures. This particle transport by a temperature gradient is called thermophoresis. The thermophoretic velocity, V_{th}, of a particle is the velocity that it achieves as a result of the thermophoretic force. The thermophoretic velocity is dependent on the temperature gradient. It is independent of particle size for particles much larger than the gas molecule mean free path (continuum-regime particle) and for particles much smaller than the gas molecule mean free path (free-molecule regime). The transport efficiency in turbulent tube flow with thermophoretic deposition, $\eta_{tube,th}$, is expressed as

$$\eta_{tube,th} = \exp\left(-\frac{\pi d L V_{th}}{Q}\right) \tag{6-57}$$

In laminar tube flow, the flow and temperature gradient conditions become more involved and no expression for the transport efficiency for laminar flow is given.

The thermophoretic velocity for particles in the continuum regime is expressed as (Friedlander 1977)

$$V_{th} = \frac{2(k_g/k_p)\, k_g\, \nabla T}{5P(1 + 2k_g/k_p)} \tag{6-58}$$

For particles in the free-molecule regime the thermophoretic velocity is expressed as (Friedlander 1977)

$$V_{th} = \frac{3\nu \nabla T}{4[1 + \pi(0.9)/8]\, T} \tag{6-59}$$

where

V_{th} = thermophoretic velocity
k_g = thermal conductivity of the gas
k_p = thermal conductivity of the particle
∇T = temperature gradient in the gas
P = pressure
T = gas temperature
ν = kinematic viscosity of the gas

The estimation of thermophoretic deposition in sample lines is not a straightforward process. The determination of the temperature gradients can be difficult. Because of the relatively low heat capacity of gases, the gas rapidly attains the tube wall temperature, causing a changing temperature gradient. The particle material thermal conductivity is not always known and, in the case of an agglomerate particle, the particle thermal conductivity is not necessarily the same as the particle material thermal conductivity because of the included voids.

Thermophoretic deposition in sampling lines can be avoided by heating the lines to the gas temperature or cooling the gas to the line temperature by dilution.

Diffusiophoretic Deposition in Sampling Lines

A particle in a nonuniform gas mixture will be acted upon by the diffusing gas molecules, diffusion arising from the concentration gradients. The force which the particle experiences is called the diffusiophoretic force. This force arises from the unequal momentum transfer to the particle from the heavier molecules on the higher-concentration side of the particle. The direction of this force is in the direction of diffusion of the heavier gas molecule. This force causes the particles to move with a velocity called the diffusiophoretic velocity, V_{dph}. This phenomenon should not be confused with Brownian diffusion of the particle. Diffusiophoresis is caused by a concentration gradient in the gas molecules. Near a condensing or an evaporating surface, an additional effect influences the force acting on the particle. Near the surface, the condensing or evaporating vapor has associated with it a concentration gradient that drives the vapor's diffusion through the gas, effecting condensation or evaporation. In order to maintain a constant gas pressure, the gas must also have a concentration gradient equal to but in the opposite direction of the vapor's gradient. The concentration gradient in the gas produces diffusion of the gas toward the surface in the case of vapor evaporation and away from the surface in the case of vapor condensation. Because the surface is not a source or a sink for the gas, an aerodynamic flow called the Stefan flow is set up to transfer the gas away from an evaporating surface and toward a condensing surface to counteract the diffusive transfer of gas. The Stefan flow exerts a force on the particle in the direction of the flow, in addition to the diffusiophoretic force acting in the direction of the diffusion of the heavier molecule (Fuchs 1964; Hinds 1982). Waldman and Schmitt (1966) and Goldsmith and May (1966) give an expression for the diffusiophoretic velocity on a particle acted on by diffusiophoretic forces and the Stefan flow:

$$V_{\mathrm{dph}} = \frac{-\sqrt{m_1}}{\gamma_1\sqrt{m_1} + \gamma_2\sqrt{m_2}} \frac{D}{\gamma_2} \nabla\gamma_1 \qquad (6\text{-}60)$$

where

m_1 = mass of a molecule of the diffusing species
m_2 = mass of a molecule of the stagnant species
γ_1 = mole fraction of the diffusing species
γ_2 = mole fraction of the stagnant species
$\nabla\gamma_1$ = gradient of the mole fraction of the diffusing species

In the case of water vapor and air, the velocity of a particle is away from an evaporating surface and toward a condensing surface. For diffusiophoresis arising from condensation in turbulent pipe flow, the transport efficiency, $\eta_{\mathrm{tube,dph}}$, of particles can be estimated from

$$\eta_{\mathrm{tube,dph}} = \exp\left(-\frac{\pi d L\, V_{\mathrm{dph}}}{Q}\right) \qquad (6\text{-}61)$$

Whitmore and Meisen (1978) produced data for diffusiophoretic deposition in turbulent flow for ammonia condensing from a number of gases. The results indicated that the particle transport efficiency was comparable to the mole fraction of the noncondensing gas. Thus, if 10% of the gas transporting particles condensed, about 10% of the particles would be deposited. These results of Whitmore and Meisen (/1978) permit a rough estimate of the transport efficiency for particles undergoing diffusiophoretic deposition based on the noncondensable mole fraction of the gas.

Conditions under which diffusiophoretic deposition would occur should be avoided in the transport of aerosol samples. This can be accomplished by avoiding condensation in sampling lines by heating or dilution.

Deposition in Chambers and Bags

Situations arise in aerosol sampling in which a sampled aerosol is temporarily stored. This can happen when a sample must be taken at a rate faster than the instrument can accommodate. For example, an aircraft sampling from a plume will often have only a few seconds of time in the plume to acquire a sample but the instrument cannot make a measurement in that short a time period. The sample is drawn into a sample chamber or bag during the short transit period and the measurement is made on the stored sample over a longer period of time. Another such sampling application is the study of aerosol size distribution evolution from gas to particle conversion and coagulation. In all these applications, the deposition of particles during their residence time in the chamber can have a definite effect on the measurement. When the time to take an aerosol measurement or a series of measurements is comparable to the time for the aerosol distribution in the chamber to change appreciably from deposition, then the deposition must be accounted for.

Two deposition mechanisms are usually considered. They are gravitational settling and diffusion. Crump and Seinfeld (1981) and Crump, Flagan, and Seinfeld (1983) have developed a model for gravitational and diffusive deposition inside stirred vessels. Other, less characterizable deposition mechanisms should be avoided. Diffusiophoresis and thermophoresis can be avoided by the sampling technique, i.e., avoid temperature gradients at the chamber wall and avoid condensing conditions. Electrostatic deposition can be avoided by the selection of an electrically conductive chamber material, such as aluminized mylar, or one that can be demonstrated not to be susceptible to electrostatic deposition. Teflon™ should not be used, for reasons outlined in the section on Electrostatic Deposition. McMurry and Raaer (1985) and McMurry and Grosjean (1985) have reported experimental data in which an enhanced loss of charged particles inside Teflon™ film chambers has been observed. They have developed a model of electrostatic deposition in chambers with static charge that explains the data well.

A model for simultaneous gravitational and diffusional losses (Crump and Seinfeld 1981) has been developed for a well-mixed vessel of arbitrary shape. For a spherical vessel, they give an expression for the wall loss coefficient, β, where the remaining fraction of particles of a given size after time t is

$$\eta_{\text{bag, grav diff}} = \exp\left(-\int_0^t \beta \, \mathrm{d}t\right) \qquad (6\text{-}62)$$

and

$$\beta = \frac{12k_e D}{\pi^2 R V_{ts}} \int_0^{\pi V_{ts}/(2\sqrt{k_e D})} \frac{x \, \mathrm{d}x}{\mathrm{e}^x - 1} + \frac{3V_{ts}}{4R} \qquad (6\text{-}63)$$

where

R = vessel radius
k_e = coefficient of eddy diffusion

The first term in Eq. 6-63 describes the losses from diffusion and includes the effect of declining eddy diffusivity (mixing level) approaching the wall. The second term in Eq. 6-63 is the contribution from gravitational settling. It is equal to the settling velocity times the ratio of the deposition area of the chamber to the volume of the chamber. The deposition area for gravitational settling is the projected area on the horizontal plane. This definition is applicable for stirred settling from any vessel shape.

The coefficient of eddy diffusion characterizes the level of mixing in the chamber. Higher values indicate a higher eddy diffusivity and, consequently, a higher loss by diffusion. A chamber with a constant throughput of gas will have a constant mixing level driven by the kinetic energy of the incoming gas, but a chamber that is initially filled and then closed off will have a mixing level that decays with time. Equation 6-62 contains the integral of β with respect to time because the coefficient of eddy diffusion, k_e, can change with time. This parameter is the only unknown in Eq. 6-63; it must be determined experimentally or estimated. Crump, Flagan, and Seinfeld (1983) indicate that k_e is proportional to the square root of the ratio of the turbulent energy dissipation rate to the kinematic viscosity of the fluid. This gives

$$k_e = CQ^{3/2}/(A^2 \nu v_c)^{1/2} \tag{6-64}$$

where

C = proportionality constant
Q = flow rate into the bag
A = area of the flow inlet
v_c = volume of the chamber

They give for their experimental 118 l spherical vessel,

$$k_e = 0.0092\,Q^{3/2} \tag{6-65}$$

where k_e is in units of s^{-1} and Q is in units of l/min. Not enough information is given to estimate a proportionality constant for Eq. 6-64. The values of k_e given by Crump, Flagan, and Seinfeld (1983) range from 0.028 to 0.068 s^{-1} and McMurry and Rader (1985) use values of 0.0064 and 0.12 s^{-1}. Conservative estimations can be performed with the higher values of k_e to assess the diffusion losses in a bag or chamber. The losses arising from settling are more straightforward.

Figure 6-15 shows the fraction of initial concentration of aerosol in a 50 cm radius spherical chamber as a function of particle size, 600 s after filling. The conditions are for air at 1 atm and 20°C. Calculations for three values of k_e that span the range mentioned above are shown and k_e is assumed constant over the 600 s time period. Gravitational deposition depends only on the size of the chamber and can be reduced by using a larger chamber. The higher values of k_e produce a higher diffusion loss. A larger bag will result in a lower diffusion loss.

In filling a bag with flexible sides, particles on the bag walls can be resuspended in the gas volume as a result of the mechanical forces on the wall during filling. Generally, larger particles are more easily resuspended. The resuspension of particles from the bag should be considered in the selection of the bag and the decision to reuse bags. In some applications, single-use bags may be prudent.

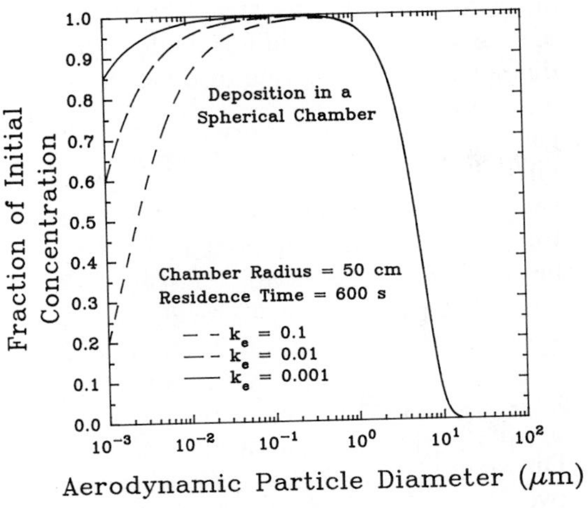

FIGURE 6-15. Plot of the Fraction of Initial Concentration, $\eta_{\text{bag, grav diff}}$, of Particles Undergoing Diffusive and Gravitational Deposition in a 50 cm Radius Spherical Chamber, 600 s after Filling, as a Function of Particle Aerodynamic Diameter at 1 atm and 20°C for Various Values of the Coefficient of Eddy Diffusion, k_e.

OTHER SAMPLING ISSUES

Sample Conditioning by Dilution

When sampling aerosols it is often necessary to reduce concentrations by dilution. Dilution may be required to bring concentrations within the range of measuring instruments, to quench chemical reactions or coagulation, or to cool a hot gas. The factors to consider in the design or selection of a diluter include the time from sample extraction to dilution (in a coagulating or chemically reacting aerosol, this time should be as short as possible), required dilution rates, particle losses in the diluter, and flow rates. Brockmann, Liu, and McMurry (1984) briefly discuss a variety of diluters and describe a diluter designed specifically for high dilution and transport of coagulating ultrafine aerosols.

An important consideration in selecting a diluter is the uncertainty with which the dilution rate is known. In diluting a sample, a dilution gas flow, Q_d, is mixed with a sample gas flow, Q_s. The total gas flow out of the diluter, Q_t, is the sum of the dilution and sample gas flows. The dilution rate, DR, is the total flow divided by the sample flow and can be expressed in terms of any of the two flows:

$$\begin{aligned} \mathrm{DR} &= Q_t/Q_s = (Q_d + Q_s)/Q_s \\ &= Q_t/(Q_t - Q_d) \end{aligned} \tag{6-66}$$

Differences in uncertainty arise, from which two flows are measured to determine the dilution rate. When Q_t and Q_s are measured, the uncertainty in the dilution rate, $U(DR)$, is related to the uncertainties in the flows, $U(Q_t)$ and $U(Q_s)$, by the relation (see the discussion of propagation of errors in Chapter 9)

$$U(\mathrm{DR}) = [U(Q_t)^2 + U(Q_s)^2]^{1/2} \tag{6-67}$$

The uncertainty is not much different when Q_d and Q_s are measured and approaches that for Q_t and Q_s for larger values of DR:

$$\begin{aligned} U(\mathrm{DR}) &= (1 - 1/\mathrm{DR}) \\ &\quad \times [U(Q_d)^2 + U(Q_s)^2]^{1/2} \end{aligned} \tag{6-68}$$

When the sample flow is one of the flows measured, the uncertainty in the dilution rate is only slightly larger than the uncertainties in the flow measurements. When Q_d and Q_t are measured, the sample flow rate is determined from the difference between the measured quantities. When this difference is of the order of the uncertainty in these measured flows, the uncertainty in the dilution rate becomes much larger than the uncertainty in either of the measured flows:

$$\begin{aligned} U(\mathrm{DR}) &= (\mathrm{DR} - 1) \\ &\quad \times [U(Q_t)^2 + U(Q_d)^2]^{1/2} \end{aligned} \tag{6-69}$$

The uncertainty in dilution rate in this situation increases with dilution rate. If the uncertainty in flow measurement is $\pm 3\%$ and the dilution rate is 10, then the uncertainties in DR for the situations described by Eqs. 6-67, 6-68, and 6-69 are $\pm 4.2\%$, $\pm 3.8\%$, and $\pm 38\%$, respectively. For higher dilutions, measurement of the sample flow is strongly recommended if the dilution rate must be known.

Measurement of the aerosol sample flow can be accomplished by drawing it through a capillary tube in laminar flow (Fuchs and Sutugin 1965; Delattre and Friedlander 1978; Brockmann, Liu, and McMurry 1984); this allows the flow to be measured. This is a well-defined system in which the particle losses can be characterized. It lends itself to high dilution rates. One must be cautious, however, to avoid plugging of the capillary.

Dilution gas can be introduced into the flow stream through holes in the wall of the containing tube or through a porous wall (Ranade, Werle, and Wasan 1976). It may also be introduced coaxially surrounding the sample gas flow. Care must be taken to avoid uneven flow, which would direct the undiluted sample flow against the walls of the sampling system that could promote uncharacterizable deposition.

The dilution gas must be well mixed with the sampled aerosol. This can be accomplished through turbulence either in the transport line or by passing the sample and dilution gas through a mixing nozzle (Brockmann, Liu, and McMurry 1984). To avoid possible particle deposition in a flow constriction, care should be taken to sheath the sample with clean gas during flow in the converging section of the nozzle or to introduce the sample flow into the throat of the nozzle. Mixing will occur in the diverging portion of the nozzle.

Plugging of Sampling Lines and Inlets

Aerosol flowing through cracks and leaks in pressurized systems has been observed to deposit in the flow paths and plug them. This has also been observed when large amounts of aerosol have flowed through pipes (Morewitz 1982). A simple model of aerosol plugging of flow paths has been developed

(Vaughn 1978; Morewitz 1982) in which the integral mass of aerosol, m, that flows through a pipe up to the point of plugging is correlated to the diameter of the pipe, d, through which it flows:

$$m = Kd^3 \tag{6-70}$$

where

m = total mass in grams of aerosol that flowed through the pipe prior to plugging
K = proportionality constant that is between 10 and 50 g/cm^3
d = pipe diameter in cm

Equation 6-70 has been referred to as the Morewitz criterion for aerosol plugging (Novick 1990). It is a very general correlation that does not depend on aerosol concentration, particle size, or flow rate.

Novick (1990) has observed situations in which the Morewitz criterion was met but no plugging occurred. Novick (1990) has examined data, including those reviewed by Morewitz (1982), and has observed that Eq. 6-70 may be a necessary condition for plugging but is not a sufficient condition. He suggests that the data support a second condition being applied to the plugging criterion, that of the violation of the Davies inertial criterion for representative sampling given in Eq. 6-31 in which the Stokes number is based on the tube diameter and the average gas velocity in the tube. It should also be noted that the presence of obstructions to flow such as discussed in the section on Deposition in Flow Constrictions can precipitate aerosol plugging.

These plugging criteria are not developed and are included to provide the reader with a means to identify situations in which plugging or high particle deposition may be encountered.

Reentrainment of Deposited Particles

Fromentin (1989) provides a discussion of the current state of research and modeling of particle reentrainment by flowing gas. Wen and Kasper (1989) point out that steady flow will not produce much reentrainment but that unsteady conditions will. These conditions are encountered in turbulent flow, with eddies impinging on the wall, and during the start-up or shutdown of flows. Reentrainment is not a well-characterized process and should be avoided by using clean sample lines and bags, minimizing deposition, elimination of mechanical shock and vibration to the sampling system, and elimination of flow transients and shocks in the sampling system.

SUMMARY AND CONCLUSIONS

Correlations for aspiration efficiency and transmission efficiency for thin-walled nozzles sampling from a flowing gas have been reported. The product of these efficiencies is the inlet efficiency. The inlet efficiency depends on the Stokes number, the ratio of ambient gas velocity to sampling gas velocity, and the sampling angle. In order to minimize loss and obtain a representative sample, the following suggestions are made:

1. Isokinetic (iso-mean-velocity) sampling (sampling velocity equal to ambient velocity and isoaxial sampling) be used whenever possible
2. Gas velocities should be large compared to the particle-settling velocity
3. Inlet diameters should be of the order of a centimeter to reduce the deposition from free-stream turbulence
4. Stokes numbers be kept as low as possible

Criteria for representative sampling from calm air are also presented.

Transport efficiency through sample lines and for residence time in storage bags is discussed. The correlations reported in the literature for deposition from gravitational settling, diffusional deposition, turbulent inertial deposition, inertial deposition in a bend, inertial deposition in an abrupt contraction, and diffusional and settling deposition in a chamber have been presented. Their dependence on the parameters of the correlations is discussed. Deposition from electrostatic, thermophoretic, and diffusiophoretic forces has been discussed and suggested practices outlined to avoid these deposition mechanisms. Resuspension of deposited material has also been discussed. Dilution as a means of sample conditioning and the uncertainties entailed in diluting a sample have been presented. These correlations allow the reader to evaluate the performance of sampling systems and to optimize their design. Because these correlations cannot address every situation and have not exhaustively covered all possible deposition mechanisms, it is strongly recommended that their use to design or evaluate a sampling system be accompanied by validating experimental calibration of the sampling system.

References

ANSI. 1969. Guide to sampling airborne radioactive materials in nuclear facilities. ANSI N13.1-1969. New York: American National Standards Institute, Inc.

Agarwal, J. and B. Y. H. Liu. 1980. A criterion for accurate aerosol sampling in calm air. *Am. Ind. Hyg. Assoc. J.* 41:191–97.

Armbruster, L. and G. Zebel. 1985. Theoretical and experimental studies for determining the aerosol sampling efficiency of annular slots. *J. Aerosol Sci.* 16(4):335–41.

Belyaev, S. P. and L. M. Levin. 1972. Investigation of aerosol aspiration by photographing particle tracks under flash illumination. *J. Aerosol Sci.* 3:127–40.

Belyaev, S. P. and L. M. Levin. 1974. Techniques for collection of representative aerosol samples. *J. Aerosol Sci.* 5:325–38.

Brockmann, J. E., B. Y. H. Liu, and P. H. McMurry. 1984. A sample extraction diluter for ultrafine aerosol sampling. *Aerosol Sci. Technol.* 4:441–51.

Brockmann, J. E., P. H. McMurry, and B. Y. H. Liu. 1982. Experimental study of simultaneous coagulation and diffusional loss of free molecule aerosols in turbulent pipe flow. *J. Colloid Interface Sci.* 88(2):522–29.

Cheng Y. S. and C. S. Wang. 1981. Motion of particles in bends of circular pipes. *Atmos. Environ.* 15:301–06.

Crane, R. I. and R. L. Evans. 1977. Inertial deposition of particles in a bent pipe. *J. Aerosol Sci.* 8:161–70

Crump, J. G., R. C. Flagan, and J. H. Seinfeld. 1983. Particle wall loss rates in vessels. *Aerosol Sci. Technol.* 3:303–09.

Crump, J. G. and J. H. Seinfeld. 1981. Turbulent deposition and gravitational sedimentation of an aerosol in a vessel of arbitrary shape. *J. Aerosol Sci.* 12(5):405–15.

Davies, C. N. 1968. The entry of aerosols into sampling tubes and heads. *Br. J. Appl. Phys.* (*J. Phys. D*) *Ser. 2* 1:921–32.

Davies, C. N. and M. Subari. 1982. Aspiration above wind velocity of aerosols with thin-walled nozzles facing and at right angles to the wind direction. *J. Aerosol Sci.* 13:59–71.

Delattre, P. and S. K. Friedlander. 1978. Aerosol coagulation and diffusion in a turbulent jet. *Ind. Engng. Chem. Fundls.* 17:189–94.

Durham, M. D. and D. A. Lundgren. 1980. Evaluation of aerosol aspiration efficiency as a function of Stokes number, velocity ratio and nozzle angle. *J. Aerosol Sci.* 11:179–88.

Fissan, H., and G. Schwientek. 1987. Sampling and transport of aerosols. *TSI J. Part. Instrum.* 2(2):3–10.

Friedlander, S. K. 1977. *Smoke, Dust, and Haze.* New York: Wiley.

Fromentin, A. 1989. Particle resuspension from a multilayer deposit by turbulent flow. PSI Bericht No. 38. Paul Scherrer Institute, Switzerland.

Fuchs, N. A. 1964. *The Mechanics of Aerosols.* Oxford: Pergamon.

Fuchs, N. A. 1975. Sampling of aerosols. *Atmos. Environ.* 9:697–707.

Fuchs, N. A., and A. G. Sutugin. 1965. Coagulation rate of highly dispersed aerosols. *J. Colloid Sci.* 20:492–500.

Goldsmith, P., and F. G. May. 1966. Diffusiophoresis and thermophoresis in water vapor systems. In *Aerosol Science*, ed. C. N. Davies, pp. 163–94. New York: Academic Press.

Gormley, P. G. and M. Kennedy. 1949. Diffusion from a stream flowing through a cylindrical tube. *Proc. Royal Irish Academy* 52A:163–69.

Grinshpun, S. A., G. N. Lipatov, and A. G. Sutugin. 1990. Sampling errors in cylindrical nozzles. *Aerosol Sci. Technol.* 12:716–40.

Hangal, S. and K. Willeke. 1990*a*. Aspiration efficiency: Unified model for all forward sampling angles. *Environ. Sci. Technol.* 24:688–91.

Hangal, S. and K. Willeke. 1990*b*. Overall efficiency of tubular inlets sampling at 0–90 degrees from horizontal aerosol flows. *Atmos. Environ.* 24A(9):2379–86.

Heyder, J. and J. Gebhart. 1977. Gravitational deposition of particles from laminar aerosol flow through inclined circular tubes. *J. Aerosol Sci.* 8:289–95.

Hinds, W. C. 1982. *Aerosol Technology.* New York: Wiley.

Holman, J. P. 1972. *Heat Transfer.* New York: McGraw-Hill.

Im, K. H. and R. K. Ahluwalia. 1989. Turbulent eddy deposition of particles on smooth and rough surfaces. *J. Aerosol Sci.* 20(4):431–36.

Jayasekera, P. N. and C. N. Davies. 1980. Aspiration below wind velocity of aerosols with sharp edged nozzles facing the wind. *J. Aerosol Sci.* 11:535–47.

John, W. and H. C. Wang. 1991. Laboratory testing methods for PM-10 samplers: Lowered effectiveness from particle loading. *Aerosol Sci. Technol.* (to be published).

John, W., W. Winklmayr, and H. C. Wang. 1991. Particle deagglomeration and reentrainment in a PM-10 sampler. *Aerosol Sci. Technol.* 14:165–76.

Kurz, J. L. and T. C. Ramey. 1988. The development, performance, and application of an advanced isokinetic stack sampling system. In *Proc. 20th DOE/NRC/Nuclear Air Cleaning Conf.* NUREG/CP-0098, pp. 847–56.

Laktionov, A. B. 1973. Aspiration of an aerosol into a vertical tube from a flow transverse to it. *Fizika Aerozoley* 7:83–87 (Translation from Russian AD-760 947, Foreign Technology Division, Wright–Patterson AFB, Dayton, OH).

Lipatov, G. N., S. A. Grinshpun, and T. I. Semenyuk. 1989. Properties of crosswise migration of particles in ducts and inner aerosol deposition. *J. Aerosol Sci.* 20(8):935–38.

Lipatov, G. N., S. A. Grinshpun, T. I. Semenyuk, and A. G. Sutugin. 1988. Secondary aspiration of aerosol particles into thin-walled nozzles facing the wind. *Atmos. Environ.* 22(8):1724–27.

Lipatov, G. N., S. A. Grinshpun, G. L. Shingaryov, and A. G. Sutugin. 1986. Aspiration of coarse aerosol by a thin-walled sampler. *J. Aerosol Sci.* 17(5):763–69.

Lipatov, G. N., T. I. Semenyuk, and S. A. Grinshpun. 1990. Aerosol migration in laminar and transition flows. *J. Aerosol Sci.* 21(Suppl. 1):S93–96.

Liu, B. Y. H. and J. K. Agarwal. 1974. Experimental observation of aerosol deposition in turbulent flow. *J. Aerosol Sci.* 5:145–55.

Liu, B. Y. H. and D. Y. H. Pui. 1981. Aerosol sampling inlets and inhalable particles. *Atmos. Environ.* 15:589–600.

Liu, B. Y. H., Z. Q. Zhang, and T. H. Kuehn. 1989. A numerical study of inertial errors in anisokinetic sampling. *J. Aerosol Sci.* 20(3):367–80.

Lundgren, D. A., M. D. Durham, and K. W. Mason. 1978. Sampling of tangential flow streams. *Am. Ind. Hyg. Assoc. J.* 39:640–44.

McFarland, A. R., N. K. Anand, C. A. Ortiz, M. E. Moore, S. H. Kim, R. E. DeOtte, Jr., and S. Somasundaram. 1988. Continuous air sampling for radioactive aerosol. In *Proc. 20th DOE/NRC Nuclear Air Cleaning Conf.* NUREG/CP-0098, pp. 834–46.

McMurry, P. H. and D. Grosjean. 1985. Gas and aerosol wall losses in Teflon film smog chambers. *Environ. Sci. Technol.* 19(12):1176–82.

McMurry, P. H. and D. J. Rader. 1985. Aerosol wall losses in electrically charged chambers. *Aerosol Sci. Technol.* 4:249–68.

Morewitz, H. A. 1982. Leakage of aerosols from containment buildings. *Health Phys.* 42(2):195–207.

Novick, V. J. 1990. Aerosol sampling and transport tube plugging criteria. In *American Association for Aerosol Research 1990 Annual Meeting Abstract Book*, p. 46, 18–22 June, Philadelphia, PA.

Okazaki, K., R. W. Wiener, and K. Willeke. 1987*a*. The combined effect of aspiration and transmission on aerosol sampling accuracy for horizontal isoaxial sampling. *Atmos. Environ.* 21(5):1181–85.

Okazaki, K., R. W. Wiener, and K. Willeke. 1987*b*. Isoaxial aerosol sampling: Nondimensional representation of overall sampling efficiency. *Environ. Sci. Technol.* 21(2):178–82.

Okazaki, K., R. W. Wiener, and K. Willeke. 1987*c*. Non-isoaxial aerosol sampling: Mechanisms controlling the overall sampling efficiency. *Environ. Sci. Technol.* 21(2):183–87.

Orr, C., Jr., and E. Y. H. Keng. 1976. Sampling and particle-size measurements. In *Handbook on Aerosols*, ed. R. Dennis. TID-26608, Technical Information Center U.S. Department of Energy.

Paulus, H. J. and R. W. Thron. 1977. Stack sampling. In *Air Pollution, Vol. 3, Measuring, Monitoring, and Surveillance of Air Pollution*, 3rd edn., ed. A. C. Stern, pp. 525–87. New York: Academic Press.

Pich, J. 1972. Theory of gravitational deposition of particles from laminar flows in channels. *J. Aerosol Sci.* 3:351–61.

Pui, D. Y. H., F. Romay-Novas, and B. Y. H. Liu. 1987. Experimental study of particle deposition in bends of circular cross section. *Aerosol Sci. Technol.* 7:301–15.

Rader, D. J. and V. A. Marple. 1988. A study of the effects of anisokinetic sampling. *Aerosol Sci. Technol.* 8(3):283–99.

Ranade, M. B., D. K. Werle, and D. T. Wasan. 1976. Aerosol transport through a porous sampling probe with transpiration air flow. *J. Colloid Interface Sci.* 56(1):42–52.

Reekis, M. W. and G. Skyrme. 1976. The dependence of particle deposition velocity on particle inertia in turbulent pipe flow. *J. Aerosol Sci.* 7:485–95.

Saffman, P. G. 1965. The lift on a small sphere in slow shear flow. *J. Fluid Mech.* 22:385–400.

Saffman, P. G. 1968. Corrigendum. *J. Fluid Mech.* 31:624.

Schlichting, H. 1968. *Boundary Layer Theory*, 6th edn. New York: McGraw-Hill.

Schwendiman, L. C., G. E. Stegen, and J. A. Glissmeyer. 1975. Rerport BNWL-SA-5138, Battelle Pacific Northwest Laboratory, Richland, WA.

Stevens, D. C. 1986. Review of aspiration coefficients of thin-walled sampling nozzle. *J. Aerosol Sci.* 17(4):729–43.

Thomas, J. W. 1958. *J. Air Pollut. Control Assoc.* 8:32.

U.S. Environmental Protection Agency. 1974. Administrative and technical aspects of source sampling for particulates. EPA-450/3-74-047.

U.S. Environmental Protection Agency. 1987. *Federal Register* 52:24634.

Vaughn, E. U. 1978. Simple model for plugging of ducts by aerosol deposits. *ANS Trans.* 28:507–08.

Vincent, J. H. 1984. A comparison between models for predicting the performances of blunt dust samplers. *Atmos. Environ.* 187(5):1033–35.

Vincent, J. H. 1987. Recent advances in aspiration theory for thin-walled and blunt aerosol sampling probes. *J. Aerosol Sci.* 18:487–98.

Vincent, J. H. 1989. *Aerosol Sampling: Science and Practice.* New York: Wiley.

Vincent, J. H., P. C. Emmett, and D. Mark. 1985. The

effects of turbulence on the entry of airborne particles into a blunt dust sampler. *Aerosol Sci. Technol.* 4:17–29.

Vincent, J. H. and H. Gibson. 1981. Sampling errors in blunt dust samplers arising from external wall loss effects. *Atmos. Environ.* 15(5):703–12.

Vincent, J. H., D. Hutson, and D. Mark. 1982. The nature of air flow near the inlets of blunt dust sampling probes. *Atmos. Environ.* 16(5):1243–49.

Vincent, J. H., D. C. Stevens, D. Mark, M. Marshall, and T. A. Smith. 1986. On the aspiration characteristics of large-diameter thin-walled aerosol sampling probes at yaw orientations with respect to the wind. *J. Aerosol Sci.* 17(2):211–24.

Waldmann, L. and K. H. Schmitt. 1966. Thermophoresis and diffusiophoresis of aerosols. In *Aerosol Science*, ed. C. N. Davies, pp. 137–62. New York: Academic Press.

Wang, C. S. 1975. Gravitational deposition of particles from laminar flows in inclined channels. *J. Aerosol Sci.* 6:191–204.

Wen, H. Y. and G. Kasper. 1989. On the kinetics of particle reentrainment from surfaces. *J. Aerosol Sci.* 20(4):483–98.

Whitmore, P. J. and A. Meisen. 1978. Diffusiophoretic particle collection under turbulent conditions. *J. Aerosol Sci.* 9:135–45.

Wiener, R. K., K. Okazaki, and K. Willeke. 1988. Influence of turbulence on aerosol sampling efficiency. *Atmos. Environ.* 22(5):917–28.

Yamano, N., and J. E. Brockmann. 1989. Aerosol sampling and transport efficiency calculation (ASTEC) and application to Surtsey/DCH aerosol sampling system. NUREG/CR-525. SAND88-1447, Sandia National Laboratories, Albuquerque, NM.

Ye, Y. and D. Y. H. Pui. 1990. Particle deposition in a tube with an abrupt contraction. *J. Aerosol Sci.* 21(1):29–40.

Zhang, Z. Q. and B. Y. H. Liu. 1989. On the empirical fitting equations for aspiration coefficients for thin-walled sampling probes. *J. Aerosol Sci.* 20(6):713–20.

7

Measurement Methods

Matti Lehtimäki

Technical Research Centre of Finland
Tampere, Finland

and

Klaus Willeke

Department of Environmental Health
University of Cincinnati
Cincinnati, OH, U.S.A.

INTRODUCTION

Most of our knowledge concerning the properties of airborne particles has been gained through experimental means. Thus, the use of appropriate measurement techniques is important in aerosol science and technology, in basic research and practical field work alike. In recent years, the pace of developing new measurement methods and instruments has been rapid and a great number of measurement techniques and instruments has become available (Willeke 1984; Pui and Liu 1989). In general, each aerosol measurement technique covers a unique range of particle concentrations, sizes, shapes, and chemical compositions. Thus, choosing the proper instrument for a particular application can be difficult. Also, reliable and accurate measurements are usually not possible without a thorough understanding of the principles and limitations of each measurement method.

This chapter introduces the basic principles of the major aerosol measurement techniques in use today. Details of the various measurement methods and of the instruments utilizing them are discussed in Part II of this book. In this chapter, different types of aerosol measurement will be discussed first, followed by an introduction of various aerosol particle collection and analysis methods. Finally, several measurement methods are presented for dynamic aerosol monitoring.

TYPES OF AEROSOL MEASUREMENT

Aerosol measurements can be classified according to the quantity to be determined. The most commonly measured aerosol quantity is particle concentration. In many applications, however, information on particle size and shape is of equal or greater importance.

Particle concentration is usually defined as the amount of particulate mass in a unit volume of air or some other gas. This quantity is measured in most practical applications, e.g., in industrial hygiene and ambient air studies, and in sampling industrial process streams. A measured value of particle concentration may refer to the total mass of particles or the particulate mass in a certain particle size range. The measurement may also be limited to the mass of a certain element or chemical compound. In some special applications the particle number concentration is equally, or more, important, e.g., when measuring fibers in ambient air or contaminant particles in clean rooms. When monitoring microorganisms in air, the biologically active material is usually recorded as the number of colony-forming units (cfu) in a unit volume of air. Radioactive aerosols are usually recorded as their activity concentration, defined as the amount of radioactive material in a unit volume of air. In aerosol research, other types of particle concentration may be relevant, among them surface area, volume, and charge concentration.

In some applications, the aim of the measurement is to determine the ratio of particle concentrations under two different conditions. For example, the removal efficiencies of air-cleaning devices and filters as well as the protection factors of respirators are determined by such concentration ratio measurements. Relative concentrations are also measured in studies on particle deposition in the human respiratory system. Aerosol measurement techniques are also used in qualitative studies, e.g., for locating leak sites in high efficiency particulate air (HEPA) filters.

In aerosol science, and in the manufacture of materials through aerosol generation and deposition techniques, the measurement of particle size and shape can be of great importance as well. When dealing with the measurement of particle size, several definitions of particle size are used. A single dimension, suitable for reporting the diameter of a spherical particle, may not appropriately represent the irregular shape of most types of aerosol particles. Therefore, an equivalent diameter is usually reported, defined as the diameter of a sphere having the same value of a specific physical property as the irregularly shaped particle being measured. As discussed in Chapter 2, the most common equivalent particle size is the aerodynamic particle diameter, defined as the diameter of a spherical unit-density particle with the same gravitational settling velocity as the particle under consideration. The definition of equivalent particle size may also be based on the mobility of the particle in an electric field, the diffusional motion of the particle or any other motion of the particle in an externally applied force field. The definition of equivalent optical particle size is based on the light scattering properties of the particle under consideration. In this case, the definition of equivalent particle size depends not only on the physical properties of the particle but also on the properties of the measurement instrument.

Concentration Levels and the Desired Size Information

The type of aerosol measurement made and the manner in which it is performed may strongly depend on the environment where the sampling is to be undertaken. The quantity to be measured as well as the particle concentration and size range also affect the decision on how to make the measurement. In some cases, the purpose of the measurement and the required accuracy may also influence the choice of proper measurement technique.

High particle concentrations may be measured in industrial process streams such as smoke stacks, where the particle mass concentration may exceed 10 g/m^3. In work environments, the mass concentration typically ranges from about 0.1 mg/m^3 to about 10 mg/m^3. The particle mass concentrations in ambient and indoor air are usually less than 0.1 mg/m^3. In highly controlled clean-room environments the particle concentrations are so low—several orders of magnitude lower than ambient concentrations—that they are recorded as number of particles per unit volume of air. When dealing with radioactive aerosols, the concentration range corresponding to maximum allowable activity concentrations extends down to 1 pg/m^3. Thus, the range of aerosol mass concentrations extends over 13 decades. Several methods and techniques are required to cover such a wide range of particle concentrations.

The diversity of aerosol measurement techniques is further increased by the wide range of particle sizes that may have to be addressed. The range of particle sizes extends over five decades from about 0.001 μm to about 100 μm. Most aerosol measurement instruments have an operational range of about one and a half decades in particle size.

Sample Collection and Analysis versus Real-Time Measurement

There are two basic approaches to the measurement of aerosol particles. The traditional approach is to sample the particles onto a collection surface and then analyze the collected particles in the laboratory. The other approach is to sample the aerosol directly into a real-time, dynamic measuring instrument.

When aerosol particles are collected for analysis at a later time, particle collection is usually achieved by deposition of the particles onto a filter. Impaction, sedimentation, electrostatic collection, thermal precipitation, and diffusion may also be used as the removal mechanisms. The major advantages of such sample collection methods are the simplicity and low cost of the sampling devices and their suitability for use under severe environmental conditions. The collected material may be subjected to different analysis techniques, another important advantage. The principal drawback of the sample collection techniques is the long time delay between sampling and the subsequent analysis which normally takes place in the laboratory. Since collected samples represent specific time intervals, the lack of time resolution in concentration and size distribution are additional drawbacks. Chemical reactions of the collected particles may bias the sample, especially when unstable aerosols are studied.

In contrast, real-time measurement methods provide a rapid data readout. The data are immediately available, a factor which is of great importance in most practical aerosol studies. Many direct reading instruments yield instantaneous or almost instantaneous information on particle concentration and size distribution. Most real-time instruments, however, are expensive and cannot be used under severe environmental conditions. Furthermore, most of them do not provide any information on the chemical composition of the particles.

Preclassification

The motion and deposition of aerosol particles in ambient air environments, in industrial processing streams, and in aerosol measurement devices are highly particle-size-dependent. The measured total aerosol mass concentration is, therefore, affected by the external air flow velocity, the sampling flow rate and the shape and dimensions of the internal surfaces containing the aerosol flow.

This particle size dependence is particularly important in health-related aerosol measurements (Lippmann 1989*a*). The aerosols that are inhaled by the nose or mouth may deposit in different physiological regions of the respiratory tract. Deposition in different regions may result in different health effects. Usually, the respiratory tract is distinguished by three distinct regions: the extrathoracic region above the larynx containing the nose and mouth, followed by the tracheobronchial region, which looks like an inverted tree, and, finally, the alveolar region, where the gas exchange occurs with the body's blood.

The penetration of particles into any of these regions is simulated by preclassifiers that remove particles ahead of the aerosol sensor in a manner similar to the particle removal occurring ahead of the physiological region of interest. This preclassification may be achieved by the use of a cyclone, impactor, filter, horizontal elutriator or any other size-selective particle removal device with appropriate removal properties (Lippmann 1989*a*; Mercer 1973).

Personal versus Area Sampling

As indicated above, a preclassifier may be used ahead of an aerosol sensor in order to relate the aerosol measurement to the purpose of the measurement. In a work environment where the particle concentration and size distribution are spatially homogeneous, an aerosol measurement may be performed by a stationary device anywhere in the work environment. However, workers are generally exposed to aerosols from localized sources, i.e., they work and move in areas with strong concentration gradients. A typical example is a welder, whose exposure is mostly due to the welding fume in the plume from the welding torch. Thus, the aerosol measurement must be made close to the worker's breathing zone, i.e., by means of personal sampling. This is generally achieved by attaching a sample collector to the worker's lapel and drawing aerosol into the collector by means of a small pump attached to the worker's belt.

Active versus Passive Sampling

Most aerosol measurements are performed by actively drawing aerosols into a sensor or onto a sampling surface by means of a pump or other air mover, as illustrated in Fig. 7-1a. In personal sampling, it is desirable to avoid the use of the pump, thus alleviating an extra burden on the worker. When studying a worker's exposure to an aerosol, the sampling pump can sometimes be avoided by using a real-time aerosol monitor such as a photometer (Armbruster and Breuer 1983). In a photometer, the instrument response depends only on the particles present in the sensing volume, independent of the air velocity through it. In this case, passive sampling can be performed, if natural air convection or the movement of the monitor transports the aerosol into the sensing volume of the instrument (Fig. 7-1b). Air velocity may, however, affect the aerosol sampling and transport efficiencies, so that active sampling may differ from passive sampling (Willeke and DeGarmo 1988).

Another example of passive sampling is shown in Fig. 7-1c. Particles, settling under the influence of gravity, will deposit onto horizontal surfaces. Settling plates are sometimes used for sampling bioaerosols in indoor air environments and hospitals. While the deposits can be used for species identifications, settling plates are discouraged for quantitative analysis because the particle collection is highly particle-size- and air-motion-dependent.

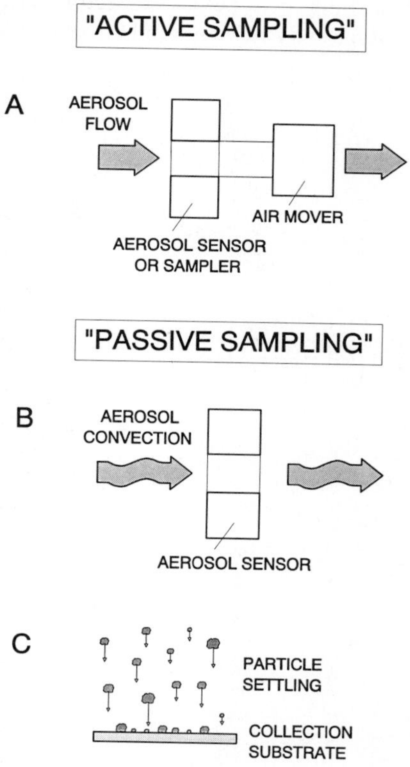

FIGURE 7-1. Active versus Passive Aerosol Sampling.

COLLECTION METHODS

Aerosol measurements may be performed by collecting the particles onto a substrate and then analyzing them, or by dynamically sensing them in the airborne state. In this section the major collection principles are introduced. Figure 7-2a shows that the collection surface may be a porous medium, e.g., a filter that passes the air or gas flow but retains all or a fraction of the particles suspended in the incident aerosol flow. Passage through such a porous medium generally has to be at a rather low flow rate and all of the aerosol flow must pass through this medium.

In many applications it is desirable not to affect the flow but to extract the particles from the aerosol flow by an externally applied force, as shown in Fig. 7-2b. The external force may be due to the gravitational pull of the earth or due to inertia when the particles are subjected to centrifugal motion in the bend of a duct or the helical path of a cyclone.

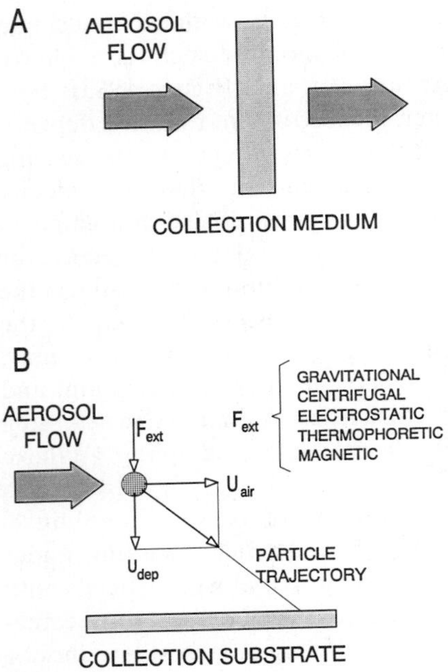

FIGURE 7-2. Aerosol Particle Collection Methods. (a) Collection by Porous Medium, (b) Collection by External Force.

Extractive forces may also be obtained by the application of electric, thermal, or magnetic force fields. Under the influence of these external forces, the particles drift towards the collection substrate along a trajectory that is the resultant of deposition velocity U_{dep} and air flow velocity U_{air}.

Filter Collection

Filtration is the most widely used method for removing particles from an aerosol flow for subsequent analysis (Lippmann 1989*b*). Particle collection in a filter is due to the simultaneous action of several collection mechanisms: inertial impaction, interception, diffusion, gravitational settling and electrostatic attraction (Hinds 1982). The relative importance of these different mechanisms depends on the particle size, density, shape, and electrical charge, the filtration flow velocity and the mechanical and electrical properties of the filter. Typical filters used for aerosol sampling consist of glass, cellulose, or plastic fibers, porous membranes, or polycarbonate pore materials. The characteristics and applications of different filter types are discussed in Part II of this book.

Inertial and Gravitational Collection

Inertia is an important particle property that is utilized in many aerosol samplers to remove particles from the aerosol flow. In a curved flow field, inertia makes the particle trajectories deviate from the flow streamlines, as shown in Fig. 7-3. In an impactor, the particle-laden air flow is pulled or pushed through a nozzle, which accelerates the air and the particles to high speed. The air jet emerging from the nozzle is deflected by the impaction surface and makes a sharp turn. Particles with sufficient inertia continue their straightforward movement towards the collection surface and impact onto the collection surface (Marple and Willeke 1979).

Inertial impaction as a collection method is utilized in different types of impactors. An impaction stage may have a single nozzle, as

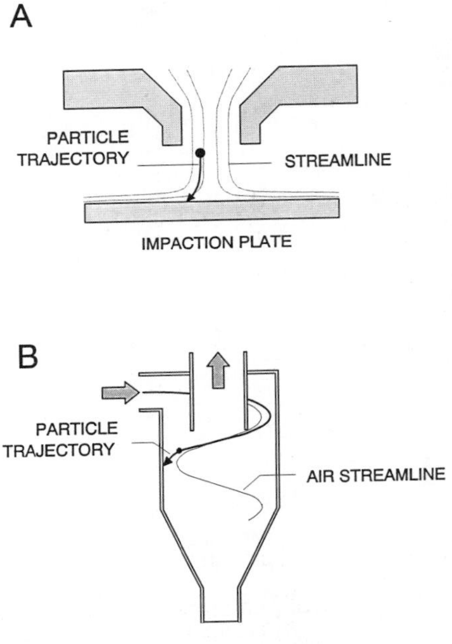

FIGURE 7-3. Inertial Collection. (a) Impactor, (b) Cyclone.

shown in Fig. 7-3a, or may have several nozzles next to each other. A typical use of a single-stage impactor is for preclassification, e.g., for removing the particles larger than the respirable fraction. A single-stage impactor may also be used as a preseparator in direct reading instruments for improving the response of the instrument to small particles or for reducing the instrument's contamination by large particles.

By combining several single-stage impactors into a multistage or cascade impactor, one may collect particles in a size-selective manner. In a cascade impactor each stage represents a specific particle size range. This is accomplished by decreasing the nozzle diameter, the number of nozzles or both in each successive stage. In operation, each stage is assumed to collect all particles of a diameter larger than its cutoff size. Because of particle collection on the previous stage, the upper limit of particle size captured by a given stage equals the cutoff size of the previous stage. Thus, each stage collects particles with diameters in a limited size range defined by the successive cutoff particle sizes.

Unfortunately, the particle behavior in a cascade impactor may be more complex than that illustrated in Fig. 7-3a. Particles may bounce or be reentrained from the impaction surface and be collected by other surfaces (Rao and Whitby 1978*a, b*). Coating the impaction surfaces with a sticky substance may improve particle collection. However, after deposition of a layer of particles, subsequent particles may not be held by the coating, as the incoming particles impact onto already deposited particles. Thus, the limited loading capacity of an impactor is a major drawback of the impaction technique.

By using several single-stage impactors with different cutoff sizes in parallel, the difference in particle mass from the different stages determines the particle mass in specific size ranges (Lee and Esmen 1983). Parallel impactors are not commonly used, however, because they take up more space and require a separate flow control for each impactor.

The inertial impaction technique is also utilized without impacting the particles onto a solid surface. Instead, the particles are impacted onto a stagnant volume of air. High-inertia particles move across the virtual surface between air flow and stagnant air volume and are collected in the latter, while the smaller particles are deflected at the virtual interface in a manner similar to solid-plate impaction. Turbulence in the stagnant air volume of this "virtual impactor" may wash the large particles back into the deflected air stream. Therefore, the air beyond the virtual interface is withdrawn at a slow rate, thus rendering this impactor a two-flow or "dichotomous" impactor. In both flows particles can be collected on a filter or analyzed in a real-time aerosol instrument. A continued suspension of the size-fractionated particles in the effluent air flows is the principal advantage of these virtual or dichotomous impactors (Loo, Jaklevic, and Goulding 1976; Chen, Yeh, and Cheng 1985).

An alternative to the inertial impactor is the cyclone in which particles are removed by centrifugal forces. As illustrated in Fig. 7-3b, the sampled aerosol moves in a helical path. Particles with high inertia are separated out by the centrifugal force while smaller particles continue with the air flow. In general, a cyclone does not yield as sharp a cutoff in particle size as an impactor. Since the cutoff by the tracheobronchial region preceding the alveolar region of the human lung is also not very sharp, a small cyclone is in common use in the U.S. as a preclassifier for sampling "respirable" aerosol fractions (Lippmann 1989*a*). The loading capacity of a cyclone is very high compared to that of an impaction stage. Thus, the cyclone is a good choice when a large amount of particles is to be sampled. Cyclones are frequently used in series to collect particle-size-classified samples, especially when measuring the particle size distributions in smoke stacks and other high-aerosol-concentration environments (Smith, Wilson, and Harris 1979).

In an inertial separator, the particle mass, size, and velocity are principal parameters constituting the driving force for removal. Particle mass alone in the gravitational force field may also be used to collect particles in a size-selective manner. In vertical and horizontal elutriators the particles are either removed or elutriated by the relative velocity between the air flow and the gravitational settling of the particles (Hinds 1982).

A vertical elutriator consists of a vertical tube or channel in which the sampled air flows upwards at a low velocity. Only particles with a gravitational-settling velocity that is less than the upward air velocity are carried out of the device. The vertical elutriator is used, e.g., for cotton dust sampling. The major drawback of the cotton dust sampler is the difficulty in achieving a precise particle size cutoff because of the uneven flow distribution in the elutriator.

The horizontal elutriator consists of a horizontal duct or a set of parallel ducts of rectangular cross section. Size-selective particle separation occurs because of the size-dependent settling in the gravitational field. The major advantages of the horizontal elutriator are its simplicity of construction, ease of modelling the size separation mechanism, and high loading capacity. In general, the horizontal elutriator is used to collect a sample of large particles or to separate large particles from the smaller ones. The main disadvantage of the horizontal elutriator is its relatively large size.

In an aerosol centrifuge (Stöber 1976; Tillery 1979), the centrifugal force is much higher than the gravitational force in an elutriator. Thus, significantly greater forces can be applied to the particles, which makes it possible to extend the measurements to lower particle size ranges. Most aerosol centrifuges collect samples with a very high size resolution. Their major drawback is their limited size range and their low sampling rate. They are not very common and their major field of application is in laboratory studies.

Other Collection Methods

The diffusional motion of aerosol particles is utilized for depositing small particles, e.g., in filters and diffusion batteries. Diffusional deposition is due to a particle's Brownian motion, which causes its path to deviate from the air streamline. The random deviations in particles' paths increase the probability of particle attachment to nearby surfaces.

Diffusion batteries separate particles according to their diffusional properties. The devices used today are mostly of the screen type, i.e., the diffusion battery consists of one or several screens perpendicular to the aerosol flow (Cheng and Yeh 1983). The screens collect smaller particles but pass the larger ones. Similarly, parallel plates or collimated hole structures have been used (Sinclair et al. 1979). Diffusion batteries are usually applied for measuring the particle size distributions in the 0.002–0.2 μm size range, but they can also be used for sampling purposes.

The thermophoretic force on a particle has also been used for sampling aerosols, especially for microscope analysis (Swift and Lippmann 1989). This force is exploited in the thermal precipitator, which employs a heated element, e.g., a wire, to create a temperature gradient in its flow channel. The particle experiences a force and, thus, a velocity component in the direction of the decreasing temperature. Since the velocity due to the thermal force is only moderately particle-size-dependent, thermal precipitation can be used over a wide particle size range. However, thermal precipitators require very low flow rates.

The electrostatic force is an effective means for collecting aerosol samples, especially in the submicrometer particle size range. The point-to-plane electrostatic precipitator (Swift and Lippmann 1989) consists of a needle electrode and a flat collection surface. The high voltage between the needle and the collection surface generates a corona discharge, which produces a high concentration of unipolar ions. The ions cause an effective particle charging and the electric field drives the charged particles towards the collection surface. An alternative to the point-to-plane precipitator is the pulsed precipitator, which utilizes a pulsed electrostatic field for collecting the charged particles. The main advantage of the latter device is sample collection without significant particle size bias (Liu, Whitby, and Yu 1967).

SAMPLE ANALYSIS METHODS

Once a sample of particles has been collected, it may be analyzed as to its physical, chemical, and biological content. The sample may be analyzed as a whole or each particle may be analyzed individually.

Physical Analysis

Gravimetry is the most widely used analysis method for collected aerosol samples. In this method, conventionally used in industrial hygiene and ambient-air studies, a known volume of aerosol is sampled by a filter and the mass increase of the filter divided by the air volume is recorded as the aerosol mass concentration. In practice, the filter is usually weighed in the laboratory before use, transported to the measurement site, exposed to an aerosol flow for a given period of time, transported back to the laboratory and reweighed. The same technique is also used for evaluating the particle deposits on the collection plates of impactors.

Since the gravimetric method records small changes in the weight of a filter or collection plate, the weighing usually has to be performed in a carefully controlled environment, so that the effects of changes in air humidity are avoided. To avoid errors due to static electricity, which may influence the reading of the balance, the electric charge on membrane and fibrous filters is usually reduced by a charge neutralizer. The proper handling of the samples during weighing and transportation is of great importance to the accuracy of the measurement.

If more than bulk analysis is needed, optical and electron microscopy of the sample are used for sizing, counting and identifying individual particles. Optical microscopy has been used extensively for particle analysis in the past. Its most widespread application today is for the analysis of sampled asbestos particles (Chapter 25). The most straightforward method is to collect the particles on a glass slide by sedimentation, electrostatic precipitation, or thermal precipitation and analyze them under the microscope. Optical microscopy may also be applied to samples collected on smooth filter surfaces. When a cellulose ester membrane filter is used, the filter may be made transparent by various techniques, such as saturating it with an immersion oil of the same refractive index as the filter material while the filter is held between a glass slide and a glass cover slip. Other examples of sample preparation are discussed in Chapter 25.

A conventional bright-field microscope can be used to count particles down to approximately 0.3–0.5 μm. With the aid of immersion oil between the microscope's objective and the glass slide, it is possible to obtain approximately 0.2 μm size resolution. A number of modifications can be made in the optical microscope in order for it to reveal special particle properties or to improve the contrast. Depending upon the modification made, the microscope may be referred to as a dark-field, polarizing, phase-contrast, interference, fluorescence, or ultraviolet microscope (Chung 1981).

Optical microscopy is essentially limited by the wavelength of the visible light, about 0.4–0.7 μm. The much shorter wavelength of radiation used in an electron microscope makes it possible to extend particle analysis to ultrafine particles. The limit of resolution for a scanning electron microscope is approximately 0.01 μm, while a transmission electron microscope's limit of resolution may be less than 0.001 μm. Electron microscope analysis, however, is made in vacuum. Thus, this technique can be applied only to stable particles, which do not evaporate or degrade when exposed to vacuum and heating by the electron beam.

The gravimetric and microscopic techniques are the most common physical analysis tools for samples of collected particles. Instead of analyzing the particle weight or number, one may, e.g., wish to record the radioactive content of the sample. This may be achieved by determining the count rate of radioactive decays in the sample. In indoor-air studies the concentrations of radon decay products are thus determined.

Chemical Analysis

An experienced microscopist, equipped with several different kinds of microscopes, can distinguish particles of many different origins, i.e., chemical compositions. More detailed information is gained by elemental analysis through, e.g., atomic absorption spectroscopy, X-ray fluorescence analysis, proton induced X-ray analysis, and neutron activation analysis (Dzubay 1977; Chung 1981).

Atomic absorption spectroscopy (AAS) is one of the most common methods for analyzing the elemental content of an aerosol sample. The typical procedure is to put the sample into solution, nebulize the liquid, and then vaporize the droplets by heating them in the airborne state. The amount of each element is determined by measuring the light absorbance in the vapor for each wavelength representing an element of interest. The major advantage of this single element method is its high sensitivity.

X-ray fluorescence (XRF) spectrometry may be used to analyze elements of atomic number 11 (Na) and higher. In this method the sample is bombarded by X-rays from an X-ray tube or a secondary target. The elements in the sample become excited and emit element-specific X-rays, which are used for identifying the elements. The intensities of the X-rays are used to quantify the elemental content.

Proton-induced X-ray emission (PIXE) analysis resembles the conventional X-ray fluorescence method. Instead of X-rays, PIXE analysis utilizes high-energy protons for exciting the elements in the sample. The major advantage of PIXE analysis is its relatively high sensitivity. The main drawback is its need of a proton accelerator.

Neutron activation analysis (NAA) is based on the activation of the sample in a flux of neutrons. It is a very sensitive method for most elements. The neutron bombardment creates radioactive isotopes which emit characteristic gamma radiation during decay. The measured gamma spectrum after activation is used to identify and quantify the elements in the sample.

While the aerosol particle generated by an industrial process may have a specific composition, ambient aerosols generally contain particles of many chemical compositions, reflecting the many sources the aerosol particles may have come from. Two samples with similar elemental compositions, as measured by the methods indicated above, may have different chemical compositions and, therefore, different effects on health, visibility, and materials. Therefore, many different methods have been developed for the analysis of different chemical compounds. The basic analytical tools for chemical analysis are gas chromatography (GC) and liquid chromatography (LC). In chromatography, the sample is injected into a carrier gas or liquid, which carries the sample into the column. The components of the mixture pass through the column at varying rates, depending on the properties of the components. Each component is then detected and quantified when it emerges from the column. The identification and analysis of complex chemical compounds is often made using another analytical instrument combined with a gas chromatograph.

Individual Particle Analysis

Bulk analysis methods provide information on the composition of the entire aerosol sample. Some of the particles may have an elemental composition and, therefore, may affect the health or material behavior in a way quite different from the one indicated by bulk analysis. In some applications information on the composition of the individual particles is, therefore, of great importance. Elemental analysis of a single particle may be made by the use of an electron microprobe in which a focused electron beam, approximately 1 μm in diameter, irradiates a spot of the sample. The wavelength or the energy of the emitted X-rays is used to identify the elements while the intensity of the X-rays is used for quantitative analysis. The electron microprobe is typically an accessory to a scanning or transmission electron microscope. The analysis of single particles can also be made with ion, laser, or scanning Auger microprobes. These are special techniques which can provide more detailed information about the composition of a given particle.

Biological Analysis

Biological methods are used when dealing with viable particles or particles containing biologically active materials. Different staining techniques may be used to differentiate various species under the microscope. When impaction methods are used to collect bioaerosols in a nutrient medium, the sample is usually incubated for a period of time, after which the colonies in the nutrient medium are counted. The number of colonies is divided by the sampled volume of air and is referred to as the concentration of colony forming units (cfu).

DYNAMIC MEASUREMENT METHODS

The measurement of particle concentration by means of sample collection and subsequent analysis usually requires that the collected sample be transported from the measurement site to a laboratory. The time lag between sampling and analysis and the reporting of the data may be longer than that desired or required. When using a dynamic measurement technique, the instrument gives the data in real time. A dynamic measurement instrument typically consists of an aerosol flow system, a particle sensor and, in some cases, a particle-size-selecting or -discriminating mechanism.

The aerosol flow system transports the particles to a sensing volume or collects the particles on a surface or filter for sensing. For instance, dynamic mass monitoring instruments collect particles on a mass-sensitive element or filter. In contrast, optical instruments typically detect particles in their airborne state without collecting them on a surface or filter, for example, by passing a laser beam through the region of interest.

Most instruments which provide information about particle size use a size-selecting mechanism to limit the particles seen by the sensor to a specific size range. The impactor, diffusion battery, electrical mobility analyzer, and time-of-flight analyzer are typical size selectors. All of these are described in detail in Part II of this book. The single-particle optical counter is an exception. It gives particle size information without size-selecting the aerosol by recording the intensities of scattered light pulses from the particles.

Mass Measurement

The most common dynamic instruments for particle mass monitoring are the piezoelectric, beta attenuation, and tapered element mass monitors. All of these devices dynamically determine the amount of aerosol particles deposited on a substrate from a known volume of air.

The principle of the piezoelectric mass monitor is shown in Fig. 7-4. A piezoelectric crystal is used as a sensitive microbalance (Sem and Daley 1979). In the example shown, electrostatic precipitation collects aerosol particles on the surface of the piezoelectric crystal. The crystal is excited in its natural frequency, which decreases with increasing mass load on its surface. Thus, the particulate mass collected on the crystal can be determined by measuring the change in the crystal's natural frequency. The relationship between the collected aerosol mass Δm (and

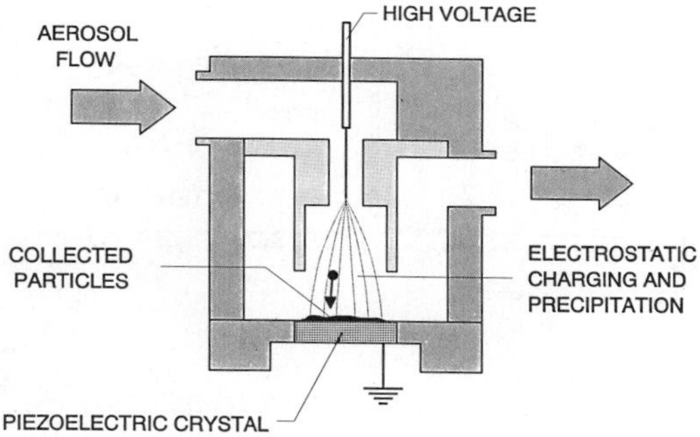

FIGURE 7-4. Piezoelectric Aerosol Mass Monitor.

particle mass concentration c_m) and the frequency shift Δf is given by

$$\Delta f = K_p \Delta m = K_p Q t c_m \qquad (7\text{-}1)$$

where K_p is the mass sensitivity coefficient of the crystal, Q is the sampling air flow rate, and t is the sampling time. The sampling time needed to collect a sufficient amount of particles depends on the particle concentration. In industrial hygiene applications, a typical sampling period is two minutes or less. The major drawback of this technique is its relatively low capacity, i.e., the crystal must be cleaned after a few measurements.

This loading limit is significantly increased in the tapered-element oscillating microbalance (TEOM®) mass monitor (Patashnick and Rupprecht 1980). The particle sample is collected on a filter mounted in a tapered oscillating element. The natural frequency of the tapered element depends on its mass and the change in the frequency is proportional to the aerosol mass collected. This technique has been used successfully for the measurement of diesel fume, ambient air and other aerosols.

The operation of the beta attenuation mass monitor (Macias and Husar 1976), illustrated in Fig. 7-5, depends on the attenuation of beta radiation in the collected aerosol sample. This attenuation of beta radiation is related to the mass of the medium between the radiation source and the detector. This mass consists of the collected particles, the filter, and the air that surrounds them. The increase of mass due to particle collection increases the attenuation of the beta rays. The relationship between the collected aerosol mass Δm and the measured beta radiation intensity I can be approximated by

$$\ln \frac{I}{I_0} \approx K_b \Delta m = K_b Q t c_m \qquad (7\text{-}2)$$

where I_0 is the radiation intensity before sampling begins and K_b is the mass sensitivity coefficient, which depends on the mass attenuation coefficient of the collected material and the geometry of the irradiation, detection, and collection system. The sensitivity of the beta attenuation method is lower than that of the piezoelectric method. Thus, the necessary sampling times are significantly longer.

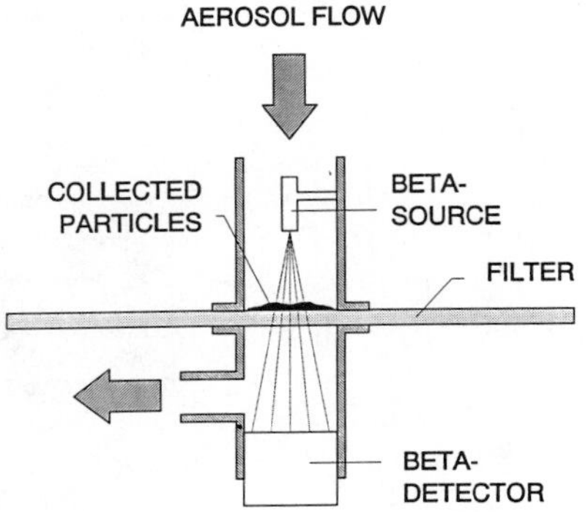

FIGURE 7-5. Beta Attenuation Aerosol Mass Monitor.

Optical Techniques

In optical aerosol measurement instruments the interaction of aerosol particles with the incident light serves as a basis for the real time measurement of particle concentration and size. Important features of optical measurement methods include high sensitivity, nearly instantaneous response and the avoidance of physical contact with the particles. The measured quantities are light absorption and scattering from a single particle or a cloud of particles.

A common technique for monitoring the concentration of aerosol particles measures the light extinction by particles in the path of a light beam, as illustrated in Fig. 7-6. Light extinction is the combination of light absorption and scattering. The relationship between the measured light intensity I, aerosol number concentration c_n and mass concentration c_m is given by

$$\ln \frac{I}{I_0} = -\frac{\pi}{4} d_p^2 Q_e L c_n$$

$$= -\frac{3 Q_e L}{2 \rho_p d_p} c_m \qquad (7\text{-}3)$$

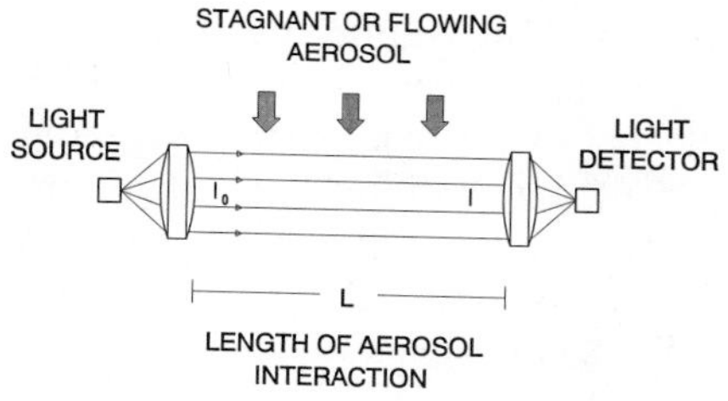

FIGURE 7-6. Light Extinction Method.

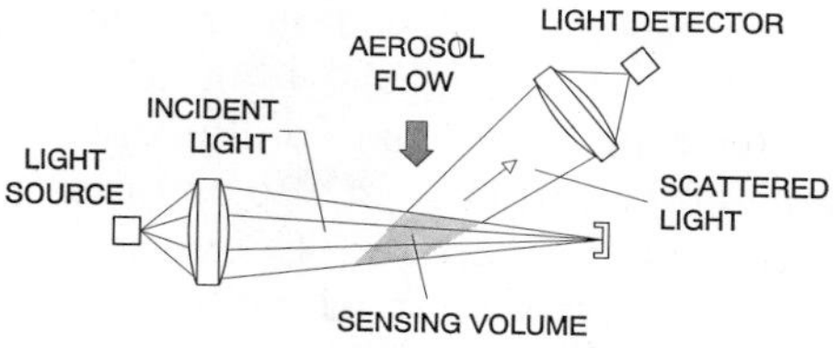

FIGURE 7-7. Light Scattering Aerosol Photometer.

where I_0 is the incident light intensity, d_p is the particle diameter, ρ_p is the particle density, Q_e is the particle extinction efficiency and L is the length of the light beam interacting with the aerosol (Hinds 1982). The extinction efficiency of a particle is a function of its refractive index, size, and shape and of the wavelength distribution of the incident light beam. For ultrafine particles Q_e is proportional to d_p^4, while for large particles ($d_p > 2$ μm) Q_e oscillates while approaching its limiting value of 2.

Equation 7-3 is valid for monodisperse aerosols. For polydisperse aerosols the combined effect of light extinction by particles of different size must be integrated over the particle size range. In practice, light extinction instruments are calibrated by means of an independent method, such as particle collection on a filter with subsequent gravimetric analysis. For example, nonabsorbing, 0.5 μm diameter particles with a refractive index of 1.5 have an extinction efficiency Q_e of 3.5 ($\lambda = 0.52$ μm) (Hinds 1982). For a path length $L = 5$ m and unit particle density $\rho_p = 1$ g/cm^3, the intensity ratio I/I_0 equals 0.9 (10% extinction) for an aerosol mass concentration of 2 mg/m^3. This simple example illustrates the relatively low sensitivity of the light extinction method. Therefore, it is used primarily in environments with high particle concentrations such as smoke stacks.

The sensitivity of an optical aerosol instrument can be significantly increased by measuring the intensity of the scattered light instead of light extinction, as shown schematically in Fig. 7-7. The intensity of the scattered light depends on the intensity of illumination I_0, particle size d_p, particle refractive index m, wavelength λ of the light and the scattering angle. Light scattering instruments collect light from a view volume over an angular range defined by the detection optics. The light flux collected by the detection optics is, therefore, an integral of the scattered-light intensities from all particles in the view volume. An important feature of the scattered-light intensity of such an instrument, referred to as a photometer, is its strong particle size dependence. For small particles ($d_p \ll \lambda$) the intensity is proportional to d_p^6. For large particles ($d_p \gg \lambda$) the intensity is approximately proportional to d_p^2. In the intermediate region ($d_p \approx \lambda$) the intensity function can undergo oscillations, depending upon the optical properties of the particles.

The photodetector of an aerosol photometer transforms the measured light into a voltage signal which is proportional to the light flux. Thus, the relationship between the output signal from an aerosol photometer and particle concentration is very complicated. For a homogeneous monodisperse aerosol the relationship between output signal U, particle number concentration c_n, and mass concentration c_m is given by

$$U = S_n(d_p)c_n = \frac{S_n(d_p)}{(\pi/6)\rho_p d_p^3} c_m = S_m(d_p)c_m \quad (7\text{-}4)$$

where $S_n(d_p)$ and $S_m(d_p)$ are the instrument number and mass sensitivities, respectively. For polydisperse aerosols, the signals generated by particles of different size must be integrated over the entire particle size range. For a cloud of aerosol particles with different optical properties, the appropriate effective sensitivity must be included in the calculations. The output signal of an aerosol photometer depends not only on the optical properties of the particles but also on the size distribution of the particles. Consequently, aerosol photometers must be calibrated by independent methods. Reliable and accurate results can be obtained only if the properties of the measured aerosol are close to the ones used for calibration. Aerosol photometers are used, e.g., to monitor industrial processes, workplace environments, and aerosol exposure levels in inhalation experiments with animals.

An aerosol photometer measures the light scattered by a cloud of particles in the sensing volume. By reducing the size of the sensing volume and introducing the aerosol through a narrow nozzle it is possible to detect scattered light pulses from single particles. A single-particle optical counter, illustrated in Fig. 7-8, can be used for the direct measurement of the particle number concentration. The number concentration is obtained by counting the number of scattered light pulses during the sample period and by dividing this number by the volume of air sampled (Willeke and Liu 1976; Knollenberg and Luehr 1976; Gebhart et al. 1976).

As discussed earlier, the intensity of the scattered light from a single particle is a strong function of particle size. The intensity of the scattered light pulse decreases steeply with decreasing particle size and finally reaches the level of background noise. The overall detected noise level depends on the amount of light scattered by the air molecules, on stray light from the internal surfaces,

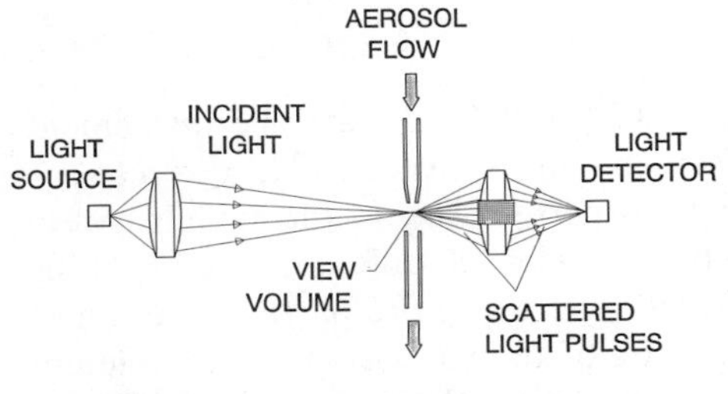

FIGURE 7-8. Single-Particle Optical Counter.

and on the electronic noise. Consequently, there is a practical lower particle size which can be detected by a single-particle counting technique. For conventional white-light instruments the lower limit is typically between 0.3 and 0.5 μm. With laser light illumination the lower limit has been extended to below 0.1 μm.

The strong dependence of the scattered-light intensity on particle size is utilized in single-particle optical counters to measure the size of the particles as they are counted. Modern optical particle counters are sophisticated instruments which produce nearly real-time information on the particle size distribution. Unfortunately, the intensity of the scattered light depends not only on the particle size but also on the optical properties of the particles passing through the view volume. Thus, it may be very difficult to relate the measured intensity distribution to the actual size distribution. In practice, the particle size indicated by a single-particle counter must be regarded as an optical equivalent particle size. The indicated particle sizes correspond to the true particle sizes only when the optical properties of the particles being measured are equal to the optical properties of the particles used for calibration (for particle sizes with a monotonic relationship between output signal and particle size). The requirement that only one particle at a time pass through the view volume limits the use of single-particle optical counter to low particle concentrations. The permissible upper concentration limit may be raised by diluting the sampled aerosol with clean air.

The lower particle size detection limit of an optical particle counter can be extended to below 0.01 μm by means of a vapor condensation technique. In condensation nucleus counters, the sampled aerosol is saturated with a vapor and then supersaturated either by adiabatic expansion or direct cooling. The supersaturation causes vapor condensation onto the particles, which leads to the rapid growth of the particles until they are sufficiently large for detection by conventional optical techniques (Sinclair and Hoopes 1975; Bricard et al. 1976; Agarwal and Sem 1980).

Since the optical sensor does not detect the original particle sizes, a condensation nucleus counter measures only the total number concentration, i.e., no particle size information is given.

The ability of optical particle counters and condensation nucleus counters to detect single particles makes them very desirable as aerosol monitors in low-concentration environments such as industrial clean rooms. The ability of the single-particle optical counter to simultaneously provide information about particle concentration and size distribution is widely used, e.g., for measuring the particle-collecting properties of filters.

Electrical Techniques

The electrical properties of aerosol particles are also used widely to measure particle concentration and size distribution while the particles are in the airborne state. Considerations for electrical detection are similar to those for optical detection in that the measurement depends on the electrical properties of the sampled particles. Electrical methods also give nearly real-time indications of the particle concentrations.

The simplest form of an electrical aerosol instrument—probably the most widely used aerosol instrument—is the ionization-type smoke detector. In a smoke detector a small radioactive source generates a measurable flow of air ions. Aerosol particles capture air ions, causing a decrease in the measured ion flow. The ionization-type smoke detector is relatively sensitive for small aerosol particles and is, therefore, an effective means for detecting a fire in its early stages.

The simplest form of an electrical aerosol instrument used in aerosol studies is the aerosol electrometer consisting of an electrometer and a filter inside a Faraday cage, as shown in Fig. 7-9. Charged particles collected by the filter generate an electric current, which is measured by the electrometer (Whitby 1976). The electric current I measured by the electrometer is given by

$$I = (c_q^+ - c_q^-)Q = c_q Q \qquad (7\text{-}5)$$

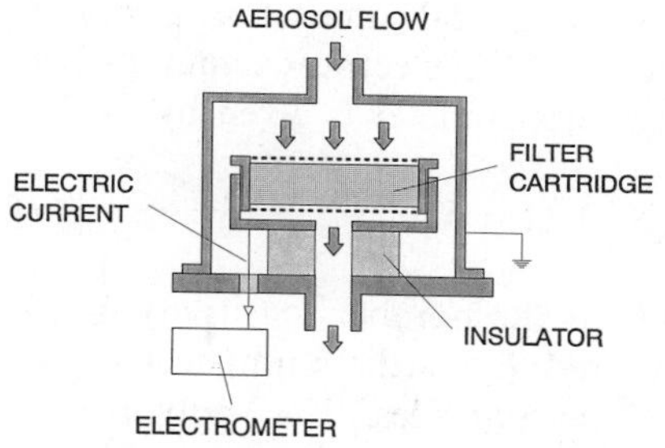

FIGURE 7-9. Aerosol Electrometer.

where Q is the sample air flow rate, c_q^+ and c_q^- are the charge concentrations of positively and negatively charged particles, respectively, and c_q is the net charge concentration of the particles. According to Gauss' law, the charge collected on the Faraday cup is the induced charge, i.e., the filter need not be a conductor. The aerosol electrometer is typically used to measure particles of unipolar charge, i.e., particles with a net charge concentration that equals the charge concentration of positively or negatively charged particles.

Unipolar charging may be achieved by diffusion charging of the aerosol, as illustrated in Fig. 7-10. A corona discharge from a high-voltage wire generates air ions that drift towards the grounded, cylindrical shell of the device. Aerosol particles passing through the annular region perpendicular to the ion drift become charged by the diffusion of air ions to them (Liu and Pui 1975). Combining the unipolar aerosol charger with the aerosol electrometer into an "electrical aerosol detector" makes it possible to measure particle

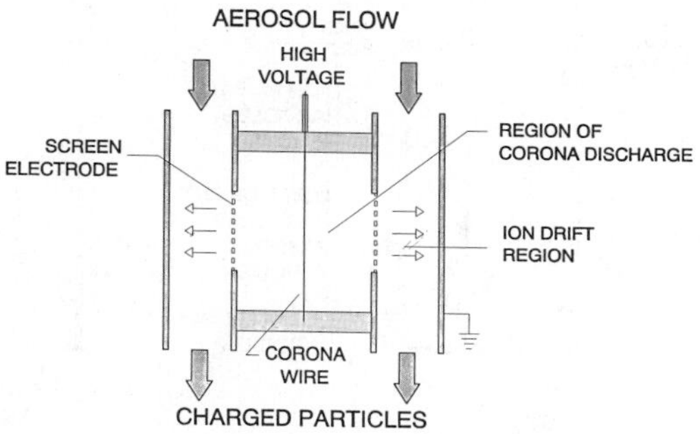

FIGURE 7-10. Unipolar Diffusion Charger.

concentrations in real time. For a monodisperse aerosol, the electric current I measured by the electrometer is given by

$$I = S_n(d_p)c_n \tag{7-6}$$

where $S_n(d_p)$ is the number sensitivity of the instrument and c_n is the number concentration of the particles. For polydisperse aerosols, the current has to be integrated over the entire particle size range.

The number sensitivity of an instrument based on unipolar diffusion charging of particles is approximately proportional to $d_p^{1.2}$ (Liu and Pui 1975). Consequently, the mass sensitivity is proportional to $d_p^{-1.8}$. The size dependencies of these sensitivities differ significantly from those of the conventional aerosol measuring instruments, i.e., particle counter (d_p^0) or mass monitor (d_p^3). Therefore, the electrical aerosol detector is not widely used outside laboratories conducting basic aerosol research.

The electric charge on aerosol particles may not only be utilized for particle detection but also for measuring the particle sizes. Charged particles placed in an electric field drift with a velocity which is a function of particle size and charge. Such a size dependence is utilized for particle sizing in the electrical mobility analyzer (Whitby 1976). Typically, charged particles are passed through a cylindrical condenser in which particle penetration is controlled by the voltage between the electrodes of the condenser. The electrical mobility analyzer operates as a low-pass filter (Fig. 7-11a) or as a band-pass filter (Fig. 7-11b), depending upon the construction of the analyzer (Knutson and Whitby 1975; Liu and Pui 1975). The low-pass device precipitates the high-mobility (smaller) particles and allows the (larger) particles with low mobility to pass through. The band-pass or "differential mobility" analyzer or classifier allows only particles within a narrow range of electrical mobility to pass through the device. In both versions, the aerosol flow is introduced in the outer annulus of the analyzer tube and clean air is introduced in the central core of the tube. The electric field between the outer and inner electrodes is used to deflect charged particles towards the inner electrode. The low-pass device detects all charged particles that have penetrated the analyzer tube, while the differential mobility analyzer detects only the charged particles extracted through the narrow mobility window in the center electrode.

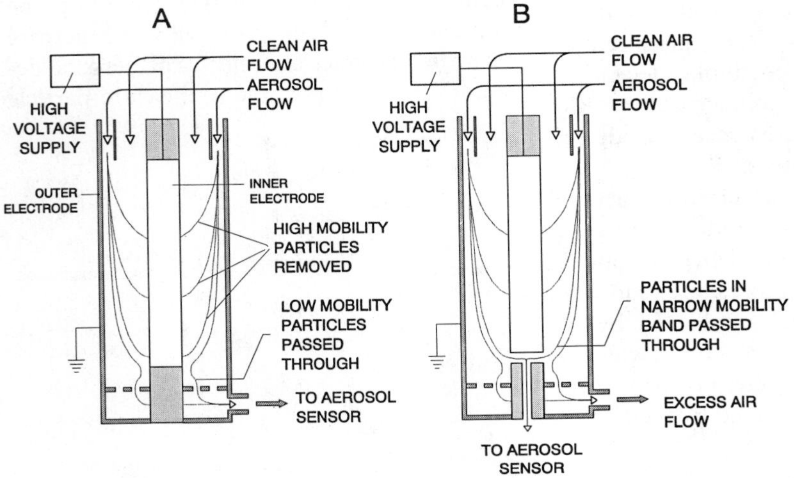

FIGURE 7-11. Electrical Mobility Analyzer. (a) Low-pass, (b) Band-pass.

Combination of Physical Measurement Techniques

Most of the dynamic aerosol monitoring techniques provide information about particle number concentration, mass concentration, light extinction, light scattering, or charge concentration. Except for the single-particle optical counter, most do not give simultaneous information about how these quantities are distributed as a function of particle size. Information on particle size distribution is, generally, obtained by combining a dynamic aerosol detection technique with a size classification technique. Some of the most commonly used instruments that combine two techniques and yield particle size information in real time are:

- Cascade impactor with piezoelectric crystals
- Time-of-flight aerodynamic sizing instrument
- Fibrous-aerosol monitor
- Diffusion battery particle size analyzer
- Differential-mobility particle sizer
- Electrical aerosol analyzer

In the first of these instruments the measurement system consists of a cascade impactor with piezoelectric crystals as impaction surfaces. The size-selected aerosol mass collected on each stage is determined by the change in natural frequency of the respective crystal (Chuan 1976).

The operation of the time-of-flight aerodynamic particle sizer is based on the detection of scattered light pulses from single particles. The major difference between this instrument and the conventional optical single-particle counter is the particle sizing technique. Instead of measuring the intensities of the scattered light pulses, the time-of-flight aerodynamic particle sizer registers the velocities of single particles exiting from an acceleration nozzle, which are related to the particle size (Wilson and Liu 1980; Remiarz et al. 1983).

The fibrous aerosol monitor (Lilienfeld, Elterman, and Baron 1979) combines optical detection with particle behavior in an oscillating high-intensity electric field. The detection of the particles occurs in an oscillating electric field, which causes the fibrous particles to rotate. Consequently, the scattered light signal measured by a fixed sensor varies with the fiber orientation. In contrast, spherical particles are unaffected by the oscillating electric field and scatter a light pulse without electric-field-induced modulation. Thus, fibrous particles are distinguished from less elongated particles.

Combining a condensation nucleus counter with a size-selective preclassifier is a widely used practice to measure the size distribution of submicrometer-sized particles (Sinclair et al. 1979). The size distribution according to the diffusion-equivalent particle size may be determined by measuring the aerosol penetration characteristics through several diffusion batteries with different effective lengths. The penetration of particles through a diffusion battery is only weakly dependent on particle size. Thus, the size resolution of this method is low.

Excellent size resolution can be achieved by using the differential-mobility analyzer as a size selective preseparator for the condensation nucleus counter (Keady, Quant, and Sem 1983). The differential-mobility analyzer separates a narrow size channel from a polydisperse aerosol (Knutson and Whitby 1975). The size-selective separation is based on the mobility of charged particles in an electric field. Thus, the particle size distribution is determined according to the electric-mobility-equivalent particle size. The particle size distribution is determined by measuring the particle penetration characteristics as a function of the voltage applied to the mobility analyzer. Inversion calculations convert the penetration characteristics to the particle size distribution. Instead of the condensation nucleus counter, an aerosol electrometer may also be used as a particle sensor, especially for ultrafine particles.

The differential-mobility particle sizer has gradually replaced the electrical aerosol analyzer (Liu and Pui 1975; Whitby 1976). In the latter instrument the aerosols are charged by unipolar diffusion and are detected by an aerosol electrometer after passage through a low-pass mobility analyzer. The electric current generated by the flow of charged particles collected by the aerosol electrometer is registered as a function of the mobility analyzer voltage and the penetration characteristics are then used to calculate the particle size distribution.

A typical feature of dynamic aerosol instruments is the operational range, which generally covers about one and a half decades in particle size. Thus, the simultaneous use of several instruments is necessary if particle size distributions are to be measured over a wide particle size range. This can be accomplished by employing, e.g., a differential-mobility particle sizer or an electrical aerosol analyzer for the ultrafine particles and optical particle counters for the larger particles. The major problem in the simultaneous use of several instruments is the matching of the size distributions when each of the instruments is based on a different physical principle and, therefore, reports a different equivalent particle size.

References

Agarwal, J. K. and G. J. Sem. 1980. Continuous flow, single-particle-counting condensation nucleus counter. *J. Aerosol Sci.* 11:343–57.

Armbruster, L. and H. Breuer. 1983. Dust monitoring and the principle of on-line dust control. In *Aerosols in the Mining and Industrial Work Environments, Vol. 3,* eds. V. A. Marple and B. Y. H. Liu, pp. 689–99. Ann Arbor: Ann Arbor Science Publishers.

Bricard, J., P. Delattre, G. Madelaine, and M. Pourprix. 1976. Detection of ultra-fine particles by means of a continuous flux condensation nuclei counter. In *Fine Particles: Aerosol Generation, Measurement, Sampling and Analysis,* ed. B. Y. H. Liu, pp. 565–80. New York: Academic Press.

Chen, B. T., H. C. Yeh, and Y. S. Cheng. 1985. A novel virtual impactor: calibration and use. *J. Aerosol Sci.* 16(4):343–54.

Cheng, Y. S. and H. C. Yeh. 1983. Performance of a screen-type diffusion battery. In *Aerosols in the Mining and Industrial Work Environments, Vol. 3,* eds. V. A. Marple and B. Y. H. Liu, pp. 1077–94. Ann Arbor: Ann Arbor Science Publishers.

Chuan, R. L. 1976. Rapid measurement of particulate size distribution in the atmosphere. In *Fine Particles: Aerosol Generation, Measurement, Sampling and Analysis,* ed. B. Y. H. Liu, pp. 763–79. New York: Academic Press.

Chung, F. H. 1981. Imaging and analysis of airborne particulates. In *Air/Particulate Instrumentation and Analysis,* ed. Paul N. Cheremisinoff, pp. 89–117. Ann Arbor: Ann Arbor Science Publishers.

Dzubay, T. G. 1977. *X-ray Fluorescence Analysis of Environmental Samples.* Ann Arbor: Ann Arbor Science Publishers.

Gebhart, J., J. Heyder, C. Roth, and W. Stahlhofen. 1976. Optical aerosol size spectrometry below and above the wavelength of light—a comparison. In *Fine Particles: Aerosol Generation, Measurement, Sampling and Analysis,* ed. B. Y. H. Liu, pp. 794–815. New York: Academic Press.

Hinds, W. C. 1982. *Aerosol Technology.* New York: Wiley.

Keady, P. B., F. R. Quant, and G. J. Sem. 1983. Differential mobility particle sizer: A new Instrument for high-resolution aerosol size distribution measurement below 1 μm. *TSI Quarterly* 9(2):3–11.

Knollenberg, R. G. and R. Luehr. 1976. Open cavity laser "active" scattering particle spectrometry from 0.05 to 5 microns. In *Fine Particles: Aerosol Generation, Measurement, Sampling and Analysis,* ed. B. Y. H. Liu, pp. 669–96. New York: Academic Press.

Knutson, E. O., and K. T. Whitby. 1975. Aerosol classification by electric mobility: Apparatus, theory, and applications, *J. Aerosol Sci.* 6:443–51.

Lee, T. C. and N. A. Esmen. 1983. Versatile parallel-stage impactor. In *Aerosols in the Mining and Industrial Work Environments, Vol. 3,* ed. V. A. Marple and B. Y. H. Liu, pp. 951–69. Ann Arbor: Ann Arbor Science Publishers.

Lilienfeld, P., P. B. Elterman, and P. Baron. 1979. Development of a prototype fibrous aerosol monitor. *Am. Ind. Hyg. Assoc. J.* 40:270–82.

Lippmann, M. 1989*a*. Size-selective health hazard sampling. In *Air Sampling Instruments,* ed. Susanne V. Hering, pp. 163–98. Cincinnati, OH: American Conference of Governmental Industrial Hygienists.

Lippmann, M. 1989*b*. Sampling aerosols by filtration. In *Air Sampling Instruments,* ed. Susanne V. Hering, pp. 305–36 Cincinnati, OH: American Conference of Governmental Industrial Hygienists.

Liu, B. Y. H. and D. Y. H. Pui. 1975. On the performance of the electrical aerosol analyzer. *J. Aerosol Sci.* 6:249–64.

Liu, B. Y. H., K. T. Whitby, and H. S. Yu. 1967. Electrostatic aerosol sampler for light and electron microscopy. *Rev. Sci. Instrum.* 38:100–02.

Loo, B. W., J. M. Jaklevic, and F. S. Goulding. 1976. Dichotomous virtual impactors for large scale monitoring of airborne particulate matter. In *Fine Particles: Aerosol Generation, Measurement, Sampling and Analysis,* ed. B. Y. H. Liu, pp. 311–50. New York: Academic Press.

Macias, E. S. and R. B. Husar. 1976. A review of atmospheric particulate mass measurement via the beta attenuation technique. In *Fine Particles: Aerosol Generation, Measurement, Sampling and Analysis*, ed. B. Y. H. Liu, pp. 535–64. New York: Academic Press.

Marple V. A. and K. Willeke. 1979. Inertial impactors. In *Aerosol Measurement*, eds. Dale A. Lundgren, et al., pp. 90–107. Gainesville: University Presses of Florida.

Mercer, T. T. 1973. *Aerosol Technology in Hazard Evaluation.* New York: Academic Press.

Patashnick, H. and G. Rupprecht. 1980. A new real time aerosol mass monitoring instrument: The TEOM. In *Proceedings: Advances in Particulate Sampling and Measurement*, ed. W. B. Smith, p. 264, EPA-600/9-80-004. U.S. Environmental Protection Agency.

Pui, D. Y. H. and B. Y. H. Liu. 1989. Advances in instrumentation for atmospheric aerosol measurement. *TSI J. Particle Instrum.*, 4(2):3–20

Rao, A. K. and K. T. Whitby. 1978*a*. Non-ideal collection characteristics of inertial impactor—I. Single-stage impactors and solid particles. *J. Aerosol Sci.* 9:77–86.

Rao, A. K. and K. T. Whitby. 1978*b*. Non-ideal collection characteristics of inertial impactor—II. Cascade impactors. *J. Aerosol Sci.* 9:87–100.

Remiarz, R. J., J. K. Agarwal, F. R. Quant, and G. J. Sem. 1983. A real-time aerodynamic particle size analyzer. In *Aerosols in the Mining and Industrial Work Environments, Vol. 3*, ed. V. A. Marple and B. Y. H. Liu, pp. 879–95. Ann Arbor: Ann Arbor Science Publishers.

Sem, G. J. and M. P. S. Daley. 1979. Performance evaluation of a new piezoelectric aerosol sensor. In *Aerosol Measurement*, eds. Dale A. Lundgren, et al., pp. 672–86. Gainesville: University Presses of Florida.

Sinclair, D., R. J. Countess, B. Y. H. Liu, and D. Y. H. Pui. 1979. Automatic analysis of submicron aerosols. In *Aerosol Measurement*, eds. Dale A. Lundgren, et al., pp. 544–63. Gainesville: University Presses of Florida.

Sinclair, D., and G. S. Hoopes. 1975. A continuous flow condensation nucleus counter. J. *Aerosol Sci.* 6:1–7.

Smith, W. B., R. R. Wilson, and D. B. Harris. 1979. A five-stage cyclone system for in situ sampling. *Environ. Sci. Technol.* 13:1387–92.

Stöber, W. 1976. Design, performance and application of spiral duct aerosol centrifuges. In *Fine Particles: Aerosol Generation, Measurement, Sampling and Analysis*, ed. B. Y. H. Liu, pp. 351–97. New York: Academic Press.

Swift, D. L. and M. Lippmann. 1989. Electrostatic and thermal precipitators. In *Air Sampling Instruments*, ed. Susanne V. Hering, pp. 387–402. Cincinnati, OH: American Conference of Governmental Industrial Hygienists.

Tillery, M. I. 1979. Aerosol centrifuges. In *Aerosol Measurement*, eds. Dale A. Lundgren et al., pp. 3–23. Gainesville: University Presses of Florida.

Whitby, K. T. 1976. Electrical measurement of aerosols. In *Fine Particles: Aerosol Generation, Measurement, Sampling and Analysis*, ed. B. Y. H. Liu, pp. 581–624. New York: Academic Press.

Willeke, K. 1984. Aerosol measurement. In *Advances in Modern Environmental Toxicology, Vol. VIII*, eds. N. A. Esmen, and M. A. Mehlman, pp. 123–51. Princeton: Princeton Scientific Publishers.

Willeke, K. and S. J. DeGarmo. 1988. Passive versus active aerosol monitoring. *Appl. Ind. Hyg.* 3:263–66.

Willeke, K. and B. Y. H. Liu. 1976. Single particle optical counter: principle and application. In *Aerosol Measurement*, eds. Dale A. Lundgren et al., pp. 697–729. Gainesville: University Presses of Florida.

Wilson, J. C. and B. Y. H. Liu. 1980. Aerodynamic particle size measurement by laser-Doppler velocimetry. *J. Aerosol Sci.* 11:139–50.

8

Factors Affecting Aerosol Measurement Quality

Paul A. Baron and William A. Heitbrink

National Institute for Occupational Safety and Health,[1]
Centers for Disease Control, Public Health Service,
Department of Health and Human Services, Cincinnati, OH, U.S.A

INTRODUCTION

Scientists involved with aerosol measurements have available to them a diversity of aerosol monitors, ranging from sample collection on a filter for later analysis to complex direct-reading instruments, which detect the airborne particles in real time and display the concentration of airborne particles as a function of aerodynamic diameter. The use of the simplest or most complex device may introduce errors in measurement and interpretation. Typically, these errors include both biases and variability. While the data from the more complex sizing instruments may make such errors evident, they also occur in the less sophisticated filter collectors. Lack of recognition of these errors may affect the interpretation of aerosol measurements. The reported information should be presented with an analysis of the correctness or accuracy of the results.

If ". . . measurement processes are to serve both the practical needs of humankind and excellence in the pursuit of new scientific knowledge, they must be endowed with an adequate level of accuracy. . . . Control, and acceptable bounds for imprecision and bias are clearly prerequisites; but scientific conventions (communication) and scientific and technological means for approaching 'the truth' must also be considered (Currie 1991)." While nomenclature provides the basis for communication of the accuracy of the measurements, the basis for developing the accuracy limits on measurements comes from experiments, assumptions, and scientific knowledge. Repeat measurements of a variable, comparison with reference methods, and the use of standard materials allow the estimation of variability and bias. Assumptions regarding the underlying basis for variability and bias, e.g., the Poisson distribution for particle counting, are often used to place limits on measurement results. Perhaps the most important factor, however, is the scientist's understanding of the processes contributing to variability and bias. This understanding comes from experience and insight into the physicochemical processes underlying the measurements. This chapter attempts to provide some of the knowledge that is

[1] Mention of product or company names does not constitute endorsement by the National Institute for Occupational Safety and Health.

useful in understanding aerosol measurements. In what follows we provide the definitions of several basic terms and, using these definitions as well as the published data, present some examples of problems that can occur in the collection, interpretation, and presentation of aerosol data.

SOME INDICATORS OF MEASUREMENT QUALITY

In reporting measurements, a scientist strives to indicate the correctness of the data. The nomenclature is an integral part of communicating this correctness. Several commonly used terms, including accuracy, precision, and bias, have been interpreted in various ways. For example, the American Society for Testing Materials (ASTM) and the American Chemical Society have established a number of definitions of these terms (ASTM 1990; Crummett et al. 1980; Keith et al. 1983; Watson, Lloy, and Mueller 1989). There are two primary schools of thought on what accuracy means: one interprets accuracy as the difference between a reference value and the average of a large number of measurements, preferably by several laboratories, while the other interprets accuracy as the difference between a reference value and a single measurement. By averaging the measurements, the first definition depends only on the measurement's bias, while the second depends both on bias and variability (ASTM 1990). The National Institute for Occupational Safety and Health (NIOSH), for instance, has developed an accuracy statement based on the second definition, stating that a laboratory evaluation should indicate that a method provides a result that is $\pm$ 25% of the correct value 95% of the time (Taylor, Kupel, and Bryant 1977). A complete evaluation protocol has been developed around this statement. There are also various definitions of bias and variability; Eisenhart, Ku, and Collé (1990) give tables of recommendations on how to present limits to bias, variability, and inaccuracy for various types of measurements. Statistical techniques for assessing the variability and bias of the data are available in many statistics texts, e.g., Box, Hunter, and Hunter (1985). There are a variety of techniques for assessing the accuracy; these are summarized by Crummett et al. (1980).

Quite often, to improve the estimate of a measured value, a measurement is repeated several times. This results in two descriptors of the data—the central tendency (e.g., the mean, mode, or median) and the spread (e.g., the standard deviation, variance, or range). The purpose of the measurement often dictates the requirements on accuracy. Generally, accuracy involves two terms: bias and variability. A bias is the difference between the central tendency of the measured value and the "true" value. The difficulty often comes in estimating the true value; it can be the result estimated from a reference instrument or technique, a value calculated using well established theory, the average of measurements from a group of competent scientists, or some combination of these.

One of the more common types of aerosol related measurements is that of particle diameter distributions. The equations in this section will be presented in terms of diameter, though they could apply to other variables just as well.

Bias

When a measurement of particle diameter d is repeated, the resultant data values are characterized by the mean $\bar{d}_p$. This is one measure of the distribution's central tendency and is calculated from the sum of the individual measured d_i values and the number of measurements n.

$$\bar{d}_p = \frac{\sum d_i}{n} \tag{8-1}$$

There are other indicators of the center of the data grouping, such as mode (the value which occurs most often); other such indicators are described in Chapter 9 and by Hinds (1982) for particle measurements. The bias is estimated from the difference between a single measurement and the true value, or between the mean of multiple measurements of the same parameter, $\bar{d}_p$, and the true value.

Small biases, e.g., $< 20\%$, often can be corrected by appropriate calculation or calibration. However, larger biases often indicate a lack of control of experimental variables and may not be completely corrected.

Variability

Variability is evaluated by making repeat measurements of a parameter and determining a measure of the spread of the results. The source of variability may be intrinsic in the measured parameter, e.g., from thermally induced noise or the random distribution of particles in a volume, or it may be due to a collection of uncontrolled biases, as in an environmental measurement affected by ambient wind speed or temperature. The difference between the two sources of variability is that the former can be predicted theoretically while the latter cannot. It is often useful to further characterize whether the variability comes from the measurement process, e.g., counting statistics, or from the process under study, e.g., aerosol generation rate.

Particle size data can often be described with a normal distribution; this is convenient because there are many statistical tests developed for the normal distribution. If a distribution is not normal, but can be normalized by some transformation, then these tests still can be used. For example, skewed number versus size data that tail off toward large particle sizes can often be normalized by taking the log of each value (lognormal distribution). The lognormal distribution is commonly used for particle size distribution data as indicated in Chapter 2. The variability in the normal distribution is commonly represented by the standard deviation σ (or its square, the variance); the estimate of σ, i.e., s can be calculated from n measurements:

$$s = \sqrt{\frac{\sum_{i=1}^{n} (\bar{d}_p - d_i)^2}{n-1}} \tag{8-2}$$

For a large number of measurements, s approaches the true standard deviation σ and the measured mean $\bar{d}_p$ approaches the true mean. For normally distributed data, approximately 68%, 95%, and 99.5% of the measurements will be within one, two, and three standard deviations from the mean, respectively.

Measurements of the particle size distributions provide a commonly used type of data. The number frequency function for a normal distribution of particles can be described using $\bar{d}_p$ and s, where the fraction of particles Δf in a small size interval Δd is given by

$$\Delta f = \frac{1}{\sqrt{2\pi}\, s} \exp\left[-\frac{(\bar{d}_p - d_i)^2}{2s^2} \right] \Delta d \tag{8-3}$$

The measured lognormal distribution is most commonly described by

$$\Delta f = \frac{1}{\sqrt{2\pi} \ln \sigma_g} \times \exp\left[-\frac{(\ln \text{CMD} - \ln d_i)^2}{2(\ln \sigma_g)^2} \right] \Delta \ln d \tag{8-4}$$

where CMD is the count median diameter (CMD is also indicated as d_{50} below); the term σ_g is called the geometric standard deviation and is always ≥ 1.

Particle count data are often described by a Poisson distribution. This distribution is the result of particles randomly distributed in space or time and, thus, arriving at random times at the detector. Poisson-distributed data have a standard deviation that is the square root of the number of particles counted. For example, if 100 particles are counted, the standard deviation of that count is $\sqrt{100}$ or 10. The Poisson distribution also can be normalized by taking the square root of the data values. For large numbers of counts, the Poisson distribution approaches the normal distribution. Particle arrival times at a detector and particles collected on a filter surface often can be described with a Poisson distribution.

Limits of Detection and Quantitation

There are several ranges into which a measurement can fall. Its value can be so small that it is indistinguishable from zero or from noise in the measurement process. There may still be useful information in this data (Helsel 1990) but, generally, these values are characterized as being below the limit of detection (LOD). At some level above the LOD, the measurement value may meet a predefined criterion of acceptability, the limit of quantitation (LOQ).

Figure 8-1 shows a range of levels for the measurement of particle diameters. At low concentrations, the measurement may be limited by false counts produced by instrumental or background noise. If we measure this background repeatedly, a distribution of values (with a mean and standard deviation) is likely to result; the LOD of the measurement can be defined as the value 3σ above the mean of the background distribution. More detailed statistical development of the LOD has been performed by Long and Winefordner (1983). One definition of LOQ is given in terms of this same background distribution, namely 10σ above the distribution mean (Keith et al. 1983). These indicators of data quality, i.e., LOD and LOQ, are commonly used in the presentation of laboratory chemical analyses. These terms may be applied to determine the useful data in particle size distributions, especially in the tails of the distributions.

AEROSOL MEASUREMENT ERRORS

Figure 8-2 summarizes some major sources of biases that may occur in an aerosol measurement. The original unsampled aerosol may range in particle size from about 0.001 μm to about 100 μm. Various portions of this range may be nondetectable with a given measurement technique. Particles smaller than about 20% of the wavelength of visible light (0.4–0.7 μm) are generally not detectable by optical means. Depending on the purpose of sampling and the type of aerosol present, different portions of this 0.001–100 μm size range may be of interest. For instance, the health scientist's concern often focuses on the 0.5–10 μm size range because the aerosol particle mass within this size range is likely to deposit in the biologically sensitive area of the lung. Measurement of such aerosols will be used as an example in some of the following discussion, parts of which have been adapted from Willeke and Baron (1990).

As the aerosol enters the sampling inlet of the aerosol measuring device, the ratio of ambient air velocity to sampling velocity, the air turbulence, as well as the size, shape, and orientation of the inlet may affect the sampling efficiency of the inlet (Vincent 1989; Okazaki, Wiener, and Willeke 1987*a*, *b*). Generally, the larger particles enter less efficiently, as illustrated in Fig. 8-2, because of properties producing inertial losses and particle settling. Various particle size preclassifiers, such as

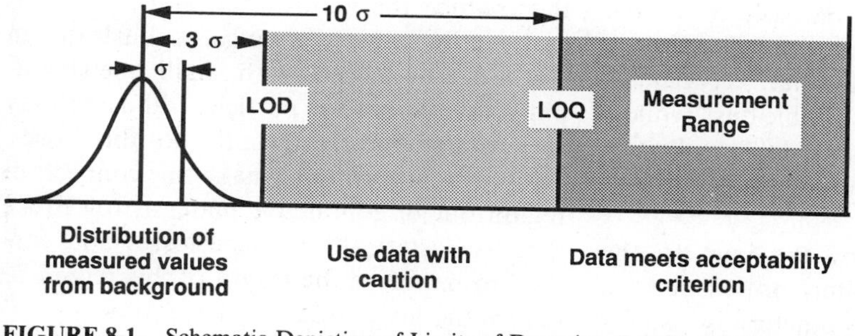

FIGURE 8-1. Schematic Depiction of Limits of Detection (LOD) and Quantitation (LOQ) for a Measurement Method Defined in Terms of the Standard Deviation of the Background Signal. The Measurement Range Covers the LOQ to the Upper Limit of the Measurement Technique (Keith et al. 1983). Other Definitions of these Terms Have Also Been Used.

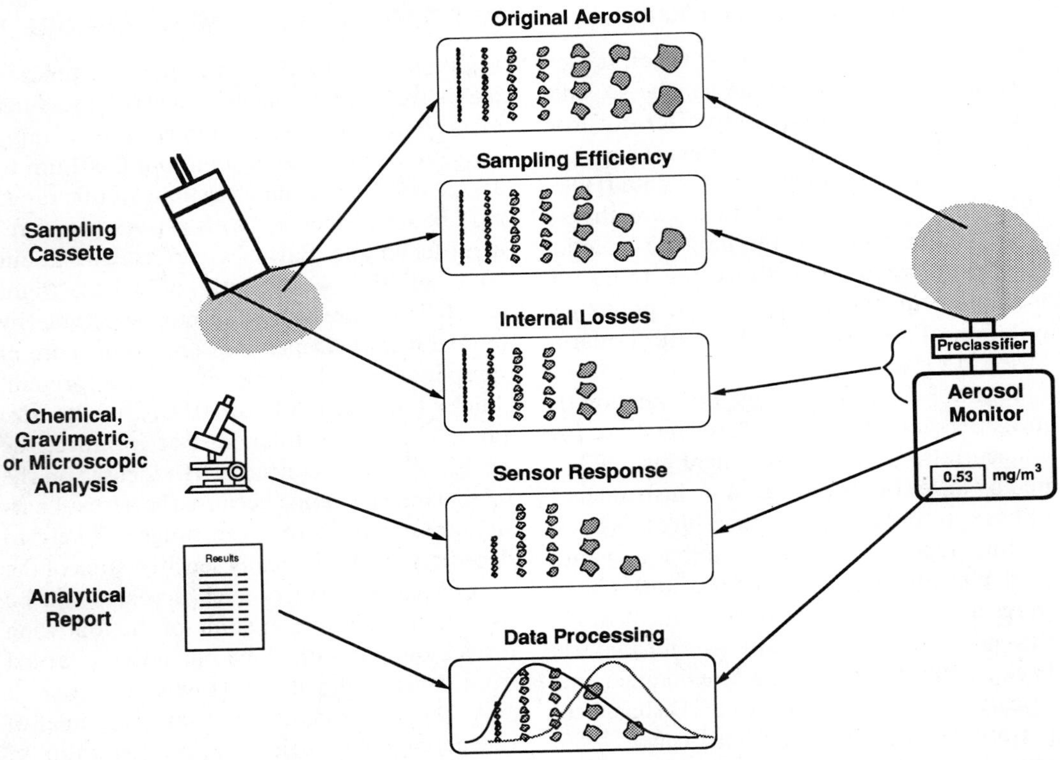

FIGURE 8-2. Schematic Representation of Some Important Biases in Aerosol Monitoring. (Figure Adapted from Willeke and Baron 1990.)

cyclones or elutriators, take advantage of these properties to impose size discrimination on sampled particles. Some of these devices are tailored to allow only a certain fraction of particles to pass through for detection. Aerosol particles reaching one specific region of health concern, i.e., the gas exchange region of the lung, is defined as respirable dust. A cyclone is generally used to measure respirable dust as defined by the American Conference of Governmental Industrial Hygienists (ACGIH) definition of respirable dust, while a horizontal elutriator is used for the British Medical Research Council (BMRC) definition (ACGIH 1984).

The section connecting the inlet to the collection device (e.g., filter) or sensor (e.g., detection region of photometer) is usually considered separately from the inlet or the point at which the aerosol enters the measurement device. For instance, in asbestos sampling a length of tubing called a cowl (Carter, Taylor, and Baron 1989) connects the sampling inlet to the collecting filter. In a direct-reading monitor, the aerosol is generally transported from the inlet to the sensor via a tube or channel. Particle losses may occur in these channels due to electrostatic attraction, impaction, or gravitational settling and further reduce the aerosol concentration, generally in the upper size range, as illustrated in Fig. 8-2. For devices with small inlets sampling submicrometer particles, diffusion may also contribute significantly to the losses. Thus, it is important to make this connection region out of conductive material to reduce the electrostatically induced losses and, further, to minimize the length of this region to reduce the losses due to other forces.

When a filter sample is analyzed under a microscope, particles smaller than the wavelength of the illuminating light may not be detected efficiently. For an electron microscope, this wavelength is much smaller than for an optical microscope. Thus, the microscope and the human eye discriminate against detection of smaller particles. Other types of analysis also may have size-dependent biases introduced during sample preparation or analysis. The sensor of a direct-reading aerosol monitor has a lower threshold below which the smaller particles remain undetected, as illustrated in Fig. 8-2. The upper size limit of detection is generally less of a problem for the sensor. However, most sensors do not respond equally to all particles of varying size and shape. This further modifies the measurement process. Often, these effects cannot be changed but must be recognized during the analysis of the data.

A further bias can occur with instruments such as optical particle sizing instruments that depend on having only one particle at a time in the detector's view volume. If more than one "coincide" in the view volume, the sensor registers only one particle, possibly of a larger size (Willeke and Liu 1976). More complex instruments may produce more complex coincidence effects, modifying the observed size distributions in unusual ways (Heitbrink, Baron, and Willeke 1990). These coincidence errors can usually be reduced by lowering the particle concentration.

Data processing involves collection, storage, and analysis of the data. If too few particles have been sampled, the displayed particle size distribution may not reflect the true size distribution because of statistical considerations. If the particles are counted as a function of particle size under the optical microscope or *in situ* by an optical sensor, the volume or mass can be calculated for each particle, thus shifting the "weighting" from a "count" distribution to a "volume" or "mass" distribution. Various assumptions in this weighting procedure can bias the resulting distribution. The assumption of particle sphericity is usually an approximation, except for droplets. Since the particle volume depends on the cube of the particle size, a few large particles outweigh many small particles. Thus, presentation of the particle size by "count" for most naturally occurring aerosol size distributions focuses on a smaller size range than the size distribution weighted by "volume" or by "mass." The number of particles in the relevant size range, therefore, statistically limits the accuracy of the recorded aerosol concentration, indicating that a sufficient number of particles must be collected in the size range of interest.

The type of display, whether it is a histogram or a cumulative plot, emphasizes different aspects of the size distribution. Finally, the method of size calibration plays an important role in the accuracy of the results. For example, if a photometer or optical single-particle counter is calibrated with particles that scatter but do not absorb light, an absorbing aerosol, such as coal dust, will be registered as having a smaller than actual particle size.

In the following sections, the aerosol size distributions used to demonstrate some of the above points were calculated using a computer spreadsheet program. This type of program allowed a rapid calculation of lognormal size distributions (using Eq. 8-4) that can exist in sampled atmospheres, as well as how these distributions might be affected by biases and variability that occur with these measurements. Published sampling and measurement efficiency data were used to modify these lognormal distributions. Note that the number concentrations calculated were based on equal-size increments on a log scale (i.e., $\Delta(\log d_p) = \text{constant}$) so that the ordinate in each graph is $\Delta N/\Delta(\log d_p)$, where N is in the units of number of particles/cm^3. Some curves were scaled to give a desired peak concentration.

The variability present in actual aerosol measurements of finite numbers of particles was simulated in some cases. Since aerosol particles arrive at a detector at random times, the count variability was described by a simulated Poisson distribution within each size increment. This variability was introduced by adding to each size increment a random number which was normally distributed (on the square root scale) about zero and had a variance equal to the particle count in that size increment.

In the following sections, some of the sources of bias and variability in measurement and interpretation are examined in more detail for some specific measurement situations. Note that this approach to calculating size distributions provides a convenient means of data analysis, both for planning experiments and for understanding published data.

Sampling and Transport

The measured size-dependent sampling efficiencies for the open- and closed-face 37 mm cassettes (Buchan, Soderholm, and Tillery 1986), both widely used in industrial hygiene sampling with filters, have been multiplied by the corresponding values of an example lognormal size distribution with a median diameter $d_{50} = 5.0\ \mu m$ and a geometric standard deviation $\sigma_g = 2.0$ (Fig. 8-3). These samplers are used for a variety of dust measurements and a smaller-diameter version of the cassette is used for asbestos exposure measurement (Carter, Taylor, and Baron 1984). Two sampling efficiency curves are calculated for an open- and a closed-face sampler hanging down, with the inlet perpendicular to a horizontally moving wind stream

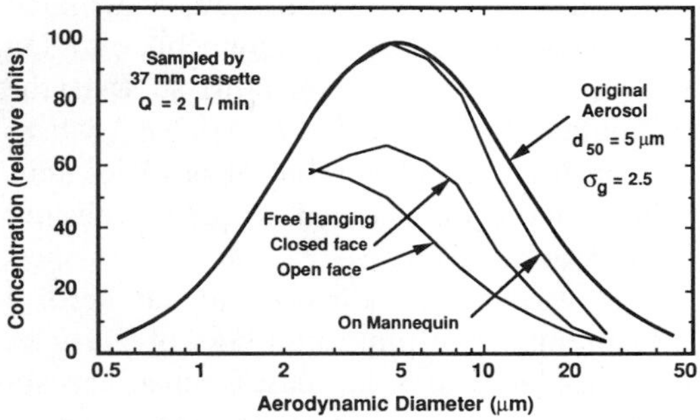

FIGURE 8-3. Sampling and Transport Biases in Several Cassette Configurations. Sampling Efficiency Data Were Taken From Buchan, Soderholm, and Tillery (1986) and Smoothed. Cassettes Hung on a Bluff Body (a Mannequin) Appear to Have Smaller Biases Than Free-Hanging Ones. (Figure Adapted from Willeke and Baron 1990.)

of 100 cm/s; the third curve was calculated from measurements with the sampler on a mannequin facing the wind under the same wind conditions. The mannequin-mounted sampler curves were nearly identical for the closed- and open-faced cassettes; so, a single average curve has been drawn for this case. It is apparent that the air flow conditions near the sampler inlet can significantly affect the collection efficiency of the sampler. The bluff mannequin body reduced the effect of wind speed on the sampler inlets. As pointed out in Chapter 6, the inlet efficiency is greater when the air flow velocity and direction in the sampler and the surrounding air are exactly or nearly matched. In Fig. 8-3, the concentration of the measured particle size distribution is reduced by varying amounts for each sampler placement compared to the true size distribution.

Electrostatic attraction to the cassette inlet and its walls reduces the amount collected on the filter, especially if the cassette is constructed of nonconducting material (Baron and Deye 1990). The loss increases with the number of electrical charges on the aerosol particles and on the sampler and decreases with increasing sampling rate. The number of charges on the aerosol particles depends on the process producing the particles, the air humidity or the amount of water on the particle surface during release and the age of the aerosol. Direct-reading aerosol monitors may have similar sampling and transport losses, depending upon the design of the inlet and the section leading to the sensor (Liu, Pui, and Szymanski 1985).

Sensor Sensitivity and Coincidence Effects

When the particles collected on a filter are analyzed by optical microscopy, many of the small particles are not detected by the microscopist, none being counted below a certain size, say 0.3 µm. The smaller particles of the original aerosol size distribution are, thus, not counted. If, in addition to inlet losses, the filter does not collect particles with 100% efficiency, the sample available for analysis may be further modified.

The combined effect of sensor response and inlet losses is illustrated for the Aerodynamic Particle Sizer (APS, TSI Inc., St. Paul, MN), a time-of-flight aerosol spectrometer that uses light scattering to detect particles. In order to illustrate the effect of a sensor's size-dependent sensitivity, a lognormal size distribution with mass median aerodynamic diameter $d_{50} = 1\ \mu m$ and $\sigma_g = 2.5$ is calculated to simulate the measured aerosol (Fig. 8-4). Based on measured efficiency curves from Blackford et al. (1988), there is a modification of the "measured" size distribution at the low end due to a lack of detector response and at the high end due to a loss of particles at the instrument inlet. Note that neither of these losses changes rapidly with particle size and that the resulting distribution appears nearly lognormal. These modifications of the shape of the distribution may result in an incorrect interpretation of the shape of the true dust distribution.

If the sensor is an optical device receiving a light-scattering signal each time a particle passes through the view volume, particle coincidence, i.e., simultaneous presence of two or more particles in the view volume, may result in the detection of a single larger particle, producing a slight shift to larger sizes and reducing the observed particle number over the entire size range. The importance of coincidence effects increases with particle number concentration.

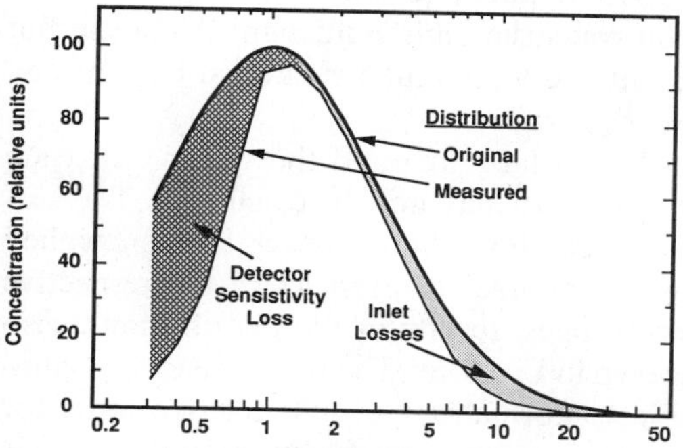

FIGURE 8-4. Sensor Bias Data for the Aerodynamic Particle Sizer (APS3300) Taken From Blackford et al. (1988). (Figure Adapted From Willeke and Baron 1990.)

In a time-of-flight device, such as the APS or the Aerosizer (Amherst Process Instruments, Inc., Hadley, MA), the time-of-flight of a particle accelerated between two path-intersecting laser beams is a measure of its aerodynamic particle size. These instruments can have coincidence losses like other optical particle-counting instruments. In addition to a loss of particle counts, these instruments produce a background of artifactual or phantom counts at all particle sizes that may overshadow the fewer correctly detected particles at the tails of the distribution (Heitbrink, Baron, and Willeke 1990). These phantom counts can be especially important if the distribution is converted to a mass distribution (a few large, phantom particles may outweigh the rest of the distribution) or if the data are used for comparison measurements (Wake 1989).

Aerosol sensors of different types may be used to measure the same parameter, such as particle aerodynamic diameter (d_a). This can provide some estimate of the biases present in the measurements. However, when the readings from different instruments result in widely disparate readings, a detailed understanding of the detection and sampling processes can be used to estimate the "best" answer. A comparison of several measurement techniques used on a grinding wheel aerosol to measure aerodynamic diameter is shown in Fig. 8-5 (O'Brien, Baron, and Willeke 1986). Filter samples were analyzed by scanning electron microscope (SEM) and real-time measurements were made with two optical particle counters (Model CI-108, Climet Instruments, San Diego; Model ASAS-X, Particle Measuring Systems, Boulder), a quartz crystal microbalance cascade impactor (Model PC-2, California Measurements, Berkeley) and an APS. The results from instruments not measuring aerodynamic diameter d_a (defined as the diameter of a unit-density particle having the same gravitational-settling speed as the particle in question) were converted to d_a. Such a conversion is generally made in health-effect studies,

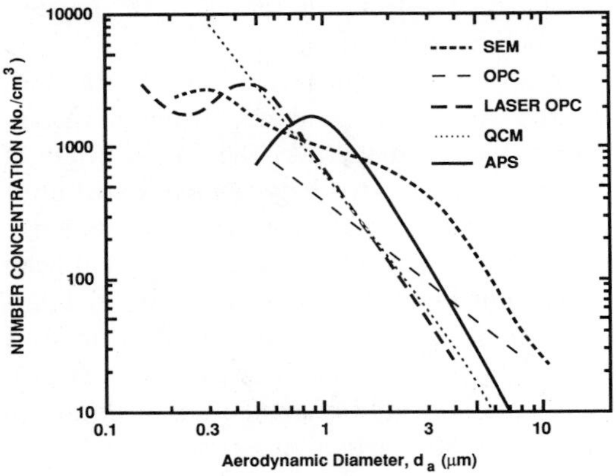

FIGURE 8-5. Measurement of Grinding Wheel Dust Using Six Different Measurement Techniques Including an SEM, Two Optical Particle Counters, a Quartz Crystal Microbalance Cascade Impactor and an APS3300. (Adapted From O'Brien, Baron, and Willeke 1986).

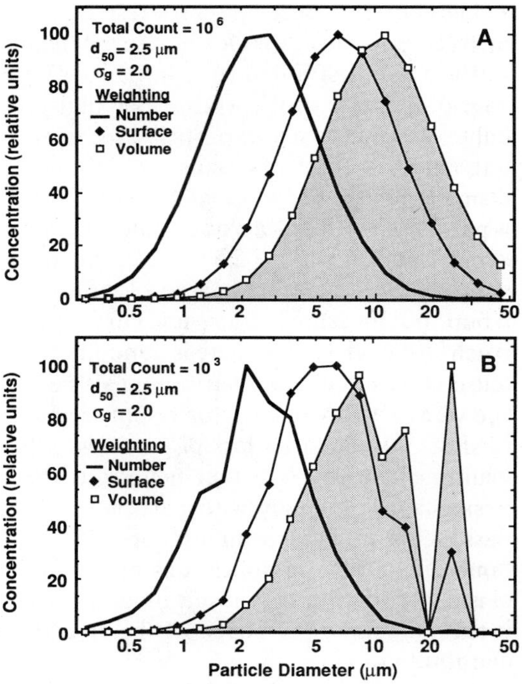

FIGURE 8-6. Variability in Volume, or Mass, Measurements. Surface Area and Volume Distributions Calculated From (a) High and (b) Low Particle Count are Given and the Curves are Normalized to the Same Height. The Volume Variability at Low Count is Due to Statistical Fluctuations, Especially in the Tail of the Number Distribution. (Adapted from Willeke and Baron 1990.)

since gravitational settling and inertial impaction of particles in the human respiratory tract in the size range of about 0.5 μm and higher are directly dependent on aerodynamic particle diameter (Hinds 1982).

One can make a best estimate of the aerosol d_a distribution based on a knowledge of the size-dependent sensitivities of the instruments and the correction factors applied to each instrument's data. With the SEM and optical particle counters, relatively large correction factors based on the assumed average particle shape, density, and refractive index may culminate in relatively unsatisfactory results. The grinding aerosol was a difficult aerosol for such a comparison because of the presence of a number of materials with widely disparate properties.

Particle Statistics

Assume that particles of an aerosol have been detected by a direct reading instrument. For a lognormal aerosol size distribution with a number median diameter of 2.5 μm and a geometric standard deviation of 2.0, the smooth number distribution curve calculated in Fig. 8-6a results from a relatively large total count of one million particles distributed in 19 size increments over the size range 0.2–45 μm. Such a high particle count is realistic for dynamic sensors, whose data acquisition systems permit multichannel analysis but might overload a filter that must be analyzed by microscopy.

The surface area and the volume for each particle size may now be calculated. The surface area and volume of each size, multiplied by the number of particles in the respective size ranges, results in the distributions, also shown in Fig. 8-6a. The peak of each distribution is normalized to 100 relative units for illustration purposes. Inclusion of the particle density would allow conversion of the volume to a mass distribution. The representation of the aerosol size distribution by any of these weightings (count, surface, or volume) results in a smooth curve because a large number of particles was used.

When the total count is reduced to 1000 (Fig. 8-6b), the number distribution curve is still recognizable as approximately lognormal, though the additional variability due to a smaller count in each size increment is apparent. Example 8-1 indicates how to perform this calculation using a spreadsheet computer program. However, the surface distribution emphasizes the larger particles, of which there are fewer. The variability in particle count for these larger particles is greater. The volume or mass distribution (highlighted by shading) emphasizes even larger particles resulting in a poorly determined curve. Conversion of a count distribution to a volume or mass distribution by counting an insufficient number of particles, may, therefore, result in considerable imprecision. Figure 8-6b illustrates and emphasizes the need for measuring a large number of particles in the particle size range of interest. Several modern real-time aerosol monitors are computer-based and offer easy conversion from one weighting to another. Such an easy conversion may tempt the user to accept numbers that may have inherent biases and high variability. Note that the variability in mass due to a small number of large particles applies also to gravimetric measurements when small samples are obtained.

Example 8-1

Calculate the number, surface, and volume values for a lognormal distribution of spherical particles with a count median diameter of 5 μm and σ_g of 2. Simulate the variability as if the entire distribution contains approximately 1000 particles.

Answer. The following equations were developed in the spreadsheet program Excel (Microsoft Corp., Bellevue, WA) and will work in other programs with few modifications. The input values and constants are listed in the first four rows Table 8-1. First we need to generate the diameters for which the lognormal distribution is produced. Column A has numbers starting at 0.25 with each following row multiplied by a constant factor, in this case 1.32, giving 19 size intervals or bins between 0.25 and 49 μm. The second column is the geometric mean of the upper and lower endpoints of each bin and is the size used to represent that bin. Thus, A2 − A1 is the first size interval and B1 is $\sqrt{A2 \times A1}$. C6 uses Eq. 8-4 to determine the concentration function in that bin or size interval:

```
C6=(LN(A7/A6)/($E$1*$E$3))
    *EXP( – ((LN(B6)
    – $E$2)^2)/(2*($E$3^2)))
```

The "$" indicates that the reference does not change in the following rows, i.e., in C7, C8, etc. The concentration function is normalized to give the appropriate number of total counts, in this case 1000:

```
D6 = C6*$E$4/$C$25
```

C25 is the sum of all the values in column C. Next, "noise" is added to the normalized particle density function to simulate the counting process. The following function produces a random number that is part of an approximately normal distribution, centered about zero with a standard deviation (Hansen 1985)

$$\sigma = \sqrt{12n}\left[\left(\frac{\sum_{i=1}^{n} \text{RAND}_i}{n}\right) - 0.5\right]$$

where n is usually chosen to be a number ≥ 3 and RAND_i is a random number between 0 and 1 that can be generated by the computer. The larger the value of n, the closer the resulting distribution will approximate a normal distribution, especially in the tails of the distribution. The Poisson distribution approaches a normal distribution for large particle counts; so, this function provides a reasonable approximation to a Poisson distribution. Poisson statistics require that the variance of the particle counts be equal

TABLE 8-1 Spreadsheet Size Distribution Calculation

Row	A	B	C	D	E	F	G
1	$d(50)$ =	5	SQRT(2*π) =	2.5066			
2	$\sigma(g)$ =	2	LN ($d(50)$) =	1.6094			
3			LN ($\sigma(g)$) =	0.6931			
4	Total Number of Particles =		1000				
5	Diameter (μm)		(Δf)	Number	No. With Random Count	Surface	Volume
6	0.2500	0.2872	3E − 05	0.03	0	0	0
7	0.3300	0.3791	0.0002	0.16	1	0.452	0.029
8	0.4356	0.5005	0.0006	0.65	1	0.787	0.066
9	0.5750	0.6606	0.0022	2.25	3	4.113	0.453
10	0.7590	0.8720	0.0067	6.69	4	9.556	1.389
11	1.0019	1.1511	0.0169	16.93	19	79.09	15.17
12	1.3225	1.5194	0.0365	36.50	30	217.6	55.10
13	1.7457	2.0056	0.0671	67.08	71	897.2	299.9
14	2.3043	2.6474	0.1049	104.96	110	2422	1069
15	3.0416	3.4946	0.1398	139.88	119	4565	2659
16	4.0149	4.6128	0.1587	158.79	154	10294	7914
17	5.2997	6.0889	0.1535	153.54	156	18170	18439
18	6.9956	8.0374	0.1264	126.45	122	24759	33167
19	9.2342	10.609	0.0887	88.71	81	28643	50646
20	12.189	14.004	0.0530	53.01	62	38200	89161
21	16.090	18.486	0.0270	26.98	19	20397	62843
22	21.238	24.401	0.0117	11.70	14	26188	106502
23	28.035	32.209	0.0043	4.32	5	16296	87482
24	37.006	42.517	0.0014	1.36	0	0	0
25	48.850		0.9995	1000	156	38200	106502

to the mean count. Thus, the standard deviation of the count in each bin is equal to the square root of the value of the density function, i.e., the count in that bin. The function is rounded to integer values as would be produced by a counting instrument:

E6 = ROUND(D6 + ((((RAND( )
+ RAND( ) + RAND( ) + RAND( )
+ RAND( ))/5) − 0.5)
*SQRT(60)*SQRT(D6)), 0)

where RAND( ) is a function that generates a random number between 0 and 1. E6 is the value for the number distribution. The surface and volume distributions are calculated from this distribution assuming spherical particles:

F6 = E6*PI( )*B6^2

G6 = E6*PI( )*B6^3/6

Finally, if one wishes to normalize the peak value of the distributions to the same value, e.g., 100, as in Fig. 8-6, three more columns, H, I, J, can be created that contain the normalized number, surface, and volume distributions, respectively. These have not been included in Table 8-1 due to space considerations. H6, I6, and J6 would contain E6*100/$E$25, F6*100/$F$25 and G6*100/$G$25, respectively. E25, F25, and G25 contain the maximum values in their respective columns. Note that the columns E, F, and G (as well as H, I, and J) will always appear somewhat different from those indicated below since the random numbers will produce different results.

Further, sampling or detection efficiencies such as those indicated in Figs. 8-2 and 8-3 can be calculated by multiplying the normalized density function (column C) by these efficiencies.

Corrections for Density and Other Physical Properties

Curve A in Fig. 8-7 shows a smooth, calculated representation of an aerosol size distribution with a median aerodynamic diameter of 5.0 μm and $\sigma_g = 2.0$. This aerodynamic particle diameter can be converted to physical particle diameter. The conversion is achieved by dividing the aerodynamic particle diameter by the square root of the particle density (see Chapter 2). A shape factor to account for nonspherical shapes also needs to be included in the conversion, but is not discussed here. For coal dust, which has a higher than unity particle density ($\rho_p \approx 1.45$ g/cm^3), the physical diameter (Curve B, Fig. 8-7) is smaller than the aerodynamic diameter. Thus, a particle of a given physical size settles in the same region of the respiratory tract as a physically larger, but less dense, particle.

Curve A in Fig. 8-7 represents the actual aerodynamic diameter distribution of a coal dust aerosol to be measured using an optical particle counter. Typically, these counters, as well as photometers, are calibrated with spherical, nonabsorbing test aerosols such as dioctyl phthalate (DOP) or polystyrene latex spheres (PSL). For example, DOP has a refractive index $m = 1.49$ (no imaginary or absorptive component). All the light received by these test particles is scattered from the particles. However, if the particles are light-absorbing, such as coal dust ($m = 1.54 - 0.5i$, with 0.5 representing the absorptive component of the refractive index), a particle of a given size scatters less light. Therefore, a coal dust particle scatters much less light and appears much smaller than a similar-sized test particle; the particle size distribution is recorded to be in a smaller size range, as illustrated by Curve C in Fig. 8-7 (Liu et al. 1974). In addition, the size correction for the absorbing particle, such as coal dust, may be strongly particle-size-dependent, further distorting the measured size distribution. While the distortion and shifting of the size distribution for coal dust is an extreme case, the assumptions involved (spherical particles, refractive index, and density values) illustrate some of the pitfalls of using optical sizing data to determine aerodynamic size.

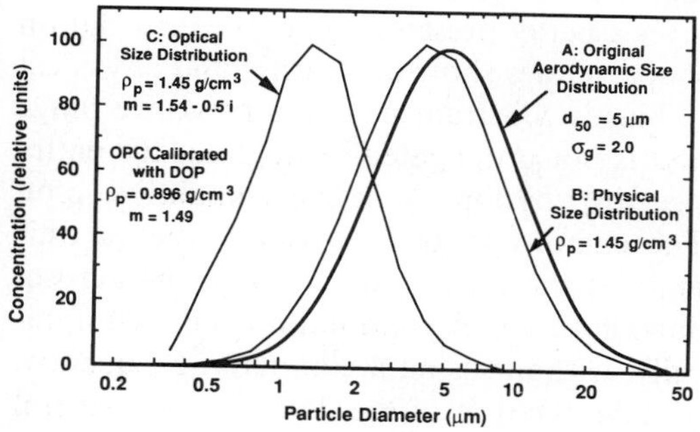

FIGURE 8-7. Measurement of Coal Dust Using Various Physical Properties of the Particles (Density, ρ_p, and Refractive Index, m). Curve A Represents the Aerodynamic Size Distribution of a Coal Dust Sample; Curve B Represents the Physical Size Distribution of that Dust (Correcting for Density); and Curve C Represents the Measurement of the Coal Dust by an Optical Particle Counter (OPC) Calibrated with Monodisperse DOP Particles (Liu et al. 1974). (Figure Adapted from Willeke and Baron 1990.)

An optical particle counter uses the optical properties of the individual particles for size discrimination. It, therefore, needs to be calibrated for the aerosol it measures. Other types of aerosol monitors use different physical properties for size discrimination. An electrical size classifier, for example, uses the electrical mobility of particles for size discrimination of submicrometer aerosols. Because the composition of the aerosol to be measured may be unknown, inadequate calibration may prevent the "size" obtained with one type of aerosol monitor from corresponding to the "size" obtained with another instrument.

Presentation of Size Distribution Data

There are several ways of presenting measured size distributions, each with advantages and disadvantages. Assume that two dusts are present in the air: Dust 1 with a geometric mean diameter of 1.5 μm and dust 2 with a geometric mean diameter of 10 μm, both with a geometric standard deviation of 2.0. Measurement of the aerosol with a direct-reading aerosol size spectrometer is calculated to produce the bimodal size distribution shown in Fig. 8-8a.

If this measurement is replotted on a cumulative plot, where the value of the ordinate indicates the number of particles less than the given size, the wavy plot of Fig. 8-8b results. Starting with the smaller particles, the curve increases with increasing particle size in an S-shaped manner. At sizes slightly larger than the mean size of dust 1, the curve levels off and then increases in slope again as the mean size of dust 2 is approached. This type of presentation is common for the results of low-resolution instruments such as a cascade impactor.

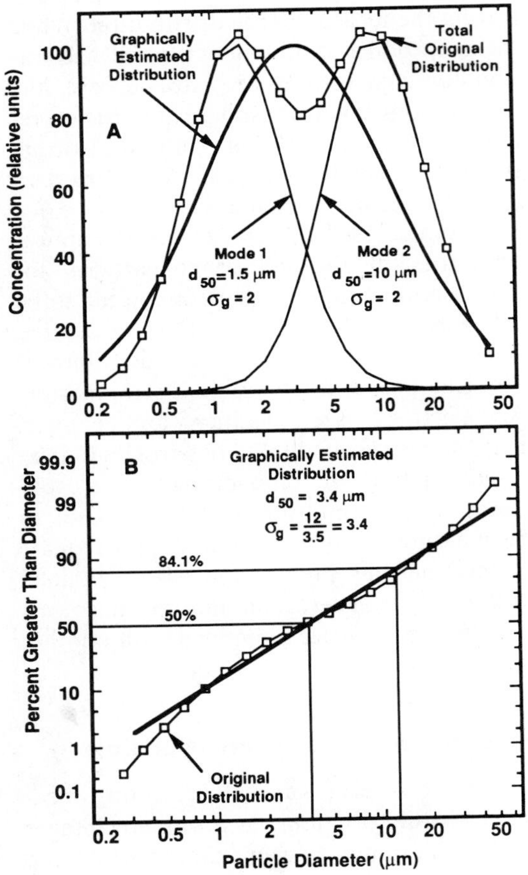

FIGURE 8-8. Representations of a Bimodal Size Distribution: Histogram versus log-Probability Plot. It is Possible to Misinterpret the log-Probability Plot of a Bimodal Distribution as Being from a Single Mode. (Figure Adapted from Willeke and Baron 1990.)

If one does not know that there are two dust modes present, one may be tempted to draw a straight line through the cumulative plot, as indicated by the heavy solid line in Fig. 8-8b. This is frequently done and justified by attributing the deviation of the data from a straight line to experimental variability. The resulting graphically estimated or "measured" aerosol, thus, has a geometric mean diameter of about 3.4 μm (corresponding to the minimum between the two dusts) and a geometric standard deviation of 3.5, indicating a single dust distribution much broader than each of the modes in the original bimodal distribution. Potentially valuable information is lost in this representation of the data, because multiple modes usually indicate different sources of aerosol.

Some statistical tests may also indicate that the cumulative data in Fig. 8-8b do not fit a single distribution. For instance, the Kolmogorov–Smirnov test (Gibson 1971) would indicate whether the measured distribution fits a specific, continuous distribution and a plot of residuals (the differences between the measured and the calculated values) qualitatively indicates whether adjacent measurements in the curve are correlated or whether the data fit the model.

Both types of representation have advantages and disadvantages. The differential plot gives a better presentation of the distribution shape: modes show up directly and any effect of bias is constrained to a narrow size range and is not propagated throughout the entire size distribution as in the cumulative plot. The cumulative plot provides a better estimate of the median diameter of the aerosol and allows easier presentation of data graphically, without using a computer. Frequently, an investigation of the data through several display techniques affords a more complete understanding of the physical meaning of the data.

Particle Size Selection

The type of aerosol monitor used depends on the purpose of sampling. The industrial hygienist generally samples from a health perspective. Because the physiological shape of the human respiratory system determines the region in which the particles will deposit, a preclassifier is frequently mounted ahead of the sensor in order to intentionally limit the particle measurement to particles reaching the physiological region of concern. A cyclone, impactor, or elutriator preclassifier separates the aerosol into respirable and nonrespirable fractions.

The respirable mass fraction can also be obtained by sensor discrimination. Size distribution results can be weighted appropriately in each size range to give the respirable dust response (Baron and Willeke 1986). Aerosol photometers are relatively inexpensive, direct-reading instruments that have a built-in size discrimination sometimes used for respirable dust measurements (Baron 1984). Such light-scattering devices monitor the scatter of light from an aerosol cloud rather than from single particles. Figure 8-9 illustrates this for a specific photometer, the TM digital μP (Hund Corp., New York, NY).

Figure 8-9 shows the calculated instrument response per unit mass concentration as a function of particle size for two kinds of aerosols with the same aerodynamic size distribution ($d_{50} = 5$ μm, $\sigma_g = 2.0$): non-light-absorbing dioctyl phthalate (DOP) droplets and dense, light-absorbing iron oxide (Fe_2O_3) particles (Armbruster 1987). The decline in response with increasing particle size above about 1 μm is common to all photometers. This decrease approximately corresponds to the classification characteristics of the ACGIH and BMRC definitions for respirable dust (ACGIH 1984), also indicated in Fig. 8-9. Complex interactions between the incident light and the particle result in similarly complex response curve patterns that differ from one type of aerosol to another.

A photometer calibrated with one type of aerosol will, therefore, generally be biased if used to measure another aerosol with a different chemistry or size distribution. This bias can be adjusted for a specific aerosol by drawing the aerosol through a filter downstream of the sensor and adjusting the sensor readout to equal the concentration measured with the filter. This procedure is valid as long as the type and size distribution of the aerosol remain unchanged. Instruments of this type can be used to make relative measurements, often providing useful real-time information, but should be used only with great care for situations requiring higher accuracy.

One approach to evaluating the accuracy of a method over a wide range of aerosol size distributions is the use of a bias map (Caplan, Doemeny, and Sorensen 1977). This involves determining the range of size distributions

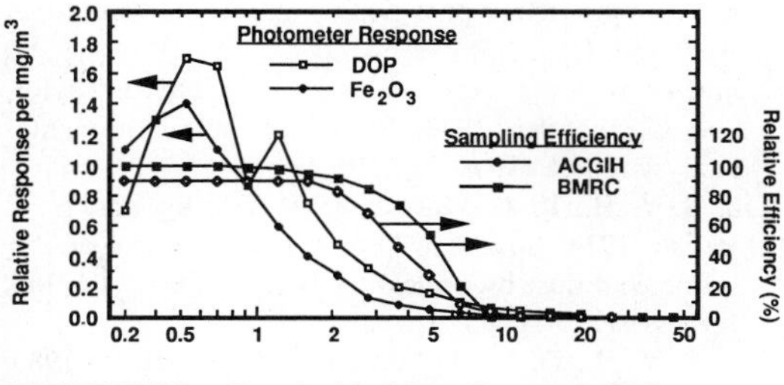

FIGURE 8-9. Respirable Mass Response Using a Photometer for Example Size Distributions of Two Materials with $d_p = 5$ μm, $\sigma_g = 2.0$. Detection Efficiency is for the TM Digital μP from the Hund Corp. Based on Measurements by Armbruster (1987). Two Definitions of Respirable Dust are Also Included. (Figure Adapted from Willeke and Baron 1990.)

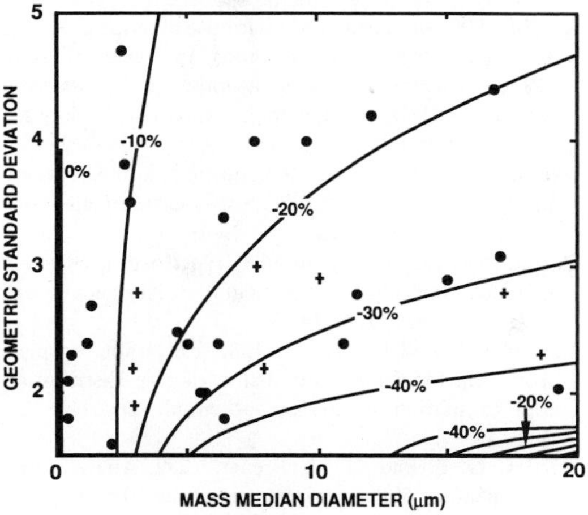

FIGURE 8-10. Bias Map Comparing Two Defined Respirable Dust Response Curves for a Range of Lognormal Size Distributions. Data Points Represent Distributions Reported by (●) Hinds and Bellin (1988) and (+) Bowman et al. (1984).

over which the measurement is expected to occur. Hinds and Bellin (1988) reviewed aerosol distributions in more than 30 workplace operations and found size distributions with σ_g ranging from 1.5 to 5 and mass median aerodynamic diameters ranging from 0.1 to 20. One can examine the bias resulting from a measurement of lognormal distributions throughout this range by a comparison of one sampler versus a standard. As an example, calculation of the bias of the ACGIH definition versus the BMRC definition (Fig. 8-9b) for each size distribution can be used to produce the bias map in Fig. 8-10. This approach has been used to evaluate the optimum flow rate through a cyclone by comparing bias maps of the cyclone relative to a respirable dust definition at different flow rates (Bartley and Breuer 1982). Hinds and Bellin used their size distribution data to estimate the effectiveness of respirators with measured size dependent leakage.

References

ACGIH. 1984. Particle size-selective sampling in the workplace. In *Annals of the ACGIH, Vol. 11*, pp. 23–100. Cincinnati, OH: American Conference of Governmental Industrial Hygienists.

ACGIH. 1990. Threshold limit values for chemical substances and physical agents. Cincinnati, OH: American Conference of Governmental Industrial Hygienists.

ASTM. 1990. Standard practice for use of terms precision and bias in ASTM test methods. In *Annual Book of ASTM Standards. General Methods and Instrumentation, Vol. 14.02.* Philadelphia: American Society for Testing Materials.

Armbruster, L. 1987. A new generation of light-scattering instruments for respirable dust measurement. *Ann. Occup. Hyg.* 31:181–93.

Baron, P. A. and G. J. Deye. 1990. Electrostatic effects in asbestos sampling I: Experimental measurements. *Am. Ind. Hyg. Assoc. J.* 51:51–62.

Baron, P. A. and K. Willeke. 1986. Respirable droplets from whirlpools: measurements of size distribution and estimation of disease potential. *Environ. Res.* 39:8–18.

Bartley, D. L. and G. M. Breuer. 1982. Analysis and optimization of the performance of the 10 mm nylone cyclone. *Am. Ind. Hyg. Assoc. J.* 43(7):520–28.

Blackford, D., A. E. Hansen, D. Y. H. Pui, P. Kinney, and G. P. Ananth. 1988. Details of recent work towards improving the performance of the TSI Aerodynamic Particle Sizer. In *Proceedings of the 2nd Annual Meeting of the Aerosol Society*, Bournemouth, U.K., March 22–24.

Box, G. E., W. G. Hunter, and J. S. Hunter. 1978. *Statistics for Experimenters.* New York: Wiley.

Buchan, R. M., S. C. Soderholm, and M. J. Tillery. 1986. Aerosol sampling efficiency of 37 mm filter cassettes. *Am. Ind. Hyg. Assoc. J.* 47:825–31.

Caplan, K. J., L. J. Doemeny, and S. D. Sorensen. 1977. Performance characteristics of the 10 mm cyclone respirable sampler: Part I—Monodisperse studies. *Amr. Ind. Hyg. Assoc. J.* 38(2):83–95.

Carter, J., D. Taylor, and P. A. Baron. 1989. Fibers, Method 7400, revision #3: 8/15/87. *NIOSH Manual of Analytical Methods*, 3rd edn., ed. P. M. Eller, (NIOSH) Pub. 84-100. Cincinnati, OH: National Institute for Occupational Safety and Health.

Crummett, W. B., F. J. Amore, D. H. Freeman, R. Libby, H. A. Laitenen, W. F. Phillips, M. M. Reddy, and J. K. Taylor. 1980. Guidelines for data acquisition and data quality evaluation in environmental chemistry. *Anal. Chem.* 52:2242–49.

Currie, L. A. 1991. In pursuit of accuracy: Nomenclature, assumptions and standards. *Pure Appl. Chem.* (in press).

Eisenhart, C., H. H. Ku, and R. Collé. 1990. Expression of the uncertainties of final measurement results: Reprints. In *Selected Publications for the EMAP Workshop.* NIST Internal Report 90-4272. Washington, DC: National Institute for Standards and Technology.

Gibson, J. D. 1971. *Nonparametric Statistical Inference.* New York: McGraw-Hill.

Hansen, A. G. 1985. Simulating the normal distribution. *BYTE.* October Issue: 137–38.

Heitbrink, W. A., P. A. Baron, and K. Willeke. 1990. Coincidence in time-of-flight aerosol spectrometers: Phantom particle creation. *Aerosol Sci. Technol.* 14(1):112–26.

Helsel, D. R. 1990. Less than obvious: Statistical treatment of data below the detection limit. *Environ. Sci. Technol.* 24(12):1766–74.

Hinds, W. C. 1982. *Aerosol Technology.* New York: Wiley.

Hinds, W. C. and P. Bellin. 1988. Effect of facial-seal leaks on protection provided by half-mask respirators. *Appl. Ind. Hyg.* 3(3):158–64.

Keith, L., W. Crummett, J. Deegan, R. Libby, J. Taylor, and G. Wentler. 1983. Principles of environmental analysis. *Anal. Chem.* 55:2210–18.

Liu, B. Y. H., V. A. Marple, K. T. Whitby, and N. J. Barsic. 1974. Size distribution measurement of airborne coal dust by optical particle counters. *Am. Ind. Hyg. Assoc. J.* 8:443–51.

Liu, B. Y. H., D. Y. H. Pui, and W. Szymanski. 1985. Effects of electric charge on sampling and filtration of aerosols. *Ann. Occup. Hyg.* 29:251–69.

Long, G. L. and J. D. Winefordner. 1983. Limit of detection: A closer look at the IUPAC definition. Anal. Chem. 55:712A–724A.

NIOSH. 1984. *Manual of Analytical Methods*, 3rd edn.

(DHHS/NIOSH Pub. No. 84-100). Cincinnati, OH: National Institute for Occupational Safety and Health.

O'Brien, D. M., P. A. Baron, and K. Willeke. 1986. Size and concentration measurement of an industrial aerosol. *Am. Ind. Hyg. Assoc. J.* 47:386–92.

Okazaki, K., R.W. Wiener, and K. Willeke. 1987*a*. Isoaxial aerosol sampling: Non-dimensional representation of overall sampling efficiency. *Environ. Sci. Technol.* 21:178–82.

Okazaki, K., R. W. Wiener, and K. Willeke. 1987*b*. Non-isoaxial aerosol sampling: Mechanisms controlling the overall sampling efficiency. *Environ. Sci. Technol.* 21:183–87.

OSHA. 1974. Occupational safety and health standards. Part 1910.93. Washington DC: Occupational Safety and Health Administration.

Taylor, D. G., R. E. Kupel, and J. M. Bryant. 1977. Documentation of the NIOSH validation tests. U.S. DHEW Publ. (NIOSH) 77-185. Washington DC: US Government Printing Office.

Vincent, J. H. 1989. *Aerosol Sampling: Science and Practice.* New York: Wiley.

Wake, D. 1989. Anomalous effects in filter penetration measurements using the aerodynamic particle sizer (APS 3300). *J. Aerosol Sci.* 20:1–7.

Watson, J. G., P. J. Lioy, and P. K. Mueller. 1989. The measurement process: Precision accuracy and validity. In *Air Sampling Instruments*, ed. S. V. Hering, pp. 51–57. Cincinnati, OH: American Conference of Governmental Industrial Hygienists.

Wiener, R. W., K. Okazaki, and K. Willeke. 1988. Influence of turbulence on aerosol sampling efficiency. *Atmos. Environ.* 22:917–28.

Willeke, K. and P. A. Baron. 1990. Sampling and interpretation errors in aerosol sampling. *Am. Ind. Hyg. Assoc. J.* 51(3):160–68.

Willeke, K. and B. Y. H. Liu. 1976. Single particle optical counter: Principles and application. In *Fine Particles: Aerosol Generation, Measurement, Sampling and Analysis*, ed. B. Y. H. Liu. New York: Academic Press.

9

Methods of Size Distribution Data Analysis and Presentation

Douglas W. Cooper

IBM Research Division
T.J. Watson Research Center
Yorktown Heights, NY, U.S.A.

INTRODUCTION

The behavior of an aerosol particle depends strongly on its size. Deposition due to gravity, diffusion, inertia, and electrostatic mechanisms depends on the particle size. Effects such as light scattering and impact on health vary with particle size. The choice of equipment and materials to reduce particle concentrations in air and other gases depends in part on the size of the particles to be captured. The equipment appropriate for sampling various aerosols, devices such as impactors and impingers and precipitators and filters, should be selected with consideration given to particle size. Even the beneficial uses of aerosols (materials fabrication, therapeutic aerosols) are more or less effective, depending upon the sizes of the particles.

Three methods of aerosol size distribution measurement are particularly common: (1) particles are sampled by an optical particle counter, and the number of particles in each of many optical equivalent diameter intervals is recorded; (2) particles are sampled by depositing them onto a surface, such as a filter, and then are sized and counted using techniques such as microscopy or light scattering; (3) particles are sampled by an impactor, and the mass of particles in each of the many aerodynamic equivalent diameter intervals is recorded. Optical particle counter and surface analysis measurements are quite useful in contamination control work, as the damage done by particles is often related to their number concentration and size. The impactor measurements are quite useful in health-related assessments, as the risk posed by particles is often related to their masses and aerodynamic diameters.

The remainder of this introduction is a brief summary of the sections to follow.

Types of Particle Size

One way to characterize particle size is on the basis of the technique used to measure it, which leads to such size types as aerodynamic diameter, optical-equivalent diameter, electrical-mobility-equivalent diameter, diffusion-equivalent diameter, etc. Another characterization is the fraction of the aerosol represented by particles in various size intervals, giving rise to the idea of a particle size distribution and various characteristic sizes, such as various means and medians.

The two most common measures of particle amount are particle count and particle mass. These data can be used to create count or mass size distributions, respectively, or—

with some loss of precision—to create other distributions (such as area). Sometimes, activity (radioactivity) or some chemical property is the basis of the measure of amount.

Particle Size Distributions

The fractions of the aerosol—by count or mass, etc.—in a set of contiguous size intervals form the particle size distribution. Particle size distributions are usually presented as frequency, histogram, or cumulative distributions, discussed below. Idealized particle size distributions include the Gaussian ("normal"), lognormal, and power-law distributions.

Concentration Distributions

The fractions of occurrence of concentrations of different magnitudes in a set of contiguous concentration intervals forms the concentration frequency distribution. Often, one assumes constant and uniform concentrations for a given situation, but there are cases where normal concentration, lognormal concentration, or Poisson count distributions are more appropriate. Concentrations can also be viewed as deterministic (rather than probabilistic) functions of time and position.

Summarizing Data with a Few Parameters

Rather than present all the particle size information, one often chooses to summarize the data with a few numbers. From the count data or the mass data, one can easily calculate a number-weighted average or a mass-weighted average, the count mean diameter or mass mean diameter, respectively.

The mean of a variable x is defined as the sum of the values of x divided by the number of values making up the sum. The standard deviation is the square root of the mean of the squared differences between the values and their mean. These definitions will be useful in much of what follows, as they are the two most common parameters used to summarize data.

The count and mass median diameters are defined as the diameters which split the total count or mass into halves. Besides such characteristic sizes as the mean or median, some measure of the spread of particle sizes is useful. Such measures include the range, which unfortunately depends on how many particles were observed, and the standard deviation.

Various mathematical methods can be used to try to match idealized distributions (such as the normal, lognormal, and power-law) with the data. These distributions can then be very succinctly described. The normal distribution requires only the mean and the standard deviation. The lognormal distribution requires only the geometric mean and the geometric standard deviation. The power-law requires only the coefficient and the exponent. Any one of these pairs of parameters completely describes an idealized approximation to the data. The parameter pairs can be obtained from simple mathematical formulas or from linear regression after the data have been transformed to approximate a linear (straight-line) relationship.

Summarizing Size Distributions Graphically

A graph can be very helpful, if it is clear. It can also be misleading. Suggestions for graphical clarity are given below. The use of cumulative distributions (total amount smaller than the indicated size) is recommended over the use of frequency distributions (number per indicated size interval). Cumulative size distributions are especially valuable if they are plotted using graph axes chosen to make them approximately straight lines.

Confidence Intervals and Error Analysis

If several different values for concentration, for example, have been measured, what is the best estimate of the true concentration? How uncertain is this estimate? The same questions can be asked about particle size, as well. One wants a "confidence interval" to bracket the best estimate of the variable. It is useful to know how much such estimates could be expected to vary from one supposedly identical situation to another. Control charts can be used to alert the user to unusual variation. One wants to know how errors in a measured variable, such as volume flow rate, become errors in a derived variable, such as concentration. This phenomenon is referred to as "propagation of error."

Testing Hypotheses with Size Distribution Data

Size and concentration distributions are valuable for summarizing data and for assisting in modeling and in making predictions. They also assist in the test of hypotheses, such as: Did the change in conditions produce a clearly different particle size distribution? Does one get clearly more particle mass or number under this situation or under that one? These questions are tests of hypotheses about the aerosols. If the data are count data, often the method of analysis chosen will be "chi-square." If the data are mass data, appropriate methods may be "Student's *t*" or "Kolmogorov–Smirnov."

Coincidence Errors

At concentration levels near or above the design limits of the instrument being used, particle counts may be lost because two or more particles are in the sensing zone simultaneously or because the electronics cannot handle such rapidly arriving signals. Such "coincidence" effects can be modeled and partially corrected for or, better, still avoided.

Data Inversion ("Deconvolution" or "Unfolding")

Particle sizing instruments do not have perfect sizing precision. There is a nonzero probability that a particle of one size will be mistakenly classified as having another. Knowing the particle size classification behavior of the instrument, one can correct for such errors. Such correction procedures are not elementary, but there is a wealth of information available on them for those who are interested in applying them. It is recommended that where possible the measurements be carried out with such instruments and under such conditions that data inversion, as it is sometimes called, may not be needed.

TYPES OF PARTICLE SIZE

We mentioned above that there are different kinds of geometric particle size for nonspherical particles. The nonspherical particle size might be its longest dimension, shortest dimension, or some combination of the two. It might be the diameter of a sphere with the same volume or the same surface area. These last two definitions are not uncommon. For spherical and nonspherical particles measured with a technique that does not give the geometric size, other equivalent diameters are often used: aerodynamic, optical, diffusion, electrical mobility, etc. These are described in Chapters 2 and 3.

PARTICLE SIZE DISTRIBUTIONS

Having decided on an appropriate measure of the particle size, which we will label "d", one is still faced with not one size but a wide range of sizes in most aerosols. Although summary statistics, such as the mean and standard deviation of particle diameter, can be used to advantage, one may want to give a detailed, perhaps approximate, description of the particle sizes. To do so, we often use either a frequency distribution or a cumulative size distribution.

Cumulative vs. Frequency Size Distributions

Table 9-1 shows a cumulative distribution, $F(d)$, a normal distribution, with a mean of $\mu = 10$ and a standard deviation of $\sigma = 1$. The first column shows the particle size at the upper end of a size interval. The next column shows the fraction of the number of particles smaller than the size indicated in the first column. This is a cumulative distribution.

TABLE 9-1 Normal (Gaussian) Distribution, with a Mean of $\mu = 10$ µm and a Standard Deviation of $\sigma = 1$ µm

Particle Size (µm)	Percentile (Percent Smaller)
$7 = \mu - 3\sigma$	0.135
$8 = \mu - 2\sigma$	2.27
$9 = \mu - 1\sigma$	15.9
$10 = \mu$	50.0
$11 = \mu + 1\sigma$	84.1
$12 = \mu + 2\sigma$	97.7
$13 = \mu + 3\sigma$	99.865

Figure 9-1 shows the cumulative size distribution from Table 9-1, plotted on linear axes. The distribution is S-shaped, "sigmoidal."

The data in Table 9-1 can be analyzed to create an approximate frequency distribution, the fractions of the aerosol in various size intervals. The most common approach is to take the cumulative distribution value at one diameter, $F(d_1)$, and subtract it from the cumulative distribution value at a larger diameter, $F(d_2)$, and divide by the difference between the two diameters:

$$f(d) \approx (F(d_2) - F(d_1))/(d_2 - d_1) \quad (9\text{-}1)$$

The value of d to which $f(d)$ corresponds is usually chosen to be the mean of d_1 and d_2:

$$d = (d_2 + d_1)/2 \quad (9\text{-}2)$$

Such an approximate frequency distribution might best be plotted as a histogram, a bar graph with the bars filling each size interval and having a height representing the frequency.

This process of finding the frequency distribution numerically is an approximation of the definition of the relationship between the frequency and cumulative distributions:

$$f(d) = \mathrm{d}F(d)/\mathrm{d}d \quad (9\text{-}3)$$

The term dd is a bit strange in appearance, but it denotes that same kind of entity that dx denotes, an infinitesimal change in or an infinitesimal interval of a variable, in this case d. The frequency distribution is the derivative of the cumulative distribution, and the derivative is the limit of the ratio of the change in F divided by the change in d as the change in d tends toward zero. The frequency distribution and the cumulative distribution contain the same information, in different form.

Figure 9-2 shows the frequency distribution that would be calculated by this method in the limit of infinitesimal size intervals. The frequency distribution is the Gaussian (or normal) distribution, discussed next.

The Normal (Gaussian) Distribution

Most scientists and engineers are at least somewhat familiar with the Gaussian (normal) frequency distribution, the "bell-shaped

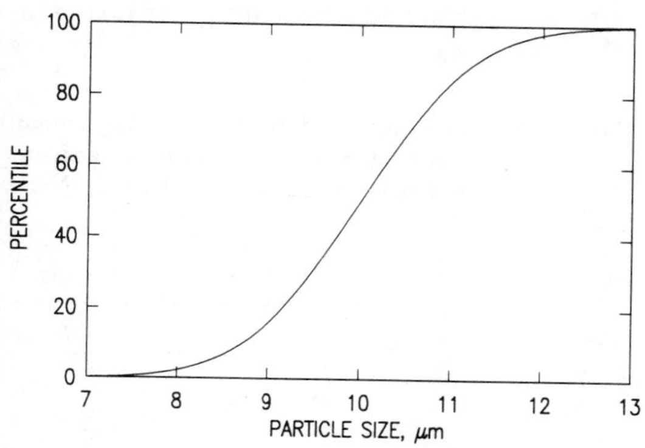

FIGURE 9-1. Cumulative Distribution, Particle Size: Gaussian (normal) Distribution on Linear Axes. (Mean is $\mu = 10$ and Standard Deviation is $\sigma = 1$.)

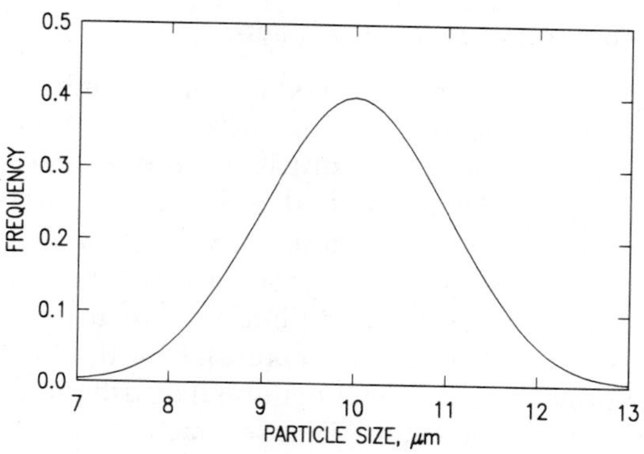

FIGURE 9-2. Frequency Distribution, Particle Size: Gaussian (normal) Distribution on Linear Axes. (Mean is $\mu = 10$ and Standard Deviation is $\sigma = 1$.)

curve" that describes many phenomena. For particles with a Gaussian size distribution, the fraction of the total (by count) that are within an infinitesimal size interval, d*d*, centered on size *d* is:

$$f(d)\mathrm{d}d = (1/\sigma\sqrt{2\pi}) \times \exp(-(d-\mu)^2/2\sigma^2)\mathrm{d}d \quad \text{(9-4)}$$

in which σ is the population standard deviation and μ is the population mean. The mean is a measure of the central tendency of the sizes and the standard deviation is a measure of their spread. Equations for the mean and standard deviation will be given below; many readers will be familiar with them already.

Normal distributions tend to arise when a multitude of small additive (+ or −) factors influence a variable that would otherwise have a single value. This is the distribution one would expect if a group of carpenters used rulers and hand-saws and tried to cut wood into 1 cm lengths. The mean would be near 1 cm and there would be about as many pieces 1.1 cm and larger as 0.9 cm and smaller. The diameters of ball bearings of the same nominal diameter would likely follow a normal distribution. This distribution is seen in aerosol science in cases where the particles are all nearly of one size, which is to say that they are nearly monodisperse, such as latex spheres used for calibration of particle measurement equipment.

The Lognormal Distribution

Lognormal distributions are more common than normal distributions in aerosol science and technology. Size distributions often have standard deviations that are large in comparison with the mean size, a situation which cannot occur for a nonnegative normal distribution. Such distributions are usually better described mathematically by the lognormal distribution. Lognormal distributions arise when multiplicative factors (both greater and less than 1) act upon a variable that would otherwise have a single magnitude.

The lognormal distribution (Aitchison and Brown 1957; Fuchs 1964; Hinds 1982; Crow and Shimizu 1988) is the distribution that results when the distribution of log(*x*) is Gaussian ("normal"). Lognormal distributions can be shown to result from the proportional breakup of large objects into smaller ones (Kolmogorov 1941; Epstein 1947) or from certain types of agglomeration of small objects into larger ones (Friedlander 1977). The lognormal distribution will result from the growth or breakup process $\mathrm{d}x(i)/\mathrm{d}t = k(i)x(i)$ when the growth constants $k(i)$ are normally distributed. The $x(i)$ are variables having to do with the sizes of the species *i*, typically the volumes of the species.

The lognormal distribution is actually a normal distribution of logarithms of particle size, $\log(d/d_0)$. The reference size, d_0, is usually 1 µm. The measure of central tendency is the median, d_{50}, which equals the geometric mean for a lognormal distribution, and the measure of spread is the geometric standard deviation, labeled σ_g. The median is the 50th percentile particle size. Common medians are the number median, area median, activity (radioactivity) median, volume median, and mass median. The geometric standard deviation (σ_g) for a lognormal distribution is the ratio of the 50th percentile size to the 16th percentile size, which for a lognormal distribution is also equal to the ratio of the 84th percentile size to the 50th percentile size. As the geometric standard deviation approaches $\sigma_g = 1.0$, the lognormal distribution approaches the normal distribution having a standard deviation of $\sigma = \sigma_g - 1$.

TABLE 9-2 Percentile Values for the Lognormal Distribution. The Median is d_{50} and the Geometric Standard Deviation is σ_g. Note the Similarity to Table 9-1

Size	Percentile
d_{50}/σ_g^3	0.135
d_{50}/σ_g^2	2.27
d_{50}/σ_g^1	15.9
d_{50}	50.0
$d_{50}\sigma_g^1$	84.1
$d_{50}\sigma_g^2$	97.7
$d_{50}\sigma_g^3$	99.865

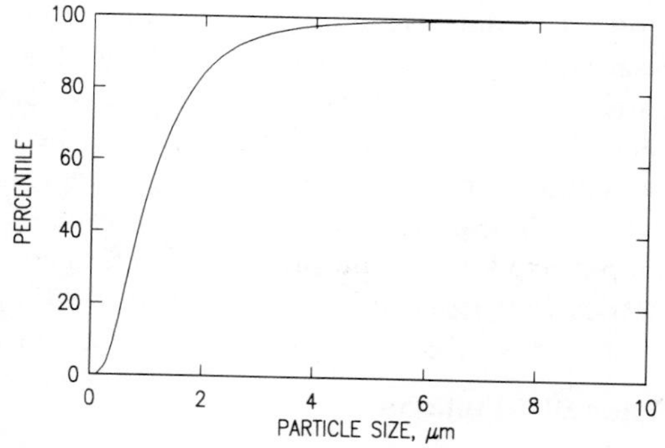

FIGURE 9-3. Cumulative Particle Size Distribution: Lognormal on Linear Axes. (Median Size is $d_{50} = 1$. Geometric Standard Deviation is $\sigma_g = 2$.)

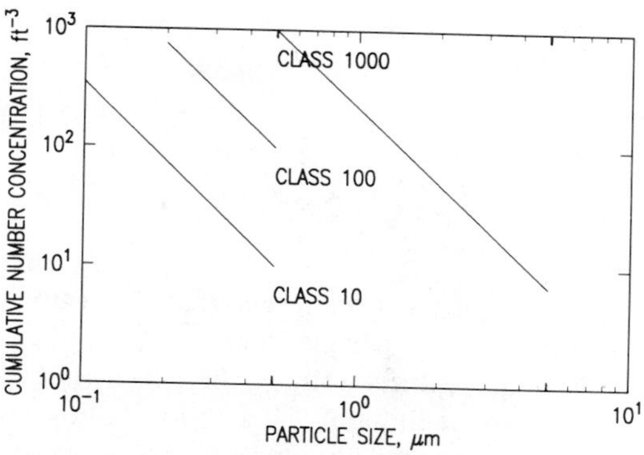

FIGURE 9-4. Cumulative Particle Size Distribution: Power Law on Log–Log Axes (Federal Standard 209D Classes 10, 100, 1000).

Table 9-2 shows the percentiles for the lognormal distribution in terms of the median and the factors of the geometric standard deviation, σ_g, from the median.

Figure 9-3 shows a cumulative lognormal distribution. The median is $\mu = 1$ and the geometric standard deviation is $\sigma_g = 2$.

The Power-Law Distribution

The power-law distribution is given by the integral of $f(d)$:

$$f(d) = ad^b \tag{9-5}$$

where b is a power less than 0. The cumulative power-law distribution is

$$F(d) = (a/(b+1))d^{(b+1)} \tag{9-6}$$

sciences, the power-law called the "Junge" distribution after an important contributor to the field.

In the contamination field, a power-law distribution is used in Federal Standard 209D (U.S. General Services Administration 1988) to characterize the typical particle size distribution in a clean room. The cumulative count size distribution is proportional to $d^{-2.2}$. The clean-room class is determined by the number of particles larger than 0.5 μm in the distribution per cubic foot of air; this value (1, 10, 100, etc.) gives the clean-room class its numerical name, so that Class 100 clean-rooms are expected to have fewer than 100 particles per cubic foot, and so on. Figure 9-4 shows the power-law cumulative particle size distribution that corresponds to Federal Standard 209D Classes 10, 100, and 1000.

The power-law distribution is just an approximation, but a useful one, and it enables one to predict easily the approximate behavior of the entire aerosol rather than just individual particles.

Example 9-1 shows the use of a power-law distribution to calculate the cumulative deposition flux due to gravitational settling of particles.

Example 9-1

How can one use a power-law distribution to predict the cumulative flux due to gravitational deposition alone?

Answer. Particle flux is the number of particles depositing per unit area per unit time. It is important in predicting the rate at which surfaces will become contaminated with particles.

The deposition velocity for a spherical particle of diameter d and material density ρ in a gas of viscosity η is

$$v(d) = g\tau = gC\rho d^2/18\eta$$

If $N(d)$ is the (cumulative) particle count concentration of particles larger than d, then

$n(d) = \mathrm{d}N/\mathrm{d}d$ is the number concentration in an infinitely small interval around the size d and

$$J(d) = \int_{d'}^{d''} n(d)v(d)\,\mathrm{d}d$$

is the flux of particles larger than d (in units such as number per cm^2 per second), where the limits of integration are the lower limit d' and an upper limit of d''. The lower limit is the smallest particle of interest, and the upper limit is the largest particle of interest.

This integration is often done numerically, once a choice of suitable upper size limit is made. However, if $n(d)$ is a power-law distribution, then

$$n(d) = ad^b$$

and if the deposition velocity can be approximated by

$$v(d) = ed^f$$

then the expression for $J(d)$ can be integrated easily as

$$J(d) = (ae/(b + f + 1))(d')^{(b+f+1)}$$

for $(b + f) < -1$.

Power-law distributions for the atmosphere use $b \simeq -4$, and those for indoor atmospheres use $b \simeq -3$, while the deposition velocity for gravitational settling is $1 < f < 2$, so that $-2 < (b + f + 1) < 0$, meaning a range of functional dependence from a flux that is almost independent of particle size to one that is almost proportional to the inverse square of particle size. When diffusion is considered, the flux sometimes increases dramatically as one includes smaller and smaller particle sizes, which have faster and faster deposition velocities due to diffusion.

A power-law distribution can be created by a process that subdivides material, with breakage rates proportional to the size of the material, which may explain why certain aerosols from wear processes are well described by this distribution. Power-law distributions also arise in the study of fractal materials, some of which are formed by agglomeration. Here again, the agglomeration rate is proportional to a power of the particle size. A fine survey of fractal geometries was presented by Mandelbrot (1977).

Other Distributions

Other distributions have been found useful in various contexts, and one should consult texts such as those by Fuchs (1964) and by Hinds (1982) for more information on them.

CONCENTRATION DISTRIBUTIONS

Concentrations are rarely uniform and constant. Instead, one finds distributions of concentration values if one measures at one location for many times or at many locations for one time or at many locations for many times. Concentration distribution information can often be summarized with the same methods as those used for particle size distributions. For example, the sample mean and the sample standard deviation are obtained for the particle count just as for the particle size.

Three such distributions are discussed next: normal, lognormal, and Poisson.

The Normal Distribution

Normal distributions of concentrations are unusual. The normal distribution arises from a large number of (positive and negative) additive effects. Sometimes, instruments that are themselves variable will be used to measure a relatively constant aerosol concentration and the resulting data will be approximately normally distributed.

The Lognormal Distribution

Lognormal distributions of concentrations are not unusual. The lognormal distribution arises from a large number of multiplicative effects. It can arise from rate equations that have variable rates, as well. The geometric mean concentration and the geometric standard deviation of the concentration values fully describe such lognormal concentration distributions.

The Poisson Distribution

The Poisson distribution, discussed next, describes the particle count data that one would obtain from sampling an aerosol that has a constant number concentration. Even though the concentration is constant, the entrance of particles into the sampler is probabilistic, leading to some variation, the smallest amount of variation encountered in practice.

Denote the particle count in the ith time interval as $n(i)$. The mean count in an interval is μ and the standard deviation of the count is σ. An interesting feature of the Poisson distribution is that the mean equals the variance: $\mu = \sigma^2$, so that the count standard deviation can be obtained from the square root of the mean count.

If one has accumulated counts over many intervals to form one long-duration sample, then the estimated standard deviation for this long-duration sample is the square root of the total count. The total count is the best estimate of the mean total count. The ratio of the standard deviation to the mean will be inversely proportional to the square root of the total count, indicating that our percentage uncertainty in the total count decreases inversely with the square root of the total count.

For the Poisson distribution, the probability of getting n counts in an interval from a population that has a mean μ in the interval is

$$P(n|\mu) = \mu^n \exp(-\mu)/n! \qquad (9\text{-}7)$$

This is the distribution one would expect when using an optical particle counter to obtain particle counts over many time intervals from a constant concentration aerosol or when inspecting a large number of equal surface areas that have been exposed to a constant concentration aerosol for the same duration.

A simple application of the Poisson distribution is the estimation of an upper limit for the true count when the interval that was sampled gave no counts. One uses

$$P(0|\mu) = \mu^0 \exp(-\mu)/0! \qquad (9\text{-}8)$$

$$P(0|\mu) = \exp(-\mu) \qquad (9\text{-}9)$$

and then notes that if the concentration had been as high as $\mu = 2.3$, there is only a 10% probability, $P(0|2.3) = 0.1$, that $n = 0$ particles would have been found. The hypothesis that $\mu = 4.6$ yields a 1% probability that $n = 0$ would be found, etc. One has 90% confidence that the true mean count for the interval is 2.3 or less and 99% confidence that it is 4.6 or less, etc.

The Poisson distribution can be used as the expected or hypothetical distribution when testing the hypothesis that the counts came from a constant, uniform count concentration distribution, using the chi-square analysis described below, for example.

Another use of the Poisson distribution is for modeling coincidence, the arrival of two or more particles (or their signals) in an interval (of space or time), discussed below.

For means much larger than 10, the Poisson distribution is very similar to the normal distribution (evaluated only at integer values). The use of the square root of the count as the variable transforms a Poisson distribution into a more nearly normal distribution (at integer values). (See Box, Hunter, and Hunter 1978.)

SUMMARIZING DATA WITH A FEW PARAMETERS

Data for Individual Particle Sizes

One kind of data to summarize with a few parameters is particle size data. Assume that each particle has a measured diameter, $d(i)$, and that the number of particles measured is N. The sample count mean particle diameter is then the sum of all the diameters divided by the number measured:

$$M = \sum d(i)/N \qquad (9\text{-}10)$$

in which we have used M to denote the sample mean.

The sample standard deviation, s, of the particle diameter is the square root of the sum of the squared differences between the particle diameters and the mean, divided by $N - 1$, one less than the number measured:

$$s = \sqrt{(\sum(d(i) - M)^2/(N - 1))} \quad (9\text{-}11)$$

The sample variance is s^2, the square of the sample standard deviation.

The sample mean, M, is an estimate of the population mean, μ, and the sample standard deviation, s, is an estimate of the population standard deviation, σ. The estimates become increasingly accurate as the number N in the sample increases. If the population has a Gaussian (normal) distribution, then it is completely described by these two parameters.

Data for the Fraction of the Aerosol in Each Interval

If the data are such that each datum represents the fraction of the aerosol (e.g., by count), $f(d(i))$, in each size interval, $\Delta d(i)$, then these equations for the sample mean diameter and the sample standard deviation become:

$$M = \sum f(d(i))d(i)\Delta d(i) \quad (9\text{-}12)$$

and

$$s = \sqrt{(\sum f(d(i))(d(i) - M)^2 \Delta d(i))} \quad (9\text{-}13)$$

Example 9-2 shows a set of hypothetical count data for some size intervals, the $f(d(i))$ calculated from these data, and the calculation of the mean and standard deviation from these data. Finally, the percentages are indicated by interval and the percentiles (cumulative percentages) shown. Percentiles are obtained by ranking the particles by sizes, then determining what percentage is smaller than or equal to the particle size of interest. If ten particle sizes were 1, 2, 3, 3, 4, 4, 5, 8, 8, 9, then the 10th percentile would be 1 and the 100th percentile would be 9, etc.

Example 9-2

How would percentiles, mean, and standard deviation be calculated from size interval data?

Answer. Assume one has the following data:

Upper Size	Interval Mid-point	Interval Count	Cum-ulat-ive Count	Per-centile
	0.25	4		
0.5			4	2
	0.75	28		
1.0			32	16
	1.50	68		
2.0			100	50
	3.00	68		
4.0			168	84
	6.00	28		
8.0			196	98
	12.00	4		
16.0			200	100

The mean particle size is obtained by multiplying the count in the size interval (0–0.5, 0.5–1.0, etc.) times the midpoint of the size interval (0.25, 0.75, etc.) for each size interval, them adding them together and dividing by the total number of particles:

$$\text{mean} = M = (4*0.25 + 28*0.75 + \cdots + 4*12)/200 = 2.72$$

Note that this can be rewritten as $(4/200)*0.25$, $(28/200)*0.75$, etc., in which case the numbers in the parentheses are the estimates of the frequency function $f(d(i))$.

The variance, the square of the standard deviation, can be calculated similarly:

$$s^2 = (4(0.25 - 2.72)^2 + 28(0.75 - 2.72)^2 + \cdots + 4(12 - 2.72)^2)/(199)$$
$$= 6.06$$

in which the $(N - 1)$ form for the sample

variance was used. The variance is 6.06, the standard deviation is 2.46. Since the mean minus two standard deviations would give a negative diameter, physically impossible, the normal distribution cannot be a good approximation. The "data" are actually from a lognormal distribution. Plotting the percentile data on lognormal probability axes would produce a straight line with a median of 2 and a geometric standard deviation of 2.

Functional Form of Distribution, *f(d)*

From the functional form of the distribution, $f(d)$, one can calculate the population mean from

$$\mu = \int f(d)(d)\,\mathrm{d}d \qquad (9\text{-}14)$$

and the population standard deviation from

$$\sigma = \sqrt{\left(\int f(d)(d-\mu)^2\,\mathrm{d}d\right)} \qquad (9\text{-}15)$$

Often, the expected value notation is used for population means:

$$E(g(d)) = \int f(d)(g(d))\,\mathrm{d}d \qquad (9\text{-}16)$$

where $E(g(d))$ is the expected value (= population mean) of $g(d)$, $f(d)$ is the frequency distribution, and $g(d)$ is a function of diameter, such as area, volume, etc.

If $f(d)$ is based on count, then $E(g(d))$ would be the count mean of $g(d)$. If $f(d)$ is based on mass, then $E(g(d))$ would be the mass mean of $g(d)$. Other possibilities exist, such as the surface mean $g(d)$ or the mass mean $g(d)$, etc. The most common means are the count (number) mean diameter and the mass mean diameter, which use $g(d) = d$.

Parameters for the Normal Distribution

As can be seen from Eq. 9-4, the normal distribution is completely determined once the mean and standard deviation are specified.

Parameters for the Lognormal Distribution

The two parameters that totally describe the lognormal distribution are the geometric mean (which, for the lognormal, equals the median) and the geometric standard deviation. These are usually obtained from count or mass distributions. The median diameters are then labeled "count median diameter" or "mass median diameter." Sometimes, one uses the surface median or activity median diameter. In all these cases, the definition of the median diameter is that half (50%) of the count or mass or surface or activity, as appropriate, is contained in particles that are smaller than the median diameter. Of course, other percentiles (such as the 90th) could be used as the basis of a descriptive diameter, and occasionally are.

The geometric standard deviation for a lognormal distribution is independent of the power of the diameter being measured. Thus, σ_g will be the same whether particles are counted or weighed.

The various diameters for a lognormal distribution are related to each other (see Example 9-3). These "Hatch–Choate" relationships for various means and medians for the lognormal distribution are presented in the text by Hinds (1982), from which we extracted the following equations, in which MMD is the mass median diameter, SMD is the surface median, and CMD is the count median diameter:

$$\mathrm{MMD} = \mathrm{CMD}\exp(3(\ln\sigma_g)^2) \qquad (9\text{-}17)$$

$$\mathrm{SMD} = \mathrm{CMD}\exp(2(\ln\sigma_g)^2) \qquad (9\text{-}18)$$

Example 9-3

What are the mass median diameter (MMD) and the surface median diameter (SMD) for a size distribution that is lognormal and has a count median diameter (CMD) of 4 and a geometric standard deviation (σ_g) of 2?

Answer. For a lognormal distribution, Eq. 9-17 predicts the mass median diameter

(MMD) to be

$$\text{MMD} = 4\exp(3(\ln 2)^2) = 4\exp(3(0.480)) = 16.90$$

and Eq. 9-18 predicts the surface mean diameter to be

$$\text{SMD} = 4\exp(2(\ln 2)^2) = 10.46$$

Note that the order of the sizes is as expected: CMD < SMD < MMD. Note also that the predictions are based on accurate data that are lognormal. This is rarely found in practice, so these predictions are usually approximate.

Often, MMD is substantially larger than CMD. For example, if $\sigma_g = 2$, MMD/CMD is $\exp(3(\ln(2)^2)) = 4.23$. $\sigma_g = 2$ is a moderate geometric standard deviation. If $\sigma_g = 3$, MMD/CMD would be 37.4, etc.

Another good reference on the Hatch–Choate relations for lognormal distributions is the book by Reist (1984).

Table 9-3 shows the various weighted means and medians for a lognormal frequency distribution (CMD = 1, $\sigma_g = 2$) based on the values given by Reist (1984). To get the area mean diameter, for example, one must transform the count or mass distribution to an area distribution, then integrate the area frequency distribution function times the particle diameter. For lognormal distributions, the Hatch–Choate relationships include such transformations.

TABLE 9-3 Various Means and Medians for a Lognormal Distribution having a Count Median of 1 and a Geometric Standard Deviation of 2 (Based on Reist 1984)

Count mode diameter (diameter at frequency peak): 0.62
Count median diameter: 1.0
Count mean diameter: 1.27
Diameter of particle having the average area: 1.62
Diameter of particle having the average mass: 2.06
Area median diameter: 2.61
Area mean diameter: 3.32
Mass or volume median diameter: 4.23
Mass or volume mean diameter: 5.37

Unfortunately, it is rare for any aerosol to be so nearly lognormal that measurements by count of the count median diameter and the geometric standard deviation can be used to estimate accurately the mass median diameter, or vice versa, except for cases of nearly monodisperse aerosols ($\sigma_g = 1$). Thus, impactor data, generally based on mass, cannot be relied upon for accurate count median diameter estimates, and optical particle counter data cannot be relied upon for accurate mass median diameter estimates, except for nearly monodisperse aerosols. This is true whether the estimates are made from the Hatch–Choate relationships or from the numerical methods of the early part of this chapter.

Linear Regression, Primarily for Power-Law Distributions

A linear regression equation is the least-squares best fit of the linear equation $y = mx + b$ to sets of $x(i)$, $y(i)$ pairs. The slope (dy/dx) of the line is m and the intercept (y at $x = 0$) is b. The least-squares best fit produces the estimates for m and b that minimize $\sum(y(i) - mx(i) - b)^2$. The equations for estimating m and b from the data are given in most statistics texts, such as Hays and Winkler (1971) or Draper and Smith (1981), and form the basis for many different computer programs, which is how virtually all regression analyses are done now.

The computer programs generally give not only the best estimates of m and b but also the uncertainty in the estimates, the standard errors of m and of b, SE(m) and SE(b). If the data were to match the assumptions of linear regression, then in 68% of the instances the confidence interval of $m \pm \text{SE}(m)$ would include the true slope and in 95% of the instances the confidence interval of $m \pm 1.96\text{SE}(m)$ would include the true slope. The same statements could be made about $b \pm \text{SE}(b)$ and $b \pm 1.96\text{SE}(b)$ and the true intercept.

The more data, and the smaller the errors in the data, the smaller the confidence limits will be for the slope and the intercept.

Example 9-4 shows excerpts from the printouts of one linear regression program and explains the terms used.

Example 9-4

How does one interpret the printout from a computer linear regression program?

Answer. Below is shown an excerpt from the printout from the General Linear Models Procedure (PROC GLM) in the SAS Institute, Inc., set of statistical programs, widely used by statisticians. The input data were the results of simulations of the detection efficiency (Y) inferred from many measurements of particles using a device of known detection efficiency (X). These data are plotted in Fig. 9-5. A few elements have been marked:

1. DEPENDENT VARIABLE: Y		
2. R-SQUARE: 0.906266		
3. C.V.: 19.9378		
4. PARAMETER:	INTERCEPT	X
ESTIMATE:	0.04525648	0.95313889
5. T FOR H0:	1.24	15.55
6. PR > |T|:	0.2279	0.0001
7. STD ERROR OF ESTIMATE	0.03661314	0.06130638

1. The dependent variable is Y and the independent variable is X.
2. The square of the correlation coefficient was $r^2 = 0.906$ ("R-SQUARE"), meaning that the correlation coefficient was $r = 0.95$. $Y = 0.9531X + 0.4525$ accounted for 90.6% of the total variance of the data.
3. The coefficient of variation (CV), the ratio of the standard deviation to the mean, was 19.9%.
4. The best estimates for the intercept (b: Y at $X = 0$) and the slope (m) were 0.045 and 0.953. Perfect calibration would have been an intercept of $b = 0.0$ and a slope of $m = 1.0$.
5. The numbers of standard errors away from zero that the best estimates represented were $t = (\text{mean} - 0)/\text{SE} = 1.24$ and 15.55.
6. The probabilities of getting estimates so different from zero by chance were 0.2279 (23%) and 0.0001, which means the true intercept was statistically indistinguishable from 0 and the true slope was almost certainly different from zero. (By "true" we mean the value for the underlying population from which the sample was drawn.)
7. The standard errors of estimate, SE(b) for the intercept, and SE(m) for the slope, were 0.0366 and 0.0613, respectively.

The assumptions underlying linear regression are rarely fulfilled exactly. These assumptions are: (1) the relationship between y and x is truly linear; (2) the measurements of $x(i)$ contain no errors; (3) the measurements of $y(i)$ contain only an additive error that is Gaussian with a mean of zero and a standard deviation that does not change. Because these assumptions are rarely fulfilled exactly, the probabilities associated with the confidence intervals are at best approximations. If the measurements of the independent variable, x, do contain an error, then the slope will be erroneously biased toward zero. The intercept will be biased in this case, too.

For a power-law distribution, the relationship becomes linear if logarithms are used:

$$f(d) = ad^b \tag{9-19a}$$

$$\log f(d) = \log a + b \log d \tag{9-19b}$$

$$F(d) = (a/(b+1))d^{b+1} \tag{9-20a}$$

$$\log F(d) = \log(a/(b+1)) + (b+1)\log d \tag{9-20b}$$

and it is easy to use a linear regression program on the logarithm of the cumulative size distribution versus the logarithm of the size. One can also simply plot $\log F(d)$ versus $\log d$, or plot $F(d)$ versus d on log–log axes, and draw the apparent best-fit line through the data. The use of a linear regression package is preferable, not so much for improved accuracy (recall that the assumptions are rarely met) as for being objective rather than subjective and, thus, being wholly reproducible by others.

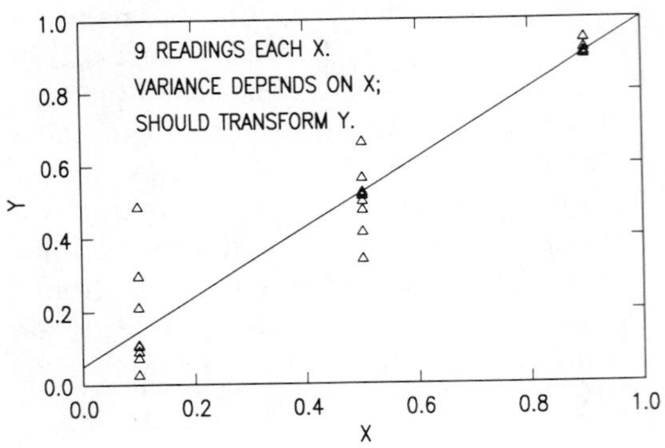

FIGURE 9-5. Example of Data with a Least-squares Linear Regression Line of Best Fit. It Would be Better to Transform *y*, Perhaps Using log *y*, So That It Has a Variance Independent of *x*.

Figure 9-5 shows the data from Example 9-4 and the best-fit straight line obtained from linear regression. Note that the assumption that the standard deviation of y is independent of x is clearly violated, because the spread in the values of y gets bigger as y gets bigger. It would be advisable to transform y to $\log(y)$ or to $\sqrt{y}$ and try regression again. $\mathrm{Log}(y)$ versus $\log(x)$ should also be tried.

Often the correlation coefficient ($-1 < r < 1$) is used as a measure of the degree to which y is linearly dependent on x. This can be misleading. The smaller the errors in y in comparison to the change in y across the range of x, the better is the correlation coefficient and the better is the fit of the line to the data. Note that data taken with a very large range in y can have large errors (in y) and still give impressive correlations ($|r|$ close to 1).

The reader is encouraged to consult statisticians or statistics texts about the fine points, such as hypothesis testing with correlation coefficients, carrying out regressions when there are errors in the independent variable (x), and how to take into account a variation in the error of y over the range of x by using regression with weights adjusted to the estimated error in y (Draper and Smith 1981).

SUMMARIZING SIZE DISTRIBUTIONS GRAPHICALLY

Although equations or tables are more useful for further analysis of the data, graphs can quickly convey the essential features of particle size distributions.

General Advice

An excellent book on the topic of graphing data is the work by Cleveland (1985). Some of his valuable suggestions are:

1. Make the legend describing the graph complete and succinct. 2. Choose symbols for the data that are easily distinguished. 3. Check carefully for errors: labels, tick marks, captions, data. 4. Make sure that the elements are large enough and clear enough for reproduction and any planned size reduction. 5. Make the data prominent. 6. Keep necessary elements only. 7. Surround the region with the scale lines, putting the tick marks outside the rectangular "data region". 8. Keep labels, captions, etc., outside the data region if possible. 9. Explain any error bars (standard deviation, standard error, etc.). 10. Choose axis scales that help minimize empty space. Zero need not always be shown, for example. 11. Choose axis scales that help readers draw conclusions (e.g. log–log where a power law, $y = ax^b$, is being investigated).

When two or more graphs are to be compared, their axes should be as similar as possible.

Cumulative vs. Frequency Representations

The familiar bell-shaped curve of the normal frequency distribution shows readily the location and width of that distribution. This frequency presentation is also convenient and clear when the distribution is made up of the sum of several distributions, for example in multimodal distributions, having more than one local maximum. A frequency distribution having more than one component may become a cumulative distribution with a shape that is not so readily interpreted by eye. To determine the fraction of the distribution that lies between two sizes, for example, the frequency distribution format is awkward, because one must integrate the area under the curve. Thus, it is generally better to use the cumulative size distribution rather than the frequency size distribution.

Perhaps the most important drawback of the frequency format is that the shape of the curve changes with the choice of the interval widths. A lognormal distribution will be skewed when plotted against the diameter, giving an asymmetric shape that has a long tail to the right (large-diameter end). Plotting a lognormal distribution against the logarithm of the diameter will give a symmetric bell-shaped curve, but then the percentage of the distribution that is between any two sizes will not be proportional to the area under the curve between these two sizes. A common mistake with frequency presentations of data is that the independent variable (particle size) is not made to progress in equal increments, such as equal diameters or equal logarithms, coupled with a failure to divide the change in cumulative distribution by the change in the independent variable, to form $\Delta F/\Delta x$. Thus, one may see a histogram presentation (bar graph with the bars extending vertically) that shows data from an instrument that has unequal sizing intervals; changing these intervals for the same aerosol would give a histogram of a different shape.

The cumulative distribution format contains the same information as the frequency distribution. The central location can be estimated from the median, the 50th percentile, and the spread can be estimated from the difference between two percentiles, such as the 16th to the 84th percentiles or the 25th to the 75th, called the "inter-quartile range" in some texts. The percentage between two sizes is readily obtained by subtraction using the cumulative distribution rather than by the more difficult process of integration (estimation of the area) using the frequency distribution.

Plotting Data and a Fitted Curve

For plotting one or a few distributions, the presentation of both the data and the best-fit curve(s) on the same graph has the advantages of completeness and clarity. Each curve helps to summarize, and the data are shown to allow others to decide on how good a summary each curve provides or to help them take the analysis further. The data should be in a tabulated form, too, not necessarily for presentation but to have them accessible. The easiest way to plot the best-fit curve is to use axes that make it into a straight line. This is discussed next.

Plotting with Linear Axes

The size data usually start as counts (or mass) in particle size intervals. By dividing by the total count (or mass), the data can be converted to segments $dF(d(i))$ of the cumulative distribution in the intervals around $d(i)$. By accumulating the segments dF, one can create the useful cumulative distribution function, $F(d)$.

If $F(d)$ plotted against d using linear axes produces a rather symmetric S-shaped curve, then it is likely that the distribution is normal or nearly so. If the curve is somewhat S-shaped but with the large-diameter part much more extended than the small-diameter part, then it is likely that the distribution is lognormal or nearly so. If the $F(d)$ curve is a straight line or if it has only one curved region rather than two, then the power-law distribution is a candidate. Plotting each of these is discussed next.

Plotting with Transformed Axes

A cumulative normal size distribution will plot as a straight line on axes that have their percentile values spaced in equal intervals of probits as the ordinate scale and micrometers (or other units of length) as the abscissa scale. Probits indicate how many standard deviations from the mean the percentile value represents for a truly normal distribution.

TABLE 9-4 Percentile Values and Corresponding Probit Values

Percentile	Probit Value
00.003	− 4.0
00.023	− 3.5
00.135	− 3.0
00.621	− 2.5
02.275	− 2.0
06.681	− 1.5
15.866	− 1.0
30.854	− 0.5
50.000	+ 0.0
69.146	+ 0.5
84.134	+ 1.0
93.319	+ 1.5
97.725	+ 2.0
99.379	+ 2.5
99.865	+ 3.0
99.977	+ 3.5
99.997	+ 4.0

The probit value for the 50th percentile is 0, and the probit values for the 16th and 84th percentile values are − 1 and 1, respectively, etc. Seventeen probit–percentile pairs are given in Table 9-4, with more values readily obtained in most statistical texts and from some pocket calculators or from computer programs.

Other probit values can be obtained in this range from the approximation formula:

$$\text{probits} = 4.9\,(F^{0.14} - (1 - F)^{0.14}) \qquad \text{(9-21)}$$

Example 9-5 illustrates conversion of percentiles into probits by using the approximation formula.

Example 9-5

How many standard deviations from the mean (probits) is the 80th percentile for a normal distribution?

Answer. Use the formula

$$\text{probits} = 4.9(F^{0.14} - (1 - F)^{0.14})$$

and substitute $F = 0.80$, giving

$$\text{probits} = 4.9(0.9692 - 0.7983) = 0.84$$

just under one standard deviation.

Plotting data from a cumulative normal distribution on probit versus linear axes will produce a straight line. However, the data will rarely be without error and perfectly normal; so, some deviation from a straight line can be expected. Another option for plotting cumulative normal curves is to use the commercially available graph paper that has the percentiles positioned according to their probit values. (This paper is sometimes called "probability paper".)

In plotting the best-fit straight line to data that are approximately normal, it is better to calculate the mean (M) and the standard deviation (s) directly from the formulas given above than to use linear regression to obtain the slope and the intercept of the curve. The probit value for points on the cumulative normal probability curve will be

$$\text{probits} = (d - M)/s \qquad \text{(9-22)}$$

Figure 9-6 shows an example of cumulative distributions plotted on a probit scale, after the particle size d has been transformed to $\log(d/d_0)$, a transformation that creates a normal distribution in $\log d$ from a lognormal

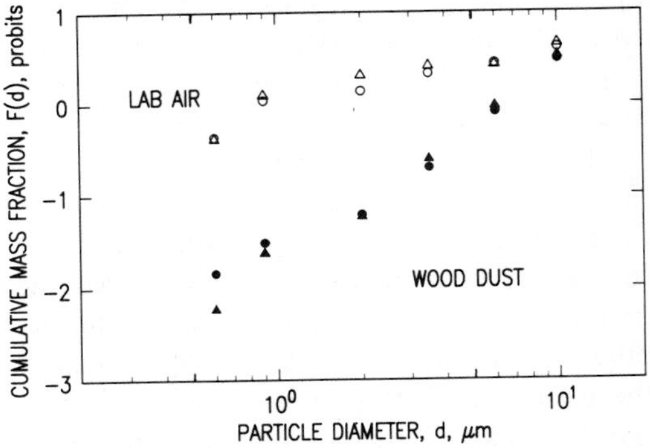

FIGURE 9-6. Cumulative Size Distribution Data (Shen and Ring 1986) Plotted as Mass Fraction Measured in Probits versus the Logarithm of the Apparent Diameter. Open Symbols are for Laboratory Air; Closed Symbols are for Wood Dust; Impactor 296A Data are Denoted with Triangles, 296B Data with Circles. Lognormal Distributions Would Plot as Straight Lines.

distribution in d. The data were taken from impactor measurements reported by Shen and Ring (1986), discussed more fully below.

The lognormal distribution independent variable is generally referred to as the logarithm of the particle size. Keep in mind that logarithms take dimensionless numbers for their arguments, so that the variable is actually $\log(d/d_0)$, where d_0 is often in micrometers, sometimes in centimeters, rarely in meters.

Commercial graph paper ("log-probability paper") is available.

Finally, the cumulative distribution for a power-law aerosol is readily plotted either as $\log F$ versus $\log(d/d_0)$ on linear axes or F versus d/d_0 on logarithmic axes.

CONFIDENCE INTERVALS AND ERROR ANALYSIS

Having made measurements, one would like an estimate of their precision, an estimate of where future measurements of the same type would fall if they were made. One can also calculate a mean from the measurements, and one would like to know how precise an estimate the sample mean is of the true mean (unknown). The measured values may be used in other equations to obtain derived values, and one would like to know how precise these derived values are. These aspects are discussed in this section.

Confidence Intervals

The confidence intervals described here are of the following type: If the data have only random errors that are nearly normal in their distribution with a standard deviation of σ, then one can estimate σ by using the sample standard deviation, s, and can estimate a confidence interval in which $P\%$ of future measurements of the same kind would fall. One can also estimate a confidence interval for the measured mean, M, which would include the real mean, μ, in $P\%$ of the instances if repeated sets of the same kind of measurement were made.

Confidence Intervals for Single Readings

If one has a large number of estimates of d, $N > 30$, from a distribution that is approximately normal, then one can calculate the sample mean and the sample standard deviation and estimate that the readings come from a normal distribution with that mean and that standard deviation. Thus, a 95% confidence interval would be the measured mean plus or minus two measured standard deviations: $M \pm 2s$. The prediction that the next value would be within $M \pm 2s$ would be correct in about 95% of the cases. Looking at the tabulation of percentiles and probit values in Table 9-4, it is clear that $M \pm 3s$ (-3 probits to $+3$ probits) would cover $99.865 - 0.135 = 99.73\%$ of the instances, etc.

With a smaller number of readings, the estimate of the mean and the estimate of the standard deviation would have more uncertainty, and the ability to predict future readings would be less.

Control Charts

Figure 9-7 shows 100 simulated observations, taken at random from a normal distribution with mean = 10 and standard deviation = 1. Lines have been drawn at ± 3 standard deviations from the mean. About 68% of the readings are within 1 standard deviation, about 95% within 2 standard deviations,

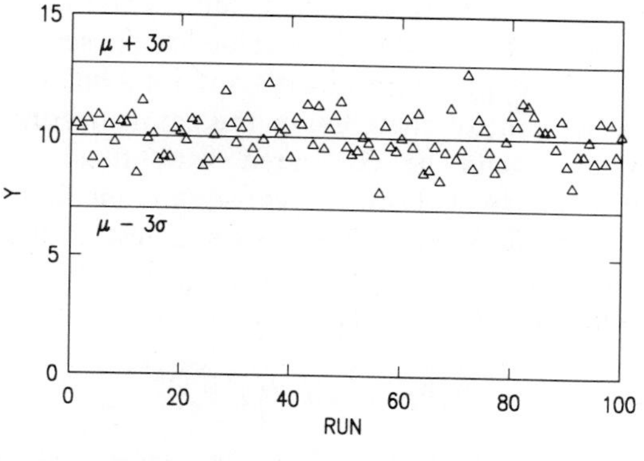

FIGURE 9-7. "Data" Drawn from a Gaussian (normal) Distribution, $\mu = 10$, $\sigma = 1$, with Lines Drawn for 3 Standard Deviations from the Mean, $\mu \pm 3\sigma$.

and none are outside 3 standard deviations. This kind of chart is a "control chart" and can be used to identify when a change in a process has occurred, by noting an unusual number of values lying outside the three-standard-deviation limits. More often, the means of many readings are plotted, rather than individual readings, and their limits are in terms of standard errors, as discussed below.

Control charts are used to keep track of the readings by charting the readings versus time. The chart has a mean marked on it, taken from a long series of readings obtained before the chart is drawn up. The chart often also has the mean plus or minus two and three standard deviations of the readings. Single readings more than three standard deviations from the mean, or two successive readings more than two standard deviations from the mean, can be used to signal a statistically significant change, one requiring investigation.

The control chart techniques available in the quality control literature are attractive. They have a long history of use and success and are good compromises between scientific exactitude and simplicity and clarity. Box, Hunter, and Hunter (1978) provided a useful summary of the traditional Shewart charts as well as the cumulative sum charts. Cumulative sum (CUSUM) methods respond to changes more rapidly than the corresponding traditional control charts. One forms cumulative sum charts by adding up the readings as they are taken, subtracting each time the expected (mean) value for the series.

Where long-term records are not available to set control chart limits, the counts in one interval can be compared with those in another by the use of chi-square methods that assume Poisson statistics, a test of the constancy of the concentration (Box, Hunter, and Hunter 1978). The statistic

$$u = (x - y)/\sqrt{(x + y)} \qquad (9\text{-}23)$$

for counts from one Poisson distribution will itself be normal with a mean of $E(u) = 0$ and a standard deviation of 1. This can be used to test whether the counts, x, from one counting interval are likely to be from the same concentration as the counts, y, from another counting interval (Cooper 1991). A value $|u| > 4$ provides strong evidence that the concentrations are not the same.

Confidence Intervals for Means

If the measurements follow a normal distribution, then the means of the measurements will have a distribution that will be nearly normal and will have a standard deviation equal to $1/\sqrt{N}$ times the standard deviation of the individual readings, where N is the number of readings used to calculate an individual mean. Even if the measurements are not quite normal, the means will be given approximately by the t-distribution (often called "Student's t" after the pseudonym of its discoverer).

For samples drawn from a normal distribution, the confidence interval for the true mean, μ, is given by the inequalities

$$M - t(N - 1, \alpha/2)s/\sqrt{N} \le \mu \le M + t(N - 1, \alpha/2)s/\sqrt{N} \qquad (9\text{-}24)$$

where $t(N - 1, \alpha/2)$ is obtained from tables of the Student's t-distribution, available in statistics texts. $N - 1$ is the number of "degrees of freedom." The value of $\alpha/2$ is half of the fraction of instances that will *not* occur within the confidence interval. If the confidence interval is a 90% confidence interval, then $\alpha/2$ is 5%. Asymmetric confidence intervals can also be obtained, such as

$$\mu \le M + t(N - 1, 0.05)s/\sqrt{N} \qquad (9\text{-}25)$$

which would cover 95% of the instances. This formula was used in the clean-room classification methodology for Federal Standard 209D (U.S. General Services Administration 1988).

Although the t-distribution is more complicated than the normal distribution, its values are tabulated widely. Furthermore, as N becomes much larger than 10, t becomes virtually a normal distribution, and the t values become the corresponding probit values. For example, for such large N values,

$$M - 2.0s/\sqrt{N} \le \mu \le M + 2.0s/\sqrt{N} \quad (9\text{-}26)$$

would be a 95% confidence interval for the mean.

The value of multiple measurements (large N) in decreasing the uncertainty (improving the precision) of the mean is evident. Regardless of the distribution that the readings come from, the mean of N readings will have a distribution that has a standard deviation (= standard error of the mean) that is $1/\sqrt{N}$ times the standard deviation of the individual readings. The $1/\sqrt{N}$ dependence means that halving the uncertainty in the mean can be achieved by going from $N = 1$ to $N = 4$ or from $N = 4$ to $N = 16$ or from $N = 16$ to $N = 64$, giving diminishing improvements as N increases.

Error Analysis: Propagation of Error

Assume that one has a set of measurements of d that have a sample mean M and a sample standard deviation s. One does not know how much a contribution to s has been made by the distribution of d and how much by the errors of measurement of the instrument used to measure d, but it is not crucial for what follows that one be able to resolve this.

"Propagation of error" concerns how the errors in measurements affect calculations based on these measurements. To carry out the analysis, it is more convenient to work with the variance, s^2, rather than the standard deviation.

The general formula for the propagation of error makes use of partial derivatives: the variance of $g(x(i))$, a function of one or more variables $x(i)$, is the sum of the products of the squares of the partial derivatives of g with respect to $x(i)$ and the variances of $x(i)$. Often, one uses just the particle diameter as a variable (so $x(i) = d$) and calculates the variance of a function $g(d)$.

The variance in $g(d)$, $s(g)^2$, is obtained from the variance in d, $s(d)^2$, as follows:

$$s(g)^2 = (\mathrm{d}g(d)/\mathrm{d}d)^2 s(d)^2 \quad (9\text{-}27)$$

If $g(d) = ad$, then $s(g)^2 = a^2 s(d)^2$, a simple result. If $g(d) = ad^b$, then the manipulation becomes somewhat more complicated, but can be simplified by using

$$s(ad^b)^2/(ad^b)^2 \approx b^2 s(d)^2/d^2 \quad (9\text{-}28a)$$

$$s(ad^b)/(ad^b) \approx bs(d)/d \quad (9\text{-}28b)$$

where the value of d to be used is the mean. If the volume is predicted from the diameter, then the relative standard deviation of the volume ($= \mu(g)/\sigma(g) \approx s(g)/M(g)$) is $b = 3$ times the relative standard deviation of the diameter. If the standard error of the mean volume is desired, then the standard error of the diameter, rather than the standard deviation, should be used in this equation. Clearly, going from the diameter to the volume, where $b = 3$, increases the uncertainty, in contrast to going from the volume to the diameter, where $b = 1/3$, which decreases the uncertainty.

An example of the use of propagation of error in analyzing experimental determination of mass concentration was presented in detail by Evans, Cooper, and Kinney (1984). Example 9-6 is based on that work.

Example 9-6

How would one apply propagation of error to the estimate of mass concentration $c = W/Qt$?

Answer (Evans, Cooper, and Kinney 1984). The equation for the concentration from a weight (mass) gain of W on a filter through which a volume flow rate Q has flowed for duration t is

$$c = W/Qt$$

The formula for the propagation of error in this situation is

$$s(c)^2 = (\mathrm{d}c/\mathrm{d}W)^2 s(W)^2 + (\mathrm{d}c/\mathrm{d}Q)^2 s(Q)^2 + (\mathrm{d}c/\mathrm{d}t)^2 s(t)^2$$

in which the derivatives are partial derivatives

with respect to the indicated independent variables.

The derivatives were evaluated at these mean values from data from filter samples:

$$W = 1.34 \times 10^5\ \mu g$$

$$Q = 1.50\ m^3/min$$

$$t = 1440\ s$$

The variances were estimated from repeated readings of the same filter mass, from comparisons of the flow meter against a standard, and from a comparison of the flow timer against a calibrated timer:

$$s(W)^2 = 3.26 \times 10^5\ (\mu g)^2$$

$$s(Q)^2 = 1.12 \times 10^{-2}\ (m^3/min)^2$$

$$s(t)^2 = 9\ (s^2)$$

Carrying out the calculation of the expected variance of c, they obtained

$$s(c)^2 \simeq 19.2$$

almost all of which was due to the uncertainty in the flow rate. This estimate of the variance produced an estimate of 4.4 μg/m³ for the uncertainty or a relative standard deviation of $4.4/62.0 = 7.1\%$ for the concentration. This compared favorably with other methods used to estimate the uncertainty.

TESTING HYPOTHESES WITH SIZE DISTRIBUTION DATA

Often, it is desired to compare one measured aerosol size distribution against another or against a theoretical distribution. One wants objective statistical measures to discriminate between two distributions for many reasons, to answer many types of questions: Are the aerosols likely to be from different sources? Has something changed during the period between the measurements of the two distributions? Do two instruments measuring the same aerosol give virtually the same results? Did different conditions produce different size distributions?

Analyzing Data from a Normal Distribution: Student's *t*

If a particle counter has been used, then the mean particle size, M_1, from one distribution can be compared with a hypothetical mean, M, by the Student's t-test, the statistic

$$t = (M_1 - M)/(s_1/\sqrt{N_1}) \qquad (9\text{-}29)$$

in which s_1 is the sample standard deviation and N_1 is the number of particles sized. This value of t is to be compared against the critical values of t at very low or very high percentiles, such as 1, 5, 10, 90, 95, 99. What is being sought is the probability that a t value of this size would have occurred if there were indeed no difference between the means. For example, if one is testing whether $M_1 = M$, then an unusually negative t value (e.g., $t \ll -1$) or an unusually positive t value (e.g., $t \gg 1$) would be a strong evidence against the hypothesis. Tables of the critical values of t for various N are available in most statistics texts.

When $N \gg 10$, the t statistic can be evaluated from the normal distribution, with $t < 0$ expected in 50% of the instances, $t < 1$ expected in 84% of the instances, $t < 2$ expected in 97.7% of the instances, etc. Usually, $|t| > 2$ would be taken as strong evidence that M_1 was not equal to M (this is the 95% significance level).

If the first distribution is to be compared not against a theoretical mean, M, but against a measured mean, M_2, then there are other formulas for Student's t-test available in the literature (e.g., Hays and Winkler 1970); one must decide whether to assume that the true underlying population distributions have equal standard deviations or not. If, as is often done, one assumes that the population standard deviations are equal (variances are equal) for the N_1 and N_2 measurements, one has

$$t = (M_1 - M_2)/\sqrt{(s^2/N_1 + s^2/N_2)} \qquad (9\text{-}30)$$

$$s^2 = ((N_1 - 1)s_1^2 + (N_2 - 1)s_2^2)/(N_1 + N_2 - 2) \qquad (9\text{-}31)$$

the latter being the "pooled estimate of variance" (Meyer 1975). The denominator of s^2, $N_1 + N_2 - 2$, is the number of degrees of freedom. If the denominator is much larger than 10, t can again be obtained from the cumulative normal distribution, as above. If the denominator is not much larger than 10, then tables of Student's t-distribution should be used with the appropriate number of degrees of freedom.

If the observations under two different conditions to be compared can be analyzed as pairs, then the differences, $z_i = x_i - y_i$, can be used as the data to be analyzed, with the mean of z tested against zero, to show some effect, or against another value, to show an effect larger than that value. In most instances, paired comparisons are to be preferred.

Correlation and Regression

There are three commonly used measures of correlation: (1) the Pearson product-moment correlation coefficient, the most common; (2) the Spearman rank correlation coefficient; and (3) the Kendall tau coefficient. Each has a somewhat different set of assumptions. (See, for example, Hays and Winkler 1970.)

Correlation analysis is of doubtful validity, however. The easiest way to see this is to note that if there were two sets of size data being compared and if we added more sizing intervals, all of which had no aerosol count or mass in them, the correlations would increase artificially, and the associated levels of probability would change correspondingly.

Analysis of Variance

Analysis of variance (ANOVA) has some dependence on the form of the underlying data distribution, but it is rather robust (Hays and Winkler 1970). The data need to come from a balanced statistical experimental design. (See standard statistics texts, such as Hays and Winkler 1970.) Transformations can be used to make data more nearly normal. (Examples are the use of the logarithm of the particle size or the square root of the count.) Multiple determinations of the distribution with a mass sensor or particle counter allow a statistical comparison, because the variability can be estimated from the replications.

Distinguishing Among Count Distributions: Chi-Square

Comparing count distributions is simpler than comparing mass distributions, because the uncertainty in the count, the standard deviation, can often be assumed to be the square root of the count (Poisson statistics). By using as the dependent variable the square root of the count, a Poisson distribution becomes more nearly normal. Furthermore, the square root transformation makes the variance for Poisson statistics approximately constant (0.25), which helps satisfy the assumptions in linear regression (Box, Hunter, and Hunter 1978).

Without making assumptions about the size distribution, one can use a chi-square analysis, after apportioning the counts into matching size intervals. Chi-square is the sum of squares of the differences between the counts, $C(i)$, in the ith interval or category, and the expected counts, $E(i)$, divided by the expected counts:

$$\chi^2 = \sum (C(i) - E(i))^2 / E(i) \qquad (9\text{-}32)$$

The values for the expected counts come from the hypothesis being tested.

The χ^2 value is to be compared with percentile values in statistical tables of χ^2 distributions, for the appropriate number of degrees of freedom. Details are available in standard statistical texts. An example of the application of χ^2 to compare size distributions is a test to see whether one size distribution of particles contaminating a surface is different from another. The data in Example 9-7 (sample A and sample B) are count data, subdivided into particle counts for diameters larger or smaller than 5 μm.

Example 9-7

A sample of aerosol A had 160 particles larger than 5 μm and 75 particles smaller than 5 μm.

A sample of aerosol B had 105 particles larger and 105 particles smaller than 5 μm. Do the data indicate the aerosols are different from each other?

Answer. This question can be answered using chi-square analysis. The chi-square analysis involves comparing the observed counts with the expected counts and calculating χ^2. An unusually large value would make it very unlikely that the aerosols are the same.

OBSERVED

	Aerosol A	Aerosol B	Totals, Fraction
$d > 5$ μm	160	105	265, 265/445 = 0.596
$d < 5$ μm	75	105	180, 180/445 = 0.404
Totals	235	210	445
Fraction	235/445 = 0.528	210/445 = 0.472	

EXPECTED

	Aerosol A	Aerosol B
$d > 5$ μm	$0.596 * 0.528 * 445 = 140.0$	$0.596 * 0.472 * 445 = 125.2$
$d < 5$ μm	$0.404 * 0.528 * 445 = 94.9$	$0.404 * 0.472 * 445 = 84.9$

If the aerosols had the same size distributions, then the counts would divide proportionately into the four cells of the two-by-two matrix, so this hypothesis gives the expected values.

CHI-SQUARE

$d > 5$ μm $(160–140.0)^2/140.0 = 2.86$

$(105–125.2)^2/125.2 = 3.26$

$d < 5$ μm $(75–94.9)^2/94.9 = 4.17$

$(105–84.9)^2/84.9 = 4.76$

$$\chi^2 = 2.86 + 3.26 + 4.17 + 4.76 = 15.05$$

Degrees of freedom

$= (\text{rows} - 1) * (\text{columns} - 1)$

$= (1) * (1) = 1$

A value of χ^2 of 15.05 or larger purely by chance is expected in less than 1/100 cases. These aerosols are almost certainly different.

The value of χ^2 is 15, a value expected in much less than 1% of the cases if the two samples were drawn from the same population. The fraction of particles larger than 5 μm in sample A is larger than that of sample B, and the difference is statistically significant.

Chi-square analysis is also often used in "goodness of fit" comparisons with hypothesized distributions. Example 9-8 shows an example of chi-square goodness of fit analysis for a distribution of particle counts compared to a hypothesized lognormal distribution.

Example 9-8

How would one apply the chi-square goodness of fit calculation to a distribution of particles by size to test whether the distribution is lognormal?

Answer. This example uses data that are 100 particles classified into intervals by size. CMD is the count median diameter. The number counted, C, is compared with the number expected, E, under the assumption that the proportions are those in a lognormal distribution. The number counted is compared with the number expected, E, using $\chi^2 = \sum(C - E)^2/E$.

Size Interval	# Expected $= E$[a]	# Counted $= C$	$\chi^2 = (C - E)^2/E$
$< \text{CMD}/\sigma_g$	16	18	0.25
CMD/σ_g–CMD	34	40	1.06
CMD–$\text{CMD} * \sigma_g$	34	30	0.47
$> \text{CMD} * \sigma_g$	16	12	1.00
Total	100	100	2.78

a. The hypothesis tested is that the data come from a lognormal distribution. The calculation of chi-square is easy. It is harder to determine the number of degrees of freedom. Here, there are four measurements, but the expected values have been chosen by matching the total number and the CMD and the σ_g to the data, using three degrees of freedom, leaving only one degree of freedom.

The total value of chi-square is $\chi^2 = 2.78$, which is borderline for one degree of freedom, being near the 90th percentile, so these data do not contradict the hypothesis of a lognormal distribution, although they cast some doubt on it.

Distinguishing Count or Mass Distributions: Kolmogorov–Smirnov

If the data can be put into a large number of intervals ($N \gg 10$ generally) that have the same boundaries, then the Kolmogorov–Smirnov (K–S) test can be quite effective. Although many aerosol sizing instruments do not have a large number of sizing channels, Heitbrink, Baron, and Willeke (1991) noted that the Aerodynamic Particle Sizer of TSI, Inc., does have many, and they used the K–S test successfully to analyze particle count data from this instrument. In other cases, the exact sizing of a large number of particles ($N \gg 10$) could be used; in that case, one could select the size intervals, with the interval end-points for the two size distributions chosen so as to provide a large number of common boundaries and comparisons.

The test involves comparing the cumulative distributions at the common end-points of each interval and calculating the maximum difference between them. The maximum difference is compared with the tabulated values for various levels of statistical significance and various numbers of intervals tested. The maximum difference values needed for statistical significance of 90% or more are differences of the order of magnitude of $1/\sqrt{N}$. For more information, see standard statistical texts, such as Hays and Winkler (1970).

Detailed Example: Reanalysis of the Data from Shen and Ring (1986)

This example demonstrates some methods for comparing size distributions, using data from the literature. It assumes that the reader has substantial familiarity with statistical techniques.

TABLE 9-5 Weights (mg) in each Impactor Size Interval for Impactor A Sampling Laboratory Air ("LA"), Impactor B Sampling Wood Dust Aerosol ("WB"), etc., from Shen and Ring (1986)

Diameter (μm)	WA	WB	LA	LB
> 10	0.68	0.86	0.066	0.068
6–10	0.47	0.60	0.021	0.015
3.5–6	0.49	0.60	0.001	0.010
2.0–3.5	0.36	0.35	0.009	0.018
0.9–2.0	0.13	0.13	0.021	0.010
0.6–0.9	0.09	0.09	0.049	0.040
0.0–0.6	0.03	0.09	0.090	0.090

Table 9-5 shows the weight data obtained from an article by Shen and Ring (1986). "WA" is a wood dust aerosol sampled by Sierra impactor 296A; "LB" is a laboratory air sampled by Sierra impactor 296B, etc. The weights were in mg and had estimated standard deviations of about 0.1 mg for the wood aerosol and 0.01 mg for the laboratory air, according to the authors.

Figure 9-6 is plotted from the data in the article of Shen and Ring. The axes are similar to log–probability axes, with cumulative mass fraction measured in probits. The log of the apparent diameter (effective cut diameter for the impactor) is the other axis. A cumulative lognormal distribution would plot as a straight line on these axes. The data certainly seem to indicate that the two distributions for the wood dust are similar to each other, the two distributions for the laboratory air are similar to each other, and the distributions for the wood dust are different from the distributions for the laboratory air. For both aerosols, sampler 296B reported a somewhat larger mass median diameter and a somewhat larger geometric standard deviation than did sampler 296A.

We analyzed the data using analysis of variance. The factors tested for were the effects of the samplers (A and B) and the effects of the aerosols (wood dust and lab air). Analysis of variance tests whether assuming there is a contribution to the mean that is associated with a factor reduces the variance by a

TABLE 9-6 Data Used in Analysis of Variance

Mass Median Diameter (MMD)

Sampler	Dust	LN(MMD)	MMD
A	Wood	1.8180	6.16
B	Wood	1.8950	6.655
A	Air	0.1988	1.22
B	Air	0.2852	1.33

Geometric Standard Deviation (GSD)

Sampler	Dust	LN(GSD)	GSD
A	Wood	1.0543	2.87
B	Wood	1.2613	3.53
A	Air	2.7973	16.4
B	Air	3.0681	21.5

(LN(X) is the natural logarithm, base e, of X.)

statistically significant degree. The values that Shen and Ring inferred for the mass median diameters were used as inputs. First, the logarithms of the mass medians were used, as shown in the top part of Table 9-6. This transformation was made to have values more closely approximating what would come from a normal distribution. A similar analysis on the logarithms of the inferred geometric standard deviations, shown in the bottom part of Table 9-6, was performed.

The ANOVA procedure was from the statistical library provided by SAS, Inc. There were two readings at each of two levels for each of two factors: aerosol, sampler. This is a balanced design. As applied to the log of the mass median diameter, the procedure indicated there was a difference between the two dusts at a better than 99.9% level of confidence and a difference between the two samplers at a better than 95% level of confidence. As applied to the log of the mass geometric standard deviation, the ANOVA procedure indicated there was a difference between the two dusts at more than a 98% level of confidence and a difference between the two samplers at a better than 90% level of confidence.

When ANOVA was applied to the untransformed mass median diameters and to the geometric standard deviations, there was no statistical significance to the differences between the samplers. Thus, using the logarithmic transformation (above), better matching the assumptions of ANOVA, produced a more sensitive test for discriminating between the samplers. ANOVA applied to the untransformed mass median diameters and geometric standard deviations indicated the dusts were different at the 98% confidence level (median diameters) and at the 90% confidence level (geometric standard deviations).

We performed Student's t-analysis on the mass median diameters and on the geometric standard deviations, and on their logarithms (using $N_1 = N_2 = 2$). Although Shen and Ring somewhat mistakenly applied the t-test to their data (Cooper 1990), Shen and Ring were correct that a t-test could be used in their situation. They had taken two samples on each of the aerosols. They mistakenly used the number of sizing intervals as the basis of their t-test, rather than the number 2 for the two samples being compared. Multiple determinations of the distribution with a mass sampler do allow statistical comparison, because the variability can be estimated from the replications. The estimates of the mass median diameter can be used as the data which are being analyzed. The logarithms of the mass median diameters give the mean logarithms, and these means can be compared: 1.818 and 1.895 for the wood dust; 0.1989 and 0.2852 for the laboratory air. The difference between the two estimates for the mean logarithm for the same aerosol is an estimate of the standard deviation of the estimate of the mean logarithm. The differences are 0.08 for the wood dust and 0.09 for the laboratory aerosol, suggesting we can pool the variances. Using Eqs. 9-30 and 9-31 for comparing means by the t-test, with $N_1 = 2$, $N_2 = 2$ for the number of samples for each mean, gave us $t = 27.9$, significant at a better than 99% level. The t-test is appropriate because the logarithms were distributed almost normally. It is not so clear that the t-test is appropriate to analyze the logarithms of the geometric standard deviations, but the test is often not very sensitive to the form of the underlying distribution. Applying it to the logarithms of the geometric standard deviations gave us $t = 10.4$, also significant at a 99% level. The t-test applied to the untransformed median diameters gave $t = 20.2$, a difference between the wood dust and the lab air at greater than the 99% confidence level. Applied to the untransformed geometric standard deviations, the t-test gave $t = 6.1$, significant at between 95% and 99%.

Our statistical analyses confirmed what "common sense" indicates when looking at the graphs of the cumulative size distributions: the wood dust and the laboratory aerosol size distribution were different in median size and in slope.

COINCIDENCE ERRORS

When two or more particles are present simultaneously in the sensing zone of an instrument, they may be counted as 1 particle (type 1 coincidence), perhaps of a different size, or as 0 particles (type 0 coincidence). Coincidence is similar to "saturation," where the pulses from the sensing zone come too rapidly for the electronics to separate. In general, coincidence causes a loss of particles, underestimating concentration, and a shift in the particle size distribution to larger sizes.

Coincidence has been treated by many authors, including Jaenicke (1972); Bader, Gordon, and Brown (1972); Julanov, Lushnikov, and Nevskii (1984, 1986); Raasch, and Umhauer (1984); and Cooper and Miller (1987). Where the particles arrive in the zone at random, the probability of having n particles in the zone can be described by the Poisson distribution. The probability that the zone is empty is $\exp(-cV)$, in which c is the concentration and V is the zone volume. The mean number will be cV. The apparent concentration (assuming coincidence type zero) is $(c)\exp(-cV)$, which is nearly c for $cV \ll 1$ and reaches a maximum at $cV = 1$; the concentration will appear to decrease if cV becomes > 1. If the coincidence causes a count in a larger size interval, then this type 1 coincidence causes less loss of count but does create spurious counts of larger particles, further distorting the size distribution (see, for example, Raasch and Umhauer 1984). For type 1 coincidence, the counting efficiency becomes $(1 - \exp(-cV))/cV$. These formulas can be adapted to correct the counts for coincidence losses.

The statistics of coincidence and time-of-arrival was explored further in a pair of papers by Julanov, Lushnikov, and Nevskii (1984, 1986), where methods for obtaining concentration estimates using the behavior of multiple nearly simultaneous arrivals were presented.

Counting errors of various types for surface scanners were analyzed by Pecen et al. (1987) for scanning-beam types and by Cooper and Miller (1987) and Cooper and Rottmann (1988) for vidicon-based types. Special precautions with design and data analysis have to be taken for scanning instruments not to overcount particles on surfaces due to multiple intersections of the scan with the same particle (Galbraith and Neukermans 1987). Such surface monitors can be calibrated with patterned surfaces, with surfaces having a known number of particles, or—by repeated counting—with surfaces having an unknown number of particles of the same size (Cooper and Neukermans 1991).

More complicated particle counting instruments, such as the Aerodynamic Particle Sizer and the Aerosizer, have more complicated coincidence effects (Heitbrink, Baron, and Willeke 1991)—see Chapter 17 for more information.

DATA INVERSION ("DECONVOLUTION" OR "UNFOLDING")

If one cannot obtain exact measurements, then perhaps one can use a knowledge of the behavior of the measurement instrument to correct the data. This is the goal of various data inversion methods that have been presented in the literature of aerosol science and technology and in the literature of many other fields as well. There is even a journal (Inverse Problems) dedicated to this topic.

Although there are instances—such as computerized axial tomography (CAT)—

where data inversion works well, there have been many papers published in the scientific literature, even just in the literature on aerosol science, criticizing previous methods and presenting a new method. There has yet to emerge a clearcut choice of method for analyzing particle size distribution data, which should be a warning to those who hope to salvage ambiguous data using mathematical magic.

An ideal instrument would have a probability of 1 of responding to particles within the nominal size boundaries of the sizing channel and a probability of 0 of responding to particles outside the nominal size interval boundaries. Figure 9-8 shows some hypothetical response functions for the lower particle size boundary of one channel of an instrument. An ideal instrument would have a step function response; more typically, one has a sigmoidal boundary; in some cases, the boundary is multivalued rather than monotonic, complicating the picture still further. The instrument would have a set of such response functions to make up the channel responses. An ideal instrument would not need to have its data subjected to data inversion. Typical instruments may profitably have their data inverted, depending on the size of the channel widths in comparison to the size interval over which the response goes from nearly 0 to nearly 1. Instruments that have multivalued responses are not good candidates for data inversion, unless the sizing intervals are chosen to segregate completely the regions of multivaluedness within distinct intervals.

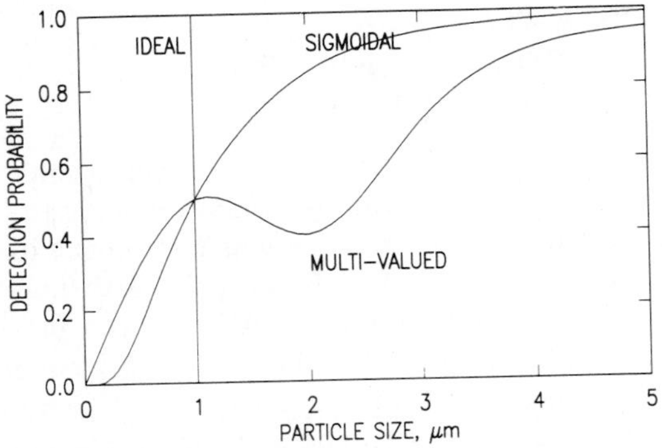

FIGURE 9-8. Some Hypothetical Response Functions for a Particle Sizing Instrument: Step Function, Sigmoidal Function, Multivalued Function.

This section outlines the topic of data inversion. As noted, much has been written on the topic. Tikhonov and Arsenin (1977), as cited by Wolfenbarger and Seinfeld (1991), summarize the difficulties of these often ill-posed problems: "Solutions will become unstable when the number of data is large. Solutions are not unique. Exact solutions often do not exist." Ill-posed, or "ill-conditioned," problems typically show large changes in the inferred solutions due to small changes, or errors, in the data.

Before discussing data inversion below, we emphasize the advice from Noble (1969): the best way to deal with ill-conditioning, the sensitivity of the results to small changes in problem formulation or in the data, is to avoid it through the choice of measuring instrument and conditions of measurement. This advice was echoed by Cooper and Wu (1990) some 20 years later. More details are given below.

What follows in this section is intended for readers with rather strong backgrounds in mathematics.

The Problem: An Integral Equation

Assume that a multichannel instrument produces a set of data that gives the fraction of the aerosol (by count or mass, typically) in each of n channels. The response $F(i)'$ can be described by the integral equation

$$F(i)' = \int K(i,d) f(d)\,\mathrm{d}d + e(i) \qquad (9\text{-}33)$$

where $f(d)$ is the particle size frequency distribution and $K(i,d)$ is the probability that a particle of size d will be counted (or weighed) in the ith channel of the instrument. The data will usually have at least some error, $e(i)$, as well, error being the degree to which $F(i)'$ does not exactly match the integral.

Although $K(i,d)$ is known, generally through calibration and perhaps through theory, $f(d)$ is unknown. The n-channel instrument, $i = 1$ to n, gives only n data points from which one tries to infer $f(d)$, which is a continuous function. The best we can expect to do is to get n elements of information on $f(d)$. "Elements" is deliberately ambiguous. One kind of element is the value of $f(d)$ at n values of $d(i)$ throughout the range of the instrument. Another kind of element is the first n moments of the $f(d)$ distribution: the mean diameter, the mean squared diameter, the mean cubed diameter, or the mean, the variance, the kurtosis, etc. Another kind of element might be the $k(i)$ coefficients of a polynomial expansion, $f(d) = k(0) + k(1)d^1 + k(2)d^2 + \cdots$. Yet another kind of element would be the Fourier expansion coefficients. The list seems boundless. The most common choice is to obtain estimates of $f(d)$ at n or fewer values of d. With only n measurements, at most only n values of $f(d)$ (or other parameters) can be measured, and an infinite number of different curves can be drawn through a finite number of points, so the unknown function is never fully determined by the data alone.

Solving by Converting to a Set of Linear Equations

A simple method to convert the integral equation into a set of linear equations in coefficients $f(d(j)) = f(j)$ is to evaluate the integral with numerical quadrature, which means to evaluate the integral by evaluating the function at a set of d values. For example, the midpoint quadrature evaluates $f(d)$ at the midpoint of the sizing intervals 1 to m and produces the approximation

$$F(i)' = K(i, d(1))f(d(1))\Delta d(1) + \cdots + K(i, d(m))f(m)\Delta d(m) + e(i) \tag{9-34}$$

for $j = 0, 1, \ldots, m$. $\Delta d(j)$ is the width of the size interval for which $f(d(j))$ is at the midpoint, in this formulation. The values of $K(i, d)$ are known, so the quadrature produces a set of n equations in m unknowns.

These equations can be written compactly as

$$F(i)' \simeq \sum K(i, j)f(j)\Delta d(j) + e(i) \tag{9-35}$$

The approximately equals sign has been used to emphasize that not only is there error in the data, but also there is error in going from the integral equation to the set of linear equations. For n equations in m unknowns, we may have a solution if the number of equations equals the number of unknowns, $m = n$; we do not have a solution if the number of equations is less than the number of unknowns, $n < m$; we may have many possible solutions if the number of equations is greater than the number of unknowns, $n > m$. Where $n > m$, more data (equations) than unknowns, we need to choose one of the solutions, generally by requiring that the chosen solution satisfy another constraint, such as being the smoothest of the alternative solutions or being the least-squares fit to the data.

The unknown values are often labeled as $x(j)$ and the known coefficients on the right-hand side are labeled $A(i, j)$. This avoids making the decision whether the user is solving for $f(j)$ or $f(j)\Delta d(j)$. The known values are often labeled as $b(i)$, and they may be part of a cumulative distribution description or part of a frequency distribution description. Regardless of the details of the description, we want to solve the set of linear equations

$$b(i) \simeq \sum A(i, j)x(j) + e(i) \tag{9-36}$$

which in vector–matrix notation becomes

$$b \simeq Ax + e \tag{9-37}$$

Some definitions are needed: A^T = transpose of A; $A(i, j)^T = A(j, i)$. $1/A$ = inverse of A; $(A)(1/A) = I$, the identity matrix. $I(i, j) = 1$ for $i = j$; $I(i, j) = 0$ otherwise.

Solving the Equations when *n* Equations Equal *m* Unknowns

If we have n equations in m unknowns and if $n = m$ and if all the equations are linearly independent of each other (i.e., none is simply equal to a constant times another), then the solution is straightforward:

$$x \simeq (1/A)b - (1/A)e \qquad (9\text{-}38)$$

the x vector is the inverse matrix operating on the data vector. We know b, but not e, so that we calculate an $x = (1/A)b$ that is in error by about $(1/A)e$.

Example 9-9 (from Cooper and Wu 1990) shows a simple two-by-two matrix from a hypothetical instrument with rather good resolution, which one can see by noting that the matrix is nearly the identity matrix (which would need no correction). The inverse matrix (labeled "$Z(j, i)$") is shown. The corrections are of the same magnitude as the off-diagonal elements of $A(i, j)$. These are rather small corrections and do not change $x(i)$ from $b(i)$ by very much nor would they multiply the error term by any large factor. Examples where much larger correction factors need be applied were given in the same article.

Example 9-9

How can one use a two-by-two response matrix to correct the data in an imperfect two-stage device?

Answer. A hypothetical instrument with two sizing channels is found to have a 5% probability of counting a particle in the wrong channel. This leads to the following matrix:

$$A(i, j) = \begin{bmatrix} 0.95 & 0.05 \\ 0.05 & 0.95 \end{bmatrix}$$

The equation $Ax = b$ becomes

$$0.95\,x(1) + 0.05\,x(2) = b(1)$$

$$0.05\,x(1) + 0.95\,x(2) = b(2)$$

The solution of the two simultaneous equations is the inverse matrix $Z = 1/A$ operating on the data vector:

$$1/A(i, j) = \begin{bmatrix} 1.056 & -0.056 \\ -0.056 & 1.056 \end{bmatrix}$$

$$x = (1/A)b$$

Thus, one would calculate a corrected $x(1)$ and $x(2)$ as

$$x(1) = 1.056 * b(1) - 0.056 * b(2)$$

$$x(2) = -0.056 * b(1) + 1.056 * b(2)$$

It is worthwhile to look at the matrix $(1/A)$ to see if it has some large elements that would greatly magnify the error e. This was discussed at greater length by Cooper and Wu (1990). A formal analysis of the errors from inverting linear equations involves a quantity called the condition number. Texts on numerical analysis or regression should be consulted, and our discussion relies on that presented by Forsythe and Moler (1967). Before defining the condition number, we need to define a norm of a matrix and a norm of a vector. Here we use the maximum norm, which is the largest absolute value for a vector, $|x| = \max x(i)$. For a matrix, the maximum norm is the total of the absolute values for the matrix row that has the largest total of absolute values:

$$|A| = \max\left\{\sum_{j=1}^{m} |A(i, j)|\right\}$$

The condition number is then

$$\text{cond}(A) = |A|\,|1/A| \qquad (9\text{-}39)$$

the product of the norm of A and the norm of the inverse of A. The error in x can be shown to be

$$|dx| \le \text{cond}(A)|e| \qquad (9\text{-}40)$$

Unfortunately, the condition number can be much larger than 1, sometimes orders of magnitude larger, so that small errors (such as fluctuations in particle counts or weight errors) can be greatly magnified.

An even simpler analysis of error can be done by noting that e is multiplied by $(1/A)$, so that the uncertainty in x could be as large as the largest element in $(1/A)$ times the largest $e(i)$, in the worst case. Large elements in the inverse matrix should be a warning sign that data inversion may produce very wrong answers.

Solving the Equations when *n* Equations Exceed *m* Unknowns

Where we have more equations than unknowns, we need an added constraint to select from the possible solutions. Generally, the constraint chosen is to minimize the sum of the squared differences between what is calculated from Ax and what was measured, b. The least-squares criterion can be shown to lead to the following equation:

$$x \simeq (1/A^{\mathrm{T}}A)(A^{\mathrm{T}}b) - (1/A^{\mathrm{T}}A)(A^{\mathrm{T}}e) \quad (9\text{-}41)$$

where, again, the error term cannot be calculated because e is unknown. The approximate solution for x is obtained by finding the matrix that is the product of A^{T} (transpose) and A, then taking its inverse and using the inverse to operate on the vector that comes from A^{T} operating on b.

This is a relatively convenient approach, because it comes down to using the readily available programs for multiple linear regression to solve the equations $Ax = b$. Simply use the multiple regression programs, giving the data, $b(i)$, and the $A(i, j)$ coefficients, and the solution will be the least-squares fit, $x(j)$.

The difficulties arise when the answers are unrealistic: negative values for some of the size distribution components or components that oscillate as particle size increases. To get around this problem, one can try to solve for fewer unknowns by reformulating the equations. For example, $x(k)$ and $x(k + 1)$ can be combined to form a new $x(k)$, for two or more of the adjacent intervals, and the set of equations solved again.

Inversion as Regression

Once it is recognized that the solution of the linear equations is simply linear regression, widely studied by statisticians and mathematicians, then a deep and broad literature opens up. There are many books on regression, and various computer programs available, such as those from the SAS Institute, Inc. (Cary, NC). An excellent source is the book by Draper and Smith (1981). Many candidate methods are revealed; for example: (1) one can choose to select only those $x(i)$ values that are statistically significantly different from zero (Student's t-test on the regression coefficients); (2) one can carry out step-wise regression, starting by keeping only the most statistically significant component or starting by dropping the least significant component; (3) to the regression equations can also be added confidence intervals on the true means for the coefficients (the $x(i)$) and on individual values.

The condition number can be orders of magnitude larger than 1. Most multiple linear regression programs will not indicate what it is. To estimate the error magnification, simply make a small change in the data for each channel, one after another, and look at the change in the computed results for x.

Some Guidelines on Reducing Sensitivity to Errors

Designing an experiment to reduce problems with data inversion can be done by calculating the condition number for various options (Cooper 1974; Jochum, Jochum, and Kowalski 1981; Yu 1983; Farzanah et al. 1985) or inspecting the inversion matrix, $(1/A^{\mathrm{T}}A)A^{\mathrm{T}}$, for particularly large coefficients (Cooper and Wu 1990).

Even without calculating an inverse, one can examine the original matrix. The best matrices to invert are those which are most nearly like the identity matrix, which has 1 on its upper left to lower right diagonal and 0 elsewhere.

Methods That Change the Response Function: Smoothing, Regularization

Wolfenbarger and Seinfeld (1990, 1991) favored the technique of "regularization," associated with Tikhonov (1963). Related books are those by Tikhonov and Arsenin (1977) and Groetsch (1984). Regularization solves a family of problems that are almost the same as the original problem, with the degree of difference specified by the investigator to reflect the uncertainty in the data. The modified problems are chosen to have stable inverses. Phillips (1962) and Twomey (1963) applied this approach to problems in which the solution was not constrained to be nonnegative but was constrained in other ways, such as the smoothest solution fitting the data within a specified degree of precision. Wolfenbarger and Seinfeld applied regularization and the constraint that the solutions be nonnegative. Wolfenbarger and Seinfeld noted, as others have, that the results can be quite misleading if the device has poor sensitivity or resolution in size ranges where the size distribution is still significant. This is one advantage for combining the results of more than one type of instrument (Wu, Cooper, and Miller 1989).

Defining a level of smoothness to rule out very unsmooth (highly oscillatory) solutions has its complications. Wahba (1985) discussed this in the context of the generalized cross-validation (GCV) inversion method. Wahba (1977) presented suggestions for handling the difficulties caused by noisy data.

Methods that Change the Data

As with any signal, the data can be decomposed by Fourier analysis into a sample from a weighted sum of sine waves, $\sin(2\pi f d)$, where f is a frequency in units of reciprocal length, such as μm^{-1}. High-frequency components are those which vary greatly with small changes in particle diameter. High-frequency components in the data become high-frequency components in the results. By averaging the data over several runs and then carrying out the inversion on the averages, some reduction in these components can be achieved. Fitting curves to the data and then inverting the values interpolated by the curves is a variation on this theme. There should be some improvement, and one is not requiring that the answer look plausible, so that if the data are still poor, an implausible answer can result and warn the user of the method. Markowski (1987) used smoothing the data to improve the Twomey (1975) nonlinear method.

Methods that Constrain the Answer

One approach to preventing unrealistic solutions is to constrain the solutions. A simple constraint is normalization: The sum of the size distribution contributions must be the total number or mass or add to 1.0, etc. Another constraint is nonnegativity, not allowing a negative solution. These two were applied, using nonlinear programming (simplex minimization) by Cooper and Spielman (1976) for the variable-slit impactor and by Kapadia (1980) and Helsper et al. (1982) for the electrical aerosol analyzer. Removing the high-frequency solution components can be used to produce smoother solutions: Twomey (1965) set lower limits on eigenvalues and used these to drop eigenvectors that contributed to the solution vector, a method related to smoothing to avoid highly oscillatory solutions that are clearly nonphysical. Similar work was presented by Baker et al. (1964) but neither they nor Twomey seem to have been aware of the other's publications. Twomey (1975) achieved nonnegativity in the solutions with his nonlinear method of inversion (see the criticism by Cooper and Wu 1990, however). Crump and Seinfeld (1982) demonstrated shortcomings in Twomey's approach and presented their own, involving constraints and "cross-validation," later found to work in some cases and not in others. Wolfenbarger and Seinfeld pointed out the weaknesses in other approaches and incorporated nonnegativity and smoothness in their regularization methods. Constraining the solution helps to reject implausible solutions, but may also hide the fact that the solutions are still wrong (e.g., good and bad solutions in Fig. 1 in Wolfenbarger and Seinfeld 1991).

A particularly constrained method is to choose to fit a preselected functional form to the data, such as a set of normal distributions

(Jaenicke 1972) or one or more lognormal distributions (Kubie 1971; Puttock 1981). Such approaches give plausible results, sometimes with answers extending beyond the range of the data, which can be misleading.

Extreme value estimation (EVE) tries to put bounds on the upper and lower values of the size distribution elements that are consistent with the data. H. Paatero of the University of Helsinki claimed some success with this approach (Paatero and Raunemaa 1989), noting that it was originally set forth by Replogle, Holcomb, and Burrus at a time when the computations were much less convenient than they are now, which may explain why it was not widely used immediately thereafter.

Further Reading

The expectation-minimization (EM) algorithm has had some success (Maher and Laird 1985) in inverting diffusion battery data, a challenging task. The method takes advantage of relations among columns as well as rows in the matrices to be inverted and of the Poisson nature of particle counts, with the errors proportional to the square root of the counts.

A recent application of deconvolution was the extraction of size distribution values from photon correlation data (Bertero et al. 1989), the response of particles undergoing Brownian motion to illumination with coherent (laser) light. The inversion is equivalent to the difficult problem of inverting the Laplace transform numerically, known to be highly sensitive to errors in the data.

Some authors have focused on rewriting the equations to estimate the various integral moments of the size distributions, the mean, variance, kurtosis, etc. Unfortunately, as shown by White (1990), some distribution types cannot be recovered from these moments.

References

Aitchison, J. and J. A. C. Brown. 1957. *The Lognormal Distribution.* Cambridge: Cambridge University Press.

Baker, H., M. R. Gordon, and O. B. Brown. 1972. Theory of coincidence and simple practical methods of coincidence count correction for optical and resistive pulse particle counters. *Rev. Sci. Instr.* 43:1407–12.

Baker, C. T. H., L. Fox, D. F. Mayers, and K. Wright. 1964. Numerical solution of Fredholm integral equations of the first kind. *Comput. J.* 7:141–48.

Bertero, M., P. Boccacci, C. De Mol, and E. R. Pike. 1989. Extraction of polydispersity information from photon correlation data. *J. Aerosol Sci.* 20(1):91–99.

Box, G. E. P., W. G. Hunter, and J. S. Hunter. 1978. *Statistics for Experimenters.* New York: Wiley.

Cleveland, W. S. 1985. *The Elements of Graphing Data.* Monterey, California: Wadsworth Advanced Books.

Cooper, D. W. 1974. The variable-slit impactor and aerosol size distribution analysis. Ph.D. Dissertation, Harvard Division of Engineering and Applied Physics, Cambridge, MA.

Cooper, D. W. 1990. On distinguishing between two aerosol size distributions. Report RC16377, IBM T.J. Watson Research Laboratory, Yorktown Heights, NY.

Cooper, D. W. 1991. Applying a simple statistical test to compare two different particle counts. *J. Aerosol Sci.* 22:773–77.

Cooper, D. W. and R. J. Miller. 1987. Analysis of coincidence losses for a monitor of particle contamination on surfaces. *J. Electrochem. Soc.* 134:2871–75.

Cooper, D. W. and A. Neukermans. 1991. Estimating an instrument's counting efficiency by repeated counts on one sample. *J. Colloid Interface Sci.* 147:98–102.

Cooper, D. W. and H. R. Rottmann. 1988. Particle sizing from disk images by counting contiguous grid squares or vidicon pixels. *J. Colloid Interface Sci.* 126(1):251–59.

Cooper, D. W. and L. A. Spielman. 1976. Data inversion using nonlinear programming with physical constraints: aerosol size distribution measurement by impactors. *Atmos. Environ.* 10:723–29.

Cooper, D. W. and J. J. Wu. 1990. The inversion matrix and error estimation in data inversion: Application to diffusion battery measurements. *J. Aerosol Sci.* 21(2):217–26.

Crow, E. L. and K. Shimizu. 1988. *Lognormal Distributions.* New York: Marcel Dekker.

Crump, J. G. and J. H. Seinfeld. 1982. A new algorithm for inversion of aerosol size distribution data. *Aerosol Sci. Technol.* 1:15–34.

Draper, N. R. and H. Smith. 1981. *Applied Regression Analysis*, 2nd edn. New York: Wiley.

Epstein, B. 1947. The mathematical description of certain breakage mechanisms leading to the logarithmic-normal distribution. *J. Franklin Inst.* 244:471–77.

Evans, J. S., D. W. Cooper, and P. Kinney. 1984. On the propagation of error in air pollution measurements. *Environ. Mon. Assess.* 4:139–53.

Farzanah, F. F., C. R. Kaplan, P. Y. Yu, J. Hong, and J. W. Gentry. 1985. Condition numbers as criteria for evaluation of atmospheric aerosol measurement techniques. *Environ. Sci. Technol.* 19(2):121–26.

Forsythe, G. E. and C. B. Moler. 1967. *Computer Solution of Linear Algebraic Systems*. Englewood Cliffs, New Jersey: Prentice-Hall.

Friedlander, S. K. 1977. *Smoke, Dust and Haze*. New York: Wiley.

Fuchs, N. A. 1964. *Mechanics of Aerosols*. New York: Pergamon.

Groetsch, C. W. 1984. *The Theory of Tikhonov Regularization for Fredholm Equations of the First Kind*. London: Pitman.

Hays, W. L. and R. L. Winkler. 1970. *Statistics*. New York: Holt, Rinehart and Winston.

Heitbrink, W., P. A. Baron, and K. Willeke. 1991. Coincidence in time-of-flight aerosol spectrometers: Phantom particle creation. *Aerosol Sci. Technol.* 14:112–26.

Helsper, C., H. Fissan, A. Kapadia, and B. Y. H. Liu. 1982. Data inversion by simplex minimization for the electrical aerosol analyzer. *Aerosol Sci. Technol.* 1:135–46.

Hinds, W. C. 1982. *Aerosol Technology*. New York: Wiley.

Jaenicke, R. 1972. The optical particle counter: Cross-sensitivity and coincidence. *J. Aerosol Sci.* 3:95–111.

Jochum, C., P. Jochum, and B. R. Kowalski. 1981. Error propagation and optimal performance in multicomponent analysis. *Anal. Chem.* 53:85–92.

Julanov, Yu. V., A. A. Lushnikov, and I. A. Nevskii. 1984. Statistics of multiple counting in aerosol counters. *J. Aerosol Sci.* 15(1):69–79.

Julanov, Yu. V., A. A. Lushnikov, and I. A. Nevskii. 1986. Statistics of multiple counting in aerosol counters—II. *J. Aerosol Sci.* 17(1):87–93.

Kapadia, A. 1980. Ph.D. Thesis, Department of Mechanical Engineering, University of Minnesota, Minneapolis, MN.

Kolmogorov, A. N. 1941. Über das logarithmisch normale Verteilungsgesetz der Dimensionen der Teilchen bei Zerstückelung, *C.R. Acad. Sci. (Doklady) URSS* XXXI:99–101.

Kubie, G. 1971. A note on the treatment of impactor data for some aerosols. *J. Aerosol Sci.* 2:23–30.

Maher, E. F. and N. M. Laird. 1985. EM algorithm reconstruction of particle size distributions from diffusion battery data. *J. Aerosol Sci.* 16:557–70.

Mandelbrot, B. 1977. *The Fractal Geometry of Nature*, New York: Freeman.

Markowski, G. R. 1987. Improving Twomey's algorithm for inversion of aerosol measurement data. *Aerosol Sci. Technol.* 7:127–41.

Meyer, S. L. 1975. *Data Analysis for Scientists and Engineers*. New York: Wiley.

Noble, B. 1969. *Applied Linear Algebra*. Englewood Cliffs, New Jersey: Prentice-Hall.

Paatero, P. and T. Raunemaa. 1989. Analysis of CO_2 thermograms by the new extreme-value estimation (EVE) deconvolution principle. *Aerosol Sci. Technol.* 10:365–69.

Pecen, J., A. Neukermans, G. Kren, and L. Galbraith. 1987. Counting errors in particulate detection on unpatterned wafers. *Solid State Technol.* 30(5):149–54.

Phillips, D. L. 1962. A technique for the numerical solution of certain integral equations of the first kind. *J. Assoc. Comput. Mach.* 9:84–97.

Puttock, J. S. 1981. Data inversion for cascade impactors: Fitting sums of lognormal distributions. *Atmos. Environ.* 15(9):1710–16.

Raasch, J. and H. Umhauer. 1984. Error in the determination of particle size distributions caused by coincidences in optical particle counters. *Part. Charact.* 1:53–58.

Reist, P. C. 1984. *Introduction to Aerosol Science*. New York: Macmillan.

Shen, A.-T. and T. A. Ring. 1986. Distinguishing between two aerosol size distributions. *Aerosol Sci. Technol.* 5:477–82.

Tikhonov, A. N. 1963. On the solution of incorrectly posed problems and the method of regularization. *Soviet Math.* 4:1035–38.

Tikhonov, A. N. and V. Y. Arsenin. 1977. *Solutions of Ill-Posed Problems*. Washington, DC: V.H. Winston and Sons.

Twomey, S. 1963. On the numerical solution of Fredholm integral equations of the first kind by the inversion of the linear system produced by quadrature. *J. Assoc. Comput. Mach.* 10:97–101.

Twomey, S. 1965. The application of numerical filtering to the solution of integral equations encountered in indirect sensing measurements. *J. Franklin Inst.* 279:95–109.

Twomey, S. 1975. Comparison of constrained linear inversion and an iterative nonlinear algorithm applied to the indirect estimation of particle size distributions. *J. Comput. Phys.* 18:188–200.

U.S. General Services Administration. 1988. Federal Standard 209D, Federal standard: Clean room and work station requirements, controlled environment. U.S. GSA, Washington, DC. Reprinted in *J. Environ. Sci.* 31(5):53–76.

Wahba, G. 1977. Practical approximate solutions to linear operator equations when the data are noisy. *SIAM J. Numer. Anal.* 14:645–67.

Wahba, G. 1985. A comparison of GCV and GML for choosing the smoothing parameter in the generalized spline smoothing problem. *Ann. Statist.* 13:1378–1402.

White, W. H. 1990. Particle size distributions that cannot be distinguished by their integral moments. *J. Colloid Interface Sci.* 135(1):297–99.

Wolfenbarger, J. K. and J. H. Seinfeld. 1990. Inversion of aerosol size distribution data. *J. Aerosol Sci.* 21:227–47.

Wolfenbarger, J. K. and J. H. Seinfeld. 1991. Estimating the variance in solutions to the aerosol data inversion problem. *Aerosol Sci. Technol.* 14:348–57.

Wu, J. J., D. W. Cooper, and R. J. Miller. 1989. Evaluation of aerosol deconvolution algorithms for determining submicron particle size distributions. *J. Aerosol Sci.* 20(4):477–82.

Yu, P. Y. 1983. Ph.D. Dissertation, University of Maryland, College Park, MD.

II

Instrumental Techniques

10

Filter Collection

K. W. Lee and Mukund Ramamurthi

Battelle Memorial Institute
505 King Avenue
Columbus, OH, U.S.A.

INTRODUCTION

Filtration is probably the most widely utilized technique for aerosol measurement, owing to its flexibility, simplicity, and economy. The central concept in aerosol filtration is the collection, through removal from the gas phase, of a *representative* sample of the aerosol on a suitable porous medium or filter. The transfer of the aerosol from a dispersed state in the air to a compact sample on the filter then facilitates the storage, transport, and sample preparation requisite for gravimetric, microscopic, microchemical, or other analytical techniques. The term "air" is used interchangeably here with "gas" to refer to the gaseous phase in which the aerosol is dispersed.

The prime objective in selecting an aerosol filtration scheme appropriate to a study is to ensure that as representative a sample of the aerosol as possible is presented to the applied analytical technique. The various aerosol filtration topics discussed in this chapter are, thus, all designed to assist the reader in developing filtration methods that would best meet this goal in specific applications. In subsequent sections of this chapter, filter sampling principles are presented, followed by a review of the various types of available aerosol measurement filters, the theoretical principles for estimating filtration efficiency and pressure drop, and the salient features and requirements of common filter analysis methods that influence filter selection.

The reader should note that the reference literature on filtration of aerosol measurement is extensive, if somewhat scattered. Several excellent and comprehensive reviews can be found, including one by Lippmann (1989*a*), in addition to numerous reports in the aerosol literature on utilizing filter methods for aerosol measurement. Subsequent chapters in this book, particularly those in Part III also provide practical examples of filter-based measurement techniques that have been perfected for various applications. Thus, the aerosol measurement practitioner would be well advised to survey the literature for references to filtration techniques utilized previously for similar applications before proceeding with a filter system design.

GENERAL PRINCIPLES OF FILTER SAMPLING

The essential components of a filter sampling system for aerosol measurement are shown in Fig. 10-1. Typically, aerosol-laden air is

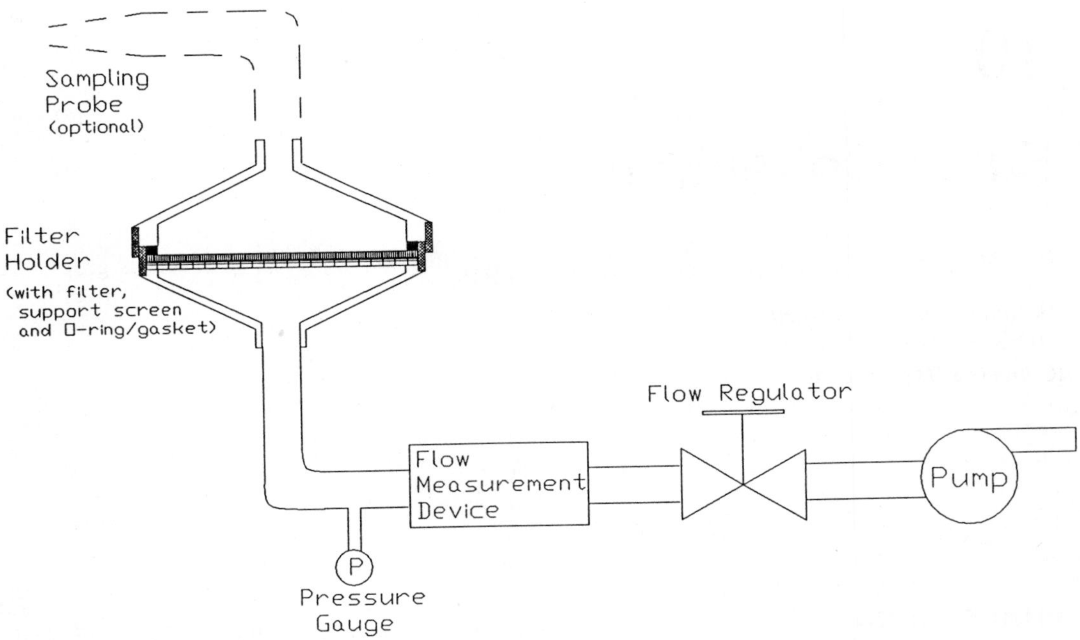

FIGURE 10-1. Schematic Example of a Filter Collection System Typically Used for Aerosol Measurement.

drawn through a sampling probe for isokinetic sampling from a flow stream into a filter holder containing an appropriately selected filter medium. Here, the aerosol is separated from the flow stream to the extent dictated by the characteristics of the filter, the air velocity through the filter, and other factors such as the particulate loading on the filter. The air drawn through the filter flows to an optional flow measurement device such as a rotameter, mass flow meter, or dry test meter, and then into a flow regulating mechanism, such as an orifice or valve coupled with an air moving device or pump. Selecting the appropriate components and the optimum order for the flow progression through the system is crucial in achieving a representative sample of the aerosol on the filter. The sampling probe upstream of the filter holder is required when sampling from moving air streams. In such applications, the volumetric sampling rate and the cross-sectional area of the sampling probe nozzle determine the air velocity through the probe inlet. The combination of an air inlet velocity equal to that of the air stream in the vicinity of the inlet, and the alignment of the probe parallel to the gas streamlines ensures isokinetic sampling of the aerosol. The importance of isokinetic sampling in assuring that a representative aerosol sample is extracted from the air stream is discussed in greater detail in Chapter 6.

The effectiveness with which the sampling probe extracts the aerosol from a moving air stream is termed the "aspiration efficiency" of the inlet. An excellent treatment of the various factors affecting inlet aspiration efficiency is provided by Davies (1968). In addition, Vincent (1989, 86–137) presents a detailed discussion of the aspiration efficiencies of thin-walled and blunt samplers in moving air streams. Sampling from still air using an inlet probe also introduces possibilities of sampling biases, that are addressed by Davies (1968), Agarwal and Liu (1980), and Vincent (1989, 144–64). In addition to the limitations imposed by aerosol dynamical behavior on inlet aspiration efficiency, particle losses due to temperature gradients between the probe inlet surfaces and the air stream, and electrostatic deposition losses due to charged probe inlet surfaces, particularly in the case of those made with plastic materials, must also be considered.

In many instances, the geometry of the sampling environment necessitates a length of connecting tubing between the inlet probe and the filter holder. The passage of the aerosol from the inlet to the filter holder is characterized by the "transport efficiency" and is influenced by a number of factors such as gravitational, diffusional, and inertial deposition onto the wall surfaces, as well as temperature-gradient-related and electrostatic wall loss effects. These issues are discussed in Chapter 6 on the sampling and transport of aerosols. It is critical that the aspiration and transport efficiencies of a filter sampling system be determined in each application to adequately characterize any distortion in the aerosol size distribution or concentration obtained on the sampling filter. Maximizing these efficiencies is particularly critical due to their inherent particle size dependence, making corrections for sampling biases extremely difficult if the aerosol size distribution is unknown. In general, sampling losses are minimized by locating the filter holder as close as possible to the sampling inlet probe, and if possible, sampling with no inlet probe or connecting tubing upstream of the filter.

Filter holders provide the means by which the filtration media can be supported, typically on a coarse wire screen or backup medium, and held in a positive seal to constrain the air drawn by the sampling pump to pass through the filter. This seal is generally obtained through the use of an O-ring or a gasket of appropriate form and material that does not damage the filter material—Teflon gaskets are increasingly popular for this application due to their inherently low adhesivity to filter surfaces. Filter holders are available for all common filter sizes in the range 13–47 mm in diameter, as well as in 8 × 10 in sizes for Hi-Volume ambient air quality sampling applications. A compilation of the wide spectrum of commercially available filter holders is presented by Lippmann (1989*a*).

Filter holders may be open-faced, in-line, or of a cassette variety, the latter commonly used in industrial hygiene sampling applications, as shown in Fig. 10-2. In general, open-faced filter holders provide an assurance of uniform filter deposition and lower sampling losses in the inlet. However, closed-face or in-line filter holders are necessary for

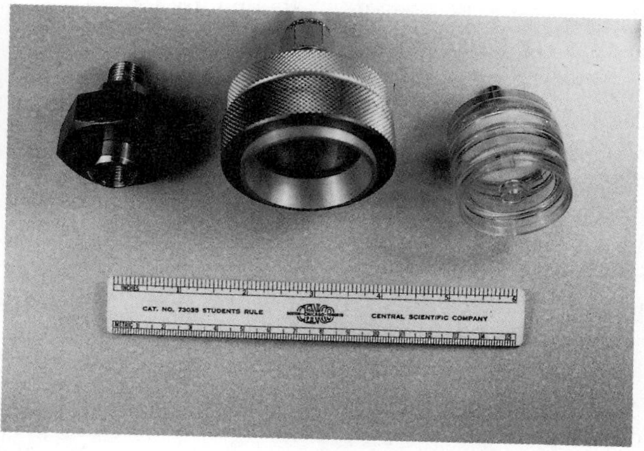

FIGURE 10-2. Examples of Filter Holders Available Commercially for Aerosol Collection (from left): (a) In-Line, 25 mm, (b) Open-Face, 47 mm, and (c) Cassette Style, 25 mm (Closed-Face Shown; Can also be Used as Open-Faced). (Photograph Courtesy: Carolyn Dye, Battelle Memorial Institute.)

probe-based sampling applications and also protect filter media from possible mechanical damage and rupture during sampling. In-line filter holders generally use a gradual expansion from the inlet to the filter surface, as well as downstream to the outlet, so as to ensure uniform air velocities over the cross section of the filter. This enables the collection of a uniform aerosol deposit on the filter surface, which is critical for subsequent analyses that may utilize only a fraction of the filter.

The use of a filter holder for aerosol collection on a filter can introduce the potential for electrostatic and diffusional aerosol deposition losses onto filter holder inlet surfaces, as well as for aerosol transformations from condensation or evaporation induced by temperature gradients present between the air stream and the filter holder. These factors must be considered in designing aerosol measurement filtration systems for specific applications. The techniques utilized in the field to overcome these potential problems include the use of a controlled-temperature enclosure for housing the filter holder, thereby minimizing temperature gradients.

The most critical function of a filter holder—that of ensuring a positive seal around the circumference of the filter—is typically the most frequent source of errors in filter sampling. Such sampling errors result from failures in the outer seal, thus permitting air to flow around the filter. Seal integrity and leak testing procedures must be employed to ensure that the sampled air is drawn only through the filter in the filter holder. Among the techniques that may be applied for this purpose are: (1) presampling positive pressurization tests conducted with the filter holder assembly capped-off appropriately, and (2) the use of a pressure gauge to monitor the vacuum pressure in the sampling line connecting the filter holder to the pump during sampling; leaks and ruptures in the filter are then easily detected from any drop in the vacuum, with normal operation indicated by a steady or gradually increasing vacuum pressure.

A wide spectrum of filters is available commercially for aerosol measurement applications, providing the user with a selection of filter materials, pore sizes, and collection characteristics, as well as a variety of shapes such as discs and sheets sized to fit commonly available filter holders. The parameters of importance in selecting a particular type of filter for an application generally include:

- collection efficiency of the filter for the aerosol size distribution to be sampled,
- pressure drop across the filter in relation to the air volumetric throughput required,
- compatibility of the filter with the sampling conditions, use and handling procedures, and analytical method to be employed. This includes issues of potential artifact formation on the filter surfaces from chemical reactions, and interferences such as would occur from the use of hygroscopic filters in the gravimetric analysis of aerosols,
- cost constraints relating to the size of the sampling effort and the number of filters required.

A more detailed discussion of the various types of filters and their collection efficiency and pressure drop characteristics is presented subsequently. Also discussed are several examples of compatibility problems typically encountered during aerosol sampling and alternatives developed to circumvent these potential problems. A comprehensive catalog of commercially available filters is provided by Lippmann (1989*a*), including a compilation of filter manufacturer trade names and corresponding filter characteristics such as pore size, thickness, pressure drop, ash content, weight per unit area, maximum operating temperature, tensile strength, and refractive index. In a subsequent section of this chapter, a useful compilation of commonly utilized filter types for various applications is presented, together with the relative advantages and limitations of each category of filters.

The use of appropriate filter handling procedures is important to assure the collection of a representative aerosol sample. The procedures range from filter seal and surface integrity testing, to minimizing interferences and artifacts from such factors as chemical reactions/transformations on the filter surface, electrostatic charging of filters, and moisture uptake in filter materials and hygroscopic aerosols. These and other procedures are discussed subsequently in connection with the requirements of filter analysis methods.

Finally, the measurement and control of the air flow rate through the filter is of as much importance as the collection of a representative aerosol sample on the filter, since the air flow rate or cumulative volume through the filter is necessary to calculate the aerosol concentration from the sample. A detailed discussion of the various alternatives available for use as flow regulating and/or measuring devices and as pumps or air movers is presented by Rubow and Furtado (1989). In addition, established scientific procedures are available for the calibration of various types of flow measurement devices as outlined, for example, by Lippmann (1989*b*). A well-designed filter collection system for aerosol measurement thus involves the collection of a representative aerosol sample on a suitable filter, combined with an accurate knowledge of the air flow rate or cumulative air volume transported through the filter.

AEROSOL MEASUREMENT FILTERS

A logical way to classify aerosol measurement filters is to divide them by their characteristic

TABLE 10-1 Summary of the Salient Characteristics of the Various Types of Filters Commonly Utilized for Aerosol Measurement

Filter Type	Characteristics
Fibrous filters	• Mat/weave of fibers with diameters of 0.1–100 μm. Cellulose or wood (paper), glass, quartz, and polymer fiber filters are available • Porosities of 60%–99%, thicknesses of 0.15–0.5 mm. • Particle collection is throughout the depth of the filter from interception, impaction, and diffusion onto fibers • High particle collection efficiencies require low air velocities. • Pressure drops are the lowest among all filters under comparable conditions
Porous-membrane filters	• Microporous membranes with tortuous pores throughout the structure • Polymer, sintered metal, and ceramic microporous filters available • Pore sizes (determined from liquid filtration) in the range 0.02–10 μm • Porosities of < 85% and thicknesses of 0.05–0.2 mm • Particle collection through attachment to microstructure elements • High collection efficiencies, but highest pressure drop among all filters
Straight-through pore filters	• Thin polycarbonate films (10 μm) with cylindrical pores perpendicular to film surface, with diameters in the range 0.1–8 μm • Porosities are low, in the range 5%–10% • Particle collection through impaction and interception near the pores and diffusion to tube walls of pores • Collection efficiencies are intermediate between fibrous and microporous membrane • Pressure drops are significantly higher than fibrous filters and comparable or higher than microporous membrane filters for equivalent collection efficiency
Granular-bed filters	• For special sampling, granules of specialty chemicals, sugar, naphthalene, sand, metal, glass beads are used • Samples are recovered by washing or volatilization • Granular bead sizes range from 200 μm to a few mm • Filtration is achieved by impaction, interception, diffusion, and gravitation • Filter porosities of 40%–60% for stationary beds • Low collection efficiency due to large granule size. To enhance diffusion low flow is used; bed depth is increased or smaller granules are used

structure. In general, filters used for aerosol sampling may be classified as fibrous filters, porous-membrane filters, straight-through pore membrane filters, and granular-bed filters, according to their structure. The terms, porous-membrane and straight-through pore membrane filters are very similar to those suggested by Hinds (1982) and discussions here closely follow his classification and the discussion provided by Lippmann (1989). A summary of the salient characteristics of each type of filter is also provided in Table 10-1, complementing the discussion that follows.

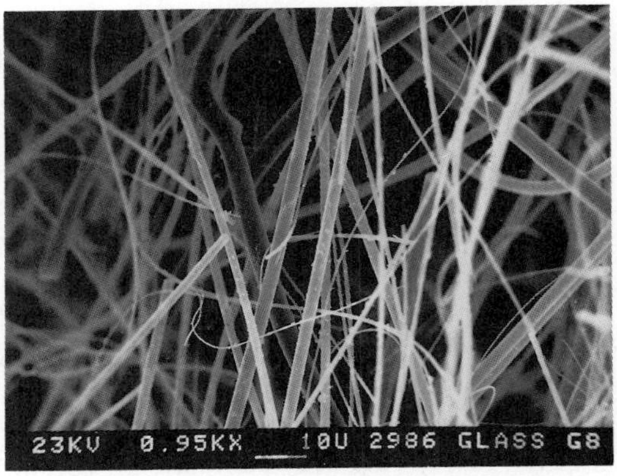

FIGURE 10-3. Electronmicrograph Showing the Typical Microstructure of a Glass Fiber Filter (Gelman Type A/E) (Scale Bar Shown on the Micrograph).

Fibrous Filters

Fibrous filters consist of a mat of individual fibers. Generally, filter porosity is relatively high ranging from about 0.6 to 0.999. Porosities of less than 0.6 are not typically found in fibrous filters because of the difficulties in effectively compressing the component fibers into a smaller thin layer. Fiber sizes range from less than 1 μm to several hundred micrometers. The range of fiber diameters for a given filter is usually very diverse, although some types of fibrous filters may consist of fibers of a uniform size. Often, the filters are fabricated using a binder material to hold individual fibers together. The mass for the binder material can be as high as 10% of the filter material. Binder-free filters are generally selected for aerosol measurement because of the artifacts and interferences caused by the presence of the organic binder in the filter. The materials used for fibrous filters include cellulose, glass, quartz, and plastic fibers. Sometimes, mixed fibers of cellulose, asbestos, and glass are also used as filters for certain low-cost qualitative sampling applications. Figure 10-3 shows the microstructure of a glass fiber filter and reveals the fibrous, mat-like nature of the filter material.

Cellulose fiber (paper) filters were used once very widely for general-purpose air sampling. Whatman filters are one of the most representative filters in this category. The filters are inexpensive, come in various sizes and have good mechanical strength and low pressure drop characteristics. Some of the critical limitations of cellulose fiber filters are their moisture sensitivity and relatively low filtration efficiency for submicrometer particles.

Glass fiber filters typically have a higher pressure drop than paper filters and often provide filtration efficiencies of greater than 99% for particles > 0.3 μm. The filters are more expensive than paper filters. However, glass fiber filters are less affected by moisture than cellulose fiber filters. Glass fiber filters are widely used as the standard filter media for high-volume air sampling. Teflon-coated glass fiber filters overcome some of the inherent inadequacies of glass fiber filters by being inert to catalyzing chemical transformations as well as by being less moisture-sensitive.

Quartz fiber filters are commonly used in high-volume air sampling applications involving subsequent chemical analyses such as atomic absorption, ion chromatography, and carbon analysis, due to their low trace contamination levels, as well as their relative inertness and ability to be baked at high temperatures to remove trace organic contaminants.

Polystyrene fiber filters have been used for sampling purposes to a limited extent. These filters have poorer mechanical strength than cellulose filters. However, their filtration efficiency is comparable to that of glass fiber filters. Other plastic materials used in filters include polyvinyl chloride and dacron. For special applications involving high temperatures and corrosive environments, filters made of stainless steel fibers have also been recently introduced.

Porous-Membrane Filters

A variety of membrane filters made of cellulose esters, polyvinyl chloride, Teflon, and sintered metals are commercially available. Membrane filters are gels formed from a colloidal solution and have a very complicated and uniform microstructure providing a tortuous or irregular air flow path. Often, the complex filter structure consists of a series of layers formed by different processes, depending upon the manufacturing technique. The pore size provided by manufacturers often do not match any of the physical filter pores or structural characteristics and are defined from liquid filtration. In general, the pressure drop and the particle collection efficiency are very high, even for particles significantly smaller than the characteristic pore size. Particles are captured by the surfaces provided by the filter structure, principally by Brownian motion and inertial impaction mechanisms. An example of a microporous membrane filter structure is shown in Fig. 10-4.

Straight-Through Pore Membrane Filters

This type of filter consists of a polycarbonate membrane with straight-through pores of a uniform size. They are very often called Nuclepore filters after their original manufacturer, Nuclepore Corp. (Pleasanton, CA), although they are also currently manufactured by Poretics Corp. (Livermore, CA). The filters are manufactured by subjecting polycarbonate membranes to neutron bombardment, followed by an etching process, that produces uniform-sized pores in the membrane. The number of pores is controlled by the bombardment time and the pore size is determined by the etching time. Capillary pore membrane filters have a very different and simpler structure compared to porous-membrane filters. They consist of a very smooth and translucent surface with straight-through capillary holes across the membrane structure. Figure 10-5 shows the microstructure of a Nuclepore filter and reveals the flat surface and uniform pores in the filter. Straight-through pore membrane filters are widely

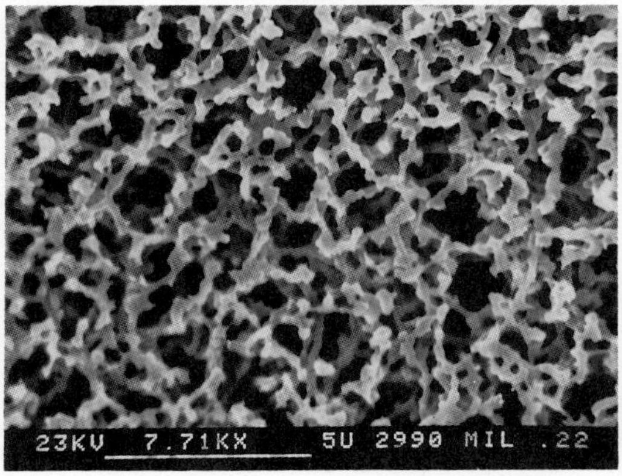

FIGURE 10-4. Electronmicrograph of a Microporous Membrane Filter (Millipore 0.22 μm Pore Size) Showing the Tortuous Flow Path and Structural Elements in the Filter (Scale Bar Shown on Micrograph).

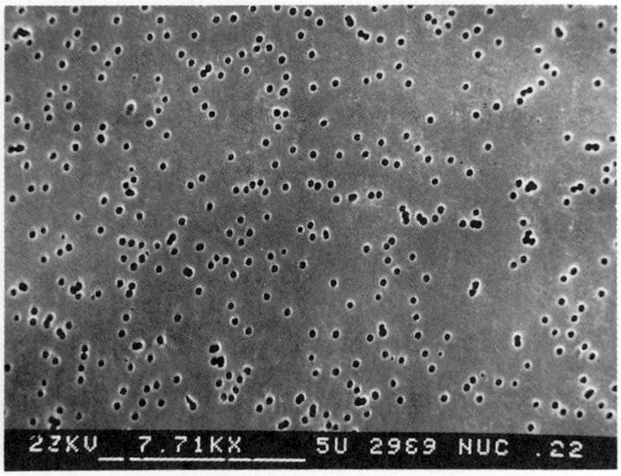

FIGURE 10-5. Electronmicrograph of a Straight-Through Pore Membrane Filter (Nuclepore 0.22 μm Pore Size) Showing the Smooth Surface and Uniform Pores in the Filter (Scale Bar Shown on Micrograph). (Micrographs Courtesy: Mark Pence, Battelle Memorial Institute.)

used for particle analysis using surface analytical techniques such as light and electron microscopy.

Recently, Mori, Emi, and Otani (1991) examined various membrane-type filters and proposed to classify both porous membrane and straight-through pore filters into five new groups according to the appearance of the filter structure. The five groups proposed are (1) random directional fiber-shaped filters, (2) uni-directional fiber-shaped filters, (3) net-shaped filters, (4) agglomerate-shaped filters, and (5) pore-shaped filters. This scheme of classification may have some utility in cataloguing the spectrum of membrane filters available.

FIGURE 10-6. Optical Microscope Photograph of Two Different Types of Granular-Bed Filtration Material. Florisil (left) and Tenax (right). (Millimeter Scale also Shown.)

Granular-Bed Filters

For specialized applications, aerosol sampling may be performed using a granular- or packed-bed filter. Filtration is accomplished by passing particle-laden air through a bed consisting of granules and recovering aerosols afterwards by extraction procedures. A major advantage of aerosol sampling by granular beds is that by selecting the proper bed media, both particulate and gaseous pollutants can be simultaneously collected (Kogan et al. 1991). Possibilities for using this method at high temperature and pressures also make this filter type attractive. Granules of activated charcoal, XAD-2, Florisil, Tenax, sugar, naphthalene, glass, sand, quartz, and metal beads have been used. XAD-2 and Tenax are polymeric adsorbent resins made of spherical granules and Florisil is a magnesia–silica material. Figure 10-6 shows two different types of granular aerosol collection media, Florisil and Tenax. Aerosols are usually recovered from the granular media for

TABLE 10-2 Commercial Sources and Typical Cost Ranges for Fibrous Aerosol Masurement Filters

Filter Material(s)	Variety of Characteristics Available	Commercial Sources	Typical Cost Range[1]
Cellulose fiber	• Fine, medium, coarse void sizes	WHA MSI MFS S&S	$10–15/100
Borosilicate glass fiber	• With organic binder • Without organic binder • With teflon coating • With cellulose content	WHA MSA GEL MIL H&V PAL NUC MFS, MSI, S&S	$10–20/100 /100 (Teflon-coated)
Quartz fiber	• Pure • With small percentage borosilicate glass	WHA MFS GEL	$40–55/100
Polymer fiber	• Polypropylene 0.45, 0.8, 1.2 μm uniform pore sizes	MSI	$20/100

1. Costs are for 47 mm diameter discs and are taken from 1990–1991 manufacturer's catalogs

TABLE 10-3 Commercial Sources and Typical Cost Ranges for Microporous Membrane Aerosol Measurement Filters

Filter Materials	Variety of Characteristics Available	Commercial Sources	Typical Cost Range[1]
Cellulose membrane	• Cellulose nitrate • Mixed ester • Cellulose acetate 0.02–8 μm pore size	MIL GEL SAR NUC MFS, MSI S&S, WHA NAL	\$50–125/100[a]
Teflon/PTFE membrane	• Pure • Polypropylene reinforced 0.2–1.0 μm pore size	Same as above	\$150–200/100[a]
Polyester/Polycarbonate/ Polypropylene membranes	0.02–15 μm pore sizes	Same as above	\$50–150/100[a]
Silver membrane	• Pure metallic silver 0.2–5 μm	POR FSI MIL NUC	\$250–300/100[b]
Nylon membrane	• Pure nylon • Laminated or impregnated with polypropylene for support 0.1–20 μm pore size	GEL MSI SAR	\$150–200/100[a]
Polyvinyl chloride membrane	• Pure PVC • PVC with acrylonitrile 0.45, 0.8, 5.0 μm pore sizes	GEL NUC	\$100–150/100[a]

1. Costs are taken from 1990–1991 manufacturers' catalogs
a. 47 mm diameter discs
b. 25 mm diameter discs

TABLE 10-4 Commercial Sources and Typical Cost Ranges for Straight-Through Pore Membrane Aerosol Measurement Filters

Filter Material	Variety of Characteristics Available	Commercial Sources	Typical Cost Range[1]
Polycarbonate	0.01–14 μm pore size (uniform with ± 15%) 6–10 μm thickness 5%–10% porosity	NUC POR	\$75–175/100
Polyester	0.1–12.0 μm pore sizes	NUC	\$75–100/100

1. Costs are for 47 mm diameter discs and are taken from 1990–1991 manufacturer's catalogs

chemical analysis by washing, volatilization, or the use of solvents.

A wide selection of fibrous and membrane filters are available commercially for use in aerosol measurement. Tables 10-2–10-4 provide a listing of a number of different filter manufacturers for each type of filter discussed, as well as a general price range indicating the relative differences in cost between the various filters.

FILTRATION THEORY

The filtration of particles by both fibrous and membrane filters has been the subject of numerous analytical, numerical, and experimental studies during recent years. As a result, the dependence of filtration efficiency on the particle size, the filter media characteristics, and the flow velocity is now well established, both qualitatively and quantitatively.

In this section, filtration mechanisms are discussed and some useful predictive equations for filter collection efficiency and the pressure drop across a filter are introduced for practical application. Filtration theory is discussed for fibrous filters and its applicability to membrane filters is described subsequently.

The starting point in characterizing fiber filtration is to consider the capture of particles by a single fiber. The single-fiber efficiency, η, is defined as the ratio of the number of particles striking the fiber to the number which would strike if the streamlines were not diverted around the fiber. If a fiber of radius R_f removes all the particles contained in a layer of thickness Y as shown in Fig. 10-7, the single fiber efficiency, η, is then defined as Y/R_f.

The overall efficiency or the total efficiency, E, of a filter composed of many fibers in a mat can be related to the single fiber efficiency as follows:

$$E = 1 - \exp\left[\frac{-4\eta\alpha L}{\pi d_f(1-\alpha)}\right] \tag{10-1}$$

where α is the solidity or packing density of the filter (1 − porosity), L the filter depth or thickness, and d_f the fiber diameter. Equation 10-1 relates the total efficiency of a filter to the single-fiber efficiency as previously defined. Usually, Eq. 10-1 is used to calculate the single fiber efficiency from the total filter efficiency, E, which can be measured experimentally.

The advantage of using the single-fiber efficiency is that it is independent of the filter thickness, L. While it is not meaningful to compare the total efficiencies of two filters of two different thicknesses, different filters can be compared on the basis of their single-fiber efficiencies. This is an important point to consider in comparing the overall filtration efficiency of filters, since a filter with a lower single-fiber efficiency may have a higher total efficiency by virtue of being thicker.

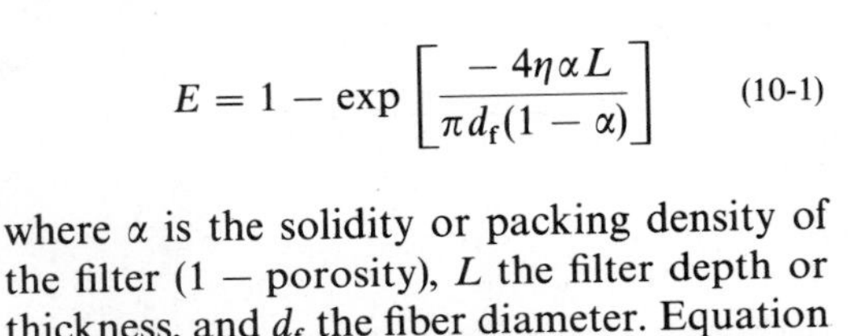

$$\eta_{total} = \frac{Y}{R_f}$$

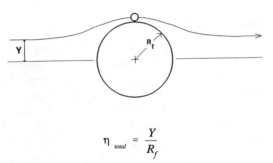

FIGURE 10-7. Definition of Single-Fiber Efficiency.

Filtration Mechanisms

As air penetrates a filter, the trajectories of particles deviate from the streamlines due to several mechanisms. As a result, particles may collide with the fiber surface and become deposited on them. The important mechanisms causing particle deposition are diffusion, inertial impaction, interception, and gravitational settling. The single-fiber efficiency, η, can then be assumed in the first approximation to be composed of the arithmetic sum of the individual efficiencies from diffusion, η_{diff}, interception η_{inter}, inertial impaction, η_{imp}, and gravitational-settling, η_{grav}, mechanisms. In addition, dendrite formation from particles collected on fibers can provide additional particle collection in filters.

Brownian Diffusion

Under normal conditions, aerosol particles undergo Brownian motion. Small particles generally do not follow the streamlines but continuously diffuse away from them. Once a particle is collected on a surface, it would adhere to it due to van der Waal's force. The particle concentration at the surface can, thus, be assumed to be zero, and the resulting concentration gradient normal to the media surface can be considered as the driving force for the diffusion of particles. Since Brownian motion generally increases with decreasing particle size, the diffusive deposition of particles is increased when the particle size is reduced. This phenomenon is illustrated in Fig. 10-8. Similarly at low flow velocities, particles can spend more time in the vicinity of the fiber surfaces, thus enhancing diffusional collection. From the convective diffusion equation describing this process, a dimensionless parameter called the Peclet

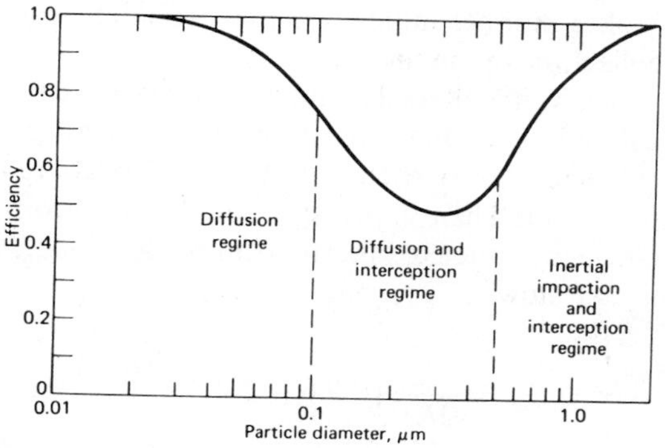

FIGURE 10-8. Schematic of Filter Efficiency vs. Particle Size Illustrating the Different Filtration Regimes.

number can be defined as

$$Pe = \frac{d_c U}{D} \tag{10-2}$$

where Pe is the Peclet number, d_c the characteristic length of collecting media, U the average air velocity inside the filter medium, and D the diffusion coefficient of the particle. For pure molecular diffusion, D can be written as

$$D = kTC/3\pi\eta d_p \tag{10-3}$$

where k is the Boltzmann constant, T the absolute temperature, η the air viscosity, d_p the particle diameter, and C_c the Cunningham slip correction. From the above discussion, particle collection by diffusion is expected to decrease with increasing Peclet number. C_c is written as

$$C_c = 1 + 2.492\frac{\lambda}{d_p} + 0.84\frac{\lambda}{d_p}\exp\left(-0.435\frac{d_p}{\lambda}\right) \tag{10-4}$$

where λ is the mean free path of the gas molecules.

The approach taken to quantify the single-fiber diffusional collection efficiency, η_{diff}, by investigators such as Friedlander (1957), Natanson (1957), and Lee and Liu (1982) is a boundary layer model that is commonly used in heat and mass transfer analysis. In early models by Friedlander and Natanson, a flow field which is an isolated cylinder model was utilized. Recent theories, such as Lee and Liu's (1982), utilize a multiple-cylinder model that takes into account the flow interference effects of neighboring fibers. Thus, these models provide a better representation of the actual flow profile in the filters. Lee and Liu's (1982) theory yields

$$\eta_{diff} = 2.58\frac{1-\alpha}{K}Pe^{-2/3} \tag{10-5}$$

where K is the hydrodynamic factor,

$$K = -\frac{1}{2}\ln\alpha - \frac{3}{4} + \alpha - \frac{\alpha^2}{4} \tag{10-6}$$

Interception

Even if the trajectory of a particle does not depart from the streamline, a particle may still be collected if the streamline brings the particle center to within one particle radius from the fiber surface. The fact that particles with a finite size can be collected even in the absence of Brownian motion and inertial impaction indicates the importance of the interception effect. One would expect the interception to be relatively independent of flow velocity for a given fiber, and this characteristic can be contrasted to the flow-dependent characteristics of diffusion and inertial impaction. The dimensionless parameter describing the interception effect is the interception parameter, R, defined as the ratio of particle diameter to fiber diameter:

$$R = \frac{d_p}{d_f} \tag{10-7}$$

If the Kuwabara flow is used, one can obtain the following expression:

$$\eta_{inter} = \frac{1+R}{2K}\left[2\ln(1+R) - 1 + \alpha + \left(\frac{1}{1+R}\right)^2\left(1-\frac{\alpha}{2}\right) - \frac{\alpha}{2}(1+R)^2\right] \tag{10-8}$$

where η_{inter} is the single-fiber efficiency due to interception. Although Eq. 10-8 is a complete expression for the interception efficiency, the form of the equation is somewhat long and it can be shown to be approximated into the following simpler form:

$$\eta_{\text{inter}} = \frac{1-\alpha}{K}\frac{R^2}{(1+R)} \tag{10-9}$$

Enhanced collection of particles by a fiber can also occur from the interception of diffusing particles and has been proposed by Stechkina, Kirsch, and Fuchs (1969). The magnitude of this additional efficiency term is of the same order as that of the errors involved in the approximation method used in the analysis. Spielman and Goren (1968) also indicate that the term is not theoretically consistent and is, consequently, not introduced here.

Inertial Impaction

The streamlines of a fluid around the fiber are curved. Particles with a finite mass and moving with the flow may not follow the streamlines exactly due to their inertia. If the curvature of a streamline is sufficiently large and the mass of a particle sufficiently high, the particle may deviate far enough from the streamline to collide with the media surface. The importance of this inertial impaction mechanism increases with increasing particle size and increasing air velocity, as shown in Fig. 10-8. Therefore, the effect of increasing air velocity on the inertial impaction of particles is contrary to that for the diffusive deposition. The inertial impaction mechanism can be studied by the use of the dimensionless Stokes number, defined as

$$Stk = \frac{C d_p^2 \rho_p U}{18 \eta d_f} \tag{10-10}$$

where ρ_p is the density of the particle. The Stokes number is the basic parameter describing the inertial impaction mechanism for particle collection in a filter. A large Stokes number implies a higher probability of collection by impaction, whereas a small Stokes number indicates a low probability of collection by impaction.

Stechkina, Kirsch, and Fuchs (1969) calculated the inertial impaction filtration efficiency for particles, using the Kuwabara flow field. Their expression for the filtration efficiency due to inertial impaction, η_{imp}, is as follows:

$$\eta_{\text{imp}} = \frac{1}{(2K)^2}[(29.6 - 28\alpha^{0.62})R^2 - 27.5R^{2.8}]Stk \tag{10-11}$$

Equation 10-11 has been used extensively for calculating the contribution by the inertial impaction mechanism.

Gravitational Settling

Particles will settle with a finite velocity in a gravitational force field. When the settling velocity is sufficiently large, the particles may deviate from the streamline. Under downward filtration conditions, this would cause an increased collection, due to gravity. When flow is upward, this mechanism causes particles to move away from the collector, resulting in a negative contribution to filtration. This mechanism is important only for particles larger than at least a few micrometers in diameter and at low flow velocities. The dimensionless parameter governing the gravitational sedimentation mechanism is

$$Gr = \frac{V_g}{U} \tag{10-12}$$

where U is the flow velocity and V_g is the settling velocity of the particle. It can be shown that the single-fiber filtration efficiency due to gravity, η_{grav}, can be approximated (Davies 1973) as

$$\eta_{\text{grav}} = \frac{Gr}{1 + Gr} \tag{10-13}$$

In filtration theories, it is common to assume that the individual filtration mechanisms discussed above are independent of each other and additive. Therefore, η, the overall single-fiber collection efficiency in Eq. 10-1 can be written as the sum of individual single-filter efficiencies contributed by the different mechanisms. This approximation has been found to serve adequately for predicting the overall collection efficiencies of fibrous filters, owing to the different ranges in particle sizes and face velocities in which the different filtration mechanisms predominate, as illustrated in Fig. 10-8.

Example 10-1

Calculate the single fiber efficiencies for a 0.5 μm diameter particle at 20°C and 1 atm due to (1) Brownian diffusion, (2) interception, (3) inertial impaction, and (4) gravitational settling for a fibrous filter having a fiber diameter of 5 μm and a solidity of 0.2 and operating at an air flow velocity of 15 cm/s. Assume the particle density is 1 g/cm³.

Answer. (1) Brownian diffusion
Using Eq. 10-3, the diffusion coefficient is calculated as

$$D = \frac{(1.38 \times 10^{-16}\,\text{dyn cm/K})\,(293\,\text{K})\,(1.33)}{3\pi\,(1.84 \times 10^{-4}\,\text{dyn s/cm}^2)\;5 \times 10^{-5}\,\text{cm}}$$

$$= 6.2 \times 10^{-7}\,\text{cm}^2/\text{s}$$

where C_c is calculated using Eq. 10-4:

$$C_c = 1 + 2.492\left(\frac{0.0653\,\mu\text{m}}{0.5\,\mu\text{m}}\right) + 0.84\left(\frac{0.0653}{0.5}\right)\exp\left(-0.435\,\frac{0.5}{0.0653}\right)$$

$$= 1.33$$

From Eq. 10-2 the Peclet number is

$$pe = \frac{(5 \times 10^{-4}\,\text{cm})\,(15\,\text{cm/s})}{6.2 \times 10^{-7}\,\text{cm}^2/\text{s}} = 1.2 \times 10^4$$

Using Eq. 10-5, the single-fiber efficiency η_{diff} is

$$\eta_{diff} = 2.58\left(\frac{1-0.2}{0.245}\right)(1.2 \times 10^4)^{-2/3}$$

$$= 0.01607$$

where

$$K = -\frac{1}{2}\ln 0.2 - \frac{3}{4} + 0.2 - \frac{(0.2)^2}{4}$$

$$= 0.245$$

(2) Interception (from Eq. 10-7)

$$R = \frac{d_p}{d_f} = \frac{0.5}{5} = 0.1$$

Therefore,

$$\eta_{inter} = \frac{1-0.2}{0.245}\,\frac{(0.1)^2}{1+0.1} = 0.02968$$

from Eq. 10-9.

(3) Inertial impaction

The Stokes number is obtained using Eq. 10-10:

$$Stk = \frac{1.33((5 \times 10^{-5})^2\,\text{cm}^2)(1\,\text{g/cm}^3)(15\,\text{cm/s})}{18(1.84 \times 10^{-4}\,\text{g/cm/s})(5 \times 10^{-4}\,\text{cm})}$$

$$= 3.01 \times 10^{-2}$$

The single-fiber efficiency due to inertial impaction is obtained using Eq. 10-11:

$$\eta_{imp} = \frac{1}{(2 \times 0.245)^2}\,[(29.6 - 28(0.2)^{0.62})0.01 - 27.5\,(0.1)^{2.8}]\;3.01 \times 10^{-2}$$

$$= 0.01862$$

(4) Gravitational settling

The settling velocity of a 0.5 μm particle is

$$V_g = \frac{\rho_p d^2 p g C}{18\eta}$$

$$= \frac{1\,\text{g/cm}^3((5 \times 10^{-5})^2\,\text{cm}^2)(980\,\text{cm/s}^2)1.33}{18 \times 1.84 \times 10^{-4}\,\text{g/cm/s}}$$

$$= 9.84 \times 10^{-4}\,\text{cm/s}$$

Then

$$Gr = \frac{V_g}{U} = \frac{9.84 \times 10^{-4}}{15} = 6.56 \times 10^{-5}$$

Therefore,

$$\eta_{\text{grav}} = \frac{Gr}{1 + Gr} = 6.56 \times 10^{-5}$$

Thus, the collection efficiency for a 0.5 μm particle has equal contributions from diffusion and inertial impaction, a bit higher from interception, but little from settling.

Loading Effects

It is well recognized and observed that the particulate collection and the pressure drop increase if high particulate concentrations are filtered for an extended period of time. This takes place because particulates accumulate on filter media and the deposited particles provide additional surfaces for collecting incoming particles. This mechanism is inherently time-dependent because the size, shape, and morphology of particle dendrites change continuously.

Payatakes (1976) treated this filtration mechanism by numerically solving sets of differential equations. Recently, Kanaoka (1989) has also developed a simple method for accounting for the dendrite filtration mechanism. According to his theory, the total particle collection efficiency, E_{loaded}, including the effects of dendrite formation, is

$$E_{\text{loaded}} = 1 - \frac{\exp(-\gamma A c U t)}{\exp(-\gamma A c U t) + \exp(AL) - 1} \tag{10-14}$$

where

$$A = \frac{4\alpha}{\pi(1-\alpha)d_f}\eta_{\text{initial}} \tag{10-15}$$

γ is the efficiency increase factor, c the particulate mass concentration, U the air velocity, t the time, and L the filter mat thickness. His theory assumes that the filtration efficiency of a single fiber increases in the following manner:

$$\eta_{\text{loaded}}/\eta_{\text{initial}} = 1 + \gamma m \tag{10-16}$$

where η_{loaded} is the filtration efficiency for the fiber which accumulated the particle mass m, and η_{initial} is the filtration efficiency of the clean fiber. γ in Eq. 10-16 was recommended by Kanaoka to be determined from experimental data.

Most Penetrating Particle Size and Minimum Efficiency

As discussed, an increase in the particle size causes increased filtration by interception and inertial impaction mechanisms, whereas a decrease in particle size enhances collection by Brownian diffusion. As a consequence, there is an intermediate particle size region where two or more mechanisms are simultaneously operating—yet none is dominant. This is the region where the particle penetration through the filter is a maximum and the filter efficiency a minimum. This is schematically illustrated in Fig. 10-8. The particle size at which the minimum efficiency occurs is termed as the most penetrating particle size. This size was previously assumed to be ≈ 0.3 μm and is the basis for the so-called DOP test method for high-efficiency particulate air (HEPA) filters.

As fibrous filtration theory has been improved in recent years, the most penetrating particle size and the corresponding minimum efficiency have been observed to vary with the type of filter and the filtration velocity. Lee and Liu (1980) derived the following equation for predicting the most penetrating particle diameter:

$$d_{p,\min} = 0.855\left[\left(\frac{K}{1-\alpha}\right)\left(\frac{\sqrt{\lambda kT}}{\eta}\right)\left(\frac{d_f^2}{U}\right)\right]^{2/9} \tag{10-17}$$

Figure 10-9 is a comparison of Eq. 10-17 with experimental data. The most penetrating particle diameter decreases with increasing flow velocity and increasing filter solidity (1 – porosity). As the filter media size increases, the most penetrating particle size increases. The corresponding minimum

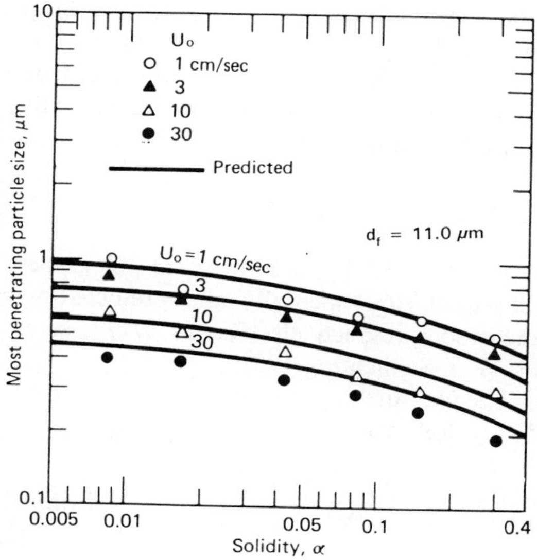

FIGURE 10-9. Comparison of Theory and Experiment for the Most Penetrating Particle Diameter.

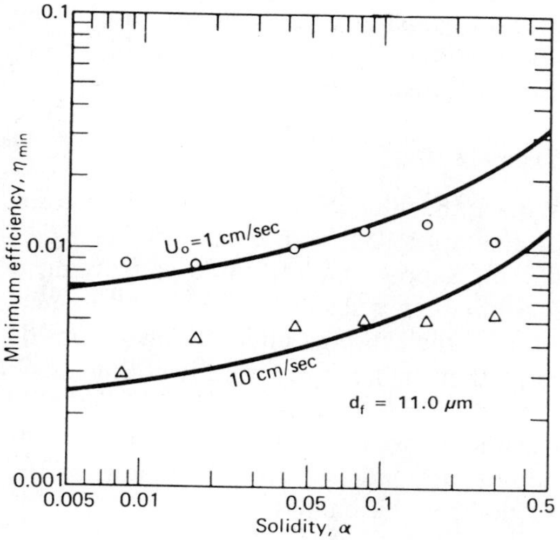

FIGURE 10-10. Comparison of Theory and Experiment for the Minimum Single Fiber Efficiency.

efficiency is given as

$$\eta_{\min} = 1.44\left[\left(\frac{1-\alpha}{K}\right)^5\left(\frac{\sqrt{\lambda}kT}{\eta}\right)^4\left(\frac{1}{U^4 d_f^{10}}\right)\right]^{1/9} \quad (10\text{-}18)$$

A comparison of the equation with experimental data for minimum efficiency is shown in Fig. 10-10.

Example 10-2

What is the particle diameter that will give minimum efficiency for the filter given in Example 10-1?

Answer. From Eq. 10-17,

$$d_{p,\min} = 0.885\left[\left(\frac{0.245}{1-0.2}\right)\right.$$
$$\times\left(\frac{\sqrt{0.0653}\sqrt{\text{cm}}\,(1.38\times10^{-16}\,\text{dyn cm/K})293\,\text{K}}{1.84\times10^{-4}\,\text{dyn s/cm}^2}\right)$$
$$\left.\times\frac{(5\times10^{-4})^2\,\text{cm}^2)}{15\,\text{cm/s}}\right]^{2/9}$$
$$= 6.64\times10^{-5}\,\text{cm} = 0.664\,\mu\text{m}$$

Membrane Filters

The predictive theory presented in this section was developed originally for fibrous filters. Studies of porous membrane filters conducted by Rubow (1981) have indicated that the filtration mechanisms for porous-membrane filters are equivalent to those for fibrous filters. Further, fibrous filtration theories were found to be applicable with the use of actual thickness and solidity, the only modification necessary being the use of an effective fiber diameter to represent the structural elements in the membrane. Similarly, a most penetrating particle size has also been shown to exist experimentally for porous-membrane filters. Thus, Eqs. 10-17 and 10-18 can also be expected to be applicable for such membrane filters with the appropriate choice of the effective fiber diameter value.

Particle collection in straight-through pore membrane filters can be estimated through the use of tube diffusion theory for diffusional collection in pore walls, and impaction and interception theory for collection near pore inlet surfaces (see, e.g., Spurny et al. 1969). Typically, the efficiency of the filters is low for particle sizes less than the pore size and greater in the size range where diffusional collection is significant (i.e., $d_p > 0.1\ \mu m$). Particles greater than the pore size are collected with high efficiency. Due to their unique, somewhat "impactor-like" cutoff filtration characteristics, Nuclepore filters have been utilized as size-selective aerosol samplers in sequential collection stages using different pore size filters (e.g., Cahill, Ashbauch, and Barone 1977; Parker et al. 1977). Heidam (1981) provides a review of aerosol fractionation by straight-through pore filters, and concludes that particle bounce can be a potential problem in using such filters in these applications.

Pressure Drop of Filters

As air passes through filter media, the filter structure causes a resistance that is a measure of air permeability or the pressure drop. A consideration of the pressure drop across filter media is important in choosing a specific filter type in a particular application. The pressure drop is easily measurable and can be used as a check on the flow fields on which deposition mechanisms are based. More importantly, the measurement of the pressure drop across filter media plays a central role in the practical estimation of filtration efficiency. Ideally, filters that exhibit a high filtration efficiency at a low pressure drop are the most desirable ones. Due to the diversity of filter types and the complicated nature of filter structure, it is difficult to describe precisely the media geometry and the corresponding flow patterns. Therefore, a prediction of the pressure drop for real filters, such as porous-membrane filters, is not straightforward. In fact, a comparison of the calculated pressure drop based on an idealized flow model with the actual measured pressure drop is used conveniently as an indication of how uniformly the media structure elements, such as fibers and pores, are arranged. A factor utilizing this concept is called the pressure drop factor, β, and is written as

$$\beta = \frac{\Delta P_{ex}}{\Delta P_{th}} \tag{10-19}$$

where ΔP_{ex} and ΔP_{th} are the pressure drop measured experimentally and predicted by the model, respectively. This pressure drop factor is then applied to the theoretically calculated filtration efficiency value as

$$E_{ex} = \beta E_{th} \tag{10-20}$$

where E_{ex} and E_{th} are the filter efficiencies measured experimentally and predicted by the model, respectively (Davies 1973; Lee et al. 1978; Schlichting 1968).

The pressure drop for fibrous filters is given by the following theoretical equation:

$$\Delta P_{th} = \frac{16\eta\alpha UL}{Kd_f^2} \tag{10-21}$$

The measured pressure drop across a filter has been found to provide an adequate approximation, through Eq. 10-21, for the effective fiber diameter, d_f, for use in the filtration efficiency theory described previously. This is particularly useful for fibrous filters composed of a highly polydisperse range of fiber filters.

Example 10-3

What is the filter efficiency for the filter given in Example 10-1 with a mat thickness of 0.1 cm? Suppose the measured pressure drop for this filter is 2.9 in (= 7.37 cm) of water. What is the expected filter efficiency, considering that the filter structure is not ideal?

Answer. Assuming the individual mechanisms are independent and additive, the single-fiber efficiency is

$$\eta = 0.01607 + 0.02968 + 0.01862 + 0.00006$$
$$= 0.06443$$

The filter efficiency is calculated using Eq. 10-1:

$$E = 1 - \exp\left[\frac{-4(0.06443)\,0.2(0.1)}{\pi(5\times10^{-4})(1-0.2)}\right]$$
$$= 0.9834 = 98.34\%$$

From Eq. 10-21, the theoretical pressure drop becomes

$$\frac{16(1.84 \times 10^{-4}\,\text{dyn s/cm}^2)(0.2 \times 15\,\text{cm/s})\;0.1\,\text{cm}}{0.245((5 \times 10^{-4})^2\,\text{cm}^2)}$$

$$= 14420\,\text{dyn/cm}^2 = 5.8\,\text{in of water}$$

The expected filter efficiency becomes

$$E_{\text{ex}} = E_{\text{th}} \frac{2.9\,\text{in of water}}{5.8\,\text{in of water}}$$

$$= 0.9834 \times 0.5 = 0.492$$

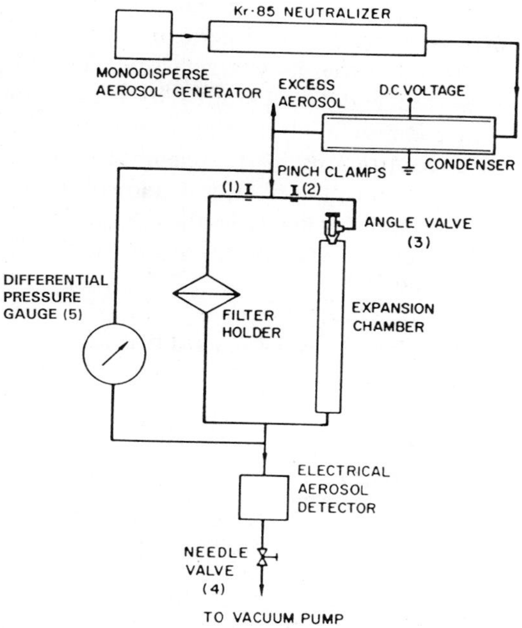

FIGURE 10-11. An Example of an Experimental System for Measuring Filter Efficiencies.

Filter Testing Method

Filter testing is important for characterizing the operational characteristics of a filter medium. Traditionally, filters were tested using the conventional dioctyl phthalate (DOP) test method. This method is described in the Military Standard (MIL-STD-282) and uses a 0.3 μm aerosol as test particles and a photometer. As discussed previously, however, particle collection efficiency depends upon the particle size and the filtration velocity. Thus, a comprehensive particle collection efficiency test needs the ability to vary particle size. Further, a number of highly efficient commercially available membrane filters exhibit considerable high pressure drops across the filter and this needs to be accommodated when the efficiency testing is performed.

In order to address these requirements, an improved filter testing method was introduced by Liu and Lee (1976). Subsequently, the method was used extensively to test various types of filters (Liu, Pui, and Rubow 1983). Figure 10-11 is a schematic diagram of the experimental setup for measuring the filtration efficiency as a function of particle size and flow velocity. The system consists of an aerosol source that is capable of providing a series of monodisperse aerosols of known size, a filter testing section, and an aerosol detector.

For the aerosol source, an atomization–condensation technique separately described by Liu and Lee (1975) is used. The technique initially atomizes DOP solution dissolved in alcohol to produce initially polydisperse particles. Subsequently, the particles become monodisperse by the vaporization–condensation method. The particle size can be varied between about 0.03 and 1.3 μm in diameter this way. The geometric standard deviation of the test particles is found to be about 1.3. These test particles are allowed to pass through a Kr-85 electrical neutralizer to avoid possible electrostatic attraction effects on filter testing results. By passing the test aerosol through the filter to be tested and then through the reference line, the amounts of the particles penetrating the paths can be compared to compute the efficiency. For testing a filter of high pressure drop, it is necessary to expand the aerosol into a low pressure in the reference line using an expansion system. In this way, the flow stream condition of both the filter tested and the reference line can be maintained the same. This can be important to ensure that the performance of the particle-detecting instrument is not affected by the different pressure condition. Consideration must also be given to particle loss in the reference sampling line and measurement for this purpose can be performed using the electrostatic charge characteristics of particles (Liu and Lee 1976) or by calibration using chemical analysis. For measuring aerosol concentrations in the reference line and downstream of the filter, any aerosol instrument suitable for detecting the aerosols in the size range of interest can be used. In the aforementioned study, an electrical aerosol detector was used.

Liu, Pui, and Rubow (1983) performed extensive filter testing for 75 different filter media and compiled the results for each filter; collection efficiencies were measured for 0.035, 0.01, 0.3, and 1.0 μm particles at four different pressure drops. Table 10-5 is an adaptation of the results of Liu, Pui, and

TABLE 10-5 List of Filters Tested and Principal Results Adapted from Liu, Pui, and Rubow (1983)

Type	Manufacturer Name	Filter	Material	Pore Size (μm)	Filter Permeability Velocity (cm/s) ($\Delta P = 1$ cmHg)	Filter Efficiency Range (%)[a]
Fibrous filter	Whatman	No. 1	Cellulose fiber	NA	6.1	49–99.96
		No. 2			3.8	63–99.97
		No. 3			2.9	89.3–99.98
		No. 4			20.6	33–99.5
		No. 5			0.86	93.1–99.99
		No. 40			3.7	77–99.99
		No. 41			16.9	43–99.5
		No. 42			0.83	92.0–99.992
	Gelman	Type A	Glass fiber	NA	11.2	99.92–>99.99
		Type A/E			15.5	99.6–>99.99
		Spectrograde			15.8	99.6–>99.99
		Microquartz			14.1	98.5–>99.99
	MSA	1106B		NA	15.8	99.5–>99.99
	Pallflex	2500 QAO	Quartz fiber	NA	41	84–99.9
		E70/2075W			36.5	84–99.95
		T60A20	Teflon-coated glass fiber		49.3	55–98.8
		(another lot)			40.6	52–99.5
		T60A25			36.5	65–99.3
		TX40H120			15.1	92.6–99.96
		(another lot)			9.0	98.9–>99.99
	Reeve Angel	934AH	Glass fiber	NA	12.5	98.9–>99.99
		(acid-treated)			20	95.0–99.96
	Whatman	GF/A	Glass fiber		14.5	99.0–>99.99
		GF/B			5.5	>99.99
		GF/C			12.8	99.6–>99.99
		EPM 1000			13.9	99.0–>99.99
	Delbag	Microsorban-98	Polystyrene	NA	13.4	98.2–>99.99
Porous membrane filters	Millipore	MF-VS	Cellulose acetate/nitrate	0.025	0.028	99.999–>99.999
		MF-VC		0.1	0.16	99.999–>99.999
		MF-PH		0.3	0.86	99.999–>99.999
		MF-HA		0.45	1.3	99.999–>99.999
		MF-AA		0.8	4.2	99.999–>99.999
		MF-RA		1.2	6.2	99.9–>99.999

TABLE 10-5 (*Continued*)

Type	Manufacturer Name	Filter	Material	Pore Size (μm)	Filter Permeability Velocity (cm/s) ($\Delta P = 1$ cmHg)	Filter Efficiency Range (%)[a]
		MF-SS		3.0	7.5	98.5–>99.999
		MF-SM		5.0	10.0	98.1–>99.9
		MF-SC		8.0	14.1	92.0–>99.9
		Polyvic-BD	Polyvinyl chloride	0.6	0.86	99.94–>99.99
		Polyvic-VS		2.0	5.07	88–>99.99
		PVC-5		5.0	11	96.7–>99.99
			Cellulose acetate			
		Celotate-EG		0.2	0.31	>99.95–>99.99
		Celotate-EH		0.5	1.07	99.989–>99.999
		Celolate-EA	Teflon	1.0	1.98	>99.99
		Mitex-LS		5.0	4.94	84–>99.99
		Mitex-LC	PTFE–polyethylene-reinforced	10.0	7.4	62–>99.99
		Fluoropore		0.2	1.31	>99.90–>99.99
		FG		0.5	2.32	>99.99
		FH		1.0	7.3	>99.99
		FA		3.0	23.5	98.2–99.98
		FS				
	Metricel	GM-6	Cellulose acetate/nitrate	0.45	1.45	>99.8–>99.99
		VM-1	Polyvinyl chloride	5.0	51.0	49–98.8
		DM-800	PVC/acrylonitrile	0.8	2.7	>99.96–>99.99
	Gelman	Gelman Teflon	Teflon	5.0	56.8	85–99.90
	Ghia	S2 37PL 02	Teflon	1.0	12.9	>99.97–>99.99
		S2 37PJ 02		2.0	23.4	99.89–>99.99
		S2 37PK 02		3.0	24.2	92–99.98
	Zefluor	P5PJ 037 50	Teflon	2.0	32.5	94.6–99.96
		P5PI 037 50		3.0	31.6	88–99.9
	Chemplast	75-F	Teflon	1.5	3	83–99.99
		75-M		1.0	6.6	54–>99.99
		75-C		1.0	32	26–999.8
	Selas Flotronics	FM0.45	Silver	0.45	1.8	93.6–99.98
		FM0.8		0.8	6.2	90–99.96
		FM1.2		1.2	9.2	73–99.7
		FM5.0		5.0	19.0	25–99.2
Straight-through membrane filter	Nuclepore	N010	Polycarbonate	0.1	0.602	99.9–>99.99
		N030		0.3	3.6	93.9–>99.99
		N040		0.4	2.9	78–>99.99
		N060		0.6	2.1	53–99.5
		N100		1.0	8.8	28–98.1
		N200		2.0	7.63	9–94.1
		N300		3.0	12	9–90.4
		N500		5.0	30.7	6–90.7
		N800		8.0	21.2	1–90.5
		N1000		10.0	95	1–46
		N1200		12.0	161	1–66

a. Filter efficiency values generally correspond to a particle diameter of 0.035–1 μm, a pressure drop range of 1–30 μmHg and a face velocity of 1–100 cm/sec

Rubow (1983) and subdivides the filters tested into various categories described previously.

FILTER SELECTION

The factors influencing the selection of a filter medium for a specific application can be numerous and varied. As enumerated earlier, the important considerations include particle collection efficiency, pressure drop through the filter at the flow required, compatibility with the analytical method to be employed, and cost constraints. In addition, constraints originating from the mechanical strength of the filter medium and compatibility with environmental sampling conditions such as temperature, pressure, humidity, and corrosiveness can also influence filter selection.

The nature and requirements of the analytical technique(s) employed to study the aerosol, following its collection on the filter, greatly influence the selection of the most appropriate filter medium. Filter analysis techniques for deposited aerosols can be divided into three general categories: gravimetric, microscopic, and microchemical. A review of these categories of analytical techniques demonstrates a number of the filter selection factors involved, as well as the potential for errors or biases and possible corrective techniques.

Gravimetric Analysis

The measurement of the increase in weight of a filter following a well-defined sampling period is the most common way to determine aerosol mass concentration. The technique requires that the filter collect the aerosol with a high efficiency (close to 100%) and that the weight increase following sampling be entirely attributable to the collected aerosol, i.e., filter weight must be independent of age and the temperature and humidity exposure conditions. Gravimetric filter analysis has been found to be most sensitive to the effects of moisture/relative humidity and static charge buildup on filter materials.

Moisture effects arise from the uptake of water vapor by the filter material and from the hygroscopicity of the aerosol sample. Filters of cellulose fibers are the most affected by water vapor uptake, with glass and cellulose quartz fiber filters being less susceptible. Water vapor uptake is lowest in Teflon membrane filters, closely followed by polycarbonate and some PVC membrane filters (Mark 1974; Demuynck 1975; Charell and Hawley

TABLE 10-6 Weight Sensitivity and Stability of a Selection of Common Filter Media to Water Vapor Uptake (Taken from Lowry and Tillery 1979)

Filter Type	Average Weight 37 mm Diameter (mg)	Weight Stability Under Standard Conditions[1] (mg)	Average Weight Change Following 24 h in a Desiccator[2] (mg)	Average Weight Change Following 24 h at 80% RH[2] (mg)
Gelman glass filter type A/E (without organic binder)	86	0.14	− 0.02	+ 0.01
MSA #457193; 5 μm pore size PVC membrane with fibrous backup filter	243	0.04	− 0.04	0
Millipore AA; 0.8-μm pore size cellulose ester membrane	53	0.13	0	+ 0.35
Millipore Teflon; 5-μm pore size Teflon membrane	108	0.02	+ 0.01	0
Selas Flotronics FM-37; 0.8-μm pore size silver membrane	460	0.03	0	+ 0.01

1. 95% confidence interval based on three measurements a day for six filters of each type over 30 days
2. Filters were equilibrated under room conditions for 24 h, weighed, desiccated, or humidified for 24 h, equilibrated under room conditions for 24 h, and reweighed

1981). Table 10-6 shows the weight stability and sensitivity to water vapor uptake of a selection of common filter media and is taken from the results of Lowry and Tillery (1979). A standard means of minimizing relative humidity interferences in gravimetric analysis involves equilibrating the filters at a constant temperature and humidity (e.g., 20°C, 50% RH) for 24 h before and after sampling. Overcoming the complicating effects of moisture uptake by hygroscopic aerosols collected on a filter is more difficult, with few approaches available other than to calibrate the weight gained under different humidity conditions using a control sample or to minimize the time lag between sampling and weight measurements.

The accumulation of static charges on a filter can result in handling difficulties, enhanced or diminished particle collection, and weighing errors in electrobalances (Engelbrecht, Cahill, and Feeney 1980). Depending upon the filter material and the manufacturing process, certain types of filters such as polycarbonate and PVC membranes can become significantly charged and result in both sampling and measurement errors. A common approach to minimize these effects is to expose the filter to a source of bipolar discharging ions such as Po-210 or Am-241 before sampling and prior to gravimetric analysis. Sampling of fibers (e.g., asbestos) has also been found to be biased by charge accumulation on plastic, nonconducting filter holders—the use of conductive filter holders has been found to alleviate this problem.

Example 10-4

Ambient air is to be sampled to collect a particulate mass of at least 10 mg for gravimetric analysis. A flow velocity of 1 ft/s (= 30.48 cm/s) will be adopted using an $8\frac{1}{2} \times 10$ in sheet quartz filter with an effective filtration area of 20×22.5 (= 450) cm². Calculate the minimum required sampling time assuming that the particle collection efficiency is 100%. The ambient average particle concentration is estimated to be $\approx 20\ \mu g/m^3$.

Answer. The flow rate through the filter is

$$450\ \text{cm}^2 \times 30.48\ \text{cm/s} = 13{,}716\ \text{cm}^3/\text{s}$$
$$= 822.96\ \text{lpm}$$
$$(29.06\ \text{cfm})$$

The particle mass collected per unit time at this flow rate is

$$(20 \times 10^{-6}\ \mu\text{g/cm}^3)\ (13{,}716\ \text{cm}^3/\text{s})$$
$$= 0.274\ \mu\text{g/s}$$

The required sampling time for collecting 10 mg is

$$10{,}000\ \mu\text{g}/(0.274\ \mu\text{g/s}) = 36{,}453\ \text{s} = 10.1\ \text{h}$$

Microscopic Analysis

Particle analysis by light or electron microscopy is frequently used to obtain information on the size, morphology, and compositional characteristics of aerosol samples. Microscopic analysis requires that the collection of the particles occur on, and as closely to, a flat filter surface as is the case with microporous and straight-through pore membrane filters. Polycarbonate, straight-through pore membrane filters are particularly well suited to microscopic applications due to their smooth, flat surface and near complete surface collection characteristics (provided pore sizes are selected appropriately).

Other surface analysis techniques that impose similar constraints on aerosol collection media include X-ray fluorescence (XRF), X-ray diffraction (XRD), and proton induced X-ray emission (PIXE) analyses for elemental and chemical species concentration measurement, and aerosol radioactivity measurement techniques. These techniques also benefit from the collection of the aerosol on or close to the surface of the filter, and impose additional constraints of minimizing both the aerosol collection surface area and the background concentration or response of the blank filter material in the analysis.

Generally, microporous and straight-through pore membrane filters are both well suited to these other filter analysis techniques. Teflon membrane filters are most commonly used for XRF analysis of filter-sampled aerosols owing to their inertness and low blank value concentrations (e.g., Chow et al. 1990). Quartz and glass fiber filters can be useful for X-ray diffraction analyses in cases of high particulate loading, while Teflon filters are superior for low loadings (Davis and Johnson 1982). Silver membranes composed of metallic silver are useful for the analysis of crystalline silica using XRD techniques due to their extremely low interference in the quartz region of the diffraction spectrum. Aerosol radioactivity analysis using alpha or beta radiation detectors usually requires a high flow rate, high collection efficiency, and the collection of particles as close to the surface as possible to minimize the absorption of radiation. Microporous cellulose ester members of pore size between 0.45 and 0.8 μm (e.g., Millipore Type AA) are commonly employed to meet these objectives. Busigin et al. (1980) provide a review of the collection characteristics of radon progeny radioactive aerosols by a variety of different filters.

The selection of filters for sampling airborne microorganisms (viruses, bacteria, and fungi) is also governed by the need to microscopically count the number of viable microorganisms or colony-forming units (cfus) collected on the filter. In these applications, the loss of some viable microorganisms may result from the dessication induced by their collection on the filter surface. Thus, collection on filter surfaces is limited to hardy microorganisms that can withstand dessication and can be transferred to suitable growth media, following collection. More detailed discussions of sampling airborne bioorganisms are presented in Chapter 21, as well as by Chatigny et al. (1989).

Microchemical Analysis

The chemical analysis of particulates collected on filter media is becoming increasingly routine in applications such as air quality monitoring. The most important factors for consideration in selecting filter media for microchemical analysis are the quantity of particulate matter required for the analysis and the minimization of (1) interferences arising from the background response of the blank filters and (2) artifact formation from chemical transformations occurring on the filter during and after sampling. The magnitude and variability of the blank filter or background trace element/chemical species concentrations of different filters can have significant importance in determining the sensitivity and the limits of detection of the analytical technique. Maenhaut (1990) reviews a number of different analytical techniques for trace atmospheric elements and discusses the blank concentration ranges for different types of filters in various analyses. A number of other important aspects in areosol sampling for microchemical analyses are also presented by Hopke (1985).

Most chemical analyses require the extraction of the particles collected on the filter following sampling in a manner suitable for input into analytical instrumentation. Filters of cellulose (Whatman-type paper), glass, Teflon-coated glass and quartz fibers are all commonly used in aerosol sampling for microchemical analysis, due to their low pressure drop characteristics that permit high-volume sampling. Cellulose paper filters are conveniently processed for aerosol extraction using incineration, ashing, or digesting in acid solution. However, cellulose paper filters suffer from low particle collection efficiency characteristics at low particulate filter loadings (see Table 10-3).

Glass, Teflon-coated glass, and quartz fiber filters have significantly higher particle collection efficiencies but must be acid-leached for recovery/extraction of the aerosols. Glass fiber filters suffer from a positive artifact mass in ambient air sampling due to their slight alkalinity that results in the *in situ* conversion of sulfur dioxide to sulfate (e.g., Coutant 1977; Rodes and Evans 1977; Stevens et al. 1978). Artifact particulate nitrate can also be formed on glass fiber filters depending on the gaseous nitric acid concentrations (Appel and Tokiwa

1981). Quartz and paper fibers do not suffer from significant sulfate artifacts in ambient air sampling, as reported by Pierson et al. (1980).

Quartz fiber filters are particularly useful in aerosol sampling for microchemical analysis due to their low water vapor uptake characteristics and low background/blank elemental concentrations. Hence, they are commonly used in ion chromatographic analysis for species such as chloride, nitrate, sulfate, potassium, and ammonium ions (e.g., Chow et al. 1990). In addition, quartz fiber filters can be baked at high temperatures to lower blank organic concentration values, allowing them to be used for collecting particulate samples that can be extracted for organic species, polycyclic aromatic hydrocarbons, alklyating aspects, and mutagenic activity (e.g., Lioy and Daisey 1983). Quartz fiber filters are also utilized in organic and elemental carbon analysis of particulate deposits on filters by a combination of flash and rapid heating of the filters and conversion of evolved carbon to CO_2 that can then be measured (e.g., Tanner, Gaffney, and Phillips 1982).

Membrane filters can also be used for microchemical analysis, although they suffer

TABLE 10-7 Compilation of Common Applications, Advantages, and Disadvantages for Various Types of Fibrous Filters

Filter Type	Typical Applications	Advantages	Limitations
Fibrous filters (general)	Air quality sampling	• Low pressure drop at high-volume sampling operation • Low cost • High particulate loading capacity	• Lower collection efficiencies for submicron particles • Particle collection occurs throughout the depth of the filter
Cellulose fiber	Typically used in limited/qualitative applications in air quality sampling	• Inexpensive • Convenient extraction of particulates	• Highly moisture-sensitive • Limited temperature range • Low particle collection efficiency • Low chemical resistance
Borosilicate glass fiber	Wide scope in air quality sampling. Used without organic binder	• Temperature resistance to ≈ 500°C • Chemically resistant to some extent	• Sulfate artifact formation due to alkalinity of fibers • Water vapor uptake can occur and must be equilibrated appropriately
Teflon-coated glass fiber	Wide range of air sampling applications–emissions analysis, gravimetric analysis, biological, and mutogenic analysis	• Low moisture uptake • Minimizes chemical transformation artifacts	• Artifact nitrate
Quartz fiber	Air sampling for chemical analysis of particulates–ion chromatography, atomic absorption, carbon analysis, PAH analysis, etc.	• Low moisture uptake • Stable to temperatures upto 800°C • Low trace contaminant levels • Can be baked to remove trace organics prior to sampling • Low artifact formation	• Friable • Artifact nitrate formation has been observed

from problems of limited particulate loading ability and the possibility of losing coarse particles during handling and transport following sampling (Dzubay and Barbour 1983). Teflon membrane filters have the advantage of low blank chemical species concentrations, as well as chemical inertness that permits sampling with minimal sulfate artifact formation. However, ammonium nitrate (nitrate salts) and nitric acid can be lost from "inert" Teflon, as well as quartz, filters through volatilization (Rodes and Evans 1977) and as a result of reactions with acidic species on the filter (Harker, Richards, and Clark 1977). Nylon filters have been utilized as an alternative for nitrate collection (Grosjean 1982; Spicer et al. 1982) although they are susceptible to sulfate artifact information (Chan, Orr, and Chung 1986).

Artifact formation in the sampling of organic aerosols on filters has also been reported in a number of studies. As discussed by Pitts and Pitts (1986), these include negative artifacts from volatilization of the more volatile particulate organics (Van Vaeck, Van Cauwenberghe, and Janssens 1984), positive artifacts from adsorption of gaseous organics on the filter (Stevens et al. 1980; Cadle, Groblicke, and Mulaya 1983) and transformations/reactions occurring with species such as O_3 and NO_2 sampled through the filter. De Raat et al. (1990) compared glass fiber, Teflon-coated glass fiber, and Teflon membrane filters for sampling of mutagens and polycyclic aromatic hydrocarbons (PAH) in ambient airborne particles and found: (1) a slightly greater mutagenicity when sampling with glass fiber filters, probably due to adsorption of gaseous PAH on the glass fibers followed by conversion to mutagens, (2) higher adsorption of volatile PAH on glass fiber filters and greater volatilization on the Teflon-coated glass fiber and Teflon membrane filters, and (3) lower concentrations of

TABLE 10-8 Compilation of Common Applications, Advantages, and Disadvantages for Various Types of Microporous Membrane Filters

Filter Type	Typical Applications	Advantages	Limitations
Membrane filters (general) (apply to all below)	Used in air sampling for surface analytical techniques, submicron particle collection	• High collection efficiency • High mechanical strength	• High pressure drop • Low particulate loading capacity, rapid clogging • Limited temperature range, typically
Cellulose mixed esters nitrate acetate, etc. and PVC membranes	Sampling of metals, cotton dust, asbestos, etc. in NIOSH standard methods	• Inexpensive among membrane filters • Low chemical resistance	• Susceptible to water vapor uptake • Operating temperature limited to 75°–130°C • Electrostatic charge buildup observed in PVC membranes
Teflon membranes	Gravimetric analysis, neutron activities analysis XRF, XRD (see text)	• Inert to chemical transformations • Extremely low moisture sensitivity • Low trace/background concentrations • Chemical resistant	• Loss of nitrates observed • Temperature range limited to $\approx$ 150°C for supported membranes and 260°C for pure PTFE membranes
Silver membranes	Organic particulate collection and analysis, e.g., benzo [a] pyrene, PAH, etc. XRD analysis of silica	• Chemical resistant • Low background interferences • High maximum operating temperature of 550°C	• Most expensive among common membrane filters

TABLE 10-9 Compilation of Common Applications, Advantages, and Disadvantages for Straight-Through Pore Membrane Filters

Filter Type	Typical Applications	Advantages	Limitations
Polycarbonate membranes	Ideal for aerosol collection for subsequent surface analytical techniques, e.g., microscopy, PIXE (see text)	• Flat, uniform surface • Nonhygroscopic • Low background/blank concentration • Surface capture characteristics • Semitransparent surface	• High pressure drop • Low particulate loading capacity • Particle size range of low collection efficiency usually exists • Susceptible to static charge buildup

TABLE 10-10 Listing of Addresses for the Commercial Sources of Filters Referenced in Chapter 10

Symbol	Source	Symbol	Source
FSI	Fisher Scientific Inc. (Fisher brand filters) 711 Forbes Avenue Pittsburgh, PA 15219	NUC	Nuclepore Corporation 7035 Commerce Circle Pleasanton, CA 94566
GEL	Gelman Instrument Company 600 S. Wagner Road Ann Arbor, MI 48106	PAL	Paliflex Production Corporation Kennedy Drive Putnam, CT 06260
H&V	Hollingsworth and Vose Company 112 Washington Street East Walpole, MA 02032	POR	Porotics Corporation 151 I Lindbergh Avenue Livermore, CA 94550
MSI	Micron Separations Inc. c/o Fisher Scientific Inc. (See FSI)	SAR	Sartorius Filters, Inc. 30940 San Clement Street Haywood, CA 94544
MFS	Micro Filtration Systems, Inc. 6800 Sierra Court Dublin, CA 94568	S&S	Schlcicher & Schnell Company 10 Optical Avenue Keene, NH 03431
MIL	Millipore Corporation Bedford, MA 01730	WHA	Whatman Reeve Angel Co. 9 Bridewell Place Clifton, NJ 07014
NAL	Nalge Company, Inc. P.O. Box 20365 Rochester, NY 14602		

the more reactive PAH species on the glass fiber filters.

The preceding discussions indicate that the requirements of filter analysis techniques and the need to minimize interferences and artifacts are often inherently conflicting and can rarely be perfectly satisfied. However, an adequate filter medium can generally be selected to meet the needs of most applications through a careful consideration of the important issues. Tables 10-7–10-9 are compilations of the common applications, advantages, and limitations of the various types of filters discussed in previous sections. The information in the tables is derived from a review of the literature and illustrates the availability of an adequate selection of filter media for the wide spectrum of applications that utilize filter sampling. Table 10-10 provides a listing of addresses for the commercial sources of filters referenced in this chapter.

References

Agarwal, J. K. and B. Y. H. Liu. 1980. A criterion for accurate aerosol sampling in calm air. *Am. Ind. Hyg. Assoc. J.* 41:191–97.

Appel, B. R. and Y. Tokiwa. 1981. Atmospheric particulate nitrate sampling errors due to reactions with particulate and gaseous strong acids. *Atmos. Environ.* 15:1087.

Cadle, S. H., P. J. Groblicki, and P. A. Mulaya. 1983. Problems in the sampling and analysis of carbon particulate. *Atmos. Environ.* 17:593.

Cahill, T. A., L. L. Ashbauch, and J. B. Barone. 1977. Analysis of respirable fractions of atmospheric particulates via sequential filtration. *J. Air Pollut. Control Assoc.* 27:675.

Chan, W. H., D. B. Orr, and D. H. S. Chung. 1986. An evaluation of artifact SO_4 formation on nylon filters under field conditions. *Atmos. Environ.* 20:2397.

Charell, P. R. and R. G. Hawley. 1981. Characteristics of water adsorption on air sampling filters. *Am. Ind. Hyg. Assoc. J.* 42:353.

Chatigny, M. A., J. M. Macher, H. A. Burge, and W. R. Soloman. 1989. Sampling airborne microorganisms and aeroallergens. In *Air Sampling Instruments for Evaluation of Atmospheric Contaminants*, 7th edn., pp. 199–220. Cincinnati, OH: American Council of Governmental Industrial Hygienists.

Chow, J. C., J. G. Watson, R. T. Egami, C. A. Frazier, Z. Lu, A. Goodrich, and A. Bird. 1990. Evaluation of regenerative-air vacuum street sweeping on geological contributions to PM_{10}. *J. Air Waste Manag. Assoc.* 40:1134.

Coutant, R. W. 1977. Effect of environmental variables on collection of atmospheric sulfate. *Environ. Sci. Technol.* 11:873.

Davies, C. N. 1968. The entry of aerosols into sampling tubes and heads. *Br. J. Appl. Phys. Ser. 2* 1:921.

Davies, C. N. 1973. *Air Filtration.* London: Academic Press.

Davis, B. L. and L. R. Johnson. 1982. On the use of various filter substrates for quantitative particulate analysis by X-ray diffraction. *Atmos. Environ.* 16:273.

Demuynck, M. 1975. Determination of irreversible absorption of water on air sampling filters. *Am. Ind. Hyg. Assoc. J.* 42:353.

Dzubay, T. G. and R. K. Baybour. 1983. A method to improve adhesion of aerosol particles on Teflon filters. *J. Air Pollut. Control Assoc.* 33:692.

Engelbrecht, D. R., T. A. Cahill, and P. J. Feeney. 1980. Electrostatic effects on gravimetric analysis of membrane filters. *J. Air Pollut. Control Assoc.* 30:391.

Friedlander, S. K. 1957. Mass and heat transfer to single spheres and cylinders at low Reynolds numbers. *A.I.Ch.E. J.* 3:43–48.

Grosjean, D. 1982. Quantitative collection of total inorganic atmospheric nitrate on nylon filters. *Anal. Lett.* 15(A9):785.

Harker, A., L. Richards, and W. Clark. 1977. Effect of atmospheric SO_2 photochemistry upon observed nitrate concentrations. *Atmos. Environ.* 11:87.

Heidam, N. Z. 1981. Review: Aerosol fractionation by sequential filtration with Nuclepore filters. *Atmos. Environ.* 15:891.

Hinds, W. D. 1982. *Aerosol Technology*. New York: Wiley.

Hopke, P. K. 1985. *Receptor Modeling in Environmental Chemistry*. New York: Wiley.

Kanaoka, C. 1989. Time dependency of air filter performance *J. Aerosol Res. Japan* 4:256–64 (in Japanese).

Kogan, V., M. R. Kuhlman, R. W. Coutant, and R. G. Lewis, 1991. Aerosol filtration by sorbent beds. (In preparation).

Lee, K. W. and B. Y. H. Liu. 1980. On the minimum efficiency and the most penetrating particle size for fibrous filters. *J. Air Pollut. Control Assoc.* 30:377–81.

Lee, K. W. and B. Y. H. Liu. 1982*a*. Experimental study of aerosol filtration by fibrous filters. *Aerosol Sci. Technol.* 1:35–46.

Lee, K. W. and B. Y. H. Liu. 1982*b*. Theoretical study of aerosol filtration by fibrous filters. *Aerosol Sci. Technol.* 1:147–61.

Lioy, P. J. and J. M. Daisey. 1983. The New Jersey project on airborne toxic elements and organic substances (ATEOS): A summary of the 1981 summer and 1981 winter studies. *J. Air Pollut. Control Assoc.* 33:649.

Lippmann, M. 1989*a*. Sampling aerosols by filtration. In *Air Sampling Instruments for Evaluation of Atmospheric Contaminants*, 7th edn., Cincinnati, OH: American Conference of Governmental Industrial Hygienists.

Lippmann, M. 1989*b*. Calibration of air sampling instruments. In *Air Sampling Instruments for Evaluation of Atmospheric Contaminants*, 7th edn., Cincinnati, OH: American Conference of Governmental Industrial Hygienists.

Liu, B. Y. H. and K. W. Lee, 1975. An aerosol generator of high stability. *Am. Ind. Hyg. Assoc. J.* 36:861–65.

Liu, B. Y. H. and K. W. Lee. 1976. Efficiency of membrane and Nuclepore filters for submicrometer aerosols. *Environ. Sci. & Technol.* 10:345–50.

Liu, B. Y. H., D. Y. H. Pui, and K. L. Rubow. 1983. Characteristics of air sampling filter media. In *Aerosols in the Mining and Industrial Work Environments*. Ann Arbor, MI: Ann Arbor Science.

Lowry, P. L. and M. I. Tillery. 1979. Filter weight stability evaluation. Los Alamos Scientific Laboratory Report No. LA-8061-MS.

Maenhaut, W. 1989. Analytical techniques for atmospheric trace elements. In *Control and Fate of Atmospheric Trace Metals*, eds. J. M. Pacyna and B. Ottar, pp. 259–301. Dordrecht: Kluwer.

Mark, D. 1974. Problems associated with the use of membrane filters for dust sampling when compositional analysis is required. *Ann. Occup. Hyg.* 17:35.

Mori, J., H. Emi, and Y. Otani. 1991. Classification of membrane gas filters and their performance evaluation. *J. Aerosol Res. Japan* 6:149–56 (in Japanese).

Natanson, G. L. 1957. Proc. Acad. Sci. USSR. *Phy. Chem. Sec.* 112:21–25.

Parker, R. D., G. H. Buzzard, T. G. Dzubay, and J. P. Bell, 1977. A two-style respirable aerosol sampler using Nuclepore filters in series. *Atmos. Environ.* 11:617.

Payatakes, A. C. 1976. Model of transient aerosol particle deposition in fibrous media with dendritic pattern. *A.I.Ch.E. J.* 23:192–202.

Pierson, W. R., W. W. Brachaczek, T. J. Korniski, T. J. Truer, and J. W. Butler, 1980. Artifact formation of sulfate, nitrate and hydrogen ion on backup filters: Allegheny mountain experiment. *J. Air Pollut. Control Assoc.* 30:34.

Pitts, B. J. F. and J. R. Pitts, Jr. 1986. *Atmospheric Chemistry: Fundametals and Experimental Techniques.* New York: Wiley.

Rodes, C. E. and G. F. Evans. 1977. Summary of LACS integrated measurements. EPA-600/4-77-034. U.S. Environmental Protection Agency, Research Triangle Park, NC.

Rubow, K. L. 1981. Submicrometer aerosol filtration characteristics of membrane filters. Ph.D. Thesis, University of Minnesota, Minneapolis, Minnesota.

Rubow, K. L. and V. C. Furtado. 1989. Air movers and samplers. In *Air Sampling Instruments for Evaluation of Atmospheric Contaminants*, 7th edn. Cincinnati, OH: American Conference of Governmental Industrial Hygienists.

Spicer, C. W. and P. M. Schumacher. 1979. Particulate nitrate: Laboratory and field studies of major sampling interferences. *Atmos. Environ.* 13:543.

Spielman, L. and S. L. Goren. 1968. Model for predicting pressure drop and filtration efficiency in fibrous media. *Environ. Sci. Technol.* 2:279–87.

Spurny, K. R., J. P. Lodge, Jr., E. R. Frank, and D. C. Sheesley. 1969. Aerosol filtration by means of Nuclepore filters: structural and filtration properties. *Environ. Sci. Technol.* 3:453.

Stechkina, I. B., A. A. Kirsch, and N. A. Fuchs. 1969. Studies in fibrous aerosol filters—IV. Calculation of aerosol deposition in model filters in the range of maximum penetration. *Ann. Occup. Hyg.* 12:1–8.

Stevens, R. K., T. G. Dzubay, G. Russwurm, and D. Rickel. 1978. Sampling and analysis of atmospheric sulfates and related species. In: *Sulfur in the Atmosphere, Proc. International Symposium,* United Nations, Dubrovnik, Yugoslavia, 7–14 September 1977. *Atmos. Environ.* 12:55.

Stevens, R. K., T. G. Dzubay, R. W. Shaw, Jr., W. A. McClenny, C. W. Lewis, and W. E. Silson. 1980. Characterization of the aerosol in the Great Smoky Mountains. *Environ. Sci. Technol.* 14:1491.

Tanner, R. L., T. S. Gaffney, and M. F. Phillips. 1982. Determination of organic and elemental carbon in atmospheric aerosol samples by thermal evolution. *Anal. Chem.* 54:1627.

Van Vaeck, L., K. Van Cauwenberghe, and J. Janssens. 1984. The gas–particle distribution of organic aerosol constituents: Measurement of the volatilization artifact in Hi-Vol cascade impactor sampling. *Atmos. Environ.* 17:900.

Vincent, J. H. 1989. *Aerosol Sampling Science and Practice.* New York: Wiley.

11

Inertial, Gravitational, Centrifugal, and Thermal Collection Techniques

Virgil A. Marple, Kenneth L. Rubow, and Bernard A. Olson

University of Minnesota
Mechanical Engineering Department
Minneapolis, MN, U.S.A.

INTRODUCTION

Inertial classification, gravitational sedimentation, centrifugation, and thermal precipitation are techniques that can be used to collect particles for subsequent analysis or for particle classification. The inertial classifiers, which include impactors, virtual impactors, and cyclones, are widely used in the sampling of particles. Settling chambers, which include centrifuges as well as gravitational settling devices, are less widely used. Thermal precipitators are seldom used.

Impactors, in particular, have been used extensively, studied both theoretically and experimentally, and are the instrument of choice for the determination of aerosol mass size distributions. Many versions of impactors are available commercially and many more have been designed, built, and used in special studies.

Virtual impactors are a more recently developed inertial classifier and possess a feature not found in conventional inertial impactors; namely, that the particles remain airborne after classification compared to the conventional impactor, where the particles are collected on a solid surface. The feature of keeping particles airborne is important for transporting the particles to other particle-analyzing instruments, or to filters, and in concentrating particles larger than the cut-size in a small fraction of the total flow.

Cyclones are also widely used as particle samplers but have been much more difficult to analyze theoretically. In general, cyclone particle cutoff characteristics are not as sharp as for impactors but still sharp enough to be useful as classifiers. They differ from impactors primarily by their ability to collect much larger quantities of particles.

Settling chambers are devices which directly measure particle terminal settling velocities, but are not commonly used. However, centrifuges, which magnify settling forces by applying centrifugal forces to the particles, are more widely used.

Thermal precipitators are also used for the collection of particles but are much less popular than inertial classifiers or centrifuges, since the particles are not classified by size. The thermal precipitator is most effective for collecting particles over a wide size range for subsequent sizing by a microscope. They do have the advantage that they collect small particles, have low pressure drops and, therefore, require only small pumps.

These classifiers are discussed in the following sections. Since inertial classifiers (especially impactors) are the most widely used of the classifiers, the major emphasis of this chapter will be placed on these classifiers.

INERTIAL CLASSIFIERS

Numerous inertial classifiers have been designed and reported in the literature with many of them being commercially available. Tables 11-1–11-4 list most of the commercially available impactors and cyclones by type and manufacturer. The following sections will discuss the classifiers in general terms, providing information on specific devices only to illustrate specific points.

Principle of Inertial Classification

The principle of inertial classification of particles is quite simple in that the particles' inertia is used in their classification. Classification is achieved in these instruments by turning the gas flow and capturing the particles with sufficient inertia to cross gas streamlines and escape the flow. Particles with less inertia will remain in the gas flow.

The simplest type of inertial classifier is a body collector, which is a body (usually cylinder or ribbon) passing through particle-laden gas. As this body moves through the gas, the gas is deflected around the body. Large particles, however, due to their inertia, are not deflected as much as small particles and will strike the surface of the body.

An excellent example of a body collector is the automobile. As the automobile passes through the air, large particles in the air will impact upon the automobile and, as passengers, we are in an excellent position to observe impaction on the windshield. Probably, the best observations are during a snow storm, when the path of the snowflakes can be observed. If the car is moving rather slowly, the snowflakes will approach the car and pass over the windshield without impaction. As the automobile increases in speed, the snowflakes will impact on the windshield. The two determining factors as to whether or not the snowflakes will impact on the windshield are the speed of the automobile and the size of each snowflake.

A similar observation is experienced when the automobile encounters flying insects. A large insect will have a trajectory undeflected by the airstream around the car and impact dramatically upon the windshield. Smaller insects will follow the airstream and not impact. However, if the body is smaller, the smaller insects will be collected. This can be seen on smaller cross-sectional-area bodies, such as the radio antenna. An inspection of the insects on the antenna will indicate that they are smaller than those collected on the automobile windshield.

The above example indicates that the velocity of the air, U, the size of the particle, d_p, and the size of the body, d_b, are three important parameters in determining whether or not a particle will be collected on the body. A dimensionless parameter, the Stokes number, defined as the ratio of the particle's stopping distance to the physical dimension of the body collector is the governing relationship as to whether or not a particle will strike the body. If the Stokes number, Stk, defined as:

$$Stk = \frac{\rho_p C_c d_p^2 U}{18 \eta d_b} \qquad (11\text{-}1)$$

where:

ρ_p = particle density
C_c = slip correction
U = relative velocity of body to air (or gas)
d_p = particle diameter
η = air (or gas) viscosity
d_b = body diameter

is larger than approximately unity, the particle will impact on the body.

Note that in the Stokes number, the three parameters discussed in the above example (U, d_p, and d_b) are included as well as properties of the gas and particles (η, C_c, and ρ_p). The Stokes number is important in all types of inertia collectors and not just body collectors. The Stokes number for a conventional impactor or a virtual impactor is the ratio of the stopping distance to the radius of a circular nozzle, or half-width of a rectangular

TABLE 11-1 Selected List of Commercially Available Impactors (Adapted from Hering 1989)

Manufacturer[1]	Sampler Name	Flow Rate (l/min)	No. of Stages	Cutpoints (range, μm)	Comments[2]
Cascade Impactors for Ambient Air Sampling					
AND	Sierra/Marple Model 210	7	10	0.16–18	a
AND	Sierra/Marple Model 260	0.3–20	6	0.5–20	b
AND	Low-Pressure Impactor	3	12	0.08–35	c
AND, GMW	One ACFM Ambient Impactor	28	8	0.4–10	
AND	Flow Sensor Ambient Impactor	28	7	0.4–6	
ATN	Low-Pressure Impactor (LPI)	1	8	0.05–4	c
BGI	May/R.E.	5	7	0.5–32	
—	Battelle Impactor, 1 l/min	1	5	0.25–4	d
—	Battelle Impactor, 12 l/min	12	6		d
HAU	Berner Impactor	30	9	0.063–16.7	
ITP	Mercer 7-Stage Impactor (02-100)	0.1	7	0.33–3.1	d
ITP	Mercer 7-Stage Impactor (02-130)	1	7	0.32–4.5	
ITP	Mercer 7-Stage Impactor (02-150)	2	7	0.25–5.0	
ITP	Mercer 7-Stage Impactor (02-170)	5	7	0.5–5.0	
ITP	Multijet CI (02-200)	10	7	0.5–8	
ITP	Multijet CI (02-220)	15	7	0.5–8	
ITP	Multijet CI (02-240)	20	7	0.5–8	
ITP	Multijet CI (02-260)	28	7	0.5–9	
MSP	MOUDI (Micro-Orifice Impactor)	30	10	0.056–18	e
PXI	Single-Orifice Impactor, Model 1CI	1	7	0.25–16	d
PXI	Single-Orifice Impactor, Model 1L-CI	1	9	0.06–16	f
CMI	Quartz Crystal Microbalance, PC-2	0.25	10	0.5–25	g
QCM	Quartz Crystal Microbalance, C-1000	0.25	10	0.5–25	g
Impactors for Ambient HiVol Samplers					
AND	HiVol Impactor, Series 65-800	1130	1	3.5	h
AND	HiVol Impactor, Series 65-800	565	4	1.1–7.0	h
AND, GMW	HiVol Impactor, Series 230	1130	4	0.49–7.2	i
AND, GMW	HiVol Impactor, Series 230	565	6	0.41–10	i
Personal Samplers					
AND, GMW	Marple Personal Sampler (Model 290)	2	8	0.5–20	a
SKC	Marple Personal Sampler	2	8	0.5–20	a
MSP	Personal Environmental Monitor	4 or 10	1	2.5	
MSP	Personal Environmental Monitor	4 or 10	1	10	
Source Test Impactors					
AND	In-Stack Air Sampler, Series 220	7	9	0.16–18	
AND	Stack Sampling Head (Mark III, IV)	3–21	8	0.4–11	a
AND	High-Capacity Stack Sampler	14	3	1.5–11	
AND	Impactor Preseparator	1	21	10	
ITP	High-Temperature, High-Pressure Impactor	16	7	0.62–8.8	
PCS	UW Source Test Cascade Impactor	28	10	0.2–20	
PCS	UW High-Capacity Source Test Impactor	28	3	1.5–11	
PCS	UW Low-Pressure Source Test Impactor	28	14	0.05–20	

1. See Table 11-4 for an explanation of manufacturer codes
2. Notes: (a) radial slot design; (b) circular jets, interchangeable nozzles; (c) four low-pressure stages; (d) one round jet per stage; (e) micro-orifice plates of 2000 jets on bottom stages; (f) two low-pressure stages added to 1 CI; (g) uses quartz crystal collection surfaces for continuous mass measurement; (h) fits on HiVol, round jets; (i) fits on HiVol, rectangular jets; (j) collection directly onto agar plates; (k) slot impactor with rotating turntable for agar plates

Table 11-2 Selected List of Commercially Available Virtual Impactors (Adapted from Hering 1989)

Manufacturer[1]	Sampler Name	Flow Rate (l/min)	No. of Stages	Cutpoints (range, μm)
AND, GMW	Dichotomous Sampler	16.7	1	2.5
BGI	Cascade Centripeter	30	3	1.2, 4, 14
MSP	High-Volume Virtual Impactor	1130	1	2.5
MSP	Microcontaminant Particle Sampler	30	1	1

1. See Table 11-4 for an explanation of manufacturer codes

Table 11-3 Selected List of Commercially Available Cyclones (Adapted from Hering 1989)

Manufacturer[1]	Cyclone Name	Flow Rate Range (l/min)	D_{50} Range (μm)
MSA, SEN, SKC	10 mm Cyclone (also called Dorr–Oliver)	0.9–5	1.8–7.0
AND, ITP	SRI V	7–28	0.3–2.0
AND, ITP	SRI IV	7–28	0.5–3.0
SEN	1/2″ HASL	8–10	2–5
AND, ITP	SRI III	14–28	1.4–2.4
—	AIHL	8–27	2.0–7.0
AND, ITP	SRI II	14–28	2.1–3.5
—	Aerotec 3/4	22–55	1.0–5.0
AND, ITP	SRI I	14–28	5.4–8.4
SEN	1″ HASL	65–350	1.0–5.0
—	BK 76	400–1100	1.0–3.0
BGI, GMW	Aerotec 2	350–500	2.5–4.0
—	BK-152	1150–2700	2.0–5.0

1. See Table 11-4 for an explanation of manufacturer codes

nozzle, and is defined as

$$Stk = \frac{\rho_p C_c d_p^2 U}{9\eta W} \qquad (11\text{-}2)$$

where

U = average air (or gas) velocity at the nozzle exit ($= Q/\pi(W/2)^2$) for a round nozzle impactor, and ($= Q/LW$) for a rectangular nozzle impactor
W = nozzle diameter (circular impactor) or nozzle width (rectangular impactor)
Q = volumetric flow rate through the nozzle
L = rectangular nozzle length

The Stokes number is a dimensionless parameter that can be used to predict whether or not a particle will impact on a body, an impaction plate, or in the collection probe of a virtual impactor, or will follow the air streamlines out of the impaction region and remain airborne. Actually, the square root of the Stokes number, $\sqrt{Stk}$, is more commonly used since it is a dimensionless particle size. A critical value of $\sqrt{Stk}$ often used to characterize inertial classifiers is $\sqrt{Stk_{50}}$. This is the value of $\sqrt{Stk}$ corresponding to d_{50}, the value of d_p collected with 50% efficiency. Thus, if the value of $\sqrt{Stk_{50}}$ is known, the value of d_{50}, corresponding to the cut-size of the impactor, can be found from

$$d_{50} = \sqrt{\frac{18\eta d_b}{\rho_p C_c U}} \sqrt{Stk_{50}} \qquad (11\text{-}3)$$

TABLE 11-4 Commercial Sources for Inertial Classifiers (Adapted from Hering 1989)

AND	Andersen Samplers, Inc. 4215 Wendell Drive Atlanta, GA 30336 (404)691-1910 or (800)241-6898	MSP	MSP Corporation 1313 Fifth Street S.E. Suite 206 Minneapolis, MN 55414 (612)379-3963
ATN	Atmospheric Technology P.O. Box 8062 Calabasas, CA 91302 (213)880-5854	MSA	Mine Safety Appliances Co. RIDC Industrial Park 121 Gamma Drive Pittsburgh, PA 15238-2919 or P.O. Box 426 Pittsburgh, PA 15230-2919
BGI	BGI Incorporated 58 Guinan Street Waltham, MA 02154 (617)891-9380	PIX	PIXE International Corp. P.O. Box 2744 Tallahassee, FL 32316 (904)222-0603
CMI	California Measurements 150 E. Montecito Avenue Sierra Madre, CA 91024 (818)355-3361	PCS	Pollution Control Systems Corp. P.O. Box 15570 Seattle, WA 98115 (206)523-7220
GMW	General Metal Works, Inc. 145 South Miami Avenue Village of Cleves, OH 45002 (513)941-2229 or (800)543-7412	QCM	QCM, Inc. P.O. Box 277 Laguna Beach, CA 92652 (714)494-9401
HAU	Hauke KG P.O. Box 63 A-4810 Gmunden, Austria (076) 12 41 33	SEN	Sensidyne Inc. 12345 Starkey Road, Suite E Largo, FL 33543 (813)530-3602
ITP	In-Tox Products 1712 Virginia NE Albuquerque, NM 87110 (505)299-1810	SKC	SKC Inc. 334 Valley View Road Eighty Four, PA 15330 (412)941-9701

for body impactors, and

$$d_{50} = \sqrt{\frac{9\eta W}{\rho_p C_c U}}\sqrt{Stk_{50}} \qquad (11\text{-}4)$$

for conventional and virtual impactors.

General Description

Inertial classifiers have been widely used for the separation of particles by their aerodynamic diameters, where the aerodynamic diameter is the diameter of a unit-density (1 g/cm^3) sphere that has the same gravitational-settling velocity as the particle in question. Four types of inertial classifiers in common use, as shown in Fig. 11-1, are body impactors, conventional impactors, virtual impactors, and cyclones. The first of these, the body impactor, is the simplest in that it consists of only a body in a moving aerosol stream onto which particles impact. The latter three of the inertial classifiers all consist of a jet of gas impinging upon a target.

A conventional impactor, in its simplest form, consists of a jet of particle-laden gas impinging upon a flat plate, with particles impacting upon the plate. Variations of the impactor include the use of either round or rectangular nozzles, single or multiple nozzles, and flat or cylindrical impaction plates.

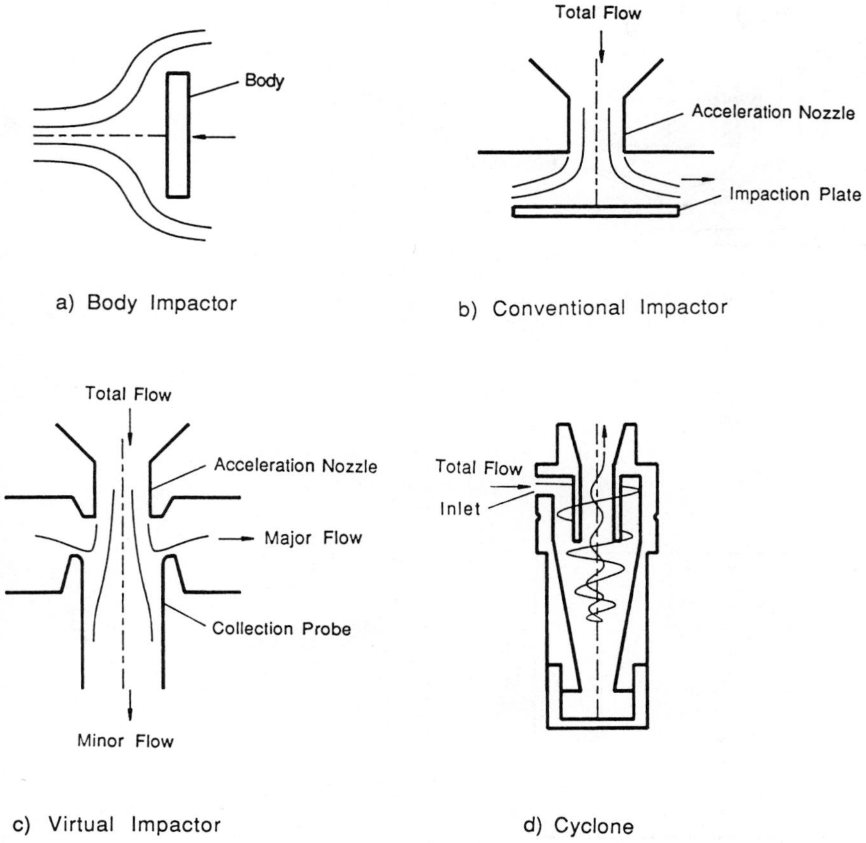

FIGURE 11-1. Four Types of Inertial Classifiers.

In a virtual impactor, the impaction plate is replaced by a collection probe slightly larger than the nozzle, with the classified particles penetrating into the collection probe. A small fraction of the gas passes through the collection probe to transport the classified particles out of the lower end of the probe. The remainder of the gas, the major portion, reverses direction in the collection probe and escapes at the upper edge.

In the cyclone, the aerosol stream is drawn through the inlet and impinges tangentially on the inner surface of a cylinder, flows in a spiral pattern down the inside of the cylinder and cone walls, reverses direction, spirals upward around the cyclone axis and exits through a centrally located tube at the upper end of the cylinder. Particles are collected on the cylindrical and conical walls by inertial forces on the particles. Clumps of impacted particles that are knocked off or dropped from the cyclone walls tend to fall to the apex of the cone, where they are collected in a cup sometimes called dust cap or grit pot.

Conventional Impactors

The most common type of impactor consists of a single jet of particle-laden gas (aerosol) impinging upon a flat plate, as shown in Fig. 11-2. Particles larger than the cut-size of the impactor will slip across the streamlines and impact upon the plate while smaller particles will follow the streamlines and not be collected. The most important impactor characteristic is the collection efficiency curve, also shown in Fig. 11-2. The collection efficiency (as a function of particle size) is defined as the fraction of particles passing

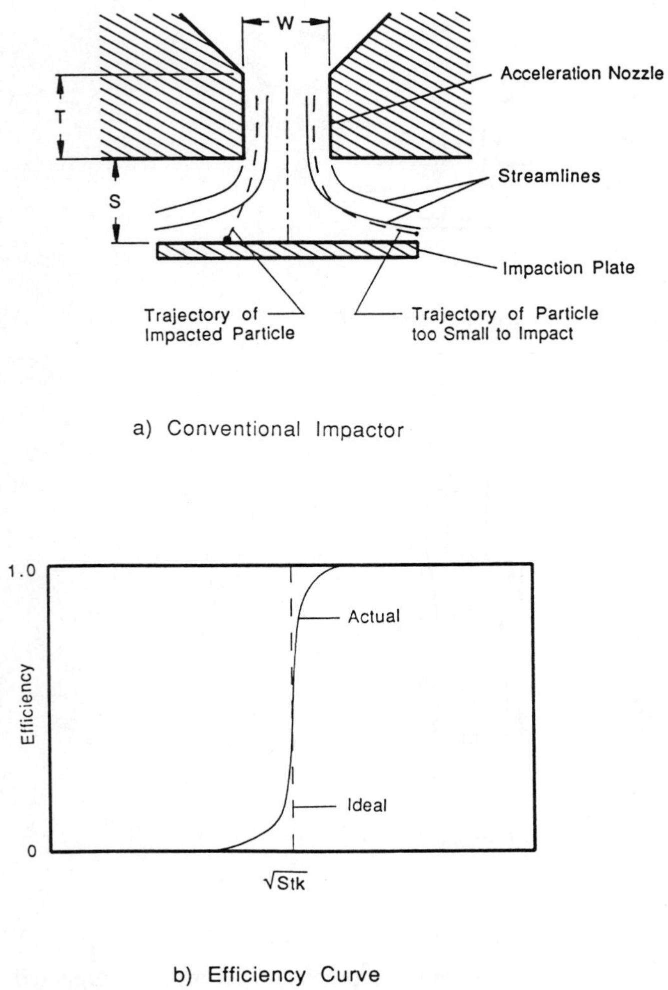

FIGURE 11-2. Schematic Diagram of a Conventional Impactor and the Corresponding Particle Collection Efficiency Curve.

through the nozzle that are collected upon the impaction plate. The ideal impactor has a perfectly sharp efficiency curve, i.e., all particles larger than the cut-size of the impactor are collected upon the plate, while all smaller particles follow the gas flow out of the impaction region.

A nozzle and impaction plate constitutes a single-stage impactor, which is useful when classifying particles into two size fractions. For example, this is the case for analyzing particles that are less than 10 μm (PM-10) or less than 2.5 μm (PM-2.5) (see Chapter 28). In these types of impactors, the particles larger than the cut-size are removed from the airstream while the smaller particles penetrate the impactor stage to be either collected on a filter, where they can be analyzed (e.g., for mass concentration, or elemental composition), or passed into some other instrument for real-time mass or number concentration measurement.

Often, it is desirable to determine the entire size distribution of the aerosol, and not just the quantity less than a certain size. In this case, a series of impactor stages are used in a cascaded fashion such that the gas passes from one stage to the next, as shown in

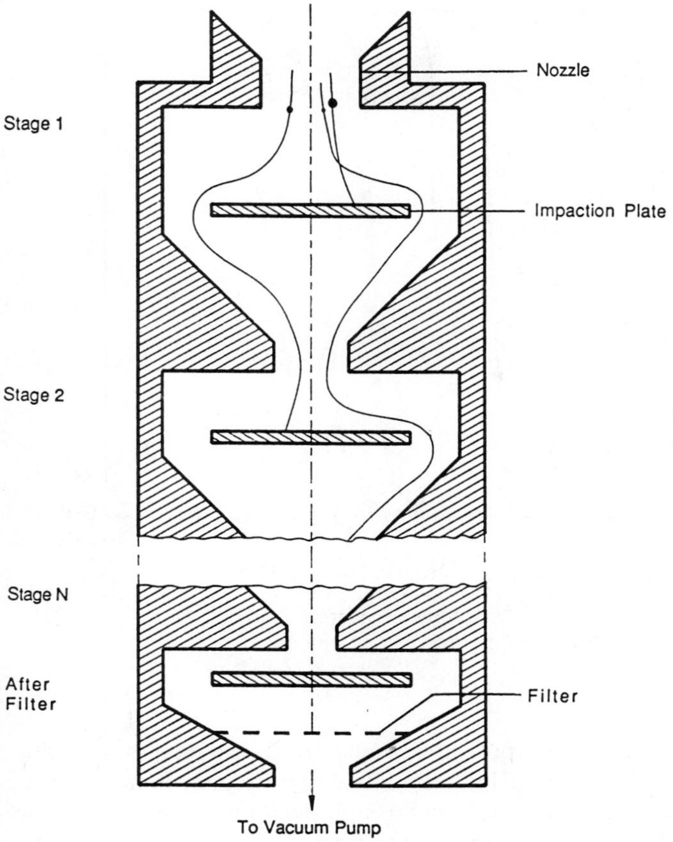

FIGURE 11-3. Schematic Diagram of Cascade Impactor.

Fig. 11-3, to remove particles in discrete size ranges (Lodge and Chan 1986). This is known as a cascade impactor and is widely used for determining size distributions of aerosols.

A cascade impactor makes use of the fact that particle collection is governed by the Stokes number. The velocity of the particle-laden gas stream is increased in successive stages, resulting in the collection of successively smaller particles in subsequent stages. For example, if a four-stage cascade impactor has cut-sizes of 10, 5, 2.5, and 1.25 μm, the first stage will collect particles larger than 10 μm, the second stage will collect particles between 5 and 10 μm, the third stage between 2.5 and 5 μm and the fourth stage between 1.25 and 2.5 μm. Particles less than 1.25 μm penetrate the final stage of the impactor and can be collected on an after-filter. The mass size distribution of the aerosol can then be determined by gravimetric analysis of the collected particles. Uncertainties in the distribution are the size of the largest particles collected on the first stage and the size of the smallest particles collected on the after-filter. These sizes may be estimated or, better, an impactor may be selected with a sufficient number of stages to span the entire size distribution so that the mass collected on the first stage and the after-filter is minimized.

As stated above, many cascade impactor designs have been built and tested. The calibration curves for no two designs will be exactly alike due to differences in design parameters (e.g., nozzle diameters, number of nozzles, sampling flow rates, etc.) and due to small influences of the boundary conditions on particle collection. However, Fig. 11-4

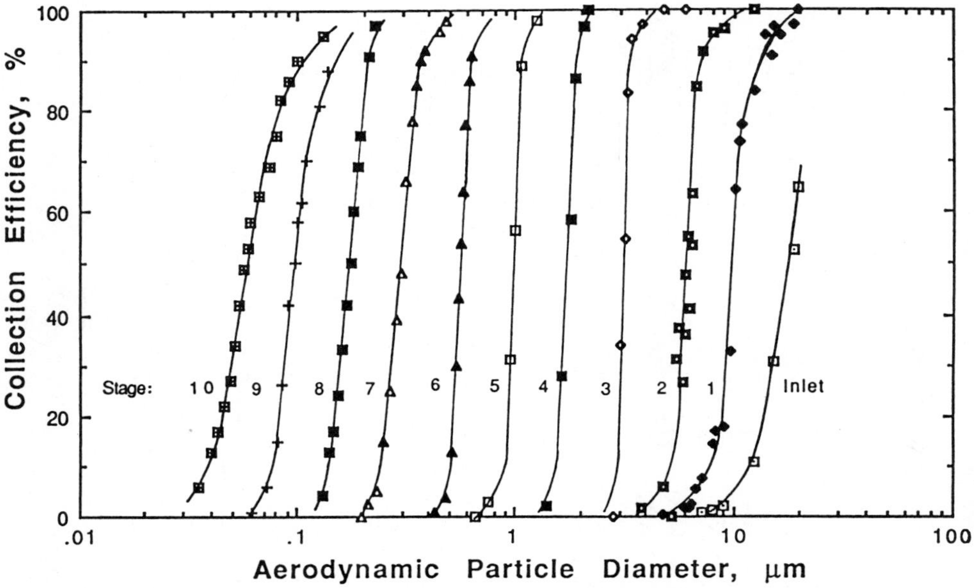

FIGURE 11-4. Particle Collection Efficiency Curves for Micro-Orifice Uniform Deposit Cascade Impactor (Marple, Rubow, and Behm 1991). (Reproduced with the Permission of Elsevier Sciences Publishing Co., Inc.)

shows a set of typical cascade impactors efficiency curves. Some impactors may have sharper cutoff characteristics and some not as sharp, but the general shape will be very similar to those shown.

An analysis of Eq. 11-4 reveals that the particle size range of an impactor can be lowered to very small sizes by either increasing the value of the slip correction, C_c (i.e., by going to low pressures in the impactor), or by decreasing the nozzle diameter, W. Impactors utilizing these techniques are known as either low-pressure impactors or micro-orifice impactors. Both types have been developed and used successfully in sampling particles down to approximately 0.05 µm diameter (Berner et al. 1979; Hering, Flagan, and Friedlander 1978; Hering and Marple 1986; Hering et al. 1979; Hillamo and Kauppinen 1991; Marple, Rubow, and Behm 1991).

There are several differences which must be considered when selecting either low-pressure or micro-orifice impactors for a particle sampling program. In a low-pressure impactor, the particles are collected by increasing the value of the slip correction at highly reduced pressures down to approximately 0.03 atm. This means that the vacuum pump drawing the flow through the impactor must be rather large, or the flow rate rather small. In addition, particles that are sensitive to evaporation at low pressures experience a reduction in size during the collection process (Biswas, Jones, and Flagen 1987). However, the low-pressure impactor is relatively simple to construct, since the nozzle diameters are similar in size to those used in conventional impactors.

In a micro-orifice impactor, the pressures are substantially larger than in low-pressure impactors, and conventional vacuum pumps can be used to obtain moderate flow rates (Marple, Rubow, and Behm 1991). Volatile particles can be more easily collected in this impactor, since the pressure drop through the entire impactor is only about 0.4 atm (Fang et al. 1991). The difficulty with the micro-orifice impactor is in its construction, since the nozzle diameters are very small (approximately 50 µm for the final stages) and the number of nozzles is large to obtain an adequate flow rate (as many as 2000 for the final stages). However, these are manufacturing problems and not problems in the use of the impactor.

Impactors are normally designed to have sharp cutoff characteristics (steep efficiency curves). However, in some cases, it is desirable to have a cutoff curve (efficiency curve) that is not sharp but rather one that follows a particular retention curve. Such is the case in designing impactors with penetration curves matching the American Conference of Governmental Industrial Hygienists (ACGIH) or the British Medical Research Council (BMRC) respirable mass criteria curves (Lippman 1989). A special class of impactors, called respirable impactors, has been designed to emulate penetration characteristics similar to these respirable curves (Marple 1978; Marple and McCormack 1983). This is accomplished by multiple-nozzle, single-stage impactors, with the nozzles having different diameters. The respirable-penetration curve is approximated by a series of steps corresponding to the number of different diameters used in the impactor stage. The fraction of the flow passing through each of the stages is proportional to the total cross-sectional area of the nozzles of a particular diameter. Using this technology, it is possible to design impactors for nearly any flow rate, with penetration characteristics that approximate any monotonically decreasing penetration curve.

Inertial impactors can be used over a wide range of conditions and have been designed for cut-sizes of 0.005 μm (Fernandez de la Mora et al. 1990) up to approximately 50 μm (Vanderpool et al. 1987) and flow rates from a few cm^3/min to several thousand m^3/min. There are, however, limitations to impactors when considering their use. Three major areas of concern are particle bounce from the collection surface, overloading of collected particle deposits on the impaction plates, and interstage losses (collection of particles on internal surfaces of the impactor other than the impaction plate).

Of the three limitations, particle bounce has received the most attention, since particles bouncing from an impaction plate will be collected on a subsequent stage that has a smaller cutpoint or on the after-filter and, thus, bias the measured size distribution or become lodged in and clog smaller nozzles of subsequent stages. The most logical approach to solving a particle bounce problem, and the technique used by many researchers, is to provide a sticky surface on the impaction plate. The mass stability and chemical composition, purity, and stability of the sticky surface are all factors in selecting the appropriate substance. Numerous types of greases and oils, including petroleum jelly, Apiezon greases (Apiezon Products Ltd., England), and silicone oils and sprays, have been used to coat the impaction plates; however, these coatings must be used correctly in order for the results to be satisfactory (Hering 1989; Rao and Whitby 1978*a*, *b*). For example, if the impaction plate has just a sticky surface, once a monolayer of collected particles forms on the surface, additional incoming particles will impact on the particles that have already been collected and particle bounce is again possible. Therefore, if a sticky surface is to be used, and a large quantity of particles collected, the sticky surface will not be adequate; a sticky substance must be used that will wick up through the particle deposit by capillary action and continually provide incoming particles with a sticky surface. As described in a later section, silicone oil works well.

The techniques that will be used for particle analysis must be considered when selecting an impaction surface. If the particles collected on the impaction plate are to be discarded, the impaction plate can be a porous material saturated with a light oil (Reischl and John 1978). This oil will wick up through the deposit and continually provide the incoming particles with a sticky impaction surface. Since these particles are to be discarded, the weight stability of the oil is not important.

If the mass size distribution for solid particles is to be determined from the deposits collected on the stages of a cascade impactor, great care must be taken in providing a sticky impaction plate surface. Since the deposits must be weighed, particle collection substrates such as metal foils, plastic films, or filters are placed upon the impaction plates so that they can be easily removed for weighing. Grease, oil, or other sticky substances are applied to the substrates. The sticky substance must have mass stability, since treated substrates will be weighed on a balance before and after sampling. In most cases, the sticky substance will be dissolved into a solvent and applied in a light film to the substrates. The treated substrates are then baked in an oven to drive off the volatiles, thereby leaving weight-stable sticky substrates onto which the particles can be collected.

The problem of deposit overloading on an impaction plate is more easy to control, since the quantity of particles collected is a function of sampling time. It is normal practice to sample from several minutes to several hours to obtain an initial size distribution. The deposits on the impaction plates are then inspected and, if any stage is overloaded, the next sampling period is shortened. Conversely, if the deposits are too light, the sampling period is extended. The major difficulty in determining the appropriate sampling time is dictated by the total quantity of particles that can be collected upon an impaction plate (before overloading occurs). This quantity varies from impactor to impactor and only experience or data from the manufacturer can provide this information.

The final area of concern is the collection of particles on surfaces other than the impaction plate. These particles are deposited on the interior surfaces of the impactor (interstage loss) and are a function of particle size, again distorting the measured size distribution. In the upper stages, where the particles are rather large, particles can be lost by impaction or turbulent deposition. However, since each stage removes particles larger than the cut-size of that stage, the interstage losses rapidly decrease as particles penetrate the upper stages. In the lower stages, particles can be lost by diffusion (a problem only for very small particles). Most of the diffusional losses occur in the final stages, where the nozzles are small and particles can diffuse to the nozzle walls.

In a properly designed cascade impactor that covers a wide range of particle sizes, one would expect to find interstage losses in the initial stages and maybe in the final stages of the impactor (if there are cut-sizes below 0.1 μm), with the losses being a minimum through the intermediate stages. Typically, particle loss varies but may be as high as 30%–40% for large particles in the inlet and first stages.

If experiments have been performed to determine the interstage loss as a function of particle size, the particle size distribution data of the impactor can be corrected. One difficulty in estimating what this correction should be is that the losses will vary as a function of the nature of the particles. For example, if the particles are liquid or sticky they will adhere to any surface with which they come in contact. A dry particle, however, may rebound when it hits a surface and remain airborne until it reaches the next stage. Thus, interstage losses would be less for a particle that bounces easily than for a sticky particle.

In summary, three major concerns of the impactor, i.e., particle bounce, overloading, and interstage losses, are a function of both the particle properties and the impactor. If the particles are sticky, particle bounce is minimized, interstage losses are maximized, and overloading is of less concern. If the particles are solid and rebound easily from a surface, particle bounce is a problem, interstage losses are decreased, and overloading is of concern.

Normally, a cascade impactor consisting of several stages is used to determine particle size distributions. However, the inertial spectrometer (INSPEC) and high-flow spectrometer (LASPEC) (Lavoro E Ambiente, Bologna, Italy), which provides particle size distribution information on one stage, have been developed (Belosi and Prodi 1987; Prodi et al. 1979, 1983; Prodi, Belosi, and Mularoni 1984). The aerosol sample flow rates are up to 0.2 and 10 l/min in the INSPEC and LASPEC, respectively. These devices employ a rectangular filter instead of an impaction plate and introduce the particles through a nozzle near one edge of the filter, as shown in

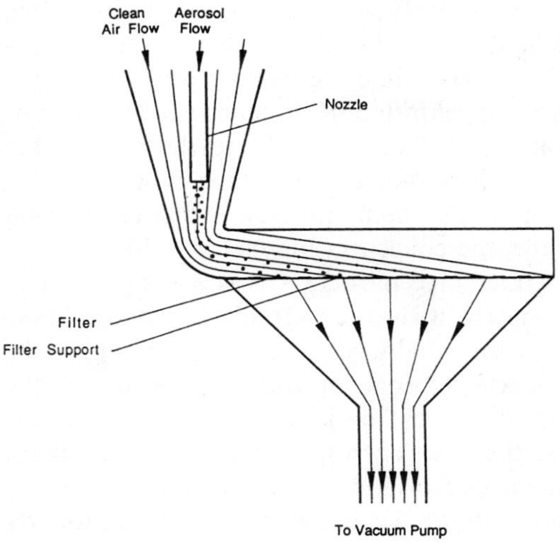

FIGURE 11-5. Schematic Diagram of an Inertial Spectrometer.

Fig. 11-5. The particles are introduced at the center of this nozzle with a sheath of clean air on both sides. The inertial properties of the particles distribute the particles in the sheath air at distances which are a function of the particles' aerodynamic diameters. Subsequently, as the particles are removed from the gas stream at the combination filter–impaction-plate, the position of the particles on the filter is a function of the aerodynamic diameter of the particles. This inertial classifier provides the entire size distribution on the filter.

It is also possible to design a spectrometer with the nozzle being at the center of a circular filter. The radial deposition of particles on the filter is a function of particle size, with the larger particles being deposited closer to the center than the smaller particles. This has been developed in a personal sampler, called a personal inertial spectrometer (PERSPEC) (Lavoro E Ambiente, Bologna, Italy), operating at a flow rate of 2 l/min (Prodi et al. 1988).

Virtual Impactors

A virtual impactor is a particle inertial classification device that is very similar to a conventional inertial impactor, with the primary difference being that the impaction plate has been replaced by a collection probe as shown in Fig. 11-6. The jet of particle-laden gas exiting the nozzle penetrates into the collection probe, where classification occurs. Large particles penetrate further into the collection probe than do small particles. A small fraction of the flow, the minor flow, is allowed to penetrate through the collection probe, carrying with it the particles larger than the cut-size of the virtual impactor. Most of the flow, the major flow, reverses in the collection probe and exits at the top of the collection probe, carrying with it the particles that are smaller than the cut-size. Therefore, both size fractions remain suspended in the gas. These two gas streams can then be directed to collection filters, into another inertial

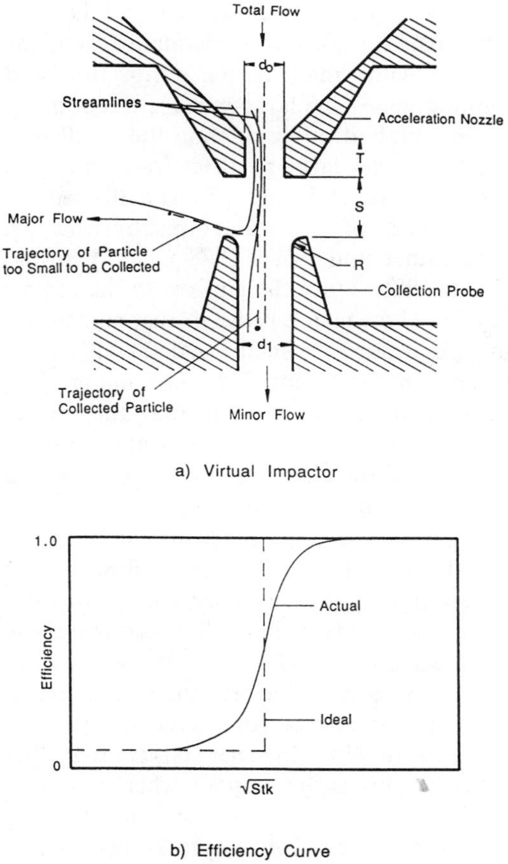

FIGURE 11-6. Schematic Diagram of a Virtual Impactor and Corresponding Particle Collection Efficiency Curve.

classification device, into another impactor stage, or into an instrument which automatically measures the concentration of particles in the gas streams. Conner (1966) described the first device to incorporate a virtual impactor with a controlled minor flow.

The particle collection efficiency curves for virtual impactors, as in the case of the conventional impactors, are quite sharp, as was shown in the theoretical study by Marple and Chien (1980). The major difference between the efficiency curves of virtual and conventional impactors is that particles smaller than the cut-size of the virtual impactor remain in both the major and minor flows. Therefore, the collection efficiency curve of particles in the collection probe asymptotically approaches the percentage of total flow which is penetrating the collection probe. For example, if the minor flow is 10% of the total flow, 10% of the particles smaller than the cut-size will remain in the minor flow and "contaminate" the large-particle fraction.

One method of eliminating the small particles in the large-particle fraction is to provide a central core of clean filtered air in the nozzle of the virtual impactor (Masuda, Hochrainer, and Stöber 1978; Chen, Yeh, and Cheng 1986). Since the air flow in the central portion of the nozzle is the air that constitutes the minor flow, the small particles will not be present in the minor flow. The major difficulty with this solution is that the virtual impactor becomes more complicated with the addition of the clean air flow. If some small particles in the large-particle fraction are not of concern, the contamination can be minimized by reducing the minor flow to the lowest value possible. In some cases this has been reduced to a value as low as 0.1% of the total flow (Xu, 1991).

Besides being an inertial classifier, a virtual impactor can also be considered as a particle concentrator for particles larger than the cut-size. This is very useful when sampling particles that are of low concentration or sampling particles into an instrument which requires concentrations greater than are normally present (Keskinen, Janka, and Lehtimäki 1987; Liebhaber, Lehtimäki, and Willeke 1991; Marple, Liu, and Olson 1989; Wu, Cooper, and Miller 1989). Since the particles larger than the cut-size of the impactor are concentrated in the minor flow, the concentration factor is equal to the ratio of the total flow rate to the minor flow rate. For example, if the minor flow is 5% of the total flow, the concentrating factor is 20.

Interstage particle loss is a major area of concern with the virtual impactor. These losses normally occur at the upper edge of the collection probe or on the backside of the nozzle plate. The losses are a maximum for particle sizes corresponding to the cutpoint of the impactor and can reach values as high as 60% if the virtual impactor is improperly designed or operated (Marple and Chien 1980). Major factors which influence these losses are the ratio of the collection probe diameter to the nozzle diameter, the shape of the collection probe inlet, the alignment of the axes of the nozzle and collection probe, the shape of the nozzle protruding through the nozzle plate, the jet Reynolds number, and the minor flow percentage. Several investigators have studied these parameters and the optimum values are still being refined (Jaenicki and Blifford 1974; McFarland, Ortiz, and Bertch 1978; Loo 1981; Chen, Yeh, and Cheng 1985, 1986; Chen and Yeh 1987; Loo and Cork 1988; Xu 1991). In general, however, the axes of the nozzle and collection probe must be aligned as close as possible, the collection probe diameter should be 30%–40% larger than the nozzle diameter, the inlet to the collection probe should be a smooth radius, the nozzle should protrude through the nozzle plate approximately two to three nozzle diameters and the minor flow should be 5%–15% of the total flow. Nearly all virtual impactors use round nozzles and collection probes; however, attempts have been made to use rectangular configurations (Forney 1976; Forney, Ravenhall, and Lee 1982).

Most virtual impactor designs have been single-stage units with one nozzle and one collection probe. However, some virtual impactor samplers have been designed with multiple nozzles and collection probes in a single stage. The purpose of this design feature is to reduce the size of the sampler and to reduce the pressure drop through the sampler (Marple, Liu, and Burton 1990; Szymanski and Liu 1989).

Body Impactors

The major problem in designing a conventional impactor for classifying large particles is that of drawing large particles into a sampling inlet. However, large particles can be classified inertially by the use of body impactors which do not involve drawing particles through an inlet. As shown in Fig. 11-7, the body impactor functions by simply sweeping an impaction surface (body) through the air, or conversely, drawing air past the body. In either case, the problems associated with the sampling of large particles is reduced.

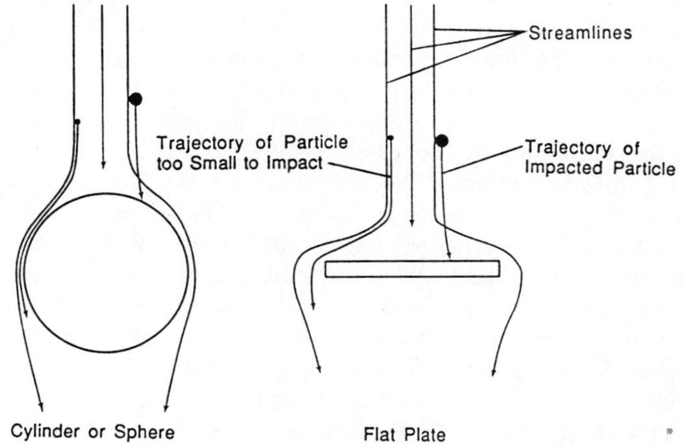

a) Body Impactors

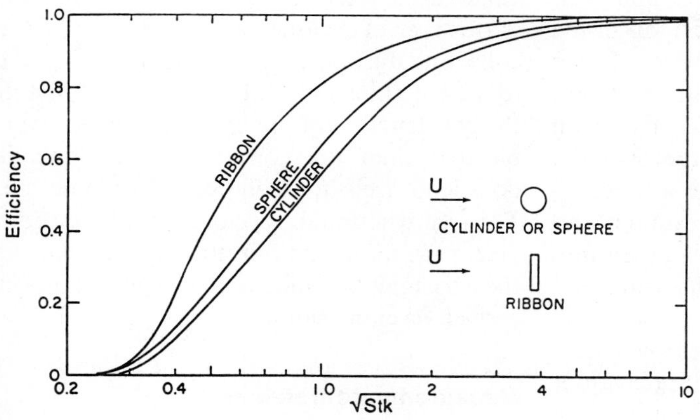

b) Efficiency Curves

FIGURE 11-7. Schematic Diagrams of Body Impactors and Corresponding Particle Collection Efficiency Curves from Golovin and Putnam (1962).

The collection efficiency of the body impactor is defined as the fraction of particles that are impacted on the body from a volume of air swept out by the body (Fig. 11-7). Particle collection curves are presented in Fig. 11-7 for particle deposition on ribbons, spheres, and cylinders. Golovin and Putnam (1962) and May and Clifford (1967) summarize particle impaction on bodies of various shapes. The cut-size is a function of the body size (dimension normal to the flow direction) and the relative velocity between the air and body. The cut-size is governed by the Stokes number equation (Eq. 11-1), where d_b is the body size and U is the velocity difference between the air and the body.

Examples of two body impactors which have been developed are the rotorod sampler and the Noll impactor. The rotorod sampler (Ted Brown Associates, Los Altos Hills, CA) is used for sampling particles larger than 15 μm and the most common application is the sampling of pollen levels in outdoor air. The Noll impactor is designed for sampling coarse atmospheric particles. It has four rotating plates of various dimensions to collect particles from 6 to 29 μm (Noll 1970; Noll et al. 1985).

Cyclones

In a cyclone, a jet of air impinges tangentially on the inner surface of a cylinder and then swirls downward in a cyclonic fashion inside the cylinder, and into a conical section. In the conical section the air reverses direction and spirals upward around the cyclone axis to an exit tube at the upper end of the cylinder. Particles larger than the cut-size are deposited on the inner surface of the cylinder and in the cone. The particles fall downward into the apex of the cone and into a dust collection cavity (grit pot).

Many cyclones have been designed in a variety of sizes for numerous applications and are very popular for collecting dust in industrial process lines. These units are normally large and have high flow rates.

Cyclones are also popular for aerosol sampling. While a variety of cyclones have been developed for respirable dust samplers, the most popular cyclone used in the United States is the 10 mm nylon cyclone which has penetration characteristics that simulate the ACGIH respirable mass criteria (Lippmann 1989).

The cut-size of the cyclone is governed by the flow rate, the size of the inlet and outlet tubes, and the size of the cylinder (e.g., Hering 1989). A rigorous theoretical analysis of a cyclone is more difficult than that of an impactor because the flow is three-dimensional and must be analyzed using a three-dimensional numerical program. Although some numerical work has been performed by the authors and others, it is not common practice to apply this technique in the design of cyclones.

Most cyclones are used as single-stage cyclones; however, cascaded versions have also been developed (Smith, Wilson, and Harris 1979; Liu and Rubow 1984). The unit developed by Smith, Wilson, and Harris for stack sampling consists of five stages with cut-sizes ranging from 0.32 to 5.4 μm at a sampling flow rate of 28.3 l/min.

Particle deposition in a cyclone is caused by the cyclonic action of the fluid in the cylinder. Most cyclones achieve this by the jet of gas impinging tangentially upon the inner surface of a cylinder. However, it is possible to achieve the cyclonic spiral flow by turning vanes at the inlet of a tube. A cascade version of this type of cyclone has been developed for collecting dust particles in a size range from 1 to 12 μm (Liu and Rubow 1984). Although the construction of this cyclone is more complicated than the conventional cyclone, it does lend itself to a compact cascade design. The conventional cyclone configuration, where the air enters tangentially and exits at the axis, makes it difficult to design a compact cascade cyclone sampler.

Measurement Strategies

Before selecting an inertial classifier, one must first decide the purpose for which the particles are being collected. If the purpose is to determine the characteristics of aerosols as a function of size, such as obtaining the mass size distribution or the chemical composition of the aerosol at various particle sizes, it is most convenient to use a cascade sampler. If only the quantity of particles less than a specified size is desired, such as is often the case for compliance sampling, then a two-stage sampler consisting of an inertial classifier followed by a filtration stage is most convenient.

A decision must also be made as to whether or not the sample needs to be time-resolved or can be an integrated sample. Normally, when samples are taken with an inertial classifier, the sample is integrated over the time period. Some impactors provide time-resolved samples on rotating impaction plates. These are the Lundgren cascade impactor (Lundgren 1971) and the Davis rotating-drum universal-size-cut monitoring (DRUM) impactor (Raabe et al. 1988).

Another important factor is the particle size range over which one needs to operate. Although inertial classifiers can classify particles in the size range of about 0.005–50 μm, a more precise size range must be known in order to select the device best suited for the test. In some cases, the size distribution of the aerosol may not be known until after the first sample has been taken. If little is known about the size distribution before the first test, a wide-range cascade sampler provides the most information.

Finally, the analysis technique to be employed for an inspection of the deposits will influence the type of sampler used for particle collection. For example, if the mass size distribution is to be determined by gravimetric analysis of the collected samples, an impactor with substrates coated with a sticky surface can be used. However, the substrates and sticky surfaces must have good mass stability. If a particle sample is to be analyzed with scanning electron microscopy (SEM), then a sticky surface is not desirable. In this case, a method must be devised whereby the particles can be collected upon a dry surface. In cases where sticky coatings cannot be used to reduce particle bounce, it is desirable to have the cut-sizes of a cascade impactor as close together as conveniently possible so that the inertia of particles impacting on the impaction plates is kept as small as possible. It has also been found that submicrometer particles do not appear to bounce as easily as supermicrometer particles. Therefore, if a sticky coating cannot be used, a greased 1 μm cut stage should follow an ungreased 1 μm stage to collect any supermicrometer particles which may have bounced to this point in the impactor. Thus, particles penetrating the 1 μm cut stage (the submicrometer particles) will not be contaminated with any supermicrometer particles that may have bounced through the upper stages.

Design Considerations

Of the various types of inertial classifiers discussed in the previous sections, impactors are most likely to be designed for specific applications. References have been provided in previous sections for the basic design of other inertial samplers. In some cases the sampling criteria may be such that a suitable commercially available impactor does not exist. This may be for reasons of size, configuration, number of stages, or the particular cut-sizes desired. For these situations it is recommended that a special impactor be designed and built for the study, based on following a few simple guidelines (Marple and Rubow 1986).

Inertial impactors have been studied extensively through theoretical analysis by numerical methods (Marple 1970; Marple and Liu 1974; Rader and Marple 1985). From these studies, numerous particle collection efficiency curves have been calculated and reported in the literature. In most cases, the theories have been compared to experimental results with good agreement. For the most part, an impactor's efficiency curve can be determined with as much accuracy from theoretical analysis as from experimental calibration.

To operate correctly, two simple design guidelines must be followed. These guidelines involve two dimensionless parameters: jet-to-plate distance divided by the nozzle diameter (S/W) and jet Reynolds number (Re_j)

(Marple 1970; Marple and Liu 1974; Marple and Willeke 1976*a,b*; Rader and Marple 1985). These guidelines were obtained from a theoretical, numerical analysis of the Navier–Stokes equations to determine the flow field and the subsequent numerical integration of particle trajectories. This process was used to determine the guidelines necessary to obtain sharp collection efficiency curves. These theoretical guidelines have been used in the design of impactors and have been shown to result in impactors with collection efficiency curves that have sharp cutoff characteristics (Marple et al. 1988).

Figure 11-8 shows the theoretically determined efficiency curves as a function of the jet-to-plate distance divided by the nozzle diameter (S/W) and the jet Reynolds number (Re_j) of the flow. Both these terms are important in that they will influence the efficiency curve and the 50% cutpoint of the impactor. For example, the position of the efficiency curve as a function of the S/W ratio is shown to be sensitive for small values of S/W. The values of Stk_{50} are relatively constant for S/W values larger than 0.5 and 1.0 for round and rectangular impactors, respectively. The S/W value should not be less than this value

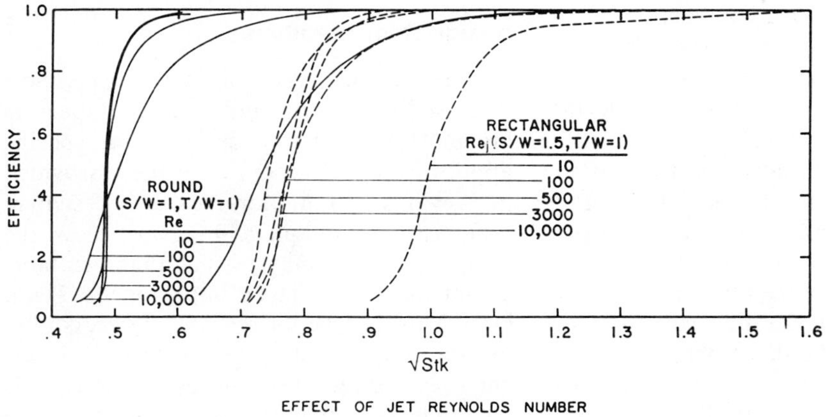

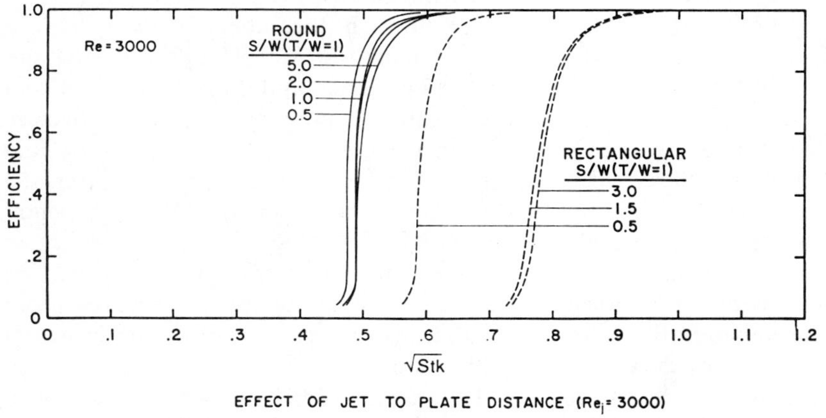

FIGURE 11-8. Theoretical Impactor Efficiency Curves for Rectangular and Round Jet Impactors Showing the Effect of the Jet-to-Plate Distance Ratio (S/W) and the Jet Reynolds Number (Re_j) (Rader and Marple 1985). (Reproduced with the Permission of the Elsevier Sciences Publishing Co., Inc.)

because small variations in S/W will then change the cut-size of the impactor. To provide a margin of safety, it is recommended that the S/W value be greater than 1.0 for round impactors and 1.5 for rectangular impactors. The upper value of S/W that can be used in an impactor is not as well known. However, impactors have operated well with S/W values as high as 5 or 10. If values larger than these are to be used, a calibration is recommended to ensure that the jet has not dissipated before impinging upon the plate.

The importance of the jet Reynolds number is more related to the sharpness of cut than to the cut-size of the impactor. If the Reynolds number is low, the gas viscous forces will be large and the velocity profile will be parabolic at the nozzle exit and in the air jet as it approaches the impaction plate. This will enhance the collection of particles near the centerline, where the smallest particles will be collected. The low velocity near the nozzle wall will require the particles in this portion of the flow to be larger in order for collection to occur. Therefore, the result of an impactor operating at a low Reynolds number is a less sharp collection efficiency curve. Theoretical analysis has shown, and experiments have verified, that the efficiency curve will be its sharpest if the Reynolds number is kept in the 500–3000 range for both round and rectangular impactors. Re_j is expressed as

$$Re_j = \frac{\rho_g W U}{\eta} \quad \text{(round)} \tag{11-5}$$

$$Re_j = \frac{2\rho_g W U}{\eta} \quad \text{(rectangular)} \tag{11-6}$$

where ρ_g is the gas density. Example 11-1 shows the calculation of the nozzle diameter and the Reynolds number for an impactor.

Example 11-1

Calculate the nozzle diameter (W) and the jet Reynolds number (Re_j) for a 10 μm cutpoint impactor. The impactor has one round nozzle and the flow rate is 30 l/min (500 cm³/s). Assume normal temperature and pressure.

Answer. The first step is to determine the required nozzle diameter for a cutpoint of 10 μm. The Stokes equation (Eq. 11-2) for a round jet impactor is

$$Stk = \frac{\rho_p C_c d_p^2 U}{9\eta W}$$

Substituting the flow rate for the average jet velocity into the Stokes equation yields

$$U = \frac{4Q}{\pi W^2}$$

and

$$Stk = \frac{4\rho_p C_c d_p^2 Q}{9\pi\eta W^3}$$

where

ρ_p = particle density (1 g/cm³)
d_p = cutpoint particle diameter (10×10^{-4} cm)
Q = volumetric flow rate (cm³/s)
C_c = Cunningham's slip correction (approximately 1)
η = gas viscosity (181×10^{-6} poise)
W = nozzle diameter (cm)

Solving the Stokes equation for the nozzle diameter gives

$$W = \sqrt[3]{\frac{4\rho_p C_c d_p^2 Q}{9\pi\eta\, Stk}}$$

The square root of the Stokes number ($\sqrt{Stk}$) corresponding to 50% collection efficiency ($\sqrt{Stk_{50}}$) can be estimated from Fig. 11-8, which shows the theoretical impactor efficiency curves for different values of the jet Reynolds number. Assume a jet Reynolds number of 3000; then the corresponding $\sqrt{Stk_{50}}$ is 0.47 and

$$W = \sqrt[3]{\frac{4 \times 1 \times 1 \times (10 \times 10^{-4})^2 \times 500}{9\pi \times (181 \times 10^{-6}) \times (0.47)^2}}$$

$$= 1.2 \text{ cm}$$

Knowing the nozzle diameter, the jet Reynolds number is calculated from Eq. 11-5

$$Re_j = \frac{\rho_g WU}{\eta}$$

Substituting the flow rate for the jet velocity yields

$$Re_j = \frac{4\rho_g Q}{\pi W\eta} = \frac{4 \times 500 \times (1.205 \times 10^{-3})}{\pi \times 1.2 \times (181 \times 10^{-6})}$$

$$= 3500$$

The jet Reynolds number is outside of, but close to, the upper limit of the suggested range of 500–3000 and, therefore, should provide sharp cutoff characteristics.

Example of Impactor Application

Impactors have been used to sample a wide variety of aerosols in a variety of studies. Since it is impossible to cover all the situations, the next few sections will describe the use of a generic impactor to solve a hypothetical problem. In each step of the process, remarks will be made as to how the step applies to the general sampling of aerosols with impactors. Examples will be given in these sections to demonstrate the type of calculations that must be made when using impactors.

The problem in this example was to determine the size distribution and mass concentration of dust particles in an industrial environment. In the initial phase of the program, it was suspected that the particle size distribution contained particles as small as 0.1 µm. The cut-size of the upper stages of the impactor had to be large enough to cover the respirable range, with the first stage of the impactor having a cutpoint of at least 10 µm.

The impactor selected for this analysis had nine stages and an after-filter. The cut-sizes of these stages are given in Table 11-5. The flow rate through the impactor was 30 l/min.

TABLE 11-5 Sample Size Distribution Data

Impactor Cutpoint (µm)	Δm (µg)	Cumulative Mass Percent Less than Indicated Size	$\frac{\Delta m}{\Delta \log(d_a)}$
18	10	99.9	392
10	100	98.9	2340
5.6	590	93.0	7410
3.2	1800	75.0	12,400
1.8	3100	44.0	11,000
1.0	2810	15.9	4920
0.56	1240	3.50	1250
0.32	305	0.45	168
0.18	42	0.03	
Filter	3	0.00	

Substrate Preparation

Since the purpose of sampling the particles was to determine the mass size distribution, the substrate had to have a stable weight and, therefore, aluminum foil substrates were selected. Filter material could also have been used, though some of these materials have a tendency to change weight with different humidity conditions. In addition, since the dust particles are solid, the application of some type of sticky substance to the substrates to reduce particle bounce was necessary. Therefore, the material used for the substrate had to be impervious to the oil so that it did not migrate through the substrate and get lost on the support of the substrate. For this application it was decided to use a silicone oil spray. A mask was prepared by cutting a hole in a clear plastic sheet just large enough to accommodate the deposited particles, placed over the foil and oil applied. The substrates were then placed in an oven at a temperature of 65°C for 90 min to evaporate the volatiles. The substrates were weighed on a microbalance, and then placed upon the impaction plates for use in the sampling program.

Sampling Time Estimations

To determine the appropriate sampling period, an estimate of the aerosol mass concentration was made to prevent overloading and bounce on the substrates. The size distribution of the aerosol was not known, so it was assumed that the aerosol was spread uniformly over the number of stages used in the impactor. The regulatory limit for the particle mass concentration in this work environment was 2 mg/m^3. Thus, it was assumed that the mass concentration was no more than this value. The nine stages of the impactor covered a size range from 0.1 to 18 μm and each stage was assumed to hold 1 mg of material. The sampling period was estimated to be 2.5 h at a sampling flow rate of 30 l/min.

Sample Analysis

After the samples were collected, the substrates were brought back to the laboratory for analysis. For this particular problem, it was necessary to determine the mass size distribution.

There were several other types of analysis that could have been performed with the substrates at this point: e.g., X-ray fluorescence analysis to determine the elemental composition, optical microscopy for inspection of the deposit, and SEM to investigate the shape and elemental composition of individual particles. The analysis of these samples with a SEM would have been difficult because of the presence of the sticky oil surface; an ungreased substrate on one of the stages would have allowed this type of analysis. If particle bounce had been too severe, this would not have resulted in good data. If grease coatings were necessary, the particles could have been washed from the substrates with a solvent and the particles separated from the solvent by filtration. These particles would then have been available for SEM analysis.

The mass concentration during one sampling period was 2.0 mg/m^3 (Example 11-2) and size distribution results tabulated in Table 11-5.

Example 11-2

The nine-stage cascade impactor was used to sample an aerosol for 167 min operating at a flow rate of 30 l/min. Table 11-5 lists the results of the gravimetric analysis. From this table, determine the total mass concentration. Plot a histogram of relative mass, i.e., $\Delta m/\Delta \log(d_a)$, versus aerodynamic diameter on semilog graph paper. Plot the cumulative mass versus aerodynamic diameter on lognormal probability graph paper. Determine the mass median diameter (MMD) and the geometric standard deviation (σ_g).

Answer. Tht total mass concentration is found by summing the Δm column in Table 11-5 and dividing that sum by the product of the flow rate and the sampling time. The total mass is 10.0 mg; therefore,

total mass concentration

$$= \frac{10.0 \text{ mg}}{(30 \text{ l/min})(167 \text{ min})(1 \text{ m}^3/1000 \text{ l})}$$

$$= 2.00 \text{ mg/m}^3$$

Figure 11-9 shows the relative mass histogram. The curve superimposed on the histogram shows that the mass distribution is lognormal. A plot of cumulative mass as a function of aerodynamic diameter is shown in Fig. 11-10. From this plot the mass median diameter and geometric standard deviation can be obtained. The mass median diameter is the diameter where 50% of the mass is collected and, from Fig. 11-10, is 2.0 μm. The geometric standard deviation is given by

$$\sigma_g = \frac{d_{84.1\%}}{d_{50\%}} = \frac{4.0}{2.0} = 2.0$$

or

$$\sigma_g = \frac{d_{50\%}}{d_{15.9\%}} = \frac{2.0}{1.0} = 2.0$$

or

$$\sigma_g = \sqrt{\frac{d_{84.1\%}}{d_{15.9\%}}} = \sqrt{\frac{4.0}{1.0}} = 2.0$$

Size Distribution Data Analysis

Particle size distribution data are normally represented as number, surface area, or volume (mass) size distributions using histograms or cumulative graphs. Since an impactor collects particles that are weighed on a balance to provide the mass of the particles in a particular size range, the mass distribution is the most common method for presenting impactor data. The data from the previous example are shown in Table 11-5 as the mass of particles in each of the size ranges represented by the stages of the impactor. It is possible to estimate the surface area and the number of particles in each of these size fractions by assuming the particles are spheres. To remove the bias of the width of the size fraction, the mass of particles is divided by the classification width. Further, since the particle size is normally represented on a log scale, it is better to have the histogram height as $\Delta m/\Delta \log d_a$. Now when the histogram is plotted as shown in Fig. 11-9, the area under the curve is representative of the percentage of mass in a particular size range. For this particular example, the mass size distribution is of interest and Fig. 11-9 shows the resulting mass size distribution.

A cumulative distribution can also be calculated. This has been performed in Table 11-5 and the results shown in Figure 11-10. Because the size distribution is made up of

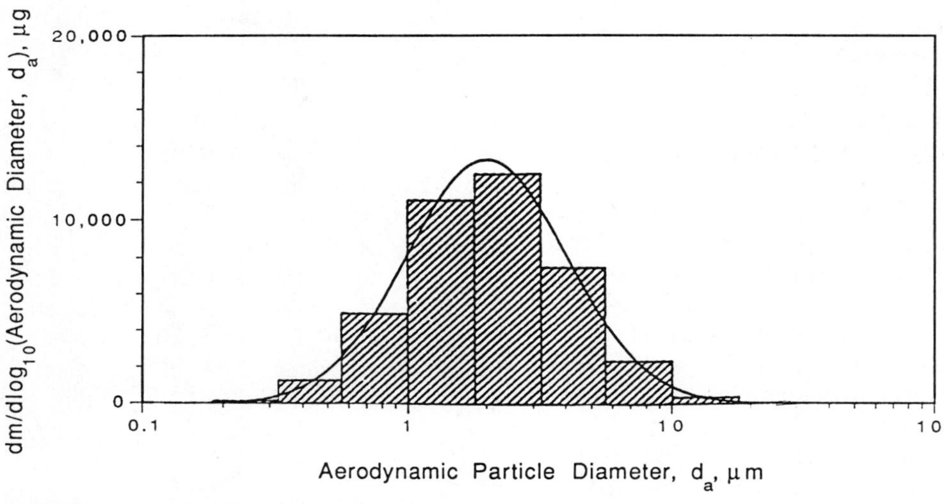

FIGURE 11-9. Relative Mass Histogram for Example 11-2.

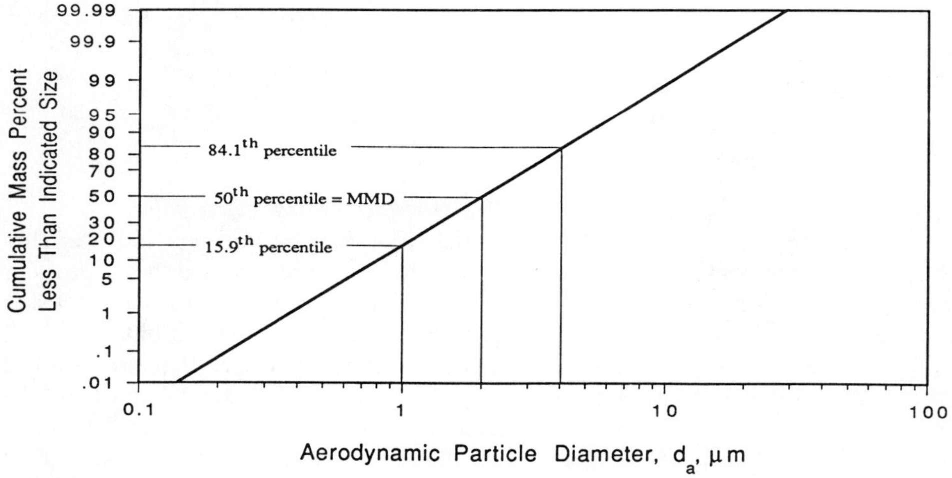

FIGURE 11-10. Lognormal Probability Plot for Example 11-2.

one distinct class of particles with a lognormal size distribution, the data lie along one straight line in the cumulative distribution.

In the above analysis it was assumed that the impactor had ideally sharp cuts at the 50% cutpoints of the stages. Since the true efficiency curve is probably S-shaped but fairly sharp, this is a good assumption, in that some particles are included from the size class above the stage and some are collected at the subsequent stage. In a rather broad distribution, these two errors tend to cancel each other and the data are a fairly good representation of the actual size distribution. However, there are techniques by which the actual shape of the efficiency curve can be incorporated into the data analysis and a more accurate description of the particle size distribution obtained. In recent years a number of these data inversion techniques have been developed (e.g., Crump and Seinfeld 1982; Markowski 1987; Rader et al. 1991; Wolfenbarger and Seinfeld 1990). Additional information on data analysis is presented in Chapter 9.

SETTLING DEVICES AND CENTRIFUGES

The definition of a particle's aerodynamic diameter is the equivalent diameter of a unit-density sphere that has the same gravitational-settling velocity as the particle in question. Therefore, a device which measures the settling velocity directly, such as a settling chamber, is a natural selection for a device to make direct measurements of a particle's aerodynamic diameter.

Settling chambers do not operate well for small particles because of low settling speeds (for example, the settling velocity of a 1 μm particle is 0.0035 cm/s). In addition, Brownian motion interferes with the settling of small particles and sets a lower limit at about 0.6 μm (Orr and Keng 1976). However, settling chambers can be used for larger particles (settling speed of 10 μm particles is 0.305 cm/s). Furthermore, because of the low settling velocities, great care must be taken for a particle to eliminate any convective air currents in the chamber.

John and Wall (1983) constructed a device to measure the settling speeds of particles in the 10–20 μm size range. This was achieved by illuminating the particles with a laser beam directed up a tube in which the particles were settling, and measuring the time taken to fall a predetermined distance.

A special form of a settling chamber is a horizontal elutriator, where an aerosol is passed slowly along a horizontal channel. Particles settle onto the bottom of the flow channel at locations dependent upon the particle size, particle density, gas velocity, and channel height. The larger aerodynamic particles settle near the entrance while the smaller particles are deposited near the exit. Two devices that incorporate this technique for particle classification are the horizontal elutriator used for respirable dust sampling and the Timbrell aerosol spectrometer.

Horizontal elutriators are used in two-stage samplers for respirable dust measurement. The nonrespirable particles deposit on horizontal, parallel plates in the elutriator, while the respirable particles penetrate through the elutriator to be either collected for subsequent gravimetric analysis or passed into a detector such as a photometer. The particle penetration characteristics of the ideal horizontal elutriator, by definition, are equivalent to the British Medical Research Council (BMRC) respirable dust criteria. The MRE gravimetric dust sampler was developed for respirable dust measurement (Wright 1954, Dunmore, Hamilton, and Smith 1964).

The physical description and operation of the Timbrell aerosol spectrometer are given in detail by Timbrell (1954, 1972). This spectrometer achieves accurate particle size classification by winnowing sampled particles in a laminar stream of clean air, where they settle along the bottom of a horizontal sedimentation chamber according to their aerodynamic diameter. The chamber is a wedge-shaped channel with microscope slides recessed into the horizontal floor. Particles ranging from 1.5 to 25 μm aerodynamic diameter can be classified at a winnowing air flow rate of 0.1 l/min. The instrument is calibrated in terms of aerodynamic particle diameter by means of spherical particles of known density. This spectrometer has been used to determine the aerodynamic size classification of spheres, particles of irregular shape, fibers, and aggregates (e.g., Griffiths and Vaughan 1986).

The difficulty in using a settling chamber or horizontal elutriators is that the force on the particles is quite small. One method to greatly increase the forces and increase the settling speed is to use a centrifuge. In a centrifuge, the air and particles are rotated at a high rotational speed and the centrifugal force is used to deposit the particles on the outer edge of an aerosol chamber.

Although several centrifuges have been developed, the one that has seen the greatest application and is currently in use today is the spiral centrifuge (Hoover, Morawietz, and Stöber 1983; Stöber 1976; Kotrappa and Light 1972; Stöber and Flachsbart 1969). One commercially available centrifuge is the Lovelace aerosol particle separator (LAPS) (In-Tox Products, Albuquerque, NM). This classifier is actually a spectrometer, in that the particles are introduced at the inner wall of a flow entering a spiral flow passage that is being rotated at a high rate of speed as shown in Fig. 11-11. At the entrance to the spiral, the particles are near the inner wall and a clean sheath air is adjacent to the outer wall. As the particles flow along the passage they are forced by centrifugal forces through the clean air and deposited on the outer edge of the spiral. The forces on the larger particles are greater and, therefore, they are deposited closer to the inlet of the spiral than are the smaller particles. Thus, the particles are deposited along the spiral passage as a function of decreasing particle size. Before sampling, a foil is placed in the outer wall of the spiral passage that is removed after sampling and the particles analyzed. The centrifuge can be calibrated with particles of known size and density, such that the aerodynamic diameter as a function of distance from the inlet is known. Once calibrated, a centrifuge can be used to obtain the aerodynamic size distribution of irregularly shaped particles, aggregate particles, and fibers (e.g., Martonen and Johnson 1990).

THERMAL PRECIPITATORS

Thermal precipitators are a class of instruments which make use of thermophoretic forces on particles to collect particles onto a sampling surface. The thermal precipitator

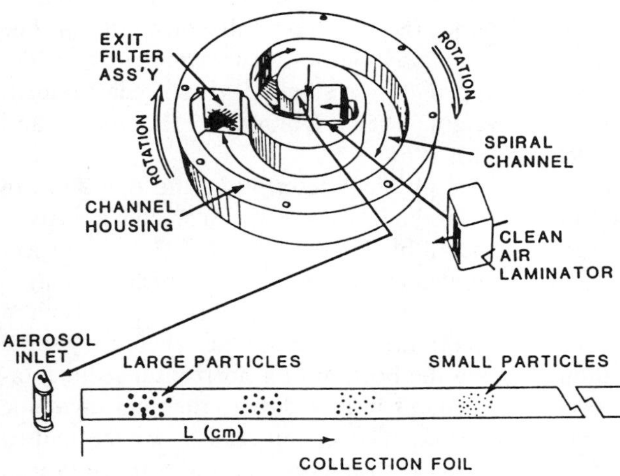

FIGURE 11-11. Schematic Diagram of Centrifuge (Cheng, Yeh, and Allen 1988). (Reproduced with the Permission of Elsevier Sciences Publishing Co., Inc.)

has been used for sampling respirable size particles in mines in Great Britain and South Africa.

The principal of operation is quite simple (Waldmann and Schmitt 1966). When a particle passes through a temperature gradient in the air, the air molecules from the warmer side of the particle strike the particle with higher energy than the air molecules on the cold side. This provides a net force in the direction of the cold surface and the particle migrates to the cold surface.

The results of this phenomenon can be experienced in everyday life. For example, where a hot water or steam radiator is adjacent to a wall, the wall behind the radiator becomes dirty. Also, in cold climates, the inside surfaces of windows in an automobile collect a film of contamination. Both these phenomena are due to the particles migrating to a cold surface when a temperature gradient exists near the surface.

By making use of the thermophoretic forces, the thermal precipitator can be simple in design. All that is necessary is to place a hot wire or filament close to a cooler surface, with particles passing between the filament and the surface. The particles then migrate to the cool surface, which is most often a microscope slide, and are collected. The particles can subsequently be analyzed by microscopic analysis.

The characteristics of thermal precipitators are that the flow rates are rather low (of the order of a few cm^3/min) and the collection efficiencies are good for particles in the submicrometer range, i.e., for particles down to 0.01 μm. The upper size limit of particles collected by this technique is of the order of 5–10 μm. With the thermal precipitator being efficient at collecting particles in this size range, it has been used most extensively for collecting respirable size particles in industrial atmospheres, especially in the mining industry. One configuration of the thermal precipitator is shown in Fig. 11-12. This shows the heated filament and the cooler surface, which contains the microscope slide.

There was a feeling that the standard thermal precipitator was rather bulky, required a fair degree of expertise to operate, and samples had to be counted under a microscope, a labor-intensive task. In the early 1950s, a modified thermal precipitator was developed, which employed a photoelectric detector to replace the optical microscope counting of the particles (Kitto and Beadle 1952; Beadle 1954). The modified thermal precipitator operated at 10 cm^3/min for sampling periods from 1 to 10 min. A horizontal elutriator was used at the inlet to remove particles that were nonrespirable and the particles deposited upon a slide. The slide would be advanced between tests and up to 11 samples were collected upon one slide. The samples on the slide were then conditioned in a desiccator and placed in a photoelectric assessor, which compared the light transmission through the deposit to the light transmission through the clean portion of the slide. The amount of light transmitted through the deposit and slide were measured by photoelectric cells and the numbers were quoted as dimensionless photoelectric readings, which could be related to the mass of particles collected.

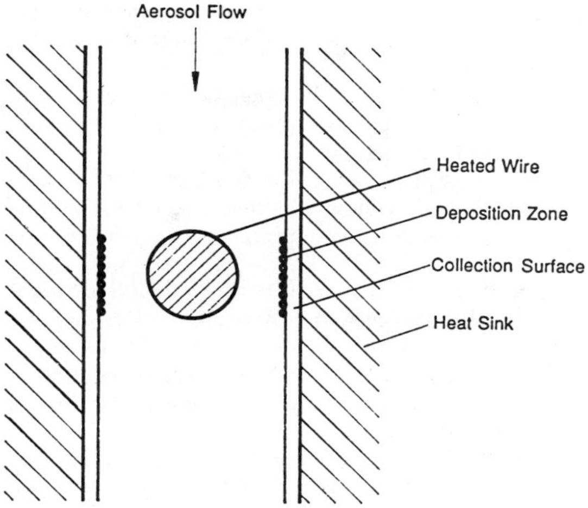

FIGURE 11-12. Cross-Sectional View of a Heated Wire and Plate Thermal Precipitator.

References

Belosi, F. and V. Prodi. 1987. Particle deposition within the inertial spectrometer. *J. Aerosol Sci.* 18:37–42.

Beadle, D. G. 1954. A photo-electric apparatus for assessing dust samples. *J. Chem. Met. Min. Soc. SA* 55:30–39.

Berner, A., C. Lürzer, F. Pohl, O. Preining, and P. Wagner. 1979. The size distribution of the urban aerosol in Vienna. *Sci. Tot. Envir.* 13:245–61.

Biswas, P., C. L. Jones, and R. C. Flagan. 1987. Distortion of size distributions by condensation and evaporation in aerosol instruments. *Aerosol Sci. Technol.* 7:231–46.

Chen, B. T. and H. C. Yeh. 1987. An improved virtual impactor: Design and performance. *J. Aerosol Sci.* 18:203–14.

Chen, B. T., H. C. Yeh, and Y. S. Cheng. 1985. A novel virtual impactor: Calibration and use. *J. Aerosol Sci.* 16:343–54.

Chen, B. T., H. C. Yeh, and Y. S. Cheng. 1986. Performance of a modified virtual impactor. *Aerosol Sci. Technol.* 5:369–76.

Cheng, Y. S., H. C. Yeh, and M. D. Allen. 1988. Dynamic shape factor of a plate-like particle. *Aerosol Sci. Technol.* 8:109–23

Conner, W. D. 1966. An inertial-type particle separator for collecting large samples. *J. Air Pollut. Control Assoc.* 16:35.

Crump, J. G. and J. H. Seinfeld. 1982. Further results of inversion on aerosol size distribution data: Higher-order Sobolev spaces and constraints. *Aerosol Sci. Technol.* 1:363–69.

Dunmore, J. H., R. J. Hamilton, and D. S. G. Smith. 1964. An instrument for the sampling of respirable dust for subsequent gravimetric assessment. *J. Sci. Instrum.* 41:669–72.

Fang, C. P., P. H. McMurry, V. A. Marple, and K. L. Rubow. 1991. Effect of flow-induced relative humidity changes on size cuts for sulfuric acid droplets in the MOUDI. *Aerosol Sci. Technol.* 14:266–77.

Fernandez de la Mora, J., S. V. Hering, N. Rao, and P. H. McMurry. 1990. Hypersonic impaction of ultrafine particles. *J. Aerosol Sci.* 21:169–87.

Forney, L. J. 1976. Aerosol fractionator for large-scale sampling. *Rev. Sci. Instrum.* 47:1264–69.

Forney, L. J., D. G. Ravenhall, and S. S. Lee. 1982. Experimental and theoretical study of a two-dimensional virtual impactor. *Environ. Sci. Technol.* 16:492–97.

Golovin, M. N. and A. A. Putnam. 1962. Inertial impaction on single elements. *Ind. Engng. Chem. Fundls.* 1:264–73.

Griffiths, W. D. and N. P. Vaughan. 1986. The aerodynamic behaviour of cylindrical and spheroidal particles when settling under gravity. *J. Aerosol Sci.* 17:53–65.

Hering, S. 1989. Inertial and gravitational collectors. In *Air Sampling Instruments*, 7th edn., ed. S. V. Hering, pp. 337–385, Cincinnati: American Conference of Governmental Industrial Hygienists.

Hering, S. V., R. C. Flagan, and S. K. Friedlander. 1978. Design and evaluation of a new low pressure impactor—1. *Environ. Sci. Technol.* 12:667–73.

Hering, S. V., S. K. Friedlander, J. J. Collins, and L. W. Richards. 1979. Design and evaluation of a new low pressure impactor—2. *Environ. Sci. Technol.* 13:184–88.

Hering, S. V. and V. A. Marple. 1986. Low pressure and micro-orifice impactors. In *Cascade Impactors: Sampling and Data Analysis*, eds. J. P. Lodge and T. L. Chan, pp. 103–127. Akron, OH: American Industrial Hygiene Association.

Hillamo, R. E. and E. I. Kauppinen. 1991. On the performance of the Berner low pressure impactor. *Aerosol Sci. Technol.* 14:33–47.

Hoover, M. D., G. Morawietz, and W. Stöber. 1983. Optimizing resolution and sampling rate in spinning duct aerosol centrifuges. *Am. Ind. Hyg. Assoc. J.* 44:131–34.

Jaenicki, R. and I. H. Blifford. 1974. Virtual impactors: A theoretical study. *Environ. Sci. Technol.* 8:648–54.

John, W. and S. M. Wall. 1983. Aerosol testing techniques for size-selective samplers. *J. Aerosol Sci.* 14:713–27.

Keskinen, J., K. Janka, and M. Lehtimäki. 1987. Virtual impactor as an accessory to optical particle counters. *Aerosol Sci. Technol.* 6:79–83.

Kitto, P. H. and D. G. Beadle. 1952. A modified form of thermal precipitator. *J. Chem. Met. Min. Soc. SA* 52:284–306.

Kotrappa, P. and M. E. Light. 1972. Design and performance of the Lovelace aerosol particle separator. *Rev. Sci. Instrum.* 43:1106–12.

Liebhaber, F. B., M. Lehtimäki, and K. Willeke. 1991. Low-cost virtual impactor for large-particle amplification in optical particle counters. *Aerosol Sci. Technol.* 15:208–13.

Lippman, M. 1989. Size-selective health hazard sampling. In *Air Sampling Instruments*, 7th edn., ed. S. V. Hering, pp. 163–98, Cincinnati: American Conference of Governmental Industrial Hygienists.

Liu, B. Y. H. and K. L. Rubow. 1984. A new axial flow cascade cyclone for size classification of airborne particulate matter. In *Aerosols: Science, Technology, and Industrial Applications of Airborne Particles*, eds. B. Y. H. Liu, D. Y. H. Pui, and H. Fissan, pp. 115–18. New York: Elsevier.

Lodge, J. P. and T. L. Chan. 1986. *Cascade impactors: Sampling and data analysis.* Akron, OH: American Industrial Hygiene Association.

Loo, B. W. 1981. High efficiency virtual impactor. United States Patent No. 4,301,002.

Loo, B. W. and C. C. Cork. 1988. Development of high efficiency virtual impactors. *Aerosol Sci. Technol.* 9:167–76.

Lundgren, D. A. 1971. Determination of particle composition and size distribution changes with time. *Atmos. Envir.* 5:645–51.

Markowski, G. R. 1987. Improving Twomey's algorithm for inversion of aerosol measurement data. *Aerosol Sci. Technol.* 7:127–41.

Marple, V. A. 1970. *A fundamental study of inertial impac-*

tors. Ph.D. Thesis, University of Minnesota, Minneapolis, MN.

Marple, V. A. 1978. Simulation of respirable penetration characteristics by inertial impaction. *J. Aerosol Sci.* 9:125–34.

Marple, V. A. and C. M. Chien. 1980. Virtual impactors: A theoretical study. *Environ. Sci. Technol.* 14:976–85.

Marple, V. A. and B. Y. H. Liu. 1974. Characteristics of laminar jet impactors. *Environ. Sci. Technol.* 8:648–54.

Marple, V. A., B. Y. H. Liu, and R. M. Burton. 1990. High-volume impactor for sampling fine and coarse particles. *J. Air Waste Manag. Assoc.* 40:762–67.

Marple, V. A., B. Y. H. Liu, and B. A. Olson. 1989. Evaluation of a cleanroom concentrating aerosol sampler. In *Proc. 35th Annual Meeting of the Institute of Environmental Sciences*, pp. 360–63.

Marple, V. A. and J. E. McCormack. 1983. Personal sampling impactor with respirable aerosol penetration characteristics. *Am. Ind. Hyg. Assoc. J.* 44:916–22.

Marple, V. A. and K. L. Rubow. 1986. Theory and design guidelines. In *Cascade Impactors: Sampling and Data Analysis*, eds. J. P. Lodge, and T. L. Chan, pp. 79–101. Akron, OH: American Industrial Hygiene Association.

Marple, V. A., K. L. Rubow, and S. M. Behm. 1991. A micro-orifice uniform deposit impactor (MOUDI). *Aerosol Sci. Technol.* 14:434–46.

Marple, V. A., K. L. Rubow, W. Turner, and J. D. Spengler. 1988. Low flow rate sharp cut impactors for indoor air sampling: Design and calibration. *J. Air Pollut. Control Assoc.* 37:1303–07.

Marple, V. A. and K. Willeke. 1976*a*. Impactor design. *Atmos. Environ.* 12:891–96.

Marple, V. A. and K. Willeke. 1976*b*. Inertial impactors: Theory, design and use. In *Fine Particles*, ed. B. Y. H. Liu, pp. 411–66. New York: Academic Press.

Martonen, T. B. and D. L. Johnson. 1990. Aerodynamic classification of fibers with aerosol centrifuges. *Part. Sci. Technol.* 8:37–63.

Masuda, H., D. Hochrainer, and W. Stöber. 1978. An improved virtual impactor for particle classification and generation of test aerosols with narrow size distributions. *J. Aerosol Sci.* 10:275–87.

May, K. R. and R. Clifford. 1967. The impaction of aerosol particles on cylinder, spheres, ribbons and discs. *Ann. Occup. Hyg.* 10:83–95.

McFarland, A. R., C. A. Ortiz, and R. W. Bertch. 1978. Particle collection characteristics of a single-stage dichotomous sampler. *Environ. Sci. Technol.* 12:679–82.

Noll, K. E. 1970. A rotary inertial impactor for sampling giant particles in the atmosphere. *Atmos. Environ.* 4:9–19.

Noll, K. E., A. Pontius, R. Frey, and M. Gould. 1985. Comparison of atmospheric coarse particles at an urban and non-urban site. *Atmos. Environ.* 19:1931–43.

Orr, C. and E. Y. H. Keng. 1976. Sampling and particle-size measurement. In *Handbook on Aerosols*, ed. R. Dennis, pp. 93–117. U.S. Energy Research and Development Adm., NTIS Publication Number TID-26608.

Prodi, V., F. Belosi, and A. Mularoni. 1984. A high flow inertial spectrometer. In *Aerosols: Science, Technology, and Industrial Applications of Airborne Particles*, eds. B. Y. H. Liu, D. Y. H. Pui, and H. J. Fissan, pp. 131–34. New York: Elsevier.

Prodi, V., F. Belosi, A. Mularoni, and P. Lucialli. 1988. PERSPEC: A personal sampler with size characterization capabilities. *Am. Ind. Hyg. Assoc. J.* 49:75–80.

Prodi, V., T. De Zaiacomo, C. Melandri, G. Tarroni, M. Formignani, P. Olivieri, L. Barilli, and G. Oberdoerster. 1983. Description and application of the inertial spectrometer. In *Aerosols in Mining and Industrial Work Environments, Vol. 3*, eds. V. A. Marple and B. Y. H. Liu, pp. 931–49. Ann Arbor: Ann Arbor Science Publications.

Prodi, V., C. Melandri, G. Tarroni, T. De Zaiacomo, M. Formignani, and D. Hochrainer. 1979. An inertial spectrometer for aerosol particles. *J. Aerosol Sci.* 10:411–19.

Raabe, O. G., D. A. Braaten, R. L. Axelbaum, S. V. Teague, and T. A. Cahill. 1988. Calibration studies of the drum impactor. *J. Aerosol Sci.* 19(2):183–95.

Rader, D. J. and V. A. Marple. 1985. Effect of ultra-stokesian drag and particle interception on impaction characteristics. *Aerosol Sci. Technol.* 4:141–56.

Rader, D. J., L. A. Mondy, J. E. Brockmann, D. A. Lucero, and K. L. Rubow. 1991. Stage response calibration of the mark III and Marple personal cascade impactors. *Aerosol Sci. Technol.* 14:365–79.

Rao, A. K. and K. T. Whitby. 1978*a*. Non-ideal collection characteristics of inertial impactors—Single stage impactors and solid particles. *J. Aerosol Sci.* 9:77–86.

Rao, A. K. and K. T. Whitby. 1978*b*. Non-ideal collection characteristics of inertial impactors—Cascade impactors. *J. Aerosol Sci.* 9:87–100.

Reischl, G. P. and W. John. 1978. The collection efficiency of impaction surfaces: A new impaction surface. *Staub-Reinhalt. Luft* 38:55.

Smith, W. R., R. R. Wilson, and D. B. Harris. 1979. A five stage cyclone system for *in situ* sampling. *Environ. Sci. Technol.* 13:1387–92.

Stöber, W. 1976. Design performance and applications of spiral duct aerosol centrifuges. In *Fine Particles, Aerosol Generation, Measurement Sampling and Analysis*, ed. B. Y. H. Liu, pp. 351–98. New York: Academic Press.

Stöber, W. and H. Flachsbart. 1969. Size-separating precipitation of aerosols in a spinning spiral duct. *Environ. Sci. Technol.* 3:1280–96.

Szymanski, W. S. and B. Y. H. Liu. 1989. An airborne particle sampler for the space shuttle. *J. Aerosol Sci.* 20:1569–72.

Timbrell, V. 1954. The terminal velocity and size of airborne dust particles. *Br. J. Appl. Phys.* 5:S86.

Timbrell, V. 1972. An aerosol spectrometer and its application. In *Assessment of Airborne Particles*, eds. T. T. Mercer, P. E. Morrow, and W. Stöber, pp. 290–330. Springfield: Charles C. Thomas.

Vanderpool, R. W., D. A. Lundgren, V. A. Marple, and

K. L. Rubow. 1987. Cocalibration of four large-particle impactors. *Aerosol Sci. Technol.* 7:177–85.

Waldmann, L. and K. H. Schmitt. 1966. Thermophoresis and diffusiophoresis of aerosols. In *Aerosol Science*, ed. C. N. Davies, pp. 137–62. New York: Academic Press.

Wolfenbarger, J. K. and J. H. Seinfeld. 1990. Inversion of aerosol size distribution data. *J. Aerosol Sci.* 21:227–47.

Wright, B. M. 1954. A size-selecting sampler for airborne dust. *Br. J. Ind. Med.* 11:284.

Wu, J. J., D. W. Cooper, and R. J. Miller. 1989. Virtual impactor aerosol concentrator for cleanroom monitoring. *J. Environ. Sci.* 5:52–56.

Xu, X. 1991. *A study of virtual impactor*. Ph.D. Thesis, University of Minnesota, Minneapolis, MN.

12

Atmospheric Sample Analysis and Sampling Artifacts

B. R. Appel

Air and Industrial Hygiene Laboratory
California Department of Health Services
2151 Berkeley Way
Berkeley, CA, U.S.A.

Dedication

Readers of this chapter will note the many references to the work of Dr. Thomas G. Dzubay, a research physicist with the U.S. EPA. His untimely death, Dec. 3, 1990, represents a loss to the scientific community of one of the most creative and productive scientists in the field of aerosol research. The author dedicates this chapter to his memory.

INTRODUCTION

The composition of atmospheric aerosols varies markedly, depending upon sampler location, the proximity of significant sources of aerosols and their gaseous precursors, and meteorology. Anthropogenic particulate matter is concentrated in fine particles (< 2.5 μm in diameter), while natural aerosols (e.g., windblown soil, sea salt, pollen, and spores) are concentrated in larger particles. Table 12-1, assembled from the results of Dzubay et al. (1982) and Stevens et al. (1984*a*), details the average concentrations in the fine and coarse (2.5–15 μm) particle fractions, obtained at a rural and at an urban location. Although generally self-explanatory, the carbon measurements require comment. Because of the complexity of the atmospheric carbonaceous material retained on sampling media, such materials are frequently characterized only as organic and elemental carbon. No standard procedure exists for these measurements, and the results from available methods can differ widely. For most air samples, total carbon closely approximates the sum of organic and elemental carbon.

The analysis of atmospheric aerosols, in the degree of detail given in Table 12-1, is usually restricted to research programs. Routine analyses, for example, those in support of current national air quality standards, require only the determination of lead in total suspended particulate matter, and the mass of suspended particles < 10 μm in aerodynamic diameter (PM-10). Studies on visibility impairment as related to atmospheric particulate matter emphasize analysis of the < 2.5 μm particle fraction for its major constituents, SO_4^{2-}, NO_3^-, NH_4^+, organic and elemental C. Analyses in support of health effects research may determine the concentrations of toxic metals (e.g., Ni, Cd, and Cr), H^+ or specific organic compounds (e.g., polyaromatic hydrocarbons).

Scope and Objective of this Review

This review surveys the analytical techniques used to determine the concentrations of the

TABLE 12-1 Average Aerosol Composition (ng/m³) for Fine and Coarse Particles at a Rural, Forested Location (Great Smoky Mountains, Tennessee) and an Urban Location (Houston, Texas)

	Smoky Mountains		Houston[1]	
	Fine[2]	Coarse[2]	Fine[2]	Coarse[2]
Total Mass	24,000 ± 3000	5600 ± 3000	42,500 ± 4250	27,200 ± 2700
SO_4^{2-}	12,000 ± 1300	NA	16,700 ± 1380	1100 ± 200
NO_3^-	300 ± 300	NA	250 ± 260	1800 ± 260
NH_4^+	2280 ± 390	NA	4300 ± 390	< 190
H^+	114	NA	67	< 1
C (organic)	2220 ± 400	1200 ± 400	NA	NA
C (elemental)	1100 ± 800	< 100	NA	NA
C (total)	3300 ± 600	1300 ± 600	7600 ± 500	3300 ± 500
Al	20 ± 18	195 ± 101	95 ± 60	1400 ± 420
Si	38 ± 10	580 ± 262	200 ± 60	3800 ± 1000
S	3744 ± 218	204 ± 187	NA	NA
Cl	< 10	7 ± 4	19 ± 6	330 ± 21
K	40 ± 3	108 ± 30	120 ± 7	180 ± 21
Ca	16 ± 1	322 ± 73	150 ± 8	3100 ± 160
Ti	< 6	18 ± 5	< 8	48 ± 14
V	< 4	< 5	NA	NA
Mn	NA	NA	13 ± 2	23 ± 3
Fe	28 ± 2	118 ± 9	170 ± 9	730 ± 40
Ni	1 ± 0.5	1 ± 0.5	3 ± 1	5 ± 1
Cu	3 ± 0.7	< 5	16 ± 2	14 ± 2
Zn	9 ± 1	< 4	102 ± 6	68 ± 5
As	2.2 ± 1	< 1	NA	NA
Se	1.4 ± 0.3	0.2 ± 0.2	NA	NA
Br	18 ± 1	5 ± 0.4	70 ± 4	39 ± 3
Pb	97 ± 5	14 ± 1	483 ± 23	127 ± 10

Source: Finlayson-Pitts and Pitts, p. 787 (1986). Reproduction with permission from Wiley

1. Samples collected during the daytime
2. Fine and coarse particles were defined in this study as having aerodynamic diameters in the ranges 0–2.5 and 2.5–15 μm, respectively

NA: not available

atmospheric aerosol constituents listed in Table 12-1. Since the relationship between measured and actual air concentrations of some aerosol constituents can be distorted by sampling errors (frequently referred to as "artifacts"), a brief discussion of such potential errors, and their influence on sampling requirements, is included. Techniques for minimizing sampling errors are reviewed in more detail elsewhere (Appel 1991).

As summarized in Table 12-2, alternative techniques are available to determine each of the species noted in Table 12-1. In general, those permitting a determination of multiple elements or ionic species are used more frequently, especially when large numbers of samples are involved. Techniques permitting analyses with little or no sample preparation are also favored. The author's goal is to assess the potential reliability and limitation of many of the procedures shown in Table 12-2 when applied to collected particulate samples. Airborne concentration measurements also impose the requirement for an accurate and precise determination of air volumes, corrected, when necessary, to standard pressure and temperature (Lumpkin 1984). However, these measurements are beyond the scope of this chapter.

Detection Limits

The instrumental detection limit (IDL) is the lowest instrumental signal which can be reliably distinguished from instrument noise.

TABLE 12-2 Analytical Alternatives for Aerosol Constituents

Species	Method[1]	Sample Preparation
PM-10 mass	Gravimetric	Equilibrate at fixed RH, temperature and neutralize charge
	β-attenuation	None
	Inertial microbalance (TEOM)	Equilibrate at 50°C
Elements	XRFA	None
	INAA	None
	PIXE	None
	ICP	Sample dissolution
	AAS	Sample dissolution
Total carbon	Combustion to CO_2 with several quantitation options	None
Organic and elemental carbon	Selective pyrolysis (and/or combustion)	None
	Light absorption for EC with OC by difference	None
SO_4^{2-}, NO_3^-, NH_4^+	IC, automated colorimetric methods, ion specific electrode (NH_4^+)	Aqueous extraction
H^+ from strong acids	Antilog of pH, Gran's titration	Extraction in 10^{-4} N $HClO_4$ protected from NH_3

1. See index for abbreviations and acronyms used.

Varying criteria for the IDL can be found (e.g., the concentration that produces a signal equivalent to twice the magnitude of the background fluctuation) (American Public Health Association 1985, 153). The method detection limit (MDL) is the smallest atmospheric concentration of a species which can be reliably distinguished from the sampling medium's contribution:

$$\text{MDL} = t\sigma/v \qquad (12\text{-}1)$$

where

t = the Student's t value for an appropriate confidence level with $n - 1$ degrees of freedom, where n is the number of blanks analyzed. U.S. Environmental Protection Agency (EPA) methods specify a 99% confidence level (one-sided t test)

σ = standard deviation of replicate analyses of blanks or samples with low concentrations of the analyte

v = air sample volume.

For example, with $n = 10$, t equals 3.3. Sampling media are subject to contamination during mounting on a sampler or in other handling. Accordingly, blanks which have been subjected to the same procedures as with the loaded media, but with no more than a brief (e.g., 30 s) operation of the sampler, "field blanks," are preferred for the determination of MDL values.

The concepts and techniques regarding detection limit determinations are explored in a volume edited by Currie (1988).

Accuracy in Chemical Analyses

The accuracy of analyses performed on atmospheric samples remains, in general, difficult to establish. Ideally, accuracy would be judged by analyzing a standard atmospheric particulate sample supported on the relevant collection medium, which offered certified values for the analytes of interest. The U.S. National Institute of Standards and Technology (NIST), formerly the National Bureau of Standards, previously provided a bulk atmospheric particulate as standard reference material (SRM) 1648, with certified analyses for nine metals (As, Cd, Cr, Cu, Ni, Zn, U, Fe, Pb), and noncertified values for 25 additional elements (Greenberg 1979). This material could be used to assess the accuracy for the certified elements using methods involving sample dissolution or extraction, or with methods such as neutron activation analysis, which can measure analyte concentrations in weighed quantities of a sample irrespective of sample distribution. (The need for a new bulk particulate standard is obvious. If sufficient requests are received by NIST, hopefully, they will respond.) The application of such SRMs to assess the accuracy of techniques such as X-ray fluorescence analysis (XRFA), which provide concentrations per unit area of sampling medium, is hampered by the large particle size of the standard. A technique for nebulizing synthetic, multielement standards for XRFA calibration is described by Giauque, Garrett, and Goda (1977). NIST SRM 1649, a bulk solid atmospheric particulate matter, although intended primarily as a reference standard for polycyclic aromatic hydrocarbons, includes non-certified values for the concentrations of leachable anions, including SO_4^{2-}, NO_3^-, and Cl^-. However, the NO_3^- and Cl^- values are probably reliable only as upper-limit values. Thin-film SRMs available from NIST, useful for XRFA, include SRM 1833 (certified for Si, K, Ti, Fe, Zn, Pb) and SRM 1832 (certified for Al, Si, Ca, V, Mn, Co, Cu).

Samples prepared in the laboratory have frequently been used to assess accuracy, although it is readily conceded that they lack much of the relevant sample matrix. To simulate particulate samples, these can be prepared by spiking a filter section with a solution containing one or more analytes of interest. NIST SRMs 2676c, 2677, and 3087, provide mixtures of varying numbers of toxic metals spiked as composite standards on cellulose ester membrane filters. These standards require analysis of the entire filter since the distribution across the surface is nonuniform. An EPA procedure (Mitchell 1988) is available to facilitate preparation of filter strips spiked with various anions and metals. Such spotting technique cannot provide a uniform distribution across a surface. However, a technique providing homogeneity within 1% for areas $\geq 25\ mm^2$ has been reported (Camp, Cooper, and Rhodes 1974). Simulated rainwater standards, SRM 2694-I and -II, provide certified values for SO_4^{2-}, NO_3^-, Cl^-, Ca^{2+}, Mg^{2+}, K^+, and H^+ at two concentration levels useful for atmospheric particulate analyses, especially those from short-term, low volume sampling. Single component metal solution SRMs are also available from NIST.

In the absence of appropriate reference materials, the degree of agreement between laboratories analyzing aliquots from the same synthetic or atmospheric samples by different methods has frequently been assessed to infer accuracy (Camp, Cooper, and Rhodes 1974; Camp et al. 1975; Hering et al. 1988; Countess 1990). With atmospheric samples, in the absence of a reference method, the interlaboratory mean values are often used as a reference in place of "true" concentrations.

SAMPLING AND STORAGE ARTIFACTS

Numerous sources of error may influence the accuracy of some of the atmospheric concentrations listed in Table 12-1. Positive and negative errors or "artifacts" are described below.

Negative Artifacts

The measurement of semivolatile materials, which can coexist in the condensed and gaseous phases, is problematic. Particle-phase materials may be lost by volatilization during collection. This has been extensively documented for NH_4NO_3 (e.g., Appel, Tokiwa, and Haik 1981), and also reported for certain polycyclic aromatic hydrocarbon (PAH) compounds (Coutant et al. 1988) during conventional filter sampling. Particulate samples collected on glass fiber filters are subject to enhanced loss of NH_4^+ because of the filter alkalinity. Extensive loss of NH_4^+, Cl^-, and NO_3^- from samples on glass and quartz fiber filters can occur during storage at room temperature (Smith, Grosjean, and Pitts 1978; Witz et al. 1990). Nitrate loss was shown to be relatively rapid, initially, with recovered nitrate decreasing proportional to the square root of time (Dunwoody 1986). Refrigeration at ca. 5°C is unable to prevent losses of some organic compounds (Clement and Karasek 1979). Although low-temperature storage is needed to minimize losses, mass determination requires 24-h storage at 15–30°C (see below). During this equilibration, in addition to water, a small loss of other semivolatile aerosol constituents is likely (McKee et al. 1972).

Chemical changes to aerosol constituents may occur both during and following collection, including neutralization of acidic species (Appel et al. 1980*b*) and oxidation of specific organic compounds (Pitts et al. 1980). Oxidation, while altering composition, would not be expected to significantly influence total or organic carbon.

Positive Artifacts

Retention of gaseous HNO_3, SO_2, and organic C by sorption or reaction with the sampling medium and/or the particulate matter thereon can result in positive errors in determining corresponding aerosol-phase concentrations. Nitric acid is retained on many types of filters (Appel 1981; Appel et al. 1984*a*), on the stages of cascade impactors (Appel et al. 1980*d*; Bondietti and Papastefanou 1989), and by previously collected particles (Appel et al. 1980*c*), leading to artifact particulate nitrate (PN). Retention is greatest on alkaline (e.g., glass fiber) and cellulose ester filters, and least on Teflon filters. Quartz filters vary markedly in their retention of nitric acid depending on composition but show little SO_2 retention. Filters with relatively low HNO_3 retention are subject to loss of NH_4NO_3 by volatilization, making nitrate results obtained with Teflon or most quartz fiber filters unreliable measures of atmospheric PN concentrations. The degree of loss is dependent on relative humidity and temperature as well as the concentrations of gaseous ammonia and nitric acid. Nitrate results obtained with glass fiber (and probably cellulose ester) filters can approximate the sum of atmospheric HNO_3 and particulate NO_3^- (Appel et al. 1981). Retention of HNO_3 on the stages of cascade impactors distorts apparent size distributions for PN. Retention of HNO_3 by previously collected particles should be more significant for longer-term (e.g., 24 h) sample collection, and should be enhanced if alkaline, coarse particles (e.g., soil dust) are collected.

Retention of SO_2 by particle sampling media leads to artifact particulate SO_4^{2-} (Coutant 1977; Appel 1981; Appel et al. 1984*a*), since the initially formed SO_3^{2-} readily undergoes oxidation. Atmospheric sulfate concentrations measured with filters of low alkalinity (e.g., Teflon and quartz fiber) are relatively reliable (Appel et al. 1984*a*). Nylon filters, useful for measurement of both gaseous and particulate NO_3^{2-}, cause partial retention of SO_2 (Cadle and Mulawa 1987), a factor likely to promote retention of gaseous NH_3 as well. Thus, their use for SO_4^{-2} and NH_4^+ measurements is suspect.

Nearly all determinations of atmospheric organic C (OC) and elemental C (EC) have been made on glass or quartz fiber filters. Alkaline filters such as glass fiber retain organic acids (e.g., benzoic acid) to a significant degree (Appel, Cheng, and Salaymeh 1990). Both quartz and glass fiber filters retain, with low efficiency, vapor-phase, polar organic compounds. Unless correction techniques are employed during sampling, particulate total and organic C measurements may be positively biased as a result. With 24-h samples, positive errors in organic C can be 15–20% (Appel, Cheng, and Salaymeh 1989). The degree to which this is offset by loss of semivolatile OC remains unclear. McDow and

Huntzicker (1990) argue that the diminished level of OC recovered from filters at higher sampling velocities reflects decreased sorption rather than increased volatilization.

Filter types should be selected based on published test results concerning the sampling errors of concern; users should carefully identify filter media by manufacturer, composition, ID number, pore, and filter size. Because of frequently inadequate quality control, individual batches of filters must be systematically sampled to measure blank values for chemical species to be determined.

MASS DETERMINATION

PM-10 mass concentrations are needed to assess compliance with the U.S. national air quality standards. Methodology for sampling, air volume measurement and gravimetric analysis is published in the U.S. Federal Register (1987) by the U.S. Environmental Protection Agency (EPA) (Lumpkin 1984; U.S. EPA 1990) and by the Intersociety Committee (ISC) (Lodge 1989, Method 501). In addition to restricting particle size, PM-10 determination requires filters with < 1% penetration of 0.3 μm DOP particles and low alkalinity (e.g., < 25 milliequivalents of H^+/g such as provided by quartz and Teflon membrane filters). However, Teflon membrane filters provide relatively high air flow resistance, especially after particle loading and, therefore, are normally used in low (10–20 lpm) and medium (ca. 100 lpm) volume samplers (see Chapter 29, for details). These require highly precise filter weighings and enhanced quality assurance efforts to obtain reliable measures of PM-10 mass.

In addition to gravimetric analysis, particle mass concentrations on filter-collected samples can be determined by β-radiation attenuation (Jaklevic et al. 1981) and by inertial oscillation (e.g., TEOM, Patashnick and Rupprecht 1990). Devices utilizing these principles can provide continuous determinations. Instruments of both types have recently been designated as equivalent to gravimetric procedures for PM-10 measurement (see Chapter 29, this volume).

Gravimetric Analysis

PM-10 samples for monitoring compliance with national ambient air quality standards are routinely collected on 8″ × 10″ quartz fiber filters with high-volume samplers equipped with PM-10 inlets. These are weighed before and after sample collection to the nearest mg or 0.1 mg. Analytical balances with oversized trays for holding such air filters are in common use. Following sampling, the weight increase is used to calculate the suspended particle mass concentration.

To achieve high accuracy and precision, the weighing area must be free of vibrations and the balance must be shielded from air currents. Other factors influencing the accuracy of aerosol mass measurements include sampling artifacts, previously discussed, leaks, the care with which filters are mounted in the sampler, loss of particles in packaging and handling loaded filters, changes in weight of the sampling medium due to loss of filter material during sampling and handling, air volume determination, especially when flow rates change with filter loading, and the changes in filter and particulate mass with RH. Static charge must be removed from the filters before weighing to minimize electrostatic attraction to balance surfaces. Potential errors due to electrostatic charge are especially significant with Teflon and other membrane filters used in low-volume samplers.

Quartz fiber exhibits greater friability compared to glass fiber filters. However, reported errors due to fiber loss were < 3 μg/m³ with very careful handling (Rehme et al. 1984; Highsmith, Bond, and Howes 1986). A quartz filter containing 5% borosilicate glass, Whatman QMA, provides markedly improved filter integrity (West 1985).

Low-volume samplers for PM-10 measurement (for example, dichotomous samplers), which provide fine (< 2.5 μm) and coarse (2.5–10 μm) particle fractions, frequently employ 37 mm diameter Teflon membrane filters weighed to the nearest 1 or 10 μg, usually with an electrobalance. Stevens (1984*b*) judged that gravimetric procedures with such filters permit mass determination with an uncertainty of $\pm 25\ \mu g$ ($\pm 1.7\ \mu g/m^3$ for a 24 h fine + coarse sample at a sampling rate of 16.7 lpm, as provided by a commercially available dichotomous sampler). Other results suggest that precision can be significantly better than this (Feeney et al. 1984). A comparison of 24 h PM-10 particle mass concentrations sampling in parallel with Teflon and quartz fiber filters in southern California showed no significant difference (Appel et al. 1984*a*). Chapter 29, this volume, includes a more detailed discussion and more extensive references relevant to PM-10 sampling.

TABLE 12-3 The Influence of Relative Humidity Change on Blank Filter Weight[1,2]

Filter Type[3]	Weight Increase ($\mu g/m^3$)[4]	
	90% RH	46% RH
Gelman microquartz (Batch 8198)	20.2	11.3
Whatman QMA (quartz)	4.6	2.0
S&S HV1 glass fiber	25.2	4.7

1. The mean weight gains experienced for five 8″ × 10″ filters on increasing RH from 18% to the final value shown, expressed in equivalent air concentrations (assuming 2500 m^3 of air sampled)
2. Calculated from data provided by K.A. Rehme, U.S. EPA (private communication, 1990)
3. See Appel et al. (1984*a*) for other data on these filters
4. Equilibration at 46% RH follows equilibration at 90% RH

Loss of Particles in Handling

A loss of particles from the coarse fractions collected on Teflon filters with dichotomous samplers was anticipated during shipping and handling; the particles are retained on the filter surface, and the coarse particle fraction lacks much of the stickier aerosol constituents (e.g., sulfates, nitrates, and carbonaceous materials) found primarily in the fine fraction. The observed losses have been variable, depending upon the sampling location and loading (Highsmith, Bond, and Howes 1986). A light oil coating decreased losses to < 5% without interfering with elemental analyses by instrumental methods (Dzubay and Barbour 1983).

Hygroscopicity of Filter Media and Particles

The use of hygroscopic filter media (e.g., cellulose and cellulose ester membrane) severely complicates particulate mass measurements. For example, a 47 mm cellulose ester filter exhibited a mass change of about 40 μg/% RH change, at 28°C between 50 and 75% RH (Cahn 1963). This imposes stringent requirements for RH control in weighing hygroscopic filters. The weights of glass, quartz, and Teflon filters are much less sensitive to RH changes over the ranges normally encountered. An EPA study (K. A. Rehme, U.S. EPA, private communication 1990) compared 8″ × 10″ glass and two types of quartz fiber filters for weight change with increasing RH (Table 12-3). Even with these relatively nonhygroscopic filters, the weight increases, expressed in $\mu g/m^3$ (assuming a 2500 m^3 air sample), clearly support the requirement for RH controls. Furthermore, hygroscopic materials in atmospheric particulate matter cause substantial weight increases at above 55% RH (Tierney and Conner 1967). The procedures for PM-10 mass measurement specify equilibration at 20–45 ± 5% RH at 15–30°C for 24 h, both before and after filter loading.

ELEMENTAL ANALYSES BY NONDESTRUCTIVE TECHNIQUES

X-ray Fluorescence Analysis (X-ray-Induced)

The theory, instrumentation, and application of X-ray induced, X-ray fluorescence analysis (XRFA) have been extensively reviewed (Birks 1969; Giauque et al. 1973; Dzubay 1977; Dzubay and Rickel 1978; Stevens 1984*b*; Van Grieken and LaBrecque 1985; Lodge 1989, 218). Detailed procedures for XRFA in filter-collected air samples are available for multielements (Cummings 1981; Lodge 1989, Method 822B), sulfur (Lodge 1989, Method 730) and lead (Moore and Witz 1980). XRFA provides both qualitative and quantitative analysis for elements with atomic numbers of about 12 and higher. Method detection limits for Teflon filter-collected samples range from 20 to 200 ng/cm^2 for 44 of 49 elements (Lodge 1989, 625). The method includes both energy-dispersive (ED) and wavelength-dispersive (WD) techniques. ED-XRFA provides a simultaneous determination of multiple elements, whereas WD-XRFA usually determines one element at a time. ED-XRFA provides a somewhat lower resolution, thereby leading to an increased potential for interferences. ED-XRFA is favored for multi-element analysis because of the availability of techniques to correct for interferences as well as the lower per-element cost.

Although described as nondestructive, the potential loss of labile elements (e.g., halides, NO_3^-, NH_4^+, carbonaceous materials) in vacuum and at elevated temperatures may invalidate samples for the analysis of such species, following XRFA.

Summary of Technique

Wavelength-dispersive instruments include an X-ray tube to irradiate the sample, a crystal to disperse the emitted X-rays according to Bragg's law, and a detector to determine the intensity of the characteristic X-rays. The detector is placed at an appropriate angle to measure the intensity at a particular wavelength. Energy-dispersive instruments often induce sample fluorescence with X-rays from a secondary target (e.g., Ti, Ni, Mo). A solid-state detector (a silicon diode containing a small amount of lithium) and a multichannel analyzer are used to disperse, detect, and record the emitted X-rays.

X-rays incident on the sample interact either by the photoelectric effect (leading to X-ray fluorescence) or by scattering, mainly from the atoms in the low atomic number constituents of the collection medium. These scattered X-rays constitute an unwanted background that defines the detection limit for the fluorescence measurement.

Sampling Requirements

The particulate sample should be deposited uniformly on the outer surface to minimize X-ray absorption by the sampling medium. The sampling medium should contribute minimal concentrations of the elements to be determined and, preferably, provide a low mass per unit area to minimize scattering the incident radiation. The contribution of an element in the sampling medium to XRFA depends on the distribution of the element in the medium.

Teflon membrane filters are frequently employed, in part, because of their low blank levels (Wagman, Bennett, and Knapp 1977). Teflo filters, made by Gelman Sciences, have a thin Teflon membrane mounted on a polyolefin ring, providing a density of about 500 $\mu g/cm^2$, excellent for XRFA. A filter of this type with 2 μm pore size provides a collection efficiency $> 99.98\%$ for particles ≥ 0.035 μm in diameter at a face velocity of 23 cm/s (Liu, Pui, and Rubow 1983). For comparison, Zefluor filters, also made by Gelman Sciences, have the Teflon membrane mounted on a porous Teflon mat. The mat adds to the total mass per unit area, increasing blank levels, and raising the minimum detection limits by about a factor of 3 (Dzubay and Stevens 1990). Furthermore, if the particles are collected on the mat side (a mistake easily made because the membrane side is not easily determined), particles are collected within the mat, rather than on the surface of the Teflon membrane. This attenuates the low-energy X-rays emitted by elements of low atomic number, causing a negative bias in their determination. Sampling with segregation into coarse and fine particle fractions is helpful in improving accuracy by minimizing self-absorption (i.e., absorption of X-rays within the same or adjacent particles of the sample) (Dzubay and Nelson 1975).

Several summaries of blank levels in filter media are available (Lodge 1989, 190; Gelman and Marshall 1975; Wagman, Bennett, and Knapp 1977), although none includes all of the filters frequently used for nondestructive analyses. Because of possible batch-to-batch variation, reported measurements of impurity levels should be considered as general guidance only. Table 12-4 lists impurity levels in six filter types. Glass and

TABLE 12-4 Content of Trace Elements in Filter Materials (ng/cm²)

Metal	Polystyrene[1] (Delbag)	Cellulose Ester (Millipore 0.45 μm)	Cellulose Paper (Whatman No. 41)	Glass Fiber	Organic Membrane	Silver Membrane
Ag	2		2			
Al	20	10	12			
As				80		
Ba	500	100	100			
Be				40	0.3	200
Bi					1	
Ca	300	250	140			
Cd					5	
Co	0.2	1	0.1		0.02	
Cr	2	14	3	80	2	60
Cu	320	40	4	20	6	20
Fe	85	300	40	4000	30	300
Hg	1	1	0.5			
Mg	1500	200	80			
Mn	2	2	0.5	400	10	30
Mo					0.1	
Ni	25	50	10	80	1	100
Pb				800	8	200
Sb	1	3	0.15	30	0.1	
Sn				50	1	
Ti	70	5	10	800	2000	200
V	0.6	0.09	0.03	30	0.1	
Zn	515	20	25	160,000	2	10

Source: Alian and Sansoni. (1985). Reproduced with the permission of Elsevier.

1. No longer commercially available.

so-called quartz fiber filters, although not providing surface deposition, can be employed for XRFA of elements of higher atomic number, for which filter absorption is of lesser importance. Whatman QMA quartz filters, distributed by the U.S. EPA for high-volume PM-10 sampling, exhibit notably higher Ba and Zn concentrations relative to other "quartz" filters (Appel et al. 1984*b*). Whatman 41 cellulose filters provide low blank levels for most elements and a low pressure drop, but relatively poor initial collection efficiency (Liu, Pui, and Rubow 1983). Partial penetration of fine-particle-related elements can be expected (Dolske, Schneider, and Sievering 1984).

Precision and Accuracy

Self-absorption of emitted X-rays results in a dependence on particle size, a factor of special concern for elements with low atomic number. In addition to sampling with particle size resolution, the use of calibration standards prepared with corresponding particle sizes can further minimize such errors. The attenuation factor due to self-absorption can be substantial, for example, 0.96 and 0.84 for 1 and 5 μm sulfur particles, respectively (Lodge 1989, 635). The overlap of the X-ray emission spectra of many elements (e.g., Mo, Bi, and Pb with the K_α X-ray line of S) also requires correction.

Numerous comparisons of methods have been made involving XRFA and other instrumental methods employing both prepared and ambient air particulate matter samples. For example, Camp et al. (1975) found, with prepared samples, an accuracy of $\pm$ 10% for XRFA for elements of atomic number ≥ 20. A comparison of water-soluble SO_4^{2-} with total S values, obtained by XRFA analyzing 400 atmospheric samples on cellulose acetate membrane filter, found a mean ratio (sulfate S)/(XRFA S) of 1.01 ± 0.01 (Appel et al. 1980*d*). These results suggest that nearly all of the atmospheric S in that study was present as SO_4^{2-}, SO_3^{2-}, or other species readily converted to sulfate during handling and SO_4^{2-} analysis.

The precision of XRFA depends on statistical counting error, drift in instrumental signal, homogeneity of the sample, and miscellaneous operational errors. The relative counting error decreases in proportion to the square root of the number of counts obtained. A single sample introduced into an instrument 10 times showed a coefficient of variation (CV) of 0.10%, employing a large accumulated count to minimize the counting error (Wagman, Bennett, and Knapp 1977).

Proton-Induced X-ray Emission Analysis

Proton-induced X-ray emission (PIXE) analysis differs from XRFA in the excitation source for producing fluorescence. It is rapid and especially well-suited for samples of a few mm^2 and thicknesses of up to approximately 10 μm. Analyses of elements ranging in atomic number from Si to Pb are readily obtained. The application of this technique to particulate samples has emphasized its capability with small, particle-sized and time-resolved samples (Nelson 1977). A comparison of PIXE and other methods applied to more typical filter samples is reported by Waetjen et al. (1983).

Summary of Technique

Bombardment of the sample with high-energy protons permits the recording of an X-ray fluorescence spectrum within a few minutes, using an energy-dispersive detector (e.g., a silicon diode containing small amounts of lithium) (Nelson 1977).

Sampling Requirements

In spite of the γ-ray background due to interactions between the proton beam and fluorine (Dzubay and Stevens 1990), Teflon filters are frequently used for samples intended for PIXE analysis. In addition to Teflon filters, Nuclepore filters have frequently been used because of their low mass per unit area (1 mg/cm^2) and low contaminant levels (Wagman, Bennett, and Knapp 1977), as well as freedom from the γ-ray background. However, relatively high collection efficiencies (e.g., $>97\%$) for fine particles with Nuclepore filters require ≤ 0.6 μm pore size filters (Liu et al. 1983). The high pressure drop of such filters severely limits sampling rates as well as filter loadings. Accurate PIXE analysis of high-volume samples for sulfur and heavier elements collected on cellulose acetate membrane filters (total particle loading of ca. 0.5 mg/cm^2, estimated filter mass 5 mg/cm^2) has also been reported (Waetjen et al. 1983). A matrix absorption correction permitted PIXE analysis on Whatman 41 cellulose filters (Kemp 1977). PIXE analysis of samples collected on a polystyrene film impaction surface is noted by Nelson (1977).

Instrumental Neutron Activation Analysis

Instrumental neutron activation analysis (INAA) of atmospheric particulate samples is, with some notable exceptions, similar to XRFA in limits of detection, and is also able to determine a large number of elements in a single sample. Similar to XRFA, this technique has often been used to characterize source "signatures," i.e., the characteristic elemental compositions of specific emission sources (e.g., Olmez and Gordon 1985). A number of reviews of the method are available (Dams et al. 1970; Alian and Sansoni 1985; Lodge 1989, 143), as well as a standard operating procedure for the analysis of atmospheric particles (Lambert 1981).

The reported advantages of INAA (Lodge 1989, 143) include: (1) the near absence of matrix effects from self-absorption and few interelement interferences, (2) the ability to analyze thick and nonhomogeneous samples, and (3) the relative nondestructiveness of the method, making the samples potentially available for some additional analyses (excluding those for labile constituents, and allowing for residual radioactivity). The disadvantages of the method include: (1) the need for a nuclear reactor and other specialized equipment and expertise,[1] (2) the 2–3 weeks required for the analysis of long-lived isotopes, and (3) the exclusion of Ni, P, and elements lighter than Na (excepting F), and a very high IDL for S. In addition, Cd is not consistently measured (Ondov et al. 1975).

Summary of Technique

A sample, usually contained in a polyethylene vial, is exposed to a high thermal neutron flux. Rapid-transfer pneumatic systems are often used to transport samples in and out of the irradiation position. Nuclear reactions occur which result in γ-ray emissions from the now radioactive sample. These are detected by a lithium-drifted germanium detector, which has a high resolution for separating the energies of the γ-rays associated with each element. Since the amount of each radionuclide produced is proportional to the mass of that element, a quantitative assay is possible. Short and long irradiation periods are used for each sample together with varying periods for radioactive decay prior to quantitation, depending on the type of sample and the elements to be determined. Irradiation periods employed, together with IDL values, are given in Lodge (1989, 143–150), Alian and Sonsoni (1985) and Ragaini et al. (1980). Some loss of labile species during INAA can be anticipated; a sample temperature of 57°C during irradiation was determined for one system (Appel et al. 1980*d*), but is found to be usually < 40°C (Lodge 1989, 148).

Sampling Requirements

As with XRFA, MDL values with INAA will frequently be limited by the variability of the blank levels in the sampling medium rather than by the IDL. The sampling media which have been employed with this method are reviewed by Alian and Sansoni (1985); Whatman 41 (cellulose), Nuclepore (polycarbonate), Millipore cellulose ester, and Delbag polystyrene (no longer commercially available) have been the most frequently employed filter media. Glass fiber filters are notable for elevated blank levels for many elements and are not often used for INAA. However, a report of INAA for 25 elements on glass fiber filter samples is instructive (Lambert and Wiltshire 1979). In it, the MDL values were calculated using Eq. 12-1, with $t = 3.3$. Less than half of 250 24 h high volume atmospheric samples in that work contained concentrations which exceeded the MDL for most elements.

Accuracy and Precision

Analysis of NIST coal and fly ash standard reference materials by four laboratories employing INAA showed excellent interlaboratory agreement (Ondov et al. 1975) . With the coal standard, the interlaboratory means showed CV values $\leq 10\%$ for 17 of 30 elements, $\leq 15\%$ for 25 of 30 elements, and $\leq 25\%$ for all 30 elements. Results with the fly ash standard were slightly poorer. Except for Ni, Zn, and As, the mean value results agreed with NIST values within the experimental uncertainties. In another interlaboratory comparison employing dried solution deposits containing known elemental concentrations, INAA agreed with the theoretical results within 11% in all cases (Camp and VanLehn 1975). The precision for replicate determinations of the NIST coal SRM by one laboratory, as measured by CV values, was 15% or better except for Co, Sr, Ti, V, and Zr (Ragaini et al. 1980). A comparison of INAA results with those by a more destructive technique with an NIST SRM is included below in the discussion of inductively coupled plasma emission spectroscopy.

ELEMENTAL ANALYSIS BY DESTRUCTIVE TECHNIQUES

Dissolution of Samples

While recognizing that the instrumental methods for elemental analysis described

[1] However, several laboratories, including the Nuclear Reactor Laboratory at MIT, have made their INAA facilities available to support outside research.

above are not nondestructive for labile constituents, it is convenient to distinguish this group of methods from those, termed "destructive," in which the sample is grossly altered. Destructive methods for elemental analyses frequently begin with sample dissolution, usually in concentrated, acid solutions at reflux temperature or, at about 100°C, using ultrasonic extraction. Such procedures "wet-ash" the samples (i.e., oxidize carbonaceous materials). For atmospheric particulate matter samples collected on nonashable (e.g., glass and quartz fiber) filters, an optimized ultrasonication technique utilized a mixture of concentrated HCl and HNO_3 (Harper et al. 1983; Lodge 1989, 365). With such filter samples, the variability in trace element concentrations extracted from the sampling medium may increase substantially the MDL. Samples suitable for total dissolution (e.g., those collected on cellulose or cellulose ester membrane filters) can be dissolved in concentrated HNO_3 at 140°C, with complete oxidation of organic material (Lodge 1989, ISC Method 822). Elements associated with silicaceous minerals may not be recovered with high efficiency by acid extractions unless the sample is finely pulverized. Alkaline fusion techniques are available for solubilizing silicon and associated elements in atmospheric particles (Appel et al. 1973). Dissolution of glass fiber filters in HF or by alkali fusion (McDonnell and Hilburn 1978) has been used to facilitate recovery of particulate constituents.

Inductively Coupled Plasma Emission Spectroscopy

Inductively coupled plasma emission spectroscopy (ICP), is notable for its speed, simplicity, accuracy, and wide dynamic range. An EPA SOP is available for acid digestion, ICP analysis and associated quality assurance of atmospheric samples on glass fiber or similar filter samples (Cummings et al. 1983; Harper et al. 1983). The latter reference describes an ICP system permitting a simultaneous measurement of up to 48 elements per sample. The technique has been reviewed by Fogg and Seeley (1984), Garbarino and Taylor (1985), and Lodge (1989, 89).

ICP response remains linear over five or more orders of magnitude change in concentration. It is suitable for the determination of most metallic elements in solution over the concentration range 10–10^5 ng/ml, and for most nonmetallic elements over the range 10^2–10^6 ng/ml. The recommended wavelengths and estimated IDL values for 48 elements are given in Cummings et al. (1983). For many metals, the IDL is more than a factor of 10 lower than that by flame atomic absorption spectroscopy.

Summary of Technique

The sample solution is nebulized into the center of a gaseous argon plasma, sustained inside an induction coil energized with a high-frequency alternating current. The analytes dissociate into their atomic form and are excited to higher energy levels.

Relaxation of the excited species back to their ground states causes emission of radiation characteristic of the elements in the sample. The emitted light intensities at characteristic wavelengths, isolated with a diffraction grating, are measured with a photomultiplier tube.

Precision and Accuracy

Harper et al. (1983) assessed both precision and accuracy for the analysis of atmospheric particulate matter samples. The accuracy of ultrasonic extraction and ICP analysis was assessed by recoveries of metals spiked in known amounts on glass fiber filter strips as well as by recoveries with NIST SRM No. 1648 (a bulk sample of urban particulate matter). Table 12-5 shows mean recovery and precision values for analyses by ICP as well as by flame atomic absorption spectroscopy (flame-AAS) and INAA. With spiked filter strips, recoveries by ICP were $\geq 88\%$, except for Cd and Cu, with CV values ranging from 3% to 8%. With SRM 1648, recoveries varied more widely, with $<25\%$ recoveries of Ti and Cr. For the remaining elements, recoveries ranged from 68% to 130%. Using spiked filters, ICP and flame-AAS results were quite

TABLE 12-5 Precision and Methods Comparison of Metals Analysis by ICP, Flame-AAS, and INAA

Element	Spiked Strips				NIST SRM No. 1648			
	Percentage Recovery		Percentage CV		Percentage Recovery		Percentage CV	
	AAS	ICP	AAS	ICP	INAA	ICP	INAA	ICP
As	106	96	7.3	2.9	105	130[a]	8.3	2.2
Ba	100	100	3.0	3.7	116	80	20	0.8
Cd	69	78	3.9	5.0	84	114[a]	32	8.5
Co	96	96	6.5	4.1	94	96	6	5.4
Cr	NA	NA	NA	NA	89	23[a]	8.3	1.4
Cu	68	69	5.1	5.0	NA	100[a]	NA	1.4
Fe	71	97	8.0	4.3	99	68[a]	6.4	1.4
Mn	86	94	8.0	5.6	NA	88	NA	1.6
Ni	88	97	7.0	4.0	126	90[a]	29	9.0
Pb	97	99	1.8	3.0	NA	95[a]	NA	1.1
Sr	86	98	4.6	5.9	NA	NA	NA	NA
Ti	NA	NA	NA	NA	NA	12	NA	5.0
V	80	92	5.7	6.3	NA	79	NA	1.9
Zn	85	88	8.8	8.4	92	97[a]	4.5	3.8

Source: Data from Harper et.al. (1983)

a. Mean percentage recoveries based on NIST-certified SRM values; other elements values are not certified by NIST
NA: Not available

similar, except for a lower recovery of Fe by AAS. With NIST SRM 1575 (Pine Needles), ICP results were within one standard deviation of the NIST certified value for all nine analytes reported (Fogg and Seeley 1984).

Comparing ICP and INAA results (Harper et al. 1983) with SRM No. 1648 for the eight elements determined by both techniques, the recovery ranges were similar, suggesting that the extraction and ICP analysis did not introduce a systematic bias to these results, except for Cr. The precision of the ICP determinations was generally much better than for INAA. The poorer precision of the INAA results for Ba and Zn reflected the variability in the filter blank results; these elements remained largely unextracted in preparing the ICP samples. Low recoveries of Cr and Ti by ICP probably reflect the inability of the mixed-acid solution to extract these soil-related elements.

Fogg and Seeley (1984) compared ICP and INAA for determining 13 elements, employing five atmospheric samples. When all data were included, the results were not significantly different. However, for Ca, Cr, Cu, and Mg, a significantly poorer agreement between methods was found compared to that for other elements.

Atomic Absorption Spectroscopy

Atomic absorption spectroscopy (AAS) is one of the most widely used methods for analysis at low- and trace-level concentrations, being applicable to about 70 elements. Notable exclusions include sulfur, carbon, and the halogens. However, the technique is subject to matrix and interelement interferences, for which correction techniques are generally available. A discussion of the principle of the method, the instrumentation employed, its advantages and disadvantages, and elimination of interferences effects are included in Lodge (1989, 83–89); detailed methods for analyzing Hg, Mo, and general trace metals in airborne materials are also included (Lodge 1989, Methods 317, 319, 822). Detailed methods are also provided by AAS instrument manufacturers (e.g., Perkin-Elmer Corp. 1982). Hwang (1972), Van Loon (1980), and Sneddon (1983) have reviewed sample preparation and analysis by AAS.

Summary of Technique

Light absorption by a medium composed of free, ground-state atoms of the element to be determined is measured at characteristic wavelengths in the range 180–900 nm. The amount of absorption increases with the concentration of atoms in the medium and, thus, with the concentration of the sample solution used in producing it (Beer's law). A hollow cathode lamp providing the emission spectrum of the analyte is typically used as an emitting source of constant intensity.

Two commonly used forms of AAS, flame- and graphite furnace-AAS, differ in the means of producing the absorbing medium. In flame-AAS, analyte solutions are nebulized into a flame, which dissociates the sample into ground-state atoms. Both air–acetylene and nitrous oxide–acetylene flames are used, depending upon the temperature needed. An air–acetylene flame provides a temperature of 2125–2400°C compared to 2600–2800°C for a nitrous oxide–acetylene flame. The higher temperature is required for more refractory elements such as B, V, Ti, and Si. In graphite furnace AAS (also referred to as electrothermal AAS, or ET-AAS), an injected sample solution is heated stepwise to dry, char, and atomize the analyte at temperatures up to 3000°C. ET-AAS provides a 50–500-fold decrease in the IDL, compared to flame AAS. (Lodge 1989, 87). However, ET-AAS is a slightly slower procedure and interference problems may be more difficult to overcome compared to flame AAS. Accordingly, it is most effectively used when sample concentrations dictate a lower IDL than provided by flame AAS. Specialized techniques are available for AAS analysis of elements which form volatile hydrides, including As, Se, Bi, Sb, and Te, as well as for Hg which can exist in an atomic state at ambient temperature. Excepting Sb, these techniques decrease the IDL by a factor of $\geq 10^3$ compared to analyses by flame AAS.

Precision and Accuracy

With flame AAS, a CV of 0.5–2% is expected for replicate determinations of analytes within the optimum analytical range, 0.1–0.5 absorbance units. The recovery of elements by flame-AAS with spiked filter strips was reported to vary from 68% to 106% (Table 12-5), as discussed above. In an interlaboratory comparison of flame-AAS analysis of atmospheric particulate lead on glass fiber filters, the interlaboratory mean result showed a CV of 3.7% (Air and Industrial Hygiene Laboratory 1973). Recovery of lead near the IDL was 101 ± 12% for trials in one laboratory. For a series of elements in water samples analyzed by numerous laboratories, the CV of the interlaboratory means ranged from 8% to 44% (American Public Health Association 1985, 156).

Interferences can result from differences in the matrix for the sample and the standard (e.g., differences in viscosity). Chemical interference results from the formation of compounds with the analyte, which are undissociated at the flame (or ET) temperature. Ionization interference results from an excitation of the analyte, depleting its ground-state concentration. Strategies to eliminate or minimize these effects are available. The method of standard additions (Lodge 1989, 617) can be used to identify such interferences as well as to obtain more accurate analyte concentrations. This method does not correct for broadband absorption and light scattering due to other sample matrix constituents. However, by employing a continuum source, background correction can be made automatically. Interferents for the AAS analysis of 24 elements found in atmospheric particulates have been tabulated (Lodge 1989, 609).

CARBON DETERMINATION

The total C (TC) present in atmospheric particles can be expressed as the sum of organic C (OC), elemental C (EC), and carbonate C (CARBC):

$$TC = OC + EC + CARBC \qquad (12\text{-}2)$$

The contribution of CARBC to TC in atmospheric aerosols is usually < 5%, at least for samples collected in urban areas (Mueller, Mosley, and Pierce 1972; Appel, Tokiwa, and

Kothny 1983). The retention of atmospheric CO_2 was judged an insignificant source of quartz fiber filter-collected carbon (Appel, Cheng, and Salaymeh 1989). Total carbon analyzers (e.g., coulometer systems, Coulometrics, Inc.) frequently include instrumentation and techniques for determining CARBC by sample acidification, and quantitation of the evolved CO_2.

Combustion of all carbon in an atmospheric sample to form CO_2, followed by a quantitation technique, permits a straightforward determination of TC, with good agreement between methods and laboratories. However, the analysis of OC and EC in atmospheric particles is hampered by the lack of appropriate reference materials and, indeed, by the lack of precise definitions of what constitutes species such as EC in atmospheric samples. As a consequence, operational definitions of OC and EC have been employed, in which these fractions are defined by a specific analytical approach. Although methods can be compared, the absolute accuracies with atmospheric samples cannot be assessed. In some instances, analysts have replaced OC and EC with terminology which reflects more closely their analytical methods, for example, "volatilizable and nonvolatilizable C" (Stevens, McClenny, and Dzubay 1982). The term "black C" (BC) has been used in place of EC by other workers (Gundel et al. 1984). Cadle and Groblicki (1982) have compared a wide assortment of techniques for the determination of EC and OC.

The present discussion describes current approaches to the analysis of TC, OC, and EC (or BC) applicable to filter-collected samples. The results of a recent interlaboratory comparison utilizing a range of analytical approaches illustrate the variability in the results. The discussion will emphasize the techniques employed to effect the discrimination between OC and EC, since this, together with the degree of conversion of OC to EC during analysis ("charring"), is thought to be the primary source of variability between the methods. The earlier techniques, relying on labor-intensive solvent extraction (Grosjean 1975; Appel, Colodny, and Wesolowski 1976), provide only the lower and upper limits for OC and EC, respectively, and will not be discussed.

Sampling Requirements

The likelihood of lesser retention of gaseous organic materials on quartz fiber compared to glass fiber filters is discussed in the Sampling and Storage Artifacts section. Quartz filters also provide lower carbon blank values (< 1 μg/cm^2), since they can be prefired at ca. 750°C, compared to < 450°C for glass fiber. The quartz filters in common use for carbon measurements include Pallflex types 2500 QAO, 2500 QAST, and 2500 QAOT (Pallflex Products Corp, Putnam, CT).

Total Carbon

Total C is determined by combustion (or decomposition) of the sample to CO_2 followed by quantitation, or, indirectly, as the sum of OC and EC, determined separately (adding CARBC determinations where significant). The direct combustion approach is expected to exhibit better precision. Pure materials (e.g., graphite and potassium acid phthalate) can be used for calibration and to assess the accuracy likely to be encountered with atmospheric samples.

The contribution of CARBC to TC may be less than quantitative, depending on analytical conditions and the specific carbonate salts present. For example, using combustion/decomposition in O_2 at 1000°C for 3 min and quantitation of CO_2 by coulometry, Na_2CO_3 and K_2CO_3 exhibited as little as 80% recovery (Hsu and Appel 1984). Total C techniques employing lower temperatures for fixed time periods may achieve even lower recoveries of carbon from the decomposition of carbonate salts.

The differences between various direct TC techniques reflect, primarily, the method of quantitation. The common techniques include determination of CO_2 by nondispersive infrared, gas chromatography with thermal conductivity detection (GC-ThC), coulometry, and by GC with flame ionization detection, following hydrogenation to methane.

Detailed methods for TC analysis are available based on GC-ThC (Belsky 1971) and coulometry (Hsu and Appel 1984). A comparison of TC results with 11 laboratories, utilizing both direct and indirect techniques, was performed using atmospheric as well as source-enriched samples (Countess 1990). For loadings of 9–380 μg C/cm^2, the CV for the interlaboratory results ranged from 4% to 14%, indicating an excellent average agreement in TC analyses, in contrast to the results for OC and EC.

Organic and Elemental Carbon

Optical Techniques

A conceptually simple approach is to measure directly the concentrations of EC and TC (and CARBC, if necessary), obtaining OC by difference:

$$OC = TC - EC - CARBC \qquad (12\text{-}3)$$

Since EC is typically only 15%–25% of TC (and CARBC $< 5\%$ of TC), the measurement of OC by difference should be reasonably precise. If it is assumed that EC is the only significant visible-light-absorbing species in atmospheric particulate matter samples, the blackness of these samples, with appropriate empirical calibration, measures EC. The EC measurement is both rapid and nondestructive. Both reflectance (Delumea, Chu, and Macias 1980) and transmittance methods (Rosen et al. 1980; Gundel et al. 1984) can be used. Optical methods for EC (or BC) are restricted to samples with EC loading less than ca. 10 μg/cm^2. The difficulty with optical methods for EC lies in the calibration technique. In one study, a difference of a factor of 4 was observed, depending on the source of EC standards used for calibrations (Appel et al. 1983).

Pyrolysis–Combustion Techniques

Based on the behavior of model compounds (e.g., graphite and individual organic compounds), the split between OC and EC has been made based on the relatively low temperature needed for pyrolysis of OC in an inert gas stream; the remaining C, assumed to be EC, is then combusted in a gas stream containing oxygen (Stevens et al. 1982; Grosjean 1984). The Grosjean procedure is notable for its use of commercially available instrumentation. Some laboratories have separated OC from EC by pyrolysis plus combustion in atmospheres containing O_2 (Cadle, Groblicki, and Mulawa 1983). The empirical nature of all such separations, and the variation introduced into the resulting OC/EC values have been well described (Cadle, Groblicki, and Mulawa 1983; Tanner, Gaffney, and Phillips 1982). Charring under pyrolysis conditions causes low OC and high EC values, and is a significant potential source of error. This error has been minimized by decreasing the pyrolysis time, and/or increasing the pyrolysis temperature (Cadle, Groblicki, and Mulawa 1983).

Thermal–Optical Techniques

Huntzicker et al. (1982) modified the pyrolysis–combustion technique to permit a correction for the degree of charring occuring with each sample. Filter reflectance as a measure of EC in the sample was monitored continuously during the analysis, together with the evolved carbon. An automated sampler, using a similar strategy, provided semicontinuous results (Turpin, Cary, and Huntzicker 1990).

Novakov (1982) described a procedure in which manually collected filter samples were subjected to a linear temperature increase in a stream of O_2, while monitoring light transmittance through the filter sample and the concentration of evolved C as CO_2. A sudden increase in transmittance at ca. 470°C and the area of the corresponding peak in CO_2 concentration served to measure the EC (or BC) in the sample. The total area of the CO_2 concentration peaks preceding the BC peak measured OC.

Accuracy and Precision of Techniques for OC and EC

The state of the art is illustrated by the results of a methods comparison with atmospheric samples (Countess 1990). Table 12-6 summarizes mean EC/(EC + OC) ratios for each

TABLE 12-6 Summary of Methods Comparison for Organic and Elemental Carbon Analysis

Method Category	No. of Participants	Ratio EC/(OC + EC)	
		Range	Mean
Pyrolysis–Combustion	7	0.10–0.29	0.20 ± 0.06
Thermal–Optical	3	0.15–0.30	0.24 ± 0.08
Optical	1	0.23	0.23

Source: Calculated from data in Countess (1990)

general approach. The observed range in the results obtained by pyrolysis–combustion as well as by thermal–optical techniques was a factor of 2 to 3. Lower EC/(EC + OC) ratios would be expected with the thermal–optical methods, since these were designed to correct for charring, in contrast to these findings. The relatively good agreement between the optical and thermal–optical methods probably has little significance, since the optical technique was calibrated against a thermal–optical method. Precision for replicate determinations by the same analyst was not a significant problem; the median CV values for all participants was 3.5%, 3.5%, and 6.7% for TC, OC, and EC, respectively. An earlier methods comparison (Groblicki et al. 1983), with similar findings, includes a detailed description of each method.

WATER-EXTRACTABLE ANION AND CATION ANALYSIS

Aqueous Extraction

The extraction of atmospheric particulate matter from glass and quartz fiber filter samples for the analysis of anions and cations (e.g., NO_3^-, SO_4^{2-}, Cl^-, NH_4^+) presents few problems; a comparison of water extraction techniques for SO_4^{2-} (Appel 1981), with samples on glass fiber filters, showed recoveries of 97%–100%, with 30 min ultrasonic extraction, 60 min extraction under reflux, or 60 min extraction in a mechanical shaker at room temperature. However, recoveries of water-soluble materials from Teflon filters are hindered because the filters are hydrophobic, and will float unless secured below the water surface. Sulfate recoveries with atmospheric samples ranged from 80% to $\geq 98\%$, depending upon the technique employed; 30 min ultrasonic extraction and mechanical shaking at room temperatures provided consistently high extraction efficiencies. Many groups employ the addition of several drops of methanol or ethanol as a wetting agent to facilitate extraction, although in one evaluation, extraction was not enhanced by such pretreatment (Appel 1981). It was noted previously that Zefluor filters (Gelman Sciences) have a porous Teflon backing, which requires very careful examination to distinguish from the membrane surface. It is likely that aqueous extraction of aerosol constituents (incorrectly) sampled onto the back filter surface would be less efficient.

Extraction of atmospheric NO_3^- from Teflon filters (Appel et al. 1980*a*) with ultrasonic extraction and mechanical shaking, as described for sulfate, gave efficiencies of $> 98\%$. However, nitric acid bound to Nylon is not efficiently recovered as NO_3^- with water extraction (Sickles et al. 1986); an alkaline medium (e.g. the dilute CO_3^{2-}–HCO_3^- buffer employed as eluent in ion chromatography) does provide efficient extraction, with the possible exception of very lightly loaded samples (Appel et al. 1988; Hering et al. 1988).

The extraction of strong acids from atmospheric particles collected on Teflon filters can be carried out in water (Appel et al. 1980*b*). However, subsequent analysis requires the elimination of dissolved CO_2 and a technique to eliminate or allow for the dissociation of weak acids. An EPA procedure for determining these strong acids, as well as anionic particle constituents, employs wetting of 47 mm Teflon filters with 100 ± 5 μl of ethanol and ultrasonic extraction of the particulate acids in 10^{-4} N $HClO_4$ (U.S. EPA 1989). The use of a pH = 4 extraction medium represses the dissociation of weak acids and minimizes the concentration of dissolved CO_2. Ultrasonic extraction efficiency of SO_4^{2-} from atmospheric samples on Teflon filters in 5×10^{-5} N $HClO_4$ (without alcohol prewetting) was reported to be $98 \pm 1\%$ and $95 \pm 2\%$ for fine and coarse particle fractions, respectively (Stevens et al. 1978). A fluted Teflon pipe resting on the outer edges of the filter was used to keep the filter submerged.

Sulfate and Nitrate

The literature abounds with methods for determining sulfate (Tanner and Newman 1976; Appel 1981) and nitrate (U.S. EPA 1982*a*) in atmospheric aerosols. The precision and accuracy, interference effects, and working ranges for various manual and automated procedures for sulfate determination have been reported (Appel 1981). Sulfate and nitrate analyses are currently dominated by the use of ion chromatography (IC), the technique discussed below. The ability of IC to determine rapidly these as well as other anions in atmospheric particulate extracts with a single sample injection has made it the pre-eminent technique for such measurements. In a recent field comparison of measurement methods for nitric acid, 13 of 15 laboratories requiring NO_3^- determinations employed IC (Herring et al. 1988). The application of IC to specific analytes in atmospheric particulate extracts, including SO_4^{2-} and NO_3^-, is discussed by numerous authors in Mulik and Sawicki (1978, 1979). Detailed methods for routine anion analysis with current IC instrumentation are available (U.S. EPA 1986; American Public Health Association 1985, 483).

The principal alternatives to IC for SO_4^{2-} and NO_3^-, at least where large numbers of samples are involved, are single analyte, automated, colorimetric methods. These can be subject to a variety of interference effects. Procedures for automated NO_3^- (Technicon Industrial Systems 1973) and SO_4^{2-} (Technicon Industrial Systems 1972) are available (Haik and Imada 1981). The results by IC are compared below to those by these automated methods.

Summary of Ion Chromatographic Technique

Ion chromatography encompasses an array of techniques varying in modes of achieving analyte separation and detection. Reviews of IC include those by Small (1981), Colenutt and Trenchard (1985), and Lodge (1989, 230). The determination of SO_4^{2-} and NO_3^- is readily achieved over at least a three-order-of-magnitude concentration range using chemically suppressed anion chromatography. This technique employs a short anion guard column (for removal of particles and organic constituents), an anion separator column, postcolumn reactor, and conductivity detector. The separator and guard column contain a high-efficiency anion exchange resin, whose permeability permits relatively high flow rates. The postcolumn reactor is an anion suppressor which may be in the form of a packed bed, fiber or micromembrane. The strongly acidic suppressor protonates the dilute carbonate–bicarbonate eluent to sharply decrease the background conductance, converting anions in the sample to highly conductive strong acids (e.g., HCl, HF, HBr, HNO_3, and HSO_4^-) for conductimetric detection. The IDL values vary with the system configuration. Using direct sample injection with a 50 μl sample loop, 10 microsiemens full scale, an IDL of 0.01 μg/ml is estimated; with sample preconcentration, IDL values decrease by a factor of 10^3 (Lodge 1989, 230).

Sampling Requirements

As discussed in the Sampling and Storage Artifacts section, sulfate concentration values obtained with quartz and Teflon filter samples can provide useful measures of atmospheric SO_4^{2-}, but NO_3^- values provide what are probably lower limits to atmospheric PN levels. Specialized collection techniques are used for more accurate measurement (Appel et al. 1981; Appel 1991).

Accuracy and Precision

Numerous evaluations have generally confirmed the excellent accuracy and precision of IC analysis for SO_4^{2-} and NO_3^-. EPA (1986) reported single-operator CV values of 2–3% for concentrations ranging from 0.8 to 3.6 μg/ml, using quality control samples for the National Acid Deposition Program; recoveries were > 97%. Lodge (1989, 540) reports a CV of < 3% at 1 μg/ml for anion peak areas. The analysis of standard solutions by 15 laboratories (American Public Health Association 1985) showed no significant bias for the interlaboratory means at the 95% confidence level, and a single-operator CV of 5–6%. However, the CV of the interlaboratory data, reflecting the scatter between laboratories, was 32% for NO_3^- at 0.5 μg/ml and 5%–8% for SO_4^{2-} over the range 0.5–44 μg/ml.

With atmospheric sample extracts, SO_4^{2-} concentrations by IC agreed within 10% with those by an automated methylthymol blue (MTB), a barium turbidimetric, and a microchemical, colorimetric method (Appel and Wehrmeister 1979). A similar comparison against the MTB method showed average agreement within about 3% (Fung et al. 1979).

Nitrate determinations by IC were compared with those by the automated, Cu–Cd reduction procedure by Fung et al. (1979). With atmospheric particulate sample extracts in the range 0.3–2.4 μg/ml, the results agreed, on an average, within about 1%. The presence of ethanol and $HClO_4$ in the extract was shown not to interfere in the IC analysis of either NO_3^- or SO_4^{2-}.

Using the anion separator column, elution conditions and the membrane suppressor recommended by Lodge (1989, 538), NO_3^- and SO_4^{2-} determinations are free of interferences due to coelution of the common anions at ambient levels. However, labile species (e.g., SO_3^{2-}) may be converted to SO_4^{2-} during sample extraction, storage, and analysis. Nitrite ion is stable to oxidation under analysis conditions. Aqueous extracts can show increasing concentrations of NO_3^- with time due to slow formation from NO_2^- and NH_4^+. The presence of Cl^- at high concentrations can interfere with the quantitation of NO_3^- by IC, a complicating factor in HNO_3 sampling with NaCl-impregnated filters.

Ammonium Ion Analysis

Ammonium ion is determined in collected samples of both gaseous NH_3 and particulate matter. Reviews of appropriate analytical techniques are available (NRC NAS 1977; Mulik, Estes, and Sawicki 1978). Both a manual and an IC method for NH_4^+ are discussed below.

Sampling and Storage Requirements

The analysis of NH_4^+ in particulate matter is most frequently done on extracts of Teflon and quartz fiber filter samples as one of a number of analytes of interest. The ease of loss of NH_4^+ from particulate matter on such filters was previously noted (see the Sampling and Storage Artifacts section). Such loss may be offset to some degree by NH_3 retention on acidic particle constituents. Accurate air concentration measurements of particulate NH_4^+ require sampling techniques which prevent NH_3 reactions with the sample and minimize the loss of NH_4^+ together with anions such as NO_3^- (e.g., Wall, John, and Ondo 1988).

Following aqueous extraction, NH_4^+ remains unstable with respect to bacteria-promoted oxidation. The extracts should be preserved by refrigeration at ca. 4°C and analyzed within 24 h. For prolonged storage of aqueous extracts, the pH should be reduced to < 2 with H_2SO_4 (assuming this has been coordinated with analyses for other analytes as necessary).

Selective Ion Electrode

Summary of Technique

The manual selective ion electrode (SIE) method for NH_4^+ is low in equipment costs and employs easily portable components. The American Public Health Association (1985, 384–386), provides a detailed method and description of the technique. The ammonia-selective electrode uses a hydrophobic, gas-permeable membrane to separate the sample extract from an internal electrode solution of NH_4Cl. Dissolved ammonia ($NH_{3(aq)}$ and NH_4^+) is converted to $NH_{3(aq)}$ by raising the pH to > 11. The $NH_{3(aq)}$ diffuses through the membrane and changes the internal solution pH. A chloride ion-selective electrode senses the fixed level of Cl^- in the internal solution, providing a reference electrode. Potentiometric measurements are made with a pH meter having an expanded millivolt scale. The OH^- concentration of the inner solution is directly proportional to the external ammonia concentration. The method is reported to be applicable in the range 0.04–1700 μg/ml (as NH_3), but responds slowly (e.g., 5–10 min) below 1 μg/ml.

Accuracy and Precision

The SIE method is subject to interference from amines, and by the metals Hg and Ag, which complex ammonia. With synthetic samples containing 0.1–1 μg/ml dissolved NH_3, the single laboratory CV for replicate measurements was 2–4%. Also with one laboratory, recoveries at 0.1 μg/ml were $> 90\%$. In an interlaboratory study with 12 laboratories using spiked aqueous samples, mean recoveries in the range 0.8–750 μg/ml were 95%–105%. However, at ≤ 0.1 μg/ml, the method was too high by a factor of 2 or more (American Public Health Association 1985, 375–377). With atmospheric particulate matter extracts, a median CV of 4.8% was observed (Appel et al. 1988). A comparison of SIE and IC NH_4^+ determination is included below.

Ion Chromatography

Summary of Technique

A detailed method (U.S. EPA 1985) was developed for common cations in rainwater using chemically suppressed ion chromatography, a method which should be equally applicable to atmospheric particulate extracts. Monovalent cations, including NH_4^+, injected from a 100 μl sample loop, are eluted with a dilute HCl solution. The cations are resolved on a cation exchange resin separator column. The suppressor, analogous to that for anions, neutralizes the eluent to reduce background conductance. The sample cations are then detected conductimetrically in their hydroxide forms. An IDL value of 0.01 μg/ml is reported for NH_4^+, with a working range of 0.03–2.0 μg/ml. U.S. EPA (1985) employs a calibration curve developed with standards at six concentrations. Mulik, Estes, and Sawicki (1978) followed each sample with a standard of similar concentration for a single-point calibration. The latter is especially useful if response factors are found to change with time.

Accuracy and Precision

Mulik, Estes, and Sawicki (1978), using techniques similar to those reported by the EPA (1985), found that aliphatic amines did not interfere with NH_4^+ determination by IC, in contrast to the SIE technique. In the range 0.09–9 μg/ml, the CV for repeated injections was 1%–3%. Using Teflon filter samples impregnated with NH_4^+ at eight levels, recoveries averaged 91% by IC, compared to 88% by SIE. For ten Teflon, dichotomous filter sample extracts of atmospheric particles, results with the two methods were highly correlated ($r = 0.98$), with average agreement within 3%.

U.S. EPA (1985) notes that the presence of Na^+ or K^+ at concentrations 10–20 times higher than that of NH_4^+ will cause unresolved peaks. Air bubbles in the system components cause baseline drift, increased signal variability, and peak distortion. Storage of solution samples at 4°C retards, but does not prevent, loss of NH_4^+. Recovery of NH_4^+ spiked into rainwater samples was 107%–114% in the range 0.2–0.5 μg/ml, with a CV decreasing from 16% to 5% with increasing concentration in this range.

Strong Particle Acidity

Atmospheric particulate matter includes acidic materials with a wide range of dissociation constants, each of which may contribute to the concentration of "free" H^+, or H_3O^+, at a given pH. Strong acids in atmospheric aerosols are frequently considered to be those with pK_a values ≤ 4, and are largely dissociated to H_3O^+ at pH ≥ 4. In addition to mineral acids, strong acids, by this definition, include such materials as citric, lactic and oxalic acids and hydrated Fe^{3+}. Acids of intermediate strength are those of p$K_a = 4$–5 (e.g., benzoic acid and the first dissociation of adipic acid), and weak acids are those with p$K_a \geq 5$. Acids of intermediate strength can have a significant degree of dissociation at pH = 4 and, therefore, represent a potential source of positive interference in measuring strong acids, as defined above.

Aside from uncertainty about the chemical nature of the strong acids, their measurement in atmospheric aerosols is difficult; an unusual degree of quality assurance is needed to obtain meaningful results. Nevertheless, the potential significance of such materials to human health has prompted a continuing effort to make and refine such measurements. No generally accepted sampling and analysis procedure has yet emerged. The present discussion describes a currently employed strategy for which a detailed method has been prepared (U.S. EPA 1989), based on the work of Koutrakis, Wolfson, and Spengler (1989). The more complex Gran's titration technique (Stevens and Dzubay 1978; Ferek et al. 1983) will not be included. Although approximate cation–anion balances support the use of this technique for particulate extracts, it has been shown to overestimate strong acid measurements in mixtures with organic acids in rainwater samples (Keene, Galloway, and Holden 1983; Keene and Galloway 1985). The methodology for measuring particulate H_2SO_4, other acid sulfates, and total strong acid has been reviewed (Lodge 1989, 715; U.S. EPA 1982*b*). A detailed method for the semi-real-time measurement of atmospheric H_2SO_4 is available (Lodge 1989, Method 713).

Summary of Technique

Fine-particle samples collected on Teflon filters are prewet with 100 µl ethanol, and extracted in 10^{-4} M $HClO_4$. On the same day as the sample extraction, the solution pH is determined to the nearest 0.01 unit, with the ionic strength increased to 0.04 M with KCl. The H_3O^+ concentration contributed by the particulate sample at pH = 4 is calculated as (− antilog solution pH), corrected by subtracting the H_3O^+ concentration of the blank extract. A technique to correct for ionic strength effects is also provided (U.S. EPA 1989).

Sampling Requirements

Measurements of strong acids in atmospheric particulate matter are subject to negative errors from (1) neutralization by NH_3 (and other alkaline gases, if present) during and subsequent to collection, (2) neutralization by reaction with basic aerosol constituents as well as with the collection medium, and (3) loss as a volatile acid following metathesis (e.g., with NaCl). Numerous papers have documented such problems and evaluated strategies to minimize them (e.g., Stevens and Dzubay 1978; Appel et al. 1980*a*; Koutrakis, Wolfson, and Spengler 1988).

The loss of strong particulate acid is minimized by removal of NH_3 ahead of the collection medium, collection on Teflon filters, restricting collection to fine (< 2.5 µm) particle fractions, and using sample handling techniques which eliminate neutralization from NH_3. Each of these strategies is used in the papers cited above.

Accuracy, Precision, and Method Detection Limit

Accuracy and precision values of ± 10% and ± 5%, respectively, are claimed for pH values in the range 4–7 (U.S. EPA 1989). However, recoveries of H_2SO_4 from (presumably blank) Teflon filters by acidity measurements were reported to be 86%–93%, following shipment to and from a field location (Koutrakis, Wolfkson, and Spengler 1988). This is comparable with the value 88 ± 8% found in an independent study, which also showed that atmospheric particulate matter on the same filter could reduce strong acid recoveries by 35–50% (Appel et al. 1980*b*).

The contribution to strong acidity measurements of benzoic ($pK_a = 4.2$) and adipic acid ($pK_{a1} = 4.4$) was assessed in trials of two-component solutions containing H_2SO_4 by titration with dilute NaOH to a pH = 4 end-point (Appel et al. 1979). Up to 20% of benzoic acid was measured as strong acid at pH = 4.0, compared to about 1% of adipic acid. Thus, acids with pK_a somewhat above 4.0 probably contribute significantly to the atmospheric strong acid measurements obtained with the EPA procedure. An MDL value of 4 nanoequiv H^+/m^3 (24 nequiv $H^+/6\ m^3$ air sample) is estimated (Koutrakis, Wolfson, and Spengler 1988). This compares to 18 nequiv H^+/m^3 (86 nequiv $H^+/5\ m^3$ air sample), based on three times the standard deviation of Teflon filter field blanks and a similar analytical procedure (Appel et al. 1987).

SUMMARY

The accurate and precise measurement of atmospheric concentrations of aerosol constituents requires attention to detail in every step of the process: sampling technique, determination of the sampled air volume, sample storage, handling, and analysis. Each must be considered in evaluating the quality of atmospheric data. The selection of analytical techniques will frequently place requirements on the sampling technique or medium. Accordingly, this chapter, although focused on

TABLE 12-7 Standard Operating Procedures for Analysis of Aerosol Constituents

Constituent	Technique[a]	Filter Medium	SOP or Method Reference No.[b]
PM-10 mass	Gravimetric	Quartz fiber, Teflon	EMD-203 and Section 2.11 of EPA Quality Assurance Handbook ISC-501[c]
Elements	ED-XRFA	Teflon	EMD-010, ISC-822B[c]
Elements	INAA	Teflon, cellulose ester, others	EMD-008
Elements	ICP	Glass, quartz fiber, others	EMD-002, ISC-303A[c]
Elements	Flame and ET-AAS	Glass, quartz fiber, others	ISC-822[c]
	Flame-AAS	Same	Standard Methods 303[d]
	ET-AAS	Same	Standard Methods 304[d]
Total carbon	Combustion and coulometry	Quartz or glass fiber	AIHL No. 94[e]
Total carbon	Combustion and GC-ThC	Quartz or glass fiber	AIHL No. 30[e]
SO_4^{2-} and NO_3^-	Ion chromatography	Quartz, Teflon, Nylon, others	Standard Methods No. 29[d], EPA Method 300.6[f]
SO_4^{2-} and NO_3^-	Autoanalyzer (MTB SO_4^{2-}, Cu–Cd reduction NO_3^-)	Same	AIHL No. 86[e] ISC-720C (SO_4^{2-})[c,g]
NH_4^+	Ion chromatography	Quartz, Teflon, others	EPA Method 300.7[f,h]
	Specific ion electrode	Same	Standard Methods 417E[d]
Strong acid	NH_3-free sampling, pH measurement	Teflon	Compendium Chapter IP-9[i]

a. See Appendix B where required
b. Except as noted below, available from: U.S. Environmental Protection Agency, Environmental Monitoring Systems Laboratory, MD-78, Research Triangle Park, NC 27711
c. Included in Lodge (1989)
d. Included in American Public Health Association (1985)
e. Available from the Air and Industrial Hygiene Laboratory, California Department of Health Services, 2151 Berkeley Way, Berkeley, CA 94704
f. U.S. Environmental Protection Agency, Environmental Monitoring and Support Laboratory, Cincinnatti, OH 45268
g. Methods 271-73W for automated NO_3^- analysis, and Method 118-71W for automated SO_4^{2-} analysis (0.3–100 ppm) are available from Technicon Instruments Corp., Tarrytown, NY
h. Procedures also available from Dionex Corp., Sunnyvale, CA
i. Included in US EPA (1989)

current analytical techniques, also provides information relevant to sampling. Table 12-7 summarizes the detailed methods (or standard operating procedures) referred to in this chapter, and their sources, for analysis by most of the techniques mentioned.

References

Air and Industrial Hygiene Laboratory, California Department of Health Services. 1973. Determination of particulate lead, Method 41 (unpublished).

Alian, A. and B. Sansoni. 1985. A review on activation analysis of air particulate matter. *J. Radioanal. Nuc. Chem.* Articles 89:191–275.

American Public Health Association. 1985. *Standard Methods for the Examination of Water and Wastewater*, 16th edn. Washington, DC: American Public Health Association.

Appel, B. R. 1981. Studies in atmospheric particulate characterization techniques. In *Air/Particulate Instrumentation and Analysis*, ed. P. N. Cheremisinoff, pp. 25–87. Ann Arbor, MI: Ann Arbor Science.

Appel, B. R. 1991. Sampling of selected labile atmospheric pollutants. In *ACS Advances in Chemistry Series: Measurement Challenges in Atmospheric Chemistry*, ed.: L. Newman. Washington, DC: American Chemical Society (in press).

Appel, B. R., W. Cheng, and F. Salaymeh. 1989. Sampling of carbonaceous particles in the atmosphere—II. *Atmos. Environ.* 23:2167–75.

Appel, B. R., W. Cheng, and F. Salaymeh. 1990. *Particulate Sampling Techniques: Measurement of Atmospheric Particulate Carbonaceous Material.* Final Report to the Electric Power Research Institute.

Appel, B. R., P. Colodny, and J. J. Wesolowski. 1976. Analysis of carbonaceous materials in Southern California atmospheric aerosols. *Environ. Sci. Technol.* 10:359–63.

Appel, B. R., E. Hoffer, M. Haik, D. Levaggi, S. Twiss, and P.K. Mueller. 1973. Determination of silicon in atmospheric particulate matter. *Environ. Anal. Chem.* 4:197–203.

Appel, B. R., E. M. Hoffer, Y. Tokiwa, M. Haik, and J. J. Wesolowski. 1980*b*. *Sampling and Analytical Problems in Air Pollution Monitoring.* Final Report to the EPA, Grant No. R806734-01-0.

Appel, B. R., Y. Tokiwa, and M. Haik. 1981. Sampling of nitrates in ambient air. *Atmos. Environ.* 15:283–89.

Appel, B. R., Y. Tokiwa, M. Haik, and E. L. Kothny. 1984*a*. Artifact particulate sulfate and nitrate formation on filter media. *Atmos. Environ.* 18:409–16.

Appel, B. R., Y. Tokiwa, J. Hsu, E. L. Kothny, E. Hahn, and J. J. Wesolowski. 1983. Visibility reduction as related to aerosol constituents. Final Report to the California Air Resources Board, Agreement No. A1-081-32. NTIS Report PB 84 243617.

Appel, B. R., Y. Tokiwa, and E. L. Kothny. 1983. Sampling of carbonaceous particles in the atmosphere. *Atmos. Environ.* 17:1787–96.

Appel, B. R., Y. Tokiwa, E. L. Kothny, V. Povard, and J. J. Wesolowski. 1984*b*. *Sampling and Analytical Problems in Air Pollution Monitoring*, Phase II, Part I. Final Report to the U.S. EPA, Cooperative Agreement CR 810798-01-0.

Appel, B. R., Y. Tokiwa, E. L. Kothny, R. Wu, and V. Povard. 1988. Evaluation of procedures for measuring atmospheric nitric acid and ammonia. *Atmos. Environ.* 22:1565–73.

Appel, B. R., Y. Tokiwa, V. Povard, and E. L. Kothny. 1987. *Determination of Acidity in Ambient Air, Phase II.* Final Report to the California Air Resources Board, Contract A4-074-32.

Appel, B. R., Y. Tokiwa, S. Wall, M. Haik, E. L. Kothny, and J. Wesolowski. 1979. *Determination of Sulfuric Acid, Total Particle-Phase Acidity and Nitric Acid in Ambient Air.* Final Report to the California Air Resources Board, Contract A6-209-30.

Appel, B. R., S. M. Wall, M. Haik, E. L. Kothny, and Y. Tokiwa. 1980*a*. Evaluation of techniques for sulfuric acid and particulate strong acidity measurements in ambient air. *Atmos. Environ.* 14:559–63.

Appel, B. R., S. M. Wall, Y. Tokiwa, and M. Haik. 1980*c*. Simultaneous nitric acid, particulate nitrate and acidity measurements in ambient air. *Atmos. Environ.* 14:549–54.

Appel, B. R. and W. J. Wehrmeister. 1979. An evaluation of sulfate analyses of atmospheric samples by ion chromatography. In *Ion Chromatographic Analysis of Environmental Pollutants*, eds. J. Mulik and E. Sawicki, Vol. II, pp. 223–33. Ann Arbor, MI: Ann Arbor Science.

Appel, B. R., J. J. Wesolowski, A. Alcocer, S. Wall, and S. Twiss. 1980*d*. Quality assurance for the chemistry of the aerosol characterization experiment. In *The Character and Origins of Smog Aerosols*, eds. G. M. Hidy, P. K. Mueller, D. Grosjean, B. R. Appel, and J. J. Wesolowski, pp. 69–105. New York: Wiley.

Belsky, T. 1971. *Carbonate and Non-Carbonate Carbon in Atmospheric Particulate Matter.* Air and Industrial Hygiene Laboratory, California Department of Health Services Method No. 30.

Birks, L. S. 1969. *X-ray Spectrochemical Analysis*, 2nd edn. New York: Interscience.

Bondietti, E. and C. Papastefanou. 1989. Large particle nitrate artifacts in the aerodynamic size distributions of ambient aerosols. *J. Aerosol Sci.* 20:667–70.

Cadle, S. and P. Groblicki. 1982. An evaluation of methods for the determination of organic and elemental carbon in particulate samples. In *Particulate Carbon-Atmospheric Life Cycle*, eds. G. T. Wolff and R. L. Klimisch, pp. 89–109. New York: Plenum.

Cadle, S., P. Groblicki, and P. Mulawa. 1983. Problems in the sampling and analysis of carbon particles. *Atmos. Environ.* 17:593–600.

Cadle, S. and P. Mulawa. 1987. The retention of SO_2 by nylon filters. *Atmos. Environ.* 21:599–603.

Cahn, L. 1963. Dynamic weight change of a membrane filter with humidity. *Mater. Res. Stand.* 3:377.

Camp, D. C., J. A. Cooper, and J. R. Rhodes. 1974. X-ray fluorescence analysis—Results of a first round intercomparison study. *X-Ray Spectrometry* 3:47–50.

Camp, D. C., A. L. VanLehn, J. Rhodes, and A. Pradzynski. 1975. Intercomparison of trace element determinations in simulated and real air particulate samples. *X-Ray Spectrometry* 4:123–37.

Clement, R. and F. Karasek, F. 1979. Sampling composition changes in sampling and analysis of organic compounds in aerosols. *Int. J. Environ. Anal. Chem.* 7:109–20.

Colenutt, B. A. and P. Trenchard. 1985. Ion chromatography and its application to environmental analysis: A review. *Environ. Pollut.* (Series B) 10:77–96.

Countess, R. J. 1990. Interlaboratory analysis of carbonaceous aerosol samples. *Aerosol Sci. Technol.* 12:114–21.

Coutant, R. W. 1977. Effects of environmental variables on collection of atmospheric sulfate. *Environ. Sci. Technol.* 11:873–78.

Coutant, R. W., L. Brown, J. Chuang, R. Riggin, and R. Lewis. 1988. Phase distribution and artifact formation in ambient air sampling for polynuclear aromatic hydrocarbons. *Atmos. Environ.* 22:403–9.

Cummings, S. L. 1981. *Standard Operating Procedure for the X-ray Fluorescence Analysis of Multielements on Dichotomous Sample Filters.* U.S. EPA Method EMSL/RTP-SOP-EMD-010.

Cummings, S. L., S. Harper, W. Loseke, and L. Pranger. 1983. *Standard Operating Procedure for the ICP-OES Determination of Trace Elements in Suspended Particulate Matter Collected on Glass-Fiber Filters.* U.S. EPA Method EMSL/RTP-SOP-EMD-002.

Currie, L. A. 1988. *Detection in Analytical Chemistry, Importance, Theory and Practice.* Washington: American Chemical Society.

Dams, R., J. A. Robbins, K. A. Rahn, and J. W. Winchester. 1970. Nondestructive neutron activation analysis of air pollution particulates. *Anal. Chem.* 42:861–67.

Delumea, R., L.-C. Chu, and E. Macias. 1980. Determination of elemental carbon component of soot in ambient aerosol samples. *Atmos. Environ.* 14:647–52.

Dolske, D., J. Schneider, and H. Sievering. 1984. Trace element pass-through for cellulose filters when used for aerosol collection. *Atmos. Environ.* 18:2557–58.

Dunwoody, C. L. 1986. Rapid nitrate loss from PM_{10} filters. *J. Air Pollut. Control Assoc.* 36:817–18.

Dzubay, T. G. 1977. *X-ray Fluorescence Analysis of Environmental Samples.* Ann Arbor, MI: Ann Arbor Science.

Dzubay, T. G. and R. K. Barbour. 1983. A method to improve the adhesion of aerosol particles on Teflon filters. *J. Air Pollut. Control Assoc.* 33:692–95.

Dzubay, T. G. and R. O. Nelson. 1975. Self absorption corrections for X-ray fluorescence analysis of aerosols. In *Advances in X-ray Analysis*, Vol. 18, eds. W. L. Pickles, C. S. Barrett, J. B. Newkirk, and C. O. Rund, pp. 619–31. New York: Plenum.

Dzubay, T. G. and D. G. Rickel. 1978. X-ray analysis of filter-collected aerosol particles. In *Electron Microscopy and X-ray Applications*, eds. P. A. Russell and A. E. Hutchings. Ann Arbor: Ann Arbor Science.

Dzubay, T. G. and R. K. Stevens. 1991. Sampling and analysis methods for ambient PM-10 aerosol. In *Receptor Modeling in Air Quality Management*, ed. P. K. Hopke. New York: Elsevier.

Dzubay, T. G., R. K. Stevens, C. Lewis, D. Hern, W. Courtney, J. Tesch, and M. Mason. 1982. Visibility and aerosol composition in Houston, Texas. *Environ. Sci. Technol.* 16:514–25.

Feeney, P., T. Cahill, J. Olivera, and R. Guidara. 1984. Gravimetric determination of mass on lightly loaded membrane filters. *J. Air Pollut. Control Assoc.* 31:376–79.

Ferek, R., A. Lazrus, P. Haagenson, and J. Winchester. 1983. Strong and weak acidity of aerosols collected over the northeastern United States. *Environ. Sci. Technol.* 17:315–24.

Finlayson-Pitts, B. J. and J. N. Pitts. 1986. *Atmospheric Chemistry: Fundamentals and Experimental Techniques*, New York: John Wiley and Sons.

Fogg, T. R. and R. Seeley. 1984. ICP-OES analysis of atmospheric aerosol particles. *American Laboratory.* December, 36–39.

Fox, D. L. and H. Jeffries. 1983. Air pollution. *Anal. Chem.* 55:233R–45R.

Fox, D. L. 1985. Air pollution. *Anal. Chem.* 57:223R–38R.

Fox, D. L. 1987. Air pollution. *Anal. Chem.* 59:280R–94R.

Fung, K. K., S. Heisler, A. Price, B. Nuesca, and P. K. Mueller. 1979. Comparison of ion chromatography and wet chemical methods for analysis of sulfate and nitrate in ambient particulate filter samples. In *Ion Chromatographic Analysis of Environmental Pollutants*, Vol. II, eds. J. Mulik and E. Sawicki, pp. 203–9. Ann Arbor, MI: Ann Arbor Science.

Garbarino, J. R. and H. E. Taylor. 1985. Recent developments and applications of inductively coupled plasma emission spectroscopy to trace elemental analysis of water. In *Trace Analysis, Vol. 4*, ed. J. F. Lawrence, pp. 186–236. New York: Academic Press.

Gelman, C. and J. C. Marshall. 1975. High purity fibrous air sampling media. *J. Am. Ind. Hyg. Assoc.* July, 512–16.

Giauque, R. D. R. B. Garrett, and L. Y. Goda. 1977. Calibration of energy-dispersive X-ray spectrometers for analysis of thin environmental samples. In *X-ray Fluorescence Analysis of Environmental Samples*, ed. T. G. Dzubay, pp. 153–64. Ann Arbor, MI. Ann Arbor Science.

Giauque, R. D., F. S. Goulding, J. M. Jaklevic, and R. H. Pehl. 1973. Trace element determination with semiconductor detector X-ray spectrometers. *Anal. Chem.* 45:671–81.

Greenberg, R. R. 1979. Trace element characterization of the NBS urban particulate matter standard reference material by instrumental neutron activation analysis. *Anal. Chem.* 51:2004–6.

Groblicki, P. J., S. Cadle, C. Ang, and P. Mulawa. 1983. *Interlaboratory Comparison of Methods for the Analysis of Organic and Elemental Carbon in Atmospheric Particulate Matter.* General Motors Research Laboratory Publication GMR-4054, ENV-152.

Grosjean, D. 1975. Solvent extraction and organic carbon determination in atmospheric particulate matter: The organic extraction–organic carbon analyzer (OE-OCA) technique. *Anal. Chem.* 47:797–805.

Grosjean, D. 1984. Particulate carbon in Los Angeles air. *Sci. Total Environ.* 32:133–45.

Gundel, L. A., R. L. Dod, H. Rosen, and T. Novakov. 1984. The relationship between optical attenuation and black carbon concentration for ambient and source particles. *Sci. Total Environ.* 36:197–202.

Haik, M. and M. Imada. 1981. *Simultaneous Determination of Sulfate and Nitrate in High-Volume Atmospheric Samples using Automated Technicon II Procedures.* Air and Industrial Hygiene Laboratory, California Department of Health Services Method No. 86.

Harper, S. L., J. Walling, D. Holland, and L. Pranger. 1983. Simplex optimization of multielement ultrasonic extraction of atmospheric particulates. *Anal. Chem.* 55:1553–57.

Herring, S. V. et al. 1988. The nitric acid shootout: Field comparison of measurement methods. *Atmos. Environ.* 22:1519–39.

Highsmith, V. R., A. E. Bond, and J. E. Howes, Jr. 1986. Particle and substrate losses from Teflon and quartz filters. *Atmos. Environ.* 20:1413–17.

Hsu, J. and B. R. Appel. 1984. *Coulometric Analysis for Total Carbon Content of Atmospheric Particulate Matter.* Method 94, Air and Industrial Hygiene Laboratory, California Department of Health Services.

Huntzicker, J., R. Johnson, J. Shah, and R. Cary. 1982. Analysis of organic and elemental carbon in ambient aerosols by a thermal–optical method. In *Particulate Carbon-Atmospheric Life Cycle*, eds. G. T. Wolff and R. L. Klimisch, pp. 79–88. New York: Plenum.

Hwang, J. Y. 1972. Trace metals in atmospheric particulates. *Anal. Chem.* 44:20A–27A.

Jaklevic, J. M., R. C. Gatti, F. S. Goulding, and B. W. Loo. 1981. A β-gauge method applied to aerosol samples. *Environ. Sci. Technol.* 15:680–86.

Keene, W., J. Galloway, and J. Holden, Jr. 1983. Measurement of weak organic acidity in precipitation from remote areas of the world. *J. Geophys. Res.* 88:5122–30.

Keene, W. and J. Galloway. 1985. Gran's titration: Inherent errors in measuring the acidity of precipitation. *Atmos. Environ.* 19:199–202.

Kemp, K. 1977. Matrix absorption corrections for PIXE analysis of urban aerosols sampled on Whatman 41 filters. In *X-ray Fluorescence Analysis of Environmental Samples*, ed. T. G. Dzubay, pp. 203–9. Ann Arbor, MI: Ann Arbor Science.

Koutrakis, P., J. Wolfson, and J. Spengler. 1988. An improved method for measuring aerosol strong acidity: Results from a nine-month study in St. Louis, Missouri and Kingston, Tennessee. *Atmos. Environ.* 22:157–62.

Lambert, J. P. 1981. *Standard Operating Procedure for the Trace Element Analysis of Dichotomous Samples by Instrumental Neutron Activation Analysis (NAA).* U.S. EPA Method EMSL/RTP-SOP-EMD-008.

Lambert, J. and F. Wiltshire. 1979. Neutron activation analysis for simultaneous determination of trace elements in ambient air collected on glass fiber filters. *Anal. Chem.* 51:1346–50.

Liu, B. Y. H., D. Y. H. Pui, and K. L. Rubow. 1983. Characterization of air sampling filter media. In *Aerosols in the Mining and Industrial Work Environment, Vol III*, eds. V. A. Marple, and B. Y. H. Liu, pp. 989–1038. Ann Arbor, MI: Ann Arbor Science.

Lodge, Jr., J. P. 1989. *Methods of Air Sampling and Analysis*, 3rd edn. Chelsea, MI: Lewis.

Lumpkin, T. A. 1984. *Standard Operating Procedure for the Andersen Model 321A PM-10 Size-Selective High Volume Sampler.* U.S. EPA Method EMSL/RTP SOP-EMD-204.

McDonnell, D. B. and J. C. Hilborn. 1978. Alkali fusion of glass fiber filters: Analysis of secondary lead emission particulate. *J. Air Pollut. Control Assoc.* 28:933.

McDow, S. R. and J. J. Huntzicker. 1990. Vapor adsorption artifact in the sampling of organic aerosol: Face velocity effects. *Atmos. Environ.* 24A: 2563–72.

McKee, H. C., R. E. Childers, O. Saenz, Jr., T. Stanley, and J. Margeson. 1972. Collaborative testing of methods to measure air pollutants—I. The high-volume method for suspended particulate matter. *J. Air Pollut. Control Assoc.* 22:342–47.

Mitchell, W. J. 1988. *Procedure for Spiking Filter Strips with Lead.* EMSL/RTP-SOP-QAD 538.

Moore, H. and S. Witz. 1980. *Determination of Lead Concentrations in Ambient Particulate Matter by Wavelength Dispersive X-Ray Fluorescence Spectrometry.* Method No. 54, California Department of Health Services, Air and Industrial Hygiene Laboratory.

Mueller, P. K., R. Mosley, and L. Pierce. 1972. Chemical composition of Pasadena aerosol by particle size and time of day—IV. Carbonate and noncarbonate carbon content. *J. Colloid-Interface Sci.* 39:235–39.

Mulik, J. D., E. Estes, and E. Sawicki. 1978. Ion chromatographic analysis of ammonium ion in ambient aerosols. In *Ion Chromatographic Analysis of Environmental Pollutants, Vol. I*, eds. E. Sawicki, J. Mulik, and E. Wittgenstein, pp. 41–51. Ann Arbor, MI: Ann Arbor Science.

Mulik, J. and E. Sawicki (ed.) 1978, 1979. *Ion Chromatographic Analysis of Environmental Pollutants, Vols. I and II.* Ann Arbor, MI: Ann Arbor Science.

National Research Council, National Academy of Sciences (NRCNAS) 1977. Ammonia. EPA Report 600/1-77-054, NTIS Document PB-278 182.

Nelson, J. W. 1977. Proton-induced aerosol analyses: Methods and samplers. In *X-ray Fluorescence Analysis of Environmental Samples*, ed. T. G. Dzubay, pp. 19–34. Ann Arbor, MI: Ann Arbor Science

Novakov, T. 1982. Soot in the atmosphere. In *Particulate Carbon-Atmospheric Life Cycle*, eds. G. T. Wolff and R. L. Klimisch, pp. 19–41. New York: Plenum.

Olmez, I. and G. E. Gordon. 1985. Rare earths: Atmospheric signatures for oil-fired power plants and refineries. *Science* 229:966–68.

Ondov, J.M., W. H. Zoller, I. Olmez, N. Aras, G. Gordon, L. Rancitelli, K. Abel, R. Filby, K. Shah, and R. Ragaini. 1975. Elemental concentrations in the National Bureau of Standards' environmental coal and fly ash standard reference materials. *Anal. Chem.* 47:1102–9.

Patashnick, H. and E. Rupprecht. Continuous PM-10 measurements using the tapered element oscillating microbalance. Paper 90-84.1 read at the Air and Waste Management Annual meeting, June 1990, Pittsburgh, PA.

Perkin-Elmer Corp. 1982. *Analytical Methods for Atomic Absorption Spectroscopy*. Norwalk, CT: Perkin-Elmer Corporation.

Pitts, J. N., Jr., D. Lokensgard, P. Ripley, K. van Cauwenberghe, L. van Vaeck, S. Schaffer, A. Thill, and W. Belser. 1980. Atmospheric epoxidation of benzo(a)pyrene by ozone: Formation of the metabolite benzo(a)pyrene-4,5 oxide. *Science* 210:1347–49.

Ragaini, R., H. R. Ralston, D. Garvis, and R. Kaifer. 1980. Instrumental neutron activation analysis techniques for measuring trace elements in California aerosols. In *The Character and Origins of Smog Aerosols*, eds. G. M. Hidy, P. K. Mueller, D. Grosjean, B. R. Appel, and J. J. Wesolowski, pp. 169–97. New York: Wiley.

Rehme, K. A., C. Smith, M. Beard, and T. Fitz-Simons. 1984. *Investigation of Filter Media for Use in the Determination of Mass Concentrations of Ambient Particulate Matter*. EPA Report 600/S4-84-048.

Rosen, H., A. D. A. Hansen, R. Dod, and T. Novakov. 1980. Soot in urban atmospheres: Determination by an optical absorption technique. *Science* 208:741–44.

Schroeder, W., M. Dobson, D. Kane, and N. Johnson. 1987. Toxic trace elements associated with airborne particulate matter: A review. *J. Air Pollut. Control. Assoc.* 37:1267–85.

Sickles, J. E., II, C. Perrino, I. Allegrini, A. Febo, M. Possanzini, and R. J. Paur. 1986. Measurement of selected inorganic air pollutants near Los Angeles using an annular denuder system. Paper read at the EPA/APCA Symposium on measurement of Toxic Air Pollutants, May 1986, Raleigh, NC.

Small, H. 1981. Applications of ion chromatography in trace analysis. In *Trace Analysis, Vol. 1*, ed. J. F. Lawrence, pp. 267–322. New York: Academic Press.

Smith, J., D. Grosjean, and J. Pitts. 1978. Observation of significant losses of particulate nitrate and ammonium from high volume glass fiber filter samples stored at room temperature. *J. Air Pollut. Control Assoc.* 28:930–33.

Sneddon, J. (1983) Collection and atomic spectroscopic measurement of metal compounds in the atmosphere: A review. *Talanta* 30(9):631–48.

Stevens, R. K. 1984. *Sampling and Analysis of Atmospheric Aerosols*, U.S. EPA Report 600/D-84-283, NTIS No. PB85-128197.

Stevens, R. K. and T. G. Dzubay. 1978. Sampling and analysis of atmospheric sulfates and related species. *Atmos. Environ.* 12:55–68.

Stevens, R. K., T. G. Dzubay, C. W. Lewis, and R. W. Shaw, Jr. 1984. Source apportionment methods applied to the determination of the origin of ambient aerosols that affect visibility in forested areas. *Atmos. Environ.* 18:261–72.

Stevens, R. K., W. A. McClenny, and T. G. Dzubay. 1982. Analytical methods to measure the carbonaceous content of aerosols. In *Particulate Carbon: Atmospheric Life Cycle*, eds. G. T. Wolff and R. L. Klimisch, pp. 111–29. New York: Plenum.

Tanner, R., J. Gaffney, and M. Phillips. 1982. Determination of organic and elemental carbon in atmospheric aerosol samples by thermal evolution. *Anal. Chem.* 54:1627–30.

Tanner, R. and L. Newman. 1976. The analysis of airborne sulfate—A critical review. *J. Air Pollut. Control Assoc.* 26:737–47.

Technicon Industrial Systems. 1972. Method 118-71W, The automated methylthymol blue method for sulfate. Tarrytown, NY.

Technicon Industrial Systems. 1973. Method 100-70W, The automated Cu–Cd reduction method for nitrate. Tarrytown, NY.

Tierney, G. and W. Conner. 1967. Hygroscopic effects on weight determinations of particulates collected on glass-fiber filters. *J. Amer. Ind. Hyg. Assoc.* July–August, 363–65.

Turpin, B., R. Cary, and J. Huntzicker. 1990. An *in situ*, time-resolved analyzer for aerosol organic and elemental carbon. *Aerosol Sci. Technol.* 12:161–71.

U.S. EPA. 1982*a*. *Air Quality Criteria for Oxides of Nitrogen*. EPA Report 600/8-82-026.

U.S. EPA. 1982*b*: *Air Quality Criteria for Particulate Matter and Sulfur Oxides, Vol. II*, pp. 3–67. EPA Report 600/8-82-029b.

U.S. EPA. 1985. Method 300.7. *Sodium, Ammonium, Potassium, Magnesium, and Calcium in Wet Deposition by Chemically Suppressed Ion Chromatography*. Cincinnati, OH: Environmental Monitoring and Support Laboratory.

U.S. EPA. 1986. Method 300.6. *Chloride, Orthophosphate, Nitrate and Sulfate in Wet Deposition by Chemically Suppressed Ion Chromatography*. Cincinnati, OH: Environmental Monitoring and Support Laboratory.

U.S. EPA. 1989. Compendium Chapter IP-9. *Determination of Reactive Acidic and Basic Gases and Particulate Matter in Indoor Air*, p. 21. Research Triangle Park, NC: Atmospheric Research and Exposure Assessment Laboratory.

U.S. EPA. 1990. *Quality Assurance Handbook for Air Pollution Measurement Systems*, Section No. 2.11. Cincinnati OH: Environmental Monitoring and Support Laboratory.

U.S. Federal Register. 1987. 40 CFR Chapter I, Appendix J, July 1. Reference Method for the Determination of Particulate Matter as PM-10.

Van Grieken, R. E. and J. J. LaBrecque. 1985. Trace analysis of environmental samples by X-ray emission spectroscopy. In *Trace Analysis. Vol. 4*, ed. J. F. Lawrence, pp. 101–83. New York: Academic Press.

Van Loon, J. C. 1980. *Analytical Atomic Absorption Spectroscopy*. New York: Academic Press.

Waetjen, U., E. Bombelka, F.-W. Richter, and H. Ries. 1983. PIXE analysis of high-volume aerosol samples and intercomparison of results with different analytical methods. *J. Aerosol Sci.* 14:305–8.

Wagman, J., R. Bennett, and K. Knapp. 1977. Simultaneous multiwavelength spectrometer for rapid elemental analysis of particulate pollutants. In *X-ray Fluorescence Analysis of Environmental Samples*, ed. T. G. Dzubay, pp. 35–56. Ann Arbor, MI: Ann Arbor Science.

Wall, S. M., W. John, J. L. Ondo. 1988. Measurement of aerosol size distributions for nitrate and major ionic species. *Atmos. Environ.* 22:1649–56.

West, L. G. 1985. A new air monitoring filter for PM_{10} collection and measurement techniques. In *Quality Assurance in Air Pollution Measurements*. Pittsburgh, PA: Air Pollution Control Association.

Witz, S., R. Eden, M. Wadley, C. Dunwoody, R. Papa, and K. Torre. 1990. Rapid loss of particulate nitrate, chloride and ammonium on quartz fiber filters during storage. *J. Air Waste Management Assoc.* 40:53–61.

13

Analysis of Individual Collected Particles

R. A. Fletcher and J. A. Small

National Institute of Standards and Technology, Gaithersburg, MD, U.S.A

INTRODUCTION

Important information about an aerosol resides in the morphology and in the chemical composition of the individual particles. Chemical, morphological, phase, and crystallographic data are often crucial for purposes of tracing formation mechanisms and possible sources of airborne particulate matter. Applications of single-particle analysis include clean-room technology, microelectronics, indoor, and outdoor air pollution, forensics and many problems related to micro-contamination. Analytical instruments and techniques must be employed to obtain compositional information at the single-particle level. Digital image processing plays an increased role in providing new information about chemical composition and morphological properties of microstructures and particles.

Single-particle analysis is defined here to mean the analysis of collected, individual particles ranging from about 5 nm to several millimeters in lateral dimension. The information obtained includes the chemical constituents that make up the particle, the morphology (size, shape), and the physical and optical properties. Two requirements must be met for single-particle analysis: the analytical technique must have sufficient spatial resolution to differentiate a single particle from the background substrate material and adjacent particles, and also sufficient sensitivity to detect at least major components of the particle's composition (picogram detection levels for micrometer-sized unit-density particles). Microanalysis, entailing the use of microscopes or microprobes, often fulfills these requirements. Microanalysis by means of microscopy originated with the advent of the light microscope, while the first microprobe work resulted from the invention of the electron microprobe (Castaing and Guinier 1950). Microanalytical developments have been motivated by the need to obtain images, and elemental and molecular (compound) information on a microscopic scale. In microanalysis, a beam of excitation radiation, which can be photons, electrons, protons, neutrons or ions, is used to bombard the sample. The interaction of the beam with the sample results in emitted radiation that is separated by a spectrometer and collected by a detector, as shown in Fig. 13-1. For microprobes, the excitation beam is focused on the particle with a spatial resolution ranging from nanometers to micrometers. To create an image, either the excitation beam is rastered across the particle or the mounted

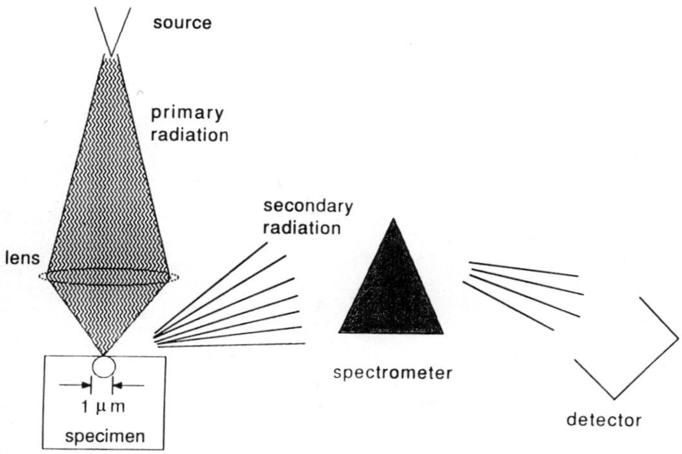

FIGURE 13-1. Schematic Representation of Generic Microanalysis Showing Excitation and Emission Radiation. The Incident Radiation can be Electrons, Photons, or Ions. The Possible Emissions are Photons in the IR, UV or VIS, Ions or X-rays.

particle is translated under the beam. Microscopes usually flood the sample with excitation radiation, but form an image by focusing the secondary radiation. Spatially resolved analysis is achieved in the microscope either by masking a portion of the secondary radiation (aperturing) or by employing a spatially sensitive detector.

Some specific examples of instruments using electron beam excitation and electron probes, scanning electron microscopes and analytical electron microscopes. The emitted radiation, back-scattered electrons or X-rays, is used for single-particle analysis. Examples of instruments that use photon radiation are the Raman microprobe (or micro-Raman), where the detected beam is frequency-shifted photons; the laser microprobe, which detects ions in a time-of-flight mass spectrometer; and the Fourier transform infrared microscope, which is based on photon absorption. Examples of instruments that use ion beams are the ion microprobe and microscope based on secondary-ion mass spectrometry (SIMS). In SIMS, either an argon, oxygen, or cesium ion beam is used to bombard the sample and either a time-of-flight mass spectrometer or, more usually, a magnetic sector or quadrupole mass spectrometer is used to detect the secondary ions generated in the interaction process.

Often, in an analysis, complementary capabilities of two or more instruments can be used in sequence on an individual particle to provide more complete particle characterization (phase, chemical, and morphological information) about that particle (Steel et al. 1984; Fletcher, Etz, and Hoeft 1990).

This chapter describes microscopes and microprobes used for the analysis of collected, individual particles. The instruments discussed are the light microscope, electron microscopes (both scanning and transmission), electron microprobes, laser, optical, and ion microprobes. The principles of operation and the instrumental capabilities are presented. The chapter also contains some basic information about sample preparation, useful for the aerosol scientist. Several excellent references which should be consulted for more detailed information include the book by Spurny (1986), the publication by Heinrich (1980), the article by Newbury (1990) and the review chapter by Grasserbauer (1978). For quick reference, Table 13-1 presents a comparison of the microanalytical techniques capable of single-particle analysis.

TABLE 13-1 Comparison of Typical Microanalytical Instrument Capabilities

Analytical Method	SEM	TEM	EPMA	LMMS	SIMS	Micro-Raman	LM	FT-IR
Excitation	Electrons	Electrons	Electrons	Photons[1]	Ions	Photons	Photons	Photons
Emission	Electrons, X-rays	Electrons, X-rays	Electrons	Ions	Ions	Photons	Photons	Photons
Quantity measured	Electrons, X-rays	Electrons, X-rays	Electrons, X-rays	m/z	m/z	Frequency shift	Light intensity	Intensity at given wave-length
Lateral resolution	5 nm	0.3 nm	0.05 μm	1 μm	1 μm, 0.1 μm (probe)	2 μm	0.5 μm	10–50 μm
Detectable elements	> Sodium	> Sodium	> Sodium	All	All	NA	NA	NA
Isotopic detection	No	No	No	Yes	Yes	NA	NA	NA
Detectability of molecular or chemical compounds	None	Inorganic	None	Organic and inorganic	Organic and inorganic	Organic and inorganic	Inorganic	Organic and inorganic
Absolute detection sensitivity	10^{-16} g	10^{-20} g	10^{-16} g	10^{-17}–10^{-19} g	10^{-19} g	10^{-12} g	NA	5×10^{-12} g
Relative sensitivity	0.1%	0.1%	0.1%	1–100 ppm	1–1000 ppm	1%	NA	1%
Sample vacuum-mounted	Yes	Yes	Yes	Yes	Yes	No	No	No
Destructive method	No	No	No	Yes	Yes	No	No	No
Surface-sensitive	No	No	No	Unknown	Yes	No	No	No
Imaging capability	Yes	Yes	Yes	Demonstrated, not widely available	Yes	Yes	Yes	Yes
Automation demonstrated	Yes	Yes	Yes	No	No	Yes	Yes	Yes
Quantitative	Yes	Yes	Yes	No	No	No	No	Semi

Source: Adapted from Wieser, Wurster, and Seiler (1980).

1. Photons in the UV to IR range.

LIGHT MICROSCOPY (LM)

Underlying Principle

Optical microscopy or light microscopy (LM) is one of the more familiar microanalytical techniques. The operating principle of light microscopes is well known. In the simplest form, the light microscope utilizes light refraction via a lens system to form enlarged images of microscopic objects. The image is focused on the detector, which can be the human eye or a camera. An object must absorb approximately 0.3% of the incident light to be visible to the eye (Dovichi and Burgi 1987).

Instrumentation

A schematic diagram of a light microscope is presented in Fig. 13-2. The important components of a light microscope are a light source, an objective lens and an ocular. The light source can be diffuse or bright and serves to illuminate the sample. The objective lens collects light that passes through the sample or is reflected from the sample surface and projects an image near the ocular. The ocular magnifies the image that is projected by the objective for the eye. Normally, the virtual image seen resides below the sample plane. There are a number of optical accessories used in conjunction with the light microscope to characterize a sample physically. Some of these capabilities will be discussed later.

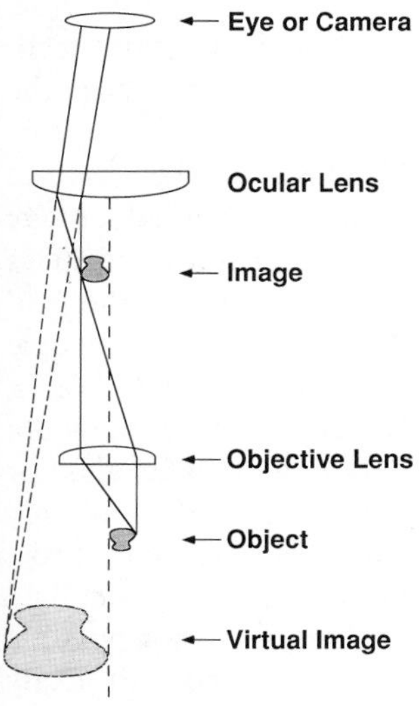

FIGURE 13-2. Schematic Diagram of a Light Microscope. Light is Transmitted Through the Sample and Focused by an Objective Lens. The Intermediate Image is then Enlarged and Transmitted to the Eye or the Detector. (Source: McCrone and Delly 1973.)

TABLE 13-2 Characteristics for Some Common Light Microscope Objectives

Magnification	Numerical Aperture	Depth of Field (μm)	Diameter of Field[1] (mm)	Resolving Power[2] (μm)
3	0.08	50	9	5
10	0.25	8	2	1.3
50	0.85	1	0.5	0.4
100	1.3	0.4	0.4	0.25

Source: Steel (1980).

1. Approximate value with 10× oculars. Wide field is now common, values may be 1.5 to 2 times larger.
2. Approximate value in green light.

Some considerations in light microscopy are: depth of field, referring to the distance beyond the plane of focus up to which the object remains in focus; magnification, quantifying the image enlargement; numerical aperture, relating the maximum light-gathering capability of the microscope objective; and resolving power, indicating the size of the smallest feature that can be discriminated. Table 13-2 presents the characteristics of various common microscope objectives (Steel 1980).

Capabilities and Applications

LM is often the first microanalytical technique used to examine a sample because it is a nondestructive approach. It is relatively easy to use as an imaging tool for many applications, but identifying a material through its optical properties can be difficult. A skilled microscopist can use the physical and optical properties of a particle (such as the size, shape, surface texture, color, refractive indices, crystallographic properties, and birefringence) to help identify a given particle and, thus, possibly its source (Grasserbauer 1978).

Two additional references containing information on the analysis of particles by LM are Steel (1980) and Friedrichs (1986). A general detailed reference for LM is given by Chamot and Mason (1958). *The Particle Atlas* (McCrone and Delly 1973) is regarded as a principal reference for the identification of particles by light microscopy.

Size and Shape Analysis

Determination of shape and size often represents the first step in single-particle analysis. Sometimes the shape will provide information about the particle type and, thus, the most probable formation mechanism of a particle. For example, fly ash normally appears under a light microscope as spheres, and dark, fractal-like (complex branched chain) structures are usually formed from a combustion source. Fibers may be asbestos or glass or may come from a wide variety of other natural or man-made sources. *The Particle Atlas* (McCrone and Delly 1973) can be used to compare the image of a sample particle to reference micrographs. Although particles can be resolved and, thus, observed at the 0.25–0.5 μm size level, shape, and size determination is most reliably done on particles larger than several micrometers in diameter (Steel 1980). Fibers are an exception to the above statement since even with a lateral dimension as small as 0.5 μm, the size and shape can be determined.

Related to particle size, particle magnification in the light microscope is limited due to the wavelength of light used (diffraction limit). A good working rule of thumb is that the maximum magnification of a light microscope is 1000 times the numerical aperture of the objective (McCrone and Delly 1973). The shape and size is useful for identifying particle origin, but some additional physical properties help to provide more definitive information about a particle's makeup.

Identification by Light Microscopy

Using the physical properties of particle shape and size and the optical properties of color, refractive index, and birefringence, the identity of an unknown particle can often be determined. Normally, optical characterization requires particles to be greater than 5–10 μm in lateral dimension (Steel 1980).

Particles that have an index of refraction different from the substrate mounting material are most easily viewed. It is the contrast between the background support and the particle that is most often important when trying to view an object by light microscopy. The contrast can be improved by a number of techniques. Figure 13-3a–d shows the same field of view of a collected airborne particle sample using different contrast enhancers. The filter material consists of mixed esters of cellulose and the sample is prepared by treating with acetone vapor to make the filter transparent (Baron and Pickford 1986). The micrographs are taken in transmitted light. Figure 13-3a illustrates the problem with viewing the filter in direct (unaided) transmission. Two large objects in the central part of the micrograph are barely visible under these conditions. When phase contrast microscopy is applied (Fig. 13-3b), the particles become quite visible; the phase shift of the light transmitted through the particles is used to enhance the contrast. Another way to increase the visibility of the particle is by differential interference contrast shown in Fig. 13-3c. This is considered a complementary technique to phase contrast, but in this case the objects take on a three-dimensional appearance. Lastly, Fig. 13-3d shows the effect of slightly uncrossed polarizers. The advantage of uncrossed polarizers is that the particles are, for the most part, still visible in the field of view and the particles made of anisotropic material stand out as illuminated objects. Crossed polarizers, on the other hand, would make all isotropic materials (such as glass or amorphous plastics) invisible and show only those particles (as luminous objects) that rotate the polarized transmitted light. A detailed discussion of these techniques is not possible here, but is given in McCrone and Delly (1973).

Several additional techniques are useful in conjunction with LM. Some microscopes are equipped with monochromatic or near-monochromatic light sources that can be used

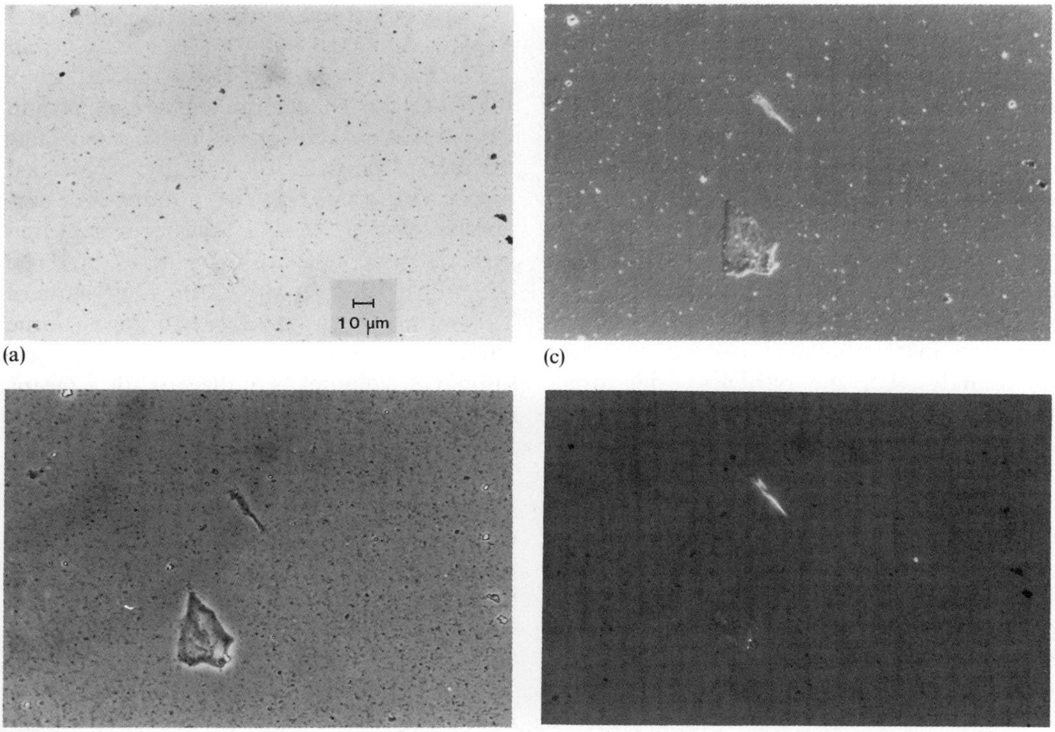

(a) (c) (b) (d)

FIGURE 13-3. Set of Four Light Micrographs Illustrating Various LM Techniques that Help to Increase Particle Visibility by Increasing Contrast and in the Last Case (3d) Help to Identify Birefringent Materials in the Sample. Frame (3a) is Straight Transmitted Light, (3b) is Transmitted Light Using Phase Contrast, (3c) Implements Differential Interference, and (3d) Shows the Effect of Slightly Uncrossed Polarizers. The Last Frame (3d) Brings out the Birefringent Material Present in the Sample as the Apparent Luminous Objects. (Courtesy of E. Steel, NIST.)

to excite fluorescence, if present, in a particle. Often the observation of fluorescence, usually excited in the UV, provides information for identifying the particle's composition. Refractive index is another parameter that can be determined and used as a powerful tool for particle identification. The refractive index measurement, with an accuracy to one part in one thousand (Grasserbauer 1978), is accomplished by immersing the particle in a series of index-matching fluids to find the matching refractive index that causes the particle to "disappear." Microchemical reactions can be used to help identify the particle composition (Seeley 1952; Chamot and Mason 1940). Analyses of these kinds require considerable experience to be useful for identifying particle composition.

The Particle Atlas can be consulted to classify a particle on the basis of its physical and optical properties. This reference contains over 600 color photomicrographs of particles from various sources and of known composition (McCrone and Delly 1973). In this reference, the types of particles are divided into four categories: (1) windblown particles such as fibers and minerals; (2) industrial particles such as abrasives, polymers, fertilizers, and cleaners; (3) combustion particles such as auto, coal-fired, and oil-fired soots; and (4) miscellaneous particles. The authors provide a step-by-step characterization procedure for classifying the particles into one of the four categories. The same reference contains scanning electron micrographs of the same 600 particles shown in the photomicrographs.

Sample Preparation and Practical Applications

In a light microscope, either transmitted or reflected light can be used. For transmitted light, the particles are mounted on a glass or quartz slide. An index or immersion oil may be applied to the sample to improve the viewing under transmitted light. An aliquot of the particles should be tested with the oil to ensure that no reaction or dissolution takes place. The oil will contaminate the particles prepared in this manner; so a subsequent microanalysis of the particles using other techniques is unlikely. Reflected light can be used to view particles collected on an aerosol filter surface, but the particles usually must be $> 1\ \mu m$. Opaque particles and particles with large indices of refraction are most easily viewed in this manner. To overcome this size limitation, some filters can be made transparent or removed entirely to allow viewing with transmitted light. Friedrichs (1986) mentions three ways to transform the filter. Each has its own advantages and disadvantages. In the first method, filters can be treated with index-matching fluid. Particles remain on the filter, but liquid particles or particles soluble in the fluid may be dissolved or possibly removed and particles with a refractive index matching the immersion oil will be difficult to see. Some "filter clearing" agents are given in Table 13-3 (Friedrichs 1986; LeGuen and Galvin 1981; Baron and Pickford 1986). The second method is to dissolve the filter using an appropriate solvent (for example, polycarbonate filters dissolve in chloroform). In the third method, the filter is ashed, leaving the refractory particles behind. In these last two approaches, the particles no longer have a filter support and must be remounted on a transparent substrate.

The number of particles collected on a substrate can be determined and related to the particle concentration in an aerosol. The number of particles is normally determined per viewing area and when a number of randomly selected viewing areas are taken together, an estimate can be made of the number of particles on the entire filter surface. This estimate can be related to the airborne particle number concentration (based on sample air volume) as in the case for asbestos number concentration (AIA 1979; Carter, Taylor, and Baron 1989).

The relevance of depth of field, especially for particle counting, is illustrated by the set of micrographs in Fig. 13-4a and b. In these micrographs, certain particles are visible and in focus, while others are difficult to see. Figure 13-4 is an example of typical LM application that might be employed by an aerosol scientist for examining a filter surface in reflectance. The particles in Fig. 13-4 were collected on a filter consisting of mixed esters of cellulose. The filter is slightly bowed in the center as a result of air flow through the filter cassette. This bowing causes a distorted planar surface for the microscopy, leading to poor particle detection due to a limited depth of focus. In Fig. 13-4a, the upper left of the field of view is in focus, but the lower right is out of focus. As the microscope focus is altered, the lower-right regions shown in Fig. 13-4b become clear images and particles in the upper left gradually become fuzzy and indistinguishable. Clearly, a shallow depth of field is problematic. The recent development of the confocal microscope provides a differ-

TABLE 13-3 Membrane Filter Clearing Agents for Light Microscopy

Filter Type	Clearing Agent	Refractive Index
Mixed esters of cellulose	1. Acetone vapor/triacetin (AIA method)	1.44–1.48
	2. Dimethyl formamide/ Euparal method	1.48
	3. Acetone vapor/Euparal method	1.48
Polycarbonate	Immersion oils or chloroform-dissolved materials	1.584 or 1.625

Sources: Friedrichs (1986), Le Guen and Galvin (1981), Baron and Pickford (1986).

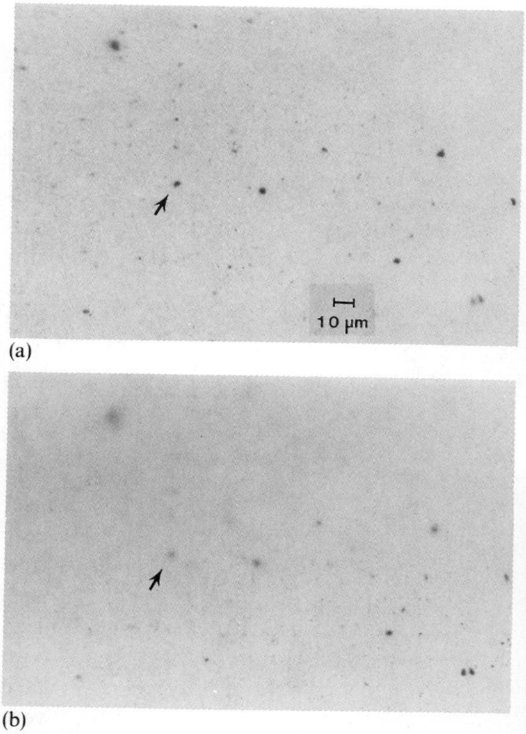

(a)

(b)

FIGURE 13-4. Light Micrograph of an Untreated Mixed Ester of Cellulose Filter Containing Collected Airborne Particles. The Magnification of the Presented Image is 620 ×. The Set of Micrographs Illustrate How the Depth of Field Influences a Viewing Volume by Selectively Bringing Certain Areas of the Filter into Focus While Leaving Other Areas Containing Particles Out of Focus. The Focused Field is Descending from the Upper Left Corner in (a) Down to the Bottom Right Corner in (b). (Courtesy of E. Steel, NIST.)

ent approach by limiting the image formation strictly to those photons scattered within the depth of field. By changing the objective to specimen distance, a series of images can be obtained as "optical sections" of particles with certain properties (e.g., fluorescence).

The depth of field usually encountered in light microscopy does not compare to that found in scanning electron micrsocopy. Figure 13-5 contains two micrographs of the same field of view of amosite asbestos. The top is a light micrograph and the bottom is an electron micrograph, both with approximately the same magnification. Note that only some of the asbestos fibers are in focus in the light micrograph, while the entire electron-generated image is in focus.

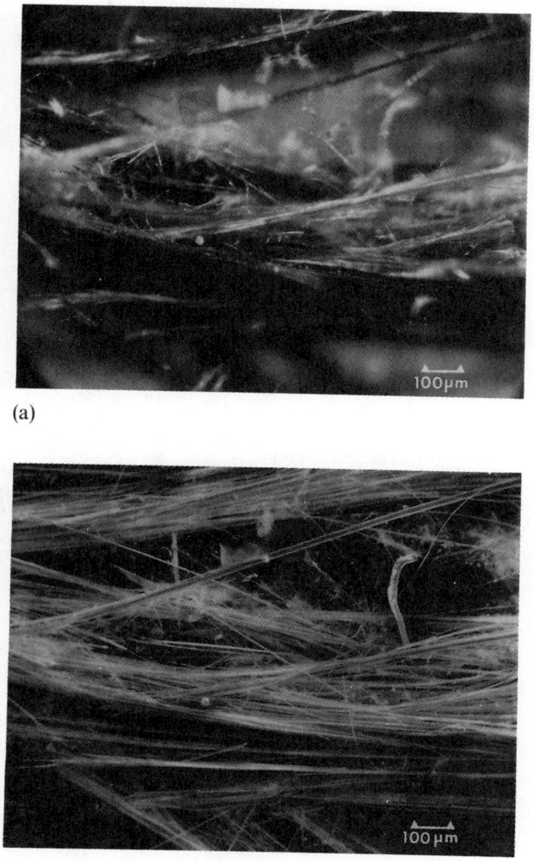

(a)

(b)

FIGURE 13-5. Two Micrographs of the Same Field of View Taken of an Amosite Asbestos Sample. The Top Light Micrograph has a Limited Depth of Field. The Depth of Field in the Bottom Electron Micograph, Taken with a Scanning Electron Microscope, Illustrates the Advantage of Using the SEM in Many Cases.

ELECTRON BEAM ANALYSIS OF PARTICLES

Principle of Electron Beam Excitation

A schematic diagram of a typical electron beam instrument and some of its analytical functions is shown in Fig. 13-6. The source of the electron beam is an emitter such as a heated tungsten filament shaped into a fine tip. The electrons emitted from the filament are formed into a beam and focused by an ion lens system onto the specimen. The electron

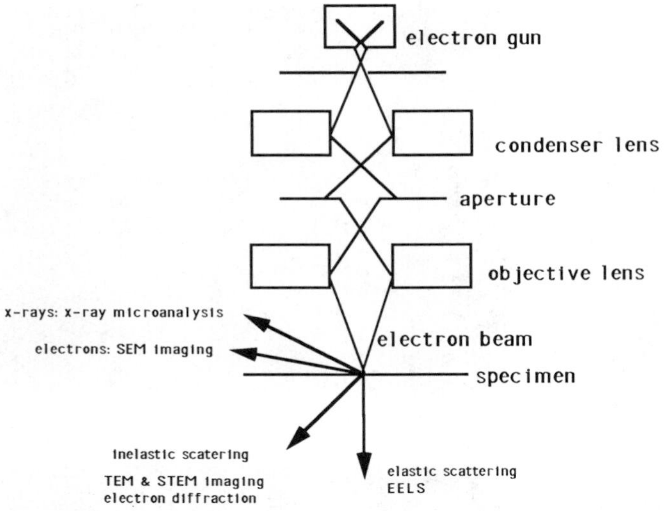

FIGURE 13-6. A Schematic Diagram of a Typical Electron Beam Instrument Showing the Electron Source and Several of the Different Analytical Signals Used in the Analysis of Particles. (Source: Goldstein et al. 1975.)

beam interacts with the atoms of the sample, resulting in the scattering of the beam electrons and the ejection of both electrons and X-ray photons from the specimen.

Capabilities

Electron Imaging

Electron imaging, which provides the analyst with particle size and morphology, is accomplished by two different methods. One of the imaging methods is used in transmission electron microscopy (TEM). An example of a TEM image of asbestos is shown in Fig. 13-7. The TEM image is observed on a phosphor screen or recorded on film and is similar to the image formed in the transmitted light microscope. The transmission microscope illuminates the sample with a static beam of high-energy (100–400 keV) electrons to image the specimen at magnifications of 1000 to about 1,000,000 times. Since the image is formed by electrons that travel through the sample, TEM is best suited for imaging small particles less than 0.5 μm in diameter.

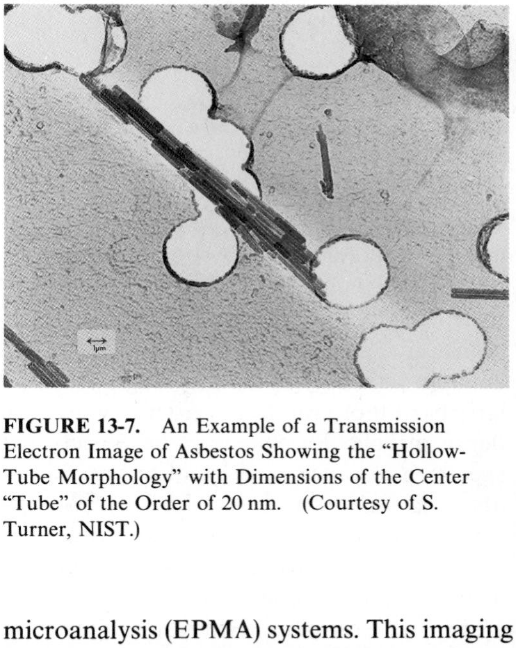

FIGURE 13-7. An Example of a Transmission Electron Image of Asbestos Showing the "Hollow-Tube Morphology" with Dimensions of the Center "Tube" of the Order of 20 nm. (Courtesy of S. Turner, NIST.)

The other type of electron imaging, an example of which is shown in Fig. 13-8, is used in conventional scanning electron microscopes (SEMs) and electron probe microanalysis (EPMA) systems. This imaging system uses a finely focused electron beam that is rastered over the sample area. The scattered electrons resulting from the elastic or inelastic interactions of the beam electrons with the sample are detected and the amplified signal is used to modulate the brightness on a CRT monitor that has a scan pattern matched to the scan pattern of the beam.

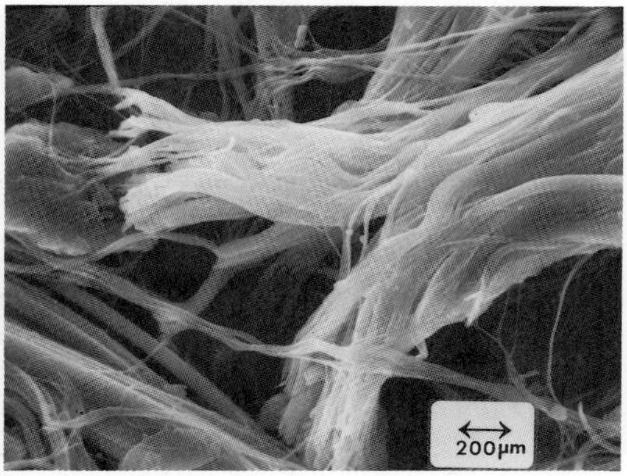

FIGURE 13-8. An Example of a Secondary Electron Image of Asbestos Taken with a Scanning Electron Microscope.

Unlike the TEM image, the image formed by the scanning instruments is an "electronic" image. In SEM and EPMA, the electron detector is generally positioned above the plane of the sample. In this configuration, the sample can be observed at magnifications of 10 to about 100,000 times and the electron images provide topographical as well as morphological information. In TEM, the electron beam can also be rastered and an electron detector placed below the sample. This configuration is used for scanning transmission electron microscopy (STEM) at magnifications up to about 300,000 times. STEM imaging produces a shadow image of the particles which can be used to determine particle morphology but not topography.

X-ray Microanalysis of Individual Particles

X-ray microanalysis is predicated on the ejection, by beam electrons, of an inner shell electron from an atom followed by the filling of these vacancies by orbital electrons and the subsequent emission of characteristic X-ray photons. Because transitions of the electrons occur between sharply defined energy levels, the X-rays are element-specific and can be used for qualitative and quantitative analysis. In addition to X-ray emission, the analysis of particles with diameters smaller than 0.1 μm can also be done in the analytical electron microscope with electron diffraction and electron energy loss spectroscopy (Cowley 1979; Joy 1979). For most inorganic applications, X-ray microanalysis provides a nondestructive microanalytical technique, where the sample remains unchanged by the analysis. The analyst must keep in mind, however, that the sample is subjected to a vacuum and any volatile components will be lost. The concepts and principles of EPMA have been discussed in detail by several authors (Heinrich 1981; Newbury et al. 1986).

Since the electron beam scatters when it interacts with the atoms of the sample and the absorption path lengths for the emitted X-rays are much longer than those for electrons, the spatial distribution of emitted X-rays and, thus, the resolution is always greater than the diameter of the electron probe. Generally, the spatial resolution of electron beam microanalysis techniques is in the range of 20 nm to approximately 2 μm, depending upon the instrument selected for the analysis, the size of the particle and other experimental conditions. The techniques can detect elements with atomic number 11 (sodium) and greater and have a lower detectability limit of about 0.1% by weight or approximately 10^{-16} g in the case of particles. Analytical results are generally reported in terms of weight fraction or weight percent. The accuracy of X-ray microanalysis applied to conventional flat, polished specimens is quite good, about 2% by weight of the measured concentration, with a precision of about 0.1% by weight of the measured concentration (Newbury et al. 1986). The accuracy and precision associated with the analysis of particles are highly dependent on particle shape, size, and composition. In general, accuracies of approximately 10%–20% by weight of the measured concentration with a precision of 5% can be expected for the quantitative analysis of particles.

The size of the particle to be analyzed often dictates the methods to be used for sample preparation as well as the instrument(s) and the analytical scheme used for an analysis. For the purpose of discussing individual

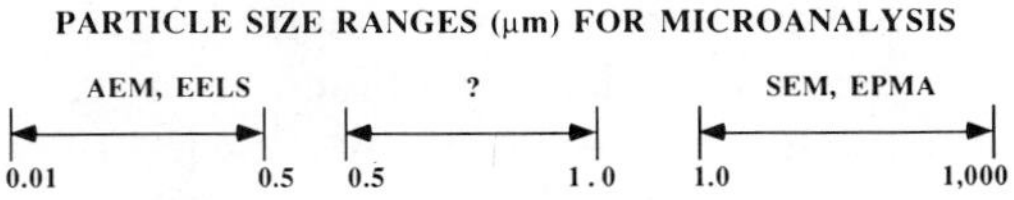

FIGURE 13-9. Particle Size Ranges for Analysis by Electron Beam Instruments Showing the Three Size Ranges of Particles Used for Analysis Methods.

particle analysis, it is appropriate to divide particles into roughly three size categories (see Fig. 13-9). The size ranges shown are only a rough guide as to the selection of a specific electron beam method. Small particles with diameters less than about 0.5 μm are best analyzed by analytical electron or scanning transmission microscopy, with electron beam accelerating potentials in excess of 80 keV. Quantitative analysis by thin-film methods is most accurate for the smallest particles and degrades as the particles become larger than about 0.1 μm in diameter. Above 0.1 μm, corrections for X-ray absorption, atomic number, and fluorescence become significant and must be incorporated into the "thin-film" analytical schemes discussed below.

At the other end of the size spectrum, the larger particles in excess of 1 μm in diameter are best analyzed in an EPMA or SEM with accelerating potentials less than about 50 keV. Quantitative analysis by conventional electron probe methods is most accurate for the largest particles, greater than about 10 μm in diameter, since they approximate a bulk target at these dimensions. Below about 10 μm, quantitative analysis degrades as the particle diameter decreases and corrections to compensate for differences between particles and bulk standards must be included in the analysis schemes.

Quantitative analysis of the particles in the intermediate size range 0.1–2 μm is difficult and can be approached in two ways. The analyst can use either AEM, in which case the particles are considered as increasing in size from a thin film, or EPMA/SEM, in which case the particles are considered as decreasing in size from a bulk sample. The remainder of this section is devoted to a discussion of the two different approaches to particle analysis.

Considerations Required for Applications

Small Particle Analysis with the Analytical Electron Microscope using X-rays and Electrons

In the quantitative analysis of particles smaller than about 0.05 μm with high-energy electrons, the effects of electron backscattering and electron energy loss in the sample can be neglected (Goldstein 1986). In addition, if the particle can be approximated as an infinitely thin film, the effects of X-ray absorption and fluorescence can also be neglected. These conditions are referred to as the "thin-film approximation" and make it possible to express the relationship between the measured and the generated X-ray intensities, for element a, as

$$I_a = I_a^* \varepsilon_a \tag{13-1}$$

where ε_a is an experimentally determined efficiency factor that is related to the overall efficiency of a particular Si–Li detector for the detection of X-rays from element a, I_a is the measured intensity and I_a^* is the generated X-ray intensity. Since the determination of particle thickness is impractical and the value of ε_a is not constant, the analysis schemes for small particles involve the measurement of the ratio of elemental concentrations (expressed in terms of weight fractions), which can be expressed by the following equation:

$$C_a/C_b = k_{ab} \cdot I_a/I_b \tag{13-2}$$

The factor k_{ab} in this equation is a sensitivity factor and is referred to in the AEM literature as the Cliff–Lorimer factor. The k_{ab} factor is related to the generation and efficiency terms for element b ratioed to those for element a (Cliff and Lorimer 1975). The analyst determines, from binary standards or first-principle calculations, a set of the k_{ab} factors for the elements of interest. For a system containing unknown concentrations of elements a, b, c, . . . the concentrations C_a, C_b, C_c, . . . can be determined from Eqs. 13-3–13-5 and the knowledge that $\sum C_i = 1$. The concentrations for the various elements can be determined if the k_{ab} factors are known (as combinations of various binaries), as given by the following equations for a ternary system:

$$C_a/C_b = k_{ab} \cdot I_a/I_b \qquad (13\text{-}3)$$

$$C_c/C_b = k_{cb} \cdot I_c/I_b \qquad (13\text{-}4)$$

$$C_a + C_b + C_c = 1 \qquad (13\text{-}5)$$

The convention in most particle schemes is to express the k_{ab} factors relative to a common matrix element such as silicon, giving $k_{a\,Si}$ factors. As the thickness of the particle increases, the electron transparency decreases, eventually reaching a level where the sample no longer conforms to the "thin-film criterion." Under these circumstances, corrections for X-ray absorption and fluorescence must be included in the analysis scheme. According to Goldstein, the limits for the failure of the thin-film criterion are an X-ray absorption > 3% and/or fluorescence > 5%. For these larger particles, Eq. 13-2 is expanded to include X-ray absorption and fluorescence. The equations for the absorption and fluorescence require an estimate of the particle thickness and a calculation of the mass absorption coefficients for the elements a and b in the specimen (Goldstein et al. 1977).

FIGURE 13-10. An Example of an Electron Diffraction Image of Chrysotile Asbestos Showing the Pattern Resulting from the Crystal Lattice Spacings. (Courtesy of S. Turner, NIST.)

Analysis with Electrons

In addition to using the characteristic X-rays to analyze small particles, there are two analytical techniques based on electron signals that can also be used for analysis, electron diffraction and electron energy loss spectrometry (EELS). Electron diffraction is used to determine the structure of crystalline particles that have thicknesses less than about 0.5 μm. This enables an unambiguous identification of the material. As the electron beam interacts with a crystalline particle, the orientation of the atoms in the crystal planes of the specimen gives rise to coherent electron scattering that is specific for the lattice spacing of that crystal (Cowley 1979). The coherent scattering is observable in the back focal plane of the microscope as intensity modulations that form an electron diffraction pattern as shown in Fig. 13-10. The spacing of the diffraction spots can be measured and translated into crystal lattice spacings that can then be referenced to an X-ray powder diffraction file or compared to a known standard.

EELS can be used for the elemental analysis of particles less than about 0.1 μm in diameter. For particles, only the elemental analysis capabilities of EELS will be discussed since the chemically dependent fine structure contained in the spectrum is difficult to interpret for unknown complex specimens. A detailed discussion of EELS is given by Egerton (1986).

In EELS, an electron energy spectrometer is used to analyze and display the energy distribution of the electrons passing through the sample. The resulting spectrum (Fig. 13-11) is a composite of three different electron–sample interactions, only one of which is used for elemental analysis. The first interaction is a result of the beam electrons passing through the sample unhindered or elastically scattered and exiting with essentially the same energy as they entered. The resulting peak, called the zero-loss peak, carries no elemental information and is the largest peak in the spectrum. The second

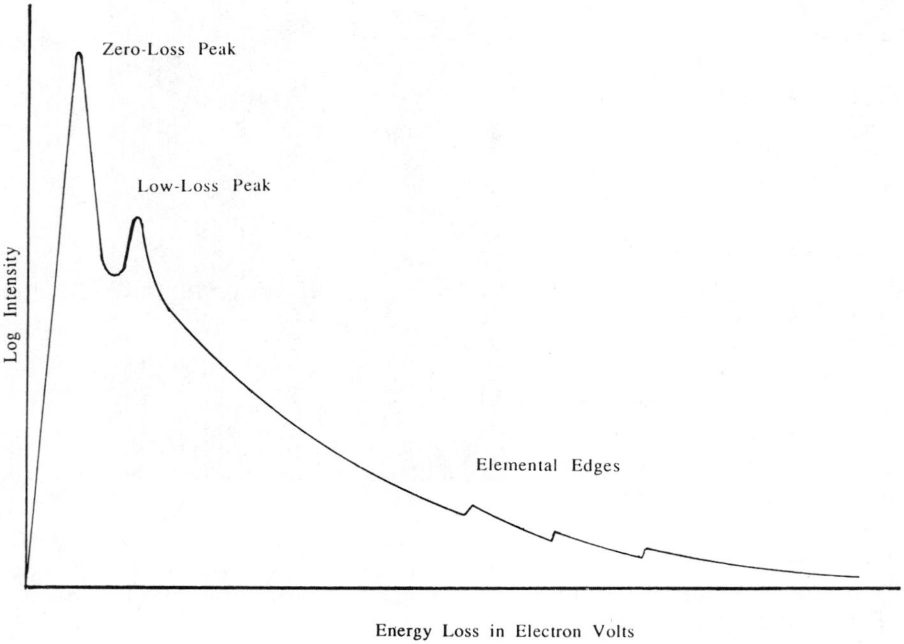

FIGURE 13-11. Diagram of An Electron Energy Loss Spectrum from the Inelastically Scattered Electrons Showing the Analytical Peaks.

interaction is the result of nonspecific inelastic collisions between the beam electrons and the coulombic field surrounding the nucleus of the sample atoms. This interaction produces what is called the low-loss peak. Following the low-loss peak is a slowly decreasing background extending (out) to higher energy loss values. Superimposed on this background are the analytical edges or lines that are a result of the inelastic collisions between beam electrons and the inner-shell electrons of the sample atoms. In the case of inelastic interactions, the beam electron ejects an inner-shell electron from the sample atom and, in the process, loses energy equal to the ionization energy of the particular electronic shell from which the electron is ejected. The resulting edges can be used to provide quantitative elemental information about the sample.

Large-Particle Analysis with the Electron Probe and Scanning Electron Microscope

In classical electron probe analysis procedures, both sample and the standard must have flat surfaces and be infinitely thick with respect to the penetration of the electron beam. The corrections for the interaction of the electron beam with the sample and the subsequent X-ray emission can be calculated from simple geometric relationships. If the effects of particle geometry are not included in a given analysis scheme, the subsequent results will be qualitative and can be used for element identification only. The errors in composition for an uncorrected analysis can be in the range of several hundred percent. It is, therefore, important to correct the X-ray data for the effects of particle shape.

In the quantitative analysis of particles, the shape and thickness of the specimen cannot be controlled or, in many cases, even measured. The difficulties in the quantitative analysis of particles or rough surfaces result from three different effects that influence the generation and measurement of X-rays from these samples (Small 1981).

The first effect is a result of the finite size (mass) of the sample. The mass effect is related to the elastic scattering of the electrons and is strongly affected by the average atomic number of the sample. This effect is important when the particle thickness is smaller than the range of the primary electron beam, so that a fraction of the beam escapes the sample before exciting X-rays. As the size of the particle decreases, the mass of material from which X-rays are excited decreases, resulting in a reduction of X-ray intensities from the particle compared to a bulk sample of identical composition. The mass effect can be demonstrated by comparing the X-ray emissions from a bulk target to the emissions from a particle of the same composition. This effect can be seen in Fig. 13-12, which depicts the Ba Lα X-ray intensity from particles (normalized to the intensity from a bulk material of the same composition) plotted as a function of particle diameter. The energy of the Ba Lα X-rays is 4.47 keV, high enough that the absorption effects are minimal and the slope of the curve reflects the mass effects. The mass effect is shown by the decrease in the Ba Lα X-ray intensity measured on particles less than 3 μm in diameter compared to the bulk. The net result of not correcting for the mass effect will be an underestimation of the composition for all elements analyzed in the particle (if the results are not normalized).

The second effect is a result of differences in X-ray absorption in particles compared to bulk standards. In the analysis of particles, the X-ray emergence angle and, therefore, the absorption path length cannot be predicted as accurately as for polished bulk specimens. For example, for the particle shown in Fig. 13-13, the X-ray intensity emitted from the particle along the A–B direction will be higher than that along the path C–D since the absorption pathlength is smaller and there is less mass for the X-ray to travel through before escaping the particle volume. The resulting elemental concentrations will, therefore, be higher than that expected for X-rays measured from the shorter pathlength and lower than that expected for X-rays measured from the longer pathlength. As expected, the magnitude of this effect is largest when there is high absorption, as is typically the case for soft X-rays from elements like Al or Si, which have energies less than 2 keV. The difference between the absorption pathlength in a particle and in a flat bulk sample can result in widely different values of the emitted X-ray intensities. The result of the varying absorption pathlengths is shown in Fig. 13-14, where Si Kα X-ray intensities from particles,

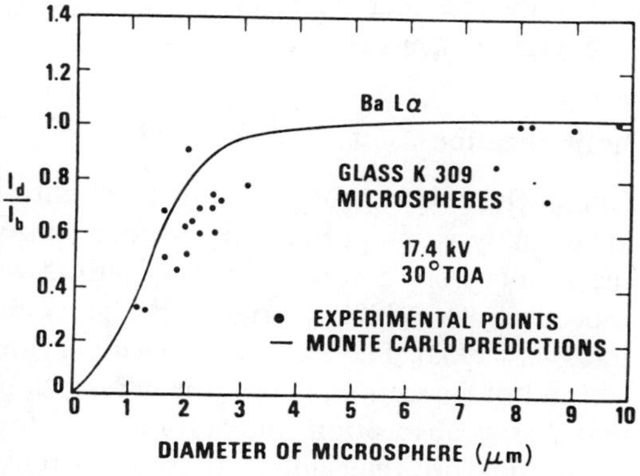

FIGURE 13-12. Plot of the Ba Lα X-ray Intensity from Particles, Normalized to the X-ray Intensity from a Bulk Target, vs. Particle Diameter. This Plot Shows the Decrease in the Particle X-ray Intensity, Compared to the Bulk Target, for the Smaller Particles Due to Their Decreased Mass. (Source: Small et al. 1978.)

X-RAY ABSORPTION IN PARTICLES

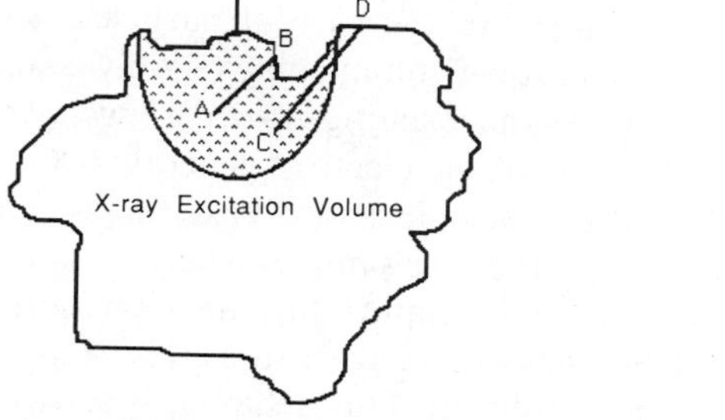

Because of particle shape x-ray absorption pathlength A-B is much shorter than C-D.

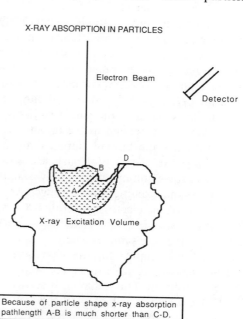

FIGURE 13-13. Schematic Diagram Showing the Random X-ray Absorption Pathlengths Encountered in Irregularly Shaped Particles.

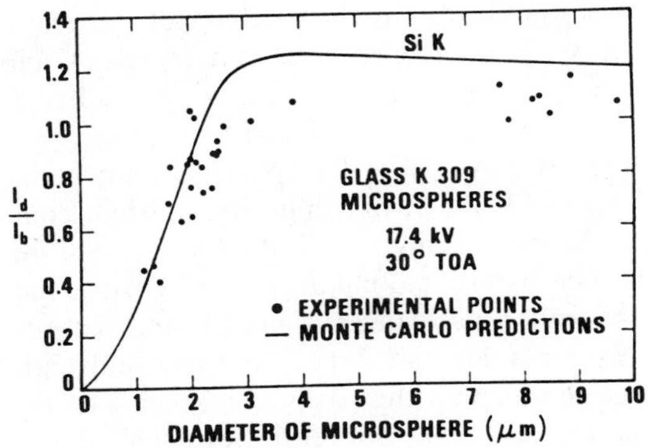

FIGURE 13-14. Plot of the Si Kα X-ray Intensities from Particles, Normalized to the X-ray Intensity from a Bulk Target, vs. Particle Diameter. This Plot Shows the Increase in X-ray Intensity for the Smaller Particles, Compared to the Bulk Material, Due to the Decreased X-ray Absorption Pathlength. (Source: Small et al. 1978.)

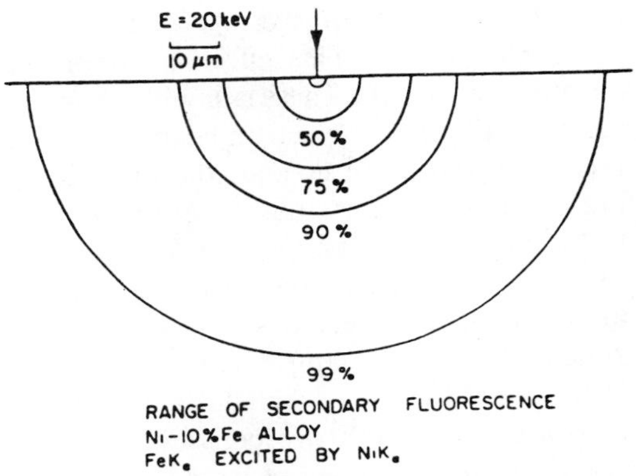

FIGURE 13-15. Diagram Showing the X-ray Range in Micrometers for Ni Kα X-rays Fluorescing Fe Kα X-rays. This Diagram Indicates that Fe X-rays are Fluoresced by Ni X-rays at a Distance Considerably Greater than the Dimensions of Microscopic Particles. (Source: Small 1981.)

normalized to the intensity from a bulk material of the same composition, are plotted as a function of particle diameter.

The third effect is caused by the secondary excitation of characteristic X-rays of interest by electron-generated X-rays, either continuum or characteristic. Because X-ray absorption coefficients in solids are relatively small compared to electron attenuation, secondary X-ray fluorescence can occur over a much larger volume or range (tens of micrometers) than the range of primary electron excitation of X-rays (1–5 μm). In bulk samples and standards, the electron-generated X-rays, for the most part, remain in the target since the samples are substantially larger than the X-ray ranges. In the case of particles, however, the particle volume may be only a small fraction of the exciting X-ray range. As a result, most of the electron-generated X-rays will exit the particle prior to exciting secondary X-rays. In those samples where secondary fluorescence is important, the deviation of particle behavior relative to a bulk standard may be significant. This behavior is shown in Fig. 13-15, which is a plot of the range for Ni Kα X-rays fluorescence of Fe Kα X-rays in a Ni–Fe alloy.

The net effects of not correcting for the low secondary excitation of X-rays in particle analysis schemes using bulk standards are:

1. An underestimate of the concentration for elements that have a significant contribution to their characteristic line from excitation by other characteristic X-rays, since the measured X-ray intensity from the particle will be reduced by the calculated value of the secondary X-rays.
2. In the case of continuum fluorescence, an underestimate of the concentrations for all elements, particularly those with higher-energy lines that are excited by the higher-energy, longer-range continuum.

Normalization

One of the simplest methods for the quantitative analysis of particles is to normalize the results from a conventional quantitative procedure to 100% (Wright, Hodge, and Langway 1963). The analyst, in selecting this method of correction, makes the assumption that X-ray absorption and fluorescence for the particle are the same as those for a bulk specimen and that the mass effect is the same for all elements. In practice, a normalization of results is effective for the correction of the mass effect since the decrease in intensity as a function of particle size is nearly the same for

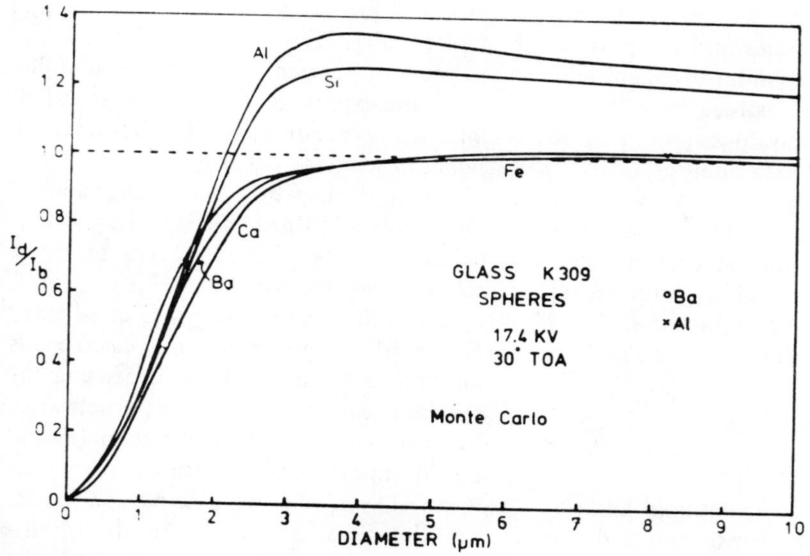

FIGURE 13-16. Plot of Calculated Monte Carlo Curves for X-ray Intensities of K-309 Glass Microspheres, Normalized to the X-ray Intensities from a Bulk Target, vs. Particle Diameter. The Plots Show that the Different Elemental Curves Merge for Particle Diameters Smaller than Approximately 2 μm.

all elements. Figure 13-16 shows that the different elemental curves merge together for particle diameters less than 2 μm. Since normalization does not compensate accurately for the absorption and fluorescence effects, the most accurate results will be obtained on particle systems that meet the following conditions:

1. Systems for which all the analytical lines for the elements are above 4 keV, where absorption effects are minimal.
2. Systems for which the analytical lines are below about 4 keV in energy; the lines for all the elements should be as close together in energy as possible so that the matrix absorption is approximately the same in all cases.
3. Systems for which there is no significant secondary fluorescence.

Table 13-4 lists the results from the analysis of lead silicate glass particles. The first set of results is taken from the analysis of the Pb Mα line at 2.3 keV that is close in energy to the Si Kα line at 1.74 keV. Since the absorption and mass corrections are similar for these two lines, the lead and silicon concentrations are in good agreement with the true values. The second set of results were determined by analyzing the Pb Lα line at

TABLE 13-4 Analysis of Lead Silicate Glass K-229 by Normalization of Quantitative Results[1]

Analysis	Si (wt%)	Percentage Error	Pb (wt%)	Percentage Error
With Pb Mα Line (Meets Conditions)				
1	0.155	+ 10.7	0.620	− 4.6
2	0.144	+ 2.9	0.643	− 1.0
3	0.136	− 2.7	0.658	+ 1.2
4	0.138	− 1.1	0.653	+ 0.5
5	0.127	− 9.0	0.675	+ 3.8
6	0.170	+ 22	0.588	− 9.5
7	0.137	− 2.5	0.657	+ 1.1
With Pb Lα Line (Does Not Meet Conditions)				
1	0.134	− 4.5	0.663	+ 1.9
2	0.177	+ 26.3	0.578	− 11.0
3	0.159	+ 14.0	0.612	− 5.8
4	0.166	+ 18.0	0.602	− 7.4
5	0.017	+ 88.0	0.894	+ 37.0
6	0.100	− 29.0	0.731	+ 12.4
7	0.157	− 12.3	0.616	− 5.2

Source: Small (1981).

1. Nominal composition: Si = 0.140; Pb = 0.650

10.6 keV. In this case, the two analytical lines have very different energies and the particle absorption effect is not similar in magnitude. As expected, the errors associated with this analysis are considerably higher than those associated with the Pb Mα analysis.

One of the major disadvantages of the normalization of results from a bulk analysis procedure is that the analyst cannot determine, by obtaining an analysis total of less than 100%, the presence of any unanalyzed elements such as those with atomic numbers less than 11.

Particle Standards

The analyst can use a conventional analysis scheme and substitute particle standards for the normal polished standards (White, Denny, and Irving 1966). In this procedure, the assumption is that the particle effects, particularly the absorption effect, will be approximately the same for the sample and standard. This assumption is reasonably valid, providing the sample and the standard are close in composition and the particle diameter is larger than about 2 μm. Below this size, any difference in size and shape between the unknown and the standard will be critical since a small change in effective diameter will result in a large change in X-ray intensity.

Geometric Modeling of Particle Shape

Geometric modeling of particle shape was developed by Armstrong and Buseck (1975). It is based on the determination of a simple geometric shape or a combination of shapes, such as square or pyramid, that define the boundaries of the particle of interest. The various particle effects are then calculated for the chosen geometric shapes defining the shape of the particle. The various effects resulting in X-ray loss in particles are corrected as follows:

1. Electron Transmission: The X-ray intensity lost as a result of electron transmission through the particle is calculated from a modified form of an expression developed by Reuter (1972) to calculate relative X-ray production in thin films. In the modified form, the expression can be used to calculate X-ray production as a function of position within a particle.
2. Electron Backscatter: The expression of Duncumb and Reed (1968) is used to calculate the loss of X-rays as a result of electron backscatter.
3. Electron Sidescatter: The loss of primary X-rays from the sidescatter of electrons is minimized by rastering or defocusing the electron beam over the entire particle area, by using an overvoltage less than 1.5 and by ratioing the concentrations.
4. X-ray Absorption: The correction for X-ray absorption sets the integration limits for the absorption pathlength to the limits determined for the geometric shape or shapes that define the overall particle. The shapes that have been included in the program include rectangular prism, tetragonal prism, right triangular prism, square pyramid, hemisphere and sphere.

The X-ray intensity from the secondary fluorescence of characteristic X-rays is calculated from the work of Armstrong and Buseck (1977), with the integration limits

TABLE 13-5 Results from the Analysis of Mineral Particles by the Geometric Modeling Method

		Composition (wt%)	
Mineral	Component	Actual	Determined
Anorthite[1]	CaO	18.9	19.0 ± 0.8
	Al_2O_3	35.7	35.6 ± 1.2
	SiO_2	44.3	44.4 ± 1.0
Rhodonite[2]	MnO	35.5	35.5 ± 1.1
	FeO	13.0	13.1 ± 0.8
	CaO	4.1	4.0 ± 0.2
	SiO_2	46.9	46.8 ± 0.8
Pyrite[3]	FeS	46.6	46.4 ± 1.0
	S	53.4	53.6 ± 1.0

Source: Armstrong (1978).

1. Results represent average of 122 analyses.
2. Results represent average of 100 analyses.
3. Results represent average of 140 analyses.

adjusted to calculate the probability of secondary fluorescence at a given location within a particle.

The results from this procedure are given in Table 13-5 for anorthite (calcium aluminum silicate), rhodonite (manganese silicate), and pyrite (iron sulfide) (Armstrong 1978). In all cases, the analyses are in good agreement with the known composition and have standard deviations less than $\pm 6\%$ of the measured concentration. Careful attention must be paid to the sizes and shapes used to define the particle boundaries if accurate concentrations are needed.

Peak-to-Background Ratios

The fourth method for the quantitative analysis of particles was developed by Small et al. (1978) and Statham and Pawley (1978). This method, an extension of the method developed for biological specimens by Hall (1968), is based on the following observation: to a first approximation, the ratio of a characteristic X-ray intensity to the continuum intensity of the same energy for a flat, infinitely thick target is equivalent to the same ratio of intensity from a particle or rough surface to its continuum intensity of the same composition.

This observation can be expressed in the following equation:

$$(I/I_{\text{con}})_{\text{particle}} = (I/I_{\text{con}})_{\text{bulk}} \qquad (13\text{-}6)$$

where I is the background-corrected peak intensity and I_{con} is the continuum intensity for the same energy window as the peak. For this procedure it is assumed that the spatial distribution for characteristic X-ray excitation is identical to the distribution for continuum X-ray excitation. As a result, the effects of particle shape and size on measured X-ray intensity will be the same for the continuum and the characteristic X-rays. It, therefore, follows that by taking the ratio of the two intensities, the particle mass and absorption effects will cancel out.

Various types of mineral particles have been analyzed with the peak-to-background method using the algorithm FRAME P (Small, Newbury, and Myklebust 1979). The results of these analyses are reported in Table 13-6 along with the result from the conventional analysis algorithm, FRAME C (Myklebust, Fiori, and Heinrich 1979). In all cases, the analyses with the peak-to-background routine are within 10% relative error of the stoichiometric values. In contrast, the errors with the conventional routine range from 7.9% for S in ZnS to 47% for Mg and Si in talc. In addition, the standard deviations for individual measurements are less for the peak-to-background routine than they are for the conventional atomic number absorption fluoresence (ZAF) algorithm (Henoc, Heinrich, and Myklebust 1973). ZAF is a quantitative analysis scheme which compares the X-ray intensities for the unknown to that for a standard.

The assumption that the generation volumes for characteristic and continuum X-rays are identical is only valid for particles greater

TABLE 13-6 Analysis of Mineral particles by the Peak-to-Background Method (FRAME P). Conventional analysis by FRAME C. The Results are the Average from 7 Analyses

	Mg (wt%)	Percentage Error	Si (wt%)	Percentage Error
Talc ($Mg_3[Si_4O_{10}](OH)_2$)				
Nominal	19.3		29.8	
Frame C	19.4	+ 0.5	29.7	− 0.3
Frame P	18.5	− 4.0	29.7	− 0.3
FeS_2	**S (wt%)**	**Percentage Error**	**Fe (wt%)**	**Percentage Error**
Nominal	53.4		46.6	
Frame C	52.7	− 1.3	47.3	+ 1.5
Frame P	52.9	− 0.9	46.4	− 0.4
ZnS_2	**S (wt%)**	**Percentage Error**	**Zn (wt%)**	**Percentage Error**
Nominal	32.9		67.1	
Frame C	35.4	+ 7.8	64.5	− 3.8
Frame P	36.0	+ 9.0	67.7	− 0.7

Source: Small (1981), Small et al. (1979)

than about 1 μm in diameter. Below this size, the anisotropic generation of the continuum results in a significantly different excitation volume for the continuum compared to the isotropically generated characteristic X-rays. This effect is shown in Fig. 13-17, where isotropic and anisotropic cross sections have been used to calculate peak-to-background ratios from K-309 glass particles normalized to bulk glass. The composition of K-309 is 7.9% Al, 18.7% Si, 10.7% Ca, 13.4% Ba, and 10.5% Fe. These plots show that the introduction of an anisotropic cross section for the continuum results in significantly higher peak-to-background ratios for the smaller particles. As a result of using peak-to-background procedures, it is necessary, for quantitative analysis of particles less than about 1 μm in size, to introduce into the analysis procedure a correction for anisotropic generation of the continuum (Small 1981).

Automated Analysis of Particles with Electron Beam Instruments

An additional method of particle analysis that makes use of both SEM and EPMA is automated analysis (Lee and Kelly 1980; Germani and Buseek 1991). In this procedure, the electron beam and specimen stage are controlled by computer and a series of particles, usually numbering from a few hundred to a few thousand, are analyzed for elemental composition and morphology. The results from automated analysis are used to infer information about the total particulate population and the sampled aerosol. In this respect automated particle analysis is different from the individual particle procedures described above. The elemental analyses are at best semiquantitative and, because of the short counting times, (often 10 s or less) do not provide information on elements with concentrations lower than about 5 wt%.

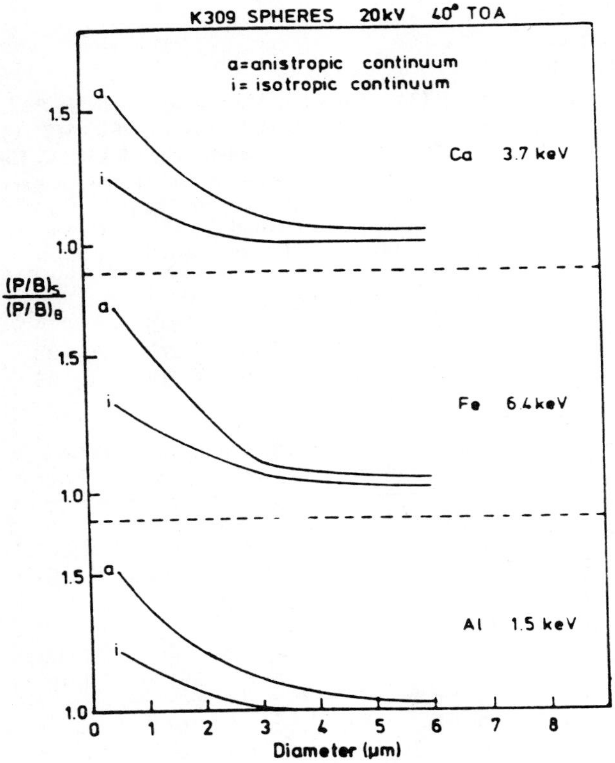

FIGURE 13-17. Plot Showing the Effects of Calculating Peak-to-Background Ratios Based on Isotropic and Anisotropic Cross Sections. (Source: Small et al. 1979.)

Once the data are collected, the different particles are divided into groups based on their composition and morphology. The groupings can be either preselected categories (discriminant analysis), or can be selected based on the data (cluster analysis) (Kaufman and Rousseeuw 1990). After grouping the data, the analyst can then estimate various population characteristics, such as mass fraction of the total sample, resulting from a given particle type.

LASER MICROPROBE MASS SPECTROMETRY (LMMS)

Underlying Principle

Laser microprobe mass spectrometry (LMMS) is a mass spectrometric technique that utilizes a focused, pulsed laser beam to generate ions by laser ablation/ionization. The process is illustrated in Fig. 13-18. The laser normally used has a short pulse length (approximately 10 ns) and a focused high irradiance (10^6–10^9 W/cm^2). The absorption of the laser beam by the sample material produces heating over a short duration. The sample is volatilized and expands (usually) into vacuum. For the lower end of the irradiance range, 10^6–10^8 W/cm^2, material is removed from the sample in a "soft" manner frequently termed laser desorption. During high irradiance, intense heating can occur, resulting in plasma formation. The sample constituents are for the most part reduced to elements, small fragments, and clusters. In both energy regimes, nonlinear optical effects are thought to occur in addition to classical absorption.

Only ions are detected in the mass spectrometer. The ionization efficiency is low, of the order of one part in one thousand. Although there are ions present, the mechanism for ionization in laser desorption is not understood. In the high irradiance mode, ionization is thought to take place by thermal and collisional processes occurring in the plasma.

Instrumentation

The laser microprobe is composed of a dual laser system coupled to a mass spectrometer

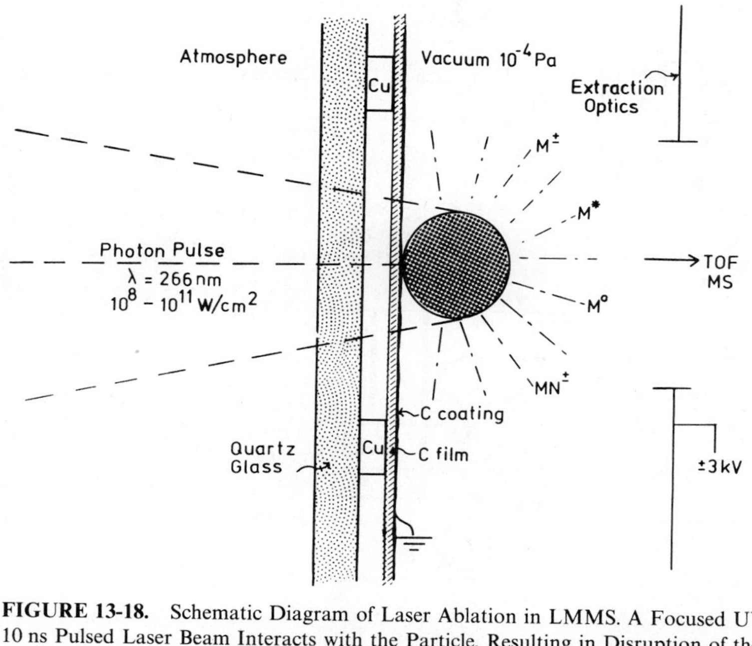

FIGURE 13-18. Schematic Diagram of Laser Ablation in LMMS. A Focused UV 10 ns Pulsed Laser Beam Interacts with the Particle, Resulting in Disruption of the Solid and Formation of Ions, Neutrals, and Metastable Species. (Source: Simons 1984.)

through an optical microscope. The sample is viewed optically and a particle is selected for analysis by aiming a co-aligned pilot laser beam spot (often from a He–Ne laser) onto the particle. The ionizing laser is a high-power pulsed laser, usually a Nd:YAG laser. The radiation from the Nd:YAG is frequency-quadrupled to provide pulsed UV radiation (266 nm) and is focused on the sample by an optical microscope objective or some other lens system. All ions generated are accelerated to uniform translational energy and mass-separated by time-of-flight (TOF) mass spectrometry. In time-of-flight mass spectrometry, flight time in the drift tube is proportional to the square root of the mass-to-charge ratio. The arrival of ion packets at the end of the drift tube is determined by a secondary electron multiplier detector.

Two geometries, reflection and transmission, are commonly used in commercial instruments. Transmission refers to having the laser beam and the mass spectrometer in-line, while reflection refers to the beam being at an acute angle with respect to the spectrometer. Figure 13-19 schematically presents only two possible examples of these two geometries. A complete description of LMMS is given by Denoyer et al. (1982), Hercules et al. (1982) and Simons (1988). Single-particle analysis is discussed by Kaufmann (1986), De Waele and Adams (1986) and by Wieser and Wurster (1986). The early work in aerosol research applications is presented by Wieser, Wurster, and Haas (1981).

Capabilities

Some of the important aspects of the LMMS instruments are nominal micrometer spatial resolution, high detection sensitivity, and detection capability for either positive and

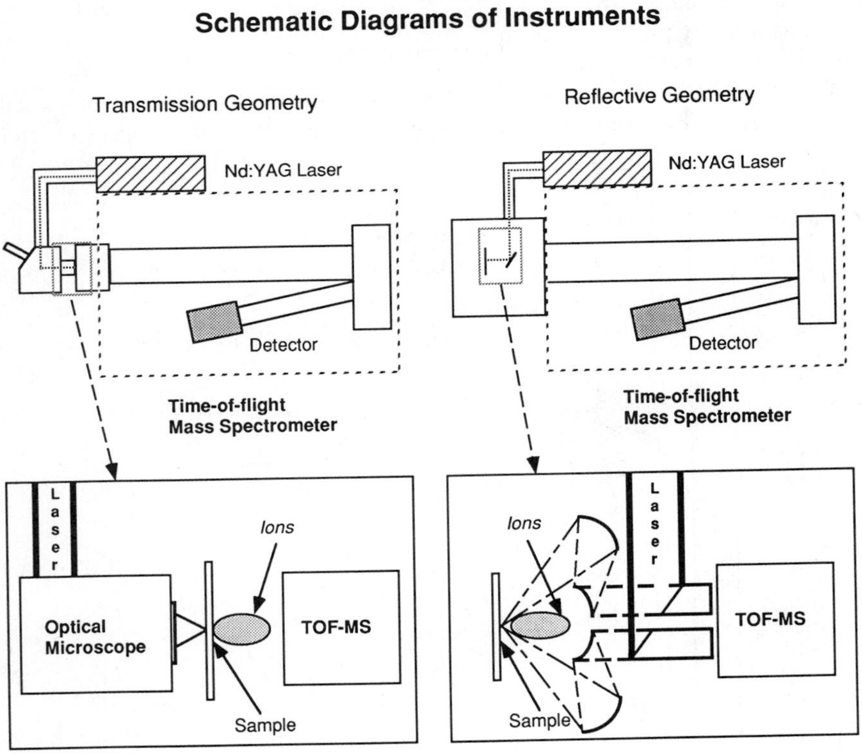

FIGURE 13-19. Schematic Diagram Showing Transmission and Reflective Geometry of Two Existing LMMS Instruments. The Transmission Geometry has Higher Spatial Resolution and is Normally Considered the Better Design for Particle Analysis. (Courtesy S. Hoeft.)

negative ions produced in the laser ionization. In addition, it is usually possible to determine the elemental, isotopic, and, in many cases, even the molecular composition of a particle. The technique is complementary to electron microscopy (Steel et al. 1984) and micro-Raman spectroscopy (Fletcher, Etz, and Hoeft 1990) and should prove to be complementary to other spectroscopies. LMMS is generally not considered quantitative even for major constituents, except with regard to isotope abundance. Matrix effects in the sample play a dominant role during the laser beam–sample interaction, making quantitation difficult. The main strength of LMMS lies in providing useful qualitative information in a rapid manner.

Elemental Analysis

One important aspect of the LMMS is that all elements are usually ionized in a single laser pulse and are detected in a single mass scan. This single, full-mass scan capability inherent to TOF mass spectrometry makes the LMMS instrument well suited for particle analysis. An example of a LMMS spectrum that displays 44 elements and their oxides present in NIST SRM 610 multielement glass is shown in Fig. 13-20 (Simons 1984). Most of these elements are present in the glass at only trace amounts (~ 500 ppm level by weight).

Detection limits for the elements vary. In most sample matrices, the alkali metals are most easily detected and are reported to have a detection limit of the order of 10^{-19} g (Kaufmann 1986). Most other elements have detection limits between 10^{-17} and 10^{-19} g, allowing the detection of minor and, in some cases, trace elements from a single particle with micrometer dimensions and picogram mass. In terms of trace analysis on bulk samples, a relative concentration of 1–100 ppm must be present for detection.

Isobaric interferences (same mass, different composition) resulting from a low mass resolution can lead to a problem in the interpretation of LMMS spectra, but can be overcome in many cases by observing the presence of isotopes or other characteristic peaks in the spectrum. For example, C_5^- and

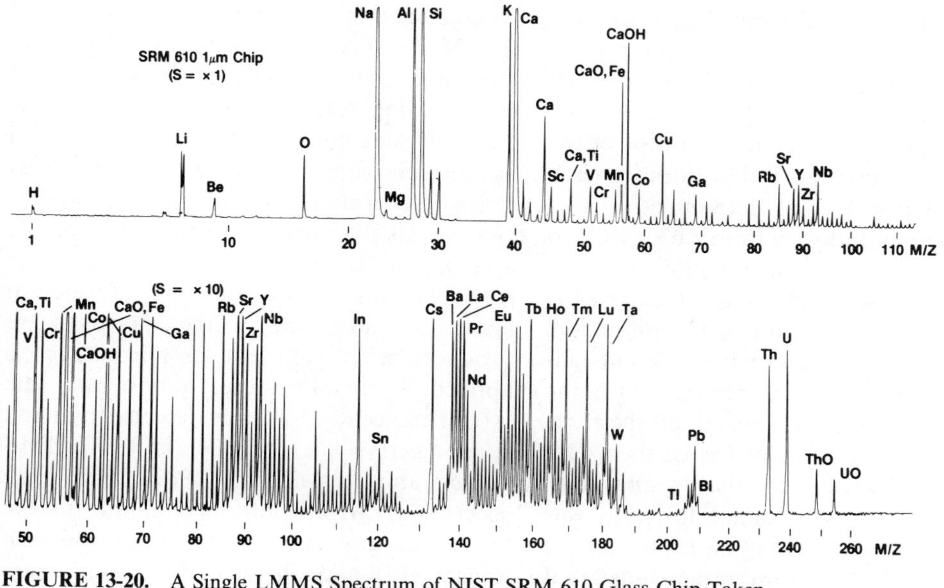

FIGURE 13-20. A Single LMMS Spectrum of NIST SRM 610 Glass Chip Taken on the Transmission Instrument. Some 44 Elements and Their Oxides are Labeled as Shown. Trace Elements, e.g., U are Present at the 400–500 ppm level. It is Characteristic of a Time-of-Flight Mass Spectrum That All Masses are shown in a Single Scan. (Source: Simons 1984.)

CO_3^- have the same m/z value of 60 and the two species would be difficult to distinguish if they are lone peaks in the mass spectrum. However, C_5^- rarely appears without associated carbon clusters, C_2^-, C_3^-, C_4^-, ...

For qualitative analysis, the instrument capabilities include sensitivity, quickness, and simplicity of analysis. Unfortunately, one disadvantage is that different matrix materials ionize with different efficiencies, leading to problems for quantitative work (Kaufmann 1986). One study that illustrates the problems involved with quantitative analysis was done by Musselman, Simons, and Linton (1988) for films, micrometer-sized spherical particles, and shards made from glass of known elemental composition. Relative sensitivity factors (RSF) of Mg, Si, Ca, Cr, Fe, Zn, Ba, and Pb were compared for the film, particles, and shards made from the same glass. A RSF of element j

$$RSF_j = \frac{I_j}{I_r} \cdot \frac{C_r}{C_j} \cdot \frac{f_r}{f_j} \qquad (13\text{-}7)$$

is defined here as a function of the concentration of the reference element, C_r, concentration of element j, C_j and their respective measured ion intensity, I_r and I_j. In Eq. 13-7, f corresponds to the isotopic fraction. Using Ba as the reference element, the ratios of the maximum to minimum RSF values for each of seven elements were determined. These ratios were found to vary from 1.63 for Ca to 5.17 for Si using all data for films, spheres, and shards. These results suggest that in the process of the laser-beam–solid interaction, the elements are ionized with different efficiencies for the three geometries since the glass composition is the same in all the three cases. When the authors restricted their data set to 1 μm spheres, 1 μm shards, and films, the RSF maximum to minimum ratio was dramatically reduced, ranging from 1.17 for Ca to 2.57 for Fe. The authors concluded that laser focus plays a role in ionization efficiency for the planar and particle geometries. It is important to utilize standards of similar geometry, size, and composition before any meaningful quantitative work can be attempted.

Isotopic abundance information for all elements is one of the strengths of mass spectrometry and supplements non-isotope-specific elemental electron microprobe capabilities. Careful isotopic measurements provide quantitative information because the isotopes experience the same matrix effects and have the same ionization properties. Accurate relative isotopic ratio measurements have been made by Lindner et al. (1986) on thin-film samples of rhenium/osmium to determine the half-life of ^{187}Re. Details of the precautions necessary to obtain such precision and accuracy are described by Simons (1983). The other major isotopic investigation by LMMS has been conducted by Schroder and Fain (1984) regarding calcium release (using ^{44}Ca) from photoreptors in the eye.

Molecular Analysis

Molecular characterization is accomplished using low irradiance (10^6–10^8 W/cm^2) directed onto the sample. Some frequently occurring anion molecular species, common to atmospheric particles, are often identified in LMMS spectra as NO_3^-, SO_4^-, HSO_4^-, PO_x^- and CO_3^-. Usually, all ions are singly charged, regardless of the natural valence state. Organic compounds are often detected as molecular ions formed through the process of laser desorption or as adducts, i.e., hydrogen or alkali metal (especially Na or K) attachment to the molecule. Characteristic fragment ions containing some information about the structure of the molecule are produced when sufficient laser irradiance is applied. It is interesting to note that the largest molecular ion ever detected by mass spectrometry, m/z 247,000, is by LMMS (Hillenkamp and Karas 1989). The mass spectra in LMMS, although unique to the technique, are similar to both electron impact (EI) and chemical ionization mass spectra. Certain types of compounds can be detected as molecular ions, such as the atmospherically important polyaromatic hydrocarbons (Mauney 1984). Conversely, organic compounds such as alkanes generate (as in EI) little or no molecular ion whatsoever. Often, the particle is not completely ablated away under low-irradiance conditions and sometimes there is no visible (by LM) damage to the particle. Van Vaeck et al. (1990) present the most recent discussion of organic compound analysis.

The mass spectra common to laser ionization mass spectrometry are often complicated. There are two approaches that can be adopted to analyze the spectra: direct identification and fingerprinting. For specific compound identification, pure reference compounds are usually analyzed and compared to the unknown. It should be pointed out that for mixtures of organic compounds, as is often the case in atmospheric samples, direct identification of even major organic components may be difficult due to the complexity of the mass spectra. Thus, fingerprinting is the only recourse. Fingerprinting is the comparison of mass peak patterns from sample particles with the peaks obtained from particles of known source, without specifically identifying the peaks. Fingerprinting can involve pattern recognition techniques, such as principal component analysis (Wieser and Wurster 1986; Currie, Fletcher, and Klouda 1987; Linder and Seydel 1989; Ro, Musselman, and Linton 1989; Odom et al. 1989).

Applications

Sample Analysis Strategy

LMMS should be reserved as the last analytical technique to be employed because in the process of analysis part or all of the sample will be destroyed. Also, the sample is vacuum-mounted in the instrument for the duration of the analysis, causing any volatile compounds present to evaporate.

Sample collection for LMMS is best accomplished directly on a substrate that can be used in the instrument. This substrate should have a mass spectrum that does not interfere with the spectrum of the sample. The reflection-designed instrument usually uses a flat substrate since ionization is achieved from the front surface. The transmission instrument requires optically transparent substrates or thin films because the beam must penetrate through the substrate to ionize the sample. A good substrate is made of quartz in the form of coverslips, which serve as windows in the instrument (see Fig. 13-18). Quartz is quite transparent to the ionizing radiation. The coverslip can be used in an impactor to collect size-segregated airborne particles. Another approach is to deposit particles using thermal precipitation onto the quartz coverslip (Cleary et al. 1992). Quartz fiber filters can be used for particle collection and subsequent LMMS. Both quartz fiber filters and quartz coverslips should be oven-heated to eliminate any possible carbon artifact from the surface. From the surface of the filter, entire quartz fibers carrying the particles can be extracted and subsequently mounted across a transmission electron microscope grid for analysis by LMMS (Fletcher 1989). As mentioned, in transmission, thin films can be used to support the particles. Formvar, a poly(vinyl formal) resin, frequently mounted on a transmission electron microscope grid, is used because it can be made into thin films that produce very low background signals in LMMS. In all these cases particles can be ablated and ionized by the laser beam individually or on a near single-particle basis with minimum background interference (with the exception of silicon and its oxides).

Environmental Analysis

LMMS is frequently used to provide information about environmental particles. Early work was done by Wieser, Wurster, and Seiler (1980) on both laboratory-generated and atmospheric aerosol particles. Mixtures and composite particles present difficult spectra to interpret, but the spectra, nevertheless, often contain information about origination or formation. For example, the coexistence of Pb and carbon in the same particle could suggest automobile exhaust. Kolaitis et al. (1989) could distinguish dimethanesulfonic acid particles collected on the lowest stages of a Battelle-type impactor from non-sea-generated particles. The authors report nitrates (derived from natural sources) forming around crystalline sea salt.

To provide more information on a sample, the analytical electron microscope (AEM) can be teamed with LMMS; LMMS supplements the AEM analysis by identifying low atomic weight elements, trace elements and molecular compounds. Sheridan and Musselman (1985) used an AEM to identify the major chemical composition and morphology and LMMS to identify low atomic weight (< Na) constituents of arctic haze aerosol. They found primarily sulfate particles. The AEM could detect the presence of sulfur in the particles and the LMMS mass spectra revealed the characteristic sulfate ion at m/z 96 and bisulfate at m/z 97. Bruynseels et al. (1988) also combined electron microscopy and LMMS to identify the chemical composition of North Sea aerosol. The authors reported the detection of nitrate particles, sulfates, sea salt, and PAHs desorbed from coal fly ash particles. Denoyer et al. (1983) and Yokozawa et al. (1987) have characterized coal fly ash for elemental composition. Surkyn et al. (1983) employed LMMS to look for elemental signatures for single-particle source identification.

An analysis of carbon soots (solid particles) and organic carbon is difficult on the single or near-single particle (~ pg) basis. Yet, in some cases the amount of carbon collected is so small that conventional bulk analysis techniques are not possible. LMMS has been used to try distinguishing soots from various combustion sources collected from controlled burns of wood, polyurethane, and heptane, as well as from atmospheric sources (Currie, Fletcher, and Klouda 1987; Fletcher and Currie 1988). Wood soot has a large amount of potassium associated with the carbon on an individual particle basis that makes it unique among the other sources studied. Efforts have been made to correlate the low mass/charge carbon cluster peak area ratios (C_4^-/C_2^-) to the amount of modern carbon present in the sample (Currie, Fletcher, and Klouda 1989).

SECONDARY-ION MASS SPECTROMETRY (SIMS)

Underlying Principle

Secondary-ion mass spectrometry (SIMS) is based on ion beam sputtering. When an ion beam interacts with a solid surface, a small fraction of the ions are reflected from the surface, but most of the ions implant themselves into the solid material transferring some of their momentum into a collisional cascade. Some of the momentum from the cascade is imparted to surface atoms or molecules, providing them with sufficient kinetic energy to escape the surface layer. The process of removing material from the top layer using ion or neutral beam techniques is called sputtering. While the primary ions may penetrate several tens of nanometers below the sample surface, the majority of the sputtered ions and neutrals are liberated from the surface. Sputtered atoms can be neutral, charged, or in a metastable state. In addition to atomic and molecular sputtered products, electrons and photons are liberated (Behrisch 1983).

Figure 13-21 illustrates the interaction of a primary ion beam with a sample surface generating secondary ions in the process. The primary ion beam is composed of one of the following: oxygen, argon, cesium, or gallium ions. These secondary ions are mass separated on the basis of their mass/charge ratio in a mass spectrometer. The surface of the sample is damaged in the process.

Instrumentation

There are two types of SIMS instruments: the ion microscope, which uses a defocused beam and specialized ion optics to form an image of the sample surface, and the ion microprobe, which uses a focused primary beam and produces ion images by sequentially rastering the beam over the sample. There are a number of commercial microprobe instruments using a highly focused liquid-metal ion source. The advantages of the microprobe instrument over the ion microscope instrument include a

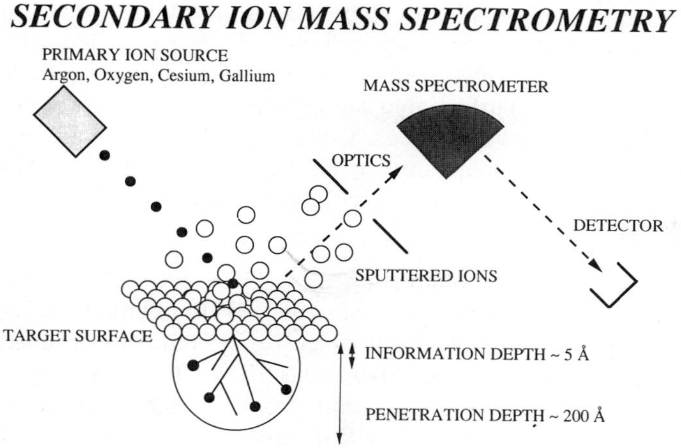

FIGURE 13-21. Schematic Diagram of the Ion Beam Sample Bombardment in SIMS Showing Approximate Depth of Interaction Between the Ions and the Sample Surface and a Representation of Ion Implantation. (Courtesy of G. Gillen, NIST.)

higher spatial resolution (of the order 0.1 μm), providing the opportunity of high spatial ion imaging, and a higher ion transmission through the use of a TOF mass spectrometer.

Different mass spectrometers are employed for ion detection. The most widely used are magnetic sector and quadrapole, but time-of-flight is beginning to be used in ion microprobes.

Capabilities

The advantages of microanalysis SIMS include high sensitivity, high spatial resolution, elemental, and isotopic analysis, molecular information, depth profiling, and ion imaging. All the elements are detectable in both types of instruments. In the case of imaging, both elemental and molecular imaging are possible on the 1 μm size level. SIMS has a large dynamic range making it applicable for trace analysis. As can be seen from Table 13-1, SIMS has detection limits comparable to or better than any technique presented. Gavrilovic (1984) estimates a yield of the order of 10^4 detectable ions from a particle with 0.1 μm diameter, assuming the particle to be completely consumed, an ionization efficiency of 1% and a secondary-ion collection rate of 10%. This is sufficient for the analysis of major elemental composition. One shortcoming in SIMS is that the combined effect of sputtering and ion production efficiency of elements varies over 5 orders of magnitude (Storms, Brown, and Stein 1977). A quantitative analysis can be done provided calibration standards of like materials are used.

Similar to LMMS, there are two modes of operation; high primary beam current (10^{-3} A/cm^2), known as dynamic SIMS, and low primary beam current (10^{-9} A/cm^2), known as static SIMS. Dynamic SIMS is a high-damage regime and is used for elemental depth profiling while static SIMS can provide the user with molecular compound information. In static SIMS, the ion dose is so low that, probabilistically, each primary ion interacts with an undamaged area of surface.

Applications to Particle Analysis

Ion Microscope

Newbury (1980) reported an example of single-particle depth profiling on a 10 μm urban dust particle, as illustrated in Fig. 13-22. The depth profile plot shows various indicated elements ratioed to ^{28}Si as a function of distance into the particle. Linton et al. (1976) showed using depth profiling that elements such as As, Pb, and Cd are associated with the surface of fly ash particles. This is not

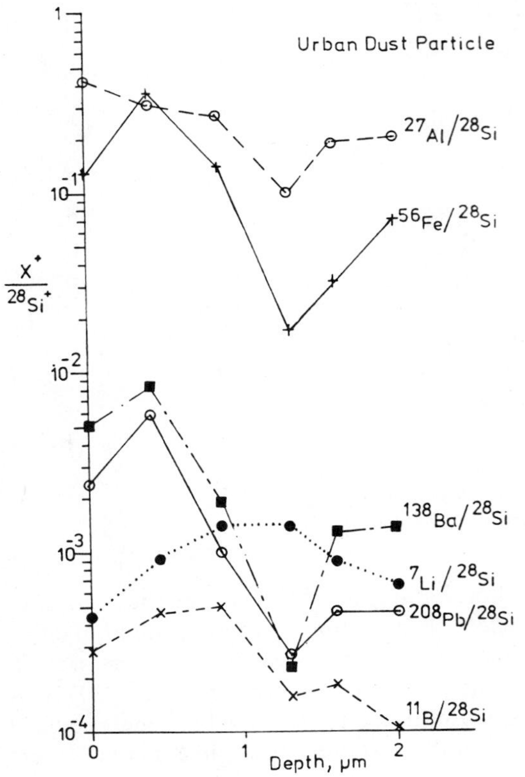

FIGURE 13-22. Depth Profiling by SIMS of a Single 10 µm Reference Particle. Essentially, the Surface Layer is Successively Sputtered Away, Generating Ions that are Separated in a Mass Spectrometer. Element Ratio Referenced to ^{28}Si. (Source: Newbury 1980.)

an example of single-particle analysis because the sample was composed of fly ash particles compressed into a disk, but the example does show the unique information derived from the surface layer of the particles found using the technique. Work on atmospheric particles is generally qualitative rather than quantitative. Klaus (1984) used a SIMS instrument to analyze aerosol particles collected onto an aluminum foil in a low-pressure impactor. Although Klaus reports that Al foils are the best substrate due to low contaminates in the foils, gold offers a higher mass/charge element and can also be very pure. Oxides of various elements, certain anions such as sulfates, nitrates, chlorides, and carbon in both graphitic and organic form were found by Klaus in some European continental aerosol plumes. A similar study was done (Klaus 1985) on automobile exhaust. Primarily, elemental constituents were reported on the particle surface analysis; no organic molecular species were found. A chapter containing many references regarding particle analysis by SIMS is presented by Klaus (1986).

Ion Microprobe

Bennett and Simons (1991) presented the results for relative sensitivity factors determined for thin 20 nm glass films and micrometer glass shards made from the same glass composition. The RSFs differed by a factor of 3–4 between film and shards.

Owari et al. (1989) presented a study using a submicrometer (of the order of 0.1 µm) liquid-metal ion source that was highly focused for imaging small structures either by secondary electrons or direct elemental secondary-ion imaging. The authors demonstrated imaging on a mix of fly ash and NaCl particles fixed onto a metal substrate with conductive silver paint. Quantitative results were reported giving relative sensitivity factors derived from NBS SRM 309 glass microspheres. Sputtering with highly focused liquid-metal ion beams appears to be a promising new research direction for particle analysis.

RAMAN MICROPROBE

Underlying Principle of Vibrational Spectroscopy

Raman scattering and infrared absorption result from vibrational and rotational processes that occur in molecules. Classical and quantum mechanical models describe the mechanism by which molecules inelastically scatter or absorb light. Classical models describe molecules as atoms (point masses) held together by chemical bonds (springs). The mass–spring systems have associated spring constants and resonant vibration frequencies. The fact that vibrational modes are distinct, exhibiting narrow bands in vibration spectra, is described quantum mechanically.

In Raman spectroscopy, the emitted radiation is inelastically scattered photons in the visible frequency range. If a beam of incident photons with energy, $h\nu$ (Planck's constant multipied by the frequency), is directed upon a material composed of molecules that have a vibrational mode at ν_m, a certain number of the scattered photons will lose energy in the vibrational process and have a new energy, $h(\nu - \nu_m)$. Additionally, if the incident photons encounter vibrating molecules with the same vibrational mode as above, some photons will gain energy $[h(\nu + \nu_m)]$. The emitted radiation that has frequencies below the incident radiation is said to be Stokes-shifted and that shifted above the incident radiation is anti-Stokes-shifted. Most of the photons are elastically scattered with no frequency shift at all (Rayleigh scattering). Photons that are inelastically scattered with the above characteristic frequency shifts are small in number, of the order of 1 part in 10^6.

Molecules that have the property of allowing an induced instantaneous dipole to be established by the incident light beam are usually strong Raman scattering species. The quantity normally used to characterize strongly Raman-active molecules are those with high molecular polarizability. Molecules with atomic members like iodine and sulfur exhibit Raman activity. Many factors that determine the scattering efficiency are discussed by Schrader (1986).

Instrumentation

A schematic diagram of a micro-Raman instrument is shown in Fig. 13-23. Monochromatic light is focused through a microscope objective on to the particle and the

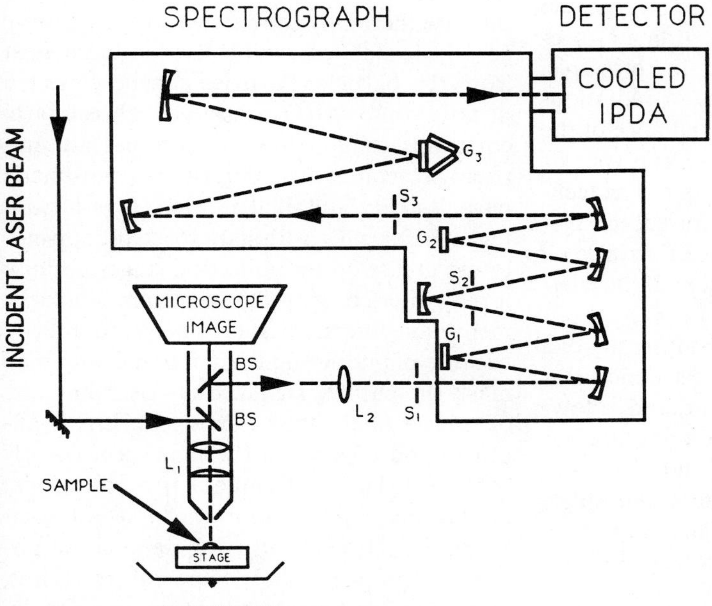

FIGURE 13-23. Schematic Diagram of a Micro-Raman Instrument Showing Excitation Radiation Coming from an Ar Ion Laser, Focused onto the Sample Through a Microscope Objective. The Beam can be Directed by Normal Microscope Translation. The Scattered Raman-Shifted Light from the Sample is Collected by the Microscope Objective and the Wavelength Separated in a Spectrograph. The Designated Components of the Spectrograph are: (G) Grating, (L) Lens, and (S) Slits. (Courtesy of E. Etz, NIST.)

scattered radiation is collected back through the microscope into a Raman spectrograph. The first stage of the spectrograph is a spectrometer to filter the Rayleigh-scattered light leaving only the Raman-shifted wavelengths. The latter stage of the spectrograph disperses the Raman-shifted light onto a multielement detector (e.g., diode array). The entire Raman spectrum is detected at once.

Capabilities

The capabilities and strengths of the micro-Raman technique are presented in some detail by Etz (1979) and Etz and Blaha (1980). The Raman microprobe offers direct molecular and structural analysis with picogram sensitivity on the single-particle level for particles 1 μm and larger in diameter. As in infrared spectroscopy, compound information may be readily obtained for crystalline materials, liquids and, to a lesser extent, for amorphous materials. Also, specific crystallographic phases can be determined by Raman spectroscopy. The technique is, in principle, nondestructive, but in the case of radiation-sensitive compounds, a high irradiance of the exciting, focused light can cause pyrolysis or photodecomposition of the sample. The technique is complementary to infrared spectroscopy, with sample fluorescence (intrinsic or from impurities) being a potential interference.

Schrader (1986) states that the intensity of a Raman shift is proportional to the concentration of the compound present and allows quantitation in many species. Quantitation is difficult from materials in the solid phase, but has been attempted for powders and single particles, employing calibration standards of closely matching materials (Grynpas, Etz, and Landis 1982).

It has been demonstrated (Etz and Blaha 1980) that reference spectra of bulk materials of the same composition match those of the same materials examined as discrete micrometer-sized particles. Therefore, the literature/library spectra of bulk materials can be utilized for compound identification of particles.

Delhaye and Dhamelincourt (1990) present a complete historical overview of the Raman microprobe, covering past and present technical developments. Other useful references are Rosasco, Etz, and Cassatt (1975), Delhaye and Dhamelincourt (1975), Rosasco (1980) and Adar (1988). Raman spectroscopy for chemical applications covering a wide range of topics, including sample preparation, biological samples, organic and inorganic compounds, is presented by Grasselli, Snavely, and Bulkin (1981).

Applications to Particle Analysis

Particle Collection and Preparation

The particles are held on a substrate having properties of low or noninterfering Raman bands and having relatively high thermal conductivity. Any material may be used as a substrate provided that it does not have interfering Raman-active bands. Substrates like Al_2O_3 are used to support the sample because they are thermally conductive and, thus, serve as a heat sink, removing laser-induced heat from the particle. Particles can be collected directly onto Al_2O_3 disks or glass/quartz coverslips using a low pressure cascade impactor to collect size segregated atmospheric particles (Etz and Blaha 1980). It is advantageous for micro-Raman spectroscopy not to use grease on the collection stages because it may interfere with the spectrum. Similarly, membrane filters are usually not used directly because of interfering Raman bands that may mask the particle signal. Analysis from glass or quartz fiber filters is possible. LiF is another good substrate for Raman spectroscopy because it has no Raman-active vibrational modes, but has one drawback—it is hygroscopic. Analysis is usually done at atmospheric pressure and temperature. Thus, many particle samples that would normally evaporate in a vacuum can be examined by micro-Raman spectroscopy.

Particle Analysis

Some of the applications to particle analysis include identification of the airborne particulates from stationary sources (Etz, Rosasco,

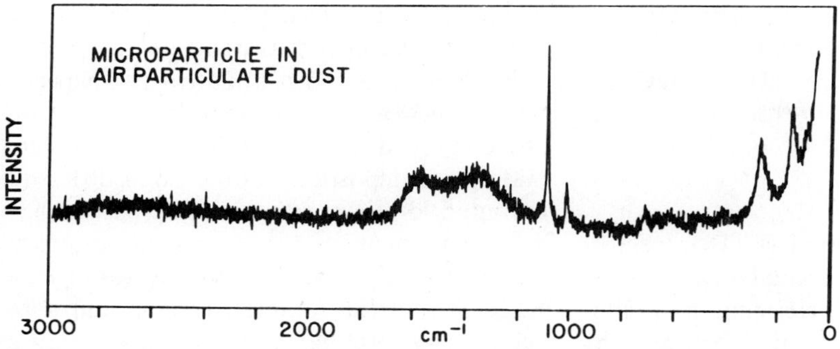

FIGURE 13-24. Micro-Raman Spectrum of an Atmospheric 10 μm Urban Dust Particle Composed Primarily of $CaCO_3$, as Evidenced by the Raman Active Bands at 1432, 1088, 714, 283, and 156 cm^{-1}. The Band at 1018 cm^{-1} Shows the Presence of $CaSO_4$ Which is Thought to Result from the Reaction of H_2SO_4 with $CaCO_3$ Found in the Particle. (Source: Blaha, Etz, and Cunningham 1979.)

and Heinrich 1978), analysis of urban dust samples and particles from oil-fired power plants (Etz, Rosasco, and Blaha 1978), graphitic carbon determination on urban particles (Blaha, Rosasco, and Etz 1978), sulfate/sulfuric acid determination (Etz, Rosasco, and Cunningham 1977), and pure reference polyaromatic hydrocarbons (Etz, Wise, and Heinrich 1979). The information that can be derived from micro-Raman on the single-particle basis is illustrated by the example of a St. Louis urban dust particle mounted on LiF substrate (Blaha, Etz, and Cunningham 1979). The Raman spectrum is shown in Fig. 13-24. The particle was identified as $CaCO_3$ from the evidence of the Raman bands at 1432, 1088, 714, 283, and 156 cm^{-1}. $CaSO_4$ was present and identified by the weak band at 1018 cm^{-1}. The broad hump-like features at approximately 1350 and 1600 cm^{-1} indicated the presence of graphitic carbon, due to carbonaceous material in or on the particle surface. The analyst proposed that $CaSO_4$ resulted from the action of H_2SO_4 with the $CaCO_3$.

INFRARED (IR) MICROSCOPY

Underlying Principle of IR Spectroscopy

IR spectroscopy is used to measure vibrational transitions in molecules occurring in the 10 to 10,000 cm^{-1} (1000–1 μm) range. The vibrational spectra are characteristic of functional groups in a given molecular compound (McMillan and Hofmeister 1988).

The model describing the vibrational processes involved in infrared spectroscopy has been discussed in the previous section. In IR spectroscopy, photons in the IR region of the spectrum are absorbed by molecular vibrational transitions. Molecules that have strong molecular dipoles are normally strong IR absorbers.

Instrumentation

The IR microscope contains an IR radiation source, optics to transfer the radiation, a spectrometer to separate light coming from or through the sample, and an IR detector. The IR microscope is an (optical) microscope that most commonly uses mid-range IR radiation, 4000–400 cm^{-1} (2.5–25 μm), to probe the sample. The incident radiation emanates from a blackbody radiation source, typically a silicon carbide "Globar" or filament. The beam-condensed incident radiation is directed at the sample using either lenses or reflecting optics. In modern IR microscopes, energy separation is achieved by an interferometer; the signal is Fourier-transformed (Katon, Sommer, and Lang 1990; McMillan and Hofmeister 1988). Absorption bands, resulting from specific IR-active molecular bonds, are usually seen in the IR spectrum as modulations in the signal intensity. One example of a typical IR detector used for mid-range wavelengths is the MCT (HgCdTe) liquid-nitrogen-cooled detector.

Capabilities

Fourier transform infrared (FTIR) spectroscopy, thus, provides an analytical method for identifying or characterizing molecular compounds. A good general reference is by Messerschmidt and Harthcock (1988). Many of the reported applications of FTIR are not examples of single-particle analysis of micrometer-sized particles, but they do show that important molecular compositional information on microsamples of material is obtainable. The technique has demonstrated its utility for a qualitative analysis of organic functional groups (for example, carbonyl or amide linkages), but direct chemical identification of complex mixtures found in atmospheric samples is often difficult. Dangler et al. (1987) point out that the technique offers fingerprint information for a source apportioning an aerosol population.

One limitation in IR microscopy, as in visible light microscopy, is that the spatial resolution is determined by the diffraction limit for the wavelength of radiation used. To obtain a spectrum of a single particle, the particle must be approximately 50 μm away from neighboring particles or objects; otherwise, there will be signal mixing due to diffraction from the particles' edges (Katon, Sommer, and Lang 1990). However, by aperturing in the image plane (a design common to most FTIR microscopes) in such a way as to define the sample area, spectra have been obtained of single objects as small as 5 μm (Messerschmidt and Reffner 1988).

Transmission or reflectance measurements are done with FTIR microscopes, depending upon the sample presentation. Transmission measurements are possible from particulate samples mounted on IR-transparent materials such as ZnSe. If the particles are mounted on metal surfaces, reflectance IR is often the desired approach. An alternative to these two methods requires transferring representative quantities of collected particle material to a cell for reflectance measurements.

FTIR offers a number of attractive aspects. Sample collection and sample preparation are combined in one step; the particulate material is deposited onto a ZnSe disk and mounted directly into the microscope for a measurement in the transmission mode (Dangler et al. 1987). Another advantage of FTIR is that the sample is not mounted in vacuum; so both high and low vapor pressure materials can be examined.

Given the breadth of IR reference spectral libraries and some of the new emerging multivariate analysis techniques being applied to FTIR, the technique provides chemical characterization for particulate samples, especially those containing organic compounds.

Applications to Particle Analysis

FTIR microscopy is a relatively new technique and, thus, only a few applications pertaining to particles have been published. The application of FTIR to airborne particle material from vehicle emissions using a low-pressure impactor for particle collection has been reported (Dangler et al. 1987; Pickle, Allen, and Pratsinis 1990; Brown et al. 1990). The authors report classes of organic and inorganic molecular constituents making up the sample. The applications of IR spectroscopy to particles have been reviewed by Allen and Palen (1989).

COMPLEMENTARY CAPABILITIES OF MICROANALYTICAL INSTRUMENTATION

There is synergism that often results from using multiple-instrument analysis; usually more information about a sample is gained by analysis with several instruments than by any single instrumental approach. The physical properties such as particle size and morphology (obtained by LM and electron microscopy characterization) are often augmented by information about the chemical composition. A complete chemical characterization may provide elemental composition (in terms of major, minor, and trace quantities in addition to isotopic abundance) and the chemical compounds that may be present. For example, elemental analysis from the electron instruments is complemented, many times, by isotopic abundance and trace analysis (mass spectrometry) and molecular composition (micro-Raman, FTIR, SIMS, and LMMS). SIMS and LMMS can further complement the elemental composition by extending the atomic number range to light elements (H–F).

A documented example of teaming two instruments together was given by Steel et al. (1984), where the capabilities of the AEM and LMMS were brought to bear on an analysis. In this work, the AEM was used to image and determine particle morphology, as well as major and minor elemental composition of submicrometer particles of zircon on a thin-film carbon substrate. AEM provided the location on the carbon substrate of the submicrometer particles for subsequent analysis by LMMS that in turn yielded trace and low-*Z* elemental composition along with isotopic abundances. The authors believed that this two-instrument approach provided a comprehensive characterization of small particles down to 0.1 μm in size.

Experimental designs applying different instruments in series or parallel are dependent upon several factors. Obviously, nondestructive analysis techniques should be attempted before analysis by methods that destroy some or all of the particles in the process of analysis (as is the case of mass spectrometry). This is a primary concern for samples with very limited numbers of particles or in the case of analysis on a single particle by several techniques.

Among the nondestructive techniques, the size of the particles in the sample dictates whether LM, SEM, or AEM can be used to obtain morphological, size, color, crystallographic properties and surface features. LM often provides the first information about particles in a sample and this information is often used to determine the course of subsequent analysis. The morphology, phase, and crystallographic nature of the particles can be an immediate clue as to which analysis technique should be pursued. A simple example is that droplets usually will not survive the vacuum environments common to many instruments, but could possibly be analyzed by one of the spectroscopies applicable at atmospheric pressures. Particles above a nominal 0.5 μm diameter can be viewed by LM and SEM, while particles smaller than 0.1 μm are best examined in the AEM. There are also considerations with regard to conductive coatings that are usually applied for the electron microscopes, the amount of total material in the sample, the composition of the sample and the information desired. For example, a sample can first be examined by light microscopy to yield size, shape, and various optical properties in the > 1 μm size range. Then, if major elemental composition is required, SEM with EDS can be done for analysis and an enlarged morphology can be imaged. Micro-Raman and FTIR provide molecular compositions by relating the vibrational spectrum to chemical bond types or, in the case of organic molecules, functional groups. The information obtained from mass spectrometry augments the vibrational bond information by often providing the molecular weight (through molecular ion identification) and additional chemical/structural information through fragment ions also present in the mass spectrum. Trace elements at concentrations down to approximately 10 ppm are often found by LMMS or SIMS in addition to the isotopic abundance. Trace elements and isotopic information may provide clues for determining particle origin.

Microanalysis utilizing multiple instruments provides a means for a comprehensive physical and chemical characterization of collected particles, which is often relevant for the aerosol, environmental, and atmospheric scientist.

Acknowledgments

The authors acknowledge the help of E. Etz, S. Hoeft, D. Newbury, E. Steel, and G. Gillen for valuable contributions to the chapter.

References

Adar, F. 1988. Developments of the Raman microprobe—Instrument and applications. *Microchem J.* 38:50–79.

Allen, D. T. and E. Palen. 1989. Recent advances in aerosol analysis by infrared spectroscopy. *J. Aerosol Sci.* 20(4):441–55.

Armstrong, J. T. 1978. Methods of quantitative analysis of individual microparticles using electron instruments. *Scanning Electron Microscopy. I*:455–67.

Armstrong, J. T. and P. R. Buseck. 1975. Methods of quantitative analysis of individual microparticles using electron microprobe: *Theoretical. Anal. Chem.* 47:2178–92.

Armstrong, J. T. and P. R. Buseck. 1977. Development of a characteristic fluorescence correction for thin films and particles. In *Proc. 12th Ann. Conf. Microbeam Anal. Soc.*, pp. 42A–42F. Boston, MA.

Asbestos International Association (AIA). 1979. Reference method for determination of airborne asbestos fibre concentration by light microscopy. Recommended Technical Method No. 1 (RTM 1). AIA Health and Safety Publication. London: AIA.

Baron, P. A. and G. C. Pickford. 1986. An asbestos sampling filter clearing procedure. *Appl. Ind. Hyg.* (1)4:169–71.

Behrisch, R. 1983. Introduction and overview. In *Sputtering by Particle Bombardment II. Topics in Applied Physics, Vol. 52*, ed. R. Behrisch. pp. 1–8. New York: Springer.

Bennett, J. and D. Simons. 1991. Relative sensitivity factors and useful yields for microfocused Ga ion beam and TOF-SIMS system. *J. Amer. Vac. Soc.* A9(3):1379–84.

Blaha, J. J., E. S. Etz, and W. C. Cunningham. 1979. Molecular analysis of microscopic samples with a Raman microprobe: Applications to particle characterization. In *Scanning Electron Microscopy/1979/I*, pp. 93–102. AMF O'Hara, IL: SEM, Inc.

Blaha, J. J., G. J. Rosaco, and E. S. Etz. 1978. Raman microprobe characterization of residual carbonaceous materials associated with urban airborne particulates. *Appl. Spectrosc.* 32(3):292–97.

Brown, S., M. C. Dangler, S. R. Burke, S. V. Hering, and D. T. Allen. 1990. Direct Fourier transform infrared analysis of size-segregated aerosols: Results from the carbonaceous species methods intercomparison study. *Aerosol Sci. Technol.* 12:172–81.

Bruynseels, F., H. Storm, R. Van Grieken, and L. V. Auwera. 1988. Characterization of North Sea aerosols by individual particle analysis. *Atmos. Environ.* 22(1):2593–602.

Carter, J., D. Taylor, and P. A. Baron. 1989. *NIOSH Manual of Analytical Methods.* Fibers, Method 7400 revision No. 3: 5/15/89. Cincinnati OH: DHHS/NIOSH.

Castaing, R. and A. Guinier. 1950. In *Proc. 1st Internat. Conf. on Electron Microscopy*, Delft, 1949, p. 60.

Chamot, E. M. and C. W. Mason. 1940. In *Handbook of Chemical Microscopy, Vol. 2.* New York: Wiley.

Chamot, E. M. and C. W. Mason. 1958. In *Handbook of Chemical Microscopy, Vol. 2.* New York: Wiley.

Cleary, T. G., G. W. Mulholland, L. K. Ives, R. A. Fletcher, and J. W. Gentry. 1992. Ultrafine combustion aerosol generator. *Aerosol Sci. Technol.* 16(3):166–70.

Cliff, G. and G. W. Lorimer. 1975. The quantitative analysis of thin specimens. *J. Microsc.* 103:203–7.

Cowley, J. 1979. Principles of image formation. In *Introduction to Analytical Electron Microscopy*, eds. J. J. Hren, J. I. Goldstein, and D. C. Joy, Chap. 1, pp. 83–117. New York: Plenum.

Currie, L. A., R. A. Fletcher, and G. A. Klouda. 1987. On the identification of carbonaceous aerosol via ^{14}C accelerator mass spectrometry and laser microprobe mass spectrometry. *Nucl. Instrum. Methods* B29:346–54.

Currie, L. A., R. A. Fletcher, and G. A. Klouda. 1989. Source apportionment of individual carbonaceous particles using 14C and laser microprobe mass spectrometry. *Aerosol Sci. Technol.* 10(2):370–78.

Dangler, M., S. Burke, S. V. Hering, and D. T. Allen. 1987. A direct FTIR method for identifying functional groups, in size segregated atmospheric aerosols. *Atmos. Environ.* 21(4):1001–4.

Delhaye, M. and P. Dhamelincourt. 1990. A perspective of the historical developments and future trends in Raman microprobe spectroscopy. In *Microbeam Analysis—1990*, eds. J. R. Michael and P. Ingram, pp. 220–27. San Francisco: San Francisco Press.

Delhaye, M. and P. Dhamelincourt. 1975. Raman microprobe and microscope with laser excitation. *J. Raman Spectrosc.* 3: 33–43.

Denoyer, E., R. Van Grieken, F. Adams, and D. F. S. Natusch. 1982. Laser microprobe mass spectrometry 1: Basic principles and performance characteristics. *Anal. Chem.* 54(1):26A–41A.

Denoyer, E., D. F. S. Natusch, P. Surkyn, and F. Adams. 1983. Laser microprobe mass analysis (LAMMA) as a tool for particle characterization: A study of coal fly ash. *Environ. Sci. Technol.* 17:457–62.

De Waele, J. K. E. and F. C. Adams. 1986. Laser-microprobe mass analysis of fibrous dusts. In *Physical and Chemical Characterization of Individual Airborne Particles*, ed. K. R. Spurny, Chap. 15, pp. 271–97. New York: Wiley.

Dovichi, N. J. and D. S. Burgi. 1987. Photothermal microscope. In *Microbeam Analysis—1987*, ed. R. H. Geiss, pp. 155–57. San Francisco: San Francisco Press.

Duncumb, P. and S. J. B. Reed. 1968. The calculation of stopping power and backscatter effects in electron probe microanalysis. In *Quantitative Electron-Probe Microanalysis*, ed. K. F. J. Heinrich, pp. 133–54. NBS Special Publication 298. Washington, DC: US Department of Commerce/National Institute of Standards and Technology.

Egerton, R. F. 1986, *Electron Energy Loss Spectroscopy in the Electron Microscope.* New York: Plenum.

Etz, E. S. 1979. Raman microprobe analysis: Principles and applications. In *Scanning Electron Microscopy Vol. 1*, pp. 67–92. AMF O'Hare, IL: SEM, Inc.

Etz, E. S. and J. Blaha. 1980. Scope and limitations of single particle analysis by Raman microprobe spectroscopy. In *Characterization of Particles*, ed. K. F. J. Heinrich, pp. 153–97. NBS Special Publication 553. Washington, DC: US Department of Commerce/National Institute of Standards and Technology.

Etz, E. S., G. J. Rosasco, and J. J. Blaha. 1978. Observation of the Raman effect from small, single particles: Its use in chemical identification of airborne particulates. In *Environmental Pollutants*, eds. T. Y. Toribara, J. R. Coleman, B. E. Dahneke, and I. Feldman. pp. 413–56. New York: Plenum.

Etz, E. S., G. J. Rosasco, and W. C. Cunningham. 1977. The chemical identification of airborne particles by laser Raman spectroscopy. In *Environmental Analysis*, pp. 295–340. New York, San Francisco, and London: Academic Press.

Etz, E. S., G. J. Rosasco, and K. F. J. Heinrich. 1978. Chemical analysis of stationary source particulate pollutants by micro-Raman spectroscopy. EPA Report, EPA-600/2-78-193. pp. 1–37. Research Triangle Park, NC: US EPA.

Etz, E., S. A. Wise, and K. F. J. Heinrich. 1979. Trace organic analysis: A new frontier in analytical chemistry. In NBS Special Publication 519, pp. 723–29. Washington, DC: US Department of Commerce/National Institute of Standards and Technology.

Fletcher, R. A. 1989. A new way to mount particulate material for laser microprobe mass analysis. *Anal. Chem.* 61:(8):914–17.

Fletcher, R. A. and L. A. Currie. 1987. Observations derived from the application of principal-component analysis to laser microprobe mass spectrometry. In *Microbeam Analysis—1987*, ed. R. H. Geiss, pp. 369–71. San Francisco: San Francisco Press.

Fletcher, R. A. and L. A. Currie. 1988. Pattern differences in laser microprobe mass spectra of negative ion carbon clusters. In *Microbeam Analysis—1988*, ed. D. E. Newbury, pp. 367–70. San Francisco: San Francisco Press.

Fletcher, R. A., E. S. Etz, and S. Hoeft. 1990. Complementary molecular information on phthalocyanine compounds derived from laser microprobe mass spectrometry and micro-Raman spectroscopy. In *Microbeam Analysis—1990*, eds. J. R. Michael and P. Ingram, pp. 89–92. San Francisco: San Francisco Press.

Friedrichs, K. H. 1986. Particle analysis by light microscopy. In *Physical and Chemical Characterization of Individual Airborne Particles*, ed. K. R. Spurny, pp. 161–72. New York: Wiley.

Gavrilovic, J. 1984. Surface analysis of small individual particles by secondary ion mass spectroscopy. In *Particle Characterization in Technology Vol. 1 Applications and Microanalysis*, ed. J. K. Beddow, pp. 3–19. Boca Raton, Fl: CRC Press.

Germani, M. S. and P. R. Buseck. 1991. Automated scanning electron microscopy for atmospheric particle analysis. *Anal. Chem.* 63:2232–37.

Goldstein, J. I. 1986. Principles of thin film X-ray microanalysis. In *Introduction to Analytical Electron Microscopy*, eds. J. J. Hren, J. I. Goldstein, and D. C. Joy, Chap. 3, pp. 83–117. New York: Plenum.

Goldstein, J. I., J. L. Costley, G. W. Lorimer, and S. J. B. Reed. 1977. Quantitative X-ray analysis in the electron microscope. In *Scanning Electron Microscopy, Vol. I*, ed. O. Johari, p. 315. Chicago: IITRI.

Goldstein, J. I., H. Yakowitz, D. E. Newbury, J. W. Colby, and J. R. Coleman. 1975. *Practical Scanning Electron Microscopy*, p. 32. New York: Plenum.

Grasselli, J. G., M. K. Snavely, and B. J. Bulkin. 1981. *Chemical Applications of Raman Spectroscopy*. New York: Wiley.

Grasserbauer, M. 1978. Characterization of individual airborne particles by light microscopy, electron and ion probe microanalysis, and electron microscopy. In *Analysis of Airborne Particles by Physical Methods*, ed. H. Malissa, pp. 125–78. West Palm Beach, Fl.: CRC Press.

Grynpas, M. D., E. S. Etz, and W. J. Landis. 1982. Studies of calcified tissues by Raman microprobe analysis. In *Microbeam Analysis—1982*, ed. K. F. J. Heinrich, pp. 333–37. San Francisco: San Francisco Press.

Hall, T. A. 1968. Some aspects of the microprobe analysis of biological specimens. In *Quantitative Electron-Probe Microanalysis*, ed. K. F. J. Heinrich. pp. 269–99. NBS Special Publication 298. Washington, DC: US Department of Commerce/National Institute of Standards and Technology.

Heinrich, K. F. J. (ed.) 1980. *Characterization of Particles*. NBS Special Publication 553. Washington, DC: US Department of Commerce/National Institute of Standards and Technology.

Heinrich, K. F. J. 1981. *Electron Beam X-ray Microanalysis*. New York: Van Nostrand Reinhold.

Henoc, J., K. F. J. Heinrich, and R. L. Myklebust. 1973. A rigorous correction procedure for quantitative electron probe microanalysis (COR 2). NBS Technical Note 769. US Department of Commerce/National Institute of Standards and Technology, Washington, DC.

Hercules, D. M., R. Day, K. Balasanmugam, T. A. Dang, and C. P. Li. 1982. Laser microprobe mass spectrometry 2. Applications to structural analysis. *Anal. Chem.* 54(2):208A–305A.

Hillenkamp, F. and M. Karas. 1989. In *Proceedings of the 37th ASMS Conference on Mass Spectrometry and Allied Topics. p.* 1168, Miami Beach, FL.

Joy, D. 1979. The basic principles of electron energy loss spectroscopy. In *Introduction to Analytical Electron Microscopy*, eds. J. J. Hren, J. I. Goldstein, and D. C. Joy, Chap. 3, pp. 83–117. New York: Plenum.

Katon, J. E., A. J. Sommer, and P. L. Lang. 1990. Infrared microspectroscopy. *Appl. Spectros. Reviews* 25(3&4):173–211.

Kaufmann, R. 1986. Laser-microprobe mass spectroscopy (LAMMA) of particulate matter. In *Physical and Chemical Characterization of Individual Airborne Particles*, ed. K. R. Spurny, Chap. 13, pp. 226–50. New York: Wiley.

Kaufman, L. and P. J. Rousseeuw. 1990. *Finding Groups in Data.* New York: Coiley.

Klaus, N. 1984. The effect of the substrate material on SIMS analysis of aerosols. *Sci. Total Environ.* 35:1–12.

Klaus, N. 1985. SIMS analysis of motor vehicle flue gas aerosols. *Sci. Total Environ.* 44:81–87.

Klaus, N. 1986. Aerosol analysis by secondary-ion mass-spectrometry. In *Physical and Chemical Characterization of Individual Airborne Particles*, ed. K. R. Spurny. Chap. 17, pp. 331–39. New York: Wiley.

Kolaitis, L. N., F. J. Bruynseels, R. E. Van Grieken, and M. O. Andreae. 1989. Determination of methanesulfonic acid and non-sea-salt in single marine aerosol particles. *Environ. Sci. Technol.* 23:236–40.

Lee, R. J. and J. F. Kelly. 1980. Applications of SEM-based automatic image analysis. In *Microbeam Analysis—1980*, ed. D. B. Wittry, pp. 13–16. San Franscisco: San Francisco Press.

Le Guen, J. M. M. and S. Galvin. 1981. Clearing and mounting techniques for the evaluation of asbestos fibres by membrane filter method. *Ann. Occup. Hyg.* 24(3):273–80.

Linder, B. and U. Seydel. 1989. Pattern recognition as a complementary tool for the evaluation of complex LAMMS data. In *Microbeam Analysis—1989*, ed. P. E. Russell, pp. 286–92. San Francisco: San Francisco Press.

Lindner, M., D. A. Leich, R. J. Borg, G. P. Russ, J. M. Bazan, D. S. Simons, and A. R. Date. 1986. Direct laboratory determination of the ^{187}Re half-life. *Nature* 320:246–48.

Linton, R. W., A. Loh, D. F. S. Natusch, C. A. Evans, and P. Williams. 1976. Surface predominance of trace elements in airborne particles. *Science* 191:852–54.

Mauney, T. 1984. Instrumental effects in LAMMA, ion kinetic energy distributions, and analysis of soot particles. Ph.D. Thesis, Colorado State University, Fort Collins, Colorado, pp. 155–69.

McCrone, W. C. and J. G. Delly. 1973. *The Particle Atlas*, Vols. 1–4, 2nd edn.. Ann Arbor, MI: Ann Arbor Science.

McMillan, F. F. and A. M. Hofmeister. 1988. Infrared and Raman Spectroscopy. In *Reviews in Mineralogy, Vol. 18 Spectroscopic Methods in Mineralogy and Geology*, ed. F. C. Hawthorne, pp. 99–159. Chelsea, MI: BookCrafters.

Messerschmidt, R. and M. Harthcock (eds.) 1988. *Infrared Microspectroscopy: Theory and Applications.* New York: Marcel Dekker.

Messerschmidt, R. and J. A. Reffner. 1988. FT-IR microscopy of biological samples: A new technique for probing cells. *Microbeam Analysis—1988*, ed. D. E. Newbury, pp. 215–18. San Francisco: San Francisco Press.

Musselman, I. H., D. S. Simons, and R. W. Linton. 1988. Effects of sample geometry on interelement quantitation in laser microprobe mass spectrometry. In *Microbeam Analysis—1988*, ed. D. E. Newbury, pp. 356–66. San Francisco: San Francisco Press.

Myklebust, R. L., C. E. Fiori, and K. F. J. Heinrich. 1979. Frame C: A compact procedure for quantitative energy-dispersive electron probe X-ray analysis. NBS Technical Note 1106. pp. 1–105. Washington, DC: US Department of Commerce/National Institute of Standards and Technology.

Newbury, D. E. 1980. Secondary ion mass spectrometry for the analysis of single particles. In *Characterization of Particles*, ed. K. F. J. Heinrich, pp. 139–52. NBS Special Publication 553. Washington, DC: US Department of Commerce/National Institute of Standards and Technology.

Newbury, D. E. 1990. Microanalysis to nanoanalysis: Measuring composition at high spatial resolution. *Nanotechnology* 1:103–30.

Newbury, D. E., D. C. Joy, P. Echlin, C. E. Fiori, and J. I. Goldstein. 1986. *Advanced Scanning Electron Microscopy and X-ray Microanalysis.* New York: Plenum.

Odom, R. W., F. Radicati di Brozolo, P. B. Harrington, and K. J. Voorhees. 1989. LAMMS: Pattern recognition and cluster ions. In *Microbeam Analysis—1989*, ed. P. E. Russell, pp. 283–85. San Francisco: San Francisco Press.

Owari, M., Satoh, H., N. Hutigami, M. Kudo, and Y. Nihei. 1989. Secondary ion mass spectrometry using a liquid metal ion source. *J. Trace and Microprobe Tech.* 7(1&2):59–85.

Pickle, T., D. T. Allen, and S. E. Pratsinis. 1990. The source and size distributions of aliphatic and carbonyl carbon in Los Angeles aerosol. *Atmos. Environ.* 24A(8):2221–28.

Reuter, W. 1972. The ionization function and its application to the electron probe analysis of thin films. In *Proc. 6th Internat. Conf. X-ray Optics and Microanalysis*, eds. G. Shinoda, K. Kohra, and T. Ichinokawa, pp. 121–30. Tokyo: Univ. Tokyo Press.

Ro, C. U., I. Musselman, and R. Linton. 1989. Molecular speciation of micro-particle: Application of pattern-recognition techniques to laser microprobe mass spectrometry. In *Microbeam Analysis—1989*, ed. P. E. Russell, pp. 293–96. San Francisco: San Francisco Press.

Rosasco, G. J. 1980. Raman microprobe spectroscopy. In *Advances in Infrared and Raman Spectroscopy, Vol. 7.* eds. R. J. H. Clark and R. E. Hester, pp. 223–82. London.

Rosasco, G. J., E. S. Etz, and W. A. Cassatt. 1975. The analysis of discrete fine particles by Raman spectroscopy. *Appl. Spectrosc.* 29(5):396–404.

Schrader, B. 1986. Micro Raman, fluorescence and scattering spectroscopy of single particle. In *Physical and Chemical Characterization of Individual Airborne Particles*, ed. K. R. Spurny, pp. 358–79. New York: Wiley.

Schroder, W. H. and G. L. Fain. 1984. Light-dependent calcium release from photoreceptors measured by laser micromass analysis. *Nature* 309(5965):268–70.

Seeley, B. K. 1952. Detection of micron and submicron chloride particles. *Anal. Chem.* 24(3):576–79.

Sheridan, P. and I. Musselman. 1985. Characterization of aircraft-collected particles present in arctic aerosol; Alaskan artic, spring 1983. *Atmos. Environ.* 19(12):2159–66.

Simons, D. S. 1983. Isotopic analysis with the laser microprobe mass analyzer. *Int. J. Mass Spectrom.* 55:15–30.

Simons, D. S. 1984. Laser microprobe mass spectrometry. In *Secondary Ion Mass Spectrometry SIMS IV*, eds. A. Benninghoven, J. Okano, R. Shimizu, and H. W. Werner, p. 158. Berlin: Springer.

Simons, D. S. 1988. Laser microprobe mass spectrometry: Description and selected applications. *Appl. Surf. Sci.* 31:103–17.

Small, J. A. 1981. Particle analysis in electron beam instruments. *Scanning Electron Microscopy*, 1:447–61.

Small, J. A., K. F. J. Heinrich, C. E. Fiori, R. L. Myklebust, D. E. Newbury, and M. F. Dilmore. 1978. The production and characterization of glass fibers and spheres for microanalysis. *Scanning Electron Microscopy* 1:445–54.

Small, J. A., K. F. J. Heinrich, D. E. Newbury, and R. L. Myklebust. 1979. Progress in the development of the peak-to-background method for the quantitative analysis of single particles with the electron probe. *Scanning Electron Microscopy* 2:807–16.

Small J. A., D. E. Newbury, and R. L. Myklebust. 1979. Analysis of particles and rough samples by FRAME P, a ZAF method incorporating peak-to-background measurements. *Microbeam Analysis—1979*, ed. D. E. Newbury, pp. 243–46. San Francisco: San Francisco Press.

Spurny, K. R. 1986. *Physical and Chemical Characterization of Individual Airborne Particles*, ed. K. R. Spurny. New York: Wiley.

Statham, P. J. and J. B. Pawley. 1978. A new method for particle X-ray micro-analysis based on peak to background measurements. *Scanning Electron Microscopy* 1:469–78.

Steel, E. B. 1980. Optical microscopy of particles. In *Characterization of Particles*, ed. K. F. J. Heinrich, pp. 5–11. NBS Special Publication 553. Washington, DC: US Department of Commerce/National Institute of Standards and Technology.

Steel, E. B., D. S. Simons, J. A. Small, and D. E. Newbury. 1984. Analysis of submicrometer particles by sequential AEM and LAMMA. In *Microbeam Analysis—1984*, eds. A. D. Romig, Jr. and J. I. Goldstein. pp. 27–29. San Francisco: San Francisco Press.

Storms, H. A., K. F. Brown, and J. D. Stein. 1977. Evaluation of cesium positive ion source for secondary ion mass spectrometry. *Anal. Chem.* 49(13):2023–30.

Surkyn, P., J. De Waele, and F. Adams. 1983. Laser microprobe mass analysis for source identification of air particulate matter. *Int. J. Environ. Anal. Chem.* 13:257–74.

Van Vaeck, L., J. Bennett, W. Lauwers, A. Vertes, and R. Gijbels. 1990. Laser microprobe mass spectrometry: Possibilities and limitations. *Mikrochim. Acta* III:283–303.

White, E. W., P. J. Denny, and S. M. Irving. 1966. *The Electron Microprobe*, eds. T. D. McKinley, K. F. J. Heinrich, and D. B. Wittry, pp. 791–804. New York: Wiley.

Wieser, P. and R. Wurster. 1986. Application of laser-microprobe mass analysis to particle collections. In *Physical and Chemical Characterization of Individual Airborne Particles*, ed. K. R. Spurny, Chap. 14, pp. 251–70. New York: Wiley.

Wieser, P., R. Wurster, and U. Haas. 1981. Applications of LAMMA in aerosol research. *Fresenius Z. Anal. Chem.* 308:260–69.

Wieser, P., R. Wurster, and H. Seiler. 1980. Identification of airborne particles by laser induced mass spectroscopy. *Atmos. Environ.* 14:485–94.

Wright, F. W., P. W. Hodge, and C. G. Langway. 1963. Studies of particles for extraterrestrial origin: 1 Chemical analysis of 118 particles. *J. Geophys. Res.* 68:5575–87.

Yokozawa, H. T. Kikuchi, K. Furuya, S. Ando, and K. Hoshino. 1987. Characterization of coal fly-ash particles by laser microprobe mass spectrometry. *Anal. Chim. Acta* 195:73–80.

14

Dynamic Mass Measurement Techniques

Kenneth Williams

U.S. Bureau of Mines
Pittsburgh, PA, U.S.A.

Chuck Fairchild

Health, Safety, and Environment Division
Los Alamos National Laboratory
Los Alamos, NM, U.S.A.

and

Joseph Jaklevic

Lawrence Berkeley Laboratory
University of California,
Berkeley, CA, U.S.A.

INTRODUCTION

This chapter describes three techniques for aerosol mass measurement, each of which could be most appropriately described as a short-term batch process. Only one of them approaches the instantaneous measurements characteristic of optical or light-scattering techniques, but all of them can determine the mass of collected aerosol samples in intervals convenient for many applications. These three techniques are the beta gauge, piezoelectric crystal, and tapered-element oscillating microbalance (TEOM®). These techniques have been used to measure a variety of aerosols. Various applications of the measurement techniques are discussed or referenced in the text.

BETA GAUGE METHOD

Measurement Principles

The beta gauge method of mass determination depends upon the near exponential decrease in the number of beta particles transmitted through a thin sample as the areal density is increased. The beta particles are emitted as a continuum energy distribution by a radioisotope source and their intensity is measured with a suitable electron counter. The method has the advantages of instrumental simplicity and ease of automation for large-scale applications. The dynamic range of sensitivity is well matched to the mass range normally of interest in aerosol monitoring in which thin membrane filters are the substrates. However, a detailed understanding of the parameters which affect the measurements is necessary in order to ensure optimal instrumental implementation and correct interpretation of results.

Figure 14-1 is a schematic diagram of a basic beta gauge instrument consisting of the radioactive source, detector, and sample. A measurement consists of determining the total flux in a continuous beta particle spectrum emitted by the radioisotopic source and transmitted through the sample. Under the proper experimental conditions, the transmitted flux (I) is related to the sample mass through the relationship (Evans 1955)

$$I = I_0 e^{-\mu x} \tag{14-1}$$

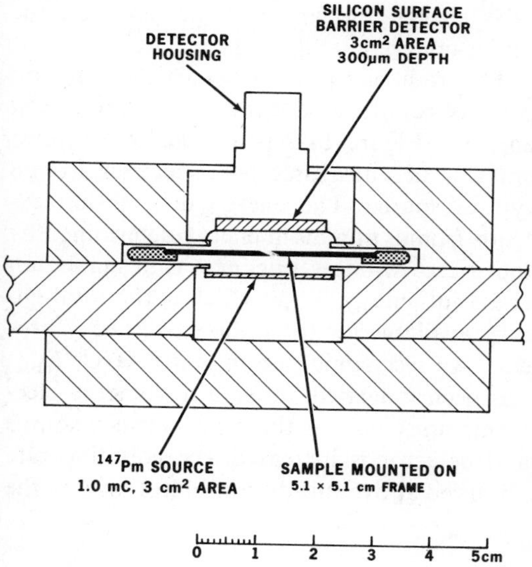

FIGURE 14-1. Schematic Diagram of a Beta-Gauge Suitable for Measuring Thin Aerosol Samples.

where I_0 is the incident flux, μ is the mass absorption coefficient for beta absorption (cm^2/g), and x is the mass thickness of the sample (g/cm^2). The mass absorption coefficient is normally determined through a calibration procedure involving the measurement of a series of known standards which bracket the mass range of interest (Jaklevic et al. 1981). The incident flux, I_0, can either be derived during the same calibration procedure or, if the interval between successive measurements is short, the value of I_0 can be made to cancel by calculating the ratio between transmitted fluxes measured with and without the particle deposit. The latter case applies for certain beta gauge designs where continuous particle deposition is monitored (Macias and Husar 1970).

Instrument Design

The optimal choice of source and detector depends upon many factors. The radioisotope source must have beta particle emission as the dominant mode of decay and exhibit half-life sufficient for the large decay corrections or frequent source replacements to be unnecessary. The source strength should provide adequate precision in counting statistics within the limitations of the detector rate-handling capabilities. Finally, as discussed in more detail below, the energy of the beta spectrum must be chosen to produce a mass absorption coefficient matched to the range of thicknesses to be measured. Table 14-1 lists a selection of radioisotopes appropriate for aerosol beta gauge applications together with relevant parameters for each (Lilienfeld 1975).

TABLE 14-1 Commonly Available Sources Suitable for Beta Attenuation Measurements

Isotope	Half-Life (years)	E_{max} (MeV)	Range (mg/cm^2) in Carbon at E_{max}	Range (mg/cm^2) in Carbon at $E_{aver} = 0.4E_{max}$
^{64}Ni	92	0.067	7.7	1.6
^{14}C	5730	0.156	32	6.6
^{147}Pm	2.62	0.225	60	13
^{85}Kr	10.76	0.67	290	77
^{36}Cl	3.1×10^4	0.712	320	84
^{204}Tl	3.8	0.765	340	94

The detector must be sensitive to the beta particles (i.e., electrons) in the energy range of interest and capable of operation at a counting rate sufficient to perform measurements in the required time interval. Beta gauges have been operated using a variety of detector types, including ionization chambers (Husar 1974), Geiger–Muller tubes (Gleason, Taylor, and Tabern 1951; Lilienfeld 1975; Klein, Ranty, and Sowa 1984), scintillation spectrometers (Jaklevic, Madden, and Wiegand 1983), and semiconductor diode detectors (Macias and Husar 1970; Jaklevic et al. 1981). Of equal importance is the choice of associated pulse processing electronics. Depending upon the type of detector used, it may be necessary to employ a preamplifier, amplifier, and pulse amplitude discriminator. Because of the level of precision required in many measurements, these analogue signal processing components must be selected and tested for long-term stability and reliability. For those laboratory applications where extreme limits of precision and accuracy are the objective, temperature stabilization of the operating environment is recommended. Most recent beta gauge systems employ semiconductor diode detectors and solid state pulse processing electronics because of their simplicity and stability of operation. The following discussion will be oriented toward this type of system although most of the comments also apply to other types of detectors.

The source–detector geometry must be maintained in a stable mechanical configuration to minimize spurious counting-rate variations. Also, it is important that the spacing be as close as possible in order that changes in atmospheric density within the gap are not interpreted as mass variations in the sample (Courtney, Shaw, and Dzubay 1982). The lower limit to the spacing is normally determined by the thickness of sample holders and the associated handling mechanisms used with automated instrumentation.

Theoretical Considerations

Studies have shown that the functional dependence of beta particle transmission expressed in Eq. 14-1 is not valid for precise experimental measurements (Jaklevic et al. 1981; Macias and Husar 1970; Heintzenberg and Winkler 1984). This result is not unexpected since the exponential behavior is not a reflection of fundamental mechanisms associated with beta particle attenuation in matter. Electrons with kinetic energies less than 1 MeV lose energy primarily through elastic collisions with atomic electrons present in the sample. As a consequence, an electron with a well-defined initial energy distribution will slow down through a series of discrete energy losses as it traverses the sample. An incident electron beam with a well-defined initial energy and direction will experience a gradual decrease in the average energy, accompanied by a spreading in the distribution of the beta ray's energy and angle of incidence on the target material.

The radioisotopic beta particle emission process results in a continuum of electron energies. Figure 14-2 is an idealized representation of a measured beta spectrum from a typical source. The energy distribution extends from a minimum energy determined by the source window thickness to a maximum endpoint energy, E_{max}, which is the total energy available for the radioactive decay process. An electronic discriminator level, E_{disc}, has been indicated above a low-energy electronic noise tail. As the mass between source and detector is increased, the counting rate observed above this discriminator level in the

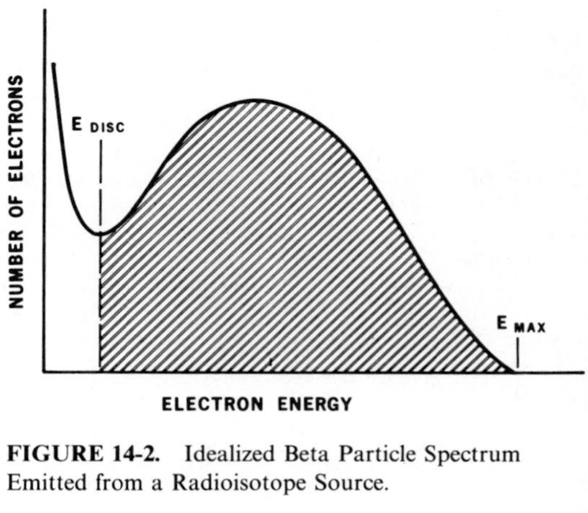

FIGURE 14-2. Idealized Beta Particle Spectrum Emitted from a Radioisotope Source.

beta gauge detector represents those electrons in the original continuum spectrum that have not been totally stopped in the sample and are still incident on the detector. This rate reflects a complex energy loss process which depends on several variables, including the average energy of the electrons in the beta spectrum and the amount of material traversed. When averaged over all these effects, the observed dependence of the counting rate on the thickness traversed is approximately exponential. Repeated measurements in a carefully defined experimental geometry using aluminum absorbers have established an approximate relationship between the mass attenuation coefficient in Eq. 14-1 and the beta spectrum endpoint energy (Gleason, Taylor, and Tabern 1951):

$$\mu\ (\mathrm{cm^2/g}) = 0.017 E_{\max}^{-1.43} \tag{14-2}$$

However, there is a considerable variation among various investigators regarding the values of the coefficients reflecting the empirical nature of these parameters (Sem and Borgos 1975).

Since the beta particle energy-loss process involves scattering from atomic electrons rather than nuclei, there is the possibility of an additional dependence of the absorption coefficient on the average atomic number of the samples. Various authors have studied this effect and an empirical relationship has been derived (Klein, Ranty, and Sowa 1984):

$$\mu\ (\mathrm{cm^2/g}) = 0.016 (Z/A)^{4/3} E_{\max}^{-1.37} \tag{14-3}$$

where Z is the atomic number and A is the atomic weight. However, the validity of this relationship seems to depend on the specific geometry of the beta gauge. A less dramatic dependence of attenuation on Z/A has been observed by Jaklevic et al. (1981) and attributed to the relative importance of the particle angular distribution in the particular source–detector arrangement employed in that study. Regardless of the details of functional dependence, it should be noted that variations associated with this effect are normally tolerable since the range of Z/A is small for all elements, with the exception of hydrogen.

The practical consequences of these theoretical observations are that, although estimates of beta gauge sensitivity can be obtained using generalized expressions, precise mass measurements require that each specific instrument is calibrated using known gravimetric standards. It is also necessary to limit the dynamic range over which one attempts to apply a given calibration using a strictly exponential approximation. To the extent that the deposited mass is normally a small fraction of the tare weight of the filter medium, this is not a severe limitation in aerosol applications.

A more complete understanding of precision and accuracy requires a detailed error analysis. If one assumes that the value of μ has been carefully determined and that the counting interval is known with complete certainty, then the precision of the mass measurement is determined by experimental variations in the determination of I_0 and I. Using Eq. 14-1, the root mean square error, $\sigma^2(x)$, in the calculated concentration, x, can be derived as (Cooper 1975)

$$\sigma^2(x) = \frac{1}{\mu^2}\left[\frac{\sigma^2(I_0)}{I_0^2} + \frac{\sigma^2(I)}{I^2}\right] \tag{14-4}$$

where I_0 and I are the incident and transmitted fluxes integrated for a fixed time interval. For the case where the errors associated with I and I_0 are the result of Poisson counting statistics only, i.e., $\sigma(I) = (I)^{1/2}$, and if the counting intervals of the two measurements are equal, the precision $\sigma(x)$ for a difference measurement varies as the inverse of the mass absorption coefficient and the inverse square root of measurement interval, as might be expected. The inverse dependence of precision on mass absorption coefficient supports the intuitive observation that lower-energy beta spectra will provide a more sensitive indicator of small mass changes in the sample. On the other hand, if the energy is too low, the exponential approximation is no longer valid since an increasing fraction of beta particles is totally stopped in the sample. Referring to

Table 14-1, a useful rule of thumb is to select a beta spectral average energy corresponding to a range which is several times the maximum thickness to be measured. For this reason, most beta gauges designed for aerosol monitoring employ either ^{14}C or ^{147}Pm as reasonable choices for use with substrates in the range 10–100 mg/cm^2 and deposits in the range 20–500 μg/cm^2 (Klein, Ranty, and Sowa 1984).

The derivation of Eq. 14-4 assumed that the value of the mass absorption coefficient was known from a previous calibration procedure. However, the value of the mass absorption coefficient is normally calculated from transmission measurements performed on a series of mass standards which bracket the anticipated range of operation of the instrument. A more complete error analysis including uncertainties in the fitting procedure is discussed by Jaklevic et al. (1981). A general conclusion of this analysis is that, although the absolute accuracy of a mass measurement is affected by such calibration errors, the precision is dominated by the variability in the determinations of I and I_0. In principle, the precision of a given mass determination can be improved by increasing the counting interval to reduce the relative error associated with Poisson counting statistics. However, one eventually reaches a limit where the variability in the measured counting rate is dominated by other effects which are not easily controlled. Sources of systematic errors can include fluctuations in atmospheric density, changes in laboratory relative humidity, which can affect the substrate mass for hygroscopic media, instabilities in the mechanical design and variability in the placement of the sample in the instrument (Courtney, Shaw, and Dzubay 1982). A major source of instrumental instability is the result of long- and short-term drift in the detector and analogue pulse processing electronics. Although modern solid state circuits can have temperature coefficients approaching 1 part in 10^4 per degree Celsius, the demands placed on precision beta gauge measurements can approach this level even in temperature-controlled environments. Because of the difficulty in controlling such sources of errors, it should be a requirement in all beta gauge measurement protocols that recalibration of the instrument be performed before each series of measurements and that replicate samples be repeatedly analyzed at the same time as the unknowns in order to monitor instrumental stability.

Potential Biases

The near exponential behavior of the beta absorption process and the variations discussed above can result in several potential measurement artifacts which should be understood. Principal among these are particle size effects, substrate inhomogeneity, and atomic number dependence.

Particle size effects result from the fact that the beta gauge transmission measurements represent an average of the absorption experienced by an aggregate of particles deposited on the filter substrate. In the limit where this deposition is a homogeneous layer of small particles whose average diameter is much less than the layer thickness, the interpretation of the results in terms of exponential absorption by a uniform deposit is valid. On the other hand, one can imagine a deposit of equivalent mass but consisting of only a few, very massive particles. In an extreme manifestation of this latter case, there could exist total absorption within a given particle. The transmission measurement in this limiting case would then reflect the fractional area covered by the particles, and interpretation in terms of an exponential absorption by a uniform deposit would be invalid.

A detailed discussion of this problem for the general case of exponential absorption is given by Cooper (1976). A similar treatment using a simplified model applied to the case of aerosol particles is given by Jaklevic et al. (1981). Figure 14-3 is a plot of the discrepancy attributed to this phenomenon for the case of a 100 μg/cm^2 deposit of unit-density particles. The ratio of the mass as determined from beta gauge measurement to the true gravimetric mass is plotted as a function of the linear

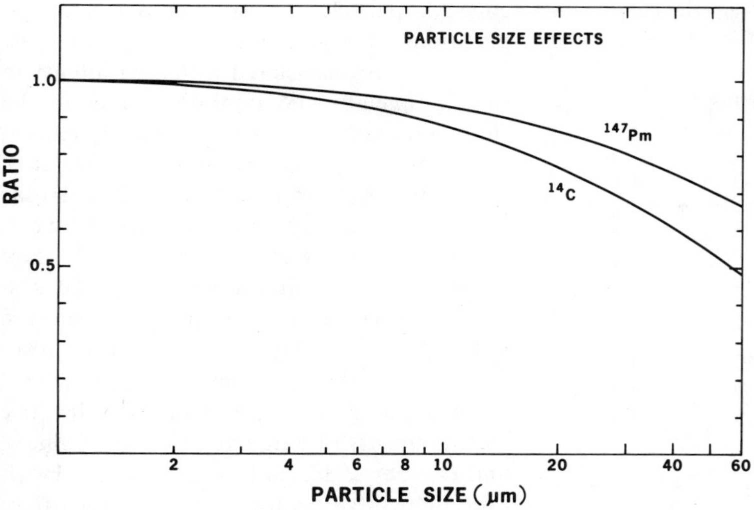

FIGURE 14-3. The Discrepancy in Beta Gauge Mass Measurements as a Function of Particle Size for the Case of Two Commonly Used Isotopes.

dimensions of particles. The discrepancy between the two increases for large particles at a rate which is greater for the source with the larger absorption coefficient.

These results indicate that one must either limit the size distribution to particles below 10 μm diameter or, if larger particles are to be analyzed, ensure that the average deposit thickness is sufficient for a statistically meaningful number of particles to be present. Similarly, for impaction samples, the deposit thickness must be uniform over the measurement area to a degree where the average of the exponential is equal to the exponential of the average.

An effect which can be explained using similar logic has to do with discrepancies caused by filter and source inhomogeneities. A microscopic examination of most membrane filter media shows that the substrate consists of a nonuniform distribution of fibers or flocculated material with relatively open spaces in between. Similarly, a radioisotope source is normally fabricated by methods which result in local inhomogeneities in the radioactivity across the face of the source. Although a point source can, in principle, alleviate this problem, there is a practical limit to the specific activity that can be concentrated in a small volume. Variations in the apparent mass of the substrate can result from random alignments between the respective inhomogeneities causing spurious high or low mass readings. The problem is exacerbated by the large mass of the substrate relative to the deposit. Since neither the source nor the substrate can be made to be perfectly spatially uniform, it is important to constrain the measurement protocol to position the filter in the instrument identically for both the initial and final weighings. This is most easily implemented in the case of large-scale automated systems and is necessary for mass measurements which aspire to achieve the limits of instrumental precision (Courtney, Shaw, and Dzubay 1982).

The atomic number dependence of the mass absorption coefficient requires that certain precautions be taken regarding the choice of calibration standards and in the interpretation of results from discrete pollution sources. Table 14-2 shows the Z/A values for a list of compounds commonly observed in ambient aerosol sampling. The mass absorption coefficients are calculated using both the Z/A dependence observed by Jaklevic et al. (1981) and those calculated from Eq. 14-3. The mass absorption coefficient for

TABLE 14-2 Effect of Atomic Number Dependence on the Measured Mass of Several Compounds

Compound	Z/A	μ (cm^2/mg)[a]	μ (cm^2/mg)[b]
$(NH_4)_2SO_4$	0.530	0.153	0.166
$NH_4H\ SO_4$	0.521	0.152	0.163
$CaSO_4 \cdot H_2O$	0.511	0.152	0.159
SiO_2	0.499	0.154	0.154
$CaCO_3$	0.500	0.154	0.154
Carbon	0.500	0.154	0.154
Fe_2O_3	0.476	0.163	0.144
NaCl	0.478	0.172	0.145
$PbSO_4$	0.429	0.193	0.126
$PbCl_2$	0.417	0.204	0.121
PbBrCl	0.415	0.206	0.120

a. From Jaklevic et al. (1981)
b. Z/A dependence calculated from Eq. 14-3. Values normalized to 0.154 for carbon to account for instrumental differences

a mixture of compounds would be the weighted sum of the respective coefficients. It is obvious that an inappropriate choice of calibration foils can affect the accuracy of the measurements if not corrected for in the data analysis. Similarly, measurements of a set of samples in which the relative contribution of diverging Z/A compounds varies widely will need special consideration in the interpretation. Although a complete correction requires that the sample composition be known, some estimate of the probable error can be obtained by observing the range of values of μ for the compounds listed in Table 14-2 and incorporating an error analysis based on Eq. 14-1. It should be noted that the errors associated with Z/A variations affect the accuracy of the measurements but not the precision and, as a consequence, have little effect on the lower limit of sensitivity of the beta gauge method.

Results and Applications

In the literature there exist descriptions of a number of beta gauge designs optimized for specific particle monitoring applications. In general, a distinction can be made between those made for continuous monitoring of discrete sources or ambient aerosols and those designed for precision laboratory analysis of collected particles. In on-line monitoring applications, it is difficult to control the systematic errors associated with a variability in the tare weight and nonuniformities in the substrate mass. Such systems usually exhibit a root mean square precision of the order of 25 μg/cm^2 (Macias and Husar 1970; Husar 1974). In laboratory-based systems where calibrations are frequently performed, tare weights are individually measured and systematic effects such as sample placement are reduced; it is possible to achieve precisions of 3 μg/cm^2 for individual mass measurements, corresponding to a precision of difference measurements of 5 μg/cm^2 (Courtney, Shaw, and Dzubay 1982; Jaklevic et al. 1981). However, there are a number of criteria other than precision and individual designs which need to be evaluated in terms of specific applications. These incude speed, convenience, cost, operating environment, and automated operation.

Beta gauge mass measurements have been incorporated into a number of studies. Loo et al. (1978) report the use of a beta gauge to

TABLE 14-3 List of Commercial Sources

Code	Address
MST	MST Measurement Systems, Inc., 327 Messner Dr., Wheeling, IL 60090, U.S.A. (Beta attenuation devices)
WAI	Wedding and Associates, Inc., P.O. Box 1756, Fort Collins, CO 80522, U.S.A. (Beta attenuation devices)
AII	Andersen Instruments Inc., 4801 Fulton Industrial Blvd., Atlanta, GA 30336, U.S.A. (Beta attenuation devices)
BCI	Berkeley Controls, Inc., 2825 Laguna Canyon Rd., Laguna Beach, CA 92652, U.S.A. (QCM cascade impactor)
CMI	California Measurements, Inc., 150 E. Montecito Ave., Sierra Madre, CA 91024, U.S.A. (QCM cascade impactor)
FTM	Femtometrics, 1721 Whittier Ave., Suite A, Costa Mesa, CA 92627, U.S.A. (SAW microbalance)
TSI	TSI, Inc., P.O. Box 64394, 500 Cardigan Road, St. Paul, MN 55164, U.S.A. (QCM aerosol mass monitor)

measure the mass of aerosols collected on cellulose ester filters as part of a large-scale sampling program in St. Louis. They report a precision of 5 μg/cm^2 for substrate masses of the order of 20 mg/cm^2. Stevens et al. (1980) have used beta gauge mass determinations for samples collected on polytetrafluoroethylene substrates with average areal densities of less than 600 μg/cm^2. Spengler and Thurston (1983) describe the use of beta gauge measurements in an extensive indoor air pollution study. Recent applications using continuous mass monitors have been described by Klein, Ranty, and Sowa (1984), and Heintzenberg and Winkler (1984). Table 14-3 lists the commercial sources of several beta attenuation instruments.

PIEZOELECTRIC CRYSTAL MEASUREMENT METHOD

Measurement Principles

Some crystalline materials, when mechanically stressed in compression or tension, respond by producing an electromotive force or voltage proportional to the stress along specific crystal planes. This phenomenon is the basis for piezoelectric strain gages. Likewise, the same material when subjected to an electric field across certain crystal planes responds by mechanically expanding or contracting. Thus, a rapidly alternating field or voltage produces physical oscillation or vibration in a piezoelectric crystal. The vibrational frequency depends on the orientation of the electric field with respect to the major axis of the crystal and, for certain modes of vibration, on the thickness and density of the crystal. Most important for their application in piezoelectric microbalances, the natural vibration frequency also depends on the mass of the crystal.

An increase of mass on sensitive planes, or areas, of a piezoelectric crystal causes the natural resonant frequency to decrease in direct proportion to the increased mass. The frequency stability at a constant mass and the sensitivity to the mass of the vibrating crystal provide its aerosol measurement capability. The sensitivity is proportional to the square of the vibrational frequency, f_0, of the crystal in its fundamental mode:

$$\Delta f = K_\alpha f_0^2 \frac{\Delta m}{A} \tag{14-5}$$

where K_α is a constant and $\Delta m/A$ is the area sensitivity as mass change per unit area producing a frequency change Δf. Sensitivities of $\sim 10^9$ Hz/g for optimized crystal geometries are typical, and a stability of ± 0.5 Hz at 10 MHz is attainable. Advantage has been taken of these properties of peizoelectric crystals to create a class of instruments sensitive to mass, or changes in mass, that are generally called quartz crystal microbalances (QCM). The piezoelectric sensor principle and its application to the QCM, as well as other types of instruments, are discussed in some depth by Ward and Buttry (1990). A review of the limitations of piezoelectric crystals in aerosol measurement is provided by Lundgren, Carter, and Daley (1976).

Piezoelectric quartz crystals used in most aerosol sampling QCMs are cut and polished on specific crystallographic planes called the AT planes. AT-cut crystals oscillate in a thickness-shear mode, that is, the oscillation occurs between two AT-cut surfaces, through the body of the crystal. In this mode, a single vibrational node is at the midplane between the two major surfaces of the thin, disk-shaped crystal. The driving electrodes are circular metallic coatings electroplated on each opposite major surface of the crystal. Oscillation occurs only in that portion of the crystal lying in the electric field between the electrode boundaries. The sensors then vibrate at a frequency proportional to the mass of the crystal and electrodes plus the mass of particles that adhere to the collecting metallic electrode surface. Particles attached to the surface of the crystal outside the electrode area have little effect on the vibrating portion of the crystal and are not detected.

Sampling Methods

A simple QCM aerosol sampler designed to measure total or respirable particle mass is

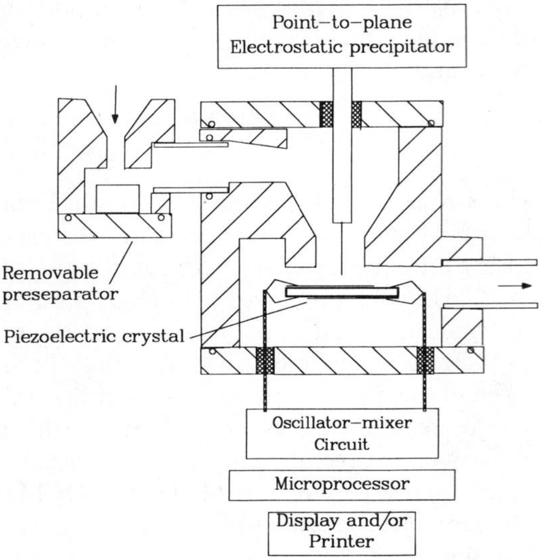

FIGURE 14-4. A Typical Sampling Arrangement for a QCM Aerosol Sampler. The Oscillator–Mixer Circuit Contains a Reference Crystal Identical to the Collection Crystal.

illustrated schematically in Fig. 14-4. The preselector at the inlet collects large particles by impaction. Respirable aerosol passes to the collection crystal, where the particles get deposited by electrostatic precipitation. The crystal (typically a 1 cm diameter by 1 mm thick disk) is suspended by posts and frames that also act as electrical conductors. Denuded air is exhausted by a pump. Both the impactor and the collection crystal are removable for cleaning, or a provision may be made for cleaning in situ.

QCM aerosol samplers may collect particles by electrostatic precipitation, inertial impaction, or other methods. In all methods, the particles are deposited onto the metal-coated quartz crystal surface. The clean crystal substrate vibrates at a natural harmonic frequency usually between 5 and 10 MHz. As material is deposited on its surface, the frequency of vibration of the crystal decreases in proportion to the mass of the deposit. A separate, matched crystal, which is protected from particle deposition but is in the same thermal and flow environment, is installed to provide a stable reference vibrational frequency. There is some evidence, however, that the reference crystal often is not in thermal equilibrium with the collection crystal (Sem and Tsurubayashi 1975). Signals produced by the two crystals are electronically mixed to provide a difference, or beat, frequency between the two. This beat frequency is an indicator of the mass of material collected. In most commercially available samplers either the beat frequency or the calculated mass can be displayed.

Potential Biases

Calibration

Initial calibration of a QCM aerosol sampler is normally performed by the manufacturer, who sets the display to correspond to a measured aerosol concentration for known changes in the frequency of vibration of the quartz crystal. After factory calibration, sample mass concentration can be calculated from the measured beat frequency, aerosol volume flow rate, mass sensitivity determined from the calibration, and sampling time. However, the instrument should be calibrated by the user against a well-characterized aerosol such as polystyrene latex (PSL) spheres or, if possible, against the expected challenge aerosols. Measurements can vary for different aerosols for several reasons.

Particle Size Effects

Both the mass collection and measurement efficiency may be affected by particle size. Collection efficiency is a function of the collection method, particle size distribution, and the substrate, or crystal surface. Collection methods such as impaction or electrostatic precipitation are discussed thoroughly elsewhere, but the collection efficiency of the crystal surfaces is of interest here. On a macroscale, the metallic electrode collecting surface is a smooth, hard surface, as are many bare impaction or other collection surfaces. Large, elastic particles may bounce, or may be reentrained from such surfaces after deposition. This is especially true in QCM impactors, where collection is due to inertial forces. An examination of the surfaces from a multistage impactor collection crystal has shown

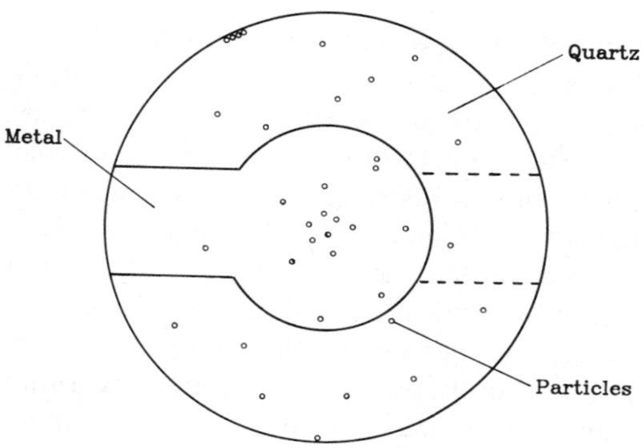

FIGURE 14-5. Typical Large, Spherical Particle Collection Pattern for Stages 1 and 2. Crystal Diameter is ~ 1 cm. (By Permission, Fairchild and Wheat 1984.)

that large, spherical particles may also migrate across the surface (Fig. 14-5, Fairchild and Wheat 1984). Bounce or migration may be eliminated or decreased by applying a soft or sticky coating to the surface. Because of the sensitivity of the piezoelectric impactor stage, the coating must be strictly controlled by following the manufacturer's instructions.

Adherence Effects

The detection of particles collected on even a coated surface may suffer another problem not observed in other samplers. Detection of the increased mass depends upon good coupling, or attachment, to the surface. Coupling is a function of the adherence of particles to the surface and can be a problem regardless of the collection technique. Simply put, the deposit mass must vibrate in unison with the collection crystal in order to be measured properly. Poor coupling generally occurs with thicker deposits. Often, its occurrence may be detected if the change in frequency is plotted as a function of time, while maintaining a uniform deposition rate. The frequency change becomes nonlinear with increase in deposit mass as the coupling decreases. Also, loss of sensitivity tends to increase with particle size above a few micrometers.

A related problem results from a variation in sensitivity of the electrode across its surface. As stated earlier, the deposit must rest on the metal electrode to be correctly sensed. Actually, the electrode itself may not be uniformly sensitive over its area because of crystal defects or electric field aberrations. This variation is corrected empirically in initial calibration but may be a source of small error if periodic recalibration is not done.

Overloading

For a relatively large collected mass, even if well attached, "saturation" of the crystal may cause severe nonlinearity. Excessive collected mass may cause a condition of gross overloading in which the vibration of the sensing crystal ceases altogether. The procedures to prevent overloading, such as periodic cleaning, are normally discussed in the instruction manuals.

Particle Losses

The QCM aerosol samplers are subject to particle losses to inlet and internal surfaces just as other aerosol samplers. Careful design can reduce these losses, but some loss is likely. Therefore, the losses should be characterized before use. Although little detailed data are available, the QCM cascade impactors presently available are particularly susceptible to wall losses. They collect a wide range of particle sizes, have a comparatively large ratio of interstage volume to flow rate (long airborne particle residence time), and have internal surface irregularities that present an opportunity for turbulence generation and particle deposition—all features conducive to wall loss.

Results and Applications

Industrial Hygiene Aerosol Sampling

At least one QCM instrument, called an aerosol mass monitor (Model 8510, TSI Inc., St. Paul, MN), is available to measure aerosols in inhalation hazard assessment. Either total dust or respirable dust concentrations common to many workplaces can be determined in a 24 s or 2 min sampling period, depending upon the expected concentration. For the measurement of respirable aerosols a single-stage impactor with a 50% effective cutoff aerodynamic diameter (d_{50}) of 3.5 μm is installed in the sampling airstream ahead of the collection crystal. Because the collection efficiency of this impactor does not match the characteristics of the ACGIH respirable dust curve (Lippmann 1978) closely, the mass monitor estimates the respirable mass concentration best for aerosols that are > 50% respirable.

In this instrument, particles are deposited on the crystal by electrostatic precipitation. A mechanism is incorporated to clean the collection crystal between samples without removing or disturbing the crystal. After the cleaning procedure, the beat frequency may be displayed to ensure that the crystal is clean and approximates its original vibrational frequency. The data reduction and the display modules are packaged together with the collection system and battery in a portable unit weighing 5 kg.

Fairchild, Tillery, and Ettinger (1980) evaluated the performance of an early version of this mass monitor against several aerosols. They reported that the mass monitor did not measure the air concentration of different aerosols equally well, and that the instrument measured the concentration of solid particles greater than 6 μm diameter poorly. It did provide a reliable measurement of small solid particle and liquid particle aerosols, particularly when multiple samples were averaged. For most industrial aerosols, the mass monitor measured the respirable concentration more accurately than the total aerosol concentration, presumably because of the better adherence and retention of the smaller particles by the sensing crystal.

The piezoelectric mass monitor, because of its mass sensitivity and short response time, allows a quick measurement of aerosol concentration, within the limitations described. It should be used primarily with respirable-size particles because it reportedly measures large particles poorly. The potential errors suggest that it be used as a survey instrument, or in cases where a number of samples can be averaged.

Near-Real-Time Cascade Impactors

Cascade impactors that operate with multiple sets of paired piezoelectric crystals—one set for each impaction stage—are available. These can be used in some environmental, industrial hygiene, clean room, and other field sampling applications, but are usually used in research applications. These impactors are inconvenient and somewhat complex for routine purposes, but they do provide a rapid indication of size distribution as well as mass concentration.

QCM impactors with ten stages have been available for a number of years (Model C-1000A, QCM Research, Laguna Beach, CA; PC-2, California Measurements, Inc., Sierra Madre, CA) and cascade impactors with fewer stages are now appearing on the market (Models MPS-3, MPS-4GL, California Measurements, Inc., Sierra Madre, CA). Figure 14-6 shows one ten-stage cascade impactor along with its separate data reduction and display module. The two modules are connected by cabling to transmit power and signals and, in some cases, to control the main valve admitting aerosol into the stages. When the valve is closed, only filtered air flows through the stages. An aerosol sample is admitted by opening the valve, usually for a measured length of time. After the valve is closed, clean air again flows into the instrument. Approximately 30 s are required for the sample to pass through all stages because of the low flow rate and large interstage volume. After a short delay, the operator may print the frequency change for each stage, the computed mass collected by each stage, and a histogram of the size distribution.

Calibration of the individual stages of this cascade impactor was performed with unit- and known density aerosols (Fairchild and Wheat 1984). They reported the stage d_{50} values to be somewhat different from the manufacturer's values. They also reported significant particle losses in the impactor, both for particles smaller than $d_a = 0.4$ μm and for particles larger than $d_a = 3$ μm.

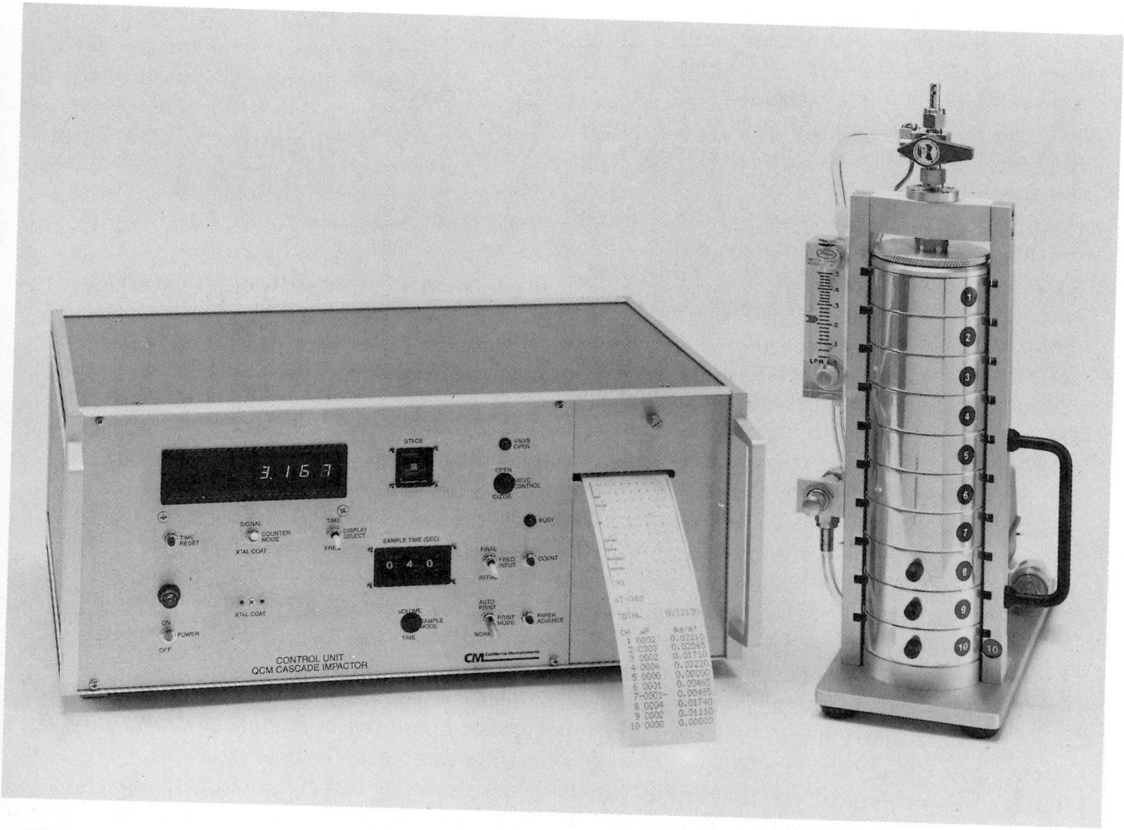

FIGURE 14-6. A QCM Cascade Impactor Showing the Impactor Stack on The Right and the Data Acquisition, Display, and Print Unit on the Left. (By Permission of Source CMI, Table 14-3.)

The cascade impaction devices have been used in atmospheric and environmental sampling applications (Rose et al. 1979; Wallace and Chuan 1977). For these applications the devices have been mounted inside aircraft and in aerodynamically shaped containers attached to aircraft wings. Karasek (1978) pointed out that piezoelectric crystal substrates are also suitable substrates in many cases for the examination of collected aerosols by light and electron microscopy.

Surface Acoustic Wave Microbalances

Another recently developed QCM aerosol instrument has a mass sensitivity much greater than samplers with AT-cut crystals. Its increase in sensitivity results from modifications to the quartz crystal and its mode of vibration. Instead of producing an electric field through the crystal thickness, the electric field is applied between electrodes on the same surface. Provided a separation of no more than several micrometers between electrodes can be achieved, this mode of excitation results in a vibrational frequency of up to 300 MHz. Theoretically, a mass sensitivity may be obtained that is three orders of magnitude greater than in the AT-cut crystal devices (Bowers and Chuan 1989). The vibrational mode along one surface is called the surface acoustic wave (SAW) mode.

Commonly, several driver electrodes are deposited in an interdigital pattern on the crystal at separations of $\sim 15\ \mu m$ between positive and negative branches. This distance produces standing-wave vibrational frequencies of about 100 MHz. Because its sensitivity is proportional to the vibrational frequency squared (Eq. 14-5), the increased sensitivity of a SAW device operating at 100 MHz over a device operating at 10 MHz is theoretically a factor of 100. Mass sensitivities in the range 10^{10}–10^{11} Hz/g appear to be attainable with the SAW devices.

As in other QCM aerosol samplers, an overloading of the active crystal is one limiting consideration for the SAW microbalances; at their oscillation frequency, saturation damping occurs at very small mass. Short aerosol collection time, very dilute air concentrations, and thin surface deposits are necessary to prevent damping the oscillations (overloading). Also, the temperature sensitivity of SAW devices is very high.

SAW microbalances are in an early stage of development and have been used in research applications only. They have been used primarily in stratospheric rockets to sample very dilute aerosols, and have been employed in the measurement of chemical vapor deposition coatings and air pollution chemical reactions (Bowers and Chuan 1989). In the last application, the crystal may be coated with a thin layer of a chemical, solid, or liquid, which then reacts with a selected gas to change the mass of material on the surface, enabling a chemist to follow the progress of the reaction. This method of sensitive, dynamic gravimetric analysis may be useful for the detection and, possibly, the identification of trace pollutants. At least one SAW microbalance (Model 200-1, Femtometrics, Costa Mesa, CA) is now available commercially (Table 14-3).

TAPERED-ELEMENT OSCILLATING MICROBALANCE METHOD

Measurement Principles

TEOM® devices are like piezoelectric crystal devices only in that mass is collected on a vibrating collection substrate, changing its frequency of oscillation. Figure 14-7 shows a typical arrangement for a TEOM® instrument. Instead of a peizoelectric crystal, the active element of any TEOM® system is a specially tapered hollow tube constructed of an elastic, glass-like material. The wide end of the tube is firmly mounted on a relatively massive base plate. The narrow end supports a replaceable collection medium such as a filter or impaction plate and is made to oscillate. Particle-laden gas streams are drawn through the collection medium, where particles are deposited. The filtered gas is then

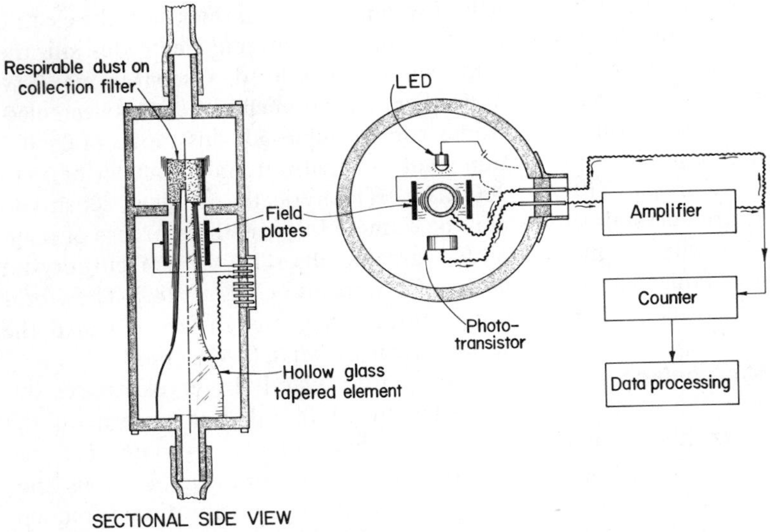

FIGURE 14-7. Typical Arrangement for TEOM® Dust Sensor.

drawn through the hollow tube, typically controlled by an automatic mass flow controller.

An electronic feedback system initiates and maintains the oscillation of the tapered element. In 1983, the U.S. Bureau of Mines (BOM) and the National Institute of Occupational Safety and Health (NIOSH) funded the development of a prototype TEOM® dust monitor for mining applications (Patashnick and Rupprecht 1983). In that particular device, a light-emitting diode (LED)–phototransistor pair aligned perpendicular to the plane of oscillation of the tapered element detects the frequency of oscillation. The light-blocking effect of the oscillating element positioned between the phototransistor and the LED modulates the output signal of the phototransistor, which is then amplified. Part of the amplified signal is applied to a conductive coating on the outside of the tapered element. In the presence of constant electric field plates, this signal provides sufficient force to keep the tapered element in oscillation. In other words, part of the amplified signal from the LED–phototransistor pair is used in an electrical feedback loop to overcome any amplitude damping of the tapered-element oscillation. The other part of the amplified signal from the LED–phototransistor pair is sent to a counter and data processing stage. Here, the oscillation frequency of the tapered element is calculated and stored in memory. The manufacturer has made several proprietary improvements to the feedback system since the early BOM/NIOSH prototype.

The equation that describes the behavior of the TEOM® system derives from the equations of motion for a simple harmonic oscillator:

$$\Delta m = K_0\left(\frac{1}{f_b^2} - \frac{1}{f_a^2}\right) \tag{14-6}$$

where Δm is the mass of the collected sample, f_b is the frequency of the oscillating element after sample collection, f_a is the frequency before sample collection, and K_0 is a constant (spring constant) unique to each tapered element. As the collection medium collects dust, the mass increases, thereby decreasing the frequency of oscillation. By measuring only the change in frequency, one can determine the gain in the mass of dust on the collection medium. Although this expression for Δm is nonlinear, it is monotonic (single-valued), independent of m, and depends only on the constant K_0. For subsequent measurements, f_b becomes f_a, a new initial frequency that reflects the total mass of the system. The new f_b after sampling will differ from f_a only because of the new mass uptake, Δm, collected during sampling.

Instrument Design

A TEOM® instrument can be tailored for a particular application. To do so, the manufacturer must know the minimum mass concentration the instrument must measure, how quickly each measurement must be made, and what sampling air flow rate will be used. To employ the highest sensitivity for a particular application, the manufacturer must consider the total mass of the tapered-element load. This mass, which in most applications is primarily the filter cartridge, must be held to a minimum to effect the maximum frequency change with a sample mass deposit. The reduction of the filter mass has practical limits that are related to both flow rate and filter life. A filter cartridge must have reasonable dimensions to sustain the desired flow before loading to a point where the flow drops to an unacceptable level. In addition to the filter mass, the tapered element has a certain amount of mass that also contributes to the total mass of the oscillating system. The limits on this mass depend on the dimensions of the tapered element. The element must have a sufficiently wide bore to allow the desired flow with minimum pressure drop and also have sufficient wall thickness to support the filter cartridge.

Potential Biases

Because TEOM® devices operate on vibrational principles like piezoelectric crystal techniques, it is instructive to examine the TEOM® technology in the light of biases that affect QCM measurements.

Calibration

Calibration for TEOM® instruments is equivalent to determining the spring constant K_0. Since K_0 is determined by the physical characteristics of the tapered element, calibration is not likely to change over a period of time. The manufacturer provides a value of K_0, but the user can easily check the value of K_0 by adding a known mass to the tapered element, measuring the change in oscillation frequency and using Eq. 14-6 to determine K_0. Using this method, Shore and Cuthbertson (1985) found that the manufacturer-supplied value for K_0 was correct within the experimental error. However, they also checked the calibration by injecting known masses of dioctyl phthalate (DOP) over the surface of the collection filter. This method suggested that the TEOM® instrument read particulate masses up to 10% lower (on an average) when the K_0 value supplied by the manufacturer was used. However, the masses of DOP used were much higher than those expected during particulate collection. They also observed that limiting injections to the center of the filter influenced the measurements. These results suggest that if one elects to verify the manufacturer-supplied K_0, aerosol depositions similar to those expected during sampling should be used.

Particle Size Effects

Since TEOM® devices typically use a filter collection medium, collection efficiency will not be significantly affected by particle size. Particles not collected by the filter medium would not represent significant mass. Particle size is likely to be more important to inlet bias of the sampling head; however, these problems are common to all sampling devices and not just to TEOM® measurement technology.

Adherence Effects

Tapered elements typically oscillate at several hundred Hz, rather than at several MHz as is typical of piezoelectric crystals. Furthermore, the collection medium is typically a filter rather than a smooth, hard surface. For these reasons, coupling of the collected aerosol is not usually an issue for TEOM® instruments. Theoretically, if sufficient mass were collected on the TEOM® filter, particles begin to flake off. In practice, however, the filters would clog before collecting particle loadings large enough to cause flaking. In fact, most TEOM® instruments provide a warning suggesting when to replace the filter.

Overloading

If the collection filter became sufficiently loaded, the added mass could conceivably damp the oscillations beyond the ability of the feedback system to sustain them. This situation, called saturation, could introduce serious error. The dynamic range of TEOM® instruments, however, is several orders of magnitude. As discussed in the previous section, filters would clog before collecting particle loadings large enough to cause saturation.

Particle Losses

As discussed under particle size effects, TEOM® instruments are subject to wall losses or particle losses because of inlet bias. A careful design of the sampling train is, as always, important. Visual wall losses were observed during a test of a prototype TEOM® device designed for personal exposure monitoring in mining environments (Williams and Vinson 1986), but these losses were later attributed to charges on the particles introduced by the test dust generation system.

Damping

If the mass of the support structure for the element is sufficiently small, the element may induce vibrations in the structure itself. The oscillation of the tapered element then will be slightly damped. This phenomenon was observed when testing the prototype TEOM® personal sampler for miners (Williams and Vinson 1986). Such damping will not be significant if the mass of the support structure is much larger than the mass of the element, or if the support structure is properly clamped to a large mass. Tests of the prototype TEOM® personal sampler for miners represent a special case; damping of this sort has not been reported for the commerical TEOM® instruments.

Results and Applications

The TEOM® technology had its beginnings at Dudley Observatory in the 1960s in conjunction with micrometeorite research (Patashnick and Hemenway 1969). A microbalance, at that time consisting of a thin quartz fiber, was designed to measure particle masses in the range 10^{-5}–10^{-11} g. Since that time, its inventors have made many improvements to the quartz fiber approach, and formed Rupprecht & Patashnick (R&P) Co., Inc. to market the device. R&P is currently the only manufacturer and vendor of TEOM® instruments.

R&P provides a variety of instruments that use the TEOM® technology, but only three are within the scope of this book. The first, the Series 1100 monitor, has been used in a variety of applications involving higher-than-ambient conditions. Many leading diesel engine manufacturers throughout the world have used the TEOM® technology in the design and development of engines. Particulate emissions from many vehicles using diesel engines must meet legislated standards in the United States. Regulations require making measurements of particulate emissions while operating the engine over a prescribed wide range of conditions. The near-real-time measurements possible with TEOM® instruments allow designers to identify those engine conditions that contribute most to emissions. In fact, vehicle-to-vehicle differences in transient particulate emissions are readily detected by the TEOM® instrument (Shore and Cuthbertson 1985). The TEOM® instrument has been demonstrated to have a one-to-one response to mass (Whitby, Johnson, and Gibbs 1985); however, care must be taken to properly account for volatile species.

Several research organizations have used the Series 1100 to measure changing concentrations in near-real-time. The BOM, in studying the health and safety hazards of materials burning in underground mine fires, used a TEOM® instrument to help study smoke characteristics. The BOM monitored the mass concentrations of smoke from wood (Egan and Litton 1986) and electrical transformer fluid (Egan 1986) fires. Newman and Steciak (1987) used the TEOM® instrument to help characterize smoke properties for fire modeling, fire detector evaluation, and assessment of the nuclear winter problem.

The Series 1200 monitor is meant primarily for research and indoor industrial applications. The Series 1400 is an industrially hardened version, better suited to remote monitoring situations. The Environmental Protection Agency has certified both instruments for PM-10 measurements of ambient air quality (U.S. Environmental Protection Agency 1989).

References

Bowers, W. D. and R. L. Chuan. 1989. Surface acoustic-wave piezoelectric crystal aerosol mass monitor. *Rev. Sci. Instr.* 60(7):1297–302.

Cooper, D. W. 1975. Statistical errors in beta absorption measurements of particulate mass concentration. *J. Air Poll. Control Assoc.* 25:1154.

Cooper, D. W. 1976. Significant relationships concerning exponential transmission or penetration. *J. Air Pollut. Control Assoc.* 26:366.

Courtney, W. J., R. W. Shaw, and T. C. Dzubay. 1982. Precision and accuracy of beta gauge for aerosol mass determinations. *Environ. Sci. Tech.* 16:236.

Egan, M. R. 1986. Transformer fluid fires in a ventilated tunnel. *BuMines* IC 9117.

Egan, M. R. and C. D. Litton. 1986. Wood crib fires in a ventilated tunnel. *BuMines* RI 9045.

Evans, R. D. 1955. *The Atomic Nucleus.* New York: McGraw-Hill.

Fairchild, C. I., M. I. Tillery, and H. J. Ettinger. 1980. An Evaluation of Fast Response Aerosol Mass Monitors. Los Alamos National Laboratory, Report LA-8220.

Fairchild, C. I. and L. D. Wheat. 1984. Calibration and evaluation of a real-time cascade impactor. *Am. Ind. Hyg. Assoc. J.* 45(4):205–11.

Gleason, G. I., J. D. Taylor, and P. L. Tabern. 1951. Absolute beta counting at defined geometries. *Nucleonics* 8:12.

Heintzenberg, J. and P. Winkler. 1984. Elemental carbon in the urban aerosol. *Sci. Tot. Environ.* 36:27.

Husar, R. B. 1974. Atmospheric particulate mass monitoring with a beta radiation detector. *Atmos. Environ.* 8:183.

Jaklevic, J. M., R. C. Gatti, F. S. Goulding, and B. W. Loo. 1981. A beta gauge method applied to aerosol samples. *Environ. Sci. Tech.* 15:680.

Jaklevic, J. M., N. W. Madden, and C. E. Wiegand. 1983. A precision beta gauge using a plastic scintillator and photomultiplier detector. *Nucl. Inst. and Methods* 214:517.

Karasek, F. W. 1978. Cascade particle analyzer. *Industr. Res./Develop.*:154–58.

Klein, F., C. Ranty, and L. Sowa. 1984. New examinations of the validity of the principle of beta radiation absorption for determinations of ambient air dust concentrations. *J. Aerosol Sci (U.K.)* 15:391.

Lilienfeld, P. 1975. Design and operation of dust measuring instrumentation based on the beta-radiation method. *Staub-Reinhalt Luft* 35:458.

Lippmann, M. 1978. Respirable dust sampling. In *Air Sampling Instruments for Evaluation of Atmospheric Contaminants*, 5th edn., pp. G1–G23, Cincinnati: American Conference of Government Industrial Hygienists.

Loo, B. W., W. R. French, R. C. Gatti, F. S. Goulding, J. M. Jaklevic, J. Llacer, and A. C. Thompson. 1978. Large-scale measurement of airborne particulate sulfur. *Atmos. Environ.* 12:759.

Lundgren, D. A., L. D. Carter, and P. S. Daley. 1976. Aerosol mass measurement using piezoelectric crystal sensors. In *Fine Particles*, ed. B. Y. H. Liu, pp. 485–510. New York: Academic Press.

Macias, E. S. and R. B. Husar. 1970. High resolution online aerosol mass measurement by the beta attenuation technique. In *Proc. 2nd Internat. Conf. on Nucl. Methods in Environ. Research*, ed. J. R. Vogt and W. Meyer, pp. 413. CONF-740701.

Newman, J. S. and J. Steciak. 1987. Characterization of particulates from diffusion flames. *Combustion and Flame* 67(1):55–64.

Patashnick, H. and C. L. Hemenway. 1969. Oscillating fiber microbalance. *Rev. Sci. Instr.* 40(8):1008–11.

Patashnick, H. and G. Rupprecht. 1983. Personal dust exposure monitor based on the tapered element oscillating microbalance. *BuMines* OFR 56–84, NTIS PB 84-173749.

Rose, W. I. Jr., R. L. Chuan, R. D. Cadle, and D. C. Wood. 1980. Small particles in volcanic eruption clouds. *Am. J. Sci.* 280:671–96.

Sem, G. J. and J. W. Borgos. 1975. An experimental investigation of the exponential attenuation of beta radiation for dust measurements. *Staub-Reinhalt Luft* 35:5.

Sem, G. J. and K. Tsurubayashi. 1975. A new mass sensor for respirable dust measurement. *Am. Ind. Hyg. Assoc. J.* 36(11):791–99.

Shore, P. R. and R. D. Cuthbertson. 1985. Application of a tapered element oscillating microbalance to continuous diesel particulate measurement. *Soc. Automotive Eng.*, Report 850, 405.

Spengler, J. D. and G. D. Thurston. 1983. Mass and elemental composition of fine and coarse particles in six U. S. cities. *J. Air Pollut. Control Assoc.* 33:1162.

Stevens, R. K., T. G. Dzubay, R. W. Shaw Jr., W. A. McClenny, C. W. Lewis, and W. E. Wilson. 1980. Characterization of the aerosol in the Great Smoky Mountains. *Environ. Sci. Technol.* 14:1491.

U.S. Environmental Protection Agency. 1989. Compendium Method IP-10B: Determination of Respirable Particulate Matter in Indoor Air Using a Continuous Particulate Monitor.

Wallace, D. and R. Chuan. 1977. A Cascade Impaction Instrument Using Quartz Crystal Microbalance Sensing Elements for Real-Time Particle Size Distribution Studies. *National Bureau of Standards Special Publication 464*.

Ward, M. D. and D. A. Buttry. 1990. *In situ* interfacial mass detection with piezoelectric transducers. *Science* 249:1000–07

Whitby, R., R. Johnson, and R. Gibbs. 1985. Second Generation TEOM Filters—Diesel Particulate Mass Comparisons between TEOM and Conventional Filtration Techniques. *Soc. Automotive Eng.*, Paper 850, 403.

Williams, K. L. and R. P. Vinson. 1986. Evaluation of the TEOM dust monitor. BuMines IC 9119.

15

Optical Direct-Reading Techniques: Light Intensity Systems

Josef Gebhart

GSF-Forschungszentrum für Umwelt und Gesundheit, GmbH, Institut für Biophysikalische Strahlenforschung, Paul-Ehrlich-Str. 20, D-6000 Frankfurt/M., Germany

INTRODUCTION

Light scattering and extinction by small particles suspended in gases are widely applied to obtain information on the concentration and size distribution of the particles. Devices based on this principle combine *in situ* measurements in the airborne state of the particles with a high degree of automation. This chapter presents measuring techniques which use the intensity of scattered light or the power of attenuated light to characterize particles and particle systems. For practical applications, two kinds of instruments exist: light scattering and extinction by single particles and light scattering and extinction by an assembly of particles. Single-particle techniques based on light scattering cover a size range from about 70 nm to more than 100 μm and are capable of measuring concentrations of less than 1 particle per liter (clean-room monitoring) to about 10^5 particles per cm^3 (aerosol research). Ensemble scattering of light is applicable for concentration measurements of particulate matter in the atmosphere from a few $\mu g/m^3$ to several hundred mg/m^3.

In the introductory section light scattering and extinction by small particles are described by applying electromagnetic theory to spherical particles of known size and refractive index. The main variables and basic assumptions of the theory are explained and computer calculations of theoretical response functions are presented, which give a general survey of the characteristics of optical systems. For a better physical understanding of the phenomena of light scattering and extinction by small particles, two approximations are introduced: dipole or Rayleigh scattering for particles much smaller than the wavelength and classical optics, including diffraction, reflection, and refraction, for particles much larger than the wavelength. With respect to ensemble techniques (aerosol photometers) and their use for concentration measurements, general relationships between photometry and gravimetry are discussed.

A special section is devoted to single-particle techniques known as optical particle counters (OPC). In an OPC, aerosol is drawn through a light beam and light flashes scattered by single particles are received by a photodetector. From the count rate of the photoelectric pulses, the number concentration, and from the pulse height, the size of the particles, is evaluated. OPCs are widely used in basic aerosol research, air pollution studies, and clean-room monitoring. Since these various kinds of applications require different specifications of an instrument, characteristic features of an OPC-like permissible range of number concentration, sensitivity, sampling flow rate, classification accuracy and size resolution are discussed on the basis of experimental results.

Aerosol photometers are useful as real-time dust monitors in industrial hygiene and for concentration measurements of atmospheric aerosols. The measuring range of a photometer is limited at high concentrations by multiple scattering and at low concentrations by the stray-light background. The noise level of sensitive instruments is determined by Rayleigh scattering from the air molecules.

Whereas theoretical and experimental calibration curves of optical systems are usually based on particles of spherical shape, most real aerosol systems consist of particles of irregular shape. Therefore, the final section deals with the influence of particle shape on scattering of light.

LIGHT SCATTERING AND EXTINCTION BY A SINGLE SPHERE

Rigorous Electromagnetic Theory

Electromagnetic theory (Mie 1908) yields the rigorous solution of light scattering and extinction by a spherical particle. The mathematical expressions have been published in extensive textbooks, for instance by van de Hulst (1957) and Kerker (1969).

Light Scattering

Let a spherical particle of diameter d_p, made up of a material of refractive index m, be situated at the origin of a system of spherical coordinates and be illuminated from the negative z-direction by a plane, linearly polarized, monochromatic wave with electric vector in the x-direction (Fig. 15-1).

The spatial distribution of the light scattered by the particle per unit solid angle can then be described by a scattering function i that depends on the scattering angle θ, the polarization angle ϕ, the particle diameter d_p, the wavelength λ and the refractive index $m = n - jk$ of the particle material (n: real part; k: imaginary part):

$$i = i(\theta, \phi, d_p, \lambda, m) \tag{15-1}$$

Particle diameter d_p and wavelength λ are connected by the dimensionless size parameter

$$\alpha = \frac{\pi d_p}{\lambda} \tag{15-2}$$

The power of light scattered by a particle per unit solid angle in direction θ is given by

$$S_\lambda(\theta, \phi, d_p, \lambda, m) = I_0 \frac{\lambda^2}{4\pi^2} i(\theta, \phi, \alpha, m) \tag{15-3}$$

where I_0 is the illumination intensity (power of light per unit area).

Various optical arrangements constructed for light scattering measurements differ in the mean scattering angle θ_0, the receiver aperture $\Delta\Omega$, and the kind of illumination. Using linearly polarized light of a plane monochromatic wave for the illumination, the power of scattered light collected by a certain optical arrangement can be calculated by the integral:

$$P_\lambda(d_p, \theta_0, \Delta\Omega, \lambda, m) = I_0 \frac{\lambda^2}{4\pi^2} \iint_{(\Delta\Omega)} i(\theta, \phi, \alpha, m) \sin\theta \, d\theta \, d\phi \tag{15-4}$$

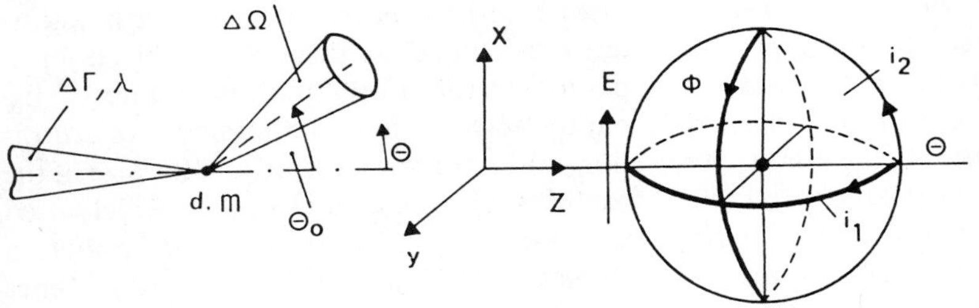

FIGURE 15-1. Parameters for Light-Scattering Calculations.

For the dependency of the scattering function i on the polarization angle one obtains

$$i = i_1 \sin^2 \phi + i_2 \cos^2 \phi \qquad (15\text{-}5)$$

where $i_1(\theta, \alpha, m)$ and $i_2(\theta, \alpha, m)$ are the two components of the scattering function in the plane perpendicular and parallel to the polarization vector E of the incident light (Fig. 15-1).

For unpolarized light one gets $i = 1/2(i_1 + i_2)$ and the scattering pattern is symmetric in rotation with respect to the axis of the illuminating beam. White light illumination additionally requires an integration over the wavelength, whereby the spectral radiation density of the light source and the spectral sensitivity of the photodetector have to be taken into account (Heyder and Gebhart 1979). If the particles are illuminated with non-collimated light a further integration over the illumination aperture has to be carried out.

For the following considerations it is useful to introduce two further quantities:

partial scattering cross-section:

$$C_s = \frac{P_\lambda}{I_0} \qquad (15\text{-}6)$$

scattering coefficient:

$$Q_s = \frac{S_\lambda}{I_0 (\pi/4) d_p^2} \qquad (15\text{-}7)$$

In this definition Q_s is the power of scattered light per unit solid angle in relation to the power $I_0(\pi/4)d_p^2$ striking the projected area of the particle.

Light Extinction

Light extinction is the attenuation of a parallel beam of light due to absorption and scattering by the particles. For a single sphere with size parameter α and refractive index m, light extinction can be described by its extinction coefficient $E_x(\alpha, m)$, which can be calculated by electromagnetic theory. The power of light $-\Delta P$ which is removed by a particle from a parallel beam of light with intensity I_0 is then given by

$$-\Delta P_\lambda = I_0 (\pi/4)\, d_p^2\, E_x(\alpha, m) \qquad (15\text{-}8)$$

Approximations

The mathematical formalism of the rigorous electromagnetic theory hardly allows a physical understanding and interpretation of the phenomena of light scattering and extinction by small particles. This can be done better by using two approximations.

Light Scattering

When the particle size is much smaller than the wavelength ($\alpha \ll 1$), the particle is subjected to an almost uniform field. The particle then oscillates like a dipole with a polarization proportional to the electric field of the incident wave. The scattering properties of such a particle can be expressed by its polarizability p, which is generally a tensor and reduces to a scalar for a homogeneous sphere. In the dipole or Rayleigh approximation and for unpolarized monochromatic light, the flux of light scattered by a spherical particle per unit solid angle is given by

$$S_\lambda(\theta, d_p, \lambda, m) = I_0 \frac{\lambda^2}{8\pi^2} \left[\frac{\pi d_p}{\lambda}\right]^6 \times \left|\frac{m^2 - 1}{m^2 + 2}\right|^2 (1 + \cos^2\theta) \qquad (15\text{-}9)$$

Equation 15-9 contains the polarizability p_s of a sphere (van de Hulst 1957):

$$p_s = 3\left[\frac{m^2 - 1}{m^2 + 2}\right] V \qquad (15\text{-}10)$$

where V is the volume of the sphere. Substituting Eq. 15-10 into Eq. 15-9 results in:

$$S_\lambda(\theta, d, \lambda, m) = I_0 \frac{\pi^2}{2\lambda^4} p_s^2 (1 + \cos^2\theta) \qquad (15\text{-}11)$$

which means that the power of scattered light is proportional to the square of the polarizability.

In the limiting case of $\alpha \gg 1$, the scattered light can be considered as consisting of three parts, which are due to the physical effects of diffraction, reflection, and refraction (Fig. 15-2). For monochromatic light, the scattering coefficient Q_s can be expressed as the sum of its components according to

$$Q_s(\theta, \alpha, m) = Q_0(\theta, \alpha) + Q_1(\theta, m) + Q_2(\theta, m) \qquad (15\text{-}12)$$

Higher-order internal reflections Q_i, $i > 2$ (rainbows) are not contained in Eq. 15-12.

$Q_0(\alpha, \theta)$ is the diffracted part of scattered light. It is independent of the optical constants of the particle material. Its angular distribution, however, depends on the size parameter α. $Q_1(\theta, m)$ is the fraction of light scattered by reflection on the surface of the particle. Its angular distribution is independent of the particle diameter but is influenced by the optical constants of the material. $Q_2(m, \theta)$ is the component scattered by two refractions by the particle surface. Its angular distribution again depends on the optical constants but not on the particle diameter.

For low scattering angles ($\theta < 10°$) the diffracted part of scattered light can be approximated by the Fraunhofer formula according to

$$Q_0(\alpha, \theta) = \frac{\alpha^2}{4\pi}\left[\frac{2I_1(\alpha \sin\theta)}{\alpha \sin\theta}\right]^2 \qquad (15\text{-}13)$$

where $I_1(\alpha, \theta)$ is a first-order Bessel function.

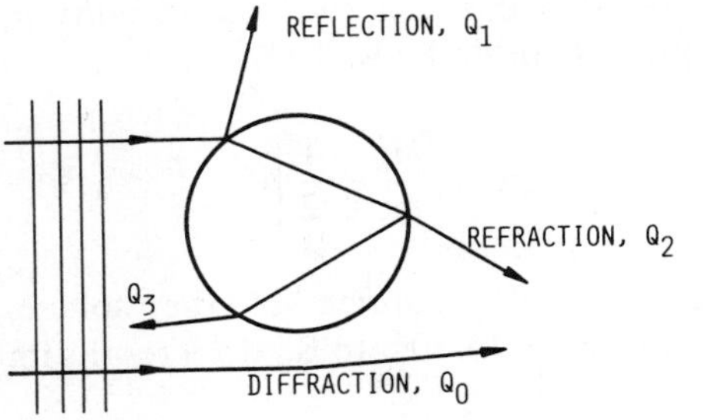

FIGURE 15-2. Decomposition of Scattered Light in Components According to Ray Optics.

The externally reflected fraction $Q_1(\theta, m)$ is resolved into the components $Q_{1,1}(\theta, m)$ and $Q_{1,2}(\theta, m)$ with polarization normal and parallel, respectively, to the plane containing the incident and reflected rays. The exact formulae for the two components are (Hodkinson 1962):

$$Q_{1,1}(\theta, m) = \frac{1}{8\pi}\left\{\frac{\sin\theta/2 - (m^2 - 1 + \sin^2\theta/2)^{1/2}}{\sin\theta/2 + (m^2 - 1 + \sin^2\theta/2)^{1/2}}\right\}^2 \qquad (15\text{-}14)$$

$$Q_{1,2}(\theta, m) = \frac{1}{8\pi}\left\{\frac{m^2\sin\theta/2 - (m^2 - 1 + \sin^2\theta/2)^{1/2}}{m^2\sin\theta/2 + (m^2 - 1 + \sin^2\theta/2)^{1/2}}\right\}^2 \qquad (15\text{-}15)$$

The fraction $Q_2(\theta, m)$ scattered by two refractions by the particle surface is given by (Hodkinson 1962)

$$Q_2(\theta, m) = \frac{2}{\pi}\left[\frac{m}{m^2 - 1}\right]^4 \times \frac{(m\cos\theta/2 - 1)^3(m - \cos\theta/2)^3}{(\cos\theta/2)(m^2 + 1 - 2m\cos\theta/2)^2} \times (1 + \sec^4\theta/2) \qquad (15\text{-}16)$$

In the final bracket the term 1 corresponds to the polarized component perpendicular to the plane of observation and the term $\sec^4\theta/2$ to the component parallel to the plane of observation.

Some angular distributions of the three fractions of scattered light can be seen in Fig. 15-3. For the reflected and the refracted light, only the components polarized perpendicular to the plane of observation are drawn in the diagram.

Since the diffraction part changes with particle diameter, two α values have been considered. The forward lobe of diffraction is limited to the angular range $\theta < \theta_{min}$, where

$$\sin\theta_{min} = \frac{5\pi}{4\alpha} \qquad (15\text{-}17)$$

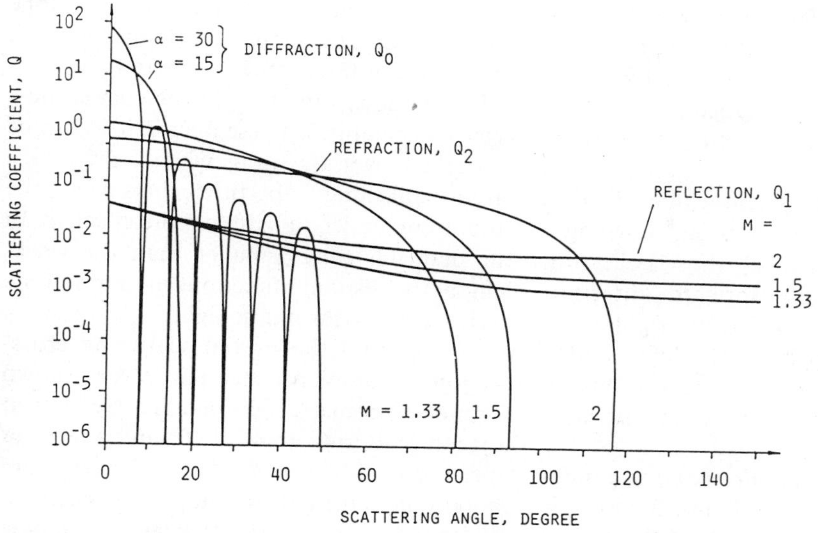

FIGURE 15-3. Angular Distribution of the Components of Scattered Light. (With Permission from Gebhart 1991.)

While the reflected component $Q_1(\theta, m)$ covers the whole angular range $0 < \theta < 180°$, the refracted part is confined to the range $\theta < \theta_{max}$, where:

$$\cos\frac{\theta_{max}}{2} = \frac{1}{m} \tag{15-18}$$

Light Extinction

The extinction coefficient in the dipole or Rayleigh approximation can be expressed by two terms, according to

$$E_x(\alpha, m) = 4\alpha \operatorname{Im}\left[-\frac{m^2-1}{m^2+2}\right] + \frac{8}{3}\alpha^4\left[\frac{m^2-1}{m+2}\right]^2 \tag{15-19}$$

where Im stands for the imaginary part. The first term represents the absorption component of extinction and the second term the scattering component.

For classical optics ($\alpha \gg 1$) the extinction coefficient has the value 2, including equal contributions of unity from light undergoing reflection, refraction, or absorption and from light being diffracted on the particle contour.

In practice, the extinction coefficient is measured to be less than 2, since some of the scattered light is collected due to the finite aperture angle ϑ of the detection unit:

$$E_x(\vartheta, \alpha, m) = 2 - 2\pi\int_0^{\vartheta} Q_s(\theta, \alpha, m)\sin\theta \, d\theta \tag{15-20}$$

Theoretical Response Functions

Although it is advisable to calibrate an optical instrument experimentally by means of test aerosols of known size and refractive index, theoretical response functions give a general survey of the characteristics of an optical system. By means of electromagnetic theory, the response functions of optical systems, which describe the power of light scattered through the collecting aperture, can be calculated as a function of the diameter of a spherical particle. The parameter of these functions is the refractive index of the particle material. Detailed calculations of this kind have been performed by Hodkinson and Greenfield (1965), Brossmann (1966), Gucker and Tuma (1968), Quenzel (1969), Oeseburg (1972), Cooke and Kerker (1975), and Heyder and Gebhart (1979). The response characteristics of coherent laser aerosol spectrometers have been calculated by Garvey and Pinnick (1983), Pettit and Peterson (1984), Soderholm and Salzman (1984), and Hinds and Kraske (1986).

Commercial optical systems can be divided roughly into instruments using low-angle scattering (diffraction lobe), those collecting scattered light in the forward direction ($\theta < 90°$) and those employing right-angle scattering. With respect to the light source one has to distinguish between polychromatic incandescent light and monochromatic laser light.

The influence of monochromatic and polychromatic light on the response function is demonstrated in Fig. 15-4, where the partial scattering cross-section c_s is plotted against the particle diameter for monochromatic and white light and a mean scattering angle of 45°. For particle diameters larger than the wavelength, light scattering can be considered as a surface effect, and outside the diffraction lobe the mean flux of scattered light depends on the square of the particle diameter.

For monochromatic light the curves show typical oscillations. These fluctuations can be smoothed over by using white light and a large collecting aperture; however, the smoothing by white light is more effective. In the diameter range smaller than the wavelength, the response functions are monotonic even for monochromatic light.

In Fig. 15-5 the partial scattering cross-section of submicrometer particles is shown for monochromatic light of a He–Ne laser at a mean scattering angle of 40° and a receiver aperture of 20°. For $d_p < \lambda/2$ all curves are monotonic and their slope indicates a d_p^6-dependence that is typical for dipole scattering, as expressed by Eq. 15-9.

Response functions for different real parts of refractive index tend to run parallel in a double-logarithmic plot, which means that they differ more or less only by a constant

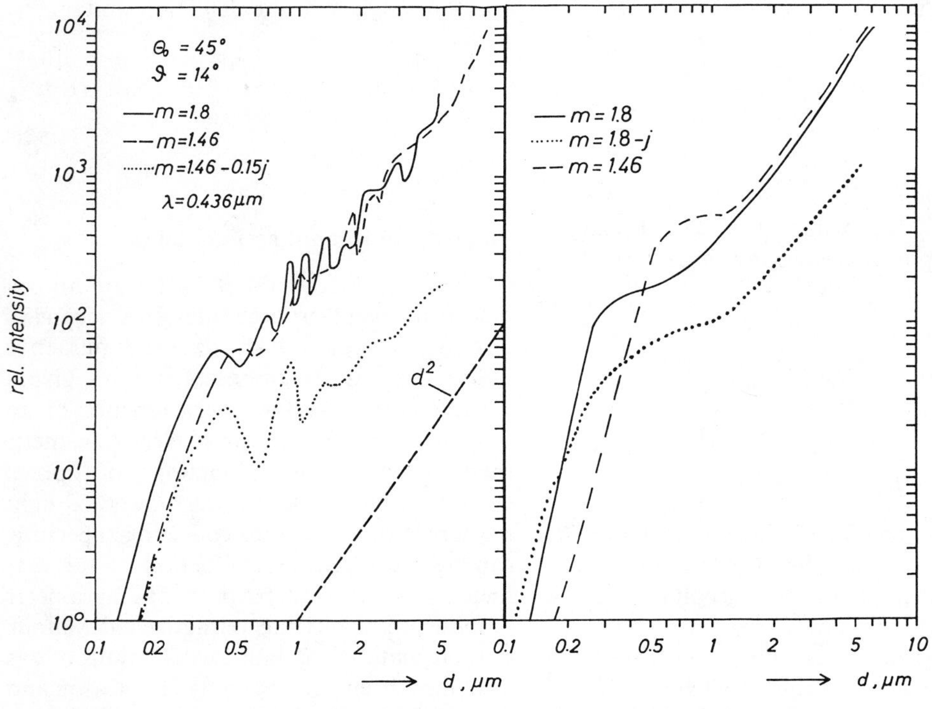

FIGURE 15-4. Theoretical Response Functions of a Forward-Scattering System Using Monochromatic (Left) and Polychromatic (Right) Illumination. (Adapted from Gebhart et al. 1976.)

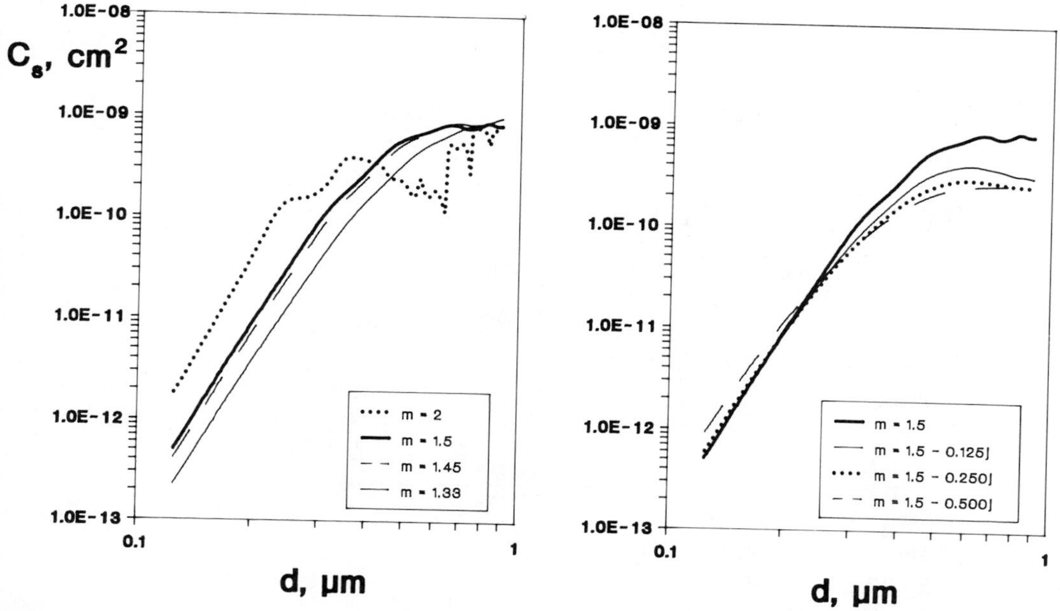

FIGURE 15-5. Theoretical Response Functions in the Submicrometer Size Range of a Forward-Scattering Laser Spectrometer. Wavelength $\lambda = 0.6328$ µm.

factor which is a function of the refractive index. For absorbing materials with a complex refractive index, $m = n - jk$, the influence of the k value is reduced considerably in the size range below the wavelength. A value of $k = 0.25$, for instance, represents the absorption properties of coal (Olaf and Robock 1961). All response curves having the same real part but different imaginary parts of refractive index cross over for particle diameters below the wavelength. High-sensitivity laser aerosol spectrometers can make use of this submicrometer part of the response function for particle size analysis.

According to classical optics, light entering the particle is scattered due to refraction. As can be seen from Fig. 15-3 this component dominates in the angular range up to about 100° except within the forward lobe of diffraction. For particles of opaque materials the fraction of light entering the particle is absorbed. Instruments which mainly collect light scattered due to refraction, therefore, show great differences in the response function between transparent and absorbing materials. Theoretical response functions which include light refracted in the forward direction are given in Fig. 15-4. Experimental calibrations of optical particle counters with light-absorbing particles have been carried out by Whitby and Vomela (1967) with particles of India ink and by Willeke and Liu (1976) with coal particles. The experimental findings confirm the theoretical predictions of Fig. 15-4 that opaque particles in the micrometer size range scatter a factor of 6–10 less light in the forward direction than the transparent particles of the same size.

Better conditions for particle size analysis can be realized if the light scattered within the forward lobe of diffraction ($\theta \leq 6°$) is collected. Within this angular range the diffraction part $Q_0(\theta, \alpha)$ dominates, and the flux of scattered light is independent of the optical properties of the particles. Theoretical response curves for low-angle scattering using a He–Ne laser are given in Fig. 15-6. The response functions belonging to absorbing particles now coincide and those of transparent spheres show periodic oscillations rather than the systematic deviations from the responses calculated for absorbing materials.

Since these oscillations originate from interferences between diffracted and refracted

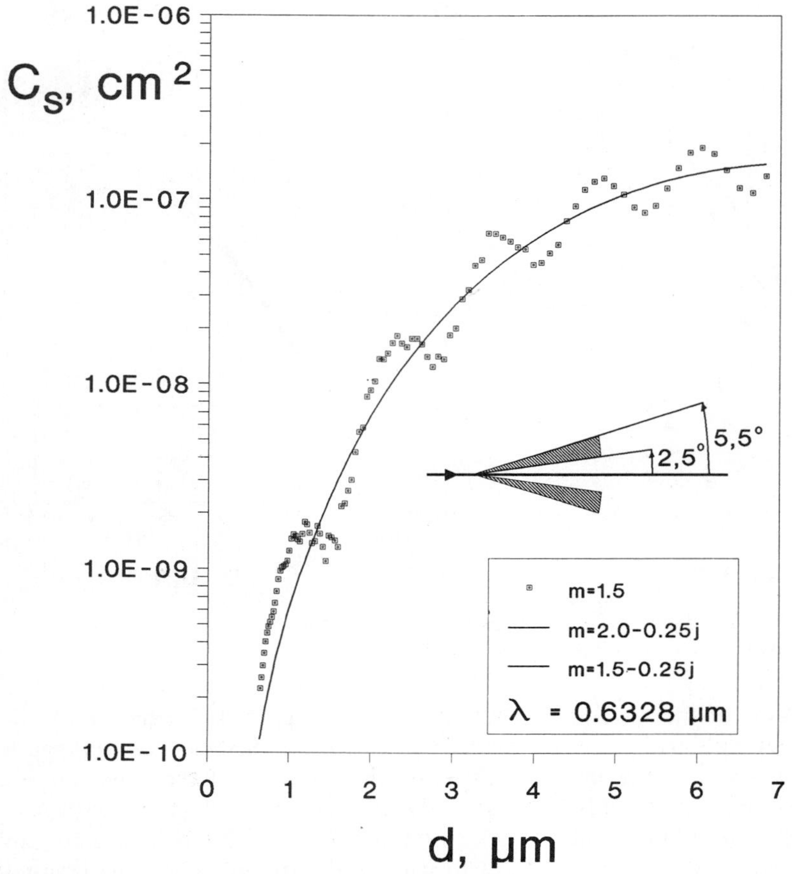

FIGURE 15-6. Theoretical Response Functions of a Low-Angle Scattering Instrument. (Adapted from Gebhart et al. 1976.)

light, they change with the real part of refractive index and can be smoothed over by using polychromatic light or particles of irregular shape. For opaque materials the refracted component is absorbed inside the particle and no interference occurs.

A laboratory instrument which uses the polychromatic light from a mercury lamp and collects light scattered within the forward lobe of diffraction has been calibrated by Gebhart et al. (1976) with particles of different optical properties. The results confirm that light scattering within the forward lobe of diffraction is nearly independent of the optical properties of the particle material.

Light extinction of a particle can be best described by its extinction coefficient $E_x(\alpha, m)$, which is obtained from electromagnetic theory. In Fig. 15-7 extinction coefficients of non-absorbing spheres of different refractive indices are plotted versus the normalized size parameter $\alpha(m - 1)$. For particle diameters of many wavelengths the extinction coefficient is seen to approach a steady value of 2. Another striking feature of the extinction curve is its oscillation, which is caused by constructive and destructive interferences of diffracted and refracted light in the forward direction (see also Fig. 15-6). As the absorption inside the particle increases, the oscillations decline in importance and settle more quickly to a value of 2. The successive cycles in the extinction curve also depend on the high symmetry of a sphere and vanish for irregular particles.

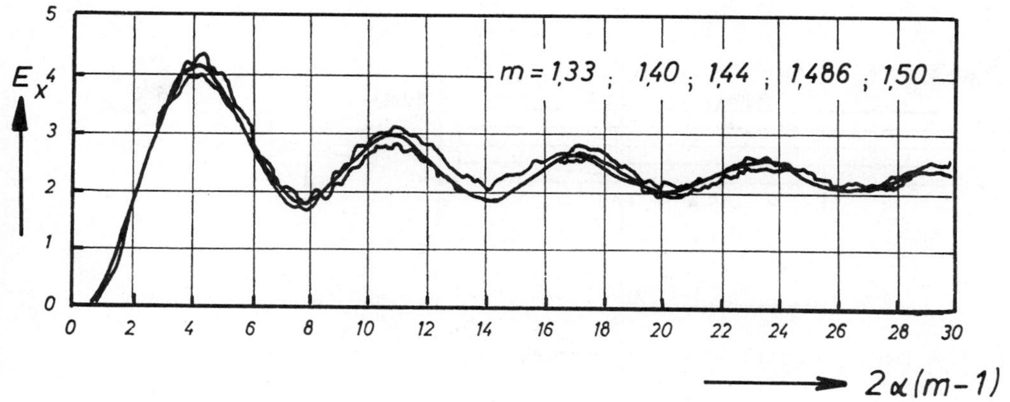

FIGURE 15-7. Extinction Coefficient of Transparent Spheres.

LIGHT SCATTERING AND EXTINCTION BY AN ASSEMBLY OF PARTICLES

Light Scattering

Let $P_\lambda(d_p, \lambda, m)$ be the flux of light scattered by a single particle into the receiver aperture of an optical system. In the presence of many particles inside the sensing volume of the photometer the resulting light flux R collected by the detector is given by

$$R = c_n \int_0^\infty f(d_p) P_\lambda(d_p, \lambda, m) \mathrm{d}d_p \qquad (15\text{-}21)$$

where c_n is the number concentration of the particles and $f(d_p)$ the probability density function of the particle size distribution. Equation 15-21 demonstrates the problem of photometric measurements of particle concentration: If the photometer response R varies, one cannot distinguish whether the number concentration c_n, the size distribution $f(d_p)$ or the optical properties m of the material have changed.

Using a test aerosol with a homogeneous chemical composition, the influence of the refractive index m can be eliminated so that only c_n and $f(d_p)$ remain as variables. In this case a linear relationship $R = \text{const} \times c_n$ can be realized if either the function $f(d_p)$ can be replaced by one particle size (monodisperse aerosol) or the function $f(d_p)$ is constant during the measurements. In these cases the photometer response is linearly correlated with the number or mass concentration of an aerosol and a simple calibration procedure can be carried out.

The simple proportionality of the scattered light flux to the number of particles inside the sensitive volume holds only if independent scattering by separate particles is considered and multiple scattering can be neglected. Estimations show that a mutual particle distance of a few diameters is a sufficient condition for independent scattering (van de Hulst 1957). The concentration level for the onset of multiple scattering depends on the particle diameter and the sensing volume of the photometer system. Examples will be given in the section on multiparticle techniques.

Light Extinction

Let a parallel beam of light with intensity I_0 and wavelength λ traverse a homogeneous mixture of particles with number concentration c_n and size distribution function $f(d_p)$. According to the Lambert–Beer law, the transmission of the light beam can then be expressed by

$$\begin{aligned} T &= I/I_0 \\ &= \exp\left\{ -c_n s \frac{\pi}{4} \int d_p^2 E_x(\alpha, m) f(d_p) \,\mathrm{d}d_p \right\} \\ &= \exp\{-Ks\} \end{aligned} \qquad (15\text{-}22)$$

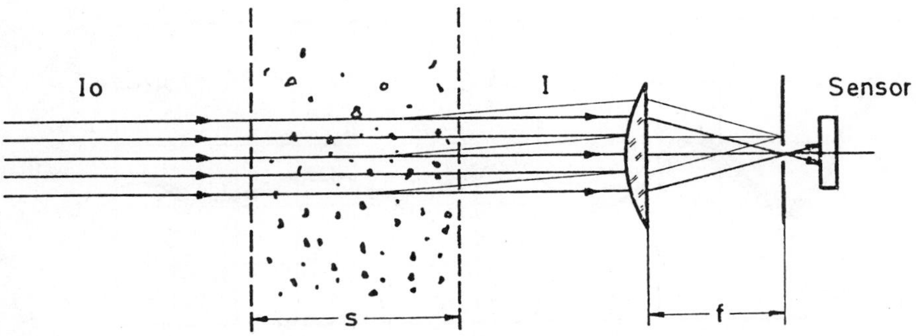

FIGURE 15-8. Set-up for Light Extinction Measurements.

where T is the transmission and s the extinction length (see Fig. 15-8).

The factor

$$K = c_n \frac{\pi}{4} \int d_p^2 E_x(\alpha, m) f(d_p) \, dd_p \qquad (15\text{-}23)$$

is called the "turbidity."

As can be seen from Eq. 15-23 the turbidity remains only a unique function of the number concentration if the size distribution $f(d_p)$ and the refractive index m of the particles do not change during the measurement.

When the particle concentration becomes very high, the particles may no longer scatter independently, and the coefficient $E_x(\alpha, m)$ may be modified. Usually, this happens at concentrations much greater than those where the Lambert–Beer law breaks down due to multiple scattering. Hodkinson (1962) performed extinction measurements on suspensions of 1.8 µm polystyrene spheres in water and showed that within the transmission range $0.37 \leq T \leq 1$ the results obeyed the Lambert–Beer law. With narrow-angle extinction techniques, where only a negligible amount of scattered light can reach the detector, the validity range of the Lambert–Beer law may be extended even to transmissions below 0.1.

Example 15-1

An aerosol generator based on controlled condensation of vapor produces uniform droplets of 3 µm diameter at a concentration of 2×10^6 cm^{-3}. The output of the generator passes a glass tube of 3 cm inner diameter. Its constancy shall be monitored by a narrow-angle extinction technique which uses a collimated laser beam. Calculate the expected transmission of the measuring system.

Answer. The relevant parameters for the turbidity K are the diameter and concentration of the particles. The extinction coefficient of 3 µm droplets in a parallel laser beam approaches 2: $K = c_n(\pi/4)d_p^2$, $E = 2 \times 10^6$ cm^{-1}) $\times$ $(\pi/4)(3 \times 10^{-4})^2$ cm^2 $\times$ $2 = 0.283$ cm^{-1}. The relevant parameters for the transmission T are the turbidity and the extinction length. From Eq. 15-22 it follows that

$$T = \exp(-Ks)$$
$$= \exp(-0.283 \text{ cm}^{-1} \times 3 \text{ cm})$$
$$= 0.428$$

Gravimetry and Photometry

Air quality standards for particulate matter in the environment are usually based on gravimetry. It is, therefore, of general interest to correlate photometer signals with the mass (volume) concentration of aerosols. The mass concentration c_m of a polydisperse aerosol is given by

$$c_m = c_n \rho_p \int_0^\infty f(d_p) \frac{\pi}{6} d_p^3 \, dd_p \qquad (15\text{-}24)$$

From the photometer response as expressed by Eq. 15-21 it follows that a linear relationship $R = \text{const.} \times c_m$ is only obtained if the flux of scattered light $P_\lambda(d_p, m, \lambda)$ collected by an optical system is a function of the particle volume. Using the electromagnetic theory of light scattering on spheres, a specific scattering function $P_{\lambda s}(d_p, m, \lambda)$ can be defined as

$$P_{\lambda s}(d_p, m, \lambda) = \frac{P_\lambda(d_p, m, \lambda)}{\rho_p(\pi/6)d_p^3} \qquad (15\text{-}25)$$

which is the flux of scattered light per unit mass (volume) concentration of aerosol. The general behavior of $P_{\lambda s}$ for a fixed m and λ is illustrated in Fig. 15-9.

In the micrometer size range, light scattering is a surface effect and $P_{\lambda s}$ decreases proportional to d_p^{-1}. For particle diameters $d_p \leq 0.5\lambda$ light scattering can be approximated by Eq. 15-9 ($S_\lambda \sim d_p^6$) and the specific scattering function increases proportional to d_p^3. At about $d_p \approx \lambda$ the specific scattering function runs through a maximum, which for dielectric spheres is additionally pronounced by resonances. Only within this narrow plateau, light scattering is nearly proportional to particle volume. Armbruster et al. (1984) calibrated a respirable dust photometer, which measures infrared light ($\lambda = 0.94$ μm) scattered by airborne particles at a mean angle of 70°, with monodisperse test aerosols of different optical properties. Their results confirm the general course of the specific scattering function as indicated in Fig. 15-9. It is interesting to note that for atmospheric aerosols and visible light, particles within the accumulation mode (0.1–1 μm) coincide with the maximum of the specific scattering function, so that particles within this size range are optically most active (Willeke and Brockmann 1977).

Light extinction, expressed by the extinction cross-section $(\pi/4)d_p^2 E_x(\alpha, m)$ of a particle, generally shows characteristics as similar to light scattering. Only for absorbing particles in the Rayleigh regime, where the first term of Eq. 15-19 dominates, does light extinction become proportional to the volume of the particles regardless of their size.

Light Scattering by Gas Molecules

For dielectric particles much smaller than the wavelength of light, dipole theory can be applied so that the power of scattered light per unit solid angle, $S_\lambda(\theta, d_p, \lambda, m)$, can be expressed by Eq. 15-11. For randomly oriented gas molecules, the mean polarizability p_m of a single molecule is connected with the macroscopic refractive index m_g of the gas, according to Born (1965)

$$c_g p_m = (m_g^2 - 1) \qquad (15\text{-}26)$$

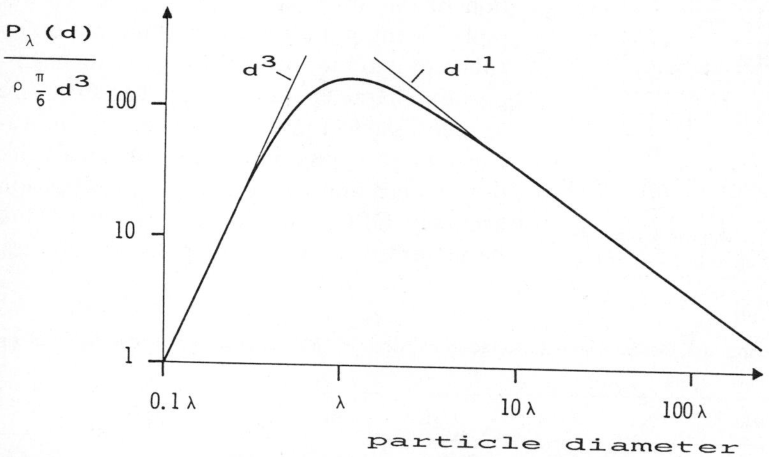

FIGURE 15-9. Schematic Diagram Illustrating the Relationship Between Photometry and Gravimetry.

where c_g is the number concentration of the molecules. Substituting Eq. 15-26 into Eq. 15-11 yields for the light scattered by a single molecule:

$$S_{\lambda,m}(\theta, d_g, \lambda, m_g) = I_0 \frac{\pi^2(m_g^2 - 1)^2}{2\lambda^4 c_g^2} \times (1 + \cos^2\theta) \quad (15\text{-}27)$$

and for the contribution of a gas volume V_g to light scattering:

$$S_{\lambda,g} = V_g I_0 \frac{\pi^2}{2\lambda^4} \frac{(m_g^2 - 1)^2}{c_g} \times (1 + \cos^2\theta) \quad (15\text{-}28)$$

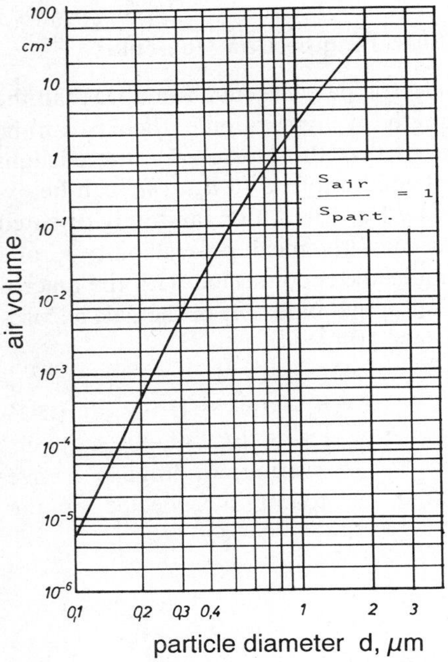

FIGURE 15-10. Rayleigh Scattering of Air in Relation to Spheres With Refractive Index $m = 1.5$.

In Fig. 15-10 Rayleigh scattering of air is compared with light scattering by single spherical particles of refractive index $m = 1.5$. Macroscopic refractive indices of some gases at $\lambda = 0.546$ μm and the corresponding fluxes of scattered light in relation to air are listed in Table 15-1.

SINGLE-PARTICLE OPTICAL COUNTERS

General Remarks

An optical particle counter (OPC) measures the size and number concentration of aerosol particles in a limited size range by means of light scattering by single particles. For this purpose a stream of aerosol is drawn through a condensed light beam. Light flashes scattered from single particles are received by a photodetector and converted into electrical pulses. From the count rate of the pulses, the number concentration, and from the pulse height, the size of the particles is derived. The light power that an individual particle scatters is a function of its size, refractive index, and shape. Particle sizing based on this principle has been known for more than 40 years. Meanwhile, the technique has been steadily developed and for about 30 years OPCs using white light illumination have been commercially available (Lieberman 1986). The invention of the laser has allowed the successful replacement of the white light illumination by coherent and monochromatic laser light. Important characteristic features of an OPC are its permissible range of number concentration, its sampling flow rate, its sensitivity (lower detection limit) and its classification accuracy. OPCs are nowadays widely used in basic aerosol research, air pollution studies,

TABLE 15-1 Rayleigh Scattering of Different Gases in Relation to Air (m_g: **Macroscopic Refractive Index of the Gas)**

Gas	Air	He	Ne	H_2	H_2O	Ar	Kr	CO_2	Xe
$(m_g - 1) \times 10^6$	292	34.89	67.25	139.6	252.7	282.3	428.7	450.5	705.5
S_g/S_{air}	1	0.015	0.053	0.228	0.75	0.934	2.15	2.38	5.84

and clean room monitoring. Since the requirements for an OPC are quite different for these different kinds of applications, the specifications of an instrument have to be adjusted to the special measurement problem.

Calibration Procedures

In the last decades many theoretical response functions of commercial OPCs have been published. Meanwhile, menu-driven programs for PCs are available to carry out light scattering calculations for spherical particles (Reist 1990). In spite of these theoretical possibilities, OPCs are usually calibrated with monodisperse test aerosols of known size and refractive index. Instruments using incandescent lamps for particle illumination have been calibrated by Whitby and Vomela (1967), Liu, Berglund, and Agarwal (1974), Liu et al. (1974*a*), Willeke and Liu (1976), and Fissan, Helsper, and Kasper (1984). Experimental calibrations of white light as well as laser light counters with various test aerosols have been carried out by Gebhart et al. (1983), Chen, Cheng, and Yeh (1984), and Liu, Szymanski, and Ann (1985). More recent papers published by Liu and Szymanski (1986), Plomb, Alderliesten, and Galjee (1986), and Szymanski and Liu (1986) report experimental work on laser particle counters.

For the production of monodisperse test aerosols several generation principles can be used. Aerosols with a high degree of monodispersity are obtained if aqueous suspensions of polystyrene latex (PSL) are nebulized, dried and drawn through the OPC (Gebhart et al. 1980). Another technique is the vibrating-orifice generator, which allows the calculation of the particle diameter to an accuracy of about 1% from the operating parameters of the generator (Berglund and Liu 1973). By varying the frequency of vibration, the liquid feed through the orifice and the concentration of the aerosol material in the volatile solvent, monodisperse aerosols from about 0.3 to 30 μm in diameter can be produced for calibration purposes. For the generation of submicrometer aerosol standards, the principle of electrostatic classification can be used (Liu and Pui 1974*b*). The aerosol material is either dissolved in a liquid or prepared as a colloidal suspension and then atomized through a jet nebulizer. After drying the mist a polydisperse aerosol consisting of solid particles or of low-volatility droplets remains. By passing the polydisperse aerosol through a differential-mobility analyzer, particles within a narrow size range are extracted according to their electrical mobility.

Optical Systems

In the following, typical optical arrangements realized in laboratory and commercial OPCs are presented.

The optical set-up of a forward-scattering instrument with converging illumination from an incandescent light source and a coaxial collecting aperture is shown in Fig. 15-11. This configuration is used, for instance, in the models PC 215, PC 245, PC 247, and PC 2102 of Hiac/Royco, Menlo Park, CA. Light from an incandescent lamp is concentrated in the sensing volume by a system of condenser lenses forming an illumination cone of 5° semi-angle. After the sensing volume the primary light beam is absorbed in a concentric light trap with a semi-angle of 16°. Light scattered by individual particles is collected through a coaxial aperture of 25° semi-angle and directed onto the cathode of a photomultiplier. Considering the illumination cone, scattered light covering an angular range from about 10° to 30° is collected by the system.

Experimental response curves of the Hiac/Royco models PC 215 and PC 245 measured

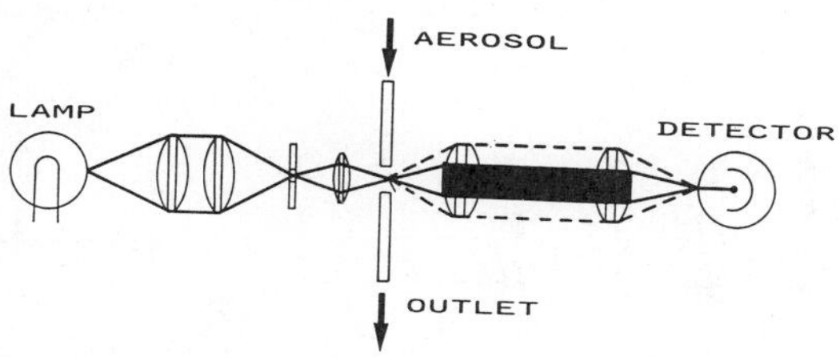

FIGURE 15-11. Optical System of a Forward-Scattering Instrument Realized, for Instance, in the Models PC 215, PC 245, and PC 247 of Hiac/Royco, Menlo Park, CA.

with PSL spheres and droplets of dioctyl phthalate (DOP) have been reported by Willeke and Liu (1976). A characteristic feature of this forward-scattering sensor is a multivalued region in the response curve between 0.7 and about 1.2 μm.

The Climet counter CI-208 (CLIMET Instr. Co., Redlands, CA) shown in Fig. 15-12 utilizes an elliptical mirror in its optical system. In the sensor, the particle sensing zone is located at the primary focal point of the elliptical mirror. High-intensity light from a quartz halogen lamp is focused on the sensing zone, where it interacts with each traversing particle.

Light scattered from each particle is collected over an angular range from 15° to 105° and is directed to a photodetector located at the secondary focal point of the ellipsoid. The sample air is surrounded by a sheath of clean filtered air, which allows a precise placement of the particles at the primary focal point. At a sample flow rate of 7000 cm^3/min a sensitivity of 0.3 μm is quoted by the manufacturer. Experimental calibrations of the Climet instrument have been carried out by Clark and Avol (1979), Mäkynen et al. (1982), and by Chen, Cheng, and Yeh (1984). A monotonically increasing response function for PSL and DOP aerosols has been found within the whole size range from 0.3 to 10 μm.

A low-angle scattering instrument developed by Gebhart et al. (1976) for the size range above the light wavelength (0.7–5 μm) is shown in Fig. 15-13. It utilizes a mercury arc lamp as a light source. A lens L_1 forms an image of the mercury arc Q on the slit D_1 which is then imaged into the sensing volume by the lens L_2. By means of a hole in the mirror M the primary beam is separated from the light scattered by the particles.

The scattered light is collected by the lens, L_3, which forms an image of the particle onto the stop D_4. Behind this stop the scattered light reaches a photomultiplier PM. The primary beam has a semi-angle of 1.5° and the receiver aperture extends from 2.5° to 5.5°. The aerosol nozzle has an inner diameter of

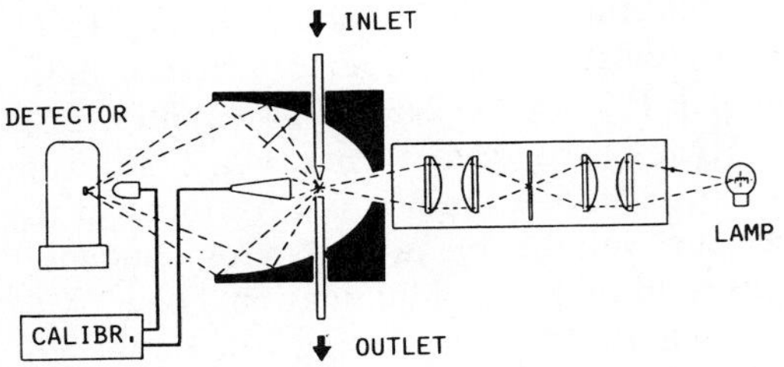

FIGURE 15-12. Optical System of the Counter CI-208 of Climet Instr. Co., Redlands, CA. (With Permission from Gebhart 1991.)

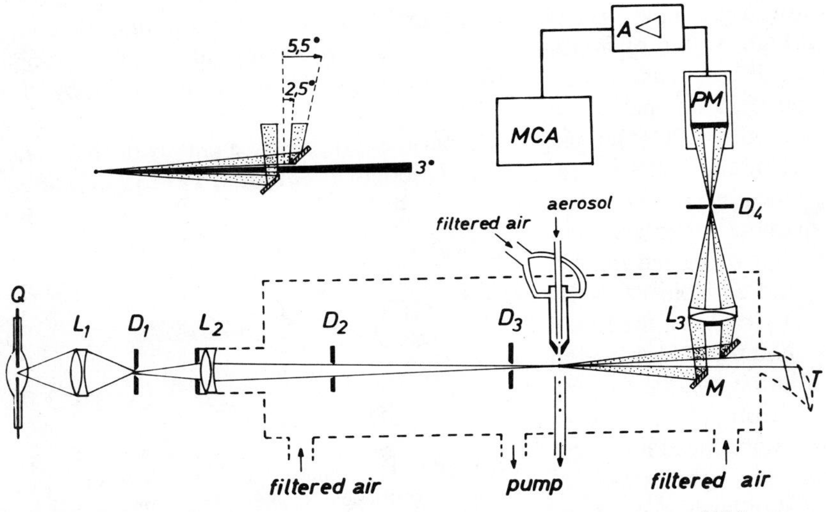

FIGURE 15-13. Optical System of a Low-Angle Scattering Instrument. (With Permission from Gebhart 1991.)

0.5 mm and by means of a clean air jacket the aerosol stream is focused onto a filament of about 0.2 mm diameter. The experimental calibration curve shown in Fig. 15-14 confirms the theoretical predictions that in an angular range where diffraction dominates, light scattering is independent of the particle material.

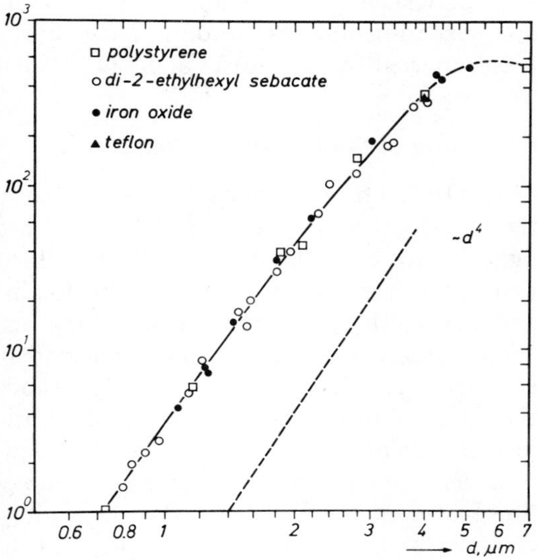

FIGURE 15-14. Experimental Response Curve of the Low-Angle Scattering Instrument. (Adapted from Gebhart et al. 1976.)

Using laser light illumination, intensities can be realized in the sensing volume that are several orders of magnitude higher than those obtained with incandescent light sources. Furthermore, a collimated laser beam needs fewer optical elements like stops and lenses, reducing the stray-light background in the chamber considerably. Three kinds of particle illumination exist with laser light: (1) the output of a laser can be focused into the sensing volume like a common incandescent light source; (2) according to Knollenberg and Luehr (1976), high illumination intensities with a low-power laser can be achieved if the sensing volume is positioned inside the laser resonator (active scattering); and (3) in the passive cavity system, the classical laser output is multireflected to create a high-intensity illumination.

A high-resolution laser aerosol spectrometer developed by Roth and Gebhart (1978) for the size range between 0.06 and 0.6 μm is shown in Fig. 15-15. The spectrometer is supplied with an argon ion laser ($\lambda = 0.5145$ μm) with an output of 2 W. The laser light is focused by an astigmatic system of lenses into the sensing volume. The light scattered by the particles is collected by a microscope objective at a mean scattering angle of 40° and is passed via a mirror to a

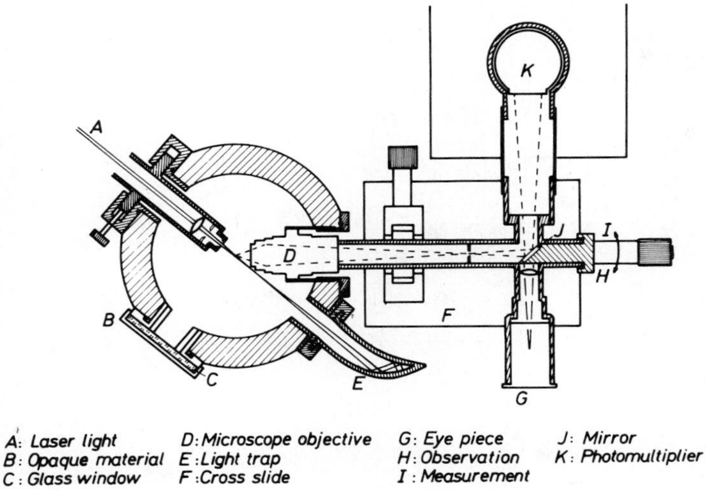

FIGURE 15-15. Laser Aerosol Spectrometer for the Submicrometer Size Range. (With Permission from Gebhart 1991.)

sensitive photomultiplier. By turning the mirror, the scattered light is reflected either on the photomultiplier cathode or into an eyepiece for observation. The sensing volume is imaged by the objective into a plane of exchangeable stops. Behind the sensing volume, the primary beam is absorbed in a light trap. The aerosol nozzle is directed perpendicular to the laser beam and to the plane of observation. The width of the aerosol stream having an original diameter of 0.2 mm is reduced by aerodynamic focusing to a small filament of about 0.03 mm width. The theoretical response functions of the spectrometer are given in Fig. 15-5 and the experimental calibration data have been published by Gebhart et al. (1976) and Roth and Gebhart (1978).

An example of an active cavity sensor is shown in Fig. 15-16. It is contained in models LAS-X, HS-LAS and LPC 101 of PMS Inc., Boulder, CO, and in the model 226/236 of Hiac/Royco, Menlo Park, CA. This sensor employs a high-Q laser cavity to achieve a high illumination intensity ($\sim$ 500 W/cm^2).

The primary collector of the scattered light is a parabolic reflector. Particles in the sample stream intersect the laser beam within the cavity at the focus of the paraboloid, which collimates the scattered light onto a 45° flat mirror. The light reflected from this mirror is refocused onto a photodetector by an aspheric lens. The whole system collects scattered light from 35° to 120° providing a 2.2 π steradian solid angle. The aerosol stream is aerodynamically focused. The theoretical response functions of this sensor have been published by Hinds and Kraske (1986) and experimental calibration data have been reported by Szymanski and Liu (1986).

Formation of the Sensing Volume

Most OPCs utilize an aerosol nozzle to blow the particles through the beam of illuminating light. In this case the sensing volume is limited by the cross-section of the aerosol stream and the width of the light beam. Using an aerosol nozzle has the advantage that under stable flow conditions the sensing volume and the sample flow rate are well-defined. In most instruments the aerosol stream is surrounded by a clean air jacket which establishes a stable flow towards the outlet of the measuring chamber and, thus, avoids stray particles in the optical system. The clean air jacket can

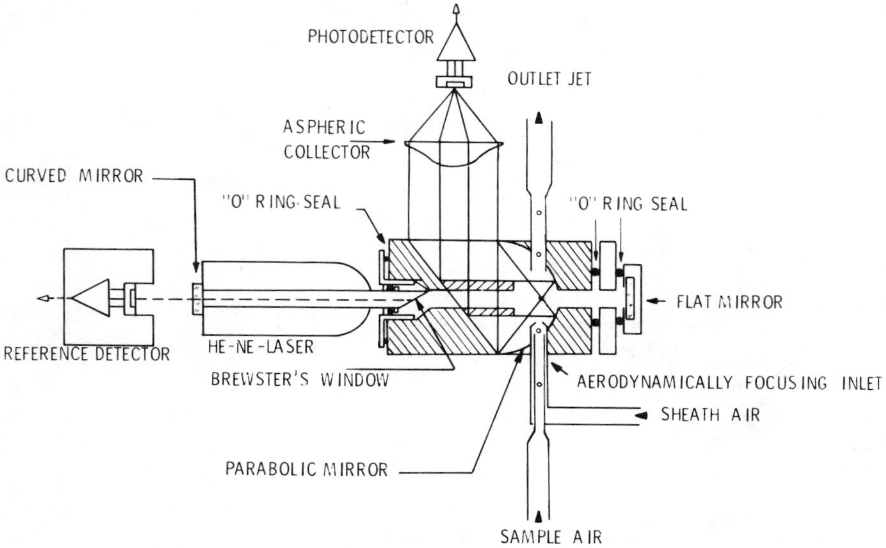

FIGURE 15-16. Set-up of an Active Scattering Laser Spectrometer Realized in the PMS Models LAS-X, HS-LAS, and LPC 101 and in the Hiac/Royco Model 226/236. (With Permission from Gebhart 1991.)

additionally be used to focus the aerosol stream aerodynamically onto a filament of less than 0.1 mm in diameter. The problem which arises in connection with an aerosol nozzle is that a sample representative of all particle sizes has to be taken out of an aerosol. Therefore, systems also have been developed that can count individual aerosol particles directly in the main stream of an aerosol.

Such an instrument is, for instance, the particle counter HC 15 of Polytec GmbH, Waldbronn, FRG, which forms its sensing volume by optical means only (Umhauer 1980). The optical system which is schematically drawn in Fig. 15-17 consists of two optical pathways, one for illumination and one for the collection of scattered light.

The lenses I and II project miniaturized images of square masks (stop I and II) into the measuring chamber, where the two optical axes cross at right angles. The images of stop I and II, which coincide with the crossing point of the two axes, form the sensing volume. Particles passing this optically defined sensitive area are illuminated with incandescent light and the light scattered by individual particles is collected at a mean scattering angle of 90°. Since an optically defined sensing volume can be made very small (about 100 μm across), high concentrations can be measured with such an instrument. The advantage of such a system is that all difficulties associated with aerosol sampling through inlets and small nozzles are avoided. There exist, however, problems in defining exactly the sensing volume and the sample flow rate. Whereas small particles are only counted when they pass the central region of the stop, bigger particles can also give rise to countable pulses at the periphery of the imaging system even if these peripheral light flashes are partially shadowed. Thus, the effective sample flow rate depends on particle size. The bigger the particles, the more peripheral light flashes contribute to a measured cumulative number distribution. This has been shown in an experimental analysis of the HC 15 counter by Helsper and Fissan (1980). The authors passed monodisperse 9 μm droplets through the sensing volume and found that about 50% of the count pulses were attenuated and simulated smaller droplets down to a detection limit of 0.3 μm.

A method to eliminate the effect of the border zone on the measured particle size distribution has been reported by Knollenberg and Luehr (1976) and Umhauer (1983). Its principle of operation is illustrated in Fig. 15-18.

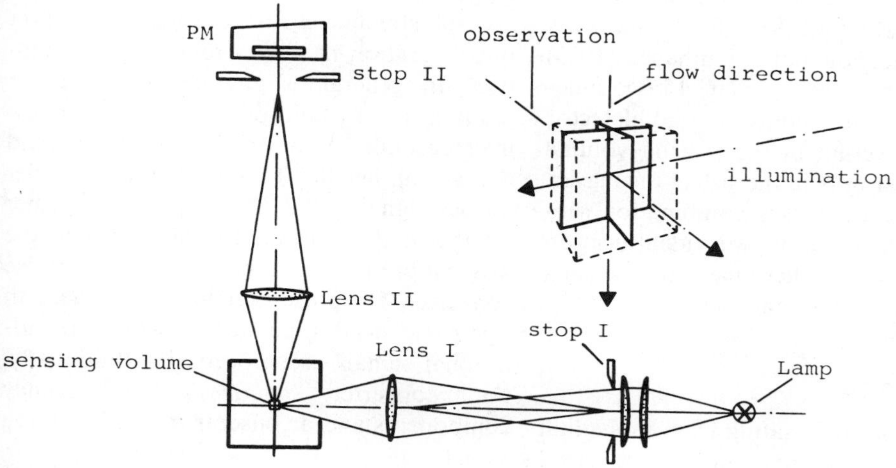

FIGURE 15-17. Arrangement of the Counter HC 15 of Polytec GmbH, Waldbronn, FRG, which forms the Sensing Volume by Optical Means Only. (With Permission from Gebhart 1991.)

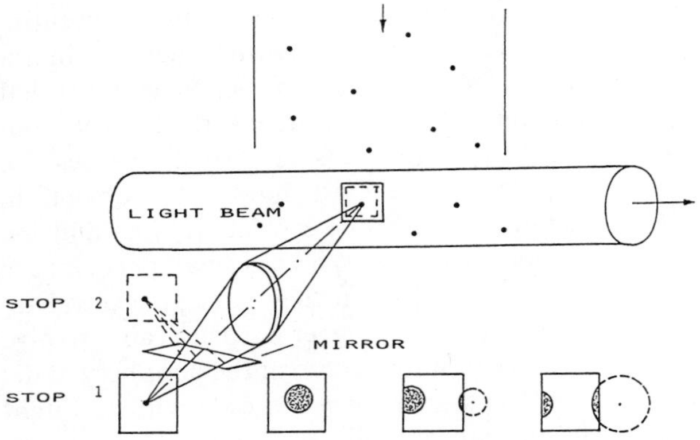

FIGURE 15-18. Elimination of Particles Traversing the Periphery of the Sensing Volume by Means of a Two-Beam System and Masked Apertures.

A beam splitter behind the collecting optics directs the scattered light through masked apertures on two independent detectors. The signals of the two detectors in conjunction with double pulse height analysis lead to a system which determines the position of a particle in the light beam and which rejects all counts originating from particles of the border zone. The model HP-LPC of PMS Inc., Boulder, CO, is a sensor of this kind that can be directly installed in a conduit for gases. Although the sensing area of this model covers only a fraction of a percent of the conduit's cross-section, a unique linear relationship between count rate and number concentration is obtained for all particle sizes above the detection limit. Umhauer (1983) constructed for the counter HC 15 two independent rectangular collector and detector units, which overlap in the sensing volume and which differ only in the size of the masks. The two different masks in combination with a double pulse analysis allow an identification of those particles which pass the sensing volume at the border zone.

Pulse Processing

The light flashes originating from individual particles in the sensing volume are converted into photocurrent pulses. Due to the fast time resolution of most photodetectors the pulse shape follows the light intensity profile inside the sensing volume. The intensity profile is approximately a Gaussian one and has a pulse width of a few to about 40 μs, depending upon the OPC used. The output pulses of the photomultiplier are usually fed to a current-to-voltage transducer, the main part of which consists of a high-frequency amplifier. Now a decision has to be made whether the pulse amplitude or its total charge (area) shall be taken as a measure of the scattered light. The charge integration method improves the resolution and the sensitivity of an OPC considerably, provided the pulse duration is very constant for all particles. A clean air jacket in conjunction with aerodynamic focusing of the aerosol stream results in such a constant pulse duration necessary for charge integration. In general, a multichannel analyzer requires a constant pulse height of a few microseconds duration in order to read the signal height. Therefore, the amplifier output signal—which is proportional either to the amplitude or to the area of the original pulse—has to be stretched in a pulse converter. The pulse stretching introduces an electronic dead time during which no additional signals are processed. Additionally, the converter removes low-frequency components and dc offset from the incoming signal.

Most OPCs have an analog output which can be connected directly to an oscilloscope. Observations of the pulse shape show that in some instruments many stray particles pass the sensing volume, producing pulses up to ten times longer than the normal ones. Since pulse times of this length exceed the recovery time of the electronics, multiple counts can originate from a single stray particle that are classified as small particles (Gebhart et al. 1983). The existence of the stray particles themselves may be explained by unstable flow conditions which prevent particles from being completely removed from the measuring chamber after their initial passage through the light beam.

Range of Number Concentration

Number concentration c_0 of the aerosol, sample flow rate Q and counting rate dN/dt of a particle counter are connected by the relation

$$\frac{dN}{dt} = c_0 Q \qquad (15\text{-}29)$$

Equation 15-29 assumes that each particle contained in the sample flow produces a single count. In practice, however, counting losses due to coincidences occur. A less than 10% loss in particle counts approximately requires (van der Meulen et al. 1980)

$$\frac{c}{c_0} = \exp\{-c_{max} Q t_r\} \geq 0.9 \qquad (15\text{-}30)$$

where t_r is the recovery time of the electronics between successive count events, including the transit time of the particle through the light beam and the pulse processing time of the multichannel analyzer. From Eq. 15-30 it can be seen that for a given recovery time t_r, measurements at high number concentrations are only possible at the expense of the sample flow rate Q and vice versa. Some maximum permissible number concentrations c_{max} derived from Eq. 15-30 for a typical recovery time of $t_r = 20$ μs are shown in Table 15-2. As can be seen, c_{max} never falls below 3.5/cm^3, which corresponds to Class 100,000 (100,000 particles per cubic foot > 0.5 μm) of the Federal Standard 209 used in contamination control (see Table 34-1 of Chapter 34 for a listing of national standards). Coincidence errors, however, are important if OPCs are applied to laboratory or environmental aerosols.

Example 15-2

A filter with an expected penetration of 0.01 for 0.5 μm particles is checked with an OPC. The OPC samples aerosol at a flow rate of 2.83 l/min = 47.16 cm^3/s and has an electronic recovery time of 12 μs. Monodisperse 0.5 μm particles with a true number concentration of $c_0 = 400$ cm^{-3} serve as the test aerosol. Calculate the experimental error due to coincidences if no dilution steps are used.

Answer. The experimental penetration is c_2/c_1, where c_1 and c_2 are the measured concentrations upstream and downstream the filter. Using Eq. 15-30 the count losses due to coincidences are,

upstream:

$$\begin{aligned}\frac{c_1}{c_0} &= \exp(-c_0 Q t_r) \\ &= \exp[(-4 \times 10^2\ \text{cm}^{-3}) \\ &\quad \times (47.16\ \text{cm}^3/\text{s}) \times (12 \times 10^{-6}\ \text{s})] \\ &= 0.797\end{aligned}$$

TABLE 15-2 Maximum Permissible Number Concentration in an OPC in Relation to the Volumetric Flow Rate Evaluated From Eq. 15-30 for an Electronic Recovery Time of 20 μs

Q (cm^3/min)	28,300	2830	283	28.3	2.83
c_{max} (cm^{-3})	11	110	1100	1.1×10^4	1.1×10^5

downstream:

$$\frac{c_2}{c_0} = \exp(-0.01 c_0 Q t_r)$$
$$= \exp[(4\ \text{cm}^{-3})$$
$$\times (47.16\ \text{cm}^3/\text{s}) \times (12 \times 10^{-6}\ \text{s})]$$
$$= 0.998 \approx 1$$

error:

$$\Delta \frac{c_2}{c_0} = \left[\frac{c_2}{c_1} - \frac{c_2}{c_0}\right] = \frac{c_2}{c_0}\left[\frac{1}{0.797} - 1\right]$$
$$= 0.25\,\frac{c_2}{c_0}$$

The lower limit c_{min} of the detectable number concentration depends on the background noise according to:

$$c_{min} > c_{ns} = \frac{\left[\frac{dN}{dt}\right]_{ns}}{Q} \qquad (15\text{-}31)$$

where $(dN/dt)_{ns}$ is the rate of noise pulses and c_{ns} an apparent number concentration produced by noise. Count noise can originate from internal particle sources, the electronics, ionizing radiation, and instabilities of the power supply. There is a general agreement that reliable concentration measurements with an OPC should start only at levels more than one order of magnitude above the background, i.e. $c_{min} \geq 10\, c_{ns}$. For the determination of the noise level of an OPC, filtered air is drawn through the instrument for several hours. Measurements of this kind have been reported by Gebhart and Roth (1986), Wen and Kasper (1986), Liu and Szymanski (1987), and Gebhart (1989*a*). Some results of background noise are presented in Table 15-3.

In most cases particles originating from the internal flow system have been identified as being responsible for the count noise. This can be concluded from the wash-out effect of the instruments and their different behavior in the positions "pump on" and "pump off." Wash-out here means that the rate of noise counts decreases with time when the instrument is supplied with particle-free air. After an exposure to environmental aerosols the instruments needed up to 15 h to obtain a stable count rate at a low level (Gebhart and Roth 1986).

In general, OPCs with incandescent light sources have somewhat lower noise levels. After completion of the wash-out most instruments are prepared to monitor Class 100 or Class 10 and in some cases even Class 1 of the Federal Standard 209 (1 cubic foot = 1 ft^3 = 28,300 cm^3). For the detection of Class 1, for instance, instruments with a sensitivity of 0.5 μm must produce less than 1 noise count per 10 ft^3, whereas laser counters with a sensitivity of 0.1 μm need a background level smaller than 3.5 counts per ft^3. In one model the count rate of noise in the first channel

TABLE 15-3 Specifications and Count Noise of a Selection of Commercial OPCs

Light Source	Model	Receiver Optics	Q (l/min)	Noise Counts per Cubic Foot: Pump on	Noise Counts per Cubic Foot: Pump off
Incandescent light	HIAC/ROYCO 4102	10°–30°	2.83	15 → 0.6	Wash-out
	HIAC/ROYCO 5000	50° range	1.0	2.4	~0
	CLIMET CI 8060	15°–150°	28.3	0.027	~0
Laser light	PMS LAS-X	35°–120°	0.3	290 → 6	Wash-out
	PMS LPC-110	35°–120°	2.83	< 0.5	~0
	HIAC/ROYCO 5100	60°–120°	28.3	120 → 0.8	Wash-out
	TSI 3755	15°–88°	2.83	< 1	
	CLIMET CI 6300	45° -135°	28.3	1–2	

Source: Gebhart and Roth (1986) and Gebhart (1989*a*)

(0.1–0.3 μm) increased with time of operation indicating that heating during the course of operation may have affected electronic noise. The high background in another instrument was caused by a leak in the housing and vanished when the instrument was put in a filtered laminar flow box (Gebhart and Roth 1986). From these experimental findings it is obvious that a certain model does not necessarily guarantee a low noise level. It is rather the status of an individual device and its history (exposure to concentrated aerosols) that determines the background noise. Therefore, instruments used for monitoring clean room Class 100 or lower should be carefully checked before the measurements.

Sensitivity and Sample Flow Rate

Manufacturers usually express the sensitivity of an OPC as the smallest particle diameter d_{min} detectable with the instrument. According to a convention of the German standard setting organization (VDI 3491, number 3), however, sensitivity should be associated with counting efficiency and the particle size with 50% counting efficiency taken as the lower detection limit d_{min}. For the determination of the counting efficiency η near the lower detection limit the measured concentration c has to be related to the concentration c_0 obtained by independent reference methods according to

$$\eta = \frac{c}{c_0} \tag{15-32}$$

During an experimental run, instruments to be checked simultaneously sample a monodisperse test aerosol from a common reservoir together with a reference instrument. Using this procedure counting efficiencies of OPCs have been measured by van der Meulen et al. (1980), Gebhart et al. (1983), Gebhart and Roth (1986), Wen and Kasper (1986), Liu and Szymanski (1987), and Gebhart (1989*a*).

A difference between c and c_0 can originate from (1) coincidence losses, (2) an incorrect sample flow rate, and (3) a decreasing sensitivity of the instrument. The effects of these parameters on counting efficiency can be separated by means of the coincidence curve of an OPC, where the ratio c/c_0 is plotted versus the reference concentration c_0. Coincidence curves of two instruments measured by Gebhart (1989*a*) are presented in Figs. 15-19 and 15-20.

Starting with test aerosols far above threshold, a single curve is obtained for all particle sizes in a semilogarithmic plot. If the extrapolation of this curve to zero concentration ($c_0 \rightarrow 0$) deviates from one, the sample flow rate of the instrument is not correct. Deviations from this curve for particle sizes approaching the lower detection limit indicate a decreasing sensitivity of the instrument. If counting efficiencies are generally evaluated by extrapolating the measured concentration ratio c/c_0 back to zero concentration, coincidence losses are eliminated and the counting efficiency reflects a specific property of an

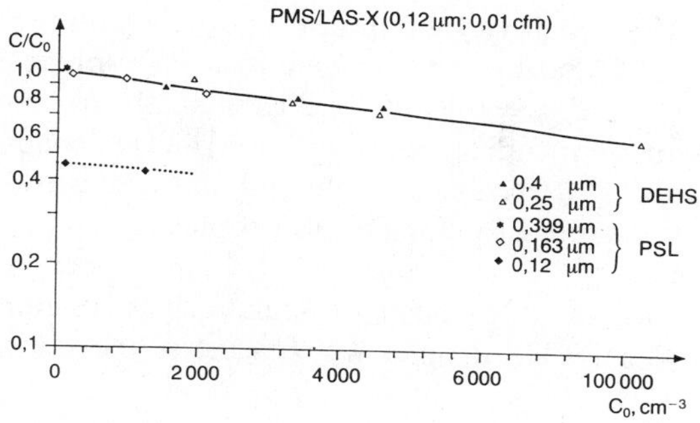

FIGURE 15-19. Experimental Coincidence Curve of the PMS Laser Spectrometer LAS-X.

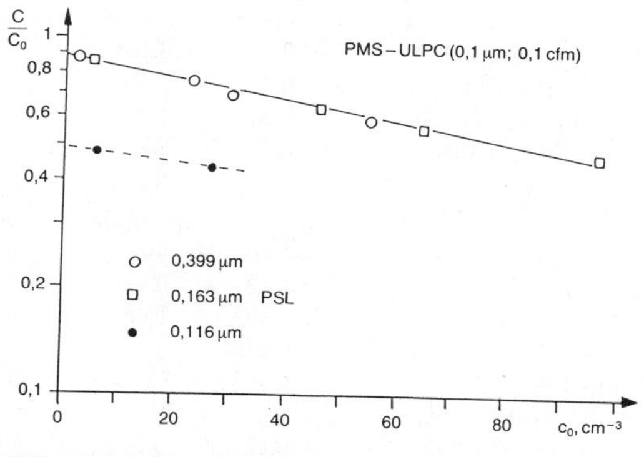

FIGURE 15-20. Experimental Coincidence Curve of the PMS Laser Counter ULPC.

instrument. Counting efficiencies of different OPCs obtained by this extrapolation technique are summarized in Table 15-4 (Gebhart 1989*a*).

In the submicrometer size range the counting efficiency decreases with decreasing particle size so that the lower detection limit is a gradual transition rather than a sharp step. The data confirm previous findings by van der Meulen et al. (1980) and Gebhart et al. (1983) that instruments using incandescent light count 0.32 μm particles with an efficiency of less than 20%. For most laser instruments, on the other hand, counting efficiencies of about 100% are achieved down to particle diameters of 0.1 μm. With the highly sensitive PMS model HS-LAS, even particles of 0.07 μm can be counted with an efficiency of about 30% at a reduced sample flow rate of 300 cm³/min.

Example 15-3

The sensitivity of a laser particle counter (LPC) has been determined with PSL spheres to be $d_{min} = 0.1$ μm. The LPC shall now be applied to droplets of di-2-ethylhexyl sebacate (DEHS). Calculate the shift in the sensitivity of the LPC for DEHS droplets.

Answer. One has to calculate the diameter of a DEHS droplet which scatters the same flux of light as a PSL sphere of 0.1 μm in diameter. In the size range of 0.1 μm the Rayleigh approximation of Eq. 15-9 can be applied, which for identical optical arrangements, reduces to

$$d_{\mathrm{PSL}}^6 \left|\frac{m^2-1}{m^2+2}\right|_{\mathrm{PSL}}^2 = d_{\mathrm{DEHS}}^6 \left|\frac{m^2-1}{m^2+2}\right|_{\mathrm{DEHS}}^2$$

PSL: $m = 1.59, \quad d = 0.1\ \mu\mathrm{m}$

DEHS: $m = 1.45$

$$d_{\mathrm{DEHS}} = d_{\mathrm{PSL}} \left[\frac{\left|\frac{m^2-1}{m^2+2}\right|_{\mathrm{PSL}}^2}{\left|\frac{m^2-1}{m^2+2}\right|_{\mathrm{DEHS}}^2} \right]^{1/6}$$

$$= 0.1\ \mu\mathrm{m} \left[\frac{\left|\frac{2.528-1}{2.528+2}\right|^2}{\left|\frac{2.103-1}{2.103+2}\right|^2} \right]^{1/6}$$

$$= 0.108\ \mu\mathrm{m}$$

In the supermicrometer size range satisfactory results are obtained. However, for particle diameters above 2 μm, inlet losses in the sampling system may again affect counting efficiency, especially if instruments with lower sampling rates are used in connection with plastic tubes. The sampling efficiency of

TABLE 15-4 Specifications and Counting Efficiency of a Selection of Commercial OPCs

Incandescent Light Model	d_{min} (μm)	Q (l/min)	Counting Efficiency at Diameter d (μm) 0.163	0.22	0.32	0.47	0.72	0.95	2.02
HIAC/ROYCO 227	0.3	0.283			0.12	0.92		1.05	
HIAC/ROYCO 4102	0.3	2.83			0.18	1.01		0.98	
HIAC/ROYCO 5000	0.3	1.0			0.18	0.89	0.91		
CLIMET CI 8060	0.3	28.3			0.15	1.02	0.98		

Laser Light Model	d_{min} (μm)	Q (l/min)	Counting Efficiency at Diameter d (μm) 0.068	0.082	0.109	0.12	0.163	0.22	0.32
PMS LAS-X	0.12	0.3				0.46	0.98		1.02
PMS LAS-X	0.9	0.06			0.98	0.98	0.99		
PMS LPC-110	0.1	28.3				0.47	0.98	1.0	
PMS HS-LAS	0.065	0.3	0.27	1.03	0.97		1.04		
HIAC/ROYCO 5120	0.2	28				0.24	0.94	0.99	1.0
HIAC/ROYCO 236	0.12	0.3					1.03	0.99	1.0

Source: Gebhart and Roth (1986) and Gebhart (1989*a*)

the particle counter Royco 245 has been investigated by Willeke and Liu (1976). Comprehensive studies of inlet characteristics and their effect on the counting efficiency of an optical sensor have been carried out by Tufto and Willeke (1982) and by Okazaki and Willeke (1987). A review of sampling and transport of aerosols has been recently given by Fissan and Schwientek (1987).

Table 15-4 indicates that the sensitivity of an OPC can obviously be increased by means of laser light illumination. On the other hand, it can be seen from this table that there exists a close connection between the sampling flow rate Q of a laser particle counter and its smallest detectable particle size $d_{\min}$. The lower the sampling flow rate, the more sensitive is the instrument. This is due to the fact that above a certain illumination intensity, light scattering by the air molecules within the sensing volume becomes the limiting factor for the sensitivity. Under these circumstances, the signal-to-noise ratio of an OPC can only be further improved by a reduction of the number of molecules which scatter light into the detector; in other words, an increase in sensitivity is only possible at the expense of the sensing volume and, in consequence, at the expense of the sampling flow rate. From Fig. 15-10 it can be seen that 1 mm^3 of air scatters about the same amount of light as a 0.22 μm particle with refractive index $m = 1.5$. This principal limitation of the sensitivity of an OPC exists regardless of the kind and intensity of illumination. The alternatives in clean room technology, therefore, seem to be either to use an instrument with high sensitivity (0.1 μm diameter particles) but a lower sampling rate or a less sensitive device with a higher sampling rate (28.3 l/min).

In a more recent development by Particle Measuring System Inc. (PMS Inc., Boulder, CO), successful attempts have been made to overcome the problem of Rayleigh scattering on gas molecules. In these models an elongated sensing volume is imaged onto a photodetector array consisting of independent elements. Owing to the facet structure of the detector unit, each element views only part of the gas molecules of the sampling flow but records the total flux of light scattered by a single traversing particle. With this segmentation technique, laser particle counters can be constructed which combine a high sensitivity (0.1 μm) with a high sampling rate of 28.3 l/min. For instance, the model LPC 110 by PMS Inc. in Table 15-4 is an instrument based on this technique.

Sizing Accuracy and Resolving Power

The sizing accuracy of an OPC describes its capability to measure the geometrical diameter of aerosol particles by classifying their photoelectric pulse heights into channels of given size intervals. For the evaluation of the sizing accuracy, cumulative size distributions of well-characterized test aerosols are compared to those measured with the OPC (Gebhart et al. 1983; Liu, Szymanski, and Ahn 1985; Szymanski and Liu 1986). The count median diameter derived from these measured size spectra in comparison to the real size of the test aerosol is then a measure of the sizing accuracy. Examples of cumulative size spectra measured with different OPCs by Gebhart et al. (1983) are shown in Fig. 15-21. The PSL test aerosols have been analyzed by electron microscopy.

Regarding the sizing accuracy, one has to distinguish between avoidable and unavoidable errors. Avoidable errors are systematic shifts on the size scale. They can be corrected by changing the linear amplification factor of the electronics. Unavoidable errors exist as long as the calibration curve has an ambiguous part or the calibration standard differs in refractive index from the aerosol to be investigated. Because of a multivalued part in the response function of the forward-scattering sensor shown in Fig. 15-11, PSL spheres of 0.95 μm in diameter are classified as 0.77 μm particles (see Fig. 15-21). An extreme example for the effect of refractive index are measurements of opaque particles (coal dust, India ink) in an OPC which applies forward scattering and has been calibrated with transparent spheres (Whitby and Vomela 1967; Willeke and Liu 1976). In this case particles of

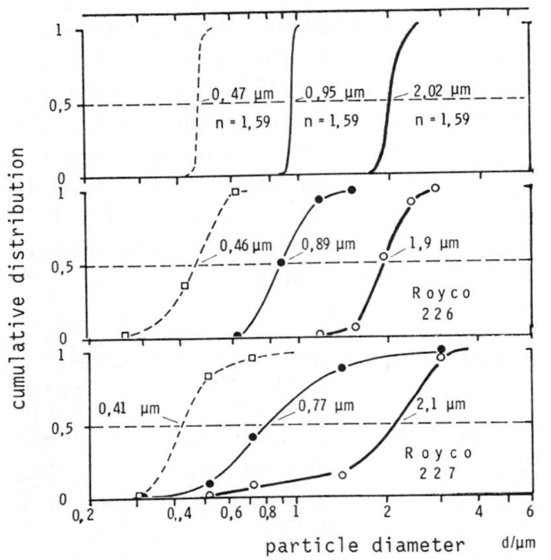

FIGURE 15-21. Cumulative Size Distributions of Different Monodisperse PSL Test Aerosols (above) in Comparison to Measurements with Different OPCs. (Adapted from Gebhart et al. 1983.)

the same geometrical size differ by about one order of magnitude in the flux of scattered light, which reduces the measured size of the particles by about a factor of 3.

To overcome these problems direct field calibrations of OPCs have been proposed. A direct aerodynamic particle size calibration of OPCs by means of inertial impactors has been performed by Marple and Rubow (1976). To obtain a single calibration point, two runs are made with the OPC. For one run an impactor is attached to the inlet of the OPC and for the other run the impactor removed. An analysis of the data from the two runs yields a calibration point corresponding to the 50% cut point size of the impactor. By changing the size of the impactor nozzle, different calibration points can be obtained.

To minimize the effect of the refractive index on the classification accuracy of the laser counter LAS-X manufactured by PMS Inc., Boulder, CO, a differential-mobility analyser (DMA, model 3071, TSI Inc. St. Paul, MN) has been used by Gebhart, Brand, and Roth (1990) to separate particle fractions of uniform electrical mobility from an unknown atmospheric aerosol. The peak of the particles carrying one charge is suitable for field calibration in terms of mobility diameters up to about 1 µm.

The resolution power of an OPC describes its capability to distinguish two monodisperse aerosols of different mean particle size. It depends on the slope of the calibration curve and on the ability of an OPC to produce uniform pulses upon exposure to a monodisperse aerosol. The main factors influencing the resolution power are the homogeneity of illumination in the sensing volume and optoelectronic noise.

Since the homogeneity of illumination is affected by the alignment of the optics, instruments of the same model may not have the same resolution due to normal manufacturing tolerances. To measure its resolution power a monodisperse aerosol of known size and standard deviation is drawn through the OPC and the instrumental broadening of the size distribution is determined. This can be done either directly by recording differential spectra (Liu, Berglund, and Agarwal 1974; Roth and Gebhart 1978) or by evaluating the slope of a cumulative size distribution (see Fig. 15-21).

A typical indication of nonuniform illumination of the particles are the recorded spectra having a sharp cutoff with respect to bigger particles and a slow decrease with respect to smaller sizes. Optoelectronic effects on the resolution power of an OPC become important only near the threshold, where the noise ripples are comparable to the particle signals. The modulation of the particle signal by quantum-statistical processes in the photodetector reduces the resolution power near threshold considerably, but has no effect for bigger particles which contain enough photons in their scattered light flashes (Liu, Berglund, and Agarwal 1974). For instruments with an optically confined sensing volume, the resolution power is additionally diminished by the effect of the border zone (Helsper and Fissan 1980).

With well-designed laser aerosol spectrometers, size resolutions comparable to that of electron microscopy can be achieved, especially in the size range below the wavelength of light (Roth and Gebhart 1978; Knollenberg 1989).

MULTIPLE-PARTICLE OPTICAL TECHNIQUES

General Considerations

Multiple-particle instruments (photometers) based on intensity measurements of scattered light are useful for concentration measurements of aerosols if certain requirements are fulfilled. For the determination of concentration ratios or relative concentrations, the composition of the aerosol (particle size distribution, refractive index) has to be constant during the experimental runs. For absolute measurements of mass concentrations the photometer has to be calibrated with the aerosol to be investigated. In both cases the instrument has to be operated in its linear range, where the number of particles in the sensing volume is linearly correlated with the photometer signal. This range of linearity is limited at high concentrations by multiple scattering and at low concentrations by the stray-light background in the chamber (see Fig. 15-22).

Stray light originates from optical elements like lenses, glass windows, and stops. Well-designed instruments are limited by Rayleigh scattering from the air molecules, resulting in nearly steady-state noise levels. By replacing air with a gas of known scattering properties, it is possible to calibrate a photometer in terms of scattering cross sections. A measure of the stray-light background is the photometer response in the presence of particle-free air. As can be seen from Fig. 15-22 a certain concentration ratio

$$\frac{c_1}{c_2} = \frac{R_1 - R_s}{R_2 - R_s} \tag{15-33}$$

is measured correctly only if the contribution of the background R_s is either negligible or subtracted from the photometer reading. In all cases where the analog signal is subjected to electronic data processing, the photometer has to be adjusted to zero response for particle-free air by means of a potentiometer.

The onset of multiple scattering depends on the specific photometer device. Example calculations for the photometer unit shown in Fig. 15-23 have been performed by Gebhart et al. (1988) based on the following assumptions: (1) a sensitive volume of 0.1 cm^3 is located in the center of an aerosol channel of 20 mm diameter and the whole channel is homogeneously filled with spherical particles of diameter d and refractive index $m = 1.45$; (2) the intensity of the illuminating beam is

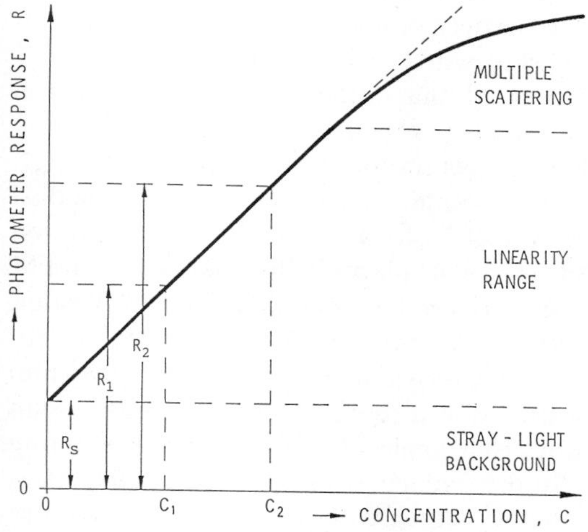

FIGURE 15-22. Linearity Range of a Light-Scattering Aerosol Photometer. (Adapted from Gebhart et al. 1988.)

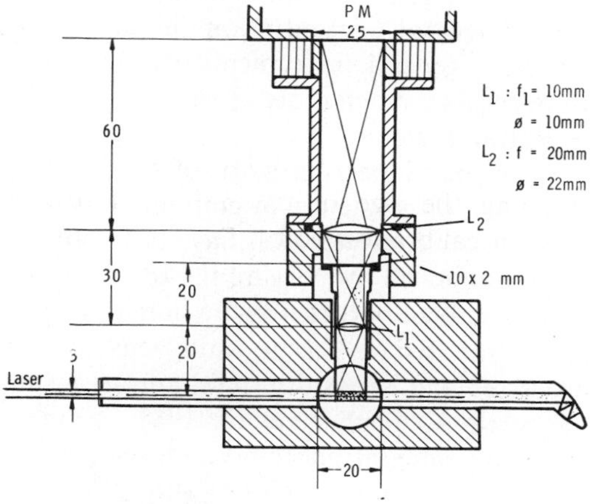

FIGURE 15-23. Laser Photometer for Aerosol Inhalation Studies.

TABLE 15-5 Onset of Multiple Scattering for the Aerosol Photometer of Fig. 15-23: (C_{MS}: Concentration Level Where the Deviation From Linearity Exceeds 10%)

d (μm)	0.26	1	2.6	10
C_{MS} (cm^{-3})	5×10^7	10^6	2×10^5	2×10^4

reduced by extinction; and (3) on its way from the particle to the receiver aperture the scattered light is attenuated by other particles inside the aerosol channel. The results in Table 15-5 show that the effect of multiple scattering depends on the concentration as well as on the diameter of the particles. The quantity C_{MS} here is defined as the concentration limit where the deviation of the photometer response from linearity exceeds 10%.

From the onset of multiple scattering in relation to the lower detection limit of 10^1–$10^2/cm^3$, it can be concluded that the linearity range of such a photometer covers at least three orders of magnitude in number concentration.

Performances in Use

Aerosol photometers in combination with inert model aerosols are nowadays applied in filter testing and in aerosol medicine. They are also useful as real-time dust monitors in industrial hygiene and for continuous recordings of aerosol concentrations in the atmosphere. In general, instruments based on light scattering are much more sensitive than light extinction systems.

Since the pioneering work of Altshuler et al. (1957), photometers in combination with monodisperse test aerosols have been used in aerosol research to measure the total deposition of aerosol particles in the human respiratory tract as well as to study gas mixing processes and deposition mechanisms inside the airways (Gebhart et al. 1988; Gebhart 1989). An integral photometer developed for aerosol inhalation studies is shown in Fig. 15-23. Its cylindrical aerosol channel can take up sample flow rates up to 1 m^3/min. An expanded parallel beam (3 mm) of a He–Ne laser traverses a system of diaphragms, crosses the aerosol channel and is absorbed in a light trap. Light scattered by the particles within the angular range of 90° ± 14° is collected by a lens L_1, whereby the illuminated particles are imaged onto a slit stop. The slit stop confines the sensitive volume (about 0.1 cm^3) and keeps away stray light from the photomultiplier. A second field lens L_2 images the aperture of lens L_1 onto the cathode of the photomultiplier. The instantaneous aerosol concentration is represented by the anode current of the photomultiplier which, after amplification, can be used for data processing. The photometer technique permits continuous recording of the aerosol concentration close to the mouth during the whole course of a breathing cycle. From such records and simultaneous measurements of respiratory volumes and flow rates the amount of aerosol in successive fractions of inspired and expired air can be evaluated. Data of this kind provide new possibilities for lung function tests (Gebhart et al. 1988).

Filter efficiencies can be checked by measuring the concentrations of uniform test aerosols upstream and downstream of a filter by means of light scattering on an assembly of particles. This technique is based on the assumption that N particles in the sensing zone of the photometer deliver the N-fold signal of one particle. A concentration–response relationship with a linearity range of at least 3 orders of magnitude is obtained. Integral photometry requires, however, a relatively high concentration upstream of the filter.

An instantaneously reading respirable dust photometer constructed for concentration levels in workplaces is the tyndallometer TM digital μP (H. Hund GmbH, D-6330 Wetzlar, FRG). The portable, battery-driven instrument measures the infrared light (0.94 μm) scattered by airborne dust particles at a mean scattering angle of 70°. Its open measuring chamber, which is insensitive to daylight, is filled with dust without elutriation of coarse particles (total dust). The photometer reading, however, is calibrated in terms of the mass concentration of respirable dust by means of a gravimetric respirable dust sampler. Experiments carried out by Armbruster et al. (1984) with monodisperse test aerosols and with real dusts confirm the instrument's principle. The photometer values obtained for real dusts can be linearly converted into mass values of respirable dust as long as ultracoarse particles are excluded. A handheld aerosol monitor HAM has been developed by PPM Inc., Knoxville, TN. It uses near-forward scattering of infrared light of 0.904 μm wavelength and its response fits the respirable fraction of aerosol. A sensitivity down to 1 $\mu g/m^3$ is quoted by the manufacturer; this makes the instrument suitable for workplace and ambient air quality monitoring. A respirable aerosol monitor RAM manufactured by GCA, Bedford, MA, allows real-time assessment of respirable dust by light scattering after aerodynamic separation of this fraction in a preselector (Lilienfeld 1983). A miniature real-time aerosol monitor MINIRAM is offered by MIE Inc., Bedford, MA. It incorporates a pulsed near-infrared light-emitting diode, a silicon detector and various collimating and filtering optics and collects light scattered over an angular range of 45°–95°. Air surrounding the instrument passes freely through its open sensing chamber and its measurement ranges cover mass concentrations from 0.01 to 100 mg/m^3.

Common mass concentration levels for respirable as well as total dusts in workplaces are in the range 0.1–100 mg/m^3, whereas typical aerosol mass concentrations in the atmosphere cover the range from 10 to about 200 $\mu g/m^3$ (Charlson, Horvath, and Pueschel 1967). Therefore, particulate pollution monitoring in the atmosphere requires aerosol photometers of high sensitivity. An instrument of this kind is the integrating nephelometer modified by Charlson et al. (1967) for air pollution and visibility studies. Its main component is a 15 cm diameter pipe of 2 m length with all optical and sampling elements contained therein. The light source is a xenon flash lamp mounted on the pipe wall and powered at a rate of 1.2 flashes/s. An opal glass in front of the lamp provides a cosine characteristic of the light source. At one end of the pipe, a photomultiplier is mounted rigidly behind a series of collimating discs. The opposite end serves as a light trap which is also constructed with discs. Sample air is drawn through the central chamber at a flow rate of about 100 l/min. The output from the phototube, a pulse of about 20 μs, is amplified and averaged over time. The device has sufficient sensitivity to distinguish gases on the basis of their Rayleigh scattering (see Table 15-1), so that different particle-free gases can be used to calibrate the instrument in terms of scattering cross sections. The results indicate that for aged atmospheric aerosols (accumulation mode), light scattering is proportional to the mass concentration of suspended particulates (Charlson, Horvath, and Pueschel 1967).

LIGHT SCATTERING BY IRREGULAR PARTICLES

General Remarks

Theoretical and experimental calibration curves of optical techniques for particle characterization are usually confined to particles of spherical shape, whereas real aerosol systems to be investigated mainly consist of particles of irregular shape. Analytical scattering solutions applicable to nonspherical particles are still in an early stage of development, so that until now limited data about special shapes have been available (Barber and Massoudi 1982). Therefore, in this section, approximations valid for particle diameters $d \ll \lambda$ and $d \gg \lambda$ are used to derive some general predictions about the influence of the particle shape on light scattering. The predictions derived from these approximations are compared with experiments on optical particle counters using irregularly shaped particles (Gebhart 1991). The experimental findings are correlated with angular scattering patterns of nonspherical particles obtained from microwave analog measurements (Zerull, Giese, and Weiss 1977) and with light scattering observations on single particles suspended in an electrodynamic balance.

Particles Smaller than the Wavelength

When the particle size is much smaller than the wavelength, the particle is subjected to an almost uniform field. In this case light scattering is mainly a volume effect and the light scattering diameter d_{sc} of such a small nonspherical particle comes very close to its volume-equivalent diameter, d_v. This can be shown theoretically by applying dipole theory (polarizability) to spheroids which are smaller than the wavelength (van de Hulst 1957; Kerker 1969). In accordance with Eqs. 15-10 and 15-11, the power of light scattered by a spheroid in relation to that by a sphere of equal volume and refractive index is given by

$$\frac{P_v^2}{P_s^2} = \left| \frac{m^2 + 2}{2L_v(m^2 - 1) + 1} \right|^2 \qquad (15\text{-}34)$$

where P_v is the polarizability of the spheroid when oscillating along its axis v and the factor L_v depends on the ratio of the axes. Some numerical results valid for a spheroid with major axis a and minor axis $b = c$ are listed in Table 15-6. For unpolarized light an average scattering ratio $(P_a^2 + P_b^2)/2P_s^2$ has to be used.

As can be seen, a ratio of the semiaxes of $a/b = 5$ results in a variation of the scattering intensity by a factor of 1.27. But due to the d^6 dependence of the scattered light on the particle diameter, the light scattering diameter d_{sc} of such a spheroid deviates only about 4% from that of a volume-equivalent sphere.

For experimental investigations agglomerates of uniform polystyrene spheres are useful objects to study the response of an OPC to nonspherical particles. Let d_{scj} be the light-scattering diameter of an agglomerate formed by j uniform spheres with diameter d_1 and refractive index m. Then a relative light-scattering diameter, F_j, of the aggregates can be defined as:

$$F_j = \frac{d_{scj}}{d_1} \qquad (15\text{-}35)$$

Relative-light scattering diameters measured with the laser aerosol spectrometer shown in Fig. 15-15 have been reported by Gebhart et al. (1976) and Gebhart (1991). Owing to the high resolving power of the instrument agglomerates up to six spheres could be distinguished. It was found that $F_j = j^{1/3}$, so that within an experimental error of 3%, d_{scj} was identical to the light-scattering diameter of a sphere of equal volume. Chen and Cheng (1984) investigated agglomerates of PSL spheres with the Royco 236 laser counter (see Fig. 15-16) and also measured a volume-equivalent response for PSL particles below 0.5 µm.

Particles Larger than the Wavelength

In the limiting case of $d \gg \lambda$, the scattered light can be considered as consisting of three components, which are due to the physical effects of diffraction, reflection, and refraction. The light-scattering components are expressed by Eq. 15-12 and illustrated in Fig. 15-3.

TABLE 15-6 Light-Scattering Diameter of Prolate Spheroids Compared to the Diameter of a Volume-Equivalent Sphere. Calculated for Unpolarized Light Using the Dipole Approximation and a Refractive Index of $m = 1.5$

Ratio of Semiaces $= a/b$	L_a	L_b	$\frac{P_a^2}{P_s^2}$	$\frac{P_b^2}{P_s^2}$	$\frac{P_a^2 + P_b^2}{2P_s^2}$	$\left(\frac{P_a^2 + P_b^2}{2P_s^2}\right)^{1/6}$
1.2	0.286	0.357	1.091	0.960	1.025	1.004
2	0.173	0.413	1.363	0.872	1.116	1.018
5	0.056	0.472	1.753	0.793	1.273	1.041

Source: Partially adapted from Gebhart (1991)

In accordance with the Fraunhofer diffraction formula, the light scattered within the forward lobe of diffraction is a function of the projected area of a particle. Outside the diffraction lobe, however, light scattering is effected by reflection and refraction on the particle surface. Therefore, specular and internal reflections can produce angular distributions which deviate considerably from the scattering diagram of a sphere of equal projected area. Consequently, the effect of the particle shape is highest if light scattered outside the diffraction part is collected through a relatively small aperture.

For irregularly shaped particles with dimensions above the wavelength of light, the effect of particle shape on light scattering is lowest if the flux of scattered light is a function of the projected area of the particle, regardless of its shape. There are two possibilities for building instruments with a projected area response: either to collect only diffracted light (low-angle scattering) or to collect all light scattered by reflection and refraction (4π-geometry scattering). The response of the low-angle scattering instrument of Fig. 15-13 to agglomerates of polystyrene spheres was investigated by Gebhart et al. (1976) and Gebhart (1991). It was found that $F_j = j^{1/2}$, so that within an experimental error of about 3%, d_{scj} was identical to the diameter of a sphere of equal projected area.

It should be mentioned in this connection that the response of a projected-area instrument to a nonspherical particle, of course, depends on its orientation in the sensing volume. It can be shown, however, by statistical considerations that even for randomly oriented agglomerates of spheres, the more frequent orientations are those that exhibit projected areas which are multiples of the cross section of a single sphere.

An instrument with a relative large receiver aperture is the Climet CI 208 counter shown in Fig. 15-12. The response of this instrument to doublet aggregates ($j = 2$) of polystyrene spheres has been investigated by Chen and Cheng (1984). Their results demonstrate that for $d_1 > 0.6$ μm the light-scattering diameter d_{scj} of such doublet aggregates is identical to the diameter of a sphere which has the same projected area as the cluster.

The results obtained from microwave analog measurements (Zerull, Giese, and Weiss 1977) show that the scattering patterns of a nonspherical particle and of a sphere of equal projected area agree quite well in the angular range where diffraction dominates. Outside the diffraction lobe, however, the intensity of light scattered by a nonspherical particle is, on an average, higher than that predicted by Mie theory and the intensity of light scattered in a fixed direction (small aperture) varies considerably with particle orientation. The experimental findings of Büttner (1983), who measured irregularly shaped particles of quartz and limestone with the right-angle scattering instrument shown in Fig. 15-17, confirm the microwave measurements. From his results it is obvious that the particles consisting of quartz and limestone scatter much more light in the 90° range than spherical particles of polystyrene and glycerin of the same Stokes diameter although the optical constants of the materials are comparable. Angular distributions of scattered laser light measured by Coletti (1984) on assemblies of isometric nonspherical particles support the experimental findings of Zerull, Giese, and Weiss (1977) and Büttner (1983).

In a recent study monosized NaCl crystals produced in a vibrating-orifice generator have been classified in the counting mode of a 2-mode laser aerosol photometer which collects light at $90° \pm 11°$ (Gebhart et al. 1988). Whereas the low-angle scattering instrument of Fig. 15-13 indicates a monodisperse NaCl aerosol at the generator outlet, the 90° aerosol photometer classifies the NaCl crystals as polydisperse. Converting the same NaCl crystals into saturated $NaCl–H_2O$ droplets, however, leads to a size distribution in the 90° aerosol photometer that can be characterized as monodisperse.

With a modified Millikan device based on the electrodynamic balance, Bottlinger and Umhauer (1989) suspended single irregular particles in an optical sensor similar to that of Fig. 15-17 and studied the signal variation for all possible orientations of the particle in the sensing volume. Based on signal spectra reflecting all orientations of a particle, the authors developed a concept to eliminate the effect of particle shape on size analysis by light scattering by single particles.

Concerning a random assembly of large irregular particles, the angular distribution of light scattered by external reflection should be about equal to that of large spheres, since there exists, on an average, a similar probability for angles of reflection. Considering light refraction by the surfaces of randomly oriented nonspherical particles, however, there is a greater possibility of high internal reflection angles inside irregular particles, which results in more total internal reflection at the expense of the refracted component scattered in the forward direction.

References

Altshuler, B., L. Yarmus, E. D. Palmes, and N. Nelson. 1957. Aerosol deposition in the human respiratory tract. *A.M.A. Arch. Ind. Health* 15:293–303.

Armbruster, L., H. Breuer, J. Gebhart, and G. Neulinger. 1984. Photometric determination of respirable dust concentration without elutriation of coarse particles. *Part. Charact.* 1:96–101.

Barber, P. W. and H. Massoudi. 1982. Recent advances in light scattering calculations for nonspherical particles. *Aerosol Sci. Technol.* 1:303–15.

Berglund, R. N. and B. Y. H. Liu. 1973. Generation of monodisperse aerosol standards. *Environ. Sci. Technol.* 7:147–53.

Born, M. 1965. *Optik.* Berlin, Heidelberg, New York: Springer.

Bottlinger. M. and H. Umhauer. 1989. Single particle light scattering size analysis: Quantification and elimination of the effect of particle shape and structure. *Part. Syst. Charact.* 6:100–09.

Brossmann, R. 1966. Die Lichtstreuung an kleinen Teilchen als Grundlage einer Teilchengrößenbestimmung. Dissertation, Karlsruhe.

Büttner, H. 1983. Kalibrierung einer Streulichtmesseinrichtung zur Partikelgrößenanalyse mit Impaktoren. *Chemie Ing. Techn.* 55:65–76.

Charlson, R. J., H. Horvath, and R. F. Pueschel. 1967. The direct measurement of atmospheric light scattering coefficient for studies of visibility and pollution. *Atmos. Environ.* 1:469–78.

Chen, B. T. and Y. S. Cheng. 1984. Optical diameters of aggregate aerosols. *J. Aerosol Sci.* 16:615–23.

Chen, B. T., Y. S. Cheng, and H. C. Yeh. 1984. Experimental response of two optical particle counters. *J. Aerosol Sci.* 15:457–64.

Clark, W. E. and E. L. Avol. 1979. In: *Aerosol Measurement*, eds. D. A. Lundgren, F. S. Harris, W. H. Marlow, M. Lippmann, W. E. Clark, and M. D. Durham, pp. 219. Gainesville: University Press of Florida.

Coletti, A. 1984. Light scattering by nonspherical particles: A laboratory study. *Aerosol Sci. Technol.* 3:39–52.

Cooke, D. D. and M. Kerker. 1975. Response calculations for light scattering aerosol particle counters. *Appl. Opt.* 14:734–39.

Fissan, H., C. Helsper, and W. Kasper. 1984. Calibration of optical particle counters with respect to particle size. *Part. Charact.* 1:108–11.

Fissan, H. and G. Schwientek. 1987. Sampling and transport of aerosols. *TSI J. Part. Instrum.* 2, 3.

Garvey, D. M. and R. G. Pinnick. 1983. Response characteristics of the particle measuring systems active aerosol spectrometer probe (ASASP-X). *Aerosol Sci. Technol.* 2:477–88.

Gebhart, J. 1989*a*. Funktionsweise und Eigenschaften Optischer Partikelzhler. *Technisches Messen tm* 56:192–203.

Gebhart, J. 1989*b*. Dosimetry of inhaled particles by means of light scattering. In *Extrapolation of Dosimetric Relationships for Inhaled Particles and Gases*, eds. J. Crapo, F. J. Miller, J. A. Graham, and A. W. Hayes, pp. 235–45. New York: Academic Press.

Gebhart, J. 1991. Response of single-particle optical counters to particles of irregular shape. *Part. Syst. Charact.* 8:40–47.

Gebhart, J., P. Blankenberg, S. Bormann, and C. Roth. 1983. Vergleichsmessungen an optischen Partikelzählern. *Staub-Reinhalt Luft.* 43:439–47.

Gebhart, J., P. Brand, and C. Roth. 1990. A combined DMPS-LAS-SYSTEM for the characterization of fine particles in gases. In *Proc. 2nd World Congress Particle Technology*, 19–22. September. Kyoto, Japan.

Gebhart, J., G. Heigwer, J. Heyder, C. Roth, and W. Stahlhofen. 1988. The use of light scattering photometry in aerosol medicine. *J. Aerosol Med.* 1:89–112.

Gebhart, J., J. Heyder, C. Roth, and W. Stahlhofen. 1976. Optical size spectrometry below and above the wavelength of light; a comparison. In *Fine Particles*, ed. B. Y. H. Liu, pp. 794–815. New York, London: Academic Press.

Gebhart, J., J. Heyder, C. Roth, and W. Stahlhofen. 1980. Herstellung und Eigenschaften von Latex-Aerosolen. *Staub-Reinhalt. Luft.* 40:1–8.

Gebhart, J. and C. Roth. 1986. Background noise and counting efficiency of single optical particle counters. In *Aerosols, Formation and Reactivity, Proc. 2nd Internat. Aerosol Conf.*, 22–26. September. Berlin. Oxford, New York: Pergamon.

Gucker, F. T. and J. Tuma. 1968. Influence of collecting lens aperture on the light-scattering diagrams from single aerosol particles. *J. Colloid Interface Sci.* 27:402–11.

Helsper, C. and H. J. Fissan. 1980. Response characteristics of a Polytec HC-15 optical particle counter. 8th

Annual Meeting Gesellschaft f. Aerosolforschung, Schmallenberg, Germany.

Heyder, J. and J. Gebhart. 1979. Optimization of response functions of light scattering instruments for size evaluation of aerosol particles. *Appl. Opt.* 18:705–11.

Hinds, W. C. and G. Kraske. 1986. Performance of PMS model LAS-X optical particle counter. *J. Aerosol Sci.* 17:67–72.

Hodkinson, J. R. 1962. Dust measurement by light scattering and absorption. Ph.D. Thesis, School of Hygiene and Tropical Medicine, London.

Hodkinson, J. R. and J. R. Greenfield. 1965. Response calculations for light-scattering aerosol counters and photometers. *Appl. Opt.* 4:1463–74.

Kerker, M. 1969. *The Scattering of Light and Other Electromagnetic Radiation.* New York, London: Academic Press.

Knollenberg, R. G. 1989. The measurement of latex particle sizes using scattering ratios in the Rayleigh scattering size range. *J. Aerosol Sci.* 20:331–45.

Knollenberg, R. G. and R. Luehr. 1976. Open cavity laser active scattering particle spectrometry from 0.05 to 5 microns. In *Fine Particles*, ed. B. Y. H. Liu, pp. 669–96. New York, London: Academic Press.

Lieberman, A. 1986. Evolution of optical airborne particle counters in the U.S.A. In *Aerosols: Formation and Reactivity, Proc. 2nd Internat. Aerosol Conf.*, 22–26, September, Berlin, pp. 590–93.

Lilienfeld, P. 1983. Current mine dust monitoring developments. In *Aerosols in the Missing and Industrial Work Environments*, eds. V. A. Marple and B. Y. H. Liu, pp. 733–57. Ann Arbor: Ann Arbor. Science Publishers.

Liu, B. Y. H., R. N. Berglund, and J. K. Agarwal. 1974. Experimental studies of optical particle counters. *Atmos. Environ.* 8:717–32.

Liu, B. Y. H., V. A. Marple, K. T. Whitby, and N. J. Barsic. 1974*a*. Size distribution measurement of airborne coal dust by optical particle counters. *Am. Ind. Hyg. Assoc. J.* 8:443–51.

Liu, B. Y. H. and D. Y. H. Pui. 1974*b*. A submicron aerosol standard and the primary absolute calibration of the condensation nucleus counter. *J. Colloid Interface Sci.* 47:155–71.

Liu, B. Y. H. and W. W. Szymanski. 1986. On sizing accuracy, counting efficiency and noise level of optical particle counters. In *Aerosols, Formation and Reactivity. Proc. 2nd Internat. Aerosol Conf.*, 22–26 September, pp. 603–06. Berlin.

Liu, B. Y. H. and W. W. Szymanski. 1987. Counting efficiency, lower detection limit and noise level of optical particle counters. In *Proc. 33rd Annual Meeting of IES*, San Jose, CA.

Liu, B. Y. H., W. W. Szymanski, and K. H. Ahn. 1985. On aerosol size distribution measurements by laser and white light optical particle counters. *J. Environ. Sci.* 28:19–24.

Marple, V. A. and K. L. Rubow. 1976. Aerodynamic particle size calibration of optical particle counters. *J. Aerosol Sci.* 7:425–33.

Mäkynen, J., J. Hakulinen, T. Kivistö, and M. Lektimäki. 1982. Optical particle counters: Response, resolution and counting efficiency. *J. Aerosol Sci.* 13:529–35.

Mie, G. 1908. Beugung an leitenden Kugeln. *Ann. Physik* 15:377–445.

Oeseburg, F. 1972. The influence of the aperture of the optical system of aerosol particle counters on the response curve. *J. Aerosol Sci.* 3:307–11.

Okazaki, K. and K. Willeke. 1987. Transmission and deposition behaviour of aerosols in sampling inlets. *Aerosol Sci. Technol.* 7:275–83.

Olaf, J. and K. Robock. 1961. Zur Theorie der Lichtstreuung an Kohle- und Bergepartikeln. *Staub* 21:495–99.

Pettit, D. R. and T. W. Peterson. 1984. Theoretical response characteristics of the coherent optical particle spectrometer. *Aerosol Sci. Technol.* 3:305–15.

Plomb, A., P. T. Alderliesten, and F. W. Galjee. 1986. On calibration and performance of laser optical particle counters. In *Aerosols, Formation and Reactivity, Proc. 2nd Internat. Aerosol Conf.*, 22–26 September, pp. 594–98. Berlin.

Quenzel, H. 1969. Influence of refractive index on the accuracy of size determination of aerosol particles with light-scattering aerosol counters. *Appl. Opt.* 8:165–69.

Reist, P. 1990. Mie theory calculations for your PC. Order: R. Enterprices, 205 Glenhill Lane, Chapel Hill, NC 27514, U.S.A.

Roth, C. and J. Gebhart. 1978. Rapid particle size analysis with an ultra-microscope. *Microscopica Acta* 81:119–29.

Soderholm, S. C. and G. C. Salzman. 1984. Laser spectrometer: Theory and practice. In *Aerosols: Science, Technology, and Industrial Applications*, eds. B. Y. H. Liu, D. Y. H. Pui, and H. J. Fissan, pp. 11–14. New York: Elsevier.

Szymanski, W. W. and B. Y. H. Liu. 1986. On the sizing accuracy of laser optical particle counters. *Part. Charact.* 3:1–7.

Tufto, P. A. and K. Willeke. 1982. Dependence of particulate sampling efficiency on inlet orientation and flow velocities. *Am. Ind. Hyg. Assoc. J.* 43:436–43.

Umhauer, H. 1980. Partikelgrößenbestimmung in Suspensionen mit Hilfe eines Streulichtzählverfahrens. *Chemie Ing. Techn.* 52:55–63.

Umhauer, H. 1983. Particle size distribution analysis by scattered light measurements using an optically defined measuring volume. *J. Aerosol Sci.* 14:765–70.

van de Hulst, H. C. 1957. *Light Scattering by Small Particles.* New York: Wiley.

van der Meulen, A., A. Plomp, F. Oeseburg, E. Buringh, R. M. van Aalst, and W. Hoevers. 1980. Intercomparison of optical particle counters under conditions of normal operation. *Atmos. Environ.* 14:495–99.

VDI-3489, Blatt 3. Messen von Partikeln; Methoden zur Characterisierung und Überwachung von Prüfaerosolen; Optischer partikelzähler. VDI, Düsseldorf, FRG.

Wen, H. Y. and G. Kasper. 1986. Counting efficiencies of six commercial particle counters. *J. Aerosol Sci.* 17:947–61.

Whitby, K. T. and R. A. Vomela. 1967. Response of single particle optical counters to nonideal particles. *Environ. Sci. Technol.* 1:801–14.

Willeke, K. and B. Y. H. Liu. 1976. Single particle optical counter: Principle and application. In *Fine Particles*, ed. B. Y. H. Liu, pp. 697–729. New York, London: Academic Press.

Willeke, K. and J. E. Brockmann. 1977. Extinction coefficients for multimodal atmospheric particle size distributions. *Atmos. Environ.* 11:995–99.

Zerull, R. H., R. H. Giese, and K. Weiss. 1977. Scattering functions of non-spherical particles vs. Mie theory. *Appl. Opt.* 16:777–79.

16

Optical Direct-Reading Techniques: *In Situ* Sensing

Daniel J. Rader and Timothy J. O'Hern

Department of Fluid and Thermal Sciences
Sandia National Laboratories
Albuquerque, NM, U.S.A.

INTRODUCTION

The characterization of airborne particles and droplets is critical in the study of a wide range of fields, including combustion, environment, industrial processing, cleanroom monitoring, and cloud and fog formation. The distributions of particle size, shape, structure, charge, and chemical composition are each important in some context. In the broadest sense, the instruments available for characterizing particles can be divided into two classes: extractive and *in situ*. In extractive sampling, a particle-laden volume of gas is removed from its environment and transported to a separate location, where the particle measurement is made. Many of the most common aerosol measurement techniques operate in this mode (see Chapter 15), as it allows the careful control of the conditions under which the measurement is made. The success of extractive techniques, however, relies on the ability to sample and transport particles without biasing the properties of interest. This condition is sometimes difficult to meet. Inlet inefficiencies, wall losses, and rapid aerosol dynamics (evaporation, condensation, coagulation) are examples of physical processes that can alter the particle size distribution. Extractive techniques can also fail when measurements need to be made in hostile environments: extremes in pressure or temperature, reactive or corrosive environments, etc. *In situ* (noninvasive) measurement techniques can overcome many of these limitations, allowing particle characterization under conditions where extractive techniques are not suitable.

Most of the *in situ* aerosol measurement techniques currently available are optically based, as the combination of optical imaging and laser illumination has provided a wealth of measurement possibilities. In fact, such a wide array of *in situ* optical techniques are available that it is necessary to limit the scope of the present review in three ways. First, we restrict discussion to instruments that are commercially available. This is a fairly strict limitation, as the laboratory development of new noninvasive particle characterization techniques is an area of active research. This decision was motivated by the desire to promote techniques which are (1) readily available to the reader (without undue investment in hardware or time), and (2) reasonably portable. Even using commercial techniques, the researcher will find that these systems are generally expensive and require significant investments in time for training. Portability is desirable as many aerosol measurements must be made in field environments.

As a second limitation, we generally consider only real-time or direct-reading equipment. This is not a serious restriction, as the majority of *in situ* techniques readily lend themselves to on-line analysis and presentation. It should be noted, of course, that subsequent (and more sophisticated) data analysis is often essential to ensure data integrity. One exception is that holography is considered. Although holograms must be first developed and later reconstructed for analysis, automated techniques are beginning to appear; so, this powerful technique has been included. Finally, we restrict discussion to techniques which provide size distribution measurements; excluded by this restriction are techniques which provide only limited information about the particle cloud. One example is a transmissometer which is capable of measuring either particle volume concentration or Sauter mean diameter (Holve and Self 1980).

In short, this chapter reviews commercially available, direct-reading, optically based, *in situ* aerosol measurement instrumentation. Among them, capabilities exist for measurement of individual particle sizes from about 0.25 to above 1000 μm, concentrations as high as $10^6/cm^3$, and speeds in the km/s range. Ensemble techniques can measure mean diameters as low as 0.01 μm. While *in situ* instruments overcome many of the limitations encountered with extractive methods, they do suffer (as a class and individually) from a wide range of new limitations. To describe these limitations, the next section provides an overview of *in situ* optical particle sizing systems. A brief section providing an introduction to light scattering theory follows. With this background in place, the chapter continues with a review of the instruments that are currently commercially available to the researcher. The individual reviews are, by necessity, short, but sufficient references are provided to help the reader to explore each method further. Although every effort has been made to include all of the available equipment, some manufacturers may have been overlooked. The chapter concludes with a section on the crucial topic of performance verification, including the issues of standards, calibration generators, and instrument comparisons.

OVERVIEW

The *in situ* measurement of particles by optical methods has been an area of active research—particularly over the last decade. Thus, many excellent reviews are available on the topic (Hirleman 1988*a*, 1983, 1984; Lefebvre 1989; Hovenac 1987). Several recent sets of proceedings contain current applications and discussions of *in situ* techniques (Hirleman, Bachalo, and Felton 1990; Hirleman 1990; Gouesbet and Grehan 1988).

It is helpful to divide optical *in situ* techniques into two general classes, based on whether they analyze single-particle events or aggregate cloud properties. Single-particle counters (SPC) generally make a size determination on one particle at a time by analyzing its scattering behavior while it passes through a well-defined (usually small) volume of high-intensity (usually laser) light. Intensity, phase, or image information in the scattered light have all been used for particle sizing. A size distribution is obtained by sizing a number of particles sufficient to ensure statistical accuracy. This class of instruments is similar, in principle, to the extractive, optical techniques covered in the previous chapter, except that the measurement volume is now located external to the instrument. SPCs generally provide a wealth of information on the counted particles, providing correlations among particle properties such as size, velocity, and time of arrival, and allowing spatial characterization of the particle field. At high number concentrations, however, single-particle counting techniques suffer from coincidence errors, which occur when more than one particle occupies the sensing volume at the same instant.

The second class of *in situ* systems, collectively called ensemble techniques, generally operate by illuminating a volume containing a large number of particles and analyzing the collective scattering. An illustrative example of an ensemble technique would be a photographic snapshot (or a hologram in three dimensions), which captures the state of a particle distribution at one instant in time. A drawback of photographic systems is that it is difficult to obtain a real-time readout of results. Real-time ensemble techniques are available which remedy this limitation, but require a mathematical inversion of the data to determine the size distribution. Ensemble techniques are well suited for measurements at high particle concentration, but become ineffective at low concentration. Generally, ensemble techniques do not provide as detailed information as SPCs, since individual particle information is lost in the averaging. Real-time ensemble techniques provide only limited spatial resolution of the particle field.

Generally, ensemble techniques measure particle concentration (number/volume), while SPC systems measure particle flux (number/area/time) (Hirleman 1988*a*). That is, ensemble techniques report the number (and sizes) of particles present in the sampling volume over the measurement time (spatial averaging), whereas SPCs report the number (and sizes) of particles *passing through* the sampling volume during the measurement time (temporal averaging). To obtain aerosol concentration, SPCs require additional particle velocity information. As Hirleman (1988*a*) points out, the distributions measured by concentration- or flux-based techniques will differ if a systematic correlation exists between particle size and velocity, as demonstrated in Example 16-1.

Example 16-1

A fluid containing a bimodal droplet distribution is rapidly accelerated prior to a measurement access window. The smaller mode consists of monodisperse droplets, with diameter $d_{p1} = 10\ \mu m$ and concentration $c_1 = 100$ particles/cm^3, which are moving with the fluid at velocity $V_1 = 10$ m/s. The larger mode consists of monodisperse droplets ($d_{p2} = 100\ \mu m$, $c_2 = 20$ particles/cm^3) which are lagging the flow at $V_2 = 2$ m/s. This aerosol is measured simultaneously by an ensemble diffraction technique (sensing volume $v = 1$ cm^3) and an SPC (sensing area $A = 1$ cm^2 normal to the flow, sample time $\Delta t = 1$ s). What is the number mean diameter measured as: (a) a spatial average by the ensemble diffraction system, (b) a temporal average by the SPC (no correction for particle velocity), and (c) a spatial average by the SPC? Assume perfect measurement by both techniques, i.e., neglect trajectory ambiguity, edge effects, etc.

Answer. (a) The ensemble technique responds to the number of particles of each size in the measurement volume. Thus, it would give a number mean of

$$\bar{d}_p = \frac{vc_1 d_{p1} + vc_2 d_{p2}}{vc_1 + vc_2}$$

$$= \frac{(1)(100)(10\ \mu m) + (1)(20)(100\ \mu m)}{(1)(100 + 20)}$$

$$= 25\ \mu m$$

(b) If no correction is made for the discrepancy in droplet velocities, the SPC would weight the mean according to the number of counts during the sample time, $n_i = A \Delta t V_i c_i$. During the sample time, the SPC would record $n_1 = 100{,}000$ counts, and $n_2 = 4000$ counts. To find the mean:

$$\bar{d}_p = \frac{n_1 d_{p1} + n_2 d_{p2}}{n_1 + n_2}$$

$$= \frac{(100{,}000)(10\ \mu m) + (4000)(100\ \mu m)}{(100{,}000 + 4000)}$$

$$= 13.5\ \mu m$$

(c) An SPC can be used to measure a spatial average concentration if the particle velocities are measured. To do this, the observed number of counts, n_i, is divided by an effective sample volume given by $A \Delta t V_i$. This calculation gives back the true concentrations ($c_1 = 100$/cm^3 and $c_2 = 20$/cm^3) given in the problem statement, and, therefore, gives the same mean diameter as in case (a).

As each SPC or ensemble particle sizing technique offers distinct strengths and weaknesses, an ideal instrument can only be defined in terms of measuring a specific set of properties for a specific aerosol in a specific environment. In this vein, Hovenac (1987) and Hirleman (1988*a*) outline an approach to *in situ* optical sizing in terms of instrument operating envelopes. The central idea is that the choice of instrument must be a two-step process: first, identify the particle properties that need to be measured and the conditions under which the measurement must be made, and second, establish that these conditions fall within the instrument's operating envelope. The final step is critical. As Hirleman (1988*a*) points out, many instruments will continue to "merrily report erroneous data and not notify the user." An instrument's operating envelope will be defined by the ability of the instrument to measure the desired property over an appropriate range to an acceptable accuracy. Hirleman (1988*a*) groups the parameters which comprise the operating envelope into three domains: particle, instrument, and environmental properties. Based on Hirleman's scheme, a general overview of the operating envelopes of *in situ* methods follows.

Particle Properties

A variety of particle properties can be of interest, including size, shape, concentration, velocity, and index of refraction. Each of these properties can be distributed among a population of particles, and the problem becomes one of measuring the related distributions. With nonspherical particles, an ambiguity arises in selecting a dimension to characterize particle size. Moreover, most measurement techniques actually infer particle size indirectly from an observation of some particle behavior (settling speed, light scattering intensity, etc.). Thus, size distribution measurements must be reported in terms of equivalent diameters: optical, aerodynamic, hydrodynamic, or electric-mobility-equivalent diameters are commonly reported. Care must be taken even when comparing among optical techniques, as the scattering behavior of the same particle will depend greatly on the details of the measurement technique used to observe it. Besides size, the particle concentration (number, surface area, or mass of particles per unit volume of gas) is also frequently of interest. A further complication arises as all particle properties can show spatial or temporal variation.

Measurement of particle size distributions demands that both particle sizing and counting be accomplished with great accuracy. High spectral resolution is required when the size distribution is itself of fundamental importance. For example, an exact description of the size distribution can be essential in understanding or predicting physical processes or in identifying origin or formation mechanisms. In some settings, however, accuracy may be less important than reproducibility. Typically, the mean size, the spread, and the shape of the distribution are all of interest. Ideally, the selected instrument's sizing range should suitably span the actual particle size range. This can complicate the characterization of wide distributions, as particle sizing over more than one order of magnitude in size is difficult to cover with one instrument in one configuration. The distribution's behavior at its tails can be important, particularly when transforming from a frequency- to a mass-weighted distribution.

A second property of interest is particle concentration: aerosol mass, area, and number per volume of gas are each of interest in some context. The ranges of concentrations which are encountered in particle measurement is impressive, ranging from less than a few particles per cubic foot in ultraclean areas to millions of particles per cubic centimeter in some industrial settings. Obviously, one technique cannot be expected to cover this entire range. In most situations, it is impractical (or impossible) to characterize every particle present; thus, it becomes necessary to infer the true aerosol properties from a measurement of some subset. Difficulties arise when the particles are only present in small numbers, as is typical at larger sizes or in clean environments. Instrument noise (phantom counts) can become important under this condition, and an effort must be made to ensure statistical significance. Instrument limitations become evident at high concentrations, as will be discussed below. Concentration measurement errors can be amplified when extrapolating volume or mass distributions from measured frequency distributions.

The particle velocity distribution can be important in understanding dispersal, transport, or flux. In some applications, the correlation between particle size and velocity is desired. Even when particle velocity is not of interest itself, it may be a limiting factor in system performance. For example, particles moving at high speeds can pose signal-processing and response-time difficulties in SPCs. If the electronics are not fast enough, the signal from a high-speed particle will broaden and its intensity peak will diminish; the result is that the particle is undersized. Size–velocity correlations can also adversely influence system performance. Generally, SPCs provide particle velocity information, while ensemble systems do not (a notable exception being pulsed photography or holography). For SPCs, a velocity measurement is typically required to infer particle concentration from the measured distribution of particle size: otherwise, faster particles will be counted preferentially in a flux-based measurement. Lower limits for particle velocity in SPCs are discussed by Hovenac (1987); upper limits are typically about 300 m/s but some systems can operate in the kilometer per second range.

Particle shape and index of refraction are less commonly of interest to the researcher, but are always important through their role in determining a particle's scattering characteristics. Some of the imaging systems discussed below are capable of recording particle shape.

Instrument Properties

An accurate determination of a particle size distribution requires that the instrument must both size and count particles accurately. Hovenac (1987) describes factors which adversely effect SPC sizing and counting performance. Although both size and count sensitivity are crucial for ensemble techniques as well, the discussion is complicated by the averaging nature of the measurement. The discussion below focuses on the measurement limits imposed by instrument features.

Perhaps the most difficult aspect of making an accurate *in situ* measurement is in defining the sample volume, as particle velocities and trajectories cannot be controlled as in sampling-type instruments (Holve 1980). This difficulty applies to both ensemble and SPC techniques, and can lead to both sizing and counting errors. For most *in situ* systems, the sample volume is determined by the intensity profile of the illuminating beam and by the geometry and characteristics of the receiving optics (apertures, stops, lenses, filters, etc.). Laser beam intensity nonuniformities within the sampling volume (in either the axial or radial directions) result in trajectory-dependent scattered intensity profiles for even monodisperse particles. For the common case of a laser beam with a Gaussian intensity profile, a particle passing through the axis of a laser beam will scatter more light than if it passed through the edge of the beam. Thus, a small particle passing through the beam axis and a large particle passing through the beam edge could give comparable scattering amplitudes ("trajectory ambiguity," Grehan and Gouesbet 1986). For intensity-based SPC techniques, such multivalued response degrades instrument accuracy. Moreover, the combination of a nonuniform beam profile and photodetector sensitivity creates the situation where the effective sample volume becomes size-dependent, e.g., small particles are detected only by passing through the central portion of the beam, whereas large particles are detected over a much larger cross section. Both ensemble and SPC *in situ* techniques can suffer this counting bias, and all SPCs require some form of sample volume correction (e.g., Holve and Self 1979*a*, *b*; Holve 1980).

One of the key parameters of interest is particle size. Several issues arise with regard to particle sizing with *in situ* techniques:

precision (repeatability), accuracy (resolution), sensitivity (lowest detectable size), and dynamic range. One requirement for sizing accuracy is a monotonic response curve (intensity or phase versus size); unfortunately, light scattering techniques are frequently multivalued due to Lorenz–Mie scattering effects (see below). Variations in particle shape and refractive index effects can dramatically affect the shape of the response curve, and will limit system accuracy unless calibrations or calculations are performed with similar particles. Many *in situ* optical systems are based on near-forward-scattering techniques, which minimize (but do not eliminate) shape and refractive index effects. The trajectory ambiguity discussed above also degrades accuracy for intensity-based techniques. All optical *in situ* techniques require that the laser beam waist be four to five times the size of the largest particle to ensure nominally uniform illumination over the particle's surface (Holve 1980). For example, Hovenac (1987) has shown for SPCs that a single large particle in a small beam could be counted as two smaller ones. Making the linear dimensions of the measurement volume much larger than the largest particles also reduces the fraction of particles that suffer edge effects (Holve and Self 1979*a*). Note that enlarging the measurement volume can increase coincidence errors, and so trade-offs must be made.

Lens imperfections, misalignment, electronic and photodetector nonlinearities, and other nonidealities can significantly degrade all aspects of system performance (Holve and Davis 1985). Beam intensity fluctuations and system misalignment transients can impair both instrument precision and accuracy. As a rule of thumb, optical and signal processing limitations generally limit the dynamic size range that can be measured (with one instrument at one setting) to about a factor of 30. Instrument noise is frequently a limiting factor in determining dynamic range, and can also influence precision, accuracy, and sensitivity.

There is always a desire for improved instrument sensitivity. For *in situ* SPCs, a lower detection limit of about 0.3 μm is typical, although sampling-type SPCs can currently detect particles to about 0.05 μm. Knollenberg (1985) describes theoretical detection limits for SPCs, and shows that the limit is dominated by background scattering from stray light or gas molecules present in the sampling volume. Interestingly, operating SPCs in a vacuum can improve instrument sensitivity by a factor of 2–6 (Knollenberg 1985). In summary, Knollenberg places the theoretical limit for SPCs operating in air or vacuum at around 0.02 μm. Some ensemble techniques offer much lower detection limits, e.g., 0.01 μm for dynamic light scattering.

High particle concentrations can also limit system performance. For example, in SPCs this can lead to coincidence, dead time, and intensity attenuation errors. Coincidence occurs when two particles occupy the measuring volume at the same time, which may be counted as a single large particle, resulting in both a sizing and counting error and consequently skewing the size distribution to larger sizes. Coincidence places an upper limit on the number concentration that can be measured without significant interference for a given system configuration. This upper limit has been shown to be proportional to the probability of interference and inversely proportional to the effective measurement volume (Holve 1980). Dead time occurs when the electronics are not ready when an event occurs because a previous event is still being analyzed; dead-time effects can reduce or skew the measured size distribution. High particle concentrations between the sample volume and the receiving optics can reduce the intensity of light scattered by the particle. The resulting error in intensity-based techniques would be to undersize all particles. In ensemble systems, multiple-particle scattering occurs at high concentrations. In this case, measurements of the size distribution become concentration-dependent. Corrections for coincidence, dead time, or multiple scattering can often be made using either hardware or software.

All of the techniques discussed in this review require sophisticated data analysis, and most require a full inversion or deconvolution to finally resolve the desired size and number distributions. Real-time ensemble instruments demonstrate the classic case of inverting a finite set of measured responses to infer an unknown distribution (Hirleman 1988*a*). For intensity-based SPC techniques, Holve (1980) has discussed the need to deconvolve the resulting intensity histograms to account for trajectory ambiguity. Although beam intensity variations have minimal effect on particle sizing with phase Doppler techniques, corrections still need to be made to account for size–velocity correlations and size-dependent sample volumes when concentration is required. The importance of proper data analysis or inversion cannot be overemphasized.

The use of exposed laser illumination by most of these systems poses eye safety concerns. Most of the systems use low-power lasers, but focusing can generate dangerous intensity levels. Proper laser safety operations are essential, and should be considered in instrument selection, location, and operation.

Environmental Properties

Refractive index gradients along the optical path can cause beam steering, with a resulting change in optical collection angles. The length of the optical path, and medium temperature and pressure gradients determine the extent of beam steering. Gas conditions (temperature, pressure, composition) also affect the gas refractive index. Laser systems are readily adaptable to high-temperature environments, as they can mitigate the influence of high thermal radiation background. There are also practical issues like optical access and window contamination that must be considered. Also, application of optical techniques in environments with high ambient light levels can lead to spurious measurements unless suitably filtered.

LIGHT SCATTERING

The field of light scattering by particles is very broad and dynamic, and a thorough presentation is well beyond the scope of this brief introduction. Instead, this section provides a limited introduction to key concepts in light scattering that are used in this chapter. For further background, the reader is referred to van de Hulst (1981), Kerker (1969, 1988), and Bohren and Huffman (1983).

Particle Light Scattering Properties

When an electromagnetic beam is projected through a particle field, some portion is transmitted through the field, while some is absorbed and some is scattered by particles in the field. Figure 16-1 is a schematic representation of light scattering from a single particle. When the scattered light has the same

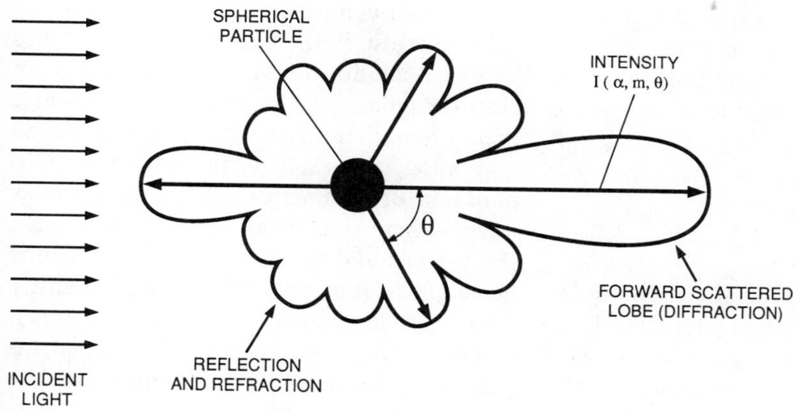

FIGURE 16-1. Single-Particle Light Scattering.

wavelength as the incident beam the scattering is called elastic scattering, while inelastic scattering describes scattering with a shift in wavelength, as is seen in Raman scattering or Doppler-shifted scattering. In general, the distribution of light scattered by a particle is a function of the particle size and shape, the incident wavelength, and the refractive indices of the particle and the surrounding medium. Both scattering and absorption characteristics of a particle can be included by describing the particle's optical properties with a complex refractive index $m = n - in'$, where the real part n describes scattered light characteristics and the imaginary part n' describes absorption. Therefore, the complex refractive index for relatively transparent particles has a very small imaginary part (absorption coefficient), while for strongly absorbing particles the imaginary part is larger, for example, $n' < 10^{-8}$ for water (Bohren and Huffman 1983) and $n' \approx 1$ for metals in the visible range (van de Hulst 1981). van de Hulst presents characteristic complex refractive indices for a number of different materials. Most handbook values of refractive index provide only the real part n, which is sufficient to calculate purely refractive properties.

Solution of the complete electromagnetic theory describing light scattering processes (Lorenz–Mie theory) is difficult, so a number of approximations have been developed which are valid within certain ranges. These ranges are determined using a dimensionless size parameter, which for a sphere is given as $\alpha = \pi d_p/\lambda$, where d_p is the diameter and λ is the incident wavelength. Note that the symbol x is also commonly used to represent the size parameter.

Rayleigh Scattering

Rayleigh scattering is typically used to describe light scattering characteristics from particles with very small values of the size parameter (diameters much smaller than the incident wavelength). Rayleigh theory is typically valid for $\alpha \ll 1$, e.g., for $\lambda = 0.6328$ μm (He–Ne laser), $d_p \ll 0.2$ μm. In the Rayleigh scattering regime, the oscillating electric field of the light wave induces an oscillating dipole in the particle, causing symmetric scattering (in the forward and backward directions) about the particle. The intensity for Rayleigh scattering is proportional to the sixth power of the particle diameter ($I \propto d_p^6$), placing severe requirements on the dynamic range of photodetectors used to measure Rayleigh-scattered light from particles of various sizes. In addition, the absolute intensity of scattered light is small, often requiring intense sources to achieve measurable Rayleigh scatter, risking particle vaporization (Hovenac 1987). As the particle size increases or as the difference between particle and medium refractive indices increases, the phase shift of light passing through particles becomes significant, the scattering becomes asymmetric, and the Rayleigh–Gans approximation must be applied to describe the scattering polar distribution.

Lorenz–Mie Theory

Particles with diameters of the same order as the incident light (say 0.1–1 μm) are too large for Rayleigh scattering, since the local electric field is no longer nearly uniform at any instant. Thus, different parts of the particle scatter portions of the incident beam with different properties. In this size range, the appropriate light scattering theory for homogeneous spheres illuminated by a plane wave is the more complex Lorenz–Mie scattering (van de Hulst 1981), which, in fact, provides an exact solution to Maxwell's equations of electromagnetic propagation. Lorenz–Mie theory is not strictly valid for light scattered by homogeneous spheres in nonuniform illumination, or for nonspherical particles. In this regime, there is a strong interaction between the particle and the incident beam. There is no simple relation between scattered intensity I and particle diameter d_p; the plot of I as a function of d_p (Fig. 16-2) is multivalued, so particles of several sizes can scatter with the same intensity. Strong dependence of scattered intensity on the refractive index further

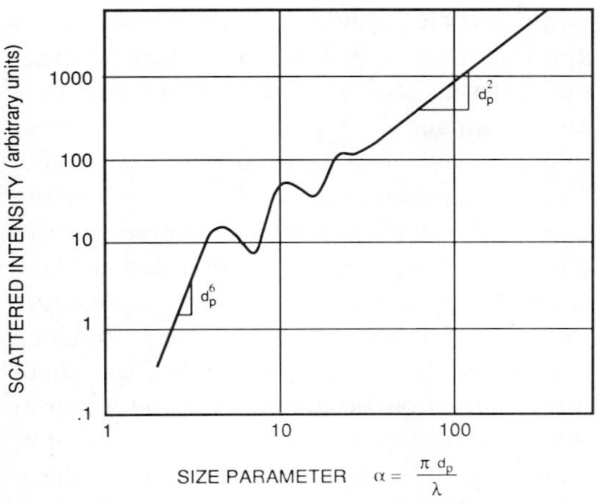

FIGURE 16-2. Lorenz–Mie Scattering Response Curve.

complicates measurements in this regime. A number of Lorenz–Mie scattering computer programs are available (e.g., Bohren and Huffman 1983).

Geometric Optics Approximation

Scattering from large particles, where $\alpha \gg 1$, can be treated using simple ray optics by considering the light wave incident on the particle to be made up of a collection of individual rays. The rays hitting the particle lead to reflection, refraction, and absorption, while those passing along the edge of the particle give rise to diffraction. The effects of each of these phenomena can be evaluated individually for particle sizes larger than the incident wavelength, unless the particle refractive index is very close to one (van de Hulst 1981).

The scattered intensity in this regime is proportional to the particle cross-sectional area ($I \propto d_p^2$), and not strongly dependent on shape or composition for aspect ratios close to one. Most geometrical optics approaches assume that the particle is illuminated by plane wave of uniform intensity and, so, are valid only for particle diameters much smaller than the beam diameter. As the particle size approaches or exceeds the incident beam diameter, corrections to standard geometrical scattering must be applied.

Diffraction (Fraunhofer and Fresnel)

For $d_p > \lambda$, diffraction begins to dominate refraction in the forward direction. For $d_p > 4$–5λ, the Lorenz–Mie theory reduces to Fraunhofer diffraction theory when limited to the near-forward direction. Diffraction from a single point particle will, in general, produce a Fresnel diffraction pattern. However, when the far-field condition is satisfied (i.e., $z \gg d_p^2/\lambda$, where z is the object to observation plane distance) the Fraunhofer diffraction pattern will be seen at the observation plane. The intensity of Fraunhofer diffraction is independent of particle refractive index, a fact that is often useful in particle sizing applications. Diffraction is dominant at near-forward scattering angles, with an intensity proportional to the particle cross-sectional area. The angular distribution of scattered light intensity is determined by the particle shape, and is inversely proportional to the diameter for spherical particles. Fraunhofer diffraction can be described theoretically using the Airy function, with the amplitude of the scattered light described by a Bessel function of the first kind.

Refraction and Reflection

Refraction in a particle can be described by Snell's law, similar to refraction at a plane refractive index interface. Reflection from a particle can be described using standard laws of reflection, with specular and diffuse reflection acting the same as for plane surfaces. The angular distribution of reflected and refracted light is independent of particle size, but depends on the complex refractive index, with higher refractive index spreading the light more strongly, and can depend strongly on particle shape (Gebhart and Anselm 1988).

Laser Optics

Coherent light interferes, while incoherent light does not. Most light sources (except lasers) generate incoherent or partially coherent light. Incoherent light is sometimes preferred for optical systems, to prevent formation of interference fringes in the image of interest, usually caused by edge diffraction. However, most optical particle sizing techniques use coherent laser light because of its intensity, monochromaticity, and coherence, needed to provide the interference key to such techniques as laser Doppler velocimetry, phase Doppler particle sizing, and holography.

Most systems use lasers with a Gaussian intensity profile (TEM_{00}). Propagation of a Gaussian beam can be described using diffraction theory (Dickson 1970), including the size and location of the focused waist, and the intensity distribution along the optical axis. When a Gaussian beam is used for light scattering measurements, a particle field is not uniformly illuminated, and, depending on the relative size of the beam and particle, the intensity may even vary over a single particle's surface. Light scattering by a single spherical particle in a Gaussian beam is discussed in Gouesbet, Maheu, and Grehan (1988), in which the theory is extended to include particles located off the beam centerline.

Multiple Scattering

The Lorenz–Mie scattering theory, and the approximations used for light scattering predictions in different size regimes, assume (among other things) single-particle light scattering. In a concentrated aerosol sample or over long path lengths, light scattered by a single particle can be rescattered by other particles in the illuminated field. The scattered light intensity measured by a photodetector will be reduced by multiple scattering. Clearly, multiple scattering must be taken into consideration in absolute intensity-based measurements, and must also be considered for diffraction-based measurements. Multiple scattering is not, however, a significant problem for phase Doppler techniques.

Nonspherical Particles

Many particles of interest in aerosol measurements are not spherical, so the detailed light scattering models must be modified to account for the particle shape. For sufficiently large particles, where the geometric optics approximation is valid, near-forward-diffraction measurements are useful because they are independent of particle shape, as mentioned above. Gebhart and Anselm (1988) discuss approximations to light scattering theory to include scattering by nonspherical particles for $d_p \ll \lambda$ and $d_p \gg \lambda$, and provide experimental comparisons using agglomerate particles and NaCl crystals. They conclude that the scattered light intensity from small nonspherical particles is within about 4% of that of an equivalent volume sphere. The light scattering characteristics of large nonspherical particles are similar to equivalent projected area spheres only when collected in the forward-diffraction lobe, or when essentially all of the reflected and refracted light is collected. Killinger and Zerull (1988) discuss scattering from nonspherical particles in the intermediate size range. They show that scattering from a 3 μm diameter rough sphere is quite close to that predicted by Lorenz–Mie theory, but that scattering from a rough, slightly nonspherical 50 μm diameter glass bead varies significantly from scattering by an equivalent sphere. They also found up to one order of magnitude variation in scattered intensity depending on nonspherical particle orientation.

Optical Imaging

In situ systems use optical configurations consisting of lenses, mirrors, filters, and apertures. All components have practical limits. An ideal lens would transform a collimated input beam to a point at the focal distance. Real lenses can cause image aberrations due to chromatic effects, astigmatism, and other effects. Physical optics shows that diffraction places limits on image resolution, so that the image of a point is a small Airy disk rather than a point. The highest achievable resolution of an optical imaging system is, therefore, the diffraction-limited resolution δ given by $\delta = 1.22\lambda l/D$, where l is the image distance and D is the limiting aperture of the system. Two points lying within a distance δ of each other will be indistinguishable in the image, as they will lie within the same Airy disk. Resolution typically limits imaging systems to examination of particles with diameters of the order 5 μm and above. Note that the resolution decreases (poorer resolving ability) linearly with measurement distance from the object. The depth of focus Δf is given by $\Delta f = 2\delta^2/\lambda$, so high resolution (small δ) leads to short depth of field, making it difficult to define the boundaries of the sampled volume, that is, to determine whether an observed particle lies inside or outside the sample volume. Since the depth of field is proportional to the square of the resolution, large particles remain in focus over a much greater distance than do smaller particles. Thompson (1984) reviews imaging techniques for particle sizing, including both coherent and incoherent light, and one- and two-lens systems, along with applications in droplet measurements.

SINGLE-PARTICLE COUNTERS: INTENSITY-BASED

This first class of instruments sizes and counts individual particles as they pass through an illuminated sample volume. As the particles pass through this region, they scatter light, which is collected over some solid angle by the receiving optics and focused onto a photodetector (Fig. 16-3). The particle size is determined by the peak intensity of the scattered light. A variety of such techniques are now available, and many reviews of the topic are available (Holve et al. 1981; Knollenberg 1979, 1981; Hovenac 1987). All of the limitations and concerns reported for SPCs in the overview apply to this class of techniques, including counting statistics at low concentrations and coincidence and dead-time effects at high concentrations. In particular, nonuniformities in the illuminating beam can result in both sizing and counting errors for this class of equipment, and some form of correction (either hardware or analytic deconvolution) is required. Intensity-based techniques are particularly sensitive to environmental features which alter either illuminating beam or scattered light intensities, such as window contamination or high particle densities between the sample volume and collection optics.

Forward-Scattering Spectrometer Probe (FSSP)

The FSSP models (Particle Measuring Systems, Inc., (PMS), Boulder, CO) are ground-

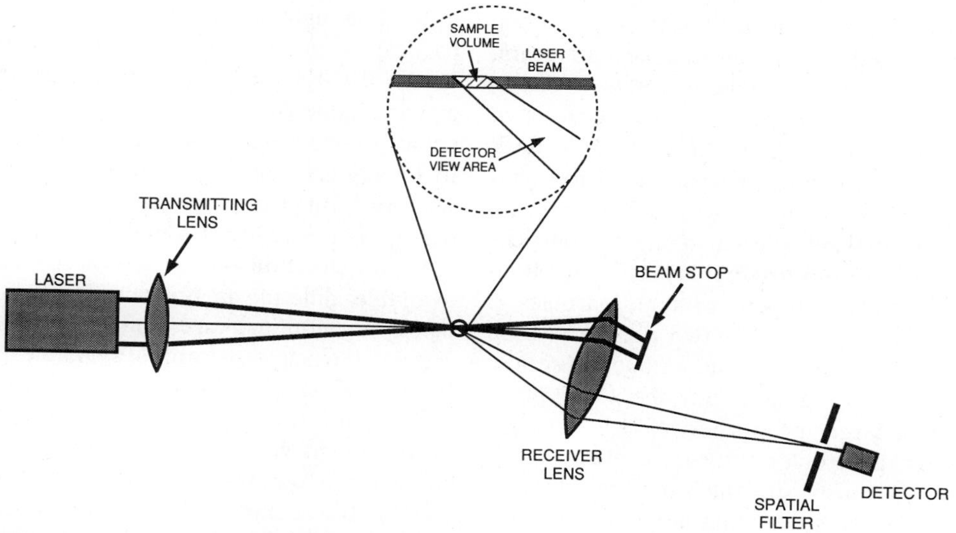

FIGURE 16-3. Example of a Single-Particle Light Scattering System.

based or aircraft-mountable probes which size particles based on the intensity of forward-scattered light as they pass through a laser illuminated sensing volume. The newer model FSSP-300 provides better sensitivity (down to 0.3 μm) and higher resolution (31 channels) over its range (0.3–20 μm) than the mechanically identical FSSP-100 (15 channels over several size ranges, such as 0.5–8.0 and 5.0–95 μm). Instrument response is up to 100 kHz. The velocity operating range for the instrument is from about 10 to 125 m/s. In the typical configuration, a particle velocity distribution is not measured by the FSSP. However, an option is available that converts the running mean particle transit time (measured to correct for edge effects) into a mean velocity. A multimode He–Ne laser beam is used to help improve the response monotonicity by diminishing the Lorenz–Mie regime oscillations. The system has been used extensively in characterizing clouds and fogs (Knollenberg 1981).

The operating principles and limitations of the FSSP have been described extensively (e.g., Knollenberg 1981). A patented dual-detector arrangement is used to size only the particles passing through a prescribed sampling volume. Briefly, the sampling volume between two probe tips is illuminated by a laser from one of the tips. When a particle enters the volume, it scatters light which is collected by optics located in the other probe tip. While a dump spot blocks the main beam, the forward-scattered light enters a beam-splitting prism and is focused onto two photodetectors (the detectors view a dark field in the absence of particles). The signal photodetector is unmasked and reports an intensity maximum used to size the particle, while the annulus detector is masked to eliminate light from in-focus, centered particles. A comparison between the two signals for each particle is used as an acceptance criterion: particles passing far from the focal plane scatter a larger proportion of light into the annular detector and are rejected. A transit time test is also performed to eliminate particles traversing the beam near an edge. This test can bias the size distribution, particularly for broad distributions with size-dependent particle velocities (Baumgardner, Cooper, and Dye 1990).

Baumgardner, Cooper, and Dye (1990) recently published a thorough review of the optical and electronic limitations of the technique, and provided an extensive bibliography of related publications. Issues addressed include sample volume, sizing, and counting uncertainties. Concentration and sizing uncertainties are found to be quite large (both about 27%). Calibrations using latex and glass spheres have been reported by Pinnick, Garvey, and Duncan (1981). Knollenberg (1981) performed an intercomparison between two supposedly identical FSSPs flown about two feet apart under an aircraft wing for over half an hour: number density readings were in excellent agreement, while water content readings varied by as much as 50% (a result of the difficulty in conversion from number to volume). As is typical of intensity-based techniques, instrument response depends on particle refractive index and shape (Baumgardner, Cooper, and Dye 1990).

Polytec Optical Aerosol Analyzers

The HC series particle sizers (Polytec GmbH, Waldbronn, Federal Republic of Germany; and Polytec Optronics, Costa Mesa, CA) detect white light scattered at 90° by single particles to measure the number distribution in the 0.4–100 μm volume-equivalent (geometric) diameter range (Umhauer 1983). The choice of white light illumination is intended to maximize the monotonicity of the scattering intensity vs. diameter response curve, and to reduce (though not eliminate) index of refraction effects. Several models are available, differing in optical geometry and, hence, in nominal size and concentration ranges. Particle size ranges available range from 0.4–22 μm (Model HC-2015) to 1.5–100 μm (Model HC-2470). The larger measurement volume required for the latter size range makes the HC-2470 more susceptible to coincidence errors, but the system is less susceptible to edge errors (Borho 1970).

Numerical correction for coincidence is possible; maximum concentration is about 10^5 particles/cm^3 for the HC-2015 and 10^3 for the HC-2470. Both models classify particles into 128 size channels, with a dynamic size range of 1:30. The velocity operating range is from 0.1 to 10 m/s (optional to 20 m/s); particle velocities are not measured.

The HC series is suited to filter efficiency testing, especially at high pressures or temperatures, and is also widely used in pharmaceutical spray sizing. The manufacturer advises that the instrument not be used for laboratory powder analysis unless material-specific calibrations are available.

Mitchell, Nichols, and van Santen (1989) investigated the performance of the Polytec HC-15/1 optical aerosol analyzer (older version of the HC-2015) with monodisperse polystyrene latex (PSL) particles with count median diameters between 0.46 and 12 μm according to British Standard BS3406 for single-particle analyzers. They also calibrated the HC-15/1 with water droplets using a single-stage impactor technique (Marple and Rubow 1976). They report a conversion factor of 2.0 ± 0.2 between water and PSL (i.e., a PSL particle of 5 μm and a water droplet of 10 μm give the same instrument response), and found that the experimental data for PSL and water agreed with Lorenz–Mie scattering calculations. Note that these calibration results demonstrate a smooth, monotonic instrument response function, although a strong index of refraction effect is observed. Fissan and Helsper (1981) used a vibrating-orifice monodisperse aerosol generator (VOMAG) to calibrate an HC-70 over a wide range of sizes of oil droplets, and also found good agreement with Lorenz–Mie calculations. For the HC-15, Fissan and Helsper (1981) found good agreement between a monodisperse VOMAG calibration and the single-stage impactor technique.

Particle Counter Sizer Velocimeter (PCSV)

The PCSV system (Insitec Measurement Systems, San Ramon, CA) is a single-particle counter that measures particle size based on the intensity of He–Ne laser light scattered in the near forward direction (see Fig. 16-3). Using near-forward-scattered (predominately diffracted) light helps reduce the particle shape and refractive index effects; thus, instrument response is mainly dependent on particle cross-sectional area. Mean particle velocity (not a velocity distribution) is determined by averaging the widths of the scattered light pulses, using an implementation (Holve 1982) of a method discussed by Hirleman (1982). The instrument's operating envelope is given by the manufacturer as: particle size between about 0.2 and 200 μm, concentration up to 10^7 particles/cm^3 for submicrometer and up to 100 ppm by volume for supermicrometer particles, and particle velocity between 0.1 and 400 m/s. A maximum particle pulse rate of 500 kHz is claimed for the system. To cover the wide range of sizes, two separate laser beams are used to form two independent measurement volumes; the narrower beam (nominal diameter of 20 μm) is used for sizing smaller particles, while the wider beam (nominal diameter of 200 μm) is used for sizing larger particles. Insitec claims an accuracy of $\pm 10\%$ and a precision of $\pm 5\%$ of the indicated size. The system is available in both bench top (PCSV-E) and fiber optic probe (PCSV-P) models. The water-cooled probe version is designed for operation in hostile environments (temperatures to 1400°C and pressures from vacuum to 10 atm) and provides gas purging to minimize window contamination. The bench top version consists of two towers supported by a common bridge, about 1 m apart with the measurement volume centrally located. Both systems contain *in situ* alignment systems to correct for beam steering in hostile environments (Holve and Annen 1984). Alignment sensitivity was explored analytically by Holve and Davis (1985).

A major feature of the PCSV system is the use of a deconvolution of the measured scattered intensity histogram to infer the size distribution (Holve and Self 1979*a*; Holve and Annen 1984; Holve and Davis 1985). The deconvolution is required due to the trajectory ambiguity. Lorenz–Mie scattering theory is used to predict the scattering response function (scattering intensity versus particle size) for the desired geometry, and has been experimentally confirmed (Holve and Self 1979*b*). Although early work characterized the intensity profile of the sample volume experimentally (Holve 1980), the current technique relies on an analytic description (Holve and Davis 1985) to generate the profile and a single-point calibration to bring predicted and observed instrument response into agreement. The validity of the sample volume analytic model was experimentally checked with monodisperse latex spheres (Holve and Davis 1985), and the accuracy of the deconvolution algorithm was established using a mixture of monodisperse oleic acid droplets (Holve and Self 1979*b*). Near-real-time output is provided via a dedicated personal computer which performs the deconvolution. Insitec provides a rotating chrome-on-glass reference reticle for instrument calibration. Calibration patterns on the reticle range from 2 to 80 μm, with either dark or clear-field formats available.

PCSV systems have been used to measure particle size distributions in coal/water slurry combustion (Holve and Annen 1984) and liquid fuel droplets and solid coal particles under combustion conditions (Holve 1980). Size distribution measurements of soda-lime glass beads in both cold and hot flows (Holve and Self 1979*b*) gave self-consistent results, showing the PCSVs ability to size particles in flames at temperatures of 1600 K.

Particle Trend Monitor (PTM)

PTM systems (High Yield Technology, Sunnyvale, CA) are intended to provide *in situ* measurement of particle contamination in semiconductor processing. In one model, a laser diode generates a beam which is projected back and forth between two mirrors, creating a "net" of light. A particle passing through the net scatters light which is collected by photodetectors to count and size (by the intensity of the scattering) the particle. A second model uses a single beam to establish the measurement volume (Borden and Larson 1989). Minimum detection is less than 0.5 μm, with the largest size detected above 10 μm. The resulting size distribution is qualitative, as no attempt is made to correct for beam nonuniformities which make the peak scattering intensity a function of both size and trajectory. As the instrument is intended only to detect variations from a baseline or trends in free particle levels, this limitation is not critical. PTM sensors are carefully constructed to allow for operation in vacuum (down to below 10^{-8} torr), and can be provided on a standard flange fitting.

LDV Signals

Laser Doppler velocimetry (LDV) is a well-established and documented technique for noninvasive measurement of particle velocities, made by measuring the Doppler-shifted frequency of light scattered by individual particles passing through a laser beam-defined measurement volume (Durst, Melling, and Whitelaw 1981). The most common LDV configuration uses crossed laser beams to define a measurement volume with typical dimensions of the order 1 mm or less. Particles passing through the measurement volume scatter light with a Doppler shift proportional to the particle speed. Speeds as high as several hundred m/s can be measured using conventional electronics. The scattered light intensity signal from each particle passage ("Doppler burst") consists of a high-frequency component superimposed on a low-frequency "pedestal" due to the Gaussian intensity distribution of the illuminating beams. After filtering the low-frequency pedestal, the remaining component (Fig. 16-4) is the Doppler frequency, directly proportional to the particle velocity. The extent of modulation of the Doppler signal (ratio of intensity of high-frequency to low-frequency component) is the signal "visibility."

A number of particle measurement techniques have been developed based on the pedestal intensity and visibility of LDV signals. Jackson (1990) provides a review of

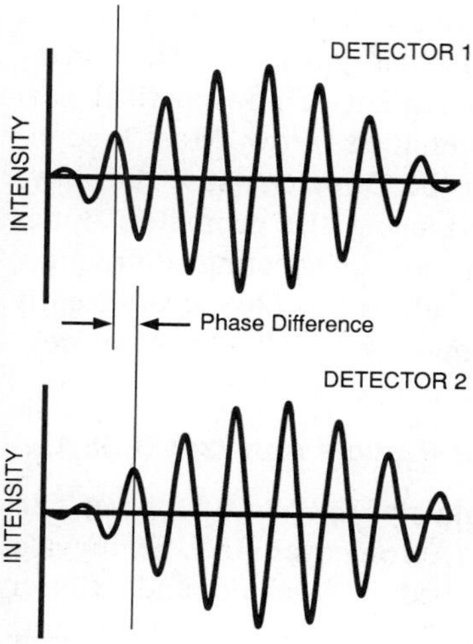

FIGURE 16-4. Characteristic Doppler Burst Signal.

interferometry-based droplet sizing techniques, including on-axis LDV visibility measurements (diffraction dominated), off-axis LDV visibility measurements (refraction dominated), and phase Doppler techniques. Such techniques are attractive since they can provide simultaneous measurement of single-particle size and velocity. Ideally, the peak intensity of the Doppler burst would be directly related to particle size. However, the Gaussian nature of the illuminating beams complicates matters due to trajectory ambiguity.

In order to avoid use of deconvolution algorithms, so that individual particle size and velocity can be determined directly, techniques have been developed which modify the incident beam intensity profile. Hess (1984) describes and demonstrates measurements made with a two-color system using a small pointing beam inside a larger Gaussian laser beam of different wavelength. The pointing beam defines a relatively uniform portion of the large Gaussian beam. Particles are only measured when detected by the pointing beam, indicating that they are in the uniform portion of the outer beam, so the scattered pedestal intensity can be recorded and the size calculated free of trajectory ambiguity. The MetroLaser PCS-100 and PCS-200 (MetroLaser, Irvine, CA) instruments use a similar technique to provide a uniform measurement region for combined LDV, particle sizing, and concentration measurements. The system can be configured to collect diffracted light in cases where the particles are of unknown shape or refractive index, and is designed to operate with no dead time, by collecting groups of particle scattering events before processing the raw data. The manufacturer-stated operating envelope includes a 0.4–6000 μm diameter range, with 2% typical resolution and a 30:1 dynamic range, and a velocity range up to several thousand m/s. Data rates up to 3×10^6 particles/s can be measured.

Grehan and Gouesbet (1986) describe a laboratory system in which the particle sizing laser beam was modified to a uniform (top-hat) profile using a Gaussian absorption filter. They state that use of an optically generated top-hat beam has several advantages over using the nearly uniform central portion of a Gaussian beam. However, true simultaneity of particle sizing and velocity measurement was not achieved with their optical system.

Another approach to using LDV signal intensity for particle size determination is by measurement of signal visibility. Lower visibility generally indicates larger scattering particles (Durst, Melling, and Whitelaw 1981). Farmer (1972) presents the theoretical basis for this technique. Takeo and Hattori (1990) apply Lorenz–Mie theory to calculate signal visibility and show that, for nonabsorbing particles, the particle size–visibility relationship is oscillatory. This finding is experimentally verified by measurement of light scattered by silicon resin particles. For absorbing particles, the visibility curves are smoother.

SINGLE-PARTICLE COUNTERS: PHASE-BASED

The phase Doppler technique is a laser Doppler velocimeter (LDV) based method for simultaneous measurement of single-particle size and velocity. This technique is not intensity-dependent like the previous group of SPC techniques, and can offer superior performance by minimizing effects such as beam attenuation or window fouling. A phase Doppler system measures the spatial and temporal frequency of the Doppler-shifted light scattered by individual particles passing through a laser-beam-crossing measurement volume. Phase Doppler systems use multiple photodetectors to sample slightly different spatial portions of the light scattered by individual particles. Figure 16-4 demonstrates high-pass-filtered Doppler bursts measured by two such detectors. The phase shift between the two signals is a measure of the scattered light spatial frequency, which can be directly related to the particle diameter, refractive index, and receiver geometry. The linear relationship between the spatial frequency of the scattered light and the particle diameter, for a fixed refractive index and receiver geometry can be shown using either Lorenz–Mie scattering theory (Saffman, Buchave, and Tanger 1984) or geometrical optics (Bachalo and Houser 1984). Particle velocity is related to the temporal frequency in the same manner as in conventional LDV. Figure 16-5 is a schematic layout of a generic phase Doppler system.

Particle sphericity is required since the phase shift is calculated for either rays refracted through spherical particles of known constant refractive index or reflected off the surface of reflective particles. Preliminary work on measurement of nonspherical particles will be presented below. Most theoretical analyses of the Doppler phase shift have been performed using the geometric optics approximation, and neglecting diffraction, since scattered light is collected at sufficiently large off-axis angles (typically 30° or more).

Phase Doppler Particle Analyzer (PDPA)

The fundamentals of the phase Doppler particle analyzer (Aerometrics, Inc., Sunnyvale, CA) are described in Bachalo and Houser (1984). It consists of a laser, transmitting optics, receiver optics package, signal processors, and data collection and analysis software, with all operations, data collection and analysis controlled by a microcomputer. The Aerometrics system can be supplied to measure one or two velocity components in addition to particle size. The manufacturer-stated operating envelope includes a 1–8000 μm diameter range, with 5% typical accuracy and a 35:1 dynamic range, and a velocity range from 1 to 200 m/s, with 1% typical accuracy. This system also calculates number density based on the number of particles passing through a calculated size-dependent measurement volume (to correct for

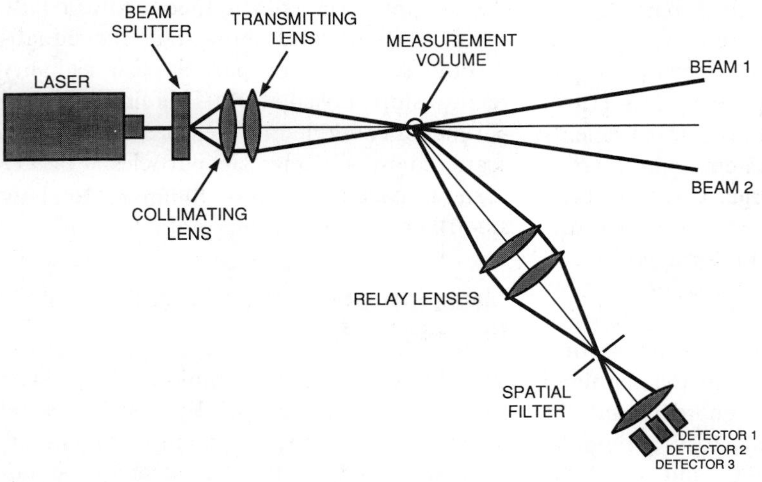

FIGURE 16-5. Layout of a Phase Doppler System.

trajectory ambiguity effects). The maximum measurable number density is $10^6/cm^3$.

There has been an abundance of recent work with this instrument, both in instrument performance characterization and in applications in industrial and research settings. Jackson (1990) presents an overview of the PDPA instrument and discusses the effect of the rotating diffraction grating (used as a combined beam splitter and frequency shifter) on PDPA concentration and velocity measurements. Variations in the spacing of etched lines on the diffraction grating can lead to fringe asymmetry in the measurement volume, leading to difficulty in the determination of measurement volume size. Imprecision in grating rotation can lead to an artificially high measured root mean square velocity. An analysis of PDPA counting errors due to data transfer dead time and coincidence losses (Oldenburg 1987) shows that the data transfer losses should be below 5% for data rates less than 2000 samples/s with direct memory access enabled, and that a new, faster Aerometrics processor improves this to less than 3% at data rates less than 70,000 samples/s. Oldenburg also discusses error due to particle coincidence and provides a correction curve for such losses up to about 15%.

Dodge, Rhodes, and Reitz (1987) performed liquid droplet measurements in sprays with the PDPA and noted that the PDPA was very sensitive to optical alignment. McDonell and Samuelson (1990) performed an evaluation of the sensitivity of PDPA measurements to operator input parameters, specifically the photomultiplier tube (PMT) gain voltage and the frequency shift level. They found that mean velocity values were insensitive to PMT voltage and frequency shift, but that fluctuating velocities were strongly dependent on PMT voltage and somewhat dependent on frequency shift, because of wobble of the rotating diffraction grating. Volume flux measurement errors were caused by inconsistencies in the determination of probe volume size, and by the strong influence of PMT voltage and the weak influence of frequency shift on particle sizing. They conclude that PDPA operations must be made with a carefully chosen standard operating procedure, and that a need exists for a reference calibration standard to guide operator selections of setup parameters for the PDPA instrument. Ceman (1990) has recently developed such a standard calibration device, a plastic disk containing a known distribution of well-characterized glass beads (see Verification section below).

Dressler and Kraemer (1990) calibrated the PDPA using a multijet droplet generator. They found PDPA measurement accuracy of about 5%. In addition, up to 5% measurement error could be caused by lens aberrations. They concluded that the PDPA system should be refocused for each different beam spacing to minimize lens spherical aberration errors. They also found that measurements were quite sensitive to the PMT gain voltage, in that a 50 V difference in gain voltage could lead to spurious data validations and artificially broadened distributions. They concluded that the lowest usable gain setting should be employed for measurements. A determination of this value can be difficult in the measurements of unknown, transient particle distributions.

Ceman (1990) performed a detailed investigation into PDPA oleic acid droplet sizing performance in the 3.5–25 μm range using a vibrating-orifice monodisperse aerosol generator and several different PDPA receiver geometries. He found that there were nonlinearities in PDPA measurements of particles below some critical diameter, in agreement with previous experiments (O'Hern, Rader, and Ceman 1989; Ceman, O'Hern, and Rader 1990) and calculations (Al-Chalabi et al. 1988) using both Lorenz–Mie theory and geometrical optics. Saffman, Buchave, and Tanger (1984) found that a large receiver lens collection solid angle tended to damp these oscillations. Recent work by Aerometrics (Sankar et al. 1990) indicates that reflections from the surface of the droplet may also contribute to the oscillations in the phase versus diameter curve in the smaller-diameter regime. Their work suggests that reflections can be minimized and linearity attained by collecting light at an angle close to the droplet

Brewster angle; however, Ceman's results show that droplet sizing oscillations can be substantially reduced by using a larger collection solid angle (shorter-focal-length receiver lens), but that working near the droplet Brewster angle provided little or no additional improvement.

Alexander et al. (1985) used the PDPA for measurement of nonspherical methanol droplets produced by a vibrating-orifice aerosol generator. The mean droplet diameters were 98 μm, with aspect ratios ranging from 0.7 to 1.4. The PDPA diameter measurements were very sensitive to aspect ratio, with the greatest error (45% oversizing) for elongated ellipsoids oriented perpendicular to the horizontal fringes in the PDPA measurement volume. Breña de la Rosa et al. (1989) have shown that the PDPA can be used to examine the shape of large nonspherical bubbles in water (1200–1800 μm diameter) by making separate measurements of the bubble major and minor diameters. This technique could possibly be extended to small particle measurements, although such an extension does not appear to be straightforward, especially for particles with rough or angular surfaces.

Recently, the PDPA has also been employed in a large number of applications. Hardalupas, Taylor, and Whitelaw (1988) used both Aerometrics and custom-made phase Doppler devices to measure solid particles transported in a gas–solid dusty jet flow, and found that asphericity of the glass beads used as seed particles in the flow limited the measurement accuracy. Bachalo, Rudoff, and Sankar (1990) presented details of PDPA operations including data analysis routines, and applied the PDPA to measurements of liquid spray droplets in both cold and combusting regimes. Droplet velocity time records showed significant spray fluctuations, leading to droplet cluster formation. Reitz (1990) applied the PDPA for droplet measurements in a pressure swirl atomizer over a wide range of operating conditions.

Particle Dynamics Analyzer (PDA)

The Particle Dynamics Analyzer (Dantec, Inc., Skovlunde, Denmark) is described theoretically in Saffman (1987) and with applications in Saffman, Buchave, and Tanger (1984). This system is similar to the Aerometrics PDPA, except that (1) a Bragg cell is used for frequency shifting instead of a rotating diffraction grating, (2) a built-in laser diode is used to generate signals for calibration of the receiving optics, and (3) signal phase is measured using a cross-correlation technique. The manufacturer-stated operating envelope for the PDA system includes a 0.5–10,000 μm size range, with 4% typical accuracy and a 40:1 dynamic range, and a velocity range to greater than 500 m/s, with 1% typical accuracy. No number density measurement capability is currently claimed.

SINGLE-PARTICLE COUNTERS: IMAGING

Determining a particle's properties by direct imaging is among the earliest techniques used in particle measurement: consider the optical (and subsequently the electron) microscope. A significant advantage is that the issues of shape and index of refraction, which complicate single-particle scattering measurements are avoided. In fact, imaging techniques provide one of the few avenues for investigating particle shape. The accuracy of single-particle imaging systems is limited by Fresnel diffraction and depth-of-field effects (Hovenac 1987). Fresnel diffraction blurs image edges and complicates sizing. The depth-of-field effect arises from its dependence on particle size, with the result that large particles remain in focus over a greater axial distance than smaller ones.

Knollenberg (1970, 1979, 1981) designed an automated, *in situ*, single-particle, optical imaging system which is commercially available (Particle Measuring Systems, Inc., Boulder, CO) as the Optical Array Imaging Probe (OAP). In this family of probes, a collimated laser beam defines a measurement volume located between two sensing tips that extend forward from the main body of the system. Receiving optics direct the beam to illuminate a linear array of photodiodes. A particle passing through the measurement volume casts a shadow on the array, resulting in a decreased signal from the individual elements that lie in the shadow. Three methods are used to analyze the resulting data from the array: 1-D, 2-D, and grey level processing. In a 1-D OAP system, the array elements are read and latched during the particle transit in a way that only provides particle size information. In a standard 2-D OAP system, the entire 2-D image of the particle is stored in high-speed memory as a series of "snapshots" of the particle during its transit. For each image, the status of each of the 32 elements in the photodiode array is recorded by one bit, indicating if the element is being shadowed or not. The great advantage of acquiring the 2-D particle images becomes apparent when measuring nonspherical particles. In the grey probe 2-D OAP system, a 64-element array is used where each element reports one of the four shadow levels. The increased sophistication of the grey probe provides twice the resolution (twice as many elements) of the standard 2-D system, as well as providing depth-of-field information.

For all of these imaging systems, both instrument resolution and sizing range depend on physical spacing of the array elements (typically 200 μm), magnification, and particle velocity. The latter requires that the user identify the expected velocity range in order to configure an OAP system. In addition, all of these systems reject particles (resulting in counting bias) which shadow elements at the edge of the array, as the fraction of the particle falling outside the array cannot be determined and, thus, precludes correct sizing. Depth-of-field rejection criteria can also be set for the grey probes, requiring that the particle must shadow at least one array element greater than a specified level to be recorded. One difficulty is that the high-speed particles may not give a sufficiently dark shadow (due to electronic limitations) to meet this acceptance criterion.

Rugged, pylon-mountable airborne versions of either the 1-D, 2-D, or grey-probe OAP configurations are available for either cloud droplet (6 cm between probe tips) or precipitation measurements (26 cm between probe tips). Particle sizing ranges and resolution depend on the particular models, which differ in the number of array elements and optical configuration used. The cloud–droplet models are suited to sizing smaller particles, with ranges such as 10–620 μm (with 10 μm resolution) or 200–6000 μm (with 200 μm resolution). The precipitation models are suited to sizing larger particles, with ranges such as 50–3100 μm (with 50 μm resolution) or 150–9300 μm (with 150 μm resolution). The resolution limits given above assume instrument operation at aircraft speeds (100 m/s); significant improvement in instrument resolution can be achieved at lower velocities. The lower limit for OAP sizing (somewhere between 1 and 10 μm according to Knollenberg 1979) results from the vanishingly small depth of field at these sizes. Particle velocities are not measured explicitly, but could be recovered by later analysis of the image sequences using the known imaging frequency.

A ground-based precipitation OAP is available in either the 1-D (droplet sizing range 200–12,400 μm with 200 μm resolution from 62 size channels) or grey probe (droplet sizing ranges 200–12,400 or 70–4340 μm with 200 or 70 μm resolution, respectively, from 62 size channels) configuration. The system is contained in a weatherized package, with the two sensing tips extending above the top (which is covered with an antisplash material). The distance between probe sensing tips is 50 cm, which provides a large sampling area for measuring the droplets in free fall.

ENSEMBLE TECHNIQUES: PARTICLE FIELD IMAGING

Imaging systems are useful since they can "freeze" the motion of a particle-laden flow, allowing later analysis and providing a permanent record of transient events. In addition, imaging techniques make no assumption of sphericity, and can be used to allow a visual examination of the shape of the individual particles, along with statistical information on the particle size distribution, concentration, velocity distribution, etc. Imaging measurements provide a primary measurement standard, since individual particle sizes are measured directly. In addition, an archival record of the particle field is formed, allowing subsequent examination by various means. However, field imaging techniques do not provide real-time analysis and imaging resolution is often degraded by dense particle fields, window effects, and other common conditions. A number of imaging techniques are available for measurement of particle fields, including photography and holography (often combined with image processing). Discussions of photographic and holographic techniques are included here because of their widespread use and unique capabilities. Several firms can provide custom design and assembly of aerosol sizing equipment using these techniques.

Photographic investigations are attractive since the technique is well established and high resolution is possible. Photography of small particles places special restrictions on the process, often requiring very short exposure times to avoid blurring of moving particles and high resolution to detect the size range of interest. The achievable resolution of photographic systems is often sufficient for particle measurements; however, the depth of field suffers. Image processing techniques promise to improve the speed and accuracy of photographic studies, and have been applied in numerous studies (e.g., Oberdier 1984; Lavergne et al. 1988).

Holography has become a fairly common technique for the study of particles with diameters larger than about 5 μm. A hologram is an interference pattern formed by the mixing of two coherent wave components: a subject wave reflected or scattered from the object or field of interest and a reference wave. Holographic reconstruction creates three-dimensional (3-D) images of the original illuminated volume, which can then be examined in detail for particle size and shape, as well as velocity and acceleration in systems with multiple-pulse illumination. The 3-D aspect allows simultaneous matching of the requirements for high resolution and good depth of field, unlike the photographic process. In addition, image premagnification (before recording) can allow examination of smaller particles, at the cost of reducing the sampled volume. A complete examination of the holographic process can be found in Collier, Burkhardt, and Lin (1971). Aerosol particle examination was historically one of the first practical applications of holography (Thompson and Ward 1966). Reviews of the use of holography for particle field measurements are given in Thompson (1974), Trolinger (1975, 1980), and Tyler and Thompson (1976). Holographic images of particle fields contain information on the size, shape, and 3-D spatial position of each individual particle comprising the field. Trolinger (1980) demonstrates that the great depth of field of a hologram gives it several orders of magnitude greater capacity than a photograph for storing volumetric information.

Both standard dual-beam and in-line or Fraunhofer holographic techniques have been applied to aerosol studies. The Fraunhofer holographic technique uses a single laser beam as both reference and subject beam during hologram recording. This is accomplished by illuminating the sample volume with a collimated coherent light beam, part of which is scattered by objects in the sample volume, while the remainder passes through the volume undiffracted, acting as a collinear reference beam. The hologram is formed by recording the interference of these two beam components on a high-resolution film. Figure 16-6A is a schematic diagram of a generic Fraunhofer holographic recording system. The in-line holographic technique is simpler and requires considerably fewer optical components than the off-axis or dual-beam method and minimizes the recording material resolution requirements. However, in-line holographic real images generally contain a high background noise component, caused by the background presence of the virtual image and out-of-focus objects. The Fraunhofer holographic technique is limited to cases in which the sample is composed of a dilute concentration of small particles in order to maintain a sufficiently strong undiffracted collinear reference beam.

Some advantages of the off-axis holographic technique are that it (1) allows the use of diffuse illumination of the sample volume,

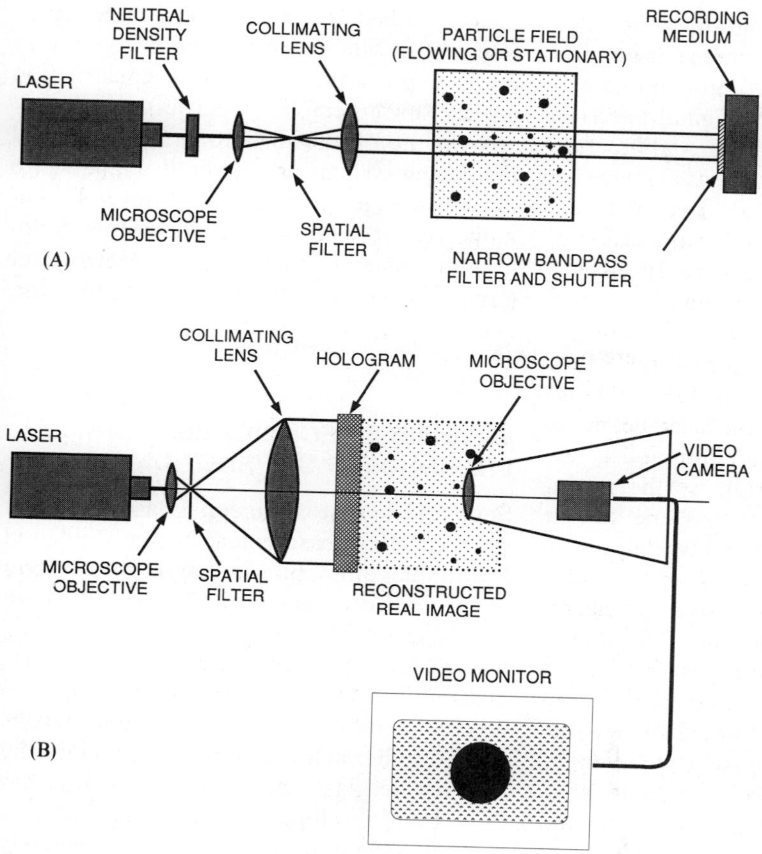

FIGURE 16-6. Layout for Fraunhofer Holography: (A) Recording System; (B) Reconstruction System.

(2) permits the use of a perfectly filtered and well-controlled reference beam, and (3) is not subject to the far-field restrictions on object size and distance from the film plane; and is not limited to dilute particle concentrations. The use of separate subject and reference beams does, however, place additional coherence requirements on the light source.

Sufficient mechanical stability is required to maintain all optical components stationary to within $\lambda/10$ during exposure, where λ is the wavelength of the illuminating beam, so pulsed lasers are typically used for examination of moving aerosols. Holographic examination of moving objects imposes a restriction on the exposure time, with the general rule of thumb for image clarity being that the object move less than 1/10 of its diameter during the exposure, so, for example, the exposure time for a 10 μm diameter object moving at 10 m/s should be less than 100 ns. High resolution generally requires that the sample volume be located close to the film plane, often requiring the use of a lens system to relay the volume image onto the film plane. The recording medium, usually a film, must have sufficient resolution to be able to record accurately the fringe pattern formed by beam interference. Spatial filtering by a pinhole aperture is generally required to eliminate beam impurities due to diffraction caused by dust particles, impurities of optical components or inherent optical noise of the laser beam.

After recording, the hologram is mounted in a reconstruction system, where it is illuminated with another coherent light source acting as the conjugate of the original reference beam. The hologram acts as a diffraction grating to form 3-D real and virtual images of the original sample volume. Use of translating stages allows a detailed examination of the reconstructed image. Figure 16-6B is a schematic diagram of an in-line reconstruction system. Unlike the in-line technique, in the off-axis reconstruction configuration the axis of the holographic images does not coincide with the reconstruction laser beam, so the viewing system is not continuously exposed to direct illumination by the reconstruction beam and the noise due to the reconstruction light source can be eliminated.

Holographic reconstruction and detailed data acquisition are very time-consuming, often requiring several "man-days" for examination of the several hundred particle images needed for statistical significance. Examinations are typically performed by a visual observation of the reconstructed images, although automated analysis techniques are a topic of current interest (e.g., Haussmann and Lauterborn 1980; Schäfer and Umhauer 1987; Chávez and Mayinger 1990). Ewan, Swithenbank, and Sarusbay (1984) and Hess and Trolinger (1985) describe unique applications of the Malvern diffractometer (see below) for evaluation of reconstructed holographic images. Instead of measuring the size distribution of a particle field, the Malvern is set up to examine a particle field hologram, yielding ensemble size distributions for each region probed by the Malvern laser beam. Use of the Malvern device could be a very useful step toward automated reconstruction (at least for ensemble properties).

While no off-the-shelf holographic aerosol measuring systems are currently available (to the knowledge of the authors), several companies can design and install custom holographic systems. Included are MetroLaser (Irvine, CA) and Physical Research, Inc. (Torrance, CA).

ENSEMBLE TECHNIQUES: FRAUNHOFER DIFFRACTION

Some of the first commercial, laser-based, *in situ* particle measurement techniques used Fraunhofer diffraction to characterize droplet sprays (Cornillault 1972; Wertheimer and Wilcock 1976; Swithenbank et al. 1977). Simply, the technique determines a particle distribution from a measurement of the ensemble diffraction pattern that results from the illumination of a particle cloud by a collimated laser beam. The technique has since been developed into a variety of commercial systems which have been extensively characterized, calibrated, and used in a wide range of particle studies. Excellent reviews of Fraunhofer diffraction techniques have recently appeared (Felton 1990; Meyer and Chigier 1986). The present discussion will provide only a brief overview of these techniques.

In typical ensemble diffraction techniques (Fig. 16-7), a laser beam is expanded and then

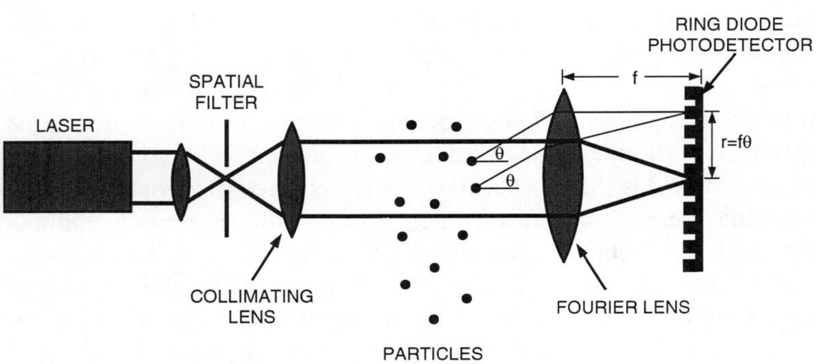

FIGURE 16-7. Layout of an Ensemble Diffraction System.

collimated into a beam, several mm in diameter, which passes through the particle cloud. Particles in the beam scatter light in all directions, although with particular efficiency in the near-forward direction. A receiving lens is used to focus both the transmitted beam and forward-scattered (predominately diffracted) light onto a detector located at the back focal plane of the lens. The transmitted light is focused to a point on the optical axis, while the diffracted light forms a series of concentric rings (Fraunhofer diffraction pattern). As the receiving lens performs a Fourier transform on the scattered light, light scattered at a given angle θ by a particle located anywhere in the illuminated sample volume will be focused at the same radial position in the transform plane. Thus, the resulting pattern is unaffected by particle location or motion.

Commercial techniques vary in the method of detecting the diffraction signature formed at the transform plane. Swithenbank et al. (1977) showed that by using a multielement set of annular, concentric photodetectors, the radial light energy distribution of the scattered light could be used to infer the size distribution. As the scattered light is most intense near the axis, the ring detectors increase in area with increasing radial distance. The theory of the annular scattering profile from diffraction is discussed elsewhere (Swithenbank et al. 1977; Felton 1990). Other techniques have been used to detect the diffraction signature (Cornillault 1972; Wertheimer and Wilcock 1976) but have not found as widespread commercial application. As a practical concern, the entire particle cloud should be within a maximum allowable distance of the receiver lens (Dodge 1984*b*). If particles are farther away than this, they may scatter light beyond the receiving lens; this effect is called "vignetting" (see Lefebvre 1989). Vignetting tends to bias the distribution towards larger sizes, as the small-particle (large-angle) signal is preferentially lost. No measurement of particle velocity is made by ensemble diffraction techniques.

The ensemble diffraction technique results in the classical inversion problem, wherein a continuous size distribution is sought which provides the best agreement with a finite set of experimental measurements (the scattered energies measured by each annular detector). A description of the inverse problem as it applies to ensemble diffraction techniques is given by Felton (1990) and Hirleman (1987). Typically, an iterative solution is used that minimizes the sum of squared errors between the predicted and measured detector responses. This method has several advantages (ensures positivity of the size distribution and guarantees stability) but requires an assumption for the size distribution (for example, Malvern systems offer a choice among a lognormal, Rosin–Rammler, or proprietary model-independent distribution). The choice of size distribution model is a critical step in the measurement process, and use of the model-independent distribution is recommended for the initial analysis of any aerosol (Meyer and Chigier 1986). Bayvel, Knight, and Robertson (1987) provide an alternative model-independent inversion algorithm. Once the size distribution is determined, the particle concentration is determined by a direct measurement of the beam attenuation through the cloud (transmitted beam intensity at the center of the detector measured before and after the aerosol is introduced).

Implicit in the standard analysis of ensemble diffraction techniques is the assumption that the particle diameter is much larger than the wavelength of the illuminating light. Thus, the lower limit for an ensemble diffraction instrument (for visible light) is about 1 μm. As an upper limit, diffraction from large particles (above several hundred micrometers) is concentrated at very small angles near the axis of the illuminating beam, making measurement difficult.

A second common assumption in the standard analysis is that each photon undergoes at most one scattering event while passing through the cloud. For high concentrations or long path lengths, the probability that a photon undergoes two or more particle interactions before reaching the detector increases. Multiple scattering spreads light over a wider angle, with the result that the size distribution appears wider and is skewed toward smaller sizes. Both experimental (Dodge 1984*a*; Felton 1990) and theoretical (Felton 1990; Hirleman 1988*b*) studies of multiple-scattering effects have been reported. Most studies agree that multiple-scattering effects can be safely ignored for obscurations (fraction of the beam scattered by the particle cloud) below about 50% (e.g., Felton 1990; Meyer and Chigier 1986). For higher obscurations many authors introduce correction factors (i.e., to distribution parameters such as mean diameter or standard deviation) and present empirical correlations to account for multiple scattering (Felton 1990; Dodge 1984*a*). Note that these correlations must be used carefully, as they depend on both the size distribution model as well as the obscuration. The recent analytic approach to multiple scattering described by Hirleman (1988*b*) was found to be in excellent agreement with the unpublished data of Dodge. Using analytic models or experimental correlations, it is possible to correct measurements at obscurations as high as 98% (Felton 1990; Hirleman 1988*b*).

Example 16-2

An ensemble diffraction based instrument measures a Rosin–Rammler droplet distribution with parameter $d_{RR} = 20.0\ \mu m$ (where d_{RR} is the drop diameter such that 63.2% of the spray volume is in drops of smaller diameter) and exponent $q = 2.0$ (where q is a measure of the spread in drop sizes) at an obscuration of 80% (OB = 0.8). What are the true parameter values ($d_{RR,0}$ and q_0) after correcting for multiple scattering effects?

Answer. For Rosin–Rammler distributions, Felton (1990) gives correlations for correcting ensemble diffraction data for multiple scattering effects. The correlations apply for obscurations between 65% and 98% and for values of the exponent between 1.2 and 3.8 with an uncertainty of $\pm 2\%$ and so may be used in this example. To correct for d_{RR}:

$$\frac{d_{RR,0}}{d_{RR}} = 1.0 + (0.036 + (0.4947(\mathrm{OB})^{8.998})q^{[1.9-3.437(\mathrm{OB})]}$$
$$= 1.0 + (0.102)(2.0)^{-0.85} = 1.057$$

To correct the exponent:

$$\frac{q_0}{q} = 1.0 + (0.035 + 0.1099(\mathrm{OB})^{8.65})q^{[0.35+1.45(\mathrm{OB})]}$$
$$= 1.0 + (0.0509)(2.0)^{1.510}$$
$$= 1.145$$

Thus, the final corrected values for the parameters are $d_{RR,0} = 21.1\ \mu m$ and $q_0 = 2.29$. Note that the observed distribution is slightly shifted to smaller sizes and is somewhat broader (smaller q) than the true distribution.

Several other general limitations of ensemble diffraction techniques should be discussed. First, the technique implicitly assumes a uniform light intensity, while the true radial intensity profile is Gaussian. As a result, the measurement volume for small particles will be smaller than for larger particles, thus biasing the measurement towards larger sizes (Meyer and Chigier 1986). Second, beam steering can result when measurements are made through regions characterized by high (particularly time varying) gradients in index of refraction. The result is similar to system misalignment, where the beam begins to illuminate the inner rings resulting in a false large particle response. Miles, Sojka, and King (1990) offer an approach to making measurements in systems with time-varying index of refraction gradients. Finally, the ensemble technique gives a line-of-sight measurement with little or no spatial resolution. Techniques for deconvolving line-of-sight data to obtain radial profiles (for axially symmetric sprays) have been discussed by Hammond (1981) and Zhu, Sun, and Chigier (1987); this process is referred to as "tomography" in the literature.

Due to the relatively long history of ensemble diffraction techniques (compared to other *in situ* optical techniques), a large literature has evolved on its performance (e.g., accuracy and precision). Calibration work has included both the use of standard reference materials (Felton 1990) suspended in test cells and a calibration reticle developed by Hirleman (Hirleman, Oechsle, and Chigier 1984; Dodge 1984*b*; Hirleman and Dodge 1985). Dodge (1984*b*) describes the importance of a calibration to correct for variations in ring detector sensitivities; using this correction, instrument-to-instrument reproducibilities were found to be better than 3% using a reticle (Hirleman and Dodge 1985). Many comparisons have been made among ensemble diffraction instruments (e.g., Dodge 1987; Dodge, Rhodes, and Reitz 1987), and between ensemble diffraction and other particle measurement techniques (see Felton 1990 for a recent review): the quality of agreement depends on the techniques and on the cloud property being compared (see the discussion below).

Although of general importance to all ensemble diffraction techniques, most of the results summarized above were obtained with Malvern Instruments systems (Malvern Instruments, Inc, Southborough, MA, U.S.A., and Malvern Instruments Ltd, Worcestershire, U.K.). Other commercially available ensemble diffraction techniques are offered by Insitec (San Ramon, CA), Compagnie Industrielle des Lasers (CILAS, Marcoussis, France), Leeds and Northrup Instruments (Microtrac Division, FL), and The Munhall Company (Worthington, OH). Typical size ranges for these systems are 0.5–2000 μm, depending on configuration.

ENSEMBLE TECHNIQUES: DYNAMIC LIGHT SCATTERING

Few techniques are available for noninvasive measurement of submicrometer diameter particles. Dynamic light scattering (DLS), also known as photon correlation spectroscopy (PCS), diffusion-broadening spectroscopy (DBS), or quasi-elastic light scattering (QELS), is an ensemble laser scattering technique for particle sizing in the 0.01–1 μm diameter range. The technique allows measurement of the mean particle size of a suspension without knowledge of particle properties (density, refractive index), and no calibration is required. In general, dynamic light scattering works by measuring the degree of spectral broadening induced in an incident laser beam by Brownian motion of particles in an illuminated sample. The mean particle diffusion coefficient is measured directly, which can then be used to determine particle sizes. PCS measures the dynamic behavior (time-dependent intensity) of the scattered light signal caused by this particle motion. Large particles produce low-frequency oscillations in the scattered signal, and small particles produce high-frequency oscillations. DLS techniques require enough particles in the measurement volume for an ensemble measurement; however, particle concentration is not measured since only ensemble light scattering is measured. Multiple scattering can lead to errors, although a new two-color cross-correlation system can minimize these errors (Drewel, Ahrens, and Schatzel 1990). In addition, limited shape information is provided on the particles, and the data analysis deconvolution is difficult for broad particle distributions.

As shown in Fig. 16-8, the optical setup of a DLS system is quite simple; a vertically polarized laser beam is focused into the sample of interest and a photomultiplier is used to collect the time-dependent light scattered at some angle by all of the particles in the beam (Weiner 1984). The signal is then processed either by a spectrum analyzer (QELS or DBS techniques) or a digital correlator (PCS). In the PCS technique, a time autocorrelation function is performed on the dynamic scattered light signal, and the characteristic decay rate of the exponential autocorrelation is used to determine the diffusion coefficient D. The effective hydrodynamic diameter of a particle is determined using the relations between D and d_p for the Knudsen number of interest, e.g., Stokes–Einstein for continuum flow, Epstein diffusion for free-molecular flow, or Stokes–Einstein with Cunningham empirical slip correction for all Knudsen numbers (Bernard 1988).

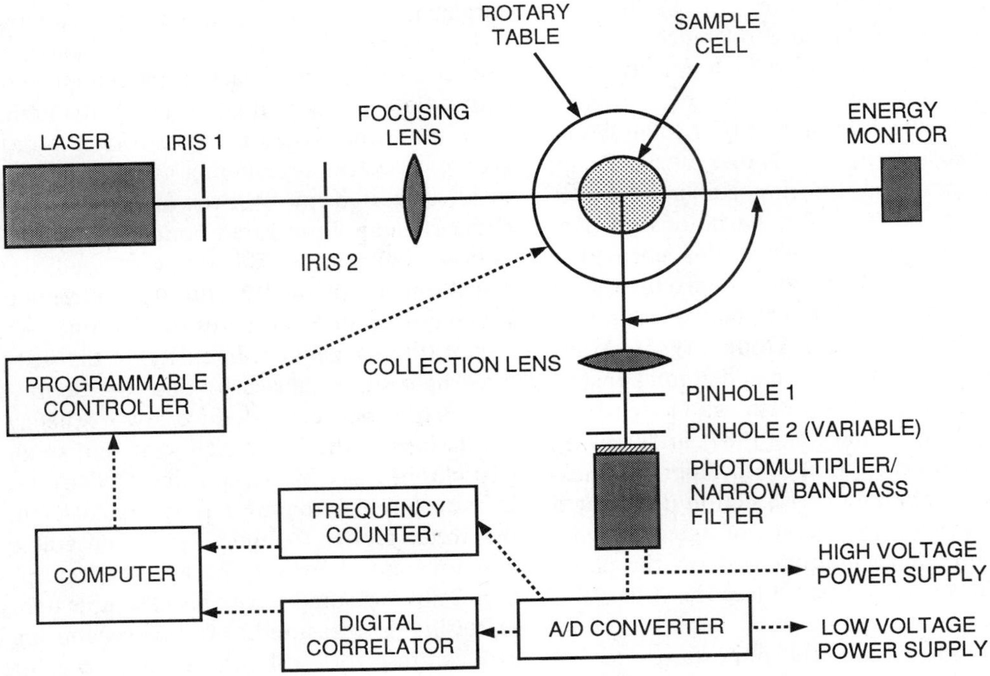

FIGURE 16-8. Photon Correlation Spectroscopy Layout.

Analysis of the single exponential curve produced by PCS of a monodisperse sample is easily achieved, but difficulties arise in polydisperse cases, where the autocorrelation is a sum of the exponentials produced by particles in various size ranges. Extraction of polydispersity information by Laplace transform inversion is discussed by Bertero et al. (1988). Successful application of PCS to a system containing an unknown distribution of particle sizes frequently requires measurements at two or more scattering angles (Bott 1987). This can be accomplished by mounting the photomultiplier on an optical rail that can rotate about the sample cell while maintaining a fixed geometry between the input laser beam and the sample.

The PCS technique has theoretical limitations on the size of the largest particles that can be measured with high precision (Martin 1987). This limit appears to be of the order of several hundred nanometers, with particle sizes above this limit achievable but with lower precision.

While commonly used for measurements of colloidal suspensions at rest, dynamic light scattering techniques have been applied to a number of aerosol measurements, often in harsh environments. Flower and Hurd (1987) used dynamic light scattering to measure silica particles in premixed flames. They measured hydrodynamic particle diameters in the 0.04–0.15 μm range, and noted that this technique is particularly useful for measurements of particles with purely real refractive indices (such as silica), where absorption techniques are invalid. Lhuissier, Nazih, and Weill (1988) applied PCS for the measurement of diesel exhaust particles of the order 0.4 μm diameter in a wind tunnel flowing at 0.8 m/s. Taylor and Sorensen (1986) present the theoretical effects of Gaussian beam properties and flow of the diffusing particles in PCS measurements, and verify experimentally using a flowing dioctylphthalate aerosol with 0.09 and 0.9 μm diameter particles.

Commercial manufacturers of PCS devices include Malvern Instruments (Southborough,

MA), Coulter Electronics (Hialeah, FL), Brookhaven Instruments (Holtsville, NY), Langley-Ford (Amherst, MA), Wyatt Technology (Santa Barbara, CA), and Particle Sizing Systems (Santa Barbara, CA). Most commercial systems include a multiangle measurement capability, and a temperature-controlled test cell to prevent convective flow from disturbing the measurements. However, use of these systems in their standard configuration typically requires extractive sampling. True *in situ* application requires system modification.

PERFORMANCE VERIFICATION

Although the *in situ* optical techniques reviewed provide a powerful resource for the measurement of particle size distributions, care must be taken to ensure that they are only used within their proper operating envelope. Thus, a means of verifying instrument performance—a standard—is essential in order to characterize instrument limits and to identify inherent instrument bias and inaccuracies. A recent review on calibration and standardization progress in Europe for *in situ* particle sizing has been given by Scarlett, Merkus, and Meesters (1990). These authors emphasize that performance verification must include a standardization of the operating procedures, as these can have a significant impact on instrument performance. Thus, a standard must include a document that prescribes a test method as well as a standard reference material that is precisely characterized (Scarlett, Merkus, and Meesters 1990). Examples of standard protocols and standard materials available for instrument calibration are discussed immediately below. In addition, a technique that allows an aerodynamic equivalent-diameter-based calibration of optical techniques is mentioned. Finally, a brief review of several instrument comparisons is provided as another method of verifying instrument performance.

Calibration

In calibration, a well-characterized aerosol or simulant (either polydisperse or monodisperse) is presented to the instrument under prescribed conditions and the instrument response is compared with the known result. For sampling instruments, it usually suffices to know the aerosol size distribution, shape, and composition (refractive index) for instrument calibration. For *in situ* techniques, however, concerns over trajectory ambiguity and coincidence require that particle trajectory (direction and velocity) and concentration also be controlled. Thus, a careful calibration of *in situ* systems becomes a difficult task. Previous efforts toward developing calibration standards can be grouped into three general areas: standard reference materials (SRM), generators, and reticles. A recent review of all the three techniques has been offered by Hemsley et al. (1988).

SRMs include suspensions or powders that have been thoroughly characterized. Polystyrene latex (PSL) monospheres (with specified mean size and standard deviation) are available as aqueous suspensions. PSL preparations are available with either the National Institute of Standards and Technology (NIST) or the Community Bureau of Reference (Bureau Communautaire de Référence, BCR) certification. A suitable airborne dispersion is typically made by spraying a dilute suspension and drying (Mitchell 1986). As an example, Mitchell, Nichols, and van Santen (1989) report a calibration of a Polytec SPC using PSL based on BS3406 (British Standards Institute 1988) for single-particle counters. Various certified powders (quartz, metals, etc.) are also available. Of particular interest are certified glass microspheres, which are available with either standard or high refractive index glass. Analytic reference particles are available from Duke Scientific Corp. (Palo Alto, CA) and Seradyn (Indianapolis, IN). Although these techniques provide a very accurate size standard, it is difficult to quantify the generated concentration.

A second approach to calibration is to use a monodisperse droplet generator. In this technique, liquid is pumped through a small orifice to generate a cylindrical liquid jet. The jet is naturally unstable and breaks up into droplets. If a periodic mechanical disturbance is applied to the jet, the breakup is found to be precisely uniform within certain frequency ranges. One droplet is formed per disturbance cycle, and so the droplet volume (diameter) can be accurately calculated from the liquid flow rate and disturbance frequency (Berglund and Liu 1973). By dissolving non-volatile materials in a volatile solvent, evaporation of the original droplets can be used to generate either liquid or solid particles of much smaller size. A single-jet monodisperse generator is available from TSI, Inc. (St. Paul, MN), while a multijet version is available from Fluid Jet Associates (Spring Valley, OH). Both single- and multiple-jet droplet generators have been used in studies to characterize the performance of the phase Doppler particle analyzer (Ceman 1990; Dressler and Kraemer 1990).

Reticles have been used extensively in the calibration of *in situ* particle counters (Hovenac 1986). Typically, a calibration reticle consists of a glass or plastic disk either with circular patterns of various sizes etched on its surface or precision spheres embedded in its matrix. The disks are typically fixed for calibrating ensemble systems, and rotated for calibrating SPCs. Circular-pattern reticles are typically used to calibrate diffraction-based instruments, as filled circles, spherical particles, or circular apertures of the same diameter give equivalent scattering patterns. Hirleman, Oechsle, and Chigier (1984) describe a calibration reticle for ensemble diffraction instruments that consists of a pattern of opaque chrome-filled circles photoetched on a glass substrate; the circles are randomly positioned and their diameters follow a Rosin–Rammler size distribution. These reticles have been used extensively in round-robin comparison of ensemble diffraction instruments (Hirleman, Oechsle, and Chigier 1984; Hirleman and Dodge 1985; Dodge 1987), and are being included as part of a standardized calibration procedure proposed to the American Society for Testing Materials (1991). These reticles are commercially available from Malvern Instruments Inc. (Southborough, MA). A similar circular-pattern reticle is available from Insitec Measurement Systems (San Ramon, CA) for calibrating their PCSV near-forward SPC systems. In this case, several tracks of monodisperse images (circle diameters from 2 to 80 μm) are etched at discrete radii from the center of the disk. A precision dc motor is used to rotate the reticle to provide a sequence of monodisperse "particles" to the sample volume. Hemsley et al. (1988) discussed the use of both precision pinholes and a circular-pattern reticle (chrome disks sputtered on glass plates) for instrument calibration.

Ceman (1990) has developed a standard calibration device for phase Doppler instruments, consisting of a plastic disk containing a known distribution of well-characterized glass beads, and has used it to make measurements with the Aerometrics PDPA. The disk can be rotated, with the PDPA measurement volume positioned at various radial locations, to provide a known velocity, size distribution, and number density calibration particle field. Several changes must be made to the standard PDPA instrument to accommodate the disk calibrations: adjustment to the transmitter angle to account for refraction through the epoxy while maintaining a fixed collection angle (see Example 16-3); correction of the effective particle refractive index used in the PDPA software to refer it to the local medium; and correction of the effective receiver lens focal length to account for changes in optical path length caused by refraction through the epoxy. His preliminary investigations show that the PDPA is able to accurately measure calibration size distributions as well as individual particle diameters in the epoxy disk.

Example 16-3

Aerosol measurements must often be made in a region physically separated from the measuring instrument by a window. For example, windows can be installed to provide optical access to an industrial flow of particle-laden gas. Consider measurements made in air ($n_1 = 1.0$), with the measurement beam passing through a 1 cm thick glass window ($n_2 = 1.4$). The beam has a 30° incident angle

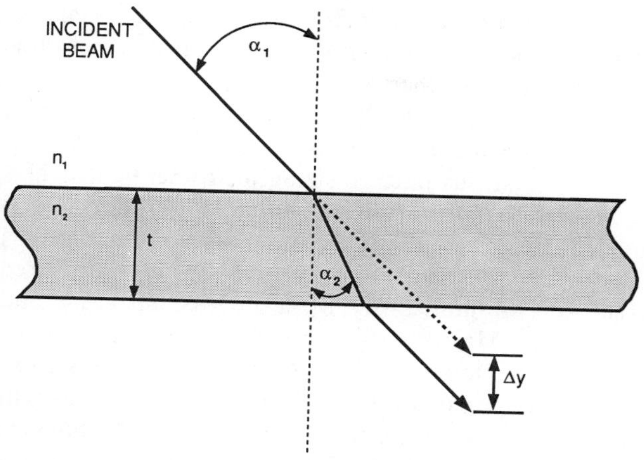

FIGURE 16-9. Measurement in an Enclosed Region.

on the window, as shown in Fig. 16-9. Find the angle of the beam (relative to its initial direction) and the apparent offset of the beam emerging from the window.

Answer. The change of angle in the window is simply given by application of Snell's law at the air–window interface:

$$n_1 \sin \alpha_1 = n_2 \sin \alpha_2$$

So, for the sample case, a 30° incident beam leads to a 20.92° beam angle inside the window. The angle inside the window is clearly independent of the window thickness. A second application of Snell's law at the inner window–air interface shows that the beam transmitted through the window recovers its original 30° angle. However, the window effectively shifts the location of any point on the beam in the measurement region (e.g., the beam crossing location for a dual beam LDV). The extent of shift is given by

$$\Delta y = t\left[1 - \frac{\cos \alpha_1}{\sqrt{(n_2/n_1)^2 - \sin^2 \alpha_1}}\right]$$

where t is the window thickness (Durst, Melling, and Whitelaw 1981). Using the given values in the formula shows that the beam is shifted by $\Delta y = 0.34$ cm by passing through the window. The reader is referred to Durst, Melling, and Whitelaw (1981) for an expression for Δy in the configuration where regions of differing refractive index exist on either side of the window, e.g., measurements in heated air or in gases other than air.

Finally, a method for an aerodynamic particle size calibration of optical particle counters (OPC) has been reported by Marple and Rubow (1976). A major advantage of this technique is providing a bridge between aerodynamic diameter (typically of interest in flowing systems) and the optical equivalent diameter reported by the systems of interest here. In their original method, an initial OPC size distribution is compared to the one taken with a single-stage inertial impactor placed on the OPC inlet. By knowing the aerodynamic cutoff of the impactor, the two size distributions can be used to establish a single calibration point for the OPC. A complete OPC calibration is obtained by using impactors with different cutoff diameters. Although this technique was originally intended for calibrating sampling-type instruments, it can also be applied to the calibration of noninvasive systems. For example, Mitchell, Nichols, and van Santen (1989) and Fissan and Helsper (1981) have calibrated a Polytec HC-15/1 with water droplets using the single-stage impactor technique. The impactor technique was selected because of the great difficulties encountered in producing small, stable water droplets by other techniques (such as with droplet breakup generators).

Comparisons

One means of evaluating instrument performance is by comparison of several instruments measuring the same aerosol, which is often simpler than providing and measuring a primary standard aerosol or simulated aerosol. However, such comparisons are not straightforward for a number of reasons: differences in instrument principles, measured statistics, definition of particle diameter, etc. Dodge (1987) discusses comparison of instruments for *in situ* measurement of spray droplets, including effects of spatial versus temporal sampling, and point versus ensemble measurements and conversions from one to the other.

With these limitations in mind, a number of researchers have compared the performance of various particle sizing instruments. Mitchell, Nichols, and van Santen (1989) found good agreement in mass distribution on a polydisperse spray (0.5–30 μm) measured simultaneously by a Polytec HC-15/1 and a Malvern 2200 laser diffractometer.

Dodge (1987) reports on an interlaboratory comparison of 17 instruments of six different types, including Malvern diffractometers (models 1800, 2200, and 2600), the Aerometrics PDPA, the PMS models FSSP-100, OAP-260X, and OAP-2D-GA1, and the Bete droplet analyzer (video imaging system). In general, the Malvern, Aerometrics, and Bete

TABLE 16-1 Listing of Selected Optical *In Situ* Particle Size Measurement Systems and Their Characteristics

Method	Commercial Source	Measure Velocity	Comments
SPC Intensity	INS	Y	PCSV (probe and bench)
SPC Intensity	HYT	N	Vacuum compatible
SPC Intensity	MET	LDV	PCS
SPC Intensity	PMS	N	FSSP, mean velocity option
SPC Intensity	POL	N	HC
SPC Phase	AEM	LDV	PDPA
SPC Phase	DAN	LDV	PDA
SPC Imaging	PMS	Y	OAP
Ensemble imaging	MET	Y	Custom holography
Ensemble imaging	PRI	Y	Custom holography
Ensemble diffraction	CIL	N	
Ensemble diffraction	INS	N	EPCS
Ensemble diffraction	LAN	N	Microtrak
Ensemble diffraction	MAL	N	Model 2600c
Ensemble diffraction	MUN	N	
Ensemble PCS	BRK	N	
Ensemble PCS	CLT	N	
Ensemble PCS	LGF	N	
Ensemble PCS	MAL	N	
Ensemble PCS	PSS	N	
Ensemble PCS	WYT	N	
Calibration	DUK	—	Standard reference particles
Calibration	FLJ	—	Multijet droplet generator
Calibration	SER	—	PSL monospheres
Calibration	TSI	—	Single-jet droplet generator

results were found to be in rough agreement for mean particle diameter, with the PMS instruments indicating larger diameters.

Presser et al. (1986) examined a spray with both the Aerometrics PDPA and an intensity deconvolution technique similar to the Insitec PCSV. They showed that the PDPA-measured mean particle sizes were considerably larger than those measured with the intensity deconvolution technique, and that both instruments gave results which showed similar trends as a function of position in the spray.

Dodge, Rhodes, and Reitz (1987) made simultaneous measurements of a spray of aircraft calibration fuel droplets (5–80 μm diameter) using the Aerometrics PDPA and the Malvern 2200, and found that both systems gave results which showed similar trends throughout the spray, but that the PDPA-measured mean droplet diameters were typically larger. They also include a discussion of the comparison of temporal averaged point measurements (PDPA) to spatial-averaged line-of-sight measurements (Malvern).

Young and Bachalo (1988) compared the Aerometrics PDPA, the Malvern 2600, and the PMS OAP-2D-GA2 performance by simultaneous measurement of falling glass beads and a droplet spray with all three instruments, and discuss calibration techniques and comparison difficulties for these instruments. In general, mean glass bead diameters measured by the three instruments were in good agreement. For spray droplet measurements, where the droplet size–velocity correlation became significant, the Malvern gave a considerably smaller mean diameter than the other two instruments. However, the PDPA

TABLE 16-2 List of Commercial Sources

Code	Address
AEM	Aerometrics, Inc., 550 Del Rey Avenue Unit A, Sunnyvale, CA, U.S.A. 94086, (408) 738-6688
BRK	Brookhaven Instruments Corporation, 750 Blue Point Road, Holtsville, NY, U.S.A. 11742, (516) 758-3200
CIL	Compagnie Industrielle des Lasers (CILAS), Route de Nozay, 91 Marcoussis, France
CLT	Coulter Scientific Instruments, P.O. Box 2145, Hialeah, FL, U.S.A. 33012-0145, (800) 526-6932
DAN	Dantec Elektronik, Tonsbakken 16-18, DK-2740, Skovlunde, Denmark, (4544) 923610
	Dantec Electronics, Inc., 777 Corporate Drive, Mahwah, NJ, U.S.A. 07430, (201) 512-0037
DUK	Duke Scientific Corp., 1135D San Antonio Road, Palo Alto, CA, U.S.A. 94303, (800) 334-3883
FLJ	Fluid Jet Associates, 1216 Waterwyck Trail, Spring Valley, OH, U.S.A. 45370, (513) 885-4882
HYT	High-Yield Technology, 762 Palomar Avenue, Sunnyvale, CA, U.S.A. 94086, (408) 730-4585
INS	Insitec Measurement Systems, 2110 Omega Road, Suite D, San Ramon, CA, U.S.A. 94583, (510) 837-1330 or (800) 995-9902
LGF	Langley-Ford, Amherst, MA, U.S.A.
LAN	Leeds and Northrup Instruments, 3000 Old Roosevelt Blvd., St. Petersburg, FL, U.S.A. 33716, (813) 573-1155
MAL	Malvern Instruments Limited, Spring Lane South, Malvern, Worcestershire WR14 1AQ, U.K. (0684) 892456
	Malvern Instruments Inc. (North America), 10 Southville Road, Southborough, MA, U.S.A. 01772, (508) 480-0200
MET	MetroLaser, 18006 Skypark Circle, Suite 108, Irvine, CA, U.S.A. 92714-6428, (714) 553-0688
MUN	The Munhall Co., 5655 N. High Street, Worthington, OH, U.S.A. 43085, (614) 888-7700
PMS	Particle Measuring Systems, Inc., 1855 South 57th Court, Boulder, CO, U.S.A. 80301, (303) 443-7100
PSS	Particle Sizing Systems, 75 Aero Camino, Suite B, Santa Barbara, CA 93117, (805) 968-1497
PRI	Physical Research, Inc., 25500 Hawthorne Blvd., Suite 2300, Torrance, CA, U.S.A. 90505-6828, (310) 378-0056
POL	Polytec GmbH, 7517 Waldbronn, Germany
	Polytec Optronics, Inc., 3001 Redhill Ave, Bldg. 5-102, Costa Mesa, CA, U.S.A. 92626, (714) 850-1835
SER	Seradyn, Inc., P.O. Box 1210, Indianapolis, IN, U.S.A. 46206, (800) 428-4007
TSI	TSI Incorporated, 500 Cardigan Road, St. Paul, MN, U.S.A. 55126, (612) 490-2833 or (800) 677-2708
WYT	Wyatt Technology Corporation, 802 East Cota St., P.O. Box 3003, Santa Barbara, CA, U.S.A. 93130, (805) 963-5904

data were shown to agree fairly well with those of the Malvern when corrected from a temporal to a spatial basis (like that of the Malvern).

On some occasions, the particle size distribution of interest will be too broad for one technique or instrument to measure. Combining different techniques can be difficult, as one is in fact performing a comparison in the overlap region (if one exists) with all the limitations discussed above. As an example, Picot et al. (1990) combined a PMS FSSP-100, a PMS OAP-260X, and a flash photography system to characterize an aerial spray nozzle, and discussed a method for incorporating data from the three systems.

CONCLUSIONS

From the extent of the previous discussion, it is easy to see that there are a wide range of optical *in situ* techniques available for measuring particle size distributions. Although this variety makes it likely that a researcher will find at least one instrument that will perform the required measurement, it is essential that a careful selection process be pursued. Particular attention is required in determining the instrument operating envelope relative to the measurement requirements in order to find the technique best suited to the application. Performance verification should be made as needed to assure data accuracy and instrument reliability. A tabulation of the various commercial techniques is given in Table 16-1, with a list of sources of this equipment provided in Table 16-2. The authors have tried to include all vendors and equipment in these lists, but realize that the field is extensive and considerable development efforts are always proceeding. We apologize for any omissions.

ACKNOWLEDGMENT

This work was performed at Sandia National Laboratories for the U.S. Department of Energy under Contract DE-AC04-76DP00789.

References

Al-Chalabi, S, A. M., Y. Hardalupas, A. R. Jones, and A. M. K. P. Taylor. 1988. Calculation of calibration curves for the phase Doppler technique: Comparison between Mie theory and geometrical optics. In *Optical Particle Sizing*, eds. G. Gouesbet and G. Grehan, pp. 107–120. New York: Plenum.

Alexander, D. R., K. J. Wiles, S. A. Schaub, and M. P. Seeman. 1985. Effects of non-spherical drops on a phase Doppler spray analyzer. In *Particle Sizing and Spray Analysis, SPIE Vol. 573,* eds. N. Chigier and G. W. Stewart, pp. 67–72. Bellingham, Washington: SPIE.

American Society for Testing Materials. 1991. Standard test method for calibration verification of laser diffraction particle sizing instruments using photomask reticles. ASTM Subcommittee E29.04, EXXX-91.

Bachalo, W. D., R. C. Rudoff, and S. V. Sankar. 1990. Time-resolved measurements of spray drop size and velocity. In *Liquid Particle Size Measurement Techniques, ASTM STP 1083, 2nd edn.*, eds. E. D. Hirleman, W. D. Bachalo, and P. G. Felton, pp. 209–24. Philadelphia: American Society for Testing and Materials.

Bachalo, W. D. and M. J. Houser. 1984. Phase Doppler spray analyzer for simultaneous measurements of drop size and velocity distributions. *Opt. Eng.* 23:583–90.

Baumgardner, D., W. A. Cooper, and J. E. Dye. 1990. Optical and electronic limitations of the forward-scattering spectrometer probe. In *Liquid Particle Size Measurement Techniques, ASTM STP 1083*, eds. E. D. Hirleman, W. D. Bachalo, and P. G. Felton, pp. 115–27. Philadelphia: American Society for Testing and Materials.

Bayvel, L. P., J. Knight, and G. Robertson. 1987. Alternative model-independent inversion programme for Malvern particle sizer. *Part. Charact.* 4:49–53.

Berglund, R. N. and B. Y. H. Liu. 1973. Generation of monodisperse aerosol standards. *Environ. Sci. Technol.* 7:147–53.

Bernard, J. M. 1988. Particle sizing in combustion systems using scattered laser light. *J. Quant. Spectroscop. Radiat. Transfer* 40:321–30.

Bertero, M., P. Boccacci, C. De Mol, and E. R. Pike. 1988. Extraction of polydispersity information in photon correlation spectroscopy. In *Optical Particle Sizing*, eds. G. Gouesbet and G. Grehan, pp. 89–98. New York: Plenum.

Bohren, C. F. and D. R. Huffman. 1983. *Absorption and Scattering of Light by Small Particles.* New York: Wiley.

Borden, P. G. and L. A. Larson. 1989. Benefits of real-time, *in situ* particle monitoring in production medium current implantation. *IEEE Trans. Semiconductor Manufacturing* 2:141–45.

Borho, K. 1970. A scattered-light measuring instrument for high dust concentrations. *Staub Reinhalt. Luft.* 30:45–49.

Bott, S. E. 1987. Submicrometer particle sizing by photon

correlation spectroscopy: Use of multiple-angle detection. In *Particle Size Distribution: Assessment and Characterization*, ed. T. Provder, pp. 74–88. Washington, DC: American Chemical Society.

Brena de la Rosa, A., S. V. Sankar, B. J. Weber, G. Wang, and W. D. Bachalo. 1989. A theoretical and experimental study of the characterization of bubbles using light scattering interferometry. In *ASME International Symposium on Cavitation Inception—1989*, eds. W. B. Morgan and B. R. Parkin, pp. 63–72. New York: ASME.

British Standards Institute 1988. British Standard Methods for Determination of Particle Size Distribution—Recommendations for Single Particle Light Interaction Methods. BS3406: Part 7.

Ceman, D. L. 1990. Phase Doppler technique: Factors affecting instrument response and novel calibration system. M.S. Thesis, The University of New Mexico, Albuquerque.

Ceman, D. L., T. J. O'Hern, and D. J. Rader. 1990. Phase Doppler droplet sizing—scattering angle effects. In *ASME Fluid Measurements and Instrumentation Forum*, eds. E. P. Rood and C. J. Blechinger, pp. 61–63. New York: ASME.

Chávez, A. and F. Mayinger. 1990. Evaluation of pulsed laser holograms of spray droplets using digital image processing. In *Proc. Second International Congress on Optical Particle Sizing*, ed. E. D. Hirleman, pp. 462–71. Arizona State University Printing Services.

Collier, R. J., C. B. Burkhardt, and L. H. Lin. 1971. *Optical Holography*. New York: Academic Press.

Cornillault, J. 1972. Particle Size Analyzer. *Appl. Opt.* 11:265–68.

Dickson, L. D. 1970. Characteristics of a propagating Gaussian beam. *Appl. Opt.* 9:1854–61.

Dodge, L. G. 1984*a*. Change of calibration of diffraction-based particle sizers in dense sprays. *Opt. Eng.* 23:626–30.

Dodge, L. G. 1984*b*. Calibration of the Malvern particle sizer. *Appl. Opt.* 23:2415–19.

Dodge, L. G. 1987. Comparison of performance of drop-sizing instruments. *Appl. Opt.* 26:1328–41.

Dodge, L. G., D. J. Rhodes, and R. D. Reitz. 1987. Drop-size measurement techniques for sprays: Comparison of Malvern laser-diffraction and aerometrics phase/Doppler. *Appl. Opt.* 26:2144–54.

Dressler, J. L. and G. O. Kraemer. 1990. A multiple drop-size drop generator for calibration of a phase-Doppler particle analyzer. In *Liquid Particle Size Measurement Techniques, ASTM STP 1083*, 2nd edn., eds. E. D. Hirleman, W. D. Bachalo, and P. G. Felton, pp. 30–44. Philadelphia: American Society for Testing and Materials.

Drewel, M., J. Ahrens, and K. Schätzel. 1990. Suppression of multiple scattering errors in particle sizing by dynamic light scattering. In *Proc. Second International Congress on Optical Particle Sizing*, ed. E. D. Hirleman, pp. 130–38. Arizona State University Printing Services.

Durst, F., A. Melling, and J. H. Whitelaw. 1981. *Principles and Practice of Laser-Doppler Anemometry*. London: Academic Press.

Ewan, B. C. R., J. Swithenbank, and C. Sorusbay. 1984. Measurement of transient spray size distributions. *Opt. Eng.* 23:620–25.

Farmer, W. M. 1972. Measurement of particle size, number density, and velocity using a laser interferometer. *Appl. Opt.* 11:2603–12.

Felton, P. G. 1990. A review of the Fraunhofer diffraction particle-sizing technique. In *Liquid Particle Size Measurement Techniques, ASTM STP 1083*, eds. E. D. Hirleman, W. D. Bachalo, and P. G. Felton, pp. 47–59. Philadelphia: American Society for Testing and Materials.

Fissan, H. J. and C. Helsper. 1981. Calibration of the polytec CH-15 and HC-70 optical particle counters. In *Aerosols in the Mining and Industrial Work Environments, Vol. 3*, eds. V. A. Marple and B. Y. H. Liu, pp. 825–31. Ann Arbor, MI: Ann Arbor Science.

Flower, W. L. and A. J. Hurd. 1987. *In situ* measurement of flame-formed silica particles using dynamic light scattering. *Appl. Opt.* 26:2236–39.

Gebhart, J. and A. Anselm. 1988. Effect of particle shape on the response of single particle optical counters. In *Optical Particle Sizing*, eds. G. Gouesbet and G. Grehan, pp. 393–409. New York: Plenum.

Gouesbet, G. and G. Grehan (eds.). 1988. *Optical Particle Sizing*, New York: Plenum.

Gouesbet, G., B. Maheu, and G. Grehan. 1988. Scattering of a Gaussian beam by a sphere using a Bromwich formulation: Case of an arbitrary location. In *Optical Particle Sizing*, eds. G. Gouesbet and G. Grehan, pp. 27–42. New York: Plenum.

Grehan, G. and G. Gouesbet. 1986. Simultaneous measurements of velocities and sizes of particles in flows using a combined system incorporating a top-hat beam technique. *Appl. Opt.* 25:3527–38.

Hammond, D. C. 1981. Deconvolution technique for line-of-sight optical scattering measurements in axisymmetric sprays. *Appl. Opt.* 20:493–99.

Hardalupas, Y., A. M. K. P. Taylor, and J. H. Whitelaw. 1988. Measurements in heavily-laden dusty jets with phase-Doppler anemometry. In *Transport Phenomena in Turbulent Flows: Theory, Experiment, and Numerical Simulation*, eds. M. Hirata and N. Kasagi, pp. 821–35. New York: Hemisphere.

Haussmann, G. and W. Lauterborn. 1980. Determination of the size and position of fast moving gas bubbles in liquids by digital 3-D image processing of hologram reconstructions. *Appl. Opt.* 19:3529–35.

Hemsley, D. J., M. L. Yeoman, C. J. Bates, and O. Hadded. 1988. The use of calibration techniques for the development and application of optical particle sizing instruments. In *Optical Particle Sizing*, eds. G. Gouesbet and G. Grehan, pp. 585–601. New York: Plenum.

Hess, C. F. 1984. Nonintrusive optical single-particle counter for measuring the size and velocity of droplets in a spray. *Appl. Opt.* 23:4375–82.

Hess, C. F. and J. D. Trolinger. 1985. Particle field

holography data reduction by Fourier transform analysis. *Opt. Eng*. 24:470–74.

Hirleman, E. D. 1982. Non-Doppler laser velocimetry: Single beam transit-time L1V. *AIAA J*. 20:86.

Hirleman, E. D. 1983. Nonintrusive laser-based particle diagnostics. In *Combustion Diagnostics by Nonintrusive Methods* (*Progress in Astronautics and Aeronautics, Vol. 92*), pp. 177–207.

Hirleman, E. D. 1984. Particle sizing by optical, nonimaging techniques. In *Liquid Particle Size Measurement Techniques, ASTM STP 848*, eds. J. M. Tishkoff, R. D. Ingebo, and J. B. Kennedy, pp. 35–60. Philadelphia: American Society for Testing and Materials.

Hirleman, E. D. 1987. Optimal scaling of the inverse Fraunhofer diffraction particle sizing problem: The linear system produced by quadrature. *Part. Charact*. 4:128–33.

Hirleman, E. D. 1988*a*. Optical techniques for particle size analysis. *Laser Topics* 10:7–10.

Hirleman, E. D. 1988*b*. Modeling of multiple scattering effects in Fraunhofer diffraction particle size analysis. *Part. Part. Syst. Charact*. 5:57–65.

Hirleman, E. D. (ed.) 1990. *Proc. Second International Congress on Optical Particle Sizing*. Arizona State University Printing Services.

Hirleman, E. D., W. D. Bachalo, and P. G. Felton (eds.). 1990. *Liquid Particle Size Measurement Techniques, ASTM STP 1083, 2nd edn*. Philadelphia: American Society for Testing and Materials.

Hirleman, E. D. and L. G. Dodge. 1985. Performance comparison of Malvern instruments laser diffraction drop size analyzers. In *Proc. ICLASS-85, Third International Conference on Liquid Atomisation and Spray Systems, Vol. 2,* p. IVA/3/1–14. London: Institute of Energy.

Hirleman, E. D., V. Oechsle, and N. A. Chigier. 1984. Response characteristics of laser diffraction particle size analyzers: Optical sample volume extent and lens effects. *Opt. Eng*. 23:610–19.

Holve, D. J. 1980. *In situ* optical particle sizing technique. *J. Energy* 4(4):176–83.

Holve, D. J. 1982. Transit timing velocimetry (TTV) for two-phase reacting flows. *Combust. Flame* 48:105–08.

Holve, D. J. and K. D. Annen. 1984. Optical particle counting, sizing, and velocimetry using intensity deconvolution. *Opt. Eng*. 23:591–603.

Holve, D. J. and G. W. Davis. 1985. Sample volume and alignment analysis for an optical particle counter sizer, and other applications. *Appl. Opt*. 24:998–1005.

Holve, D. J. and S. A. Self. 1979*a*. Optical particle sizing for *in situ* measurements Part 1. *Appl. Opt*. 18:1632–45.

Holve, D. J. and S. A. Self. 1979*b*. Optical particle sizing for *in situ* measurements Part 2. *Appl. Opt*. 18:1646–52.

Holve, D. and S. Self. 1980. Optical measurements of mean particle size in coal-fired MHD flows. *Combustion and Flame* 37:211–214.

Holve, D. J., D. Tichenor, J. C. F. Wang, and D. R. Hardesty. 1981. Design criteria and recent developments of optical single particle counters for fossil fuel systems. *Opt. Eng*. 20:529–39.

Hovenac, E. A. 1986. Use of rotating reticles for calibration of single particle counters. In *ICALEO '86 Proc., Vol. 58, Flow and Particle Diagnostics*. pp. 129–34. Arlington, VA: Laser Institute of America.

Hovenac, E. A. 1987. Performance and operating envelope of imaging and scattering particle sizing instruments. NASA CR-180859, National Aeronautics and Space Administration, Lewis Research Center, Cleveland, OH.

Jackson, T. A. 1990. Droplet sizing interferometry. In *Liquid Particle Size Measurement Techniques, ASTM STP 1083*, 2nd edn. eds. E. D. Hirleman, W. D. Bachalo, and P. G. Felton, pp. 151–69. Philadelphia: American Society for Testing and Materials.

Kerker, M. 1969. *The Scattering of Light and Other Electromagnetic Radiation*. New York: Academic Press.

Kerker, M. (ed.). 1988. *Selected Papers on Light Scattering, SPIE Vol. 951*. Bellingham, Washington: SPIE.

Killinger, R. T. and R. H. Zerull. 1988. Effects of shape and orientation to be considered for optical particle sizing. In *Optical Particle Sizing*, eds. G. Gouesbet and G. Grehan, pp. 419–29. New York: Plenum.

Knollenberg, R. G. 1970. The optical array: An alternative to scattering or extinction for airborne particle size determination. *J. Appl. Meteor*. 9:86–103.

Knollenberg, R. G. 1979. Single particle light scattering spectrometers. In *Aerosol Measurement*, eds. D. A. Lundgren, F. S. Harris, W. H. Marlow, M. Lippmann, W. E. Clark, and M. D. Durham, pp. 271–93. Gainsville: University Press of Florida.

Knollenberg, R. G. 1981. Techniques for probing cloud microstructure. In *Clouds: Their Formation, Optical Properties, and Effects*, eds. P. V. Hobbs and A. Deepak, pp. 15–89. New York: Academic Press.

Knollenberg, R. G. 1985. The measurement of particle sizes below 0.1 micrometers. *J. Environ. Sci*. 28:32–47.

Lavergne, G., Y. Biscos, F. Bismes, and P. Hebrard. 1988. Optical particle sizing: Digital video image processing application. In *Optical Particle Sizing*, eds. G. Gouesbet and G. Grehan, pp. 471–82. New York: Plenum.

Lefebvre, A. H. 1989. *Atomization and Sprays*. New York: Hemisphere.

Lhuissier, N., A. Nazih, and M. E. Weill. 1988. Measurement of diesel exhaust particles in a dilution tunnel by photon correlation spectroscopy. In *Photon Correlation Techniques and Applications, Proc. OSA Topical Meeting, Vol. 1*, eds. J. B. Abbiss and A. E. Smart, pp. 84–89.

Marple, V. A. and K. L. Rubow. 1976. Aerodynamic particle size calibration of optical particle counters. *J. Aerosol Sci*. 7:425–33.

Martin, J. E. 1987. Slow aggregation of colloidal silica. *Phys. Rev. A* 36:3415–26.

McDonell, V. G. and S. Samuelson. 1990. Sensitivity assessment of a phase-Doppler interferometer to user-controlled settings. In *Liquid Particle Size Measurement Techniques, ASTM STP 1083*, 2nd edn., eds. E. D.

Hirleman, W. D. Bachalo, and P. G. Felton, pp. 170–89. Philadelphia: American Society for Testing and Materials.

Meyer, P. and N. Chigier. 1986. Dropsize measurements using a Malvern 2200 particle sizer. *Atom. Spray Technol.* 2:261–98.

Miles, B. H., P. E. Sojka, and G. B. King. 1990. Malvern particle size measurements in media with time varying index of refraction gradients. *Appl. Opt.* 29:4563–73.

Mitchell, J. P. 1986. Aerosol generation for instrument calibration. UK Atomic Energy Report AEEW-R 2092.

Mitchell, J. P., A. L. Nichols, and A. van Santen. 1989. The characterization of water-droplet aerosols by Polytec optical particle analysers. *Part. Part. Syst. Charact.* 6:119–23.

Oberdier, L. M. 1984. An instrumentation system to automate the analysis of fuel-spray images using computer vision. In *Liquid Particle Size Measurement Techniques, ASTM STP 848*, eds. J. M. Tishkoff, R. D. Ingebo, and J. B. Kennedy, pp. 123–36. Philadelphia: American Society for Testing and Materials.

O'Hern, T. J., D. J. Rader, and D. L. Ceman. 1989. Droplet sizing calibration of the phase Doppler particle analyzer. In *ASME Fluid Measurements and Instrumentation Forum*, eds. E. P. Rood and C. J. Blechinger, pp. 49–57. New York: American Society of Mechanical Engineers.

Oldenburg, J. R. 1987. Counting error analysis of the phase/Doppler particle analyzer. NASA TM-100231, National Aeronautics and Space Administration, Lewis Research Center, Cleveland, OH.

Pinnick, R. G., D. M. Garvey, and L. D. Duncan. 1981. Calibration of Knollenberg FSSP light-scattering counters for measurement of cloud droplets. *J. Appl. Meteor.* 20:1049–51.

Picot, J. C., M. W. van Vliet, N. J. Payne, and D. D. Kristmanson. 1990. Characterization of aerial spray nozzles with laser light-scattering and imaging probes and flash photography. In *Liquid Particle Size Measurement Techniques, ASTM STP 1083*, eds. E. D. Hirleman, W. D. Bachalo, and P. G. Felton, pp. 142–50. Philadelphia: American Society for Testing and Materials.

Presser, C., A. K. Gupta, R. J. Santoro, and H. G. Semerjian. 1986. Laser diagnostics for characterization of fuel sprays. In *ICALEO '86 Proc., Vol. 58, Flow and Particle Diagnostics*, pp. 160–67. Toledo, OH: Laser Institute of America.

Reitz, R. D. 1990. Effect of vaporization and turbulence on spray drop-size and velocity distributions. In *Liquid Particle Size Measurement Techniques, ASTM STP 1083*, 2nd edn., eds. E. D. Hirleman, W. D. Bachalo, and P. G. Felton, pp. 225–37. Philadelphia: American Society for Testing and Materials.

Saffman, M. 1987. Optical particle sizing using the phase of LDA signals. *Dantec Inform. Measurement Analysis*, 5:8–13.

Saffman, M., P. Buchave, and H. Tanger. 1984. Simultaneous measurement of size, concentration and velocity of spherical particles by a laser Doppler method. In *Proc. Second International Symposium on Applications of Laser Anemometry to Fluid Mechanics*, eds. R. J. Adrian, D. F. G. Durão, F. Durst, H. Mishina, and J. H. Whitelaw, pp. 85–103. Lisbon: Ladoan.

Sankar, S. V., B. J. Weber, D. Y. Kamemoto, and W. D. Bachalo. 1990. Sizing fine particles with the phase Doppler interferometric technique. In *Proc. Second International Congress on Optical Particle Sizing*, ed. E. D. Hirleman, pp. 277–87. Arizona State University Printing Services.

Scarlett, B., H. G. Merkus, and G. M. H. Meesters. 1990. European progress on calibration and standardization for particle sizing. In *Liquid Particle Size Measurement Techniques, ASTM STP 1083*, 2nd edn., eds. E. D. Hirleman, W. D. Bachalo, and P. G. Felton, pp. 9–18. Philadelphia: American Society for Testing and Materials.

Schäfer, M. and H. Umhauer. 1987. Realization of a concept for the complete evaluation of double pulse holograms of particulate phases in flows. *Part. Charact.* 4:166–74.

Swithenbank, J., J. M. Beer, D. S. Taylor, D. Abbot, and G. C. McCreath. 1977. A laser diagnostic technique for the measurement of droplet and particle size distribution. *Experimental Diagnostics in Gas Phase Combustion Systems, Progress in Astronautics and Aeronautics* 53:421–47.

Takeo, T. and H. Hattori. 1990. Visibility analysis of laser Doppler anemometry for spherical particles smaller than several light wavelengths. *J. Appl. Phys.* (Japanese) 29:419–26.

Taylor, T. W. and C. M. Sorensen. 1986. Gaussian beam effects on the photon correlation spectrum from a flowing Brownian motion system. *Appl. Opt.* 25:2421–26.

Thompson, B. J. 1984. Droplet characteristics with conventional and holographic imaging techniques. In *Liquid Particle Size Measurement Techniques, ASTM STP 848*, eds. J. M. Tishkoff, R. D. Ingebo, and J. B. Kennedy, pp. 111–22. Philadelphia: American Society for Testing and Materials.

Thompson, B. J. 1974. Holographic particle sizing techniques. *J. Phys. E: Sci. Inst.* 7:781–88.

Thompson, B. J. and J. H. Ward. 1966. Particle sizing—the first direct use of holography. *Scientific Res.* 1:37–40.

Trolinger, J. D. 1975. Particle field holography. *Opt. Eng.* 14:383–92.

Trolinger, J. D. 1980. Analysis of holographic diagnostic systems. *Opt. Eng.* 19:722–26.

Tyler, G. A. and B. J. Thompson. 1976. Fraunhofer holography applied to particle size analysis—a reassessment. *Optica Acta* 23:685–700.

Umhauer, H. 1983. Particle size distribution analysis by scattered light measurements using an optically defined measuring volume. *J. Aerosol Sci.* 14:765–70.

van de Hulst, H. C. 1981. *Light Scattering by Small Particles*. New York: Dover.

Weiner, B. B. 1984. Particle sizing using photon

correlation spectroscopy. In *Modern Methods of Particle Size Analysis*. eds. H. G. Barth, pp. 93–116. New York: Wiley.

Wertheimer, A. L. and W. L. Wilcock. 1976. Light scattering measurements of particle distributions. *Appl. Opt.* 15:1616–20.

Young, B. W. and W. D. Bachalo. 1988. The direct comparison of three "in-flight" droplet sizing techniques for pesticide spray research. In *Optical Particle Sizing*, eds. G. Gouesbet and G. Grehan, pp. 483–97. New York: Plenum.

Zhu, H. M., T. Y. Sun, and N. Chigier. 1987. Tomographical transformation of Malvern spray measurements. *Atom. Spray Technol.* 3:89–105.

17

Direct-Reading Techniques Using Optical Particle Detection

Paul A. Baron

National Institute for Occupational Safty and Health[1]
Centers for Disease Control, Public Health Service
Department of Health and Human Services, Cincinnati, OH, U.S.A.

M. K. Mazumder

University of Arkansas at Little Rock
Department of Electronics and Instrumentation, Little Rock, AR, U.S.A.

and

Y. S. Cheng

Inhalation Toxicology Research Institute, Albuquerque, NM, U.S.A.

Introduction

Optical direct-reading particle counting techniques have the advantage of rapid, continuous, nondestructive particle detection. However, the amount of light scattered may not be directly related to the property that one wishes to measure. By combining the advantages of optical detection techniques with the manipulation of particle motion, several instruments have been developed that detect more specific properties of aerosol particles. The aerodynamic size of particles is used to describe the behavior of particles in gravitational settling, filtration, respiratory deposition, sampling systems, etc.

Measurement of the aerodynamic size at one time could only be achieved by manually observing the settling velocity of individual particles. Subsequently, impactors allowed the measurement of size distributions on a routine basis, although gravimetric and/or chemical analysis still had to be carried out in the laboratory. With the advent of new technology (e.g., lasers and microcomputers), real-time measurements became possible. Several instruments were developed to measure the aerodynamic size as rapidly and accurately as possible. These included the electric single-particle aerodynamic relaxation time analyzer (E-SPART; Hosakawa Micron International, Osaka, Japan; distributed by Micron Powder Systems, Summit, NJ), the Aerodynamic Particle Sizer (APS; TSI, Inc., St. Paul, MN) and the Aerosizer (Amherst Process Instruments, Hadley, MA). While the latter two

1. Mention of company or product names does not constitute endorsement by the National Institute for Occupational Safety and Health.

instruments allow a rapid determination of size distributions, they measure particle behavior largely outside the Stokes regime and the recorded size must be corrected to give an accurate aerodynamic size of individual particles.

While the aerodynamic diameter describes the inertial properties, the electrostatic charge influences the electrodynamic behavior of the particle in transport processes. Both aerodynamic diameter and electrostatic charge measurements on individual particles are needed in many electrodynamic processes; some examples are electrophotography and laser printing, electrostatic powder coating, electrostatic precipitation, electrostatically enhanced fabric filtration, and electrostatic beneficiation of minerals and coal. The E-SPART is capable of measuring particle charge as well as aerodynamic diameter.

Airborne asbestos fiber measurements went through a similar progression over time in that, originally, relative crude measurements of concentration were made by collection with midget impingers and microscope counting of all large particles. Filter collection with microscopic analysis was developed so that only fibers were detected. Finally, the development of the fibrous aerosol monitor (model FAM-1, MIE Inc., Bedford, MA) allowed continuous, real-time detection of airborne fibers. The FAM-1 was designed to give results close to those of the phase contrast light microscope method (see Chapter 25).

These more sophisticated instruments provide more specific data about aerosols; however, because of the complexity of their detection and analysis systems, they may also have various limitations and subtle problems associated with the interpretation of the data. The following sections present a discussion of these instruments.

ELECTRIC SINGLE-PARTICLE AERODYNAMIC RELAXATION TIME ANALYZER

Measurement Principles

The E-SPART analyzer can be operated in several modes, the first of which is the original

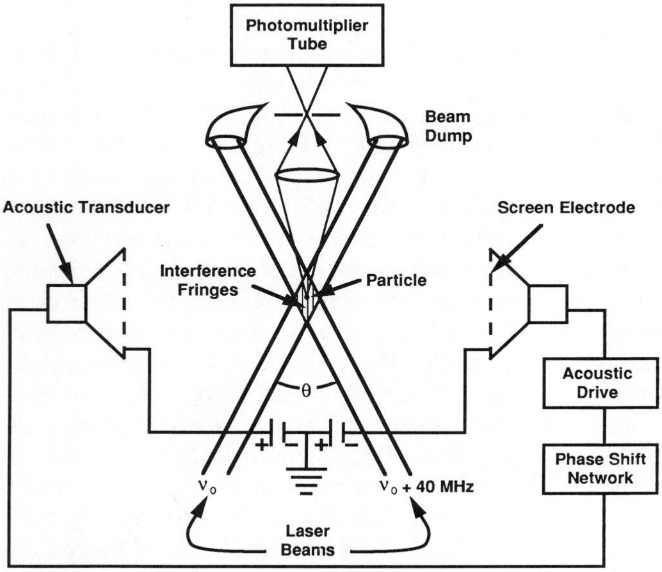

FIGURE 17-1. Aerosol Relaxation Chamber Showing the Geometrical Configuration of Acoustic Transducers and Electrodes for Applying Acoustic and Electric Fields in the E-SPART Analyzer. One of the Illuminating Laser Beams has been Shifted by 40 MHz.

SPART mode (Mazumder and Kirsch 1977; Mazumder et al. 1979). The SPART analyzer determines the aerodynamic diameter by subjecting particles to an acoustic field of frequency f and measuring the response of these particles to the acoustic excitation. In its typical sampling configuration, the aerosol is sampled in a laminar air flow moving vertically downward through the SPART's sensing volume (Fig. 17-1). The acoustic field induces an oscillatory velocity component to the particle motion in the horizontal direction. The inertia of each particle causes a phase lag ϕ in the particle's periodic motion with respect to the acoustic field. This phase lag ϕ is related to the relaxation time of the particle τ_p, which is a function of the aerodynamic diameter d_a of the particle. The SPART analyzer employs a differential laser Doppler velocimeter (LDV) to measure the oscillatory velocity component (in the horizontal direction) of the particle, and a microphone to measure the acoustic field. The phase lag of the particle motion with respect to the acoustic field driving the particle is converted to aerodynamic diameter using a microcomputer. The microcomputer stores the aerodynamic size data for the particles sampled and provides the measured size distribution. Although the response of the SPART can be calculated theoretically, it is calibrated with monodisperse latex particles because some instrumental parameters are difficult to measure.

In the E-SPART analyzer, an electrical particle acceleration field is used. There are two configurations: (1) a dc electric field superimposed upon the acoustic field (Mazumder, Ware, and Hood 1983), and (2) an ac electric field replacing the acoustic field (Renninger, Mazumder, and Testerman 1981). In the first configuration (i.e., an acoustic E-SPART analyzer), the horizontal motion of a charged particle is caused by the superposition of two fields: (a) the acoustic field forcing the particle in an oscillatory motion and (b) a dc electric field inducing a migration velocity component, which depends on the polarity and magnitude of the electrical charge q of the particle and the field strength. In this configuration, d_a is measured for either electrically charged or uncharged particles by determining the phase lag ϕ. Electrical charge q is determined from the measured electrical migration velocity V_e and the aerodynamic diameter d_a. The direction of V_e, which is also measured, provides the polarity of q.

In the second E-SPART configuration, no acoustic field is used and the particles are subjected to an ac electric field. Therefore, the measurement process is applicable only to electrically charged particles. If a particle is charged, it experiences an oscillatory motion caused by the applied ac electric field. This oscillatory velocity component of the particle will have a phase lag ϕ with respect to the applied electric field and the measurement of ϕ again allows a determination of d_a. The amplitude of the oscillatory velocity component of the particle is directly proportional to the electric charge on the particle. Thus, from the measurement of the velocity amplitude V_p and the phase lag ϕ, the electrostatic charge for individual particles can be calculated. There is a phase shift of 180° for particles of opposite polarity and this 180° phase shift is detected to determine the polarity of the electrical charge. Both the phase lag and the amplitude information are obtained by the LDV and the associated signal processing electronics.

Particle Motion: External Oscillating Force

The oscillation of a particle experiencing a sinusoidal force F_e in a gaseous medium was derived by Stokes and can be represented in the Stokes regime by the following equation (Fuchs 1964, 80):

$$m_r \frac{dv_p}{dt} + \frac{v_p}{B_r} - F_e = 0 \qquad (17\text{-}1)$$

where m_r is the effective mass of the particle, B_r is the effective mobility of the particle, v_p is the time-dependent (instantaneous) particle velocity, and F_e is the external force. Since particle velocity is not constant, Stokes showed that m_r and B_r can be replaced by

$$m_r = m_p + \frac{9m'\xi}{4} \tag{17-2}$$

$$B_r = \left(\frac{3\pi\eta d_p}{C_c} + \frac{9m'\omega\xi}{4}\right)^{-1} \tag{17-3}$$

where m_p is the particle mass, η is the gas viscosity, C_c is the slip correction factor, m' is the mass of air displaced by the particle, ω is the angular frequency of oscillation ($2\pi f$),

$$\xi = \frac{2}{d_p\sqrt{2\nu/\omega}} \tag{17-4}$$

ν is the kinematic viscosity (η/ρ_g), and ρ_g is the gas density. Equation 17-1 is a simplified version of the original Stokes equation for $m_p \gg m'$. When the frequency approaches zero, i.e., the velocity approaches a constant, ξ approaches zero and the effective particle mass can be replaced by m_p, B_r is reduced to the mobility B and

$$F_r = -3\pi\eta d_p(V_p - U_g)/C_c \tag{17-5}$$

where V_p is the particle velocity and U_g is the steady-state gas velocity. This is the more familiar form of the Stokes equation.

Particle Motion: Acoustic Field

Equation 17-1 describes the motion of a particle under the influence of an external field, e.g., a charged particle in an ac electric field. When there is no external field, but the medium itself is oscillating, e.g., the particle experiences an acoustic excitation, the time-dependent gas velocity u_g can be expressed as

$$u_g = U_g \sin(\omega t) \tag{17-6}$$

where U_g is the maximum gas velocity. The parameters measured by the instrument are the velocity amplitude ratio

$$\frac{V_p}{U_g} = \sqrt{\frac{1 + 3\xi + \frac{9}{2}\xi^2 + \frac{9}{2}\xi^3 + \frac{9}{4}\xi^4}{\alpha^2 + 3\alpha\xi + \frac{9}{2}\xi^2 + \frac{9}{2}\xi^3 + \frac{9}{4}\xi^4}} \tag{17-7}$$

and the phase lag of the particle behind the air motion, given by

$$(\phi - \theta) = \tan^{-1}\left[\frac{3(\alpha - 1)\xi(\xi + 1)}{2(\alpha + \frac{3}{2}\alpha\xi + \frac{3}{2}\xi + \frac{9}{2}\xi^2 + \frac{9}{2}\xi^3 + \frac{9}{4}\xi^4)}\right] \tag{17-8}$$

where

$$\theta = \tan^{-1}\left[\frac{\frac{2}{3} + \xi}{\xi(\xi + 1)}\right] \tag{17-9}$$

and

$$\alpha = \frac{2\rho_p}{3\rho_g} \tag{17-10}$$

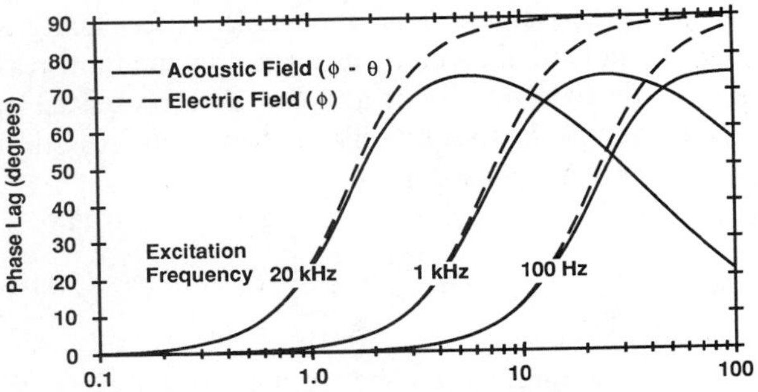

FIGURE 17-2. Phase Lag of the Particle Motion with Respect to an Acoustic Excitation Field (Eq. 17-8) and Electrostatic Excitation Field (Eq. 17-12) Plotted as a Function of Aerodynamic Diameter for Several Drive Frequencies.

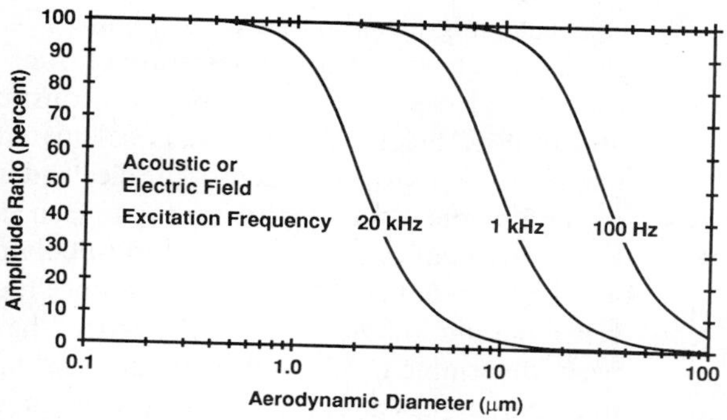

FIGURE 17-3. Amplitude Ratio of the Particle Motion with Respect to an Electrostatic Field Drive Plotted as a Function of Aerodynamic Diameter for Several Drive Frequencies for the E-SPART (Eq. 17-7). The Simpler Equation (Eq. 17-13) Gives Similar Curves throughout the Entire Range of Sizes.

The right-hand sides of Eqs. 17-7 and 17-8 are described by the terms α, ξ, and v, which are functions of the particle and gas properties. Figures 17-2 and 17-3 show $(\phi - \theta)$ and V_p/U_g, respectively, plotted as a function of aerodynamic diameter for several acoustic drive frequencies.

If the inertial terms caused by the acceleration of the particle in the medium are neglected, then the particle motion can be represented using the particle relaxation time τ_p

$$\tau_p = \frac{\rho_0 d_a^2 C_c}{18\eta} \tag{17-11}$$

where ρ_0 is unit density (1 g/cm^3). The phase lag ϕ of the particle behind the air motion in the acoustic field is given by

$$\phi = \tan^{-1}(\omega\tau_p) \tag{17-12}$$

where $\omega/2\pi$ is the acoustic frequency. The phase lag calculated using this equation is also plotted in Fig. 17-2. Under this condition, the ratio of the amplitude of particle velocity V_p to the amplitude of the gas motion U_g due to the acoustic field is given by

$$\frac{V_p}{U_g} = \frac{1}{\sqrt{1 + \omega^2\tau_p^2}} \tag{17-13}$$

Equations 17-12 and 17-11 can also be applied when the particle motion is induced by an external field, such as with a charged particle in an ac electric field. These equations, as well as Eqs. 17-7 and 17-8, indicate that the measurement of either the phase lag of the particle motion relative to the gas motion or the velocity–amplitude ratio of the particle in the acoustic field is sufficient to determine τ_p or d_a.

In the case of acoustic excitation, there are two major forces acting on the particle beside gravitational field: (1) the viscous drag force and (2) the force caused by the pressure gradient in the medium. The first one is caused by the fluid resistance due to the viscosity of the medium and the second is due to the inertial resistance and the effective fluid resistance depending on the product $\omega\tau_p$. For small values of this product, the fluid resistance is primarily viscous and for large values it is inertial. For sizing the aerodynamic diameter, the product $\omega\tau_p$ can vary from 0.01 to 100. In the range 0.01–2, the resistance can be approximated by the viscous drag (Eq. 17-12) and when $\omega\tau_p > 2$, both viscous and inertial resistance need to be considered (Eq. 17-8). Note that for $d_a \leq 100$ μm, particle Reynolds number Re_p is less than 0.1.

The value of $\omega\tau_p = 2$ corresponds to $\phi = 63.5°$. For $\phi \leq 63.5°$, Eq. 17-12 gives the results for an acoustic excitation within 15% of Eq. 17-8. For larger values of phase lag, the two curves are quite different (Fig. 17-2).

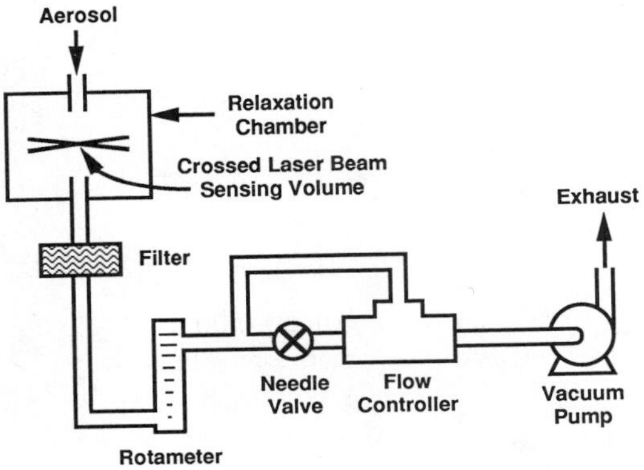

FIGURE 17-4. A Schematic Diagram of the Aerosol Sampling System Used in the E-SPART Analyzer.

However, the amplitude ratio curves given by Eqs. 17-13 and 17-7 are nearly identical (within less than 1%, Fig. 17-3) and, therefore, the simplified Eq. 17-13 holds over the entire range of $\omega\tau_p$.

To operate the analyzer over a wide size range, it is necessary to use two frequencies of excitation either in tandem inside a single relaxation chamber, or simultaneously in two relaxation chambers connected in series. For example, a prototype E-SPART analyzer has been operated at two frequencies, 24 kHz (for 0.3–4.0 μm) and 1.0 kHz (for 2.0–20.0 μm), using two chambers connected in series. Integration of the experimental data can be accomplished with the appropriate software. The aerosol sampling system with only one relaxation chamber used in the E-SPART analyzer, is shown in Fig. 17-4.

Particle Motion: dc Electric Field

When placed in a constant electric field E, the electrostatic force on a charged particle can be expressed as $F_e = qE$, where q is the particle charge. A particle of diameter d_a with n elementary charges will move with an electrical migration velocity V_e given by (Chapter 3)

$$V_e = \frac{neEC_c}{3\pi\eta d_a} \tag{17-14}$$

where e is the elementary charge. As shown in Fig. 17-3, the acoustic velocity component $V_p\sin(\omega t - \phi + \theta)$ is superimposed on this electrical migration velocity V_e. Figure 17-1 shows the geometrical configuration of transducers and electrodes for applying acoustic and electric fields. The field E is calculated from the voltage applied across the electrodes divided by the distance between them. Hence, a measurement of V_e can be used to calculate n, the number of elementary charges, once the aerodynamic diameter of the particle has been determined. The software performing the charge measurement reads the voltage (which is adjustable) applied across the electrodes and computes the field E for determining the magnitude of the charge q (ne) for each particle. The analyzer also recognizes the direction of V_e, which depends on the polarity of the charge q of the particle. Thus, from the direction and magnitude of V_e, the computer can record both polarity and magnitude of particle charge.

Particle Motion: ac Electric Field

The phase lag measurement technique can also be used for a charged particle in an ac E-SPART analyzer with the same electrode configuration as that shown in Fig. 17-1, with no acoustic field applied. An electrical sinusoidal voltage $V_0\sin(\omega t)$ is applied across the two electrodes. In this process, it is necessary that the particles be electrically charged in order to make size and charge measurements. When a charged particle transits the LDV sensing volume, the particle will experience an oscillatory electric field, $E_0\sin(\omega t)$, and a zero gas velocity. When the time $t \gg \tau_p$, ϕ has the same expression as Eq. 17-12 and the amplitude ratio is

$$\frac{V_p}{E_0} = \frac{qC_c}{3\pi\eta d_a\sqrt{1 + \omega^2\tau_p^2}} \tag{17-15}$$

Equations 17-12 and 17-15 indicate that for a charged particle, d_a can be determined from the measured value of the phase lag or the amplitude ratio as with the acoustical SPART analyzer.

In the case of electric excitation, there are two major forces acting on the particle besides the gravitational field: (1) the coulombic force and (2) the viscous drag force. For $Re_p < 0.1$, which is the case for many practical applications, the inertial resistance of the fluid can be neglected and the phase lag is given by Eq. 17-12. Unlike in the case of acoustic excitation, there is no pressure gradient force and, hence, there is no foldover in the phase lag relationship (Fig. 17-2). The amplitude ratio, in the case of either electric or acoustic excitation, is given by Eq. 17-13, without significant error. For the cases where $Re_p > 1$, appropriate correction will be needed to compute the viscous and inertial resistance forces acting on the particle for accurate size and charge measurements.

In the E-SPART analyzer, the measurement of d_a is independent of the driving field amplitude E_0 and the magnitude of the particle charge q. Once d_a is determined from ϕ, the analyzer then calculates the electrical mobility (qB) or the electrostatic charge q of the particle from Eq. 17-15. The phase lag measurement technique is independent of the amplitude of the driving force as long as the particle amplitude is sufficiently large for accurate measurement of ϕ.

E-SPART Analyzer

The E-SPART analyzer consists of four components: (1) a dual-beam, frequency-biased laser Doppler velocimeter, (2) a relaxation cell, (3) an electronic signal and data processing system, and (4) a personal computer. The LDV measures the particle velocity. The sensing volume of the LDV is formed by the intersection of the two laser beams and is located between the electrodes as shown in Fig. 17-1. As a particle passes through the sensing volume in the direction normal to the plane containing the two converging laser beams, the particle experiences an acoustic and/or an electric field. The LDV detects only the horizontal velocity component of the particle. It does not detect the vertically downward sampling velocity. However, the duration of the LDV signal burst is the residence time of the particle within the sensing volume and it is inversely proportional to the sampling velocity. The residence time must be long enough to measure $\phi - \theta$ and V_e or ϕ and V_p/E_0. The residence time is discussed further in the Aerosol Sampling section below.

A helium–neon laser (632 nm) or an argon ion laser (488 nm) is used as the monochromatic light source for the LDV. The choice of laser (He–Ne or Ar^+) and the output power (10 mW for He–Ne, or 50–500 mW for Ar^+ laser) depends upon the application. Two output laser beams are derived by passing the laser beam through an acoustooptic cell (Bragg cell) modulator. The output beams have nearly equal power, but one of the beams is shifted in frequency by 40 MHz by the modulator (Fig. 17-1). The two beams intersect at the focal volume within the relaxation cell. Light scattered from aerosol particles passing through the sensing volume is collected by the receiving lenses and focused to a pinhole directly in front of a photomultiplier tube. The output of the photomultiplier is an electrical signal which represents the Doppler burst containing the particle motion information.

Aerosol Sampling

Figure 17-4 shows a schematic diagram of the flow control system used for the E-SPART analyzer. The aerosol sample is drawn into the relaxation chamber by using a vacuum pump. A differential flow controller is used to maintain a constant rate of sampling flow approximately 0.5 l/min through the relaxation chamber. The sampling rate through the LDV sensing volume is a few milliliters per minute, depending on the specific optical configuration and the diameter of the particle. If the length of the sensing volume is L and the sampling velocity of the aerosol particles is V_z, then the maximum residence time of the particle in the sensing volume will be L/V_z. This residence time is set equal to NT, where N is the number of acoustic cycles and T is the time period of the acoustic or electric excitation. Under this condition, an aerosol particle passing vertically downward through the center of the sensing volume will undergo periodic motion for N excitation cycles.

Phase lag ϕ measurements on each individual cycle for this particle can be performed by the E-SPART analyzer and the average value of ϕ over N cycles is used to determine d_a.

The choice of N depends upon three factors: (1) the size resolution desired, (2) the response time of the signal processing electronics, and (3) the particle concentration. Typically, N is set between 3 and 8 by adjusting the sampling velocity V_z. Since the time period T depends on the frequency f of the acoustic or electric excitation drive, the residence time for particles is varied depending upon f and the size measurements. All the particles may not pass through the center of the sensing volume, resulting in a shorter sensing time; however, the particle must stay in the volume for at least one cycle to be measured.

It is essential that the flow field is maintained laminar as the aerosol sample passes through and around the LDV sensing volume. Velocity components of the particle in the x-direction in absence of any acoustic and electrical excitation should be less than the particle's Brownian motion.

Signal and Data Processing Electronics

The E-SPART signal processing electronics is organized into five functional sections: (1) a receiver containing the RF amplifier, mixer, and demodulator, (2) the signal conditioning circuitry, (3) the size and charge measurement circuitry, (4) a direct memory access board for interfacing with the computer, and 5) a personal computer.

The instantaneous Doppler signal frequency generated by a particle traversing the sensing volume in the absence of any excitation, is given by f_0, which is the LDV bias frequency (40 MHz) as determined by the Bragg cell. When a particle experiences an acoustic excitation of frequency f, and a dc electric field, the Doppler frequency f_D is

$$f_D = f_0 + \Delta f + \frac{2}{\lambda} V_p \sin(\omega t - \phi)\sin(\Theta/2) \qquad (17\text{-}16)$$

where Θ is the intersection angle of the two laser beams, λ is the laser radiation wavelength and ω is $2\pi f$. It is assumed that ϕ is $< 63.5°$ in Eq. 17-16. A similar equation can be used when ac electric field excitation is used. The carrier frequency shift Δf is

$$\Delta f = \frac{2neEC_c}{3\pi\eta d_a \lambda}\sin(\Theta/2) \qquad (17\text{-}17)$$

where n is the number of electronic charges on the particles and E is the dc electric field.

In an acoustic E-SPART, the phase lag ϕ of the particle motion is determined from the time interval between the zero crossings of the acoustic field and the resultant particle motion (Fig. 17-5). The d_a for the particle is computed from f as follows:

$$d_a = \sqrt{\frac{9\eta\tan\phi}{\pi f \rho_0 C_c}} \qquad (17\text{-}18)$$

and from Eq. 17-17,

$$ne = q = \frac{3\pi\Delta f \eta d_a \lambda}{2EC_c \sin(\Theta/2)} \qquad (17\text{-}19)$$

In the case of the ac E-SPART, an ac electric field $E_0\sin(\omega t)$ replaces the dc electric field and the acoustic excitation. The d_a is determined from ϕ, and q is calculated from the ratio of the amplitude of the particle motion (V_p) to the amplitude of the ac electric field E_0, as shown in Eq. 17-15. The maximum value of

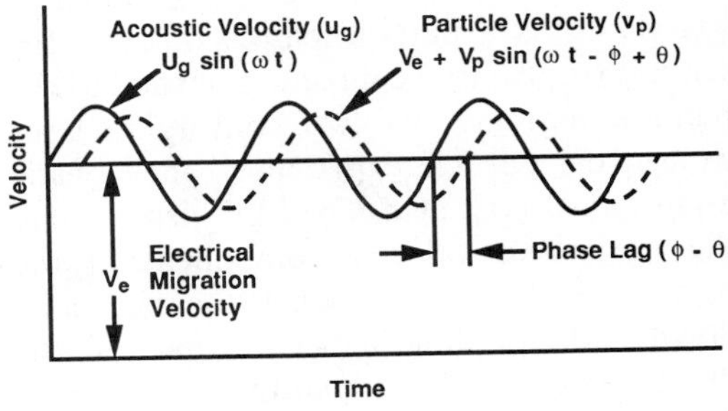

FIGURE 17-5. Wave Forms of the Motion of Charged Particles within the Sensing Volume of the E-SPART Analyzer.

the frequency deviation is:

$$\Delta f_0 = 2V_p \sin(\Theta/2)/\lambda \tag{17-20}$$

and

$$ne = q$$

$$= \frac{3\pi|\Delta f_0|\eta d_a \lambda\sqrt{1+\omega^2\tau_p^2}}{2EC_c \sin(\Theta/2)} \tag{17-21}$$

The frequency deviation Δf_0, with respect to f_0, can be either positive or negative, depending upon the polarity of the charge. The magnitude of the charge is determined from $|\Delta f_0|$.

Instrument Calibration

Size Calibration

While it is possible to calculate the aerodynamic diameter and the particle charge if the instrumental parameters are measured accurately and the particle equation of motion is known, the uncertainty in some of these parameters normally necessitates a calibration of the instrument. Once this calibration is performed, the size calibration should be consistent for the entire range.

To perform the calibration, polystyrene latex (PSL) spheres of known size are aerosolized by nebulizing a suspension of PSL particles in water. The aerosol stream is then dried, passed through a diffusion dryer and ducted through the E-SPART sensing volume. An offset count is then adjusted so that the measured size channel correlates with the actual size of the PSL particles. The calibration should be made under known conditions of temperature, humidity, and other instrument parameters upon which the measured value of the phase lag may depend. For calibrating the instrument for particle sizes larger than 10 μm, a vibrating-orifice aerosol generator (Model 3450, TSI, Inc., St. Paul, MN) is used to produce droplets of known size of a known density liquid, e.g., bis-ethylhexyl sebacate (BES).

Charge Calibration

Since the electrical field inside the sensing volume can be readily calculated from the distance between the electrodes and the applied voltage across the electrodes, the need for charge calibration in the E-SPART analyzer is less stringent compared to the size calibration. It is also difficult to generate particles with known size and charge for calibration purposes. To check the charge calibration of the E-SPART analyzer, a differential-mobility analyzer (DMA) was used to classify a PSL aerosol for a known charge and then the classified aerosol particle was used to test the validity of the measured values of the charge (Mazumder et al. 1991).

The E-SPART data showed close agreement with the particle charge classified by the DMA. For example, the measured charge of 0.8 μm diameter PSL aerosol was within ± 2 electronic charges when classified for 50 electronic charges per particle.

Instrument Operation

The computer used for operating the E-SPART analyzer provides various options. Features include selection of sampling time, maximum count per channel, and data storage and retrieval routines. In a typical operational mode, the instrument provides the aerodynamic size distribution of the sampled particles, the electrostatic charge distribution for any of the size channels selected by the operator, and a table showing the particle count vs. the magnitude of the electrostatic charge for that given size channel. Tables are separated, one for positively charged particles and the other for negatively charged particles. For a given size, the computer calculates the net average electrostatic charge for all the particles counted in that particular size channel and the charge-to-mass ratio for that particular size. The operator can choose any size channel for measuring charge distributions.

A summary of data can be requested, giving the total number of particles counted with the average charge tabulated for a given size channel for both positive and negative charged particles. The software can provide plots of the size distribution of the aerosol in

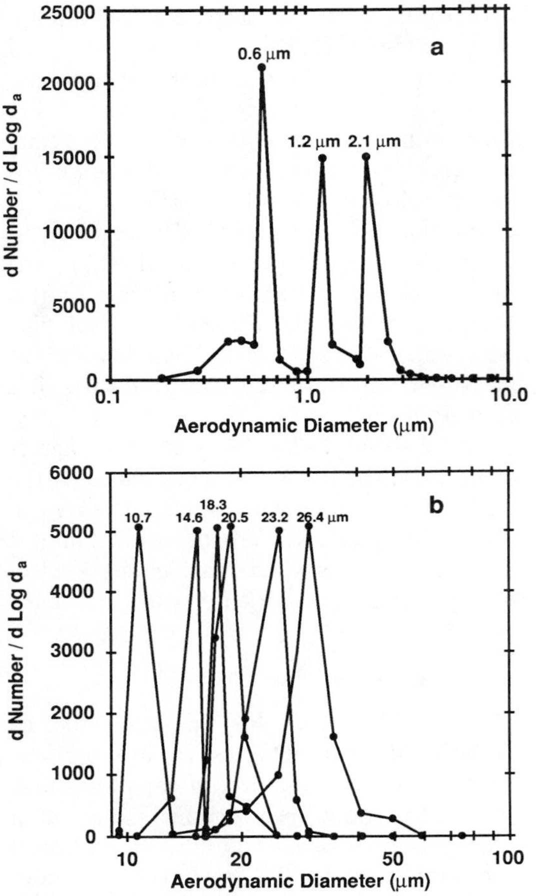

FIGURE 17-6. (a) Size Frequency Distribution $dN/d \log d_a$ for a Mixture of PSL Spheres with Diameters of 0.6, 1.2, and 2.1 μm (Acoustic E-SPART). (b) Size Frequency Distribution $dN/d \log d_a$ for Monodisperse BES Droplet Aerosols of Different Sizes (ac E-SPART). The Data were Obtained to Give Equal Peak Concentrations.

terms of number (Figure 17-6a, b), cumulative number, volume, cumulative volume, as well as statistics including count median diameter, mass median diameter, and geometric standard deviation.

For each particle, the aerodynamic diameter (d_a) and the charge (q) are determined in the E-SPART analyzer and the average value of the charge-to-mass ratio computed. For a spherical particle of diameter d_p and specific gravity ρ_p, we can write an approximate relationship

$$d_p^2 \rho_p = d_a^2 \rho_0 \tag{17-22}$$

The mass m_p of the particle can be computed from the measured value of d_a, if ρ_0 is known. Thus,

$$m_p = \pi \rho_0^{3/2} d_a^3 / 6 \rho_p^{1/2} \tag{17-23}$$

For each size channel $(d_a)_i$, $i = 1$–32, the particle count is stored as n_i; m_p for each channel is approximately $\pi n_i \rho_0^{3/2} d_a^3 / 6 \rho_p^{1/2}$. The total mass of the particle sample is given by summing over the i channels:

$$m = \sum (m_p)_i = \frac{\pi \rho_0^{3/2}}{6 \rho_p^{1/2}} \sum n_i (d_a)_i^3 \tag{17-24}$$

for each size channel, $(d_a)_i$; the total count n_i is also stored in the charge channels. The sums are performed over all 32 channels. The number of charged particles, n_i, is equal to $n^0 + n_i^+ + n_i^-$, where n^0, n^+, and n^- represent the total number of particles with zero, positive, and negative charges with diameter $(d_a)_i$, respectively. The software provides computations and plots of n_i^+, n^0, n_i^- vs. charge-to-mass (q/m) ratio for any channel $(d_a)_i$, as shown in Fig. 17-7. A three-dimensional plot

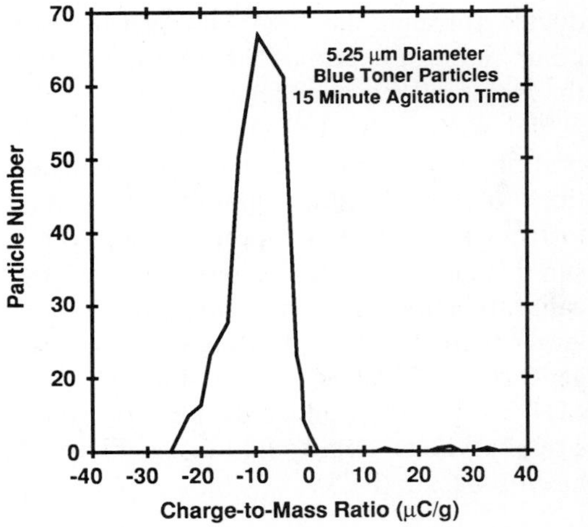

FIGURE 17-7. Electrostatic Charge Distribution of a Sample of Blue Toner Showing the Variation of Charge Density for a Selected Aerodynamic Diameter.

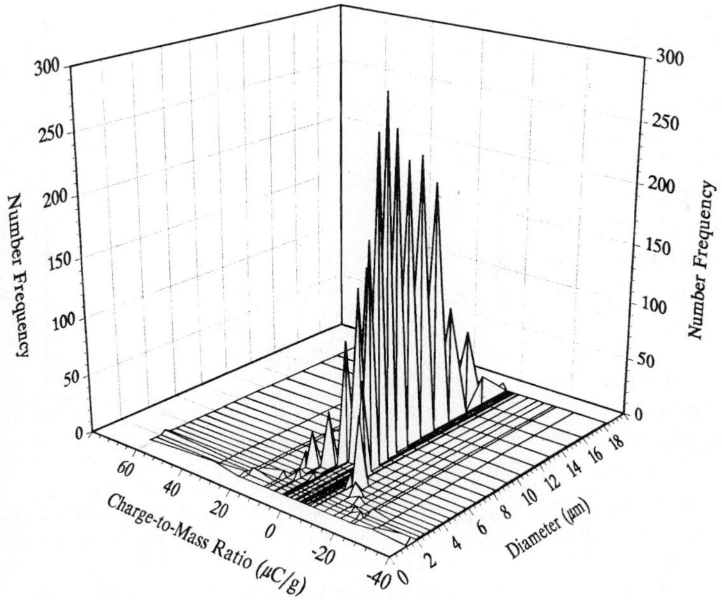

FIGURE 17-8. A 3-D Representation of the Variation of Particle Count and Charge Density as a Function of Aerodynamic Diameter.

of number vs. charge-to-mass ratio and d_a is also available (Fig. 17-8).

Size Resolution

Particle size is measured by determining the phase lag ϕ between the particle motion and the driving force (acoustic or electric). In practice, a time interval Δt is measured and the relationship between the phase lag ϕ and Δt is given by $\Delta t = \phi/\omega$. The signal processing electronics determines Δt by generating a phase comparator pulse with the duration Δt and then counting the time period of that pulse using a counter of frequency f_c. To obtain good resolution, the counter frequency f_c is made many times larger than the excitation frequency f. The number of counts n_c for a given time interval Δt can be written as $n_c = \Delta t f_c$. Since the maximum value of phase shift is 90°, the maximum count will occur for a 90° phase shift. As shown in Fig. 17-2, the variation of phase shift with respect to aerodynamic diameter, $d\phi/d(d_a)$, depends on the frequency of operation. For particles with the Cunningham correction factor equal to 1, i.e., $d_a > 2$ μm, maximum resolution (Renninger, Mazumder, and Testerman 1981) is obtained when $d\phi/d(d_a)$ is maximum, which gives a phase angle of $\phi = 30°$ or $\omega\tau_p = 1/\sqrt{3}$ for maximum resolution. In terms of channel number, the greatest resolution occurs when n_c is approximately equal to 1/3 of the maximum count n_m.

Measurements have shown that a resolution of $\pm 1\%$ can be obtained when the instrument is operated under controlled ambient conditions, so that there is no variation of temperature and the constituents of the gas phase in the aerosol. For example, it was possible to easily distinguish particle size differences between singlets, doublets, and triplets of 1.01 μm PSL particles.

Charge Resolution

The resolution in measuring the particle charge also depends on the parameters of the instrument and the magnitude of the electric field. High-resolution charge measurement was needed to study the Boltzmann distribution of the particle charge. In such applications, a high electric field (dc) was used in the acoustic E-SPART and the charge distribution was resolved within ± 1 electronic charge, with the average charge varying from 0 to 20 electronic charges for 1.0 μm diameter particles. For highly charged particles, the resolution of the instrument is difficult to determine, since there is no readily available process to control q accurately, nor any other instrument to measure the exact value based on a single particle.

Range of Operation

The E-SPART analyzer can be operated in the range 0.3–75 μm in aerodynamic diameter. The commercial instrument is normally operated at a single frequency. However, in order to cover this entire range, it is necessary to modify the analyzer to operate with at least two different frequencies of excitation, 25 and 0.5 kHz, in tandem or simultaneously. For high-resolution sizing of particles, it is necessary to operate the analyzer at three frequencies: 25, 2.0 and 0.5 kHz. The maximum particle count rate (10 particles/s to 2000 particles/s) depends on the frequency of operation.

The desired range of measurement of electrostatic charge on each particle is from zero charge to its saturation value, with positive or negative polarity. Charge distribution of PSL particles with Boltzmann equilibrium charges have been measured. Figure 17-9 shows the average values of the charge-to-mass ratio of a toner sample experimentally measured as a function of d_a and the calculated Gaussian limit for maximum q/m as a function of diameter. The particles were tribo-charged. The saturation charge-to-mass ratio for tribo-charged, dielectric solid particles varies inversely with particle diameter. The data show that the analyzer can measure particles with saturation charge.

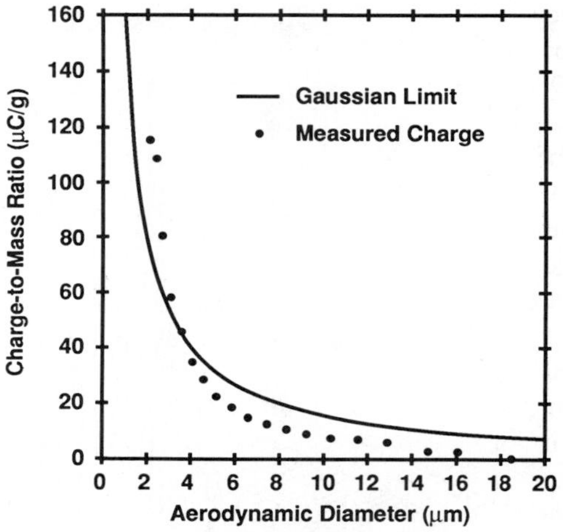

FIGURE 17-9. The Charge Distribution of a Tribo-Charged Toner Sample as Measured by an E-SPART Analyzer. The Solid Line Shows the Saturation Charge Calculated from the Gaussian Limit. The Experimental Data Show that the E-SPART Analyzer can Measure Particle Charge Near the Saturation Limit.

Precision and Accuracy

The basic principle applied in the E-SPART analyzer can provide absolute measurements of particle size and electric charge if the physical parameters involved are accurately known. For example, the aerodynamic diameter depends on the viscosity of the gas in which the particles are suspended. Since viscosity is independent of pressure, the size measurement can be performed at different ambient pressures. However, if the temperature or the constituents of the gas change, the viscosity will change and, therefore, the measured value of the aerodynamic diameter will be related to the properties of the gas in which the particles are suspended. This is an advantage of the E-SPART analyzer for *in situ* measurements. However, if there are uncertain variations in the ambient conditions from sample to sample, such as changes in temperature, the instrument's operation will be affected adversely. When the instrument is operated at a high acoustic frequency (25 kHz), it is important to keep a constant temperature in the relaxation cell so that the phase offset value does not change. The constraint is less severe when the acoustic E-SPART is operated at a frequency of 1 kHz or lower, or when an ac excitation is used.

AERODYNAMIC PARTICLE SIZER

The development of the APS (APS 3310, TSI, Inc. St. Paul, MN) was based on a particle acceleration nozzle and laser Doppler detection system constructed by Wilson and Liu (1980). In this study particles were introduced into the center of an accelerating nozzle. Small particles followed the motion of the air closely while larger particles lagged behind, causing an increase in relative velocity between air and particles. This increase in relative velocity is analogous to the increase in settling velocity with particle d_a. Wilson and Liu indicate that the particle velocity is a function of d_a as long as Re_p stays small (within the Stokes regime, $Re_p < 0.1$). As Re_p increases, apparent particle size becomes a function of particle density and shape as well as d_a. In addition, there is a tradeoff in size resolution and nozzle velocity. Thus, at high nozzle velocities, particle motion is more non-Stokesian (less accurate aerodynamic sizing) but particle sizing is more rapid. Based on similar principles, TSI, Inc., developed the APS with support from the National Institute for Occupational Safety and Health (NIOSH) (Agarwal and Remiarz 1981).

The APS sizes particles by measuring their velocity relative to the air velocity within an acceleration nozzle. This velocity is compared with a calibration curve established using monodisperse spheres.

Instrument Description

The APS consists of a sensor unit containing the sampling system, detector, preliminary processing electronics and internal flow rate indicator, and a computer. The computer receives the data from the sensor unit about once per second and updates the calculated aerodynamic size distribution. One version of the software that collects and displays the data comes with the instrument; more sophisticated software providing near-real-time display is available as an option. While the entire unit is sufficiently portable and rugged for it to be used for field measurements (Baron 1986; Baron and Willeke 1986; Szewczyk et al. 1991), it is more suited to laboratory environments.

Aerosol is introduced to the inlet of the APS at a flow rate of 5 l/min. Four l/min of this flow is removed and passed through a filter and reintroduced as sheath air. The remaining 1 l/min aerosol flow is fed through a focusing nozzle, recombined with the sheath air and accelerated through the final nozzle (Fig. 17-10). The pressure below the nozzle is approximately 100 mmHg (Chen, Cheng, and Yeh 1985). The sheath and total flow are controlled by valves and monitored with thermal mass flow meters.

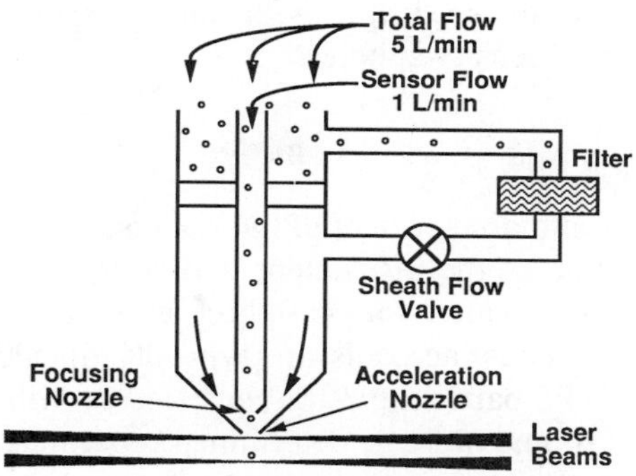

FIGURE 17-10. Schematic Diagram of Aerodynamic Particle Sizer (Model APS 3300) Nozzle and Laser Velocimeter.

At the exit of the acceleration nozzle, each particle passes through two laser beams. The light scattered from the particle causes two pulses to be detected by a photomultiplier and the time lag between the two pulses is recorded. Because larger particles have not accelerated to the air velocity in the sensing zone, they are represented by larger time lags. The time-lag data is stored in an accumulator in bins representing equal time intervals. Two sets of data are stored: one by the small-particle processor (SPP) in 4 ns bin intervals and the other by the large-particle processor (LPP) in bin intervals of 66.7 ns. The software gives the option of using just the SPP for particle distributions in the range 0.5 to 15.9 μm, while the LPP can be used to extend that range up to 30 μm. The LPP has special anticoincidence circuitry that virtually eliminates excess counts due to coincidence. When both processors are used, the two sets of data are linearly combined in the range from 5.7 to 15.9 μm. The final aerodynamic size is determined by calibrating the accumulator spectrum with monodisperse, near-unit-density spheres.

Sample Inlet

The 2 cm diameter inlet of the APS is located at the top of the instrument and is not conveniently located for sampling moving air directly. Thus, aerosols are typically ducted to the APS with external tubing. Losses within this tubing must be determined separately. Within the inlet, the air is split between the inner inlet (measured aerosol flow) and the sheath flow, as indicated in Fig. 17-11. The gas velocity at the measured flow inlet is higher than the velocity at the APS inlet, i.e., superisokinetic. This sampling arrangement produces some oversampling of larger particles to compensate for losses within the inner nozzle tube. Calibration curves have been developed by the manufacturer and are provided as part of the computer software, so that compensation for sampling losses at this point can be achieved.

Sample dilution systems have been provided as additional options for the APS to reduce problems with particle coincidence in the sensor. Penetration curves for these dilutors were measured by the manufacturer and are provided as part of the software to correct size distributions. At 15 μm the loss within the dilutor is near 50% and increases rapidly with increasing d_a; corrections of this magnitude make the data in the larger-particle channels questionable.

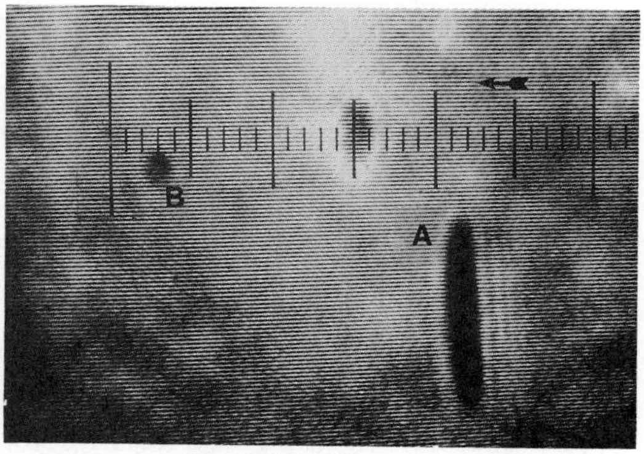

FIGURE 17-11. Picture of Droplets in the High-Velocity Air Jet just beyond the APS Nozzle Taken with a High-Speed Laser Imaging System, Showing a Droplet Flattened by the Drag Force. The Nozzle Tip is about 200 μm to the Right of Droplet A with the Air and the Droplets Moving to the Left. The Scale Markers are Approximately 5 μm Apart. The Larger Droplet A Has an Extreme 10 × 60 μm Ellipsoidal Shape while the Smaller Droplet B is about 8 × 10 μm and is Only Slightly Flattened.

At the bottom of the inner inlet, a nozzle constricts the flow and focuses the aerosol in the center of the acceleration flow. The inner walls of the focusing nozzle form a 60° angle with the direction of flow. Impaction may produce particle accumulation on this nozzle surface, further restricting the penetration of the inner inlet to 50% for about 8 μm oil droplets (Kinney 1990). Kinney also evaluated modifications to this nozzle and found that a smaller nozzle angle (2° or 8°) produces less internal loss but decreases the resolution of the APS (Kinney 1990). The amount of particle loss for the 60° inlet nozzle can be estimated using an equation developed for particle deposition efficiency η in a tube with a 90° contraction:

$$\eta = [1 - \exp(1.721 - 8.557x + 2.227x^2)]^2 \tag{17-25}$$

where

$$x = \sqrt{Stk}/(D_i/D_n)^{0.31} \tag{17-26}$$

Stk is the Stokes number in the inlet tube, D_i is the diameter of the inlet tube, and D_n is the diameter of the nozzle (Ye and Pui 1990). Impaction of liquid particles will follow the deposition efficiency described by Eq. 17-25, while solid particles may bounce and exhibit lower deposition.

Laser Velocimeter Sensor

Aerosol passing through the inner nozzle is combined with the sheath flow and focused into the center of the acceleration nozzle. The air flow conditions in the nozzle region have been modeled and agree well with experimental measurements (Ananth and Wilson 1988). The air velocity reaches approximately

150 m/s at the exit of the acceleration nozzle. The flow conditions affecting the particle acceleration depend on the nozzle dimensions as well as the spacing of the nozzles. A helium–neon laser beam is split into two parallel, flattened beams that intersect the particle path 200–500 μm from the acceleration nozzle. The distance of these beams from the acceleration nozzle also affects the measured particle velocity. These various dimensions are difficult to control precisely during instrument manufacture. Thus, the calibration of each APS instrument is slightly different.

A calcite crystal is used to split the laser beams. The resulting beams are polarized at 90° to one another and may also have a slightly different intensity. Compactly shaped particles passing through these beams will produce two pulses of approximately the same intensity. However, if the particles are anisotropic, the light scattering intensity as the particle passes through each beam may be slightly different. When the particle is small enough to produce a scattering pulse near the threshold limit of the photomultiplier detector/electronics system, there is significant probability that only one of these pulses of light from the particle will be detected (Heitbrink, Baron, and Willeke 1991). This probability of incomplete, single pulse detection has implications for the production of "phantom" coincidence particles, as discussed below.

Detection and Data Analysis

As each particle passes through the two laser beams, the pulses are detected by a photomultiplier and the time between the pulses is recorded. Two high-speed data accumulator systems are used in the detector module: a small-particle processor (SPP) and a large-particle processor (LPP). The SPP collects the time-of-flight data in an accumulator in increments of 4 ns and the LPP collects data in increments of 66.7 ns. These data are passed from the detection module to an IBM (or IBM-compatible) personal computer for transformation to size distribution according to a stored calibration curve, as well as for any further manipulation or storage as desired. The SPP covers the aerodynamic diameter range of 0.5 to 15.9 μm and the LPP covers 5 to 30 μm. In the overlap range, the data from the two processors are blended together, proportionately increasing the LPP contribution with increasing size. The treatment of the overlap range will be discussed further in the section Coincidence Effects. Once a number distribution has been measured, various other differential and cumulative distributions can be calculated similarly as described for the E-SPART.

Calibration

Monodisperse latex spheres are typically used for calibration of the full size range of the APS if it is to be used for measuring solid particles. Latex spheres smaller than about 5 μm can readily be generated by nebulizing a water suspension of the spheres. Note that while isopropyl alcohol suspensions of latex spheres may be easier to generate and dry, the alcohol slowly dissolves in the spheres and will cause a slight increase in size after a period of time. Larger calibration particles can be generated dry from a surface by suction, as with the small-scale powder disperser (Model 3433, TSI, Inc., St. Paul, MN). Since latex particles are available only in specific discrete sizes, the calibration curve is completed using a spline function to fit the calibration points.

The calibration of each APS instrument is unique due to variations in the nozzle sizes, spacing, and laser beam locations. However, once the calibration of the APS has been completed in air at ambient pressure, calibration for other gas viscosities and pressures can be achieved as described by Rader et al. (1990). The gas velocity U_g in the nozzle can be calculated from the Bernoulli equation for compressible flow:

$$U_g = \left[\frac{2RT}{M}\ln\left(\frac{P}{P - \Delta P}\right)\right]^{1/2} \qquad (17\text{-}27)$$

where R is the universal gas constant, T is the

absolute temperature, M is the gas molecular weight, P is the ambient pressure, and ΔP is the pressure drop across the nozzle. ΔP is measured by the flow transducer in the APS. The particle velocity V_p can be determined from

$$V_p = \frac{U_g t_{min}}{t} \tag{17-28}$$

where t is the transit time of the particle between the laser beams and t_{min} is the minimum transit time for small particles observable in the APS accumulator. Plotting the ratio V_p/U_g as a function of Stokes number results in a universal response curve. This means that the check on the APS size response under the same or new pressure or viscosity conditions can be achieved by always setting U_g to the same value.

Other monodisperse particles, such as those generated from the vibrating-orifice monodisperse aerosol generator (model 3450, TSI, Inc., St. Paul, MN) can also be used for calibration. However, it was found that oil droplets generated in this fashion distorted due to the high acceleration and, therefore, exhibited a smaller aerodynamic diameter than predicted for a spherical shape (Baron 1986). Figure 17-12 shows two dioctyl phthalate droplets detected with a laser imaging system (Laser Holography, Inc., Mammoth Lakes, CA) just past the tip of the acceleration nozzle. Air motion is directed from right to left, causing the deformation of the larger droplet into an extreme oblate shape and increasing the droplet drag. Unless the calibration is used for measuring similar oil droplets, only solid particles should be used for the calibration of the APS.

Non-Stokesian Corrections

The acceleration in the nozzle produces Reynolds numbers outside the Stokes regime, as indicated in Table 17-1 for particles in the APS size range. Thus, the measured size is dependent on other factors besides the aerodynamic size, including gas density, viscosity, particle density, and particle shape. Using the approach of Wang and John (1987), correction factors for the measured size of compact particles can be calculated if the particle density, gas density, and gas viscosity are known. These calculations have been validated by the Navier–Stokes calculation of the flow field in the APS nozzle by Ananth and Wilson (1988). The following equations (Rader et al. 1990) are iterated until no further significant change occurs:

$$\sqrt{Stk_2} = \sqrt{Stk_1}\left[\frac{6 + R_2^{2/3}}{6 + R_1^{2/3}}\right]^{1/2} \tag{17-29}$$

$$R_i = \xi_i^{3/2}\sqrt{Stk_i}|U_g - V_p| \tag{17-30}$$

$$\xi_i = \left(\frac{18\rho_{gi}^2 S}{\rho_{pi}\mu_i U_g}\right)^{1/3} \tag{17-31}$$

where subscript 1 refers to calibration conditions with unit-density spheres, subscript 2 refers to measurement conditions, Stk is the Stokes number, and S $(= U_g t_{min})$ is the

TABLE 17-1 Particle Properties in the APS Nozzle

Particle Diameter (μm)	Relative Velocity (cm/s)	Particle Reynolds Number	Weber Number (Oil Droplets)
0.5	40	0.013	2.9×10^{-6}
1.0	1750	1.16	0.0113
3.0	6490	12.9	0.468
10.0	10600	69.6	4.13
15.0	11500	114.0	7.36
20.0	12300	163.0	11.2

Source. Baron (1986)

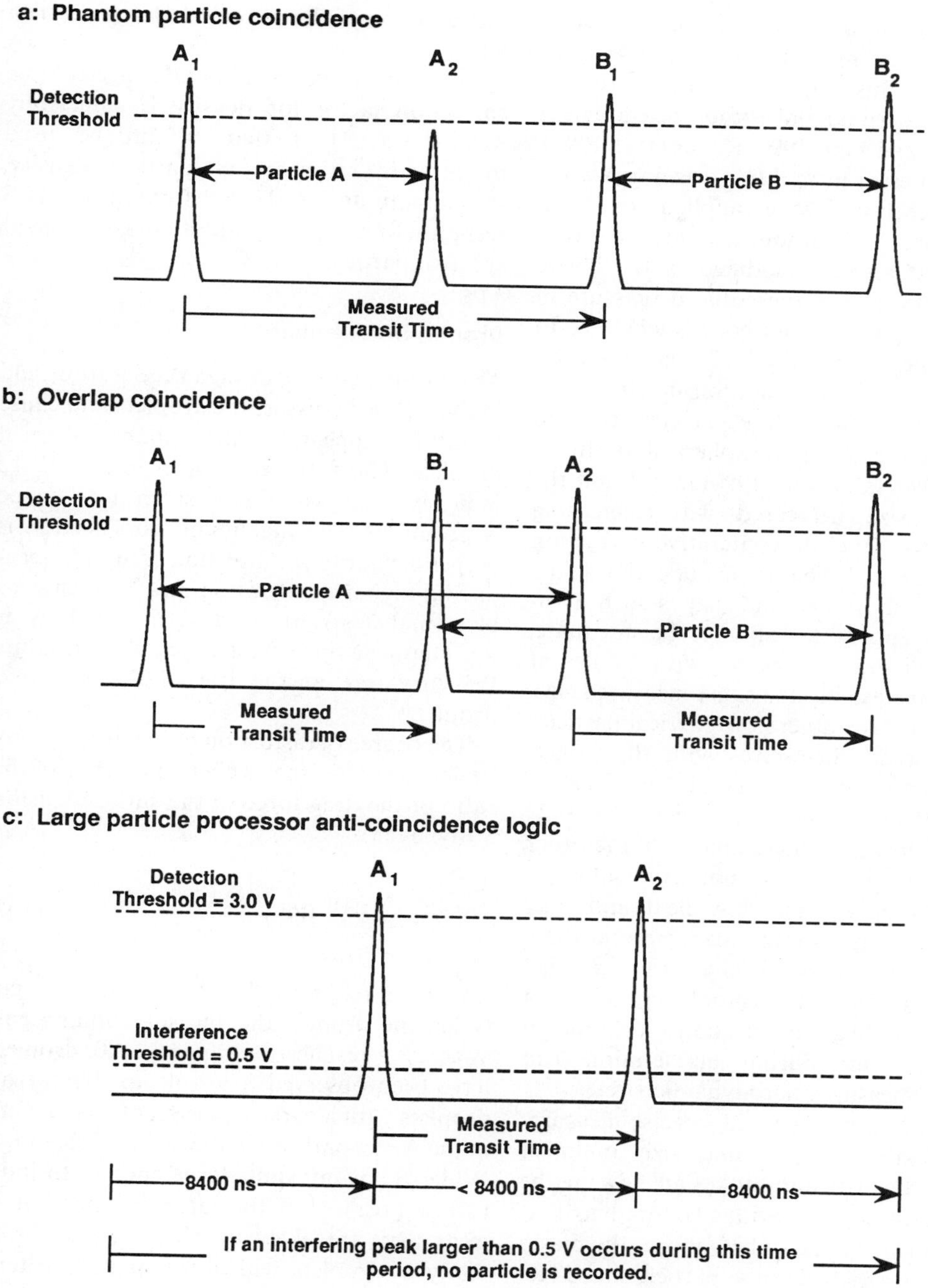

FIGURE 17-12. Coincidence Scenarios in the SPP and the LPP in the APS: (a) A Single Detected Pulse Triggers the Timer, While a Pulse from a Second Particle Produces a Measured Transit Time That May Indicate a Particle of Random Size; (b) Overlap of the Pulses from Two Coincident Particles Produces Two Smaller Detected Particles; (c) A Particle Detected by the LPP Must Have Pulses Larger than 3 V. The LPP Anticoincidence Circuitry Prevents Detection of a Particle When an Interfering Pulse Occurs within 8400 ns before or after the Evaluated Pulse Pair. (Adapted from Heitbrink, Baron, and Willeke 1991.)

distance between the laser beams. Measurements were made in argon and N_2O to confirm that this approach improved the accuracy of aerodynamic size measurement (Lee, Kim, and Han 1990; Rader et al. 1990). The largest error in d_a (12%) was noted when these corrections were applied to large (30 μm) particles in argon. The slip-correction factor must also be modified in the above equation because of the reduced pressure in the nozzle and computer code is available to perform these corrections (Wang and John 1989). The high acceleration in the APS nozzle may also cause inaccuracies in measuring the d_a of nonspherical particles. Cheng, Chen, and Yeh (1990) found that the measured size decreased with increasing shape factor. The above iterative correction was further modified to include the shape factor. For more extreme shapes such as fibers, this approach may not be adequate. Identical-diameter fibers with different lengths gave exactly the same measured size. Fibers, as well as other nonspherical particles tend to orient themselves with their maximum cross section oriented perpendicular to the flow (Clift, Grace, and Weber 1978, 142). However, larger fibers (diameter of the order of 10 μm diameter) may not have sufficient time to orient in the flow field and may produce a measured size intermediate between the perpendicular and parallel orientation. Thus, the initial conditions of the particle (e.g., orientation, location in the flow field) during acceleration can affect the measured aerodynamic size.

The APS allows a rapid, precise measurement of the aerodynamic size of most particles. Due to non-Stokesian flow in the acceleration nozzle, various factors bias that measurement. As described above, the biases caused by particle density, particle shape factor, gas viscosity, and gas density are sufficiently well understood for the corrections to measured size to be made. The size of these biases is often of the order of 25% or less. Thus, for many purposes, an estimated value of the particle density can yield sufficient accuracy in the corrected aerodynamic size. An exception may occur when the dynamic shape factor of the particle is large, as with fibers.

Chen and coworkers (1990) suggested the correction factor for density is sufficiently well characterized that it can be used to provide estimates of particle density. Brockmann and Rader (1990) used the APS response to measure shape factors for several types of particles.

Droplet Deformation

As indicated above, the high acceleration field in the APS will distort the droplets and cause them to appear smaller than spherical droplets. The flattened droplet in Fig. 17-12 indicates that an oblate spheroidal shape is produced. The maximum cross section is perpendicular to the direction of motion, producing higher acceleration and the behavior of a smaller spherical particle. The shape is not elliptical since the surface tension limits the curvature at the rim of the distorted droplet.

The degree of distortion of the droplet may be indicated by the Weber number We, the ratio of the drag force to the surface tension σ (dyne/cm):

$$We = \frac{\rho_g V_p^2 d_p}{\sigma} \tag{17-32}$$

where V_p refers to the relative velocity between the droplet and the surrounding gas. When We reaches the range 12–20, droplets have been observed to break up. We for oil droplets with a surface tension of 33 dyne/cm in the APS nozzle are indicated in Table 17-1. While We is not quite large enough to indicate a breakup of the larger droplets, it is clear they are likely to distort. However, a study of droplets with different surface tension and viscosity indicates that the surface tension determines the limiting shape of the droplet, while the viscosity determines the rate at which the droplet distorts (Griffiths, Iles, and Vaughan 1986). At high acceleration as in the APS nozzle, small droplets may distort to their maximum extent, while large, high-viscosity droplets may not distort completely by the time they reach the detection region. In addition, the degree of distortion is dependent on the precise acceleration history and, therefore, will vary from instrument to instrument. Water droplets, which have a relatively high surface tension, do not appear to be distorted noticeably in the acceleration field (Baron 1986).

Coincidence Effects

An accurate detection of a particle in the time-of-flight detection system of the APS requires full detection of two pulses from the same particle. The coincidence results from such a system are more complex than those of a standard optical particle counter, where two coincident particles produce one, somewhat larger, measured particle. Several APS coincidence scenarios are presented in Fig. 17-12. Coincidence between two particles can result in two smaller particles (Fig. 17-12a), one smaller particle, one randomly sized particle (Fig. 17-12b), or no particles detected. The relative frequency of these possible coincidence results and the effect on the measured distribution depends on the shape of the size distribution, the particle concentration, as well as which of the two signal processors (small-particle processor, SPP, or large particle processor, LPP) is used to detect the particles. The LPP is designed to completely eliminate coincidence counts in the large particle range, where coincidence can produce relatively large changes in apparent concentration (Fig. 17-12c). The coincidence effects in the APS system have been modeled mathematically as well as with a Monte Carlo calculation (Heitbrink, Baron, and Willeke 1991).

Particle counting systems such as the APS are prone to detection problems when more than one particle is present, or coincident, in the detection volume at the same time. The number of coincidence events in a measured distribution can be estimated from the difference between the actual concentration C_a and the measured concentration C_m:

$$C_a - C_m = C_a[1 - \exp(-C_a Q t)] \qquad (17\text{-}33)$$

where Q is the flow rate through the detection volume and t is the residence time of the particle in the detection volume (Willeke and Liu 1976). This equation can be used to estimate the coincidence loss to the peak of a monodisperse size distribution. Obtaining an accurate coincidence level for most size distributions is more difficult because t depends on particle size. For instance, the concentration level giving 1% coincidence in the SPP for 0.8, 3, and 10 μm particles is 558, 387, and 234 particles/cm^3, respectively. For the LPP, a 1% coincidence level is predicted for 10, 20, and 29 μm particles at concentrations of 55, 48, and 43 particles/cm^3 (TSI 1987). By using Eq. 17-33 for variously sized particles, the upper limit to the number of coincidence events can be estimated for broader size distributions.

C_a may also be difficult to estimate for many distributions where many of the particles detected by the SPP are smaller than 1 μm. These particles may be only partially detected, resulting in single detected pulses that contribute to coincidence events (Fig. 17-12a), but not to the observed small-particle concentration. These coincidence events result in a randomly sized "phantom" particle. The result is a nearly constant background of these coincidence-induced phantom particles (Heitbrink, Baron, and Willeke 1991).

The phantom-particle background produced by the SPP is, therefore, dependent on the number of particles near the pulse height detection limit of the sensor as well as the concentration of fully detected particles. This background becomes important in size regions where relatively few real particles are detected. Thus, when particle number distributions are converted to mass distributions, a few large phantom particles can bias the calculated mass and unrealistically skew the distribution (Baron 1986). Another situation arises when two particle distributions are being compared, such as the distributions before and after a filter to measure penetration efficiency (Wake 1989). In the size range where phantom-particle concentration is more than a few percent of the real-particle concentration, the ratio of the before and after distributions will be inaccurate.

The SPP, thus, tends to produce overestimates of particle concentration near the tail of a distribution, such as often occurs at large particle sizes. On the other hand, the large-particle processor (LPP) is designed to completely eliminate phantom particles (Fig. 17-12c). Therefore, coincidence results solely in a loss of LPP-detected particles. The difference between the LPP and the SPP concentration in the overlap range can give a hint of the magnitude of coincidence effects. The SPP coincidence can sometimes be reduced by lowering the photomultiplier gain. For size distributions skewed to a small particle size, this reduces the number of small particles detected, thus reducing the phantom-particle creation. If a region of the size spectrum is known not to contain any real particles, the detected particles can be assumed to be due to coincidence. The average detected particle number per channel in this region can be subtracted from the entire distribution to obtain a more accurate distribution (Heitbrink and Baron 1992).

Resolution and Accuracy

As noted above, the APS measures particles largely outside the Stokes regime and requires corrections to the data to provide an accurate d_a. Measurement of spherical particles can be corrected by taking into account particle density and, if necessary, changes in sampled gas density and viscosity. Liquid particles can also be accurately measured if the APS is calibrated with the same liquid. The resolution for spherical particles is high. For example, Remiarz et al. (1983) found geometric standard deviations in the range of 1.0058 (6.8 μm oil) to 1.025 (0.8 μm latex) for monodisperse particles. Due to non-Stokesian behavior and variations in the acceleration flow field experienced by particles approaching the detection region, resolution and accuracy may be diminished for nonspherical particles. Marshall, Mitchell, and Griffiths (1991) found that the d_a of particles with a shape factor of 1.19 was underestimated by 25% in the APS.

AEROSIZER

Measurement Principles

The Aerosizer (Amherst Process Instruments, Amherst, MA) is also based on acceleration of particles and time-of-flight principles. The idea of accelerating particles in a sonic expansion flow and measuring the terminal velocity was first proposed and demonstrated in laboratory prototype instruments by Dahneke and his coworkers (Dahneke, 1973; Dahneke and Padliya 1977; Cheng and Dahneke, 1979). The Aerosizer is the commercial product of this aerosol beam research. Figure 17-13 depicts the schematic of the particle detection section of the device. The aerosol enters the inner capillary tube, surrounded by particle-free sheath air in the outer tube. Air molecules and particles are accelerated through a convergent nozzle (0.75 mm diameter) with a 15° half-angle and are delivered into a partially evacuated chamber. The nozzle acts as a critical orifice and controls the flow rate through the nozzle. The ratio of pressure in the chamber to that in the ambient air is much smaller than 0.53; therefore, the air at the nozzle exit, U_g, attains sonic velocity:

$$U_g = \sqrt{\frac{\gamma R T}{M}} \tag{17-34}$$

where γ is the ratio of specific heat capacities (1.40 for air), R is the gas constant, T is the temperature in K, and M is the molecular weight (28.96 for air).

The air continues to expand in the chamber with a supersonic free-jet flow. The flow field in the convergent nozzle and the supersonic expansion have been described previously (Dahneke and Cheng 1979). From numerical calculations, it has been shown that the particle attains a terminal velocity soon after exiting the nozzle (Dahneke and Cheng 1979). At a distance of five nozzle diameters downstream of the nozzle, the calculated axial velocity is within about 2% of that at a distance of 50 diameters downstream.

The time of flight of a particle is measured with two laser beams located close to the

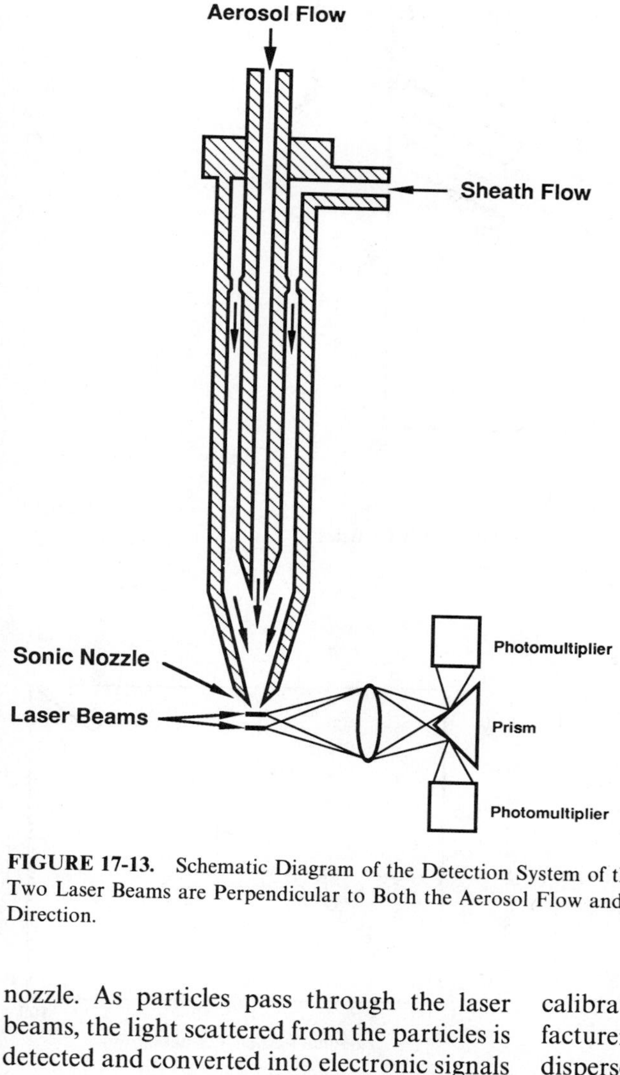

FIGURE 17-13. Schematic Diagram of the Detection System of the Aerosizer. The Two Laser Beams are Perpendicular to Both the Aerosol Flow and Detection Direction.

nozzle. As particles pass through the laser beams, the light scattered from the particles is detected and converted into electronic signals by two photomultiplier tubes. The distance between the two laser beams is about 1 mm, with the spread of each individual beam being about 20–30 μm. One photomultiplier detects light as the particle passes through the first beam, whereas the other photomultiplier detects the light from the second beam. The time between these two events (the time of flight) is measured and recorded.

The terminal velocity and time of flight are functions of particle diameter, density, and shape. The calibration curves (Fig. 17-14) provided by the manufacturer are based on theoretical calculations for solid spherical particles. The experimental validation of the calibration curves is provided by the manufacturer using a limited number of monodisperse spherical particles of polystyrene latex (density = 1.05 g/cm^3) and glass (density = 2.45 g/cm^3). A more extensive calibration of the instrument using similar particles in the size range 1–100 μm obtained at the Inhalation Toxicology Research Institute is shown in Fig. 17-15. This figure shows a good agreement between the calibration curves and the experimental data for densities of 1.05 and 2.45 g/cm^3 when the measured values are corrected for density. However, there are no data for more dense particles and the exact responses for nonspherical particles have not yet been investigated for this instrument. Because of supersonic flow conditions, the Reynolds number of micrometer size par-

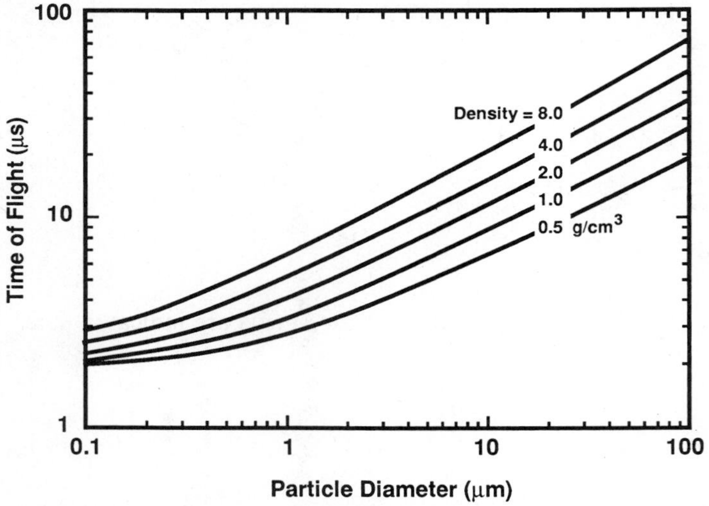

FIGURE 17-14. Calculated Aerosizer Calibration Curves for Spherical Particles of Different Densities. (Courtesy of Amherst Process Instruments, Inc.)

ticles in the sensing zone is outside of the Stokes regime (Reynolds number $\gg 0.05$) and the terminal velocity is not a unique function of aerodynamic particle size (which combines the particle diameter, density, and shape factor into a single function). Until more studies on the performance of the instrument are conducted, accurate determination of the particle aerodynamic size distribution is valid only for spherical particles of known density.

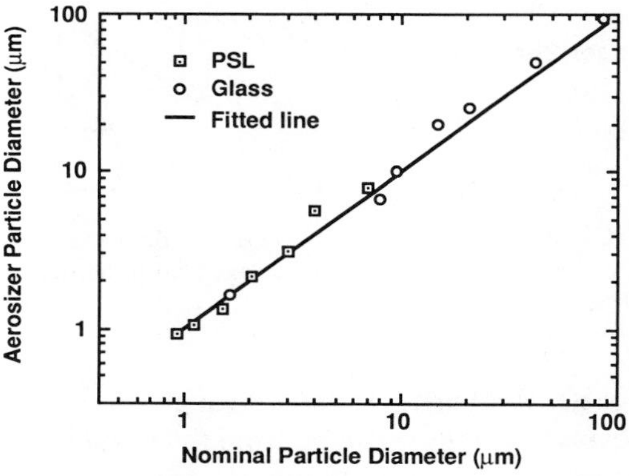

FIGURE 17-15. Aerosizer Calibration Data for Latex Particles (Density = 1.05) and Glass Beads (Density = 2.45).

Particle Coincidence

The aerosizer uses two separate detectors and clocks to measure the pulse arrival. This allows a differentiation between the starting and ending pulse for each particle. However, rather than just associating two adjacent pulses with each other, the instrument considers all pulses over some time period and associates each individual pulse with all other pulses. This creates a background of uncorrelated pulse pairs, or "phantom particles", and a spectrum of correlated pulse pairs, or real particles. The uncorrelated pulse pairs create a uniform background, the mean of which is subtracted from the spectrum.

Coincidence causes particles to be subtracted from the spectrum and added to the uncorrelated background, much the same way that the coincidence-produced phantom particles add a continuous background in the APS spectrum. In addition, the noise produced by the uncorrelated pulse pairs remains in the spectrum after the mean is subtracted, resulting in additional noise on the particle spectrum as well as a low-level noise background.

Instrument Capabilities

The applicable size range of the aerosizer is between 0.5 and 200 μm, and in this size range response curves for solid spherical particles have been confirmed (Fig. 17-16). The instrument is capable of determining particle size distributions for aerosol concentrations between 1 and 1000 particles/cm^3. However, the coincidence errors increase with particle size; therefore, the effective aerosol concentration limits decrease with increasing particle size. The unit comes with a dry-powder disperser. This is very useful for dispersing the dry powder for the size measurement, especially for powders of larger sizes (> 10 μm) which have very low transmission efficiency in aerosol generation and sampling systems.

Other Information

The aerosizer consists of three basic units: (1) the sensor, which houses the sampling inlet, the aerosol beam assembly, a vacuum chamber, and a 5 mW He–Ne laser; (2) a diaphragm vacuum pump, and (3) a personal computer with software to control the operation and to analyze the size distribution. The size distribution data from the aerosizer is normalized so that absolute concentration information is not available. This precludes making comparison measurements, e.g., before and after a filter.

The sensor (48 cm × 25 cm × 27 cm) has a mass of 10 kg and the pump (48 cm × 30 cm × 29 cm) has a mass of about 27 kg. The sampling flow rate is 5.3 l/min calculated from the critical flow conditions. At this time, the calibration of the instrument has been verified only for solid spherical particles of density 1–2.5 g/cm^3. For other particles, including liquid droplets and nonspherical particles, the responses are largely unknown, and it is advisable to perform a calibration with the material of interest, if an accurate size distribution is needed.

COMPARISON OF AERODYNAMIC SIZING INSTRUMENTS

The E-SPART, APS, and Aerosizer are all based on the acceleration of a particle in an aerosol beam and particle velocity measurement for size determination. They all attempt to provide an indication of aerodynamic particle size. However, there are substantial differences in the design and operation of these instruments. The APS has been in commercial production the longest, of the three instruments, and has had more published evaluations of its operation than the other two. Some of the similarities and differences between the instruments are indicated here.

The E-SPART operates at low particle velocities (< 50 cm/s) and approximately ambient pressures; the APS accelerates particles at less than sonic velocities with about 100 mmHg pressure differential between the ambient air and sensing volume (Chen, Cheng, and Yeh 1985), and the Aerosizer is operated under supersonic conditions with much higher pressure drop (> 700 mm Hg). Thus, the E-SPART operates largely within the Stokes regime, while the APS operates largely outside the Stokes regime, with the aerosizer even further outside. Thus, correction factors often must be applied to obtain accurate aerodynamic particle diameters. Correction factors for particle density and gas viscosity can be calculated for the APS and the Aerosizer and have been validated for the APS. Coincidence levels limit the measurable concentration in all these instruments, with a decreasing concentration limit in the following order: Aerosizer, APS, and E-SPART.

The E-SPART is capable of providing both aerodynamic size as well as particle charge information, with graphics capability for presenting the data. The other two instruments provide only size measurement, but also have similar data presentation capabilities.

The APS and Aerosizer are relatively mobile and can be used in field situations, though they are primarily laboratory instruments. The current version of the E-SPART is considerably larger and is less amenable to measurements outside the laboratory.

FIBROUS AEROSOL MONITOR (FAM)

The measurement of fibers is useful for controlling the health risk due to exposure to airborne asbestos fibers and other similar materials. Currently, the most common method for estimating such exposure is personal filter cassette sampling followed by phase contrast light microscope counting of fibers. This method, exemplified by NIOSH Method 7400 (Carter, Taylor, and Baron 1984), often requires sampling times of hours and analysis times of 5–20 min (see Chapter 26). The fibrous aerosol monitor (model FAM-1, MIE Inc., Bedford, MA) was developed under joint sponsorship of NIOSH, the Bureau of Mines, and the Environmental Protection Agency (Lilienfeld, Elterman, and Baron 1979) to supplement this technique with real-time indication of fiber concentration.

The FAM-1 preferentially detects fibers by aligning the fibers in an oscillating electric field, illuminating the fibers with a laser beam and detecting the resulting pulses of light with photomultiplier. Compact particles do not align in the oscillating field and, thus, do not produce pulses synchronous with the oscillating electric field.

Fiber Alignment

Particles in the sensing zone are subjected to a constant 3000 V/cm electric field perpendicular to both the laser beam and the detector axis. A smaller ac field is applied at 45° to this field to oscillate the electric field in the ϕ direction (Fig. 17-16). With this field rocking motion, fibers aligned parallel to the electric field vector spend more time scattering light into the detector, thus improving the sensitivity, than if the field exhibited a full 360° rotation (used in the prototype version).

Conductive fibers are predicted to align in the FAM with less than a 0.1° lag behind the rotating field (Lilienfeld 1985). Nonconductive fibers are expected to align very poorly, if at all. Conductivity on the scale of fibers that are a few micrometers or tens of micrometers in length depends on surface conductivity as well as bulk conductivity. For instance, glass fibers have a very low bulk conductivity, but can align in the electric field of the FAM at humidities above about RH 30%. Thus, adsorbed water can provide enough surface

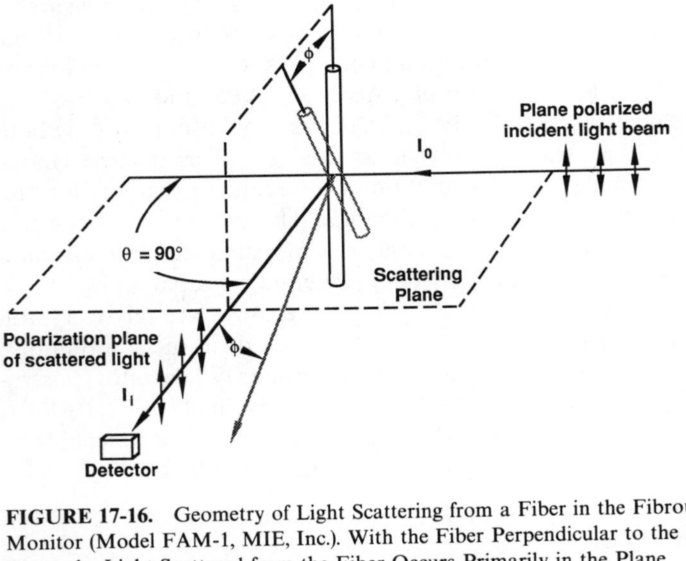

FIGURE 17-16. Geometry of Light Scattering from a Fiber in the Fibrous Aerosol Monitor (Model FAM-1, MIE, Inc.). With the Fiber Perpendicular to the Laser Beam, the Light Scattered from the Fiber Occurs Primarily in the Plane Perpendicular to Both the Fiber and the Laser Beam and Results in a Pulse of Light Reaching the Detector as the Fiber Rotates. (Adapted from Lilienfeld 1987.)

conductivity to allow complete alignment. Asbestos fibers adsorb water more readily and align at much lower humidities.

Brownian motion can cause fibers to be randomly displaced from a fully aligned condition. At room temperature, the average angular deflection from alignment θ_B can be calculated from

$$\sin\theta_B \cong \frac{1}{\pi E}\sqrt{\frac{kT(\sin(2\beta)-1)}{2\varepsilon_0 L^3}} \qquad (17\text{-}35)$$

where E is the electric field, k is the Boltzmann constant, T is the absolute temperature, β is the fiber length-to-width (aspect) ratio, ε_0 is the permittivity of free space, and L is the fiber length. The deflection is a strong function of fiber length and, while fibers longer than 5 μm are predicted to align well in the FAM, shorter fibers may not hold alignment sufficiently well to be detected.

Sensing System

The FAM uses two fiber properties to exclude detection of other types of particles: the alignment of conductive fibers with their long axis parallel to an electric field and the peculiar light scattering pattern produced by fibers. A fiber in the FAM sensing zone is oriented by an electric field such that its major axis is always perpendicular to the incident light beam (Fig. 17-16). Under these conditions, the light scatters primarily into a plane that is perpendicular to the fiber axis and contains the light beam. The detector is situated to receive the maximum scattered light when the fiber is also perpendicular to the detector fiber axis. With very long fibers, this plane of scattering is quite thin. With a smaller length-to-diameter ratio, or aspect ratio, the scattering plane becomes broader, eventually approaching a uniform pattern as a function of ϕ for spherical particles. As with compact particles, more light is scattered in the forward direction (small θ) than in the backward direction and this proportion increases with fiber diameter. As with scattering from compact particles, the scattering intensity and angle dependence is a function of refractive index. The 90° detection angle in the FAM allows multiple detection of the same particle as it moves down the detection volume, improving sensitivity and specificity. In addition, for small-diameter fibers the scattered light tends to be plane-polarized parallel to the fiber axis. To increase the amount of scattered light from these fibers relative to that from nonaligned compact particles, a polarizing filter is placed in front of the photomultiplier detector.

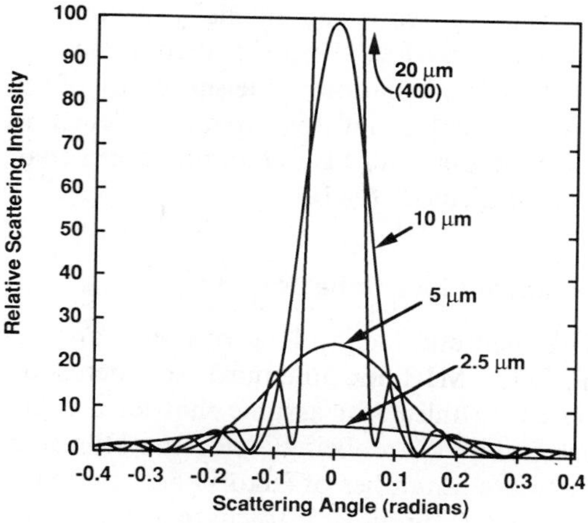

FIGURE 17-17. Light Scattering Pulses Calculated for Several Fiber Lengths for a Single Fiber Diameter (Eq. 17-36). (Adapted from Lilienfeld 1987.)

A detector is placed at a right angle to the incident beam in the scattering plane, as indicated in Fig. 17-16. As a fiber rotates in the ϕ direction, a pulse of light is detected. The intensity profile of this pulse for small values of ϕ is indicated by

$$I \propto L^2\left[\frac{\sin(\pi L\varphi/\lambda)}{\pi L\varphi/\lambda}\right]^2 \qquad (17\text{-}36)$$

where I is the scattering intensity, L is the fiber length, and λ is the light wavelength (0.6328 μm) (Lilienfeld 1987). The sharpness of the scattering pattern indicated by this function depends on the fiber length. Pulse sharpness is used as a discrimination in the FAM to limit the detection to fibers longer than 5 μm. The curves in Fig. 17-17 indicate the relative scattering intensity profiles as calculated by Eq. 17-36 for fibers of several lengths. A calculation of the sensitivity of the FAM indicates that it is capable of detecting 0.075 μm diameter fibers that are longer than 5 μm (Lilienfeld 1987).

Nonideal Fiber Behavior

The theoretical considerations that are the basis for the fiber alignment and detection system in the FAM assume that ideal cylindrical fibers are being detected. However, there are a number of factors that can make fiber detection difficult to characterize theoretically. For instance, fibers may not have the ideal cylindrical or even an ellipsoidal shape. Asbestos fibers are typically bundles of fibrils that may have splayed ends, noncylindrical cross section, curvature, attached particles, etc. Fibers may exist as clumps of multiple fibers floating in the air. Different fiber types have different refractive index and, hence, different scattering efficiencies. Fibers with curvature or with other particles attached may not align at 90° to the detector axis to give optimum detection geometry.

The fibers of one of the most common types of asbestos, chrysotile, often exhibit curvature. This curvature was found to result in lower sensitivity in the FAM prototype (Lilienfeld, Elterman, and Baron 1979). However, by applying a sufficiently high voltage field in the detection volume the fibers can be straightened, so that they produce responses similar to more ideal fibers (Lilienfeld 1985).

The laser beam does not uniformly illuminate the detection volume. The profile of the beam is Gaussian, indicating that fibers at the edge of the beam will scatter less light than fibers in the center of the beam. Large-diameter fibers may settle too far to be completely detected. In addition, charged fibers may drift away from the beam due to the high electric field. All these factors will affect the sensitivity and response of the instrument. Thus, as with other light scattering instruments, a calibration procedure is required to make the instrument respond in a manner equivalent to light microscopic detection.

Instrument Characteristics

Sampling System

The air flow in the FAM is controlled by a diaphragm pump to 2 l/min and a rotameter provides a visual indication of this flow. A cassette filter in front of the rotameter collects fibers that have been detected by the FAM. An adjustment to the pump voltage allows a calibration of the flow rate. The aerosol enters a 1.27 cm diameter tube inlet and passes through two right-angle bends. This arrangement is required by laser safety regulations to prevent direct viewing of the laser beam. The aerosol flow continues down the 1 cm diameter sensing tube to the sensing volume and then out the other end of the sensing tube (Fig. 17-18). A minor amount of clean-air flow from the pump through the beam orifices keeps the Brewster angle windows (used to prevent backscatter of light) and orifices clean.

An additional option is available with the FAM that uses a virtual impactor to remove particles larger than 10 μm aerodynamic diameter from the sampled air stream. The impactor is driven entirely by the flow into the FAM inlet. The jet pressure in the impactor forces 10% of the flow with the large particles into the minor-flow orifice, through a glass fiber filter and a rotameter. This 10% is thus impacted, filtered, and returned to the major flow entering the FAM. This arrangement has the advantage that an additional pump to control the minor flow is not required. With 10% of the small particles as well as the large particles removed, the final fiber count must be increased to adjust for this 10% loss.

The sensing volume is far enough downstream from the inlet for the laminar flow to be established. Under these conditions, the fibers are expected to align parallel to the flow until they reach the sensing zone. As indicated in Fig. 17-19, the laser beam shines down the center of the sensing tube; only those fibers in the center, illuminated by the beam, are detected. In addition, fibers must stay in the beam for a sufficient time to be recorded as a fiber. Thus, any fibers that settle or otherwise deviate more than a fraction of the beam

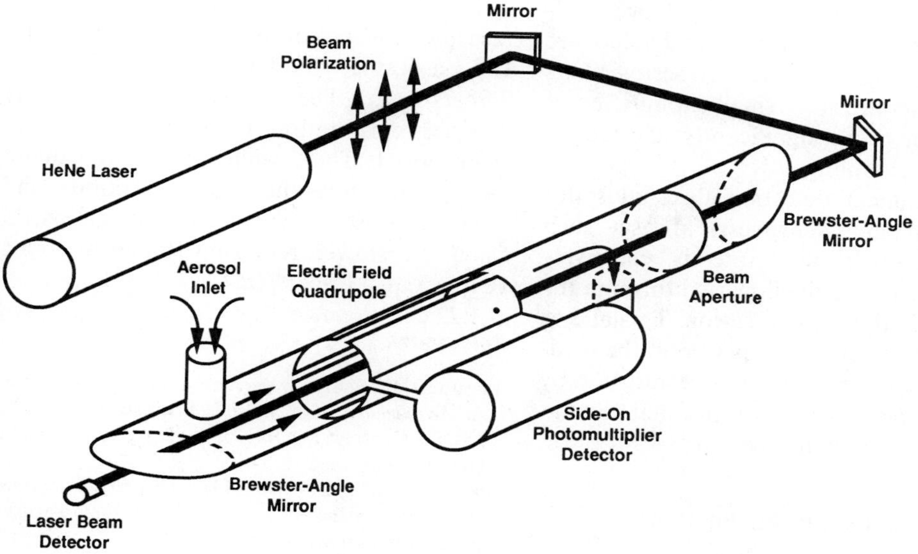

FIGURE 17-18. Sensing System of the FAM-1, Including the Light Scattering System, Fiber Alignment System and Air Flow Pattern. (Adapted from Lilienfeld 1987.)

radius (about 0.7 mm) in a straight trajectory, in the 0.1 s it takes the fiber to traverse the sensing zone, will not be recorded as fibers.

As with other particle counters, the FAM exhibits coincidence errors at high fiber concentrations. Coincidence of two fibers in the detection volume results in only one fiber being counted. The coincidence level can be estimated using Eq. 17-33 with a detection volume (Qt) of 0.0083 cm^3. This indicates a coincidence loss of about 15% at a measured concentration of 15 fibers/cm^3.

Instrument Dimensions and Controls

The FAM is a small suitcase-sized instrument with a mass of 11.4 kg and dimensions 53 cm × 35 cm × 20 cm. It is normally operated from 115 V ac power, though a 7 kg battery pack option is available, that allows up to 3 h of operation. The readout is by liquid crystal display in fibers/cm^3. An additional option allows recording of cumulative fiber count continuously on a strip chart recorder.

The instrument controls include dials for setting the pulse amplitude and pulse sharpness thresholds, a selector for sampling periods of 1, 10, 100, or 1000 min or for continuous measurement, and an analog output meter. This meter can be selected to indicate beam alignment, power supply voltage level, average photomultiplier output (an indicator of aerosol concentration), and fiber sensor output (an indicator of fiber peak concentration). Two LEDs provide visual indication of too high an aerosol concentration (from total light scattered) and too high a fiber concentration.

Calibration

The original aim in the development of the FAM was to provide a real time direct-reading monitor capable of giving results equivalent to those provided by the reference method using filter sample collection and phase contrast light microscope analysis (e.g., NIOSH Method 7400, as described in Chapter 25). Since the instrument response cannot be predicted theoretically, it must be calibrated by a side-by-side comparison with the reference method. Each instrument is calibrated in this fashion by the manufacturer with an amosite aerosol produced from a vibrating fluidized-bed generator. During calibration, the instrument and filter inlets are placed close together. Most asbestos fibers from the generator have small enough Stokes numbers not to be significantly affected by anisokinetic sampling.

The instrument detection thresholds (for pulse height and sharpness) are set such that the measurement rate is effectively 10 cm^3/min. This is the flow rate through the laser beam in the sensing region. This allows the fiber count number displayed in the readout to give the correct fiber concentration by appropriate placement of the decimal point at the end of the sampling period.

FAM Evaluations and Application

A number of studies have been performed on the FAM though most of these have not been in-depth evaluations. In general, these studies indicated that the FAM provides comparable results to the reference method. An evaluation was carried out by Iles and Shenton-Taylor (1986) that indicated good correlation of laboratory measurements with the reference phase contrast method, but field measurements gave much poorer agreement. The FAM manufacturer indicated that this study may have suffered from poor reliability of the instrument. It was found that instruments produced in the first several years of production were notoriously unreliable, primarily because of the ease of laser misalignment. Improvements in the instrument ruggedness have largely eliminated these and other stability problems. However, no further detailed studies of the FAM have been carried out since the mid-1980s.

The FAM has come into fairly extensive use in recent years as a monitor for asbestos removal operations. The FAM provides real-time indication of fiber concentrations outside these sites to ensure the integrity of the containment system. Since the filter sampling typically takes an hour or more, with filter analysis adding at least another 30 min, the FAM can provide feedback in minutes to prevent unnecessary exposure to potentially high concentrations of asbestos aerosols.

The FAM was originally designed to measure concentrations at levels near the U.S. asbestos standard in 1978, namely 2 fibers/cm^3. The current standard is 0.2 fibers/cm^3 and a level half that value is proposed. Thus, while the FAM can be used to measure these concentrations, the time response at such concentrations is somewhat slower, requiring 10 min to count 10 fibers at 0.1 fibers/cm^3.

References

Agarwal, J. K., and R. J. Remiarz. 1981. *Development of an aerodynamic particle size analyzer.* USDHEW-NIOSH Contract Report No. 210-80-0800. Cincinnati OH: NIOSH.

Ananth, G. and J. C. Wilson. 1988. Theoretical analysis of the performance of the TSI aerodynamic particle sizer. *J. Aerosol Sci.* 9:189–99.

Baron, P. A. 1986. Calibration and use of the aerodynamic particle sizer (APS 3300). *Aerosol Sci. Technol.* 5(1): 55–67.

Baron, P. A. and K. Willeke. 1986. Respirable droplets from whirlpools: Measurements of size distribution and estimation of disease potential. *Environ. Res.* 39:8–18.

Brockmann, J. E. and D. J. Rader. 1990. APS response to nonspherical particles and experimental determination of dynamic shape factor. *Aerosol Sci. Technol.* 13:162–72.

Carter, J., D. Taylor, and P. A. Baron. 1984. Fibers, Method 7400 Revision no. 3:5/15/89. *NIOSH Manual of Analytical Methods.* Cincinnati OH: DHHS (NIOSH).

Chen, B., Y. S. Cheng, and H. C. Yeh. 1985. Performance of a TSI Aerodynamic Particle Sizer. *Aerosol Sci. Technol.* 4:89–97.

Chen, B. T., Y. S. Cheng, and H. C. Yeh. 1990. A study of density effect and droplet deformation in the TSI aerodynamic particle sizer. *Aerosol Sci. Technol.* 12:278–85.

Cheng, Y. S., B. T. Chen, and H. C. Yeh. 1990. Behavior of isometric nonspherical aerosol particles in the aerodynamic particle sizer. *J. Aerosol Sci.* 21(5):701–10.

Cheng, Y. S. and B. E. Dahneke. 1979. Properties of continuum source particle beams. II. Beams generated in capillary expansions. *J. Aerosol Sci.* 10:363–68.

Clift, R., J. R. Grace, and M. E. Weber. 1978. *Bubbles, Drops and Particles.* New York: Academic Press.

Dahneke, B. 1973. Aerosol beam spectrometry. *Nature Physical Sci.* 244:54–55.

Dahneke, B. E. and Y. S. Cheng. 1979. Properties of continuum source particle beams. I. Calculation methods and results. *J. Aerosol Sci.* 10:257–74.

Dahneke, B. and D. Padliya. 1977. Nozzle-inlet design for aerosol beam. *Instruments in Rarefied Gas Dynamics,* 51, Part II, pp. 1163–72.

Fuchs, N. A. 1964. *Mechanics of Aerosols*. New York: Pergaman.

Griffiths, W. D., P. J. Iles, and N. P. Vaughan. 1986. The behaviour of liquid droplet aerosols in an APS3300. *J. Aerosol Sci.* 17(6): 921–30.

Heitbrink, W. A. and P. A. Baron. 1992. An approach to evaluating and correcting Aerodynamic Particle Sizer measurements for phantom particle count creation. *Am. Ind. Hyg. Assoc.* 53(7):427–431.

Heitbrink, W. A., P. A. Baron, and K. Willeke. 1991. Coincidence in time-of-flight spectrometers: phantom particle creation. *Aerosol Sci. Technol.* 14:112–26.

Iles, P. J. and T. Shenton-Taylor. 1986. Comparison of a fibrous aerosol monitor (FAM) with the membrane filter method for measuring airborne asbestos concentrations. *Ann. Occup. Hyg.* 30(1):77–87.

Kinney, P. D. 1990. Inlet efficiency study for the TSI aerodynamic particle sizer. M.S. Thesis. Department of Mechanical Engineering, University of Minnesota.

Lee, K. W., J. C. Kim, and D. S. Han. 1990. Effects of gas density and viscosity on response of aerodynamic particle sizer. *Aerosol Sci. Technol.* 13:203–12.

Lilienfeld, P. 1985. Rotational electrodynamics of airborne fibers. *J. Aerosol Sci.* 16(4):315–22.

Lilienfeld, P. 1987. Light scattering from oscillating fibers at normal incidence. *J. Aerosol Sci.* 18(4):389.

Lilienfeld, P., P. Elterman, and P. Baron. 1979. Development of a prototype fibrous aerosol monitor. *Am. Ind. Hyg. Assoc. J.* 40(4):270–82.

Marshall, I. A., J. P. Mitchell, and W. D. Griffiths. 1991. The behaviour of regular-shaped non-spherical particles in a TSI Aerodynamic Particle Sizer. *J. Aerosol Sci.* 22(1):73–89.

Mazumder, M. K. and K. J. Kirsch. 1977. Single particle aerodynamic relaxation time analyzer. *Rev. Sci. Instrum.*. 48(4):622.

Mazumder, M. K., R. E. Ware, and W. G. Hood. 1983. Simultaneous measurements of aerodynamic diameter and electrostatic charge on single-particle basis. In *Measurements of Suspended Particles by Quasi-Elastic Light Scattering*. ed. B. Dahneke. New York: Wiley.

Mazumder, M. K., R. E. Ware, J. D. Wilson, R. G. Renninger, F. C. Hiller, P. C. McLeod, R. W. Raible, and M. K. Testerman. 1979. SPART analyzer: Its application to aerodynamic size distribution measurements. *J. Aerosol Sci.* 10:561–69.

Mazumder, M. K., R. E. Ware, T. Yokoyama, B. J. Rubin, and D. Kamp. 1991. Measurement of particle size and electrostatic charge distributions on toners using E-SPART analyzer. *IEEE Trans. Ind. Appl.* 27(4):611–19.

Rader, D. J., J. E. Brockmann, D. L. Ceman, and D. A. Lucero. 1990. A method to employ the APS factory calibration under different operating conditions. *Aerosol Sci. Technol.* 13(4):514–21.

Remiarz, R. J., J. K. Agarwal, F. R. Quant, and G. J. Sem. 1983. Real-time aerodynamic particle size analyzer. In *Aerosols in the Mining and Industrial Work Environments. Vol. 3*, eds V. A. Marple and B. Y. H. Liu. Ann Arbor, MI: Ann Arbor Science.

Renninger, R. G., M. K. Mazumder, and M. K. Testerman. 1981. Particle sizing by electrical single particle aerodynamic relaxation time analyzer. *Rev. Sci. Instrum.* 52(2):242.

Szewczyk, K. W., M. Lehtimäki, M.-J. Pan, U. Krishnan and K. Willeke. 1991. Measurement of the change in aerosol size distribution with bacterial growth in a pilot scale fermenter. *Biotech. Bioeng.* (to be published).

TSI, Inc. 1987. *Model APS33B System Aerodynamic Particle Sizer Instruction Manual*. TSI, Inc. 500 Cardigan Road, P.O. Box 64394, St. Paul, MN.

Wake, D. 1989. Anomalous effects in filter penetration measurements using the aerodynamic particle sizer (APS 3300). *J. Aerosol Sci.* 20(1):13–17.

Wang, H. and W. John. 1989. A simple iteration procedure to correct for the density effect in the aerodynamic particle sizer. *Aerosol Sci. Technol.* 10:501–5.

Wang, J. C. and W. John. 1987. Particle density correction for the aerodynamic particle sizer. *Aerosol Sci. Technol.* 6:191–98.

Willeke, K. and B. Y. H. Liu. 1976. Single particle optical counter: Principles and application. In *Fine Particles: Aerosol Generation. Measurement, Sampling and Analysis*, ed. B. Liu. New York: Academic Press.

Wilson, J. C. and B. Y. H. Liu. 1980. Aerodynamic particle size measurement by laser-Doppler velocimetry. *J. Aerosol Sci.* 11(2):139–50.

Ye, Y. and D. Y. H. Pui. 1990. Particle deposition in a tube with an abrupt contraction. *J. Aerosol Sci.* 21(1):29–40.

18

Electrical Techniques[1]

Hsu-Chi Yeh

Lovelace Inhalation Toxicology Research Institute
P. O. Box 5890
Albuquerque, NM, U.S.A.

INTRODUCTION

The electrostatic charge on particles is an important property of aerosols and influences the aerosol behavior (Whitby and Liu 1966). Most aerosol particles, whether produced naturally or by man-made processes, carry some electrostatic charge. Depending upon the nature of the aerosol state, some particles may be highly charged. For charged particles under the influence of an external electric field and those that are highly charged, the electrical force can be very large as compared to other forces, such as gravity. In most cases, the electrical force on particles is undesirable, because it enhances particle deposition onto the surfaces (causing contamination) or modifies particle movement and makes its behavior difficult to analyze. However, under well-controlled conditions, this electrical force can be used to design many important types of aerosol control equipment, and aerosol sampling and measuring instruments. In this chapter, we shall discuss how the electrical properties of aerosols can be used for aerosol measurement.

PARTICLE CHARGING

Charging Mechanisms

There are several mechanisms by which aerosol particles can acquire charge: static electrification (Hinds 1982), charging by small ions (diffusion charging and field charging), thermionic charging, and self-charging (for radioactive aerosols). Charging by small ions requires the production of either bipolar or unipolar ions and is often used to produce controlled-charge aerosols.

Static electrification causes particles to become charged as they are separated from bulk material or other surfaces. Thus, particles are usually charged by this mechanism during formation, resuspension, or high-velocity transport. Static electrification can be further subdivided into several mechanisms. These include (1) electrolytic charging, resulting from separation of high dielectric liquids from solid surfaces; (2) spray charging, resulting from the disruption of charged liquid surfaces; (3) contact charging, resulting from the separation of dry particles from solid surfaces due to differences in electrochemical potential

1. Research supported by the U.S. Department of Energy, Office of Health and Environmental Research under Contract No. DE-AC04-76EV01013.

of the materials (Liu 1970; Yeh, Carpenter, and Cheng 1988); and (4) induced charging, such as by applying a high voltage to the spray or jet nozzle (Reischl, John, and Devor 1977). In general, the particle charge acquired by static electrification is difficult to predict and in most cases is undesirable.

Aerosol particles can acquire charge by collision with small ions. Based on the nature of the ions (bipolar or unipolar) and whether an external electric field is applied, charging by ions can be further divided into two charging mechanisms: diffusion charging and field charging. These charging mechanisms have been widely studied and were used in developing aerosol measuring instruments based on aerosol electrical properties. A more detailed discussion will be presented in the next section.

Thermionic charging of particles is primarily caused by thermionic emission of either electrons or ions when particles are heated to a sufficiently high temperature (Yeh and Cheng 1980). For a radioactive aerosol, particles may gain charges due to a self-charging mechanism by the loss of either alpha or beta particles and the ejection of valence electrons, or the release of charged fragments from atoms of particles during alpha, beta, or gamma radiation (Yeh et al. 1976; Yeh, Newton, and Teague 1978).

Diffusion and Field Charging

When aerosol particles are exposed to unipolar ions, the particles become charged due to collisions between the ions and the particles. In the absence of an applied electric field, particle charging is primarily due to collisions caused by Brownian motion of the ions and particles and is referred to as diffusion charging. Field charging is the charging by unipolar ions in the presence of a strong electric field. The ions move rapidly in the direction of the applied electric field, and this increases the frequency of collisions between the ions and particles. Therefore, field charging can achieve a much higher particle charge status than diffusion charging for the same ion concentration. Diffusion charging has been used in aerosol sizing instruments, such as the electrical aerosol analyzer (EAA), whereas field charging has been used primarily in aerosol sampling or collection devices, such as electrostatic precipitators.

Many theoretical and experimental studies on diffusion charging have been reported (Bricard 1948; Fuchs 1947; Liu and Pui 1977; Marlow and Brock 1975; Pui, Fruin, and McMurry 1988; White 1951). It has been experimentally shown that, for the continuum regime with $Kn_i \ll 1$ (Knudsen number based on ion mean free path $= Kn_i = 2\lambda_i/d_p$, where λ_i = mean free path of the ions, d_p = particle radius), Fuchs and Bricard's theory applies for particle diameters larger than about 0.1 μm (Liu and Pui 1977; Kirsch and Zagnit'ko 1981). For the free-molecule regime with $Kn_i \gg 1$, Pui, Fruin, and McMurry (1988) showed that the theory of Marlow and Brock (1975) best predicts charging rates in the ultrafine particle size range. In the transition regime, where $Kn_i \approx 1$, the approaches of Fuchs (1963) and Marlow and Brock (1975) can be applied. Except for the classical diffusion charging theory, equations developed by Fuchs, Bricard, or Marlow and Brock do not have a closed form, and numerical integration is required. To estimate the order of magnitude of the number of charges, n, acquired by a particle of diameter d_p, White's (1951) equation can be used:

$$n = \frac{d_p kT}{2e^2} \ln\left(1 + \frac{d_p \bar{c} \pi e^2 N_i t}{2kT}\right) \tag{18-1}$$

where

k = Boltzmann's constant (1.38×10^{-16} dyn cm/K)
T = absolute temperature (K)
e = elementary unit of charge (4.8×10^{-10} statC)
$\bar{c}$ = mean thermal speed of the ions ($\approx 2.4 \times 10^4$ cm/s)
N_i = ion concentration (ions/cm³) and
t = charging time (s)

If charging takes place when a strong external electric field is applied, the motion of the ions will be confined along the electric field lines. Assuming that diffusion charging can be neglected, the following field charging equation has been derived (White 1951):

$$n = \frac{3\varepsilon}{\varepsilon + 2} \frac{Ed_p^2}{4e} \frac{\pi N_i e Z_i t}{(\pi N_i e Z_i t + 1)} \quad (18\text{-}2)$$

where E = applied electric field (statV/cm), ε = dielectric constant of particle, and Z_i = mobility of ions ($\approx 450\ \text{cm}^2/\text{statV s}$). Equation 18-2 suggests that a limited charge (indicated by the first two factors) will be reached when sufficient charging time is applied ($t \to \infty$). This limited charge is referred to as saturation charge (n_s). It depends on the material of the particle, particle size, and the field strength, and is independent of ion concentration N_i.

When particles are charged under moderate electric field intensities, both the field and diffusion mechanisms of particle charging should be considered. Cochet (1961) derived the following equation, which agreed with his experimental data:

$$n = \left[\left(1 + \frac{Kn_i}{2}\right)^2 + \frac{2}{1 + Kn_i}\left(\frac{\varepsilon - 1}{\varepsilon + 2}\right)\right] \times \frac{Ed_p^2}{4e} \frac{\pi N_i e Z_i t}{(\pi N_i e Z_i t + 1)} \quad (18\text{-}3)$$

Smith and McDonald (1976) and Liu and Kapadia (1978) presented results of numerical solutions to the basic equation governing the diffusion of ions in an electric field. Sato (1987) suggested that the sum of the charges calculated using the classical diffusional and field theories independently (White 1951) agreed best with his experimental data.

BEHAVIOR OF CHARGED PARTICLES

Charged Part1cles Without External Electric Field

The force between two point charges is governed by Coulomb's law. For two particles with n_1 and n_2 unit charges, the force, in CGS units, can be written as

$$F = \frac{n_1 n_2 e^2}{r^2} \quad (18\text{-}4)$$

where r (cm) is the distance between the particles. Depending upon the polarity of the charge carried by the particles, the force can be attractive (opposite polarity) or repulsive (same polarity). The coulombic force decreases rapidly with separation distance between particles. For high-concentration aerosols, this attractive or repulsive force will cause particles to coagulate or disperse, thus affecting particle losses or size distribution within a confined space or during particle transport through a duct.

In addition to the coulombic force, the image force sometimes plays an important role. A charged particle can be attracted to an uncharged object or surfaces at close range by image forces, because the charged particle induces an equal and opposite charge in the uncharged object or surfaces and creates its own field for attraction. The image force is weaker than the coulombic force.

Charged Particles With External Electric Field

When a particle carrying n charges is placed in an electric field of intensity E, it experiences an electrostatic force of

$$F_{elec} = neE \quad (18\text{-}5)$$

For particle motion in the Stokes regime, the terminal velocity or drift, V_{elec}, when the electrostatic force and drag force are equal, is:

$$V_{elec} = \frac{neEC_c}{3\pi\eta d_p} \quad (18\text{-}6)$$

where η is the viscosity of air and C_c is the slip correction factor. Equation 18-6 can be written in terms of particle mechanical mobility, B (Eq. 3-14), as

$$V_{elec} = neEB \quad (18\text{-}7)$$

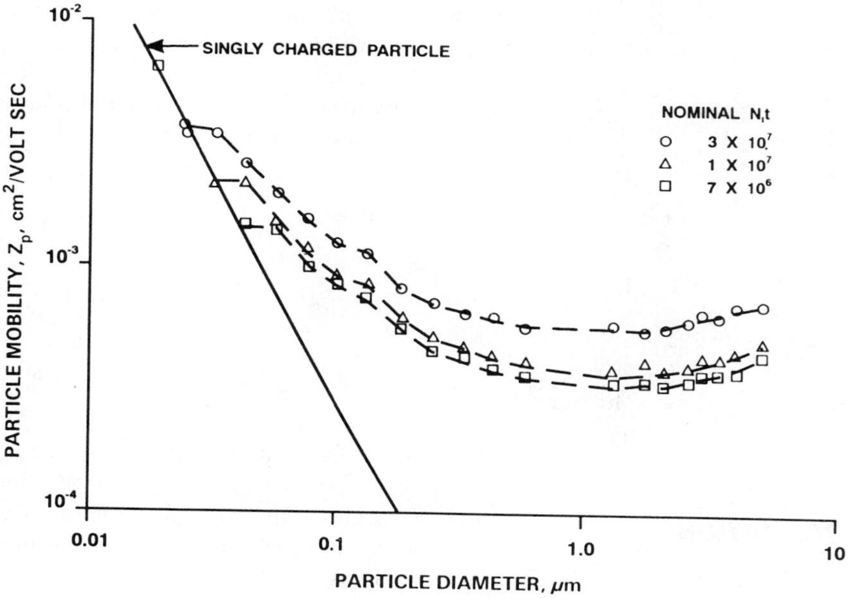

FIGURE 18-1. Measured Particle Electrical Mobility as a Function of Particle Diameter for a Singly Charged Particle and for Unipolar Diffusion-Charged Particles at various $N_i t$ Values. (Modified from Pui, 1976.)

Equation 18-6 shows that the V_{elec} is proportional to the intensity of the applied electric field. The proportionality constant is, by definition, the electrical mobility of the particle, Z_p (cm^2/statV s), i.e.,

$$Z_p = \frac{V_{elec}}{E} = \frac{neC_c}{3\pi\eta d_p} = neB \qquad (18\text{-}8)$$

Figure 18-1 shows the particle electrical mobility as a function of particle diameter for a singly charged particle and for unipolar diffusion-charged particles at various $N_i t$ values. For singly charged particles, particle electrical mobility decreases as particle size increases. However, for unipolar diffusion charging, there is a particle size with minimum electrical mobility.

Example 18-1

A unit-density spherical particle with diameter of 2.0 µm, carrying 100 excess electrons, is placed between two parallel plates 2 cm apart, with one plate grounded while a high voltage of 2000 volt is applied to the other plate. (1) What is its electrical mobility? (2) What is the electrical drift velocity V_{elec} of the particle? (3) How does this drift velocity compare to gravitational terminal velocity?

Answer. From Eq. 18-8, the electrical mobility is

$$\begin{aligned} Z_p &= neC_c/3\pi\eta d_p \\ &= (100 \times 4.8 \times 10^{-10} \times 1.08)/ \\ &\quad [3\pi(1.81 \times 10^{-4})(2 \times 10^{-4})] \\ &= 0.15 \text{ cm}^2/\text{statV s}. \end{aligned}$$

The drift velocity can also be obtained from Eq. 18-8, after rearrangement:

$$\begin{aligned} V_{elec} = Z_p E &= 0.15[2000/(300 \times 2)] \\ &= 0.51 \text{ cm/s} \end{aligned}$$

The gravitational terminal velocity can be

obtained from Eq. 3-23:

$$V_{\text{grav}} = \rho_p d_p^2 g C_c / 18\eta$$
$$= 1.0 \times (2.0 \times 10^{-4})(980)(1.08)/$$
$$[18(1.81 \times 10^{-4})]$$
$$= 0.013 \text{ cm/s}$$

Therefore,

$$V_{\text{elec}}/V_{\text{grav}} = 39.2$$

and the electrical drift velocity is much larger than the gravitational-settling velocity.

CHARGE NEUTRALIZATION

Bipolar Ion Sources

In principle, the minimum charge of an aerosol particle is zero. However, this condition is rarely achieved because of the random charging of aerosol particles by ions that exist in the air. In the normal atmosphere, bipolar ions are produced by cosmic rays and other radioactive elements in the atmosphere. This air concentration of ions is generally of the order of 500 to 1000 ion-pairs/cm^3 of air. When aerosol particles are exposed to bipolar ion atmospheres, after a sufficiently long time, the particles will come to a state of charge equilibrium with the ionic atmosphere due to frequent collisions between the particles and ions. This equilibrium charge state is called the Boltzmann equilibrium or residual charge distribution, and can be written as

$$f_n = \frac{\exp(-n^2/2\sigma^2)}{\sum \exp(-n^2/2\sigma^2)} \qquad (18\text{-}9)$$

where

$$\sigma^2 = \frac{d_p kT}{2e^2} \qquad (18\text{-}10)$$

and f_n is the fraction of particles of size d_p having n elementary units of charge. For particles larger than 0.05 μm diameter, Eq. 18-9 can be approximated by the normal Gaussian distribution:

$$f_n = \frac{\exp(-n^2/2\sigma^2)}{\sqrt{2\pi\sigma^2}} \qquad (18\text{-}11)$$

Equilibrium bipolar charge distributions of aerosols predicted by Boltzmann's law have been experimentally (Liu and Pui 1974*a*) and numerically (Takahashi 1971) verified for particle diameters larger than 0.02 μm and 0.1 μm, respectively. However, a study by Reischl, Scheibel, and Porstendörfer (1983) indicated that for particle diameters less than

TABLE 18-1 Equilibrium Charge Distributions Based on Boltzmann's Law and Fuchs' Theory (1963)

		Particle Diameter, d_p (μm)						
Charge Ratio	Theory	0.002	0.006	0.02	0.06	0.2	0.6	2.0
$n_1/2n_0$	Fuchs	0.0056	0.029	0.159	0.45	0.77	0.91	0.97
	Boltzmann	4.3E−13	7.6E−05	0.058	0.39	0.75	0.91	0.97
$n_2/2n_0$	Fuchs	0	0	0.0056	0.043	0.36	0.71	0.89
	Boltzmann	3.5E−50	3.3E−17	1.1E−05	0.022	0.32	0.68	0.89
$n_3/2n_0$	Fuchs	0	0	0	0.0081	0.102	0.46	0.78
	Boltzmann	5.2E−112	8.0E−38	7.4E−12	0.0002	0.077	0.43	0.77
$n_4/2n_0$	Fuchs	0	0	0	0	0.021	0.25	0.64
	Boltzmann	1.4E−198	1.1E−66	1.6E−20	2.5E−07	0.011	0.22	0.63

n_0 = number of particles with zero charge
n_1 = number of particles with either +1 or −1 charge
n_2 = number of particles with either +2 or −2 charge
n_3 = number of particles with either +3 or −3 charge
n_4 = number of particles with either +4 or −4 charge

0.05 μm, Boltzmann's law underestimates the charged fraction, and Fuchs' (1963) theory should be used. Fuchs' equations were complicated and required numerical integration. Table 18-1 lists the equilibrium charge distributions for particle radius less than 1.0 μm, according to Boltzmann's law and Fuchs' theory.

^{85}Kr Aerosol Discharger

Aged aerosols will eventually reach charge equilibrium, due to random collisions between the particles and ions in the air. In the free atmosphere, the equilibrating process will take approximately 30 min. However, by increasing the ion concentration using a radioactive source or a bipolar ion generator, the equilibration time can be greatly shortened. Several methods can be used to increase ion concentration; for example, using a bipolar ion generator (Whitby 1961) or various types of radioactive sources such as ^{90}Sr, ^{90}Y, ^{85}Kr, ^{241}Am, ^{210}Po, or ^{3}H (Whitby and Liu 1968; Thomas and Rimberg 1967; Mercer and Chow 1968; Cooper and Reist 1973).

Among these aerosol dischargers or neutralizers (sometimes called deionizers), the ^{85}Kr aerosol discharger is one of the most commonly used. Cooper and Reist (1973) described a general analysis of neutralizing charged aerosols with radioactive sources, and Liu and Pui (1974*b*) presented a detailed theoretical and experimental study of electrical neutralization of aerosols using a ^{85}Kr discharger. Yeh (1976) extended the theory of Liu and Pui to include self-charging, radioactive aerosols. Teague, Yeh, and Newton (1978) described the design criteria and fabrication method to construct an ^{85}Kr discharger (Fig. 18-2). Note that, as a rule of thumb, to adequately discharge aerosols to near-Boltzmann equilibrium, the $N_i t$ product of the discharger should be of the order of 6×10^6 (ions/cm^3 s), where N_i is the ion concentration, and t is the neutralization time. The ion concentration, N_i, is dependent on the strength of the ion source used, whereas

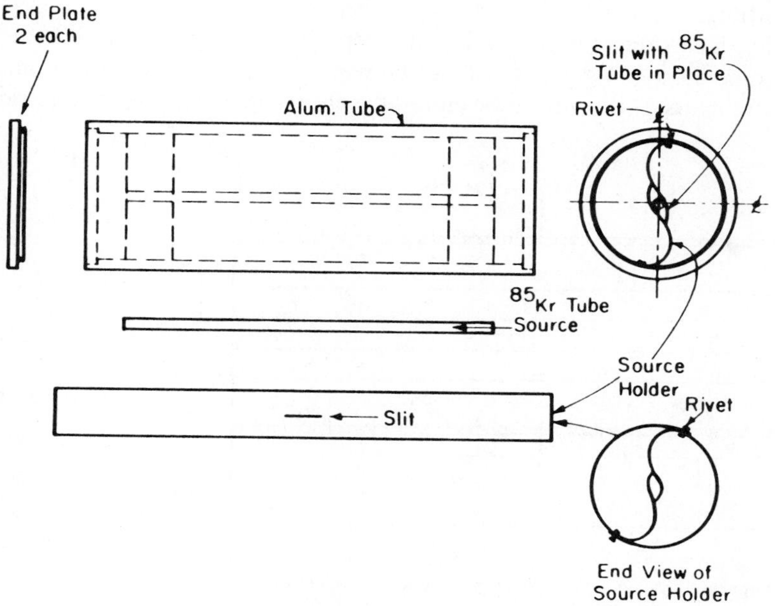

FIGURE 18-2. Cross-Sectional View of ^{85}Kr Aerosol Discharger Showing Dimensional Details and Radioactive Source Positioner. (Modified from Teaque et al. 1978.)

the charging time t can be adjusted by changing the flow rate through the discharger.

CHARGE DISTRIBUTION MEASUREMENT

Several methods can be used to determine the charge distribution of an aerosol. Most of these methods are based on the measurement of electrical mobility distribution of monodisperse particles. One exception to this is the measurement of ultrafine particles less than 0.01 μm; where particles will either carry a single charge or zero charge under charge equilibrium conditions, a direct measurement of charged and uncharged fractions can be made. Instruments used for measuring electrical mobility include the Millikan Cell (Millikan 1935; Sano, Fikitan, and Sakata 1953; Ivanov et al. 1974; Arnold 1979), parallel-plate electrical mobility spectrometer (Megaw and Wells 1969; Maltoni et al. 1973; Takahashi and Kudo 1973; Yeh et al. 1976; Porstendörfer et al. 1984), and concentric electrical mobility analyzer (e.g., electrical aerosol analyzer and differential-mobility analyzer) (Whitby and Clark 1966; Knutson and Whitby 1975*b*). The Millikan cell method is based on a single-particle measurement; as such, it is tedious and time-consuming for charge distribution measurements. It is primarily used as a research tool and is not commercially available or suitable for practitioner use.

Figure 18-3 depicts a schematic diagram of a parallel-plate electrical mobility spectrometer, showing a particle trajectory with respect to flow streamlines within the spectrometer. The aerosol is drawn into the device as a thin ribbon through a narrow-slot nozzle extending across the spectrometer, midway between the plates. Clean filtered air flows through the laminator and shields the thin aerosol stream on all sides. As the aerosol and clean air flow through the channel, particles move across the clean air under electrostatic forces. Those particles with electrical mobility greater than Z_{pL} are deposited onto the plates, which have a electric potential applied between them. Particles with $Z_p < Z_{pL}$ will penetrate the spectrometer. The Z_{pL} is determined by:

$$Z_{pL} = \frac{(\psi_c - \psi_0)}{EL} = \frac{Q}{2EL} = \frac{2Uh^2}{VL} \tag{18-12}$$

where ψ_c and ψ_0 are the values of the stream functions at the points occupied by the particle at the center of the aerosol inlet nozzle and at the deposit site, respectively, L is the distance between the aerosol inlet nozzle and the end of the electrode plate, E is the electric

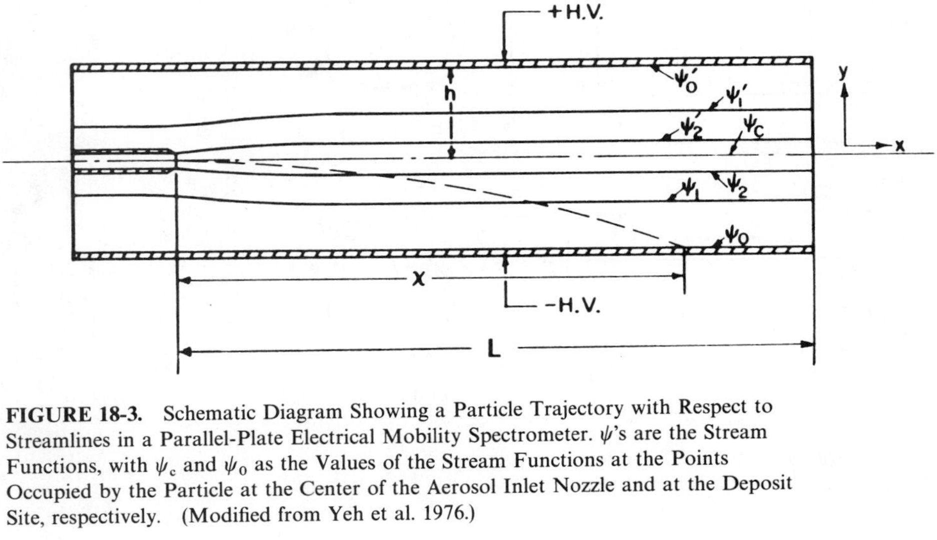

FIGURE 18-3. Schematic Diagram Showing a Particle Trajectory with Respect to Streamlines in a Parallel-Plate Electrical Mobility Spectrometer. ψ's are the Stream Functions, with ψ_c and ψ_0 as the Values of the Stream Functions at the Points Occupied by the Particle at the Center of the Aerosol Inlet Nozzle and at the Deposit Site, respectively. (Modified from Yeh et al. 1976.)

field between the plates ($= V/2h$), U is the mean flow velocity, h is half the inter-plate distance, and V is the potential difference between the plates. By varying the voltage V and measuring the corresponding particle penetration, the electrical mobility distribution can be obtained. Then, from Eq. 18-8, the corresponding charge distribution can be calculated for a given monodisperse aerosol.

A variation to the basic design of Fig. 18-3 has been reported by Johnston (1983). In his design, the outlet flow of the electrical mobility spectrometer was split into two equal-flow outlets; thus, both the positively and negatively charged particles can be measured directly by an optical particle counter. This arrangement improves the speed of measurement and data analysis. It has been used to measure the electrostatic charge of laboratory aerosols and for workplace aerosols (Johnston, Vincent, and Jones 1985, 1987).

Figure 18-4 shows two types of electrical mobility analyzers, similar to the parallel plate electrical mobility spectrometer but with a cylindrical configuration. The aerosol is introduced into the mobility analyzer through an outer annulus with a core of particle-free, clean air. An adjustable voltage is applied to the inner electrode with the outer cylinder grounded. As shown in Fig. 18-4a, particles with a sufficiently high electrical mobility are collected, while those with lower mobilities penetrate the mobility analyzer and are sensed by an aerosol sensor (e.g., electrometer or particle counter). The mean electrical mobility, Z_p, of the particles precipitated near the lower end of the collector rod is given by (Liu and Pui 1975)

$$Z_p = \frac{(Q_{tm} - 0.5Q_{am})\ln(r_2/r_1)}{2\pi VL} \tag{18-13}$$

where Q_{tm} and Q_{am} are the total flowrate and aerosol flowrates in the mobility analyzer, r_1 and r_2 are the inner and outer electrode radii of the analyzer, L is the length of the inner electrode, and V is the voltage on the collector rod. Again, for a given monodisperse aerosol, by varying the voltage V and measuring the corresponding particle penetration, the electrical mobility distribution and, thus, the charge distribution can be obtained. This type of mobility analyzer is sometimes referred to as an integrated-mobility analyzer, because any particle with mobility less than Z_p in Eq. 18-13 will penetrate the analyzer.

A variation to the basic design, called a differential-mobility analyzer (DMA), is

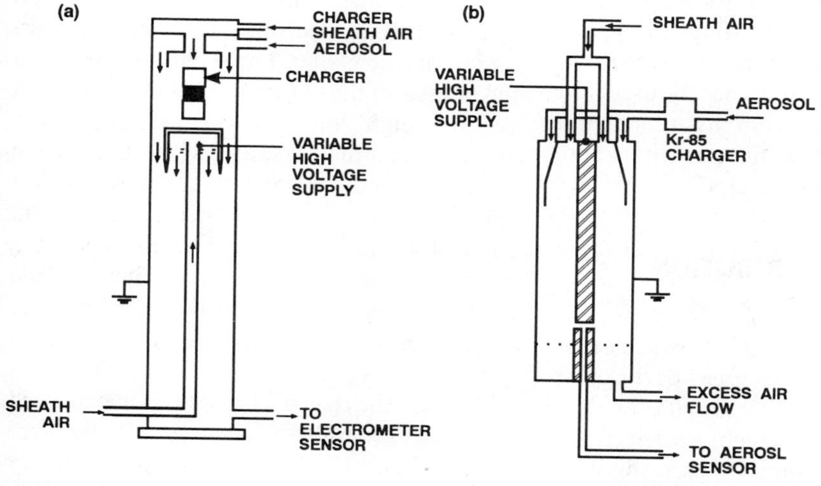

FIGURE 18-4. Schematic Diagrams of (a) an Integrated Electrical Mobility Analyzer and (b) a Differential-Mobility Analyzer.

shown in Fig. 18-4b. As shown in the figure, an extraction slit has been cut in the center rod. Therefore, particles with higher electrical mobility will be collected on the upper portion of the rod and those with lower mobility will be carried along with the major outlet flow. Only those particles with a narrow range of electrical mobility will pass through the narrow slit and will be carried out by the minor sampling flow to the aerosol sensor. The mean electrical mobility of the particles, Z_p, extracted through the slit, is given by (Knutson and Whitby 1975*a*)

$$Z_p = \frac{Q_t - (Q_s + Q_a)/2] \ln (r_2/r_1)}{2\pi VL} \quad (18\text{-}14)$$

with the width of the mobility spread,

$$\Delta Z_p = \frac{(Q_s + Q_a) \ln(r_2/r_1)}{2\pi VL} \quad (18\text{-}15)$$

where Q_s and Q_a are the sample and aerosol flows, and Q_t is the total flow in the analyzer. The DMA has higher resolution than the integrated-mobility analyzer due to its design. However, it may require an aerosol sensor with higher sensitivity for detection, because only a narrow mobility fraction is extracted. Generally, in order to measure particle charge distributions, monodisperse or size-classified aerosols should be used as input to the mobility analyzer. This is because, for a given electrical mobility, there can be different particle sizes with different numbers of electrostatic charges associated with them. If users are interested only in determining the charge fraction, then a simple condenser with a suitable aerosol sensor can be used.

AEROSOL SIZE DISTRIBUTION MEASUREMENT

Equation 18-7 indicates that the electrostatic drift velocity of a charged particle can be much higher than the gravitational or inertial velocities under high electric fields. The electrostatic precipitator uses this property for aerosol sampling and air cleaning. In addition to measuring aerosol charge distribution, the electrical mobility analyzer can also be used to measure the aerosol size distribution. For this purpose, it generally requires charging the aerosol under well-controlled charging conditions prior to mobility measurement and conversion to size distribution.

Electrostatic Precipitator

Two basic steps are involved in the electrostatic precipitator (ESP): (1) charging the particles and (2) subjecting them to a strong electric field, so that their electrical drift will cause them to deposit on the collecting substrate for subsequent examination by light or electron microscopy. Particles are usually charged by field charging, using a corona discharge. Normally, the unipolar ion concentration and the electric field strength are such that the particles will reach their saturation charge in a very short time. Depending on the design of the precipitating region, the ESP can be classified into single-stage and two-stage precipitators.

The point-to-plane ESP is a typical single-stage precipitator and is used for sampling aerosols for electron microscope studies. Figure 18-5 shows a schematic diagram of a point-to-plane ESP of the Rochester design (Morrow and Mercer 1964). The corona and precipitating fields are formed between a corona needle (the point) and a flat collection surface with an electron microscope grid sitting on it (the plane). The separation between the point and the plane is about 0.5–0.7 cm. A negative high voltage of several kilovolts is applied to the needle with the plane grounded, and the ion current is between 1 and 6 μA. The sampling flow rate varies from 50 to 1000 cm^3/min. In general, depending on the design of an ESP, the point-to-plane distance ranges from a few millimeters to a few centimeters, flow rates from 0.05 to 5 l/min, and voltages from 2 to 15 kV.

Because the electrical mobility of a charged particle is generally a function of its size (see Eqs. 18-2 and 18-8), and the simple geometry of the point-to-plane ESP, where charging and collection are essentially taking place in

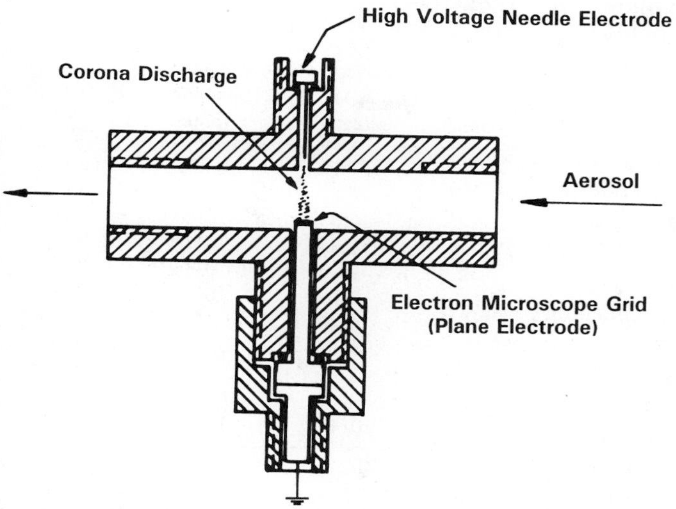

FIGURE 18-5. Schematic Diagram of a Point-to-Plane Electrostatic Precipitator of Rochester Design. (Modified from Cheng, Yeh, and Kanapilly 1981.)

the same area in a short aerosol transit time, the collection efficiency may be a function of particle size. Thus, data obtained with such samples may be biased with particle size. Cheng, Yeh, and Kanapilly (1981) indicated that, for the ESP as shown in Fig. 18-5, collection efficiencies decrease as particle size decreases or aerosol flow rate increases. They suggested that, with flow rates below 100 cm^3/min, ion currents of 2–3 μA, and with a sharp needle, the representative samples could be obtained for particle sizes larger than 0.03 μm with efficiency around 85%.

Liu, Whitby, and Yu (1967) developed a pulsed precipitating electrostatic aerosol sampler in an attempt to overcome the bias in collection efficiency due to particle size effect. This is a two-stage ESP. The particles are first charged in the charging section and subsequently collected in the precipitation section, where high-voltage precipitating pulses are applied at a rate of approximately 13 cycles/min. The charged aerosol is allowed to flow into the precipitation section with the field off for 3 s, and is then subjected to a strong electric field for 1.5 s, when the particles are deposited onto the plate. Because only those particles that are present in the precipitation section at the beginning of a pulse are deposited onto the collecting surface, the deposition is relatively uniform and without size bias over a portion of this collecting area.

Samples are collected for electron or optical microscopy by placing electron microscope grids or glass slides over the ESP collecting surface. For optical microscopy, the Porton eyepiece graticule can be used to size the particles. For electron microscopy, electron micrographs of an aerosol sample can be obtained and particles sized using the Zeiss particle size analyzer (Raabe 1971) or an image analyzer. Details of particle sizing by microscopy will not be presented here (see Chapter 13).

Electrical Aerosol Analyzer

The electrical aerosol analyzer (EAA) is based on the "diffusion charging mobility analysis" principle. It was first developed at the Particle Laboratory, University of Minnesota (Whitby and Clark 1966) and subsequently developed into a commercial instrument (Model 3030, TSI Inc., St. Paul, MN). This instrument consists of three sections: a diffusion charger, a mobility analyzer, and an electrometer current sensor. The mobility analyzer section is similar to that shown in Fig. 18-4a. The mean electrical mobility, Z_p, of the particles precipitated near the lower end of the collector rod was given in Eq. 18-13. The relationship between particle size, particle charge, and Z_p was given by Eq. 18-8.

For controlled diffusion charging, a one-to-one relationship exists between particle size and particle charge for submicrometer-sized spherical particles, i.e., $n_p = n_p(d_p)$. Experimental relationships under different charging conditions have been published (Liu and Pui 1975). The results under the charging condition of $N_i t = 1.0 \times 10^7$ ions/cm^3 s; (Fig. 18-1) have been used in the design of the EAA (TSI 1977). A set of standard operating conditions has been suggested, and a standard data reduction table is given for manual data analysis. A PC program has also been written, incorporating the standard data reduction table for automatic operation and data reduction. Under standard operating conditions, the instrument's volumetric flow rates are as follows: $Q_{tm} = 50$ l/min, $Q_{am} = 5$ l/min (4 l/min from aerosol sample and 1 l/min for diffusion charger), and 45 l/min of clean sheath air. Details of the operation and maintenance procedures can be found in "Operating and Service Manual, Model 3030 Electrical Aerosol Size Analyzer" (TSI 1977).

Example 18-2

The design of the TSI EAA 3030 was based on the following criteria:

$$N_i t = 1.0 \times 10^7 \text{ (ions/cm}^3\text{ s)}$$

$$r_1 = 1.151 \text{ cm}$$

$$r_2 = 2.985 \text{ cm}$$

$$L = 30.48 \text{ cm}$$

$$Q_t = 50 \text{ l/min}$$

$$Q_{am} = 5 \text{ l/min}$$

Table 18-E1 lists part of the data reduction table for the EAA, indicating the (1) channel number, (2) collector rod voltage, and (3) the corresponding channel boundary particle size when operated under the suggested standard operating conditions. What is their corresponding electrical mobility at each channel?

Answer. The mobility equation of the EAA is expressed by Eq. 18-13. For channel 1 with the applied voltage of 19 V, the corresponding Z_p is

$$\begin{aligned} Z_p &= (Q_t - 0.5Q_s)[\ln(r_2/r_1)]/2\pi VL \\ &= [(50 - 0.5 \times 5)(1000/60)] \\ &\quad \times [\ln(2.985/1.151)]/[2\pi(30.48)V] \\ &= 3.94/V \\ &= 3.94/(19 \text{ V}/300 \text{ V/statV}) \\ &= 62.2 \text{ cm}^2/\text{statV s} \end{aligned} \quad (18\text{-E1})$$

Repeat these calculations using Eq. 18-E1 with corresponding V, and the results are listed in column 4 of Table 18-E1.

TABLE 18-E1 Calculated Z_p for Each Channel (Voltage Step) Particle Size

Channel	Collector Rod Voltage (V)	Channel Boundary (μm)	Z_p at 1 atm (cm^2/statV s)
	19	0.0032	62.2
1			
	59	0.0056	20.0
2			
	186	0.0100	6.35
3			
	588	0.0178	2.01
4			
	1870	0.0316	0.632
5			
	2600	0.0562	0.455
6			
	4440	0.100	0.266
7			
	6600	0.178	0.179
8			
	8380	0.316	0.141
9			
	9600	0.562	0.123
10			
	10,600	1.000	0.111

Two precautions in using the EAA should be mentioned here. First, as indicated in Fig. 18-1, there exists a minimum electrical mobility versus particle size for a given diffusion charging condition. For the charging condition used in the EAA ($N_i t = 1 \times 10^7$ ions/cm^3 s), this minimum lies between 1.0 and 2.0 μm particle diameter. This means that the one-to-one relationship between particle size and particle electrical mobility exists only for particles either smaller than about 1.0 μm or larger than about 2.0 μm. The design of the EAA is primarily for sizing submicrometer-sized aerosols. Therefore, to avoid errors in electrical mobility measurements caused by contributions from particles larger than 1.0 μm, an aerosol pre-classifier, such as an impactor, should be used if one suspects that the aerosol to be measured may contain such particles.

The second point of caution concerns the operating conditions. As described previously, the "data reduction table" was derived from calibration under the suggested standard operating conditions that included designated volumetric flow rates, charging condition, preprogrammed step-voltages on the collector rod, and operating temperature and pressure. Any deviation from these standard operating conditions [e.g., using the EAA aboard a research aircraft (Liu, Whitby, and Pui 1974) or at high-elevation locations, such as in the Rocky Mountain region], can lead to possible error in the measurement results, unless adjustments are made in the data reduction procedure or operating parameters. Yeh, Cheng, and Kanapilly (1981) and Yeh and Cheng (1982) discussed this problem and suggested two alternate schemes to overcome it. These involve adjusting the data reduction procedures or operating parameters. Briefly, the first scheme involved adjusting the instrument channel boundary particle sizes and the conversion constant of measured current to the number of particles while keeping the volumetric flow rate and other operating conditions as suggested by the manufacturer. The second scheme involved adjusting the applied voltage steps to the center rod while keeping the channel boundary particle sizes and the rest of the data analysis package intact.

Example 18-3

The EAA in Example 18-2 will be used in a Rocky Mountain area where the local atmospheric pressure is equal to 0.8 atm. The instrument will be operated according to suggested operating conditions, i.e., the same charging condition and the same volumetric flow rates for Q_{tm} and Q_{am}. Assuming that under the same charging condition (i.e., same $N_i t$), the particle charge for a given particle size is independent of pressure when $P \leq 1$ atm, what would be the Z_p corresponding to each channel boundary size?

Answer. Let Z_p and Z_{p0} denote the electrical mobility at $P = 0.8$ atm and $P_0 = 1$ atm, respectively; then from Eq. 18-8, one has

$$Z_p/Z_{p0} = (neC_c/3\pi\eta d_p)/(n_o eC_{c0}/3\pi\eta d_p)$$
$$= C_c/C_{c0}$$

or

$$Z_p = (C_c/C_{c0})\, Z_{p0},$$

TABLE 18-E2 Calculated Electrical Mobility vs. Channel Boundary Particle Sizes for the EAA Under a Given Charging Condition ($N_i t = 1 \times 10^7$ ions/cm³ s) at Two Different Atmospheric Pressures

Size (μm)	C_∞ at 1 atm	C_c at 0.8 atm	Z_p at 1 atm (cm²/statV s)	Z_p at 0.8 atm (cm²/statV s)
0.0032	70.7763	88.3177	6.2199E + 01	7.7615E + 01
0.0056	40.6410	50.6631	2.0031E + 01	2.4970E + 01
0.0100	22.9677	28.5773	6.3537E + 00	7.9055E + 00
0.0178	13.1214	16.2682	2.0099E + 00	2.4919E + 00
0.0316	7.6257	9.3907	6.3198E − 01	7.7825E − 01
0.0562	4.5500	5.5305	4.5453E − 01	5.5248E − 01
0.1000	2.8591	3.3934	2.6617E − 01	3.1591E − 01
0.1780	1.9576	2.2381	1.7906E − 01	2.0472E − 01
0.3160	1.5000	1.6411	1.4103E − 01	1.5429E − 01
0.5620	1.2709	1.3416	1.2311E − 01	1.2996E − 01
1.0000	1.1512	1.1891	1.1149E − 01	1.1516E − 01

where the slip correction, C_c and C_{c0}, can be calculated from Eq. 3-8 or Eq. 3-9 for $P = 0.8$ atm (81 kPa) or $P_0 = 1$ atm (101.33 kPa) for each particle size. For $d_p = 0.0032$ µm, Eq. 3-9 gives

$$C_c = 88.3 \quad \text{at } P = 0.8 \text{ atm}$$

$$C_{c0} = 70.8 \quad \text{at } P_0 = 1.0 \text{ atm}$$

Therefore, from the above equation,

$$Z_p = (88.3/70.8)(62.2) = 77.6 \text{ cm}^2/\text{statV s}$$

Repeat the same procedure for the rest of the particle sizes listed in Table 18-E1; the results are listed in Table 18-E2 and Fig. 18-E1. Because the EAA measures the mobility, for each voltage step, the penetration cutoff fraction is for mobility less than a given mobility corresponding to the applied voltage rather than a given particle size. Therefore, one should note that the so-called channel boundary particle sizes have changed under different operating atmospheric pressure. For example, as shown in Fig. 18-E1, the original channel boundary size of 0.0316 µm has $Z_p = 0.63$ cm²/statV s. For the same Z_p at 0.8 atm, the size becomes 0.046 µm, a difference of about 45%.

Differential-Mobility Analyzer

As with the EAA, the differential-mobility analyzer (DMA) is based on the known charge state of the particles and subsequent mobility analysis. Although the idea of differential mobility is not new (Rohmann 1923), the current interest in the DMA was kindled at the Particle Technology Laboratory of the University of Minnesota. Knutson and Whitby (1975*a*, *b*) and Knutson (1976) gave a detailed description of the DMA. Initially, the DMA gained its fame in aerosol science and technology mainly as an aerosol classifier for producing submicrometer-sized monodisperse aerosols (Liu and Pui 1974*c*; Liu 1974). With the advancement in ultrafine particle detection, such as the improved sensitivity of an electrometer or a condensation nucleus counter (CNC), the advantages of

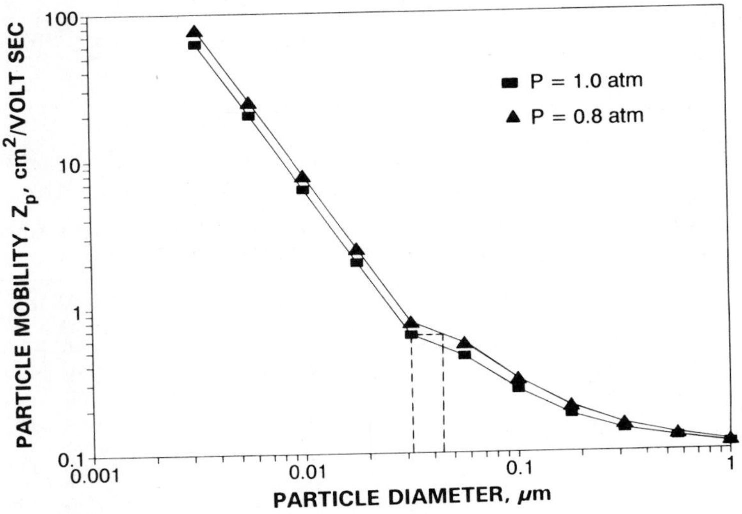

FIGURE 18-E1 Particle Electrical Mobility and Size Relationship for the EAA under a Given Charging Condition ($N_i t = 1 \times 10^7$ ions/cm³ s) with Pressure as a Parameter. As an Example, As Shown by the Dashed Vertical Lines, the Original Channel Boundary Size of 0.032 µm has $Z_p = 0.63$ cm²/statV s, Whereas for the Same Z_p at 0.80 atm, the Size becomes 0.046 µm, a Difference of About 45%.

using the DMA method for particle size analysis have been recognized (Hoppel 1978; Fissan, Helsper, and Thielen 1983; ten Brink et al. 1983; Keady, Quant, and Sem 1984; Kousaka, Okuyama, and Adachi 1985; Reischl 1991). Furthermore, the application of the DMA can be greatly enhanced when two DMAs are used in a tandem fashion. This arrangement has been used to study the size distribution of ultrafine particles (Haaf 1978) and the growth or evaporation of droplets (Rader and McMurry 1986).

However, there are two key differences between the EAA and the DMA. Unlike the EAA, where unipolar diffusion charging is applied, the DMA is based on the equilibrium charge state of the particles. This equilibrium charge state, sometimes called the Boltzmann equilibrium, can be obtained by exposing the particles to a bipolar ion atmosphere, such as the ^{85}Kr charge neutralizer. The second deviation from the EAA is that, instead of measuring the "integrated band of mobility", the DMA measures the extracted "narrow band" of mobility (Fig. 18-4b). This feature, plus the fact that the equilibrium charge distribution is much more stable than the charge distribution obtained from unipolar diffusion charging contributes to the better resolution of the DMA as compared to the EAA. Furthermore, the theoretical basis for predicting the equilibrium charge distribution is much better understood and more accurate than the diffusion charging theory. The data inversion computer program can use these theoretical equations instead of relying on calibrations that may have to be repeated each time when operating conditions are changed.

Limitations

Table 18-2 lists some of the commercially available aerosol measurement instruments that are based on the electrical technique. In addition to these commercial instruments, various designs of aerosol measurement instruments are available from numerous investigators.

Although the electrical technique is very useful in aerosol measurement, there are some limitations. In the following sections, limitations of each type of instrument will be briefly discussed.

Electrostatic Precipitator

1. It is well known that it is very difficult to charge ultrafine particles less than 0.01 μm in size. Therefore, a large fraction of ultrafine particles might not be charged, and the sample obtained might be biased toward larger particles.
2. As indicated by Cheng, Yeh, and Kanapilly (1981), the precipitation efficiency for the point-to-plane ESP is a function of particle size and flowrates. To minimize these effects, they suggested that the point-to-plane ESP should be operated with flow rates below 100 cm^3/min, ion

TABLE 18-2 Lists of Typical Commercially Available Aerosol Measurement Instruments which are Based on the Electrical Technique

Name of Instrument	Manufacturer	Remark
Point-to-plane electrostatic precipitator	In-Tox products, Albuquerque, NM	Requires EM for sizing
Model 3100 electrostatic aerosol sampler	TSI Inc., St. Paul, MN	Requires EM for sizing
Model 3932 differential-mobility particle sizing system (DMPS)	TSI Inc., St. Paul, MN	0.01–1.0 μm Either with a CNC or an electrometer
Electrical aerosol analyzing system (EAA)	TSI Inc., St. Paul, MN	0.01–1.0 μm
Electrical mobility spectrometer (EMS VIE-06)	Univ. of Vienna	Requires a CNC or an electrometer
HAUKE-AERAS submicron aerosol analyzing system SAAS 3/150	Hauke KG, Gmunden, Austria	0.003–0.15 μm

currents of 2–3 μA, and with a sharp needle.

Electrical Aerosol Analyzer (EAA)

1. As indicated in Fig. 18-1 and discussed previously, there exists a minimum electrical mobility for a given diffusion charging condition as a function of particle size. The design of the EAA is primarily for sizing submicrometer-sized aerosols. Therefore, to avoid the error in electrical mobility measurements caused by contributions from particles larger than 1.0 μm, an aerosol preclassifier, such as an impactor, should be used if one suspects that the aerosol to be measured may contain such particles.
2. Because the current diffusion charging (or combined field and diffusion charging) theory cannot predict accurately the charge distribution of an aerosol under various charging conditions, the data analysis scheme primarily relies on experimental data for given charging conditions and calibration data on instrument sensitivity under given operating conditions. As described previously, the "data reduction table" was derived from calibration under the suggested standard operating conditions that include designated volumetric flow rates, charging conditions, preprogrammed step-voltages on the collector rod, and operating temperature and pressure. Any deviation from these standard operating conditions will lead to possible error in the measurement results, unless adjustments are made in the data reduction procedure or operating parameters. If one has to operate the EAA under other than the manufacturer's conditions (e.g., different atmospheric pressure), suggestions made by Yeh, Cheng, and Kanapilly (1981) and Yeh and Cheng (1982) should be considered.
3. Because the charging phenomenon results in a charge distribution, rather than in single-value charges, the instrument sensitivities from calibration for various voltage steps are being used in the data inversion scheme. The implication of this is that the calculated geometric standard deviation (σ_g) of the aerosol distribution is, generally, larger than 1.3 even if the σ_g of the input aerosol is less than 1.3. Therefore, the EAA may predict the correct median particle size, but the σ_g may be in error for a narrowly distributed aerosol.
4. The charging data and the instrument sensitivity used are valid only for spherical particles. For particles shaped very close to a sphere, the potential error is small. However, for particles that deviate greatly from spherical shape, the error can be large. For example, the charge distribution on fiber aerosols is not well understood; therefore, EAA might not be a good instrument to use for sizing fibers.
5. For radioactive aerosols, especially those with high specific radioactivity, the measured electrical mobility may be different from a nonradioactive aerosol because of the potential self-charging due to radioactive decay.

Differential-Mobility Analyzer (DMA)

1. Because a similar principle is applied to the EAA and the DMA, limitations 4 and 5 listed under the EAA also apply to the DMA.
2. Because only a small number of particles with essentially one charge per particle are detected at each voltage step, the detecting limit of the accompanying CNC or electrometer will determine the lower limit of measurement.
3. For aerosol size measurement, the DMA requires the use of a neutralizer to bring particle charge distribution to Boltzmann equilibrium. Radioactive sources are commonly used as neutralizers; this includes the α-source (such as ^{241}Am or ^{210}Po) and the β-source (such as ^{85}Kr). It has been reported that the radiolytic processes can cause the formation of ultrafine particles (Leong et al. 1983; Winklmayr et al. 1990), especially for α-emitters. This may cause artifacts in a size measurement.

References

Arnold, S. 1979. Determination of particle mass and charge by one electron differentials. *J. Aerosol Sci.* 10:49–53.

Bricard, J. 1948. Sur l'equilibre ionique de la basse atmosphere. *C. R. Acad. Sci.* (Paris) 226:1536–38.

Cheng, Y. S., H. C. Yeh, and G. M. Kanapilly. 1981. Collection efficiencies of a point-to-plane electrostatic precipitator. *AIHAJ* 42:605–10.

Cochet, R. 1961. *Colloque Internationale-La Physique des Forces Electrostatiques et Leurs Applications*, pp. 331–38. Paris: Centre National de la Recherche Scientifique.

Cooper, D. W. and P. C. Reist. 1973. Neutralizing charged aerosols with radioactive sources. *J. Colloid Interface Sci.* 45:17–26.

Fissan, H. J., C. Helsper, and H. J. Thielen. 1983. Determination of particle size distributions by means of an electrostatic classifier. *J. Aerosol Sci.* 14:354–57.

Fuchs, N. A. 1947. The charges on the particles of aerocolloids. *Investiya Acad. Nauk. SSSR, Ser. Geogr. Geofiz.* 11:341.

Fuchs, N. A. 1963. On the stationary charge distribution on aerosol particles in a bipolar ionic atmosphere. *Geofis. Pura Appl.* 56:185–93.

Haaf, W. 1978. An electrostatic tandem mobility analyzer for size distribution measurements in the Aitken range of atmospheric pure-air aerosols. *Proc Gesellschaft für* Aerosolforschung, pp. 165–73, Wien, Austria, 26–28 September.

Hinds, W. C. 1982. *Aerosol Technology: Properties, Behavior, and Measurement of Airborne Particles*, pp. 292, New York: Wiley.

Hoppel, W. A. 1978. Determination of the aerosol size distribution from the mobility distribution of the charged fraction of aerosols. *J. Aerosol Sci.* 9:41–54.

Ivanov, V. D., V. N. Kirichenko, B. V. Shan'gin, and V. M. Berezhnoi. 1974. Method and apparatus for the investigation of the electrical charging of beta-active 'hot' aerosol particles. *Colloid J. USSR* 36:427–30.

Johnston, A. M. 1983. A semi-automatic method for the assessment of electric charge carried by airborne dust. *J. Aerosol Sci.* 14:643–55.

Johnston, A. M., J. H. Vincent, and A. D. Jones. 1985. Measurements of electric charge for workplace aerosols. *Ann. Occup. Hyg.* 29:271–84.

Johnston, A. M., J. H. Vincent, and A. D. Jones. 1987. Electrical charge characteristics of dry aerosols dispersed by a number of laboratory mechanical dispersers. *Aerosol Sci. Technol.* 6:115–27.

Keady, P. B., F. R. Quant, and G. J. Sem. 1984. Automated differential mobility particle sizer. In *Aerosols*, eds. B. Y. H. Liu, D. Y. H. Pui, and H. J. Fissan, pp. 71–74. New York: Elsevier.

Kirsch, A. A. and A. V. Zagnit'ko. 1981. Diffusion charging of submicrometer aerosol particles by unipclar ions. *J. Colloid Interface Sci.* 80:111–17.

Knutson, E. O. and K. T. Whitby. 1975*a*. Aerosol classification by electric mobility: Apparatus, theory, and applications. *J. Aerosol Sci.* 6:443–51.

Knutson, E. O. and K. T. Whitby. 1975*b*. Accurate measurement of aerosol electric mobility moments. *J. Aerosol Sci.* 6:453–60.

Knutson, E. O. 1976. Extended electric mobility method for measuring aerosol particle size and concentration. In *Fine Particles*, ed. B. Y. H. Liu, pp. 739–62. New York: Academic Press.

Kousaka, Y., K. Okuyama, and M. Adachi. 1985. Determination of particle size distribution of ultra-fine aerosols using a differential mobility analyzer. *Aerosol Sci. Technol.* 4:209–25.

Leong, K. H., P. K. Hopke, J. J. Stukel, and H. C. Wang. 1983. Radiolytic condensation nuclei in aerosol neutralizers. *J. Aerosol Sci.* 14:23–7.

Liu, B. Y. H., K. T. Whitby, and H. S. Yu. 1967. Electrostatic aerosol sampler for light electron microscopy. *Rev. Sci. Inst.* 38:100–02.

Liu, B. Y. H. 1970. Electrical properties of aerosols—Principles and applications. Particle Laboratory Publication No. 142, University of Minnesota.

Liu, B. Y. H. 1974. Laboratory generation of particulates with emphasis on submicron aerosols. *JAPCA* 24:1170–72.

Liu, B. Y. H. and D. Y. H. Pui. 1974*a*. Equilibrium bipolar charge distribution of aerosols. *J. Colloid Interface Sci.* 49:305–12.

Liu, B. Y. H. and D. Y. H. Pui. 1974*b*. Electrical neutralization of aerosols. *J. Aerosol Sci.* 5:465–72.

Liu, B. Y. H. and D. Y. H. Pui. 1974*c*. A submicron aerosol standard and the primary, absolute calibration of the condensation nuclei counter. *J. Colloid Interface Sci.* 47:155–71.

Liu, B. Y. H. K. T. Whitby, and D. Y. H. Pui. 1974. A portable electrical analyzer for size distribution measurement of submicron aerosols. *JAPCA* 24:1067–72.

Liu, B. Y. H. and D. Y. H. Pui. 1975. On the performance of the electrical aerosol analyzer. *J. Aerosol Sci.* 6:249–64.

Liu, B. Y. H. and D. Y. H. Pui. 1977. On unipolar diffusion charging of aerosols in the continuum regime. *J. Colloid Interface Sci.* 58:142–49.

Liu, B. Y. H. and A. Kapadia. 1978. Combined field and diffusion charging of aerosol particles. *J. Aerosol Sci.* 9:227–42.

Maltoni, G. G., C. Melandri, V. Prodi, G. Tarroni, A. De Zaiacomo, G. F. Bompane, and M. Formignani. 1973. An improved parallel plate mobility analyzer for aerosol particles. *J. Aerosol Sci.* 4:447–55.

Marlow, W. H. and J. R. Brock. 1975. Unipolar charging of small aerosol particles. *J. Colloid Interface Sci.* 50:32–8.

Megaw, W. J. and A. C. Wells. 1969. A high resolution charge and mobility spectrometer for radioactive submicrometre aerosols. *J. Sci. Instrum.* (*J. Phys. E.*) *Series 2* 2:1013–16.

Mercer, T. T. and H. Y. Chow. 1968. Impaction from rectangular jets. *J. Colloid Interface Sci.* 27:75–83.

Millikan, R. A. 1935. Electrons (+ and −), protons, photons, neutrons, and cosmic rays, *The Exact Evaluation of e*, chapter V, pp. 90–124. Chicago: The University of Chicago Press.

Morrow, P. E. and T. T. Mercer. 1964. A point-to-plane electrostatic precipitator for particle size sampling. *AIHAJ* 25:8–14.

Porstendörfer, J., A. Hussin, H. G. Scheibel, and K. H. Becker. 1984. Bipolar diffusion charging of aerosol particles—II. Influence of the concentration ratio of positive and negative ions on the charge distribution. *J. Aerosol Sci.* 15:47–56.

Pui, D. Y. H. 1976. Experimental study of diffusion charging of aerosols. Ph.D. Thesis, University of Minnesota, Minnesota.

Pui, D. Y. H., S. Fruin, and P. H. McMurry. 1988. Unipolar diffusion charging of ultrafine aerosols. *AS&T* 8:173–87.

Raabe, O. G. 1971. Particle size analysis utilizing grouped data and the log-normal distribution. *J. Aerosol Sci.* 2:289–303.

Rader, D. J. and P. H. McMurry. 1986. Application of the tandem differential mobility analyzer to studies of droplet growth or evaporation. *J. Aerosol Sci.* 17:771–87.

Reischl, G., W. John, and W. Devor. 1977. Uniform electrical charging of monodisperse aerosols. *J. Aerosol Sci.* 8:55–65.

Reischl, G. P., H. G. Scheibel, and J. Porstendörfer. 1983. The bipolar charging of aerosols: Experimental results in the size range below 20-nm particle diameter. *J. Colloid Interface Sci.* 91:272–75.

Reischl, G. P. 1991. Measurement of ambient aerosols by the differential mobility analyzer method: Concepts and realization criteria for the size range between 2 and 500 nm. *Aerosol Sci. Technol.* 14:5–24.

Rohmann, H. 1923. Method of size measurement for suspended particles. *Z. Phys.* 17:253–65.

Sano, I., F. Fikitan, and S. Sakata. 1953. Colloid chemistry of tobacco smokes, distribution of size and charge of the particles. *J. Chem. Soc. Japan; Pure Chem. Sec.* 74:664–68.

Sato, T. 1987. Charging process of fine particles in unipolar ion flow. *Trans. I.E.E. Japan* 107:155–61.

Smith, W. B. and J. R. McDonald. 1976. Development of a theory for the charging of particles by unipolar ions. *J. Aerosol Sci.* 7:151–66.

Takahashi, K. 1971. Numerical verification of Boltzmann's distribution for electrical charge of aerosol particles. *J. Colloid Interface Sci.* 35:508–10.

Takahashi, K. and A. Kudo. 1973. Electrical charging of aerosol particles by bipolar ions in flow type charging vessels. *J. Aerosol Sci.* 4:209–16.

Teague, S. V., H. C. Yeh, and G. J. Newton. 1978. Fabrication and use of krypton-85 aerosol discharge devices. *Health Phys.* 35:392–95.

ten Brink, H. M., A. Plomp, H. Spoelstra, and J. F. van de Vate. 1983. A high-resolution electrical mobility aerosol spectrometer (MAS). *J. Aerosol Sci.* 5:589–97.

Thomas, J. W. and D. Rimberg. 1967. A simple method for measuring the average charge on a monodisperse aerosol. *Staub* (English) 27:18–22.

TSI. 1977. Operating and service manual: Model 3030 electrical aerosol size analyzer. TSI Inc., St. Paul, MN.

Whitby, K. T. 1961. Generator for producing high concentrations of small ions. *Rev. Sci. Instrum.* 32:1351–55.

Whitby, K. T. and B. Y. H. Liu. 1966. The electrical behaviour of aerosols. In *Aerosol Science*, ed. C. N. Davies, pp.59–86. New York: Academic Press.

Whitby, K. T. and W. E. Clark. 1966. Electric aerosol particle counting and size distribution measuring system for the 0.015 to 1 μ size range. *Tellus* 18:573–86.

Whitby, K. T. and B. Y. H. Liu. 1968. Polystyrene aerosols—electrical charge and residue size distribution. *Atmos. Environ.* 2:103–16.

White, H. J. 1951. Particle charging in electrostatic precipitation. *AIEE Trans.* 70:1186–91.

Winklmayr, W., M. Ramamurthi, R. Strydom, and P. K. Hopke. 1990. Size distribution measurements of ultrafine aerosols, $d_p > 1.8$ nm, formed by radiolysis in a diameter measurement analyzer aerosol charger. *Aerosol Sci. Technol.* 13:394–98.

Yeh, H. C., G. J. Newton, O. G. Raabe, and D. R. Boor. 1976. Self-charging of ^{198}Au-labeled monodisperse gold aerosols studied with a miniature electrical mobility spectrometer. *J. Aerosol Sci.* 7:245–53.

Yeh, H. C. 1976. A theoretical study of electrical discharging of self-charging aerosols. *J. Aerosol Sci.* 7:343–49.

Yeh, H. C., G. J. Newton, and S. V. Teague. 1978. Charge distribution on plutonium-containing aerosols produced in mixed-oxide reactor fuel fabrication and the laboratory. *Health Phys.* 35:500–03.

Yeh, H. C. and Y. S. Cheng. 1980. An experimental study of the effect of temperature upon aerosol charge state. *Environ. Sci. Technol.* 14:726–29.

Yeh, H. C., Y. S. Cheng, and G. M. Kanapilly. 1981. Electrical aerosol analyzer: Data reduction for high altitude or reduced pressure. *Atmos. Environ.* 15:713–18.

Yeh, H. C. and Y. S. Cheng. 1982. Electrical aerosol analyzer: An alternate method for use at high altitude or reduced pressure. *Atmos. Environ.* 16:1269–70.

Yeh, H. C., R. L. Carpenter, and Y. S. Cheng. 1988. Electrostatic charge of aerosol particles from a fluidized bed aerosol generator. *J. Aerosol Sci.* 19:147–51.

19

Condensation Detection and Diffusion Size Separation Techniques[1]

Yung-Sung Cheng

Inhalation Toxicology Research Institute
Lovelace Biomedical and Environmental Research Institute
P.O. Box 5890
Albuquerque, NM, U.S.A.

INTRODUCTION

The diffusion and condensation techniques described in this chapter are used primarily for the measurement of ultrafine particles smaller than 0.2 μm in diameter. Condensation techniques have been used in studying atmospheric aerosols since the time of Aitken (1888). The sizes of atmospheric nuclei measured by Aitken were between 0.02 and 0.2 μm (Aitken nuclei). The Aitken nuclei and smaller particles could not be seen under an optical microscope; so he constructed an instrument to increase the particle size by saturating the atmosphere with water vapors; subsequent cooling of the atmosphere caused condensation. The particles grew to larger than 1.0 μm in size and could then be counted in an optical microscope. From this early technique have evolved two types of instruments: (1) condensation nuclei counters (CNCs) for determination of the number concentration of ultrafine particles, and (2) cloud chambers for measurement of high-energy charged particles at atomic and subatomic levels. This chapter will concentrate on the CNC. Improvements on the CNC design have continued over the last century, and now several types of CNCs have been developed. They vary in the method by which that supersaturation is induced and in the method of particle detection.

The diffusion technique is used to collect ultrafine particles and vapors, and to determine the size distribution of ultrafine particles. The technique was first conceived following the observation that losses of atmospheric nuclei in tubes were related to their diffusion coefficients (Nolan and Guerrini 1935). Mathematical equations for diffusion losses in rectangular or circular tubes were subsequently derived (Nolan and Nolan 1938; Gormley and Kennedy 1949). This enabled an accurate determination of diffusion coefficients and submicrometer particle sizes from measurement of particle penetration through these tubes.

Diffusion samplers separate particles or vapors by differential diffusion mobilities. Two types of diffusion samplers are often used in air sampling. A diffusion battery can be used to measure the size distribution of submicrometer particles, and a diffusion denuder is designed to separate and collect gases or

1. Supported by the Office of Health and Environmental Research of the U.S. Department of Energy under Contract. No. DE-AC04-76EV01013.

vapors from airborne particles. Some diffusion denuders can also be used to determine the diffusion coefficient and, therefore, the size of a gas or vapor. Diffusion samplers include tubes of different shapes and stacks with fine-mesh screens of well-defined characteristics.

Both diffusion and condensation devices are used for the measurement of ultrafine aerosol particles. A diffusion battery is often used with a CNC in a sampling train to determine both the concentration and particle size distribution. This chapter describes the operating principle, theory, design, applications, and data analyses of the CNCs, diffusion batteries, and diffusion denuders.

CONDENSATION THEORY

A CNC is an instrument for detecting ultrafine particles. The principle on which the condensation technique is based involves three processes: (1) supersaturation of water or other working fluids, (2) growth of particles by condensation of vapors, and (3) detection of particles.

Supersaturation

Ultrafine particles, including Aitken's nuclei, are too small to be detected by an optical microscope, which has a detection limit of about 0.1 μm. In a CNC, the aerosol is first saturated with a vapor of water or alcohol and subsequently cooled to induce a supersaturated condition, i.e., the vapor pressure is higher than the saturation pressure at a given temperature. Vapor molecules condense on particles to form liquid droplets of larger sizes. For bulk liquids, the vapor pressure at the saturation or equilibrium condition (p_s in mmHg) as a function of temperature (T, in degrees Kelvin) is shown by the following equation:

$$\log_{10} p_s = a - \frac{b}{T - c} \qquad (19\text{-}1)$$

For some compounds, a slightly different equation gives the best fit to the data:

$$\log_{10} p_s = \frac{-52.3\,b}{T} + c \qquad (19\text{-}2)$$

Table 19-1 Vapor Pressures of CNC Working Fluids

Fluid	Equation	*a*	*b*	*c*
Water	19-1	8.108	1750	38
Methanol	19-1	7.879	1233	45
Ethanol	19-1	8.045	1554	50.2
n-butanol	19-2		46.78	9.136
Dibutyl phthalate	19-1	16.27	5099	109

Table 19-1 lists the constants (a, b, and c) for some working fluids used in the CNCs, including water and alcohols. The saturation vapor pressure (p_s) is defined as the equilibrium partial vapor pressure for a flat liquid surface. For liquid droplets in an aerosol system, the partial pressure required to maintain the equilibrium is greater than that for a flat surface. This is called the Kelvin effect. The relationship between the saturation vapor pressure (p_d) on the droplet surface and particle diameter (d_p) can be expressed as the following function (see, for example, Hinds 1982):

$$\frac{p_d}{p_s} = \exp\left(\frac{4 v \gamma}{R T d_p}\right) \qquad (19\text{-}3)$$

where γ is the surface tension, v the molar volume of the liquid, and R the gas constant. Figure 19-1 plots the saturation ratio at equilibrium ($S_R = p_d/p_s$) as a function of the particle size for a water droplet. As the diameter becomes smaller, the saturation ratio needed to induce condensation becomes larger. At a given level of saturation ratio in a CNC (S_R between 1.5 and 3), the Kelvin diameter, d^*, as defined by Eq. 19-3 by setting $d^* = d_p$, is the minimum size to initiate the condensation. Particles larger than the Kelvin diameter will grow, while those smaller than d^* will be too small for vapor condensation. Note that Eq. 19-3 is derived for vapor condensed on liquid droplets of the same material or on

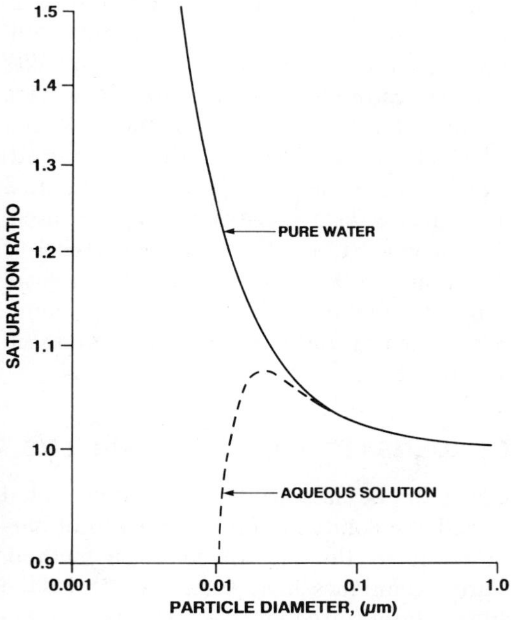

FIGURE 19-1. Saturated Ratio of Water as a Function of Droplet Size at 20°C. The Solid Line is the Pure Water Droplet and the Dashed Line is the Solution Particle.

insoluble particles with wettable surface properties for the working fluid.

For aerosol systems consisting of nonvolatile solute particles and a volatile solvent, such as hygroscopic particles in existence with water vapor, the saturation vapor pressure is lowered, because there are fewer solvent molecules in the surface layers than in the case of a pure solvent. For an ideal solution, the reduction of vapor pressure should be proportional to concentration (Raoult's law). In a solution droplet, the vapor pressure can be expressed in a form similar to the Kelvin equation (Eq. 19-3) (Tang 1976; Friedlander 1977):

$$\frac{p_d}{p_s} = \delta m_f \exp\left(\frac{4\gamma v_1}{RTd_p}\right) \tag{19-4}$$

where δ, v_1, and m_f are the activity coefficient, partial molar volume of the solvent, and mole fraction of the solute, respectively. Thus, there are two competing effects for solution droplets. The Kelvin effect tends to increase the vapor pressure, whereas the solute tends to reduce it. For an ideal solution ($\delta = 1$), Eq. 19-4 can be approximated in the following form (Friedlander 1977):

$$\frac{p_d}{p_s} = \exp\left(\frac{4\gamma v_1}{RTd_p} - \frac{6n_2 v_1}{\pi d_p^3}\right) \tag{19-5}$$

where n_2 is the number of moles of the solute. Figure 19-1 compares the vapor pressure for a pure water droplet to that of a solution particle, showing that a solution particle may be stable even at vapor pressures below saturation. In a CNC using water as the working fluid, hygroscopic particles, such as sodium chloride and other salts, may have lower detection limits because they are soluble in water.

If a particle carries an electrostatic charge, the vapor pressure at the surface is also reduced, and so is the saturation ratio required for the condensation (Scheibel and Porstendörfer 1986):

$$\frac{p_d}{p_s} = \exp\left(\frac{4v}{RT}\left[\frac{\gamma}{d_p} - \frac{q^2(1 - 1/\varepsilon)}{2\pi d_p^4}\right]\right) \tag{19-6}$$

where q is the electrostatic charge, and ε the dielectric constant of the droplet.

Droplet Growth

The droplet growth by condensation can be calculated by the following equation (Sutugin and Fuchs 1965; Zhang and Liu 1990):

$$d_p \frac{\mathrm{d}}{\mathrm{d}t}(d_p) = \frac{4D_a v}{R}\left[\frac{p}{T} - \frac{p_d}{T_d}\right] f(Kn) \tag{19-7}$$

where D_a is the diffusion coefficient of the condensing vapor, p and T are the vapor pressure and temperature in the surrounding gas far away from the particle, and v is the molar volume. The Fuchs correction, $f(Kn)$, is important for particles smaller than 0.1 μm:

$$f(Kn) = \frac{1 + Kn}{1 + 1.71\,Kn + 1.333\,Kn^2} \tag{19-8}$$

where Kn ($= 2\lambda/d_p$) is the Knudsen number

and λ the mean free path of the gas medium. Furthermore, p_d and T_d are the vapor pressure and temperature at the surface. For the Kelvin effect, p_d can be calculated from Eq. 19-3. The temperature on the droplet surface, T_d, includes the effect of the droplet temperature increase due to the latent heat of condensation and can be estimated as follows (Hinds 1982):

$$T_d = T + \frac{LMD_a}{k_v R}\left(\frac{p}{T} - \frac{p_d}{T_d}\right) \qquad (19\text{-}9)$$

where L is the latent heat of the working fluid and k_v is the thermal conductivity of the carrier gas. Equation 19-7 describes the growth of a homogeneous liquid droplet or the growth of insoluble particles with wettable surfaces. The growth of soluble particles is described by Friedlander (1977).

Droplet Detection

In a CNC, particles grow to a near-uniform size between 5 and 15 µm, large enough to be detected by optical means. Aitken (1888) used an optical microscope to count particles collected on slides for determining the concentration. Later on, Pollak and others (Pollak and Daly 1958; Jaenicke and Kanter 1976) mounted a camera on the CNC to photograph the particles. The Aitken and photographic CNCs are sometimes referred to as the absolute CNCs, because the number concentration is determined by direct counting. However, these CNCs are manually operated and they have been used primarily as calibration references for newer types of CNCs, in which the particles are detected by photoelectrical means. Early photoelectric CNCs, such as the Pollak counter (Pollak and Metnieks 1959), used the light extinction method, whereas recent models use light scattering techniques. The detection system consists of a light source and a photodetector. In a light extinction device, the transmission of light is detected, whereas in newer CNCs, the scattered light from the aerosol is detected in the photodetector. The signals from individual particles can be identified and counted, or the intensity of the scattered light is used as an indication of the particle concentration. The response curve of a photoelectric CNC must be calibrated. Early calibrations were performed with the photographic or Aitken CNCs. In recent studies, positively charged, monodisperse aerosol particles generated from an electrostatic classifier have been used; the concentration of a charged aerosol is determined with an electrometer. The aerosol concentration measured by the CNC is number concentration, conventionally expressed as particles/cm^3.

CONDENSATION NUCLEI COUNTERS

CNCs can be classified by the technique used to activate condensation and growth of particles, or by the aerosol detection method. Three techniques have been used to induce supersaturation conditions: (1) adiabatic expansion, (2) conductive cooling, and (3) mixing of hot and cold air streams. The aerosol detection systems include photography, light extinction, and light scattering, as described in the Condensation Theory section. Both photographic and light extinction methods are associated exclusively with CNCs having expansion chambers, whereas the light scattering detection method is incorporated in to more recent CNCs, with condensation provided by thermal systems. This section describes the CNCs according to the supersaturation technique used, with emphasis on the commercially available instruments (also listed in Table 19-2).

Description of Condensation Nuclei Counter

Three types of CNCs based on the technique used to induce supersaturation conditions are described here.

Expansion-Type CNC

An expansion-type CNC consists of a humidifier, an expansion chamber, and a detector. Water is the working fluid for this type of CNC. The aerosol stream is first humidified to reach the saturation of water vapor at

Table 19-2 Commercially Available CNCs and Diffusion Batteries

Company	Model	Type of Saturation	Type of Detection	Detection Limit (nm)	Company Address
BGI	Pollak Counter	Expansion/ overpressure	Light extinction	2.8	58 Guinan St. Waltham, MA 02154 (617) 891-9380
Gardner		Expansion/ overpressure	Light scattering	3.8	3643 Carmen Rd. Schenetady, NY (518) 355-2330
Environmental One	Rich 200	Expansion/ underpressure	Light scattering		2773 Balltown Schenetady, NY (518) 346-6161
TSI	3020/3022	Conductive	Light scattering	4	500 Cardigan Rd. St. Paul, MN 55160 (612) 483-0900
TSI	3025	Conductive	Light scattering	2	
TSI	3076	Conductive	Light scattering	10	

room temperature, and then the aerosol in an expansion chamber is suddenly cooled by volume expansion or pressure release. At lower temperatures the chamber becomes supersaturated with water vapor, which then condenses on particles. The expansion-type CNC was first designed by Aitken (1888) and was modified later by others. Several commercial instruments are based on these modifications dating back to the 1950s and 1960s, including the Pollak counter (Pollak and Metnieks 1959) and the Rich counters (Rich 1955, 1961), the Environmental One and the Gardner counters (Hogan and Gardner 1968), and the GE counters (Skala 1963; Haberl 1979).

In the original Aitken counter, the air volume was expanded by using a piston; in other models, the pressure inside the expansion chamber is reduced to cause the expansion. In an overpressure system, the chamber is pressurized by pumping air into the chamber using a hand pump (Pollak counter; Pollak and Metnieks 1959) or using bellows (Gardner counter; Hogan and Gardner 1968), and then a valve is opened to release the chamber pressure to the ambient value. In the Environmental One CNC (Model 200, Environmental One, Schenetady, NY) (Rich 1961; Skala 1963), underpressure systems are used in which air in the chamber under ambient pressure is released to an evacuated chamber. The volume or pressure expansion ratios as defined in the following equations are determined by the geometry of the expansion chamber:

$$\text{pressure expansion ratio} = \frac{P_i}{P_f} \tag{19-10}$$

$$\text{volume expansion ratio} = \frac{v_f}{v_i} \tag{19-11}$$

where P_i and P_f are pressures at the beginning and the end stage of expansion, v_i and v_f denote volumes at the beginning and end stage of the expansion. The expansion ratios in a CNC determine the supersaturation ratio of vapors. Usually, the dry adiabatic expansion of ideal gases is assumed for air and water vapor, because during the short time of expansion, there is little time for heat transfer. Under these assumptions, the two expansion ratios are related by

$$\frac{P_i}{P_f} = \left(\frac{v_f}{v_i}\right)^{\gamma} \tag{19-12}$$

where γ is the specific heat ratio, and is 1.4 for air. Because both the expansion ratios are greater than 1, Eq. 19-12 shows that the pressure expansion ratio is larger than the volume expansion ratio. The pressure expansion ratio of CNCs ranges from 1.1 to 1.5 (Miller and Bodhaine 1982).

After expansion, the chamber temperature at the onset of condensation (T_f) can be related to the initial temperature (T_i) before expansion and the expansion ratios (Miller and Bodhaine 1982):

$$\frac{T_f}{T_i} = \left(\frac{P_f}{P_i}\right)^{(\gamma - 1/\gamma)} = \left(\frac{v_i}{v_f}\right)^{\gamma - 1} \qquad (19\text{-}13)$$

At this point, the water vapor in the system becomes supersaturated; the saturation ratio is defined as the ratio of the partial pressure of water, $p(T_f)$, to the saturated water pressure, $p_s(T_f)$:

$$S_R = \frac{p(T_f)}{p_s(T_f)} \qquad (19\text{-}14)$$

Assuming that the air is initially saturated with water, the partial pressure of water after expansion can then be calculated similar to Eq. 19-12:

$$\frac{p(T_f)}{p_s(T_i)} = \left(\frac{v_i}{v_f}\right)^{\gamma} = \frac{P_f}{P_i} \qquad (19\text{-}15)$$

Substituting Eq. 19-15 into Eq. 19-14, we obtain the saturation ratio in terms of saturated vapor pressures and the pressure expansion ratio:

$$S_R = \frac{p_s(T_i)}{p_s(T_f)}\left(\frac{P_f}{P_i}\right) \qquad (19\text{-}16)$$

An expansion-type CNC is normally operated in a cyclic fashion. The chamber is first filled with aerosols; the expansion occurs, the aerosol particles grow, and the concentration is determined. In photographic CNCs (Pollak and Daly 1958; Jaenicke and Kanter 1976; Scheibel and Porstendörfer 1986), additional steps are necessary to develop photographs and to count the particles for concentration. Some of the early CNCs, including the Pollak counter and other photographic CNCs, used an overpressure system with slow manual operation. They were used as primary standards for other CNCs, because the counting method was considered as a direct measurement of particle concentration. The Environmental One and Gardner CNCs (Model CN, Gardner Associated, Schenectady, NY) (Skala 1963) use underpressure systems, and the expansion cycles are controlled by rotary valves. The sampling rates are 1–5 cycles/s, providing near-continuous operation.

Example 19-1

An expansion-type CNC with water as the working fluid has a pressure expansion ratio of 1.21. If the CNC is operated at a room temperature of 20°C and 740 mmHg pressure, what is the minimum particle diameter which can grow in the CNC (Kelvin diameter)?

Answer. The following are the steps to solve the problem:

(1) Estimate the temperature after expansion, T_f, from Eq. 19-13:

$$T_f = T_i\left(\frac{P_f}{P_i}\right)^{(\gamma - 1)/\gamma} = 293.2\left(\frac{1}{1.21}\right)^{0.4/1.4}$$
$$= 277.7\ \text{K} = 4.5°\text{C}$$

(2) Calculate the saturated water pressure, $p_s(T_i)$ and $p_s(T_f)$, from Eq. 19-1:

$$p_s(T_i) = 10^{[8.108 - 1750/(293.1 - 38)]}$$
$$= 17.7\ \text{mmHg}$$

Similarly, $p_s(T_f) = 6.33$ mmHg

(3) Estimate the supersaturation ratio using Eq. 19-25:

$$S_R = \frac{17.7}{6.33} \times \frac{1}{1.21} = 2.31$$

(4) Calculate the Kelvin diameter using Eq. 19-3 and surface tension of 75 dyn/cm for

water:

$$d^* = \frac{(4 \times 75 \text{ dyn/cm})(18 \text{ cm}^3/\text{gmol})}{(8.31 \times 10^7 \text{ dyn/cm/gmol/K})(277.7 \text{ K} \times \ln 2.31)}$$

$$= 2.79 \times 10^{-7} \text{ cm} = 2.79 \text{ nm}$$

Figure 19-2 shows a schematic diagram of a Pollak counter (Pollak and Metnieks 1959). The counter includes a vertical expansion chamber or fog tube (2.5 cm inner diameter and 60 cm long) with a water-saturated ceramic lining, a light source at the top, a photodetector at the bottom, and a hand pump. Thus, the fog tube combines the functions of humidification and expansion and also provides the light path for photoelectrical detection. The aerosol is first allowed to flush through the chamber, and then the inlet and the exit to the chamber are closed. Next, the fog tube is pressurized to 160 mmHg over the ambient pressure by operating the hand pump. After a delay to allow the air to become saturated, the initial light intensity is taken. Then the expansion valve is opened for air to expand to the ambient pressure. A second reading is taken, and the ratio of the final reading to the initial reading is the light transmission. The light transmission was calibrated against the aerosol concentration measured simultaneously from a photographic CNC, as listed in the tables of Pollak and Metnieks (1959). These tables have been checked more recently using a monodisperse aerosol generated from electrical aerosol analyzers (Liu et al. 1975; Sinclair 1984).

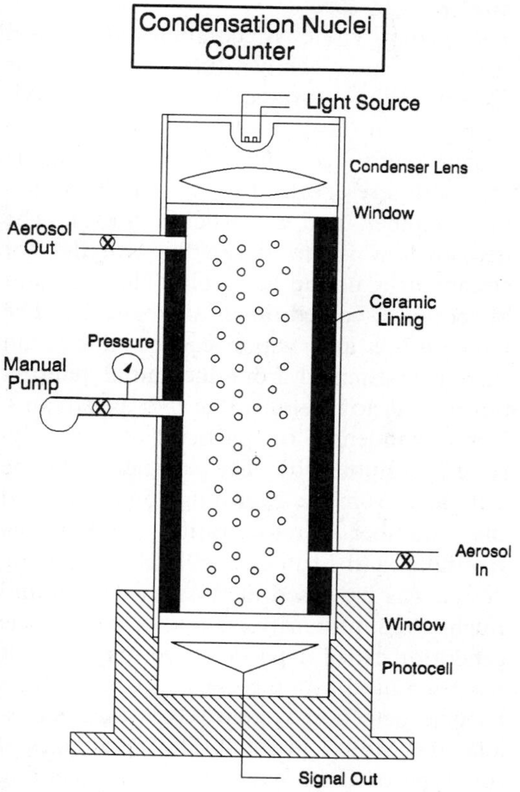

FIGURE 19-2. Schematic Diagram of the Pollak Counter.

Conductive-Cooling-Type CNC

A major disadvantage of the expansion-type CNC is that the flow is cyclic and, therefore, incompatible with the requirement of steady-state flow when the CNC is used to measure concentration from a diffusion battery or electrical mobility analyzer. The continuous-flow CNCs are based on the principle of thermal cooling to induce the supersaturation of a working fluid with a steady flow in the system. The conductive-cooling-type CNC was first designed by Sinclair and Hoopes (1975*b*), Bricard et al. (1976), and Agarwal and Sem (1980). As shown in Fig. 19-3, the conductive-cooling CNC consists of a saturator, condenser, and particle detector. The aerosol passes through an alcohol reservoir kept at an elevated temperature. The residence time is such that the aerosol will be saturated with the working fluid at the set temperature. Then the aerosol enters a condenser tube, which is kept at a lower temperature by cooling the wall. In the condenser, gas cooling takes place by conduction and convection, which leads to supersaturation in the cooled aerosol stream. Particles then grow by condensation to form droplets. The concentration of alcohol vapor and temperature are functions of the location in the condenser tube and are determined by numerically solving the heat and mass transfer equations (Ahn and Liu 1990; Zhang and Liu 1990).

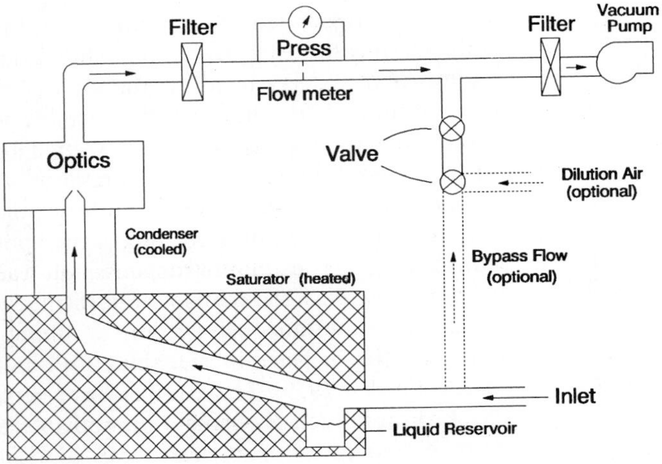

FIGURE 19-3. Schematic Diagram of a TSI Model 3022 CNC.

Figure 19-3 shows a continuous-flow CNC described by Agarwal and Sem (1980) and Keady, Quant, and Sem (1988) (Model 3022, TSI, St. Paul, MN). The saturator with butanol is kept at 35°C, and the condenser tube is maintained at 10°C. The aerosol flow rate is 300 ml/min. Under the normal operating condition at ambient pressure, the saturation ratio is calculated as a function of location (Zhang and Liu 1990). The highest saturation ratio is at the center of the tube mainly because of the high vapor concentration. The vapor concentration and saturation ratio decrease as they approach the tube wall. Because of the distribution of the vapor concentration, not all particles are activated for growth, resulting in a decreased counting efficiency for particles smaller than 10 nm. Particles that grow to a size of about 12 μm are detected with a light scattering system. The photodetector is operated as a single-particle counter for any concentration lower than 1000 particles/cm^3 and as a photometer for higher concentrations. To increase the counting efficiency of nanometer-sized particles, Wilson and Blackshear (1983) modified the condenser tube so that aerosol is introduced in the center surrounded with clean sheath air saturated with alcohol, where the supersaturation ratio is at the maximum. Based on similar concepts, continuous-flow CNCs with lower detection limits were designed by Stolzenburg and McMurry (1986) and Keady, Quant, and Sem (1988).

Mixing-Type CNC

The mixing-type CNC combines an aerosol stream at room temperature and a hot air stream with saturated vapor of dibutylphthalate or dioctyl sebacate with a high boiling point. The CNC consists of a saturator with a reservoir of working fluid, a mixing chamber, and a particle detector. The aerosol flow is mixed rapidly with the hot stream in a nozzle assembly. The resultant mixed air is cooled down adiabatically. The temperature and vapor concentration can then be estimated from the initial temperatures and flow rates of the two air streams. Vapor condenses on particles at a steady-state continuous flow. The particle size can be estimated from the operating conditions and the number concentration (Okuyama, Kousaka, and Motouchi 1984). This mixing device was first used in Russia (Sutugin and Fuchs 1965) primarily as an ultrafine particle generator called a particle size magnifier. It has been used as a CNC by incorporating a particle detection system (Okuyama, Kousada, and Motouchi 1984). The advantages of this type of CNC over the thermal cooling CNC are: (1) higher aerosol flow rates (0.5–2 l/min) can be used, and (2) minimum diffusional losses of aerosol particles occur, because of the short aerosol delivery distance, since the aerosol stream does not pass through the saturator. However, there are no commercial mixing-type CNCs available in the U.S.

THEORIES OF THE DIFFUSION TECHNIQUE

Mathematical expressions relating collection or penetration of vapors and particles through cylindrical and rectangular tubes and screens have been derived. These expressions can be used to calculate diffusion coefficients or particle sizes from experimental measurements through diffusion samplers.

These mathematical expressions were derived from the convective diffusion equation describing the concentration profile (c) in various geometries and flow profiles:

$$\frac{D_{\mathrm{p}}}{r}\frac{\partial}{\partial r}\left(r\frac{\partial c}{\partial r}\right) = U(r)\frac{\partial c}{\partial z} \qquad (19\text{-}17)$$

where D_{p} is the particle diffusion coefficient, r the radial direction, z the axial direction, and $U(r)$ the velocity profile in the axial direction. Several assumptions were made in the derivation of Eq. 19-17: (1) the concentration is in a steady-state condition; (2) the flow field in the device is a fully developed laminar flow; (3) the effect of diffusion in the direction of flow is neglected; (4) no production or reaction of the gas or aerosol occurs in the device; and (5) the sticking coefficient of the gas or particle is 100% on the collection surface (walls or screens). Diffusion devices can be classified as tube (channel)-type or screen-type. Each type has different flow profiles. Solutions to Eq. 19-17 for different types of diffusion samplers are summarized in the following section.

Tube (Channel)-Type

Penetration (P) of particles or gases due to the diffusional mechanism has been derived for channels of different geometries, including cylindrical, rectangular, disk, and annular shapes. The general solution of Eq. 19-17 can be expressed as a series of exponential functions:

$$P = \sum_{n=1}^{\infty} A_n \exp(-\beta_n \mu) \qquad (19\text{-}18)$$

where μ is the dimensionless argument relating the diffusion coefficient, channel length, and flow rate, and β_n are the eigenvalues. Convergence of Eq. 19-18 depends on the magnitude of μ. For larger values of μ (low penetration), few terms are needed for convergence, whereas at smaller values of μ (high penetration), many terms are required. For low penetration, alternative equations have been derived. Specific equations for each channel-type are described below.

Cylindrical Tubes

Penetration through a circular tube (Fig. 19-4) at a flow rate Q for particles with a diffusion coefficient D has been derived by several investigators as a function of the parameter μ defined as (Gormley and Kennedy 1949; Sideman, Luss, and Peck 1965; Davis and Parkins 1970; Tan and Hsu 1971; Lekhtmakher 1971; Bowen, Levine, and Epstein 1976)

$$\mu = \frac{\pi D L}{Q} \qquad (19\text{-}19)$$

The numerical solution obtained by Bowen, Levine, and Epstein (1976) for μ between 1×10^{-7} and 1 is most accurate. Results obtained by Davis and Parkins, Tan and Hsu, and Sideman, Luss, and Peck, and Lekhtmakher agree substantially with those of Bowen, Levine, and Epstein (1976). By comparison of the various expressions, the following analytical solutions have the accuracy of four significant numbers as compared to Bowen's result in the entire range of μ:

$$\begin{aligned} P = {} & 0.81905\exp(-3.6568\,\mu) \\ & + 0.09753\exp(-22.305\,\mu) \\ & + 0.0325\exp(-56.961\,\mu) \\ & + 0.01544\exp(-107.62\,\mu) \qquad (19\text{-}20) \\ & \text{for } \mu > 0.02 \end{aligned}$$

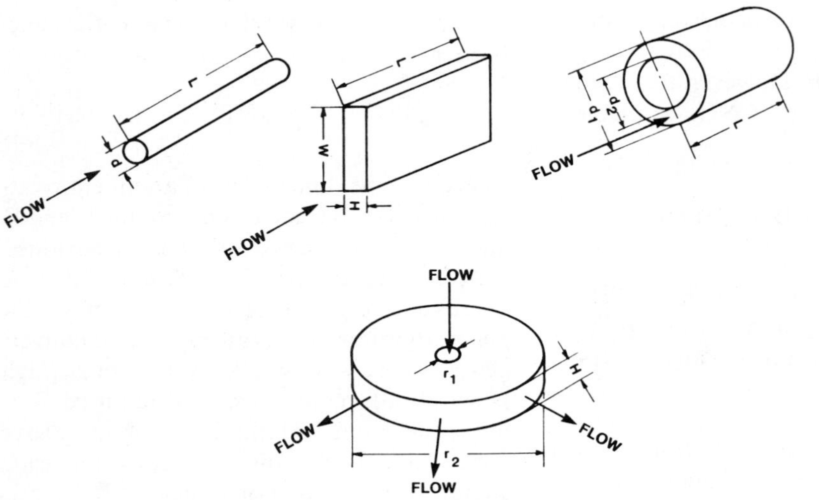

FIGURE 19-4. Schematic Diagram of Different Shapes of Tubes.

$$P = 1.0 - 2.5638\,\mu^{2/3} + 1.2\,\mu + 0.1767\,\mu^{4/3} \quad \text{for } \mu \leq 0.02 \qquad (19\text{-}21)$$

The formula for small values of μ is taken from Gormley and Kennedy (1949), Newman (1969), and Ingham (1975).

Rectangular Channels and Parallel Circular Plates

Particle penetration through a parallel narrow rectangular tube (Fig. 19-4) of width W and separation H, where $H \ll W$, has been derived as a function of μ defined as $8DLW/3QH$ (Gormley as cited by Nolan and Nolan 1938; DeMarcus and Thomas 1952; Sideman, Luss, and Peck 1965; Mercer and Mercer 1970; Tan and Thomas 1972; Bowen, Levine, and Epstein 1976). The same equation can be used to calculate penetration through parallel circular plates (Fig. 19-4), where the diffusion parameter μ is defined as $8\pi D(r_2^2 - r_1^2)/3QH$, where r_2 and r_1 are outer and inner radii of the disks (Mercer and Mercer 1970). The most accurate solution was given by Tan and Thomas (1972) and Bowen, Levine, and Epstein (1976); other investigators agreed substantially with their results. Using the first four terms of the solution given by Tan and Thomas, the penetration can be calculated to the accuracy of four significant numbers as compared to the numerical solution of Bower, Levine, and Epstein (1976):

$$P = 0.9104\exp(-2.8278\,\mu) + 0.0531\exp(-32.147\,\mu) + 0.01528\exp(-93.475\,\mu) + 0.00681\exp(-186.805\,\mu) \quad \text{for } \mu > 0.05 \qquad (19\text{-}22)$$

$$P = 1 - 1.526\,\mu^{2/3} + 0.15\,\mu + 0.0342\,\mu^{4/3} \quad \text{for } \mu \leq 0.05 \qquad (19\text{-}23)$$

The formula for small values of μ is given by Ingham (1976). Kennedy (quoted by Nolan and Kenny 1953) derived a similar formula with different coefficients, but the results are different by only 1%.

Annular Tubes

A theoretical equation has not been derived for diffusional losses through an annular tube (Fig. 19-5). The following empirical equation for the annular tube was derived from a sorption study with SO_2 (Possanzini, Febo,

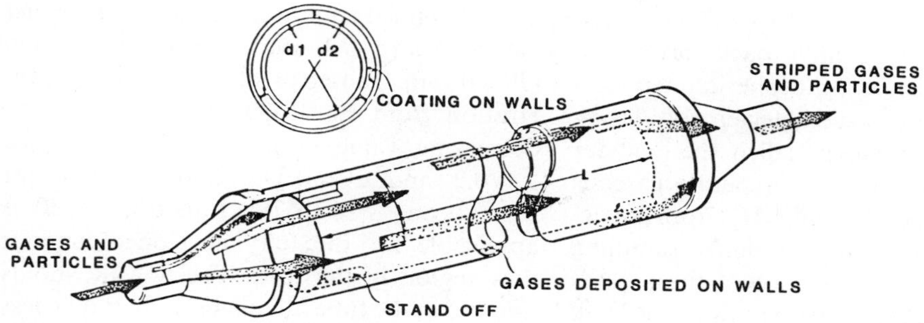

FIGURE 19-5. Schematic Diagram of an Annular Denuder. (Reprinted from Stevens (1986) with the Permission of Lewis Publishers.)

and Aliberti 1983):

$$P = (0.82 \pm 0.10)\exp(-22.53 \pm 1.22\mu) \quad (19\text{-}24)$$

where μ is defined as $\pi DL(d_1 + d_2)/4Q(d_2 - d_1)$, and d_2 and d_1 are outer and inner diameters, respectively. Equation 19-24 is valid only for annular tubes, where μ is large and penetration through the device is less than 10%.

Screen-Type

Aerosol penetration through a stack of fine-mesh screens with circular fibers of uniform diameter and arrangement has been derived (Cheng and Yeh 1980; Cheng, Keating, and Kanapilly 1980; Yeh, Cheng, and Orman 1982). A stack of fine-mesh screens simulates a fan model filter (Kirsch and Fuchs 1968; Kirsch and Stechkina 1978) in terms of flow resistance and aerosol deposition characteristics (Cheng, Yeh, and Brinsko 1985). The theoretical penetration was derived based on the aerosol filtration in the fan model filter:

$$P = \exp\left[-B_f n\left(2.7Pe^{-2/3} + \frac{1}{\kappa}R^2 + \frac{1.24}{\kappa^{1/2}}Pe^{-1/2}R^{2/3}\right)\right] \quad (19\text{-}25)$$

$$B_f = \frac{4\alpha h}{\pi(1-\alpha)d_f} \quad (19\text{-}26)$$

where n is the number of screens, d_f the fiber diameter, h the thickness of a single screen, α the solid volume fraction of the screen, and κ the hydrodynamic factor of the screen:

$$\kappa = -0.5\ln(2\alpha/\pi) + (2\alpha/\pi) - 0.75 - 0.25(2\alpha/\pi)^2 \quad (19\text{-}27)$$

$R = d_p/d_f$, the interception parameter, Pe is the Peclet number:

$$Pe = Ud_f/D \quad (19\text{-}28)$$

and U is the velocity entering the screen. Equation 19-25 includes the diffusional and interceptional losses of aerosol on screens and is valid for particles up to 1 μm in size (Cheng, Yeh, and Brinsko 1985). For particles larger than 1 μm, inertial impaction becomes an important mechanism, and Eq. 19-25 may not be adequate. For smaller particles ($d_p < 0.01$ μm), diffusional deposition is the dominant mechanism, and Eq. 19-25 is simplified to

$$P = \exp(-2.7B_f nPe^{-2/3}) \quad (19\text{-}29)$$

DIFFUSION DENUDERS

Gas or vapor molecules diffuse rapidly to the wall of a diffusion sampler and adsorb onto the wall coated with material suitable for collecting the gas. Diffusion tubes have been used to measure diffusion coefficients of several gases in the air (Thomas 1955; Fish and Durham 1971; Ferm 1979; Durham, Spiller, and Ellestad 1987). Since 1980, diffusion denuders followed by a filter pack have been developed to sample atmospheric nitric acid vapors and nitrate particulate aerosols. Using this sampling technique, called the denuder difference method, one can separate gaseous species, such as HNO_3 and NH_3, from particulate nitrates and, thus, minimize sampling artifacts due to the presence of these gases (Stevens et al. 1978; Appel, Tokiwa, and Haik 1981; Shaw et al. 1982; Forrest et al. 1982; Ferm 1986; Stevens 1986). Diffusion denuders are also used to monitor vapors, such as formaldehyde, chlorinated organics, and tetraalkyl lead, in the ambient air or work environments (Johnson et al. 1985; Cecchini, Febo, and Possanzini 1985; Febo, DiPalo, and Possanzini 1986). Some personal samplers have also been developed for industrial hygiene use (DeSantis and Perrino 1986; Gunderson and Anderson 1987).

Description of Diffusion Denuders

Two types of diffusion denuders have been designed: the cylindrical tube and annular tube.

Cylindrical Denuders

In cylindrical denuders, a single cylindrical glass or teflon tube is often used for collecting gases or vapors. The diameter and length of the tube and the sampling flow rate are designed to have greater than 99% collection efficiency. For example, a glass tube of 3 mm ID and 35 cm long would have over 99% efficiency for ammonia ($D_a = 2.47 \times 10^{-5}\ m^2/s$) at 3 l/min (Ferm 1979). For higher sampling flow rates, parallel tube assemblies have been designed (Stevens et al. 1978; Forrest et al. 1982), consisting of 16 glass tubes 5 mm ID and 30 cm long. The sampling flow rate was 50 l/min, and the collection efficiency for ammonia was over 99% (Stevens et al. 1978).

Penetration through the tube-type denuders can be estimated by taking only the first term of Eq. 19-20:

$$P = 0.819 \exp(-3.66\mu) \quad (19\text{-}30)$$

This simplified equation is accurate at higher values of μ (> 0.4) and at lower penetration ($P < 0.190$). The error of the estimated penetration from Eq. 19-30 increases with the decreasing value of μ (-0.25% error for $\mu = 0.2$ and $P = 0.395$, and -1.8% for $\mu = 0.1$ and $P = 0.578$). Equation 19-30 is applicable for the fully developed laminar flow region in the tube. The flow Reynolds number in the tube (diameter d) should be less than 2300 for laminar flow:

$$Re_f = \frac{4\rho_f Q}{\eta \pi d} < 2300 \quad (19\text{-}31)$$

In the entrance of the tube, the flow is in a transition region from plug flow to fully developed laminar flow. The length of entrance, L_e, is defined by the following equation and should be minimized:

$$L_e = 0.035\, d\, Re_f \quad (19\text{-}32)$$

Annular Denuders

Higher sampling flow rates are desirable, especially for sampling trains consisting of denuders and filters or dichtomous samplers (Shaw et al. 1982). An annular tube denuder was recently designed for this purpose (Possanzini, Febo, and Aliberti 1983). It consists of two coaxial cylinders, with the inner one sealed at both ends, so that air is forced to pass through the annular space (Fig. 19-5). The collection efficiency of the annular tube can be estimated from Eq. 19-24 for lower Reynolds numbers ($Re < 2300$), defined as

$$Re_f = \frac{4\rho_f Q}{\eta \pi (d_1 + d_2)} \quad (19\text{-}33)$$

Comparing the performance of the cylindrical and annular denuders in removing a gas from an air stream, a typical annular denuder ($d_2 = 3.3$ cm and $d_1 = 3.0$ cm) is possible by equating Eqs. 19-24 and 19-30. It can be shown that

$$\left.\frac{Q}{L}\right|_{\text{annular}} = 31.5 \left.\frac{Q}{L}\right|_{\text{cylinder}} \quad (19\text{-}34)$$

This relationship shows that for equal sampling time and tube length, the annular denuder can operate at 30 times the flow rate of the cylindrical denuder and still have the same removal efficiency. Also, the Reynolds number would still indicate laminar flow conditions for the annular tube system. A multichannel annular diffusion denuder has been tested and used in ambient air sampling (Johnson et al. 1985).

Compact Coil Denuder

A compact coil denuder consisting of a 1.0 cm inner diameter and 95 cm long (L) glass tube bent into a three-turn helical coil with a 10 cm diameter (Fig. 19-6) has been designed (Pui et al. 1990). The heat and mass transfer rates to the tube wall in a curved tube are much higher than that in a straight tube operated under the same conditions (Mori and Nakayama 1967*a, b*). This denuder is operated at 10 l/min (Q) with a Reynolds number of 1400. The penetration through the denuder can be expressed as (Pui et al. 1990):

$$P = 0.82 \exp\left(-\frac{\pi L D}{Q} Sh \right) \tag{19-35}$$

$$Sh = \frac{0.864}{\delta} De^{1/2}(1 + 2.35\, De^{-1/2}) \tag{19-36}$$

where Sh is the Sherwood number, De the Dean number ($= Re_f/$[coil radius/tube radius]), equal to the flow Reynolds number divided by the square root of the curvature, and δ is the thickness ratio of the concentration boundary layer to the momentum boundary layer. This unit has 99.3% collection efficiency for SO_2, with less than 6% particle losses for particles between 0.015 and 2.5 μm in diameter. It is compact and easy to operate.

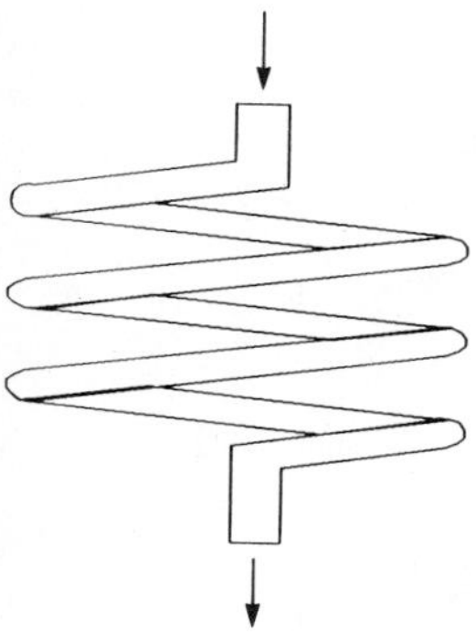

FIGURE 19-6. Schematic Diagram of a Compact Coil Denuder.

Transition-Flow Denuder

Both cylindrical and annular denuders are operated under laminar flow conditions, and they are designed to remove all gases of interest from the aerosol stream. In passing through such denuders, particle evaporation may increase the concentration of some gases, especially in the case of the decomposition of NH_4NO_3 into HNO_3 and NH_3 gases. To avoid biases due to evaporation of particles, one approach to sampling such gases is to collect only a known fraction of gases in the denuder and then calculate the gas concentration.

A transition-flow denuder was designed by Durham et al. (1986) to permit higher sampling flow. The cylindrical denuder has an inside diameter of 0.95 cm with a 6 cm distance of the first active surface to allow the development of a stable flow profile. The denuder section is lined with a 3.2 cm long nylon sheet. By assuming complete mixing in the active section, the penetration can be expressed as

$$P = \exp\left(-\frac{2\pi\alpha L}{Q} \right) \tag{19-37}$$

$$\alpha = \frac{r D_a}{\delta} \tag{19-38}$$

where L is the length of the active surface, Q is the flow rate, r the radius of the pipe, and δ the boundary layer thickness, which is a function of the flow Reynolds number. The penetration must be determined empirically. Operating the denuder at 16.1 l/min ($Re = 2500$), Durham et al. (1986) obtained a penetration of 0.911 for HNO_3.

Scrubber-Type Diffusion Denuders

Most diffusion denuders have a solid coating on the wall to collect gaseous species, and the coating substrates are washed after sampling for analysis. For continuous analysis of gas species, diffusion scrubbers are used, where the absorbent or solvent in the liquid form are continuously flowing along the tube wall, and the analyte can be analyzed in real time (Dasgupta 1984; Dasgupta et al. 1988). A tubular scrubber is made by inserting a membrane tube into the glass tube to form a jacket between the glass wall and the membrane (Fig. 19-7). Porous membranes, such as PTFE and polypropyline, allow gases but not particles to permeate and dissolve in solution, which flows continuously through the jacket. The collection characteristics of this type of diffusion scrubber should be similar to the tube diffusion denuder. Another type of diffusion denuder consists of a tube with a membrane tube at the center, in which air flows through the annular space, while the solvent passes through the membrane tube.

Coating Substrates

Absorbent material can be coated onto the tube wall of a denuder to collect the gas of interest from the air stream. Table 19-3 lists substrates for removal of some gases as reported in the literature. Some materials absorb more than one gaseous species. For example, sodium carbonate can absorb acidic gaseous species found in the ambient air, including HCl, HNO_2, HNO_3, and SO_2. The method of application of material to the tube wall depends largely on the nature of the material. Most materials are first dissolved and then applied to the tube wall. Solvents are allowed to evaporate, leaving the absorbent on the glass tube wall. In some cases the glass denuder wall has been etched by sandblasting the surface to increase the capacity of walls to support the denuding chemical substrate (Possanzini, Febo, and Aliberti 1983). Absorbent paper impregnated with liquid or solution substrate, such as oleic acid, has been used to line the inside of the denuder wall (Thomas 1955). A nylon sheet has also been used as a liner (Durham et al. 1986). Anodized aluminum surfaces have recently been found to be a good absorbing surface for nitric acid. Annular denuders made of anodized aluminum do not need coating (John 1987). Tenax or silica gel powder is more difficult to apply; however, these materials adhere to the glass wall coated with silicon grease (Johnson et al. 1985; Gunderson and Anderson 1987).

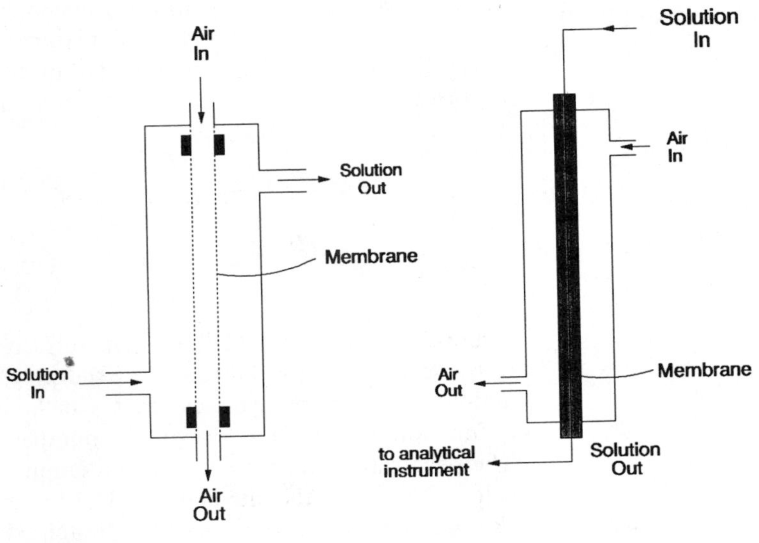

FIGURE 19-7. Schematic Diagram of Diffusion Scrubbers.

Table 19-3 Materials for Absorbing Gases in the Diffusion Denuder

Coating Material	Gas Absorbed	References
Oxalic acid	NH_3, aniline	Ferm (1979), DeSantis and Perrino (1986)
Oleic acid	SO_3	Thomas (1955)
uH_3PO_3	NH_3	Stevens et al. (1978)
K_2CO_3	SO_2, H_2S	Durham, Wilson, and Bailey (1978)
Na_2CO_3	SO_2, HCI, HNO_3, HNO_2	Forrest et al. (1982)
$CuSO_4$	NH_3	Thomas (1955)
PbO_2	SO_2, H_2S	Durham, Wilson, and Bailey (1978)
WO_3	NH_3, HNO_3	Braman, Shelley, and McClenny (1982)
MgO	HNO_3	Stevens et al. (1978)
NaF	HNO_3	Slanina et al. (1981)
NaOH and guaiacol	NO_2	Buttini, DiPalo, and Possanzini (1987)
Bisulfite-triethanolamine	Formaldehyde	Cecchini, Febo, and Possanzini (1985)
Nylon sheet	SO_2, HNO_3	Durham et al. (1986)
Tenax powder	Chlorinated organics	Johnson et al. (1985)
Silica gel	Aniline	Gunderson and Anderson (1987)
ICI	Tetraalkyl lead	Febo, DiPalo, and Possanzini (1986)

Sampling Trains

When sampling ambient or working atmospheres, it is sometimes necessary to collect gas species and particulate materials separately. In this case, a sampling train consisting of diffusion denuders and a filter pack has been used. A more complex system, as the one shown in Fig. 19-8, including a cyclone preclassifier, two Na_2CO_3-coated annular denuders, and a filter pack with a teflon and a nylon filter, has been used to collect acidic gases (HNO_3, HNO_2, SO_2, and HCl) separately from nitrate and sulfate particles (Stevens 1986). The first denuder removes gases quantitatively, whereas the second accounts for the interference from particulate material deposited on the wall under the assumption that particle deposition on each denuder is the same (Febo, DiPalo, and Possanzini 1986). The denuders are placed vertically to avoid particle deposition on the walls by sedimentation. A diffusion scrubber can be connected to an ion chromatograph or other analytical instruments for real-time analysis of gaseous species (Lindgren and Dasgupta 1988; Dasgupta et al. 1988).

DIFFUSION BATTERIES

Diffusion batteries were originally developed to measure the diffusion coefficient of particles less than 0.1 mm in diameter. They have since been used for the determination of particle size distributions by converting the diffusion coefficient to the particle size. Diffusion batteries are one type among only a few

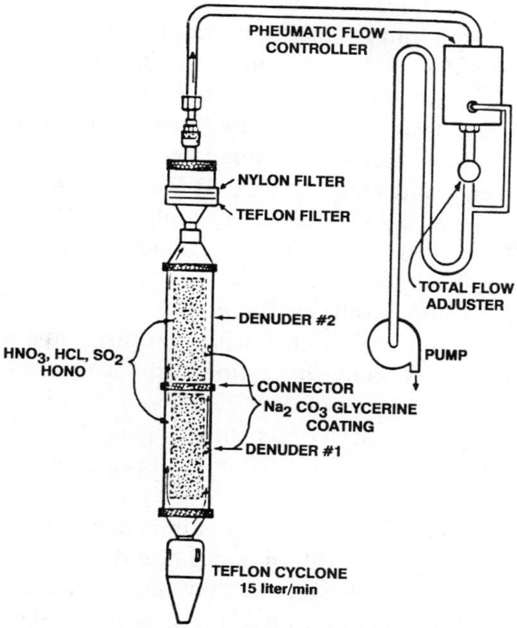

FIGURE 19-8. An Ambient Acidic Aerosol Sampler Consisting of a Precutter, Two Annular Denuders, and a Filter Pack. (Reprinted from Stevens (1986) with the Permission of Lewis Publishers.)

instruments that are applicable in measuring ultrafine particles between 0.1 mm and about 1 nm, corresponding to the size of molecular clusters. In this section, various designs of the instrument, detection of particles, and methods of data analysis will be discussed.

Description of Diffusion Batteries

Several types of diffusion batteries have been designed. Those based on rectangular channels and parallel circular plates are single-stage diffusion batteries. Cylindrical-tube and screen-type diffusion batteries usually have several stages.

Rectangular Channel

Rectangular-channel diffusion batteries usually consist of many rectangular plates forming parallel channels of equal width. These plates are separated by spacers and glued to a container with an airtight seal. For example, a diffusion battery consisting of 20 parallel channels (0.01 cm wide, 12.7 cm high, and 47.3 cm long) made from graphite plates has been designed for a 1 l/min sampling flow rate (Thomas 1955). Other instruments have been made from aluminum or glass plates with similar construction (Nolan and Nolan 1938; Nolan and Doherty 1950; Pollak, O'Conner, and Metnieks 1956; Megaw and Wiffen 1963; Rich 1966). Each channel should be parallel and have the same width. Deviation of channel width results in a nonuniform flow rate through each channel, which, in turn, causes the deviation of penetration from the theoretical prediction of Eqs. 19-22 and 19-23. A diffusion battery of ten single channels, each separately housed in a box, has been designed by Pollak and Metnieks (1959).

A single-stage diffusion battery can be used to measure the diffusion coefficient of monodisperse aerosols at one flow rate. When it is used to measure polydisperse aerosols, such as those found in the ambient air, several measurements taken at different flow rates are necessary to determine the distribution of diffusion coefficients.

Parallel Disks

A diffusion sampler has been designed by Kotrappa, Bhanti, and Dhandayutham (1975). It is based on the diffusional losses of particles from a fluid flowing radially inward between two coaxial, parallel, or circular plates as originally proposed by Mercer and Mercer (1970). Stainless steel plates (3.77 cm diameter) with a central hole of 0.2 cm diameter in the upper plate are the collecting substrate. Separation between the plates is 0.225 cm. An absolute filter is used to collect the material penetrating the device. This sampler has been used to determine the diffusion coefficient of radon decay products, which have diffusion coefficients of the order of 0.05 cm^2/s. The amount of radioactivity collected at the plates and absolute filter was determined, and the diffusion coefficient calculated from Eq. 19-22, simplified to contain only the first term:

$$P = 0.9104 \exp\left(-2.8278 \frac{8\pi D_p (r_2^2 - r_1^2)}{3QH}\right) \tag{19-39}$$

Cylindrical Tubes

Tube-type diffusion batteries made of cylindrical tubes usually consist of a cluster of thin-walled tubes with diameters less than 0.1 cm inner diameter. Large equivalent length (actual length × number of tubes) is required for the measurement of particle size, because particles have much smaller diffusion coefficients than gas molecules. Several cluster tube diffusion batteries have been designed by Sinclair (1969), Breslin, Guggenheim, and George (1971), and Scheibel and Porstendörfer (1984). Figure 19-9 shows a schematic diagram of a tube-type diffusion battery as reported by Scheibel and Porstendörfer (1984). Three diffusion batteries with 100, 484, and 1000 single tubes were used with lengths of 5.0, 9.3, and 39.03 cm, respectively. Tube-type diffusion batteries use materials that are commercially available and are also easier to construct than the parallel-plate diffusion battery. A lightweight material

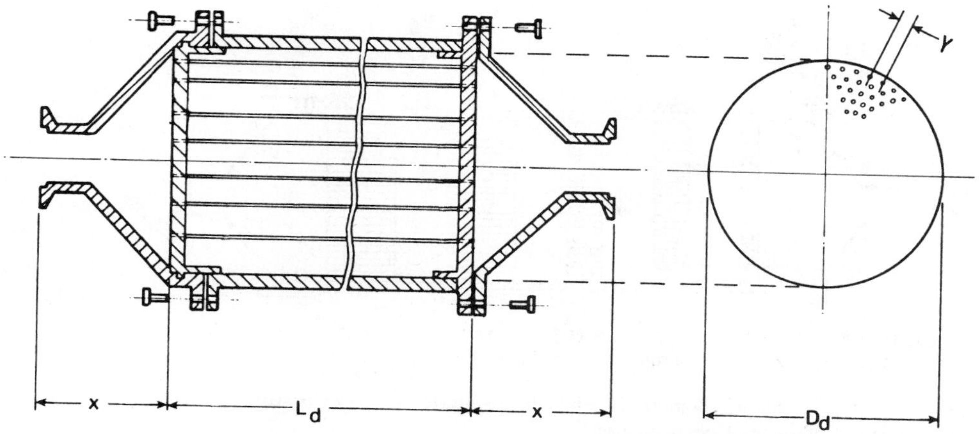

D_d = 10 cm
x = 7 cm

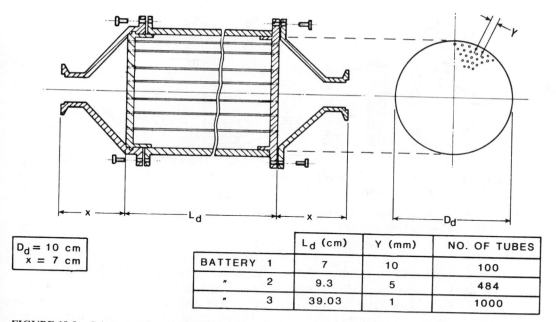

	L_d (cm)	Y (mm)	NO. OF TUBES
BATTERY 1	7	10	100
" 2	9.3	5	484
" 3	39.03	1	1000

FIGURE 19-9. Schematic Diagram of a Cluster-Tube Diffusion Battery. (Reprinted from Scheibel and Porstendörfer (1984).)

such as aluminum, is usually used; however, this type of diffusion battery is still heavy, bulky, and expensive. Most cluster tube diffusion batteries consist of one to three stages (Breslin, Guggenheim, and George 1971; Scheibel and Porstendörfer 1984), although an eight-stage diffusion battery has been constructed (Sinclair 1969).

Compact diffusion batteries with many stages have been designed by using collimated hole structures (CHS). The CHS are disks containing a large number of near-circular holes. Figure 19-10 shows a $1\frac{3}{4}$ in diameter CHS disk made from stainless steel containing 14,500 holes of 0.009 in diameter (Brunswick Co., Chicago, IL). With a thickness of 1/8–1 in, the equivalent length ranged from 46 to 369 m. A portable 11-stage diffusion battery has been designed with CHS elements (Sinclair 1972). The total length is 60 cm, and the equivalent length is 5094 m. Figure 19-11 shows a schematic diagram of a five-stage diffusion battery made from CHS elements. A multiple-stage diffusion battery is required to measure the size distribution of a polydisperse aerosol. The development of a multiple-stage CHS diffusion battery opens up the possibility of routine measurements of submicrometer aerosols. Other CHS disks made from glass capillary tubes of 25–50 mm in diameter and a thickness of 0.5–2.0 mm are

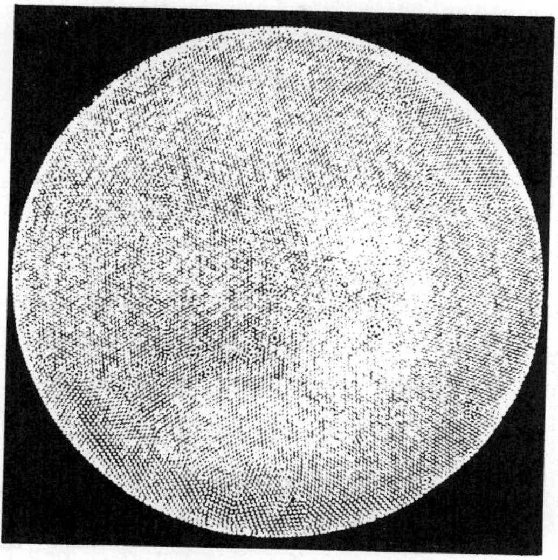

FIGURE 19-10. A Stainless Steel Collimated Hole Structure Disc.

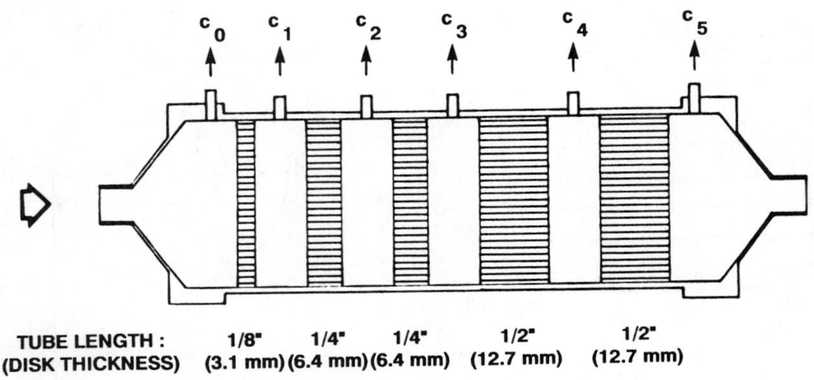

FIGURE 19-11. Schematic Diagram of a Five-Stage Diffusion Battery Consisting of a Stainless Steel Collimated Hole Structure.

also available (Galileo Electro Optical Corp., Struburg, MA). A six-stage CHS diffusion battery made from glass has also been designed (Brown, Beyer, and Gentry 1984).

Screen-Type

Diffusion batteries using stacks of filters as the cell material have been used by Sinclair and Hinchliffe (1972) and Twomey and Zalabsky (1981). This material is lightweight and inexpensive to build. However, commercial fiber or membrane filters are not ideal materials because of the nonuniformity in the fiber diameter and porosity. Aerosol penetration through the filter may not be consistent and could not be predicted accurately by filtration theory. Sinclair and Hoopes (1975*a*) designed a ten-stage unit using stainless steel 635-mesh screens of uniform diameter, opening and thickness (Fig. 19-12). The designed flow rate ranged from 4 to 6 l/min. Stacks of these well-defined screens simulate a fan model both in geometry and in flow resistance (Cheng, Yeh, and Brinsko 1985). Penetration through screens can be predicted by the fan model filtration theory (Eq. 19-25) (Cheng and Yeh 1980; Cheng, Keating, and Kanapilly 1980). Subsequently, this unit has become commercially available (Model 3040, TSI Inc., St. Paul, MN). Other types of screens have also been tested and found useful (Yeh, Cheng, and Orman 1982; Cheng, Yeh, and Brinsko 1985). Table 19-4 lists characteristics of the different screens as shown in Fig. 19-13. Screen-type diffusion batteries are compact in size and simple to construct. Screens can be cleaned and replaced easily when they are contaminated or worn out.

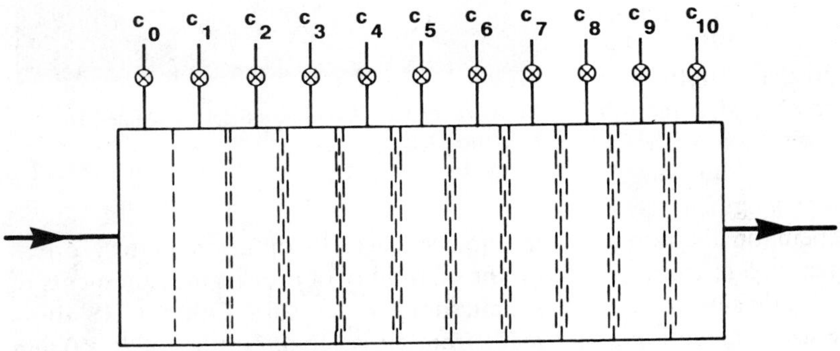

FIGURE 19-12. Schematic Diagram of a Ten-Stage Screen-Type Diffusion Battery.

Table 19-4 Characteristic Dimensions and Constants for Various Types of Screens in Screen-Type Diffusion Batteries

Weave	Square	Square	Square	Twill	Twill
Screen diameter (mm)	55.9	40.6	25.4	25.4	20
Screen thickness (mm)	122	96.3	57.1	63.5	50
Solid volume fraction	0.345	0.244	0.230	0.292	0.313
a	1.677	0.8969	0.9021	0.180	1.450
k	0.216	0.330	0.352	0.269	0.246

FIGURE 19-13. Photomicrograph of Stainless Mesh Screens.

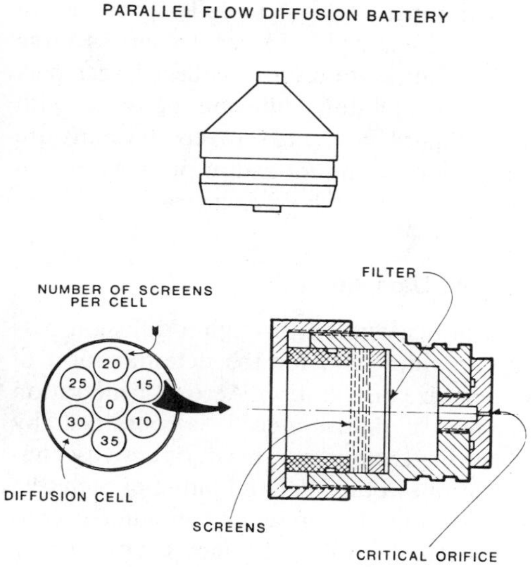

FIGURE 19-14. Schematic Diagram of a Parallel Flow Diffusion Battery.

Most multistage diffusion batteries described here are arranged in a series so that the aerosol concentration decreases continuously through the cells. Aerosol penetration is usually detected by a CNC. Based on parallel flow and mass collection principles, a parallel-flow diffusion battery (PFDB) has been designed (Cheng et al. 1984). This unit measures the penetration by mass or radioactivity without a particle detecting unit. It is also more useful to detect unstable aerosols with fluctuating size and concentration. A schematic diagram of the PFDB is shown in Fig. 19-14. It consists of a conical cap and a collection section containing seven cells. Each diffusion cell contains a different number of stainless steel 200-mesh screens followed by a 25 mm Zefluor filter (Gelman, Ann Arbor, MI). The seven cells contain typically 0–35 screens. Critical orifices provide a 2 l/min flow rate through each cell, resulting in a total flow rate of 14 l/min. Gravimetric determination of collected filter samples from each cell provides the direct mass penetration as a function of screen number for the determination of aerosol size distribution, thereby eliminating the sometimes inaccurate conversion of number to mass.

Screen diffusion batteries have been used routinely to determine the activity size distribution of radon progeny. Single screen and a filter have been used to estimate the unattached fraction. The screen and flow rate with a 50% penetration in the screen about 4–10 nm are usually used. The radioactivities collected on both the screen and the filter are counted, and the amount of activity collected on the screen is assumed to be the unattached radon progeny. The activity size distribution of the radon progeny can be measured with either a graded-screen diffusion battery (Holub, Knutson, and Soloman 1988) or a parallel-flow diffusion battery (Reineking and Porstendörfer 1986). A graded diffusion battery consists of several stages, each with a different type screen. The screens with lower mesh number are used to collect particles in the nm size range, whereas screens of large mesh number are used to collect larger particles. Parallel-flow diffusion batteries with much higher flow rates (over 30 l/min) are used to collect indoor radon progeny, which usually have very low concentrations.

Use and Data Analysis

Aerosol penetration through a diffusion battery provides data for the determination of particle size distribution. Aerosol penetration through a diffusion cell is obtained by measuring the number, mass, or activity concentrations at the inlet and outlet of each cell. A CNC is used to measure the number concentration. Figure 19-15 shows a schematic diagram of a system including a diffusion battery, automatic switching valve, and a CNC. With the automatic sampling system, it takes 3 min to complete an 11-channel measurement.

For radioactive aerosols, penetration based on activity can be obtained by collecting samples at the diffusion cell and a backup filter at the end of the diffusion battery. The single-stage parallel disk diffusion sampler (Kotrappa, Bhanti, and Dhandayutham 1975) and screen diffusion batteries have been used for this purpose. Screens can be counted directly for radioactivity (Reineking and Porstendörfer 1986). Penetration based on the mass can be obtained by using a parallel flow diffusion battery.

Data Analysis

Monodisperse Aerosol

Particle size distributions are calculated from penetration data obtained from the diffusion battery measurement. For a monodisperse aerosol, the diffusion coefficient, D_p, can be calculated directly from the corresponding Eqs. 19-20–19-25. The particle size is then calculated from the following relationship:

$$D_p = \frac{kTC_c(d_p)}{3\pi\eta d_p} \tag{19-40}$$

$$C_c(d_p) = 1 + \frac{2\lambda}{d_p}\left[1.142 + 0.558\exp\left(-0.999\frac{d_p}{2\lambda}\right)\right] \tag{19-41}$$

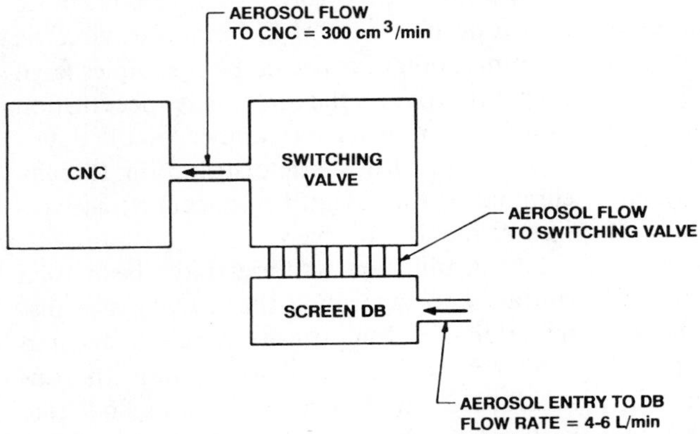

FIGURE 19-15. Schematic Diagram of a Diffusion Battery, Automatic Switch valve, and Condensation Nuclei Counter.

where k is the Boltzmann constant (1.38×10^{-16} erg/K), T the absolute temperature (K), C the Cunningham slip correction factor, and λ the mean free path of air (0.673 μm at 23°C and 760 mmHg). With monodisperse particles, measurements from a single-stage diffusion battery are sufficient, and measurements from multiple-stage devices should improve the accuracy.

Example 19-2

What is the aerosol penetration through a single stage tube diffusion battery (10 cm long, 100 tubes, and 4 l/min flow) for a 0.01 μm particle diameter, under an ambient pressure of 760 mmHg and 23°C?

Answer. The following are the steps to solve the problem:

(1) Calculate the diffusion coefficient under the ambient condition using Eq. 19-40 by first calculating the slip correction factor C_c (Eq. 19-41):

$$C_c = 1 + \frac{2 \times 0.0673}{0.01}\left[1.142 + 0.558 \exp\left(-0.999 \times \frac{0.01}{2 \times 0.0673}\right)\right]$$

$$= 23.3$$

$$D_p = \frac{(1.38 \times 10^{-16}) \times 296.2 \times 23.3}{3 \times 3.14(1.81 \times 10^{-4}) \times 0.01 \times 10^{-4}}$$

$$= 5.58 \times 10^{-4}\ \text{cm}^2/\text{s}$$

(2) Calculate the penetration through a circular tube using Eqs. 9-19 and 9-20:

$$\mu = \frac{3.14(5.58 \times 10^{-4}\ \text{cm}^2/\text{s})(10\ \text{cm})}{\frac{4}{100} \times \frac{1000}{60}\ \text{cm}^3/\text{s}}$$

$$= 0.0263$$

$$P = 0.819 \exp(-3.65 \times 0.0263) + 0.0975 \exp(-22.3 \times 0.0263) + 0.0325 \exp(-57.0 \times 0.0263) + 0.01544 \exp(-107 \times 0.0263)$$

$$= 0.806$$

Polydisperse Aerosols

Most aerosols in ambient environments and workplaces have polydisperse size distributions and the method described in the previous section does not apply. Three penetration data points are the minimum required, but more points will improve the accuracy of the size determination. Both graphical and numerical inversion methods have been developed for the size determination from penetration data.

Fuchs, Stechkina, and Starosselskii (1962) have generated a family of penetration curves for rectangular-channel diffusion batteries, assuming the aerosol size distribution is lognormally distributed. Mercer and Greene (1974) have provided curves representing the penetration of aerosols in both cylindrical and rectangular channels as functions of a diffusion parameter and the geometrical standard deviation. Once the data are properly aligned with one of the curves, this method gives a rough estimate of the mean and geometric standard deviation of the diffusion coefficient. Similar curves have been derived for screen-type diffusion batteries (Lee, Connick, and Gieseke 1981). This method does not apply to aerosols which do not follow lognormal size distributions.

Sinclair (1969) used a graphical "stripping" method to estimate the particle size distribution from penetration data through a multistage cylindrical-type diffusion battery. A family of penetration curves is calculated for monodisperse particles over a range of equivalent length. The experimental penetration data are plotted on a different sheet of paper of the same scale. The experimental curve is matched against the theoretical curves and the one having the best fit at the right hand of experimental curves (i.e., where penetration is least) is subtracted, leaving a new experimental curve. The process is repeated until the original experimental curve is entirely eliminated. Particle size and fractions of each size in the original aerosol are indicated by the matched theoretical curves and their intercepts with the ordinate of the graph. A similar method has been applied to the screen-type diffusion battery (Sinclair et al.

1979). This method does not assume a certain size distribution and, thus, is more useful. However, the results for both graphical methods depend on judgement in matching curves.

More consistent results can be obtained by using numerical inversion methods. In a diffusion battery, aerosol penetration through stage i, P_i, can be expressed mathematically as the integration of the aerosol penetration equation for monodisperse aerosol $P_i(x)$:

$$P_i = \int_0^\infty P_i(x) f(x)\,\mathrm{d}x \qquad (19\text{-}42)$$

where $f(x)$ is the size distribution and $P_i(x)$ the aerosol penetration of size x in stage i. Each observed penetration for stages $i = 1$–n can be expressed in the form of Eq. 19-42. Several numerical inversion methods have been developed to obtain the aerosol size distribution, $f(x)$. Raabe (1978) developed a nonlinear least-squares regression to solve Eq. 19-42 under the assumption of a lognormal distribution for $f(x)$. A similar method was used by Soderholm (1979) for diffusion battery data analysis. A nonlinear iterative method was proposed by Twomey (1975) and was applied to diffusion batteries by Knutson and Sinclair (1979). A modification of Twomey's method was used for data analysis of screen-type diffusion batteries (Kapadia 1980; Cheng and Yeh 1984). Recently, an expectation maximization algorithm was developed for the screen-type diffusion battery and appeared to work as well as or better than the least-squares regression and Twomey's method (Maher and Laird 1985). See also the discussion in Chapter 9.

CONCLUSIONS

Diffusion batteries and condensation nuclei counters are useful instruments in measuring the aerosol concentration and size distribution of ultrafine particles. The limit of detection in a CNC is about 2–5 nm, thus, the lower limit of usefulness of a condensation nuclei counter includes large molecular clusters. However, diffusion denuders and diffusion batteries can still be used to collect and separate particles and gases smaller than a few nanometers. Other techniques, including radioactivity detection and analytical chemical techniques, can be used to identify and quantitate materials collected on diffusion devices. These techniques are important in studying particles between the size range of gases and aerosols, a region where interesting phenomena can occur.

References

Agarwal, J. K. and G. J. Sem. 1980. Continuous flow, single-particle-counting condensation nucleus counter. *J. Aerosol Sci.* 11:343–57.

Ahn, K. H. and B. Y. H. Liu. 1990. Particle activation and droplet growth processes in condensation nucleus counter—I. Theoretical background. *J. Aerosol Sci.* 21:263–76.

Aitken, J. 1888. On the number of dust particles in the atmosphere. *Proc. Roy. Soc. Edinburgh* 35.

Appel, B. R., Y. Tokiwa, and M. Haik. 1981. Sampling of nitrates in ambient air. *Atmos. Environ.* 15:283–89.

Bowen, B. D., S. Levine, and N. Esptein. 1976. Fine particle deposition in laminar flow through parallel-plate and cylindrical channels. *J. Colloid Interface Sci.* 54:375–90.

Braman, R. S., T. Shelley, and W. A. McClenny. 1982. Tungstic acid for preconcentration and determination of gaseous and particulate ammonia and nitric acid in ambient air. *Anal. Chem.* 54:358–64.

Breslin, A. J., S. F. Guggenheim, and A. C. George. 1971. Compact high efficiency diffusion batteries. *Staub-Rein. Luft* 31(8):1–5.

Bricard, J., P. Delattre, G. Madelaine, and M. Pourprix. 1976. Detection of ultra-fine particles by means of a continuous flux condensation nuclei counter. In *Fine Particles*, ed. B. Y. H. Liu, pp. 566–80. New York: Academic Press.

Brown, K. E., J. Beyer, and J. W. Gentry. 1984. Calibration and design of diffusion batteries for ultrafine aerosols. *J. Aerosol Sci.* 15:133–45.

Buttiini, P., V. Di Palo, and M. Possanzini. 1987. Coupling of denuder and ion chromatographic techniques for NO_2 trace level determination in air. *Sci. Tot. Environ.* 61:59–72.

Cecchini, F., A. Febo, and M. Possanzini. 1985. High efficiency annular denuder for formaldehyde monitoring. *Anal. Lett.* 18:681–93.

Cheng, Y. S., J. A. Keating, and G. M. Kanapilly. 1980. Theory and calibration of a screen-type diffusion battery. *J. Aerosol Sci.* 11:549–56.

Cheng, Y. S. and H. C. Yeh. 1980. Theory of a screen-type diffusion battery. *J. Aerosol Sci.* 11:313–20.

Cheng, Y. S. and H. C. Yeh. 1984. Analysis of screen diffusion battery data. *Am. Ind. Hyg. Assoc. J.* 45:556–61.

Cheng, Y. S., H. C. Yeh, and K. J. Brinsko. 1985. Use of

wire screens as a fan model filter. *Aerosol Sci. Technol.* 4:165–74.

Cheng, Y. S., H. C. Yeh, J. L. Mauderly, and B. V. Mokler. 1984. Characterization of diesel exhaust in a chronic inhalation study. *Am. Ind. Hyg. Assoc. J.* 45:547–55.

Dasgupta, P. K. 1984. A diffusion scrubber for the collection of atmospheric gases. *Atmos. Environ.* 18:1593–99.

Dasgupta, P. K., S. Dong, H. Hwang, H. C. Yang, and G. Zhang. 1988. Continuous liquid-phase fluorometry coupled to a diffusion scrubber for the real-time determination of atmospheric formaldehyde hydrogren peroxide and sulfur dioxide. *Atmos. Environ.* 22:946–63.

Davis, H. R. and G. V. Parkins. 1970. Mass transfer from small capillaries with wall resistance in the laminar flow regime. *Appl. Sci. Res.* 22:20–30.

DeMarcus, W. and J. W. Thomas. 1952. Theory of a diffusion battery. Report ORNL-1413, Oak Ridge National Laboratory.

DeSantis, F. and C. Perrino. 1986. Personal sampling of aniline in working site by using high efficiency annular denuders. *Ann. Chimica* 76:355–64.

Durham, J. L., T. G. Ellestad, L. Stockburger, K. T. Knapp, and L. L. Spiller. 1986. A transition-flow reactor tube for measuring trace gas concentrations. *J. Air Pollut. Control Assoc.* 36:1228–32.

Durham, J. L., L. L. Spiller, and T. G. Ellestad. 1987. Nitric acid–nitrate aerosol measurements by a diffusion denuder, a performace evaluation. *Atmos. Environ.* 21:589–98.

Durham, J. L., W. E. Wilson, and E. B. Bailey. 1978. Application of an SO_2 denuder for continuous measurement of sulfur in submicrometric aerosols. *Atmos. Environ.* 12:883–86.

Febo, A., V. DiPalo, and M. Possanzini. 1986. The determination of tetraalkyl lead air by a denuder diffusion technique. *Sci. Tot. Environ.* 48:187–94.

Ferm, M. 1979. Method for determination of atmospheric ammonia. *Atmos. Environ.* 13:1385–93.

Ferm, M. 1986. A Na_2CO_3-coated denuder and filter for determination of gaseous HNO_3 and particulate NO in the atmosphere. *Atmos. Environ.* 20:1193–1201.

Fish, B. R. and J. L. Durham. 1971. Diffusion coefficient of SO_2 in air. *Environ. Lett.* 2:13–21.

Forrest, J., D. J. Spandau, R. L. Tanner, and L. Newman. 1982. Determination of atmospheric nitrate and nitric acid employing a diffusion denuder with a filter pack. *Atmos. Environ.* 16:1473–85.

Friedlander, S. K. 1977. *Smoke, Dust and Haze.* New York: Wiley.

Fuchs, N. A., I. B. Stechkina, and V. I. Starosselskii. 1962. On the determination of particle size distribution in polydisperse aerosols by the diffusion method. *Br. J. Appl. Phy.* 13:280–81.

Gormley, P. G. and M. Kennedy. 1949. Diffusion from a stream flowing through a cylindrical tube. *Proc. Roy. Irish Academy* A52:163–69.

Gunderson, E. C. and C. C. Anderson. 1987. Collection device for separating airborne vapor and particulates. *Am. Ind. Hyg. Assoc. J.* 48:634–38.

Haberl, J. B. 1979. General Electric condensation nuclei counters. In *Aerosol Measurement*, eds. D. A. Lundgren, F. S. Harris, W. H. Marlow, M. Lippmann, W. E. Clark, and M. D. Durham, pp. 568–73. Gainsville, FL: University Press of Florida.

Hinds, W. C. 1982. *Aerosol Technology, Properties, Behaviors, and Measurement of Airborne Particles.* New York: Wiley.

Hogan, A. W. and G. Gardner. 1968. A nucleus counter of increased sensitivity. *J. Rech. Atmos.* 3:59–61.

Holub, R. F., E. O. Knutson, and S. Soloman. 1988. Tests of the graded wire screen technique for measuring the amount and size distribution of unattached radon progeny. *Rad. Protect. Dosim.* 24:265–68.

Ingham, D. B. 1975. Diffusion of disintegration products of radioactive gases in circular and flat channels. *J. Aerosol Sci.* 6:395–402.

Ingham, D. B. 1976. Simultaneous diffusion and sedimentation of aerosol particles in rectangular tubes. *J. Aerosol Sci.* 7:373–80.

Jaenicke, R. K. and H. J. Kanter. 1976. Direct condensation nuclei counter with automatic photographic recording, and general problems of absolute counters. *J. Appl. Meteorol.* 15:620–32.

John, W. 1987. Personal communication.

Johnson, N. D., S. C. Barton, G. H. S. Thomas, D. A. Lane, and W. H. Schroeder. 1985. Development of gas/particle fractionating sampler of chlorinated organics. 78th Annual Meeting of Air Pollution Control Assocation, Detroit, MI.

Kapadia, A. 1980. Data reduction techniques for aerosol size distribution measurement instruments. Ph.D. Thesis, University of Minnesota.

Keady, P. B., F. R. Quant, and G. J. Sem. 1988. Two new condensation particle counters: Design and performance. Paper read at AAAR Annual Meeting, 10–14 October 1988, Chapel Hill, NC.

Kirsch, A. A. and N. A. Fuchs. 1968. Studies of fibrous aerosol filters—III. Diffusional deposition of aerosols in fibrous filters. *Ann. Occup. Hyg.* 11:299–304.

Kirsch, A. A. and I. B. Stechkina. 1978. The theory of aerosol filtration with fibrous filter. In *Fundamentals of Aerosol Science*, ed. D. T. Shaw, pp. 165–56. New York: Wiley.

Knutson, E. O. and D. Sinclair. 1979. Experience in sampling urban aerosols with the Sinclair diffusion battery and nucleus counter. In *Proc. Advances in Particle Sampling and Measurement*, ed. W. B. Smith, pp. 98–120. EPA 600/7-79-065.

Kotrappa, K., D. P. Bhanti, and R. Dhandayutham. 1975. Diffusion sampler useful for measuring diffusion coefficients and unattached fraction of radon and thoron decay products. *Health Phys.* 29:155–62.

Lee, K. W., P. A. Connick, and J. A. Gieseke. 1981. Extension of the screen-type diffusion battery. *J. Aerosol Sci.* 12:385–86.

Lekhtmakher, S. O. 1971. Effect of Peclet number on the precipitation of particles from a laminar flow. *J. Engrg. Phys.* 20:400–2.

Lindgren, P. F. and P. K. Dasgupta. 1988. Measurement

of atmospheric sulfur dioxide by diffusion scrubber coupled ion chromatography. *Anal. Chem.* 61:19–24.

Liu, B. Y. H., D. Y. H. Pui, A. W. Hogan, and T. A. Rich. 1975. Calibration of the Pollak counter with monodisperse aerosols. *J. Appl. Meteorol.* 14:46–51.

Maher, E. F. and N. M. Laird. 1985. EM algorithm reconstruction of particle size distributions from diffusion battery data. *J. Aerosol Sci.* 16:557–70.

Megaw, W. J. and R. D. Wiffen. 1963. Measurement of the diffusion coefficient of homogeneous and other nuclei. *J. Rech. Atmos.* 1:113–25.

Mercer, T. T. and T. D. Greene. 1974. Interpretation of diffusion battery data. *J. Aerosol Sci.* 5:251–55.

Mercer, T. T. and R. L. Mercer. 1970. Diffusional deposition from a fluid flowing radially between concentric, parallel, circular plates. *J. Aerosol Sci.* 1:279–85.

Miller, S. W. and B. A. Bodhaine. 1982. Supersaturation and expansion ratios in condensation nuclei counters: An historical perspective. *J. Aerosol Sci.* 13:481–90.

Mori, Y. and W. Nakayama. 1967*a*. Study on forced convective heat transfer in curved pipes. *Int. J. Heat Mass Transfer* 10:37–59.

Mori, Y. and W. Nakayama. 1967*b*. Study on forced convective heat transfer in curved pipes. *Int. J. Heat Mass Transfer* 10:681–95.

Newman, J. 1969. Extension of the Leveque solution. *J. Heat Transfer* 91:177–78.

Nolan, P. J. and D. J. Doherty. 1950. Size and charge distribution of atmospheric condensation nuclei. *Proc. Roy. Irish Academy* 53A:163–79.

Nolan, J. J. and V. H. Guerrini. 1935. The diffusion coefficient of condensation nuclei and velocity of fall in air of atmospheric nuclei. *Proc. Roy. Irish Academy* 43:5–24.

Nolan, P. J. and P. J. Kennedy. 1953. Anomalous loss of condensation nuclei in rubber tubing. *J. Atmos. Terrestrial Phys.* 3:181–85.

Nolan, J. J. and P. J. Nolan. 1938. Diffusion and fall of atmospheric condensation nuclei. *Proc. Roy. Irish Academy* A45:47–63.

Okuyama, K., Y. Kousaka, and T. Motouchi. 1984. Condensational growth of ultrafine aerosol particles in a new particle size magnifier. *Aerosol Sci. Technol.* 3:353–66.

Pollak, L. M. and J. Daly. 1958. An improved model of the condensation nucleus counter with stereo-photomicrographic recording. *Geofis. Pura Applicata* 41:211–16.

Pollak, L. W. and A. L. Metnieks. 1959. New calibration of photo-electric nucleus counters. *Geofis. Pura Applicata* 43:285–301.

Pollak, L. W., T. C. O'Conner, and A. L. Metnieks. 1956. On the determination of the diffusion coefficient of condensation nuclei using the static and dynamic methods. *Geofis. Pura Applicata* 34:177–94.

Possanzini, M., A. Febo, and A. Aliberti. 1983. New design of a high-performance denuder for the sampling of atmospheric pollutants. *Atmos. Environ.* 17:2605–10.

Pui, D. Y. H., C. W. Lewis, C. J. Tsai, and B. Y. H. Liu. 1990. A compact coiled denuder for atmospheric sampling. *Environ. Sci. Technol.* 24:307–12.

Raabe, O. G. 1978. A general method for fitting size distributions to multi-component aerosol data using weighted least-squares. *Environ. Sci. Technol.* 12:1162–67.

Reineking, A. and J. Porstendörfer. 1986. High-volume screen diffusion batteries and α-spectroscopy for measurement of the radon daughter activity size distributions in the environment. *J. Aerosol Sci.* 17:873–80.

Rich, T. A. 1955. A photo-electric nucleus counter with size discrimination. *Geofis. Pura Applicata* 31:60–65.

Rich, T. A. 1961. A continuous recorder for condensation nuclei. *Geofis. Pura Applicata* 50:46–52.

Rich, T. A. 1966. Apparatus and method for measuring the size of aerosols. *J. Rech. Atmos.* 2:79–85.

Scheibel, H. G. and J. Porstendörfer. 1986. Counting efficiency and detection limit of condensation nuclei counters for submicrometer aerosols. I. Theoretical evaluation of the influence of heterogeneous nucleation and wall losses. *J. Colloid Interface Sci.* 109:261–74.

Scheibel, H. G. and J. Porstendörfer. 1984. Penetration measurements for tube and screen-type diffusion batteries in the ultrafine particle size range. *J. Aerosol Sci.* 15:673–82.

Shaw, R. W., R. K. Stevens, J. Bowermaster, J. W. Tesch, and E. Tew. 1982. Measurements of atmospheric nitrate and nitric acid; the denuder difference experiment. *Atmos. Environ.* 16:845–53.

Sideman, S., D. Luss, and R. E. Peck. 1965. Heat transfer in laminar flow in circular and flat conduits with (constant) surface resistance. *Appl. Sci. Res.* A14:157–71.

Sinclair, D. 1969. Measurement and production of submicron aerosols. In *Proc. 7th Conference on Condensation and Ice Nuclei*, pp. 132–37.

Sinclair, D. 1972. A portable diffusion battery. *Am. Ind. Hyg. Assoc. J.* 33:729–35.

Sinclair, D. 1984. Intrinsic calibration of the Pollak Counter-A revision. *Aerosol Sci. Technol.* 3:125–34.

Sinclair, D., R. J. Countess, B. Y. H. Liu, and D. Y. H. Pui. 1979. Automatic analysis of submicron aerosols. In *Aerosol Measurement*, eds. W. E. Clark and M. D. Durham, pp. 544–63. Gainsville, FL: University Press of Florida.

Sinclair, D. and L. Hinchliffe. 1972. Production and measurement of submicron aerosols. In *Assessment of Airborne Particles*, eds. T. T. Mercer et al., pp. 182–99. Springfiled, IL: Thomas.

Sinclair, D. and G. S. Hoopes. 1975*a*. A novel form of diffusion battery. *Am. Ind. Hyg. Assoc. J.* 36:39–42.

Sinclair, D. and G. S. Hoopes. 1975*b*. A continuous flow condensation nucleus counter. *J. Aerosol Sci.* 6:1–7.

Skala, G. F. 1963. A new instrument for the continuous measurement of condensation nuclei. *Anal. Chem.* 35:702–6.

Slanina, J., L. V. Lamoen-Doornebal, W. A. Lingerak, and W. Meilof. 1981. Application of a thermo-denuder analyzer to the determination of H_2SO_4, HNO_3 and NH_3 in air. *Int. J. Environ. Anal. Chem.* 9:59–70.

Soderholm, S. C. 1979. Analysis of diffusion battery data. *J. Aerosol Sci.* 10:163–75.

Stevens, R. K. 1986. Modern methods to measure air pollutants. In *Aerosols: Research, Risk Assessment and Control Strategies*, ed. S. D. Lee, pp. 69–95. MI: Lewis.

Stevens, R. K., T. G. Dzubay, G. Russwurm, and D. Rickel. 1978. Sampling and analysis of atmospheric sulfates and related species. *Atmos. Environ.* 12:55–68.

Stolzenburg, M. R. and P. H. McMurry. 1986. Counting efficiency of an ultrafine aerosol condensation nucleus counter. Theory and experiment. In *Aerosol: Formation and Reactivity*, pp. 786–89. New York: Pergamon.

Sutugin, A. G. and N. A. Fuchs. 1965. Coagulation rate of highly dispersed aerosols. *J. Colloid Sci.* 20:492–500.

Tan, C. W. and C. J. Hsu. 1971. Diffusion of aerosols in laminar flow in a cylindrical tube. *J. Aerosol Sci.* 2:117–24.

Tan, C. W. and J. W. Thomas. 1972. Aerosol penetration through a parallel-plate diffusion battery. *J. Aerosol Sci.* 3:39–43.

Tang, I. K. 1976. Phase transformation and growth of aerosol particles composed of mixed salts. *J. Aerosol Sci.* 7:361–72.

Thomas, J. W. 1955. The diffusion battery method for aerosol particle size determination. *J. Colloid Sci.* 10:246–55.

Twomey, S. A. and R. A. Zalabsky. 1981. Multifilter technique for examination of the size distribution of the natural aerosol in the submicrometer size range. *Environ. Sci. Technol.* 15:177–84.

Twomey, S. 1975. Comparison of constrained linear inversion and an alternative nonlinear algorithm applied to the indirect estimation of particle size distribution. *J. Comput. Phys.* 18:188–200.

Wilson J. C. and E. D. Blackshear. 1983. The function and response of an improved stratospheric condensation nucleus counter. *J. Geophys. Res.* 88:6781–85.

Yeh, H. C., Y. S. Cheng, and M. M. Orman. 1982. Evaluation of various types of wire screens as diffusion battery cells. *J. Colloid Interface Sci.* 86:12–16.

Zhang, Z. Q. and B. Y. H. Liu. 1990. Dependence of the performance of TSI 3020 condensation nucleus counter on pressure, flow rate and temperature. *Aerosol Sci. Technol.* 13:493–504.

20

Electrodynamic Levitation of Particles

E. James Davis

Department of Chemical Engineering, BF-10
University of Washington
Seattle, WA, U.S.A.

INTRODUCTION

Numerous characteristics and properties of single aerosol particles can be measured by suspending the particle in one or more laser beams. Detection of the light scattered by an aerosol droplet provides information on the size and refractive index of the material, and Raman and fluorescence scattering from a microparticle or microdroplet provide chemical information. Small particles can be levitated stably by applying a variety of phenomena including electrostatic, electrodynamic, radiometric, magnetic, acoustic, and aerodynamic forces. Brandt (1989) briefly reviewed the principles of these and other levitation methods, but in this chapter the focus is on the use of electrodynamic levitation to measure basic aerosol particle properties. It should be pointed out, however, that aerosol particle levitation by light pressure is a useful technique for microparticle Raman spectroscopy (for example, see Thurn and Kiefer 1985).

Electrodynamic levitation is an outgrowth of the classical Millikan oil drop experiment, in which a dc electrical field was used to balance the gravitational force on a charged droplet. But there is an important difference between electrodynamic levitation and Millikan's (1911) electrostatic technique in that an ac electrical field is superposed on the dc field to provide radial stability. Stable levitation in a purely electrostatic field is not possible without the introduction of a radial electrical field to center the particle and some type of feedback control system to rebalance the particle when it is perturbed from the null point of the balance. By means of the ac field the electrodynamic balance provides strong restoring forces when perturbations from the null point arise, and this maintains the particle at the center of the device. The ac field exerts no time-average force (averaged over one cycle) on the particle, so the dc field is used to balance the gravitational force and any other vertical forces on the particle.

The range of operation of the electrodynamic balance is remarkable, for masses ranging from about 10 μg (10^{-5} g) to that of a single electron can be trapped. Electrodynamic trapping at radio frequencies is the basis for the 1989 Nobel Prize in Physics awarded to Paul and Dehmelt for their work on atomic spectroscopy. Many investigators have used the device to study aerosols since Davis and Ray (1980) introduced it to the aerosol community, for the instrument, coupled with light-scattering capabilities, is particularly well-adapted to the size range

0.05–50 μm. Because a microparticle can be held tenaciously in an electrodynamic balance it is possible to flow a gas around the levitated particle to study aerodynamic drag, convective heat and mass transfer, and gas/particle chemical reactions. The details of such measurements are examined here.

LEVITATION PRINCIPLES

To illustrate the primary function of either an electrostatic or electrodynamic balance let us consider a Millikan condenser consisting of two flat plates (electrodes) of infinite extent (no edge effects) separated by a distance $2z_0$. Let us suppose that the particle to be suspended is negatively charged with coulombic charge q and has mass m. The charge is the product of the number n of elementary charges and the charge e on the electron (where $e = -1.602 \times 10^{-19}$ C in the SI system or -4.803×10^{-10} statC or esu in the cgs system). If we apply a positive potential $+V_{dc}$ to the top electrode and negative potential $-V_{dc}$ to the bottom electrode, the uniform dc field E_{dc} is given by

$$E_{dc} = -\frac{\left[+V_{dc} - (-V_{dc})\right]}{2z_0} = -\frac{V_{dc}}{z_0} \tag{20-1}$$

Because this electrical field is uniform in the space between the electrodes, the particle can be suspended anywhere in that space provided that the electrostatic force ($F_e = qE_{dc}$) on it balances the gravitational force, assuming that there are no other vertical forces on the object. A force balance yields

$$-q\frac{V_{dc}}{z_0} = mg \tag{20-2}$$

where g is the gravitational acceleration constant. Thus, for constant charge, the voltage required to levitate the particle is proportional to the particle mass. This is the basis for using the device as an analytical balance for microparticles.

Example 20-1

A Millikan condenser equipped with temperature and humidity control is used to levitate a 1 μm diameter water droplet at 20°C and atmospheric pressure. If the electrode spacing is 2 cm and if the charge on the droplet is equivalent to 100 electrons, what potential difference between the upper and lower plates is required?

Answer. The charge on one electron is -1.602×10^{-19} C or -4.803×10^{-10} esu, so the droplet charge is

$$\begin{aligned} q &= -4.803 \times 10^{-10}(100) \\ &= -4.803 \times 10^{-8} \text{ esu} \end{aligned}$$

The droplet mass is

$$\begin{aligned} m &= \pi \rho_p d_p^3/6 \\ &= \pi(1.0 \text{ g/cm}^3)(1.0 \times 10^{-4} \text{ cm})^3/6 \\ &= 5.24 \times 10^{-13} \text{ g} \\ &= 0.524 \text{ pg} \end{aligned}$$

The required potential difference is

$$\begin{aligned} \Delta V_{dc} &= V_{dc} - (-V_{dc}) \\ &= 2V_{dc} = -2z_0 m_p g/q \\ &= -2(1.0 \text{ cm})(5.24 \times 10^{-12} \text{g}) \\ &\quad \times (981 \text{ cm/s}^2)/(-4.803 \\ &\quad \times 10^{-8} \text{ esu}) \\ &= 2.14 \times 10^{-2} \text{ statV} \\ &= 2.14 \times 10^{-2} \text{ statV} \\ &\quad \times (299.8 \text{ V/statV}) \\ &= 6.41 \text{ V} \end{aligned}$$

Various modifications to Millikan's electrostatic balance have been introduced to control the position of the particle. One commercial version is the Science Spectrum Differential II Spectrometer (Wyatt and Phillips 1972). The dc field is generated by an electrified pin and a lower plate, which are mounted in a light-scattering chamber. The vertical position of the suspended particle is controlled by means of an electro-optic feedback control system. With this electrode configuration there exists a "potential well" which provides a weak radial restoring force when the particle drifts from the vertical axis of the chamber. The device has been used primarily for light-scattering measurements, for it is not well-suited to force measurements. The electrodynamic balance has gained a wider acceptance than electrostatic levitators.

The electrodynamic balance provides a strong lateral restoring force by means of the additional ac field. A frequently used electrodynamic balance design has the bihyperboloidal electrode configuration introduced by Wuerker, Langmuir, and Shelton (1959), and Fig. 20-1 shows one of these bihyperboloidal devices in the author's laboratory (Davis 1987). The dc electric field is generated by applying dc potentials to the upper and lower endcaps, and an ac potential V_{ac} is applied to the ring electrode. With this configuration the dc field is not uniform, but the dc field at the center of the chamber ($r = 0$ and $z = 0$) may be approximated by

$$E_{dc} = -C_0 \frac{V_{dc}}{z_0} \tag{20-3}$$

Here C_0 is a geometrical constant, the balance constant, which depends on the shapes of the electrodes and is affected by the existence of holes drilled through the electrodes for optical ports or for gas flow. Several investigators have computed C_0 for a variety of configurations (Philip, Gelbard, and Arnold 1983; Davis 1985; Sloane and Elmoursi 1987), and Davis (1985) showed how it can be determined experimentally. For the configuration of Fig. 20-1, $C_0 \approx 0.8$.

The balance shown in Fig. 20-1 is equipped with an electro-optic feedback controller to adjust the dc voltage as the particle changes mass and/or charge, and it is also designed for flow experiments. Metered gas enters through a hole in the bottom electrode to form a laminar jet in which the particle is suspended. The aerodynamic drag on the particle can be varied by adjusting the gas flow rate. Detectors of the scattered light are also mounted on the ring electrode to measure the optical properties of spherical particles. A 512-element linear photodiode array is used to record the intensity of the scattered light as a function of angle in the horizontal plane (the phase function), and a photomultiplier tube is used to record the "optical resonance spectrum" at a fixed angle. Phase functions and

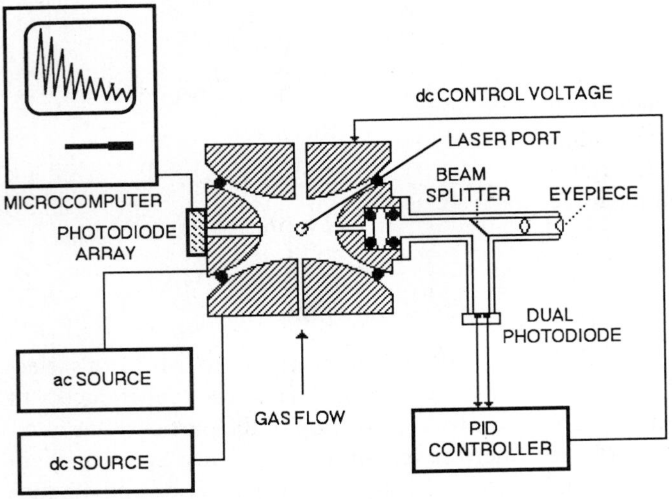

FIGURE 20-1. The Bihyperboloidal Electrodynamic Balance Configuration with Electro-optic Proportional/Integral/Differential (PID) Feedback Control. (Source: Davis 1987.)

resonance spectra are very sensitive to the size and refractive index of a dielectric sphere.

Electrode Configurations

The "quadrupole" shown in Fig. 20-1 is only one of many electrode configurations that have been used for electrodynamic levitation of aerosol particles. Figure 20-2 shows simpler designs that are equally effective. The earliest is that of Straubel (see Straubel and Straubel 1987) (Fig. 20-2b), which consists of a Millikan condenser with an ac disk inserted at the midplane between the dc endcaps. As shown in the figure, the lower electrode can be replaced by a conducting liquid pool. Weiss-Wrana (1983) introduced the double ac ring configuration so that scattered light can be detected in the horizontal plane, and Ray and Souyri (1989) simplified the device even further by superposing the ac and dc potentials on two parallel rings. The balance constant C_0 is different for each configuration, and Davis, Buehler, and Ward (1990) analyzed the electrical fields and developed scaling laws for determining C_0 for such devices. Richardson and Spann (1984) used a trap with hemispherical mesh dc electrodes and a cylindrical ac electrode, shown in Fig. 20-2a.

Stability Characteristics

The operating characteristics of the electrodynamic balance are governed by the equation of particle motion, particularly the vertical component of the vectorial equation of motion. The vertical forces can include aerodynamic drag, gravity, electrostatic and phoretic forces, and a force balance yields

$$m\frac{d^2z}{dt^2} = -K_d\left(\frac{dz}{dt} - U_g\right) + F_z - mg + qE_{dc,z} + qE_{ac,z} \tag{20-4}$$

where z is the vertical distance between the particle and the midplane of the balance, U_g is

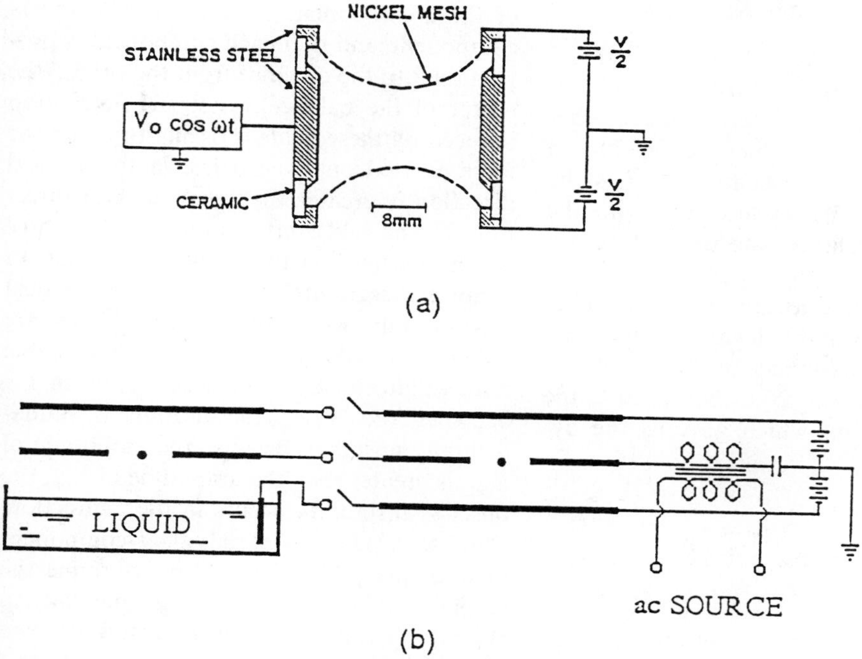

FIGURE 20-2. Electrodynamic Balance Configurations with (a) Hemispherical dc Electrodes (Source: Richardson and Spann 1984) and (b) Disk Electrodes (Source: Straubel and Straubel 1987.)

the uniform gas velocity upstream of the particle, F_z is any vertical force other than gravity, drag, and electrostatic, $E_{ac,z}$ and $E_{dc,z}$ are the vertical components of the ac and dc electrical field vectors, respectively, and K_d is an aerodynamic drag parameter. For Stokes flow around a sphere in the continuum regime K_d is given by

$$K_d = 3\pi d_p \eta \tag{20-5}$$

where d_p is the particle diameter and η is the gas viscosity.

The first criterion for stable particle trapping is that the dc electrical field must be adjusted to balance all time-invariant vertical forces (or at least balance all slowly varying forces). Thus, one must adjust the dc potential to satisfy the force balance

$$qE_{dc,z} = mg - K_d U_g - F_z \tag{20-6}$$

If $E_{dc,z}$ is given by Eq. 20-3, Eq. 20-6 reduces to

$$-C_0 q \frac{V_{dc}}{z_0} = mg - K_d U_g - F_z \tag{20-7}$$

In the absence of gas flow through the chamber and if there are no external forces other than gravity, Eq. 20-7 reduces to Eq. 20-2 in the case that $C_0 = 1$. Equation 20-7 is the basis of any force measurement involving the electrodynamic balance, and we shall return to it repeatedly.

Equation 20-7 is a necessary but not sufficient condition for stable levitation of a particle because the dynamics of the system are important. Using Eq. 20-6 in Eq. 20-4, the equation of motion which governs the dynamics becomes

$$m\frac{d^2z}{dt^2} = -K_d \frac{dz}{dt} + qE_{ac,z} \tag{20-8}$$

This result applies to any of the electrode configurations used for electrodynamic levitation, but before it can be applied to determine the operating parameters of the balance one must determine $E_{ac,z}$ for the particular configuration of interest. In general, $E_{ac,z}$ is a function of the radial and axial coordinates of the particle, and mathematical expressions for this component of the electrical field have been worked out for a few electrode configurations. In the absence of a dc bias voltage on the ring electrode of the bihyperboloidal configuration of Fig. 20-1, the radial and vertical components of the ac field are given by the following approximations valid to first-order in r and z:

$$E_{ac,r} = -\frac{V_{ac}}{z_0^2} r \cos \omega t \quad \text{and}$$

$$E_{ac,z} = \frac{2V_{ac}}{z_0^2} z \cos \omega t$$

where $V_{ac} \cos \omega t$ is the ac potential applied to the ring electrode, $\omega = 2\pi f$, and f is the frequency of the ac source. Here $2z_0$ is the minimum distance between the endcap electrodes. Note that the radial and vertical components are out of phase, that the amplitude of the z-component is twice that of the r-component, and that each component is proportional to the distance from the origin (the center of the balance). The particle motion induced by these fields is oscillatory, and the force exerted on the particle in the vertical direction is greater than that in the r-direction. The ac field exerts no time-average force on the particle, so the dc field is needed to balance the gravitational and other vertical forces. Only when the vertical forces are balanced by the dc field is there a stable nonoscillatory state. Furthermore, when V_{ac} becomes too large, the motion is fundamentally unstable. Because the amplitude of $E_{ac,z}$ is greater than the amplitude of $E_{ac,r}$ the onset of instability occurs in the z-direction. Thus, we need consider only the z-component of the equation of motion to determine the second stability criterion. Experimentally, a violent vertical oscillation is observed when V_{ac} is too large.

The solutions for $E_{ac,z}$ for various electrode configurations all lead to a stability equation of the form

$$m\frac{d^2z}{dt^2} + K_d\frac{dz}{dt} - \frac{qV_{ac}}{G}z\cos\omega t = 0 \qquad (20\text{-}9)$$

where G is a geometrical parameter which involves the size and shape of the ac electrode(s). For bihyperboloidal electrodes considered above, $G = z_0^2/2$. Davis, Buehler, and Ward (1990) derived expressions for G for double-ring balances, and Müller (1960) determined solutions for $E_{ac,z}$ for a variety of designs, including that of Straubel.

It is important to note that Eq. 20-9 does not have stable solutions for all values of ac amplitude V_{ac} and ac frequency f. The stability of the system is governed by an aerodynamic drag parameter $A = K_d/m\omega$ and an ac field strength parameter $B = 2qV_{ac}/m\omega^2G$. For the bihyperboloidal electrode configuration of Fig. 20-1, the field strength parameter is defined by

$$B = \frac{4qV_{ac}}{z_0^2 m\omega^2} \qquad (20\text{-}10)$$

Using an approximate solution technique to examine the stability of Eq. 20-9, Müller determined that a particle is unstable if it satisfies the following inequality:

$$B > \sqrt{0.8245 + 4A^2} \qquad (20\text{-}11)$$

More rigorous analyses by Frickel, Shaffer, and Stamatoff (1978) and by Davis (1985) demonstrate that there are multiple regions of stability. Table 20-1 lists the coordinates of the "marginal stability curve" for the first unstable region computed by the more rigorous approach, and Fig. 20-3 shows the first stability envelope and the lower boundary of the second envelope. The marginal stability curve is the boundary between stable and unstable regions. The figure indicates that a particle can be stably trapped when a point (A, B) lies outside of the unstable region delineated by the stability envelope shown. Not shown in the figure are other stability envelopes which lie above that plotted. Müller's approximation for the lower part of the stability envelope is found to be satisfactory for $A < 1$.

Note that if the field strength parameter B is too large, a particle cannot be trapped. But B can be decreased either by decreasing V_{ac} or by increasing the ac frequency. For that reason, it is desirable to provide both amplitude and frequency control of the ac field in the design of an electrodynamic balance. The stability characteristics of the device make it possible to reject particles of an undesired size by altering V_{ac} and f to drive them into unstable oscillation, which will cause them to

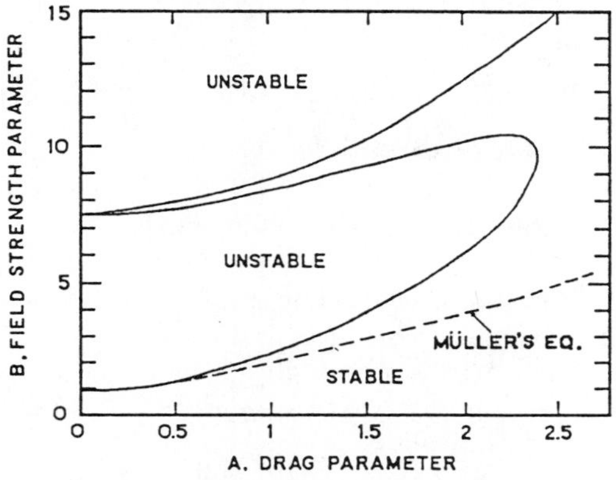

FIGURE 20-3. The First Stability Envelope for the Electrodynamic Balance and Müller's (1960) Approximation for the Stability Criterion.

TABLE 20-1 The First Two Marginal Stability Curves for the Electrodynamic Balance

Region I															
A	0.2	0.4	0.6	1.0	1.4	1.8	2.0	2.3	2.0	1.6	1.4	1.0	0.8	0.4	0.2
B	0.99	1.20	1.51	2.37	3.54	5.09	6.06	8.01	10.2	9.51	9.13	8.41	8.11	7.67	7.56
Region II															
A	0.6	1.0	1.4	1.8	2.0	2.4	2.6	3.0	3.4	3.8	4.09	4.0	3.8	3.3	3.2
B	8.03	8.82	9.97	11.5	12.4	14.4	15.6	18.3	21.6	25.6	28.4	32.5	32.5	30.8	30.4

hit the electrodes and be removed from the free space of the chamber.

Example 20-2

It is desired to levitate a 2.0 μm diameter polystyrene latex sphere ($\rho_p = 1.01$ g/cm^3) in a bihyperboloidal electrodynamic balance. The minimum distance between the endcap electrodes is $2z_0 = 1.0$ cm, and the ac source is a transformer which supplies an adjustable voltage at a fixed frequency of 10 kHz. Suppose that the particle is negatively charged with a charge equivalent to 1000 electrons. If the balance chamber contains air at 20°C and atmospheric pressure, what is the maximum ac voltage that can be applied to levitate the droplet? Furthermore, if the balance constant is given by $C_0 = 0.80$ and there is no air flow through the chamber and no phoretic forces acting on the particle, what dc voltage is required for stable suspension?

Answer. The particle mass is

$$m = \pi\rho_p d_p^3/6$$
$$= \pi(1.01\ \text{g/cm}^3)(2.0 \times 10^{-4}\ \text{cm})^3/6$$
$$= 4.23 \times 10^{-12}\ \text{g}$$

The drag parameter for the particle is

$$A = 3\pi d_p \eta/m\omega = 1.5 d_p \eta/mf$$
$$= 1.5(2.0 \times 10^{-4}\ \text{cm}) \times (1.813 \times 10^{-4}\ \text{g/cm s})/(4.23 \times 10^{-12}\ \text{g})(10^4\ \text{Hz})$$
$$= 1.286$$

From Fig. 20-3 or from Table 20-1, with $A = 1.286$, the point (A, B) lying on the stability envelope gives $B = 3.2$. Thus, if B is greater than 3.2, the particle will be unstable. From the definition of B the maximum ac voltage is

$$V_{ac} = Bm(z_0\omega)^2/4q = 3.2(4.23 \times 10^{-12}\text{g}) \times [(0.5\ \text{cm})(2\pi \times 10^4\ \text{Hz})]^2/(4)(10^3 \times 4.803 \times 10^{-10}\ \text{esu})$$
$$= 6.95 \times 10^3\ \text{statV}$$
$$= 6.95 \times 10^3\ \text{statV}(299.8\ \text{V/statV})$$
$$= 2.08 \times 10^6\ \text{V}$$

This high voltage indicates that the particle can be stably levitated at voltages easily attained in the laboratory. A typically suitable ac voltage is 1000 V.

Answer. The dc voltage required is obtained using Eq. 20-7 with $U_g = F_z = 0$. Thus,

$$V_{dc} = -mgz_0/C_0 q$$
$$= (4.23 \times 10^{-12}\text{g})(981\ \text{cm/s}^2)(0.5\ \text{cm})/(0.80)(10^3 \times 4.803 \times 10^{-10}\ \text{esu})$$
$$= 0.00540\ \text{statV}$$
$$= 0.00540\ \text{statV}(299.8\ \text{V/statV})$$
$$= 1.62\ \text{V}$$

MEASUREMENT TECHNIQUES

Equation 20-7 is the basis of charge, mass, and force measurements with the electrodynamic balance. For a sphere of known density the mass can also be determined by measuring the size using light-scattering techniques.

Mass and Charge

Changes in particle mass are readily monitored by recording the dc levitation voltage, provided that the charge remains constant. Ionizing radiation produces charge loss (Ward and Davis 1989) as does photoemission of electrons due to ultraviolet radiation (Arnold and Hessel 1985). In addition, droplet explosions occur when an evaporating charged droplet approaches the Rayleigh limit of charge (Taflin, Ward, and Davis 1989), and charge is lost when the droplet breaks up. At elevated temperatures ($T > 1000$ K) thermionic emission leads to catastrophic charge loss (Bar-Ziv et al. 1989). In the absence of these phenomena, however, a particle generally retains its charge while undergoing evaporation, condensation, and chemical reaction. In this event, with $U_g = F_z = 0$, Eq. 20-7 reduces to direct proportionality between V_{dc} and particle mass, that is,

$$m = -\frac{C_0 q}{g z_0} V_{dc} \tag{20-12}$$

For a sphere we may write $m = 4\pi a^3 \rho_p/3$, and then if the radius is measured by light scattering, Eq. 20-12 can be used to calculate the charge. Absolute mass measurement can be performed by the method of Arnold (1979), who used a UV light source to alter the charge on a particle in a controlled manner. Let V_n be the dc voltage required for levitating a particle of fixed mass with charge $q_n = en$ equivalent to n electrons. Here e is the charge on the electron. If UV illumination is used to emit one electron from the surface, then V_{n-1} will be the levitation voltage corresponding to charge state $q_{n-1} = (n-1)e$. Using Eq. 20-12 to relate the voltage and charge before and after electron emission, one obtains

$$q_n - q_{n-1} = ne - (n-1)e = e$$
$$= \frac{mgz_0}{C_0}\left(\frac{1}{V_{n-1}} - \frac{1}{V_n}\right) \tag{20-13}$$

Thus, from a knowledge of the geometrical constants e, g, C_0, and z_0 and from two measurements of the dc voltage, one can compute the absolute mass of the particle. Since it is possible for two or more charges to be emitted simultaneously, the experiment should be repeated to yield a series of steps when plotted as levitation voltage versus time. A step involving two electron emissions can then be clearly distinguished from a single emission.

Example 20-3

An electrodynamic balance with $z_0 = 1.25$ cm and $C_0 = 0.75$ is used to measure the absolute mass of a solid particle by the one-electron differential method. Prior to illumination with a UV source the dc levitation voltage was found to be 1200 V (4.003 statV) and after illumination for a short period the voltage increased to 1220 V (4.069 statV). What is the mass of the particle? What is the initial charge on the particle?

Answer. Using Eq. 20-13, one obtains

$$m = eC_0 V_n V_{n-1}/[gz_0(V_n - V_{n-1})]$$
$$= (-4.803 \times 10^{-10}\ \text{esu})(0.75)$$
$$\times (4.003\ \text{statV})(4.069\ \text{statV})/$$
$$[(981\ \text{cm/s}^2)(1.25\ \text{cm})$$
$$\times (4.003 - 4.069\ \text{statV})]$$
$$= 7.25 \times 10^{-11}\text{g} = 72.5\ \text{pg}$$

Answer. The charge on the particle is obtained from Eq. 21-12:

$$q_n = -mgz_0/V_n C_0$$
$$= -(7.25 \times 10^{-11}\ \text{g})(981\ \text{cm/s}^2)$$
$$\times (1.25\ \text{cm})/[(4.003\ \text{statV})(0.75)]$$
$$= -2.96 \times 10^{-8}\ \text{esu}$$

The number of elementary charges is

$$n = q_n/e = (-2.96 \times 10^{-8}\ \text{esu})/$$
$$(-4.803 \times 10^{-10}\ \text{esu/electron})$$
$$= 61.7 \approx 62$$

An alternate procedure is available for determining the charge on an evaporating droplet. Taflin, Ward, and Davis (1989) used such a procedure to measure the charge before and after droplet explosion, which occurs when the electrical stress produced by the increasing surface charge density balances surface tension forces. The method is based on a rearrangement of Eq. 20-7 for a sphere in the creeping flow continuum regime. In this case a gas flow is used to sweep vapor from the balance chamber, and the gas flow produces an aerodynamic drag on a slowly evaporating droplet. In the absence of phoretic and other such external forces, Eq. 20-7 may be written

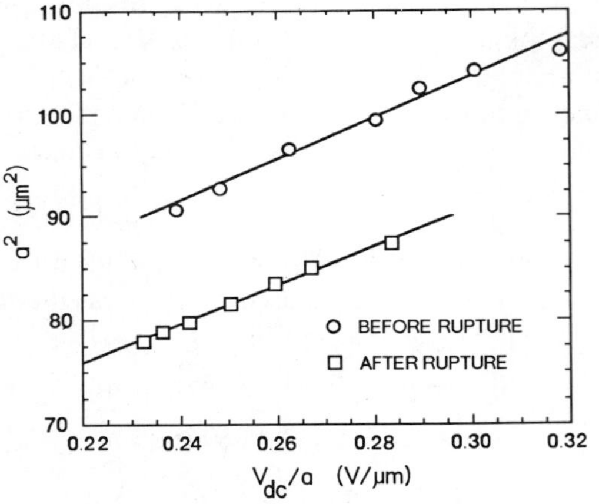

FIGURE 20-4. Evaporation Data for Dodecanol (Source: Bridges 1990.)

in the following form for a sphere with density ρ_p:

$$-C_0 q \frac{V_{dc}}{z_0} = \frac{4}{3}\pi a^3 \rho_p g - 6\pi a \eta U_g \quad (20\text{-}14)$$

This equation may be rearranged to yield

$$a^2 = -\frac{3C_0 q}{4\pi \rho_p g z_0}\frac{V_{dc}}{a} + \frac{9\eta U_g}{2\rho_p g} \quad (20\text{-}15)$$

Thus, a plot of a^2 versus V_{dc}/a for an evaporating droplet should yield a straight line with slope equal to $-3C_0 q/4\pi\rho_p g z_0$ and intercept equal to $9\eta U_g/2\rho_p g$. Figure 20-4 shows such data obtained in the author's laboratory by Bridges (1990) for a droplet of 1-dodecanol evaporating in air at 22°C and atmospheric pressure. The droplet radius was measured by the light-scattering methods described below. There is a slight change in slope subsequent to the droplet fission, indicating that charge was lost.

Example 20-4

For the electrodynamic balance used to obtain the data presented in Fig. 20-3, $C_0 = 1.0$ and $2z_0 = 3.08$ cm. The density of 1-dodecanol is 0.820 g/cm³. Calculate the charge on the droplet prior to droplet rupture using the data of Fig. 20-3.

Answer. Using the upper data in Fig. 20-3, the slope of the graph of a^2 versus V_{dc}/a is estimated to be 201 $\mu m^3/V$. In the cgs system this is

$$201\ \mu m^3/V = (201\ \mu m^3)(10^{-4}\ cm/\mu m)^3 \times (299.8\ V/statV) = 6.03 \times 10^{-8}\ cm^3/statV$$

From Eq. 21-15 the slope is

$$-3C_0 q/4\pi\rho_p g z_0 = -(3)(1.0)q/[4\pi(0.820\ g/cm^3) \times (981\ cm/s^2)(1.54\ cm)] = -1.93 \times 10^{-4} q$$

Solving for the droplet charge,

$$q = -6.03 \times 10^{-8}/1.93 \times 10^{-4} = -3.13 \times 10^{-4} statC$$

The number of elementary charges is

$$n = q/e = \frac{(-3.13 \times 10^{-4}\ statC)}{(-4.803 \times 10^{-10}\ statC/electron)} = 6.51 \times 10^5$$

Light-Scattering Size

In the last century Lord Rayleigh developed expressions for the intensity of light scattered by small spheres. Mie (1908) and Debye (1909) developed more general solutions of the Maxwell equations for electromagnetic scattering from a dielectric sphere, and those solutions reduce to Rayleigh's result when the sphere is small compared to the wavelength λ_0 of the incident light. Details of the so-called Mie theory are beyond the scope of this Chapter, but a cursory review of the results of the Mie theory are needed to interpret data.

Consider an electromagnetic wave propagating in the z-direction which impinges on a dielectric sphere with complex refractive index $N = n + ik$ and radius a. For a sphere which is transparent for wavelength λ_0, $k = 0$.

The theory shows that the intensity of the scattered light, observed in the horizontal plane, is a function of scattering angle θ, of the ratio $m = n/n_0$, where n_0 is the refractive index of the surrounding medium (essentially unity for air), and of the nondimensional light-scattering size $\alpha = 2\pi a/\lambda_0$. The polar angle θ is measured from the z-axis. The scattered intensity also depends on the polarization of the incident beam.

Several types of experiments can be used to determine the light-scattering size of a sphere, and these include a measurement of (i) the polarization ratio, (ii) the phase function, and (iii) the optical resonance spectrum. The first method involves a measurement of the scattered light at some fixed angle θ using two polarization states of the incident beam, usually vertically polarized and horizontally polarized light. The ratio of the measured intensities is sensitive to size and refractive index, and the Mie theory is applied to interpret the results. The so-called phase function is a measurement of the scattered light as a function of scattering angle, and its structure depends on α and m. Optical resonances or morphological resonances, which are measured at a fixed angle as α varies, are extremely sensitive to size and refractive index. Physically, morphological resonances can be interpreted in terms of constructive and destructive interferences of internal waves and, mathematically, they correspond to poles in the mathematical equations of the Mie theory. Figure 20-5 is a schematic diagram of the apparatus used by Bridges (1990)

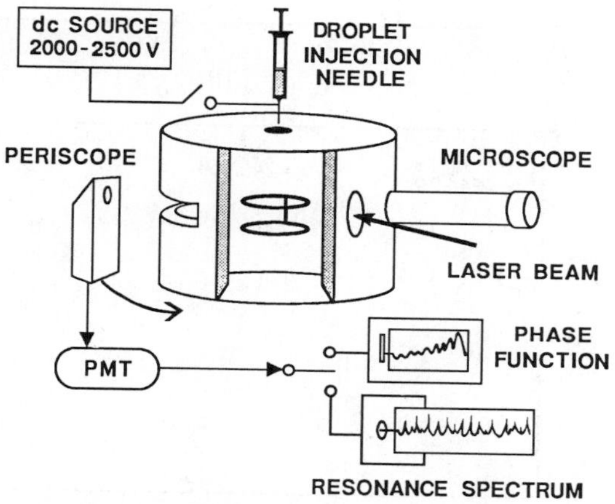

FIGURE 20-5. The Apparatus used for Droplet Evaporation Experiments in the Author's Laboratory. (Source: Bridges 1990.)

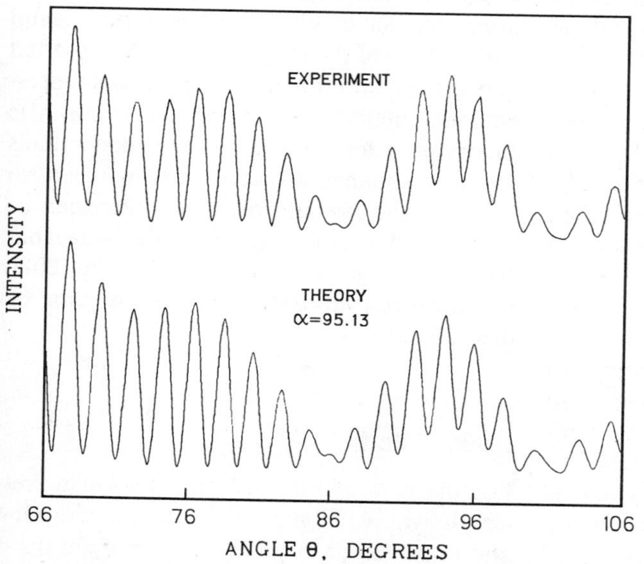

FIGURE 20-6. A Comparison Between Experimental and Theoretical Phase Functions for a Dioctyl Phthalate Droplet. (Source: Ray et al. 1990.)

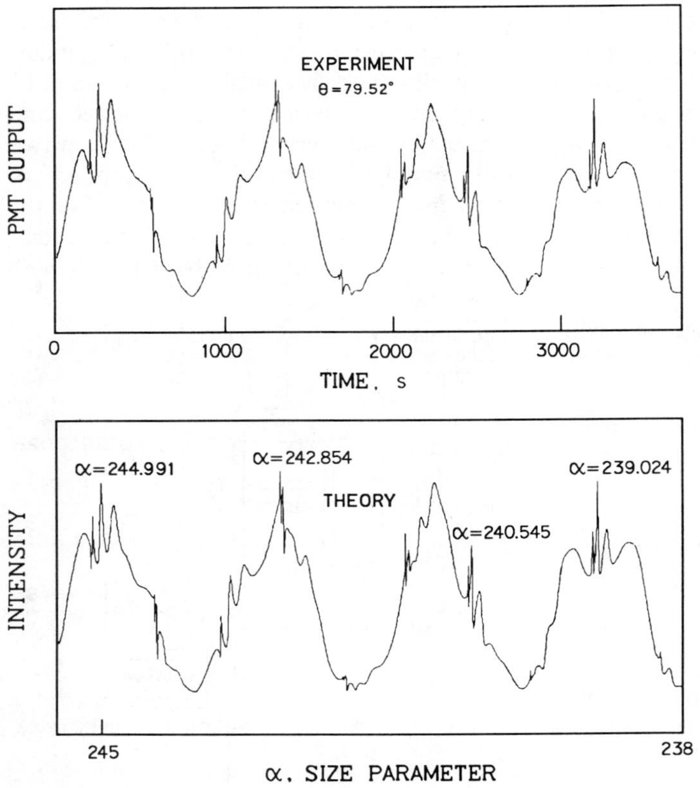

FIGURE 20-7. A Comparison Between Experimental and Theoretical Resonance Spectra for a Glycerol Droplet Evaporating in Air at 25°C. (Source: Ray et al. 1991.)

to record phase functions and optical resonance spectra. To record the former signal, he used a rotating periscope connected to a photomultiplier tube (PMT) and an x–y recorder. The periscope was driven by a stepper motor, and the voltage pulse used to drive the stepper motor served as a measure of the scattering angle. The PMT signal provided the intensity data. With the periscope set at a fixed angle, the optical resonance spectrum was recorded on an oscillograph.

Ray et al. (1991) used a similar apparatus to record the phase function for a droplet of dioctyl phthalate presented in Fig. 20-6. Also shown in the figure is phase function computed from the Mie theory for $\alpha = 95.13$ ($a =$ μm) and $m = 1.4860$. The agreement is excellent and permits one to determine the size to two parts in 10^4. Even greater precision is attainable using resonance spectra, and Fig. 20-7 displays experimental and theoretical spectra obtained for a slowly evaporating droplet of glycerol. The experimental data consist of the output of the PMT plotted versus time, and the theoretical results represent the intensity computed from the Mie theory as a function of α. The finest details of the resonance structure are matched so well that the size and refractive index can be determined to one part in 10^5. Note that by comparing the two spectra in Fig. 20-7, the droplet evaporation rate da/dt can be determined.

Example 20-5

For the data of Fig. 20-7 the incident wavelength was 632.8 nm. Estimate the evaporation rate (da/dt in μm/s) of the droplet during the time period covered by the experimental data ($0 \leq t \leq 3700$ s).

Answer. The peaks identified by $\alpha_1 = 244.991$ and $\alpha_4 = 239.024$ correspond to droplet radii given by

$$a_1 = \alpha_1 \lambda_0/2\pi$$
$$= (244.991)(632.8 \text{ nm})/2\pi$$
$$= 24674 \text{ nm} = 24.674 \ \mu\text{m}$$

and

$$a_4 = \alpha_4 \lambda_0/2\pi$$
$$= (239.024)(632.8 \text{ nm})/2\pi$$
$$= 24073 \text{ nm} = 24.073 \ \mu\text{m}$$

The time increment Δt corresponding to these two peaks is estimated from the experimental data to be

$$\Delta t = 2962 \text{ s}$$

Thus,

$$da/dt \approx (a_4 - a_1)/\Delta t$$
$$= (24.073 - 24.674 \ \mu\text{m})/(2962 \text{ s})$$
$$= -2.029 \times 10^{-4} \ \mu\text{m/s}$$

Drag and other Forces

Equation 20-7 applies to the measurement of the aerodynamic drag force or any other vertical force acting on a levitated particle. The drag force measurement is accomplished either by means of a laminar jet formed by introducing the gas through a hole in the bottom electrode (Davis et al. 1987) or by using flow through mesh electrodes (Allen, Buehler, and Davis 1990). Let V_0 be the dc voltage required for levitation when only gravitational and electrostatic forces are applied. In this event Eq. 20-7 may be rearranged to yield

$$-\frac{C_0 q}{z_0} = \frac{mg}{V_0} \qquad (20\text{-}16)$$

When a drag force is exerted on the particle, Eq. 20-7 becomes

$$-C_0 q \frac{V_{dc}}{z_0} = mg - F_d \qquad (20\text{-}17)$$

where F_d is the drag force. Thus, combining Eqs. 20-16 and 20-17, one obtains a simple relationship between the two voltages and the ratio of the drag force to the particle weight:

$$\frac{F_d}{mg} = 1 - \frac{V_{dc}}{V_0} = \frac{\Delta V}{V_0} \qquad (20\text{-}18)$$

where $\Delta V = V_0 - V_{dc}$.

If the drag force is varied by changing the upward gas flow through the balance chamber, one can balance the gravitational force with the drag force, at which point $V_{dc} = 0$. This principle can be used to calibrate the flow system of the balance to determine the velocity as a function of flow rate.

Example 20-6

A spherical silicate particle with a diameter of 12.5 μm is trapped in an electrodynamic balance of the type shown in Fig. 20-1. The silicate particle has a density of 2.86 g/cm³. Nitrogen at 20°C and atmospheric pressure ($\eta = 1.75 \times 10^{-4}$ g/cm s and $\rho_g = 1.16 \times 10^{-3}$ g/cm³) is introduced into the chamber and flows out through a hole in the top electrode. The flow rate is adjusted until no dc voltage is required to suspend the particle. What is the gas velocity "seen" by the particle when the particle weight is balanced by the drag force?

Answer. Assuming Stokes flow, which must be verified *a posteriori*, we may equate the drag force and the weight as follows:

$$F_d = 6\pi a \eta U_g = mg = 4\pi a^3 \rho_p g/3$$

Thus, solving for the gas velocity,

$$U_g = 2a^2 \rho_p g/9\eta$$
$$= \frac{2(6.25 \times 10^{-4} \text{ cm})^2 (2.86 \text{ g/cm}^3)(981 \text{ cm/s}^2)}{[9(1.75 \times 10^{-4} \text{ g/cm s})]}$$
$$= 1.39 \text{ cm/s}$$

This calculated velocity is based on the assumption of Stokes flow around the particle, which is valid for $Re_f < 0.1$, where the Reynolds number is defined by $Re_f = d_p \rho_g U_g / \eta$. We must verify that the Reynolds number is small. Thus,

$$Re_f = 2(6.25 \times 10^{-4}\ \text{cm}) \times (1.16 \times 10^{-3}\ \text{g/cm}^3)(1.39\ \text{cm/s}) / (1.75 \times 10^{-4}\ \text{g/cm s}) = 0.0115$$

Since $Re_f < 0.1$, the use of the Stokes law is valid here.

Any other vertical force can be measured in a similar fashion. For example, Lin and Campillo (1985) obtained the photophoretic force on ammonium sulfate crystalline particles in this way, and Fig. 20-8 shows their results plotted as $\Delta V/V_0$ versus the chamber pressure.

EVAPORATION/CONDENSATION

Many investigators have used electrostatic and electrodynamic balances for evaporation and condensation measurements in the continuum ($Kn \ll 1$) and the free-molecule ($Kn \gg 1$) regimes. Here the Knudsen number Kn is the ratio of the mean free path of the gas-phase molecules to the radius of the particle. In the continuum regime, isothermal diffusion-controlled quasi-steady-state evaporation or condensation processes for a sphere are described by the equation derived by Maxwell:

$$\frac{da}{dt} = -\frac{D_{ij}}{\rho_p a}(C_a - C_\infty) \tag{20-19}$$

where D_{ij} is the diffusivity for vapor i in the surrounding gas j, C_a is the vapor concentration in equilibrium with the particle surface (at $r = a$), and C_∞ is the vapor concentration far from the interface. For an ideal vapor-gas mixture we may write the concentration in terms of the partial pressure of the vapor p_i:

$$C_i = \frac{M_i p_i}{RT} \tag{20-20}$$

where M_i is the molecular weight of the

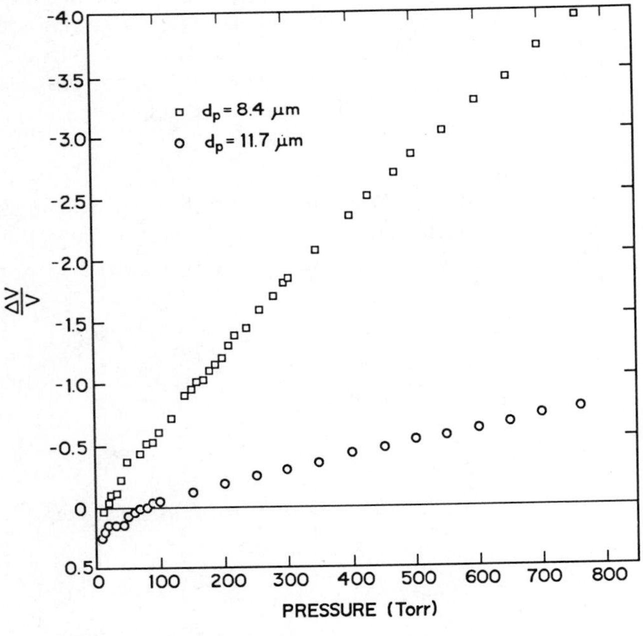

FIGURE 20-8. Photophoretic Force Data for Two Different Particles of Crystalline Ammonium Sulfate. (Source: Lin and Campillo 1985.)

vapor, R is the ideal gas constant, and T is the temperature at the radial position in question.

In the free-molecule regime, da/dt for evaporation into a vapor-free space may be written as

$$\frac{da}{dt} = -\varepsilon\frac{c_i^* M_i p_i^0}{4\rho_p RT} \qquad (20\text{-}21)$$

where ε is the evaporation coefficient, p_i^0 is the vapor pressure of the vapor at the surface temperature, and c_i^* is the average velocity of the vapor molecules, given by

$$c_i^* = \sqrt{\frac{8RT}{\pi M_i}} \qquad (20\text{-}22)$$

A similar equation applies for condensation if we replace the vapor pressure by the partial pressure far from the interface and replace ε by a condensation coefficient. The intermediate or Knudsen aerosol regime is more difficult to describe (see Hidy and Brock 1970), and a variety of semi-theoretical equations and theoretical solutions of the Boltzmann transport equation have been proposed.

For low-volatility single-component droplets at atmospheric pressure, evaporation proceeds at very nearly the temperature of the surrounding gas. But for water droplets and other relatively high vapor pressure chemicals, the interfacial temperature is less than the surrounding temperature because heat must be transferred to the interface at a rate sufficient to provide the latent heat of vaporization. This results in a radial temperature gradient that is not negligible (Taflin et al. 1988). Another complication which arises when rapid evaporation occurs is the Stefan flow (see Hidy and Brock 1970). We shall focus on low-vapor-pressure species here.

Diffusivities and Vapor Pressures

If a low-vapor-pressure-levitated droplet evaporates into a surrounding gas containing no accumulated vapor ($C_\infty = 0$), Eq. 20-19 may be integrated, using Eq. 20-20, to yield

$$a^2 = a_0^2 - \frac{2M_i D_{ij} p_i^0(T_a)}{\rho_p RT_a}(t - t_0) \qquad (20\text{-}23)$$

where a_0 is the droplet radius at time t_0, and T_a is the interfacial temperature. This equation indicates that if the square of the radius is plotted versus time, the data should fall on a straight line with slope S_{ij} given by $S_{ij} = 2M_i D_{ij} p_i^0(T_a)/\rho_p RT_a$. If the vapor pressure is known, the diffusion coefficient can be determined from the measured slope; conversely, if the diffusion coefficient is known, the vapor pressure can be determined. Davis and Ray (1977) and Ray, Davis, and Ravindran (1979) showed how both the vapor pressure and diffusivity can be determined by obtaining droplet evaporation rate data at several different temperatures.

Example 20-7

The glycerol evaporation data presented in Fig. 20-7 were obtained for a glycerol droplet levitated in air at 25°C and at atmospheric pressure. Calculate the slope S_{ij} of the data considered as a graph of a^2 versus time. Using the method of Fuller, Schettler, and Giddings (1966) for estimating gas-phase diffusivities, the diffusion coefficient for glycerol vapor in air at the conditions of the experiment is calculated to be 9.53×10^{-2} cm²/s. For glycerol, $M_i = 92.11$ and $\rho_p = 1.2613$ g/cm³. Estimate the vapor pressure of glycerol at 25°C based on the data of Fig. 20-7.

Answer. Using the data of Example 20-5 for the time interval $\Delta t = 2962$ s, we have $a_1^2 = (24.674)^2 = 608.80\ \mu m^2$ and $a_4^2 = (24.073)^2 = 579.51\ \mu m^2$. Thus, the slope S_{ij} for this time increment is

$$\begin{aligned} S_{ij} &= (a_4^2 - a_1^2)/\Delta t \\ &= (579.51 - 608.80)/2962 \\ &= -0.009889\ \mu m^2/s \\ &= (-0.009889\ \mu m^2/s) \\ &\quad \times (10^{-4}\ cm/\mu m)^2 \\ &= -9.889 \times 10^{-11}\ cm^2/s \end{aligned}$$

Answer. From the definition of S_{ij} and the properties of glycerol, one obtains

$$S_{ij} = -2M_iD_{ij}p_i^0(T_a)/\rho_pRT_a$$

$$= \frac{-2(92.11\ \text{g/mol})(9.53\times10^{-2}\ \text{cm}^2/\text{s})p_i^0(T_a)}{(1.2613\ \text{g/cm}^3)(8.314\times10^7\ \text{dyn cm/mol K})(298\ \text{K})}$$

$$= -5.618\times10^{-10}p_i^0(T_a)\ \text{cm}^4/\text{dyn s}$$

Equating this result with the slope determined from the graphical data, we have

$$p_i^0(T_a) = \frac{(-9.889\times10^{-11}\ \text{cm}^2/\text{s})}{(-5.618\times10^{-10}\ \text{cm}^4/\text{dyn s})}$$

$$= 0.176\ \text{dyn/cm}^2$$

$$= (0.176\ \text{dyn/cm}^2)(0.1\ \text{dyn/cm}^2\ \text{Pa})$$

$$= 0.0176\ \text{Pa}$$

This is in reasonable agreement with data reported in the literature. Smith and Srivastava (1986) reported a value of 0.0198 Pa for glycerol at 297 K, and Ross and Heideger's (1962) data yielded 0.0227 Pa at 297 K. These comparisons suggest that the estimated diffusivity is too high.

Accommodation Coefficient

Evaporation and condensation measurements at low pressures can be used to determine evaporation and condensation coefficients and the thermal accommodation coefficient. Chang and Davis (1976) used an electrostatic balance to study evaporation in the Knudsen regime, Ray, Lee, and Tilley (1988) measured evaporation rates at large Knudsen numbers with an electrodynamic balance, and Sageev et al. (1986) measured water condensation on aqueous solution droplets in the Knudsen regime by electrodynamic levitation. Rubel and Gentry (1984*a*) used electrodynamic suspension to measure the evaporation coefficient for water evaporating from a phosphoric acid droplet in the presence of hexadecanol vapor. The hexadecanol acted as a surfactant which could suppress evaporation.

Sageev and coworkers heated a droplet with an infrared pulse and then followed the relaxation of the droplet via light scattering as it reequilibrated by condensation of water on it. In principle, this dynamic technique yields the thermal accommodation coefficient α_T and the condensation coefficient β_M, but the investigators found the results to be insensitive to β_M. For the condensation of water on $(NH_4)_2SO_4$ solution droplets, they found the results to be consistent with $\alpha_T = 1.0$. The evaporation rate data of Chang and Davis and of Ray and coworkers for low-volatility dioctyl phthalate are consistent with an evaporation coefficient of unity. Rubel and Gentry showed that the evaporation coefficient is near unity until a monolayer is formed on the droplet surface, and then the evaporation rate falls dramatically. Taflin et al. (1988) saw the same effect on the evaporation of sodium dodecyl sulfate solutions at atmospheric pressure.

A direct measurement of da/dt at low pressures and the application of Eq. 20-21 permits one to determine the evaporation coefficient, provided that the vapor pressure is known. For low-volatility chemicals this is a relatively straightforward process, but if rapid evaporation occurs, the issue of thermal accommodation arises. The latent heat required for the evaporation process must be provided by energy transfer from the surrounding gas molecules.

Example 20-8

A dioctyl phthalate (DOP) droplet was evaporated in a chamber at 298.8 K. To operate in the free-molecule regime, the pressure was maintained at 0.1 mmHg by initially evacuating the chamber and then back-filling with nitrogen to operate at 0.1 mmHg. The molecular weight of DOP is 390.6 and its density is 0.981 g/cm^3. From light-scattering measurements it was found that $da/dt = -0.00510\ \mu\text{m/min}$ $(-8.50\times10^{-9}\text{cm/s})$. From an equation, proposed by Ray, Davis, and Ravindran (1979), the vapor pressure of DOP at 298.8 K is estimated to be 1.76 $\times 10^{-7}$ mmHg (2.35×10^{-4} dyn/cm^2). Calculate the evaporation coefficient for DOP from this limited data.

Answer. The average velocity of DOP molecules is

$$c_i^* = [8(8.314 \times 10^7 \text{ erg/mol K})(298.8 \text{ K})/\pi(390.6 \text{ g/mol})]^{1/2}$$
$$= 12{,}730 \text{ cm/s}$$

Solving for ε, using Eq. 21-21,

$$\varepsilon = -4(da/dt)\rho_p RT/c_i^* M_i p_i^0$$
$$= 4(8.50 \times 10^{-9} \text{ cm/s})(0.981 \text{ g/cm}^3) \times (8.314 \times 10^7 \text{ erg/mol K})(298.8 \text{ K})/(12{,}730 \text{ cm/s})(390.6 \text{ g/mol}) \times (2.35 \times 10^{-4} \text{ dyn/cm}^2)$$
$$= 0.709$$

Thus, one may state that ε is of order unity, based on this calculation for a single data point.

CHEMICAL REACTIONS

Electrodynamic suspension in a flow field makes it possible to use the device to study chemical reactions between a reactive vapor or gas and the particle. Rubel and Gentry (1984*b*) performed such experiments using a phosphoric acid droplet levitated in an NH_3/air flow. They followed the reaction by the weight changes in the particle. Taflin and Davis (1990) used electrodynamic levitation to study the reaction between bromine vapor and a droplet of 1-octadecene, using both weight changes and optical resonance data to measure the reaction rate. Since morphological resonances are extremely sensitive to the refractive index of the dielectric sphere, changes in the refractive index can be determined by recording the resonance spectrum during chemical reaction. Figure 20-9 shows the resonance spectrum and the corresponding dc levitation voltage obtained by Taflin (1988) for a droplet of 1-octadecene reacting with bromine vapor at 22°C. There is a signficant change in weight, as indicated by the voltage increase, and the resonance structure indicates that, initially, there was a rapid uptake of bromine, followed by a slowing of the absorption and reaction rate as the irreversible reaction neared completion. The refractive index of 1-octadecene is 1.4448 and that of the product, dibromooctadecane, is 1.4805, so the change in refractive index could be extracted from the resonance spectrum. Table 20-2 lists the droplet suspension voltage as a function of time for the data presented in Fig. 20-9.

A photochemically activated polymerization of a levitated aerosol particle was examined by Ward et al. (1987). This was accomplished by adding a trace amount of a fluorophore to acrylamide monomer, activating the polymerization with UV light, and following the change in fluorescence as polymerization proceeded. The fluorophore added, auramine-O, has the property that it

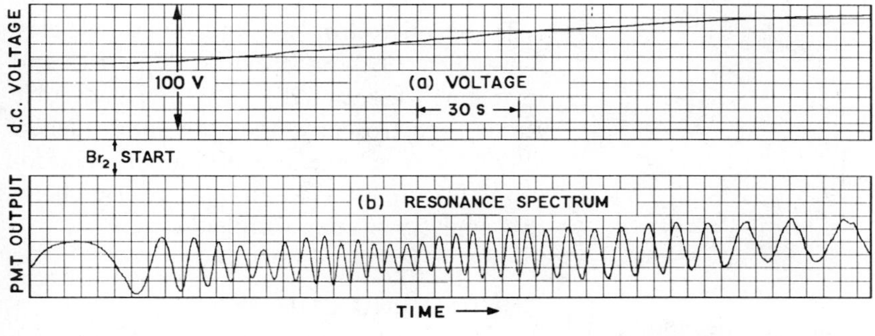

FIGURE 20-9. Oscillograph Tracings for the Chemical Reaction Between Bromine Vapor and a Droplet of 1-octadecene Showing (a) the dc Voltage as a Function of Time and (b) the Morphological Resonance Spectrum. (Source: Taflin 1988.)

TABLE 20-2 The Levitation Voltage as a Function of Time for the Bromination of a Droplet of 1-octadecane

Time(s)	0	30	60	90	120	150	180	210	∞
dc voltage	55.0	58.1	63.6	72.0	80.3	83.0	87.6	90.5	91.0

Source. Taflin 1988

does not fluoresce in low-viscosity media, but as the viscosity increases fluorescence increases.

Weight change measurements are quite adequate for chemical reactions which involve signficant changes in molecular weight, but when there is little change in molecular weight due to chemical reaction, the gravimetric method is not accurate. For the polymerization studied by Ward and his coworkers no weight change occurred apart from some evaporation of the monomer. Because of the large change in the molecular weight associated with the bromination of 1-octadecene, electrodynamic weight change measurements are quite adequate for studying this reaction.

Example 20-9

Using the data in Table 20-2, calculate the fractional conversion of 1-octadecene ($M_{OCT} = 252.49$) to dibromooctadecane ($M_{DBO} = 412.30$) 120 s after the start of bromine addition.

Answer. The reaction is

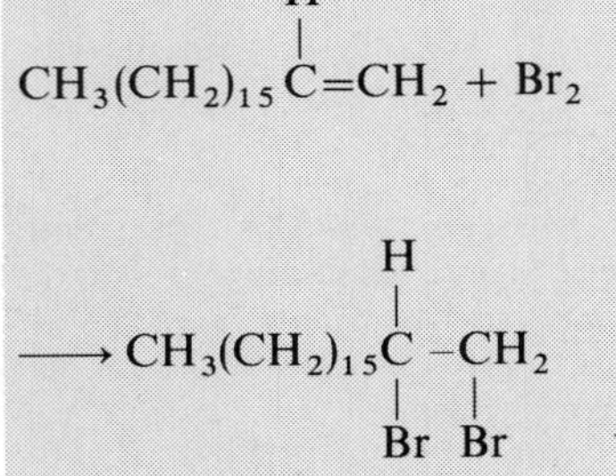

Thus, one mole of 1-octadecene reacts with one mole of bromine to form one mole of dibromooctadecane. Since the dc voltage is directly proportional to the droplet mass, the ratio of the initial mass m_0 to the mass at the completion of the reaction m_∞ should equal the molecular weight ratio. Comparing these ratios, we have

$$M_{DBO}/M_{OCT} = 412.30/252.49 = 1.633 \quad \text{and}$$
$$m_\infty/m_0 = V_\infty/V_0 = 91.0/55.0 = 1.655$$

These ratios are in reasonably good agreement. At $t = 120$ s, the dc voltage is 80.3 V, so the fractional conversion to the product is given by

$$\begin{aligned} x &= (m - m_0)/(m_\infty - m_0) \\ &= (V - V_0)/(V_\infty - V_0) \\ &= (80.3 - 55.0)/(91.0 - 55.0) \\ &= 0.703 \end{aligned}$$

or 70.3% conversion.

Although refractive index changes can be detected with high precision by means of morphological resonance measurements, it is very difficult to interpret resonance data if both the droplet size and the refractive index change. Furthermore, the refractive index is not a unique function of composition when the droplet contains more than two components. These problems have led to the use of Raman spectroscopy for the chemical analysis of microdroplets and microparticles. Schweiger (1990) reviewed the principles and problems associated with microparticle Raman spectroscopy, and Davis and Buehler (1990) explored the use of Raman spectroscopy for qualitative and quantitative chemical analysis of microdroplets.

CONCLUDING COMMENTS

This review has touched on only a few applications of aerosol particle levitation. Many other types of experiments have been carried out with the technique, including solution thermodynamics studies, infrared and photophoretic spectroscopic analyses, thermogravimetric analysis of char particles, and a variety of photothermal and radiometric measurements. New uses appear continually.

References

Allen, T. M., M. F. Buehler, and E. J. Davis. 1990. Radiometric effects on absorbing microspheres. *J. Colloid Interface Sci.* 145:343–56.

Arnold, S. 1979. Determination of particle mass and charge by one electron differentials. *J. Aerosol Sci.* 10:49–53.

Arnold, S. and N. Hessel. 1985. Photoemission from single electrodynamically levitated microparticles. *Rev. Sci. Instrum.* 56:2066–69.

Bar-Ziv, E., D. B. Jones, R. E. Spjut, D. R. Dudek, A. F. Sarofim, and J. P. Longwell. 1989. Measurement of combustion kinetics of a single char particle in an electrodynamic thermogravimetric analyzer. *Combustion Flame* 75:81–106.

Brandt, E. H. 1989. Levitation in physics. *Science* 243:349–55.

Bridges, M. A. 1990. Measurement of surface tension using an electrodynamic balance. M.S. Thesis, University of Washington.

Chang, R. and E. J. Davis. 1976. Knudsen aerosol evaporation. *J. Colloid Interface Sci.* 54:352–63.

Davis, E. J. 1985. Electrodynamic balance stability characteristics and applications to the study of aerocolloidal particles. *Langmuir* 1:379–87.

Davis, E. J. 1987. The picobalance for single microparticle measurements. *ISA Trans.* 26:1–5.

Davis, E. J. and M. F. Buehler. 1990. Chemical reactions with single microparticles. *MRS Bulletin* 15:26–33.

Davis, E. J., M. F. Buehler, and T. L. Ward. 1990. The double-ring electrodynamic balance for microparticle characterization. *Rev. Sci. Instrum.* 61:1281–88.

Davis, E. J. and A. K. Ray. 1977. Determination of diffusion coefficients by submicron droplet evaporation. *J. Chem. Phys.* 67:414–19.

Davis, E. J., S. H. Zhang, J. H. Fulton, and R. Periasamy. 1987. Measurement of the aerodynamic drag force on single aerosol particles. *Aerosol Sci. Technol.* 6:273–87.

Debye, P. 1909. Light pressure on spheres of any material. *Ann. Physik* 30:57–136.

Frickel, R. H., R. E. Shaffer, and J. B. Stamatoff. 1978. Chambers for the electrodynamic containment of charged aerosol particles, Report No. ARCSL-TR-77041, Chemical Systems Laboratory, Aberdeen Proving Ground, MD.

Fuller, E. N., P. D. Schettler, and J. C. Giddings. 1966. A new method for prediction of binary gas-phase diffusion coefficients. *Ind. Engng. Chem.* 58:19–27.

Hidy, G. M. and J. R. Brock. 1970. *The Dynamics of Aerocolloidal Systems.* New York: Pergamon.

Lin, H.-B. and A. J. Campillo. 1985. Photothermal aerosol absorption spectroscopy. *Appl. Opt.* 24:422–33.

Mie, G. 1908. Beiträge zur optik trüber medien speziell kolloidaler metallösungen. *Ann. Physik* 25:377–445.

Millikan, R. A. 1911. Isolation of an ion, a precision measurement of its charge, and the correction of Stokes's law. *Phys. Rev. Series 1* 32:349–97.

Müller, A. Von. 1960. Theoretische untersuchungen über das verhalten geladener teilchen in sattelpunkten electrischer wechselfelder. *Ann. Physik* 6:206–12.

Philip, M. A., F. Gelbard, and S. Arnold. 1983. An absolute method for aerosol particle mass and charge measurement, *J. Colloid Interface Sci.* 91:507–15.

Ray, A. K., E. J. Davis, and P. Ravindran. 1979. Determination of ultra-low vapor pressures by submicron droplet evaporation. *J. Chem. Phys.* 71:582–87.

Ray, A. K., J. Lee, and H. L. Tilley. 1988. Direct measurement of evaporation rates of single droplets at large Knudsen numbers. *Langmuir* 4:631–37.

Ray, A. K. and A. Souyri. 1989. Paper 296, 63rd Colloid and Surface Sci. Symp., Seattle.

Ray, A. K., A. Souyri, E. J. Davis, and T. M. Allen. 1991. The precision of light scattering techniques for measuring optical parameters of microspheres. *Appl. Opt.* 30:3974–83.

Richardson, C. B. and J. F. Spann. 1984. Measurement of the water cycle in a levitated ammonium sulfate particle. *J. Aerosol Sci.* 15:563–71.

Ross, G. R. and W. J. Heideger. 1962. Vapor pressure of glycerol. *J. Chem. Eng. Data* 7:505–9.

Rubel, G. O. and J. W. Gentry. 1984*a*. Measurement of the kinetics of solution droplets in the presence of adsorbed monolayers: determination of water accommodation coefficients. *J. Phys. Chem.* 88:3142–48.

Rubel, G. O. and J. W. Gentry. 1984*b*. Investigation of the reaction between single aerosol acid droplets and ammonia gas. *J. Aerosol Sci.* 15:661–71.

Sageev, G., R. C. Flagan, J. H. Seinfeld, and S. Arnold. 1986. Condensation rate of water on aqueous droplets in the transition regime. *J. Colloid Interface Sci.* 113:421–29.

Schweiger, G. 1990. Raman scattering on single aerosol particles and on flowing aerosols: A review. *J. Aerosol Sci.* 21:483–509.

Sloane, C. S. and A. A. Elmoursi. 1987. Characterization of an electrodynamic balance for suspending charged droplets. In *Conf. Record of the 1987 IEEE Industrial Applications Meeting, Part II*, pp. 1568–77. New York: IEEE Publ. Services.

Smith, B. D. and R. Srivastava. 1986. *Thermodynamic Data for Pure Compounds.* New York: Elsevier.

Straubel, E. and H. Straubel. 1987. Electro-optical measurement of chemical and physical changes taking place in an individual aerosol particle. In *Physical and Chemical Characterization of Individual Airborne Particles*, ed. K. R. Spurny. New York: Wiley.

Taflin, D. C. 1988. Interpretation of microdroplet mass

transfer phenomena by optical resonance spectroscopy. Ph.D. Dissertation, University of Washington.

Taflin, D. C. and E. J. Davis. 1990. A study of aerosol chemical reactions by optical resonance spectroscopy. *J. Aerosol Sci.* 21:73–86.

Taflin, D. C., T. L. Ward, and E. J. Davis. 1989. Electrified droplet fission and the Rayleigh limit. *Langmuir* 5: 376–84.

Taflin, D. C., S. H. Zhang, T. Allen, and E. J. Davis. 1988. Measurement of droplet interfacial phenomena by light-scattering techniques. *A.I.Ch.E. J.* 34:1310–20.

Thurn, R. and W. Kiefer. 1984. Raman-microsampling technique applying optical levitation by radiation pressure. *Appl. Spectrosc.* 38:78–83.

Ward, T. L. and E. J. Davis. 1989. Electrodynamic radioactivity detector for microparticles. *Rev. Sci. Instrum.* 60:414–21.

Ward, T. L., S. H. Zhang, T. Allen, and E. J. Davis. 1987. Photochemical polymerization of acrylamide aerosol particles. *J. Colloid Interface Sci.* 118:343–55.

Weiss-Wrana, K. 1983. Optical properties of interplanetary dust: Comparison with light scattering by larger meteoritic and terrestrial grains. *Astron. Astrophys.* 126:240–50.

Wuerker, R. F., H. Shelton, and R. V. Langmuir. 1959. Electrodynamic containment of charged particles. *J. Appl. Phys.* 30:342–49.

Wyatt, P. J. and D. T. Phillips. 1972. A new instrument for the study of individual aerosol particles. *J. Colloid Interface Sci.* 39:47–51.

21

Bioaerosol Sampling

Aino Nevalainen[1], Klaus Willeke,
Frank Liebhaber[2], and Jozef Pastuszka[3]

Department of Environmental Health,
University of Cincinnati,
Cincinnati, OH, U.S.A

Harriet Burge

University of Michigan,
Ann Arbor, MI, U.S.A

and

Eva Henningson

National Defence Research Establishment
S-901 82 Umeå, Sweden

INTRODUCTION

Bioaerosols are particles of variable biological origin, e.g., pollen, fungal spores or fragments of fungal mycelium, bacterial cells, viruses, protozoa, excreta or fragments of insects, skin scales or hair of mammals, or other components, residues or products of organisms, such as bacterial lipopolysaccharides, i.e., endotoxins, or fungal mycotoxins. Generally, the collection of bioaerosol particles is based on the same sampling principles as those for non-biological aerosols. However, ensuring the survival or biological activity of the bioaerosol particles during and after collection is an important concern which differs from physical aerosol particle sampling. Furthermore, sample handling and storage, as well as the analysis of the collected aerosol, are considerably different from general particle sampling.

General Characteristics

Bioaerosols have specific characteristics that influence their sampling. From the sampling perspective, bioaerosol particles may be:

1. Single spores, pollen grains, bacterial cells, or viruses
2. Aggregates of several spores or cells
3. Fragments of spores and cells or their metabolic products
4. Biological material carried by other, non-biological particles.

1. On leave from the National Public Health Institute, Kuopio, Finland
2. On leave from the U.S. Air Force Air University
3. On leave from the Institute of Environmental Protection, Katowice, Poland

Many bioaerosol particles, such as fungal spores and pollen, are designed by nature for transmission to other areas and to stay alive during the transmission. Therefore, this type of hardy bioaerosol is resistant to environmental stresses, such as ultraviolet light, cold, heat, dryness, and toxic gases, and to sampling stresses, such as shear forces acting on them during sampling. Vegetative cells of bacteria are easily damaged and their viability may be compromised by both environmental and sampling stresses (Cox 1987). Some bioaerosol particles are bits and pieces of biological origin that are shed by an ecosystem. Bacteria emitted from aerated wastewater are residues of evaporated droplets. Many of the bacteria shed by humans, for example, are residues of the skin. They probably stay viable while airborne because they are adapted to the dry conditions of their environment and protected by their original substrate, i.e., the skin scale. Otherwise, most vegetative cells of bacteria are prone to damage as a result of becoming airborne.

Microbial cells in an aerosol may be viable or nonviable. The definition of the viability of a cell is not explicit (Roszak and Colwell 1987), but, generally, viable cells are able to reproduce or they have metabolic activity. Nonviable organisms are unable to reproduce or they are dead. In addition, the nutritional requirements of many environmental microorganisms are not known and not all microbes can be grown on laboratory media. From a natural soil or water sample, for example, it is typically possible to culture less than 1% of the microbes present (Atlas 1988), and this may be true for airborne microbes as well. Bioaerosols that are not whole cells, such as endotoxins, mycotoxins, or various allergens, may also be present. Therefore, depending upon the detection method, the results are commonly expressed as colony-forming units (cfu), when viable organisms are determined, or as cells, spores or pollen grains per unit volume of air, when the viability of the organisms is not determined. The results of chemical analyses, e.g., for endotoxins, are expressed as ng/m^3 or mg/m^3.

The density of microbial cells is somewhat variable, depending upon the degree of cell hydration, the reserve materials and lipid content of the cell (Doetsch and Cook 1974). They consist mainly of water, about 70% of their weight. The rest of the cell material consists of macromolecules, such as nucleic acids, proteins, lipids, carbohydrates and their combinations. The density of microbial cells has been reported to be 1.07–1.09 g/cm^3 (Hamer 1985), 0.9–1.3 g/cm^3 (Orr and Gordon 1956), 1.09–1.24 g/cm^3 (Bratbak and Dundas 1984) or 1.5 g/cm^3 (May 1966).

Bioaerosol particles cover a wide size range. Viruses are the smallest potentially living particles, about 0.02–0.3 μm in length. Bacteria and fungal spores cover a size range from about 0.3 up to 100 μm. Pollen, algae, protozoa, and dander are several tens to hundreds of micrometers in diameter. When microbial cells or spores are carried by other materials or when they are present as aggregates, their migration and deposition depend on the overall size of the whole unit.

Although most of the bioaerosols are harmless constituents of normal environments, some bioaerosol particles may be infectious agents or allergens, or they may carry toxic or irritant components or metabolites. To be infectious, an organism must be viable, but to cause allergenic or toxic effects viability is not a prerequisite. Thus, dead cells as well as cell residues may affect human health. Host factors, including the genetic and environmental factors affecting the individual immunologic response, also play an important role.

Biological characteristics may be utilized in controlling bioaerosol sources. Control measures can be targeted on environmental factors that regulate the growth of microbes, e.g., temperature and moisture.

BIOAEROSOL TYPES

Bacteria

Bacteria are generally found in all soils, waters, plants, and animals. In air, bacteria may occur as vegetative cells or endospores.

They may be carried by other particles, such as water droplet residues, plant materials, or the skin fragments of animals. Bacteria are single-cell microorganisms with a size range of 0.5-30 μm. Their shape varies from spherical to rod-shaped, spiral or filamentous. Many spherical bacteria occur as pairs, tetrads or clusters, e.g., *Micrococcus* and *Staphylococcus*, or as chains, e.g., *Streptococcus*. The rod-shaped bacteria may occur as single cells or in chains, e.g., *Lactobacillus*. Bacteria can be divided into two groups based on the ability of the cell wall to retain crystal violet. Gram-positive bacteria, such as *Bacillus* and *Staphylococcus*, retain the dye while gram-negative bacteria, such as *Pseudomonas* and *Legionella*, do not retain it.

Bacterial cells that are actively metabolizing and dividing into new cells are called vegetative cells. Endospores are formed within the vegetative cells of certain bacterial genera, e.g., *Bacillus* and *Thermoactinomyces*. Bacterial endospores are dormant forms of cells and very resistant to cold, heat, radiation, and other environmental stresses. The sizes of bacterial spores range from 0.5 to 3 μm. They are easily carried away by air currents because gravitational settling is fairly insignificant for particles of this size range.

Pathogenic bacteria are known to cause disease in humans, animals, and plants. Pathogens are often specific and cause disease only to a certain species of animal or plant. Most animal and plant pathogens are different from human pathogens. Knowledge about human pathogens is the major focus of clinical microbiology.

Environmental or saprophytic bacteria are found everywhere and their nutritional and temperature requirements vary. Only a fraction of environmental bacteria appear to have been identified and characterized so far. Some of the environmental bacteria are opportunistic pathogens, i.e., they may attack an individual having a weakened immunological response. Some species in the well-known genus *Legionella* are opportunistic pathogens. *Legionella* species are common in natural waters but, until recently, were difficult to culture in the laboratory. For reasons presently not known, some *Legionella* species find an ecological niche in man-made warm water systems, where they may multiply and, if aerosolized, cause serious disease in exposed people (Keleti and Shapiro 1988).

Actinomycetes are a group of soil bacteria that resemble fungi in their growth morphology. Their spores may contribute significantly to occupational exposure in agricultural work situations (Lacey and Crook 1989). In occupational and environmental hygiene, important genera of actinomycetes are *Streptomyces*, which gives soil its characteristic odor, and the thermophilic genera *Thermoactinomyces* and *Faenia*. These organisms cause hypersensitivity pneumonitis. Actinomycete spores may be found in office or residential buildings which have excessive microbial growth due to moisture accumulation within the structure or inside the heating, ventilation and air-conditioning (HVAC) system.

In air, bacteria may occur alone or may be carried by other particles. Bacteria tend to grow in colonies in their natural habitats, such as water and soil, and on different surfaces such as biofilms. Therefore, whenever they become aerosolized, they often occur as aggregates or microcolonies attached to other materials (Eduard et al. 1990). For example, the skin scales of mammals, which are abundant in indoor air, usually contain colonies of bacteria, e.g., *Micrococcus* and *Staphylococcus* (Lundholm 1982; Noble 1975).

Bacterial endotoxins are lipopolysaccharides specific to the cell wall of gram-negative bacteria. Endotoxins are heat-resistant and chemically stable and they maintain their biological activity after the bacterial cell is no longer viable. Endotoxins cause acute toxic effects that include fever, malaise and decrements in pulmonary function. They can be abundant in agricultural environments, some industries and in humidified indoor air, and may contribute to symptoms of humidifier fever (Rylander and Haglind 1984; Jacobs 1989).

Fungi

The fungi are also omnipresent microorganisms. They are responsible for most of the aerobic decay of natural organic material (Kendrick 1985). The terms mold and mildew refer to visible fungal growth on surfaces. Fungi can be unicellular, e.g., yeasts, but are usually multicellular, forming long chains of cells called hyphae which, in mass, are called mycelium. Fungi are classified into groups based on their methods of spore production. Fungal spores are usually the primary dissemination for the organism and are well adapted to airborne transport. The size range of fungal spores, 0.5–30 μm or sometimes even larger, allows for their transport by winds to long distances. They are often resistant to various environmental stresses such as dryness, cold, heat, and ultraviolet radiation.

Some fungi can utilize living plant materials and cause crop diseases of great economic importance. A few fungi, e.g., *Histoplasma*, *Blastomyces*, and *Coccidioides*, readily invade the living animal tissue and cause infectious disease. Others are opportunistic pathogens, e.g., *Aspergillus* and *Cryptococcus*, and cause infection only in people with impaired immunity. However, most fungi are saprophytic, i.e., they utilize and grow on any nonliving organic material, provided adequate moisture is present. The moisture levels required for fungal growth are often quite low. Sufficient moisture can be absorbed from air by some organic materials, e.g., human skin scales, if the relative humidity exceeds 70%.

Most fungal aerosols can cause allergic reactions and diseases, such as asthma, allergic rhinitis, or, in some cases, hypersensitivity pneumonitis. Studies of these allergens usually focus on fungal spores, although fragments of mycelium and metabolites can also become airborne. Although a few genera, such as *Cladosporium*, *Alternaria*, basidiospores, and ascospores, dominate the outdoor aerosols over most of the world, there is some geographical variation. In areas of seasonal variation, the levels of fungal spores are highest in summer and fall and lowest in winter.

Massive exposure to fungal spores may occur in farming and food handling occupations as well as in some industries (Kotimaa, Oksanen, and Koskela 1991; Rylander 1986; Kotimaa 1990). In office and residential indoor environments, the outdoor air is an important source of fungal spores. The detection of airborne spores resulting from growth occurring on indoor substrates can be difficult in the presence of normal background levels of outdoor bioaerosol.

Mycotoxins are toxic chemicals produced by some fungi. Several species of the genus *Aspergillus* produce carcinogenic toxins. These toxins have been studied primarily with respect to ingestion. There is some indication that aerosol exposure is a hazard as well. Another category of mycotoxins, the trichothecenes, are produced by fungi such as *Fusarium*, and *Stachybotrys*. These can cause serious acute effects, including headaches, dizziness, and immunosuppression, and have been isolated from the air (Sorenson et al. 1987).

Viruses

Viruses differ from other microorganisms because they can reproduce only inside a host cell. Therefore, they are intracellular parasites and never grow on nonliving substrates. Viruses infect either bacteria, plants, animals, or humans and they are mostly very specific within these groups. Viruses that replicate only within bacterial cells are called bacteriophages. Viruses are the smallest of all microorganisms, 0.02–0.3 μm. They consist of only one type of nucleic acid, either RNA or DNA, and are, therefore, not able to generate genetic information without a complete host cell. Viruses are surrounded by a protein layer called a capsid.

Viruses can be transmitted through air in the absence of the host cell (Gerone et al. 1966). Although individually small, viruses travel in air as droplet nuclei carried by other materials such as respiratory secretions. Thus, their particle sizes may vary greatly and depend on many factors, including relative humidity.

Viruses can cause infectious diseases, e.g., influenza, measles, and chicken pox, and the viral sources are almost always other infected humans. There is epidemiological evidence for the transmission of viral infections inside and between buildings (Riley, Murphy, and Riley 1978; Donaldson 1978).

Pollen

Pollen grains are produced by plants to transmit the "male" genetic material to "female" flower structures. Many ornamental plants produce pollen that is transported by insects. However, many trees, grasses, and weeds produce pollen that is adapted for airborne dispersal. To be successfully transmitted, this kind of pollen is usually produced in large amounts. For example, one shoot of hemp may produce over 10^9 pollen grains (Faegri and Iversen 1989). Airborne pollen types are resistant to environmental stresses such as desiccation, temperature, and light; hence, they tend to resist sampling stresses. Pollen from different plants varies in size, surface structure and, to some extent, shape. Size range is approximately 10–100 μm, with many grain types being between 25 and 50 μm. Many kinds of pollen contain important allergens. Patterns of prevalence for airborne pollen vary with geography and climate. For example, ragweed pollen is considered to be one of the most important allergens in North America, while birch pollen is considered the most important in Northern Europe.

Algae

Algae are primitive plants that occur naturally in water. Individual cells can become airborne with water droplets (Brown, Larson, and Bold 1964). They have been isolated in house dust (Bernstein and Safferman 1970) and there is evidence that airborne algae may play the role of a source of respiratory hypersensitivity reactions (Bernstein and Safferman 1973).

House Dust Mites

House dust mites, e.g., *Dermatophagoides pteronyssinus* and *D. farinae*, are common allergens in residential environments. They live in mattresses, carpets, and upholstered furniture. The allergens derive from their excreta, which may easily become aerosolized, although particles are probably relatively large, about 10 μm. Airborne antigens can be collected using breathing zone filtration. However, samples are usually collected directly from the source, e.g., mattresses and carpets, with a vacuum cleaner. The amount of dust mite allergens in the dust sample is determined indirectly based on the colorimetric indication of guanine, which is a component of dust mite excreta (von Bischoff 1989). Other methods use monoclonal antibodies specific for chemically purified mite allergens. A specialized knowledge is needed to visually identify the mites in samples.

Other arthropod material, e.g., cockroach excreta as well as mammalian and avian amorphous material, such as serum, dander, or urine of cats, dogs, and birds, are also allergenic. Both source samples and air samples have been analyzed for these allergens using specific immunoassay techniques.

SOURCES OF BIOAEROSOLS

Many indoor bioaerosols originate outdoors. The surfaces of living and dead plants are probably the most important sources of airborne fungal spores and bacteria. Therefore, the mechanical movement of plants, as in many farming activities, or soil, e.g., farming or construction, generates bioaerosols together with other dust. Actinomycetes apparently become airborne from soil (Atlas and Bartha 1987). All natural waters as well as anthropogenic waters, such as sewage lagoons and cooling water systems, contain large numbers of microorganisms. For example, gram-negative bacteria, actinomycetes, and algae are common constituents of water ecosystems. Therefore, water or liquid droplets resulting from rain, splashes, or bubbling processes may contain bioaerosols which may remain airborne after the water evaporates.

Strong sources of bioaerosols may exist in many work situations, when organic material is handled, such as plants, hay, straw, wood chips, cereal grains, tobacco, cotton, organic waste, or wastewater. In agricultural and horticultural occupations, exposure to fungal and actinomycete spores may be severe (Clark, Rylander, and Larsson 1983; Lacey and Crook 1988; Kotimaa 1990).

In nonindustrial situations, specific bioaerosol sources may develop due to microbial growth in a building's heating, ventilation and air-conditioning systems or in the structure itself. Generally, the prerequisite for microbial growth is excessive and accessible moisture. Standing water is always a good reservoir for microbial growth and, therefore, a potential source of microbial aerosols when disturbed (Burge, Solomon, and Boise 1980; Keleti and Shapiro 1987; ACGIH 1989). In addition to water, microorganisms only need minute amounts of nutrients, which are available in the water or building materials, such as cellulose, wood, or concrete. Therefore, a source of spores or other bioaerosol material may develop wherever water is leaking or condensing inside a building.

In nonindustrial indoor environments, the most important source of airborne bacteria is the presence of humans. Air with a high concentration of human bacteria is not necessarily a health hazard, but indicates the presence of many humans, their physical activity, or inefficient ventilation. In such cases, pathogenic bacteria or viruses may also occur in air, although their presence may be difficult to verify by conventional bioaerosol sampling techniques. Therefore, the concentration in air of normal human skin bacteria is often used as an indicator of indoor air quality.

GENERAL SAMPLING CONSIDERATIONS

The purpose of bioaerosol sampling is most often to verify and quantify the presence of bioaerosols for exposure assessment, or to identify their source for control. Dose–response relationships are poorly known and, therefore, exposure guidelines have not been established for acceptable healthful levels of any bioaerosol. Bioaerosol concentrations have timely variations of several orders of magnitude. For example, concentrations of the order of 10–10^3 cfu/m^3 can be found in homes or occupational environments with moderate sources, e.g., tobacco processing, sanitary landfills, or biotechnical industries (Macher, Huang, and Flores 1991; Martinez et al. 1988; Rahkonen et al. 1990; Reponen, Nevalainen, and Raunemaa 1989; Verhoeff et al. 1990). Lower concentrations of $\leq 10^2$ cfu/m^3 can be found in well-ventilated facilities without significant sources, such as offices, laboratories, clean rooms, and operating rooms in hospitals. High concentrations of bioaerosols, with peak concentrations from 10^4 to as high as 10^{10} cfu/m^3, can occur in environments such as textile mills, sawmills, some agricultural exposure situations and in seriously contaminated homes and offices (Lacey and Crook 1988; Kotimaa 1990; Eduard et al. 1990). In most of these environments, the concentrations vary considerably in time and space. This is partly because bioaerosol sources do not necessarily generate particles continuously. For example, spore production and spore release from fungal mycelium may occur in bursts under certain air humidity and velocity conditions.

Sampling Strategy

No single sampling method can collect, identify, and quantify all of the bioaerosol components existing in a particular environment. Therefore, a source inventory is important and useful. It may include a preliminary microbiological analysis of the water reservoirs and a contact sample from a surface with assumed fungal growth. In industrial exposure situations, the type and location of sources are frequently evident. In nonindustrial environments, the sources are often less obvious and, therefore, difficult to sample.

Sampling Efficiency of Bioaerosol Samplers

The overall sampling efficacy of a bioaerosol sampler can be divided into three components:

1. The inlet sampling efficiency is a function of the sampler inlet's ability to extract particles from the ambient environment without bias with regard to the particle size, shape, or aerodynamic behavior.
2. The removal efficiency is determined by the sampler's ability to remove the particles from the airstream and deposit them into or onto the collection medium.
3. The biological aspect of sampling efficacy is the sampling and removal of biological particles without altering their viability or biological activity, and to provide the proper conditions for the organisms to form colonies or to be otherwise detected.

The physical and biological parameters of sampling performance must be separated in order to quantify their effects. Presently, little is documented with regard to the extent to which these parameters affect sampling.

None of the presently available samplers for culturable bioaerosols can be considered as a reference method, although the all-glass liquid impinger (AGI-30) and the six-stage Andersen impactor have been suggested for that purpose (Brachman et al. 1964). Few of the currently available samplers have been adequately characterized as to their sampling performance. The results of reported field comparison studies are not easily comparable to each other, partly because the samplers, sampling times, and sampled volumes have varied within and between studies, and partly due to the different operational principles and parameters of each of the samplers. Table 21-1 presents some of the comparison studies performed under field conditions. It can be seen that certain patterns apparently exist: some samplers are more efficient collectors of viable bioaerosol particles than others.

Bioaerosol particles must be collected from the ambient air in an unbiased way. To be efficient over a broad particle size range, aspiration sampling should occur under isokinetic conditions. This may, however, cause different types of sampling bias, such as inlet losses of larger particles. The aspects of isokinetic sampling are further discussed in Chapter 6.

The physics of particle removal from air, as well as the principles of good sample collection, is common to all airborne particles. Therefore, physical principles can be applied to the sampling of bioaerosols. These principles determine the amount of sample collected and the sampling time needed for a proper analysis.

TABLE 21-1 Results of Field Comparison Studies Performed with Different Bioaerosol Samplers[1]

Bioaerosol Samplers Compared[2]	Environment Studied	Bioaerosols Studied	Reference
AND 6-stage ~ N6	Office environment	Fungal spores	Jones et al. (1985)
AND 6-stage > AND 2-stage	Wastewater facilities	Bacteria	Gillespie et al. (1981)
AND 6-stage > STA	Cotton mill	Bacteria	Lundholm (1982)
AND 6-stage > AGI	Cotton mill	Bacteria	Lundholm (1982)
AND 6-stage > AGI	Greenhouse	Bacteria	Buttner and Stetzenbach (1991)
AND 2-stage ~ May 3-stage	Wastewater irrigation	Bacteria	Zimmerman, Reist, and Turner (1987)
AND 8-stage > AND 2-stage	Class rooms	Bacteria	Curtis, Balsbaugh, and Drummond (1978)
RCS > STA	Laboratory environment	Bacteria	Placencia et al. (1982)
RCS > STA	Hospital, low concentrations	Bacteria	Groeschel (1980)
RCS < STA	Hospital, high concentrations	Bacteria	Groeschel (1980)
RCS ~ STA ~ N6 > SAS	Occupational settings	Fungal spores	Smid et al. (1989)
N6 > STA > RCS > SAS[3]	Homes	Fungal spores	Verhoeff et al. (1989)

1. AND = Andersen impactor; N6 = Andersen 6th stage alone; STA = Slit-to-agar sampler; AGI = All-glass impinger; RCS = Reuter centrifugal sampler; SAS = surface air system sampler
2. Approximately equal collection efficiency is expressed by ~ and higher collection efficiency is expressed by >
3. Results obtained with malt extract agar; results somewhat variable with other media

COLLECTION PROCESS[1]

Principles of Bioaerosol Particle Collection

Bioaerosol sampling involves separating the particle trajectory from the air streamline trajectory. To achieve this, different physical forces are applied as illustrated in Fig. 21-1. In Fig. 21-1a, the inertia of the particle forces its impaction onto a solid or semisolid impaction surface, usually either a culture medium, or an adhesive surface that can be examined microscopically. Applications of this principle are cascade impactors with one, two, or more stages (e.g., the Andersen and SAS culture plate impactors) and slit samplers (e.g., Burkard spore traps, Casella, New Brunswick and Mattson-Garvin culture plate samplers; more information is given further below).

Figure 21-1b illustrates a virtual impactor, which is also based on the inertial behavior of particles. Size separation of particles occurs when the small low-inertia particles follow the air flow and the large high-inertia particles cross the "virtual" impaction surface and are collected or sensed below this interface. An experimental sampler of this kind has been used to collect outdoor air antigens (Kauppinen et al. 1989).

The principle of particle separation by centrifugal force is shown in Fig. 21-1c. It also uses the inertial behavior of the particle, but in a radial geometry. The Reuter centrifugal sampler (RCS) is of this type.

When filtration is the collection mechanism, as shown in Fig. 21-1d, inertial forces are also responsible for separating the particles from the airstream. However, in concert

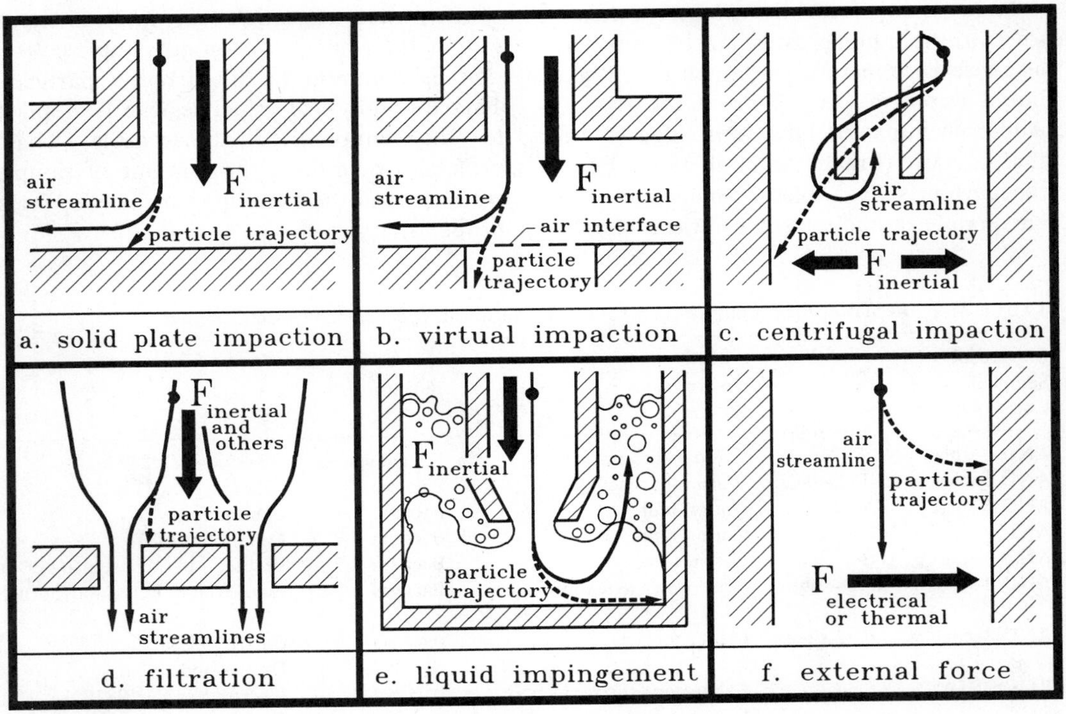

FIGURE 21-1. Mechanisms of Particle Removal from Air. (Adapted from Nevalainen et al. 1992.)

---

1. The concepts and data for collection efficiency and sampling time were developed under Contract No. 90-16A between the Center of Indoor Air Research and the University of Cincinnati (K. Willeke)

with inertial impaction, other mechanisms such as interception, diffusion, and electrostatic attraction contribute to the deposition of particles onto the filter material.

Liquid impingement (Fig. 21-1e) is a method that mainly uses inertial forces to collect particles, but also uses diffusion within the bubbles to enhance particle collection. Several versions of liquid impingers are currently available (e.g., AGI-4 and AGI-30 impingers) including a preseparating unit and a multistage impinger (May 1966).

Particles may also be removed from the airstream by externally applied forces, such as electrical forces on charged particles, or thermal forces on an aerosol flow which has a thermal gradient perpendicular to its flow. This principle is illustrated in Fig. 21-1f. An electrostatic sampler has been used for virus sampling (Gerone et al. 1966).

Because of the wide size range of the bioaerosol particles, aspiration is the preferred method for sample collection. An exception to this is the rotating-arm device, e.g., the Rotorod sampler, that is commonly used to sample pollen.

Settling plates rely on gravity to cause particles to fall onto a culture plate. The gravitational settling of particles is highly particle-size-dependent and is strongly influenced by the air motion in the surrounding air. The colony-forming units seen on the settling plate or spores on a settling slide can, therefore, not be directly related to the concentrations of biological particles in the air.

Particles can bounce when they strike the impaction surface or other previously collected particles. The latter often occurs when the sample is overloaded. The bouncing of larger particles can lead to undersampling in single-stage impactors when the particles bounce up and away. Particle bounce can affect the reported size distribution in cascade impactors when large particles bounce to the next, smaller-size collection stage. Collection efficiency, especially for larger particles, gradually falls as the sampling surface becomes loaded. This effect has not been accurately examined for any of the commonly used bioaerosol samplers. However, it is a common practice to apply a thin film of grease to the impaction surface to reduce the bounce effect and, hence, to increase the collection efficiency. The bouncing phenomenon is discussed in Chapter 11.

Particle Removal by Inertial Impaction

Inertial impaction is the most widely used mechanism of particle removal in bioaerosol samplers. The impaction process depends on the particles' inertial properties, such as size, density, and velocity, and on the impactor's physical parameters, such as inlet-nozzle dimensions and airflow paths. How these parameters interact in the impaction zone is

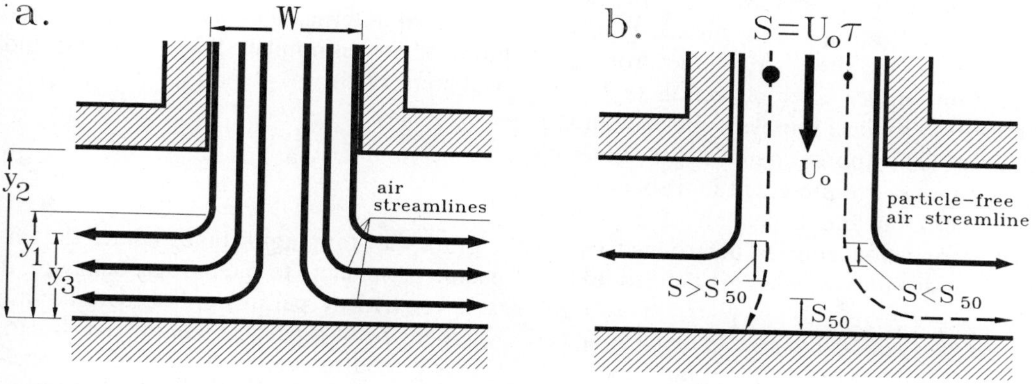

FIGURE 21-2. Particle Removal by Inertial Impaction. (a) Critical Parameters of the Impaction Zone. (b) Stopping Distance and Its Importance in Particle Removal. (Adapted from Nevalainen et al. 1992.)

illustrated in Fig. 21-2. Aerosol particles follow the air as it flows into and through the impactor's inlet nozzle, which may be round or rectangular.

The size of the nozzle is given by W, which is either the diameter of a round nozzle or the width of a rectangular one. To relate W to the performance characteristics of the impactor in a simple manner, the concept of "hydraulic diameter" has been used (Willeke and McFeters 1975). The hydraulic diameter, D_h, for air or liquid flows in a conduit is defined by

$$D_h = \frac{4A}{P} \tag{21-1}$$

where A is the cross-sectional area of the flow and P is the perimeter "wetted" by the flow, or the boundary of the cross-sectional area (Daugherty, Franzini, and Finnemore 1985). In an impactor, P is the circumference of the nozzle interior. For a round nozzle, W equals D_h. For a rectangular nozzle, if the ratio of width over length (W/L) is much less than one, the hydraulic diameter approximately equals twice the nozzle width, $2W$.

In the impaction zone illustration of Fig. 21-2a, the removal of particles occurs where the particle trajectory significantly deviates from the air streamline. The impactor airflow, the shape and size of the nozzle, and the distance from the impaction surface where the flow significantly starts deflecting (y_1) determine the minimum distance between the nozzle and the impaction surface (y_2), the distance of the free streamline from the impaction surface after deflection (y_3), and the aerosol particle removal efficiency of the impactor. The approximate integer values of y_1, y_2, and y_3 are presented in Table 21-2.

Deceleration and collection of a particle in the impaction zone is determined by its stopping distance, S, which is defined as

$$S = V_0 \tau \simeq U_0 \tau \tag{21-2}$$

where V_0 is the initial particle velocity (usually assumed to be equal to the incoming flow velocity, U_0) and τ is the relaxation time. The latter relates to the time required by a particle to adjust or "relax" itself to a new set of forces or conditions (Hinds 1982). It depends on particle size, expressed by the diameter, d_p, particle density, ρ_p, and the viscosity, η, of the fluid medium the particle is traveling in:

$$\tau = \frac{\rho_p d_p^2 C_c}{18\eta} \tag{21-3}$$

C_c is the Cunningham correction factor (see Chapter 3), which is only significant for submicrometer-sized particles. To relate the travel of the particle to its confined flow within an impactor, the nondimensional Stokes number, Stk, is used:

$$Stk = \frac{S}{W/2} \tag{21-4}$$

Expressed in terms of hydraulic diameter, the modified Stokes number, Stk', is (Willeke and McFeters 1975):

$$Stk' = \frac{S}{D_h/2} \tag{21-5}$$

Since the stopping distance relates the impactor parameters to the particle behavior, it is a convenient parameter for determining whether or not a particle will impact. It can, thus, be used to predict the collection efficiency of an impactor. The spatial relationship of a particle's stopping distance to its potential impaction is illustrated in Fig. 21-2b. For

TABLE 21-2 Approximate Values of Critical Impaction Parameters

Parameter	Nozzle Shape: Round	Nozzle Shape: Rectangular
W	D_h	$D_h/2$
y_1	$\sim W/2 = D_h/2$	$\sim W = D_h/2$
y_2	$> D_h/2$	$> D_h/2$
Optimal y_2	$> 1.0W$	$> 1.5W$
y_3	$\sim 1/10W$	$\sim 1/2W$
Stk_{50}	$\sim 1/4$	$\sim 1/2$
Stk'_{50}	$\sim 1/4$	$\sim 1/4$
S_{50}	$\sim 1/8W$	$\sim 1/4W$

each streamline, there is a certain stopping distance for which all the particles larger than a certain diameter, d, are collected. On another streamline further away from the centerline of the incoming flow, a particle may need a longer stopping distance to be collected. Thus, a particle must have a higher inertia to be collected from the further streamline. For the entire flow field, S_{50} designates the stopping distance for which 50% of the particles are collected and 50% pass through the impactor. Similarly, the "cutoff size", d_{50}, designates the particle diameter for 50% removal. Since most impaction stages have very sharp cutoff characteristics, almost all the particles larger than that size are collected. Therefore, d_{50} is generally assumed to be the size above which all particles larger than that size are collected. d_{50} is an important characteristic of any bioaerosol sampler; however, it is only known for a few samplers. In order to evaluate the sampling performances of different samplers, d_{50} can be numerically estimated (Willeke and McFeters 1975; Marple and Willeke 1976):

$$d_{50} = \sqrt{\frac{9\eta W Stk_{50}}{\rho_p U_0 C_c}} \propto \sqrt{Stk_{50}} \propto \sqrt{Stk'_{50}} \propto \sqrt{S_{50}} \qquad (21\text{-}6)$$

Approximate values of Stk_{50}, Stk'_{50}, and S_{50} that can be used for calculating d_{50} are given in Table 21-2.

Numerical Estimation of Cutoff Sizes

The design and performance characteristics of five commonly used air samplers, designed to collect various bioaerosols, are compared in Table 21-3 and Fig. 21-3.

The five samplers range in volumetric flow rate, Q, from 10 to 180 l/min and have a single impactor nozzle or several in parallel. For the six-stage Andersen impactor, the first and sixth stages have been treated as single impactors. The 50% stopping distances (S_{50}) and the cutoff diameters (d_{50}) for these bioaerosol samplers have been calculated using Eq. 21-6 and values for the viscosity of air, $\eta = 1.81 \times 10^{-5}$ Pa s at normal air temperature and pressure (20°C, 1 atm), and the approximate density of bioaerosol particles, $\rho_p = 1.0$ g/cm^3.

The aerodynamic diameter of a particle expresses the diameter of a spherical particle, density 1 g/cm^3, that has the same settling velocity as the particle in question. For bioaerosols with a density close to unity and a shape close to spherical, the physical and aerodynamic diameters are approximately equal.

Figure 21-3 graphically shows how the calculated stopping distances for the five samplers of Table 21-3 change with aerodynamic particle size. The stopping distance curves terminate at the calculated S_{50}. For comparison purposes, the cutoff sizes available in the literature are also included in Fig. 21-3. Also included are the size ranges of viruses,

TABLE 21-3 Sampling Parameters and Calculated and Reported Cutoff sizes for Selected Bioaerosol Samplers[1]

				Nozzle						d_{50} (μm)	
Bioaerosol Sampler	Collection Medium	Q (l/min)	U_0 (m/s)	Shape	W (mm)	L (mm)	A[2] (mm^2)	Number	S_{50} (mm)	Calculated	Reported
AGI-30	Liquid	12.5	265.2	Round	1.0		0.79	1	0.125	0.31	
AND-I	Nutrient	28.3	1.08	Round	1.18		1.09	400	0.148	6.61	7.0 (1)
AND-VI	Nutrient	28.3	24.02	Round	0.25		0.05	400	0.031	0.57	0.65 (1)
BURK	Adhesive	10.0	11.90	Rectangle	1.0	14	14	1	0.25	2.52	
SAS	Nutrient	180	17.34	Round	1.0		0.79	219	0.125	1.45	1.9 (2)
MK-II	Nutrient	30.0	51.42	Rectangle	0.35	28	9.8	1	0.087	0.67	

1. See Fig. 21-3 and Tables 21-1 and 21-5 for key to sampler abbreviations
2. Area per nozzle
Reference (1) Andersen (1958); (2) Lach (1985)

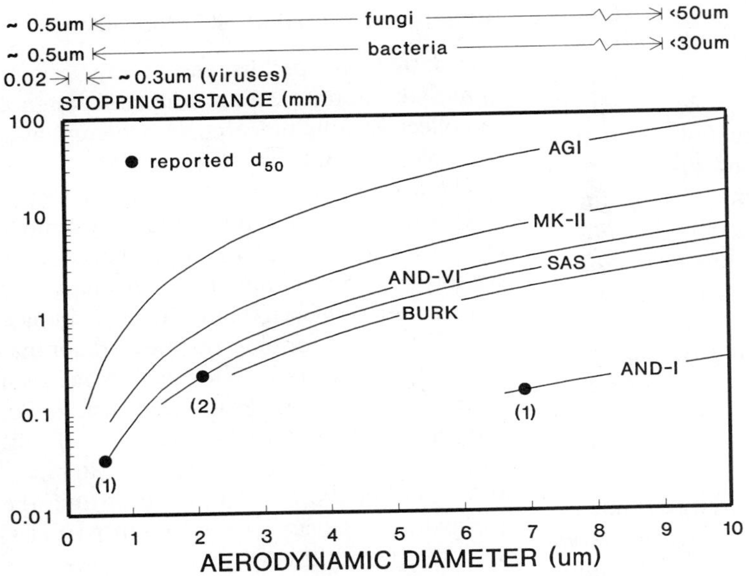

FIGURE 21-3. Stopping Distance as a Function of the Aerodynamic Diameter of the Collected Particle for Selected Bioaerosol Samplers. Reported Cutoff Sizes: (1) Andersen (1958) (2) Lach (1985). (Adapted from Nevalainen et al. 1991.)

bacteria, and fungi, although these bioaerosols may also occur in larger particles, as agglomerates or carried by non-biological materials. The calculated d_{50} values, based on the approximate values of Table 21-2, are somewhat smaller than the reported, experimentally determined cutoff sizes. As can be seen in Fig. 21-3, the cutoff sizes for the various samplers or sampler stages range from less than 0.5 μm to over 5 μm. Therefore, only the sampler with the smallest cutoff size may collect larger viruses unless they are carried by other particles. A virus that is attached to a carrier may be collected by any sampler whose cutoff size is smaller than the size of the particle to which the virus is attached. The differences in the cutoff sizes of the various samplers may at least partly explain the differences between sampler performances that have been previously reported.

Stopping distances and cutoff sizes can be determined for other inertial samplers as well, although in some situations, it may be more difficult. For instance, in a spiral sampler using centrifugal motion, the linear impaction parameters mentioned earlier cannot be applied to centrifugal impaction in the same manner, because the stopping distance increases as the particle moves towards the wall. Macher and First (1983) reported an experimentally determined cutoff size of 3.8 μm for the Reuter centrifugal sampler.

Example 21-1

An Andersen N6 sampler (or the 6th stage of a six-stage cascade impactor) is operated at a volumetric flow rate of 20 l/min instead of the required 28.3 l/min (1 cfm). This is a single-stage sampler with 400 nozzles of 0.25 mm diameter each, i.e., 400 jets of aerosol flow are directed towards the nutrient medium below them.

(a) Calculate the cutoff diameter, d_{50}, for the reduced flow. What is the smallest particle collected?
(b) The N6 is operated at the required flow rate of 28.3 l/min, but 40 of its 400 nozzles are plugged. What is the smallest particle collected?

Answer. (a) The air velocity and, therefore, the approximate particle velocity in each of the n jets of cross-sectional area A_j is

$$U_0 = \frac{Q}{nA_j} = \frac{Q}{n\left(\frac{\pi W^2}{4}\right)}$$

$$= \frac{\left(20 \times 10^3 \frac{cm^3}{min}\right)\left(\frac{min}{60\ s}\right)}{400\left[\frac{\pi(0.025\ cm)^2}{4}\right]}$$

$$= 1697\ \frac{cm}{s}$$

To determine the cutoff diameter, d_{50}, use Eq. 21-6. At normal pressure and air temperature (1 atm and 20°C), the viscosity of air equals 1.81×10^{-4} poise (1.81×10^{-4} g/cm s). The density of bioaerosol particles ranges from 0.9 to 1.5 g/cm^3 and is assumed to equal 1 g/cm^3 for this example. For the round nozzles the value of Stk_{50} is approximately 1/4 (see Table 21-2):

$$d_{50} = \sqrt{\frac{9\eta W Stk_{50}}{\rho_p U_0 C_c}}$$

$$= \sqrt{\frac{9 \times 1.81 \times (10^{-4}\ g/cm\ s)(0.025\ cm) \times 0.25}{(1\ g/cm^3)(1697\ cm\ s) \times 1}}$$

$$= 7.74 \times 10^{-5}\ cm = 0.77/\mu m$$

Therefore, a 0.77 μm particle is approximately the smallest one collected. In the above calculation, the Cunningham correction factor, C_c, is assumed to be one. It is actually somewhat larger than one for the calculated d_{50}. Through an iteration procedure, the calculated d_{50} is, therefore, somewhat smaller than indicated. At the recommended flow rate of 28.3 l/min, d_{50} is 0.57 μm (see Table 21-3).

(b) If 40 of the nozzles are plugged, the air velocity through the 360 remaining nozzles at a flow rate of 28.3 l/min is 2668 cm/s. For this velocity and $C_c = 1$ the new cutoff size is $d_{50} = 0.62$ μm. For this particle size and $C_c > 1$, the actual cutoff size is even smaller.

COLLECTION TIME

An essential part of the sampling strategy is to define the sample collection times. Bioaerosol concentrations vary greatly over time, which is graphically shown in Part I of Fig. 21-4, where c is the instantaneous concentration of the bioaerosol. Typically, periods of low concentrations, i.e., t_1–t_2, are followed by periods of high and peaking concentrations, i.e., t_3–t_4, and vice versa. These large fluctuations are best represented on a logarithmic concentration scale as shown. An average concentration, c_a, of 1000 bioaerosol particles/m^3 of air is presented as an example. Bioaerosol concentrations of this order of magnitude are common in outdoor air and many indoor situations. The ambient concentrations rarely remain stable within a narrow concentration range unless the time period is relatively short, i.e., minutes, or the atmosphere is undisturbed, such as in an unventilated, nonpopulated, closed room.

Sampling during periods of changing concentration must be sufficiently long, or many short samples must be combined to properly represent the average environmental concentration. Part II of Fig. 21-4 reflects how the concentration varies in air volume v being sampled during the sampling period from starting time, t_s, through the finish, t_f. The volume of air equals the product of sampling flow rate, Q, and sampling time, t:

$$v = Qt \qquad (21\text{-}7)$$

The sampling process actually integrates the instantaneous concentrations within the sampled period with respect to time. Hence, after impaction, Part III, the number of particles collected, N, can be expressed by the average concentration, c_a, times the sampled volume, v, or the sampler's flow rate, Q, multiplied by the sampling time, t:

$$N = c_a Qt \qquad (21\text{-}8)$$

Surface Density of Collected Particles

Part IV of the sampling process illustration shows the bioaerosol particles being collected

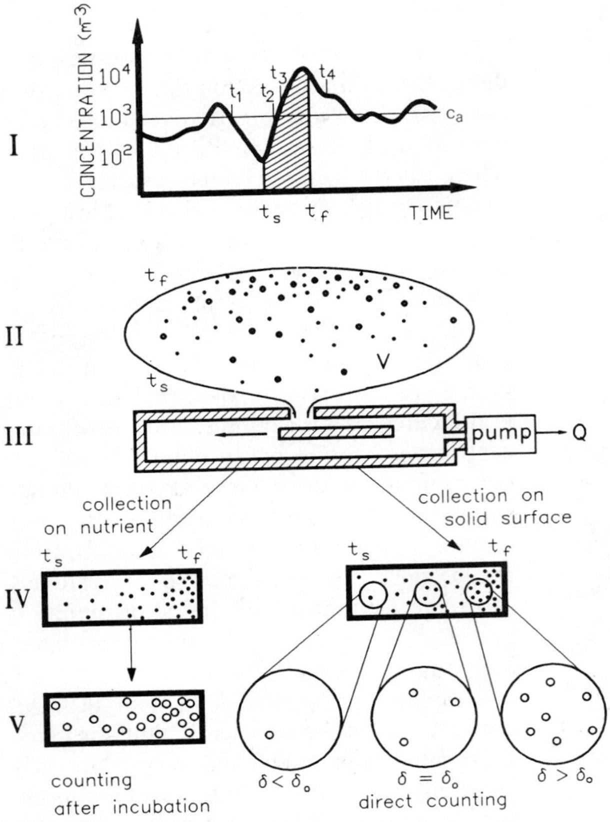

FIGURE 21-4. The Process of Bioaerosol Sampling. (Adapted from Nevalainen et al. 1992.)

on a nutrient or solid surface which moves to the left during sampling, resulting in varying numbers of particles per unit area according to the changing concentration in the sampled air volume. The number of objects on the surface of a viewing area, whether they be microbial colonies on a Petri dish or microscopic particles on an adhesive surface, will be referred to as the "surface density" of a sample. This surface density, annotated as δ, is linearly related to the concentration of the airborne particles, the sampling airflow rate and the sampling time. The number of particles collected in time t is $c_a Qt$. The surface density, δ, on the collection area is, therefore

$$\delta = \frac{N}{A} = \frac{c_a Q}{A} t \tag{21-9}$$

Part V of Fig. 3 presents the postcollection phase of sampling, where the collected material is analyzed. For example, this can occur either immediately upon collection by viewing the collected particles directly with a microscope, or after an incubation period when the colonies are sufficiently developed to be viewed and identified with the unaided eye. Viewing, counting, and identifying the particles collected in the optimal sample, either optically or otherwise, is facilitated by an appropriate surface density, δ_0. If the sample surface density is very low, $\delta \ll \delta_0$, both sampling and counting errors will be high with respect to the levels, and the reported data may not accurately reflect the true airborne concentration. If the sample surface density is very high, $\delta \gg \delta_0$, on a microscope slide, the particles may be too close to individually count and identify and the bioaerosol particles may be covered and masked by dust particles. If the sample density is too high on a nutrient surface, the collected organisms may grow together or inhibit each other's growth. This is especially important for fungus spores which often release substances that inhibit germination of adjacent spores. Nonbiological particles that impact on nutrient surfaces along with the organisms may not cause viewing problems, but may inhibit growth.

Optimal Sampling Time

All these factors relate to sampling time. Since ambient concentrations and sampler flow rates are normally beyond the control of the investigator, the usual way to mitigate the problems of nonrepresentative or overloaded samples is to control the sampling period.

An optimal sample should integrate the temporal changes in the sample population it is attempting to represent. This ideal sample is a direct result of an optimal surface density for the particular sampling method. This, in turn, can be referenced to an arbitrarily chosen bioaerosol concentration because, under normal sampling conditions, the ambient concentration is not known. By inverting Eq. 21-9, the sampling time can be calculated for each sampler by using the desired value of surface density and assuming the order of magnitude of the airborne bioaerosol concentration:

$$t = \frac{\delta A}{c_a Q} \tag{21-10}$$

Optimal Sampling Times for Different Samplers

The optimal sampling time for a given ambient concentration will be different for each sampler according to the sampler's sampling flow rate and collection surface area. For purposes of this discussion, we will assume an ambient concentration of 10^3 bioaerosol particles/m^3 of air and a sample collection efficiency of 100%. We will also assume that $\delta_{macro} = 1$ particle/cm^2 is an ideal number of colonies on a culture plate and that $\delta_{micro} = 10^4$ particles/cm^2, which equals three particles per microscopic field of 200 μm diameter, is the minimum number of bioaerosol particles on a slide sample. Using these assumptions and Eq. 21-10, we have calculated optimal sampling times for five commonly used bioaerosol samplers. For these calculations, the density of the countable objects on the sampling surface is used as the surface density, δ. The area of the sampling surface, A, is defined differently for various samplers. For slit samplers using an adhesive impaction surface, e.g., the Burkard samplers, A is approximately the area of the slit nozzle. For slit-to-agar impactors and sieve-type impactors, such as the Andersen impactor with 400 jets or the SAS sampler with its 219 jets, the actual sampling surface is the sum of the areas of the nozzles. However, in culture plate samplers the sampled particles are allowed to develop into colonies prior to viewing and, therefore, the area of the sampling surface, A, is the area of the nutrient plate. For the Andersen six-stage impactor, it is assumed that bioaerosol particles impact onto each of the six stages. Therefore, the total area of all six stages is used in calculating the sampling time. This assumption is valid when the particle size distribution allows for even particle deposition among the stages. When the sixth stage is used as a separate sampler, annotated here as the AND-VI and often referred to as the N-6, only the surface area of one Petri dish is used.

It can be seen in Fig. 21-5 that for $c_a = 10^3$ particles/m^3, to obtain the same surface density, the sampling time for the surface air system sampler (SAS) is 8 s, while the Burkard personal air sampler has a sampling time of 140 min. This represents a difference of three orders of magnitude. Lengthening the time of sampling for culture plate devices may allow for better sample representation, but it is likely to cause overloading, leading to counting errors and inhibition effects. Therefore, a short-period culture plate sampler is best suited for environments where a very low

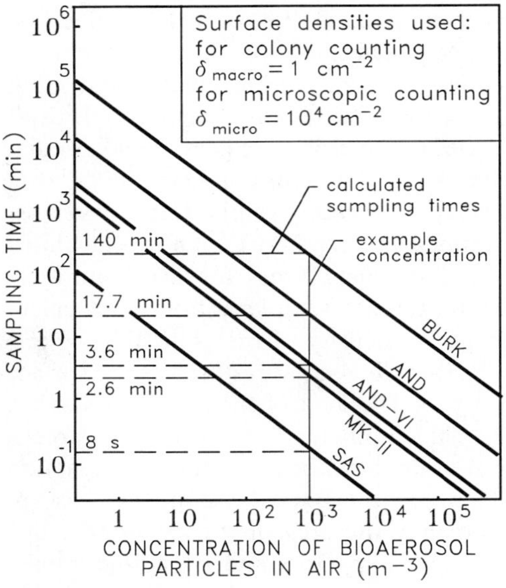

FIGURE 21-5. Collection Times for Selected Bioaerosol Samplers. (Adapted from Nevalainen et al. 1992.)

ambient concentration, e.g., $\leq 10^2$ cfu/m^3, allows longer sampling periods, or numerous short, sequential samples.

The calculated sampling time for the Burkard personal sampler at 10^3 bioaerosol particles/m^3 is 140 min, but in most environments this would result in a sample grossly overloaded with nonbiological particles. Therefore, the assumed or expected values (c, δ) can be modified in order to obtain a sampling time more in line with realistic expectations. This means that this sampler may be used in environments where the concentration of bioaerosol particles is higher, e.g., 10^4/m^3, and when the sampling time is shorter, e.g., 14 min. Otherwise, a compromise in the desired surface density can be made. For example, for one-tenth of the surface density previously used, i.e., $\delta_{\text{micro}} = 10^3$ particles/cm^2 = 0.3 particles per field, the sampling time is reduced to one-tenth of the original value, $t = 14$ min. This may not be a very representative sample, but it is within the range this sampler is commonly used. Using this sampler for a short period of time may produce sampling bias due to fluctuations in the bioaerosol concentration or because of statistically low surface densities.

Example 21-2

The surface air system (SAS) sampler is used to collect bioaerosols at an expected concentration level of 500 particles/m^3. The microbiologist who is going to analyze the results would like to see 50 colonies on each 55 mm Rodac plate (nutrient dish) sample. The SAS sampler has 219 sampling jets and a flow rate of 180 l/min. Calculate the optimal sampling time.

Answer. The area for the 55 mm Rodac plate equals 23.8 cm^2. Thus, the surface density for counting colonies per plate (Eq. 21-9) is:

$$\delta = \frac{N}{A} = \frac{50 \text{ particles}}{23.8 \text{ cm}^2} = 2.1 \frac{\text{particles}}{\text{cm}^2}$$

To determine the optimal sampling time, use Eq. 21-10:

$$t = \frac{\delta A}{c_a Q}$$

$$= \frac{(2.1 \text{ particles/cm}^2)(23.8 \text{ cm}^2)}{(500 \text{ particles/m}^3)(180 \times 10^{-3} \text{ m}^3\text{/min})}$$

$$= 0.55 \text{ min} = 33 \text{ s}$$

This sampling time may be too short for an environment with variable concentrations of bioaerosols. Several consecutive samples should be taken or a different sampler with a larger sampling period should be chosen (see Fig. 21-5).

Example 21-3

You want to sample in a hospital. Cleaning of the postoperative patient room will take 20 min. Previous reports indicate an average bacterial aerosol concentration of $\sim 10^2$ cfu/m^3. Which sampler would you choose to sample the 20 min work situation?

Answer. By looking at Fig. 21-5, the concentration of 10^2 bioaerosol particles/m^3 can be located on the *x*-axis and followed up to the 20 min level on the *y*-axis. At this intersection, MK-II or the Andersen N6 are indicated as suitable samplers at a surface density $\delta_{macro} = 1\ cm^{-2}$.

SELECTION OF SAMPLER

Sampler selection depends on the kinds and levels of bioaerosols of interest in the environment to be sampled. For example, when evaluating allergens, total rather than viable counts are important. Also, when selecting a cultural sampler one must consider the vulnerability of the bioaerosol to the sampling forces since some samplers affect viability more than others. The most commonly used instruments are discussed briefly in order to provide the user with the basis for deciding which sampler to use and how it operates. Table 21-4 gives a list of the presently available samplers and their suitability to various sampling purposes.

The names and addresses of the manufacturers of the instruments are given in Table 21-5. Further information about the samplers is available in the text by Chatigny et al. (1989).

The Andersen six-stage sampler has been characterized in many studies and it allows a characterization of bioaerosols in specific size ranges. There are also two-stage and one-stage (N6) models. When the sample concentrations are high, several bioaerosol particles may deposit below an impaction nozzle. Only one of the organisms may grow, preventing others from developing into visible colonies. In some cases, two or more organisms may grow together into a single colony. When a high percentage of the impaction spots below the nozzles of a multiple-nozzle impaction stage develops colonies, the actual concentration of bioaerosol particles can be statistically calculated by the positive-hole conversion method described by Andersen (1958) and Macher (1989).

The "surface air system" (SAS) samplers are one-stage multiple-hole impactors commercially available in three models, which allow portability and consecutive sampling. Contact plates (55 mm diameter) filled with nutrient medium are used in these samplers.

TABLE 21-4 Optimal Sampling Times for Commonly used Bioaerosol Samplers

Sampler	Sampling Flow Rate (l/min)	Sampling Time (min)	Remarks
1. Slit-to-agar[1]			Some of these samplers have
100 mm dish	30	2.6	different speeds of rotation
150 mm dish	700	0.3	Petri dish; therefore, dictate time
2. Impactor[1]			
SAS	180	0.14	Results should be "positive hole"
AND 6-stage	28.3	17.7	corrected
N6	28.3	3.0	
3. Centrifugal sampler[2]			
RCS	40	0.5–8	Impaction onto plastic strips
4. Slit sampler[3]			
Nonmoving surface	10	140	Direct microscopic counting
Moving surface	10–28.3	1 h–7days	
5. Impinger			
AGI-4	12.5	1–30	Collectate can be serially diluted
AGI-30	12.5	1–30	
6. Filter			
37 mm cassette	1–2	5–60	For hardy bioaerosols

1. Optimal sampling time is calculated for $\delta = 1$ colony/cm^2 and $c = 1000$ bioaerosol particles/m^3
2. Sampling times preset
3. Optimal sampling time is calculated for $\delta = 10^4$ particles/cm^2 and $c = 1000$ bioaerosol particles/m^3

TABLE 21-5 Company Information About Bioaerosol Samplers

Sampler	Company Address	Telephone
Andersen impactor 1-, 2-, or 6-stage	Andersen Samplers Inc., 4215-C Wendell Drive, Atlanta, GA 30336, U.S.A.	(800) 241-6898 (404) 241-6898
Mattson–Garvin slit sampler	Barramundi Corp., P.O. Drawer 4529 Homosassa Springs, FL 32647, U.S.A.	(904) 628-0200
Burkard sampler and multi-stage liquid impinger	Burkard Manufacturing Co. Ltd., Woodcock Hill, Rickmansworth Hertfordshire WD3 1PJ, U.K.	(0923) 773134/5
Casella sampler	BGI Inc., 58 Guinan St., Waltham, MA 02154, U.S.A.	(617) 891-9380
New Brunswick sampler	New Brunswick Scientific Co., Inc., Box 4005, 44 Talmadge Rd., Edison, NJ 08818, U.S.A.	(800) 631-5417 (201) 287-1200
SAS sampler	Spiral Systems Instruments, 4853 Cordell Avenue, Suite A-10 Bethesda, MD 20814, U.S.A.	(301) 657-1620
RCS sampler	Biotest Diagnostics Corp., 66 Ford Road, Suite 131, Denville, NJ 07834, U.S.A.	(800) 522-0090 (201) 625-1300
Rotorod sampler	Sampling Technologies, Inc., 26338 Esperanza Drive, Los Altos Hills, CA 94022, U.S.A.	(415) 941-1232
All-glass impinger	Ace Glass Inc., P.O. Box 688, Vineland, NJ 08360, U.S.A.	(609) 692-3333
Multi-stage liquid impinger	A.W. Dixon, 30 Anerly Station Road, London S.E. 20, U.K.	

These samplers also require statistical adjustments for multiple impactions.

There are several models of slit impactors. Some rotate a Petri dish below the inlet nozzle; others impact particles upon a microscope slide or tape. The rotating culture plate impactors are especially useful for determining temporal changes in viable bioaerosol concentrations. This feature can be used to implicate a specific source of bioaerosol when used before, during, and after an emission episode. The Burkard recording spore traps provide time discrimination in units as little as one hour or as long as one week.

The Reuter centrifugal sampler (RCS) is portable and inconspicuous and, therefore, causes minimal disturbance to a room's occupants. However, the results obtained from these samplers must be interpreted carefully since the d_{50} is larger than for the other inertial impactors. Furthermore, the device cannot be easily calibrated, except the model RCS Plus which has a built-in calibration system.

Filtration is an easy-to-use method to sample robust or microscopically identifiable bioaerosols in heavily contaminated environments (Palmgren et al. 1986; Eduard et al. 1990). It is probably not a suitable method for evaluating the levels of vegetative cells due to its desiccating effects.

The liquid of an impinger-type sampler can be used for serial dilutions and subsequent analysis of viable organisms. Collection in a liquid can also be used for endotoxin determinations (Milton et al. 1990), as well as for immunologic, genetic, and viral analyses. Liquid impingement is not an efficient method for collection of hydrophobic particles, such as many of the fungal spores.

Calibration

No matter how efficient a sampler is, unless the air flow through the device is properly calibrated, reliable bioaerosol quantitation will not be possible. In addition to precisely knowing the volume of air sampled in order to quantify the concentrations, accurately calibrated airflow rates are especially critical for impactors, which are designed to operate at specific airflow rates. Improperly set airflow rates will alter an impactor's cutoff size and, thus, its ability to collect an air sample that represents the ambient bioaerosol size distribution. When sampling with impactors, the distance from the inlet to the surface of the collection medium must also be correct. If the impactor is equipped with a movable stage, as in slit-to-agar instruments, the adjustment of the medium-to-nozzle distance is fairly easy. If the sampler is not equipped with such an alignment device, the nutrient medium must be poured into the Petri dish to a predetermined depth. Table 21-2 gives the minimum and optimal medium-to-nozzle distances (y_2).

Selection of the Sampling Medium

When sampling microbes by culture methods, the proper sampling medium must be chosen. Most growth media contain sources of carbon, nitrogen, phosphate, sulphate, iron, magnesium, sodium, potassium, and chloride ions. By varying the composition of the medium, or using still more specific nutrients, selected microbes can be cultivated or excluded.

For air sampling purposes, a nonselective medium is most often used. Many media provide the basic nutrients needed for the growth of the most common environmental microbes. Many broad-spectrum media are commercially available, e.g., tryptone–glucose–yeast agar, nutrient agar, or casein–soy–peptone agar for bacteria and malt extract agar for fungi. In a comparative study by Smid et al. (1989), almost similar results were obtained with malt extract agar, dichloran glycerol agar, and dichloran rose-bengal chloramphenicol agar.

Selective media are available for many types of microorganisms, e.g., gram-negative bacteria, and xerophilic fungi which grow on media with low water content. Fungi and bacteria may prevent each other's growth. Therefore, it is sometimes beneficial to use a fungicide in the bacterial medium and an antibiotic against bacteria in a fungal medium. A commonly used fungicide for this purpose is cycloheximide (~ 500 mg/l). Streptomycin or chlortetracycline (~ 40 mg/l) are used as antibiotics. Growth suppressants, such as rose-bengal, may be used in fungal media to limit the size of the fastest growing colonies (Morring Sorenson, and Attfield 1983). However, growth suppressants may prevent the growth of some organisms of interest and, therefore, their use is still controversial among bioaerosol researchers. An example of these controversial properties is the phototoxicity of rose-bengal (Larson and Berenbaum 1988).

Incubation Temperature

Bioaerosol samples for viable colony counts are usually incubated at room temperature (21°C–25°C), or at 28°C or 35°C in an incubator. These temperatures may favor the growth of some microbes more than others, but generally, most environmental strains grow well at temperatures below 30°C. A temperature of 35°C is generally used when monitoring food and water. As a general rule, it is important to incubate all the samples, including the reference samples, at the same temperature. Exceptions to this are selective media, for which the needed temperature is usually given by the manufacturer. If thermophilic organisms are of interest, they are usually differentiated from other species by using a much higher incubation temperature, e.g., 55°C.

Contamination

To avoid contaminating and, therefore, compromising bioaerosol samples, the principles of aseptic techniques must be followed. Aseptic technique means that a systematic practice is maintained to prevent undesired microbes and spores from contaminating the sample. In the case of ambient air sampling, for example, both the microbes on the human skin and within the respiratory system are among the nondesirable particles.

All surfaces, including washed hands, contain bacteria and spores unless they are specifically sterilized. Sterilization means total destruction of both cells and spores. This can be done by autoclaving or flaming the objects or materials in question. For example, all microbial culture media are sterilized before use. Not all objects can be sterilized and, therefore, disinfection is used to remove the majority of microbes from the surfaces and materials. Disinfection treatment with an oxidizing chemical or alcohol destroys pathogenic organisms and most other vegetative cells. Although it does not destroy all spores, disinfecting sampling equipment is usually enough to prevent significant contamination of the air sample.

Other than settling dishes, most samplers are reusable and should be thoroughly cleaned prior to each use and decontaminated either by autoclaving or by disinfection via a chemical soak or wipe-down. Special care should be given to samplers that have convoluted inlets and pathways leading to the collection media; contaminating organisms and debris can accumulate within the sampler making it difficult to disinfect. This creates the potential for compromised successive samples. When mounting the nutrient medium dishes or slides into samplers, avoid touching the nutrient surfaces by hand or other nonsterile objects. This includes falling droplets and settling dust, e.g., from inside a ventilation system. A set of control nutrient dishes should be present during the same sampling period and be incubated with the samples in order to confirm the sterility of the medium. Once used, the Petri dishes should be sealed and transported gently with the sampling surface facing down.

Sample Analysis

There are different ways to detect and quantify the collected bioaerosol particles. Directly counting the bioaerosol particles with a microscope may be appropriate for an experienced microscopist counting large particles, but small bioaerosol particles are difficult to identify and are easily masked by other particles that occur in much larger quantities. Direct counting is used for pollen grains and fungal spores, but bacterial cells are not visible with a light microscope unless stained. All the particles containing biological material can be detected by staining their nucleic acids with a fluorescent stain, e.g., acridine orange, and counting them with an epifluorescence microscope.

When microorganisms are cultured into countable colonies, the incubation conditions and the medium should be suitable for the organisms of interest, ideally all viable microbes. In most cases, this may be difficult because of the vast variation in the growth needs of the different organisms present. In this type of sample analysis, the absolute number of bacterial cells or fungal spores in the sampled air cannot be obtained because the airborne particles may have been aggregates of two or more cells or spores (Eduard et al. 1990). The results are given in units of cfu/m^3, i.e., colony-forming units in a cubic meter of air.

Identification of fungal colonies is based on the morphology of the colonies, spores, and hyphae as well as on different physiological tests. A number of handbooks are available that allow the identification of fungi to a specific genus (e.g., Kendrick 1985). The identification of fungal species is an elaborate task which calls for specific expertise in environmental mycology.

Identification of bacterial colonies is based mainly on the morphology and staining properties of the cells and different physiological and biochemical tests. Easy-to-use test kits are available that identify with a certain probability when supported by morphological characteristics. Identification of environmental bacteria calls for specified experience because these organisms differ from clinically significant species. The principles of identification and classification are presented in bacteriological handbooks, e.g., Krieg et al. (1984) or Truper and Kramer (1981). In most cases, a general classification of the colony types, e.g., gram-negative rods or gram-positive cocci, or genus identification, e.g., *Staphylococcus* or *Pseudomonas*, provides enough information to draw conclusions from the sampling results. However, in some cases, species identification is needed. For example, it may be important to know whether pathogenic *Staphylococcus aureus* or *Pseudomonas aeruginosa* is present. This may require assistance from bacteriological laboratories.

Antigen and toxin analyses are still research techniques and in the event they are needed, arrangements must be made with appropriate research laboratories.

Data Analysis and Interpretation

There are no guidelines for acceptable or harmful levels of bioaerosols. Therefore, it is necessary to decide in advance the criteria that will be used to determine whether or not an environment is contaminated. For this purpose, reference data about the range of concentrations and the airborne microbial flora occurring in the sampled environments should be collected with the sampler and the analysis method to be used. An unusual exposure situation exists if the concentrations in the studied environment differ from other, symptom-free or noncomplaint environments by more than an order of magnitude when sampled with the same method. In non-industrial indoor environments, the levels should be lower than those outdoors. Furthermore, the composition of the species in the environment under study may differ from the control environments.

References

ACGIH. 1989. *Guidelines for the Assessment of Bioaerosols in the Indoor Environment*. Cincinnati, OH: American Conference of Governmental Industrial Hygienists.

Andersen, A. A. 1958. New sampler for the collection, sizing and enumeration of viable airborne particles. *J. Bacteriol.* 76:471–84.

Atlas, R. M. 1988. *Microbiology, Fundamentals and Applications*. New York: Macmillan.

Atlas, R. M. and R. Bartha. 1987. *Microbial Ecology, Fundamentals and Applications*. Reading, MA: Addison–Wesley.

Bernstein, I. L. and R. S. Safferman. 1970. Viable algae in house dust. *Nature* 227:851–52.

Bernstein, I. L. and R. S. Safferman. 1973. Clinical sensitivity to green algae demonstrated by nasal challenge and *in vitro* tests of immediate hypersensitivity. *J. Aller. Clin. Immunol.* 51:22–8.

von Bischoff, E. 1989. Sources of pollution of indoor air by mite allergen containing house dust. *Environ. Int.* 15:181–92.

Brachman, P. S., R. Ehrlich, H. F. Eichenwald, V. J. Gabelli, T. W. Kethley, S. H. Madin, J. R. Maltman, G. Middlebrook, J. D. Morton, I. H. Silver, and E. K. Wolfe. 1964. Standard sampler for assay of airborne microorganisms. *Science* 144:1295.

Bratbak, G. and I. Dundas. 1984. Bacterial dry matter content and biomass estimations. *Appl. Environ. Microbiol.* 48:755–57.

Brown, R. M. Jr., D. A. Larson, and H. C. Bold. 1964. Airborne algae: Their abundance and heterogeneity. *Science* 143:583.

Burge, H., W. R. Solomon, and J. R. Boise. 1980. Microbial prevalence in domestic humidifiers. *Appl. Environ. Microbiol.* 39:840–44.

Buttner, M. P. and L. D. Stetzenbach. 1991. Evaluation of four aerobiological sampling methods for the retrieval of aerosolized *Pseudomonas syringae*. *Appl. Environ. Microbiol.* 57:1268–70.

Chatigny, M. A., J. M. Macher, H. A. Burge, and W. R. Solomon. 1989. Sampling airborne microorganisms and aeroallergens. In *Air Sampling Instruments for Evaluation of Atmospheric Contaminants*, ed. S. V. Hering, pp. 199–220. Cincinnati, OH: American Conference of Governmental Industrial Hygienists.

Clark, S., R. Rylander, and L. Larsson. 1983. Airborne bacteria, endotoxin and fungi in dust in poultry and swine confinement buildings. *Am. Ind. Hyg. Assoc. J.* 44:537–41.

Cox, C.S. 1987. *The Aerobiological Pathway of Microorganisms*. Chichester: Wiley.

Curtis, S. E., R. K. Balsbaugh, and J. G. Drummond. 1978. Comparison of Andersen eight-stage and two-stage viable air samplers. *Appl. Environ. Microbiol.* 35:208–9.

Daugherty, R. L., J. B. Franzini, and E. J. Finnemore. 1985. *Fluid Mechanics with Engineering Applications*. New York: McGraw-Hill.

Doetsch, R. N. and T. M. Cook. 1973. *Introduction to Bacteria and Their Ecobiology*. Baltimore: University Park Press.

Donaldson, A. I. 1978. Factors influencing the dispersal, survival and deposition of airborne pathogens of farm animals. *Vet. Bull.* 48:83–94.

Eduard, W., J. Lacey, K. Karlsson, U. Palmgren, G. Ström, and G. Blomquist. 1990. Evaluation of methods for enumerating microorganisms in filter samples from highly contaminated occupational environments. *Am. Ind. Hyg. Assoc. J.* 51:427–36.

Faegri, K. and J. Iversen. 1989. *Textbook of Pollen Analysis*. IV edn., eds. K. Faegri, P. E. Kaland, and K. Krzywinski. Chichester: Wiley.

Gerone, P. J., R. B. Couch, G. V. Keefer, R. G. Douglas, E. B. Derrenbacher, and V. Knight. 1966. Assessment of experimental and natural viral aerosols. *Bacteriol. Rev.* 30:576–84.

Gillespie, V. L., C. S. Clark, H. S. Bjornson, S. J. Samuels, and J. W. Holland. 1981. A comparison of two-stage and six-stage Andersen impactors for viable aerosols. *Am. Ind. Hyg. Assoc. J.* 42:858–64.

Groeschel, D. H. M. 1980. Air sampling in hospitals. *Ann. N.Y. Acad. Sci.* 353:230–40.

Hamer, G. 1985. Chemical engineering and biotechnology. In *Biotechnology, Principles and Applications*, eds. I. J. Higgins, D. J. Best, and J. Jones. Oxford: Blackwell.

Hinds, W. C. 1982. *Aerosol Technology.* New York: Wiley.

Jacobs, R. R. 1989. Airborne endotoxins: An association with occupational lung disease. *Appl. Ind. Hyg.* 4:50–6.

Jones, W. J., K. Morring, P. Morey, and W. Sorenson. 1985. Evaluation of the Andersen viable impactor for single stage sampling. *Am. Ind. Hyg. Assoc. J.* 46:294–98.

Kauppinen, E. I., A. V. K. Jäppinen, R. E. Hillamo, A. Rantio-Lehtimäki, and A. Koivikko. 1989. A static particle sized selective bioaerosol sampler for the ambient atmosphere. *J. Aerosol Sci.* 20:829–38.

Keleti, G. and M. A. Shapiro. 1987. *Legionella* and the environment. *CRC Critical Rev. Environ. Control* 17:133–85.

Kendrick, B. 1985. *The Fifth Kingdom.* Waterloo, Ontario, Canada: Mycologue.

Kotimaa, M. 1990. Occupational exposure to spores in the handling of wood chips. *Grana* 29:153–56.

Kotimaa, M. H., L. Oksanen, and P. Koskela. 1991. Feeding and bedding materials as sources of microbial exposure on dairy farms. *Scand. J. Work Environ. Health* 17:117–22.

Krieg, N. R., P. H. A. Sneath, N. S. Mair, M. E. Sharpe, and J. G. Holt (eds.). 1984. *Bergey's Manual of Systematic Bacteriology.* Baltimore: Williams & Wilkins.

Lacey, J. and B. Crook. 1988. Fungal and actinomycete spores as pollutants of the workplace and occupational allergens. *Ann. Occup. Hyg.* 32:515–33.

Lach, V. 1985. Performance of the surface air system air samplers. *J. Hosp. Inf.* 6:102–7.

Larson, R. A. and M. R. Berenbaum. 1988. Environmental phototoxicity. *Environ. Sci. Technol.* 22:354–60.

Lundholm, M. 1982. Comparison of methods for quantitative determinations of airborne bacteria and evaluation of total viable counts. *Appl. Environ. Microbiol.* 44:179–83.

Macher, J. M. 1989. Positive-hole correction of multiple-jet impactors for collecting viable microorganisms. *Am. Ind. Hyg. Assoc. J.* 50:561–68.

Macher, J. M. and M. W. First. 1983. Reuter centrifugal air sampler: Measurement of effective airflow rate and collection efficiency. *Appl. Environ. Microbiol.* 45:1960–62.

Macher, J. M., F.-Y. Huang, and M. Flores. 1991. A two-year study of microbiological indoor air quality in a new apartment. *Arch. Environ. Health* 46:25–9.

Marple, V. A. and K. Willeke. 1976. Impactor design. *Atmos. Environ.* 10:891–96.

Martinez, F. K., J. W. Sheehy, J. H. Jones, and L. B. Cusick. 1988. Microbial containment in conventional fermentation processes. *Appl. Ind. Hyg.* 3:177–81.

May, K. R. 1966. Multistage liquid impinger. *Bacteriol. Rev.* 30:559–70.

Milton, D. K., R. J. Gere, H. A. Feldman, and I. A. Greaves. 1990. Endotoxin measurement: Aerosol sampling and application of a new Limulus method. *Am. Ind. Hyg. Assoc. J.* 51:331–37.

Morring, K. L., W. G. Sorenson, and M. D. Attfield. 1983. Sampling for airborne fungi: A statistical comparison of media. *Am. Ind. Hyg. Assoc. J.* 44:662–64.

Nevalainen, A., J. Pastuszka, F. Liebhaber, and K. Willeke. 1992. Performance of bioaerosol samplers: Collection characteristics and sampler design considerations. *Atmos. Environ.* 26(A):531–40.

Noble, W. C. 1975. Dispersal of skin microorganisms. *Br. J. Dermatol.* 93:477–85.

Orr, C. Jr. and M. T. Gordon. 1956. The density and size of airborne *Serratia marcescens. J. Bacteriol.* 71:315–17.

Palmgren, U., G. Ström, G. Blomquist, and P. Malmberg. 1986. Collection of airborne microorganisms on Nuclepore filters, estimation and analysis —CAMNEA method. *J. Appl. Bacteriol.* 61:401–6.

Placencia, A. M., J. T. Peeler, G. S. Oxborrow, and J. W. Danielson. 1982. Comparison of bacterial recovery by Reuter centrifugal air sampler and slit-to-agar sampler. *Appl. Environ. Microbiol.* 44:512–13.

Rahkonen, P., M. Ettala, M. Laukkanen, and M. Salkinoja-Salonen. 1990. Airborne microbes and endotoxins in the work environment of two sanitary landfills in Finland. *Aerosol Sci. Technol.* 13:505–13.

Reponen, T., A. Nevalainen, and T. Raunemaa, T. 1989. Bioaerosol and particle mass levels and ventilation in Finnish homes. *Environ. Int.* 15:203–8.

Riley, R. L., G. Murphy, and R. L. Riley. 1978. Airborne spread of measles in a suburban elementary school. *Am. J. Epidemiol.* 107:421–32.

Roszak, D. B. and R. R. Colwell. 1987. Survival strategies of bacteria in the natural environment. *Microbiol. Rev.* 51:365–79.

Rylander, R. 1986. Lung diseases caused by organic dusts in the farm environment. *Am. J. Ind. Med.* 10:221–27.

Rylander, R. and P. Haglind. 1984. Airborne endotoxins and humidifier disease. *Clin. Allergy* 14:109–12.

Smid, T., E. Schokkin, J. S. M. Boleij, and D. Heederik. 1989. Enumeration of viable fungi in occupational environments: A comparison of samplers and media. *Am. Ind. Hyg. Assoc. J.* 50:235–39.

Sorenson, W. G., D. G. Frazer, B. B. Jarvis, J. Simpson, and V. A. Robinson. 1987. Trichothecene mycotoxins in aerosolized conidia of *Stachybotrys atra. Appl. Environ. Microbiol.* 53:1370–75.

Truper, H.G. and J. Kramer. 1981. Principles of characterization of prokaryotes. In *The Prokaryotes*, eds. M. P. Starr, H. Stolp, H. G. Truper, A. Balows, and H. Schlegel. Berlin: Springer.

Verhoeff, A. P., J. H. van Wijnen, J. S. M. Boleij, B. Brunekreef, E. S. van Reenen-Hoekstra, and R. A. Samson. 1990. Enumeration and identification of airborne viable mould propagules in houses. *Allergy* 45:275–84.

Willeke, K. and J. J. McFeters. 1975. The influence of flow entry and collecting surface on the impaction efficiency of inertial impactors. *J. Colloid Interface Sci.* 53:121–27.

Zimmerman, N. J., P. C. Reist, and A. G. Turner. 1987. Comparison of two biological aerosol sampling methods. *Appl. Environ. Microbiol.* 53:99–104.

22

Instrument Calibration

Bean T. Chen

Inhalation Toxicology Research Institute
Lovelace Biomedical and Environmental Research Institute
P.O. Box 5890, Albuquerque, NM, U.S.A.

INTRODUCTION

Most knowledge concerning aerosol properties has been obtained by experimental means using aerosol instruments. These instruments can be categorized as: (1) the collection-and-analysis instruments, such as a cascade impactor, the Atkin-type condensation nuclei counter, or a filter sampler, that are designed to remove particles from gas streams for direct measurements; and (2) the real-time, direct-reading instruments, such as an optical particle counter, photoelectric condensation nuclei counter, or photometers, that may provide high sensitivity, good accuracy, and quick response. Ideally, instrument response can be theoretically computed based on equations and procedures described in the previous chapters. However, practical considerations, such as compactness, portability of the instrument, and convenience of operation, may influence the design of the instrument. As a result of the design, the sampling and counting efficiencies of the instrument may vary, and a theoretical prediction of the instrument response based on the ideal conditions may not be fulfilled. For example, although 50% effective cutoff diameters and collection efficiencies for an impactor stage can be computed, the phenomena of particle bounce, reentrainment, electrostatic charge effects, and wall losses do occur (Cheng and Yeh 1979). Therefore, the measured impaction efficiency may be different from the theoretically determined collection efficiency and, thus, experimental calibration is essential.

Most instruments are usually calibrated and evaluated by the manufacturer or the inventor before being used by others. For an instrument intended to collect and analyze an aerosol, the collection efficiency and wall loss are generally determined in the calibration. For a real-time, direct-reading instrument, calibration establishes the relationship between an instrument's response (e.g., electronic signal or channel number) and the value of the property (e.g., particle size, number concentration, or mass concentration) being measured. However, the operating conditions and the parameters used during the original calibration can vary from those under which the eventual user operates. As a result, the original calibration data may not apply, and the user must calibrate the instrument to operate it with confidence. In general, a reliable and accurate calibration requires (1) a proper selection of a desired test aerosol, (2) a complete understanding of the principles and procedures of operation, (3) a thorough investigation of relevant parameters, and (4) a sufficient knowledge of the capabilities and limitations of the instrument.

In the last two decades, the developments in monodisperse aerosol generation and classification, along with the progress in electron microscopy and imaging analysis, have made instrument calibration easier and more reproducible. This chapter reviews the calibration techniques relevant to aerosol measurement devices, such as sizing instruments, condensation nuclei counters, and mass monitors. The generation methods for test aerosols and important parameters in instrument calibration are emphasized. Also reviewed are the calibration and use of flow-monitoring devices that play an integral role in aerosol sampling and instrument calibration.

DIRECT MEASUREMENT AND PRIMARY STANDARDS

Although aerosol instruments can be categorized based on either the particle properties characterized (inertia, gravitational, optical, diffusional, thermal, or electrical) or the measuring techniques used (real-time or sample collection, personal or area, passive or active), the measured parameters are either particle size, number concentration, or mass concentration. For direct measurements of these parameters, an aerosol sample is traditionally collected, using a filter sampler or an electrostatic precipitator, and then analyzed. Particle size distribution and number concentration can be determined by examining the sample with an optical or electron microscope. A microscopic observation of aerosol particles from the sample permits a direct measurement of the particle size. This is in contrast to indirect methods, such as sedimentation, impaction, mobility analysis, or light scattering, where the particle size is estimated from the measurement of a property related to size. A microscope can also be used to count particles in the sample. Number concentration is then obtained by dividing the total number of particles collected by the gas volume sampled. The most common way to determine aerosol mass concentration is to pass a known volume of an aerosol through a high-efficiency filter and determine the increase in the mass of the filter due to the collected aerosol particles. The mass can be determined using either an analytical balance (0.01 mg precision) or an electronic balance (0.1 μg–0.1 mg precision). Mass concentration is obtained by dividing the increased mass by the gas volume sampled. These methods provide direct measurements of aerosol size distribution, number concentration, and mass concentration and are considered as absolute standards for instrument calibration. For example, an Aitken-type condensation nuclei counter (CNC) is traditionally considered as an absolute standard for other CNCs because it allows direct counting of the settled droplets through a microscope.

It is, however, time-consuming and sometimes impractical to calibrate instruments using direct methods. Therefore, a primary standard that has been well calibrated with the absolute standard is normally used for routine instrument calibration. For example, the commercially available monodisperse polystyrene latex spheres with a known size are used as primary standards for calibrating sizing instruments and the measurement of electric charge on the aerosol classified by a differential-mobility analyzer is used by several researchers as a primary standard for the CNCs. For mass concentration measurements, the simultaneous filter samples followed by gravimetric measurements are the most common calibration standards for real-time mass monitors. The trend is that a well-calibrated, real-time instrument with a direct readout is more widely used as a primary calibration standard than one which provides a direct measurement.

GENERAL CONSIDERATIONS

Instrument calibration is essential to a successful measurement of aerosol properties in a sampling environment. Practically, one should consider not only the techniques and data analysis used during calibration, but also the information on the sampling environment and the aerosol properties to be measured.

Sampling Environment

The type of aerosol instrument selected and the manner in which it is calibrated may strongly depend on the environment in which the aerosol is to be sampled. In general, one should first attempt to identify the aerosol sources in the environment and decide what information is needed and for what purpose before selecting the parameters and the test aerosol for calibration.

Parameters to be Investigated

After gathering enough information on the sampling environment, one should select a set of parameters to be investigated during the instrument calibration. These parameters should be chosen depending on the measuring property in the instrument as well as on the aerosol properties of interest in the sampling environment. Generally, two types of parameters are considered in instrument calibration: one is related to the instrument operation and the other is related to the test aerosol properties. For example, the volumetric flow rate, pressure drop, and light source intensity are operating parameters, while particle size, shape, and intrinsic properties of the suspending gas medium are aerosol parameters. The parameters selected can be different between two instruments. To calibrate an aerodynamic sizing device, the effects of particle density, velocity, and ambient pressure on the instrument response are important, while, in an optical particle counter (OPC), the particle refractive index, wavelength of the light source, and collection angles of scattered light are the important parameters.

Selection of Test Aerosols

A proper selection of the test aerosols is essential to instrument calibration. Because most instruments have responses strongly dependent on the physical and chemical properties of the aerosol particles, the calibration curve of an instrument is usually only valid for the test aerosol. For an aerosol whose physical and chemical properties are different from those of the test aerosol, the data interpretation based on the calibration could be misleading (Willeke and Baron 1990). For example, one would underestimate the size distribution of a carbon black aerosol by using the calibration curve of an OPC obtained from polystyrene latex spheres because the carbon particles cannot scatter as much light as the polystyrene latex spheres can. Ideally, an aerosol that has physical and chemical properties (e.g., size, shape, density, refractive index, dielectric constant, and thermal conductivity) similar to those of the aerosol to be measured should be selected as the test aerosol during calibration. For general calibration purposes, a monodisperse aerosol with spherical particles is often used.

Calibration Practices

System setups and sampling techniques should be considered during calibration. The integrity of the system should be maintained by avoiding possible leakage and blockage of flow. The techniques and criteria of isokinetic sampling and sampling from still air should be followed, especially for aerosol particles greater than 5 μm in aerodynamic diameter. For each data point, at least three calibrations are required to provide statistically valid information.

Data Analysis

A calibration curve that contains the relationship between the instrument's responses and the values of a certain aerosol property is established after calibration. In the case of an instrument that directly indicates a value of the measured property, calibration provides an adjustment (or a correction factor) to the indicated value. In addition, the resolution and sensitivity of the instrument should be examined and analyzed. The rule of thumb is to conduct the calibration based on all important parameters, to assemble all the data, and express that data in a generalized mathematical equation, relating the instrument response to a single parameter (Chen, Cheng, and Yeh 1985; Zhang and Liu 1990).

For data analysis, instrument manufacturers sometimes provide a built-in algorithm whose properties, accuracy, and limitations are often unknown to the user. Unless a user understands the algorithm, the analysis should be developed based only on the raw calibration data, without any manipulation by the built-in algorithm.

CALIBRATION APPARATUS AND PROCEDURES

Figure 22-1a is a schematic diagram of a typical calibration apparatus for aerosol instruments. It includes an aerosol generator, aerosol conditioning devices (e.g., diffusion

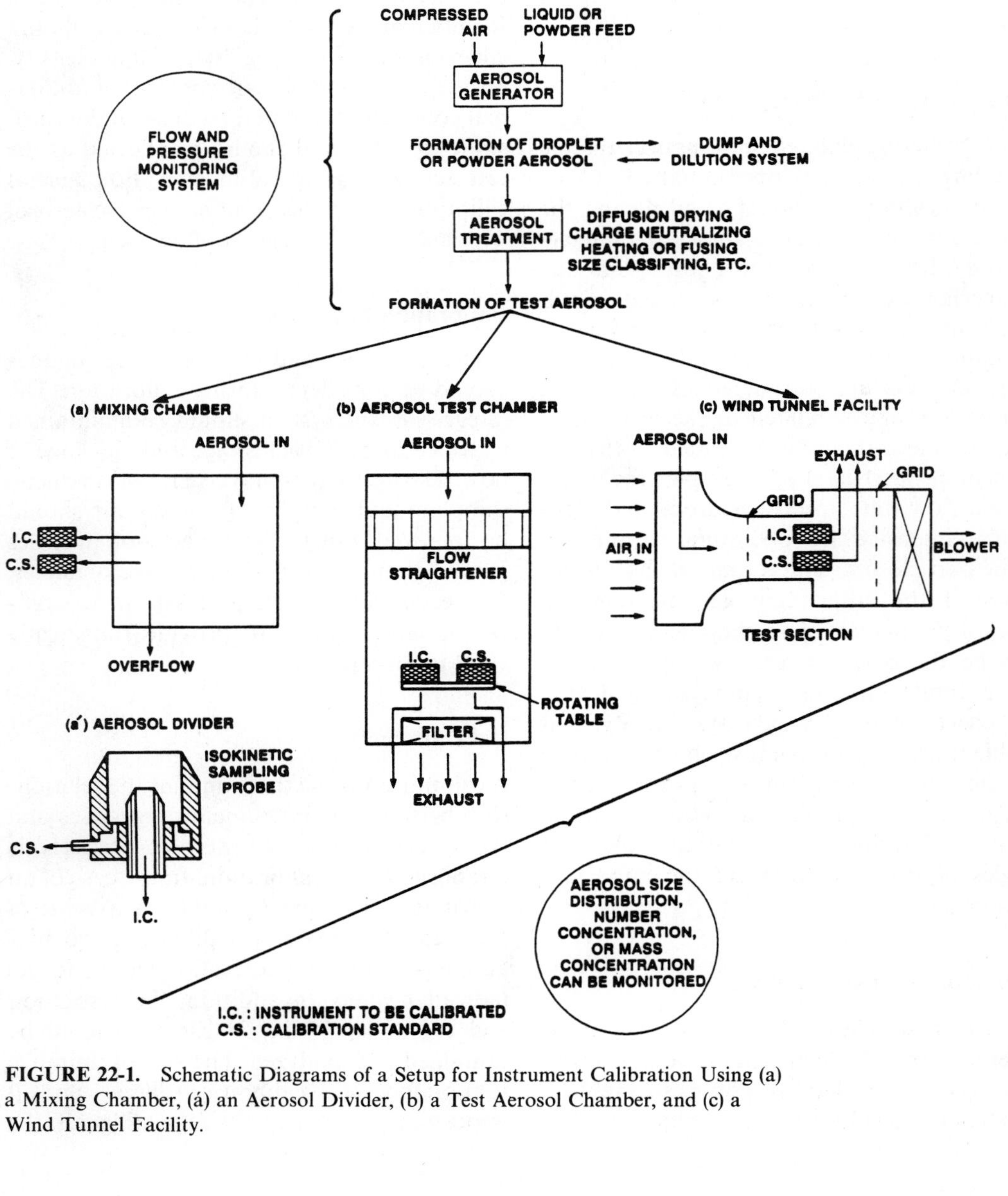

FIGURE 22-1. Schematic Diagrams of a Setup for Instrument Calibration Using (a) a Mixing Chamber, (á) an Aerosol Divider, (b) a Test Aerosol Chamber, and (c) a Wind Tunnel Facility.

dryer, charge neutralizer, aerosol classifier, aerosol concentrator, and dilution air supply), a mixing chamber, pressure and air flow monitoring equipment, the instrument to be calibrated, and a primary calibration standard for direct measurement. The aerosol produced from the generator can be monodisperse or polydisperse, solid or liquid, wet or dry, charged or noncharged, or spherical or nonspherical (described later). Generally, this aerosol requires several steps of conditioning before use. For an aerosol containing volatile vapors or water droplets, a diffusion dryer with desiccant and/or charcoal is commonly used to produce a dry aerosol. In some cases, a heat treatment using a high-temperature furnace is required for the production of test aerosol (Kanapilly, Raabe, and Newton 1970; Chen, Cheng, and Yeh 1990). The heat treatment involves either sintering or fusing the particles to reach the desired particle morphology and chemical form, or initiating particle evaporation and subsequent condensation to produce monodisperse particles. Because aerosol particles are usually charged by static electrification during formation, a neutralizer containing a bipolar ion source (e.g., ^{85}Kr) is often used in the aerosol treatment. This reduces the number of charges on the particles and results in an aerosol with charge equilibrium (John 1980). In addition, a size-classifying device is often used in the aerosol treatment to segregate particles of similar size or of a desired size fraction (Liu and Pui 1974; Chen, Yeh, and Rivero 1988; Romay-Novas and Pui 1988). Furthermore, a concentrator or a dilutor is often used to adjust the aerosol concentration (Barr et al. 1983; Yeh and Carpenter 1983).

The desired test aerosol can be used to calibrate instruments in several ways. The simplest way, as shown in Fig. 22-1a, is to introduce the test aerosol into a mixing chamber in which the aerosol is uniformly distributed and sampled by both the instrument to be calibrated and the calibration standard. Pressure in the chamber and flow rate through the instrument are monitored. A sampling device, such as a filter sampler or an electrostatic precipitator, is often used to collect reference samples for direct measurement. To ensure that both the calibration device and the instrument to be calibrated have compatible aerosol samples, an aerosol divider is used as a common sampling port for calibrating a mass monitor (Marple and Rubow 1978). In the aerosol divider, the flow is split isokinetically into two streams: one passes directly into the instrument to be calibrated and the other flows through the calibration standard (Fig. 22-1a′).

Another way of calibrating an instrument is to introduce the test aerosol into a test aerosol chamber (Fig. 22-1b) which contains the subject instrument (Marple and Rubow 1983). This chamber provides a quiescent atmosphere in which the entire instrument can be exposed to the aerosol as in the real sampling environment. The test aerosol is introduced at the top of the chamber and uniformly distributed in the section where the instrument is set on a rotating table. Rotation provides a means for collecting samples at various locations to reduce any effects due to possible temporal and spatial variations in aerosol concentration.

Both the mixing chamber and the aerosol test chamber are used when the instrument to be calibrated is operated with low or zero ambient-wind velocity. To evaluate the collection efficiency and internal loss of an atmospheric sampler or a personal sampler, a wind tunnel facility (Fig. 22-1c) is generally the site for instrument calibration (Prandtl 1952). The sampler to be calibrated or evaluated is located inside the tunnel. The wind tunnel provides a wide range of wind velocity (up to 40 km/h) and different orientation angles to simulate different atmospheric conditions. The wind velocity, flow uniformity, and turbulence are monitored using flow-monitoring devices (described later) to ensure a uniform aerosol concentration at the sampling zone and a minimal turbulent effect on sampling efficiency and internal loss. During calibration of the test sampler, an isokinetic sampler is generally used to collect reference samples.

Before any calibration, a standard operating protocol should be prepared. First, the manual for the instrument to be calibrated should be read carefully. However, the manual only serves as an initial guide to instrument operation; developing a successful calibration procedure is the responsibility of the user. For example, a laser-operated optical particle counter tends to produce oscillated responses when the particles are larger than the wavelength of the laser beam; however, the calibration curve provided by the manufacturer seldom reveals this phenomenon (Chen, Cheng, and Yeh 1984). Unless the user calibrates the instrument using monodisperse aerosols of many sizes, this oscillated scattering phenomenon will not be revealed experimentally.

In addition, the integrity of the whole system should be checked prior to calibration. The integrity can be quantified from a series of pressure measurements on a sealed system that has initially been brought to a pressure slightly above or below the ambient pressure (Mokler and White 1983). Any possible leakage can be discovered by various methods. The simplest one is to pressurize the system slightly and then put soapy water on the surface of the system to detect the leakage. Freon can also be injected into the system to detect the location of the leakage.

TEST AEROSOL GENERATION

In Chapter 5, several test aerosol generators have been described. The test aerosols contain either monodisperse or polydisperse, spherical or nonspherical, solid, or liquid particles (Mercer 1973; Raabe 1976; Hinds 1982; Lippmann 1989). The characteristics of an ideal generator are a constant and reproducible output of monodisperse and stable aerosol particles, whose size and concentration can be easily controlled. For general instrument calibration, the test aerosol often contains monodisperse, spherical particles. To calibrate an instrument for a specific environment, the test aerosol should have physical and chemical properties similar to those of the aerosol of interest. In addition, one should consider the environment in which the instrument is to be operated when selecting the test aerosol. For example, if the instrument is to be operated in a high-temperature environment, the desired test aerosol could be a refractory metal oxide, such as cerium oxide, because of its thermal stability and chemical inertness. Generally, as long as the desired aerosol to be used for instrument calibration can be determined, the method of generation is not a problem. Table 22-1 lists the test aerosols frequently used for instrument calibration. Monodisperse aerosols containing spherical particles are the most widely used. Particles with nonspherical shapes are sometimes used in calibration to study the possible effect of shape on the instrument response. Polydisperse dust particles have also been used in calibrating dust monitors. This is important, because most real aerosols contain nonspherical particles of different sizes.

The size distribution and concentration of a test aerosol depend on both the characteristics of the generator and the feed material. The information given in this section is intended as a guide for the selection of appropriate generation techniques. The actual size distribution in each application should always be measured directly with the appropriate instruments.

Monodisperse Aerosols with Spherical Particles

The methods of producing monodisperse aerosols with spherical particles have been reviewed by Fuchs and Sutugin (1966), Mercer (1973), and Raabe (1976). These methods include the atomization of a suspension of monodisperse particles, the formation of uniform droplets by dispersion of liquid jets with periodic vibration or a spinning disc (top), and the growth of uniform particles or droplets by controlled condensation.

Atomization of Suspensions of Monodisperse Particles

The simplest way of producing monodisperse aerosols is by air-blast-nebulizing a dilute

TABLE 22-1 List of Test Aerosols and Generation Methods Used for Instrument Calibration

Test Aerosol	Particle Morphology	Size Range[1]		Density (g/cm^3)	Refractive Index	Generation Method	Aerosol Output ($particles/cm^3$)
		VMD (μm)	σ_g				
PSL (PVT)	Spherical, solid	0.01–30	≤ 1.02	1.05 (1.027)	1.58	Nebulization	$< 10^4$
Fluorescent uranine	Irregular, solid	< 8	1.4–3	1.53	—	Nebulization	$< 10^9$
Dioctyl phthalate	Spherical, liquid	0.5–40	≤ 1.1	0.99	1.49	Vibrating atomization	$< 10^5$
Oleic acid	Spherical, liquid	0.5–40	≤ 1.1	0.89	1.46	Vibrating atomization	$< 10^5$
Ammonium fluorescein	Spherical, solid	0.5–50	≤ 1.1	1.35	—	Vibrating atomization	$< 10^5$
Fused ferric oxide	Spherical, solid	0.2–10	≤ 1.1	2.3	—	Spinning-disc (top) atomization	$< 10^7$
Fused aluminosilicate	Spherical, solid	0.2–10	≤ 1.1	3.5	—	Spinning-disc (top) atomization	$< 10^7$
Fused cerium oxide	Spherical, solid	0.2–10	≤ 1.1	4.33	—	Spinning-disc (top) atomization	$< 10^7$
Sodium chloride	Irregular, solid	0.002–0.3	≤ 1.2	2.17	1.54	Evaporation/condensation	$< 10^6$
Silver	Irregular, solid	0.002–0.3	≤ 1.2	10.5	0.54	Evaporation/condensation	$< 10^6$
Coal dust	Irregular, solid	≈ 3.3	≈ 3.2	1.45	1.54–0.5i	Dry-powder dispersion	$< 30\ mg/m^3$
Arizona road dust	Irregular, solid	≈ 3.8	≈ 3.0	2.61	—	Dry-powder dispersion	$< 30\ mg/m^3$

1. Aerosol treatment of drying, charge neutralization, and size classification is generally used

liquid suspension containing monodisperse polystyrene (PSL) or polyvinyl toluene (PVT) latex spheres. These spheres are commercially available in a size range from 0.01 to 30 μm (Duke Scientific, Palo Alto, CA; Dyno Industrier A. S., Lillestrom, Norway; Japan Synthetic Rubber, Tokyo, Japan; Polysciences, Warrington, PA; Seragen Diagnostics, Indianapolis, IN; 3M, Minneapolis, MN). PSL particles of different sizes have also been concurrently produced in an aerosol to obtain more than one data point per experimental run. Monodisperse latex particles containing fluorescent dye or radiolabeled isotopes are also used in calibrations when quantitative measurements by fluorometric or radiometric techniques are needed (Newton et al. 1980; Chen et al. 1991).

Three problems arise in the generation of these latex particles: particle size measurement, formation of aggregates, and existence of residual particles. The diameters of the particles reported by the manufacturer can be different from those measured by researchers using an electron microscope, because the particles tend to evaporate and shrink due to electron irradiation or to increase in size due to the absorption of a contaminant under irradiation. Porstendörfer and Heyder (1972) and Yamada, Miyamoto, and Koizumi (1985) recommend a measurement of these particles by an electron microscope, but the size calibration should be conducted under the conditions of no particle shadowing, minimal electron beam intensity, and short exposure time.

The second problem in generating latex particles is the formation of aggregate latex particles in the aerosol. The percentage of aggregates can be reduced by diluting the suspension. Assuming that the probability of the number of particles in an atomized droplet can be described by Poisson statistics, and that the droplet size distribution can be approximated by a lognormal distribution, Raabe (1968) derived the following equation to calculate the latex dilution factor, Y, necessary to give a desired single ratio, R, which is the number of droplets containing single particles relative to the total number of droplets containing particles:

$$Y = F(\mathrm{VMD})^3 \exp(4.5 \ln^2 \sigma_g) \times [1 - 0.5\exp(\ln^2 \sigma_g)]/(1 - R)d_p^3 \quad (22\text{-}1)$$

where F is the volumetric fraction of individual particles of diameter d_p in the original latex suspension, and VMD and σ_g are the volume median diameter and the geometric standard deviation of the droplet size distribution, respectively. The values of VMD and σ_g of commonly used air-blast atomizers are listed in Table 22-2. This equation is limited to values of $\sigma_g < 2.1$ and $R > 0.9$.

Example 22-1

A bottle of 1 μm PSL suspension containing a 10% solid is being used to produce a test aerosol containing at least 95% singlets. What is the dilution factor required in this suspension if the Retec X-70/N nebulizer is used and operated at 20 psig?

Answer. Using Eq. 22-1

$$Y = F(\mathrm{VMD})^3 \exp(4.5 \ln^2 \sigma_g) \times (1 - 0.5\exp(\ln^2 \sigma_g))/(1 - R)d_p^3$$

$$F = 10\% = 0.1$$

$$R = 95\% = 0.95$$

$$d_p = 1\ \mu\mathrm{m}$$

Based on size distribution data given in Table 22-2,

$$\mathrm{VMD} = 5.7\ \mu\mathrm{m}$$

$$\sigma_g = 1.8$$

$$Y = (0.1)(5.7)^3 \exp(4.5 \ln^2(1.8)) \times (1 - 0.5\exp(\ln^2(1.8)))/(1 - 0.95)(1)^3$$

$$= (0.1)(5.7)^3 \exp(1.555) \times (1 - 0.5\exp(0.345))/(0.05)$$

$$= (0.1)(5.7)^3(4.734)(0.294)/(0.05)$$

$$= 514.9$$

TABLE 22-2 Operating Parameters of Air-Blast and Ultrasonic Nebulizers

Nebulizer	Operating Conditions: Orifice Diameter (mm)	Operating Conditions: Air Pressure (psig)	Operating Conditions: Frequency (mHz)	Flow[1] Rate (l/min)	Aerosol Output (μl/l)	Droplet Size Distribution: VMD (μm)	Droplet Size Distribution: σ_g	Commercial Source
Airblast Type								
Collison	0.35	15		2.0	8.8	2.5–3.0	—	BGI, Inc.,
		25		2.7	7.7	1.9–2.0	—	Waltham, MA
DeVilbiss[2] D-40	0.84	15		12.4	15.5	4.2	1.8	The DeVilbiss Co.,
		30		20.9	12.1	2.8	1.9	Somerset, PA
DeVilbiss D-45	0.76	15		9.4	23.2	4.0	—	The DeVilbiss Co.,
		30		14.5	22.9	3.4	—	Somerset, PA
Lovelace	0.26	20		1.5	40	5.8	1.8	In-Tox Products,
		50		2.3	27	2.6	2.3	Albuquerque, NM
Retec X-70/N	0.46	20		5.0	46	5.7	1.8	Cavitron Corp.,
		50		9.7	47	3.2	2.2	Portland, OR
Ultrasonic Type								
DeVilbiss 880 (2)[a]			1.35	41.0	54	5.7	1.5	The DeVilbiss Co.,
(4)[a]			1.35	41.0	150	6.9	1.6	Somerset, PA
Mistogen[d]			1.40	24.7	62	6.5	1.4	The Timeter Group, Lancaster, PA
Monaghan 650 (5)[a]			1.65	8.0	16	5.1	1.6–2.0	Monaghan Co.,
(10)[a]			1.65	8.0	50	5.2	1.6–2.0	Littleton, CO

1. Output per orifice. 2. Vent closed. a. Power settings. d. With reservoir.

A dilution factor of at least 515 is needed to produce an aerosol containing 95% of singlet PSL particles.

The third problem arises when nonlatex residual particles are present in the aerosol as a result of the impurities and surfactant in the liquid suspension. The nonlatex particles could either distort the size of an individual particle, if they attach to it, or become individual particles with a size different from that of the latex particles in the aerosol. To reduce the impurity in the diluting water, a system that provides deionized and double-distilled water is normally used. To remove the surfactant from the suspension, a procedure of diluting, centrifuging, and discarding the supernate is generally followed.

Vibrating Orifice and Spinning-Disc (Top) Atomizers

The generators that produce monodisperse droplets can also produce monodisperse aerosols. There are two methods available. The vibrating-orifice atomizer creates a jet of liquid through an orifice and breaks the jet into droplets of a uniform diameter by applying certain disturbing frequencies (Fulwyler, Glascock, and Hiebert 1969; Raabe and Newton 1970; Berglund and Liu 1973). The stability, monodispersity, and diameter of the droplets depend on the diameter of the orifice, the density and surface tension of the liquid, the velocity of the jet, and the disturbing frequency. The advantage of this method is that the diameter of the droplet, d_d (μm), and the aerosol particle, d_p (μm), can be determined by the following equations (Berglund and Liu 1973):

$$d_d = 10^4(Q_l/10\pi f)^{1/3} \quad (22\text{-}2)$$

$$d_p = C^{1/3}\, d_d \quad (22\text{-}3)$$

where Q_l is the liquid feed rate in cm^3/min, f is the vibrating frequency in Hz, and C is the volumetric concentration of the solute in the solution. Unfortunately, the output aerosol concentrations are low, about 30–400 cm^{-3}, depending upon the primary droplet size. In addition, liquid suspension can clog the orifice and may not be used as feeding material. In the case of producing monodisperse solid aerosols (e.g., ammonium fluorescein), the drying process for suspended droplets is crucial to the surface smoothness of the final particles. The inclusion of small amounts of oil can improve the sphericity by preventing crystallization. However, the density may be affected if the particle dries too fast.

Example 22-2

A vibrating orifice atomizer is used to produce monodisperse ammonium fluorescein particles for instrument calibration. The solution is prepared by dissolving 20 g of fluorescein powder in 1 l of 0.1 N ammonium hydroxide. The atomizer is operated at a liquid feed rate of 0.18 cm^3/min. The orifice is 20 μm in diameter and is oscillated at 55 kHz. What are the diameters of the initial droplets and the final ammonium fluorescein particles after the drying process?

Answer. Using Eqs. 22-2 and 22-3:

$$d_d = 10^4(Q_l/10\pi f)^{1/3}$$

$$d_p = C^{1/3} d_d$$

$$Q_l = 0.18 \text{ cm}^3/\text{min}$$

$$f = 55{,}000 \text{ Hz}$$

Density of ammonium fluorescein

$$= 1.35 \text{ g/cm}^3 \text{ (Table 22-1)}$$

$$C = (20/1) \text{ g/l}$$

$$= (20/1.35) \times 10^{-3} \text{ l/l}$$

$$= 1.48 \times 10^{-2}$$

Droplet size

$$d_d = 10^4(0.18/(10(\pi)55{,}000))^{1/3}$$

$$= 47.1 \text{ μm}$$

Particle size

$$d_p = (1.48 \times 10^{-2})^{1/3}(47.1)$$

$$= 11.6 \text{ μm}$$

The second method of producing monodisperse droplets is by the spinning-disc (top) atomizer, in which a liquid jet is fed at a constant rate onto the center of a rotating disc (top). The liquid spreads over the disc (top) surface in a thin film, accumulating at the rim until the centrifugal force acting to discharge it exceeds the capillary force acting to hold it together, and a droplet is thrown off. The droplet size, d_d, depends on the disc (top) diameter, d_s (in μm), and the rotating speed, ω_s (in rpm), as follows:

$$d_d = (W\gamma / \rho_L \omega_s^2 d_s)^{1/2} \quad (22\text{-}4)$$

where γ and ρ_L are the surface tension and the density of the liquid, and W is a constant. The application of this process has been investigated by Walton and Prewett (1949) and May (1949) using an air-driven spinning top, and by Whitby, Lundgren, and Peterson (1965) and Lippmann and Albert (1967) using a motor-driven spinning disc. Unlike a vibrating-orifice atomizer, aqueous suspensions, as well as solutions, can be used. A disadvantage of this method is that undesired satellite droplets are frequently formed and must be removed from the useful aerosol produced by the primary droplets. In addition, the constant W (Eq. 22-4) varies with the instrument and the feed material used, and the droplet size and the final particle size cannot be easily calculated as for the vibrating orifice atomizer.

Controlled Condensation Techniques

Condensation is also a method that produces monodisperse aerosols for calibration purposes. In this method, the heated vapor of a substance that is normally liquid or solid at room temperature is mixed with the nuclei, on which it condenses when it passes in laminar flow through a cooling zone. If the condensation process is diffusion-controlled, the surface area of the growing droplet will increase at a constant rate, producing a particle having a diameter d_t at time t related to the initial diameter, d_0, of the nucleus by

$$d_t^2 = d_0^2 + bt \quad (22\text{-}5)$$

where b is a constant, related to the concentration and diffusivity of the vapor and to the temperature. If bt is the same for all particles and much larger than d_0, the diameter of the nucleus has little effect on the final diameter of the particle, so that an aerosol containing monodisperse particles is produced. In practice, uniform temperature profile, sufficient vapor concentration, and sufficient residence time in the condensation region are the key controls, and a constant nuclei concentration provides a stable aerosol concentration (Sinclair and LaMer 1949; Rapaport and Weinstock 1955; Prodi 1972; Liu and Lee 1975; Tu 1982). Particle sizes from 0.03 to greater than 2 μm with a σ_g of 1.2–1.3 can be produced in this way. The number concentration can be as high as 10^7 particles/cm^3.

Monodisperse Aerosols with Nonspherical Particles

The effects of particle shape on instrument response are important, especially for the sizing instruments in which the measured properties are generally dependent on particle shape. Information concerning the effects of shape on instrument response can be obtained by using monodisperse aerosols of nonspherical particles during calibration. One way of generating these aerosols is to nebulize the liquid suspension containing monodisperse nonspherical particles. Various techniques have been used to produce monodisperse particles of highly uniform particle size and shape. Matijevic (1985) produced inorganic and polymer colloid particles of cubic, spindle, and rhombohedral shapes by chemical reactions. Fiber-like particles of a narrow size range were also produced using different methods (Esman et al. 1980; Loo, Cork, and Madden 1982; Vaughan 1990; Hoover et al. 1990). The vibrating-orifice and spinning-disc (top) aerosol generators described above can also be used to generate irregularly shaped particles, such as crystalline sodium chloride particles. Although the generators produce spherical droplets, the crystal form of the solid particles becomes the shape of the final aerosol after drying the liquid vapor.

In addition, naturally occurring materials, such as fungal spores, pollens, and bacteria or the fortuitous occurrence of multiplets of spheres, are also frequently used as test aerosols of nonspherical particles (Corn and Esmen 1976; Adams, Wennerstrom, and Mazumder 1985). The aerosols of fungal spores and pollens are commonly generated using the dry powder dispersion technique (described later).

Polydisperse Aerosols

Polydisperse aerosols are seldom used as test aerosols to calibrate sizing instruments; however, some polydisperse aerosols, such as coal dust and Arizona road dust, are frequently used in calibrating dust monitors, which provide information on the mass concentration of total and/or respiratory dust. There are two ways to generate polydisperse aerosols: wet droplet dispersion and dry powder dispersion.

Wet Dispersion

The simplest way to disperse a droplet aerosol is by wet nebulization. Two types of nebulizers are often used to produce aerosols. Air-blast nebulizers (Mercer et al. 1968) use compressed air (15–50 psig, 1 psig = 6.87 $\times 10^4$ dyn/cm^2) to draw bulk liquid from a reservoir as a result of the Bernoulli effect. The high-velocity air breaks up the liquid into droplets and then suspends the droplets as part of the aerosol. Droplets produced from this method have a VMD of 1–10 μm and a σ_g of 1.4–2.5 (Table 22-2). The aerosol size distribution can be modified by varying the pressure in the compressed air or the dilution ratio in the solution. One problem arises when the bulk liquid contains a volatile solvent that evaporates rapidly after droplet formation. The continuous loss of solvent increases the solute concentration in the reservoir and causes the particle size to increase gradually with time. This problem can be circumvented by circulating the solution through a large reservoir, presaturating the supply air, and cooling the nebulizer.

In the ultrasonic nebulizer, the mechanical energy necessary to atomize a liquid comes from a piezoelectric crystal vibrating under the influence of an alternating electric field produced by an electronic high-frequency oscillator. The vibrations are transmitted through a coupling fluid to a nebulizer cup containing the solution to be aerosolized. At a certain frequency (1.3–1.7 mHz), a heavy mist appears above the liquid surface of the cup. The diameter of the droplets making up the mist is related to the wavelength of the capillary waves, which decreases with increasing frequency of the ultrasonic vibrations. Normally, the VMD is 5–10 μm, with a σ_g of 1.4–2.0 (Table 22-2).

Aerosol particles with chemical properties different from those of the liquid feed material can be produced through wet dispersion by using suitable gas-phase reactions, such as polymerization or oxidation. The production of spherical particles of insoluble oxides and aluminosilicate particles with entrapped radionuclides has been described by Kanapilly, Raabe, and Newton (1970) and Newton et al. (1980).

Dry Dispersion

Numerous techniques have been described by Hinds (1980) to disperse dust or fiber particles (Table 22-3). Basically, the techniques consist of two steps: (1) a means of continuously delivering powder into the disperser at a constant rate; and (2) a means of dispersing the powder to form an aerosol. However, dispersibility of a powder depends on powder material, particle size, particle shape, and moisture content. There is no standard generation method for powder dispersion.

One advantage of this dry dispersion method is that the aerosol generated from this technique has a similar size distribution as the single powder particles when they are suspended in the air. In addition, because no solvent or heat treatment is required, the physical and chemical form of the material is preserved. A problem common to dry-dispersion aerosols is the buildup of charge on

TABLE 22-3 Operating Parameters of Dry-Powder Dispersers

	Wright Dust Feed	Fluidized Bed	NBS II Dust Generator	Small-Scale Powder Disperser	Jet-O Mizer Model 00
Type of Operation	Scraping the packed plug and dispersing it with air	Feeding the power to the bed on a conveyor and air-fluidizing it	Using metering gear to deliver the powder and air-dispersing it	Using a rotating plate to deliver the powder and dispersing it with venturi suction	Using venturi suction to feed the powder into a fluid energy mill in which centrifugal force and air velocity are used to break up the agglomerate and to disperse the powder
Air flow rate, l/min	8.5–40	5–20	50–85	12–21	14–113
Feed flow rate, mm^3/min	0.24–210	1.2–3.6	1200–50,000	0.9–2.5	2000–30,000
Output mass concentration, g/m^3 ($\rho = 1$ g/cm^3)	0.012–11.5	0.13–4.0	15–200	0.0003–0.04	10–1500
Recommended size range, μm	0.2–10	0.5–40	1–100	1–50	0.2–30
Source	BGI, Inc.	TSI, Inc.	BGI, Inc.	TSI, Inc.	Fluid Energy Process and Equipment Co.

particles as they touch and separate from a surface in the generator. This charging varies the output concentration due to a variable loss to the wall of the system. The problem can be solved by passing the aerosol through a chamber containing a bipolar ion source as described before.

Size Classification of Polydisperse Aerosols

Although polydisperse aerosols may be used for instrument calibration or to simulate the actual use of equipment under controlled laboratory conditions, they can also be classified according to size in order to provide an aerosol with a narrow size range for instrument calibration. For particles smaller than 0.1 μm, Liu and Pui (1974) developed a differential-electrical-mobility analyzer to classify aerosol particles of the same electrical mobility. Because most classified particles are singly charged, the aerosol produced is monodisperse. This classification technique has been used to produce a submicrometer-sized aerosol standard in calibrating CNCs, and diffusion batteries, and in determining particle deposition in human nasal and oral casts (Liu et al. 1975; Scheibel and Porstendörfer 1984; Cheng et al. 1990). For particles greater than 1 μm, the size-classifying technique based on the aerodynamic property is generally used. Two virtual impactors can be placed in a series to segregate the desired fraction of the input aerosol for use in instrument calibration (Chen, Yeh, and Rivero 1988; Pilacinski et al. 1990). To classify aerosols in the 0.1–1.0 μm range, a technique that involves both the mobility analyzer and a single-stage micro-orifice impactor has been used (Romay-Novas and Pui 1988). The above techniques are also used for reducing the undesired particles, such as PSL aggregates from an air nebulizer or satellite particles by the spinning-disc generator.

All the devices and techniques described above classify aerosol particles when the particles are still in the airborne state. Other instruments, such as elutriators, spectrometers, cascade impactors, and cascade cyclones, classify particles by means of collecting size-classified particles on a substrate and, then, resuspending the size-classified particles. For example, a spiral centrifuge collects aerodynamically classified particles on aluminum foil; a resuspension of the particles caught on a narrow segment of the foil can be used to produce monodisperse aerosols (Kotrappa and Moss 1971). The disadvantage of all size-classifying techniques is that only a small quantity of particles is produced.

Test Aerosols with Tagging Materials

For some applications, particle detection is often facilitated by incorporating dye or radioisotope tags in the particles during their production. For example, test aerosols composed of fluorescent dye can be analyzed in solutions containing as little as 10^{-10} g/m^3. Radiolabeling techniques have been used in many forms and can usually be detected at extremely low concentrations (Newton et al. 1980).

CALIBRATION OF FLOW, PRESSURE, AND VELOCITY

Accurate measurement of gas flow rate, pressure, and velocity is an integral part of aerosol research, as well as an indication of successful instrument calibration (Mercer 1973; Hinds 1982; Lippmann 1989). Various instruments (Table 22-4) and techniques involved in the measurement of these parameters are discussed in this section.

Flow Rate Measurement

Two types of meters are conventionally used in an aerosol system to measure the flow rate: the variable-pressure head meters, such as orifice or venturi meters, and the variable-area meters, such as rotameters (Table 22-4). Both operate on the principle of conservation of energy. Specifically, they use Bernoulli's theorem for the exchange of potential energy for kinetic energy and/or friction heat. Each consists of a flow restriction that causes an increase in gas velocity and kinetic energy

TABLE 22-4 Instruments that Measure Flow Rate, Volume, Pressure, and Velocity of Gases[1]

Quantity Measured	Instrument	Range	Standard
Volume	Spirometer	1 l–1 m^3	Primary standard for flow rate calibration
	Bubble flow meter	1 cm^3–10 l	
	Piston-displacement meter[2]	1 cm^3–12 l	
	Dry-gas meter	Unlimited[3]	Secondary standard for flow rate calibration
	Wet-gas meter	Unlimited[4]	
Volumetric Flow Rate	Venturi meter[5]	1 l/s–100 m^3/s	
	Orifice meter[5]	1 cm^3/s–100 m^3/s	
	Rotameter[5]	0.01 cm^3/s–50 l/s	
Pressure Differential	Manometer	0–2 atm	Calibration standard
	Pressure gauge[5]	0–20 atm	
	Pressure transducer[5]	0–220 atm	
Velocity	Pitot tube	>5 m/s	Calibration standard
	Hot-wire anemometer[5]	5 cm/s–40 m/s	

1. Commercial sources of these instruments can be found in Lippmann (1989)
2. Mercury-sealed
3. Range for flow rate calibration is 5–150 l/min
4. Range for flow rate calibration is 0.5–230 l/min
5. Frequent calibration against a standard is needed

and, therefore, requires a corresponding decrease in potential energy, i.e., static pressure. The flow rate can be calculated by knowing the pressure drop, the cross-sectional area upstream and at the constriction, the density of the gas, and the discharge coefficient. Taking into account the flow constriction and frictional effects, the discharge coefficient is the ratio of the actual flow rate to the ideal flow rate and is dependent on the design of the flow restriction.

The variable-head meter determines the average flow rate by measuring the pressure differential across a calibrated resistance in the flow stream. The venturi meter (Fig. 22-2a) is designed to streamline the constriction throat in the flow stream to minimize energy loss. The volumetric flow rate, Q, referred to ambient air, is given by

$$Q = (kA_2/\rho_a)[2(P_1 - P_2)\rho_1/(1 - (A_2/A_1)^2)]^{1/2} \qquad (22\text{-}6)$$

where k is the discharge coefficient ($=0.985$ for standard conditions), ρ_a the ambient-air density, P the pressure, and A the cross-sectional area. Subscripts 1 and 2 denote the position upstream and at the throat, respectively.

A simpler form of a variable-head meter is the orifice meter (Fig. 22-2b), in which a thin plate with a sharp-edged circular orifice is inserted at the center of the flow to converge the streamline to a shape similar to that for a venturi meter. Although a large energy loss takes place in the orifice meter, a result of sudden area increase after the gas passes through the flow restriction, the meter is widely used because of its ease of installation and low cost. Because the cross-sectional area of the narrowest flow region cannot be measured directly, it is customary to replace A_2 in Eq. 22-6 with the orifice area A_o. The discharge coefficient k for an orifice meter depends on the orifice design and is generally obtained by calibrating it against a reliable reference instrument. Both venturi and orifice meters must be cleaned to maintain their integrity at the constriction when used in an aerosol system. For a constant flow control of filter sampling, a type of orifice meter, called a critical orifice meter, is widely used downstream of the filter. The orifice is usually small enough to provide a downstream pressure P_2

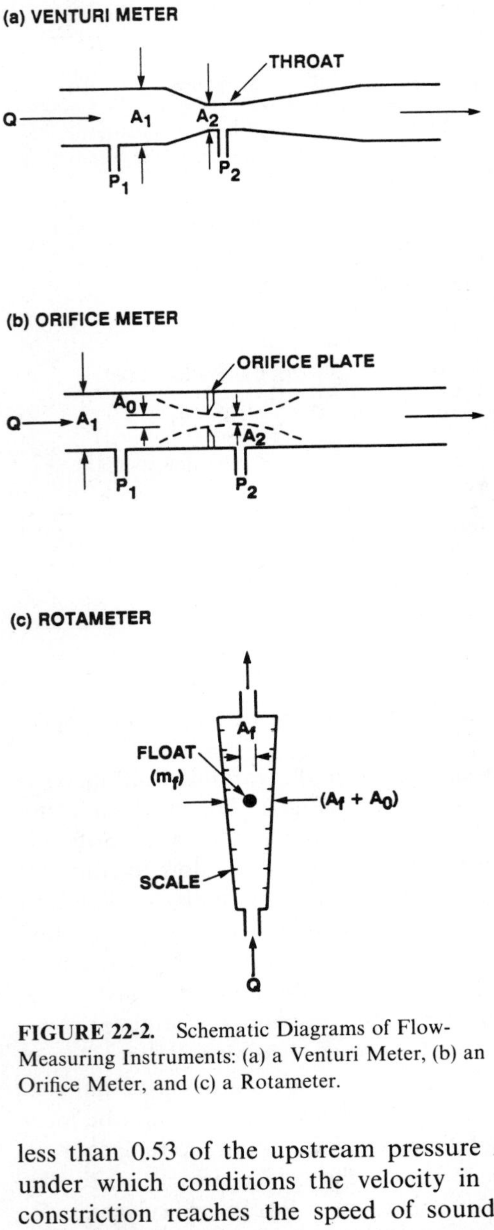

FIGURE 22-2. Schematic Diagrams of Flow-Measuring Instruments: (a) a Venturi Meter, (b) an Orifice Meter, and (c) a Rotameter.

less than 0.53 of the upstream pressure P_1, under which conditions the velocity in the constriction reaches the speed of sound; a further reduction in the downstream pressure does not increase the velocity through the system. For a critical orifice, the flow rate, Q, is proportional to the upstream pressure and is expressed as

$$Q = \frac{0.58\,kA_o}{\rho_a}(\gamma\rho_1 P_1)^{1/2} \qquad (22\text{-}7)$$

where γ is the ratio of specific heats and is 1.4 for air. A critical orifice is normally operated downstream of a filter so that the integrity of the orifice can be maintained.

Example 22-3

A critical orifice with 0.4 mm diameter is fabricated for air sampling purposes and used downstream of a filter. The flow rate is measured to be 1 l/min when the upstream pressure is close to the ambient pressure (760 mmHg, 20°C).

1. What size of orifice must be fabricated if the sampling flow rate is 2 l/min (assuming that the downstream pressure is still less than 0.53 of the upstream pressure)?
2. What is the sampling flow rate when this orifice is used in Albuquerque, NM (625 mmHg)?
3. When the filter is loaded and the pressure gauge at the upstream of the orifice reads − 10 cmH_2O, what is the sampling flow rate?

Answer. Using Eq. 22-7:

$$Q = \frac{0.58kA_o}{\rho_a}(\gamma\rho_1 P_1)^{1/2}$$

1. $Q \propto A_o$. If the sampling flow rate is doubled, then the orifice diameter needs to be $(2)^{1/2}$ times of the original size, i.e., $(0.4) \times (2)^{1/2} = 0.57$ mm.
2. $Q \propto \frac{1}{\rho_a}(\rho_1 P_1)^{1/2}$. Because $\rho_1 = \rho_a$ and $P_1 \propto \rho_1$, the sampling flow rate remains unchanged (1 l/min).
3. $Q \propto (\rho_1 P_1)^{1/2} = P_1$ $(P_1 \propto \rho_1)$. Because − 10 cmH_2O = − 7.35 mmHg, the sampling flow rate is $[1 \times (760 - 7.35/760)] = 0.99$ l/min.

Unlike the variable-head meters, where the pressure drop fluctuates with flow rate, the variable-area meter changes the orifice area with flow to maintain a constant pressure drop. The most common type of the variable-area meter is the rotameter (Fig. 22-2c). It consists of a "float" that moves up and down within a vertical tapered tube which is larger at the top than at the bottom. The gas flows upward, causing the float to rise until the pressure drop across the annular area (A_o) between the float and the tube wall is just sufficient to support the float. The height of the float indicates the flow rate, which is

$$Q = \frac{kA_o}{\rho_a}(2m_f g \rho_g / A_f)^{1/2} \tag{22-8}$$

where m_f and A_f are the mass and the cross-sectional area of the float, ρ_g is the gas density in the tube, and g is the acceleration due to gravity. This equation can be obtained from Eq. 22-6 by replacing the pressure drop with $m_f g/A_f$ and assuming $A_2 \ll$ tube diameter. The float position is usually calibrated by marks on the tube in terms of the flow rate at ambient pressure. For operation at a gas pressure (density) different from that used during calibration, the true flow rate must be corrected. According to Eq. 22-8, the true flow rate is proportional to $(\rho_g)^{1/2}/\rho_a$ or $(P_g)^{1/2}/P_a$, where P_g and P_a are the gas pressure in the rotameter and at the ambient, respectively. For a rotameter calibrated at an ambient pressure, $P_{a,c}$, and used at a different ambient pressure, $P_{a,i}$, the true volumetric flow rate Q_i at a fixed float position is given by

$$Q_i = Q_c(P_{a,c}/P_{a,i})^{1/2} \tag{22-9}$$

where Q_c is the flow rate indicated by the rotameter during calibration. For a rotameter calibrated and used at the same ambient pressure, the measured flow rate depends on the gas density, ρ_g, (pressure, P_g), in the tube. For example, if the rotameter is operated at a different gas density in the tube than that used during calibration (e.g., when the rotameter is located downstream of a filter or an impactor during aerosol sampling), then the actual volumetric flow rate at a fixed float position is given by

$$Q_i = Q_c(\rho_{g,i}/\rho_{g,c})^{1/2}$$
$$= Q_c(P_{g,i}/P_{g,c})^{1/2} \tag{22-10}$$

where the subscripts i and c refer to the actual condition in the rotameter and the condition during calibration, respectively. Normally, $P_{g,c}$ is the ambient pressure, P_a, and $P_{g,i}$ is $(P_a - \Delta P)$, where ΔP is the gauge pressure downstream of the sampler. The rotameter flow rate ranges from 0.01 cm^3/s to 50 l/s, and the rotameter is generally calibrated using a calibrated dry gas meter, bubble flow meter, or spirometer.

Example 22-4

A rotameter is calibrated by the manufacturer at sea level (760 mmHg) and used in Albuquerque, NM (625 mmHg). What percentage of error in flow rate will result if the rotameter is not recalibrated? Assuming that this rotameter is recalibrated and then used downstream of a filter to measure the sampling flow rate, what will be the true flow rate if the flow rate indicated by the recalibrated rotameter is 5 l/min and the pressure drop is 10 in H_2O?

Answer. Using Eq. 22-9

$$Q_i = Q_c(P_{a,c}/P_{a,i})^{1/2}$$
$$P_{a,c} = 760 \text{ mmHg}$$
$$P_{a,i} = 625 \text{ mmHg}$$
$$Q_c/Q_i = (625/760)^{1/2} = 0.907$$

$$\text{Percentage error} = (Q_i - Q_c/Q_i)\,100$$
$$= (1 - Q_c/Q_i)100$$
$$= 9.3\%$$

Using Eq. 22-10

$$Q_i = Q_c(P_{g,i}/P_{g,c})^{1/2}$$
$$P_{g,i} = P_a - \Delta P$$
$$= 625 - (10 \times 25.4/13.6)$$
$$= 625 - 18.7 = 606.3 \text{ mmHg}$$
$$P_{g,c} = P_a = 625 \text{ mmHg}$$
$$Q_c = 5 \text{ l/min}$$

$$\text{True flow rate} \quad Q_i = 5(606.3/625)^{1/2}$$
$$= 4.92 \text{ l/min}$$

Flow meters must be calibrated, and the calibration standards measure gas volume directly over time (Table 22-4). A spirometer (Fig. 22-3a) is a cylindrical bell with its open end under a liquid seal. The bell is supported by a cord and is balanced by a counterweight. In operation, the volume of air entering the spirometer is determined by calculating the distance raised in the bell times the cross-sectional area of the bell. The spirometer can measure a volume as high as 1 m^3.

For a smaller volume, a soap bubble meter (Fig. 22-3b) is a widely used primary standard. A bubble film is created in a tube from a reservoir of soapy water and acts as a nearly frictionless piston as the air passes through the tube. The distances of the bubble displaced along the tube and the cross-sectional area of the tube are used to determine the volume of the air entering the bubble meter. Several automated bubble meters that incorporate bubble-detecting sensors, automatic timing, and readout of flow rate are commercially available and widely used as calibration standards for flow rates up to 40 l/min. Similar to the bubble meter is a device based on the mercury-sealed piston-displacement technique (Fig. 22-3c). It is also being used as a primary standard in the laboratory.

A dry-gas meter is also often used to calibrate rotameters and orifice meters, although it must be calibrated against a primary standard, such as spirometer. The dry-gas meter

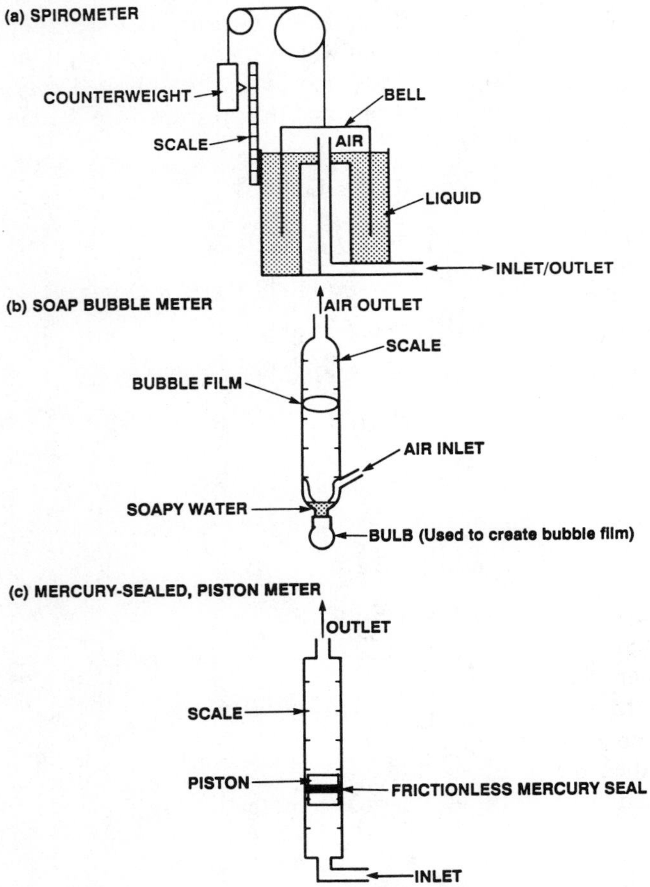

FIGURE 22-3. Schematic Diagrams of Instruments Used to Measure Gas Volume: (a) a Spirometer, (b) a Soap Bubble Meter, and (c) a Mercury-Sealed, Piston Meter.

(Table 22-4) contains two bellows that are alternately filled and emptied by the metered gas. The movement of the bellows controls the action of mechanical valves that direct the flow and operates a cycle-counting device that registers the total volume of gas passing through the meter. In operation, one inlet of the meter is always open to ambient pressure because the housing of the device cannot support a very large pressure drop. In addition, at least ten revolutions are recommended for each measurement to avoid the effect of nonlinear strokes. In a wet test meter, gas flows into a rotating system of chambers that connects to a revolution counter. The water level seals the chambers and acts as a valve to direct the flow to the proper chamber. The measured volume from this meter must be corrected for water vapor content. There is, however, no limit on the total gas volume measured.

Pressure Measurement

Three types of pressure-sensing devices are commonly used to determine the pressure differential between two points in a system or the gauge pressure: manometers, mechanical gauges, and pressure transducers (Table 22-4).

The liquid-filled manometer (Fig. 22-4a) consists of a glass or plastic tube sized to allow the height of the fluid level to balance against the incoming pressure. The manometer expresses the pressure differential, ΔP

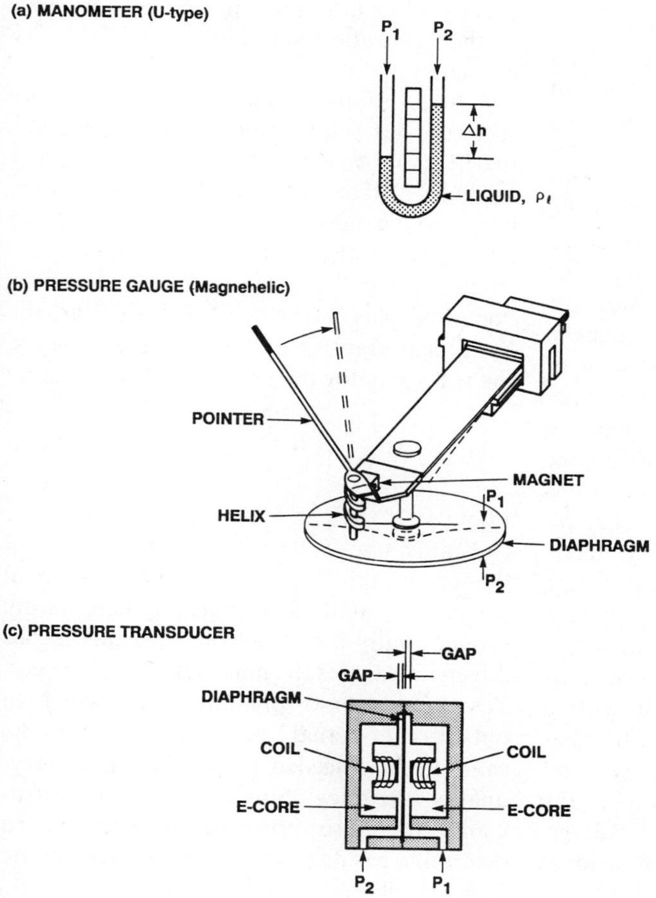

FIGURE 22-4. Schematic Diagrams of Instruments Used to Measure Gas Pressure: (a) a Manometer, (b) a Pressure Gauge, and (c) a Pressure Transducer.

($= P_1 - P_2$), by measuring the difference in liquid column height, Δh. For a given liquid, the pressure difference can be determined by the relationship

$$\Delta P = \rho_L \Delta h g \tag{22-11}$$

where ρ_L is the density of the liquid, and g is the acceleration due to gravity. For different applications, there are three types of manometers: U-tube, well-type, and inclined-type. Normally, a manometer does not require any calibration and can be used as a pressure standard. Corrections, however, may be applied if one is concerned about changes in the specific gravity of the fluid due to changes in ambient pressure, scale expansion, and tube expansion.

The mechanical pressure gauge (Fig. 22-4b) is widely used as a pressure sensor of an aerosol system both in the laboratory and in the field. It normally consists of a metal or plastic housing which contains a diaphragm assembly. The diaphragm movement due to the pressure differential is transferred to the dial indicator mechanism. The pressure gauge provides an accurate reading as a percentage of the full-scale range of the device. The most common pressure gauge is the Magnehelic (Dwyer Inst.), with a range from 0.01 in of water to 100 psig. In this device, the diaphragm transmits the effect of pressure to an indicator by means of magnetic linkage without direct physical contact to ensure the accuracy and sensitivity of the instrument.

The pressure transducer is used to determine the pressure differential across an orifice meter and to provide a quick digital output of measurement. In this device (Fig. 22-4c), a diaphragm is embedded between two coils that connect to an electronic unit with a display of voltage. The deflection of the diaphragm changes the gap between the coil and the diaphragm and, as a result, changes the magnetic reluctance and the output voltage. Although the output voltage is proportional to the pressure applied to the transducer, frequent calibrations against manometers are required.

Velocity Measurement

The measurement of the local gas velocity in a duct is critical for proper isokinetic sampling and for calibrating flow-measuring devices. The most widely used device for measuring velocity is the pitot tube (Fig. 22-5a), that directly measures the velocity pressure in a moving gas flow and then relates the pressure to the velocity, V. The pressure, normally measured using a manometer, is expressed as a difference in liquid column height, Δh. Using Bernoulli's theorem, the velocity is

$$V = (2\rho_L g \Delta h / \rho_g)^{1/2} \tag{22-12}$$

where ρ_L and ρ_g are the liquid density in the manometer and the gas density in the duct, respectively, and g is the acceleration due to gravity. The pitot tube is normally considered as the calibration standard for gas velocity measurement.

A hot-wire anemometer (Fig. 22-5b) measures the gas velocity by sensing the convective heat loss on a hot wire when the gas flows through it. The wire is heated electrically, heat loss changes the temperature, and the resistance of the wire that is sensed electronically can be converted to a meter or digital display of velocity. Considering the basic heat transfer in thermal anemometry, the voltage of the output signal, V, is given by

$$V^2 = (A + B\,U^n)\,(T_s - T) \tag{22-13}$$

where U is the gas velocity, A, B, and n are constants, and T_s and T are temperatures of the sensor and the gas. A and B are empirical constants, usually determined by calibration; n is normally 0.3–0.8. Note that this device directly measures the mass velocity of the gas. Temperature and pressure are needed to obtain the actual velocity, and periodic calibration is needed to provide laboratory-quality accuracy. Similarly, in the instruments that use hot-wire anemometry to determine gas flow rate, the actual volumetric flow rate should be calibrated.

If one is interested in the velocity of aerosol particles (not the gas medium), a laser

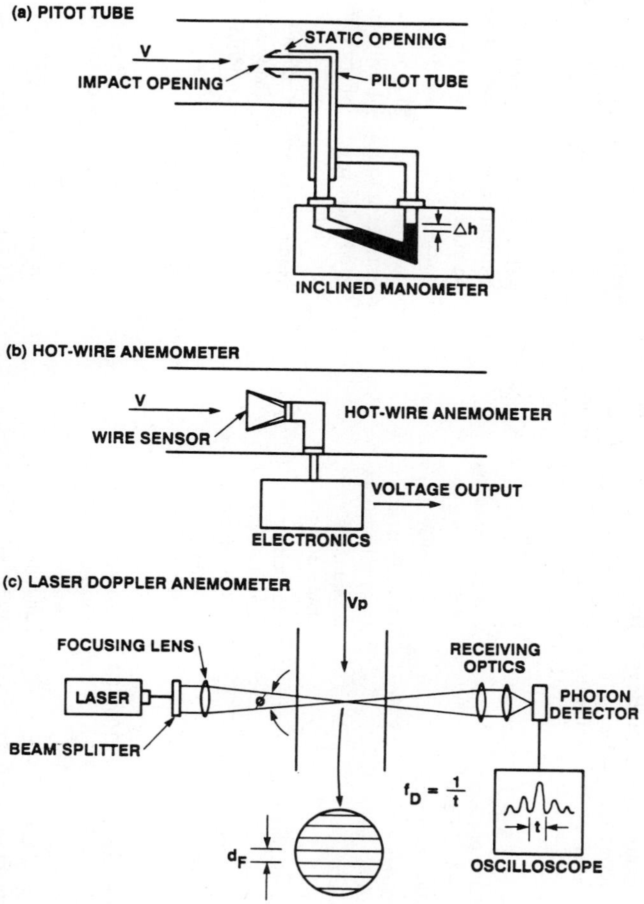

FIGURE 22-5. Schematic Diagrams of Instruments Used to Measure Gas Velocity: (a) a Pitot Tube, (b) a Hot-Wire Anemometer, and (c) a Laser Doppler Anemometer.

Doppler anemometer can be used (Fig. 22-5c). This device does not require the insertion of a sensing probe as used in both the pitot tube and hot-wire anemometer. It uses two laser beams to form an interference pattern with a fringe spacing d_F. As a particle travels through the fringe pattern, its scattered-light intensity provides a shift in the detected frequency, f_D, a phenomenon known as the Doppler effect. The velocity of the particle (V_p) can be obtained by

$$V_p = d_F f_D \tag{22-14}$$

$$d_F = \lambda/2\sin(\phi/2) \tag{22-15}$$

where λ is the wavelength of the laser, and ϕ is the angle between the beams. Although the device is delicate and only suitable for laboratory measurement, this method provides rapid-response, on-line measurement related to particle velocity.

INSTRUMENT CALIBRATIONS

In this section, several aerosol instruments that determine particle size, number concentration, and mass concentration are briefly described. Their calibration standards and important parameters are summarized in Table 22-5.

TABLE 22-5 Summary of Direct Measurements and Primary Standards of Aerosol Instruments and Important Parameters to be Considered in Instrument Calibration

Instrument	Operating Principle	Measured Quantity	Important Aerosol Parameter	Important Instrument Parameter	Particle Size Range (μm)	Direct Calibration Standard	Primary Calibration Standard	Main Advantage	Main Disadvantage
Size Measurement									
Cascade impactor	Particle inertial impaction	Mass	Size, shape, density	Flow rate, gas medium, physical dimension in and around the nozzle	0.05–30	Monodisperse spherical particles with a known size and density	—	Aerodynamic size distribution based on mass concentration	Internal loss, particle bounce and re-entrainment
Aerodynamic particle sizing instrument	Time of flight	Velocity	Size, shape, density, rigidity	Flow rate, pressure, gas medium	0.5–20	Monodisperse spherical particles with a known size, shape, and density	—	Real-time instrument with good sensitivity and resolution	Coincidence, density, and shape effects
Optical particle counter	Interaction between particle and light	Scattered light intensity	Size, shape, refractive index	Wavelength of the light source, range of scattering angles, sensitivity of detector	0.3–15	Monodisperse spherical particles with a known size and refractive index	—	Noninvasive, real-time, *in situ* measurement; also good for number concentration measurement	Calibration changes with the material
Electrical mobility analyzer	Size classification based on electric mobility	Electric charge or particle count	Size, shape, dielectric constant, humidity	Flow rate, charging mechanism, electric field strength	0.001–0.1	Monodisperse,[1] spherical particles with a known size and dielectric constant	—	Suitable for particles smaller than 0.1 μm	Multiple charges on the particle

Diffusion battery	Particle diffusional deposition	Particle count or mass	Size, shape, number concentration	Flow rate, temperature, deposition surface	0.001–0.1	Monodisperse,[1] spherical particles with a known size	—	Suitable for particles smaller than 0.1 μm	Unsuitable for large particles or particles with a large aspect ratio
Number Concentration Measurement									
Condensation nuclei counter	Vapor condenses on particles and makes them detectable	Particle count	Size, number concentration, hygroscopicity	Flow rate, saturation ratio, temperature gradient	0.001–0.5	Atkin counter with a microscopic and direct measurement	Pollak counter, photographic counter, or electrically classified monodisperse aerosol	Suitable for concentration measurement of submicrometer particles	Size-dependent counting efficiency
Mass Concentration Measurement									
Photometer	Interaction between particle and light	Total light scattering	Size, shape, refractive index, density	Wavelength of the light source, range of scattering angles, sensitivity of detector	0.3–1.5	Gravimetric measurement of filter samples	—	Real-time, continuous readout	Calibration changes with material type
β-attenuation monitor	Mass-dependent absorption of β-radiation	Mass	Size	Uniformity of particle deposit	1–15	Gravimetric measurement of filter samples	—	Real-time measurement	Low sensitivity
Quartz crystal mass balance	Mass-dependent resonant frequency of the crystal	Mass	Size	Sensitivity of the sensor	0.02–10	Gravimetric measurement of filter samples	—	Real-time measurement	Frequent sensor cleaning

1 For particle size smaller than 0.01 μ/m, the electrically classified monodisperse aerosols are used calibration standard

Particle-Sizing Instruments

Most instruments used in particle size analysis actually measure some physical property of the particle rather than measure a simple linear geometric dimension. Particle size is then set to the diameter of a spherical reference particle that possesses the same physical property in the same amount. This can be an aerodynamic, optical, electrical, or diffusional property of the particle (Chen et al. 1989).

Aerodynamic sizing instruments, including collection-and-analysis devices (cascade impactors, elutriators, aerosol centrifuges, and cascade cyclones) and real-time analyzers (particle relaxation size analyzers), are generally designed to measure particle sizes between 0.2 and 25 μm. The collection-and-analysis devices measure masses of particles separated by their aerodynamic properties in different force fields (Chapter 11). The real-time analyzers determine particle size, as well as velocity, as particles pass through a sensing zone that can be a Doppler interference fringe pattern or a two-laser-beam arrangement (Chapters 16 and 17). Generally, parameters such as particle size, flow rate (velocity), density, and intrinsic gas properties can affect the collection efficiency or instrument response (Stöber 1976; Marple and Willeke 1976; Chen, Cheng, and Yeh 1985; Baron 1986; Hering 1989). In addition, loading capacities and wall losses in each instrument should be fully examined. Without the careful analysis, the results of wall losses can cause incorrect data interpretation.

The optical particle counter is a real-time instrument that uses single-particle light scattering techniques to measure aerosol size distribution (0.1–10 μm) and number concentration (Chapter 15). Important parameters of this device include the size, shape, orientation, and refractive index of the aerosol particles to be detected, as well as the wavelength of the light source, the range of scattering angles, and the sensitivity of the photodetector in the instrument. Particle size and refractive index are the two more important variables in OPC calibrations, and different values of these variables are often used in the Mie scattering equations to predict theoretically the relative response of an OPC. However, the coincidence error due to multiple particles in the sensing zone should be minimized by using a low particle concentration in calibration. The results from cross-calibrations and theoretical predictions indicate that, for a particle smaller than the wavelength of the light source, an OPC with a forward scattering collection responds independently of the particle shape and composition. However, for a large particle, an OPC with a polychromatic light source and a concentric light collecting system tends to have a smooth response curve independent of the particle shape and orientation (Hodkinson 1966; Willeke and Liu 1976; Gebhart et al. 1976; Chen et al. 1989).

Electrical mobility analyzers (Chapter 18) and diffusion batteries (Chapter 19) are sizing instruments based on the electrical and diffusional properties of submicrometer particles (< 0.5 μm). Important parameters for the electrical mobility analyzer are the geometric diameter and dielectric constant of the particle, and flow rate and charging mechanism in the instrument (Liu and Pui 1974; Pui and Liu 1979; Yeh, Cheng, and Kanapilly 1983). The flow rate, temperature, and particle size, as well as the geometric dimensions of the diffusion surfaces (e.g., screen wire diameter and tube length), are important for diffusion batteries (Cheng 1989).

Condensation Nuclei Counters

Particle number concentration can be determined by sampling particles through a high-efficiency membrane filter and examining the filter with an optical or electron microscope. This is, however, time-consuming and not reliable for particles smaller than 0.005 μm. For submicrometer aerosol particles, the number concentration is generally determined by using CNCs (Chapter 19). In a CNC, supersaturation of water or alcohol vapor occurs in a cloud chamber to initiate vapor condensation on the particle surface. Droplets grow to a micrometer in diameter regardless of their initial size and are detected by microscopic, photographic, or photoelectric methods. Several primary standards have been used for CNC calibrations. The Pollak counter and the photographic-type counters have been the standards against which others are normally calibrated (Liu et al. 1975; Jaenicke and Kanter 1976; Winters, Barnard, and Hogan 1977; Podzimek, Carstens, and Yue 1982; Sinclair 1984). Submicrometer aerosols produced from a differential-mobility analyzer have also been used as a standard (Liu and Pui 1974). Two types of parameters are important in a CNC calibration: particle size, number concentration, and hydrophilic and hydrophobic properties of the aerosol; and vapor saturation ratio, temperature gradient, and flow rate in the device. Cross-calibrations among different instruments and techniques have indicated that, as long as all the important parameters are considered, the results will agree among all the instruments and techniques involved (Liu et al. 1982).

Mass Concentration Monitors

The most common way to determine aerosol mass concentration is to determine both the mass collected on a filter and the gas volume sampled (Chapter 10). This direct, gravimetric approach is usually achieved with a filter of high collection efficiency and low pressure drop, e.g., a glass-fiber filter. Several real-time monitors have been developed to determine aerosol mass concentration: photometers, quartz crystal mass balances, and beta attenuation mass monitors. These monitors can provide either the total mass concentration or only the respirable mass concentration by separating the nonrespirable particles with an impactor, cyclone, or horizontal elutriator.

Photometers measure by detecting light scattering or light extinction from many particles at one time (different from an OPC, in which no more than one particle is detected at a time) (Chapter 15). Important parameters for a photometer are the same as those for an OPC (e.g., particle size, shape, and refractive index). Quartz crystal (piezoelectric) mass (QCM) balances measure the mass collected on the crystal by detecting the change in the oscillating resonant frequency of the crystal due to increased mass (Chapter 14). Beta attenuation monitors (BAM) determine aerosol concentration by collecting mass on a Mylar film or other substances and measuring the mass of collected material by beta ray attenuation (Chapter 14). Collection efficiency, detection sensitivity (frequency change and beta attenuation for QCM and BAM, respectively), and mass-loading capacity are the most important parameters for these instruments. Several instrument calibrations and comparisons have been done by Kuusisto (1983), Marple and Rubow (1984), Smith, Baron, and Murdock (1987), and Baron (1988). In general, for each sampling environment, a filter sample should be taken in parallel to real-time mass monitors for reference standards.

SUMMARY AND CONCLUSIONS

Because the accuracy of measuring aerosols depends on the precision of the aerosol instrument, extreme care should be taken in performing all calibration procedures. The following comments summarize this chapter and the philosophy of aerosol instrument calibration:

1. Standard devices should be used with care and attention to detail.
2. All standard materials, instruments, and procedures should be checked periodically to determine their stability and operating conditions.
3. A newly acquired instrument should be calibrated. The calibration curve and built-in software provided by the manufacturer may not be directly applicable to the user, especially if the ambient pressure, temperature, and wind velocity in the sampling environment are very different from those used during the manufacturer's calibration. In addition, recalibration is required when a different gas medium is used, because gas density and viscosity play an important role in instrument performance.

4. A device should be calibrated after it has been changed or repaired by the manufacturer, subjected to use, mishandled or damaged, or at any time when there is a question as to its accuracy.
5. The operation and limitation of an instrument should be understood before attempting to calibrate it. One should use a procedure or setup that does not change the characteristics of the instrument or standard within the operating range required.
6. Prior to calibration, a series of important parameters to be investigated during calibration should be determined. These parameters should include both the operating parameters in the instrument and the variables of the aerosol properties.
7. Proper test aerosols that have similar physical and chemical properties to the aerosol to be measured should be selected for instrument calibration. Depending upon the instrument, the output of the test aerosol should be checked throughout the calibration to ensure its consistency in size, number concentration, and mass concentration.
8. When in doubt about calibration procedures or data, assure their validity before proceeding to the next operation.
9. All sampling and calibration connections should be as short and free of constrictions and resistance as possible. To ensure that both the calibration device and the instrument to be calibrated have compatible aerosol samples, the sampling lines to both instruments should be identical. This is especially important when the aerosol particles are greater than 10 μm or smaller than 0.1 μm. Leakage should be checked and avoided.
10. Extreme care should be exercised in reading scales, timing, adjusting, leveling, and in all other operations involved.
11. Sufficient time should be allowed for the instrument to be warmed up, flow equilibrium to be established, inertia to be overcome, and conditions to stabilize.
12. Enough data should be obtained to give confidence in the calibration curve for a given parameter. Each calibration point should be made up of at least three readings to ensure statistical confidence in the measurement.
13. A permanent record of all procedures, data, and results should be maintained. This record should include trial runs, known faulty data with appropriate comments, instrument identification, barometric pressure, temperature, flow rate, properties of the test aerosol, and the name of the operator.
14. When a calibration data point differs from previous records, the cause of change should be determined before accepting new data or repeating the procedure.
15. Calibration curves and factors should be properly identified as to conditions of calibration, the device calibrated and what it is calibrated against, the unit involved, the range and precision of calibration data, and who performed the procedure. Often it is convenient to indicate where the original data are filed and to attach a tag to the instrument indicating the above information.

References

Adams, A. J., D. E. Wennerstrom, and M. K. Mazumder. 1985. Use of bacteria as model nonspherical aerosol particles. *J. Aerosol Sci.* 16:193–200.

Baron, P. A. 1986. Calibration and use of the aerodynamic particle sizer (APS 3300). *Aerosol Sci. Technol.* 5:55–67.

Baron, P. A. 1988. Modern real-time aerosol samplers. *Appl. Ind. Hyg.* 3:97–103.

Barr, E. B., M. D. Hoover, G. M. Kanapilly, H. C. Yeh, and S. J. Rothenberg. 1983. Aerosol concentrator: Design, construction, calibration, and use. *Aerosol Sci. Technol.* 2:437–42.

Berglund, R. N. and B. Y. H. Liu. 1973. Generation of monodisperse aerosol standards. *Environ. Sci. Technol.* 7:147–53.

Chen, B. T., Y. S. Cheng, and H. C. Yeh. 1984. Experimental responses of two optical particle counters. *J. Aerosol Sci.* 15:457–64.

Chen, B. T., Y. S. Cheng, and H. C. Yeh. 1985. Performance of a TSI aerodynamic particle sizer. *Aerosol Sci. Technol.* 4:89–97.

Chen, B. T., Y. S. Cheng, and H. C. Yeh. 1990. A study of density effect and droplet deformation in the TSI aerodynamic particle sizer. *Aerosol Sci. Technol.* 12:278–85.

Chen, B. T., Y. S. Cheng, H. C. Yeh, W. E. Bechtold, and G. L. Finch. 1991. Test of the size resolution and sizing accuracy of the Lovelace parallel-flow diffusion battery. *Am. Ind. Hyg. Assoc. J.* 52:75–80.

Chen, B. T., H. C. Yeh, Y. S. Cheng, and G. J. Newton. 1989. Particle size analyzer for air quality studies. In *Encyclopedia of Environmental Control Technology, Vol. II*, ed. P. N. Cheremisinoff, pp. 453–514. Houston, TX: Gulf Publishers.

Chen, B. T., H. C. Yeh, and M. A. Rivero. 1988. Use of two virtual impactors in series as an aerosol generator. *J. Aerosol Sci.* 19:137–46.

Cheng, Y. S. 1989. Diffusion batteries and denuders. In *Air Sampling Instruments*, ed. S. V. Hering, pp. 405–19. Cincinnati, OH: ACGIH.

Cheng, Y. S., Y. Yamada, H. C. Yeh, and D. L. Swift. 1990. Deposition of ultrafine aerosols in a human oral cast. *Aerosol Sci. Technol.* 12:1075–81.

Cheng, Y. S. and H. C. Yeh. 1979. Particle bounce in cascade impactors. *Environ. Sci. Technol.* 13:1392–96.

Corn, M. and N. A. Esmen. 1976. Aerosol generation. In *Handbook on Aerosols*, ed. R. Dennis, Publ. TID-26608, pp. 9–39. Springfield, VA: National Technical Information Service, U.S. Dept. of Commerce.

Esmen, N. A., R. A. Kahn, D. LaPietra, and E. D. McGovern. 1980. Generation of monodisperse fibrous glass aerosols. *Am. Ind. Hyg. Assoc. J.* 41:175–79.

Fuchs, N. A. and A. G. Sutugin. 1966. Generation and use of monodisperse aerosols. In *Aerosol Sci*, ed. C. N. Davies, pp. 1–30. New York: Academic Press.

Fulwyler, M. J., R. B. Glascock, and R. D. Hiebert. 1969. Device which separates minute particles according to electronically sensed volume. *Rev. Sci. Instrum.* 40:42–48.

Gebhart, J., J. Heyder, C. Roth, and W. Stahlhofen. 1976. Optical aerosol size spectrometry below and above the wavelength of light—A comparison. In *Fine Particles: Aerosol Generation, Measurement, Sampling, and Analysis*, ed. B. Y. H. Liu, pp. 793–815. New York: Academic Press.

Hering, S. 1989. Inertial and gravitational collectors. In *Air Sampling Instruments*, ed. S. V. Hering, pp. 337–85. Cincinnati, OH: ACGIH.

Hinds, W. 1980. Dry dispersion aerosol generator. In *Generation of Aerosols and Facilities for Exposure Experiments*, ed. K. Willeke, pp. 171–88. Ann Arbor, MI: Ann Arbor Science Publishers.

Hinds, W. 1982. *Aerosol Technology*. New York: Wiley.

Hodkinson, J. R. 1966. The optical measurement of aerosols. In *Aerosol Science*, ed. C. N. Davies, pp. 287–357. New York: Academic Press.

Hoover, M. D., S. A. Casalnuovo, P. J. Lipowicz, H. C. Yeh, R. W. Hanson, and A. J. Hurd. 1990. A method for producing non-spherical monodisperse particles using integrated circuit fabrication techniques. *J. Aerosol Sci.* 21:569–75.

Jaenicke, R. and H. J. Kanter. 1976. Direct condensation nuclei counter with automatic photographic recording, and general problems of absolute counters. *J. Appl. Meteor.* 15:620–32.

John, W. 1980. Particle charge effects. In *Generation of Aerosols and Facilities for Exposure Experiments*, ed. K. Willeke, pp. 141–51. Ann Arbor, MI: Ann Arbor Science Publishers.

Kanapilly, G. M., O. G. Raabe, and G. J. Newton. 1970. A new method for the generation of aerosols of insoluble particles. *J. Aerosol Sci.* 1:313–23.

Kotrappa, P. and O. R. Moss. 1971. Production of relatively monodisperse aerosols for inhalation experiments by aerosol centrifugation. *Health Phys.* 21:531–35.

Kuusisto, P. 1983. Evaluation of the direct reading instruments for the measurement of aerosols. *Am. Ind. Hyg. Assoc. J.* 44:863–74.

Lippmann, M. 1989. Calibration of air sampling instruments. In *Air Sampling Instruments*, ed. S. V. Hering, pp. 73–109. Cincinnati, OH: ACGIH.

Lippmann, M. and R. E. Albert. 1967. A compact electric-motor driven spinning disc aerosol generator. *Am. Ind. Hyg. Assoc. J.* 28:501–6.

Liu, B. Y. H. and K. W. Lee. 1975. An aerosol generator of high stability. *Am. Ind. Hyg. Assoc. J.* 36:861–65.

Liu, B. Y. H. and D. Y. H. Pui. 1974. A submicron aerosol standard and the primary, absolute calibration of the condensation nuclei counter. *J. Colloid Interface Sci.* 47:155–71.

Liu, B. Y. H., D. Y. H. Pui, A. W. Hogan, and T. A. Rich. 1975. Calibration of the Pollak counter with monodisperse aerosols. *J. Appl. Meteor.* 14:46–51.

Liu, B. Y. H., D. Y. H. Pui, R. L. McKenzie, J. K. Agarwal, R. Jaenicke, F. G. Pohl, O. Prening, G. Reischl, W. Szymanski, and P. E. Wagner. 1982. Intercomparison of different absolute instruments for measurement of aerosol number concentration. *J. Aerosol Sci.* 13:429–50.

Loo, B. W., C. P. Cork, and N. W. Madden. 1982. A laser-based monodisperse carbon fiber generator. *J. Aerosol Sci.* 13:241–48.

Marple, V. A. and K. L. Rubow. 1978. An evaluation of the GCA respirable dust monitor 101-1. *Am. Ind. Hyg. Assoc. J.* 39:17–25.

Marple, V. A. and K. L. Rubow. 1983. An aerosol chamber for instrument evaluation and calibration. *Am. Ind. Hyg. Assoc. J.* 44:361–67.

Marple, V. A. and K. L. Rubow. 1984. *Respirable Dust Measurement*. A Mining Research Contract Report, Bureau of Mines, U.S. Dept. of the Interior.

Marple, V. A. and K. Willeke. 1976. Inertial impactors: Theory, design, and use. In *Fine Particles: Aerosol Generation, Measurement, Sampling, and Analysis*, ed. B. Y. H. Liu, pp. 411–46. New York: Academic Press.

Matijevic, E. 1985. Production of monodisperse colloidal particles. *Ann. Rev. Mater. Sci.* 15:483–516.

May, K. R. 1949. An improved spinning top homogeneous spray apparatus. *J. Appl. Phys.* 20:932–38.

Mercer, T. T. 1973. *Aerosol Technology in Hazard Evaluation*. New York: Academic Press.

Mercer, T. T., M. I. Tillery, and H. Y. Chow. 1968.

Operating characteristics of some compressed-air nebulizers. *Am. Ind. Hyg. Assoc. J.* 29:66–78.

Mokler, B. V. and R. K. White. 1983. Quantitative standard for exposure chamber integrity. *Am. Ind. Hyg. Assoc. J.* 44:292–95.

Newton, G. J., G. M. Kanapilly, B. B. Boecker, and O. G. Raabe. 1980. Radioactive labeling of aerosols: Generation methods and characteristics. In *Generation of Aerosols and Facilities for Exposure Experiments*, ed. K. Willeke, pp. 399–425. Ann Arbor, MI: Ann Arbor Science Publishers.

Pilacinski, W., J. Ruuskanen, C. C. Chen, M. J. Pan, and K. Willeke. 1990. Size-fractionating aerosol generator. *Aerosol Sci. Technol.* 13:450–58.

Podzimek, J., J. C. Carstens, and P. C. Yue. 1982. Comparison of several Aitken nuclei counters. *Atmos. Environ.* 16:1–11.

Porstendörfer, J. and J. Heyder. 1972. Size distribution of latex particles. *J. Aerosol Sci.* 3:141–48.

Prandtl, L. 1952. *Essentials of Fluid Dynamics*, pp. 247–49, 306–11. London: Blackie & Son.

Prodi, V. 1972. A condensation aerosol generator for solid monodisperse particles. In *Assessment of Airborne Particles*, eds. T. T. Mercer, P. E. Morrow, and W. Stöber, pp. 169–81. Springfield, IL: C. C. Thomas.

Pui, D. Y. H. and B. Y. H. Liu. 1979. Electrical aerosol analyzer: Calibration and performance. In *Aerosol Measurement*, eds. D. A. Lundgren et al., pp. 384–99. Gainesville, FL: University Presses of Florida.

Raabe, O. G. 1968. The dilution of monodisperse suspensions for aerosolization. *Am. Ind. Hyg. Assoc. J.* 29:439–43.

Raabe, O. G. 1976. The generation of fine particles. In *Fine Particles: Aerosol Generation, Measurement, Sampling, and Analysis*, ed. B. Y. H. Liu, pp. 57–110. New York: Academic Press.

Raabe, O. G. and Newton, G. L. 1970. Development of techniques for generating monodisperse aerosols with the Fulwyler droplet generator. In *Fussion Product Inhalation Program Annual Report, LF-43*, pp. 13–17.

Rapaport, E. and S. E. Weinstock. 1955. A generator for homogeneous aerosols. *Experientia* 11:363–64.

Romay-Novas, F. J. and D. Y. H. Pui. 1988. Generation of monodisperse aerosols in the 0.1–1.0 μm diameter range using a mobility classification–inertial impaction technique. *Aerosol Sci. Technol.* 9:123–31.

Scheibel, H. G. and J. Porstendörfer. 1984. Penetration measurements for tube and screen-type diffusion batteries in the ultrafine particle size range. *J. Aerosol Sci.* 15:673–82.

Sinclair, D. 1984. Intrinsic calibration of the Pollak counter—A revision. *Aerosol Sci. Technol.* 3:125–34.

Sinclair, D. and V. K. LaMer. 1949. Light scattering as a measure of particle size in aerosols. *Chem. Rev.* 44:245–67.

Smith, J. P., P. A. Baron, and D. J. Murdock. 1987. Response characteristics of scattered light aerosol sensors used for control monitoring. *Am. Ind. Hyg. Assoc. J.* 48:219–29.

Stöber, W. 1976. Design, performance and applications of spiral duct aerosol centrifuges. In *Fine Particles: Aerosol Generation, Measurement, Sampling, and Analysis*, ed. B. Y. H. Liu, pp. 351–97. New York: Academic Press.

Tu, K. W. 1982. A condensation aerosol generator system for monodisperse aerosols of different physicochemical properties. *J. Aerosol Sci.* 13:363–71.

Vaughan, N. P. 1990. The generation of monodisperse fibers of caffeine. *J. Aerosol Sci.* 21:453–62.

Walton, W. H. and W. C. Prewett. 1949. The production of sprays and mists of uniform drop size by means of spinning disc type sprayers. *Proc. Phys. Soc.* B62:341–50.

Whitby, K. T., D. A. Lundgren, and C. M. Peterson. 1965. Homogeneous aerosol generator. *Int. J. Air Water Pollut.* 9:263–77.

Willeke, K. and P. A. Baron. 1990. Sampling and interpretation errors in aerosol monitoring. *Am. Ind. Hyg. Assoc. J.* 51:160–68.

Willeke, K. and B. Y. H. Liu. 1976. Single particle optical counter: Principle and application. In *Fine Particles: Aerosol Generation, Measurement, Sampling, and Analysis*, ed. B. Y. H. Liu, pp. 697–729. New York: Academic Press.

Winters, W., S. Barnard, and A. Hogan. 1977. A portable, photo-recording Aitken counter. *J. Appl. Meteor.* 16:992–96.

Yamada, Y., K. Miyamoto, and A. Koizumi. 1985. Size determination of latex particles by electron microscopy. *Aerosol Sci. Technol.* 4:227–32.

Yeh, H. C. and R. L. Carpenter. 1983. Evaluation of an in-line dilutor for submicron aerosols. *Am. Ind. Hyg. Assoc. J.* 44:358–60.

Yeh, H. C., Y. S. Cheng, and G. M. Kanapilly. 1983. Use of the electrical aerosol analyzer at reduced pressure. In *Aerosols in the Mining and Industrial Work Environments*, eds. V. A. Marple and B. Y. H. Liu, pp. 1117–33. Ann Arbor, MI: Ann Arbor Science Publishers.

Zhang, Z. Q. and B. Y. H. Liu. 1990. Dependence of the performance of TSI 3020 condensation nucleus counter on pressure, flow rate, and temperature. *Aerosol Sci. Technol.* 13:493–504.

23

Data Acquisition and Analysis

Dennis O'Brien

National Institute for Occupational Safety and Health[1]
Centers for Disease Control
Public Health Service
U.S. Department of Health and Human Services
Cincinnati, OH, U.S.A.

INTRODUCTION

In determining the properties of an aerosol, some physical characteristic of the aerosol, such as its concentration or size, is associated by the measuring device with an output. This output usually takes the form of another physical parameter such as the position of a needle on a meter. An output of this type is an example of analog data. Analog data are represented by physical quantities such as time, temperature, length, weight, and voltage, and can assume any value within defined upper and lower limits. Analog data are continuous in the mathematical sense. The signal from a galvanometer is an example of an analog output. Digital data, on the other hand, are not continuous, but discrete. Digital data are represented by numbers and can only assume fixed values or steps within the defined lower and upper limits. The numbers stored in a personal computer are in digital form. Although the discussion presented in this chapter is largely from an industrial hygiene perspective, these techniques are generally applicable to other fields in aerosol measurement.

Manual Observation vs. Automated Recording

If a physical parameter changes very slowly and if the observer has ample time, a pencil and paper provide a convenient and inexpensive means of recording the parameter. Many aerosol measurements integrate the variable of interest (e.g., concentration) over the sampling period, which may range from minutes to days. For these types of measurements, manual data collection, with subsequent manual entry into a permanent log or computer record, may be satisfactory. Other aerosol measurements provide a real-time output signal that changes too rapidly for manual data collection. Formerly, these types of measurements were recorded as analog data in the form of a tracing using a strip chart recorder. Data analysis required that the information be manually read from the chart and analyzed. Currently, various analog-to-digital recording devices allow vastly greater and more efficient data collection than manual techniques. These devices electronically read the data from the measuring instrument and store it in a form compatible

1. Any mention of company names or products does not constitute an endorsement by the National Institute for Occupational Safety and Health

with a personal computer (Gressel, Heitbrink, and McGlothlin 1988; Praml and Hartmann 1988; Cecala, McClelland, and Jankowski 1989; O'Brien et al. 1989).

Types of Instrument Output

Most aerosol instruments have either one of two types of signal output: a string of voltage pulses or a dc voltage signal. The output of particle sizing or counting instruments is typically a string of randomly spaced voltage pulses, with the pulse height proportional to the size of the particle measured. Other types of particle sizing instruments may have different signals that are measured. For example, in a time-of-flight spectrometer, the time between pulses is measured rather than the pulse height. However, most of the general principles of acquisition and analysis of data are the same. The output of aerosol photometers and other concentration measuring devices is usually a dc voltage signal, typically of the order of 1–10 V, full scale. The output may be an amplified and conditioned signal connected directly to the instrument meter or it may be a raw signal connected directly to a sensor. If the output is connected to the meter, the signal is generally linear with the measured variable, with a zero voltage corresponding to a zero measured variable. If the output is connected to a sensor, the output signal may be offset with respect to zero and may be nonlinear, requiring the application of a calibration function.

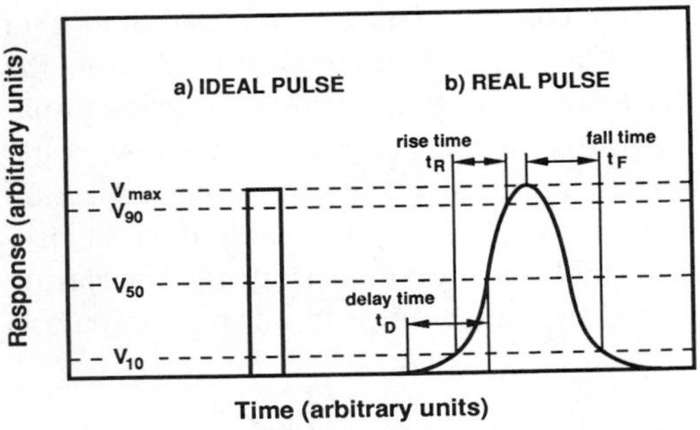

FIGURE 23-1. Pulse Parameters.

RECORDING AND ANALYSIS OF PULSES

Many modern particle sizing instruments are purchased as a complete package: particle sensor, pulse analyzer, personal computer, and software (Agarwal, Remiarz, and Nelson 1981). While a knowledge of the individual components of the system is not necessary to operate these instruments, a basic understanding may assist in the interpretation of the results obtained.

Definitions

An ideal pulse consists of a momentary positive or negative voltage "spike", as represented in Fig. 23-1a. Particle sizing instruments rarely produce ideal pulses because a finite time is required for both the measurement and the transport of the particle into and out of the detection zone. Optical particle counters are particle sizing instruments which produce pulses as output. Figure 23-1b shows a hypothetical output from the photomultiplier tube of an optical particle counter. This output can be described in terms of the following characteristics: the *delay time*, t_D, is the time required for the pulse to rise to one-half of its maximum value, V_{max}; the *rise time*, t_R, is the time required for the pulse to increase from 10% to 90% of V_{max}; and, the decay, or *fall time*, t_F, is defined as the time required for the pulse to drop from V_{max} to 10% of its maximum voltage.

Pulse Analyzer

The simplest form of a pulse analyzer is the pulse counter, which as the name implies simply counts the input pulses. In order for the pulses to be counted, they must exceed some minimum threshold (to screen out noise) and must not exceed some maximum value. In addition, the shape of the pulse must be recognizable by the counter. Most pulse counters and pulse height analyzers are designed to accept the pulses produced by nuclear detection devices. Nuclear pulses may be of much shorter duration (< 2 μs) than the pulses produced by particle counting and sizing instruments (4–140 μs reported by Maykynen et al. 1982). Therefore, it may be necessary to use a pulse-shaping amplifier, a preamplifier, or an attenuator to produce a pulse compatible with the counter/analyzer. In addition, each counter has an electronic dead time associated with each pulse. During this dead time, any incoming pulse is not counted. This can result in an observed counting rate which is lower than the true counting rate, if the recommended maximum rate is exceeded. These counting losses, and not necessarily those due to coincidence (for a discussion of coincidence effects, refer to Chapters 9 and 17), limit the maximum aerosol concentration that can be measured (Graedel 1974). The effect of electronic dead time on the counting efficiency of a particle counter can be described by the following equation (adapted from Maykynen et al. 1982):

$$\frac{n_i}{n_t} = e^{-n_t Q T} \tag{23-1}$$

where

n_t = true concentration
n_i = indicated concentration
Q = volumetric flow rate
t = electronic dead time

Simple pulse height analysis can be performed using a pulse counter with a variable threshold. This variable threshold is sometimes called a base line discriminator. It only passes pulses above a certain amplitude. If the distribution of pulse sizes remains constant, a cumulative (integrated) distribution can be obtained by repeated sampling during equal-length periods for different discriminator settings. Typical single-particle optical counters incorporate pulse counters into the instrument. These counters use variable baseline discriminators to measure particles (pulses) greater than 0.3, 0.5, 0.7 μm, etc.

Single-Channel Analyzer

A single-channel analyzer permits an evaluation of the differential distribution of pulse heights. It contains a second discriminator and an anticoincidence circuit. Only pulses with amplitudes greater than the base-line value and lower than the second discriminator set point are allowed to pass to the counting circuits. The second discriminator is tied to the base-line, so that a constant-width voltage "window" or "channel" is open to the pulse counter. The window can be stepped through the voltage spectrum so that the entire (differential) spectrum can be obtained.

Multichannel Analyzer

Multichannel analyzers perform the same function as single-channel analyzers, except that multiple channels are counted simultaneously. Most multichannel analyzers allow a selection of the number of channels or "bins" into which the pulses are sorted. In most analyzers the number of bins is usually some power of 2, with a choice ranging from about 256 to 4096. The increased resolution offered by increasing the number of bins is offset by the increased memory and data storage requirements. A hypothetical pulse height analysis is shown in Fig. 23-2. In the figure, a series of 13 randomly spaced voltage pulses "accumulate" in "bins" that correspond to discrete size ranges. The output of the multichannel analyzer may be a screen, a printer, or a storage device. Many instruments incorporate multichannel analyzers into the instrument package. Individual multichannel analyzers are available as stand-alone systems, or as cards that can be installed in personal computer systems.

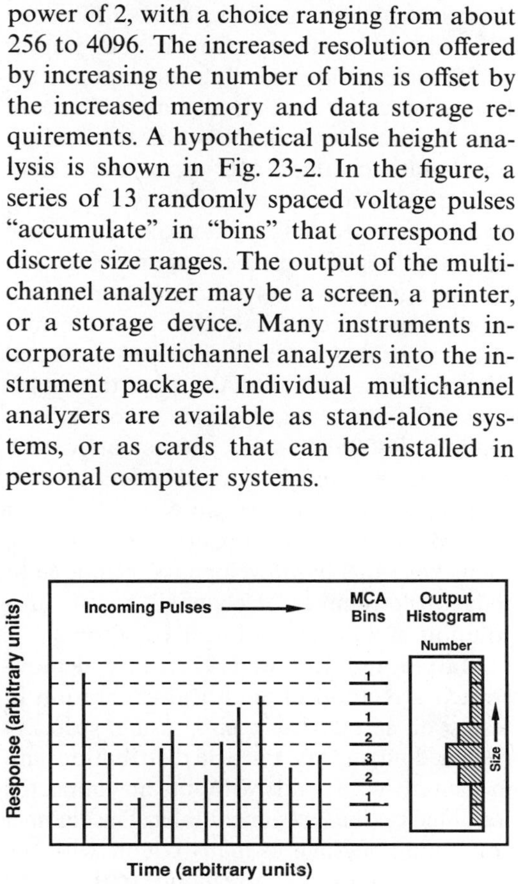

FIGURE 23-2. Pulse Height Analysis Process.

Calibration

To provide useful information, the output of the pulse height analyzer must be calibrated. In the case of particle sizing instruments, a series of monodisperse aerosols, such as polystyrene latex spheres, is used to calibrate the channels or bins. A data set is obtained containing particle size and the corresponding channel number (or voltage range). A calibration curve can then be generated using a curve-fitting routine such as a cubic spline. Particle size is determined in terms of the equivalent calibration particle, with size computed from the calibration curve.

The size resolution afforded by the calibration is limited by the particle measuring device, not the analysis system. For example, consider the case of single-particle optical counters. The response characteristics of these instruments can be calculated from the Mie scattering theory. Cooke and Kerker (1975) have performed these calculations for five commercial particle counters. Some of these response curves show points of inflection (multiple values) over a portion of their size range. The selection of the width of reported individual size intervals should be such that regions of the response function having multiple values are incorporated into single intervals; otherwise these instrumental characteristics may be interpreted as actual particle size variations. Although the multichannel analyzer may provide a voltage resolution of 4096 channels, the actual particle size resolution may be several orders of magnitude less. In order to be more realistic about the capabilities of the measuring instruments, Jaenicke (1972) has developed a technique to reduce the number of channels based on a solution of a system of linear equations.

Marple and Rubow (1976) have demonstrated a method of empirical calibration in terms of aerodynamic size, using specially designed impactors. Particle distributions are measured with and without an impactor mounted on the instrument inlet. The channel containing one-half as many counts with the impactor as without corresponds to the cutoff size of the impactor. Various calibration points can be obtained by using different nozzles to change the cutoff size of the impactor. Marple and Rubow (1978) also explain the use of an inertial impactor technique for calibration. They note that, by using an average calibration curve, the uncertainty in particle size at any particle counter output voltage will generally be less than about 30%. Chen, Yeh, and Tang (1990) report the use of a Lovelace two-stage virtual impactor for aerodynamic classification of monodisperse aerosols for calibration of a Royco 226 optical particle counter. They used this technique with aerosols of fly ash, oil shale, triphenyl phosphate, talc, alumina, polystyrene latex spheres, and glass microspheres. A comparison of optical equivalent and aerodynamic diameters indicated that the optical counter could overestimate or underestimate aerodynamic size, using calibration based on polystyrene latex spheres. The authors concluded that a two-stage virtual impactor is useful for such calibrations. Calibration can also be performed using monodisperse aerosols generated from atomizing latex spheres, vibrating-orifice generators, spinning-disk generators, and other techniques. A more detailed discussion of calibration techniques can be found in Chapter 22.

Spreadsheet Analysis

Multichannel data acquisition systems are typically tailored for nuclear applications, e.g., gamma ray spectroscopy; thus, the type of analyses that can be performed are of limited use to the aerosol scientist. Many multichannel analysis systems produce data files in the American Standard Code for Information Interchange (ASCII) format which allows ease of manipulation by the user, especially if the data are imported into an electronic spreadsheet. Spreadsheets are versatile tools that can be tailored for the analysis of multichannel data. A spreadsheet is an electronic work sheet composed of "rows" and "columns". The intersection of each row and column is the "cell". The cell may contain labels, data values, or formulae. Spreadsheets allow the user to copy, move, format, graph, search, and analyze data contained in selected ranges of cells. A file imported from the multichannel analyzer will appear as two or more columns of data. One column of data will contain the bin (channel) number, which is proportional to the pulse height, and is related to particle size. The other column will contain the number of pulses counted in each channel.

Channel numbers are converted into particle size data in the spreadsheet by summing the counts contained in a specified range of channels (cells in the spreadsheet) and assigning this sum to a cell which corresponds to a given size interval. For lognormal distributions the size interval is a geometric progression: the upper limit of the size interval is some multiple of the lower limit. This process is repeated for each channel. If the sampling efficiency of the instrument is known, a corrected size distribution can be calculated. Dividing each entry by the sample volume yields a size distribution by number concentration (particles per sampled volume). By assuming a spherical shape (or applying shape factors) and uniform density, the size distributions by surface and mass can be calculated. Cumulative size distribution can be determined from the number, surface, or mass distributions by summing all cells corresponding to sizes smaller than the indicated cell. Many other analyses can be easily performed, including the removal of background counts, calculation of counting statistics, normalization, spectra comparisons (e.g., for filter penetration studies), and application of particle removal/deposition criteria.

ANALOG-TO-DIGITAL CONVERSION OF DC VOLTAGE DATA

Aerosol photometers and many other real-time instruments have analog output capabilities. This usually takes the form of a dc voltage signal, typically of the order of 1–10 V, full scale. In the past, this analog output has been used to drive a strip chart recorder. However, in order to perform data analysis with a computer, the data from the strip chart recorder needed to be manually read and keyed into the computer, a rather tedious process. With the advent of the personal computer, the real-time analog output from these monitors can now be stored digitally, allowing the data to be transferred to the computer with just a few steps.

Real-time monitoring can be used in industrial hygiene to relate some measurement of concentration (exposure) to some work activity or piece of equipment, to determine some dynamic property of the work environment, or to contrast concentration differences caused by the introduction of some control method. Implicit in these purposes is the measurement of other variables affecting the concentration measurement. These variables may be qualitative, i.e., the performance of a specific work task, or they may be quantitative, i.e., the rate at which a material is being used in a particular industrial process. Quantitative process variables may be recorded using the same data acquisition system as that used for the concentration measurements, or they may be recorded manually in a notebook. Qualitative variables may also be measured by the use of video recording. This allows an after-the-fact review of the experiment or measured events. The only requirements are that the video recording system needs an on-screen clock or timer that can be synchronized with the clock of the data recording device, and that the resolution of the clock or timer be compatible with the activities measured.

Types of Recording Devices

Data recording devices generally fall into two different categories: portable data loggers and computer-based analog-to-digital (A/D) convertor systems. Portable data loggers store the data in a built-in bank of memory. They may record data from one or more sensors (concentration, pressure, temperature, etc.) into respective channels (not to be confused with the channels of a pulse height analyzer). After data collection, the data are transferred to (interrogated by) a personal computer, typically through the computer's communication (usually the RS-232) port. How this is accomplished varies with the manufacturer, some using available communications programs, others requiring proprietary software written specifically for their data logger. Computer-based systems may consist of either an internally mounted card or an external, separately housed A/D unit. Internal systems are capable of recording from multiple inputs (typically 8 or 16); external systems are capable of recording up to several hundred inputs. The computer-based systems store the data directly into the computer memory or onto the disk drive. These systems require software for control of the A/D convertor. While some users prefer to write their own control software, several commercial packages are available which are capable of controlling A/D boards made by several manufacturers. Many boards contain digital-to-analog functions as well, so that automatic control of the instrument or experiment is possible.

Resolution

Resolution refers to the ability to distinguish different input signals. A/D convertors will either have a fixed input range, 0–2 V, for example, or the working range can be chosen with hardware switches or control software. This working range is then broken down into intervals. The resolution of an analog-to-digital convertor is usually given in bits. An 8-bit data logger with a working range of 0–2 V will break that range into 256 intervals, each interval corresponding to about 0.008 V. The number of intervals, n, is determined by the following equation:

$$n = 2^{\text{Bits}} \tag{23-2}$$

Therefore, for the 8-bit convertor, 0–2 V working range example, in order for the data recording device to differentiate between two voltage readings, the difference in readings must be greater than 0.008 V. The magnitude of the voltage interval is calculated from the following equation:

$$V_i = \frac{V_u - V_l}{n} \tag{23-3}$$

where

V_i = interval, V
V_u = upper working range, V
V_l = lower working range, V
n = number of intervals

Data loggers are typically 8-bit units while computer-based A/D convertors are available from 8 to 16 bits.

Example 23-1

An aerosol photometer is operated in the range of 0–100 mg/m^3. The analog output of the instrument is 0–2 V, which is connected to an 8-bit data logger with an input range of 0–10 V. What is the resolution of the recorded measurements in mg/m^3?

Answer. From Eq. 23-2, the A/D convertor divides its 10 V input into $2^8 = 256$ intervals. Since the output of the instrument is only 2 V, it makes use of only 2/10 (51) of the available 256 intervals. Since 51 intervals correspond to a full-scale reading of 100 mg/m^3, each interval (the resolution) represents approximately 2 mg/m^3.

Sampling Rate

The process by which a series of digital values are assigned to a continuous analog signal is called sampling. The analog-to-digital convertor observes the input signal at regular intervals and assigns a discrete value of the measured variable to this time. The assigned value depends on the resolution of the A/D convertor as described above. The highest-frequency, f_H, component contained in the input signal is called the *Nyquist frequency*. The minimum rate at which this signal can be sampled, $2f_H$, is called the *Nyquist rate*. Sampling at lower than this rate can produce a phenomenon known as "aliasing", where a high frequency signal is disguised as a lower-frequency signal. Figure 23-3 illustrates this phenomena.

Spreadsheet Analysis

Software created for data acquisition systems is often limited in the types of analyses that can be performed. Generally, these packages offer integration, averaging, determination of minima and maxima, and graphic output. For additional analyses it is desirable to convert the data into a form recognizable by electronic spreadsheets.

Importing Data

Concentration data are typically contained in data logger files, which can only be read by proprietary data logger software. The software may contain utilities that can be used to create compatible files (e.g., ASCII format) that can be imported by the spreadsheet. Since all but the shortest of test runs can create an extraordinarily large spreadsheet, the data logger software often allows the user the selection of a time interval between spreadsheet entries. Individual instruments vary in the speed of their response; the time interval selected should not be less than this response time.

The imported file will appear as two or more columns of data. One column of data will contain time (either elapsed time or clock time). The other column(s) will contain the output signal(s) of the measuring instrument(s).

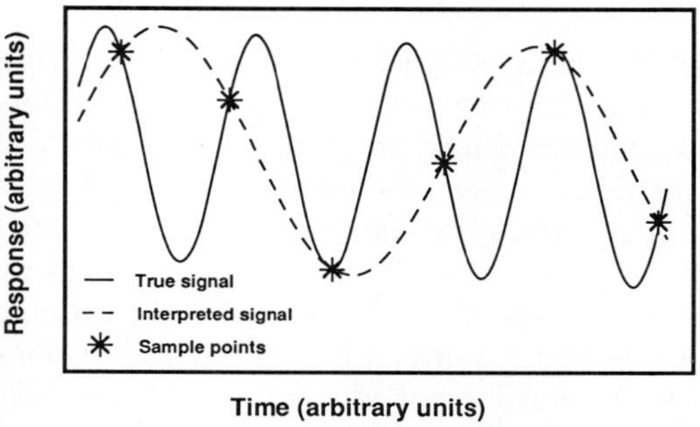

FIGURE 23-3. "Aliasing" Caused by Excessive Sampling Period. Sampling at Too Low a Sampling Rate Results in a False Interpretation of the "True" Signal.

Instrument and Transportation Time Lag

Instruments will vary in their ability to respond to a step change in the measured variable. The time required for the output of an instrument to reach 63% ($1/e$) of its response to a step change in input is termed as its *time constant* (Willard, Merritt, and Dean 1974). Some instruments have an adjustable time constant, higher settings having the effect of damping meter and output fluctuations in order to permit easier reading.

TABLE 23-1 Commercial Sources for Data Acquisition and Analysis

Type	Commercial Source	Remarks
Multichannel analyzer	CAN	Several models, stand-alone and PC-based
Multichannel analyzer	ORT	Several models, PC-based
Multichannel analyzer	TEN	Several models, stand-alone and PC-based
Analog-to-digital cards for PC	CYB	
Analog-to-digital cards for PC	DAT	
Analog-to-digital cards for PC	KTH	
Data acquisition software	AST	"ASYST"—software for the control of A/D boards and data analysis
Data acquisition software	LAB	"Labtech Notebook"—software for the control of A/D boards and data analysis
Data acquisition software	LOT	"Acquire"—software for the control of A/D boards and data analysis
Data logger	MET	
Data logger	RUS	
Spreadsheet software	LOT	Lotus 1-2-3
Spreadsheet software	BOR	Quattro-Pro
Spreadsheet software	XCL	Excel

TABLE 23-2 List of Commercial Sources for Data Acquisition and Analysis

Code	Address
AST	Asyst Software Technologies, Inc., 100 Corporate Woods, Rochester, NY 14623. Tel. (800) 348-0033
BOR	Borland International, 4585 Scotts Valley Drive, Scotts Valley, CA 95066. Tel. (800) 255-8008
CAN	Canberra Industries, Inc., One State Street, Meriden, CN 06450. Tel. (203) 238-2351
CYB	Cyborg, 94 Bridge St., Newton, MA 02158. Tel. (617) 964-9020
DAT	Data Translation, Inc., 100 Locke Drive, Marlboro, MA 01752. Tel. (508) 481-3700
KTH	Keithley Data Acquisition and Control, Keithley Instruments, Inc., 28775 Aurora Rd., Cleveland, OH 44139. Tel. (216) 248-0400
LAB	Laboratory Technologies Corporation, 400 Research Drive, Wilmington, MA 01887. Tel. (508) 657-5400
LOT	Lotus Development Corporation, 55 Cambridge Parkway, Cambridge, MA 02142. Tel. (800) 232-1662
MET	Metrasonics, Inc., Box 23075, Rochester, NY 14692. Tel. (716) 334-7300
ORT	EG&G ORTEC, 100 Midland Road, Oak Ridge, TN. Tel. (615) 482-4411
RUS	Rustrak Instruments, Rte 2 & Middle Rd., East Greenwich, RI 02818. Tel. (800) 332-3202
TEN	Tennelec/Nucleus Inc., 601 Oak Ridge Turnpike, Oak Ridge, TN. Tel. (615) 483-8405
XCL	Microsoft, 16011 NE 36th Way, Redmond WA 98073

In addition to the delay caused by the time constant of the instrument, there will be an additional delay caused by the environmental transport of the contaminant to the instrument inlet. This transport delay may vary, depending on the type of operation being performed and its physical arrangement. It may not necessarily remain constant in time. Instrument and environmental response may affect the independence of the measurements. This topic is discussed later in this chapter in the Autocorrelation section.

In performing analyses using a spreadsheet, the concentration measurements need to be "slipped" with respect to the indicated time to allow for instrument response and environmental transport, so that the concentration measurements will correspond to the causal activities. This synchronization is basically a trial-and-error procedure.

Data Coding

As noted earlier, one purpose of real-time monitoring is to relate some measurement of concentration to some causative variables. Quantitative variables may be imported into the spreadsheet from electronically obtained data files or they may be manually entered based on careful note taking. Qualitative measurements can be obtained by a review of video recordings of the measured events. The qualitative variables are entered into the spreadsheet as a column of labels corresponding to the activity (other variables) occurring during that time interval (row). The qualitative variables also may be entered into the spreadsheet as a column of dummy variables (1's and 0's) corresponding to "yes" or "no" (or vice versa) to the activity represented by that column. A dummy or indicator variable is one that assumes a finite number of values for the purpose of identifying different categories of interest (Kleinbaum and Kupper 1978).

Database Functions

Database functions, available in many spreadsheets, can provide a quick and easy method of analysis. These functions allow the user to identify a range of cells as a database. This database can then be searched for data which meet certain criteria. These criteria can be either quantitative or qualitative variables. The data can be extracted, or simple statistical parameters can be calculated (e.g., average, count). The computation of descriptive statistics can be automated by setting up a "data table" using a combination of the database statistical functions, the database query capabilities, and the data table function built into most of the current spreadsheet programs. The relative importance of each activity can be ascertained by determining the integrated concentration (the concentration–time product).

Example 23-2

One worker uses two different tools (A and B) in the course of performing his job tasks. Dust exposure was determined using an aerosol photometer, the data recorded with a data logger, and the work period was videotaped. The data was down-loaded from the logger into the spreadsheet of Fig. 23-5. What is the average dust exposure when using each tool?

Answer. The videotape is reviewed and the tool use is entered into the spreadsheet for each time–concentration pair as the descriptive variables "A" and "B". The database range B5 . . . C15 is assigned the name, "NAME". The average exposure is determined for the use of each tool using the database functions @davg using the criteria G19 . . . G20 (tool "A") and G24 . . . G25 (tool "B"). The resulting average dust exposure is 1.1 mg/m^3 when using tool "A" and 3.6 mg/m^3 when using tool "B".

Regression Analysis

Multiple regression provides a more sophisticated method of analyzing real-time data. Multiple regression determines if one variable is dependent on two or more others. Typically, the dependent variable (the Y-range) is the contaminant concentration, but it also may be a concentration difference, or a mathematical function of the concentration. The independent variables (the X-range) are the quantitative process variables or dummy variables described in the Data coding section. The regression routine calculates the coefficients and the intercept of the best-fitting straight line. A typical regression equation is of the following form:

$$\text{concentration} = c_0 + c_1 * x_1 + c_2 * x_2 + \cdots + c_n * x_n \qquad (24\text{-}4)$$

where

c_0 = constant or intercept of regression line

	A	B	C	D	E	F	G
1	One worker uses two tools (A & B) to perform his job.						
2	Which tool causes the greatest exposure?						
3							
4							
5	Time	Conc.	Tool				
6	0.0	1.0	A				
7	1.0	1.0	A				
8	2.0	1.2	A				
9	3.0	1.3	A				
10	4.0	1.0	A		<-- Database "NAME"		
11	5.0	4.0	B				
12	6.0	4.0	B				
13	7.0	3.0	B				
14	8.0	2.0	B				
15	9.0	5.0	B				
16							
17							
18							
19		Exposure when using tool "A":				Tool	<-- Criterion
20						A	
21			@DAVG(NAME,0,G19..G20) -->			1.1	
22							
23							
24		Exposure when using tool "B":				Tool	<-- Criterion
25						B	
26			@DAVG(NAME,0,G24..G25) -->			3.6	
27							
28							

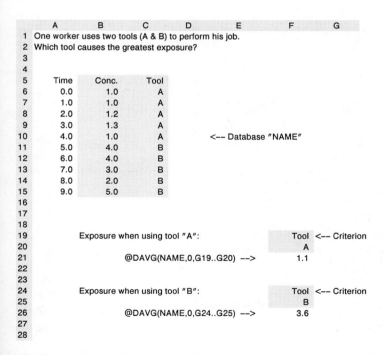

FIGURE 23-5. Example 1 Spreadsheet for Time/Task Analysis Using Database Functions.

$c_1 – c_n$ = coefficients of regression
$x_1 – x_n$ = values of independent variables

Example 23-3

Using the data of Example 23-2 and regression analysis, what is the average dust exposure when using each tool?

Answer. The videotape is reviewed as before and the tool use is entered into the spreadsheet of Fig. 23-6 for each time–concentration pair as the dummy variables 0 and 1, corresponding to use of the tools "A" and "B", respectively. Using the spreadsheet functions to perform a simple linear regression of the data, the regression equation "EXPOSURE + 2.5 * TOOL + 1.1" is obtained. Substituting the dummy variables into this equation results in an average dust exposure of 1.1 mg/m^3 when using tool "A" and 3.6 mg/m^3 when using tool "B" in the spreadsheet of Fig. 23-6. The 95% confidence limit for the regression coefficients can be calculated by multiplying the "Std Err Coef" calculated by the spreadsheet by 1.96 and adding/subtracting the result to the coefficient. (Note: the factor 1.96 is valid only for the large number of data points ($\gg 10$) typically encountered when analyzing electronically acquired data.) In the example spreadsheet, coefficient c_0 corresponds to exposure when

	A	B	C	D	E	F	G
1	One worker uses two tools (A & B) to perform his job.						
2	Which tool causes the greatest exposure?						
3							
4		Measured		Predicted			
5	Time	Conc.	Tool	Conc.	Residual		
6		(Y)	(X)				
7	0.0	1.0	0.0	1.1	−0.1		
8	1.0	1.0	0.0	1.1	−0.1		
9	2.0	1.2	0.0	1.1	0.1		
10	3.0	1.3	0.0	1.1	0.2		
11	4.0	1.0	0.0	1.1	−0.1		
12	5.0	4.0	1.0	3.6	0.4		
13	6.0	4.0	1.0	3.6	0.4		
14	7.0	3.0	1.0	3.6	−0.6		
15	8.0	2.0	1.0	3.6	−1.6		
16	9.0	5.0	1.0	3.6	1.4		
17							
18		MODEL:	Y	= C1 * X + Constant			
19			EXPOSURE = 2.5 * TOOL + 1.1				
20							
21			Regression Output:				
22		Constant			1.1		
23		Std Err of Y Est			0.81		
24		R Squared			0.75		
25		No. of Observations			10		
26		Degrees of Freedom			8		
27							
28		X Coefficient(s)		2.5			
29		Std Err of Coef.		0.51			
30							
31							
32		Exposure when using tool "A":				For "A", TOOL (X) = 0	
33			EXPOSURE(A) = 2.5 * 0 + 1.1				
34			EXPOSURE(A) = 1.1				
35							
36							
37		Exposure when using tool "B":				For "B", TOOL (X) = 1	
38			EXPOSURE(B) = 2.5 * 1 + 1.1				
39			EXPOSURE(B) = 3.6				
40							

FIGURE 23-6. Example 2 Spreadsheet for Time/Task Analysis Using Linear Regression.

using tool "A". Coefficient c_1 corresponds to the difference in exposures when using the two tools.

Autocorrelation

Real-time data do not consist of simple sets of independent measurements. The measuring instrument, the electronics, and the environment do not respond immediately to changes in the independent variables. The precise response of the environment and the instrument/electronics package to an external stimulus is not known. Therefore, each data point obtained may be some function of the preceding measurement(s). To determine the degree of this dependence or autocorrelation, the regression equation is used to calculate a predicted value of the dependent variable (e.g., dust concentration). The predicted value (e.g., dust concentration as a function of the qualitative variables) minus the observed value (e.g., measured concentration) yields the residual. If the real-time data are independent, the residual values should be random numbers. To test for time dependence, the residuals can be copied to an empty section of the spreadsheet. The residuals can then be recopied to adjoining columns but offset by

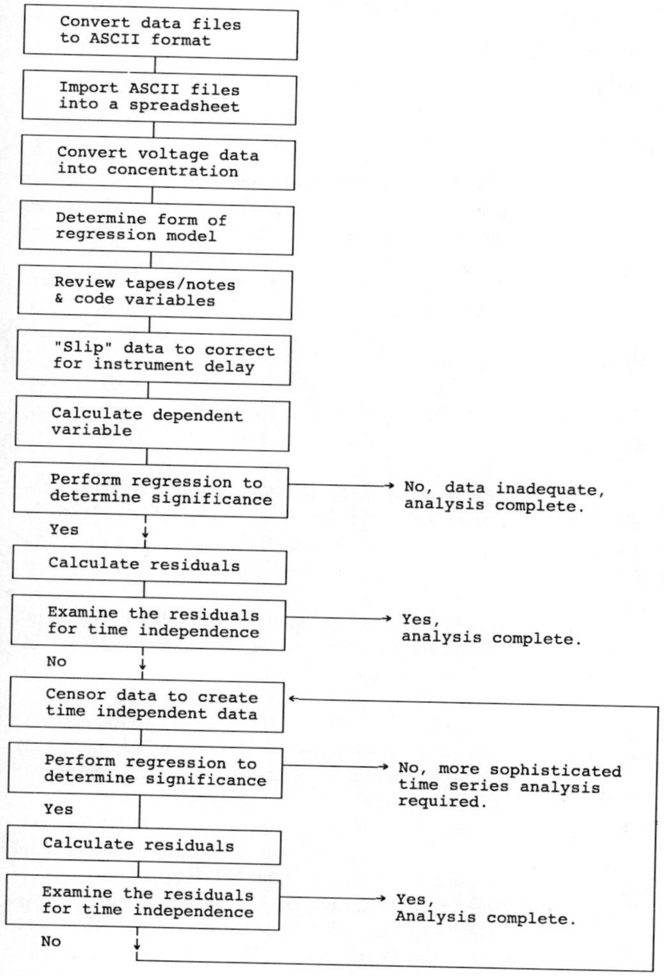

FIGURE 23-4. Flow Diagram for Spreadsheet Analysis of Real-Time Concentration Data.

one, two, and three readings corresponding to delays of 1, 2, and 3 time intervals. Regression analyses can then be performed on these sets of residuals to determine autocorrelation. If the data set demonstrates that each reading is dependent only on the reading immediately preceding it, then the time dependence can be removed from the original data set by eliminating every other data point, then performing a new regression on the reduced, time-independent data set. Similar data cleansing can be performed if a time dependence exists for readings separated by 2 or 3 time intervals. The database functions can be used on the "cleansed" data to determine the relative contributions to the concentration.

Measurements collected at equally spaced intervals are known as *time series*. Autocorrelation is the phenomenon that distinguishes time series from other branches of statistical analysis. The above procedure of data censoring may result in the elimination of too much data. In that case, the use of more sophisticated analytical procedures will be required, such as those outlined by Brocklebank and Dickey (1986).

Figure 23-4 summarizes the steps used in the spreadsheet analysis of real-time data. First, the data logger/acquisition system file(s) is converted to ASCII format, and imported into a spreadsheet. Multiple files can be combined into a single spreadsheet if necessary. Within the spreadsheet, voltage data can be transformed into units of concentration, using a calibration factor (if desired). Next, videotapes or field notes are reviewed and these variables are coded into the spreadsheet. The data are then examined for instrument delay and corrected as necessary. Multiple regression is performed to determine the significance of the relationship, and to calculate the predicted differences and the residuals. The residuals (the difference between the predicted value and the dependent variable) are examined for time independence. Finally, a time-dependent data set is created by cleansing the data, and multiple regression is performed to determine if the relationship is still significant.

FUTURE TRENDS

The trends in the personal computer and electronics industries have been to create smaller, lighter, and less expensive instruments, yet with increased sophistication and ease of use. Thus, manufacturers of laboratory instruments increasingly have been assembling packaged systems, whereby data acquisition and analysis are integrated into the system design. Portable aerosol instruments are currently lagging behind their gas and vapor monitoring counterparts in the degree of integration of data recording and analysis capabilities; future instruments should see an incorporation of computer-transferable, automated data recording.

References

Agarwal, J. K., R. J. Remiarz, and P. A. Nelson. 1981. An instrument for real time aerodynamic particle size analysis using laser velocimetry. In *Proc. Inhalation Toxicology and Technology Symposium*, ed. B. K. J. Leong, Ann Arbor, MI: Ann Arbor Science Publishers.

Brocklebank, J. C. and D. A. Dickey. 1986. *SAS System for Forecasting Time Series*. Cary, North Carolina: SAS Institute.

Cecala, A. B., J. J. McClelland, and R. A. Jankowski. 1988. Substantial time savings achieved through computer dust analysis. *Appl. Ind. Hyg.* 3:203–6.

Chen, B. T., H. C. Yeh, and J. A. Tang. 1990 A new technique to calibrate optical particle counters aerodynamically. *Am. Ind. Hyg. Assoc. J.* 51(1):32–35.

Cooke, D. and M. Kerker. 1975. Response calculations for light-scattering aerosol particle counters. *Appl. Opt.* 14:734–39.

Graedel, T. E. 1974. Channel width determination and electronic pulse processing losses in optical particle counters. *J. Aerosol Sci.* 5:125–31.

Gressel, M. G. and W. A. Heitbrink (eds.). 1991. *A Manual for Analyzing Workplace Exposures Using Real-Time Monitoring Techniques*. Cincinnati, OH. National Institute for Occupational Safety and Health (in preparation).

Gressel, M. G., W. A. Heitbrink, and J. D. McGlothlin. 1988. Advantages of real time data acquisition for exposure assessment. *Appl. Ind. Hyg.* 3:316.

Jaenicke, R. 1972. The optical particle counter: Cross-sensitivity and coincidence. *J. Aerosol Sci.* 30:95–111.

Kleinbaum, D. G. and L. L. Kupper. 1978. *Applied Regression Analysis and Other Multivariable Methods*. North Scituate, MA. Duxbury Press.

Marple, V. A. and K. L. Rubow. 1976. Aerodynamic particle size calibration of optical particle counters. *J. Aerosol Sci.* 7:425–33.

Marple, V. A. and K. L. Rubow. 1978. A portable optical particle counter system for measuring dust aerosols. *Am. Ind. Hyg. Assoc. J.* 39:210–18.

Maykynen, J., J. Hakulinen, T. Kivisto, and M. Lehtimaki. 1982. Optical particle counters: Response, resolution, and counting efficiency. *J. Aerosol Sci.* 13:529–35.

O'Brien, D. M., T. J. Fischbach, T. C. Cooper, W. F. Todd, M. G. Gressel, and K. F. Martinez. 1989. Acquisition and spreadsheet analysis of real time dust exposure data: A case study. *Appl. Ind. Hyg.* 4:238–43.

Praml, G. J. and A. L. Hartmann. 1988. Personal full shift fine dust registration: A handy measuring system for field application. *J. Aerosol Sci.* 19:1449–51.

Willard, H. W., L. L. Merritt, and J. A. Dean. 1974. *Instrumental Methods of Analysis.* New York: Van Nostrand.

III

Applications

24

Industrial Hygiene

Paul A. Jensen and Dennis O'Brien

National Institute for Occupational Safety and Health[1]
Centers for Disease Control
Public Health Service
U.S. Department of Health and Human Services
Cincinnati, OH, U.S.A.

INTRODUCTION

Definition of Industrial Hygiene

Industrial hygiene is both a science and an art. Industrial hygiene encompasses the total realm of control, including the recognition and evaluation of those factors of the environment emanating from the place of work which may cause illness, lack of well being, or discomfort either among workers or among the community as a whole. In short, industrial hygiene can be defined as the identification, evaluation, and control of occupational health hazards. Industrial hygiene aerosol sampling differs from general aerosol measurements in several ways. First, in evaluating the industrial environment, the aerosol contaminant is almost always known, or it can be deduced from a knowledge of the process or product in use. Only in limited cases is the sampling performed to identify an unknown contaminant. Second, industrial hygiene sampling today is mainly performed by the use of portable, battery-operated air sampling equipment. This equipment is usually worn by the worker for the duration of the work shift to define the exposure of that individual. This is known as personal breathing zone air sampling. The personal breathing zone is that air that would most nearly represent the air inhaled by the worker and is usually considered to be either less than about 30 cm from the workers mouth or a distance of four to six feet from the floor. Lastly, the concentrations encountered in industrial hygiene are usually higher than those encountered in community air sampling but much lower than that potentially present in monitoring the effluent of a stack or other process vent.

Historical Perspectives

Air sampling equipment for the measurement of aerosols has evolved over the past several decades in the direction of miniaturization and automation. First (1988) cites the example of the Greenberg–Smith impinger, which was the workhorse of aerosol sampling in the 1920s. This device was too large and heavy to permit its use as a personal aerosol sampling instrument. It was supplanted a decade later by the smaller and lighter midget impinger which required a pump and motor only one-tenth the size of the earlier equipment. For aerosols, he notes, the midget impinger has been largely replaced by the use of

1. Any mention of commercial names or products does not constitute an endorsement by the National Institute for Occupational Safety and Health.

filter media contained in disposable cassettes with airflow provided by portable, battery-operated pumps. Electronic controls permit the automated start-up and shut-off of the pump, record the sampling time, monitor pump performance, and control the pump speed to compensate for the increase in the pressure drop across the filter as the contaminant is collected. The increased miniaturization and automation has not been limited to personal sampling equipment. In the past, real-time (direct-reading) aerosol sampling instruments have been limited by their size and power requirements to measurements made in a fixed location in the work environment. Portable real-time instruments have been developed to measure aerosol concentration and/or particle size distribution. These instruments can be carried into the work site and measurements can be made in the breathing zone of the worker. Even smaller aerosol measuring instruments have been developed which can be worn by the worker. These instruments may have data recording and averaging capabilities. They can also provide a warning to the worker if concentrations exceed a preset alarm condition so that the worker can leave the hazardous environment and action can be taken to alleviate the hazard.

The evolution of industrial hygiene measurement practice from fixed-location (area) monitoring to personal sampling arises from the general inadequacy of area sampling for monitoring worker exposures. Area monitoring does not yield adequate exposure estimates for the individual worker because of temporal and spatial variation of aerosol contaminants in the work environment. Industrial aerosol mass and size distributions must not be assumed to be the same in all locations or at different times. Often, the individual's own work activities are the source of aerosol contamination. A dust-producing source located at arm's length results in a rapidly decreasing concentration with increasing distance from the worker. Tebbens (1973) notes that it is the likelihood of large temporal and spatial errors which led to the concept of personal sampling, that is, attaching the sensing element directly to the worker.

PURPOSES OF SAMPLING

Industrial hygiene sampling is performed for a wide variety of reasons: to determine if a risk to worker health exists; to comply with regulatory requirements; to monitor continued performance of good work practices and engineering controls; to gather data for use in epidemiological studies; and to identify highly hazardous jobs or specific work tasks so that corrective actions may be taken. The above reasons imply that what is measured and how the measurement is made is relevant to the biological effects of the hazardous material.

Toxicological Considerations

Contaminants enter the body by skin absorption, ingestion, and inhalation. Inhalation is the most common route of entry for industrial dusts (although for certain contaminants, most notably lead, ingestion can also play an important role). The effects of inhaled aerosols depend on the specific chemical species, the concentration, the duration of the exposure, and the site of deposition within the respiratory system. Particle size is the most important factor affecting the location of deposition.

The health effects produced by exposure to a toxic aerosol vary based on the duration of exposure. Acute effects occur within a few minutes or hours of exposure. In some instances, the acute effects may be easily reversed; in others, irreparable harm may be done to the worker. For example, an aerosol of an irritant such as anhydrous ammonia and water has reversible, acute effects at low levels of exposure and possible death due to respiratory arrest at high levels (Rom and Barkman 1983). Chronic effects may take years or a lifetime to develop and are not usually reversible. For example, death due to lung cancer may not occur until 5–35 years after the onset of work with asbestos (Rom and Barkman 1983).

Local effects occur when inhaled contaminants produce a toxic effect at the site of contact. An example of a local effect within the respiratory system is silicosis, a disease of the lung resulting from inhalation of crystalline silica (Jones 1983). A toxic effect, known as a systemic effect, may also be caused at sites within the body far removed from the point of entry. An example of a systemic effect is the toxic consequence for the blood-forming organs and the nervous system resulting from an overexposure to lead fumes (Fischbein 1983). The chemical and biological properties of the aerosol determine if the effect produced will be at the site of deposition or at some other location within the body. For example, the solubility of aerosol particles within the fluids of the lung may determine the availability of the chemical to the bloodstream. Particle size distribution affects not only the site of deposition but also the chemical reactivity, since the total surface area of a given mass of dust increases as the particle size decreases (Hinds 1982). Since the toxic effect depends on the site of deposition, the aerosol measurement techniques and the control technologies must target the particle sizes producing the undesirable effect.

Deposition of Particles in the Respiratory System

The determination of the health consequences of exposure to an aerosol requires an analysis of the inhalation and deposition of the aerosol within the human respiratory system. The toxic action of an aerosol may be related to the number of particles, their surface area, or the mass deposited. Many occupational diseases are associated with the deposition of particles within a certain region of the respiratory tract. The aerosol properties associated with the location of deposition in the respiratory system are particle size and density. The parameter most closely associated with this regional deposition is the aerodynamic diameter, d_a, defined as the diameter of a sphere of unit density possessing the same terminal settling velocity as the particle in question.

Many particles present in the air do not enter the respiratory tract because of the size-selective sampling of the nose and the mouth. Still others are removed in the upper respiratory tract. The concept of size-selective air sampling calls for measurement of particles in industrial aerosols of the size that are associated with a specific health effect. This concept can be traced to proposals from the British Medical Research Council and the U.S. Atomic Energy Commission in the 1950s defining respirable dust (Morrow 1964). More recently, the International Standards Organization (ISO) (Ogden 1983) and the American Conference of Governmental Industrial Hygienists (ACGIH) (ACGIH 1985, 1991) have proposed definitions of dust size ranges for use in sampling industrial aerosols. For chemical substances present in inhaled air as suspensions of solid particles or droplets, the potential hazard depends on the particle size as well as on mass concentration. The ACGIH has defined three particle-size-selective threshold limit values (PSS TLVs): inhalable particulate mass TLVs (IPM TLVs) for those materials that are hazardous when deposited anywhere in the respiratory tract; thoracic particulate mass TLVs (TPM TLVs) for those materials that are hazardous when deposited anywhere within the lung airways and the gas-exchange region; and, respirable particulate mass TLVs (RPM TLVs) for those materials that are hazardous when deposited anywhere in the gas-exchange region.

These three particulate mass fractions are quantitatively defined by ACGIH (1991) as follows:

1. Inhalable particulate mass (IPM) consists of those particles that are captured according to the following collection efficiency regardless of sampler orientation with respect to wind direction:

$$SI(d_a) = 50\% \times (1 + \exp(-0.06 d_a)) \quad \text{for } 0 < d_a \leq 100\ \mu\text{m} \qquad (24\text{-}1)$$

where $SI(d_a)$ is the collection efficiency in percent and d_a is the aerodynamic diameter in μm.

2. Thoracic particulate mass (TPM) consists of those particles that are captured according to the following collection efficiency:

$$ST(d_a) = SI(d_a) \times (1 - F(x)),$$
$$x = \frac{\ln(d_a/\Gamma)}{\ln(\Sigma)} \tag{24-2}$$

where $ST(d_a)$ is the collection efficiency in percent, the mass median aerodynamic diameter, Γ, is 11.64 μm, the geometric standard deviation, Σ, is 1.5, and $F(x)$ is the cumulative probability function of a standardized lognormal variable, x.

3. Respirable particulate mass (RPM) consists of those particles that are captured according to the following collection efficiency:

$$SR(d_a) = SI(d_a) \times (1 - F(x)),$$
$$x = \frac{\ln(d_a/\Gamma)}{\ln(\Sigma)} \tag{24-3}$$

where $SR(d_a)$ is the collection efficiency in percent, the mass median aerodynamic diameter, Γ, is 4.25 μm, the geometric standard deviation, Σ, is 1.5, and $F(x)$ is the cumulative probability function of a standardized lognormal variable, x.

These performance criteria are described graphically in Fig. 24-1.

An international convention for particle-size-selective sampling was proposed, which requires the resolution of the differences between the firmly established British Mine Research Council (BMRC) and ACGIH particle size-selective sampling in the workplace (Soderholm 1989; Lippmann 1989*a*). The proposed definitions differ approximately equally from the BMRC and old ACGIH definitions and is at least as defensible when compared to available human data. Much international confusion will be avoided once ISO adopts the RPM, TPM, and IPM definitions recommended by Soderholm (1989) and adopted by ACGIH (Soderholm 1991). The ACGIH and ISO RPM, TPM, and IPM collection efficiency curves and the BMRC RPM collection efficiency curve are shown graphically in Fig. 24-1.

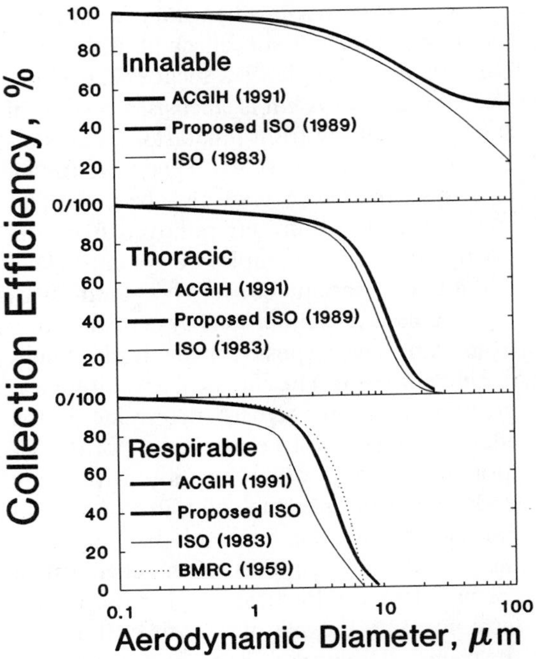

FIGURE 24-1. American Conference of Governmental Industrial Hygienists (ACGIH), British Medical Research Council (BMRC), and International Organization for Standardization (ISO) Size-Selective Sampling Criteria.

Occupational Exposure Limits

Industrial hygienists employ various criteria for the evaluation of the hazards posed by workplace exposures. In the United States, the primary sources of occupational exposure limits for the workplace are: (1) National Institute for Occupational Safety and Health (NIOSH) recommended exposure limits (RELs), (2) the U.S. Department of Labor (OSHA and MSHA) permissible exposure limits (PELs), and (3) the American Conference of Governmental Industrial Hygienists' (ACGIH) Threshold Limit Values® (TLVs®). These criteria cite the levels of exposure to which workers may be exposed. NIOSH

RELs are time-weighted average (TWA) concentrations for up to a 10 h workday during a 40 h workweek (NIOSH 1990). OSHA PELs are TWA concentrations that must not be exceeded during any 8 h work shift of a 40 h workweek (U.S. DOL 1986). The AGGIH TLVs are 8 h TWA concentrations for a normal 8 h workday and 40 h workweek, to which nearly all workers may be exposed, day after day, without adverse effects (ACGIH 1991). Even below these various exposure limits, a small percentage of workers may experience adverse health effects because of individual susceptibility, a pre-existing medical condition, and/or a hypersensitivity (allergy). In addition, some materials may act in combination with other workplace exposures to produce undesirable health effects even if the occupational exposures to individual contaminants are controlled at the level set by the evaluation criterion. For example, gases such as the oxides of nitrogen and sulfur dioxide may adsorb on dust particles and produce health effects at levels normally considered safe. Also, some substances are absorbed by direct contact with the skin and mucous membranes, and, thus, potentially increase the overall exposure. Unfortunately, the terms "health effects" and "lung deposition" are often used interchangeably. In addition to lung deposition, the toxicology of the particulate must be taken into account. The ACGIH is examining their occupational exposure limits for chemical substances which exist as dusts, mists, or fumes to determine which size range is of health concern and, therefore, which size range should be measured.

Measurement of Concentration

The measurement of the concentration of an industrial aerosol requires both collection and analysis. The collection phase involves the transport of a representative sample of the workplace aerosol into the sampling device, separation of the particles from the air stream, and an accurate measurement of the volume of air sampled. In traditional aerosol sampling, the analysis is performed in a laboratory on the accumulated sample. A large-enough air volume must have been sampled in order to have a sufficient quantity of contaminant for analysis. In real-time aerosol sampling (with direct-reading instruments) the analysis is either performed continuously or in a series of short, consecutive samples or "batches". Many instruments do not separate the dust particles from the air stream but measure the particles *in situ*. All instruments have a limited range of concentration and often a limited size range, which they are able to measure.

Measurement of Particle Size

Particle size distribution can be described by the number, surface, or volume (mass) of particles as a function of particle diameter. The diameter can in turn be given on the basis of a calibration aerosol of known particle size distribution or the actual aerosol measured (e.g., as a projected area or aerodynamic diameter). The aerodynamic diameter is particularly useful in industrial hygiene when considering the deposition of particles in the lung, the fate of particles in aerosol samplers, and the removal efficiency in air pollution control equipment. Size distribution may also vary spatially. Larger particles may be found near a contaminant source but may not be transported far.

In industrial hygiene practice, coarse estimates of particle size are often determined by identifying the process of generation. For example, most of the airborne copper produced by a grinding operation is supermicrometer-sized and may be classified as a "dust", while most of the copper generated during the melting of an alloy is submicrometer-sized and may be termed a "fume" (Tossavainen 1976; Walworth 1945).

To supplement aerosol sampling, bulk samples of settled dust may be collected from work surfaces, or samples may be retrieved from dust collectors attached to local exhaust systems. In addition, bulk samples may assist the industrial hygienist in choosing an appropriate method for sampling the chemical species of the aerosol (NIOSH 1973). These samples may be analyzed by optical or electron microscopy, or resuspended and analyzed by techniques such as sedimentation, light scattering, or time-of-flight spectrometry. Unfortunately, since these bulk samples have been size-selectively deposited, they may not be representative of the particle size distribution of the aerosol present in the work environment (NIOSH 1973).

Industrial hygienists have a wide variety of techniques at their disposal for determining the particle size of aerosols. Instruments usually do not measure particle diameter directly, but infer the particle size indirectly using differing physical principles: particle inertia (time-of-flight spectrometer and cascade impactor), light scattering (single-particle optical counter), and projected-area microscopy (optical and electron microscope). Many of these instruments are limited to ranges spanning roughly one to one- and one-half decades of particle size (Pui and Liu 1988). Since industrial aerosols may span a wide size range, and the industrial hygienist may only be interested in a limited portion of the industrial aerosol, erroneous conclusions regarding the size distribution can easily be drawn, particularly for particles near the lower and upper particle size ranges of the instrument. This is primarily due to detection problems and sampling efficiency near the upper and lower size limits of an instrument. Small particles may not be detected by the instrument with the same efficiency as large particles. Large particles may not be collected with the same efficiency as small particles (Pui and Liu 1988).

Spatial and Temporal Resolution

Exposure estimate problems in industrial hygiene monitoring arise when the sampling device cannot be attached directly to the worker, the contaminant concentration is not constant in space, the worker's location (or breathing zone) is not constant in time, or the contaminant generation rate is not constant in time (Bhaskar, Ramani, and Jankowski 1988; Whitney 1972). In addition, side-by-side sampling has shown significant variation in the measured mass concentration (Raynor, Ogden, and Hayes 1975).

Present personal sampling and analytical methods for most industrial dusts are intended for long sampling periods, i.e., full-shift sampling (8–10 h). While generally adequate for exposure assessment, these methods are of limited use in the evaluation of potential emission points or in the study of the temporal or time-related characteristics of a process (Gressel, Heitbrink, and McGlothlin 1988).

Sampling Strategies

As a matter of convention, exposure measurements for chronic hazards are usually taken for the duration of a single work shift. A TWA mass concentration (c_m) refers to the average mass concentration of a substance during a normal 8–10 h workday. A TWA c_m can be determined from a single full-shift sample, or it can be calculated from a series of consecutive samples (Leidel, Busch, and Lynch 1977). The following equation can be used to calculate the TWA from multiple consecutive samples of the same environment:

$$c_m = \frac{\sum_{i=1}^{n} c_{mi} \times t_i}{\sum_{i=1}^{n} t_i} = \frac{\sum_{i=1}^{n} c_{mi} \times t_i}{t} \qquad (24\text{-}4)$$

where t_i is the duration of sample i in minutes, c_{mi} is the mass concentration of sample i in mg/m^3, and t is the total sampling time in minutes.

For purposes of determination of compliance with occupational exposure limits, it is generally desirable to sample the workers assumed to be at maximum risk. Where the maximum-risk employees cannot be ascertained, employees should be selected at random.

To determine compliance with an occupational exposure limit (OEL), Leidel, Busch, and Lynch (1977) recommend calculating the 95% one-sided lower confidence limit (LCL) and the 95% one-sided upper confidence limit (UCL).

LCL and UCL are computed as follows:

$$\mathrm{LCL}(95\%) = \chi - t_\alpha \times \mathrm{CV_T}$$
$$\mathrm{UCL}(95\%) = \chi + t_\alpha \times \mathrm{CV_T} \tag{24-5}$$
$$\chi = \frac{C_\mathrm{m}}{\mathrm{OEL}}$$

where $t_\alpha = 1.645$ when $\alpha = 0.95$, and $\mathrm{CV_T}$ is the coefficient of variation for the sampling/analytical method.

If LCL and χ are above unity, then the exposure is classified noncompliant. If UCL and χ are below unity, then the exposure is classified as compliant. Finally, if unity lies between LCL and χ, or between UCL and χ, then the exposure is classified as possible overexposure.

Some substances have short-term exposure limits (STEL) or ceiling values that are intended to supplement the TWA where there are recognized toxic effects from high short-term exposures. Samples of materials which represent acute hazards are collected for shorter time periods, typically 15 min or less, and are collected during periods when the concentration of contaminant is expected to be highest. Other STELs are for carcinogens; these substances do not have an 8 h TWA. Real-time instruments are particularly useful for identifying the periods of peak exposure.

Example 24-1

What is the time-weighted average exposure calculated from the following data? Is the worker's exposure in compliance with an occupational exposure limit of 50 μg/m³? Three consecutive air samples for lead are collected on filter cassettes in the breathing zone of a worker in a brass foundry with the following results:

TABLE 24-1 Example Filter Cassette Sampling Data for a Worker in a Brass Foundry

Sample Sequence Number	Sample Time (min)	Sample Flow Rate (l/min)	Sample Volume (l)	Mass Collected (μg)	Mass Concentration (μg/m³)
001	60	2.0	120	30	250
002	180	2.0	360	36	100
003	240	2.0	480	5	10

Answer. The TWA mass concentration for the work shift is calculated from Eq. 25-4:

$$c_{\mathrm{mTWA}} = \left[\frac{(250\ \mu\mathrm{g/m^3} \times 60\ \mathrm{min}) + (100\ \mu\mathrm{g/m^3} \times 180\ \mathrm{min}) + (10\ \mu\mathrm{g/m^3} \times 240\ \mathrm{min})}{60\ \mathrm{min} + 180\ \mathrm{min} + 240\ \mathrm{min}}\right]$$
$$= 74\ \mu\mathrm{g/m^3}$$

The reader should note that this 8 h TWA mass concentration is greater than the occupational exposure limit of 50 μg/m³. To verify noncompliance, the one-sided lower confidence limit (LCL) must be calculated using Eq. 25-5 and compared to unity. If the industrial hygienist used NIOSH Method 7105

(Eller 1989), CV_T equals 0.068:

$$\chi = \frac{74\ \mu g/m^3}{50\ \mu g/m^3} = 1.48$$

$$\text{LCL (95\%)} = 1.48 - (1.645 \times 0.68)$$

$$= 1.37$$

Because LCL is greater than unity, the exposure average is classified as noncompliant and appropriate action must be taken to reduce the worker's exposure to lead.

TRADITIONAL SAMPLING METHODS

Aerosol measurement requires the transport of a representative sample of the workplace aerosol into the sampling device, separation of the particles from the air stream, and an accurate measurement of the volume of air sampled. The following sections deal with the transport and measurement of the aerosol sample, the separation and collection of the dust particles, and factors which may cause a bias of the sample.

Air Mover and Sampler Calibration

Present-day personal sampling devices usually rely on either diaphragm or piston-type pumps to draw air. The pump is connected to a direct current (dc) motor, supplied by a battery pack of rechargeable nickel–cadmium cells. The flow rates of pumps vary among manufacturers, but most will provide flows of 1–3 l/min against pressures of 6.25 kPa (25 in of water) for periods up to 8 h. Some pumps incorporate a rubber bladder to minimize flow pulsations, which may adversely affect the performance of size-separating devices (e.g., respirable dust samplers) (Bartley et al. 1984).

Olin, Sem, and Christenson (1971) defined the ideal instrument for measuring the mass concentration of an industrial aerosol as having four characteristics. First, the instrument should be automatic, reliable, and able to be operated unattended. Second, instantaneous measurements or an average measurement over a short time (e.g., 5–10 min) should be an option. Third, the analog readout signal should be capable of remote transmission and recording, reading directly in units of $\mu g/m^3$. Fourth, the detection of mass concentration should be direct, rather than by a correlation of parameters such as visibility with mass concentration.

Many sampling pumps incorporate a visual indication of flow rate. Usually, this is a rotameter which is affected by the fluid's specific gravity (density of gas/density of the air). Specific gravity is a function of the pressure in the rotameter. Thus, the rotameter should be calibrated for the conditions of use with the same sampling train (e.g., cyclone, cassette, filter media, cassette fittings, and connecting tubing) as that used in the field (McCammon 1989). The observed flow rate (Q_r) is then converted to flow rate (Q_n) at "normal" temperature and barometric pressure (Matheson 1983). During use at the sampling site, the temperature and barometric pressure should be recorded at the sampling site so that Q_n can be converted to the actual flow rate (Q_s) through the sampler or air mover. The following correction factors should be applied to the rotameter readings (Matheson 1983):

$$Q_n = Q_r \times \sqrt{\frac{SG_r}{SG_n} \times \frac{T_r}{T_n} \times \frac{P_n}{P_r}}$$

$$Q_n = Q_r \times \sqrt{\frac{SG_r}{SG_n} \times \frac{T_r}{T_n} \times \frac{P_n}{P_r}} \times \sqrt{\frac{SG_n}{SG_s} \times \frac{T_n}{T_s} \times \frac{P_s}{P_n}} \tag{24-6}$$

$$Q_s = Q_n \times \sqrt{\frac{SG_n}{SG_s} \times \frac{T_n}{T_s} \times \frac{P_s}{P_n}}$$

for conditions r, n, and s, where r are the conditions when the rotameter was calibrated, n are the conditions of normal temperature and pressure, and s are the conditions of actual sampling; SG is the specific gravity (unitless), T is the absolute temperature in °K, and P is the absolute pressure in kPa.

The actual volume of air sampled at the work site is determined using Q_s and sampling time and is used in calculating the exposure (c_m). The accuracy of determination of the concentration of a toxic substance in air is no greater than the accuracy with which the air volume is measured. Therefore, an accurate calibration of the airflow rate through the sampling train is necessary. The frequency of calibration depends on the use, abuse, and handling to which the pump is subjected. Primary standards, such as a spirometer or soap bubble meter, are recommended for calibration. The soap bubble meter is often used for field calibration. Figure 24-2 shows the field calibration setup for (A) a cyclone assembly having a cassette with a personal sampling pump and (B) a cassette with a personal sampling pump. McCammon (1989) and Lippmann (1989*b*) provide detailed instructions for the calibration of personal sampling pumps.

Flow decreases as the filter loading increases. Modern sampling pumps have flow sensors which change pump speed to compensate for the increased pressure drop and maintain constant volumetric flow rate. Older sampling pumps lack these controls, and, therefore, require periodic manual adjustments (usually of a flow control valve) to maintain a constant airflow rate (Lippmann 1989*c*).

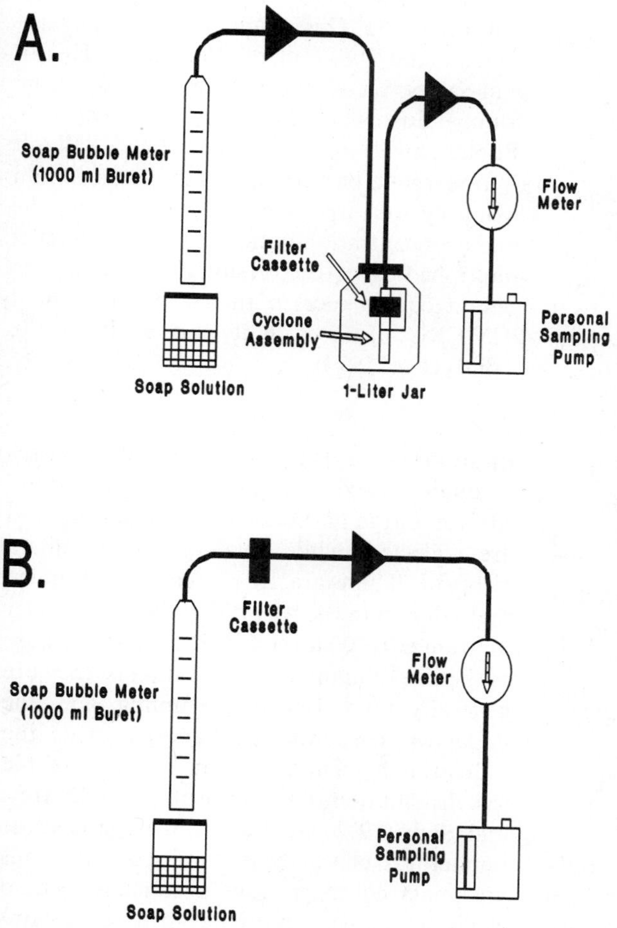

FIGURE 24-2. Arrangement for Flow Rate Calibration of Sampling Pumps with (A) Cyclone Assembly with a Filter Cassette and (B) Filter Cassette.

Preclassifiers

Preclassifiers are devices used to alter the size distribution of the airborne dust presented to the collection device. The preseparation of the industrial aerosol is performed so that the measured concentration reflects airborne particles of the size associated with a particular health effect or with any effect or the particle-size-selective TLV. Three types of devices are used as preclassifiers in industrial hygiene: cyclone, elutriator, and impactor (John 1984).

Cyclone

A cyclone removes particles larger than a desired size range from an air stream by particle inertia. Aerosol enters the cyclone through a tangential inlet and follows a cycloidal path before exiting through an axial opening. Large particles are forced toward the walls of the cyclone. Excess deposits on the walls fall into a dust pot. The smaller, respirable particles remain in the cyclone's air stream. The 10-mm diameter nylon cyclone is the most common preclassifier used in the United States. When operated at 1.7 l/min, the removal characteristics of this device approximately follow the ACGIH criteria for respirable dust. By design, this device is highly efficient for the removal of larger particles (i.e., greater than 10 μm) and is not efficient for particles below about 2 μm (Lippmann 1989*a*). The 10-mm nylon cyclone followed by a filter cassette is used to measure pneumoconiosis-producing dusts (Hering 1989; Carsey, Shulman, and Lorberau 1987). In general, a full-shift sample is required to obtain a quantifiable mass of particulate, but the minimum sampling time can be calculated using Eq. 24-4. Larger cyclones (i.e., 0.5–1 in) can be used with stationary aerosol samplers operated at higher flow rates to collect a larger-volume air sample (Pasceri 1973).

Elutriator

An elutriator removes large particles from the air stream by particle settling (Wright 1954). Two configurations are used: vertical and horizontal. Elutriators are too large in size to be used with personal sampling devices. Cylindrical vertical elutriators are used for the sampling of cotton dust (Wesley et al. 1978; NIOSH 1975). Particles with terminal settling velocities less than the upward flow of aerosol are collected; large particles and lint are removed from the air stream. Penetration curves for the vertical elutriator have been measured by Claasen (1981) and Robert (1979). Horizontal elutriators consist of a series of horizontal plates, stacked at equal spacings. Particles are removed from the air stream as they settle onto the plates (Hering 1989). The Mining Research Establishment (MRE) elutriator, a gravimetric respirable dust sampler, incorporates a horizontal elutriator which meets the British Medical Research Council (BMRC) criteria for respirable dust (Lynch 1970; Ogden, Birkett, and Gibson 1978). This instrument is important because the U.S. Federal Coal Mine Health and Safety Act of 1969 standard of 2.0 mg/m^3 for coal dust is based on this instrument (U.S. Public Law 91-173 1969). Since the MRE unit is too large to be a personal sampler, a 10 mm nylon cyclone and a filter are used, and the mass concentration obtained from the filter is multiplied by a conversion factor (1.38) to obtain equivalence to the MRE instrument (CMCRC 1980; Knight and Lichti 1970; Tomb et al. 1973).

Impactor

In an impactor, the sample air stream passes through a nozzle and impacts on a collection surface. Large particles strike the surface and are collected, while small particles remain airborne. There are, at present, no samplers that fall within the ACGIH IPM criteria over the range 0–100 μm (ACGIH 1985). Because of the size of many of these samplers, they are generally used for area sampling. Cascade impactors (i.e., multiple stages) provide the industrial hygienist with an estimate of the size distribution of the aerosol. Chan, D'Arcy, and Siak (1990) used a seven-stage cascade impactor to characterize the size distributions and mass concentrations of machining fluid aerosols in an automotive transmission manufacturing plant. The industrial aerosols were found to be polydisperse and generally bimodally distributed with no distinct seasonal variation. Recently, smaller versions of impactors have been developed and tested in the breathing zone of workers for respirable dust sampling (Baron 1983; Marple and McCormack 1983; Rubow et al. 1987). Macher and Hansson (1987) modified a personal impactor to collect bioaerosols. This method proved to be quite labor intensive, and the modifications resulted in a loss of collection efficiency over the personal dust sampler.

Collection Techniques

Most aerosol sampling methods require that the dust particles be separated from the air stream for analysis. Filtration, impaction, and precipitation are three typical methods of accomplishing this separation in industrial hygiene.

Filtration

The separation of particles from an aerosol sample is most commonly achieved by the use of filtration. Filter media are available in both fibrous (typically glass) and membranous forms. A common misconception is that aerosol filters work like microscopic sieves in which only particles smaller than the holes can get through. Particle removal occurs by collision and attachment of particles to the surface of the fibers or membrane. Sampling filter media may have pore sizes of 0.01–10 μm. The efficiency of fibrous filters is a function of the face velocity. For particles less than 1 μm, the overall efficiency decreases with increasing face velocity. For particles greater than approximately 1 μm, the filter collection efficiency is greater than 99%. The overall efficiency of membrane filters is approximately 100% for particles larger than the pore size. Filters for use in industrial hygiene are usually supplied as disks of either 25 or 37 mm diameter. Since the pressure drop across the filter increases with the air velocity through the filter, the use of a 37-mm diameter filter for a given volumetric flow rate results in a lower pressure drop and is generally recommended. The use of the smaller (25 mm) filter concentrates the deposit of the contaminant into a smaller total area, thus increasing the area density of particles. Thus, the 25-mm diameter filters are recommended when the analysis is by microscopy (Beckett 1980; Hunsaker et al. 1988). Membrane filters may be used singularly or in combination with a preclassifier to eliminate large particles and analyzed gravimetrically, chemically, or microscopically (Harris and Maguire 1968, Wesley et al. 1978; Carsey, Shulman, and Lorberau 1987).

Membrane filters are manufactured in a variety of pore sizes from polymers such as cellulose ester, polyvinyl chloride, and polycarbonate. Because membrane filters are thin, they lack rigidity and must be used with a support pad. The choice of a filter medium depends on the contaminant of interest and the requirements of the analytical technique. For gravimetric analysis, nonhygroscopic materials such as glass fibers, silver, or polyvinyl chloride membranes should be selected. For analysis by microscopy, or for acid digestion for the analysis of metals, cellulose ester membranes are the usual choice. When several analyses are to be performed on the same filter, some compromises may be necessary (NIOSH 1973). For instance, if a sample is to be analyzed for metals as well as for total mass, a polyvinyl chloride filter may be selected, even though it may be more difficult to digest than a cellulose ester membrane, which cannot be used for gravimetric analysis because it is hygroscopic (Eller 1989). Filtration can also be used for the determination of certain fungi and endospore-forming bacteria. Desiccation-resistant organisms can be washed from the surface of smooth-surface polycarbonate filters and can then be cultured, or the filter may be examined microscopically (Wolf et al. 1959; Fields et al. 1974; Lundholm 1982; Palmgren et al. 1986).

Filters are often held in disposable plastic filter cassettes during aerosol sampling. Some of the styles used in the U.S. are shown in Fig. 24-3. With the exception of the coal mine respirable dust sampler (D), these styles are available for both 25- and 37-mm diameter filters (Hinds 1982; Treaftis et al. 1984). The

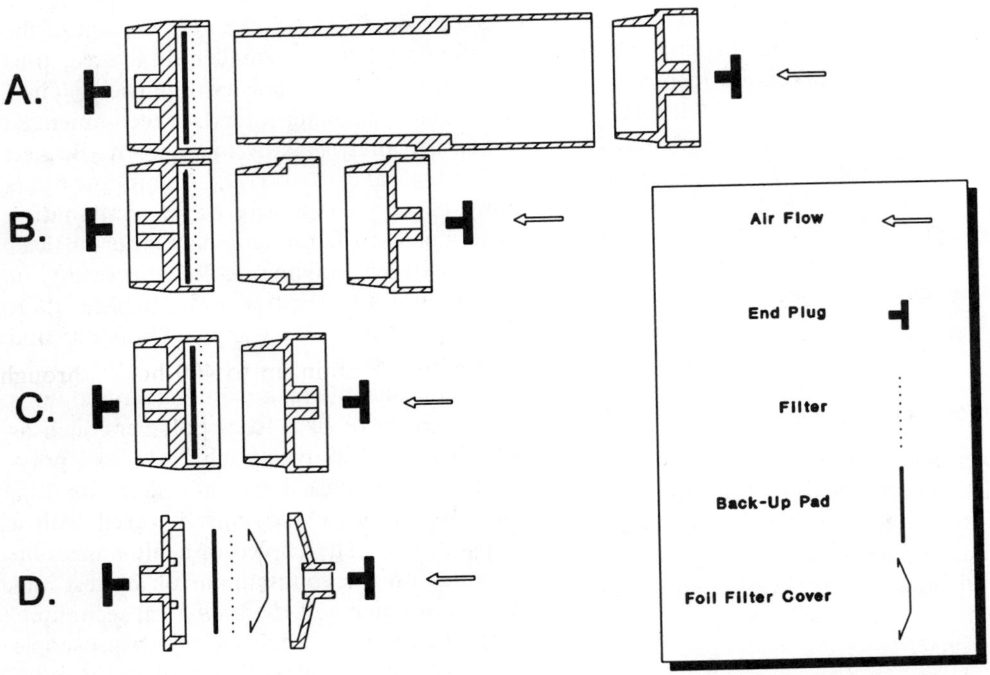

FIGURE 24-3. Filter Cassettes Used in Industrial Hygiene: A. 3-piece Cassette with Extended Cowl for Open-Faced Sampling; B. 3-piece Cassette for Open or Close-Faced Sampling; C. 2-piece Cassette for Close-Faced Sampling and for Use with 10 mm Nylon Cyclone; and D. Coal Mine Dust Cassette for Use with 10 mm Nylon Cyclone. Note that the Foil Filter Cover is Crimped onto the Filter and the Two are Weighed as a Unit.

two-piece cassette (C) is intended for closed-face use only. Because of size limitations, the two-piece cassette (C) and the coal mine dust sampler (D) are the only cassettes that fit into the apparatus with the 10-mm nylon cyclone. The three-piece cassette (B) can be used either open or closed-face. Closed-face sampling is performed by removing the end plug and is used when the particulates will be analyzed chemically and/or gravimetrically. Open-face sampling is performed by removing the end plug and the plastic cover and is used when the particulate must be uniformly deposited (i.e., for microscopic analysis). If used in the open-face mode then, in order to protect the filter, the plastic cover should be retained for use when sampling is concluded. Wide variations between open- and closed-face sample pairs have been documented in environmental sampling as compared to the small variations when on a mannequin (Beaulieu et al. 1980). The closed-face cassette reduces accidental contamination by large amounts of nonairborne and rapidly settling dust and the possibility of tampering. It is for this reason that Doemeny and Smith (1981) recommend the use of a closed-face cassette in the majority of filter sampling methods. To promote a uniform deposition of particles on the filter, extended cowls (A) are also used when the sample is analyzed by microscopy. These cowls should be made of conductive materials to minimize the accumulation of electrical charge and subsequent undesirable electrostatic effects (Baron 1987; Baron and Deye 1990). In addition, the cowl is used to reduce the direct impaction of large particles on the filter and to protect the filter from fingers. As with the standard three-piece cassette, the plastic cover and plug should be saved to prevent contamination of the sample in subsequent storage and handling. All plastic cassettes should be securely assembled and sealed with a cellulose shrink band or tape around the seams of the cassette to prevent leakage past the filter.

Impaction

Impaction is used to separate a particle from a gas stream, based on its inertia. As discussed in the section on preclassifiers, an impactor consists of a nozzle and a target. Impactors are usually classified by their cutoff diameter, i.e., the particle diameter at which 50% passes through the device (Hinds 1982). Thus, impactors should be selected so that the minimum size particle expected to be present will be collected. The midget impinger is an example of an impactor. Once the workhorse of dust sampling, the midget impinger still finds use in the collection of reactive aerosols (e.g., polyisocyanates), where chemical solutions are used to dissolve and form stable compounds with the reactive species (Moseley 1985; Skarping, Smith, and Dalene 1985). Impingers are also useful for the collection of viable aerosols (White et al. 1975; Lembke et al. 1981; Henningson et al. 1988).

Cascade impactors consist of a stack of impaction stages: each stage consists of one or more nozzles and a target or substrate. The nozzles may take the form of holes or slots. The target may consist of a greased plate, filter material, or nutrient media contained in Petri dishes (for the growth of microorganisms). Each succeeding stage collects smaller particles than the one preceding it. A filter may be used as the final stage so that all particles not impacted on the previous stages are collected. The target may be weighed to determine the collected mass, or it may be washed and the wash solution analyzed chemically. Filter media used as the target material offer the advantage of a lower tare weight; filter media can also be digested for chemical analysis of the deposited particles. Filters also induce more particle bounce than greased or oiled plates although fibrous filters usually have the same cut diameter. Although personal cascade impactors are available, these devices are not as widely used in personal sampling as filters (Baron 1983; Marple and McCormack 1983; Rubow et al. 1987).

Industrial hygienists are more likely to employ stationary cascade impactors or individual impactors used in survey instruments, either as the primary collection mechanism, or as a preseparator (for example, to remove nonrespirable particles from the sampled air stream). Impactors are also used for the collection of airborne microorganisms. These impactors contain up to 400 holes, through which the air jets impact onto nutrient media with one or more bacterial or fungal colonies forming at some of the impaction sites. Since more than one particle may have been deposited at a given site, correction factors are needed to adjust for the possibility of multiple organisms collecting at each site (Macher 1989).

Precipitation

Precipitation uses an external force field, typically either thermal or electrical, to separate particles from a sampled air stream. Historically, these devices were used as survey instruments; some of these devices allowed collection directly on a microscope slide (Poppoff 1954; Alfheim and Lindskog 1984).

Sampling Bias

Commonly used cassette samplers mounted on the lapel do not necessarily measure either the total aerosol concentration or the inhalable fraction; performance varies with the specific sampler and the ambient-air velocity. The fraction of aerosol which enters the nose and/or mouth has been referred to as the "inspirable" fraction. More recently, however, there has been a general international agreement that the term should be "inhalable" (Vincent and Mark 1990). Ogden (1983) reports that a sampler similar to that of a closed-face 37-mm cassette operated at 2 l/min has a sampling efficiency of 50% for a 15 μm particle, and 10% for a 35 μm particle when the wearer faces a wind of 1 m/s. The superceded ACGIH performance criteria include sampling efficiencies of about 70% and

50% for 15 and 35 μm particles, respectively (ACGIH 1991). Vincent and Mark (1990) studied a variety of aerosol samplers in relation to the ISO particulate-size-selective criteria recommendations previously mentioned (ISO 1983). They found that an aluminum 25-mm open-face cassette, 25-mm closed-face cassette, and a 25-mm closed-face cassette with seven 4-mm holes drilled in the front cover, matched the 1983 ISO Inhalable Particulate Mass curve quite well for low wind speed (0.5 μm/s) and for particles with aerodynamic diameters up to about 15 μm. Several European samplers (both personal and area) show promise as inhalable dust samplers (Vincent and Mark 1990).

Electrostatic effects can also result in sample bias. Baron (1987) found that aerosol charge is capable of producing a significant sampling error in asbestos fiber sampling. Cornett et al. (1989) noted fiberglass deposition on the inside surface of 50-mm conductive cowls used in conjunction with 25-mm mixed cellulose ester filters. When the extension cowls were washed, an overall increase of 16.5% was observed in the weight of fibers deposited. In 88% of the cases, fibers on the cowl were visible to the naked eye. Blackford, Harris, and Revell (1985) presented data on the reduction of dust losses within the cassette of the SIMPEDS personal dust sampler. Hunsaker et al. (1988) showed that under conditions of minimal electrostatic deposition onto cassette walls, the NIOSH 7400 method exhibits a nonuniform distribution of fibers within a 60° wedge. It is conceivable that the selection of counting fields in particular regions of the filter allow a distinction between compliance and noncompliance.

METHODS OF ANALYSIS

Microscopy

Microscopy ranks as one of the important tools in the analysis of aerosols. Historically, microscopy was used to count particles collected in impinger solutions. Today, microscopy is the principal technique used for the differentiation of fibers from other dust particles. Optical microscopy is limited to the detection of particles larger than about 0.3 μm. Scanning electron microscopy (SEM) and transmission electron microscopy (TEM) allow the measurement of smaller particles (Middleton 1982; Warner 1988; Stewart 1988). In addition, electron microscopy allows elemental analysis on a particle by particle basis. Depending on the specific techniqtue used, particles may be analyzed directly on the filter, or a sample preparation procedure may be employed to resuspend the particles for redeposition on a specific substrate.

In direct analysis of filter samples, particles are counted in f discrete, randomly selected microscope fields, each of known area a. The number of particles on the filter, N, and the number concentration of particles in air, c_n, are determined by the following equations:

$$N = n\left(\frac{A}{fa}\right)$$

$$c_{\mathrm{n}} = \left(\frac{n}{Qt}\right)\left(\frac{A}{fa}\right) = \frac{N}{Qt} \qquad (24\text{-}7)$$

where n is the total number of particles counted, Q the volumetric flow rate, t the sampling time, and A the effective filter area (i.e., the area available for collection and not obscured by the filter holder).

Because this technique assumes that particles are distributed randomly on the filter, care must be exercised in sampling to ensure uniform deposition by the use of appropriate collection devices, e.g., open-faced cassettes using conductive cowls (Cornett et al. 1989). Hunsaker et al. (1988) suggests developing a standardized method of field selection for asbestos fiber counting, such as counting lines oriented radially or strictly within the inner portions of the filter, because of the nonuniform distribution of fibers within a 60° wedge.

The precision of the particle count depends on the total number of particles counted. Particle counting (for total counts greater than about 16 particles) obeys Poisson statistics: the variance of this distribution is equal to the mean. Under ideal conditions, the relative standard deviation (RSD) in percent is, therefore, given as

$$\text{RSD} = n^{-1/2} \times 100\% \qquad (24\text{-}8)$$

where n is the particle count. For example, if n was increased from 16 to 100 particles, the RSD would be reduced from 25% to 10%. For reliable analysis, particles deposited on the filter can be increased by using a higher flow rate in sampling, which is limited only by particle overlap on the filter. Counting statistics can also be enhanced by counting more fields. Reducing the magnification lowers resolution and also results in an increasing area of the filter and, hence, the number of particles counted (if the particles remain visible). Equation 24-8 applies both to the relative standard deviation for individual size intervals as well as the total particle count. Because industrial aerosols usually follow a lognormal size distribution (i.e., fewer large particles than small particles), it is often advantageous to perform the size analysis at two or more levels of magnification, compiling the particle distribution in discrete ranges.

Chemical and Gravimetric

In chemical and gravimetric analysis of filter samples, the total mass (or total mass of the specific chemical) is measured. The time-weighted average mass concentration c_m is determined by dividing the total mass collected, m, by the volume of air sampled, Qt:

$$c_m = \frac{m}{Qt} \qquad (24\text{-}9)$$

Gravimetric analysis is performed by weighing the filter before sampling (the tare weight), and again after sampling to determine the weight gain. As mentioned earlier, not all filter media are suitable for gravimetric analysis because of the tendency of some materials to absorb water (Sass-Kortsak, Tracy, and Purdham 1989). An accurate gravimetric determination requires strict sample conditioning procedures. Filters are usually equilibrated at the relative humidity of the weighing room (e.g., 50% RH). Gravimetric sensitivity is limited not only by the sensitivity of the balance but also by the tare weight of the filter relative to the sample weight and by the weight stability of the filter. (Roach 1973; Vaughan, Milligan, and Ogden 1989).

Chemical analysis usually requires that the filter be ashed so that the sample can be presented to an instrument for analysis. In the case of metals, the filter is destroyed using an acid, and the resulting solution is analyzed by atomic absorption spectrometry (Owen, Delaney, and Neff 1951). Because of the differing requirements of various analytical procedures, often separate filters are required for the sampling of multicomponent work atmospheres (Eller 1989). If the industrial hygienist has an estimate of mass concentration, Eq. 24-4 can be rearranged to solve for either Q or t such that the filter will not be overloaded or underloaded.

The industrial hygienist should note that, according to Soderholm (1988), aerosols are inherently unstable. Four types of processes associated with aerosol instability are coagulation, deposition of particles on surfaces, deposition and evaporation of vapor from surfaces, and vapor/particle interactions. Coagulation occurs as particles collide and stick together, resulting in a decrease in particle number concentration, but no change in the mass concentration. Particles are deposited either gravitationally, inertially, or diffusionally on surfaces. Deposition of vapor and evaporation of high vapor pressure mists are primarily functions of the diffusion coefficients and the thickness of the diffusion boundary layer. Finally, particle/vapor interactions involve mass transfer due to mass imbalances of substances on the surface (Soderholm 1988).

Example 24-2

What would be the minimum sample time to measure silica at a suspected concentration of 100 μg/m^3, if the minimum quantifiable mass is 50 μg per filter?

Answer. Since silica is a hazard to the lung, the respirable fraction is measured using a 10-mm nylon cyclone followed by a filter cassette and a sampling pump operating at 1.7 l/min. Equation 24-9 can then be rearranged to solve for time t:

$$t = \frac{m}{Qc_{\rm m}}$$

$$= \left[\frac{50\ \mu\text{g}}{1.7\ \text{l/m} \times 100\ \mu\text{g/m}^3 \times 10^{-3}\ \text{m}^3/\text{l}}\right]$$

$$= 294\ \text{min}$$

Biological

Methods for the determination of viable organisms rest on the ability of the organism to remain viable and reproduce. Organisms are collected in impingers, on filters, on microscope slides, or on nutrient media in Petri plates (Chatigny et al. 1989). After sampling, quantifying airborne microorganisms is accomplished by diluting and plating the collection fluid or by using a membrane filtration plating technique when the expected microbial load is low (i.e., < 60 cells per 20 ml of impinger fluid) (Blomquist, Palmgren, and Strom 1987; Morey 1990). In the latter method, the membrane filter is placed on a sterile pad saturated with nutrient broth in a sterile Petri dish. The plate is incubated at the appropriate temperature and the colonies counted. Microorganisms collected on membrane filters can either be eluted with sterile buffered water or plated directly on a sterile pad saturated with nutrient broth in a sterile Petri dish (U.S. EPA 1978). These filters or impingers can be used over a wide range of airborne concentrations. More commonly, the organisms are collected directly on nutrient media used as collection targets in impactors (Chatigny et al. 1989). The nutrient media are incubated and the number of colony-forming units (CFUs) are counted by the naked eye or under low magnification. CFUs are visible groups of cells that initiated from a single cell. Nutrient media are available from commercial supply houses for the cultivation, isolation, and differentiation of specific microorganisms (Baron and Finegold 1990). Differential media support the growth of various species, while providing an environment that makes it easier to distinguish among different organisms. Selective media contain agents that prevent or inhibit the growth of certain organisms, while permitting the growth of others. Individual colonies can be isolated and identified by morphological and biochemical characteristics. More details on the measurement of bioaerosols can be found in Chapter 21.

REAL-TIME MEASUREMENT

Real-time aerosol instruments are used for a variety of tasks in industrial hygiene. Baron (1988, 1989) and Pui and Liu (1988) summarized the advances in particle measuring and sampling instruments. Aerosol photometers and piezoelectric instruments can be used as survey instruments. Aerosol photometers (and, more recently, condensation nuclei counters) have been used in respirator fit testing (Willeke, Ayer, and Blanchard 1981; NIOSH 1987; Ernstberger, Gall, and Turok 1988). Aerosol photometers have been used in conjunction with video recording as a powerful study tool to identify operations or activities causing exposure (Gressel et al. 1987; Gressel, Heitbrink, and McGlothlin 1988).

Concentration Measurements

Piezoelectric devices, such as TSI Model 3800 (TSI, Inc., St. Paul, MN), are available for the measurement of respirable aerosols (Sem, Tsurubayashi, and Homma 1977; Swift and Lippmann 1989). Particles which pass a preclassifier (< 3.5 μm) are deposited onto a quartz collection crystal by electrostatic precipitation. Particles collected on the surface increase the mass of the crystal and thereby cause a shift in the oscillation frequency of the crystal. This shift is directly proportional to the mass loading. Since piezoelectric devices respond directly to mass, no correction for the particle density is required. These devices are not continuous samplers but collect samples for discrete periods. Sem, Tsurubayashi, and Homma (1977) found that for piezoelectric devices calibrated with welding smoke, field and laboratory data showed good agreement (± 10%) with parallel filter samples of ten aerosols, including electric arc welding fumes, asbestos mill dust, oil mist, walnut shell abrasive dust, powdered metal dust, and cotton dust. The piezoelectric device consistently measured tobacco smoke 15% low. Experimental data also showed that sensor loading must not exceed certain limits. For most aerosols, the limit is 4 mg-min/m^3 (e.g., 2 min with a concentration of 2 mg/m^3 or 24 s with a concentration of 10 mg/m^3) (Sem, Tsurubayashi, and Homma 1977).

Aerosol photometers are versatile direct-reading instruments for indicating aerosol concentration. The measurement is based on the amount of light scattered by the aerosol particles. Light scatter is a function of the refractive index, density, and shape of the particles sampled. In environments of high relative humidity, the instrument will detect water vapor as well as the particles (Dimmick 1961). A beam of light inside the instrument is focused onto a view volume, through which the air stream passes. The air stream can be pumped through the sensitive volume or allowed to passively circulate through it. The amount of light scattered is measured by a photosensitive detector. The concentration measured is equivalent to the concentration of a calibration aerosol. The optical characteristics of common aerosol photometers are such that they are most sensitive to respirable aerosols. As a first approximation, the aerosol photometers respond roughly to particle volume, so the instrument readings can be corrected for particle density by multiplying by the ratio of the actual particle density to the density of the factory calibration aerosol. Most photometers have an analog output which can be electronically recorded. Light scattering photometers have been used to characterize penetration through HEPA filters (Biermann and Bergman 1988), nuisance dust exposure (Gressel et al. 1987), and welding fume exposure (Glinsmann and Rosenthal 1985).

Particle Size Measurement

Real-time cascade impactors are available using quartz crystals as the target material. They consist of several inertial impactors of decreasing cutoff diameter (Carpenter and Brenchley 1972; Fairchild and Wheat 1984). The collection surface in each stage is a greased quartz collection crystal coupled to a reference crystal located outside of the sample stream. High-frequency oscillations of the two crystals are mixed, resulting in a beat signal of lower frequency. Particles collected on each stage increase the mass of the measurement crystal and thereby cause a shift in the oscillation frequency of the crystal. This results in a shift in the beat frequency directly proportional to the mass loading. O'Brien, Baron, and Willeke (1986) used such a device for measuring the size distribution and concentration of the aerosol produced during the high speed grinding of gray iron castings.

Single-particle optical counters (OPC) size particles based on the amount of light scattered by individual particles. A beam of light inside the instrument is focused onto a view volume, through which the particles pass one at a time. The amount of light scattered by each particle is measured by a photosensitive detector. The particle size distribution is then determined from an analysis of the photodetector output by a multichannel (pulse height) analyzer. The size indicated is equivalent to the size of a calibration aerosol, typically latex spheres. The estimation of the aerodynamic diameter of an unknown aerosol from the diameter obtained from an optical particle counter (scattered-light-equivalent) is difficult as it is dependent on the shape and complex index of refraction of the unknown particles and on instrumental characteristics such as the geometry of the optical system and the photodetector sensitivity. One approach to determination of the aerodynamic size with single-particle optical counters (Marple and Rubow 1976) involves an empirical calibration using specially designed impactors. Particle distributions are measured with and without an impactor mounted on the counter inlet. The counter channel containing one-half as many counts with the impactor as without corresponds to the cutoff size of the impactor. Various calibration points can be obtained by using different nozzles to change the cutoff size of the impactor. Crouch (1987) and O'Brien, Baron, and Willeke (1987) used OPCs to measure the size distribution and concentration of airborne particles in aerosols from grinding gray iron castings. Skillern (1971) used three different light-scattering aerosol photometers for testing high-efficiency filters and clean-room atmospheres, but frequent checking by optical or electron microscopy was necessary to identify the actual particle size distribution of the aerosol.

Time-of-flight aerosol spectrometers, such as the APS Model 33B and the Aerosizer MACH 2 (TSI, Inc., St. Paul, MN, and Amherst Process Industry, Inc., Amherst, MA), share some characteristics of optical particle counters: particles are counted one at a time, using a light source and a photomultiplier tube. The lower detection limits of both instruments are limited by the amount of light scattered by an individual particle. Although OPCs and time-of-flight aerosol spectrometers are both based on light scattering, the measured quantity is different. The time-of-flight aerosol spectrometer sizes particles by measuring the transit time between two planes of laser light as the particles leave an accelerating flow field. This time is proportional to the aerodynamic size (Remiarz et al. 1983).

The maximum concentration that can be measured by both optical particle counters and time-of-flight spectrometers is limited by coincidence—the simultaneous presence of two or more particles in the viewing volume. The coincidence of two particles in an OPC causes a single larger particle to be counted (Willeke and Liu 1976); coincidence in time-of-flight aerosol spectrometers results in randomly sized particles (Remiarz et al. 1983); and phantom particles are created when particle concentrations exceed 100 particles/cm^3 (Heitbrink, Baron, and Willeke 1991). These instruments incorporate circuitry or software logic to reduce this effect.

O'Brien, Baron, and Willeke (1986) compared four methods of aerosol measurement. Samples of an industrial aerosol from a grinding operation of gray iron castings were analyzed by a scattering aerosol spectrometer (OPC), and a time-of-flight spectrophotometer (APS), a piezoelectric microbalance cascade impactor (QCM), and a scanning electron microscope (SEM). Plots of number concentration and mass concentration versus aerodynamic diameter show large differences among the instruments. For particle sizes less than 0.3 μm, OPCs indicated an increase in the number of particles with decreasing particle size, while SEM showed the opposite trend. The QCM was of questionable accuracy below 0.5 μm. The counting efficiency of the APS dropped off below 1.0 μm and dipped at about 8.0 μm, due to internal circuitry and logic software. The QCM underestimated the aerosol concentration by a factor of about 2. The slopes of the particle size distribution between 1.0 and 7.0 μm obtained on the APS, OPC, and QCM were similar, while the SEM slope was almost linear on the mass-weighted plot. It was concluded that a combination of instruments can be used to get an estimate of the true particle size distribution.

Surveys/Calibration

Most real-time devices are used as survey instruments. Aerosol photometers and piezoelectric instruments can be used to identify and prioritize potential sources of exposure to dust. Since these instruments are not specific for individual chemical species and the period of measurement is necessarily brief, they may not reflect actual exposures measured by long-term sampling techniques. These measurements can be used to identify areas or operations potentially causing exposures.

Since aerosol photometers do not measure mass directly, they should be calibrated before use. Ideally, this can be accomplished by placing the instrument in a calibration chamber, where an aerosol generator produces a known concentration of the contaminant of interest (Pui and Liu 1988). Realistically, very few industrial hygienists have access to such chambers, so field calibration is necessary. In field calibration, the instrument is placed in the work environment, and a parallel measurement is made by a reference method (e.g., membrane filter). The output of the instrument is recorded for the identical sampling period of the reference. After the sampling is completed, the instrument output is averaged and compared to the results obtained by the reference method. The ratio of the reference method to the instrument average is the calibration factor. Some real-time aerosol instruments have available calibration accessories to facilitate field calibration (e.g., a reference scatter device) (Skillern 1971; Gero and Tomb 1988). In using the calibration factor, one assumes that the particle size distribution and composition do not change appreciably between calibration and sampling conditions.

Respirator Fit Testing

Another application of aerosol measurement involves fit-testing a respirator. Because each person's face is unique, one style or size of respirator will not provide the optimum protection for all workers. In an attempt to optimize the selection of a respirator for an individual worker, various test schemes have been developed to measure respirator fit. The simplest procedures involve the use of amyl acetate (banana oil), saccharin, or an irritant smoke to challenge the respirator and rely on the subjective response of the wearer to evaluate the fit (Marsh 1984*a*, 1989*b*; Myers 1986). Two quantitative fit-test methods commonly used challenge respirators with either a liquid aerosol in the di-2-ethylhexyl phthalate test method or a solid aerosol in the sodium chloride test method. Air samples are continuously drawn off from the probe into the mask and fed to a photometer for challenge aerosol detection. Protection factors are determined by calculating the challenge aerosol concentration outside the mask to concentration sampled inside the mask (Kolesar et al. 1982; Lowry et al. 1977).

Field quantitative fit-test procedures have evolved which utilize an aerosol generator (typically corn oil), a portable exposure chamber, and an aerosol photometer. The respirator is modified with a fitting which permits sampling within the face piece. The photometer is connected to both the exposure chamber and the respirator. The ratio of the aerosol concentration in the chamber to that within the respirator is used as an index of fit. A recent improvement in quantitative fit testing (Willeke, Ayer, and Blanchard 1981) does away with the aerosol generator and exposure chamber. Since the concentration of submicrometer-sized particles in the atmosphere is relatively stable (for the duration of a test), ambient aerosol is used as the challenge atmosphere and the concentration of particles in the 0.02–1 μm size range is measured using a miniaturized condensation nucleus counter.

Task/Exposure Analysis

One purpose of real-time aerosol monitoring is to relate some measurement of exposure to some work activity or to a particular piece of equipment, to determine some dynamic property of the work environment or to contrast exposure differences caused by the introduction of a control measure (Gressel, Heitbrink, and McGlothlin 1988). The aerosol measurement is correlated with some variable affecting the exposure. Aerosol photometers have been used in conjunction with video recording as a powerful study tool to identify operations or activities causing exposure. The concentration data can be recorded and entered into a spreadsheet for manipulation and analysis. Observational data can likewise be entered into the spreadsheet. The qualitative data may originate from careful note taking or they may be obtained after-the-fact by an observation of time-superimposed videotapes. This permits a calculation of the contribution of the various factors to the exposure scenario, allowing the identification and subsequent control of the offending tasks (Gressel, Heitbrink, and McGlothlin 1988).

FUTURE TRENDS

The establishment of criteria for size-selective sampling by the American Conference of

Governmental Industrial Hygienists may lead to changes in conventional sampler technology. The implementation of these new samplers by industrial hygienists will depend on the adoption of particle size-specific occupational exposure limits by the ACGIH and various governmental agencies. An increasing awareness of sample loss caused by electrostatic forces will increase the use of samplers made with conductive materials. The trend toward miniaturization and automation of sampling equipment will likely continue, especially for real-time instruments. Task/exposure analysis, largely a research tool in the 1980s, should be adopted by more practicing industrial hygienists in the 1990s. This will spur the demand and subsequent development of still smaller instruments with built-in data recording capabilities. The lack of specificity of real-time monitors will remain a problem in the foreseeable future, with little progress likely for real-time chemical analysis.

References

Alfheim, I. and A. Lindskog. 1984. A comparison between different high volume sampling systems for collecting ambient airborne particles for mutagenicity testing and for analysis of organic compounds. *Sci. Tot. Environ.* 34:203–22.

American Conference of Governmental Industrial Hygienists (ACGIH). 1985. Particle size-selective sampling in the workplace. Report of the ACGIH Technical Committee on Air Sampling Procedures. Cincinnati: American Conference of Governmental Industrial Hygienists. (This report was also published in *ACGIH Transactions—1984, Ann. Am. Conf. Ind. Hyg.* 11:23–100, and parts were summarized in Phalen, R. F., W. C. Hinds, W. John, P. J. Lioy, M. Lippmann, M. A. McCawley, O. G. Raabe, S. C. Soderholm, and B. O. Stuart. 1986. Rationale and recommendations for particle size selective sampling in the workplace. *Appl. Ind. Hyg.* 1:3–14.

American Conference of Governmental Industrial Hygienists (ACGIH). 1991. Threshold limit values for chemical substances and physical agents in the workroom environment with intended changes for 1990–1991. Cincinnati: American Conference of Governmental Industrial Hygienists.

Baron, E. J. and S. M. Finegold. 1990. *Bailey and Scott's Diagnostic Microbiology*, 8th edn., pp. 81–141. Philadelphia: Mosby Company.

Baron, P. A. 1983. Sampler evaluation with an aerodynamic particle sizer. In *Aerosols in the Mining and Industrial Work Environments, Vol. 3, Instrumentation.* eds. V. A. Marple and B. Y. H. Liu, pp. 861–77. Ann Arbor: Ann Arbor Science.

Baron, P. A. 1987. Overloading and electrostatic effects in asbestos fiber sampling. *Am. Ind. Hyg. Assoc. J.* 48:A709–10.

Baron, P. A. 1988. Modern real-time aerosol samplers. *Appl. Ind. Hyg.* 3:97–103.

Baron, P. A. 1989. Aerosol photometers for respirable dust measurements. In *NIOSH Manual of Analytical Methods*, 3rd edn., ed. P. M. Eller, DHHS (NIOSH) Publication No. 84-100, Cincinnati, OH, with first (1985), second (1987), and third (1989) supplements.

Baron, P. A. and G. J. Deye. 1990. Electrostatic effects in asbestos sampling I and II: Experimental measurements and comparison of theory and experiment. *Am. Ind. Hyg. Assoc. J.* 51:51–69.

Bartley, D. L., G. M. Breuer, P. A. Baron, and J. D. Bowman. 1984. Pump fluctuations and their effect on cyclone performance. *Am. Ind. Hyg. Assoc. J.* 45:10–18.

Beaulieu, H. J., A. V. Fidino, K. L. B. Arlington, and R. M. Buchan. 1980. Comparison of aerosol sampling techniques: "Open" versus "closed-face" filter cassettes. *Am. Ind. Hyg. Assoc. J.* 41:758–65

Beckett, S. T. 1980. The effects of sampling practice on the measured concentration of airborne asbestos. *Ann. Occup. Hyg.* 23:259–72.

Bhaskar, R., R. V. Ramani, and R. A. Jankowski. 1988. Experimental studies on dust dispersion in mine airways. *Mining Eng.* 40:191–95.

Biermann, A. H. and W. Bergman. 1988. Filter penetration measurements using a condensation nuclei counter and an aerosol photometer. *J. Aerosol. Sci.* 19:471–83.

Blackford, D. B., G. W. Harris, and G. Revell. 1985. The reduction of dust losses within the cassette of the SIMPEDS personal dust sampler. *Ann. Occup. Hyg.* 29:169–80.

Blomquist, G., U. Palmgren, and G. Strom. 1987. Methodological aspects of measurement of exposure to mould. *Eur. J. Respir. Dis.* 71:29–36.

Carpenter, T. E. and D. L. Brenchley. 1972. A piezoelectric cascade impactor for aerosol monitoring. *Am. Ind. Hyg. Assoc. J.* 33:503–10.

Carsey, T. P., S. Shulman, and C. D. Lorberau. 1987. An investigation of the performance of the 10-mm nylon cyclone. *Appl. Ind. Hyg.* 2:47–52.

Chan, T. L., J. B. D'Arcy, and J. Siak. 1990. Size characteristics of machining fluid aerosols in an industrial metalworking environment. *Appl. Occup. Environ. Hyg.* 5:162–70.

Chatigny, M. A., J. M. Macher, H. A. Burge, and W. R. Solomon. 1989. Sampling airborne microorganisms and aeroallergens. In *Air Sampling Instruments for Evaluation of Atmospheric Contaminants*, 7th edn., ed. S. V. Hering, pp. 199–220. Cincinnati: American Conference of Governmental Industrial Hygienists.

Claasen, B. 1981. Experimental determination of the vertical elutriator particle transfer efficiency. *Am. Ind. Hyg. Assoc. J.* 42:305–9.

Committee on Measurement and Control of Respirable Dust (CMCRC). 1980. *Measurement and Control of Respirable Dust in Mines.* National Materials Advisory Board, Commission on Sociotechnical Systems, National Research Council. NMAB-363. Washington, DC: National Academy of Sciences.

Cornett, M. J., C. Rice, V. S. Hertzberg, and J. E. Lockey. 1989. Assessment of fiber deposition on the conductive sampling cowl in the refractory ceramic fiber industry. *Appl. Ind. Hyg.* 4:201–4.

Crouch, K. G. 1987. Aerosols and wind generated by hand grinders: Experimental. *Am. Ind. Hyg. Assoc. J.* 48:15–22.

Dimmick, R. L. 1961. A light-scatter probe for aerosol studies. *Am. Ind. Hyg. Assoc. J.* 22:80–82.

Doemeny, L. J. and J. P. Smith. 1981. Letter to the forum. *Am. Ind. Hyg. Assoc. J.* 42:A22.

Eller, P. M. 1989. *NIOSH Manual of Analytical Methods*, 3rd edn., DHHS (NIOSH) Publication No. 84-100, Cincinnati, OH, with first (1985), second (1987), and third (1989) supplements.

Ernstberger, H. G., R. B. Gall, and C. W. Turok. 1988. Experiments supporting the use of ambient aerosols for quantitative respirator fit testing. *Am. Ind. Hyg. Assoc. J.* 49:613–19.

Fairchild, C. I. and L. D. Wheat. 1984. Calibration and evaluation of a real-time cascade impactor. *Am. Ind. Hyg. Assoc. J.* 45:205–11.

Fields, N. D., G. S. Oxborrow, J. R. Puleo, and C. M. Herring. 1974. Evaluation of membrane filter field monitors for microbiological air sampling. *Appl. Microbiol.* 27:517–20.

First, M. W. 1988. Sampling and analysis of air contaminants: An overview. *Appl. Ind. Hyg.* 3:F20–27.

Fischbein, A. 1983. Environmental and occupational lead exposure. In *Environmental and Occupational Medicine*, ed. W. N. Rom, pp. 433–47. Boston: Little, Brown and Company.

Gero, A. J. and T. F. Tomb. 1988. MINIRAM calibration differences. *Appl. Ind. Hyg.* 3:110–14.

Glinsmann, P. W. and F. S. Rosenthal. 1985. Evaluation of an aerosol photometer for monitoring welding fume levels in a shipyard. *Am. Ind. Hyg. Assoc. J.* 46:391–95.

Gressel, M. G., W. A. Heitbrink, J. D. McGlothlin, and T. J. Fischbach. 1987. Real-time, integrated, and ergonomic analysis of dust exposure during manual materials handling. *Appl. Ind. Hyg.* 2:108–13.

Gressel, M. G., W. A. Heitbrink, and J. D. McGlothlin. 1988. Advantages of real-time data acquisition for exposure assessment. *Appl. Ind. Hyg.* 3:316.

Harris, G. W. and B. A. Maguire. 1968. Gravimetric dust sampling instrument (SIMPEDS): Preliminary results. *Ann. Occup. Hyg.* 11:195–201.

Heitbrink, W. A., P. A. Baron, and K. Willeke. 1991. Coincidence in time-of-flight aerosol spectrometers: Phantom particle creation. *Aerosol. Sci. Technol.* 14:112–26.

Henningson, E. W., I. Fangmark, E. Larsson, and L. E. Wikstrom. 1988. Collection efficiency of liquid samplers for microbiological aerosols. *J. Aerosol. Sci.* 19:911–14.

Hering, S. V. 1989. Inertial and gravitational collectors. In *Air Sampling Instruments for Evaluation of Atmospheric Contaminants*, 7th edn., ed. S. V. Hering, pp. 337–86. Cincinnati: American Conference of Governmental Industrial Hygienists.

Hinds, W. C. 1982. *Aerosol Technology.* New York: Wiley.

Hunsaker, H. A., R. M. Buchan, T. J. Keefe, and W. E. Marlatt. 1988. Characterization of asbestos fiber distribution on membrane filters from 25- and 37-mm sampling cassettes. *Appl. Ind. Hyg.* 3:284–90.

International Organization for Standardization (ISO). 1983. Air quality—Particle size fraction definitions for health-related sampling. Tech. Rep. ISO/TR 7708-1983. Geneva: International Organization for Standardization.

John, W. 1984. Sampler efficiencies: Respirable mass fraction. *Ann. ACGIH* 11:81–84.

Jones, R. N. 1983. Silicosis. In *Environmental and Occupational Medicine*, ed. W. N. Rom, pp. 197–206. Boston: Little, Brown and Company.

Knight, G. and K. Lichti. 1970. Comparison of cyclone and horizontal elutriator size selectors. *Am. Ind. Hyg. Assoc. J.* 31:437–41.

Kolesar, E. S., D. J. Cosgrove, C. M. DeLaBarre, and C. F. Theis. 1982. Comparison of respirator protection factors measured by two quantitative fit test methods. *Aviation Space Environ. Med.* 53:1116–22.

Leidel, N. A., K. A. Busch, and J. R. Lynch. 1977. *Occupational Exposure Sampling Strategy Manual*, DHEW (NIOSH) Publication No. 77-173, Cincinnati, OH: NIOSH.

Lembke, L. L., R. N. Kniseley, R. C. Van-Nostrand, and M. D. Hale. 1981. Precision of the all-glass impinger and the Andersen microbial impactor for air sampling in solid-waste handling facilities. *Appl. Environ. Microbiol.* 42:222–25.

Lippmann, M. 1989*a*. Size-selective health hazard sampling. In *Air Sampling Instruments for Evaluation of Atmospheric Contaminants*, 7th edn., ed. S. V. Hering, pp. 163–98. Cincinnati: American Conference of Governmental Industrial Hygienists.

Lippmann, M. 1989*b*. Calibration of air sampling instruments. In *Air Sampling Instruments for Evaluation of Atmospheric Contaminants*, 7th edn., ed. S. V. Hering, pp. 73–110. Cincinnati: American Conference of Governmental Industrial Hygienists.

Lippmann, M. 1989*c*. Sampling Aerosols by Filtration. In *Air Sampling Instruments for Evaluation of Atmospheric Contaminants*, 7th edn., ed. S. V. Hering, pp. 305–36. Cincinnati: American Conference of Governmental Industrial Hygienists.

Lowry, P. L., C. P. Richards, L. A. Geoffrion, S. K.

Tasuda, L. D. Wheat, and J. M. Bustos. 1977. *Respiratory Studies for the National Institute for Occupational Safety and Health, January 1–December 31, 1977*. Publication No. 78-161. Cincinnati, OH: U.S. Department of Health, Education, and Welfare, DHEW (NIOSH).

Lundholm, M. 1982. Comparison methods for quantitative determinations of airborne bacteria and evaluation of total viable counts. *Appl. Environ. Microbiol.* 44:179–83.

Lynch, J. R. 1970. Evaluation of size-selective presamplers: I. Theoretical cyclone and elutriator relationships. *Am. Ind. Hyg. Assoc. J.* 31:548–51.

Macher, J. M. 1989. Positive-hole correction of multiple-jet impactors for collecting viable microorganisms. *Am. Ind. Hyg. Assoc. J.* 50:561–68.

Macher, J. M. and H. C. Hansson. 1987. Personal size-separating impactor for sampling microbiological aerosols. *Am. Ind. Hyg. Assoc. J.* 48:652–55.

Marple, V. A. and K. L. Rubow. 1976. Aerodynamic particle size calibration of optical particle counters. *J. Aerosol Sci.* 7:425–33.

Marple, V. A. and J. E. McCormack. 1983. Personal sampling impactor with respirable aerosol penetration characteristics. *Am. Ind. Hyg. Assoc. J.* 44:916–22.

Marsh, J. L. 1984*a*. Evaluation of irritant smoke qualitative fitting test for respirators. *Am. Ind. Hyg. Assoc. J.* 45:245–49.

Marsh, J. L. 1984*b*. Evaluation of saccharin qualitative fitting test for respirators. *Am. Ind. Hyg. Assoc. J.* 45:371–76.

Matheson Gas Products, Inc. 1983. *Guide to Safe Handling of Compressed Gases*. Secausus, NJ: Matheson Gas Products, Inc.

McCammon, C. S. 1989. Considerations for sampling airborne substances. In *NIOSH Manual of Analytical Methods*, 3rd edn., ed. P. M. Eller, DHHS (NIOSH) Publication No. 84-100, Cincinnati, OH, with first (1985), second (1987), and third (1989) supplements.

Middleton, A. P. 1982. Visibility of fine fibers of asbestos during outine electron microscopical analysis. *Ann. Occup. Hyg.* 25:53–62.

Morrow, P. E. 1964. Evaluation of inhalation hazards based upon the respirable dust concept and the philosophy and application of selective sampling. *Am. Ind. Hyg. Assoc. J.* 25:213–36.

Moseley, C. L. 1985. Solvent exposure from liquid sampling media. *Am. Ind. Hyg. Assoc. J.* 46:B10–14.

Morey, P. R. 1990. Practical aspects of sampling for organic dusts and microorganisms. *Am. J. Ind. Med.* 18:273–78.

Myers, W. R. 1986. Quantitative fit testing of respirators: Past, present future. In *Fluid Filtration: Gas, Vol. I*, ed. R. R. Raber, pp. 181–92. Publication No. ASTM STP 975, Philadelphia: American Society of Testing and Materials.

National Institute for Occupational Safety and Health (NIOSH). 1990. *NIOSH Pocket Guide to Chemical Hazards*. DHHS (NIOSH) Publication No. 90-117, Cincinnati, OH: U.S. Department of Health and Human Services.

National Institute for Occupational Safety and Health (NIOSH). 1987. *NIOSH Guide to Industrial Respiratory Protection*. DHHS (NIOSH) Publication No. 87-116, Cincinnati, OH: U.S. Department of Health and Human Services.

National Institute for Occupational Safety and Health (NIOSH). 1975. *Criteria for a Recommended Standard to Cotton Dust*. DHEW (NIOSH) Publication No. 75-118, Cincinnati, OH: U.S. Department of Health, Education, and Welfare.

National Institute for Occupational Safety and Health (NIOSH). 1973. *The Industrial Environment—Its Evaluation and Control*. DHEW (NIOSH) Publication No. 74-117, Cincinnati, OH: U.S. Department of Health, Education, and Welfare.

O'Brien, D., P. Baron, and K. Willeke. 1986. Size and concentration measurement of an industrial aerosol. *Am. Ind. Hyg. Assoc. J.* 47:386–92.

O'Brien, D., P. Baron, and K. Willeke. 1987. Respirable dust control in grinding gray iron castings. *Am. Ind. Hyg. Assoc. J.* 48:181–87.

Ogden, T. L. 1983. Inhalable, inspirable and total dust. In *Aerosols in the Mining and Industrial Work Environments. Vol. 1*, eds. V. A. Marple and B. Y. H. Liu, pp. 119–38. Ann Arbor: Ann Arbor Science Publishers.

Ogden, T. L., J. L. Birkett, and H. Gibson. 1978. Large-particle entry efficiencies of the MRE 113A gravimetric dust sampler. *Ann. Occup. Hyg.* 21:251–63.

Olin, J. G., G. J. Sem, and D. L. Christenson. 1971. Piezoelectric–electrostatic aerosol mass concentration monitor. *Am. Ind. Hyg. Assoc. J.* 32:209–20.

Owen, L. E., J. C. Delaney, and C. M. Neff. 1951. The spectrochemical analysis of air-borne dusts for beryllium. *Ind. Hyg. Quart.* 12:112–14.

Palmgren, U., G. Strom, G. Blomquist, and P. Malmberg. 1986. Collection of airborne micro-organisms on Nuclepore filters, estimation and analysis—CAMNEA method. *J. Appl. Bact.* 61:401–6.

Pasceri, R. E. 1973. Respirable dust content in ambient air. *Environ. Sci. Technol.* 7:623–27.

Poppoff, I. G. 1954. Dust sampling by thermal precipitation. *Am. Ind. Hyg. Assoc. J.* 15:145–51.

Pui, D. Y. H. and B. Y. H. Liu. 1988. Advances in instrumentation for atmospheric aerosol measurement. *Physica Scripta* 37:252–69.

Raynor, G. S., E. C. Ogden, and J. V. Hayes. 1975. Spatial variability in airborne pollen concentrations. *J. Allergy Clin. Immunol.* 55:195–202.

Remiarz, R. J., J. K. Agarwal, F. R. Quant, and G. J. Sem. 1983. Real-time aerodynamic particle size analyzer. In *Aerosols in the Mining and Industrial Work Environments. Vol. 3*, eds. V. A. Marple and B. Y. H. Liu, pp. 879–95. Ann Arbor: Ann Arbor Science Publishers.

Roach, S. A. 1973. Sampling air for particulates. In *The Industrial Environment—Its Evaluation and Control*, 3rd edn., DHEW (NIOSH) Publication No. 74-117, pp. 139–53.

Robert, K. 1979. Cotton dust sampling efficiency of the vertical elutriator. *Am. Ind. Hyg. Assoc. J.* 40:535–42.

Rom, W. N. 1983. Asbestos and related fibers. In *Environ-*

mental and Occupational Medicine, ed. Rom, W. N., pp. 157–82. Boston: Little, Brown and Company.

Rom, W. N. and H. Barkman. 1983. Respiratory irritants. In *Environmental and Occupational Medicine*, ed. Rom, W. N., pp. 273–83. Boston: Little, Brown and Company.

Rubow, K. L., V. A. Marple, J. Olin, and M. A. McCawley. 1987. A personal cascade impactor: Design, evaluation and calibration. *Am. Ind. Hyg. Assoc. J.* 48:532–38.

Sass-Kortsak, A. M., C. Tracey, and J. Purdham. 1989. Filter preparation techniques for dust exposure determination by gravimetric analysis. *Appl. Ind. Hyg.* 4:222–26.

Sem, G. J., K. Tsurubayashi, and K. Homma. 1977. Performance of the piezoelectric microbalance respirable aerosol sensor. *Am. Ind. Hyg. Assoc. J.* 38:580–88.

Skarping, G., B. E. F. Smith, and M. Dalene. 1985. Trace analysis of amines and isocyanates using glass capillary gas chromatography and selective detection. V. Direct determination of isocyanates using nitrogen-selective and electron-capture detection. *J. Chromat.* 331:331–38.

Skillern, C. P. 1971. Problems using Mie scattering photometers for in-place HEPA filter tests and aerosol studies. *Am. Ind. Hyg. Assoc. J.* 32:96–103.

Soderholm, S. C. 1988. Aerosol instabilities. *Appl. Ind. Hyg.* 3:35–40.

Soderholm, S. C. 1989. Proposed international conventions for particle size-selective sampling. *Ann. Occup. Hyg.* 33:301–20.

Soderholm, S. C. 1991. Why change ACGIH's definition of respirable dust? *Appl. Occup. Environ. Hyg.* 6:248–50.

Stewart, I. M. 1988. Asbestos—Analytical techniques. *Appl. Ind. Hyg.* 3:F24–26.

Swift, D. L. and M. Lippmann. 1989. Electrostatic and thermal precipitators. In *Air Sampling Instruments for Evaluation of Atmospheric Contaminants*, 7th edn., ed. S. V. Hering, pp. 387–404. Cincinnati: American Conference of Governmental Industrial Hygienists.

Tebbens, B. D. 1973. Personal dosimetry versus environmental monitoring. *J. Occup. Med.* 15:639–41.

Tomb, T. F., H. M. Treaftis, R. L. Mundell, and P. S. Parobeck. 1973. Report of Investigations 7772, Bureau of Mines. Pittsburgh: U.S. Department of the Interior.

Tossanainen, A. 1976. Metal fumes in foundries. *Scand. J. Work Environ. Health* 4:42–49.

Treaftis, H. N., A. J. Gero, P. M. Kacsmar, and T. F. Tomb. 1984. Comparison of mass concentrations determined with personal respirable coal dust samplers operating at 1.2 liters per minute and the Casella 113A gravimetric sampler (MRE). *Am. Ind. Hyg. Assoc. J.* 45:826–32.

U.S. Department of Labor, Occupational Health and Safety Administration (USDOL). 1986. *Code of Federal Regulations.* Title 29, §1910.1000. Washington, DC: U.S. Government Printing Office.

U.S. Environmental Protection Agency (USEPA). 1978. *Microbiological Methods for Monitoring the Environment*, eds. R. Bordner, J. Winter, and P. Scarpino. Publication No. EPA-600/8-78-017, Cincinnati, OH: U.S. Environmental Protection Agency.

U.S. Public Law 91-173. 1969. *Federal Coal Mine Health and Safety Act of 1969*, 91st Congress (10 December 1969).

Vaughan, N. P., B. B. Milligan, and T. L. Ogden. 1989. Filter weighing reproducibility and the gravimetric detection limit. *Ann. Occup. Hyg.* 33:331–37.

Vincent, J. H. and D. Mark. 1990. Entry characteristics of practical workplace aerosol samplers in relation to the ISO recommendations. *Ann. Occup. Hyg.* 34:249–62.

Walworth, H. T. 1945. Health hazards in some non-ferrous metal smelters. *I.H. Supp., Ind. Med.* 6:367–72.

Warner, M. 1988. Optical and electron microscopy can be used to determine asbestos in ambient air. *Anal. Chem.* 60:395A–96A.

Wesley, R. A., J. D. Hatcher, O. L. McCaskill, and J. B. Cocke. 1978. Dust levels and particle-size distributions in high-capacity cotton gins. *Am. Ind. Hyg. Assoc. J.* 39:368–77.

White, L. A., D. J. Hadley, J. E. Davids, and R. I. Naylor. 1975. Improved large-volume sampler for the collection of bacterial cells from aerosol. *Appl. Microbiol.* 29:335–39.

Whitney, C. 1972. The spatial inference problem. Contract No. HSM-99-72-18, Cambridge, Massachusetts: Charles Stark Draper Laboratory.

Willeke, K., H. E. Ayer, and J. D. Blanchard. 1981. New methods for quantitative respirator fit testing with aerosols. *Am. Ind. Hyg. Assoc. J.* 42:121–25.

Willeke, K. and B. Y. H. Liu. 1976. Single particle optical counter: Principle and Application. In *Fine Particles: Aerosol Generation, Measurement, Sampling, and Analysis*, ed. B. Y. H. Liu, pp. 697–729. New York: Academic Press.

Wolf, H. W., P. Skaliy, L. B. Hall, M. M. Harris, H. M. Decker, L. M. Buchanan, and C. M. Dahlren. 1959. *Sampling Microbiological Aerosols*, Public Health Monograph No. 60. Washington, DC: U.S. Department of Health, Education, and Welfare, Public Health Service.

Wright, B. M. 1954. A size-selecting sampler for airborne dust. *Br. J. Ind. Med.* 11:284–88.

25

Measurement of Asbestos and Other Fibers

Paul A. Baron

National Institute for Occupational Safety and Health[1]
Centers for Disease Control, Public Health Service
Department of Health and Human Services, Cincinnati, OH, U.S.A.

Introduction

The term fiber has been applied to a wide variety of particles having an elongated shape, i.e., one particle dimension significantly greater than the other two. Because of this elongation, fibers can have aerodynamic and other properties quite different from more compact particles. Certain fibers have several unique properties that make them not only useful from a commercial standpoint, but also important from a health standpoint. Asbestos, for instance, includes six commercial fibrous minerals that have high tensile strength, chemical resistance, and heat resistance. These properties have made asbestos useful in a variety of products, including friction materials, high-temperature insulating materials, acoustic insulating materials, fire proof cloth and rope, and floor tiles. While the bulk materials in these products consist primarily of macroscopic-sized fibers, many of them can release long, thin fibers into the air.

A variety of materials can be considered fibers from an aerosol behavior standpoint. Besides asbestos, other mineral fibers exist in nature. Several materials, including glass and mineral slags, have been melted and spun into fibers. Ceramic materials have similarly been spun into fibers, as well as grown by chemical and vapor crystallization. Carbon and graphite fibers are produced commercially for high-strength products. Organic fibers, such as cotton, wood, and other cellulosic materials, are widely present in the environment, both from commercially produced materials as well as natural sources. Besides cylindrical particles that have relatively high strength, chains of particles also may behave as fibers and can serve as models for some aerodynamic properties. Some organic materials can be crystallized into well-defined fibrous shapes and can be used to test theories of fiber aerodynamic behavior.

Asbestos fiber aerosols have been closely associated with several diseases, such as asbestosis (a fibrosis of the lung), mesothelioma (cancer of the lining around the lung), and lung cancer (NIOSH 1976*a*). Thus, the fibers that can enter the respiratory system are of greatest concern. In addition, the high rate of

1. Mention of product or company names does not constitute endorsement by the National Institute for Occupational Safety and Health.

disease prompted the statement that there "is no evidence for a threshold or a 'safe' level of asbestos exposure" (NIOSH 1990*a*). The seriousness of the diseases has driven measurement technology to provide maximum sensitivity and accuracy for measuring asbestos aerosols. Other airborne fibers may have one or more of the same physical and chemical properties as asbestos. In some cases, human exposures and/or animal studies suggested the disease potential of these fibers. Thus, there is concern regarding the health effects of fibers other than asbestos. While the commercial properties of mineral fibers have created a store of knowledge about physical and geologic properties, it is the health concerns that have largely driven the technology for detecting and quantifying airborne concentrations of fibers. Thus, much of the fiber aerosol research and measurement capability relates to the ability of microscopic-sized fibers to enter and deposit in the human respiratory system.

The dimensions of fibers in aerosols can cover a wide range. For asbestos, diameters can be as small as 0.025 μm (Langer, Mackler, and Pooley 1974), while lengths can be several hundred micrometers. The dimension distributions depend on the fiber type as well as on how the fibers were comminuted from the bulk material. The magnitude of disparity between length and diameter often makes it difficult to make accurate size distribution measurements. Several protocols, using various types of microscopes, have been developed to deal with fiber distribution measurement. Other types of instruments, primarily using light scattering properties, have been developed to characterize fibers. However, these instruments usually provide only an approximate indication of fiber dimensions.

Note that the following discussion, except where otherwise indicated, will deal largely with measurement of aerosolized fibers, generally visible only with a microscope, and not with the macroscopic or bulk properties of the fibrous material. Since asbestos has been the most intensely studied type of fiber, many comments will relate to this material. Many issues regarding asbestos mineralogy, health effects, and measurement techniques are discussed in a review by Walton (1982); further reviews in Environmental Health Perspectives are introduced by Langer, Mackler, and Pooley (1974) and Dement (1990). Additional topics are presented in the books by Selikoff and Hammond (1979), Rajhans and Sullivan (1981), Michaels and Chissick (1979), Chissick and Derricott (1983), and Holt (1987). Similar reviews have been carried out for man-made fibers (NIOSH 1976*a*, 1977; IARC 1988).

FIBER SHAPE

The behavior of fibers suspended in a gas is a function of the fiber dimensions. Assuming either a cylindrical or prolate spheroidal shape, these dimensions can be defined by two parameters, length and diameter. A third parameter β is often invoked to indicate the fibrosity or aspect ratio, i.e., the ratio of the length to the diameter. However, often, real fibers meet neither the ideal cylindrical assumption nor the prolate spheroid shape assumption. Glass or mineral fibers are often nearly cylindrical, but even these fibers frequently display curvature along their length as well as bulbous or jagged ends. Asbestos fibers are formed from a unique crystal habitat in which the bulk mineral has slip planes in two directions, but only rarely in the third. This results in a propensity to produce particles that can split longitudinally to produce thinner and thinner fibers, ultimately resulting in fibrils about 0.025–0.05 μm diameter. Thus, while some asbestos fibers exhibit a nearly ideal cylindrical shape, others may have various combinations and degrees of splayed ends, curvature, splitting, noncircular circumference, etc. For instance, the magnetically aligned chrysotile fibers in Fig. 25-1 show many of these characteristics. In spite of these possible variations in shape, fibers are still most often characterized just by length and diameter.

Distributions of natural fibers are rarely monodisperse in diameter and even more rarely in length. This has made it difficult to provide adequate calibration for instruments

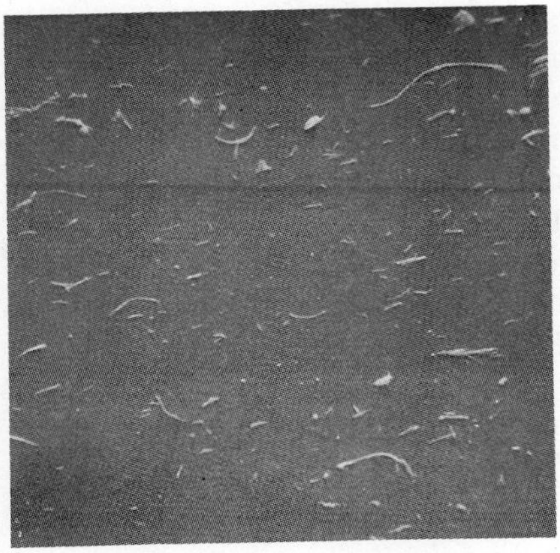

FIGURE 25-1. Scanning Electron Micrograph (1500×) of an Aqueous Sample of Magnetically Aligned UICC Canadian Chrysotile Collected on a 0.1 μm Pore Size Filter. (Source: Timbrell 1973.)

that attempt to measure fibers as well as to perform measurements of fiber toxicity as a function of fiber dimension. Distributions of fibers can often be described by a two-dimensional (length L and diameter W) lognormal distribution (Cheng 1986; Schneider, Holst, and Skotte 1983), i.e., $\ln L$ and $\ln W$ are each distributed normally. The probability density function is given by

$$f(L, W) = \frac{1}{2\pi\sigma_W\sigma_L\sqrt{1-\tau^2}\,LW} \times \exp\left[-\frac{A^2 + B^2 - 2\tau AB}{2(1-\tau^2)}\right] \tag{25-1}$$

where

$$A = (\ln W - \mu_W)/\sigma_W$$

$$B = (\ln L - \mu_L)/\sigma_L$$

μ_i and σ_i^2 are the mean and variance, respectively, of the natural logarithm of L and W and τ is the correlation between $\ln L$ and $\ln W$. The five parameters μ_L, μ_W, σ_L, σ_W, and τ are needed to define completely a two-dimensional size distribution. The two-dimensional lognormal size distribution has the properties that the marginal and the conditional distributions are lognormal (Holst and Schneider 1985). The former property indicates that the length and diameter distributions are each, separately, lognormal. The latter indicates that functions of length and diameter of the form kW^pL^q, where k, p, and q are constants, are also lognormal. Such functions include the aspect ratio, surface area, volume, and aerodynamic diameter. Deviations from lognormality can sometimes be attributed to artifacts in sampling or analysis or to multiple aerosol generation sources.

Many fiber distributions reported in the literature include the length and diameter means and variances but, unfortunately, not τ. However, if the original data are reported in a table as a function of both length and diameter, the correlation term can be estimated (Cheng 1986). Most fiber distributions have positive τ, suggesting that diameter often increases with length.

There have been measurements of a variety of fibrous aerosols. Table 25-1 lists the results of some examples. Some of these materials have been generated for toxicity studies, some have been measured in environmental studies, while others have been generated as calibration materials.

FIBER BEHAVIOR

Translational Motion

As with other aerosol particles, fiber dimensions can cover a relatively wide range: the smaller fibers are affected primarily by diffusional forces, while the larger ones are primarily affected by flow shear, inertial, and gravitational forces. Fiber behavior has been observed and theoretically calculated for several fibrous shapes, including prolate ellipsoids, cylinders, and chains of spheres. The motion of these various shapes, generally, differs only slightly.

Fiber behavior differs, depending upon whether the major axis is oriented parallel to or perpendicular to the direction of the motion relative to the surrounding gas (Fig. 25-2a and b). The drag on a fiber is greatest when it is oriented perpendicular to the flow of the surrounding gas. Fiber behavior is

TABLE 25-1 Examples of Measured Fiber Size Distributions

Material	Diameter (μm)	σ_g	Length (μm)	σ_g	MMAD (μm)	σ_g	Measurement Technique
Chromoglycic acid[1]	0.205	1.58	2.09	1.83			SEM
					0.65	1.88	Cascade impactor
Sugar cane silicate[2]	0.3–1.5*		3.5–65*				TEM
Caffeine[3]	1.13	1.08	5.55	1.12			SEM
					2.1	1.1	Sedimentation
Ceramic fibers[4]							
Sample a	0.5		10.1				TEM
Sample b	0.66		8.3				TEM
Sample c	0.98		22.8				TEM
Chrysotile[5]							
Preform ring	0.13	2.15	1.6	2.7			TEM
Yarn dressing	0.08	1.92	1.0	2.4			TEM
Cure press	0.13	1.94	1.5	2.2			TEM
Crocidolite[6]							
Mine/Mill**	0.08–0.10	1.86–2.08	0.98–1.25	2.30–2.55			TEM
Manufacturing	0.04	1.58	0.54	2.32			TEM
Fibrous glass[7]							
Code 100	0.12	1.8†	2.7	2.2†			TEM
Code 110	1.8	1.7†	26	2.0†			TEM
Iron oxide chains[8]							
	0.059	1.1	1††	2.0			TEM
					0.32	1.11	Centrifuge

Source. 1. Chan and Gonda (1989); 2. Boeniger et al. (1988); 3. Vaughan (1990); 4. Rood (1988); 5. Rood and Scott (1989); 5. Pinkerton et al. (1983); 6. Hwang and Gibbs (1981); 7. Timbrell (1974); 8. Kaspar and Shaw (1983)

* These values represent the range of particle sizes rather than the median diameters

** These values represent the range of several measurements that produced similar results

† Estimated from data in reference

†† Estimated from mean chain length of 22 primary particles

often described in terms of a combination of the two orientations. While the difference in drag between the two orientations is typically about 15–30%, it can be difficult to determine the contribution of each orientation in experimental systems. At low Reynolds number (Re_p), fiber orientation will be stable (discounting Brownian rotation) and not change due to translational motion, e.g., during gravitational settling (Gallily 1971). In addition, fibers settling in still air will not settle exactly in the direction of the gravitational force, but will drift somewhat due to orientation (Weiss, Cohen, and Gallily 1978). Larger fibers, with Re_p greater than about 0.01, will settle with their major axis oriented perpendicular to the direction of motion (Fig. 25-2a). With increasing Re_p ($Re_p > 100$), longer fibers ($\beta > 20$) are still stable in the perpendicular orientation, but there is an increasing trend toward instability (Clift, Grace, and Weber 1978, 154).

The aerodynamic diameter d_a of a prolate spheroid has been calculated from

$$d_a = d_f \sqrt{\frac{\rho_f \beta}{\rho_0 \chi}} \qquad (25\text{-}2)$$

by using the numerical shape factor χ of a prolate ellipsoid of revolution (Fuchs 1964, 37); d_f is the physical fiber diameter, ρ_f is the fiber density and ρ_0 is unit density. A cylinder with the same diameter and length as a prolate ellipsoid has 3/2 times greater volume and mass. Therefore, for cylinders with the same axial dimensions, the right-hand side of Eq. 25-2 must be multiplied by $(3/2)^{1/3}$ or $(3/2)^{1/2}$ to obtain the equivalent-volume diameter or equivalent-weight diameter, respectively

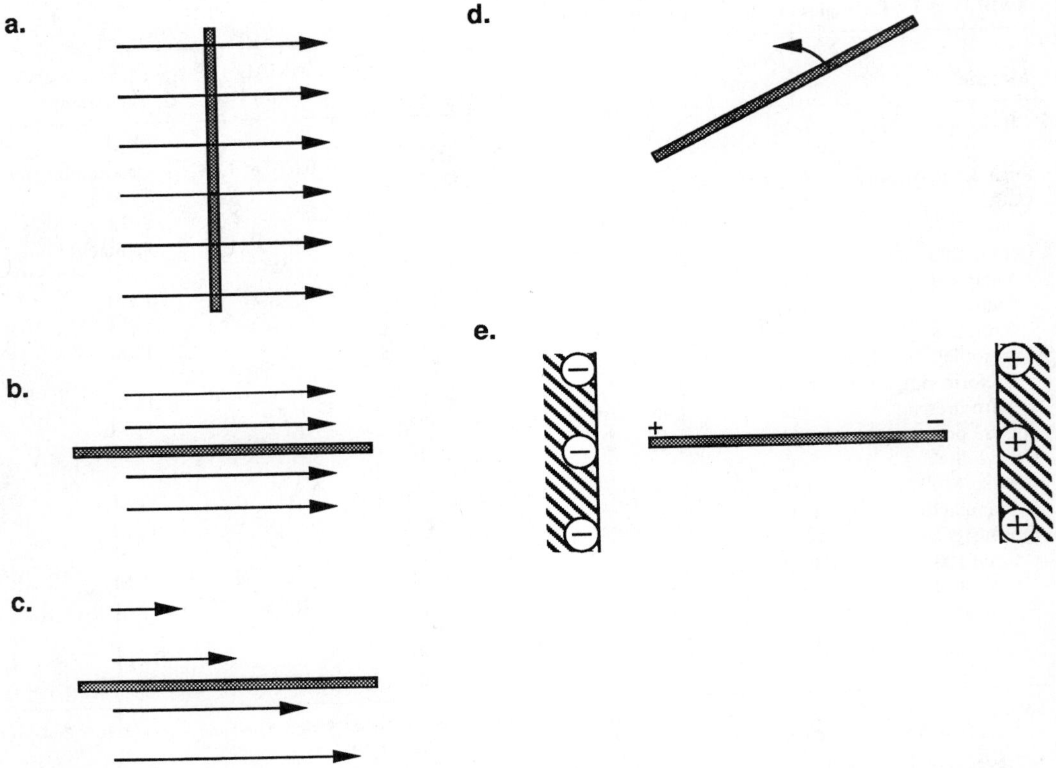

FIGURE 25-2. Fiber Alignment in Various Force Fields. (a) Fiber Aligned Perpendicular to Relative Gas Motion. This is the Preferred Orientation During Gravitational Settling and Acceleration at $0.01 < Re < 100$ in the Absence of Other Forces. (b) Fiber Parallel to Relative Gas Motion. Fiber Motion is Often Treated as a Combination of Cases a and b. (c) Fiber is Readily Oriented Parallel to, or at Some Small Angle to, the Direction of Shear Flow in the Suspending Gas Medium. (d) Small Fibers Governed by Diffusional Forces may Exhibit Completely Random Orientation. (e) Conductive Fibers are Aligned Parallel to an Electric Field. Many Fibers are also Aligned in a Magnetic Field, Usually Parallel to the Field Lines, Though They may be Aligned Perpendicular to the Field or, for Some Materials, at Some Intermediate Angle.

(Griffiths and Vaughan 1986). For motion parallel and perpendicular to the fiber major axis, the respective shape factors $\chi_{\parallel}$ and $\chi_{\perp}$ are (Stöber 1972; Kasper 1982):

$$\chi_{\parallel} = \frac{4(\beta^2-1)}{3} \bigg/ \left\{ \frac{2\beta^2-1}{\sqrt{\beta^2-1}} \ln(\beta+\sqrt{\beta^2-1}) - \beta \right\} \tag{25-3}$$

$$\chi_{\perp} = \frac{8(\beta^2-1)}{3} \bigg/ \left\{ \frac{2\beta^2-3}{\sqrt{\beta^2-1}} \ln(\beta+\sqrt{\beta^2-1}) + \beta \right\} \tag{25-4}$$

(see Fig. 25-2a and b). Note that a dynamic shape factor χ_d is also defined that is related to the numerical shape factor for prolate spheroids by $\chi = \chi_d \beta^{-1/3}$. The dynamic shape factor is applied when the equivalent-volume diameter of the particle is used rather than the physical diameter.

An alternate approach for directly calculating the aerodynamic diameter of cylinders (Cox 1970) gives similar results:

$$d_{a,\parallel} = d_f \sqrt{\frac{9\rho_f}{4\rho_0}\,[\ln(2\beta) - 0.807]} \tag{25-5}$$

and

$$d_{a,\perp} = d_f \sqrt{\frac{9\rho_f}{8\rho_0}\,[\ln(2\beta) + 0.193]} \quad (25\text{-}6)$$

Others have also provided formulae for prolate ellipsoids and cylinders (Gonda and Khalik 1985; Prodi et al. 1982).

If fibers are not preferentially oriented by a drag force or any other alignment force, the orientation may be completely random. Then, a single average shape factor $\bar{\chi}$ that is a function of the two shape factors noted above may be used, i.e.,

$$\frac{1}{\bar{\chi}} = \frac{1}{3\chi_{\parallel}} + \frac{2}{3\chi_{\perp}} \quad (25\text{-}7)$$

In the presence of air gradients, the fiber will experience a torque until the fiber is oriented parallel to the direction of the shear force (Fig. 25-2c). Thus, a fiber settling in a horizontal laminar flow will tend to be oriented horizontally (parallel to the shear). However, the fiber will experience a periodic instability and perform a "flip". This instability is a function of fiber dimensions as well as the flow gradients. Under such conditions, the aerodynamic diameter is not strictly an inherent property of the particle and depends on the experimental conditions of measurement (Gallily and Eisner 1979).

Inertial separation is commonly used for particle separation and sizing, e.g., in impactors and cyclones. In such systems where flow conditions are rapidly changing, the fiber mechanics are governed by initial orientation and flow relaxation time besides the usual parameters observed for spherical particles (Gallily et al. 1986). For instance, fibers with large rotational inertia (especially long fibers) may not orient completely or may over-rotate in passing through a nozzle. Fiber behavior under such a situation may be only approximately defined by the Stokes number or other nondimensionalized parameters.

The experimental measurement of fiber deposition has been carried out in horizontal elutriators (Gallily and Eisner 1979; Griffiths and Vaughan 1986; Iles 1990), centrifuges (Stöber, Flachsbart, and Hochrainer 1970; Martonen and Johnson 1990), impactors (Burke and Esmen 1978; Prodi et al. 1982), and cyclones (Fairchild et al. 1976; Iles 1990) for a variety of fiber types.

The extended shape also means that interception during translational motion plays a larger role in fiber deposition than for compact particles. However, the alignment of fibers by shear flow often reduces the effect of length on interception.

Rotational Motion

The rotational mobility B_r of a high aspect ratio ellipsoid can be approximated by (Lilienfeld 1985)

$$B_r = \frac{3[2\ln(2\beta - 1)]}{2\pi\eta L^3} \quad (25\text{-}8)$$

where η is the viscosity of the gas. Note that the rotational mobility is a strong inverse function of fiber length. Similarly, the rotational diffusion coefficient D_r for fibers is also a strong function of fiber length:

$$D_r = \frac{3kT}{\pi\eta\beta L^3}(\ln 2\beta - \delta) \quad (25\text{-}9)$$

where k is the Boltzmann constant, T is the temperature and δ is 1.4 for aspect ratios β larger than ten. Rotational mobility can be estimated by measuring the rate of relaxation after removal of an electrostatic alignment force (Cheng et al. 1991).

Behavior in the Transition Regime

Under molecular bombardment, fibers can exhibit both rotational diffusion as well as translational diffusion. Such fibers are likely to be randomly oriented (Fig. 25-2d). As for fibers in the Stokes regime, it is often convenient to separate the translational motion of fibers into motion in which the major axis is parallel to the direction of motion and another in which the major axis is perpendicular to the translational motion. The diffusion of fibers is described by the diffusion coefficient

D_f (cm^2/s):

$$D_f = BkT = \frac{kT}{f} = \frac{kTC_f}{f^0} \quad (25\text{-}10$$

where B is the fiber mobility (dyn cm/s), and f^0 is the drag per unit velocity of the fiber in the continuum regime and f is the drag per unit velocity of the fiber corrected for slip by the fiber slip correction factor C_f. A theory for the slip correction factors for nonspherical particles is described by Dahneke (1973a, b, c; 1982).

Fiber diffusional behavior is usually treated as a modification of spherical particle diffusion using particle shape factors (Asgharian, Yu, and Gradon 1988). This approach has agreed well with the experimental diffusion coefficient measurement of fibers with mean diameters of 0.24–0.38 μm (Gentry et al. 1983). Diffusional coefficients of much smaller fibers have also been measured (Gentry, Spurny, and Soulen 1988), which show higher diffusion coefficients than expected.

As with the stagnant flow conditions for fibers in the continuum regime, fibers are expected to be randomly oriented unless affected by shear flow or other forces. Again, the longer the fiber, the more likely it is to be oriented by such forces.

Several studies have estimated the effects of various deposition mechanisms (diffusion, impaction, interception) to determine overall particle deposition in filters (Fu, Cheng, and Shaw 1990) and lung airways (Asgharian and Yu 1989; Balashazy, Martonen, and Hofmann 1990).

Charging

Theories for unipolar diffusion charging (Laframboise and Chang 1977) and bipolar diffusion charging (Wen, Reischl, and Kasper 1984) of fibers have been developed. Unipolar charging of fibers causes the charge of long, thin fibers to increase dramatically, though the electrical mobility of such fibers changes more slowly with aspect ratio (Yu, Wang, and Gentry 1987). Such a variation of mobility with fiber aspect ratios may allow the separation of fibers of different lengths.

Electric Field Effects

A fiber may be aligned in an electric field by an induced dipole in the fiber. This requires that charges in the fiber be separated so that the polarity is opposite to that of the surrounding electric field as shown in Fig. 25-2e. The charge separation from conduction is usually greater than that from polarization of the material. For charge separation to occur, the fiber must be sufficiently conductive so that the charges can migrate the length of the fiber in a reasonable time. Aerosol particles, even those consisting of a normally nonconducting material, can often be considered conductive because of their low capacitance and small dimensions (Fuchs 1964; Lilienfeld 1985). Surface impurities can also contribute to a particle's conductivity. Thus, an electric field of sufficient strength (1000–5000 V/cm) can overcome diffusional randomization and flow shear forces to align most types of fibers, including relatively nonconductive ones. For instance, electrostatically aligned zinc oxide fibers were used to modulate microwave radiation (Tolles, Sanders, and Fritz 1974).

When fibers and compact particles of the same aerodynamic diameter are charged under the same conditions, the fibers may have higher mobility than compact particles. Field studies of work environments suggested that fibers carried a charge proportional to fiber length (Johnston, Vincent, and Jones 1985). Other studies indicated that unipolar, charged particles can be separated according to aspect ratio (Griffiths, Kenny, and Chase 1985; Yu, Wang, and Gentry 1987).

Dielectrophoresis has also been investigated for separating fibers of different lengths (Lipowicz and Yeh 1989). Uncharged, conductive fibers may be separated according to length in a nonuniform electric field. Since the electrical mobility of charged fibers is generally higher than the dielectric mobility, such a separation must be carried out on fibers with low charge, in an ac electric field, or both.

An electrostatic enhancement of fiber deposition in lungs (conductive tubing) has been observed (Jones, Johnston, and Vincent 1983). Calculations support such an enhancement of sedimenting charged fibers (Chen and Yu 1990).

Magnetic Field Effects

If a suspension of fibers in a liquid or gas is subjected to a magnetic field, fibers with sufficient magnetic susceptibility will align at some angle to the field. Usually this angle is either 0° or 90°; some amphibole asbestos samples have fibers aligned at both angles. Timbrell (1975) developed a technique for preparing permanently aligned samples by allowing a suspension of fibers in 0.5% celloidin/amyl acetate to dry in a 5–10,000 G magnetic field. Several fiber types have been aligned by Timbrell (1972, 1973), including carbon fibers and the various types of asbestos. Fibers of silicon carbide, silicon nitride, and tungsten-cored boron did not align in similar fields.

Figures 25-3–25-5 contain the images of fibers magnetically aligned on a slide surface with light scattering patterns from magnetically aligned liquid suspensions of the same types of fibers. The direction of the magnetic field is shown in the figures. The scattering pattern has the main laser beam in the center, with the plane of scattering radiating in opposite directions. In Fig. 25-3, monodisperse diameter carbon fibers are all aligned parallel to the field so that a well-defined scattering pattern perpendicular to the field is produced. The crocidolite fibers in Fig. 25-4 are aligned the same way, but are not monodisperse. In other cases, the fibers are aligned perpendicular or both perpendicular and parallel to the magnetic field (e.g., Fig. 25-5), the latter resulting in two planes of scattering. A synthetic fluoroamphibole was observed to align at ±65° to the magnetic field direction. The degree and direction of alignment has not been adequately explained; however, it appears to be more a function of the mineralogical source of the material rather than of the primary crystal structure.

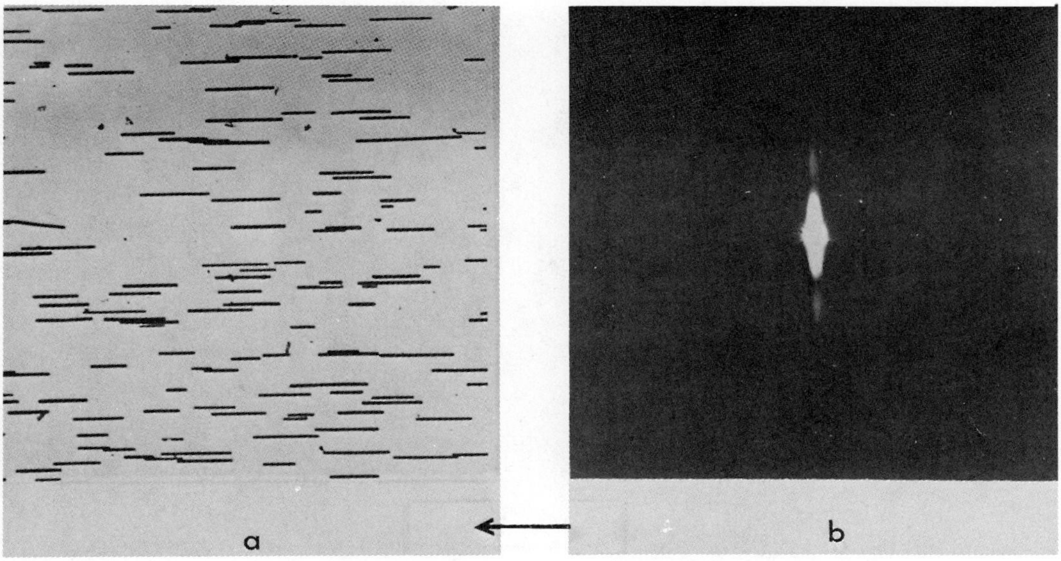

FIGURE 25-3. (a) Phase Contrast Microscope (PCM) Image of Magnetically Aligned Carbon Fibers Suspended in Celloidin on a Glass Slide. (b) Light Scattering Pattern from the Same Fibers in Aqueous Suspension. The Direction of the Magnetic Field is Indicated by the Arrow. Note the Monodisperse Diameter of the Fibers, Reflected in the Sharply Defined Scattering Pattern. (Source: Timbrell 1973.)

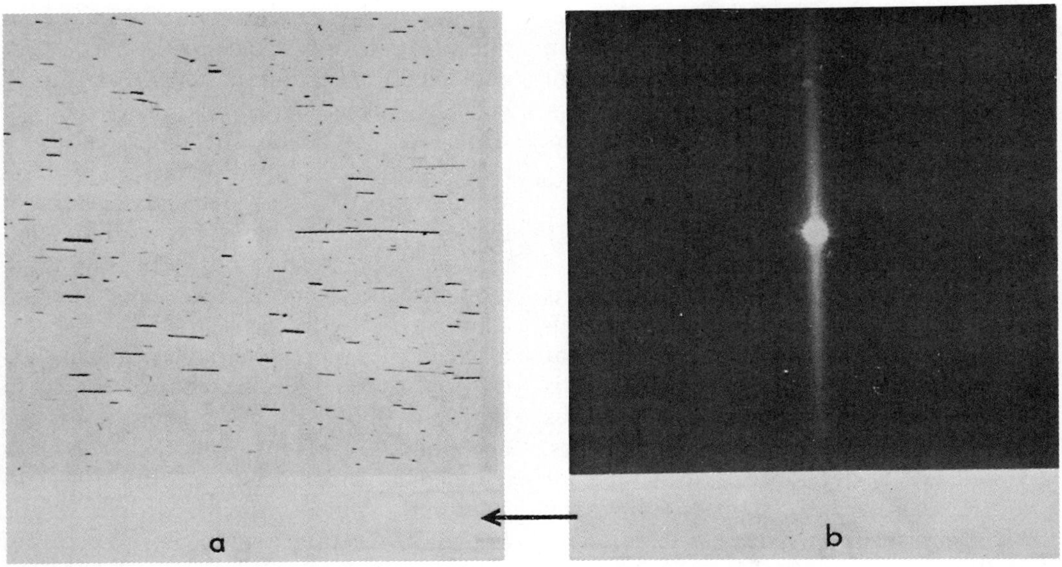

FIGURE 25-4. (a) PCM Image of Magnetically Aligned UICC Crocidolite Fibers Suspended in Celloidin on a Glass Slide. (b) Light Scattering Pattern from the Same Fibers in Aqueous suspension. The Direction of the Magnetic Field is Indicated by the Arrow. (Source: Timbrell 1973.)

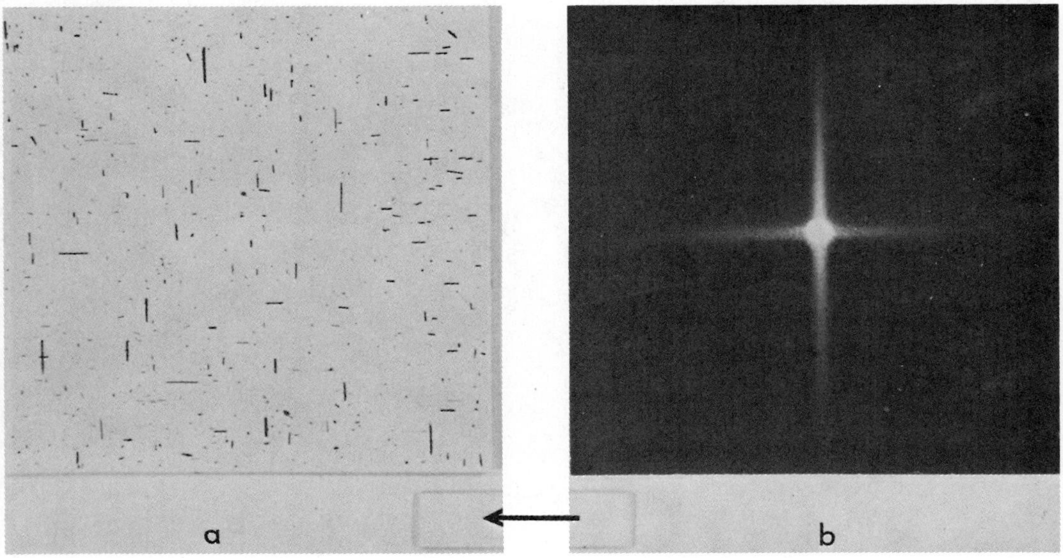

FIGURE 25-5. (a) PCM Image of Magnetically Aligned UICC Amosite Fibers Suspended in Celloidin on a Glass Slide. (b) Light Scattering Pattern from the Same Fibers in Aqueous Suspension. The Direction of the Magnetic Field is Indicated by the Arrow. (Source: Timbrell 1973.)

Thus, Ugandan tremolite was observed to align perpendicular to the magnetic field, while Zululand tremolite aligned parallel to the field (Timbrell 1973).

Light Scattering

If a flashlight beam were shone on a glass rod such that the light beam was perpendicular to the rod axis, one would expect the refracted and scattered light to be dispersed into a plane perpendicular to the rod's axis. Light scattered and refracted from microscopic fibers produces a similarly unique pattern. Such scattering patterns from magnetically aligned fibers in an aqueous suspension are given in Figs. 25-3–25-5. If the fiber is not perpendicular to the light beam, the light scattering pattern is not as conveniently constrained to a plane, becoming a cone of light. The details of light scattering from infinitely long cylinders, ellipsoids, and several other regular elongated shapes can be described by the Mie theory (Kerker 1969; Van de Hulst 1957). As with scattering from spherical and compact particles, the scattering from fibers with diameters larger than the light wavelength is concentrated in the forward direction. The scattering from smaller-diameter fibers is less, but more uniform in all directions around the fiber axis. In addition, the scattered light tends to be polarized in the direction parallel to the fiber axis.

The unique, planar scattering pattern for right-angle illumination has been the basis of several useful fiber detection techniques. As noted above, the fiber must be held perpendicular to the light-beam axis to obtain the characteristic fiber pattern. A fiber collected on a surface such as a glass slide will generally be parallel to the slide surface, thus allowing a beam perpendicular to the slide surface to produce the narrow, planar scattering pattern from a single fiber. To obtain a characteristic scattering pattern from a group of fibers, they must be aligned using some force, such as magnetic, electric, or shear flow.

Figures 25-3–25-5 show the scattering patterns for several types of fibers. Note the well-defined scattering pattern for monodisperse carbon fibers (Fig. 25-3) while the broad distribution of diameters for the other fiber types produces a more diffuse pattern.

LABORATORY FIBER GENERATION

Fibers are more difficult to generate than compact particles because of their tendency to intertwine when in contact with each other. This tendency is the basis for some commercial properties of fibers, e.g., the formation of rope and felt. Various types of fibers can be generated in various concentration ranges for instrument calibration, analytical method validation, and quality assurance, and toxicology studies. Various generation mechanisms have been used to produce well-dispersed fibers.

Nebulization of liquid suspensions has been used for generating relatively short fibers at low concentrations. Fibers larger than the nebulized droplet diameter may not be generated efficiently and, if the concentration is too high, more than one fiber may be present in a droplet, resulting in fiber agglomerates.

Several researchers have used high speed chopping of a packed fiber plug. Timbrell, Hyett, and Skidmore (1968) developed a version of this approach using a household coffee grinder. A specially ground reference material, Union Internationale Contre le Cancer (UICC) asbestos (Timbrell, Gilson, and Webster 1968), was packed uniformly in a syringe body and pushed slowly into the rotating blades of the grinder. The consumption rate of asbestos was 0.6–1.0 g/h, though not all of this was generated as an aerosol. The dispersion appeared largely to be single fibers, with relatively few clumps or flocculates, though no size distribution was reported. A generator with similar operating principle and dimensions was constructed using more durable materials, including tungsten carbide blades and a stainless steel body (Fairchild et al. 1976). This device was used for both chrysotile and fibrous glass. Although there were relatively few clumps of chrysotile by number, the mass in these clumps accounted for the largest fraction of the mass distribution. Airborne concentrations of 6–8 mg/m^3 were achieved, though the feed rate could be lowered to one-fourth or increased to ten times the value used for this measurement. Fibrous glass aerosols were more successfully generated with this device (Fairchild et al. 1978).

Fluidized-bed generators have also been used for creating fibrous aerosols. Fluidized beds consist of two phases: a powder phase, containing one or more components, and an air phase passing through the powder. The powder is made to act like a boiling fluid by passing air through it or by applying vibrational or acoustic energy. The powder may consist solely of the dust to be generated or it may consist of the dust plus particles that are too large to be carried away by the airflow. These powder particles separate the dust particles from each other and tend to break apart agglomerates.

A two-component fluidized bed was used for inhalation exposure experiments with fibrous glass (Carpenter et al. 1981) and crocidolite (Griffis et al. 1983). The bed consisted of a stainless steel powder mixed with the fibers as a slurry and then dried. Air passing through the bed fluidized the bed and released the fibers, initially at a high rate, then decreasing exponentially. A similar air-fluidized bed with bronze powder as the fluidizing powder was used for generating multiple filter samples of chrysotile for a quality assurance program (Baron and Deye 1987).

Charge must be reduced on generated aerosols to produce uniform air concentrations and consistent measurements. This can be accomplished by charge neutralizers (see Chapter 19) or by modifying the generation conditions. It was found that fluidized beds produced highly charged fibers when operated with dry air; the charge level dropped about tenfold when the relative humidity of the air was increased to about 15% (Baron and Deye 1989*a*).

A two-component fluidized-bed generator with a screw feed system to refresh the bed continually with premixed powder was found to produce a constant output concentration (Tanaka and Akiyama 1984). This generator, using glass beads as the large-particle fluidizing component, was found to produce a constant output concentration 6 mg/m^3 fibrous glass (from a glass fiber filter) for one week (Tanaka and Akiyama 1987). A similar system using stainless steel beads and asbestos fibers was developed by Sussman, Gearhart, and Lippmann (1985).

A one-component fluidized bed developed by Spurny (Spurny, Boose, and Hochrainer 1975; Spurny 1980) used vibrational energy to assist bed fluidization. The output of the bed for several types of asbestos was constant with time and the fiber size was controlled by the vibration frequency and amplitude. A fluidized-bed generator of this type, using mechanical vibration, is commercially available (PALAS Co., Karlsruhe).

Chains of iron oxide particles have been generated using a laminar carbon monoxide flame (Kasper, Shon, and Shaw 1980). This generation system created a high concentration of relatively monodisperse ($\sigma_g \approx 1.4$), nearly spherical particles that, under appropriate conditions, agglomerated into chains with little branching. Similar chains have been observed in other flame systems. An example of particle chain size distribution is noted in Table 25-1. These chains can be used for investigating fiber diffusion, alignment, and aerodynamic size, especially in the transition regime (Kasper and Shaw 1983).

FIBER HEALTH EFFECTS

While asbestos fibers have many useful commercial properties, there has been much concern regarding their ability to cause disease. There are three primary diseases that have been attributed to asbestos fiber exposure: asbestosis (a fibrosis or scarring of the lung tissue), mesothelioma (a cancer of the pleura or peritoneum), and lung cancer. Other cancers, e.g., of the gastrointestinal tract, also have been attributed to asbestos exposure. In spite of an abundance of research into disease mechanisms of asbestos fibers, the etiologies of these diseases are still not well understood. Fiber shape appears to play a major role although other properties, such as fiber chemistry and solubility in body fluids, clearly, are also important.

Fiber diameter must play a role in disease, since the aerodynamic properties resulting in respiratory system deposition are strongly dependent on diameter (Timbrell 1982). In general, fibers must be smaller than about 2–3 μm diameter to reach the thoracic region and thinner still to reach the air exchange regions of the respiratory system (Stöber, Flachsbart, and Hochrainer 1970). It has been hypothesized that short fibers have much less disease potential because macrophages in the lung can engulf these particles and remove them from the lung with relative ease. Longer fibers cannot be completely engulfed by these cells and, therefore, tend to remain in the lung much longer (Holt 1987). Timbrell (1982) found that clearance occurred for fibers up to 17 μm long. Besides shape, the chemical properties of fibers also appear to play a role. Some fibers, especially glass, have been found to dissolve in lung tissue over an extended period, reducing their potential for disease (Johnson, Griffiths, and Hill 1984; Law, Bunn, and Hesterberg 1990). Chrysotile fibers longer than 5 μm were found to increase in number in lung tissue, apparently due to a longitudinal splitting of the fibers (Bellmann et al. 1986). *In vitro* studies have suggested that also the surface properties of fibers can affect cell toxicity (Light and Wei 1977).

An extrapolation of these properties indicates that long (especially those > 5 μm long), thin (< 3 μm diameter) insoluble fibers may have significant disease potential. Several researchers have postulated more specific size ranges as causing the various diseases (Pott 1978; Lippmann 1988; Timbrell 1990).

FIBER REGULATIONS

The potentially severe health effects of asbestos fiber exposure have prompted several regulatory and health research organizations to publish regulations and guidelines for controlling the airborne concentrations of asbestos fibers. Since the health-based data indicate that disease at current exposure concentrations is primarily related to fiber number, most regulatory air concentration measurements are based on asbestos fiber number concentration rather than on mass concentration. In the United States, for example, the Occupational Safety and Health Administration (OSHA) provides regulations for exposure to hazardous agents in industrial and other workplace settings. The OSHA regulations require that workers are not to be exposed to more than 0.2 asbestos fibers/cm^3 averaged over an 8 h period or more than 1.0 fiber/cm^3 over 30 min as measured using the filter collection/phase contrast microscope (PCM) method (Occupational Safety and Health Administration 1986). The Mine Safety and Health Administration (MSHA) regulates exposures in mines and mills, and limits miner exposure to 2.0 fibers/cm^3 for an 8 h average and 10 fibers/cm^3 for a 15 min average (Mine Safety and Health Administration 1988).

The Environmental Protection Agency (EPA) regulates the environmental levels of pollutants. Apart from prohibiting visible emissions, EPA has not implemented limits for environmental concentrations. However, to protect children from being exposed to asbestos in schools, the EPA has mandated procedures for removing and measuring asbestos in schools (Environmental Protection Agency 1987). The EPA has defined asbestos-containing material (ACM) as material containing more than 1% asbestos, to be measured using polarized light microscopy. After the removal of ACM in schools, the EPA requires that the airborne asbestos concentration in the cleaned area be no greater than that outside the area. The measurement is conducted using five air samples inside and five outside for the comparison. An analysis by transmission electron microscope (TEM) is required for monitoring the completion of all asbestos removal operations, except that the PCM can be used when removing small amounts of asbestos. Guidance documents describing the methods for controlling asbestos in buildings (Environmental Protection

Agency 1985*a*, 1990*a*) and for the measurement of asbestos after removal (Environmental Protection Agency 1985*b*) have been provided.

The Consumer Product Safety Commission (CPSC) provides guidance to manufacturers regarding the material content and potential hazards of commercial products. One such product to be targeted was hairdryers, prompting measurements of their emissions (Geraci et al. 1979). Individual state agencies set regulations that are often more stringent than those of the national agencies (Abbot 1990).

The National Institute for Occupational Safety and Health (NIOSH) recommends health standards to OSHA and MSHA. NIOSH (1990*a*) has indicated that there is no safe airborne concentration of any asbestos fibers. Based on this, NIOSH urged that the goal be to eliminate, or reduce to the lowest possible levels, exposures to asbestos fibers and recommended an exposure guideline of 0.1 fibers/cm^3, based on practical limitations of PCM measurements.

Regulation of other fibers, e.g., fibrous glass and mineral wool, has generally dealt with these materials as nuisance dust. However, this may change since fibers other than asbestos have been demonstrated to have disease potential in humans and animals. For instance, erionite (a fibrous zeolite) has been associated with human mesotheliomas (Baris 1980) and several man-made fibers have produced disease in animal exposure studies (Pott et al. 1987; Smith et al. 1987).

ASBESTOS TERMINOLOGY

Asbestos is a term applied to several commercial minerals exploited for their useful properties, largely due to their tendency to produce long, thin fibers. The development of asbestos crystal structure occurs primarily along one crystal axis. This results in a structure consisting of fibrils (the smallest-diameter fibers, 0.025–0.05 μm diameter) bundled together. When subjected to comminution, asbestos normally breaks down into particles (fibers) that have high aspect ratios both on macroscopic and microscopic scales; the mineralogical term for this condition is asbestiform (American Society for Testing Materials 1982). Other minerals with the same chemistry and crystal structure but without the unequal crystal development tend to produce more regular particles (termed cleavage fragments), though some of these particles may also be elongated. Although these cleavage fragments have a length/width distribution different from the asbestiform fibers (Virta et al. 1983), individual elongated cleavage fragments are often indistinguishable from asbestiform fibers when measured with commonly available techniques.

The asbestiform minerals that have been regulated include the following six: chrysotile (serpentine), amosite (cummingtonite–grunerite), crocidolite (glaucophane–riebeckite), tremolite asbestos, actinolite asbestos, and anthophyllite asbestos. The nonasbestiform mineral names for the first three materials are provided in parentheses. The latter three have the term asbestos attached because nonasbestiform varieties of these minerals have the same name. The latter five types of asbestos are classified as amphiboles. There are other asbestiform or fibrous minerals, but these are relatively rare and, except for attapulgite and wollastonite, have not been exploited commercially (Zumwalde and Dement 1977).

Health-related regulations, based on available microscope measurement methods, have specified asbestos fibers as particles with the elemental composition (from X-ray analysis) and crystal structure (from electron diffraction) appropriate to asbestos and with a length greater than 5 μm and an aspect ratio greater than 3:1 (Occupational Safety and Health Administration 1986; NIOSH 1990*a*). Thus, the health-related regulatory definitions of asbestos fibers are based on measurements that may include particles that are cleavage fragments and not necessarily asbestiform. This definition is in contrast to the mineralogical definition relating to the asbestiform crystal structure, which often produces particle distributions with higher mean aspect ratios (Kelse and Thompson 1990; Wylie 1979).

The term "asbestos structure" is used for reporting asbestos air concentration under the Asbestos Hazard Emergency Relief Act (AHERA) regulations (Environmental Protection Agency 1987). An asbestos structure is any particle (fiber, bundle, cluster, or matrix) consisting of or containing an observed fiber segment. This fiber segment must be longer than 0.5 μm, with an aspect ratio of 5:1 or greater, and be identified as asbestos by elemental analysis and electron diffraction. This definition is used to provide a sensitive method for assessing cleanliness after asbestos removal has been completed.

MEASUREMENT TECHNIQUES

There are two main classes of measurement techniques for fibers: microscopic observation of individual fibers and light-scattering-based instruments. Other instruments described in this book also can detect fiber aerosols, but, since they are not specific for fibers, will not be considered here. Microscopic techniques require the collection of samples, most often filter samples, that are returned to the laboratory for preparation and analysis by a microscopist. Four principal types of microscopes are used for fiber detection and analysis: the phase contrast light microscope (PCM), the polarized light microscope (PLM), the scanning electron microscope (SEM), and the transmission electron microscope (TEM). A review of microscope techniques for workplace and environmental asbestos measurements is given by Chatfield (1986). Descriptions of various light and electron microscopic techniques as well as pictures of many different types of fibers and particles are given in the seven-volume *Particle Atlas* (McCrone and Delly 1973).

Sample analysis can take place at various levels of complexity, e.g., counting fibers within prescribed size limits, determining the fiber size distribution, and measurement of the complete size distribution, as well as identifying individual fiber types. The first is usually applied to establish compliance with regulations and may require some qualitative analysis as well as a simple counting of fibers. The latter types of analysis are usually reserved for research studies or environmental assessments where the fiber sources are unknown.

Sample preparation is extremely important for microscopic analysis, since the view of the fibers and other particles is largely two-dimensional. Thus, the particles must be uniformly distributed at optimum concentration or loading over the sample surface. If the loading is too high, fibers and particles will overlap and be difficult to analyze; if the loading is too low, it will take too long to find a useful number of fibers (Iles and Johnston 1983; Peck, Serocki, and Dicker 1986).

The Environmental Protection Agency (1989) has evaluated asbestos sample preparation techniques. Most analyses of fibrous aerosol samples are performed on filter samples that are prepared without disturbing the location of the fibers on the filter surface (direct-transfer sample preparation). This approach has been taken because asbestos fibers can break up when suspended in a liquid, increasing the fiber number. When sampling other fiber types that do not break apart, or when sampling problems dictate the dilution of the collected sample, a liquid suspension and redeposition of the fibers may be performed (indirect-transfer sample preparation). Indirect sample preparation has the advantages of allowing the removal of some interfering particles as well as providing a more uniformly deposited and optimally loaded sample.

Bulk sample analysis by polarized light microscopy is often used in conjunction with air sampling to find potential sources of airborne fibers. X-ray diffraction (Abell 1984) and other bulk analysis techniques have also been used for asbestos and other fibers; however, these techniques are not discussed here since they are not specific for fibers and usually are not sufficiently sensitive for aerosol sample analysis.

Phase Contrast Microscope (PCM)

The phase contrast microscope (PCM) technique uses an optical microscope selected for the detection of small-diameter fibers. It is primarily used to measure fiber number concentrations, but generally cannot be used to distinguish between different types of fibers. PCM-based methods are inexpensive and readily available, lending themselves to measuring asbestos fiber concentrations in places known to contain asbestos fibers. PCM only provides an index of asbestos fiber concentration since it cannot be used to detect those fibers thinner than about 0.25 μm. The exact limit of detection is proportional to the difference in refractive index between the fiber and the surrounding medium (Rooker, Vaughan, and Le Guen 1982). For high refractive index fibers, such as crocidolite, fibers somewhat thinner than 0.25 μm may be detected (Kenny, Rood, and Blight 1987). While phase contrast optics improves the ability to detect thin fibers, it also reduces the resolution to approximately 0.6 μm diameter; i.e., thin fibers are detected but appear wider than they really are. Thus, the PCM is not useful for measuring the diameters of thin fibers.

Several fiber detection methods have been published based on the PCM (Asbestos International Association 1979; Carter, Taylor, and Baron 1989*a*; National Health and Medical Research Council 1976; Leidel et al. 1979; World Health Organization 1981).

Sampling

Current sampling methods for asbestos fibers employ a 25 mm sampling cassette with a 50 mm long cylindrical, conductive cowl attached to the inlet. The cowl is primarily intended to reduce contact contamination by personnel who might be wearing the sampler. The collection medium is a 0.45, 0.8, or 1.2 μm pore size, mixed cellulose ester, membrane filter. The smaller the pore size, the greater the pressure drop across the filter and the greater the fraction of fibers depositing on the top surface of the filter. Thus, 0.45 μm filters are used for high-volume area samples analyzed by TEM, while the others are used for personal samples (taken with a battery-powered pump) analyzed by PCM. Sampling periods are chosen to obtain optimum fiber loading on the filter, usually 100–1300 fiber/mm^2. Fiber loadings above this range result in undercounting, while samples below this range tend to result in overcounting (Cherrie, Jones, and Johnston 1986). Non-fibrous particulate present at high concentrations will obscure the fibers and cause undercounting.

Much of the sampling for asbestos fibers is performed to establish compliance with a regulatory standard or exposure guideline. The number of samples required to show compliance or noncompliance have been outlined and is based on the number of samples as well as on the confidence limits around each analysis result (Leidel, Busch, and Lynch 1977).

As with other types of sampling, unit (or 100%) sampling efficiency is needed to obtain samples that represent the environment. Since the sampler diameter and flow rate are constrained by method specifications and sample size requirements, respectively, sampling is rarely conducted isokinetically. For asbestos fibers, this is generally not a problem since most of the fibers posing a health risk have a small enough Stokes number that sampling efficiency is close to 100%. Sampling in stagnant air and at several air velocities, no differences in fiber concentration were found at flow rates up to 16 l/min (Johnston, Jones, and Vincent 1982). However, for fibers with larger median diameters (e.g., > 3 μm), such as glass, graphite, and cellulose fibers, significant undersampling or oversampling may occur under anisokinetic conditions. Measurements of < 5 μm aerodynamic diameter particles under anisokinetic conditions suggest that the 25 mm cowled sampler is accurate within about ± 10% at wind speeds up to 0.2 m/s, while at 0.5 m/s it may undersample or oversample as much as ± 25% (Fletcher et al. 1989).

Besides optimum loading, the filter deposit must be uniform, since only a small portion of the filter is observed during analysis. Various sampling forces, e.g., electrostatic, inertial, gravitational, can affect the trajectory of fibers entering the sampler and depositing on the filter. Anisokinetic sampling allows some fibers to impact or settle on the cowl surface and results in noticeable distortion of the filter deposit uniformity. Cornett et al. (1989)

reported that a significant portion of glass fibers is collected on the inner walls of the cowl and not on the filter. Breysse et al. (1990) developed a technique for determining the fraction of fibers on the cowl on the filter. Whether these cowl-deposited fibers represent "losses" and should be included in the analysis has yet to be established.

Measurements of asbestos fibers produced in industrial settings suggest that these fibers are more highly charged than particles with a compact shape and, thus, are more mobile in an electric field (Johnston, Vincent, and Jones 1985). The sampling cassette may carry a charge, in spite of being conductive, since it is electrically isolated in many sampling situations. An electric field produced by a charged cassette can produce biased sampling as well as a nonuniform deposition of fibers on the filter (Baron and Deye 1989*a*, *b*; Johnston, Vincent, and Jones 1985).

Sample Preparation

A direct sample preparation technique is used for PCM analysis. The filter is made optically clear and a liquid or resin with a refractive index close to that of the filter is used to fill the space between the filter and a glass cover slip. The cover slip is required by the design of the microscope objective. Several techniques have been used for clearing filters and surrounding the sampled particles in a medium of refractive index 1.48. The three most commonly used techniques for sample preparation in recent years are the dimethyl phthalate–diethyl oxalate (DMP–DEO) method (Leidel et al. 1979), the acetone–triacetin method (Carter, Taylor, and Baron 1989*b*), and the dimethyl formamide (DMF)–Euparal method (Le Guen and Galvin 1981).

Samples prepared by the DMP–DEO method are temporary, since the filter breaks up and the mixture crystallizes after about two days. The acetone–triacetin method has the advantage of providing samples lasting for approximately six months to two years (Shenton-Taylor and Ogden 1986), as well as being quick and easy to perform (Baron and Pickford 1986). The relative permanence of the samples makes it easier to perform sample recounts for quality assurance programs. This method also provides slightly better sample clarity than the DMP–DEO method. The DMF–Euparal prepared samples are stable for many years, since Euparal is a resin. Although having a slightly higher refractive index, these preparations also have the best clarity of the three (Le Guen and Galvin 1981). The DMF mixture has a higher toxicity than the other materials and the preparation should be conducted in a well-ventilated area (NIOSH 1990*b*).

Some materials, such as fibrous glass, may have a refractive index too close to that of the prepared filter to allow good contrast. An alternate preparation technique involves collapsing the filter and etching the surface with a low-temperature oxygen-plasma (Rendall and Schoeman 1985). Since the fibers are surrounded by air, resulting in a larger difference in the refractive index, higher and presumably more accurate counts can be obtained.

Fiber Counting Procedures

Typical counting rules are listed in NIOSH Method 7400 (Carter, Taylor, and Baron 1989*b*). Particles longer than 5 μm, with an aspect ratio greater than 3:1 are counted. Other commonly used counting rules differ slightly in that neither fibers with diameters greater than 3 μm nor fibers attached to particles with diameters greater than 3 μm are counted.

The graticule depicted in Fig. 25-6 defines the area observed in the microscope field in which fibers are counted (Walton and Beckett 1977). This graticule is adapted specifically for each microscope so that it will present a 100 μm diameter field to the microscopist. A fiber within the graticule field is counted as one, and a fiber with one end within the field is counted as one-half fiber; all others are not counted (see the examples in the figure). Fields are blindly chosen (to prevent biased field selection) from areas on the filter surface, generally along a radius of the filter. A minimum of 20 fields is evaluated. Measurement continues until 100 fibers have been counted,

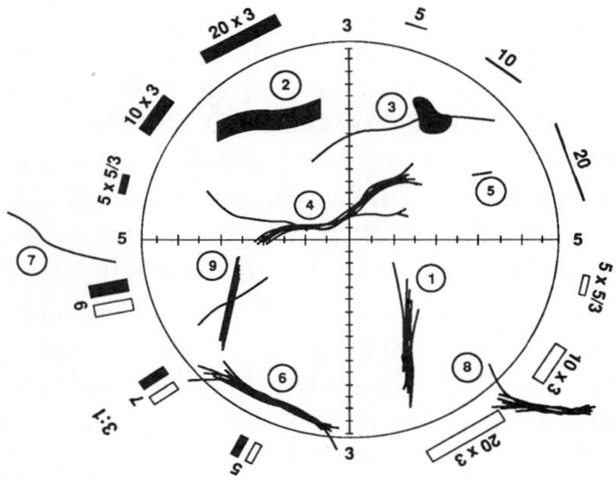

FIGURE 25-6. Walton–Beckett Light Microscope Graticule for Counting Fibers. Particles 1, 2, 3, and 4 are counted as one fiber; 5, 6, and 7 are not counted; 8 is counted as one-half fiber; and 9 is counted as two fibers.

or 100 fields have been evaluated, whichever comes first.

The fiber count observed on each filter sample is corrected for background counts performed on blank filters. The airborne concentration C (fibers/cm^3) is calculated, based on the number of fibers F observed in N microscope fields:

$$C = \frac{(F/N\,A_g)\,A_f}{1000\,V} \tag{25-11}$$

where each graticule field A_g has a nominal area of 7.85×10^{-3} mm^2, the sampling filter has a collection area A_f (approximately 385 mm^2 for a 25 mm diameter filter) and the sampled volume V is measured in liters. F is adjusted for blank filter counts.

An alternate approach to counting fibers, especially at low concentrations, is to count voids, that is, the number of fields that contain no fibers (Attfield and Beckett 1983). If the fibers are distributed on the filter according to the Poisson distribution (uniformly random), the number of void fields can be directly related to the fiber concentration per field (c_f) on the filter:

$$c_f = \ln\,(n/E) \tag{25-12}$$

where n is the number of fields observed and E is the number of void fields. Sahle and Larsson (1989) found that for lightly loaded samples, this approach is faster than conventional counting by almost a factor of two and also may be more precise.

Measurement Accuracy

The accuracy of various fiber counting techniques is poor when compared to other analytical methods. For instance, most methods in the NIOSH Manual of Analytical Methods state an overall accuracy (combined variability and bias) of better than 25% (NIOSH 1984). This requires that the variability of a method be less than about 12%. Under optimum analysis conditions (uniform sample deposit, no background dust interference, optimum loading), a relative standard deviation of 0.10 (or 10%) is predicted for the fiber counting method. Thus, the accuracy (including bias and variability) can be no better than this level.

In fact, other sources of variability and bias can occur. Some of these are due to the small sample size observed as part of the measurement procedure. For instance, fibers may be nonuniformly distributed in the sample due to inertial, electrostatic, or other sampling influences. Since the analysis assumes a uniform distribution on the filter, taking a small portion for microscopic analysis can, therefore, result in significant variations in the reported concentration. In addition, one microscopist may introduce biases relative to other microscopists due to decisions regarding which particles to count as fibers. When comparisons are made between groups of microscopists, these biases may appear as an increased variability in the overall results.

Comparisons of the PCM with other microscopic techniques have been made (Marconi, Menichini, and Paoletti 1984). Because of differences in sensitivity, resolution and type of illumination, different microscopic techniques will often produce different results. However, if the analysis of fibers by a high-resolution technique, e.g., electron microscopy, is limited to those fibers visible by PCM, good agreement can be obtained (Marconi, Menichini, and Paoletti 1984; Taylor et al. 1984).

The use of established analytical procedures for fiber count analysis is extremely important. This is the only way that results from one laboratory can be compared reliably with those from other laboratories. Microscopist training, proper equipment, and established quality control procedures are all important components of proper laboratory practice. To ensure a uniformity of application of these analytical procedures, both within-laboratory and interlaboratory sample exchanges are necessary (Abell, Shulman, and Baron 1989; Ogden et al. 1986). The usual technique for determining analytical biases, that is, the comparison with a reference method, does not work because an alternative fiber counting technique that measures the "true" fiber concentration does not exist. Thus, the final test of fiber counting accuracy is the comparison of one's results with those of a group of competent laboratories.

Several formal programs of sample exchange have been established for PCM. These include the American Industrial Hygiene Association's (AIHA) Proficiency Analytical Testing (PAT) program (Groff, Schlecht, and Shulman 1991), U.K.'s Regular Inter-Laboratory Counting Exchange (RICE) program (Crawford and Cowie 1984) and the International Asbestos Fibre Regular Interchange Counting Arrangement (AFRICA) program (Institute of Occupational Medicine, Edinburgh). For a laboratory performing PCM analyses to establish compliance with OSHA asbestos fiber exposure level regulations, regular sample exchanges with other laboratories are required.

The measurement of fiber concentrations for comparative use within a single study may not need all the components of a complete quality assurance scheme. For instance, if a study is intended only to provide relative fiber concentrations to show differences with time or location, interlaboratory exchange of samples may not be necessary. However, the use of published counting and sample preparation procedures, as well as the performance of blind repeat analyses, are important for establishing analytical confidence limits.

Application

The PCM is primarily used for fiber counting to provide an index of asbestos fiber exposure in workplaces in which asbestos is known to be present. The current U.S. regulations for determining airborne asbestos fiber exposure in industrial workplaces (Occupational Safety and Health Administration 1986), mines and mills (Mine Safety and Health Administration 1988), and, sometimes, for determining acceptable levels after asbestos abatement in schools (Environmental Protection Agency 1987) are based on PCM analysis.

Although the morphology observed under the PCM allows some discrimination between fiber types, it is not specific enough to allow positive identification of asbestos or other fibers. The PCM is used also for the measurement of man-made mineral fibers (MMMF), e.g., mineral wool and fibrous glass (NIOSH 1977). It can be used with other analytical techniques, such as PLM, SEM, or TEM, for specific fiber types. Such an approach allows compliance with regulations and guidelines as well as comparison with the epidemiological and toxicological data that have been obtained for various types of fiber exposure.

For instance, a TEM method, NIOSH Method 7402 (Carter, Taylor, and Baron 1989*b*), provides qualitative analysis for asbestos fibers. In this method, all fibers thicker than 0.25 μm (presumed optically visible) are analyzed by electron diffraction and energy dispersive X-ray analysis (EDXA). Rather than using the TEM count directly, the asbestos fiber fraction (relative to the total number of fibers) is applied to the PCM count on the same sample. This approach is taken because Taylor et al. (1984) found that the TEM count alone had higher variability than the combined TEM/PCM result.

Polarized Light Microscope (PLM)

PLM is often used to determine the percentage of asbestos or other fibres in bulk materials that can potentially release an aerosol. The EPA (1987) has defined asbestos-containing material (ACM) as material containing more than 1% asbestos. PLM permits the identification of fiber type by an observation of the fiber morphology, light diffraction, and optical rotation properties relative to the surrounding medium. Under carefully controlled conditions, an estimate of percentage of asbestos may be obtained.

Several PLM techniques are used for identifying fiber type as well as quantifying the percentage of fibrous material (usually asbestos) in a sample (Klinger et al. 1989; Middleton 1979; McCrone, McCrone, and Delly 1978). These techniques depend on the refractive index and other optical properties of individual particles. Many of these PLM techniques require a visual observation of color in the fiber and become less reliable for fibers thinner than about 1 μm (Vaughan, Rooker, and Le Guen 1981).

Sampling

Several procedures have been suggested for obtaining representative bulk samples of ACM in a fashion that prevents unnecessary exposure to asbestos aerosol (Jankovic 1985). The material should be wetted or sealed during sample removal. A small coring device, such as a cork borer, can be used to obtain a sample from the full depth of the material. At least three samples per 1000 ft^3 of ACM should be taken. The sample should be placed in a well-sealed, rugged container. Finally, the sampled area should be repaired or sealed to minimize further fiber release.

Sample Preparation and Analysis

Sample preparation for a PLM analysis involves grinding the material to the optimum particle size range (1–15 μm diameter) and dispersing the particles in a liquid of known refractive index on a glass slide. Some ACM, such as vinyl asbestos floor tiles, may require dissolution or ashing of the matrix material so that the fibers are separated and are visible in the microscope. Before and after preparation, the sample is observed with a stereomicroscope at 10–100× magnification to evaluate sample uniformity and observe whether fibrous material is present.

Some materials that interfere with accurate fiber identification either by their similarity or by covering up the fibers can be removed by physical treatment of the sample. For instance, organic materials, such as cellulose fibers or diesel soot can be removed by low-temperature oxygen-plasma ashing (Baron and Platek 1990). Leather fibers and chrysotile have a similar appearance and refractive index. The leather can be removed by ashing at 400°C (Churchyard and Copeland 1988).

Morphology of fibers can give some indication of fiber type. For instance, it assists in differentiation of chrysotile (curly fibers) from the amphibole asbestos (straight fibers, especially when shorter than 50 μm). Asbestos fibers often have frayed or split ends. Glass or mineral wool fibers are typically straight or slightly curved with fractured or sometimes bulbous ends; diameters are usually in the 5–15 μm range. Many plant fibers are flattened and twisted, with diameters 5–20 μm. Note that it is not recommended to base identification solely on morphology.

Crossed polarizing filters in the microscope can be used to indicate whether a fiber is isotropic (isometric or amorphous) or anisotropic (uniaxial or biaxial crystal structure). A measurement of the angle at which these fibers disappear (the extinction angles, a function of the crystal structure) narrows the possible identity of the fibers.

When a transparent fiber has a larger refractive index than its surrounding medium, the bright halo (Becke line) around that fiber appears to move into it as the microscope focus is raised; when the fiber has a smaller refractive index, the Becke line moves out of it. This technique can be used to bracket the fiber refractive index.

Dispersion, or refractive index change with wavelength, of a fiber can be used for identification. When particles are placed in a liquid whose dispersion is different from that of the particle, the particle may exhibit a color caused by the refraction of light. This technique requires the use of special "dispersion staining" optics. By using several refractive index liquids in series, the refractive index and the dispersion of the fiber can be established and compared with those of standard materials or published data (McCrone 1980).

Once the sample has been prepared in the appropriate refractive index liquid, specific fiber types, e.g., asbestos, can be identified and the percentage of fibers estimated. Two approaches are typically used: visual comparison with prepared reference slides or pictures and point counting. When attempting to estimate whether a material is ACM (i.e., $> 1\%$ asbestos), the visual comparison technique is adequate when more than about 10% of the particles observed are asbestos. Point counting is used for these lower concentration samples to provide higher accuracy (Environmental Protection Agency 1990*b*). It involves observing 400 randomly selected "points" (identified with a reticle crosshair) in the sample. The number of points containing asbestos is divided by the total number of points observed to give the percentage of asbestos. A combination of these approaches balances the analysis time and accuracy of the results (Webber et al. 1990).

PLM also can be used for qualitative analysis of air sample filters by collapsing the filter and using low-temperature plasma etching of the surface to expose the fibers. Various refractive index liquids can then be placed on the etched surface to surround the fibers, allowing techniques noted above to be used (Vaughan, Rooker, and Le Guen 1981). The smallest fibers that can be identified by this method are about 1 μm in diameter.

Accuracy

PLM analysis is primarily used for qualitative identification of fiber type. Accurate identification of asbestos and other fibers requires a proper training in the crystallographic properties of particles as well as training and familiarization with the PLM. As with fiber counting, a laboratory quality assurance program is necessary to ensure consistently accurate results. The National Voluntary Laboratory Accreditation Program (NVLAP) operated by the National Institute for Standards and Technology (NIST) inspects laboratories for proper practice as well as provides unknown samples four times a year to check their performance in fiber identification. Under a predecessor to this program, approximately 350 laboratories correctly classified 98.5% of the samples as asbestos and correctly identified the specific asbestos types in approximately 97% of the samples. A blind test of 51 laboratories resulted in 97.5% correct classifications and 79.1% correct identifications (Parris and Starner 1986).

PLM has been cast in a quantitative measurement role by the EPA requirement of determining whether a school building material is ACM. Many variables, including particle size, density, and shape, are not adequately controlled or measured in the analysis and contribute to errors in the percentage mass estimate. Thus, PLM analysis is at best a semi-quantitative technique. Performance evaluation of laboratories' ability to estimate the percentage of asbestos is also planned under the NVLAP program.

Scanning Electron Microscope (SEM)

The SEM is manufactured in a variety of models, ranging from inexpensive devices that have relatively low resolution and can only display particle images, to more sophisticated devices that rival the TEM in resolution and can also provide elemental analysis of particles. Thus, the SEM provides analyses intermediate between those of the light microscope and the TEM, giving elemental analysis and higher-resolution images than the light microscope. Under routine fiber counting conditions, fibers thinner than about 0.1 μm are not generally visible with the SEM, nor can the internal structure of fibers be seen. Additionally, a definitive determination of fiber type by electron diffraction is not available with the SEM (Middleton 1982; Steen et al. 1983). However, in the "photographic mode", a high-quality SEM can be used to detect small fibers quite effectively. The photographs can be used to measure fiber size distributions (Platek et al. 1985).

Sample Collection and Preparation

A sample that is to be used directly in the SEM must have the fibers exposed on the sampling substrate. Samples are typically either collected on the smooth-surfaced polycarbonate capillary pore filter or deposited on a smooth substrate by impaction or electrostatic precipitation. Samples that are taken with a membrane filter can also be used if the filter is collapsed and the filter material is etched away by low-temperature plasma ashing. The sample must generally be coated with a conductive layer to reduce charge accumulation from the electron beam in the SEM. Gold and carbon are common coating materials. Gold allows a thinner layer to be effective, resulting in somewhat higher resolution, but interferes with EDXA of the fibers. Once a sample is coated, it can be placed directly in the SEM for analysis. Relatively large samples, each up to several cm in diameter, can be analyzed.

Application

For the most part, the SEM has been used for fiber detection and analysis as an intermediate between TEM and PCM analysis. Generally, the SEM has not been recommended for measuring asbestos, especially in environmental situations, because of the inability to positively identify fibers by electron diffraction as well as the lack of instrument standardization. For occupational exposures, where the fibers present are generally of a known type, PCM has been preferred as the less expensive and more available method. A SEM method was developed by the Asbestos International Association (1984). A SEM method was recommended by the World Health Organization for sizing man-made mineral fibers (World Health Organization 1985).

However, the SEM does permit the microscopist (1) to obtain higher resolution and sensitivity than the light microscope, (2) to obtain qualitative information about fiber type from elemental analysis, and (3) to reduce sample preparation and analysis costs vs. the TEM. In research studies of known fiber types when analysis time is not important, photographic images from the SEM can be analyzed to obtain size distributions including all fiber sizes (Platek et al. 1985). The SEM field of view is larger than that for the TEM and, thus, long fibers may be sized more accurately.

Transmission Electron Microscopy (TEM)

TEM allows the most definitive analysis of individual fibers: particle shape can be observed and measured, elemental composition can be determined by EDXA, and crystal structure can be deduced from electron diffraction patterns (Langer, Mackler, and Pooley 1974). TEM has sufficiently high resolution and sensitivity to allow observation of the smallest fiber sizes. However, TEM analysis is the most expensive one described in this chapter because of instrument cost and complexity, high operator expertise, and complex sample preparation. Furthermore, quantitative fiber concentration measurements have relatively poor reproducibility (Environmental Protection Agency 1985*b*).

Sampling

Currently, post-asbestos abatement clearance monitoring is perhaps the most common use of TEM analysis. The measurement is intended to indicate that a location is acceptably clean so that it can be reoccupied. The EPA (1987) AHERA regulations require the use of 25 mm conductively cowled cassettes with 0.45 μm pore size, mixed cellulose ester filters. Polycarbonate capillary pore filters also have been used in the past; however, they have fallen into disfavor because of erratic contamination of the filter medium with asbestos fibers (Powers 1986).

Sampling requirements for TEM analysis are similar to those of other microscopic analyses, although the high magnification afforded by TEM allows a somewhat higher filter loading than that for PCM. For direct-transfer sample analysis, the sample must be optimally loaded, the sampling efficiency should be as close to 100% as possible, and the filter deposit uniform. For indirect-transfer sample analysis, only the sampling efficiency is important. Mixed cellulose ester filters are commonly used, but may not be optimal because short, thin fibers can deposit deep within the filter and not be completely transferred to the TEM sample grid. Other filter materials have also been used, e.g., vinyl copolymer. Note that vinyl and polycarbonate filters tend to retain more nonuniform surface charges than cellulosic filters because of their lower conductivity. These charges may cause nonuniform particle deposition.

The AHERA regulations also require the use of "aggressive sampling". This is a procedure in which high-velocity air blowers (e.g., leaf blowers) are used to release dust particles from the surfaces in a room so that they can be sampled. At the same time, fans keep the dust suspended and mixed with the air. An ac line operated pump is used to sample at flow rates up to 10 l/min. The sampling cassette is placed at a height of 1.5–2 m from the floor in a downward-facing position. Sampling times are typically 2–10 h.

Sample Preparation

Sample preparation, always an important part of any microscopy, is especially important for the TEM. The potentially high levels of magnification dictate that the sample be relatively small (the grid holding the sample is 3 mm diameter), thin (the electron beam must be able to penetrate the particles to perform electron diffraction), and evenly dispersed (the grid should accurately represent the entire sample). Two general approaches have been taken in preparing samples: direct transfer (the carbon film contains the particles in the same location and orientation as when they were originally deposited on the filter or substrate) and indirect transfer (collected particles are dispersed as a liquid suspension and redeposited on a filter for grid preparation) (Environmental Protection Agency 1989). Many of the following procedures are described in the EPA (1987) AHERA method.

As with other measurement techniques, the purpose of the analysis is important in determining the approach to take. Sometimes, the fibers may be fragile (agglomerate chains, asbestos fibers) and minimal manipulation of the sample is necessary to prevent breakup. In such cases, direct transfer of the particulate to the electron microscope grid is preferred. The direct-transfer technique may have difficulty including all the particles collected on a filter (especially membrane filters) in the carbon membrane/grid sample due to varying particle thickness and collection below the filter surface. This technique is the one most widely adopted for environmental asbestos measurements.

For other analyses where number concentration is not important or the fibers are more sturdy, an indirect-transfer technique may have advantages. The collected sample is resuspended in a liquid, mixed (usually by ultrasonication) and an appropriate-sized aliquot is redeposited on a capillary pore filter for preparation of the carbon film. The indirect-transfer method homogenizes the sample and ensures that the final preparation has the appropriate concentration as well as good uniformity. This approach may cause an apparent increase in particle concentration because resuspension tends to separate particles, especially fibers, that were attached in the aerosol phase. One study found 15.5 times as many fibers with the indirect approach as with the direct approach (Huang and Wang 1983). However, it has been suggested that this increase is primarily due to shorter (< 1 μm) fibers being separated from larger particles (Chatfield 1983) and that the number of long (> 5 μm) fibers is not changed significantly as long as the ultrasonication is gentle (Chatfield 1986).

Complete separation of fibers into fibrils by extensive ultrasonication was used as a technique for asbestos fiber mass determination since it eliminated the variability of fiber diameter, gave a more uniform dispersion of fibers in the sample, and increased the number of fibers available for counting (Selikoff, Nicholson, and Langer 1972).

Figure 25-7 provides an indication of the major steps involved in the direct-transfer preparation of the TEM sample. The membrane filter is collapsed using vapors of a solvent such as acetone. It has been suggested that acetone vapor collapse of the filter causes

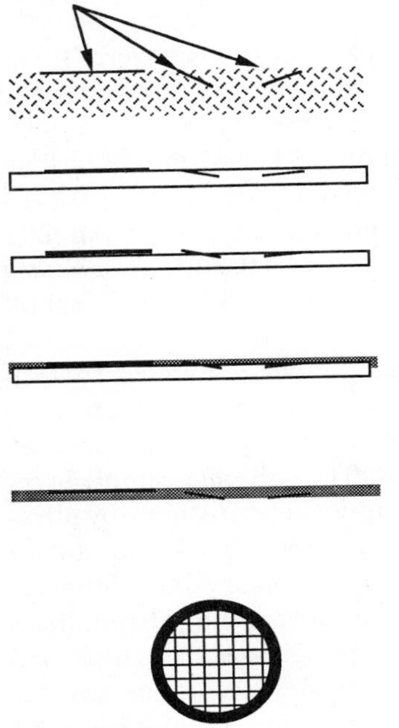

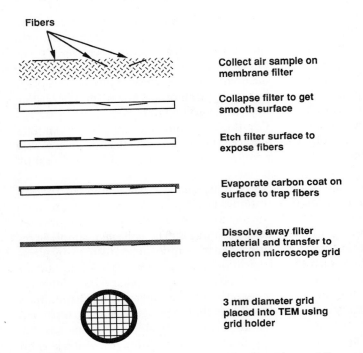

FIGURE 25-7. Principal Steps Used in Transmission Electron Microscope Sample Preparation by the Direct Transfer Method.

fibers to be washed into the body of the filter. Thus, the carbon film does not capture all these fibers for transfer to the electron microscope grid (Chatfield 1986). For this reason, the DMF technique described above for PCM samples, combined with the use of a 0.1 μm pore size filter, may provide an improved TEM sample (Burdett and Rood 1983; Chatfield 1986). To expose more of the fibers, especially those that may lie below the top surface of the collapsed filter, a low-temperature ashing step is often performed. Carbon is then vaporized onto the surface, generally from a range of angles to produce a uniform coating. The carbon film must be of the proper thickness to provide a stable film on the grid; thinner films will crack and break apart while thicker films prevent the electron beam from providing good contrast images of the fibers. Thicker films may also obscure the electron diffraction patterns of very small fibers.

A copper grid is placed on the carbon-coated surface and the filter is dissolved away using an appropriate solvent. This step is important to complete because incomplete removal of the filter degrades imaging by the TEM severely. The removal must also be gentle to prevent a breakup of the carbon film. Complete filter removal may take 4–24 h.

Several grid preparations are generally made to ensure that enough grid openings have the carbon film intact to provide an adequate area for analysis. Observing more than one grid also reduces the probability of biased results due to local variations of fiber deposition on the filter.

Analysis

The TEM operates by focusing a high-energy (10–1000 keV) electron beam on an area of the sample. The beam passes through the sample, which is usually supported on a grid by a thin, nearly transparent, carbon film. By controlling the magnetic focusing optics, the image of the sample can be observed on a phosphorescent viewing screen at various magnifications ranging from about 200 × to 1,000,000 × . For asbestos fibril detection,

10,000–20,000 × magnification is used. Particles absorb more of the electron beam energy than the supporting carbon film and appear as dark objects on a light background.

Particle shape is often an important indicator of the type or origin of fiber. Larger asbestiform fibers can usually be distinguished by their fibrillar structure, especially at the ends where the fibrils tend to splay apart. Chrysotile fibrils, when observed at high magnification, usually display a tubular structure. These various shape characteristics are often sufficient when the presence of certain fiber types has been established. However, in samples of unknown source or composition, further qualitative analysis of each fiber is usually necessary.

Besides observing particle shape, the TEM allows the use of electron diffraction, which indicates crystal structure, and EDXA, which indicates the elemental composition of particles. Diffraction and elemental composition patterns from reference materials are used to confirm the presence of specific types of fibers. It often happens that for a given fiber, the diffraction pattern and/or the elemental composition pattern may not be completely detected or may change over the length of the fiber. This may be due to fiber thickness, interfering materials, variations in fiber composition or twisting of the crystallographic axis along the fiber length. Thus, a full confirmation of the identities of all fibers is often not possible.

Applications

The TEM is a microchemical instrument that can be used for analyzing collected samples of aerosols as well as bulk samples. It can provide the most definitive qualitative information regarding fiber type as well as quantitative information on fiber number. In addition, the fiber dimensions, composition, and number concentration information can be used to estimate the mass concentration.

Accuracy

A detailed study by Steel and Small (1985) of parameters affecting the TEM analysis accuracy showed that instrument capabilities, including mechanical stage, imaging and contrast, and diffraction, can produce order-of-magnitude errors if they are not up to the demands of asbestos analysis. By a careful comparison between several counters and microscopes using well-characterized chrysotile samples, it was also found that a careful and experienced analyst could attain accuracies of 90% in finding all asbestos fibrils in a sample. The accuracy degraded significantly for fibrils shorter than 1 μm and analysts had a 50% or less chance of finding those shorter than 0.5 μm. In an interlaboratory comparison on prepared grid samples, Turner and Steel (1991) found that the average result reported by the laboratories was 0.67 of the NIST-verified count. Forty percent of the fibers shorter than 1 μm were missed; half as many longer than 1 μm were missed.

Repeat counts by a single analyst of laboratory-generated amosite fiber samples gave a relative standard deviation of 0.27 (Taylor et al. 1984). The interlaboratory relative standard deviation for repeat counts on the same sample has been estimated to be approximately 1.0 (Environmental Protection Agency 1985*b*).

Recently, a laboratory accreditation program has been established by the National Institute for Standards and Technology under the National Voluntary Laboratory Accreditation Program (NVLAP). Data regarding laboratory performance in this program have not yet been published. Quality assurance guidelines for laboratories participating in this program are available (Berner et al. 1990).

AUTOMATED FIBER ANALYSIS TECHNIQUES

IMAGE ANALYSIS

There have been several attempts to improve the accuracy and speed and to reduce the subjectivity of microscope analysis by using automated image analysis of asbestos fiber samples (Whisnant 1975). Automated image analysis involves taking the image from a microscope, digitizing the image, and using a computer to evaluate the number and size of objects present in the image. For fiber analysis, this may involve counting the number of fibers present or obtaining a size distribution of the fibers.

Fiber counting has been achieved with reasonable success for PCM analysis of asbestos fibers. The Manchester Asbestos Program (MAP) was developed at the Manchester University (U.K.) with support from the Health and Safety Executive (Kenny 1984). The MAP (Applied Imaging, Tyne & Wear, U.K.) operates in a semiautomated mode, with the analyst selecting fields and focusing the microscope.

Evaluations of this program indicated that it provided a good correlation with manual counting (by microscopist) in most cases. The precision of the results from MAP was better than that produced by manual counting (Baron and Shulman 1987; Kenny 1988). Some difficulties the program had included breaking fiber images apart and counting them as more than one, not detecting very thin fibers, detecting edges of large particles as fibers, and detecting chains of particles as fibers. A second version of the program was developed to deal with samples in which most of the particles were not asbestos, such as samples from abatement sites (Kenny 1988). The MAP program is used as the reference for a U.K. quality assurance program (Ogden et al. 1986).

Many image analysis systems have available software for counting larger, well-defined fibers. For instance, they work reasonably well for fibrous glass insulation or synthetic organic textile fibers. Such programs will not work as well as the MAP for asbestos fiber images because of the difficulty of detecting fibers of various diameter and curvature, some barely visible, some overlapping, in the presence of a noisy background containing particles of different shapes and sizes. However, with increased computer power and improved image analysis techniques, more accurate automated fiber counting may be possible in the future.

Image analysis systems are integrated with some SEMs, since it is possible to apply direct computer control of the electron beam. Such systems work well for compact particles and are capable of determining size distribution and elemental analysis of about 200 particles per hour (Stettler, Platek, and Groth 1983). A system based on principles similar to those of the MAP for fiber detection was developed for an SEM (Stott and Meranger 1984) but has not been commercialized.

An image analysis system can be attached to a TEM by attaching an imaging camera below the observation screen. The image can then be acquired by the computer system and analyzed in much the same way as an optical image. However, many image processing and analysis functions are integrally tied to the magnification of the image. Therefore, any change in the magnification may require extensive reprogramming of the software.

Direct-Reading Fiber Measurement

An optical particle counter has been used for the detection of chrysotile fibers in a textile plant where asbestos was the primary aerosol contaminant (Rickards 1978). Other direct-reading monitors also may be used in situations where the fibers are the major constituent in the aerosol, as in laboratory studies.

A fibrous aerosol monitor was developed for a more specific measurement of asbestos fibers in the presence of compact particles (Lilienfeld, Elterman, and Baron 1979). A commercial version of the Fibrous Aerosol Monitor (FAM-1, MIE Inc., Bedford, MA) was produced and has undergone continuous improvement, primarily in ruggedness for field use, over the past ten years. A new version with computer control has recently been produced (FM-7400, MIE Inc., Bedford, MA). This instrument detects fibers via a combined electrostatic alignment/optical scattering technique. Fibers are aligned in a plane perpendicular to a laser light beam and then oscillated back and forth within this plane. These fibers will scatter the light primarily in a direction perpendicular to the fiber's principal axis. Thus, as the fibers oscillate, pulses of light are observed at a photomultiplier detector, which views a direction at right angles both to the laser beam and to the fiber axis. These pulses can be related to fiber size and shape. See Chapter 17 for a further discussion of the FAM-1 operation principles.

The FAM-1 depends on light scattering patterns for the detection of fibers and is, therefore, not specific for fiber type, e.g., asbestos. The instrument is calibrated with asbestos fibers and compared with filter samples counted using NIOSH Method 7400. Several evaluations show that data from the FAM-1 correlate reasonably well with PCM-based measurements (Lilienfeld 1986), though one study found that the FAM-1 was not much better for monitoring fibers in workplaces than an optical particle counter (Iles and Shenton-Taylor 1986). Some instrumental problems noted in that study may have been due to reliability problems associated with early production instruments. Currently, the FAM-1 is frequently used to monitor control systems that are installed to reduce asbestos exposures at asbestos abatement sites.

Another direct-reading fiber monitor (Hygenius, Inc. Mississauga, Ont.) has recently been marketed that also uses electrostatic alignment and light scattering detection principles. However, the light scattering patterns are detected with a solid state imaging array rather than a photomultiplier.

OTHER MEASUREMENT TECHNIQUES

An asbestos fiber analysis system, the M-88 Fiber Analyzer (Vickers Instruments, York), was developed based on magnetic alignment/light scattering detection of asbestos fibers on filter samples. Fibers were floated from the filter surface with a solvent, magnetically aligned, and then trapped in an aligned position when the solvent evaporated. The integrated scattering from a large portion of the filter was analyzed and compared to reference standards to give a quantitative indication of fiber count. Evaluations of this instrument found that the calibration varied significantly with fiber source and size distribution (Jones and Gale 1982; Abell, Molina, and Shulman 1984). However, even these observed correlations between instrument response and fiber concentration disappeared at low filter loadings (Verrill 1982). Production of the instrument subsequently ceased.

Asbestos and other fibers have been stained with fluorescent dyes to enhance their visibility in a light microscope (Benarie 1983). While this technique may be used as a rapid screening technique, it is not sufficiently specific for asbestos to provide unequivocal identification of fiber type.

References

Abbot, S. H. 1990. State regulatory watch. Asbestos Issues 3(12):16–25.

Abell, M. T. 1984. Chrysotile asbestos. Method 9000. 2/15/84. In NIOSH Manual of Analytical Methods. Cincinnati, OH: DHHS/NIOSH.

Abell, M. T., D. Molina, and S. Shulman. 1984. *Laboratory Evaluation of the M-88 Rapid Fibre Counter*. Division of Physical Sciences and Engineering, National Institute for Occupational Safety and Health Internal Report.

Abell, M. T., S. A. Shulman, and P. A. Baron. 1989. The quality of fiber count data. *Appl. Ind. Hyg.* 4(11):273–85.

American Society for Testing Materials 1982. *Definitions for Asbestos and Other Health-Related Silicates*. ASTM STP 834:1–213. Philadelphia: American Society for Testing Materials.

Asbestos International Association. 1979. Airborne Asbestos Fiber Concentrations at Workplaces by Light Microscopy *(Membrane Filter Method)*. AIA Health and Safety Publication RTM1. Paris: Asbestos International Association.

Asbestos International Association. 1984. *Method for the Determination of Airborne Asbestos Fibres by Scanning Electron Microscopy*. AIA Health and Safety Publication RTM2. Paris: Asbestos International Association.

Asgharian, B. and C. P. Yu. 1989. A simplified model of interceptional deposition of fibers at airway bifurcations. *Aerosol Sci. Technol.* 11:80–88.

Asgharian, B., C. P. Yu, and L. Gradon. 1988. Diffusion of fibers in a tubular flow. *Aerosol Sci. Technol.* 9(3):213–19.

Attfield, M. D. and S. T. Beckett. 1983. Void counting in assessing membrane filter samples of asbestos fibres. *Ann. Occup. Hyg.* 27:273–82.

Balshazy, I., T. B. Martonen, and W. Hofmann. 1990. Fiber deposition in airway bifurcations. *J. Aerosol Med.* 3(4):243–60.

Baris, Y. 1980. The clinical and radiological aspects of 185 cases of malignant pleural mesothelioma. In *Biological Effects of Mineral Fibres, Vol. 2*, ed. J. C. Wagner, pp. 937–47. IARC Scientific Publications, No. 30. Lyon: International Agency for Research on Cancer.

Baron, P. A. and G. J. Deye. 1987. Generation of replicate asbestos aerosol samples for quality assurance. *Appl. Ind. Hyg.* 2(3):114–18.

Baron, P. A. and G. J. Deye. 1989*a*. Electrostatic effects in asbestos sampling I: Experimental measurements. *Am. Ind. Hyg. Assoc. J.* 51:51–62.

Baron, P. A. and G. J. Deye. 1989*b*. Electrostatic effects in asbestos sampling II: Comparison of theory and experiment. *Amer. Ind. Hyg. Assoc. J.* 51:63–69.

Baron, P. A. and G. C. Pickford. 1986. An asbestos sample filter clearing procedure. *Applied Ind. Hyg.* 1(4):169–71.

Baron, P. A. and S. F. Platek. 1990. NIOSH method 7402-asbestos fibers (Revision #1)—Low temperature ashing of filter samples. *Amer. Ind. Hyg. Assoc. J.* 51:A730–31.

Baron, P. A. and S. A. Shulman. 1987. Evaluation of the Magiscan image analyzer for asbestos fiber counting. *Am. Ind. Hyg. Assoc. J.* 48(1):39–46.

Bellmann, B. H. Konig, H. Mühle, and F. Pott. 1986. Chemical durability of asbestos and of man-made mineral fibers *in vivo*. *J. Aerosol Sci.* 17:341–45.

Benarie, M. 1983. Identification of asbestos fibers by fluorochrome staining. In *Aerosols in the Mining and Industrial Work Environment*. Ann Arbor: Ann Arbor Science.

Berner, T., E. Chatfield, J. Chesson, and J. Rench. 1990. *Transmission Electron Microscopy Asbestos Laboratories: Quality Assurance Guidelines*. EPA/5-90-002.

Boeniger, M., M. Hawkins, P. Marsin, and R. Newman. 1988. Occupational exposure to silicate fibres and PAHs during sugar-cane harvesting. *Ann. Occup. Hyg.* 32(2):153–69.

Breysse, P. N., C. H. Rice, P. Auborg, M. J. Komoroski, M. Kalinowski, R. Versen, J. Woodson, R. Carlton, and P. L. S. Lees. 1990. Cowl rinsing procedure for airborne fiber sampling. *Appl. Ind. Hyg.* 5(9):619–22.

Burdett, G. J. and A. P. Rood. 1983. A membrane-filter, direct-tranfer technique for the analysis of asbestos fibers or other inorganic particles by transmission electron microscopy. *Environ. Sci. Technol.* 17(11):643–48.

Burke, W. A. and N. A. Esmen. 1978. The inertial behavior of fibers. *Am. Ind. Hyg. Assoc. J.* 39:400–5.

Carpenter, R. L., J. A. Pickrell, B. V. Mokler, H. C. Yeh, and P. B. DeNee. 1981. Generation of respirable glass fiber aerosols using a fluidized bed aerosol generator. *Am. Ind. Hyg. Assoc. J.* 42:777–84.

Carter, J., D. Taylor, and P. A. Baron. 1989*a*. Asbestos fibers, Method 7402 revision no. 1: 5/15/89. In *NIOSH Manual of Analytical Methods*. Cincinnati, OH: DHHS/NIOSH.

Carter, J., D. Taylor, and P. A. Baron. 1989*b*. Fibers, Method 7400 revision no. 3:5/15/89. In *NIOSH Manual of Analytical Methods*. Cincinnati, OH: DHHS/NIOSH.

Chan, H. K. and I. Gonda. 1989. Aerodynamic properties of elongated particles of chromoglycic acid. *J. Aerosol Sci.* 20(2):157–68.

Chatfield, E. J. 1983. Measurement of asbestos fibre concentrations in ambient atmospheres. Mississauga, Ontario, Canada: Ontario Research Foundation.

Chatfield, E. J. 1986. Asbestos measurements in workplaces and ambient atmospheres. In *Electron Microscopy in Forensic, Occupational and Environmental Health Sciences*. New York: Plenum.

Chen, Y. K. and C. P. Yu. 1990. Sedimentation of charged fibers from a two-dimensional channel flow. *Aerosol Sci. Technol.* 12:786–92.

Cheng, Y.-S. 1986. Bivariate lognormal distribution for characterizing asbestos fiber aerosols. *Aerosol Sci. Technol.* 5:359–68.

Cheng, M. T., G. W. Xie, M. Yang, and D. T. Shaw. 1991. Experimental characterization of chain-aggregate aerosol by electrooptic scattering. *Aerosol Sci. Technol.* 14:74–81.

Cherrie, J., A. D. Jones, and A. M. Johnston. 1986. The influence of fiber density on the assessment of fiber concentration using the membrane filter method. *Am. Ind. Hyg. Assoc. J.* 47:465–74.

Chissick, S. S. and R. Derricott (eds.) 1983. *Asbestos: Properties, Applications and Hazards, Vol. 2*. Chichester: Wiley.

Churchyard, M. P. and G. K. E. Copeland. 1988. Is it really chrysotile? *Ann. Occup. Hyg.* 32(4):545–47.

Clift, R., J. R. Grace, and M. E. Weber. 1978. *Bubbles, Drops and Particles*. New York: Academic Press.

Cornett, M. J., C. H. Rice, V. Herzberg, and J. Lockey. 1989. Assessment of fiber deposition on the conductive cowl in the refractory ceramic fiber industry. *Appl. Ind. Hyg.* 4(8):201–4.

Cox, R. G. 1970. The motion of long slender bodies in a viscous fluid I: General theory. *J. Fluid Mech.* 44(4):791–810.

Crawford, N. P. and A. J. Cowie. 1984. Quality control of asbestos fibre counts in the United Kingdom—the present position. *Ann. Occup. Hyg.* 28:391–98.

Dahneke, B. E. 1973*a*. Slip correction factors for nonspherical bodies—I. Introduction and continuum flow. *J. Aerosol Sci.* 4:139–45.

Dahneke, B. E. 1973*b*. Slip correction factors for nonspherical bodies—II. Free molecule flow. *J. Aerosol Sci.* 4:147–61.

Dahneke, B. E. 1973*c*. Slip correction factors for nonspherical bodies—III. The form of the general law. *J. Aerosol Sci.* 4:163–70.

Dahneke, B. E. 1982. Viscous resistance of straight-chain aggregates of uniform spheres. *Aerosol Sci. Technol.* 1:179–85.

Dement, J. 1990. Overview: Workshop on fiber toxicology research needs. *Environ. Health Perspec.* 88:261–68.

Environmental Protection Agency. 1985*a*. *Guidance for*

Controlling Asbestos-Containing Materials in Buildings. EPA 560/5-85-024.

Environmental Protection Agency. 1985*b*. *Measuring Airborne Asbestos Following an Abatement Action.* EPA EPA 600/4-85-049.

Environmental Protection Agency. 1987. *Asbestos-Containing Materials in Schools. Federal Register.* 40 CFR Part 763. Washington, DC: Government Printing Office.

Environmental Protection Agency. 1989. *Comparison of Airborne Asbestos Levels Determined by Transmission Electron Microscopy (TEM) Using Direct and Indirect Transfer Techniques.* EPA 560/5-89-004

Environmental Protection Agency. 1990*a*. *Managing Asbestos in Place: A Building Owners Guide to Operations and Maintenance Programs for Asbestos Containing Materials.* EPA 20T-2003.

Environmental Protection Agency. 1990*b*. National emission standards for hazardous air pollutants; asbestos NESHAP revision; final rule. 20 November 1990. 40 CFR Part 61. Washington, DC: Government Printing Office.

Fairchild, C. I., L. W. Ortiz, H. J. Ettinger, and M. I. Tillery. 1976. *Aerosol Research and Development Related to Health Hazard Analysis.* Los Alamos National Laboratory Progress Report LA-6277-PR.

Fairchild, C. I., L. W. Ortiz, M. I. Tillery, and H. J. Ettinger. 1978. *Aerosol Research and Development Related to Health Hazard Analysis.* Los Alamos National Laboratory Progress Report LA-7380-PR.

Fletcher, R. A., E. B. Steel, M. Beard, C. C. Wang, and J. W. Gentry. 1989. Uniformity of particle deposition for indoor air sampling under anisokinetic conditions. *J. Aerosol Sci.* 20(8):1593–96.

Fu, T.-H., M.-T. Cheng, and D. Shaw. 1990. Filtration of chain aggregate aerosols by model screen filter. *Aerosol Sci. Technol.* 13:151–61.

Fuchs, N. A. 1964. *The Mechanics of Aerosols.* Oxford: Pergamon.

Gallily, I. 1971. On the drag experienced by a spheroidal, small particle in a gravitational and electrostatic field. *J. Colloid Interface Sci.* 36(3):325–39.

Gallily, I. and A. D. Eisner. 1979. On the orderly nature of the motion of nonspherical aerosol particles I. Deposition from a laminar flow. *J. Colloid Interface Sci.* 68(2):320–37.

Gallily, I., D. Schiby, A. H. Cohen, W. Hollönder, D. Schless, and W. Stöber. 1986. On the inertial separation of nonspherical aerosol particles from laminar flows. I. The cylindrical case. *Aerosol Sci. Technol.* 5:267–86.

Gentry, J. W., K. R. Spurny, J. Schörmann, and H. Opiela. 1983. Measurement of the diffusion coefficient of asbestos fibers. In *Aerosols in the Mining and Industrial Work Environments.* Ann Arbor: Ann Arbor Science Publishers.

Gentry, J. W., K. R. Spurny, and S. A. Soulen. 1988. Measurements of the diffusion coefficients of ultrathin asbestos fibers. *J. Aerosol Sci.* 19(7):1041–44.

Geraci, C. L., P. A. Baron, J. W. Carter, and D. L. Smith. 1979. Testing of hair dryer emissions. Report of Interagency Agreement NIOSH IA-79-29. National Institute for Occupational Safety and Health, Cincinnati, OH.

Gonda, I. and A. F. A. E. Khalik. 1985. On the calculation of aerodynamic diameters of fibers. *Aerosol Sci. Technol.* 4:233.

Griffis, L. C., J. A. Pickrell, R. L. Carpenter, R. K. Wolff, S. J. Allen, and K. Y. Yerkes. 1983. Deposition of crocidolite asbestos and glass microfibers inhaled by the beagle dog. *Am. Ind. Hyg. Assoc. J.* 44:216–22.

Griffiths, W. D., L. C. Kenny, and S. T. Chase. 1985. The electrostatic separation of fibres and compact particles. *Ann. Occup. Hyg.* 16(3):229–43.

Griffiths, W. D. and N. P. Vaughan. 1986. The aerodynamic behaviour of cylindrical and spheroidal particles when settling under gravity. *J. Aerosol Sci.* 17(1):53–65.

Groff, J. H., P. C. Schlecht, and S. Shulman. 1991. Laboratory reports and rating criteria for the proficiency analytical testing (PAT) program. DHHS (NIOSH) Publication No. 91-102. Cincinnati, OH: NIOSH.

Holst, E. and T. Schneider. 1985. Fibre size characterization and size analysis using general and bivariate log-normal distributions. *J. Aerosol Sci.* 5:407–13.

Holt, P. F. 1987. *Dust and Disease.* New York: Wiley.

Huang, C. Y. and Z. M. Wang. 1983. Comparison of methods of assessing asbestos fiber concentrations. *Arch. Environ. Health* 38(1):5–10.

Hwang, C. Y. and G. W. Gibbs. 1981. The dimensions of airborne asbestos fibres—I. Crocidolite from Kuruman area, Cape Province, South Africa. *Ann. Occup. Hyg.* 24(1):23–41.

IARC. 1988. IARC Monographs of the evaluation of the carcionogenic risk of chemicals in humans, Vol. 43, *Man-Made Mineral Fibers and Radon*, pp. 33–171. Lyon, France: International Agency for Research on Cancer.

Iles, P. J. 1990. Size selection of fibres by cyclone and horizontal elutriator. *J. Aerosol Sci.* 21(6):745–60.

Iles, P. J. and A. M. Johnston. 1983. Problems of asbestos fibre counting in the presence of fibre–fibre and particle–fibre overlap. *Ann. Occup. Hyg.* 27(4):389–403.

Iles, P. J. and T. Shenton-Taylor. 1986. Comparison of a fibrous aerosol monitor (FAM) with the membrane filter method for measuring airborne asbestos concentrations. *Ann. Occup. Hyg.* 30(1):77–87.

Jankovic, J. T. 1985. Asbestos bulk sampling procedure. *Am. Ind. Hyg. Assoc. J.* 46(2):B8–9.

Johnson, N. F., D. M. Griffiths, and R. J. Hill. 1984. Size distribution following long term inhalation of MMMF. In *Biological Effects of Man-Made Mineral Fibers, Vol. 2*, pp. 102–25. Copenhagen: World Health Organization.

Johnston, A. M., A. D. Jones, and J. H. Vincent. 1982. The influence of external aerodynamic factors on the measurement of the airborne concentration of asbestos fibres by the membrane filter method. *Ann. Occup. Hyg.* 25(3):309–16.

Johnston, A. M., J. H. Vincent, and A. D. Jones. 1985. Measurements of electric charge for workplace aerosols. *Ann. Occup. Hyg.* 29:271–84.

Jones, A. D. and R. W. Gale. 1982. Industrial trials with the Vickers M88 rapid asbestos fibre counter. *Ann Occup. Hyg.* 25(1):39–51.

Jones, A. D., A. M. Johnston, and J. H. Vincent. 1983. Static electrification of airborne asbestos dust. In *Aerosols in the Mining and Industrial Work Environment.* Ann Arbor: Ann Arbor Science.

Kasper, G. 1982. Dynamics and measurement of smokes. *Aerosol Sci. Technol.* 1:187–99.

Kasper, G. and D. T. Shaw. 1983. Comparative size distribution measurements on chain aggregates. *Aerosol Sci. Technol.* 2:369–81.

Kasper, G., S. N. Shon, and D. T. Shaw. 1980. Controlled formation of chain aggregates from very small metal oxide particles. *Am. Ind. Hyg. Assoc. J.* 41:288–96.

Kelse, J. W. and C. S. Thompson. 1990. The regulatory and mineralogical definitions of asbestos and their impact on amphibole dust analysis. *Am. Ind. Hyg. Assoc. J.* 50:613–22.

Kenny, L. C. 1984. Asbestos fibre counting by image analysis—The performance of the Manchester Asbestos Program on Magiscan. *Ann. Occup. Hyg.* 28:401–15.

Kenny, L. C. 1988. Automated analysis of asbestos clearance samples. *Ann. Occup. Hyg.* 32(1):115–28.

Kenny, L. C., A. P. Rood, and B. J. N. Blight. 1987. A direct measurement of the visibility of amosite asbestos fibres by phase contrast optical microscopy. *Ann. Occup. Hyg.* 31(2):261–64.

Kerker, M. 1969. *The Scattering of Light and Other Electromagnetic Radiation.* New York: Academic Press.

Klinger, P. A., K. R. Nicholson, F. J. Hearl, and J. T. Jankovic. 1989. Asbestos (bulk). Method 9002. 5/15/89. In *NIOSH Manual of Analytical Methods.* Cincinnati, OH: DHHS/NIOSH.

Laframboise, J. G. and J.-S. Chang. 1977. Theory of charge deposition on charged aerosol particles of arbitrary shape. *J. Aerosol Sci.* 8:331–38.

Langer, A. M., A. D. Mackler, and F. D. Pooley. 1974. Electron microscopical investigation of asbestos fibers. *Environ. Health Perspect.* 9:63–80.

Law, B. D., W. B. Bunn, and T. W. Hesterberg. 1990. Solubility of polymeric organic fibers and manmade vitreous fibers in Gambles solution. *Inhal. Toxicol.* 2:321–39.

Le Guen, J. M. and S. Galvin. 1981. Clearing and mounting techniques for the evaluation of asbestos fibres by the membrane filter method. *Ann. Occup. Hyg.* 24(3):273–80.

Leidel, N. A., S. G. Bayer, R. D. Zumwalde, and K. A. Busch. 1979. USPHS/NIOSH membrane filter method for evaluating airborne asbestos fibers. DHEW (NIOSH) Publication No. 79-127. Washington, DC: U.S. Government Printing Office.

Leidel, N. A., K. A. Busch, and J. R. Lynch. 1977. *Occupational Exposure Sampling Strategy Manual.* DHEW (NIOSH) Publication No. 77-173.

Light, W. G. and E. T. Wei. 1977. Surface charge and asbestos toxicity. *Nature* 265:537–39.

Lilienfeld, P. 1985. Rotational electrodynamics of airborne fibers. *J. Aerosol Sci.* 16(4):315–22.

Lilienfeld, P. 1986. Low concentration airborne asbestos monitoring with the GCA FAM-1. In *Aerosols: Formation and Reactivity*, pp. 1020–23. Berlin: Pergamon Journals.

Lilienfeld, P., P. Elterman, and P. Baron. 1979. Development of a prototype fibrous aerosol monitor. *Am. Ind. Hyg. Assoc. J.* 40(4):270–82.

Lipowicz, P. J. and H. C. Yeh. 1989. Fiber dielectrophoresis. *Aerosol Sci. Technol.* 11:206–12.

Lippmann, M. 1988. Asbestos exposure indices. *Environ. Res.* 46:86–106.

Marconi, A., E. Menichini, and L. Paoletti. 1984. A comparison of light microscopy and transmission electron microscopy results in the evaluation of the occupational exposure to airborne chrysotile fibres. *Ann. Occup. Hyg.* 28:321–31.

Martonen, T. B. and D. L. Johnson. 1990. Aerodynamic classification of fibers with aerosol centrifuges. *Part. Sci. Technol.* 8:37–53.

McCrone, W. C. 1980. *The Asbestos Particle Atlas.* Ann Arbor: Ann Arbor Science.

McCrone, W. C. and J. G. Delly. 1973. *The Particle Atlas*, 2nd edn., 7 Volumes. Ann Arbor: Ann Arbor Science.

McCrone, W. C., L. B. McCrone, and J. G. Delly. 1978. *Polarized Light Microscopy.* Ann Arbor: Ann Arbor Science.

Michaels, L. and S. S. Chissick (eds). 1979. *Asbestos: Properties, Applications and Hazards.* Chichester: Wiley.

Middleton, A. P. 1979. The identification of asbestos in solid materials. In *Asbestos: Properties, Applications and Hazards*, eds. L. Michaels and S. S. Chissick. Chichester: Wiley.

Middleton, A. P. 1982. Visibility of fine fibres of asbestos during routine electron microscopical analysis. *Ann. Occup. Hyg.* 25(1):53–62.

Mine Safety and Health Administration. 1988. *Exposure Limits for Airborne Contaminants.* CRF Part 56.5001. Washington, DC: U.S. Government Printing Office.

National Health and Medical Research Council. 1976. *Membrane Filter Method for Estimating Airborne Asbestos Dust.* Australian Department of Health.

NIOSH. 1976*a*. Revised recommended asbestos standard. DHEW (NIOSH) Publication No. 77-169. Cincinnati: National Institute for Occupational Safety and Health.

NIOSH. 1976*b*. Occupational exposure to fibrous glass: Proceedings of a symposium. DHEW (NIOSH) Publication No. 76-151. Washington, DC: US Government Printing Office.

NIOSH. 1977. Criteria for a recommended standard: Occupational exposure to fibrous glass. DHEW (NIOSH) Publication No. 77-152. Washington, DC: US Government Printing Office.

NIOSH. 1984. NIOSH Manual of Analytical Methods. 3rd edn., 2 Vol. DHHS (NIOSH) Publication No. 84-100. Cincinnati: National Institute for Occupational Safety and Health.

NIOSH. 1990*a*. Testimony of NIOSH on occupational exposure to asbestos, tremolite, anthophyllite and actinolite. 29CFR Parts 1910 and 1926, 9 May 1990.

NIOSH. 1990*b*. Preventing adverse health effects from exposure to dimethylformamide (DMF). DHHS (NIOSH) Publication No. 90-105. Cincinnati: National Institute for Occupational Safety and Health.

Occupational Safety and Health Administration. 1986. *Occupational Exposure to Asbestos, Tremolite, Anthophyllite, and Actinolite Asbestos; Final Rules*. 29 CFR Part 1910.1001 and 1926. Washington, DC: U.S. Government Printing Office.

Ogden, T. L., T. Shenton-Taylor, J. W. Cherrie, N. P. Crawford, S. Moorcroft, M. J. Duggan, P. A. Jackson, and R. D. Treble. 1986. Within-laboratory quality control of asbestos counting. *Ann. Occup. Hyg.* 30(4):411–25.

Parris, M. L. and K. Starner. 1986. *Asbestos Containing Materials in School Buildings: Bulk Sample Analysis Quality Assurance Program*. EPA/600/4-86/028.

Peck, A. S., J. J. Serocki, and L. C. Dicker. 1986. Sample density and the quantitative capabilities of PCM analysis for the measurement of airborne asbestos. *Am. Ind. Hyg. Assoc. J.* 47(4):A230–34.

Pinkerton, K. E., A. R. Brody, D. A. McLaurin, B. Adkins Jr., R. W. O'Connor, P. C. Pratt, and J. D. Crapo. 1983. Characterization of three types of chrysotile asbestos after aerosolization. *Environ. Res.* 31:32–53.

Platek, S. F., D. H. Groth, C. E. Ulrich, L. E. Stettler, M. S. Finnell, and M. Stoll. 1985. Chronic inhalation of short asbestos fibers. *Fund. Appl. Tox.* 5:327–40.

Pott, F. 1978. Some aspects on the dosimetry of the carcinogenic potency of asbestos and other fibrous dusts. *Staub-Reinhalt. Luft.* 38:486.

Pott, F., U. Ziem, F. J. Reiffer, F. Huth, H. Ernst, and U. Mohr. 1987. Carcinogenicity studies of fibers, metal compounds and some other dusts in rats. *Exp. Pathology* 32:129–52.

Powers, T. J. 1986. Filter blank contamination in asbestos abatement monitoring procedures: Proceedings of a peer review workshop, 24–25 April 1986. Cincinnati: Environmental Protection Agency.

Prodi, V., T. D. Zaiacomo, D. Hochrainer, and K. Spurny. 1982. Fibre collection and measurement with the inertial spectrometer. *J. Aerosol Sci.* 13:49–58.

Rajhans, G. S. and J. L. Sullivan. 1981. *Asbestos Sampling and Analysis*. Ann Arbor: Ann Arbor Science Publishers.

Rendall, R. E. G. and J. J. Schoeman. 1985. A membrane filter technique for glass fibres. *Ann. Occup. Hyg.* 29(1):101–8.

Rickards, A. L. 1978. The routine monitoring of airborne asbestos in an occupational environment. *Ann. Occup. Hyg.* 21(3):315–22.

Rood, A. P. 1988. Size distribution of airborne ceramic fibres as determined by transmission electron microscopy. *Ann. Occup. Hyg.* 32(2):237–40.

Rood, A. P. and R. M. Scott. 1989. Size distributions of chrysotile asbestos in a friction products factory as determined by transmission electron microscopy. *Ann. Occup. Hyg.* 33(4):583–90.

Rooker, S. J., N. P. Vaughan, and J. M. Le Guen. 1982. On the visibility of fibers by phase contrast microscopy. *Am. Ind. Hyg. Assoc. J.* 43:505–15.

Sahle, W. and G. Larsson. 1989. The usefulness of void-counting for fibre concentration estimation by optical phase contrast microscopy. *Ann. Occup. Hyg.* 33(1):97–111.

Schneider, T., E. Holst, and J. Skotte. 1983. Size distribution of airborne fibres generated from man-made mineral fiber products. *Ann. Occup. Hyg.* 27(2):157–71.

Selikoff, I. J. and E. C. Hammond. 1979. *Health Hazards of Asbestos Exposure*. New York: New York Academy of Sciences.

Selikoff, I. J., W. J. Nicholson, and A. M. Langer. 1972. Asbestos air pollution. *Arch. Environ. Health* 25:1–13.

Shenton-Taylor, T. and T. L. Ogden. 1986. Permanence of membrane filter clearing and mounting methods for asbestos measurement. *Microscope* 34:161–72.

Smith, D. M., L. W. Ortiz, R. F. Archuleta, and N. F. Johnson. 1987. Long-term health effects in hamsters and rats exposed chronically to man-made vitreous fibers. *Ann. Occup. Hyg.* 31(4B):731–54.

Spurny, K. R. 1980. Fiber generation and length classification. In *Generation of Aerosols and Facilities for Exposure Experiments*, pp. 257–98. Ann Arbor: Ann Arbor Science.

Spurny, K., C. Boose, and D. Hochrainer. 1975. Zerstäubung von asbestfasern in einem fliessbett-aerosolgenerator. *Staub-Reinhalt. Luft.* 35(12):440–45.

Steel, E. B. and J. A. Small. 1985. Accuracy of transmission electron microscopy for the analysis of asbestos in ambient environments. *Anal. Chem.* 57:209–13.

Steen, D., M. P. Guillemin, P. Buffat, and G. Litzistorf. 1983. Determination of asbestos fibres in air: Transmission electron microscopy as a reference method. *Atmos. Environ.* 17(11):2285–97.

Stettler, L. E., S. F. Platek, and D. H. Groth. 1983. Particle analysis by scanning electron microscopy/energy dispersive X-ray/image analysis. In *Aerosols in the Mining and Industrial Work Environments, Vol. 3*. Ann Arbor: Ann Arbor Science.

Stöber, W. 1972. Dynamic shape factors of nonspherical aerosol particles. In *Assessment of Airborne Particles*, eds. T. T. Mercer, P. E. Morrow, and W. Stöber, pp. 249–89. Springfield, IL: C. C. Thomas.

Stöber, W., H. Flachsbart, and D. Hochrainer. 1970. The aerodynamic diameter of latex aggregates and asbestos fibers. *Staub-Reinhalt. Luft.* 30(7):1–12.

Stott, W. R. and J. C. Meranger. 1984. Automated fiber

counting in the scanning electron microscope. *Scanning Electron Microsc.* 1984(II):583–88.

Sussman, R. G., J. M. Gearhart, and M. Lippmann. 1985. A variable feed rate mechanism for fluidized bed generators. *Am. Ind. Hyg. Assoc. J.* 46:24–27.

Tanaka, I. and T. Akiyama. 1984. A new dust generator for inhalation toxicity studies. *Ann. Occup. Hyg.* 28(2):157–62.

Tanaka, I. and T. Akiyama. 1987. Fibrous particles generator for inhalation toxicity studies. *Ann. Occup. Hyg.* 31(3):401–3.

Taylor, D. G., P. A. Baron, S. A. Shulman, and J. W. Carter. 1984. Identification and counting of asbestos fibers. *Am. Ind. Hyg. Assoc. J.* 45: 84–88.

Timbrell, V. 1972. Alignment of carbon and other man-made fibers by magnetic fields. *J. Appl. Phys.* 43(11):4839–40.

Timbrell, V. 1973. Desired characteristics of fibres for biological experiments. In *Fibres for Biological Experiments.* ed. P. V. Pelnar, p. 89. Montreal: Institute of Occupational and Environmental Health.

Timbrell, V. 1974. Aerodynamic considerations and other aspects of glass fiber. In *Occupational Exposure to Fibrous Glass.* DHEW Publication No. (NIOSH) 76-151, pp. 33–50.

Timbrell, V. 1975. Alignment of respirable asbestos fibres by magnetic fields. *Ann Occup. Hyg.* 18:299–311.

Timbrell, V. 1982. Deposition and retention of fibres in the human lung. *Ann. Occup. Hyg.* 26:347–69.

Timbrell, V. 1990. Review of the significance of fibre size in fibre-related lung disease: A centrifuge cell for preparing accurate microscope-evaluation specimens from slurries used in inoculation studies. *Ann. Occup. Hyg.* 33:483–505.

Timbrell, V., J. C. Gilson, and I. Webster. 1968. Preparation of UICC reference asbestos materials. *Int. J. Cancer.* 3:406.

Timbrell, V., A. W. Hyett, and J. W. Skidmore. 1968. A simple dispenser for generating dust clouds from standard reference samples of asbestos. *Ann. Occup. Hyg.* 11:273–81.

Tolles, W. M., R. A. Sanders, and G. W. Fritz. 1974. Dielectric response of anisotropic polarized particles observed with microwaves: A new method of characterizing properties of nonspherical particles in suspension. *J. Appl. Phys.* 45(9):3777–83.

Turner, S. and E. B. Steel. 1991. Accuracy of transmission electron microscopy analysis of asbestos on filters: Interlaboratory study. *Anal. Chem.* 63:868–72.

Van de Hulst, H. C. 1957. *Light Scattering by Small Particles.* New York: Wiley.

Vaughan, N. P. 1990. The generation of monodisperse fibres of caffeine. *J. Aerosol Sci.* 21(3):453–62.

Vaughan, N. P., S. J. Rooker, and J. M. Le Guen. 1981. *In situ* identification of asbestos fibres collected on membrane filters for counting. *Ann. Occup. Hyg.* 24(3):281–90.

Verrill, K. J. 1983. Vickers Instruments M88 rapid asbestos fiber monitor. (Letter to the Editor) *Ann. Occup. Hyg.* 27(1):111.

Virta, R. L., K. B. Shedd, A. G. Wylie, and J. G. Snyder. 1983. Size and shape characteristics of amphibole asbestos (amosite) and amphibole cleavage fragments (actinolite, cummingtonite) collected on occupational air monitoring filters. In *Aerosols in the Mining and Industrial Work Environments.* Ann Arbor: Ann Arbor Science.

Walton, W. H. 1982. The nature, hazards, and assessment of occupational exposure to airborne asbestos dust: A review. *Ann. Occup. Hyg.* 25:115–247.

Walton, W. H. and S. T. Beckett. 1977. A microscope eyepiece graticule for the evaluation of fibrous dusts. *Ann. Occup. Hyg.* 20:19–23.

Webber, J. S., R. J. Janulis, L. J. Carhart, and M. B. Gillespie. 1990. Quantitating asbestos content in friable bulk sample: Development of a stratified point-counting method. *Am. Ind. Hyg. Assoc. J.* 51(8):447–52.

Weiss, M. A., A.-H. Cohen, and I. Gallily. 1978. On the stochastic nature of the motion of nonspherical aerosol particles. II. The overall drift angle in sedimentation. *J. Aerosol Sci.* 9:527–41.

Wen, H. Y., G. P. Reischl, and G. Kasper. 1984. Bipolar diffusion charging of fibrous aerosol particles—I. Charging theory. *J. Aerosol Sci.* 15(2):89–101.

Whisnant, R. A. 1975. Evaluation of image analysis equipment applied to asbestos fiber counting. Contract Report 210-75-0080/5. Cincinnati: U.S. Department of Health Education and Welfare, National Institute for Occupational Safety and Health.

World Health Organization. 1981. *Methods of Monitoring and Evaluating Airborne Man-Made Mineral Fibres.* EURO Reports and Studies 48. Copenhagen: World Health Organization.

World Health Organization. 1985. *Reference Methods for Measuring Airborne Man-Made Mineral Fibres (MMMF).* Copenhagen: World Health Organization.

Wylie, A. G. 1979. Fiber length and aspect ratio of some selected asbestos samples. In *Health Hazards of Asbestos Exposure.* New York: Annals of the New York Academy of Science.

Yu, P. Y., C. C. Wang, and J. W. Gentry. 1987. Experimental measurement of the rate of unipolar charging of actinolite fibers. *J. Aerosol Sci.* 18(1):73–85.

Zumwalde, R. D. and J. M. Dement. 1977. *Review and Evaluation of Analytical Methods for Environmental Studies of Fibrous Particulate Exposures.* DHEW (NIOSH) Publication No. 77-204.

26

Mine Aerosol Measurement

B. K. Cantrell, K. L. Williams, W. F. Watts, Jr., and R. A. Jankowski

U.S. Department of the Interior, Bureau of Mines[1]
Washington, DC, U.S.A.

INTRODUCTION

Exposure to mineral aerosol is an occupational health hazard in mining and mineral processing industries because of the risk of developing pneumoconiosis. Agricola (1556) described this hazard for metal mining in Carpathia. He described shortness of breath and consumption, conditions which are now associated with asthma and emphysema. The mortality of miners under these conditions was such that some women in the area had lost seven husbands to "consumption". Ramazzini (1964) associated such conditions with the presence of small particles of stone or sand in the lung. By the end of the 19th century, several respiratory diseases were known to affect miners, including silicosis and coal workers' pneumoconiosis (Fletcher 1948; Seaton et al. 1981). By the mid-20th century, it was clear that the risk of simple pneumoconiosis is associated with a miner's cumulative exposure to mine aerosol in the respirable range and that prevention lies in reducing exposure through regulation (Seaton 1986).

In the United States, the regulation of mine worker exposure to airborne mine dust is the responsibility of the United States Department of Labor, Mine Safety and Health Administration (MSHA). MSHA receives technical assistance from the Department of the Interior, Bureau of Mines (BOM) and the National Institute for Occupational Safety and Health (NIOSH). NIOSH also provides recommended exposure limits (RELs) for contaminants to MSHA. MSHA's regulation of mines includes the establishment of permissible exposure limits (PELs) based on these RELs, approval of mine operator dust control plans designed to meet the PELs, periodic mine inspections to determine if mines are complying with their dust plans, and a respirable dust measurement program to provide samples for comparison with the PELs (MSHA 1991).

This chapter summarizes aerosol measurement technology currently used in the U.S. mining industry. Because of the regulatory aspect, the application of aerosol measurement technology has evolved in two areas: compliance measurements, which support regulatory programs, and research measurements, which aid in the development of new compliance measurement techniques

1. References made in this article to specific products do not imply endorsement by the Bureau of Mines.

and assist in determining the fundamental properties of mine aerosol.

MINE AEROSOL SOURCES

Before discussing measurement techniques used in the mining industry, a digression is necessary to explore the sources and characteristics of mine aerosol as it occurs in the underground environment. This focuses more on aerosol sources in underground coal mines because they are more numerous than metal and nonmetal mines by approximately six to one.

The Coal Mining Process

There are two types of underground bituminous coal mine plans used in the United States: room-and-pillar, used with continuous mining systems and conventional mining; and longwall mining. In the room-and-pillar plan, narrow entries, typically 5 to 6 m wide, are driven into the coal seam and interconnected at 24 to 60 m intervals by crosscut entries (Fig. 26-1). Coal pillars are left in place to support the mine roof, but may be partially removed prior to abandoning that section of the mine. In longwall operations a large block or panel of coal 150 by 1500 m is usually developed with room-and-pillar entries on all sides Fig. 26-2. The entire panel is then extracted using a longwall mining system, with the mine roof allowed to collapse following mining, forming a "GOB" area. Each of these plans uses different equipment and procedures and thus represents different sources of respirable dust generation.

The conventional mining process consists of the following procedures: (1) undercutting a 3 m deep slot across the bottom of the coal face to allow the coal to break easily when blasted; (2) drilling holes into the coal face for the explosives; (3) loading the explosives into

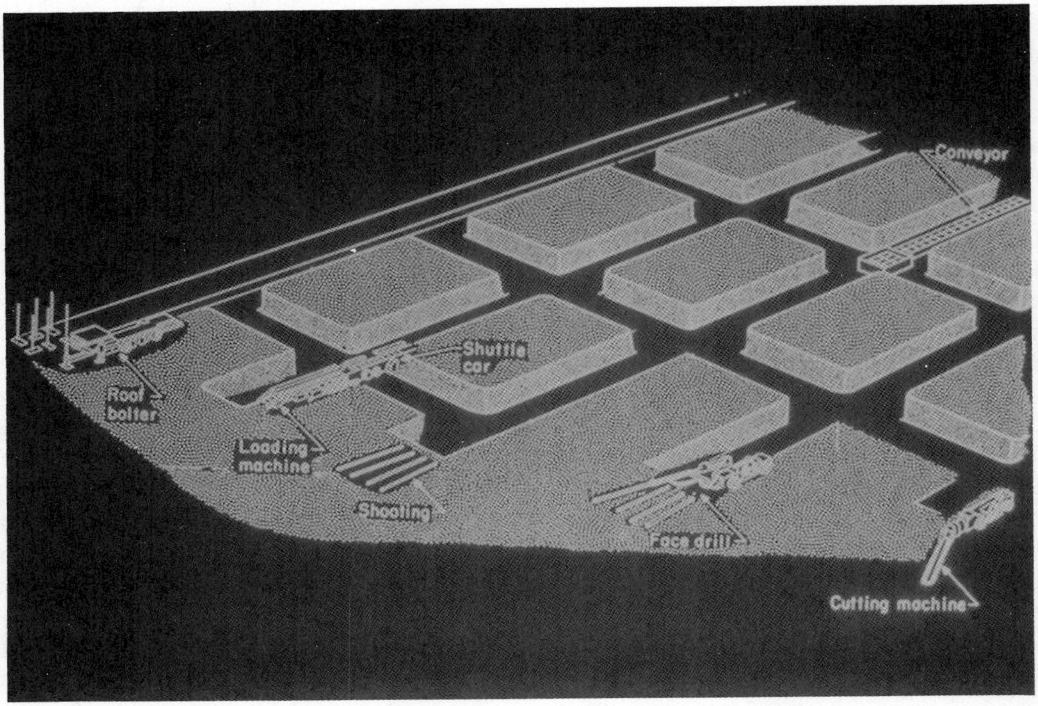

FIGURE 26-1. Typical Room and Pillar Mine Plan Showing a Conventional Mining Operation. (Source: Bureau of Mines.)

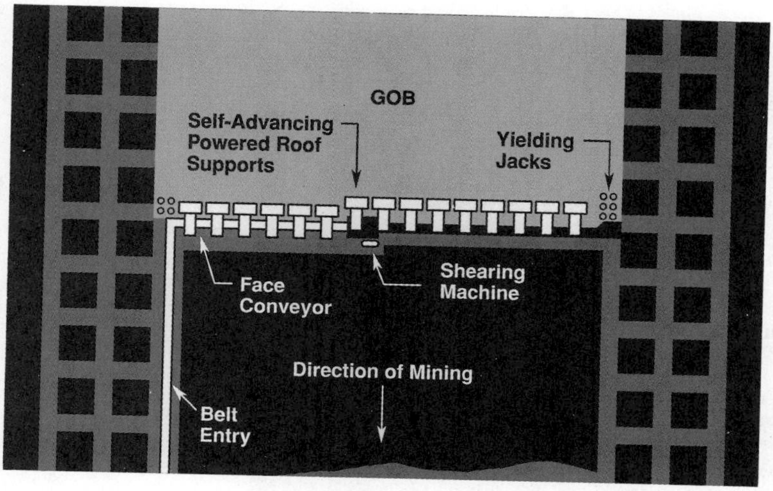

FIGURE 26-2. Typical Longwall Mine Plan Showing a Top View of the Mining Operation. (Source: Bureau of Mines.)

FIGURE 26-3. Continuous Miner. (Source: Bureau of Mines.)

the holes and blasting down the coal face; (4) using a loading machine to transfer the blasted coal into a haulage vehicle, shuttle car, or a bridge conveyor system; (5) transporting the coal from the face area to material transfer facilities used to remove coal from the mine, usually a belt transport system; and, (6) installing roof bolts, timbers, or other roof control measures in the entry just mined. This mining process accounts for approximately 7% of current underground coal production.

In continuous mining, a continuous mining machine, shown in Fig. 26-3, cuts coal from the face and loads it into the shuttle cars, thus replacing the first four procedures listed above. The shuttle cars transport the coal from the face area to the material transfer system. The continuous mining machine is

FIGURE 26-4. Longwall Shear. (Source: Bureau of Mines.)

then moved to a new entry while roof bolts or timbers are installed to control the mine roof in the entry just mined. The continuous mining method is the most prevalent underground coal extraction process, accounting for 60% of underground coal mined (Organiscak 1989).

Longwall mining is the most efficient and productive of the coal extraction methods. Coal is typically removed from a longwall panel in 1 m wide slices, using a double-drum shearing machine. The shearer, shown in Fig. 26-4, consists of two cutting heads, mounted on ranging arms at each end of a frame body. The shearer is self-propelled along a track parallel to the coal face. The cut coal is conveyed by the drums onto a chain conveyor running parallel to the shearer track. The chain conveyor transports the coal from the face area to the material transfer system. The immediate roof over the chain conveyor and shearing machine is supported by large hydraulic jacks and shields, each approximately $1\frac{1}{2}$ m wide. These supports are installed flush against one another, along the entire length of the longwall face. After the shearing machine passes, the supports are advanced, allowing the roof to collapse. Longwall mining currently accounts for approximately 33% of U.S. underground coal production (Jankowski and Hake 1989).

Dust Sources in Underground Coal Mines

Each extraction method generates respirable dust aerosol from a variety of sources. Concentrations of respirable dust aerosol in the mine environment are dependent on a number of factors. These include ventilation, dust controls such as sprays, the type of coal, and the rate at which coal is removed. The mechanisms for removing coal from a seam in continuous and longwall mining are similar; however, the removal rate and energy used in longwall mining are higher. The energy of coal removal by blasting, associated with conventional mining, is lower than that associated with either continuous or longwall mining. Concentrations of respirable dust measured on longwalls are the highest, while the conventional method generates the lowest levels.

Sources of respirable dust generation during the conventional extraction process include: coal transport, drilling, cutting, blasting, and loading (Jankowski and Hake 1989). Here, blasting accounts for 34% and loading for 46% of the dust, while cutting and drilling account for the remainder. Sources of respirable dust generation during the continuous mining process include coal transport, cutting, and drilling. The continuous mining machine accounts for 73% of the dust and roof bolting accounts for 25%. Sources of respirable dust generation during the longwall mining process include coal transport out of the mine, face coal transport, support movement, and shearer cutting. Shearer cutting and fragment reduction by the conveyor belt loader account for 81%, support movement for 13%, and face coal transport for 6% of the dust. Dust sources common to all methods of underground coal extraction include conveyor belts, belt transfer points, and haulage roads.

Metal and Nonmetal Mining Processes

The conventional mining procedures described in the previous section are used in over 90% of metal and nonmetal mines (Organiscak 1989). Some minerals, such as trona and potash, which are soft or occur in strata having a high methane content, may be mined using continuous mining techniques but these are in the minority. Sources of aerosols in mining activities in the face area include: drilling, continuous-mining machine operations, conventional mining methods, etc.; materials handling procedures such as the use of fragment reduction equipment on conveyor belts; and diesel-powered equipment necessary for material transfer tasks. Diesel-powered haulage vehicles are almost universally used in noncoal mining operations. Secondary sources of aerosol include dust entrainment during load–haul–dump operations and dust reentrainment from floor and walls caused by mine traffic.

Diesel Exhaust Aerosol

In addition to being exposed to dust, a miner working in an underground mine with diesel-powered equipment is exposed to a wide array of pollutants from the diesel exhaust. These include CO, CO_2, NO, NO_2, SO_2, diesel exhaust aerosol, and a variety of aerosol-associated and gas-phase hydrocarbon compounds. Diesel exhaust aerosol is of particular concern because it is almost entirely respirable in size, with more than 90% of the particles, by mass, having an aerodynamic diameter less than 1.0 μm (Cantrell 1987). This means that the aerosol can penetrate to the deepest regions of the lungs and, if retained, cause or contribute to the development of obstructive or restrictive lung disease (Watts 1987).

PHYSICAL CHARACTERISTICS OF MINE AEROSOL

Aerosols from mine sources have particle size distributions that are determined by both the mining method and the source of the aerosol (Cantrell et al. 1987). A size distribution summarizing the physical characteristics of mine aerosols is shown in Fig. 26-5. The shape of the aerosol size distribution is influenced by the different sources contributing to the aerosol. This figure displays some of these sources and the physical mechanisms, such as condensation and coagulation, that transfer aerosol mass from one size to another. It should be noted that these mechanisms and the general shape of the distribution are not unique to mine aerosols.

There are three distinct aerosol size ranges identifiable by features in measured mine aerosol size distributions. The smallest of these, from 0.001 to 0.08 μm, is the Aitken nuclei range, which contains primary aerosol from combustion sources, such as diesel engines, and secondary aerosols or chain aggregates, formed by coagulation of primary aerosols. The next size range, from 0.08 to approximately 1.0 μm, is termed the accumulation range. It contains emissions in this size range plus aerosol accumulated by mass transfer through coagulation and condensation processes from the Aitken nuclei range. The last range, 1.0 to approximately 40 μm, is termed the coarse particle range. Aerosols within this size fraction generally result from mechanical processes such as rock fracture and bulk material handling. Mineral dust

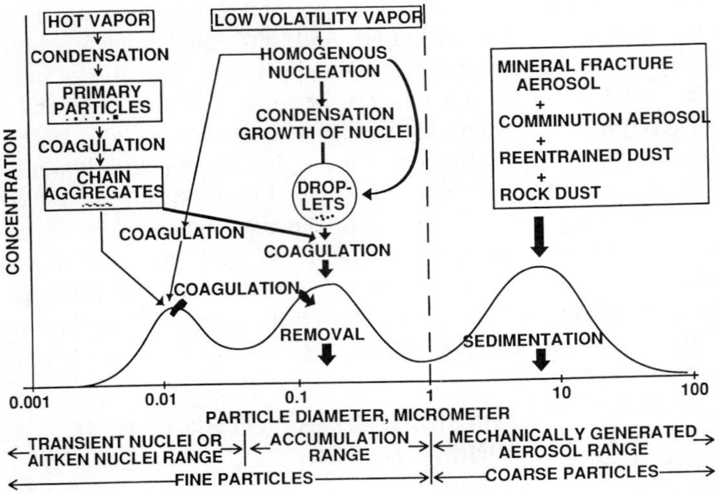

FIGURE 26-5. Size Distribution Summarizing the General Physical Characteristics of a Mine Aerosol. (Source: Cantrell et al. 1987.)

aerosol reentrained by mine haulage vehicles during the load–haul–dump cycle is an example of an in-mine emission that will contribute aerosol to this size range.

For convenience, the Aitken nuclei and the accumulation ranges are combined in a single "fine" particle range. A division is usually made between this range and the "coarse" particle range at 1.0 μm. This distinction is possible because sources of aerosol in the two ranges are usually different, and the coarse particle range contains very little mass transferred from the accumulation range by coagulation.

In each of the ranges mentioned, the size distribution of mine aerosol can exhibit a maximum, or mode, which takes its name from the size range in which it occurs. Hence, the maximum in the accumulation range is termed the accumulation mode. Figure 26-6 presents a typical size distribution of aerosol mass concentration measured in a haulage entry of a diesel-equipped coal mine (Cantrell and Rubow 1990). Here the modal character of the size distribution is discernible even though the nuclei mode is suppressed compared to the accumulation mode. In contrast, Fig. 26-7 shows a mass size distribution measured in the haulage way of an all-electric equipped coal mine. Here the accumulation

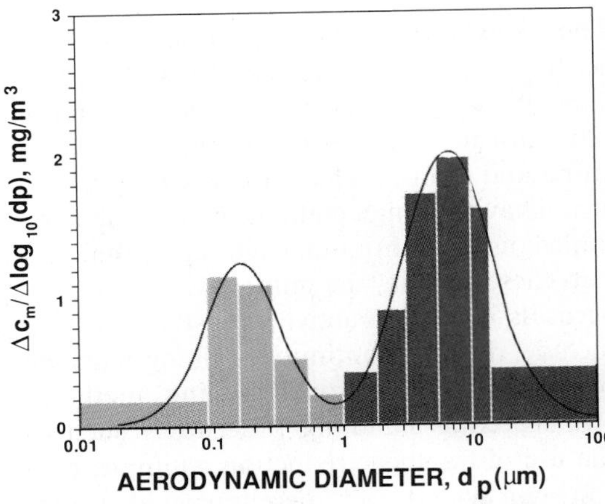

FIGURE 26-6. Mass Size Distribution of Mine Aerosol in a Diesel-Equipped Mine. (Source: Cantrell and Rubow 1990.)

mode is much smaller than the coarse particle mode. Taken together, the figures indicate that diesel aerosols can make a strong contribution to accumulation mode aerosol in a diesel-equipped mine.

MEASUREMENT TECHNOLOGY

Compliance measurements are primarily intended for use in the enforcement of PELs

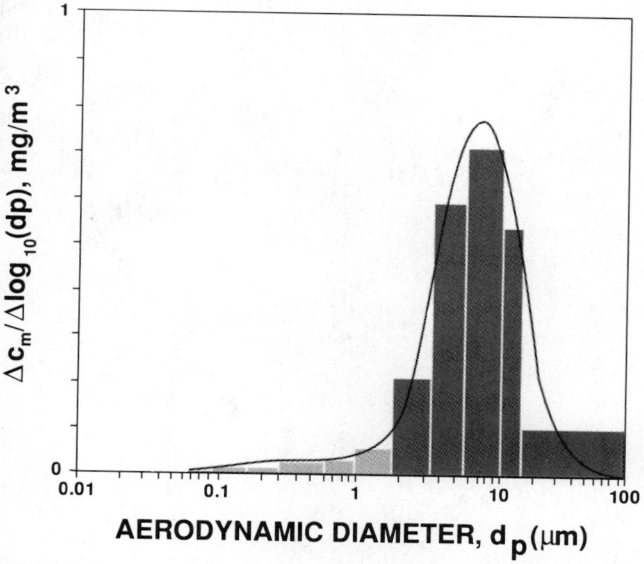

FIGURE 26-7. Mass Size Distribution of Mine Aerosol in an All-Electric Equipped Mine. (Source: Cantrell and Rubow 1990.)

established by MSHA. Regulatory requirements determine the sampling strategy and instrumentation used for such measurements. Research aerosol measurements are used to develop new aerosol instrumentation, to define new occupational health hazards, and to expand the knowledge regarding mine aerosols. They draw on all of the technology available to the aerosol community. Consequently, compliance and research aerosol measurements are treated here separately. Also, the following discussion focuses on aerosol measurements in U.S. mines. For a discussion of measurement practices in the European Community, see Vincent et al. (1988).

Compliance Measurements and the Regulatory Environment

Regulatory Requirements—Metal and Nonmetal

MSHA regulates practices affecting health and safety in metal and nonmetal mines and mills under the authority of the Federal Mine Safety and Health Act of 1977 (U.S. Congress 1977*a*). The specific regulations are found in the Code of Federal Regulations, Title 30 (MSHA 1991). The occupational health standards established by MSHA were adopted from the 1973 recommended threshold limit values of the American Conference of Governmental Industrial Hygienists (ACGIH 1973). Compliance with these regulations is determined by the collection of environmental samples by MSHA inspectors. Aerosol-related contaminants that are regulated include total dust, quartz, asbestos, radon daughters, and welding fumes (MSHA 1990).

Examples of sampling and analysis procedures for one of these regulated aerosol contaminants are those for respirable dust containing more than 1% quartz. A sample is collected using the personal respirable dust sampler in Fig. 26-8 (MSA Co., Pittsburgh, PA).[1] Sample air is first passed through a Dorr-Oliver nylon cyclone preclassifier at a flow rate of 1.7 l/min to remove the nonrespirable fraction of sampled dust. Respirable dust is then collected on a filter that is analyzed gravimetrically to determine mass concentration. The filter deposit is also analyzed for quartz content using X-ray diffraction (MESA 1975). The measured mass concentration is compared to the PEL determined from the quartz content

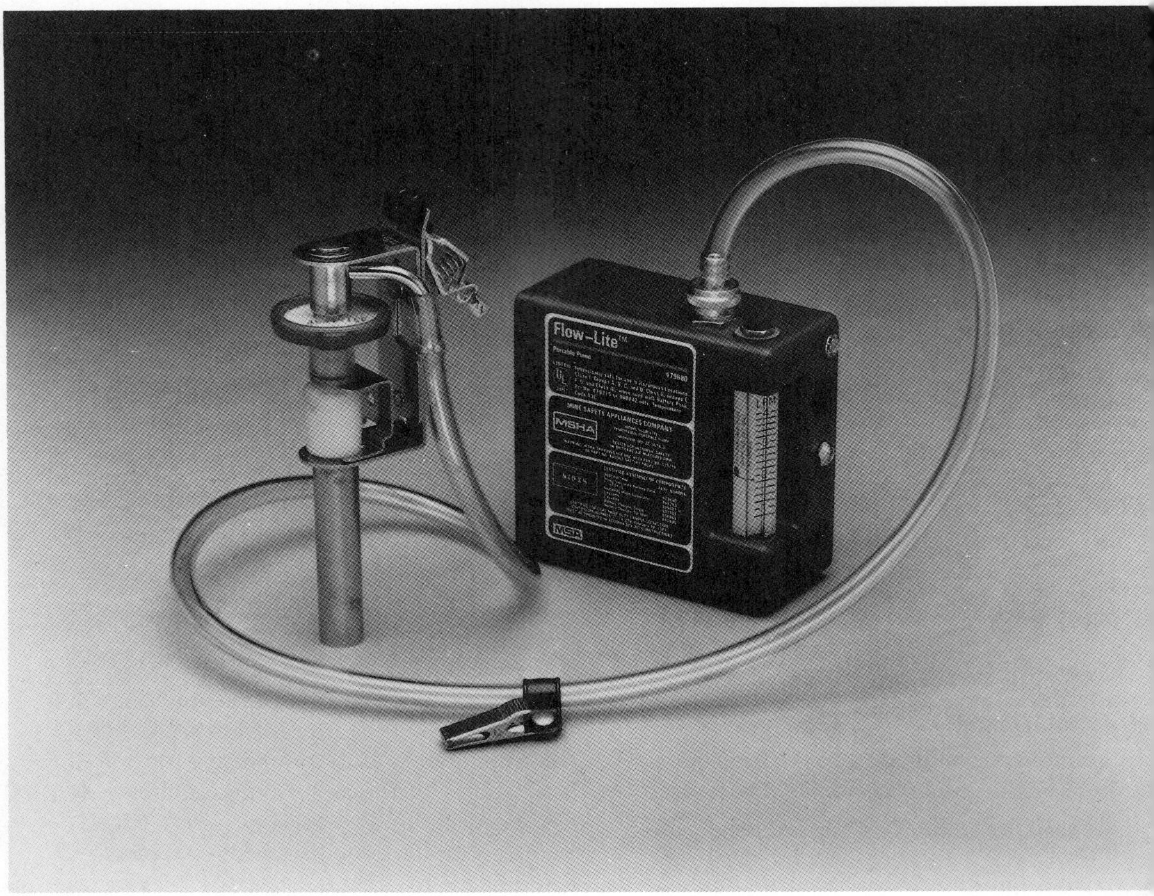

FIGURE 26-8. Personal Respirable-Dust Monitor. (Courtesy of MSA Co.)

of the respirable dust by

$$PEL_{quartz} = \frac{10\ \text{mg/m}^3}{\text{percent respirable quartz} + 2} \tag{26-1}$$

For a given exposure level the magnitude of the toxicity is proportional to the quartz content (ACGIH 1980). The factor 2 in the denominator of the PEL formula ensures that dust exposures will not be excessively high when the quartz content is less than 5%, and limits respirable dust concentrations to 5 mg/m^3 when no quartz is present in the sample. MSHA has proposed a revision of many of the existing health regulations (MSHA 1989*a*). Included in these revisions is a proposed change in the PEL for respirable quartz. The proposed PEL, which is undergoing review, is 100 μg/m^3 of respirable quartz.

Between 1985 and 1990, MSHA inspectors have collected more than 21,900 respirable dust samples containing quartz in metal and nonmetal mines and mills. A summary of average respirable dust concentrations measured in this period for selected commodities is given in Table 26-1 (Watts 1992). This analysis is described by Watts and Parker (1987). Quoted are the number of samples, arithmetic mean with its standard deviation, and the percent quartz based on the arithmetic mean. Since the distribution of measured quartz concentrations is lognormal (Watts et al. 1984), the geometric mean and standard deviation are also given.

TABLE 26-1 Respirable quartz statistics for selected commodities, 1985 through 1990

Commodity	Samples (Number)	Concentration (μm/m³) AM	ASD	GM	GSD	Quartz (%) AM
Metal:						
Copper	321	99	135	58	3	10.2
Gold	1541	107	210	55	3	11.5
Iron	529	65	99	35	3	11.8
Lead and zinc	118	41	41	29	2	6.1
Molybdenum	120	61	62	41	3	10.4
Silver	369	109	225	55	3	12.2
Uranium	205	68	101	40	3	8.0
Nonmetal:						
Common clay	886	61	83	39	3	7.5
Fire clay	119	123	198	67	3	14.5
Nonmetal, not elsewhere classified	250	130	149	80	3	32.8
Phosphate rock	157	43	73	23	3	7.9
Sand and gravel	6327	86	185	42	3	14.7
Stone:						
Cement	184	41	68	22	3	5.6
Granite:						
Crushed	1442	68	102	41	3	11.8
Dimension	351	154	240	75	3	13.5
Limestone: Crushed	5027	47	115	27	3	6.0
Sandstone:						
Crushed	1341	116	201	65	3	24.6
Dimension	107	98	99	65	3	24.7
Stone: Crushed	645	100	220	44	3	12.1
Traprock: Crushed	420	66	102	35	3	8.5

Source: Watts 1992

AM—Arithmetic mean, ASD—Arithmetic standard deviation, GM—Geometric mean, GSD—Geometric standard deviation

Regulatory Requirements—Coal

Respirable coal mine dust measurements are made to determine compliance with an MSHA-approved dust control plan that is based on established PELs (MSHA 1989*b*). In 1970 a mandatory total respirable dust PEL of 3.0 mg/m³ was established for underground coal mines in the Federal Coal Mine Health and Safety Act of 1969. An effective PEL of 100 μg/m³ was established for quartz in 1971 (MSHA 1971). The respirable dust standard was subsequently lowered in 1972 to 2.0 mg/m³. Mandatory dust standards for surface work areas of underground coal mines and surface mines also became effective in 1972. These regulations were continued under the Federal Mine Safety and Health Act of 1977 (U.S. Congress 1977*b*), which amended the 1969 Coal Act and merged coal and noncoal regulations into one law. In the 1969 Act, "concentration of respirable dust" was defined as that measured using a Mining Research Establishment (MRE) parallel-plate elutriator (Casella, model 113A, U.K.) sampling instrument (Fig. 26-9), or such an equivalent concentration measured with another device. This instrument was designed to have a sampling efficiency equivalent to the respirable response curve specified by the British Medical Research Council (BMRC) (Lippman 1989). The 1977 Act changed the definition of "concentration of respirable dust" to the "average concentration of respirable dust measured with a device approved by the Secretary (of Labor) and the Secretary of Health Education and Welfare".

The personal respirable dust sampler illustrated in Fig. 26-8 is also approved for

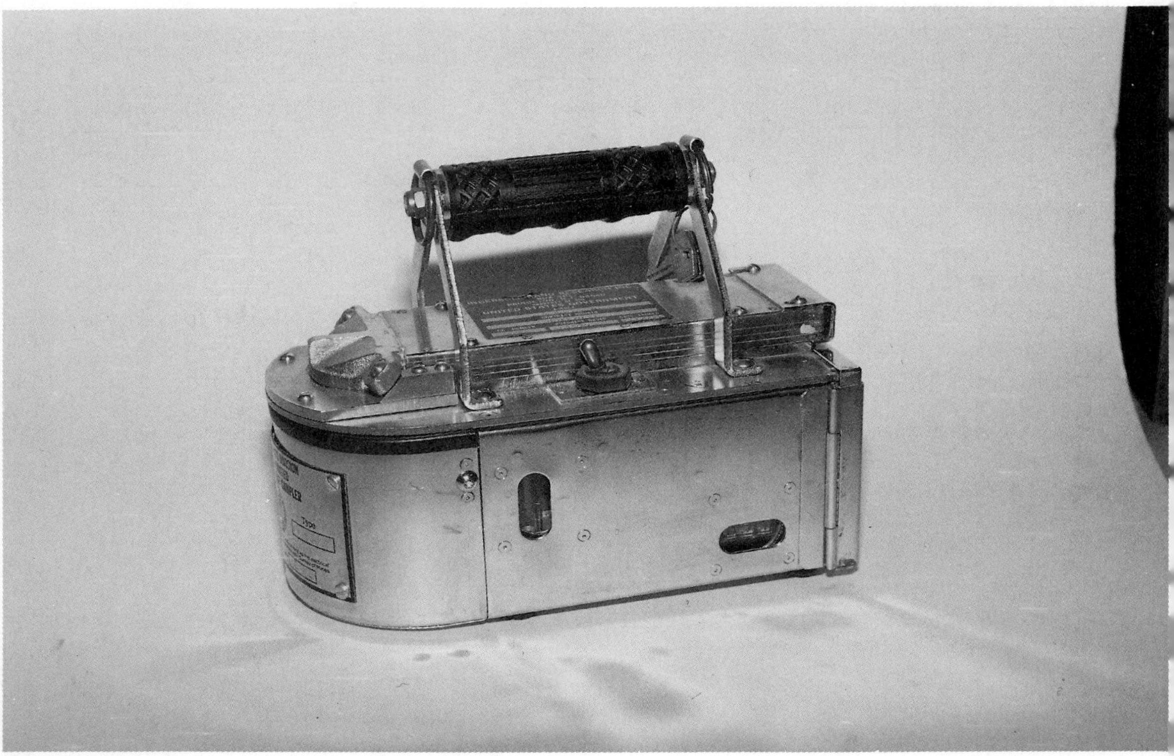

FIGURE 26-9. Casella, Model 113a, MRE Respirable Dust Monitor. (Cantrell et al. 1987)

measuring respirable coal mine dust (MSHA 1989*b*). The sampling rate used for coal, however, is 2 l/min (Tomb and Raymond 1970). Sample analysis for total respirable dust is gravimetric (Raymond, Tomb, and Parobeck 1987). Analysis for quartz is by Fourier transform infrared spectrometry (Ainsworth, Parobeck, and Tomb 1989). The measurements are converted to equivalent MRE concentrations by multiplying the measured concentrations by an accommodation factor of 1.38 (Treaftis et al. 1984). The difference between the Dorr–Oliver and BMRC sampling efficiency curves is evident from Fig. 26-10 (Caplan, Doemeny, and Sorenson 1977*a*, *b*; Lippman 1989). Specific regulations detailing the methods for collecting respirable dust samples are found in the Code of Federal Regulations, Title 30 (MSHA 1991).

A mine is in noncompliance with its dust plan if the arithmetic average concentration of five consecutive respirable dust samples is in excess of the applicable standard (MSHA 1989*b*). If the percentage quartz is less than 5%, the standard is 2.0 mg/m^3. If the percentage of quartz is greater than 5%, the applicable standard is 10 divided by percentage quartz. In underground coal mines, samples are collected on workers with a designated occupation, usually a mining machine operator. MSHA is required to inspect all underground coal mines four times a year and the mine operators are required to sample bimonthly. The results of these inspections, for designated occupations in longwall and continuous miner sections during the period 1985–1990, are shown in Fig. 26-11 (Watts 1992). Shown are the arithmetic-mean concentrations of respirable dust measured by both MSHA inspector and mine operators. Considering typical arithmetic standard deviations for the means of $\pm$ 1.1 mg m^3 for the

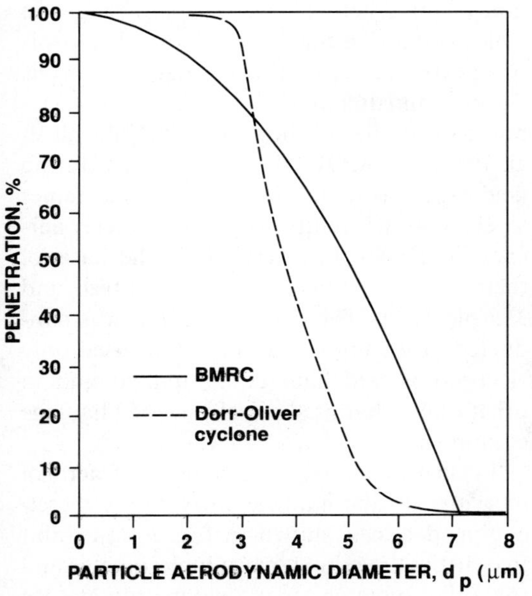

FIGURE 26-10. Respirable Aerosol Sampling Criteria; BMRC, Dorr–Oliver Cyclone. (Source: Caplan, Doemeny, and Sorenson 1977*a, b*; Lippman 1989.)

continuous miner and $\pm$ 1.3 mg/m^3 for longwall sections, these concentrations have remained essentially constant over these six years.

Diesel Exhaust Aerosol

NIOSH (1988) has recommended that "whole diesel exhaust be regarded as a "potential occupational carcinogen", as defined in the Cancer Policy of the Occupational Safety and Health Administration." NIOSH further stated that "though the excess risk of cancer in diesel-exhaust-exposed workers has not been quantitatively estimated, it is logical to assume that reduction in exposure to diesel exhaust in the workplace would reduce the excess risk." The International Agency for Research on Cancer (1989) has also classified diesel exhaust as "probably carcinogenic to humans." Additionally, the Mine Safety and Health Administration (MSHA 1988) received a recommendation from an advisory committee to establish a diesel exhaust aerosol standard and to establish regulations to

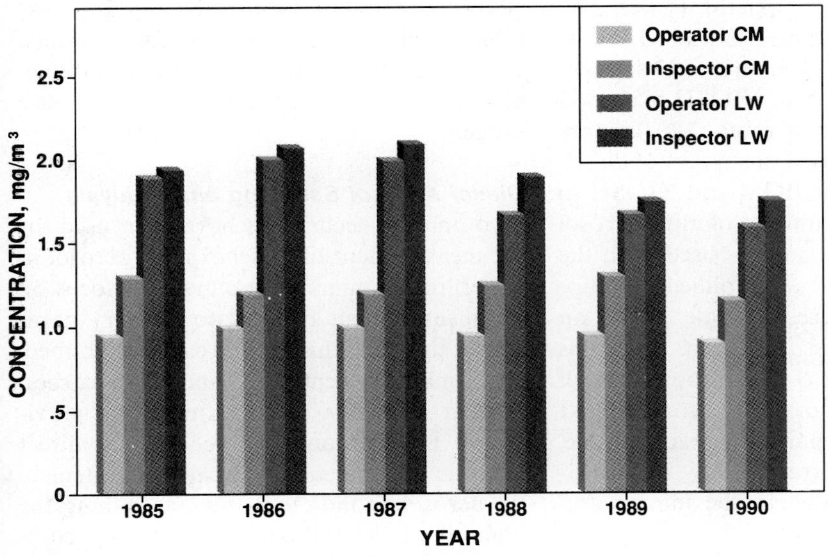

FIGURE 26-11. Average Coal Dust Aerosol Concentrations Determined for Designated Occupations by MSHA Inspectors and the Mine Operators, 1985–1990. CM and LW Refer to Continuous Miner and Longwall Sections, Respectively. (Source: Watts 1992.)

minimize exposure to all diesel pollutants in underground coal mines. MSHA is proceeding with the development of a PEL for diesel exhaust aerosol (MSHA 1992).

Research Aerosol Measurements

Research measurements are used to characterize the mine aerosol's mass and specific components such as quartz, trace elements, and carbon as a function of aerosol size. Recent emphasis on the potential health hazard associated with exposure to diesel exhaust aerosol has focused attention on specific techniques to measure these aerosols. The primary technique used to collect samples in coal mines for all of these measurements is size-selective sampling using inertial impaction. Recently, several aerosol sensor techniques were examined for possible use in a continuous monitor for respirable mine aerosol. These included light scattering and direct mass measurement.

Size-Selective Sampling

A size-selective sampling technique that has been useful for in-mine measurement of both diesel and mineral dust aerosol employs the Marple personal impactor (series 290, Anderson Samplers Inc., Atlanta, GA) discussed in Chapter 11. The series 290 sampler was originally designed for NIOSH as a wood dust sampler by Rubow et al. (1987). More recently, it has been used in surveys of diesel-equipped mines by the BOM and NIOSH to measure the size distribution of mine aerosol (NIOSH 1987). These surveys used both the cascade impactor and a simplified dichotomous sampler that uses a single impaction stage with a cut point of 1.5 μm. Estimates were made of average concentration levels of respirable diesel aerosol for the working shift using the sub-1.0 μm portion of each sample. The average concentration for submicrometer aerosol generated in the mining sections was 0.7 ± 0.3 mg/m^3.

The micro-orifice, uniform-deposit impactor (MOUDI) (model 100, MSP Inc., Minneapolis, MN), discussed in Chapter 11, has also been used to measure the size distribution of mine aerosol over the size ranges in which respirable coal dust and diesel aerosols are expected to predominate (Marple, Rubow, and Behm 1991). In addition to laboratory studies of these aerosols (Marple et al. 1986), the MOUDI has been used during field experiments in underground coal mines to evaluate its ability to separate diesel aerosol from coal dust aerosol on the basis of their size distributions (Rubow, Cantrell, and Marple 1990). The field evaluation was conducted in underground mines that used only electric-powered haulage equipment and in other mines that used diesel-powered haulage equipment.

Typical mass size distributions of aerosol measured in the haulage entry of the diesel-equipped mines, shown in Fig. 26-6, exhibit two distinct maxima: one submicrometer and the other greater than one micrometer in size. These measurements indicate that more than 90% of diesel exhaust aerosols in the diesel-equipped coal mines studied were submicrometer in size (Cantrell 1987). The diesel-associated submicrometer aerosol accounted for approximately 40–60% of the respirable-aerosol mass concentration. In contrast, aerosol size measurements in the all-electric coal mines, typified by Fig. 26-7, exhibited very small submicrometer maxima. Less than 10% of the measured respirable-aerosol mass was in the submicrometer size range.

Diesel Aerosol Sampling and Analysis

Two analysis techniques have been used for the measurement of diesel exhaust aerosol in underground mines. These methods focus on the measurement of the two primary parameters by which this aerosol can be described: (1) its mass concentration and (2) its carbon content. The first is measured using gravimetric analysis and the second, by direct analysis of the elementary carbon content of the aerosol or indirectly, by determining the fraction of the sample removable by combustion. In all cases, size-selective sampling is used to provide a sample that contains most of the diesel-associated portion of the sampled respirable aerosol.

A personal diesel exhaust aerosol sampler based on size-selective sampling, developed for use in underground coal mines (Rubow et al. 1990), is shown in Fig. 26-12. It has three stages and employs inertial impaction for separating and collecting the diesel and mineral dust fractions of the sampled respirable aerosol. The first stage is an inertial preclassifier that separates and collects the larger, nonrespirable aerosol. The preclassifier used in this design is a 10-mm Dorr–Oliver cyclone. Its second stage is a four-nozzle impactor with a 50% cut point of 0.8 μm aerodynamic diameter. The third stage, which is a filter, collects the respirable aerosol less than 0.8 μm aerodynamic diameter in size. The sampler operates at a sampling flow rate of 2 l/min and is designed to be compatible with commercial personal sampling pumps.

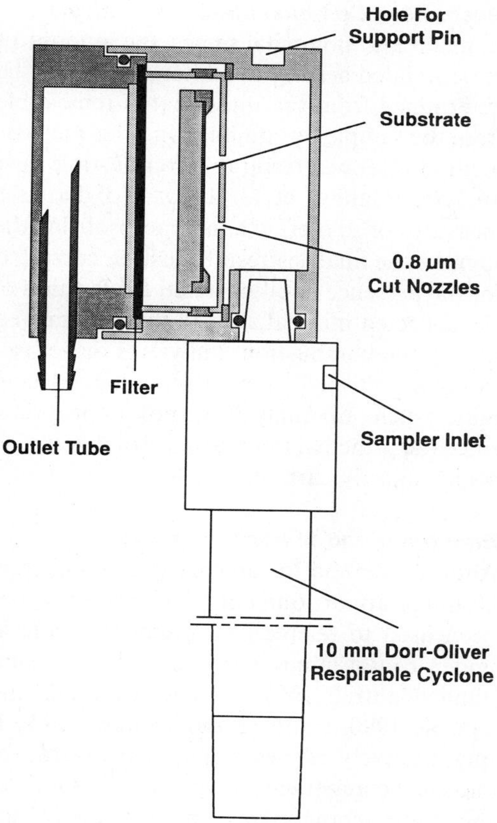

FIGURE 26-12. Personal Diesel Exhaust Aerosol Sampler. (Source: Rubow et al.1990.)

Gravimetric Analysis

Coupled with gravimetric analysis, the personal diesel exhaust aerosol sampler can provide measurements of diesel exhaust aerosol concentrations in coal mines, under worst-case sampling conditions, which are accurate to within 50% for concentration levels greater than 0.3 mg/m^3 (Cantrell and Rubow 1991). During field evaluation tests, the sampler was used to make numerous aerosol concentration measurements in underground coal mines that use diesel haulage equipment. Figure 26-13 summarizes respirable-aerosol concentrations in these mines, determined from area samples collected in the haulage entries, on coal shuttle cars (Ramcar, Dresser Ind., Jeffrey Div., Columbus, OH), and in the ventilation return entry (Cantrell et al. 1991). Figure 26-14 summarizes the diesel exhaust aerosol concentrations measured using the same samples.

The haulage, shuttle car, and return locations have similar distributions for the diesel exhaust aerosol portion of the respirable total. This implies that diesel exhaust aerosol concentrations are uniform regardless of where they are measured in the section. The total respirable-aerosol concentration levels are different, depending upon where the samples are taken. The highest is in the ventilation return and the lowest is in the haulage

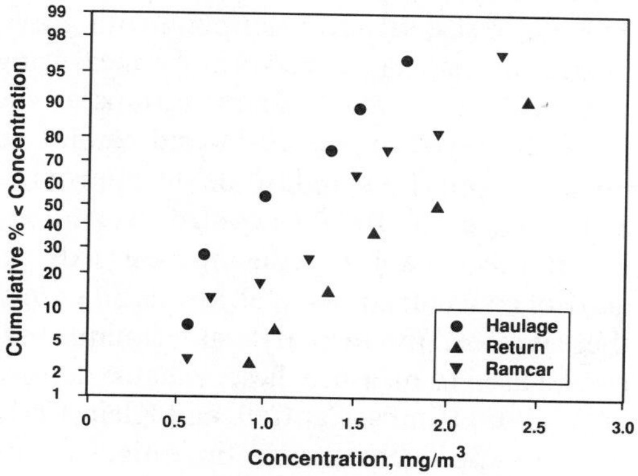

FIGURE 26-13. Cumulative Frequency Plot of Respirable-Aerosol Concentrations in Continuous Mining Operations Using Diesel-Equipped Haulage Vehicles. (Source: Cantrell et al. 1991.)

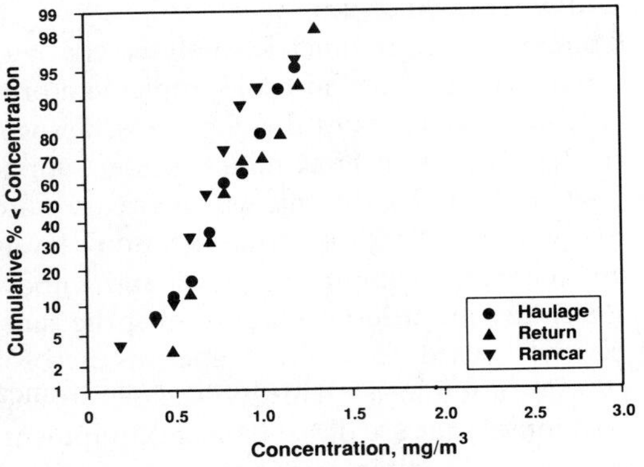

FIGURE 26-14. Cumulative Frequency Plot of Respirable Diesel Aerosol Concentrations in Continuous Mining Operations Using Diesel-Equipped Haulage Vehicles. (Source: Cantrell et al. 1991.)

way. These results imply that exposure to respirable mineral dust is location- and, hence, occupation-dependent.

The median diesel exhaust aerosol concentration determined with the personal sampler at the haulage location for the three mines surveyed was 0.8 ± 0.2 mg/m^3. Diesel aerosol contributed 65% of the respirable aerosol at this location. It is clear from these data that diesel exhaust aerosol contributes significantly to respirable coal mine aerosol concentrations in mines using diesel haulage and that reductions in such concentrations can improve mine air quality.

Elemental Carbon Analysis

The use of size-selective sampling with gravimetric analysis is intended only for measuring diesel exhaust aerosol concentrations at the levels displayed in Fig. 26-14 and cannot be used to support a standard below approximately 300 μg/m^3. Because control technology is currently capable of reducing diesel exhaust aerosol concentrations to as low as 50 μg/m^3, there is a need for an analytical technique that can be used to measure diesel exhaust aerosol at this level (Ambs, Cantrell, and Klein 1991).

One approach, being investigated by NIOSH, uses the elementary carbon content of an aerosol sample as a surrogate for the diesel exhaust aerosol portion. An evolved gas analysis technique called the "thermal–optical method" is being evaluated for this determination (Birch 1992). This method is a modification of a technique developed for the analysis of atmospheric aerosol (Cadle and Groblicki 1980; Johnson et al. 1981). It operates by converting the aerosol carbon to carbon dioxide in a furnace as the temperature is varied over the range 0°C–750°C. An optical feature corrects for pyrolitically generated elemental carbon, or "char", which is formed during the analysis of some samples. A measurement of the evolved gas can be used to quantify volatile carbon, carbonate carbon, and graphitic carbon in the aerosol sample at very low levels. Such methods are in the research stage and have yet to be validated for use in the underground mining environment. They promise, however, to provide sensitive techniques for measuring diesel exhaust aerosol.

Respirable Combustible Dust Analysis

In metal and nonmetal mines, the amount of diesel-related aerosol in a filter sample can be determined from the mass that is removable from the sample by combustion. This analysis method is termed respirable combustible dust analysis (Gangal et al. 1990). To derive a measure of diesel exhaust aerosol in the sample, the analysis result must be corrected for the presence of oil mist and the fraction of the collected mineral aerosol that is removed during the combustion analysis. For aerosol samples collected in a diesel-equipped coal mine, such an analysis is not appropriate since the principal mineral aerosol involved is also primarily carbon.

Elemental and Mineral Analysis

Another method for analyzing diesel aerosol is source apportionment. This technique has been used to referee measurements made in underground mines using size-selective sampling (Cantrell 1987; Rubow, Cantrell, and Marple 1990; Cantrell and Rubow 1990). It quantitatively relates elemental, mineral, or chemical components in an aerosol sample to the same components in sources of the aerosol. Using this relationship, the contribution of each source to the sampled aerosol can be determined. For diesel-equipped coal mines the primary sources of aerosol are coal, rock dust used as a fire retardant on the mine walls, and diesel exhaust aerosol emissions. One specific analytical technique used to apportion collected aerosol among these sources is chemical mass balance model analysis (Watson 1984; Henry et al. 1984; Davis 1984). The model is expressed as

$$c_{ei} = \sum_{j=1}^{p} a_{ij} S_j \qquad (26\text{-}2)$$

Here, c_{ei} is the mass concentration of the ith elemental, mineral, or chemical component of the sample in ug/m^3; a_{ij} is the fractional amount of component i in emissions from source j; S_j is the total contribution of source j to the sample; and p is the number of sources. Apportionment of the source is achieved by first characterizing the aerosol sources, obtaining values for a_{ij}, then analyzing the aerosol in the sample for the same components, and finally solving for S_j. A least-squares regression analysis is used to determine the S_j of the overdetermined system of equations expressed by Eq. 26-2.

Real-Time Measurement

Accurate, real-time measurement of mine dust aerosol is an unrealized goal (Lilienfeld, Stern, and Tiani 1983). Although many techniques have been investigated and several instruments are commercially available, few are capable of realizing the accuracy of filter sampling techniques and none are available as compliance monitors. The following discussion summarizes two of the more promising technologies.

Light Scattering Instruments

Often called photometers or nephelometers, light scattering instruments use a light source to illuminate a sample aerosol and a light sensor to measure the scattered light, which can be related to the mass concentration of the dust. These aerosol monitors were characterized in the laboratory for different dusts (Kuusisto 1983; Marple and Rubow 1981, 1984; Keeton 1979; Rubow and Marple 1983; Williams and Timko 1984). Except for the work by Keeton, none of the laboratory research has involved diesel exhaust particulate. In all cases the relationship of the instruments' response to dust concentration is not simple but depends on particle size, particle composition, and on instrument design and manufacturing differences. Usually, these instruments must be calibrated for the specific dust being monitored, although this is not necessary for cases where only relative measures are needed and the particle properties do not change significantly during the measurement period.

Most of the mining community's experience with real-time, photometric instruments is with the RAM-1 and MINIRAM devices (MIE Inc., Billerica, MA), which were initially designed for use in the mine environment. Intrinsically safe versions of these instruments are certified by MSHA for use in the potentially explosive atmospheres of underground coal and gassy noncoal mines. They are normally operated with the Dorr–Oliver 10 mm cyclone to remove water droplets and ensure a respirable dust sample. The RAM, shown in Fig. 26-15, is a light scattering aerosol monitor of the nephelometric type. The instrument uses a pulsed Gallium arsenide light-emitting source that generates a narrow-band emission centered at 880 nm. Radiation scattered by airborne particles in the view volume is collected over an angular range of approximately 45°–95° from the forward direction by means of a silicon light detector. The MINIRAM employs the same sample geometry and optical arrangement.

In-mine experience with these instruments for mineral dust aerosol measurements is extensive (Tomb, Treaftis, and Gero 1984) but little published data are available for diesel aerosol. McCawley (NIOSH 1987) obtained good correlations ($r^2 = 0.92$) between in-mine MINIRAM measurements and those obtained using a gravimetric sampler fitted with an intake impactor to restrict the penetration of particles larger than 1.0 μm. Analysis of the collected sample was accomplished using RCD measurements. This successful comparison between diesel aerosol

FIGURE 26-15. MIE RAM-1 Real-Time Respirable-Aerosol Monitor Equipped with a Dorr–Oliver Cyclone Sample Inlet. (Source: Olson and Veith 1987.)

and the response of a RAM-type instrument holds promise that it might be used to monitor diesel aerosol.

Response of the RAM-1 and MINIRAM to diesel aerosol in the laboratory has been reported by Zeller (1987*a*, *b*). Figure 26-16a, b shows the recorded response during the tests of two MINIRAMs and three RAMs for diesel exhaust aerosol from a Caterpillar 3304 engine operated at different combinations of steady-state speeds and loads. The exhaust was diluted about 25:1 with clean air prior to measurement. The mass concentrations are obtained using simultaneously collected filter samples. Data shown in Fig. 26-16a exhibit a different response for each of the instruments. This individual-instrument bias is normal and was observed for rock and coal dusts (Marple and Rubow 1981, 1984). Figure 26-16b shows these same data adjusted for instrument bias by multiplying the instrument responses by a value which is constant for each instrument. The data now lie on a line with only moderate scatter attributed to imprecision or random error. The manufacturer provides an internal adjustment for each instrument that achieves the same effect.

Light scattering technology remains an attractive technique for respirable coal mine dust monitoring because it is mechanically simple and provides almost instantaneous readings. Such readings can be averaged over any appropriate interval, stored, or presented in a variety of formats. Unfortunately, light scattering instruments do not measure the dust mass concentration directly. Significant changes in dust particle characteristics such as shape, size, surface properties, and density

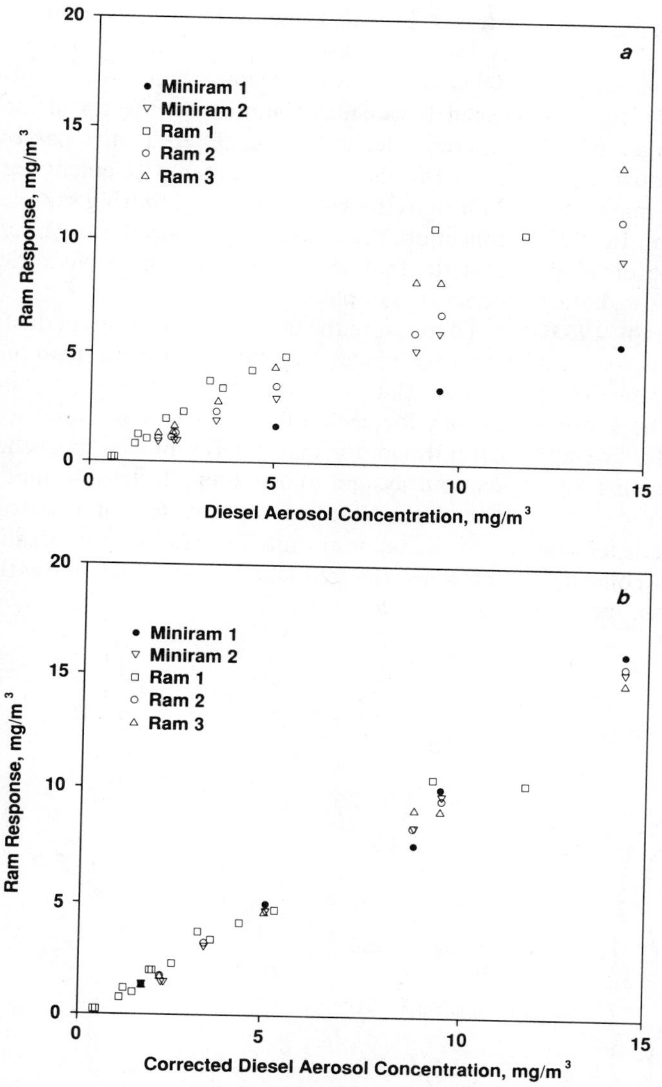

FIGURE 26-16. Nephelometer, RAM-1 and MINIRAM, Response to Diesel Aerosol: (a) actual Instrument Response; (b) Same Data Adjusted for Instrument Bias. (Source: Cantrell et al. 1987.)

can affect the instrument's correlation with mass concentration and will require a recalibration of the instrument (Williams and Timko 1984).

Tapered-Element Oscillating Microbalance

The tapered-element oscillating microbalance (TEOM), described in Chapter 14, has found only limited use in the mining industry. These instruments are usually not certified to be explosion-proof or intrinsically safe for use in explosive atmospheres. Electrical equipment used in underground mines must be approved by MSHA. Nonapproved equipment may sometimes be used in intake airways away from the coal face but cannot be used at the coal face or in return airways, where methane/air mixtures can reach explosive levels. Also, as of this writing, no regulation requires the use of real-time dust monitoring in underground coal mines.

Despite the limited application of the TEOM in the mining industry, the technology offers one notable advantage: direct measurement of dust mass. Since dust exposure standards are based on dust mass, this attribute of the TEOM is significant. In 1983, the BOM and NIOSH funded the development of a prototype TEOM personal dust monitor (Patashnick and Rupprecht 1983). The prototype monitor developed is a system configured for end-of-shift measurements. It is not a real-time monitor, but uses oscillating-microbalance technology to "weigh" the collection filter before and after dust sampling.

Figure 26-17 shows the special canister that houses the tapered element and collection filter. During dust sample collection, the canister is connected to a Dorr–Oliver nylon cyclone to remove nonrespirable dust from the sample stream. Respirable aerosol is collected by a filter mounted on the tip of the tapered element. The filtered air passes through the hollow tapered element at 2 l/min to the sampling pump. During sample collection, the canister functions like any filter cassette that might be used in gravimetric personal samplers.

To measure the mass concentration of dust, the user places the canister into a readout system that records the characteristic frequency for each vibrating element. Once accomplished, the canister is removed from the readout system and is then electrically inactive. After sampling, the canister is reinserted into the readout unit. The readout unit again measures the frequency of each element with its dust-laden filter. The change in frequency

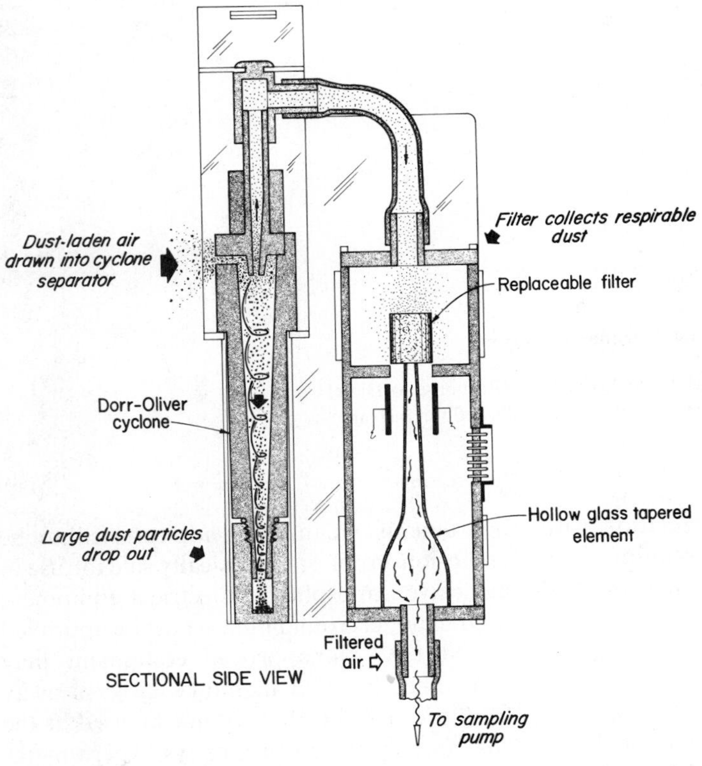

FIGURE 26-17. Canister Housing TEOM Tapered Element and Collection Filter. (Source: Cantrell et al. 1987.)

is a function of the collected aerosol mass. Its conversion to mass is discussed in Chapter 14.

The BOM has evaluated this prototype system in the laboratory for both end-of-shift and near-real-time applications (Williams and Vinson 1986). The effective standard deviation of repeated measurements was 1.6 μg. The tests at controlled temperature and humidity showed less than 20 μg of drift during an 8 h shift. When subjected to a somewhat abrupt change in relative humidity (46–78%), the TEOM dust monitor measured a total increase of less than 40 μg over a period of 70 h. For an end-of-shift measurement system, the measurement errors from humidity can be adequately controlled by conditioning the filter before presampling and postsampling frequency analysis. In general, the dust monitor showed excellent agreement with measurements made using personal respirable dust samplers. The coefficient of variation of the monitor measurements was slightly higher than that of the personal sampler measurements, although less than 10% for most cases.

References

Agricola, G. 1556. *De re Metallica.* Translated by M. C. Hoover, and L. H. Hoover. New York: Dover.

Ainsworth, S. M., P. S. Parobeck, and T. F. Tomb. 1989. Determining the quartz content of respirable dust by FTIR. MSHA IR 1169, Arlington, VA.

American Conference of Governmental Industrial Hygienists. 1973. *TLVs—Threshold Limit Values for Chemical Substances in Workroom Air.* Cincinnati, OH: ACGIH.

American Conference of Governmental Industrial Hygienists. 1980. *Documentation of the Threshold Limit Values,* 4th edn., pp. 364–65. Cincinnati, OH: ACGIH.

Ambs, J. L., B. K. Cantrell, and A. D. Klein. 1991. Evaluation of a disposable diesel exhaust filter. In *Advances in Filtration and Separation Technology, Vol. 4,* ed. K. L. Rubow, pp. 287–91. Am. Fil. Soc.

Birch, M. E. 1992. *A Sampling and Analytical Method for Airborne Diesel-Exhaust Particles.* Presented at the 203rd National Meeting of the American Chemical Society, Division of Health and Safety, Symposium on Industrial Hygiene Chemistry, San Francisco. Washington DC: American Chemical Society.

Cadle, S. H. and J. Groblicki. 1980. An evaluation of methods for determination of organic and elemental carbon in particulate samples. General Motors Report GMR-3452, ENV #86.

Cantrell, B. K. 1987. Source apportionment analysis applied to mine dust aerosols: Coal dust and diesel emissions aerosol measurement. In *Proc. 3rd Mine Ventilation Symp.,* SME, pp. 495–501.

Cantrell, B. K. and K. L. Rubow. 1990. Mineral dust and diesel exhaust aerosol measurements in underground metal and nonmetal mines. In *Proc. VIIth International Pneumoconioses Conf.,* NIOSH Pub. No. 90-108, pp. 651–55. NIOSH.

Cantrell, B. K. and K. L. Rubow. 1991. Development of personal diesel aerosol sampler design and performance criteria. *Mining Engineering Mag.* February 1991 Issue: 232–36.

Cantrell, B. K., K. L. Rubow, W. F. Watts, and D. H. Carlson. 1991. Pollutant levels in underground coal mines using diesel equipment. SME Preprint No. 91-35, Soc. for Mining, Metallurgy, and Exploration, Inc., Littleton, CO.

Cantrell, B. K., H. W. Zeller, K. L. Williams, and J. Cocalis. 1987. Monitoring and measurement of in-mine aerosol: Diesel emissions. BuMines IC 9141, pp. 18–40. Washington, DC.

Caplan, K. J., L. J. Doemeny, and S. D. Sorenson. 1977*a*. Performance characteristics of the 10-mm respirable mass sampler: Part I—Monodisperse studies. *Am. Ind. Hyg. Assoc. J.* 38:83–95.

Caplan, K. J., L. J. Doemeny, and S. D. Sorenson. 1977*b*. Performance characteristics of the 10-mm respirable mass sampler: Part II—Coal dust studies. *Am. Ind. Hyg. Assoc. J.* 38:162–73.

Davis, B. L. 1984. X-ray diffraction analysis and source apportionment of Denver aerosol. *Atmos. Environ.* 18(10):2197–208.

Fletcher, C. M. 1948. Pneumoconiosis of coal-miners. *Br. Med. J.* 1:1015–65.

Gangal, M., J. Ebersole, J. Vallieres, and D. Dainty. 1990. Laboratory study of current (1990/91) SOOT/RCD sampling methodology for the mine environment. Mining Research Laboratory, Canada Centre for Mineral and Energy Technology (CANMET). Ottawa, Canada.

Henry, R. C., C. W. Lewis, P. K. Hopke, and H. J. Williamson. 1984. Review of receptor model fundamentals. *Atmos. Environ.* 18(8):1507–15.

International Agency for Research on Cancer. 1989. *IARC Monographs on the Evaluation of Carcinogenic Risks to Humans—Diesel and Gasoline Engine Exhausts and Some Nitroarenes. Vol. 46,* p. 458. Lyon: IARC.

Jankowski, R. A. and J. Hake. 1989. Dust sources and controls for high production longwall faces. In *Proceedings of Longwall U.S.A.,* Pittsburgh, PA.

Johnson, R. L., J. J. Shaw, R. A. Cary, J. J. Huntzicker. 1981. An automated thermal-optical method for the analysis of carbonaceous aerosol. In *Atmospheric Aerosol: Source/Air Quality Relationships,* eds. W. S. Macias and P. K. Hopke, ACS Symposium Series, No. 167, pp. 223–33. Washington, DC: Am. Chem. Soc.

Keeton, S. C. 1979. Carbon particulate measurements in a diesel engine. Sandia Laboratories Publication SAND 79-8210.

Kuusisto, P. 1983. Evaluation of the direct reading instruments for the measurement of aerosols. *Am. Ind. Hyg. Assoc. J.* 44(11):863–74.

Lilienfeld, P., R. Stern, and G. Tiani. 1983. Continuous Respirable Dust Monitoring System (CRDMS). BuMines OFR *204-83*; NTIS 84-126473. Washington, DC: Office of the Federal Register, National Archives and Records Administration.

Lippman, M. 1989. Size-selective health hazard sampling. In *Air Sampling Instruments*, 7th edn., ed. S. V. Hering, pp. 163–98.

Marple, V. A., D. B. Kittleson, K. L. Rubow, and C. Fang. 1986. Methods for the selective sampling of diesel particulates in mine dust aerosol. BuMines OFR 44-87, NTIS PB 88-130810, Washington, DC.

Marple, V. A. and K. L. Rubow. 1981. Instruments and techniques for dynamic particle size measurement of coal dust. BuMines OFR 173-83; NTIS PB 83-262360, Washington, DC.

Marple, V. A. and K. L. Rubow. 1984. Respirable dust measurement. BuMines OFR 92-85, NTIS PB 85-245843, Washington, DC.

Marple, V. A., K. L. Rubow, and S. M. Behm. 1991. A micro-orifice uniform deposit impactor (MOUDI): Description, calibration, and use. *Aerosol Sci. Technol.* 14:434.

MESA. 1975. The determination of free silica in airborne dust collected on membrane filters. U.S. Dept. of the Interior, Mining Enforcement and Safety Administration, IR No. 1021, MSHA, Arlington, VA.

Mine Safety and Health Administration. 1971. Code of Federal Regulations (CFR), Title 30—Mineral Resources; Chapter 1—Mine Safety and Health Administration, Department of Labor, Rev. of 10 March 1971.

Mine Safety and Health Administration. 1988. Report of the Mine Safety and Health Administration Advisory Committee on Standards and Regulations for Diesel-Powered Equipment in Underground Coal Mines. Report to the Secretary of Labor, July 1988, Arlington, VA.

Mine Safety and Health Administration. 1989*a*. Air quality, chemical substances, and respiratory protection standards; proposed rule. *Federal Register* 54(166):35760–85.

Mine Safety and Health Administration. 1989*b*. Coal mine health inspection procedures. In *MSHA Handbook Series, Handbook No. 89-V-1*, Chapter 1. Arlington, VA: MSHA.

Mine Safety and Health Administration. 1990. Metal and nonmetal mine health inspection procedures. In *MSHA Handbook Series, Handbook No. PH 90-IV-4*, Chapter 1. Arlington, VA: MSHA.

Mine Safety and Health Administration. 1991. Code of Federal Regulations (CFR), Title 30—Mineral Resources; Chapter 1—Mine Safety and Health Administration, Department of Labor, Rev. of 1 July 1991, pp. 3–705.

Mine Safety and Health Administration. 1992. Permissible exposure limit for diesel particulate; proposed rule. *Federal Register* 57(3):500–3.

National Institute of Occupational Safety and Health. 1987. Diesel particulate measurement techniques applied to ventilation control strategies in underground coal mines. BuMines contract No. JO 145006. For information contact Mr. M. McCawley NIOSH, Morgantown, WV.

National Institute of Occupational Safety and Health. 1988. Carcinogenic effects of exposure to diesel exhaust. In *Current Intelligence Bulletin 50*, Dept. Health and Human Services (NIOSH) Publ. 88-116. NIOSH.

Olson, K. S. and D. L. Veith. 1987. Fugitive dust control for haulage roads and tailing basins. BuMines RI 9069, Washington, DC.

Organiscak, J. A. 1989. Respirable dust generation, comparison of continuous and conventional mining methods when excavating rock. BuMines RI 9233, Washington, DC.

Patashnick, H. and G. Rupprecht. 1983. Personal dust exposure monitor based on the tapered element oscillating microbalance (contract HO308106, Rupprecht & Patashnick Co., Inc.). BuMines OFR 56-84; NTIS PB 84-173749, Washington, DC.

Rahn, K. A. 1976. Chemical composition of the atmospheric aerosol. University of Rhode Island, Technical Report, Narragansett, RI.

Ramazzini, B. 1964. *De Morbis Artificum Diatriba (1713)*. Translated by W.C. Wright. New York: Hafner.

Raymond, L. D., T. F. Tomb, and P. S. Parobeck. 1987. Respirable coal mine dust sample processing. MSHA IR 1156, Arlington, VA.

Rubow, K. L., B. K. Cantrell, and V. A. Marple. 1990. Measurement of coal dust and diesel exhaust aerosols in underground mines. In *Proc. VIIth International Pneumoconioses Conf.*, NIOSH Pub. No. 90-108, pp. 645–50. NIOSH.

Rubow, K. L. and V. A. Marple. 1983. An instrument evaluation chamber: Calibration of commercial photometers. In *Aerosols in the Mining and Industrial Work Environment, Vol III*, eds. V. A. Marple and B. Y. H. Liu, pp. 777–98. Ann Arbor, MI: Ann Arbor Science Publishers.

Rubow, K. L., V. A. Marple, J. Olin, and M. A. McCawley. 1987. A personal cascade impactor: Design, evaluation and calibration. *Am. Ind. Hyg. Assoc. J.* 48:532–38.

Rubow, K. L., V. A. Marple, Y. Tao, and D. Liu. 1990. Design and evaluation of a personal diesel aerosol sampler for underground coal mines. SME Preprint No. 90-132. Littleton, CO, Soc. for Mining, Metallurgy, and Exploration, Inc.

Seaton, A. 1986. Respiratory disease in miners. *Proc. Ann. Am. Conf. Gov. Ind. Hyg.* 14:21–25.

Seaton, A., J. A. Dick, J. Dodgson, and M. Jacobsen. 1981. Quartz and pneumoconiosis in coal-miners. *Lancet* 1:1272.

Tomb, T. F. and L. D. Raymond. 1970. Evaluation of the penetration characteristics of a horizontal plate elutriator and of a 10-mm nylon cyclone elutriator. BuMines RI 7367, Washington, DC.

Tomb, T. F., H. N. Treaftis, and A. J. Gero. 1981. Instantaneous dust exposure monitors. *Environ Internat.* 5:85–96.

Treaftis, H. N., A. J. Gero, P. M. Kacsmar, and T. F. Tomb. 1984. Comparison of mass concentrations determined with personal respirable coal mine dust samplers operating at 1.2 liters per minute and the Casella 113A gravimetric sampler (MRE). *Am. Ind. Hyg. Assoc. J.* 45(12):826–32.

U.S. Congress. 1977*a*. The Federal Mine Safety and Health Act of 1977. Public Law 91-173, as amended by Public Law 95-164, Nov. 9, 1983, Stat. 803.

U.S. Congress. 1977*b*. The Federal Mine Safety and Health Act of 1977. Public Law 91-173, as amended by Public Law 95-164, Nov. 9, 1991, Stat. 1291 and 1299.

Vincent, J. H., D. Mark, W. A. Witherspoon, and I. Parker. 1988. Joint investigations of new generations of dust sampling instruments. Final Report on CEC Res. Contract 7260-03/14/08.

Watson, J. G. 1984. Overview of receptor model principles. *J. Air Pollut. Control Assoc.* 34(6):619–23.

Watts, W. F., Jr. 1987. Industrial hygiene issues arising from the use of diesel equipment in underground mines. BuMines IC 9141, Washington, DC, pp. 4–8.

Watts, W. F., Jr. 1992. Dust exposure data for the period 1985–1990; collected by MSHA and complied by the U.S. Department of the Interior, Bureau of Mines, "Mine Inspection Data Analysis System". Available from W. F. Watts, Jr., Bureau of Mines, Minneapolis, MN.

Watts, W. F., Jr., and D. R. Parker. 1987. Respirable dust levels in coal, metal, and nonmetal mines. BuMines IC 9125, Washington, DC.

Watts, W. F., Jr., D. R. Parker, R. L. Johnson, and K. L. Jensen. 1984. Analysis of data on respirable quartz dust samples collected in metal and nonmetal mines and mills. BuMines IC 8967, Washington, DC.

Williams, K. L. and R. J. Timko. 1984. Performance evaluation of a real-time aerosol monitor. BuMines IC 8968, Washington, DC.

Williams, K. L. and R. P. Vinson. 1986. Evaluation of the TEOM dust monitor. BuMines IC 9119, Washington, DC.

Zeller, H. W. 1987*a*. Effects of a barium-based fuel additive on diesel exhaust particulate. BuMines RI 9090, Washington, DC.

Zeller, H. W. 1987*b*. Effects of fuel additives on emissions. BuMines IC 9141, Washington, DC, pp. 79–93.

27

Practical Aspects of Particle Measurement in Combustion Gases

David S. Ensor

Research Triangle Institute
Research Triangle Park, North Carolina, U.S.A.

INTRODUCTION

Scope

Particle measurement requires the application of microanalysis techniques in atmospheres that are often dirty and corrosive. This chapter reviews the practical considerations of measuring aerosol mass concentration and particle size distributions in combustion gas streams upstream and downstream of control equipment.

The logistics of a sampling program depend on the size of a facility and its location. A small pilot plant or laboratory may be convenient to test for combustion aerosols. On the other hand, a large power plant may require testing very close to bends and flow disruptions, necessitating the use of many sample points because of possible particle stratification. Hoists and scaffolding may be required to complete such testing. In tests calling for coordinated measurements at several locations within a facility, work crews may be separated by long distances and may need to communicate by radio or temporary telephone systems.

Safety Considerations

Hot combustion gases can cause burns and may be toxic. In addition, heated sampling probes and exposed hardware can cause serious burns. Besides the dangers of harmful gases and exposed hardware, sources with high humidities or entrained water droplets often require the use of electrical heaters, which carry an inherent potential for shocks. Also, testing may involve the use of platforms and scaffolds to test inaccessible sections of the facility. This equipment may require assembly and awkward maneuvering under adverse conditions. Effort should be made at the test site to protect individuals from falling materials and from falls.

Often, laboratory instrumentation is used under conditions never contemplated by the developers, possibly causing failure of the test equipment and unforeseen hazards. Sampling conditions may have high positive or negative pressures or high temperatures, requiring custom equipment for safe sampling. And, the processes themselves may be subject to sudden releases or explosions.

Field crews need to be trained in first aid and in the use of all necessary safety equipment, including respirators, face shields, goggles, safety hats, climbing belts, protective boots, protective gloves, protective clothing, and ear protectors. All hazards, such as electric shock, toxic gases, and falling objects, need to be identified and precautions taken to minimize their dangers. Under extreme conditions of cold, heat, and moisture, work becomes even more dangerous, and extra precautions should be taken.

COMBUSTION AND CONTROL DEVICES

Combustion and associated control equipment have been studied extensively for certain industries over the last few years. The reader should consult summaries such as Flagan and Friedlander (1978) for a description of the basic mechanisms of ash coalesce/breakup and volatilization and condensation. McElroy et al. (1982) described the particle-size-dependent chemistry and concentration from six different electric utility power plants and showed particle-dependent enrichment and the effects of baghouses and electrostatic precipitators.

Control Equipment

The careful evaluation of particulate control equipment is a demanding task. Test procedures have been described by Smith, Cushing, and McCain (1977), Smith, Cavanaugh, and Wilson (1978), McDonald and Dean (1980), Markowski et al. (1980), Cushing and Smith (1978), and Ensor et al. (1979). Shendrikar and Ensor (1986) summarized the additional constraints required when measuring the particle-size-dependent chemistry of combustion gases and control device efficiency. The following must be considered when testing particulate control equipment:

- *Wide range in concentration.* The inlet and outlet concentrations of high-efficiency control equipment will vary widely. The inlet concentration can be orders of magnitude greater than the outlet concentration. When the purpose of the test is to measure the particle-size-dependent efficiency, matching the test method and the measurement period is difficult.
- *Stratification of particles in the ducts.* In large power plants, the ducts are very large, and aerosol stratification will result from particle sedimentation. Often, the sample locations are near flow disruptions and leaks from the atmosphere may bias the measurements. Therefore, the concentration and size distributions may vary greatly depending upon the location in the duct.
- *Control equipment cycling.* Electrostatic precipitators remove collected particles from the plates by rapping them with mechanical hammers. A rap liberates a puff of particles about 6 μm in diameter. The rapping puff has been experimentally found to be quite similar in particle size in all units (Smith, Cushing, and McCain 1977). The rappers are adjusted for minimum emissions and the frequency of rapping depends on the location of the rappers in the electrostatic precipitator.

  Fabric filter baghouses clean the bags either by shaking, reverse air flow, or pulses of air, all of which create a puff of particles. These process cycles need to be documented and considered during the test plan.

  Elaborate schemes have been developed to measure the emissions during plate rapping or bag cleaning. Two complete sets of sampling probes have been used: one between puffs, and the other during puffs (Smith, Cushing, and McCain 1977). The probes are turned downstream when not in use. Another approach is to use an extractive sampling system with a real-time instrument.
- *Entrained water.* Ducts downstream of a wet scrubber may contain water droplets. Usually, the test equipment must be heated above the dew point. The sampling inlets may need to be extended and heated to allow sufficient residence time to evaporate the water droplets.
- *Unexpected sources of particles.* Some processes, such as water sprays to cool the gases, may generate particles. Ensor and Harmon (1980) found that evaporation of water droplets in a particulate scrubber generated submicrometer particles in proportion to the concentration of dissolved solids.

- *Air infiltration.* Many ducts have subatmospheric pressure. Outside air may leak into the duct from holes or equipment such as air heaters.
- *Acidic gases.* Many combustion processes contain acidic gases (e.g., sulfur dioxide, sulfur trioxide, oxides of nitrogen, or hydrochloric acid) as products of combustion. These, in addition to being corrosive to equipment, may complicate the test program. Cascade impactors with a significant pressure drop may create supersaturated conditions and cause condensation (Biswas, Jones, and Flagan 1987). In addition, the collected particles on filters and impaction stages may have artifactual weight increases because of chemical reactions (Lundgren and Balfour 1980).

SUPPORTING MEASUREMENTS

Table 27-1 summarizes the supporting measurements. This summary is by no means comprehensive but is intended to illustrate test requirements. Process conditions should be documented to allow a reproduction of conditions. Process measurements include fuel analysis, quantity, and load.

Depending upon the sampling system, the velocity must be known in order to determine isokinetic sampling conditions. Velocity (times concentration) is also used to determine the rate of material emitted from a process. The determination of velocity is described in EPA Methods 1 and 2 (U.S. EPA 1977*a*, *b*).

For sampling particle-laden streams, a type-S pitot tube is used to measure gas velocity because it is rugged and its probe is resistant to plugging. A temperature probe is often attached to the pitot tube. The number of sampling points required for velocity and mass depends on the proximity of the sampling point to upstream and downstream flow distributions.

TABLE 27-1 Supporting Measurements

	Comments
Process	
Fuel	Standard process instrumentation and bulk samples
Load	Plant instrumentation
Duct	
Velocity	EPA Methods 1 and 2
Temperature	Thermometer or thermocouple
Pressure	Manometer or pressure gauge
Gas	
Oxygen, CO_2, CO	EPA Method 3
Water	Part of EPA Methods 5 and 17
SO_2, NO, CO, SO_3	Either manual method or instrumentation
Mass concentration	EPA Methods 5 and 17
Control equipment	
Pressure drop	Plant instrumentation
Energy	Process calculation
Process flows	Plant instrumentation and process calculation

MASS MEASUREMENT

Particle mass emissions are measured by obtaining a filter sample from the flue gas. Two different approaches are used: external filtration and in-stack filtration. EPA Method 5 uses an external filtration approach and is illustrated in Fig. 27-1 (U.S. EPA 1977*c*). The flue gas is sampled with an isokinetic nozzle and is piped through a heated tube and a heated chamber (maintained at $120 \pm 14°C$) containing the filter. Immediately following the filter, impingers capture water and condensable material. The in-stack filter technique, called EPA Method 17, is shown in Fig. 27-2 (U.S. EPA 1978). The filter is near the sample nozzle and is kept at the duct temperature.

EPA Method 17 is much simpler to use than EPA Method 5; it employs a flexible hose, which allows one person to hold and move the probe simultaneously. Positioning Method 5 requires a roller-and-rail assembly because the probe and the filter box are connected. In addition, the inside of the tube in the long probe must be brushed carefully to quantitatively recover particulate matter that may have deposited during the test. Data taken with the two test methods may differ

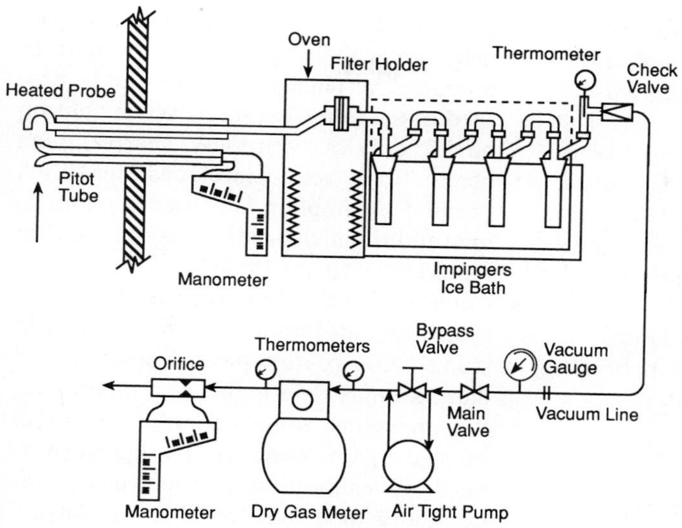

FIGURE 27-1. EPA Method 5 Particulate-Sampling Train (U.S. EPA 1977).

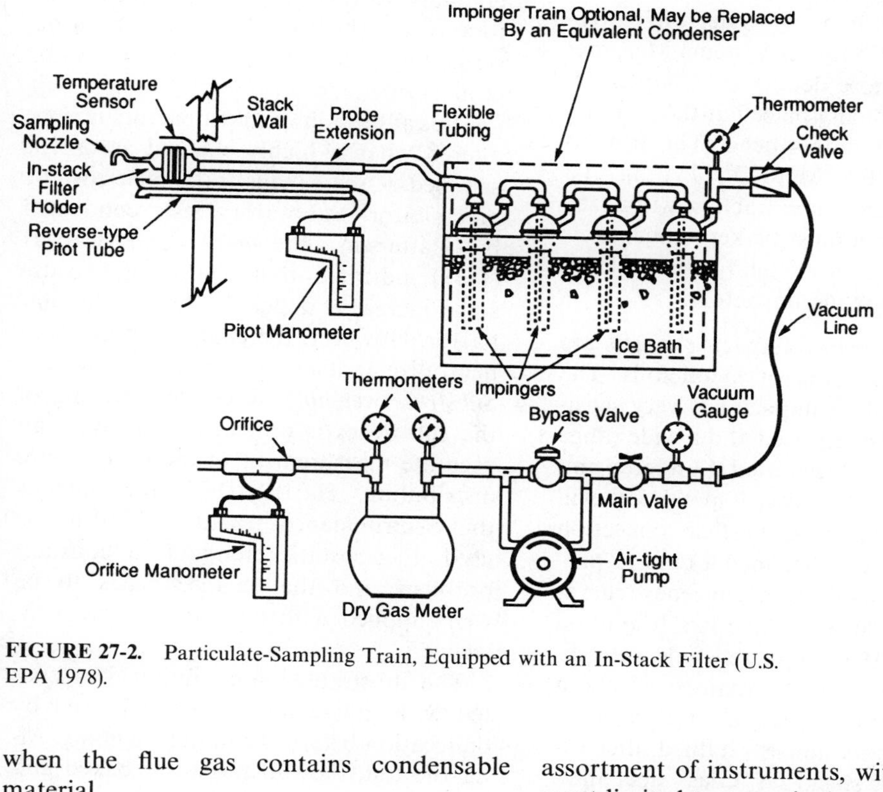

FIGURE 27-2. Particulate-Sampling Train, Equipped with an In-Stack Filter (U.S. EPA 1978).

when the flue gas contains condensable material.

PARTICLE SIZE DISTRIBUTION

Measurement of the particle size distribution in combustion effluents usually employs an assortment of instruments, with each instrument limited to a particular range of particle sizes. The cascade impactor has been used for over 20 years to obtain particle size distributions (Pilat, Ensor, and Bosch 1970). Impactors require great skill to use but can be used to obtain particle measurements from 0.3

to about 10 μm. Low-pressure impactors can be used to collect particles as small as 0.05 μm (Pilat et al. 1978; Vanderpool, Lundgren, and Kerch 1987).

In the mid-1970s, in an effort to obtain particle size distributions for particle control, a number of extractive sampling systems were developed. The first systems consisted of condensation nuclei counters combined with diffusion batteries (Smith et al. 1975). The dilution systems were later adapted to electrical aerosol size analyzers (Sem 1979).

Cascade Impaction

Cascade impactors, described in more detail by Marple and Willeke (1979) and in Chapter 11, have been developed specifically for sampling flue gas. Impactors developed for flue gas are manufactured by Anderson Instruments, Inc., and Pollution Systems Measurement, Inc. For more details see Chapter 11. An impactor is usually placed at the end of a pipe with a long radius bend. The train is similar to an EPA Method 17 train. To maintain a consistent size fraction within the impactors, the flow must be kept constant.

The following issues should be considered when selecting cascade impactors:

- *Collection substrates.* Impactor inserts are typically filter media or metal foils. The performance of an impactor is very sensitive to the substrate used and a wide range of materials have been tried (Lundgren and Balfour 1980). A greased metal foil will produce nearly ideal particle collection curves. However, filter media tend to produce particle collection efficiency curves that are broader and have less resolution (Smith et al. 1975). Greased foils are subject to temperature limitations of about 500°F. The best greases are high-vacuum because the adhesion is retained after exposure to hot stack gas. Glass fiber filter media are subject to chemical reactions with acidic gases and may produce anomalous weight gains (Felix et al. 1977).
- *Wall losses.* The first stages have the largest losses. In a laboratory evaluation, Smith et al. (1975) found that wall losses for particles larger than 5 μm were important. In practice, the material on the wall is brushed to the following stage. When "button hook" nozzles with very sharp tubing bends, that were developed for EPA Method 5 sampling, are used for impactors, additional deposition may be found in the nozzle (Knapp 1980).
- *Calibration.* Effort has been made to calibrate cascade impactors used in field studies on a routine basis. Corrosion and damage may affect the performance of the impactor. Some variation exists because the jet holes tend to be slightly oversized. A calibration method using one stage at a time was described by Calvert, Lake, and Parker (1976); the method uses polystyrene aerosol and optical particle counters. Examples of whole impactor calibrations with dye aerosol have been described by Cushing et al. (1976) and Rader et al. (1991).

  The calibration of the impactors may be altered when a highly charged aerosol is sampled at the exhaust of an electrostatic precipitator. Laboratory tests conducted by Patterson, Riersgard, and Calvert (1978) indicated that charge may cause some increased deposition within the impactor. However, the charge is difficult to neutralize at stack temperatures.
- *Substrate weighing.* Accurate weighing of the substrates is critical to obtaining an accurate measurement of the particle size distribution. Harris (1977) recommends that electrobalances sensitive to 0.01 mg be used. Electrostatic charge is a constant problem, and the balance needs to be well supplied with fresh radioactive neutralizers.

  The substrates are conditioned or coated with grease and baked, followed by desiccation before the initial weighing. After the test, the substrates are baked and desiccated until reweighed. Hygroscopic deposits on the substrates can cause large weight gains even in a laboratory environment. The author has placed the analytical balance in a dry box. Samples should not be conditioned at constant elevated humidity because the humidity in a stack environment is usually less than 5% RH.

- *Particle bounce and blowoff.* The impaction process is a tradeoff between impaction of the particle and reentrainment of deposited material. The fraction of particles blown from the surface depends on the properties of the particles and the surface. Also, the blowoff is suspected to increase as the loading increases (Harris 1977). Therefore, the generally accepted maximum loading for an impactor stage is a few milligrams. Markowski (1987) suggests that duplicate stages be used to detect particle reentrainment and to obtain information for correction of the final data.

Extractive Sampling

Extractive sampling with dilution is very useful for measuring aerosols from either hot or concentrated sources. The diluted gas can be introduced with instruments developed for sampling particles under ambient conditions. Two objectives usually guide the development of dilution systems:

- Quenching the aerosol and measuring under conditions with only the concentration altered, with the objective of quantifying the particle size collection efficiency of control equipment.
- Simulating a specific process, such as a smoke plume entering a cold atmosphere. The condensible inorganic and organic components will condense and coagulate, as would be expected upon discharge to the atmosphere.

The resulting particle size distribution may be quite different from that found with undiluted gas because of the condensation of semivolatile materials.

A number of different approaches to dilution have been reported: the injector approach (Heinsohn, Davis, and Knapp 1980), the mixing-orifice approach (Ensor and Hooper 1977), the multijet mixing chamber (Smith et al. 1975), the porous-wall diluter approach (Ranade et al. 1976; Ensor et al. 1981), and the mixing-tee approach (Hildemann, Cass, and Markowski 1989). Heinsohn, Davis, and Knapp (1980) and Hildemann, Cass, and Markowski (1989) developed smoke plume simulators; the others sampled stack gas with a minimum of modification.

The porous-wall diluter seems to exhibit less wall loss than the other methods and to have the widest range of dilution ratios. Ranade, Werle, and Wasan (1976) observed that aerosol is pushed to the center of the porous tube with the dilution air.

Several factors must be considered when planning the construction of a dilution system to be applied to a wide range of sources. First among these considerations is the ability to adjust dilution ratios over a wide range of values. The consistency of the particle size distribution as a function of dilution ratio is one way to test for condensation artifacts. Second, the flow of undiluted aerosol needs to be monitored and measured. This allows a quantification of the dilution ratio. Third, the equipment should allow an adjustment to obtain clean-air background checks of the dilution system and the supporting instrumentation during a use. Sometimes aerosol deposits from previous use will vaporize, causing confusing results.

An example of a porous-wall dilution system is shown in Fig. 27-3 (Ensor et al. 1981). The first stage of the diluter is inserted through a sample port. A venturi flowmeter was used to measure the gas velocity with the pressure-drop gauge mounted outside the duct. For particle sizes less than 2 μm, it was found that loading of the system caused by large particles could be prevented by pointing the probe nozzle downstream. A venturi was selected for minimum modification of the aerosol. The clean-air background conditions also can be obtained by increasing the dilution pressure and reversing the flow through the venturi, blowing the clean air into the duct. The direction of flow is indicated by a reversal in the pressure drop at the gauge. The dilution is controlled by the flow controller and the bleed of outside air into the system.

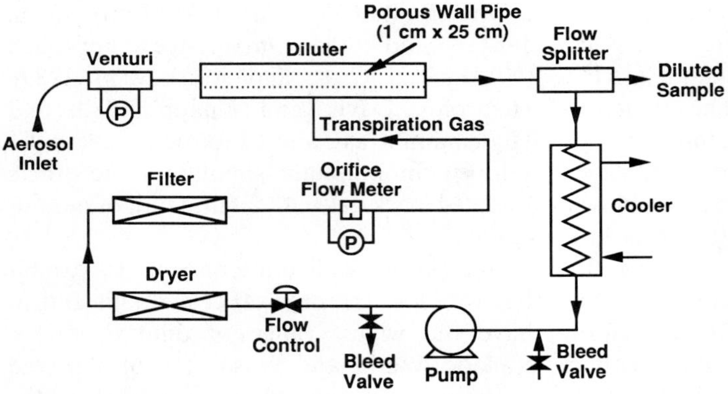

FIGURE 27-3. Flow Diagram of the Dilution System for Flue Gas Sampling.

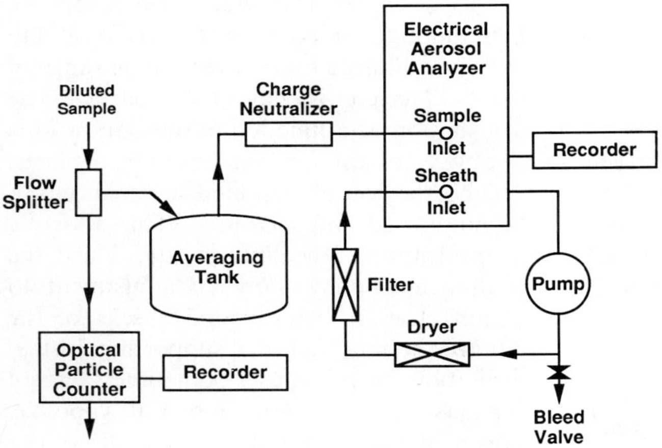

FIGURE 27-4. Sample Flow Diagram of Particle Size Measurement with an Electrical Aerosol Analyzer and an Optical Particle Counter.

The flow splitter contains a nozzle that samples isokinetically a fraction of the diluted gas stream. The system is constructed in four porous-wall modular dilution units, which are connected in series, allowing a wide range of dilution ratios. The remainder of the system is shown in Fig. 27-4. The flow is split between an optical particle counter and an electrical aerosol analyzer. An averaging tank is used to reduce the concentration fluctuation to be compatible with the cycle time of the instrument. The system was calibrated, as shown in Fig. 27-5, with an optical particle counter.

Quality Assurance

The collection of high-quality data under field conditions requires attention to the details and the use of standard procedures for quantifying data quality. Tatsch (1982) identifies the following quality assurance procedures:

- Project planning
- Adequate preparation by calibration and procedure development in preliminary tests
- Record keeping and documentation
- Chain-of-custody of samples and data sheets

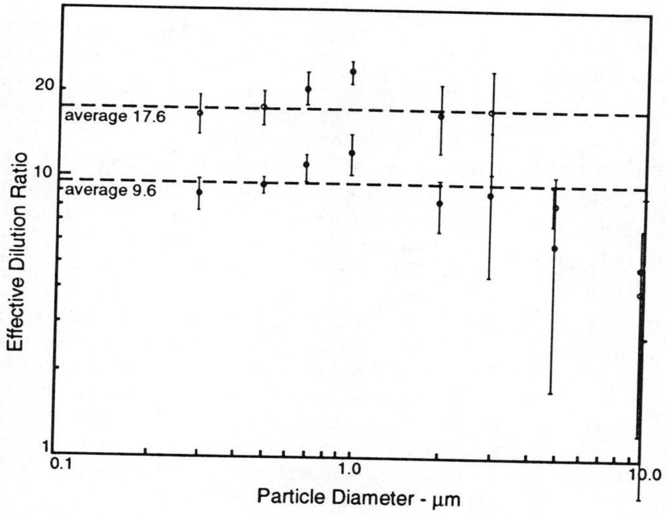

FIGURE 27-5. Performance of the Diluter in Fig. 27-3 for Source Measurement (Average of Four Diluters). The Effective Dilution Ratio (Dotted Line) is that Obtained from the Air and Flow Rates by Compulation. The Dilution Ratio Obtained by Measuring the Concentration of Particles at the Inlet and Exhaust is Shown by the Data Points with 1 Standard Deviation Error Bars. This Shows the Accuracy of Dilution and the Wall Losses in the Porous Wall Dilution System.

- Devising tests for artifact formation such as system leak tests, blank impactor tests with filtered stack gas, various dilution ratios during extractive sampling
- Near-real-time data reduction to detect errors and artifacts
- Control samples to isolate problems in handling
- Calibration of equipment
- Establishing data quality goals in terms of data recovery and variability

IMPACTOR DATA REDUCTION

In the reduction of particle size distribution data, three factors must be considered:

- Cascade impactor data are obtained with different impactors with different size cuts and a limited number of points.
- Submicrometer particle measurements are made with instruments having a broad cutoff.
- How these data can be combined.

A widely used method for cascade impactor data reduction was devised by Calvert and reported by Ensor et al. (1975). The procedure was automated by Markowski and Ensor (1977), McCain et al. (1978), and Durham et al. (1982) for microcomputers. The procedure is as follows:

1. The particle diameter corresponding to 50% collection is computed for each stage in the impactor, using the appropriate temperature, pressure, and gaseous molecular weight.
2. A cumulative smaller-than-mass concentration curve is plotted as a function of geometric mean diameter.
3. The curve is then fitted with a spline function.
4. At selected particle diameters, the slope of the curve is computed to obtain a differential curve, $\Delta M/\Delta d$.
5. The particle-dependent penetration is the computed ratio of the outlet $\Delta M/\Delta d$ to the inlet $\Delta M/\Delta d$ for selected particle diameters. A computation of the particle-size-dependent penetration is preferred to that of efficiency, because a consistent number of significant figures can be retained and

the mean and standard deviation can be computed easily. This method can be applied to data from different kinds of cascade impactors.

The reduction of extractive system data involves correcting the particle size distributions for the dilution ratios. One persistent problem is the comparison of particle size distributions taken with different instrumentation. For a further discussion of size distribution data deconvolution see Chapter 9.

References

Anderson Instruments, Inc., 4801 Fulton Industrial Blvd., Atlanta, GA 30336.

Biswas, P., C. L. Jones, and R. C. Flagan. 1987. Distortion of size distributions by condensation and evaporation in aerosol instruments. *Aerosol Sci. Technol.* 7:231–46.

Calvert, S., C. Lake, and R. Parker. 1976. *Cascade Impactor Calibration Guidelines*, EPA-600/2-76-118, Washington, DC: U.S. Environmental Protection Agency.

Cushing, K. M., C. E. Lacey, J. D. McCain, and W. B. Smith. 1976. *Particulate Sizing Techniques for Control Device Evaluation: Cascade Impactor Calibrations*, EPA-600/2-76-280. U.S. Environmental Protection Agency.

Cushing, K. M. and W. B. Smith. 1978. *Procedures Manual for Fabric Filter Evaluation*, EPA-600/7-78-113. Washington, DC: U.S. Environmental Protection Agency.

Durham, M., S. Tegtmeyer, K. Wasmundt, and L. E. Sparks. 1982. A microcomputer-based cascade-impactor data-reduction system. In *Proc. 3rd Symposium on the Transfer and Utilization of Particulate Control Technology. Vol. 4*, pp. 285–96, EPA-600/9-82-005d. U.S. Environmental Protection Agency.

Ensor, D. S. et al. 1975. *Evaluation of a Particulate Scrubber on a Coal-Fired Utility Boiler*. EPA-600/2-75-074. Research Triangle Park, NC: U.S. Environmental Protection Agency.

Ensor, D. S. and D. L. Harmon. 1980. Evaluation of the Ceilcote ionizing wet scrubber. In *Proc. 2nd Symposium on the Transfer and Utilization of Particulate Control Technology*. EPA-600/9-80-039c. Research Triangle Park, NC: U.S. Environmental Protection Agency.

Ensor, D. S. and R. G. Hooper. 1977. The Meteorology Research, Inc., extractive sampling system for submicron particles. In *Proceedings of the Seminar on In-Stack Sizing for Particulate Control Device Evaluation*, PB-266-103, EPA-600/2-77-060. Research Triangle Park, NC: U.S. Environmental Protection Agency.

Ensor, D. S., R. G. Hooper, G. R. Markowski, and R. C. Carr. 1979. Evaluation of performance and particle size dependent efficiency of baghouses. In *Proceedings Advances in Particle Sampling and Measurement*, pp. 314–36. EPA-600/7-79-065. Asheville, NC: U.S. Environmental Protection Agency.

Ensor, D. S., M. B. Ranade, P. A. Lawless, D. W. VanOsdell, and A. S. Damle. 1981. Extractive sampling with dilution probes. In *Proceedings Particulate Sampling and Characterization*. pp. 125–47. Morgantown, WV: U.S. Department of Energy Contractor's Workshop.

Felix, L. G., G. I. Clinard, G. E. Lacey, and J. D. McCain. 1977. *Interial Cascade Impactor Substrate Media for Flue Gas Sampling*, EPA-600/7-77-060, Washington, DC: U.S. Environmental Protection Agency.

Flagan, R. G. and S. K. Friedlander. 1978. Particle formation in pulverized coal combustion—A review. In *Recent Developments in Aerosol Science*, ed. D. T. Shaw, pp. 25–59. New York: Wiley.

Harris, D. B. 1977. *Procedures for Cascade Impactor Calibration and Operation in Process Streams*, EPA-600/2-77-004. Washington, DC: U.S. Environmental Protection Agency.

Heinsohn, R. J., J. W. Davis, and K. T. Knapp. 1980. Dilution source sampling system. In *Proceedings Advances in Particle Sampling and Measurement*, pp. 107–29. EPA-600/9-80-004. Washington, DC: U.S. Environmental Protection Agency.

Hildemann, L. M., G. R. Cass, and G. R. Markowski. 1989. A dilution stack sampler for collection of organic aerosol emission: Design, characterization and field tests. *Aerosol Sci. Technol.* 10:193–204.

Knapp, T. 1980. The effects of nozzle losses on impactor sampling. In *Proceedings Advances in Particle Sampling and Measurement*, pp. 101–106, EPA-600/9-80-004. Research Triangle Park, NC: U.S. Environmental Protection Agency.

Lundgren, D. A. and W. D. Balfour. 1980. *Use and Limitations of Instack Impactors*. EPA-600/2-80-048. Research Triangle Park, NC: U.S. Environmental Protection Agency.

Markowski, G. R. 1987. On identifying and correcting for reentrainment in cascade impactor measurement. *Aerosol Sci. and Technol.* 7:143–59.

Markowski, G. R. and D. S. Ensor. 1977. A procedure for computing particle size dependent efficiency for control devices from cascade impactor data. Paper presented at the 70th Annual Meeting of the Air Pollution Control Association, June, at Toronto, Canada.

Markowski, G. R., D. S. Ensor, M. E. Drehsen, and A. D. Shendrikar. 1980. *Fine Particle Sampling, Analysis and Data Reduction Procedures, and Manual for Low Pressure Impactor and Electrical Aerosol Analyzer*, EPRI Project 1410-3. Palo Alto, CA: Electric Power Research Institute.

Marple, V. A. and K. Willeke. 1979. Inertial impactors. In *Aerosol Measurement*, eds. D. A. Lundgren, et al., pp. 90–107. Gainesville, FL: University Presses of Florida.

McCain, J. D., B. Clinard, L. Felix, and J. Johnson. 1978. A data reduction system for cascade impactors. In *Proceedings Advances in Particle Sampling Measurement*, EPA-600/7-79-065. Asheville, NC: U.S. Environmental Protection Agency.

McDonald, J. R. and A. H. Dean. 1980. *A Manual for the Use of Electrostatic Precipitators to Collect Fly Ash Particles*, p. 746, EPA-600/8-80-025. Washington, DC: U.S. Environmental Protection Agency.

McElroy, M. W., R. C. Carr, D. S. Ensor, and G. R. Markowski. 1982. Size distribution of fine particles from coal combustion. *Science* 215:13–19.

Patterson, R. G., P. Riersgard, and C. Calvert. 1978. *Effects of Charged Particles on Cascade Impactor Calibrations*, EPA-600/7-78-195. Washington, DC: U.S. Environmental Protection Agency.

Pilat, M. J., D. S. Ensor, and J. C. Bosch. 1970. Source test cascade impactor. *Atmos. Environ.* 4:671–79.

Pilat, M. J., G. A. Raemhild, E. B. Powell, G. M. Fiorette, and D. F. Myer. 1978. *Development of Cascade Impactor Systems for Sampling 0.02 to 20-micron Diameter Particles, Vol. 1*, FP-844. Palo Alto, CA: Electric Power Research Institute.

Pollution Control Systems, Inc., P.O. Box 15570, Seattle, WA 98115.

Rader, D. J., L. A. Mondy, J. E. Brockmann, D. A. Lucero, and K. L. Rubow. 1991. Stage response calibration of the Mark III and Marple personal cascade impactors. *Aerosol Sci. Technol.* 14:380–87.

Ranade, M. B., D. K. Werle, and D. T. Wasan. 1976. Aerosol transport through a porous sampling probe with transpiration air flow. *J. Colloid Interface Sci.* 56:42.

Sem, G. J. 1979. Electrical aerosol analyzer: Operation, maintenance, and application. In *Aerosol Measurement*, eds. D. A. Lundgren, et al., pp. 400–30, Gainesville, FL: University Presses of Florida.

Shendrikar, A. D. and D. S. Ensor. 1986. Sampling and measurement of trace element emissions from particulate control devices. In *Toxic Metals in the Atmosphere*. eds. J. O. Nriagu and C. I. Davidson, pp. 53–111. New York: Wiley.

Smith, W. B., P. R. Cavanaugh, and R. R. Wilson. 1978. *Technical Manual: A Survey of Equipment and Methods for Particulate Sampling in Industrial Process Streams*, EPA-600/7-78-043. Washington, DC: U.S. Environmental Protection Agency.

Smith, W. B., K. M. Cushing, G. E. Lacey, and J. D. McCain. 1975. *Particle Sizing Techniques for Control Device Evaluation*, EPA-650/2-74-102-a. Washington, DC: U.S. Environmental Protection Agency.

Smith, W. B., K. M. Cushing, and J. D. McCain. 1977. *Procedures Manual for Electrostatic Precipitator Evaluation*, EPA-600/7-77-059. Washington, DC: U.S. Environmental Protection Agency.

Tatsch, C. E. 1982. Quality assurance for particle-sizing measurements. In *Proc. 3rd Symposium on the Transfer and Utilization of Particulate Control Technology, Vol. 4*, EPA-600/9-82-005d. U.S. Environmental Protection Agency.

U.S. Environmental Protection Agency. 1977*a*. EPA Method 1: Sample and velocity traverses for stationary sources. *Federal Register* 42(160): 41755.

U.S. Environmental Protection Agency. 1977*b*. EPA Method 2: Determination of stack gas and volumetric flowrate. *Federal Register* 42(160):41758.

U.S. Environmental Protection Agency. 1977*c*. EPA Method 5: Determination of particulate emissions from stationary sources. *Federal Register* 42(160): 41776.

U.S. Environmental Protection Agency. 1978. EPA Method 17: Determination of particulate emissions from stationary sources (in-stack filtration method). *Federal Register* 43(37).

Vanderpool, R. W., D. A. Lundgren, and P. E. Kerch. 1987. Design and calibration of an in-stack low-pressure impactor. *Aerosol Sci. Technol.* 12:143–59.

28

Ambient Air Sampling

John G. Watson and Judith C. Chow

Desert Research Institute
University and Community College System of Nevada
5625 Fox Avenue,
Reno, NV, U.S.A.

INTRODUCTION

Ambient aerosol sampling is performed for a variety of purposes. A sampling system designed for one purpose does not necessarily meet the needs for other or additional purposes. Recent experience has shown, however, that sampling components which were developed, tested and applied in different types of air pollution studies can be practically assembled in different ways to meet multiple sampling goals.

The objectives of this chapter are: (1) to outline the generic requirements of ambient aerosol sampling systems; (2) to identify the components which can meet these requirements; and (3) to describe the existing particle sampling systems which have been assembled and which have applied these components to real-world problems. The major emphasis of this discussion concerns sampling systems which acquire aerosol deposits on filter media that are prepared and analyzed in a laboratory setting. Chapter 12 discusses analytical methods that can be applied to these deposits. Owing to the brevity of this chapter, a substantial number of references are identified which provide more quantitative information and more complete explanations than are offered here.

SAMPLING SYSTEM REQUIREMENTS

Ambient sampling of suspended particulate matter is commonly undertaken to: (1) determine compliance with air quality standards; (2) evaluate the extent and causes of deposition and visibility impairment; (3) enhance an understanding of the chemical and physical properties of atmospheric pollution; and (4) apportion the chemical constituents of suspended particulate matter to their emitting sources.

The current U.S. National Ambient Air Quality Standard for suspended particulate matter applies to the mass of suspended particles less than 10 μm aerodynamic diameter (PM_{10}). This standard is exceeded when the mass of these particles is greater than 150 μg/m^3 over a 24 h period more than three times in three years or when an annual average of 24 h samples exceeds 50 μg/m^3 (Federal Register 1987*a*). The Federal Register (1987*b*) specifies requirements which must be met by any system used to determine compliance with the PM_{10} standard. These include: (1) wind tunnel testing of sampler inlets; (2) sampling efficiency and alkalinity standards of filter media; (3) stability of sample flow rates; and (4) precision of gravimetric analysis.

To evaluate visibility reduction, it is necessary to measure the mass and chemical composition of particles with aerodynamic diameters less than ~ 2.5 μm ($PM_{2.5}$), since particles in this range scatter light most efficiently (Hinds 1982). The major constituents of $PM_{2.5}$ which scatter and absorb light are sulfates, nitrates, ammonium, organic carbon, elemental carbon, and crustal species (e.g., Al, Si, Fe, Ca, Ti) (Sloane et al. 1990). Each of these chemicals has a different effect on light extinction, and each one needs to be quantified. A simple mass measurement is insufficient to determine the contributions to light extinction. Samples need to be taken over time periods which correspond to visible haze (daylight hours), thereby requiring multiple samples of less than 24 h duration.

Research is being conducted to determine the concentrations of acidic aerosols which affect human health (Koutrakis et al. 1988) and forests (Ashbaugh, Chow, and Watson 1991) and to understand the chemical formation mechanisms and equilibrium of inorganic (e.g., Pilinus and Seinfeld 1988) and organic (e.g., Grosjean and Seinfeld 1989) species. For certain purposes, the measurement of chemical concentrations in finely divided size fractions is needed (e.g., John et al. 1990; Sloane et al. 1991). Diurnal changes in ambient aerosol concentrations have also been required for certain research studies.

Source apportionment models (also termed "receptor models") use the chemical and physical characteristics of gases and particles measured at source and receptor to both identify the presence of and to quantify source contributions to pollutants measured at the receptor. The temporal and spatial variability of these concentrations can also be incorporated into receptor models to further specify the origins and contributions of species measured at a receptor. To distinguish the contributions of one source from another, the chemical and physical characteristics must be such that: (1) they are present in different proportions in different source emissions; (2) these proportions remain relatively constant for each source type; and (3) changes in these proportions between source and receptor are negligible or can be approximated. Source apportionment of PM_{10} has been commonly applied to develop emissions reduction strategies using software (Watson et al. 1990), applications and validation protocols (Watson, Chow, and Pace 1991), and model reconciliation protocols (Pace 1991) issued by the U.S. Environmental Protection Agency (EPA).

The general ambient aerosol sampling requirements for all of these study objectives are: (1) well-defined size fractions; (2) filter media which are compatible with the intended analysis methods; (3) sampled air volumes which are stable and provide sufficient deposit for the desired analyses without overloading the filter; (4) sampling surfaces which do not react with the measured species; and (5) available, cost-effective, and practical sampling hardware and operating procedures. No single ambient aerosol sampling system exists which meets the needs of every study. However, different sampling components which have been developed for different purposes can be assembled to attain a large variety of study objectives. These components are: (1) one or more size-selective inlets; (2) sampling surfaces; (3) substrates; (4) substrate holders; and (5) flow movers and controllers. A number of options are available within each of these categories.

SAMPLING INLETS

Sampling inlets are intended to remove particles which exceed a specified aerodynamic particle diameter. These inlets are characterized by sampling effectiveness curves which show the fraction of spherical particles of unit density penetrating through the inlet to the filter surface (Watson et al. 1983). These curves are measured by presenting known concentrations of particles of selected aerodynamic diameters to the inlet at different wind velocities in a wind tunnel (Federal Register 1987*b*). This sampling effectiveness is characterized by a 50% cut-point (d_{50}), the diameter at which half of the particles in the ambient air pass through the inlet, and a slope (or geometric standard deviation),

TABLE 28-1 Summary of the Size-Selective Inlets for Ambient Aerosol Samplers

Inlet Identifier & Manufacturer	d_{50} (μm); Slope[c]	Flow Rate (lpm)	Operating Principle	Sampling Effectiveness Reference	PM_{10} Reference or Equivalence Number	Comments
High-Volume						
SA[a] or GMW[b] Model 320	15; 1.5	1,130	Impactor	McFarland, Ortiz, and Rodes (1980)	N.A.	Single stage, no greased shim.
SA or GMW Model 1200	9.7; 1.40	1,130	Impactor	McFarland and Ortiz (1987)	RFPS-1287-063 FR[d] 12/1/87, p. 45683	Single-stage with greased shim (body hinged)
SA or GMW Model 321-A	10.2; 1.45	1,130	Impactor	McFarland and Ortiz (1984)	RFPS-1287-065 FR 12/1/87, p. 45683 (named 321-C)	Single-stage with greased shim
SA or GMW Model 321-B	9.7; 1.40	1,130	Impactor	McFarland and Ortiz (1987), VanOsdell and Chen (1990), Woods, Chen, and Ranade (1986)	RFPS-1287-064 FR 12/1/87 p. 45683	Two-stage with greased shim
Wedding IP_{10} (PM_{10})	9.6; 1.37	1,130	Cyclonic Flow	Wedding and Weigand (1985), Woods, Chen, and Ranade (1986)	RFPS-1087-062 FR 10/6/87 p. 3736	
GMW Wedding PM_{10}	8.8	1,130	Cyclonic Flow	Woods, Chen, and Ranade (1986)	None	
Medium-Volume						
SA-254 Medium Volume (PM_{10})	10; 1.6[c]	113	Impactor	Olin and Bohn (1983)	RFPS-0389-071 FR 3/24/87, p. 12273)	Designated for Oregon only
Wedding Medium Flow	9.5; 1.12	113	Cyclonic Flow	Wedding et al. (1983)	None	

Bendix 240 Cyclone	2.5; 0.82	113	Cyclonic Flow	Chan and Lippman (1977), Mueller et al. (1983)	None	
Low-Volume						
SA-246	10.2; 1.41	16.7	Virtual Impactor	McFarland, Ortizs, and Bertch (1984), VanOsdell and Chen (1990)	RFPS-0789-073 FR 7/27/89, p. 31247 EQPM-0990-076 FR 9/18/90, p. 38387 EQPM-1090-079, FR 10/29/90, p. 43406	Used on dichotomous sample, beta attenuation monitor, and tapered element oscillating microbalance
Wedding TP_{10}	9.8	16.67	Cyclonic Flow	Wedding, Weigand, and Carney (1982)	EQPM-0391-081 FR 3/5/91, p. 9216	
Bendix Unico 18	2.5	18	Cyclonic Flow	Chan and Lippmann (1977)	None	
AIHL Cyclone	2.5	22	Cyclonic Flow	John and Reischl (1980)	None	
SA 244 and 245		15	Virtual Impactor	McFarland, Ortiz, and Bertch (1978), Olin (1978)	None	
AIHL Cyclone	2.2	30	Cyclonic Flow	None	None	
Stacked Filter Unit	2–3	10	Selective Filtration	John et al. (1983*a*, *b*)	None	Cahill et al. (1990)

a. Sierra–Andersen
b. General Metal Works
c. $c_{slope} = (\sqrt{d_{84}/d_{16}})$, as defined in text
d. FR = Federal Register

which is the square root of the ratio of the diameter of particles excluded by the inlet with an 84% efficiency (d_{84}) to the diameter removed with a 16% efficiency (d_{16}). A slope of 1 indicates a step function, while a slope which exceeds 2 does not provide a definitive size cut. Slopes of 1.3–1.5 are considered to provide well-defined particle size fractions.

Acceptable size-selective inlets must also possess particle transmission characteristics which are independent of wind speed and wind direction. The sampling efficiency of the rectangular peaked-roof inlet of the high-volume (hivol) sampler for total suspended particulate (TSP) matter was found to have a large variability in sampling effectiveness in response to these environmental variables (Wedding, McFarland, and Cermak 1977; McFarland, Ortiz, and Rodes 1980).

Hering (1989) provides a comprehensive list of commercially available size-selective inlets and photographs of several of them. Table 28-1 identifies inlets which have been commonly used in ambient-air pollution studies, especially those corresponding to PM_{10} and $PM_{2.5}$ size ranges. Where available, both the 50% cut-points and slopes are given with citations for the tests which were applied to measure them. Inlets passing the Federal Register (1987*b*) requirements for PM_{10} are identified by EPA method number. In some cases there is more than one reference method number because the same inlet has been coupled to different sampling components. The 50% cut-point (d_{50}) and slope ($\sqrt{d_{84}/d_{16}}$) apply to the listed flow rates. These flow rates fall into ranges appropriate for high-volume (~ 1000 lpm), medium-volume (~ 100 lpm), and low-volume (~ 10–20 lpm) sampling systems. The medium- and high-volume inlets are especially useful when samples are taken in parallel on several substrates, since flow rates can be kept high enough to obtain an adequate deposit for analysis.

Ambient sampling inlets operate on the principles of direct impaction, virtual impaction, cyclonic flow, selective filtration, and elutriation. All the direct impaction systems consist of a set of circular jets positioned above an impaction plate. The impactor dimensions are selected to allow the particles which exceed the desired cut-point to strike the plate, and those which are less than the cut-point to follow the airstream which passes the plate. John and Wang (1991) proved that reentrainment and disaggregation of particles occurs on the impaction surface of high-volume impactor inlets, and these surfaces are often oiled or greased to assure specified sampling effectiveness. The Sierra–Andersen SA-1200 inlet has a hinged lid which allows easy access for cleaning and greasing. The SA-246 inlet can be unscrewed for access to the impaction plate. The SA-254 inlet requires the removal and replacement of a dozen small screws for cleaning and is not as convenient as the other inlets.

The virtual impactor operates on a similar principle, with the exception that the impaction surface is replaced by an opening which directs the larger particles to another sampling substrate. This principle is used effectively in the Sierra–Andersen dichotomous sampler, described below, to separate particles into $PM_{2.5}$ and coarse (PM_{10} minus $PM_{2.5}$) particle size fractions. Four small screws must be removed to disassemble the virtual impactor for cleaning.

Cyclonic flow inlets use an impeller to impart a circular motion to air entering the inlet. This air enters a cylindrical tube oriented perpendicular to the impellers and the centripetal force imparted to the particles in the airstream moves them toward the walls of this tube. Those particles reaching the tube wall adhere to it, often with the help of an oil or grease coating. The Bendix cyclones have a small collection area at the bottom of the cyclone and direct the air to be sampled out of the top. Heavy particles which do not adhere to the cyclone wall settle into this collection area, which can be accessed for cleaning by removal of a plastic cap. The Wedding IP_{10} inlet has an access port through which a cleaning brush may be inserted.

Selective filtration takes advantage of the uniform pore size and known sampling effectiveness of etched polycarbonate filters manufactured by Nuclepore Corporation. Eight-micrometer pore size filters collect particles by interception and impaction in the vicinity of the pores to provide 50% cut-points for particles between 2 and 3 μm at flow rates of ~ 10 lpm. A ~ 0.4 μm pore size filter is placed behind the 8 μm filter to collect the transmitted particles. Cahill et al. (1990) observed the reentrainment of large, dry particles from the front filter and developed a greasing method to reduce this artifact. The sampling effectiveness curve for this method has a broad slope, and the method does not provide as distinctive a separation between $PM_{2.5}$ and larger size ranges as the direct impaction, virtual impaction, or cyclonic inlets. The method is much less expensive than other methods, however, and has been used in situations where simplicity and expense are major concerns.

Elutriator inlets draw air into a stilled-air chamber surrounding an open duct which leads to the filter. When the upward velocity due to flow through the inlet exceeds the particle settling velocity, that particle penetrates the inlet. When the settling velocity exceeds the upward velocity, the particle is not transmitted. This type of inlet was originally mated to the virtual impactor dichotomous sampler to provide a 15 μm cut-point. Wind tunnel tests (Wedding et al. 1980) found the cut-point to be highly dependent on wind speed, and this inlet was later replaced by the SA-246 PM_{10} direct impaction inlet.

Mathai et al. (1990), in evaluating the differences between several samplers used in western U.S. visibility studies, concluded that most differences between measurements from collocated sampling systems can be attributed to differences in inlet characteristics. These differences caused major controversies with respect to the development of PM_{10} reference methods in the early 1980s. Direct impaction and cyclonic high-volume samplers from different manufacturers were operated side by side and 30%–50% differences in PM_{10} mass were found (e.g., Rodes et al. 1985). Several different versions of these inlets were constructed with different dimensions, different numbers of stages, and different impaction surfaces. Extensive wind tunnel tests (Woods, Chen, and Ranade 1986) were conducted on clean and dirty, greased and ungreased high-volume PM_{10} inlets. These showed a positive shift in d_{50} for the impactors and a negative shift for the cyclones with increasing intervals between inlet maintenance. These results imply that frequent inlet cleaning is essential to obtain a well-defined size fraction.

Another source of confusion was introduced by the EPA tolerance on the PM_{10} cut-point of 10 ± 0.5 μm (Federal Register 1987*b*). The Wedding high-volume cyclonic inlet had a d_{50} of 9.6 μm, while the original Sierra–Andersen high-volume direct impaction inlet had a d_{50} of 10.2 μm. This difference made the cyclonic inlet more attractive to potential purchasers, since sampling with a lower d_{50} decreases the PM_{10} measured and minimizes the probability of violating the standard. Sierra–Andersen redesigned the inlet to provide a lower d_{50} to meet this competition and provided new impactor jets for its original inlet.

Differences in d_{50} impaction plate greasing and in cleaning schedule are rarely specified in published reports of sampler comparisons, and differences between measurements by different PM_{10} reference methods are difficult to explain. Wedding et al. (1988) and Hoffman et al. (1988) provide the most recent data for currently used PM_{10} inlets. These comparisons show that the Wedding high-volume samplers measure consistently less PM_{10} than do the Sierra–Andersen high-volume PM_{10} samplers.

SAMPLING SURFACES

Certain sampling surfaces absorb or react with gases and particles, thereby preventing their collection on sampling substrates (Hering et al. 1988; John et al. 1986). This is especially the case for nitric acid vapor which sticks to nearly everything. The removal of nitric acid in an inlet or sampling duct can change the gas–particle equilibrium of particulate ammonium nitrate, causing this substance to dissociate into ammonia and nitric acid gases. This is also true for certain volatile organic species.

In the case of denuder-type sampling systems (Stevens et al. 1990), a surface which *does* absorb nitric acid is desired. Denuders are constructed with dimensions such that over 90% of selected gases will diffuse to the sampler surface while less than 10% of the particles will deposit on this surface. The surface can be coated with an absorbent, which is then extracted and analyzed for the desired gaseous species. In the case of all other sampling components, an inert surface which *does not* act as a sink for atmospheric constituents is desired.

John et al. (1986) have tested different materials with respect to their affinity for nitric acid. These studies indicate that surfaces coated with perfluoroalkoxy (PFA) Teflon can pass nitric acid with 80%–100% efficiency. These researchers have also determined that the aluminum surfaces common to many samplers and inlets have an almost infinite capacity for absorbing nitric acid vapor while transmitting particles with high efficiency. Prior to use in sampling, PFA Teflon surfaces should be washed with a dilute solution of nitric acid to season them.

Little has been published concerning the affinity of other gaseous components to sampler surfaces. Additional tests need to be conducted for volatile organic particles. Most sampling surfaces are aluminum (which is usually oxidized), stainless steel, or plastic (polycarbonate or polyvinyl chloride). The plastic surfaces can acquire an electrical charge which might attract suspended particles to them, though the dimensions of most ambient sampling systems are sufficiently large for this attraction to be negligible (Rogers, Watson, and Mathai 1989).

FILTER MEDIA

The choice of filter media results from a compromise among the following filter attributes: (1) mechanical stability; (2) chemical stability; (3) particle or gas sampling efficiency; (4) flow resistance; (5) loading capacity; (6) blank values; (7) artifact formation; (8) compatibility with analysis methods; and (9) cost and availability. Chapter 12 discusses the chemical characteristics of different filter media. Lippmann (1989) provides a comprehensive list of commercially available filter types, their sampling efficiencies, and their manufacturers.

U.S. EPA filter requirements for PM_{10} sampling specify 0.3 μm DOP (dioctyl phthalate) sampling efficiency in excess of 99%, weight losses or gains due to mechanical or chemical instability of less than a 5 $\mu g/m^3$ equivalent, and an alkalinity of less than 25 microequivalents/g to minimize the absorption of sulfur dioxide (SO_2) and nitrogen oxides (NO_x). These are only the minimal requirements for samples which require chemical analyses. The most commonly used filter media for atmospheric particle and gas sampling are cellulose fiber, glass fiber, Teflon-coated glass fiber, Teflon membrane, etched polycarbonate membrane, quartz fiber, and nylon membrane. None of these materials is perfect for all purposes.

Cellulose fiber filters consist of a tightly woven paper mat. These filters meet the requirements in most categories with the exception of sampling efficiency and water vapor artifacts. Sampling efficiencies below 50% in the submicrometer region have been observed, but these are highly dependent on the filter weave. Cellulose fiber is hygroscopic and requires a precise relative humidity control in the filter-processing environment to obtain accurate mass measurements. This substrate has low elemental blanks and is commonly used for elemental and ionic analyses of the deposit, but it is not suitable for carbon analysis. Cellulose fiber filters can be impregnated with gas-absorbing compounds and located behind more efficient particle-collecting filters. This allows gases such as sulfur dioxide, nitrogen dioxide, and ammonia to be measured with suspended particles. The most commonly used cellulose fiber filter is Whatman 41.

Glass fiber filters consist of a tightly woven mat of borosilicate glass filaments. These filters meet the requirements in most categories with the exception of artifact formation and blank levels. Sampling efficiency is very high for all particle sizes. The high alkalinity of these substrates causes sulfur dioxide, nitrogen oxides, and gaseous nitric acid to be adsorbed (Coutant 1977; Spicer and Schumacher 1977). Blank levels for most elements of interest are extremely high and variable (Witz, Smith, and Moore 1983). Particulate nitrate and ammonium losses have been observed when these samples are stored at room temperature for long periods, but this is also true of deposits on other filter media (Witz et al. 1990). Glass fiber filters absorb organic carbon vapors which are measured as particulate carbon during analysis. The most commonly used glass fiber filters are Gelman Type A/E and Whatman EP2000.

Teflon-coated glass fiber filters impregnate a Teflon slurry onto a loosely woven glass fiber mat. These filters meet the requirements in all categories except blank element and carbon levels. Although a small nitric acid artifact has been observed (Mueller et al. 1983), it is tolerable in most situations. These filters are excellent for ion analyses but not for carbon analyses owing to their Teflon coating. The most commonly used Teflon-coated glass fiber filters are Pallflex TX40HI20 and T60A20.

Teflon membrane filters consist of a porous Teflon sheet which is either stretched across a plastic ring or supported by a loosely woven Teflon mat. These filters meet the requirements in all categories except flow resistance and carbon blank levels. Because of their low porosity, it is not usually possible to attain the flow rates needed by the size-selective inlets in high-volume sampling, though it is possible to obtain flow rates required for low-volume and medium-volume inlets. These filters cannot be analyzed for carbon because of its presence in the filter material, though they have very low blank levels for ions and elements. Most nondestructive multielemental analysis methods use Teflon membrane filters. The deposit of particles on the filter surface makes these substrates especially amenable to X-ray fluorescence (XRF) and proton-induced X-ray emission (PIXE) analyses. Gelman 1.0, 2.0, and 3.0 μm pore size Teflon filters, which are made of polytetrafluoroethylene (PTFE) Teflon stretched across a polymethylpentane ring, are the most commonly used Teflon membrane filters. Gelman Zefluor filters consist of PTFE Teflon mounted on a woven PTFE mat. The Zefluor filters are less desirable because their larger mass density decreases the XRF and PIXE sensitivity, and because the similar appearance of both sides causes them to be mounted upside down with the particles being drawn through the mat rather than onto the surface of the membrane.

Etched polycarbonate membrane filters are constructed from a thin polycarbonate sheet through which pores of uniform diameter have been produced by radioactive particle penetration and chemical etching. These filters have low sampling efficiencies (< 80%), even for small pore sizes (Liu and Lee 1976; Buzzard and Bell 1980). This low efficiency is used as an advantage when making size-specific measurements. Polycarbonate membrane filters have low elemental blank levels and are appropriate for elemental and ion analysis. They are the best filter media for single-particle analysis by electron microscopy, but they cannot be submitted to carbon analysis owing to their composition. The filters hold an electrostatic charge which influences mass measurements unless substantial effort is invested in discharging them with a small radioactive source (Engelbrecht, Cahill, and Feeney 1980). Electrostatic discharging is a good practice for all filter media, even though others do not retain as much charge as the Nuclepore membranes. The Nuclepore 8.0 and 0.4 μm filters are most commonly used in ambient aerosol sampling. While the 0.2 μm pore size filter provides a higher sampling efficiency, its higher flow resistance requires excessive vacuum for a reasonable flow rate.

Quartz fiber filters consist of a tightly woven mat of quartz filaments. These filters meet the requirements in most categories and have artifact properties which are significantly lower than those for glass fiber filters, though quartz substrates adsorb hydrocarbon gases during sampling (Eatough et al. 1990; McDow and Huntzicker 1990). They should be baked at ~800°C prior to sampling to remove adsorbed organic vapors. Blank levels are high and variable for several elements (especially aluminum and silicon), though newer formulations are cleaner than earlier formulations. These filters are widely used for ion and carbon analyses. The greatest drawback of quartz fiber filters is their fragility; they require extremely careful handling for accurate mass measurements. The Whatman QM/A quartz fiber filter contains a 5% borosilicate glass binder which minimizes its friability. This filter is often used in high-volume PM_{10} samplers for mass measurements. The Pallflex 2500 QAT-UP filter is pure quartz and undergoes a distilled-water washing (thus, the "UP", or "ultrapure" designation). This filter should not be used for mass analyses, but it is the best choice for carbon measurements.

Nylon membrane filters consist of thin sheets of porous nylon. Nylon filters are used almost exclusively for the collection of nitric acid. These filters were not originally manufactured for this purpose, however, and there is a substantial difference among the properties of filters from different manufacturers. Nylon filters have high flow resistance, which increases rapidly with filter loading. These filters passively absorb nitric acid, and their blank nitrate levels can be high, depending upon how long they have been exposed to an acid-rich environment. They should be washed in a sodium bicarbonate solution followed by distilled water (Watson et al. 1991*b*) prior to use in the field. Schleicher and Schuell Grade 66 and Gelman Nylasorb are the nylon membrane filters most commonly used for ambient-air sampling.

All filter batches may be contaminated, and a sample from each batch of filters (1 out of 50–100 filters) should be submitted to the intended chemical analyses prior to use in a field study. Although not yet reported in the literature, recent studies have found elevated levels of lead, calcium, and nitrate in batches of blank Teflon membrane filters.

FILTER HOLDERS

Filter substrates must be protected from contamination prior to, during, and after sampling. This is accomplished by placing the filter in a filter holder. These holders must: (1) mate to the sampler and to the flow system without leaks; (2) be composed of inert materials which do not absorb acidic gases; (3) allow a uniformly distributed deposit to be collected; (4) have a low pressure drop across the empty holder; (5) accommodate the sizes of commonly available air sampling filters (e.g., 37 or 47 mm); and (6) be durable and reasonably priced.

Most filter holders are configured as in-line or open-faced. In-line holders often concentrate the particles in the center of the substrate, and this will bias the results if analyses are performed on portions of the filter. Fujita et al. (1989) show differences as high as 600% between chemical measurements in the middle and at the edges of filters sampled with in-line filter holders. Openfaced filter holders are better choices for ambient aerosol sampling systems.

Lippmann (1989) shows specifications and photographs for a large variety of commercially available filter holders. The holders most commonly used for ambient aerosol sampling are: (1) Gelman stainless steel; (2) Nuclepore polycarbonate plastic; (3) Savillex PFA Teflon; and (4) Sierra–Andersen polyethylene rings.

The Gelman stainless-steel filter holder accommodates 28 and 47 mm filters and, though it has an open-faced adapter, it is sometimes used with an in-line coupler. As noted above, this in-line holder results in a spot in the center of the filter. These filter holders cost several hundred dollars apiece, and filters are generally loaded into them in the field rather than in a laboratory environment. These filter holders cannot be used to sample nitric acid.

The Nuclepore polycarbonate plastic filter holders accommodate 25, 37, and 47 mm diameter filters. Each one costs a few tens of dollars, making it possible to purchase many of them for laboratory loading and unloading. These holders can be modified by widening the outlet hole to reduce flow resistance, using multiple extender sections for filter stacking, and replacing the rubber O-ring with a Viton O-ring to minimize carbon adsorption from the rubber. The polycarbonate material has not been tested for inertness with respect to nitric acid.

The Savillex PFA Teflon 47 mm filter holder is made of injection-molded PFA Teflon, which was previously noted as having the least inclination to absorb nitric acid. These filter holders have a tapered extender section (called a receptacle) which can be mated to a sampler plenum with an O-ring in a retainer ring. Several grids and grid rings can be stacked within the holder to obtain series filtration. The cost of these filter holder combinations is a few tens of dollars, low enough to allow many to be purchased for laboratory loading.

The Sierra–Andersen polyethylene filter holders are two-piece rings which accommodate 37 mm diameter filters. They are used exclusively with the Sierra–Andersen 240 series of virtual impactor samplers. They are inexpensive (< \$10 apiece) and can be loaded in the laboratory for shipment to field sites. Their affinity for nitric acid is unknown.

FLOW MOVEMENT AND CONTROL

Air is passed through the sampling substrates by means of a vacuum created by a pump. Rubow and Furtado (1989) describe the commercially available air pumps, their capacities and operating principles. Rogers, Watson, and Mathai (1989) have found that a 3/4 HP carbon vane pump is sufficient to draw in excess of 120 lpm through a Teflon membrane filter. Smaller pumps can be used for lower flow rates and filter media with lower resistances.

Regardless of the pump used, the quantity of air per unit time must be precisely measured and controlled to determine particle concentrations and to maintain the size-selective properties of the sampling inlet. Four general methods of flow control are used in fine particle sampling systems: (1) manual volumetric; (2) automatic mass; (3) differential pressure volumetric; and (4) critical orifice volumetric.

Manual control is accomplished when the operator initializes a setting, such as a valve adjustment, and then relies on the known and constant functioning of sampler components, such as pumps and tubing, to maintain flows within specifications. Flow rates which are set manually generally change over a sampling period as the collection substrate loads up and presents a higher flow resistance. For most filter loadings ($< 200\ \mu g/m^3$), the flow will not change by more than 10% during sampling and the average of flow rates measured before and after sampling provides an accurate estimate of the actual flow.

Automatic mass flow controllers use thermal anemometers, which measure the heat transfer between two points in the gas stream. To a first approximation, the heat transfer is proportional to the flux of gas molecules between the two points and, hence, the mass flow controller is able to sense the flux of mass. Mass flow controllers require compensating circuitry to avoid errors due to absolute temperature variations of the gas itself as well as the controller sensing probe. Wedding (1985) estimates potential differences in excess of 10% between mass and volumetric measurements of the same flow rates, depending on temperature and pressure variations.

Differential pressure volumetric flow control maintains a constant pressure across an orifice (usually a valve which can be adjusted for a specified flow rate) by a diaphragm-controlled valve located between the filter and the orifice (Chow et al. 1992). The diaphragm is controlled by the pressure between the orifice and the pump. When this pressure increases (as when filters load up), the diaphragm opens the valve and allows more air to pass.

A critical orifice consists of a small circular opening between the filter and the pump. When the downstream pressure at the minimum flow area downstream of the orifice is less than 53% of the upstream pressure, the air velocity attains the speed of sound and it will remain constant, regardless of increased flow resistance. Critical orifices provide very stable flow rates, but they require large pumps and low flow rates (typically less than 20 lpm with commonly available pumps) to maintain the high pressure differences. Wedding et al. (1987) have developed a "critical throat", which uses a diffuser arrangement allowing recovery of over 90% of the energy which is normally expended in back pressure behind a critical orifice. This design allows higher flow rates to be obtained with a given pump.

SAMPLING SYSTEMS

Rubow and Furtado (1989), Perry (1989), and Hering (1989) describe commercially available systems for ambient aerosol sampling. The most widely used of these are the high-volume PM_{10} samplers sold by Wedding and Associates and Sierra–Andersen that are designated reference methods for PM_{10}, as shown in Table 28-1. These samplers use a low-pressure blower to draw air through 8 × 10 in (20.3 × 25.4 cm) fiber filters. The peaked-roof dust cover which was formerly used to measure TSP is replaced with one of the inlets described in Table 28-1. The Wedding units use critical throat flow control, while the Sierra–Andersen units offer a choice between mass flow control or critical throat control. The procedures for these samplers are well established (e.g., Watson et al. 1989). As noted above, frequent inlet cleaning is necessary for an accurate size sampling by these units.

The Sierra–Andersen low-volume dichotomous sampler is also commercially available as a PM_{10} reference sampler, and this unit is often used with appropriate filter media when elemental, ionic, and carbon analyses are desired. This sampler uses a virtual impactor to separate the $PM_{2.5}$ and coarse particle size fractions. Flow rates are controlled by a differential pressure regulator. Ten percent of the fine particles are sampled on the coarse particle filters, and corrections must be made (Evans and Ryan 1983) to the coarse particle measurements to compensate. Several dichotomous samplers may be collocated to accommodate different filter types. John, Wall, and Ondo (1988) describe how dichotomous samplers can be adapted for nitric acid sampling. The inlet and virtual impactor should be disassembled and thoroughly cleaned on a regular schedule. The virtual impactor can be assembled in a reverse orientation, with the impactor jet over the fine particle filter rather than over the coarse particle filter, and care must be taken to reassemble this unit correctly.

In addition to these commercially available samplers, several other sampling systems have been developed using the components described above for specific research studies. A few of these are compared in Table 28-2 to provide examples of how different components can be assembled into an ambient aerosol sampling system.

The stacked filter unit (SFU) used in the National Park Service Visibility Network (Cahill et al. 1986) applies the etched polycarbonate membrane filter size separation method described above with a nominal sample duration of 72 h. Flow rates are manually set and recorded before and after sampling. This sampler operated reliably at remote locations for over a decade, and a large database of elemental concentrations from PIXE analyses has been derived from its samples.

The sequential filter sampler (SFS) was originally designed in the late 1970s for use in the sulfate regional experiment (SURE) (Mueller et al. 1983) and the Portland aerosol characterization study (PACS) (Watson 1979) and has been applied in over a dozen subsequent studies. A version of this sampler has since received PM_{10} reference method status. The SFS consists of an aluminum plenum to which a SA-254 PM_{10} or a Bendix 240 $PM_{2.5}$ inlet is attached. Up to 12 sampling ports within the plenum are controlled by solenoid valves which divert flow from one channel to the next by means of a programmable timer. These ports accept filters which have been preloaded into open-faced 47 mm Nuclepore filter holders. The sample flow can be divided for simultaneous collection on two or more

... Sampling Systems Assembled for Major Studies

Sampling System	Inlet and Particle Size (µm)	Flow Rate (lpm)	Sampling Surface	Filter Holders	Filter Media	Features	Descriptive Reference
Stacked Filter Unit (SFU)	Selective Nuclepore Filters (2–3 µm)	10	Polycarbonate	Nuclepore	47 mm Nuclepore 8.0 µm 25 mm Teflon		Cahill et al. (1990)
Sequential Filter Sampler (SFS)	SA-254 Medium-Volume Inlet (PM_{10})	20 out of 113	Anodized Aluminum	Nuclepore open-face	47 mm Teflon, quartz cellulose, nylon	Option to add nitric acid denuders in the sampling stream	Watson et al. (1991*b*)
	Bendix 240 Cyclone ($PM_{2.5}$)	20 out of 113	Anodized Aluminum	Nuclepore open-face	47 mm Teflon, quartz, impregnated cellulose, nylon		
WRAQS	SA-320 High-Volume Inlet (PM_{15})	113 out of 1,130	Stainless Steel	Nuclepore in-line	47 mm Teflon, quartz		Tombach et al. (1987)
	Bendix 240 Cyclone	113	Stainless Steel	Nuclepore in-line	47 mm Teflon, quartz		
SCISAS	SA-320 High-Volume Inlet (PM_{15})	113 out of 1,130	Aluminum	Nuclepore open-face	47mm Teflon, quartz	Up to 6 sequential samples	Rogers, Watson, and Mahai (1989)
	Bendix 240 Cyclone ($PM_{2.5}$)	113 out of 1,130	Stainless Steel	Nuclepore open-face	47 mm Teflon, quartz		
IMPROVE	Wedding Low-Volume Inlet (PM_{10})	18	Stainless Steel	Nuclepore	25 mm Teflon, quartz 47 mm nylon	Nitric acid denuders can be placed in inlet line	Eldred et al. (1990)
	Modified AIHL Cyclone ($PM_{2.5}$)	24	Aluminum	Nuclepore	25 mm Teflon, quartz 47 mm nylon		
California Institute of Technology	SA-246 Low-Volume Inlet (PM_{10})	16.7	Stainless Steel	Gelman in-line	37 mm Teflon, quartz		Solomon et al. (1989)
	AIHL Cyclone	22	Stainless Steel	Gelman in-line	37 mm Teflon, quartz, nylon, impregnated cellulose		
SCAQS	SA-254 Medium-Volume Inlet (PM_{10})	35 out of 113	Stainless Steel	Gelman in-line	47 mm Teflon, quartz	Option to add 20 cm flow homogenizer	Fitz et al. (1989)
	Bendix 240 Cyclone ($PM_{2.5}$)	35 out of 113	Stainless Steel	Gelman in-line	47 mm Teflon, quartz, nylon, impregnated cellulose	Option to add 20 cm flow homogenizer	
CADMP	SA-245 Medium-Volume Inlet (PM_{10})	20 out of 113	Anodized Aluminum	Savillex open-face	47 mm Teflon, impregnated cellulose fiber	Nitric acid denuders. Two sequential samples	Chow et al. (1991)
	Bendix 240 Cyclone (Teflon-coated)	20 out of 113	PFA Teflon	Savillex open-face	47 mm Teflon, quartz, nylon		

filter media. The differential-pressure volumetric flow controller has been adapted to split the flow between filters and maintain a constant flow rate despite filter loading.

The Western Regional air quality sampler (WRAQS) was developed for sampling of low concentrations related to visibility impairment in the southwest U.S. desert. The WRAQS draws air through the high-volume SA-320 PM_{15} inlet into a large-diameter polyvinyl chloride sampling duct. Samples are drawn isokinetically from this duct directly onto filters for a PM_{15} sample, and through Bendix 240 cyclones for $PM_{2.5}$ size fractions. The high flow rate through these filters was needed to obtain enough sample for chemical analyses from the pristine environments in which WRAQS was operated. Flow rates are manually controlled. The WRAQS uses in-line filter holders and Tombach et al. (1987) document how the resulting deposit inhomogeneities affect the accuracy of the chemical analysis performed on sections of the filter.

The size-classifying isokinetic sequential aerosol sampler (SCISAS) combines elements of the WRAQS sampler and the SFS to provide for up to six sequential samples in two size ranges on two filter media for the *S*ubregional *C*ooperative *E*lectric Utility, Department of Defense, *N*ational Park Service, and *E*PA *S*tudy on visibility (SCENES) (Mueller et al. 1986). The inlets and sample extraction methods are similar to those of the WRAQS, with the addition of solenoid valves to allow up to six samples to be taken without operator intervention. This sequential feature was required owing to the SCENES need for daily sampling without a daily visit to change samples. Flow rates are manually controlled.

The California Institute of Technology (CIT) and the Southern California Air Quality Study (SCAQS) samplers are of similar design, though inlets and flow rates differ. Both samplers were intended to measure PM_{10}, $PM_{2.5}$, and gaseous components of the aerosol on substrates for chemical analysis. The SCAQS sampler took samples of 4–6 h duration, and required higher flow rates to obtain sufficient deposit for analysis. These systems draw air through several filters in parallel and in series using Gelman in-line filter holders and use critical orifice flow control.

The Interagency Monitoring of Protected Visual Environments (IMPROVE) sampler was developed to replace the SFU at U.S. National Parks and Monuments in support of regional visibility assessment. This sampler consists of different modules controlled by a common sample switching system. Each module can be tailored to a specific type of measurement. The standard configuration acquires one PM_{10} sample for gravimetric and PIXE analysis and three $PM_{2.5}$ samples (one on Teflon membrane for gravimetric and PIXE analysis, one on Teflon membrane for ion analysis, and one on quartz fiber for carbon analysis). Up to four sequential samples can be taken without operator intervention by using a timer to control solenoid valves. Flow rates are controlled by critical orifice.

The California Acid Deposition Monitoring Program (CADMP) sampler was assembled to measure species related to dry acidic deposition in urban and forested regions of California (Watson et al. 1991*b*). It consists of two units with conical plena, one with a PM_{10} inlet and the other with a Teflon-coated $PM_{2.5}$ inlet. Two samples are taken simultaneously out of each plenum to obtain deposits of suspended particles and acidic gases. The 20 lpm flow rates are controlled by differential-pressure regulation. Denuded and nondenuded sample streams are included for nitric acid measurements, and a solenoid switching system takes unattended daytime and nighttime samples.

The sampling systems listed in Table 28-2 are relatively well standardized, though noncommercial. Several annular denuder sampling systems and cascade impactor systems are also being developed and applied in research studies.

As noted above, annular denuder systems are designed for the measurement of acidic species. These systems use Teflon-coated inlets, mass flow controllers, and in-line filter holders. Nitric acid, sulfur dioxide, and other gases are absorbed on the inner surfaces of the denuder inlet and are removed by washing with an extraction solution. These systems are still undergoing design changes. The system developed by Koutrakis et al. (1988) has been applied in a nationwide network and has developed a degree of standardization.

Several cascade impactors have been developed to partition PM_{10} into smaller size ranges on substrates suitable for chemical analysis. Each of these impaction methods uses low-pressure stages to attain the lower particle cut-points. The low-pressure impactor (LPI) (Hering et al. 1979), the Davis Rotating drum Unit for Monitoring (DRUM) (Raabe et al. 1988), the Berner impactor (Berner et al. 1979), and the Micro Orifice Uniform Deposit Impactor (MOUDI) (Marple, Liu, and Kuhlmey 1981) have been applied in ambient aerosol chemistry and visibility studies.

The low flow rate (1 lpm) and nonuniform deposits, as well as substantial handling of the substrates, make the LPI impractical for many studies, though it has been coupled with high-sensitivity sulfer analysis methods, and can be used for this element in many environments. The DRUM is a Lundgren-style impactor sampling at 30 lpm with low-pressure stages which acquires particle deposits in nine size ranges from 0.07 to 8.5 μm. Mylar impaction substrates are mounted on cylinders which rotate below the impaction jet so that particle deposits can be monitored as a function of time and monitored by PIXE. The MOUDI and Berner cascade impactors are different in design but similar in function. These impactors operate at 30 lpm and have been used to acquire carbon and ion deposits on substrates which are amenable to chemical analysis by thermal–optical carbon analysis on aluminum foils and sulfate and nitrate analyses on Teflon substrates in at least eight size ranges from 0.03 to 15 μm.

SELECTING A SAMPLING SYSTEM

This chapter was not intended to identify every existing or every conceivable ambient aerosol sampling system. It has concentrated on sampler components and sampler configurations in common use for which a body of knowledge and confidence has been established. Nevertheless, the practitioner is presented with a myriad of choices when confronted with the challenge of designing, installing, operating, and using the data from a sampling network to accomplish a specific purpose. A few simple steps can be followed to aid in the network design.

The first step is to decide what the specific objectives of monitoring are. Four types of monitoring goals were given as examples in the first part of this chapter. The specific objectives should be as comprehensive as possible. For example, it is a common practice to field a network of high-volume PM_{10} samplers for the immediate objective of determining compliance with air quality standards. If one or more of these samples exceeds a standard, then the objectives broaden to include source apportionment. Sampling which meets the first objective is often inadequate for the source apportionment objective.

The second step is to determine the particle size fractions, chemical analyses, sampling frequencies, and sample durations needed to address the objectives. More frequent samples, or samples taken at remote locations, may require a sequential sampling feature. Shorter sample durations may require a larger flow rate to obtain an adequate sample deposit for analysis. The types of analyses and size fractions desired affect the number of sampling ports and different filter media needed.

The third step is to calculate the expected amount of deposit on each filter for each chemical species and compare it with typical detection limits for the analyses being considered. Chapter 12 provides typical concentrations and detection limits which can be normalized to flow rates and filter sizes of different sampling systems. In general, urban samples acquire adequate deposits for analysis with flow rates as low as ~ 20 lpm for as low as 4 h durations. Samples at nonurban sites may require >100 lpm flow rates for 24 h durations to obtain an adequate deposit.

TABLE 28-3 Commercial Sources for Filters and Filter Holders

Andersen Samplers, Inc. 4215 Wendell Drive Atlanta, GA 30336 (404) 691-1910 (800) 241 6898	Savillex Corporation 6133 Baker Road Minnetonka, MN 55345 (612) 935-4100
Gelman Instrument Company 600 South Wagner Road Ann Arbor, MI 48106 (313) 665-0651	Schleicher and Schuell, Inc. 543 Washington Street Keene, NH 03431 (800) 245-4029
General Metal Works, Inc. 145 South Miami Avenue Village of Cleves, OH 45002 (513) 941-2229 (800) 543-7412	Wedding & Associates, Inc. P.O. Box 1756 Fort Collins, CO 80552 (303) 221-0678 (800) FOR-PM10
Nuclepore Corporation 7035 Commerce Circle Pleasanton, CA 94566 (415) 462-2230	Whatman, Inc. 9 Bridewell Place Clifton, NJ 07014 (201) 773-5800
Pallflex Production Corporation Kennedy Drive Putnam, CT 06260 (203) 928-7761	

The analytical laboratory should be involved at the sampler design stage to assure a compatibility between sampling methods, analysis methods, filter media, and lower quantifiable limits.

The fourth step is to create, adapt, or purchase the sampling system which provides the most cost-effective and reliable means of meeting the monitoring needs. Table 28-3 provides the names and addresses of companies which offer many of the items cited here as off-the-shelf products. It is also worthwhile to contact the authors of cited references for the noncommercial sampling systems, since these investigators may be in a position to loan, rent, sell, or build one of these units to meet a special need. If no existing unit can fulfill all the requirements, then it will be necessary to assemble different sampling components into a new configuration which will meet those needs.

The final step is to create or adapt an operating procedure which specifies the methods and schedules for inlet cleaning, filter transport and handling, calibration and performance tests, and record keeping. All the sampling systems discussed here have such written procedures which can serve as guides for specific procedures.

CONCLUSIONS

Ambient aerosol sampling systems have evolved in order to meet different monitoring needs. Sampler inlets, monitoring surfaces, filter media, filter holders, flow movers, and operating procedures have been developed and tested. These components have been integrated into various sampler configurations for application in dozens of studies. Using these components and the lessons learned from previous studies, it is possible to construct sampling systems customized to meet specific objectives without extensive original development and testing.

References

Ashbaugh, L., J. C. Chow, and J. G. Watson. 1991. Atmospheric acidity data quality and measurement characteristics for California's Acid Deposition Monitoring Program. Paper 91-89.7, presented at the 84th Annual Meeting, Vancouver, British Columbia, Canada. Pittsburgh: Air and Waste Management Association.

Berner, A., C. H. Lurzer, L. Pohl, O. Preining, and P. Wagner. 1979. The size distribution of the urban aerosol in Vienna. *Sci. Total Environ.* 13:245–61.

Buzzard, G. H. and J. P. Bell. 1980. Experimental filtration efficiencies for large pore Nuclepore filters. *J. Aerosol Sci.* 11:435.

Cahill, T. A., R. A. Eldred, and P. J. Feeney. 1986. Particulate monitoring and data analysis for the National Park Service, 1982–1985. Air Quality Group, University of California, Davis, CA.

Cahill, T. A., R. A. Eldred, P. J. Feeney, P. J. Beveridge, and L. K. Wilkinson. 1990. The stacked filter unit revisited. In *Transactions, Visibility and Fine Particles*, ed. C. V. Mathai, p. 213. Pittsburgh: Air and Waste Management Association.

Chan, T. and M. Lippman. 1977. Particle collection efficiencies of sampling cyclones: An empirical theory. *Environ. Sci. Technol.* 11:377–81.

Chow, J. C., J. G. Watson, J. L. Bowen, A. W. Gertler, C. A. Frazier, K. K. Fung, and L. Ashbaugh. 1992. A sampling system for reactive species in the western U.S. *ACS Symposium Series*, ed. E. Winegar. Washington, DC: American Chemical Society, in press.

Coutant, R. W. 1977. Effect of environmental variables on collection of atmospheric sulfate. *Environ. Sci. Technol.* 11:873.

Eatough, D. J., N. Aghdaie, M. Cottam, T. Gammon, L. D. Hansen, E. A. Lewis, and R. J. Farber. 1990. Loss of semi-volatile organic compounds from particles during sampling on filters. In *Transactions, Visibility and Fine Particles*, ed. C. V. Mathai, p. 146. Pittsburgh: Air and Waste Management Association.

Eldred, R. A., T. A. Cahill, L. K. Wilkinson, P. J. Feeney, J. C. Chow, and W. C. Malm. 1990. Measurement of fine particles and their components in the NPS/IMPROVE network. In *Transactions: Visibility and Fine Particles*, ed. C. V. Mathai, pp. 187–96. Pittsburgh: Air and Waste Management Association.

Engelbrecht, D. R., T. A. Cahill, and P. J. Feeney. 1980. Electrostatic effects on gravimetric analysis of membrane filters. *J. APCA* 30:391–92.

Evans, J. S. and P. B. Ryan. 1983. Statistical uncertainties in aerosol mass concentrations measured by virtual impactors. *Aerosol Sci. Technol.* 2:531–36.

Federal Register. 1987*a*. Revisions to the national ambient air quality standards for particulate matter: 40 CFR Part 50. *Federal Register* 52:24634.

Federal Register. 1987*b*. Ambient air monitoring reference and equivalent methods: 40 CFR Part 53. *Federal Register* 52:24724.

Fitz, D., M. Chan, G. Cass, D. Lawson, and L. Ashbaugh. 1989. A multi-component size-classifying aerosol and gas sampler for ambient air monitoring. Presented at the 82nd Annual Meeting, Anaheim, CA. Pittsburgh: Air and Waste Management Association.

Grosjean, D. and J. H. Seinfeld. 1989. Parameterization of the formation potential of secondary organic aerosols. *Atmos. Environ.* 23:1733.

Hering, S. V. 1989. Inertial and gravitational collectors. In *Air Sampling Instruments for Evaluation of Atmospheric Contaminants*, ed. S. V. Hering, 7th edn., pp. 337–85. Cincinnati, OH: American Conference of Governmental Industrial Hygienists.

Hering, S. V., R. C. Flagan, and S. K. Friedlander. 1979. Design and evaluation of new low-pressure impactor I. *Environ. Sci. Technol.* 12:667.

Hering, S. V., D. R. Lawson, I. Allegrini, A. Febo, C. Perrino, M. Possanzini, J. E. Sickles II, K. G. Anlauf, A. Wiebe, B. R. Appel, W. John, J. Ondo, S. Wall, R. S. Braman, R. Sutton, G. R. Cass, P. A. Solomon, D. J. Eatough, N. L. Eatough, E. C. Ellis, D. Grosjean, B. B. Hicks, J. D. Womack, J. Horrocks, K. T. Knapp, T. G. Ellstad, R. J. Paur, W. J. Mitchell, M. Pleasant, E. Peake, A. MacLean, W. R. Pierson, W. Brachaczek, H. I. Schiff, G. I. Mackay, C. W. Spicer, D. H. Stedman, A. M. Winer, H. W. Biermann, and E. C. Tuazon. 1988. The nitric acid shootout: Field comparison of measurement methods. *Atmos. Environ.* 22:1519–39.

Hinds, W. C. 1982. *Aerosol Technology: Properties, Behavior, and Measurement of Airborne Particles.* New York: Wiley.

Hoffman, A. J., L. J. Purdue, K. A. Rehme, and D. M. Holland. 1988. 1987 PM_{10} sampler intercomparison study. In *Transactions, PM_{10}: Implementation of Standards*, eds. C. V. Mathai and D. H. Stonefield, pp. 138–49. Pittsburgh: Air Pollution Control Association.

John, W., S. Hering, G. Reischel, G. Sasaki, and S. Goren. 1983*a*. Anomalous filtration of solid particles by Nuclepore filters. *Atmos. Environ.* 17:373.

John, W., S. Hering, G. Reischl, and G. V. Sasaki. 1983*b*. Characteristics of Nuclepore filters with large pore size—II. Filtration properties. *Atmos. Environ.* 17:373.

John, W. and G. Reischl. 1980. A cyclone for size-selective sampling of ambient air. *J. APCA* 30:872.

John, W., S. M. Wall, and J. L. Ondo. 1988. A new method for nitric acid and nitrate aerosol measurement using the dichotomous sampler. *Atmos. Environ.* 22:1627–35.

John, W., S. M. Wall, J. L. Ondo, and H. C. Wang. 1986. Dry deposition of acidic gases and particles. Air and Industrial Hygiene Laboratory, California Department of Health Services, Berkeley, CA.

John, W., S. M. Wall, J. L. Ondo, and W. Winklmayr. 1990. Modes in the size distributions of atmospheric inorganic aerosol. *Atmos. Environ.* 24A:2349–59.

John, W. and H. -C. Wang. 1991. Laboratory testing method for PM-10 samplers: Lowered effectiveness from particle loading. *Aerosol Sci. Technol.* 14:93.

Koutrakis, P., J. M. Wolfson, J. L. Slater, M. Brauer, and J. D. Spengler. 1988. Evaluation of an annular denuder/filter pack system to collect acidic aerosols and gases. *Environ. Sci. Technol.* 22:1463.

Lippmann, M. 1989. Sampling aerosols by filtration. In *Air Sampling Instruments for Evaluation of Atmospheric Contaminants*, ed. S. V. Hering, 7th edn., pp. 306–36.

Cincinnati, OH: American Conference of Governmental Industrial Hygienists.

Liu, B. Y. and K. W. Lee. 1976. Efficiency of membrane and Nuclepore filters for submicrometer aerosols. *Environ. Sci. Technol.* 10:345.

Marple, V. A., B. Y. H. Liu, and G. A. Kuhlmey. 1981. A uniform deposit impactor. *J. Aerosol Sci.* 12:333.

Mathai, C. V., J. G. Watson, C. F. Rogers, J. C. Chow, I. Tombach, J. Zwicker, T. Cahill, P. Feeney, R. Eldred, M. Pitchford, and P. K. Mueller. 1990. Intercomparison of ambient aerosol samplers used in western visibility and air quality studies. *Environ. Sci. Technol.* 24:1090–99.

McDow, S. R. and J. J. Huntzicker. 1990. Vapor adsorption artifact in the sampling of organic aerosol: Face velocity effects. *Atmos. Environ.* 24A:2563–72.

McFarland, A. R. and C. A. Ortiz. 1984. Characterization of Sierra–Andersen PM_{10} inlet model 246b. Texas A&M Air Quality Laboratory Report 4716/02/02/84/ARM, College Station, TX.

McFarland, A. R. and C. A. Ortiz. 1987. Aerosol sampling characteristics of the Sierra–Andersen model 321b PM_{10} inlet. Texas A&M Air Quality Report 4716/02/08/87/ARM, College Station, TX.

McFarland, A. R., C. A. Ortiz, and R. W. Bertch. 1978. Particle collection characteristics of a single-stage dichotomous sampler. *Environ. Sci. Technol.* 12:679.

McFarland, A. R., C. A. Ortiz, and R. W. Bertch. 1984. A 10 μm cutpoint size selective inlet for hivol samplers. *J. APCA* 34:544.

McFarland, A. R., C. A. Ortiz and C. E. Rodes. 1980. Characterization of sampling systems. In *Proceedings of the Technical Basis for A Size Specific Particulate Standard*, p. 59. Pittsburgh: Air Pollution Control Association.

Mueller, P. K., D. A. Hansen, and J. G. Watson. 1986. The *S*ubregional *C*ooperative *E*lectric Utility, Department of Defense, *N*ational Park Service, and *E*nvironmental Protection Agency *S*tudy (SCENES) on visibility: An overview. Report NO. EA-4664-SR, Electric Power Research Institute, Palo Alto, CA.

Mueller, P. K., G. M. Hidy, J. G. Watson, R. L. Baskett, K. K. Fung, R. C. Henry, T. F. Lavery, and K. K. Warren. 1983. The sulfate regional experiment: Report of findings, Vols. 1–3. EPRI Report EA-1901, Electric Power Research Institute, Palo Alto, CA.

Olin, J. G. 1978. A new virtual impactor (dichotomous sampler) for fine particle air monitoring. Presented at the 71st Annual Meeting, Air Pollution Control Association, Pittsburgh, PA.

Olin, J. G. and R. R. Bohn. 1983. A new PM_{10} medium flow sampler. Presented at the 76th Annual Meeting, Atlanta, GA. Pittsburgh: Air Pollution Control Association.

Pace, T. G. 1991. Receptor modeling in the context of ambient air quality standard for particulate matter. In *Receptor Modeling for Air Quality Management, Vol. 7, Data Handling in Science and Technology*, ed. P. K. Hopke, Chap. 8, pp. 255–97. Amsterdam: Elsevier.

Perry, W. H. 1989. Sequential and tape samplers—Unattended sampling. In *Air Sampling Instruments for Evaluation of Atmospheric Contaminants*, ed. S. V. Hering, 7th edn., pp. 291–303. Cincinnati, OH: American Conference of Governmental Industrial Hygienists.

Pilinis, C. and J. H. Seinfeld. 1988. Development and evaluation of an eulerian photochemical gas–aerosol model. *Atmos. Environ.* 22:1985–2001.

Raabe, O. G., D. A. Braaten, R. L. Axelbaum, S. V. Teague, and T. A. Cahill. 1988. Calibration studies of the DRUM impactor. *J. Aerosol Sci.* 19:183–95.

Rodes, C. E., D. M. Holland, L. J. Purdue, and K. A. Rehme. 1985. A field comparison of PM_{10} inlets at four locations. *J. APCA* 35:45.

Rogers, C. F., J. G. Watson, and C. V. Mathai. 1989. Design and testing of a new size classifying isokinetic sequential aerosol sampler. *J. APCA* 39:1569–76.

Rubow, K. L. and V. C. Furtado. 1989. Air movers and samplers. In *Air Sampling Instruments for Evaluation of Atmospheric Contaminants*, ed. S. V. Hering, pp. 241–74. Cincinnati, OH: American Conference of Governmental Industrial Hygienists.

Sloane, C. S., J. G. Watson, J. C. Chow, L. Pritchett, and L. W. Richards. 1990. Size distribution and optical properties of the Denver Brown Cloud. In *Transactions, Visibility and Fine Particles*, ed. C. V. Mathai, pp. 384–93. Pittsburgh: Air & Waste Management Association.

Sloane, C. S., J. Watson, J. Chow, L. Pritchett, and L. W. Richards. 1991. Size-segregated fine particle measurements by chemical species and their impact on visibility impairment in Denver. *Atmos. Environ.* 25A:1013–24.

Solomon, P. A., T. Fall, L. Salmon, G. R. Cass, H. A. Gray, and A. Davison. 1989. Chemical characteristics of PM_{10} aerosols collected in the Los Angeles area. *J. APCA* 39:154–63.

Spicer, C. W. and P. M. Schumacher. 1977. Interference in sampling atmospheric particulate nitrate. *Atmos. Environ.* 11:873.

Stevens, R. K., L. J. Purdue, H. M. Barnes, R. P. Ward, J. O. Baugh, J. P. Bell, H. Sauren, J. E. Sickles II, and L. L. Hodson. 1990. Annular denuders and visibility studies. In *Transactions, Visibility and Fine Particles*. ed. C. V. Mathai, p. 122. Pittsburgh: Air and Waste Management Association.

Tombach, I. H., D. W. Allard, R. L. Drake, and R. C. Lewis. 1987. Western regional air quality studies. Visibility and air quality measurements: 1981–1982. Report EA-4903, prepared for Electric Power Research Institute, Palo Alto, CA, by AeroVironment, Inc., Monrovia, CA.

VanOsdell, D. W. and F. L. Chen. 1990. Wind tunnel test report no. 28: Test of the Sierra–Andersen 246B dichotomous sampler inlet at 2, 8, and 24 km/h. Research Triangle Institute Report No. 432U-3999-93-28, Research Triangle Park, NC.

Watson, J. G. 1979. Chemical element balance receptor

model methodology for assessing the sources of fine and total particulate matter. Ph.D. Dissertation, Oregon Graduate Center, Beaverton, OR. Ann Arbor, MI: University Microfilms International.

Watson, J. G., J. C. Chow, J. J. Shah, and T. G. Pace. 1983. The effect of sampling inlets on the PM_{10} and PM_{15} to TSP concentration ratios. *J. APCA* 33:114.

Watson, J. G., N. F. Robinson, J. C. Chow, R. C. Henry, B. M. Kim, T. G. Pace, E. L. Meyer, and Q. Nguyen. 1990. The USEPA/DRI chemical mass balance receptor model, CMB 7.0. *Environ. Software* 5(1):38–49.

Watson, J. G., J. L. Bowen, J. C. Chow, C. F. Rogers, M. G. Ruby, M. J. Rood, and R. T. Egami. 1989. Method 501: High volume measurement of size classified suspended particulate matter. In *Methods of Air Sampling and Analysis*, ed. J. P. Lodge, 3rd edn., pp. 427–39. Chelsea, MI: Lewis Publishers.

Watson, J. G., J. C. Chow, L. W. Richards, D. L. Haase, C. McDade, D. L. Dietrich, D. Moon, and C. Sloane. 1991*a*. The 1989–90 Phoenix Urban Haze Study, Vol. II: The apportionment of light extinction to sources. Final report. DRI Document No. 8931.5F1, prepared for Arizona Department of Environmental Quality, Phoenix, AZ, by Desert Research Institute, Reno, NV.

Watson, J. G., J. C. Chow, R. T. Egami, J. L. Bowen, C. A. Frazier, A. W. Gertler, D. H. Lowenthal, and K. K. Fung. 1991*b*. Measurements of dry deposition parameters for the California Acid Deposition Monitoring Program. Final report. DRI Document No. 8068.2F1, prepared under Contract No. A6-076-32 for California Air Resources Board, Sacramento, CA, by Desert Research Institute, Reno, NV.

Watson, J. G., J. C. Chow, and T. G. Pace. 1991. Chemical mass balance. In *Receptor Modeling for Air Quality Management, Vol. 7, Data Handling in Science and Technology*, ed. P. K. Hopke, Chap. 4, pp. 83–116. Amsterdam: Elsevier.

Wedding, J. B. 1985. Errors in sampling ambient concentrations employing setpoint temperature compensated mass flow transducers. *Atmos. Environ.* 19:1219.

Wedding, J. B., T. C. Carney, M. W. Litgotke, and R. E. Baumgardner. 1983. The Wedding ambient aerosol sampling inlet for an intermediate flowrate (4 CFM) sampler. *Environ. Sci. Technol.* 17:379.

Wedding, J. B., A. R. McFarland, and J. E. Cermak. 1977. Large particle collection characteristics of ambient aerosol samplers. *Environ. Sci. Technol.* 4:387.

Wedding, J. B. and M. A. Weigand. 1985. The Wedding ambient aerosol sampling inlet (D_{50} = 10 μm) for the high volume sampler. *Atmos. Environ.* 19:535.

Wedding, J. B., M. A. Weigand, and T. C. Carney. 1982. A 10 μm cutpoint inlet for the dichotomous sampler. *Environ. Sci. Technol.* 16:602.

Wedding, J. B., M. Weigand, W. John, and S. Wall. 1980. Sampling effectiveness of the inlet to dichotomous sampler. *Environ. Sci. Technol.* 14:1367.

Wedding, J. B., M. A. Weigand, Y. J. Kim, and J. P. Lodge, Jr. 1988. PM-10 data: Collocated Andersen and Wedding samplers—Partial convergence. In *Transactions, PM_{10}: Implementation of Standards*, eds. C. V. Mathai, and D. H. Stonefield, pp. 104–19. Pittsburgh: Air Pollution Control Association.

Wedding, J. B., M. A. Weigand, Y. J. Kim, D. L. Swift, and J. P. Lodge. 1987. A critical flow device for accurate PM_{10} sampling and correct indication of PM_{10} dosage to the thoracic region of the respiratory tract. *J. APCA* 37:254–58.

Witz, S., M. M. Smith, and A. B. Moore, Jr. 1983. Comparative performance of glass fiber hi-vol filters. *J. APCA* 33:988.

Witz, S., R. W. Eden, M. W. Wadley, C. Dunwoody, R. P. Papa, and K. J. Torre. 1990. Rapid loss of particulate nitrate, chloride and ammonium on quartz fiber filters during storage. *J. APCA* 40:53–61.

Woods, M. C., F. Chen, and M. B. Ranade. 1986. Wind tunnel test reports 14 through 25. Research Triangle Institute, Research Triangle Park, NC.

29

Fugitive Dust Emissions

Chatten Cowherd, Jr.

Midwest Research Institute
Kansas City, Missouri, U.S.A.

INTRODUCTION

The initial air pollution control strategies designed to attain national ambient-air quality standards (NAAQS 1990) in the United States focused almost entirely on ducted sources of emissions to the atmosphere. Fugitive emissions, i.e., those air pollutants that enter the atmosphere without first passing through a confined flow stream, were thought to be relatively insignificant in terms of air quality impact.

However, with the lack of success in achieving the NAAQS for particulate matter, it became evident that fugitive particulate emissions had been underestimated in importance. This included the contributions of commonplace surface dust entrainment processes, such as wind erosion and vehicle travel on unpaved roads, referred to as fugitive dust sources.

In 1987, the basis for the primary (health-related) particulate NAAQS was changed from total suspended particulate matter (particles smaller than about 30 μm in aerodynamic diameter) to PM-10 (particles smaller than 10 μm in aerodynamic diameter). Because fugitive dust emissions were viewed as consisting of mostly coarse particles, their importance was projected to be significantly reduced.

Once again, however, many areas of the country with significant fugitive dust impacts were unable to attain the new PM-10 standards and were classified as particulate "nonattainment areas". As a result, the Clean Air Act Amendments of 1990 (Section 190) called for more stringent control measures for fugitive dust sources within any of the 72 PM-10 nonattainment areas where such control is needed to ensure attainment.

Fugitive dust sources may also contribute to localized air quality problems involving hazardous air pollutants. Examples of such cases are hazardous-waste treatment, storage and disposal facilities (TSDFs), and abandoned waste dumps or industrial deposition areas where contaminated soils are subject to disturbance by wind or machinery. Remediation plans must address potential air pathway exposure to area residents from soil excavation and other mechanical disturbances that entail direct emissions or increase the wind erodibility of the contaminated materials.

This chapter addresses the techniques used to estimate particulate emission rates from fugitive dust sources before and after control application. Specific topics include the characteristics of fugitive dust sources, field techniques for emission quantification, and available models for emission estimation and control performance estimation.

SOURCE CHARACTERIZATION

The sources of fugitive particulate emissions may be separated into two broad categories: process sources and open dust sources. Process sources of fugitive emissions are those associated with industrial operations that alter the chemical or physical characteristics of a feed material. Open dust sources are those that generate fugitive emissions of solid particles by the forces of wind or machinery acting on exposed materials. Open dust sources are also referred to as fugitive dust sources.

Open dust sources include industrial sources of particulate emissions associated with the open transport, storage, and transfer of raw, intermediate, and waste aggregate materials, and nonindustrial sources such as unpaved roads and parking lots, paved streets and highways, heavy construction activities, and agricultural tilling. Generic categories of open dust sources are listed in Table 29-1 (Cowherd and Kinsey 1986).

Factors Affecting Mechanical Dust Generation

Mechanically generated emissions from open dust sources exhibit a high degree of variability from one site to another, and emissions at any one site tend to fluctuate widely. The site characteristics which cause these variations may be grouped into two categories: measures of energy expended by machinery interacting with the surface (e.g., the speed of a vehicle traveling over an unpaved surface), and properties of the exposed surface material (e.g., the content of suspendable fines in the surface material and its moisture content). These site characteristics are further discussed below.

TABLE 29-1 Generic Categories of Open Dust Sources

1. *Unpaved Travel Surfaces*
   - Roads
   - Parking lots and staging areas
   - Storage piles
2. *Paved Travel Surfaces*
   - Streets and highways
   - Parking lots and staging areas
3. *Exposed Areas (wind erosion)*
   - Storage piles
   - Bare ground areas
4. *Materials Handling*
   - Batch drop (dumping)
   - Continuous drop (conveyor transfer, stacking)
   - Pushing (dozing, grading, scraping)
   - Tilling

Surface Material Texture

The dry-particle size distribution of the exposed soil or surface material determines its susceptibility to mechanical entrainment. The upper size limit for particles that can become suspended has been estimated at about 75 μm in aerodynamic diameter (Cowherd et al. 1974). Conveniently, 75 μm in physical diameter is also the smallest particle size for which size analysis by dry sieving is practical (ASTM 1984). Particles passing a 200-mesh screen (74 μm opening) on dry sieving are termed "silt" by highway officials. Note that for fugitive dust particles, the physical diameter and aerodynamic diameter are roughly equivalent because of the offsetting effects of higher density and irregular shape.

Surface Material Moisture

Dust emissions are known to be strongly dependent on the moisture level of the mechanically disturbed material (Cowherd et al. 1974). Water acts as a dust suppressant by forming cohesive moisture films among the discrete grains of surface material. In turn, the moisture level depends on the moisture added by natural precipitation, the moisture removed by evaporation, and moisture movement beneath the surface. The evaporation rate depends on the degree of air movement over the surface, material texture and mineralogy, and crust presence. The moisture-holding capacity of the air is also important, and it correlates strongly with the surface temperature. Vehicle traffic intensifies the drying process primarily by increasing air movement over the surface.

Mechanical Equipment Characteristics

In addition to the material properties discussed above, it is clear that the physical and mechanical characteristics of materials handling and transport equipment also affect dust emission levels. For example, visual observation suggests (and field studies have confirmed) that vehicle emissions per unit of unpaved road length increase with increasing vehicle speed (Cowherd et al. 1974). For traffic on unpaved roads, studies have also shown positive correlations between emissions and (a) vehicle weight and (b) number of wheels per vehicle (Cowherd, Bohn, and Cuscino 1979). Similarly, dust emissions from materials-handling operations have been found to increase with increasing wind speed and drop distance.

Factors Affecting Windblown Dust Generation

Wind-generated emissions from open dust sources also exhibit a high degree of variability from one site to another, and emissions at any one site tend to fluctuate widely. The site characteristics which cause these variations may be grouped into two categories: measures of energy expended by wind interacting with the erodible surface (e.g., wind speed) and properties of the exposed surface material (e.g., content of suspendable fines in the surface material and its moisture content or, for a crusted surface, the strength of the crust). These site characteristics are discussed further below.

Surface Material Texture

As in the case of mechanical entrainment, the dry-particle size distribution of the exposed soil or surface material determines its susceptibility to wind erosion. Wind forces move soil particles by three transport modes: saltation, surface creep, and suspension. Saltation describes particles, ranging in diameter from about 75 to 500 μm, that are readily lifted from the surface and jump or bounce within a layer close to the air–surface interface. Particles transported by surface creep range in diameter from about 500 to 1000 μm. These large particles move very close to the ground, propelled by wind stress and by the impact of small particles transported by saltation. Particles smaller than about 75 μm in diameter move by suspension and tend to follow air motions. As stated above, the upper size limit of silt particles (75 μm in physical diameter) is roughly the smallest particle size for which size analysis by dry sieving is practical. The threshold wind speed for the onset of saltation, which drives the wind erosion process, is also dependent on soil texture, with 100–150 μm particles having the lowest threshold speed. Saltation provides energy for the release of particles in the PM-10 size range that typically are bound by surface forces to larger clusters.

Nonerodible Elements

Nonerodible elements, such as clumps of grass or stones (larger than about 1 cm in diameter) on the surface, consume part of the shear stress of the wind which otherwise would be transferred to erodible soil. Surfaces impregnated with a large density of nonerodible elements behave as having a "limited reservoir" of erodible particles, even if the material protected by nonerodible elements is itself highly erodible. Wind-generated emissions from such surfaces decay sharply with time, as the particle reservoir is depleted. Surfaces covered by unbroken grass are virtually nonerodible.

Surface Material Moisture

Dust emissions from wind erosion are known to be strongly dependent on the moisture level of the erodible material (Woodruff and Siddoway 1965). The mechanism of moisture mitigation is the same as that described above for mechanical entrainment.

Crust Formation

Following the wetting of a soil or other surface material, fine particles will move to form a surface crust. The surface crust acts to hold in soil moisture and resist erosion. The degree of protection that is afforded by a soil crust to the underlying soil may by measured by the modulus of rupture and thickness of the crust

(Cowherd et al. 1985). This modulus of rupture is roughly a measure of the hardness of the crust. Exposed soil that lacks a surface crust (e.g., a disturbed soil or a very sandy soil) is much more susceptible to wind erosion.

Frequency of Mechanical Disturbance

Emissions generated by wind erosion are also dependent on the frequency of disturbance of the erodible surface. A disturbance is defined as an action which results in the exposure of fresh surface material. This would occur whenever a layer of aggregate material is either added to or removed from the surface. The disturbance of an exposed area may also result from the turning of surface material to a depth exceeding the size of the largest material present. Each time that a surface is disturbed, its erosion potential is increased by destroying the mitigative effects of crusts, vegetation, and friable nonerodible elements and by exposing new surface fines. Although vehicular traffic alters the surface by pulverizing surface material, this effect probably does not restore the full erosion potential, except for surfaces that are protected only by a friable crust. In this case, breaking of the crust over the area of the tire/surface contact once again exposes the erodible material beneath.

Wind Speed

Under high wind conditions that trigger wind erosion by exceeding the threshold velocity, the wind speed profile near the erodible surface is found to follow a logarithmic distribution (Gillette 1978):

$$u(z) = \frac{u^*}{0.4} \ln \frac{z}{z_0} \quad (z > z_0) \tag{29-1}$$

where

u = wind speed, cm/s
u^* = friction velocity, cm/s
z = height above test surface, cm
z_0 = roughness height, cm
0.4 = von Karman's constant, dimensionless

The friction velocity (u^*) is a measure of wind shear stress on the erodible surface, as determined from the slope of the logarithmic velocity profile. The roughness height (z_0) is a measure of the roughness of the exposed surface as determined from the y-intercept of the velocity profile (i.e., the height at which the wind speed is zero) on a logarithmic-linear graph. These parameters are illustrated in Fig. 29-1.

Agricultural scientists have established that total soil loss by continuous wind erosion of highly erodible fields is dependent roughly on the cube of wind speed above the threshold velocity (Woodruff and Siddoway 1965). More recent work has shown that the loss of particles in suspension mode follows a similar dependence. Soils protected by nonerodible elements or crusts exhibit a weaker dependence of suspended particulate emissions on wind speed (Cowherd 1988).

Wind Gusts

Although mean atmospheric wind speeds in many areas of the country are not sufficient to initiate wind erosion from "limited-reservoir" surfaces, wind gusts may quickly deplete a substantial portion of the erosion potential of surfaces having a "limited reservoir" of erodible particles. In addition, because the erosion potential (mass of particles constituting the "limited reservoir") increases with increasing wind speed, estimated emissions should be related to the gusts of highest magnitude.

The routinely measured meteorological variable which best reflects the magnitude of wind gusts is the "fastest mile of wind" (U.S. WS 1988). The quantity represents the wind speed corresponding to the whole mile of wind movement which has passed by the 1-mile contact anemometer in the least amount of time. Daily measurements of the fastest mile are presented in the monthly Local Climatological Data (LCD) summaries for weather stations throughout the United States (NEDS 1992). The duration of the fastest mile, typically about 1–2 min (for fastest miles of 30–60 mph, respectively), matches well with the half-life (i.e., the time required to remove one-half the erodible particles on the

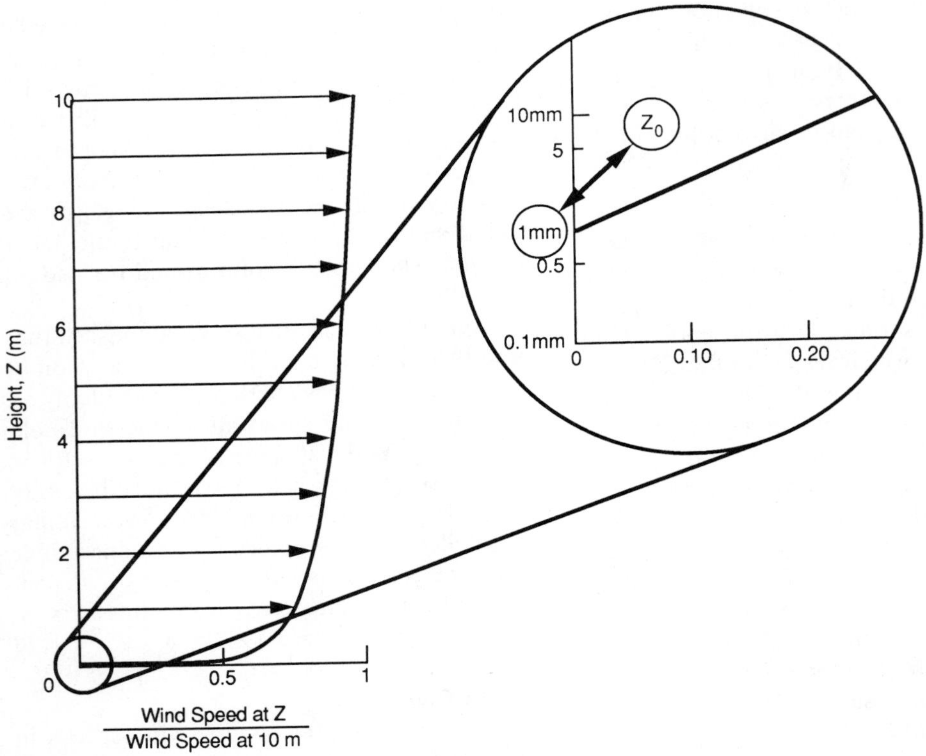

FIGURE 29-1. Logarithmic Wind-Speed Profile.

surface) of the erosion process. It should be noted, however, that instantaneous peak wind speeds can significantly exceed the daily fastest mile.

The strong time decay of erosion rate is due not only to the limited availability of erodible particles but also to the efficiency of short-duration wind gusts in depleting the reservoir. Furthermore, because the threshold wind speed must be exceeded to trigger the possibility of substantial wind erosion, the dependence of erosion potential on wind speed cannot be represented by any simple linear function. For this reason, the use of an average wind speed to calculate an average emission rate is inappropriate.

Wind Accessibility

If the erodible material lies on the surface of a large relatively flat pile or exposed area with little penetration into the surface wind layer, then the material is uniformly accessible to the wind. If this is not the case, it is necessary to divide the erodible area into subareas representing different degrees of exposure to wind. For example, the results of physical modeling show that the frontal face of an elevated pile is exposed to surface wind speeds of the same order as the approach wind speed upwind of the pile at a height matching the top of the pile (Billings-Stunder and Arya 1988); on the other hand, the leeward face of the pile is exposed to much lower wind speeds. An illustration of this nonuniform wind accessibility is presented later.

Emission Calculation Procedure

A calculation of the estimated emission rate for a given source requires data on source extent, uncontrolled emission factor, and control efficiency. The mathematical expression for this calculation is as follows:

$$R = Me\,(1 - c) \qquad (29\text{-}2)$$

where

R = estimated mass emission rate in the specified particle size range
M = source extent
e = uncontrolled emission factor in the specified particle size range, i.e., mass of uncontrolled emissions per unit of source extent
c = fractional efficiency of control

The source extent is the appropriate measure of source size or the level of activity which is used to scale the uncontrolled emission factor to the particular source in question. For process sources of fugitive particulate emissions, the source extent is usually the production rate (i.e., the mass of product per unit time). Similarly, the source extent of an open dust source entailing a batch or continuous drop operation is the rate of mass throughput.

For other categories of open dust sources, the source extent is related to the area of the exposed surface which is disturbed by either wind or mechanical forces. In the case of wind erosion, the source extent is simply the area of erodible surface. For emissions generated by mechanical disturbance, the source extent is also the surface area (or volume) of the material from which the emissions emanate. For vehicle travel, the disturbed surface area is the travel length times the average daily traffic (ADT) count, with each vehicle having a disturbance width equal to the width of a travel lane.

A compendium of emission factors (referred to as AP-42) is maintained by the U.S. Environmental Protection Agency (1985). Section 11.2 of AP-42 contains the emission factors for fugitive dust sources. These "uncontrolled" emission factors typically incorporate the effects of natural mitigation (e.g., rainfall).

If an anthropogenic control measure (e.g., treating the surface with a chemical binder which forms an artificial crust) is applied to the source, the uncontrolled emission factor in Eq. 29-2 must be multiplied by an additional term to reflect the resulting fractional control.

In broad terms, anthropogenic control measures can be considered as either continuous or periodic, as the following examples illustrate:

Continuous Controls	*Periodic Controls*
Wet suppression at conveyor transfer points	Watering or chemical treatment of unpaved roads
Enclosures/wind fences around storage piles	Sweeping of paved travel surfaces
Continuous vegetation of exposed areas	Chemical stabilization of exposed areas

The major difference between the two types of controls is related to the time dependency of performance. For continuous controls, efficiency is essentially constant with respect to time. On the other hand, the efficiency associated with periodic controls tends to decrease (decay) with time after application. An example is the chemical treatment of an unpaved road; immediately after application, the road surface is thoroughly wetted and complete control is assumed. A generally high level of control is observed for approximately one week. Thereafter, the efficiency tends to decrease until the next chemical application (at which time the cycle repeats but with some residual effects from the previous application).

In order to quantify the performance of a specific periodic control, two measures of control efficiency are required. The first is "instantaneous" control and is defined by

$$c(t) = \left(1 - \frac{e_c(t)}{e_u}\right) \times 100\% \qquad \text{(29-3)}$$

where

$c(t)$ = instantaneous control efficiency (percent)
$e_c(t)$ = instantaneous emission factor for the controlled source
e_u = uncontrolled emission factor
t = time after control application

The other important measure of periodic control performance is average efficiency, defined as

$$C(T) = \frac{1}{T}\int_0^T c(t)\,dt \qquad (29\text{-}4)$$

where

$C(T)$ = average control efficiency during the period ending at time T after application (percent)

$c(t)$ = instantaneous control efficiency at time t after application (percent)

T = time period over which the average control efficiency is referenced

The average control efficiency values are needed to estimate the emission reductions due to periodic applications.

In order to estimate the emissions of a specific contaminant that is present within the disturbed surface material, the emission rate should be adjusted by the fraction of contaminant in the specified airborne-particle size range (Cowherd et al. 1985). However, as an approximation, the fraction of contaminant in the silt portion of the disturbed surface material may be used to adjust the uncontrolled emission factor and, in turn, the calculated emission rate, so that it applies to the contaminant of interest.

EMISSION QUANTIFICATION TECHNIQUES

Fugitive dust emission rates and particle size distributions are difficult to quantify because of the diffuse and variable nature of such sources and the wide range of particle sizes, including particles which deposit immediately adjacent to the source. Standard source testing methods, which are designed for application to confined flows under steady-state, forced-flow conditions, are not suitable for the measurement of fugitive emissions unless the plume can be drawn into a forced-flow system. The available source testing methods for fugitive dust sources are described in the following paragraphs.

Mechanical Entrainment Processes

For the field measurement of fugitive mass emissions from mechanical entrainment processes, four basic techniques have been defined:

1. The quasi-stack method involves capturing the entire particulate emissions stream with enclosures or hoods and applying conventional source-testing techniques to the confined flow (Kolnsberg et al. 1976).
2. The roof monitor method involves the measurement of particulate concentrations and airflows across well-defined building openings such as roof monitors, ceiling vents, and windows, followed by a calculation of the particulate mass flux exiting the building (Kenson and Barlett 1976).
3. The upwind–downwind method involves the measurement of upwind and downwind particulate concentrations, utilizing ground-based samplers under known meteorological conditions, followed by a calculation of the source strength (mass emission rate) with atmospheric dispersion equations (Kolnsberg 1976).
4. The exposure-profiling method involves simultaneous, multipoint measurements of particulate concentration and wind speed over the effective cross section of the plume, followed by a calculation of the net particulate mass flux through integration of the plume profiles (Cowherd et al. 1974).

Because it is usually impractical to enclose open dust sources or to capture the entire emissions plume, only the upwind–downwind and exposure-profiling methods are suitable for the measurement of particulate emissions from most open dust sources. These two methods are discussed separately below.

Upwind–Downwind Method

The upwind–downwind method involves the measurement of airborne particulate concentrations both upwind and downwind of the pollutant source. The number of upwind sampling instruments depends on the degree of isolation of the source operation of concern (i.e., the absence of interference from other sources upwind). Increasing the number of downwind instruments improves the reliability in determining the emission rate by providing better plume definition. In order to reasonably define the plume emanating from a point source, instruments need to be located at a minimum of two downwind distances and three crosswind distances. The same sampling requirements pertain to line sources except that measurement need not be made at multiple crosswind distances. A basic upwind–downwind array is shown in Fig. 29-2.

Net downwind (i.e., downwind minus upwind) concentrations are used as input to atmospheric dispersion equations (normally of the Gaussian type) to back-calculate the particulate emission rate (i.e., source strength) required to generate the pollutant concentrations measured. Emission factors are obtained by dividing the calculated emission rate by the source extent. A number of meteorological parameters must be concurrently recorded for input to this dispersion equation. As a minimum, the wind direction and speed must be recorded on-site.

While the upwind–downwind method is applicable to virtually all types of sources, it has significant limitations with regard to the development of source-specific emission factors. The major limitations are as follows:

1. In attempting to quantify a large area source, overlapping plumes from upwind (background) sources may preclude the determination of the specific contribution of the area source.
2. Because of the impracticality of adjusting the locations of the sampling array for shifts in wind direction during sampling, it may be questionable to assume that the plume position is fixed in the application of the dispersion model.
3. The usual assumption that an area source is uniformly emitting may not allow for a realistic representation of spatial variation in source activity.

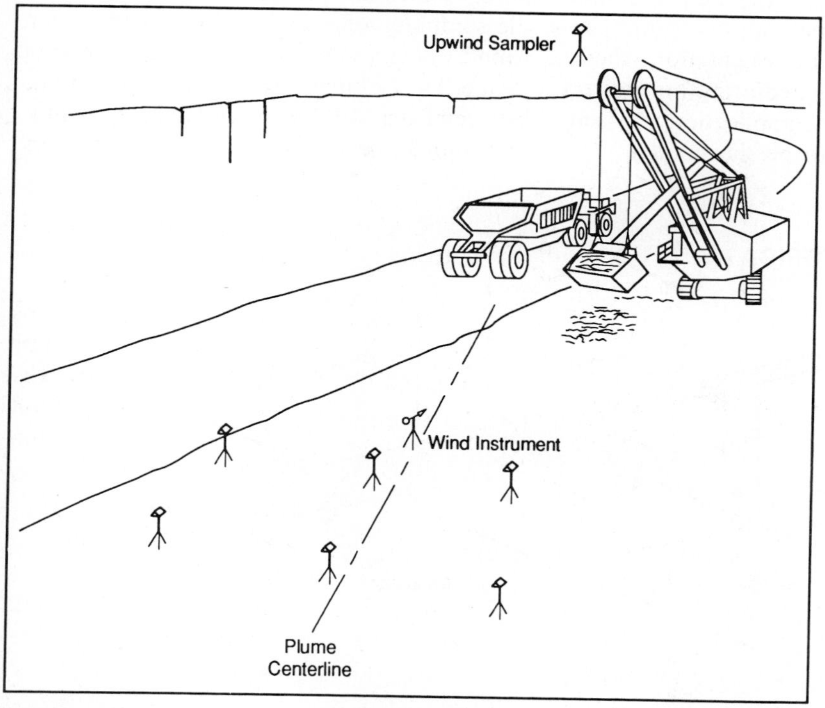

FIGURE 29-2. Upwind–Downwind Sampling Method.

4. The typical use of an uncalibrated atmospheric dispersion model introduces the possibility of substantial error. According to Turner (1970), the error in the calculated emission rate can be as much as a factor of three, even if the stringent requirement of unobstructed dispersion from a simplified source configuration is met (e.g., constant emission rate from a single point).

Exposure-Profiling Method

As an alternative to conventional upwind–downwind sampling, the exposure-profiling technique utilizes the isokinetic profiling concept, which is the basis for conventional ducted source testing (EPA Method 5), except that, in the case of exposure-profiling, the ambient wind directs the plume to the sampling array. The passage of airborne particulate matter immediately downwind of the source is measured directly by means of a simultaneous, multipoint sampling of particulate concentration and wind velocity over the effective cross section of the fugitive emissions plume. Unlike the conventional upwind–downwind method, exposure-profiling uses a mass-balance calculation scheme (Fig. 29-3) rather than requiring an indirect calculation through the application of a generalized atmospheric dispersion model.

For the measurement of nonbuoyant fugitive emissions using exposure profiling, sampling heads are distributed over a vertical network positioned just downwind (usually about 5 m) from the source. Particulate sampling heads should be symmetrically distributed over the concentrated portion of the plume containing at least 80% of the total mass flux. A vertical line grid of at least three samplers is sufficient for the measurement of emissions from line or moving point sources (Fig. 29-4), while a two-dimensional array of at least five samplers is required for quantification of the fixed virtual point source of emissions (Fig. 29-5). In the profiler head configuration shown in Fig. 29-4, air is drawn through a 5 × 5 cm directional intake into a settling chamber (trapping particles larger than about 50 μm in aerodynamic diameter) and then upward through a horizontally positioned glass fiber filter. At least one upwind sampler must be operated to measure the background concentration, and wind speed must be measured concurrently on-site.

To achieve isokinetic sampling, the sampling intakes are pointed into the wind and the sampling velocity matched to the mean wind velocity approaching the sampling intake. The isokinetic velocity ratio should lie between 0.8 and 1.2, and the sampler orientation should be adjusted whenever it differs

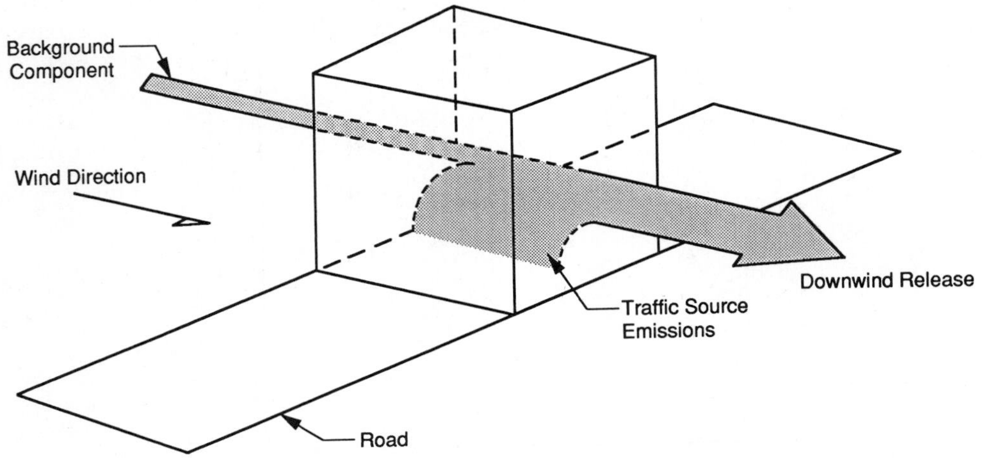

FIGURE 29-3. Particulate Mass Balance—Roadway.

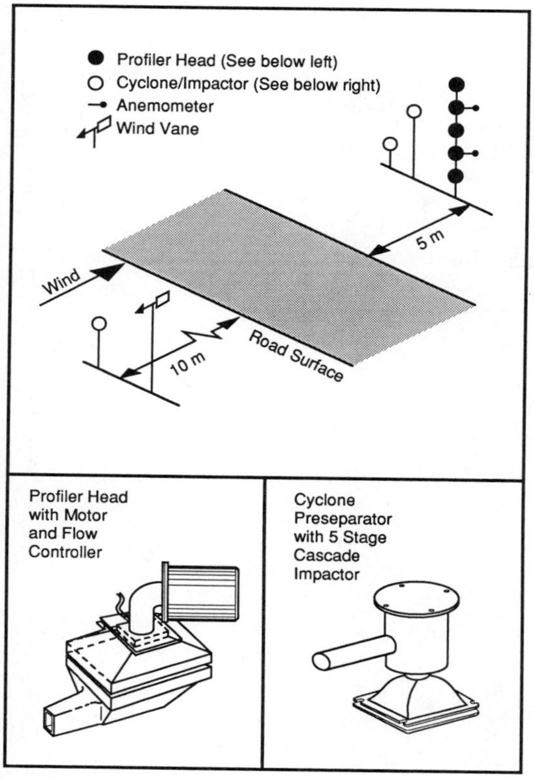

FIGURE 29-4. Exposure Profiling Method—Roadway.

from the mean wind direction by more than 30°. A minimum wind velocity of 1.8 m/s (4 mph) is usually required to assure acceptable consistency in wind speed and direction.

The particulate emission rate is obtained by a spatial integration of the distributed measurements of exposure (accumulated mass flux), which is the product of mass concentration and wind speed:

$$R = \int_A C(h, w)u(h, w)\,\mathrm{d}h\,\mathrm{d}w \qquad (29\text{-}5)$$

where

R = emission rate, g/s
C = net particulate concentration, $\mathrm{g/m^3}$
u = wind speed, m/s
h = vertical distance coordinate, m
w = lateral distance coordinate, m
A = effective cross-sectional area of plume, $\mathrm{m^2}$

Usually, a numerical integration scheme is used to calculate the emission rate.

Regardless of which method is used, isokinetic sampling is required for a representative collection of particles larger than about 5 μm in aerodynamic diameter. Biases may result from differences between the sampling intake direction and the direction of the wind. Because of natural fluctuations in wind speed and direction, some anisokinetic sampling effects will always be encountered. In estimating the magnitudes of anisokinetic sampling errors, an aerodynamic particle diameter of about 12 μm is assumed because this approximates the mass median diameter of a high-volume catch near a fugitive dust source.

To hold sampling bias to an acceptable minimum, the angle α between the mean wind direction and the direction of the sampling axis should not exceed 30°. For $\alpha = 30°$, the sampling error is about 10% for particles of 12 μm aerodynamic diameter (Watson 1954). The restriction on horizontal wind direction fluctuation (i.e., standard deviation σ_α less than 22.5°) excludes sampling under Stability Class A, which is characterized by large horizontal wind meander and low wind speeds (Turner 1970).

In the wind speed range of 4–20 mph, the sampling rate can be readily adjusted and matched to the corresponding mean wind speed. An isokinetic flow ratio (IFR = ratio of sampling intake speed to approach wind speed) of less than 0.8 or greater than 1.2 may lead to large concentration errors. For particles of 12 μm diameter, it has been shown that the sampling error is less than about 5% for an IFR between 0.8 and 1.2 (Watson 1954). If these wind conditions are met, background concentrations usually amount to less than 5% of downwind concentrations (for uncontrolled open dust sources).

Wind Erosion

The two wind erosion source testing methods of interest are as follows:

1. The upwind–downwind method involves the measurement of upwind and downwind particulate concentrations, utilizing

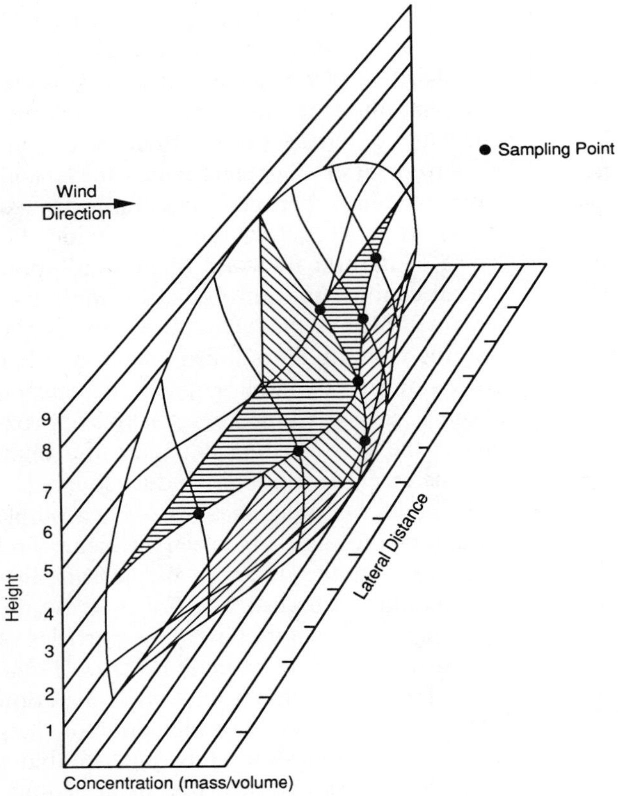

FIGURE 29-5. Exposure Profile for Fixed-Point Source.

ground-based samplers under known meteorological conditions, followed by a calculation of the source strength (mass emission rate) using atmospheric dispersion equations (Kolnsberg 1976).

2. The wind tunnel method involves the use of a portable open-floored wind tunnel for *in situ* measurement of emissions from representative surfaces under predetermined wind conditions (Cuscino, Muleski, and Cowherd 1983).

Each of these methods will be discussed below.

Upwind–Downwind Method

The upwind–downwind method and its limitations for quantification of mechanically generated emissions were discussed above. The upwind–downwind method is burdened with practical difficulties for the study of wind erosion, in that the onset of erosion and its intensity is beyond the control of the investigator. As an illustration of this point, two published emission factors for coal pile wind erosion were developed from upwind–downwind sampling under light wind conditions (Blackwood and Wachter 1977; PEDCo 1978). In addition, background (upwind) particulate concentrations tend to be high during erosion events, making source isolation very difficult.

Wind Tunnel Method

An alternative approach to this problem entails the use of a portable open-floored wind tunnel for *in situ* measurement of emissions from representative test surfaces under predetermined wind conditions. In effect, the experimental problem is divided into two parts: determination of the relationship between the rate of windblown dust emissions

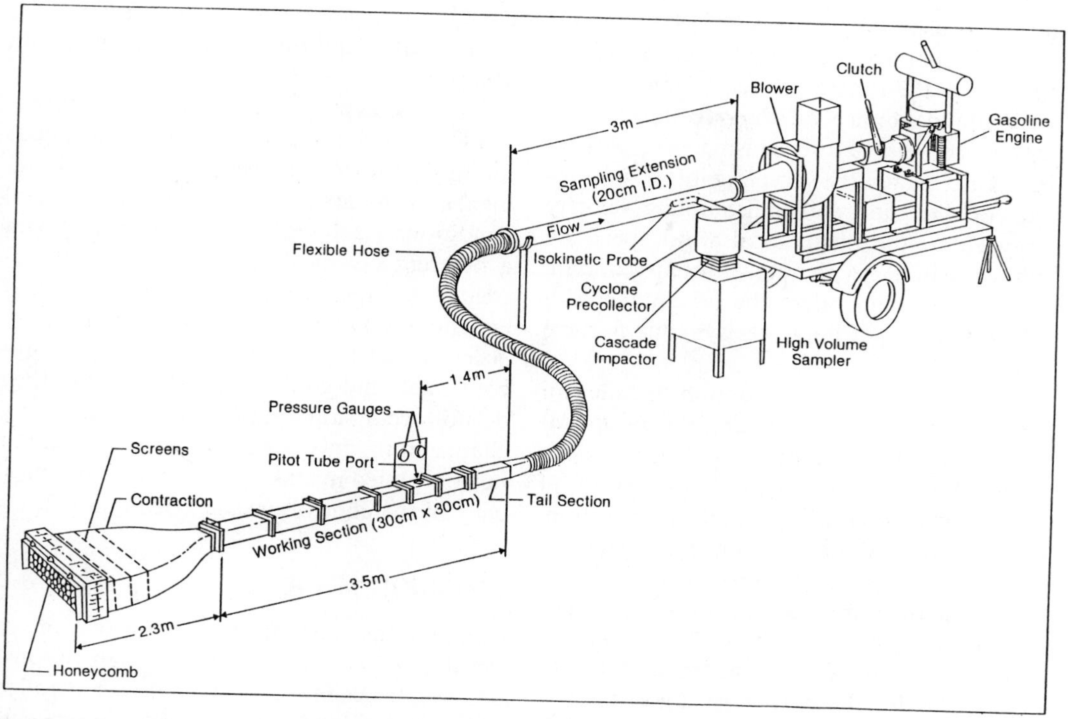

FIGURE 29-6. Portable Open-Floored Wind Tunnel Equipment.

and the physical parameters which enter into the wind erosion process; and analysis of wind flow patterns around storage piles or other non-uniformly exposed surfaces.

The most common version of the wind tunnel method utilizes a pull-through wind tunnel with an open-floored test section placed directly over the surface to be tested. An example device of this type is shown in Fig. 29-6. Air is drawn through the tunnel at controlled velocities. The exit air stream from the test section passes through a circular duct fitted with a directional sampling probe at the downstream end. Air is drawn isokinetically through the probe by a high-volume sampling train.

The wind tunnel method incorporates the essential features of Method 5 stack sampling (U.S. EPA 1977). The one prime difference, the use of single-point sampling, is justified by the high turbulence levels in the sampling module. The measurement uncertainty inherent in this method is of the same order as that in Method 5, which has been subjected to extensive collaborative testing by EPA.

The wind tunnel method relies on a straightforward mass-balance technique for the calculation of emission rate. By sampling under light ambient wind conditions, background interferences from upwind erosion sources can be avoided. Although a portable wind tunnel does not generate the larger scales of turbulent motion found in the atmosphere, the turbulent boundary layer formed within the tunnel simulates the smaller scales of atmospheric turbulence. It is the smaller-scale turbulence which penetrates the wind flow in direct contact with the erodible surface and contributes to the particle entrainment mechanisms (Gillette 1978).

Particle Sizing

Concurrent with the measurement of mass emissions, the aerodynamic particle size distribution should be characterized. For this purpose, a high-volume cyclone/cascade impactor featuring isokinetic sample collection has been used (Cowherd et al. 1986). A cyclone preseparator (or other device) is needed to remove the coarse particles, which otherwise would bounce off the substrate stages within the impactor, causing fine-particle bias. Once again, the sampling intake is pointed into the wind and the sampling velocity adjusted to the mean local wind speed by fitting the intake with a nozzle of appropriate size. This system offers the advantage of a direct determination of aerodynamic particle size.

Another particle sizing option includes an analysis of the particulate deposit by optical or electron microscopy. Disadvantages include (a) potential artificial disaggregation of particle clusters during sample preparation, and (b) uncertainties in converting physical size data to equivalent aerodynamic diameters. In a collaborative field test of the exposure-profiling method, the cyclone/impactor method was judged to be more suitable than microscopy for the particle sizing of fugitive dust emissions (McCain, Pyle, and McGrillis 1985).

Control Efficiency Estimation

Field evaluation of the control efficiency requires that the study design include not only adequate emission measurement techniques but also a proven "control application plan". In the past, two major types of plans have been used:

Type 1: Controlled and uncontrolled emission measurements are obtained simultaneously.

Type 2: Uncontrolled tests are performed initially, followed by controlled tests.

In order to ensure comparability between the operating characteristics of the controlled and uncontrolled sources, many evaluations are forced to employ Type-2 plans. An example would be a wet suppression system used on a primary crusher. One important exception to this, however, is unpaved-road dust control. In this instance, testing under a Type-1 plan is conducted on two or more contiguous road segments. One segment is left untreated and the others are treated with separate dust suppressants.

Under a Type-2 plan for testing unpaved-road dust controls, uncontrolled testing is initially performed on one or more road segments, generally under worst-case (dry) conditions. Each segment is then treated with a different chemical; no segment is left untreated as a reference. A normalization of emissions may be required to allow for differences in vehicle characteristics during the uncontrolled and controlled tests because they do not occur simultaneously. For example, a change in the mix of vehicle types should not be interpreted mistakenly as part of the efficiency of the control measure being tested.

EMISSION MODELS

Early in the EPA field testing program to develop emission factors for fugitive dust sources, it became evident that uncontrolled emissions within a single generic source category may vary over two (or more) orders of magnitude as a result of variations in source conditions (equipment characteristics, material properties, and climatic parameters). Therefore, it would not be feasible to represent an entire generic source category in terms of a single-valued emission factor, as traditionally used by the EPA to describe average emissions from a narrowly defined ducted source operation. In other words, it would take a large matrix of single-valued factors to adequately represent an entire generic fugitive dust source category. In order to account for emissions variability, therefore, the approach was taken that fugitive dust emission factors be constructed as mathematical equations for sources grouped by the dust generation mechanisms. The emission factor equation for each source category would contain multiplicative correction parameter terms that explain much of the variance in observed emission factor values on the basis of variances in specific source parameters. Such factors would be applicable to a wide range of source conditions, limited only by the extent of experimental verification.

For example, the use of the silt content as a measure of the dust generation potential of a material acted on by the forces of wind or machinery proved to be an important step in extending the applicability of the emission factor equations to a wide variety of aggregate materials of industrial importance.

As a result of the successful development of emission factor equations for open dust sources, in 1975 EPA published a new section of AP-42 (Section 11.2) dealing with fugitive dust sources on a generic basis (U.S. EPA 1975). The original source categories included unpaved roads, agricultural tilling, and aggregate storage piles.

Because of the increased rate of open dust source test data accumulation after 1975 and the development of improved emission factor equations, EPA published a major update and expansion of Section 11.2 (Supplement 14) in May 1983 (U.S. EPA 1983). This included improved emission factor equations for unpaved roads, agricultural tilling, industrial paved roads, aggregate storage piles, and materials handling (including batch- and continuous-drop operations). Also with each of the new equations was provided a set of particle size multipliers for adjusting the calculated emission factors to specific particle size fractions.

According to EPA's current Section 11.2 of AP-42 (U.S. EPA 1985), the fugitive particulate emission factor equations "retain their assigned quality ratings if applied within the ranges of source conditions that were tested in developing the equations." These ranges of source conditions are provided with each emission factor equation. There is no restriction as to the composition of the emitting aggregate material as long as its silt and moisture contents are available for comparison to the tested source conditions.

For illustrative purposes, two example emission factor models from AP-42 are presented below.

Vehicle Traffic on Unpaved Roads

For the purpose of estimating emissions, the AP-42 emission factor equation applicable to vehicle traffic on unpaved roads takes source characteristics (such as average vehicle weight and road surface texture) into consideration (Cowherd, Bohn, and Cuscino 1979):

$$e = 0.61\left(\frac{s}{12}\right)\left(\frac{S}{48}\right)\left(\frac{W}{2.7}\right)^{0.7}\left(\frac{w}{4}\right)^{0.5} \times \frac{(365 - p)}{365} \text{ (kg/VKT)} \tag{29-6}$$

$$e = 2.1\left(\frac{s}{12}\right)\left(\frac{S}{30}\right)\left(\frac{W}{3}\right)^{0.7}\left(\frac{w}{4}\right)^{0.5} \times \frac{(365 - p)}{365} \text{ (lb/VMT)}$$

where

- e = PM-10 emission factor in units stated (VKT denotes vehicle-kilometers traveled and VMT denotes vehicle-miles traveled)
- s = silt content of road surface material, percent (ASTM-C-136)
- S = mean vehicle speed, km/h (mile/h)
- W = mean vehicle weight, Mg (ton)
- w = mean number of wheels (dimensionless)
- p = number of days with ≥ 0.254 mm (0.01 in) of precipitation per year

Using the scheme given in AP-42, the above equation is rated "A", *assuming the availability of site-specific correction parameter data.* The 95% confidence interval for this equation is a factor of 1.46, as compared to a factor of 5.20 when an average emission factor value is used in place of the predictive equation (Cuscino, Cowherd, and Bohn 1981). The denominators in each of the multiplicative terms of the equation constitute normalizing default values, in case no site-specific correction parameter data are available.

The number of wet days per year, p, for the geographical area of interest should be determined from local climatic data. Maps giving similar data on a monthly basis are available from the National Climatic Center at Asheville, North Carolina.

The AP-42 predictive emission factor equation for unpaved roads, which is routinely used for emission inventorying purposes, is based on source tests conducted under dry conditions. Extrapolation to annual average uncontrolled emission estimates (including natural mitigation) is accomplished by assuming that emissions are occurring at the estimated rate on days without measurable precipitation and, conversely, are absent on days with measurable precipitation. This assumption has not been verified in a rigorous manner; however, experience with hundreds of field tests indicates that it is a reasonable assumption if the source operates on a fairly "continuous" basis (Cowherd, Muleski, and Kinsey 1988).

Industrial Wind Erosion

The emission factor for wind-generated particulate emissions from mixtures of erodible and nonerodible surface material subject to disturbance may be expressed in units of g/m^2 yr as follows (Cowherd 1988):

$$e = 0.5 \sum_{i=1}^{N} P_i \qquad (29\text{-}7)$$

where

e = PM-10 emission factor, g/m^2 yr
N = number of disturbances per year
P_i = erosion potential corresponding to the observed (or probable) fastest mile of wind for the ith period between disturbances, g/m^2

In calculating emission factors, each area of an erodible surface that is subject to a different frequency of disturbance should be treated separately. For a surface disturbed daily, $N = 365$/yr, and for a surface disturbance once every 6 months, $N = 2$/yr.

The erosion potential function for a dry, exposed surface has the following form:

$$P = 58(u^* - u_t^*)^2 + 25(u^* - u_t^*)$$
$$P = 0 \text{ for } u^* \leq u_t^* \qquad (29\text{-}8)$$

where

u^* = friction velocity (m/s)
u_t^* = threshold friction velocity (m/s)

Because of the nonlinear form of the erosion potential function, each erosion event must be treated separately.

Equations 29-7 and 29-8 apply only to dry, exposed materials with limited erosion potential. The resulting calculation is valid only for a time period as long as or longer than the period between disturbances. Calculated emissions represent intermittent events and should not be input directly into dispersion models that assume steady-state emission rates.

For uncrusted surfaces, the threshold friction velocity is best estimated from the dry aggregate structure of the soil. A simple hand sieving test of surface soil [adapted from a laboratory procedure published by Chepil (1952)] can be used to determine the mode of the surface aggregate size distribution by an inspection of relative sieve catch amounts. The threshold friction velocity for erosion can be determined from the mode of the aggregate size distribution, as described by Gillette (1980). Threshold friction velocities for several surface types have been determined by field measurements with a portable wind tunnel (Cowherd 1988; Gillette 1980; Muleski 1985; Nickling and Gillies 1986).

The fastest mile of wind for the periods between disturbances may be obtained from the monthly LCD summaries for the nearest reporting weather station that is representative of the site in question, available from the National Climatic Center. These summaries report actual fastest-mile values for each day of a given month. Because the erosion potential is a highly nonlinear function of the fastest mile, mean values of the fastest mile are inappropriate. The anemometer heights of reporting weather stations are found in Changery (1978), and should be corrected to a 10 m reference height using Eq. 29-1.

To convert the fastest mile of wind (u^+) from a reference anemometer height of 10 m to the equivalent friction velocity (u^*), the logarithmic wind speed profile may be used to yield the following equation:

$$u^* = 0.053\, u_{10}^{+} \qquad (29\text{-}9)$$

where

u^* = friction velocity (m/s)
u_{10}^{+} = fastest mile of the reference anemometer for a period between disturbances (m/s)

This assumes a typical roughness height of 0.5 cm for open terrain. Equation 29-9 is restricted to large relatively flat piles or exposed areas with little penetration into the surface wind layer.

If the pile significantly penetrates the surface wind layer (i.e., with a height-to-base ratio exceeding 0.2), it is necessary to divide the pile area into subareas representing different degrees of exposure to wind. The results of physical modeling show that the frontal face of an elevated pile is exposed to wind speeds of the same order as the approach wind speed at the top of the pile.

For two representative pile shapes (conical pile and oval pile with flat-top, 37° side slope), the ratios of surface wind speed (u_s) to approach wind speed (u_r) have been derived from wind tunnel studies (Billings-Stunder and Arya 1988). The results are shown in Fig. 29-7, corresponding to an actual pile height of 11 m, a reference (upwind) anemometer

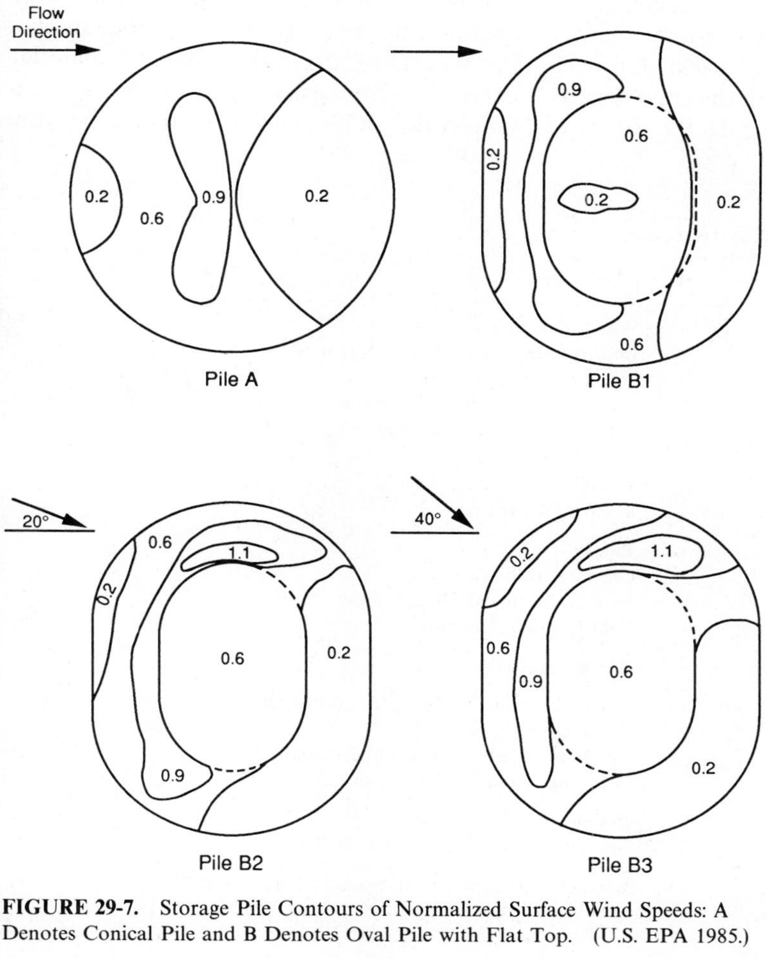

FIGURE 29-7. Storage Pile Contours of Normalized Surface Wind Speeds: A Denotes Conical Pile and B Denotes Oval Pile with Flat Top. (U.S. EPA 1985.)

height of 10 m, and a pile surface roughness height (z_0) of 0.5 cm. The measured surface winds correspond to a height of 25 cm above the surface. The profiles of u_s/u_r can be used to estimate the surface friction velocity distribution around similarly shaped piles, using the procedure described in AP-42.

EMISSION CONTROL OPTIONS

Typically, there are several options for the control of fugitive particulate emissions from any given source. This is clear from Eq. 29-2, used to calculate the emission rate. Because the uncontrolled-emission rate is the product of the source extent and the uncontrolled-emission factor, a reduction in either of these two variables produces a proportional reduction in the uncontrolled-emission rate.

Although the reduction of source extent results in a highly predictable reduction in the uncontrolled-emission rate, such an approach usually requires a change in the source activity. Frequently, a reduction in the extent of one source may necessitate an increase in the extent of another, as in the shifting of vehicle traffic from an unpaved road to a paved road.

In the case of open sources, the reduction in the uncontrolled-emission factor may be achieved by adjusted work practices. The degree of the reduction of the uncontrolled-emission factor can be estimated from the known dependence of the factor on source conditions that are subject to alteration. For open dust sources, this information is embodied in the predictive emission factor equations for fugitive dust sources as presented in Section 11.2 of AP-42. The reduction of source extent and the incorporation of adjusted work practices which reduce the amount of exposed dust-producing material are preventive measures for the control of fugitive dust emissions.

Add-on controls can also be applied to reduce emissions by reducing the amount (areal extent) of dust-producing material, other than by cleanup operations. For example, the elimination of mud/dirt carryout

TABLE 29-2 Controls for Fugitive Dust Sources

Source category	Control action
Paved roads	Water flushing/sweeping Improvements in sanding/salting applications and materials Truck covering Prevention of track-on/wash-on: • Construction site measures • Curb installation • Shoulder stabilization • Storm water drainage
Unpaved roads	Paving Chemical stabilization Surface improvement (graveling) Vehicle speed reduction
Storage piles (transfer operations)	Wet suppression
Construction/demolition	Paving permanent roads early in project Truck covering Access apron construction and cleaning Watering of graveled travel surfaces
Open area wind erosion	Revegetation Limitation of off-road vehicle traffic
Agricultural tilling	Land conservation practices under Food Security Act

onto paved roads at construction and demolition sites is a cost-effective preventive measure. On the other hand, mitigative measures involve the periodic removal of dust-producing material. Examples of mitigative measures include: cleanup of spillage on travel surfaces (paved and unpaved) and cleanup of material spillage at conveyor transfer points. Mitigative measures tend to be less favorable from a cost-effectiveness standpoint.

Periodically applied control techniques for open dust sources begin to decay in efficiency almost immediately after implementation. The most extreme example of this is the watering of unpaved roads, where the efficiency decays from nearly 100% to 0% in a matter of hours (or minutes). The control efficiency for broom sweeping and flushing applied in combination on a paved road may decay to 0% in 1 or 2 days. Chemical dust suppressants applied to unpaved roads can yield control efficiencies that will decay to 0% in several months.

Consequently, a single-value control efficiency is usually not adequate to describe the performance of most intermittent-control techniques for open dust sources. The "time-weighted average" control efficiency must be reported along with the time period over which the value applies. For continuous-control systems (e.g., wet suppression for continuous-drop materials transfer), a single control efficiency is usually appropriate.

Table 29-2 lists fugitive dust control measures that have been judged to be generally cost-effective for application to metropolitan areas unable to meet PM-10 standards. The most highly developed performance models available apply to application of chemical suppressants on unpaved roads. These models relate the expected instantaneous control efficiency to the application parameters (application intensity and dilution ratio) and to the number of vehicle passes (rather than time) following the application. More details on available dust control measure performance and cost are presented by Cowherd, Muleski, and Kinsey (1988) and Cowherd (1991).

References

ASTM. 1984. American Society of Testing and Materials, Subcommittee C09.03.05. Standard method for sieve analysis of fine and coarse aggregates. Method C-136,84a, Philadelphia, PA.

Billings-Stunder, B. J. and S. P. S. Arya. 1988. Windbreak effectiveness for storage pile fugitive dust control: A wind tunnel study. *J. Air Pollut. Control Assoc.* 38:135–43.

Blackwood, T. R. and R. A. Wachter. 1977. *Source Assessment: Coal Storage Piles.* EPA Contract No. 68-02-1874. Cincinnati, OH: U.S. EPA.

CAA. 1990. Clean Air Amendments of 1990, Public Law 101-549, 15 November 1990.

Chepil, W. S. 1952. Improved rotary sieve for measuring state and stability of dry soil structure. *Soil Science Society of America Proceedings* 16:113–17.

Cowherd, C. Jr. 1988. A refined scheme for calculation of wind generated PM-10 emissions from storage piles. In *Proceedings: APCA/EPA Conference on PM-10: Implementation of Standards.* San Francisco, CA: U.S. EPA.

Cowherd, C. Jr. 1991. *Best Available Control Measures (BACM) for Fugitive Dust Sources.* Revised Draft Guidance Document submitted to Air Quality Management Division. Research Triangle Park, NC: U.S. EPA.

Cowherd, C. Jr., K. Axetell Jr., C. M. (Guenther) Maxwell, and G. A. Jutze. 1974. *Development of Emission Factors for Fugitive Dust Sources,* EPA Publication No. EPA-450/3-74/037, NTIS Publication No. PB-238 262.

Cowherd, C. Jr., R. Bohn, and T. Cuscino Jr. 1979. *Iron and Steel Plant Open Source Fugitive Emission Evaluation,* EPA-600/2-79/103, NTIS Publication No. PB-299 385.

Cowherd, C. Jr., J. S. Kinsey, D. D. Wallace, M. A. Grelinger, T. A. Cuscino, and R. M. Neulicht. 1986. *Identification, Assessment, and Control of Fugitive Particulate Emissions,* EPA-600/8-86-023: NTIS Publication No. PB86-230083.

Cowherd, C. Jr., G. E. Muleski, P. J. Englehart, and D. A. Gillette. 1985. *Rapid Assessment of Exposure to Particulate Emissions from Surface Contamination Sites,* EPA/600/8-85/002. Washington, DC: U.S. EPA.

Cowherd, C. Jr., G. E. Muleski, and J. S. Kinsey. 1988. *Control of Open Fugitive Dust Sources,* EPA-450/3-88-008. Research Triangle Park, NC: U.S. EPA.

Cuscino, T. A. Jr., C. Cowherd Jr., and R. Bohn. 1981. Fugitive emission control of open dust sources. *Proceedings: Symposium on Iron and Steel Pollution Abatement Technology for 1980, Philadelphia, Pennsylvania, November 1980,* EPA-600/9-81/017, 71-84, NTIS Publication No. PB81-244808.

Cuscino, T. Jr., G. E. Muleski, and C. Cowherd Jr. 1983. *Iron and Steel Plant Open Source Fugitive Emission Evaluation.* EPA-600/2-83-110, NTIS Publication No. PB84-110568.

Cuscino, T. Jr., G. E. Muleski, and C. Cowherd Jr. 1983. *Iron and Steel Plant Open Source Fugitive Emission Control Evaluation.* EPA-600/2-83-110. Research Triangle Park, NC: U.S. EPA.

Gillette, D. A. 1978*a*. A wind tunnel simulation of the erosion of soil. *Atmos. Environ.* (Great Britain). 12:1735–43.

Gillette, D. A. 1978*b*. Tests with a portable wind tunnel for determining wind erosion threshold velocities. *Atmos. Environ.* 12:2309.

Gillette, D. A. 1980. Threshold velocities for input of soil particles into the air by desert soils. *J. Geophys. Res.* 54(C10):5621–30.

Kenson, R. E. and P. T. Bartlett. 1976. *Technical Manual for the Measurement of Fugitive Emissions: Roof Monitor Sampling Method for Industrial Fugitive Emissions,* EPA-600/2-76-089b, NTIS Publication No. PB257847.

Kolnsberg, H. J. 1976. *Technical Manual for the Measurement of Fugitive Emissions: Upwind/Downwind Sampling Method for Industrial Fugitive Emissions.* EPA-600/2-76-089a, NTIS Publication No. PB253092.

Kolnsberg, H. J. et al. 1976. *Technical Manual for the Measurement of Fugitive Emissions: Quasi-Stack Sampling Method for Industrial Fugitive Emissions,* EPA-600/2-76-089c, NTIS Publication No. PB257848.

McCain, J. D., B. E. Pyle, and R. C. McCrillis. 1985. Comparative study of open source particulate emission measurement techniques. In *Proceedings of the Air Pollution Control Association Annual Meeting.* Pittsburgh PA: APCA.

Muleski, G. E. 1985. *Coal Yard Wind Erosion Measurements. Midwest Research Institute Final Report.* Industrial client.

NAAQS. 1990. CAA § 109(b), 42 U.S.C. § 7409(b) (1990), Public Law 101-549, 15 November 1990.

NEDS. 1992. National Environmental Data Service. Local climatological data annual and monthly summaries. Asheville, NC: National Climatic Center.

Nickling, W. G. and J. A. Gillies. 1986. *Evaluation of Aerosol Production Potential of Type Surfaces in Arizona.* Submitted to Engineering-Science, Arcadia, California, for EPA Contract No. 68-02-388.

PEDCo Environmental. 1978. *Survey of Fugitive Dust from Coal Mines,* EPA-908/1-78-003. Denver, CO: U.S. EPA.

Turner, D. B. 1970. *Workbook of Atmospheric Dispersion Estimates (AP-26).* Research Triangle Park, NC: U.S. EPA.

U.S. EPA. 1975. *Compilation of Air Pollution Emission Factors, AP-42,* 3rd edn. Research Triangle Park, NC: U.S. EPA.

U.S. EPA. 1977. Standards of performance for new stationary sources, revision to reference methods 1-8. *Federal Register.* 18 August 1977, Part II.

U.S. EPA. 1983. *Compilation of Air Pollution Emission Factors, AP-42,* 3rd edn., Supplement 14. Research Triangle Park, NC: U.S. EPA.

U.S. EPA. 1985/1986/1988/1990. *Compilation of Air Pollution Emission Factors, AP-42,* 4th edn. Research Triangle Park, NC: U.S. EPA. Supplement A, October 1986. Supplement B, September 1988. Supplement C, September 1990.

U.S. WS. 1988. U.S. Weather Service. Manual of Surface Observations. Washington, DC.

Watson, H. H. 1954. Errors due to anisokinetic sampling of aerosols. *Am. Ind. Hyg. Assoc. Quart.* 15:21.

Woodruff, N. P. and F. H. Siddoway. 1965. A wind erosion equation. *Soil Sci. Soc. Am. Proc.* 29(5):602–8.

30

Indoor Aerosols and Aerosol Exposure[1]

Russell W. Wiener

Aerosol Physics and Methods Branch
Atmospheric Research and Exposure Assessment Laboratory
U.S. Environmental Protection Agency, MD-77
Research Triangle Park, NC, U.S.A.

and

Charles E. Rodes

Department of Environmental Sciences and Engineering
University of North Carolina, CB7400
Chapel Hill, NC, U.S.A.

INTRODUCTION

This chapter provides an overview of indoor aerosols and issues related to assessing human exposure to such aerosols. The methods and equipment available for measuring exposures to indoor aerosols are described in modest detail. In addition, the various types of studies (surveys, field studies, models) used to assess population exposure levels are described. The potential limitations of physical sampling and modeling are also discussed.

The study of aerosols in indoor air and the assessment of human exposure to aerosols are relatively recent activities. The terms *indoor air* and *exposure assessment* in this chapter refer primarily to nonindustrial settings, such as homes, offices, and public-access buildings (e.g., museums, airport terminals, retail stores). Although many occupational settings are "indoors", the aerosol concentrations and constituents, airflow regimes, and turbulence levels pose related, but different, aerosol measurement constraints.

Until recently, it was commonly believed that the quality of indoor air was superior to that of the outdoor (ambient) air nearby. However, even though ambient air quality may have improved, the indoor air quality (IAQ) has not. Several factors have influenced the apparent deterioration of indoor air quality: life-styles have changed; building construction techniques have changed; and people have become more concerned about environmental tobacco smoke (ETS). All these factors have combined to increase the relative importance of studying indoor exposure.

Changes in people's life-styles have introduced new contaminants into the indoor

1. The information in this document has been funded wholly or in part by the United States Environmental Protection Agency. It has been subjected to Agency review and approved for publication. Any mention of trade names or commercial products does not constitute an endorsement or recommendation for use.

environment, including synthetic fibers, residues from spray propellants, deodorants, pesticides, and combustion particles from kerosene space heaters. In addition, the proportion of time spent indoors has increased, partly because of television. The average adult now spends 61%–78% of the day indoors (U.S. Environmental Protection Agency (EPA) 1989; Szalai 1972; Jenkins, Phillips, and Mulberg 1990). It is, therefore, very likely that people will be exposed to aerosol particles in their homes, offices, etc. Approximately 96 million people cook with gas stoves for an average of 4 h/day, and 7 million individuals are exposed to kerosene heater fumes for an average of 2 h/day. No estimates are available on the number of "sick" (polluted) homes with faulty or unvented combustion appliances (U.S. EPA 1987, 1.8).

The energy crisis of the 1970s brought about measures to conserve energy. Buildings were made tighter; replacement air was reduced; and heating, ventilation, and air-conditioning (HVAC) systems were centrally controlled. The resulting reduction of the infiltration rate of air from outdoors to indoors has decreased the potential for diluting indoor pollutants.

Dust particles indoors appear to contribute significantly to the problem. Some of the earliest data on human exposure to aerosols in residential settings were related to pesticides adsorbed on dust particles (Starr et al. 1974). In addition, Binder et al. (1976) determined that a person's daily dosage of aerosols is more strongly influenced by exposure to indoor, rather than outdoor, pollutants. Van Houdt and Boleij (1984) reported that indoor airborne dust in this study was more mutagenic than that found outdoors. Spengler et al. (1985) showed that personal exposures to respirable suspended particles for 24 h (mass median diameter $\leq \approx 3.5\ \mu m$) are consistently higher than the time-weighted average concentrations predicted using 24 h indoor samples from stationary monitors.

Since the mid-1960s, many researchers have demonstrated the contributions of ETS to IAQ and human exposure to aerosols (National Research Council 1986*a*, *b*; Repace and Lowrey 1980, 1983; Repace 1987). EPA's Total Exposure Assessment Methodology (TEAM) program has recently begun to study the representative metropolitan population's exposure to aerosols (Ozkaynak et al. 1990; Wiener 1989; Wiener, Spengler, and Pellizzari 1990), and EPA's Nonoccupational Pesticides Exposure Study (NOPES) is specifically addressing residential pesticides (Immerman and Schaum 1990; Bond, Lewis, and Immerman 1990). The Harvard Six-City Study was one of the earliest studies to investigate the relationships between ambient aerosol exposure and indoor aerosol exposure (Sexton, Spengler, and Treitman 1984*b*; Spengler et al. 1981; Samfield 1985). Although some IAQ studies have been concerned with aerosol issues other than human health (e.g., Nazaroff, Salmon, and Cass (1990) studied aerosol fates in museums), the great majority of IAQ and exposure studies have focused on a health response.

Indoor aerosol measurements are made by using a sampling device in a fixed location or by using personal samplers carried by subjects in the environment of study. Personal environmental monitors (PEMs) are small, self-contained, battery-powered sampling systems that can be transported by the individual (see Fig. 30-1). These relatively unobtrusive samplers are used to assess the time-integrated exposure of an individual to aerosols, without influencing his or her normal activity patterns.

Exposure occurs when an aerosol particle reaches the entrance to a person's nose or mouth (for respiratory deposition) or the skin (for dermal adsorption) (Lioy 1990; Ott 1990). The best measurements for the assessment of exposure to aerosols typically require that a sampler be placed within a person's breathing zone to characterize the inhaled air. Alternatively, each microenvironment (e.g., a room, work area) that the participant visits can be monitored using a surrogate microenvironmental monitor (MEM) or a fixed-location sampler (see Fig. 30-2). (The individual locations can then represent cells in an exposure model.) A *microenvironment* is a relatively small area—in which the aerosol distribu-

(a)

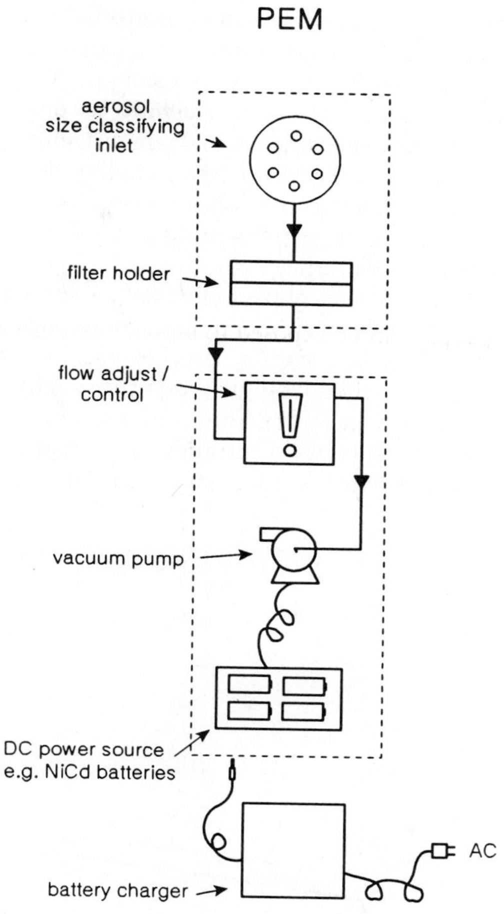

(b)

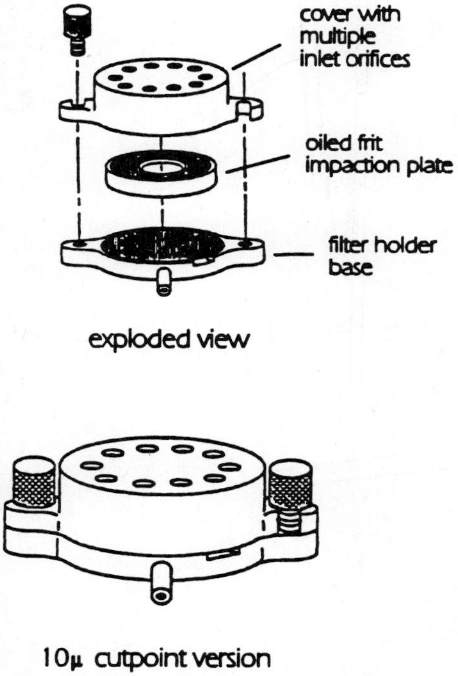

(c)

FIGURE 30-1. Schematic Diagram and Setup for a Personal Exposure Monitor (PEM): (a) Diagram of a Person Wearing PEM; (b) Schematic Diagram of a Typical PEM; and (c) Diagram of an MSP Personal Environmental Monitoring Impactor. (Adapted from U.S. EPA 1990.)

tions are assumed to be homogeneous or of a well-described distribution—that can be fully represented by an individual sample. A *representative sample* is then defined as one in which the concentration and size distribution of particles collected by the sampling device are equivalent to the concentration and size distribution of particles present in the atmosphere of concern.

The exposure of an individual to aerosols in both industrial and residential settings is related to the breathing process (e.g., inhalation rate, tidal volume, respiratory physiology, etc.), in combination with the aerosol's characteristics (concentration, size distribution, morphology, chemistry, etc.). The aerodynamic size of aerosol particles provides a direct relationship to the deposition of such particles in a human being's respiratory system. The respiratory system's removal processes of impaction and settling are defined in terms of this diameter, while interception and diffusion are determined by the particle's physical dimensions. One or more of these removal processes may operate simultaneously. Depending upon the location of particles in the individual's respiratory system, the action of the removal processes may produce different aerosol deposition patterns. For example, particles greater than about 10 μm in aerodynamic diameter are deposited on the surfaces of the upper portion of the

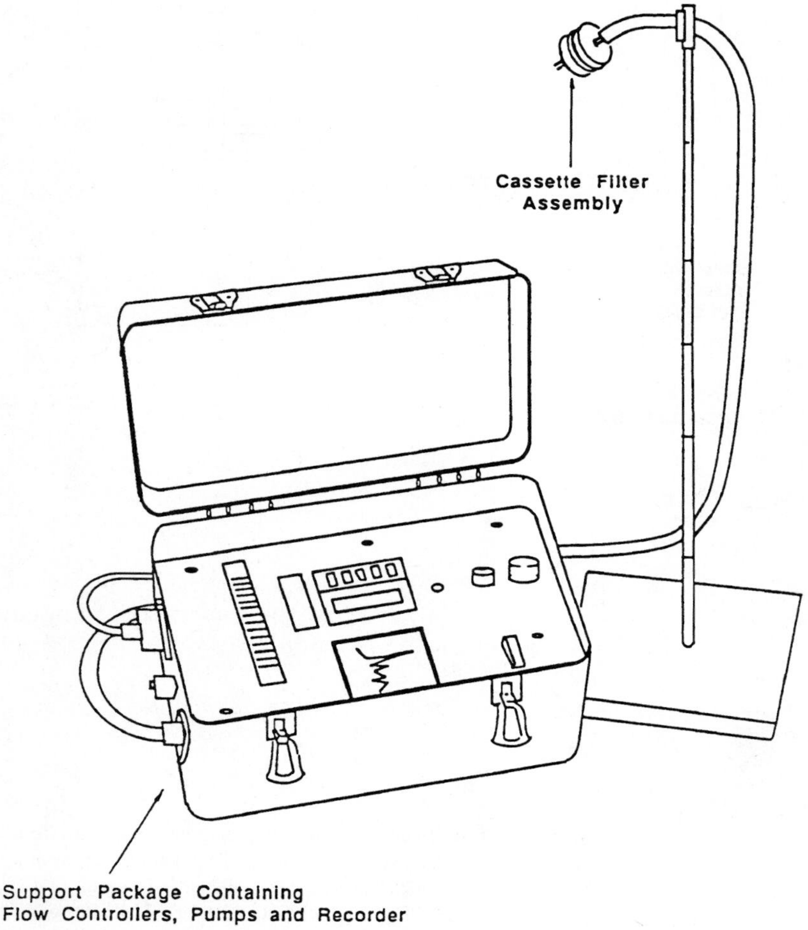

(a)

FIGURE 30-2 (a)

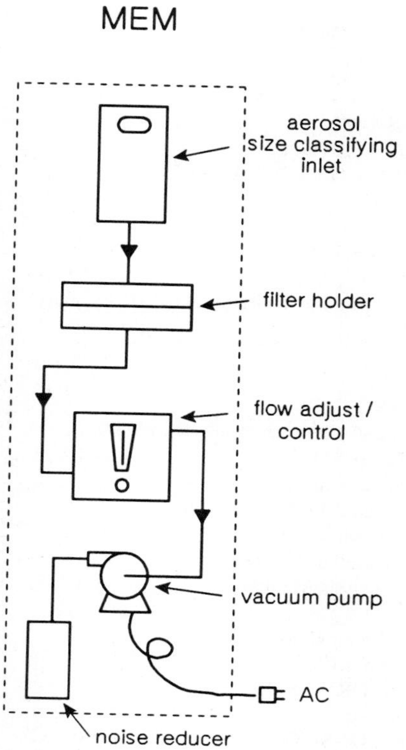

(b)

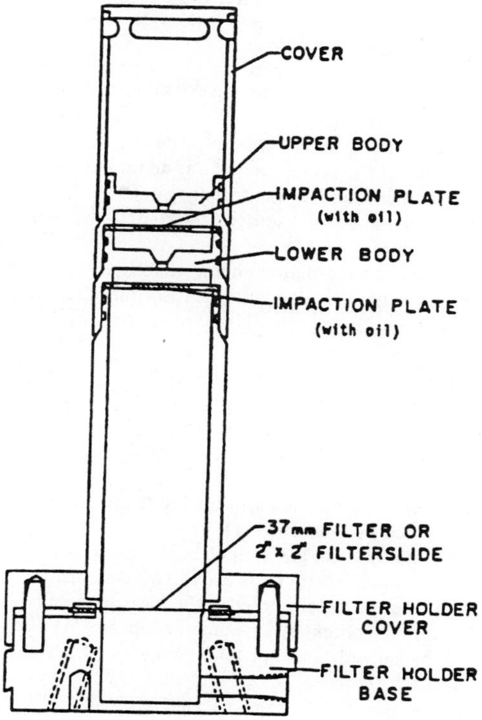

(c)

FIGURE 30-2. Schematic Diagram for a Microenvironmental Monitor (MEM): (a) Diagram of Sampling Setup for MEM; (b) Schematic Diagram of a Typical MEM; and (c) Diagram of an Indoor Sampler Impactor. (Adapted from U.S. EPA 1990.)

respiratory tract, but small particles, less than 2 μm in aerodynamic diameter, are removed primarily in the gas-exchange (alveolar) region.

To characterize indoor air aerosols in detail (e.g., surface area distribution, condensation nuclei count, electrical mobility, etc.), one would need to use the measurement principles described elsewhere in this book. The focus of this chapter is on describing the measurements of integrated aerosol masses and individual species that may be related to human health responses.

The purposes for making such integrated measurements are: (1) to characterize a microenvironment's spatial and/or temporal concentrations, (2) to characterize a population's aerosol exposure by making breathing-zone measurements, or (3) to conduct statistically based population exposure studies. Categories (1) and (2) may be studied jointly and, therefore, are considered in this chapter in the Defined-Focus Studies section. To be able to extrapolate limited data to the expected aerosol exposure of larger populations, one must consider probability-based sampling designs. These are described briefly in the Population Exposure Assessment Studies section.

After determining the study category, several key questions should be reviewed to better define the study design and methodology (see Table 30-1).

SOURCES

The term *indoor aerosols* is not restricted to aerosols generated solely indoors. The indoor

TABLE 30-1 Questions for the Design Indoor and Exposure Aerosol Sampling Studies

Questions for Consideration	Reference Sections
1. What specific aerosols are expected and which ones are to be studied?	SOURCES SPECIFIC AEROSOLS
2. What are the expected concentrations?	SAMPLING CONSIDERATIONS
3. What are the expected spatial and temporal distribution gradients and concentrations?	MODELING
4. What are the aerosol sources, their strengths, and locations?	Emissions from Indoor Sources
5. What are the ventilation characteristics of the microenvironments that may affect aerosol sampling (e.g. velocities, turbulence levels, recirculation patterns, etc.)?	Transport, Ventilation, Mixing, and Air Exchange
6. Can spatial homogeneity be assumed?	MODELING
7. Will speciation beyond mass be needed?	Elemental and Chemical Analysis Methods
8. Is aerosol morphology important?	SPECIFIC AEROSOLS
9. Are viable biological aerosols a concern?	Sick Building Syndrome Bioaerosols and Viable Sampling
10. Are there suitable locations for microenvironmental (fixed location) monitoring?	Personal vs. Microenvironmental Sampling Physical Sampling Instrumentation
11. Are suitably precise and unobtrusive personal monitors available for the desired measurement?	Physical Analysis Methods
12. What are the uncertainties in the measurements?	Other Sampling Considerations
13. What is the potential for human exposure?	*Population Statistics*
14. Do the aerosols affect people? (What are the health hazards?)	REGULATORY ASPECTS
15. Who is affected? (Is there a sensitive population?)	DESIGN CONSIDERATIONS FOR AEROSOL EXPOSURE STUDIES
16. What is the distribution of exposure for the population at risk?	Population Exposure Assessment Studies

aerosols most often of concern are tobacco smoke; combustion products from wood stoves and fireplaces, kerosene heaters, and gas ranges; residues from personal care and household cleaning products; pesticides; asbestos and other man-made fibrous aerosols; radon decay products; and biological aerosols (e.g., pollen, spores, bacteria).

Principal outdoor aerosol contaminants include suspended particles from fugitive dusts, soil, sea salt, reentrained road dust; combustion products from stationary and mobile sources (e.g., power plants and cars), forest fires, wood and crop burning; secondary aerosols from photochemical processes; acid aerosols; and biological aerosols.

Activities that generate aerosol particles include woodworking, walking on carpets (causes particle resuspension), shedding of hair and skin detritus, cooking, vacuuming, dusting, and using aerosol sprays (e.g., paints, pesticides, cleaning agents). Carpets serve as a sink for suspended particulate matter, semivolatile chemicals, and soil particles brought into the home by foot traffic (tracking).

People are typically exposed to ambient aerosols when they spend time outdoors; however, they may be exposed indoors due to the infiltration of ambient aerosols inside (Colome, Spengler, and McCarthy 1982; Dockery and Spengler 1981*b*). Vu Duc and Favez (1981) found lead indoors, for which 60% of the total mass was submicrometer-sized particles, and attributed it to automobile exhaust. Polycyclic aromatic hydrocarbons (PAHs) were also found on submicrometer-sized particles (Vu Duc and Favez 1981) and were attributed to the infiltration of combustion aerosols. In fact, PAH concentrations are often higher indoors than outdoors (Pellizzari et al. 1983; Wallace et al. 1986; Wallace 1987; Chuang, Hannan, and

Wilson 1987; Wilson et al. 1985; National Research Council 1986*b*).

PHYSICAL AND CHEMICAL PROPERTIES, PARTICLE SIZE, AND HEALTH EFFECTS

Aerosols are complex chemically, biologically, and physically, and their characteristics directly influence both IAQ and exposure measurements. Numerous physical, biological, and chemical properties of aerosols are listed in Table 30-2.

Individual aerosol particles can be liquid or solid; organic or inorganic; viable or nonviable; spherical, fibrous, or irregular; acidic or basic; etc. Particles may adsorb materials as they are transported (e.g., radon decay products, pesticides, other organics from home products). Any particle mass collected reflects a heterogeneous mixture of material —material that has chemical properties ranging from inert to highly toxic, pH values from acidic to basic, and various degrees of reactivity. The particle size also varies, as shown by the following study:

> Particles studied indoors have primarily been those in the fine mode (< 2.5 μm), primarily resulting from cigarette smoking, or as emission[s] from combustion appliances. Coarse-mode (2.5–10 μm) particles resulting from reentrainment of fibers, dust, house dust mite fecal pellets, animal and human dander, and mold spores and fragments probably constitute the second most common form of indoor particle pollution (U.S. EPA 1987, 2.58).

Particles resulting from combustion processes often contain an organic fraction (that is extractable by solvents) and an elemental carbon fraction. The soluble fraction may contain compounds known to be mutagenic in short-term bioassays and tumorigenic in animals. Constituents adsorbed or attached to a specific-sized particle have pulmonary deposition sites and deposition efficiencies that are different from the same constituents existing in vapor phase. A particle's size, hygroscopicity, density, and shape influence its deposition, bioavailability, retention, and pharmacokinetics. Synergistic effects also occur. For example, benzo[a]pyrene's carcinogenic potential is enhanced when it is coprecipitated with ferric oxide (which alone is not a carcinogen) and instilled into the lungs of animals (U.S. EPA 1987, 2.60; Wolff et al. 1989).

Both viable (biologically active) and nonviable indoor air contaminants can induce allergic responses—ranging from lethal diseases to mild allergies—and can induce hypersensitivity in susceptible individuals. Death may result from aerosol exposure to *Legionella pneumophila* or from carcinomas induced by asbestos or radon exposures (McDade et al. 1977). Bioaerosols elicit an array of allergic reactions in sensitive individuals, ranging from hypersensitivity pneumonitis to mild nasal discomfort. Other

TABLE 30-2 Properties of Aerosol Particles

Physical	Chemical	Biological
density	phase distribution	viability
number (count)	stability reactivity	colony-forming potential
mass	organic/inorganic	reproductive requirements
surface area	solubility (aqueous)	pathogenicity
size (aerodynamic and physical diameter	acidity/alkalinity	carcinogenicity
size distribution	constituency	mutagenicity
shape		toxicity
degree of agglomeration		

indoor aerosol pollutants, such as ETS, can be irritants and can cause coughing and eye irritation.

SAMPLING CONSIDERATIONS

For any aerosol study, sampling methods can be separated into two groups: physical and chemical. Physical sampling methods provide information regarding particle size, number, and mass concentration; chemical sampling methods provide information on phase distribution, elemental constituents, and chemical constituents. Analyses of specific compounds in the aerosol may be needed to determine a specific pollutant or marker compounds/elements to characterize likely sources of the pollutant.

Physical Analysis Methods

The type of sampling method selected depends largely on the physical characteristics of the aerosol being studied. Different types of instruments provide different sizing information, based on the physical parameter being measured. Factors such as particle mass, number, surface area distribution, and the influence of each on respiratory deposition influence the type of measurement method selected. An analytical method is selected based on factors such as the need to collect a physical sample for subsequent analysis, the need to restrict sampling to a particular size fraction, or the need to place the sampler appropriately (indoor vs. outdoor). The time period necessary for sample integration may range from real-time (optical measurements) to longer time intervals (necessary to obtain an adequate sample).

For any given method, the physical factor that is actually determining the measurement may change the results perceived. Impactors size particles aerodynamically, by the interaction of the inertial force of the particle and the viscous force of the fluid. Particle elutriators may use gravity or electrostatic forces to separate the aerosols. Optical particle counters and photometers are based on light scattering by particles, according to physical size and concentration. Even though the same aerosol may be sampled by different techniques, the particle size distribution measured may be very different if the techniques are based on measurements of different aerosol properties—such as optical diameter, inertial mass, or electrical mobility. This difference may be reconciled if suitable calibrations are performed and compensations are made for the measurement methods.

Personal and microenvironmental inertial impactors, virtual impactors, and cyclones have been developed that are suitable for work in indoor air or exposure studies. These include research prototype samplers developed by the University of Minnesota, the Harvard School of Public Health, EPA, and the National Institute of Standards and Technology (Fletcher 1984; Fletcher and Bright 1983; Rubow et al. 1985; Spengler et al. 1981, 1985, 1989; Turner et al. 1979). Some specific samplers available are the Harvard/EPRI (Electric Power Research Institute) 1.7 lpm cyclone sampler (Dockery and Spengler 1981a), the EPA-NBS sampler (McKenzie et al. 1982; Bright and Fletcher 1983); the MS&T® Personal Impactor (Buckley et al. 1991); and the Personal Environmental Monitoring Impactor (Marple et al. 1988, 1989; Hering 1989). Commercially available devices are available from a number of companies (see Tables 30-3 and 30-4), including MSP, Andersen, Air Diagnostics, and BGI (Hering 1989; Nagda, Rector, and Koontz 1987; Fletcher 1984).

The inlets of samplers used indoors can be designed to mimic specific locations in the respiratory system where aerosols deposit. Examples of this type of sampler design include respirable cyclones with 3.5 μm cutpoints (Lippmann 1983) that mimic the American Conference of Governmental Industrial Hygienists (ACGIH) curve and ambient PM_{10} (particulate matter ≥ 1 μm aerodynamic diameter) samplers (Wedding, Weigand, and Kim 1985; McFarland and Ortiz 1983; Marple et al. 1987; Hering 1989) that mimic EPA's inhalable-particle curve.

The shape of the PM_{10} sampling efficiency curve is obtained using combinations of aerosol separation principles, including

TABLE 30-3 Commercial Sources for Microenvironmental (MEM) and Personal (PEM) Impactors*

Type	Name	Commercial Source	Remarks
Impactor	MST Personal Sampler	ADI	4-l/min with 2.5- or 10-μm cut sizes.
Impactor	Indoor Sampler	ADI	10-l/min with 2.5- or 10-μm cut sizes Includes pump, noise damping and flow control.
Impactor	290 Marple Personal Cascade Impactor	AND	2-l/min, 8 stages, from 0.6- to 20-μm cut sizes.
Impactor	RM1	BGI	25-mm open face filter holder.
Impactor	PERSPEC	BGI	4-l/min, Personal Aerosol Spectrometer.
Impactor	Marple MEM Sampler	MSP and ESI	10-l/min dual flow 2.5- and 10-μm cut sizes. Includes pump, noise damping and flow control.
Impactor	Personal Environmental Monitoring Impactor	MSP	4-l/min with 2.5- or 10-μm sizes.
Cyclone	Gravimetric Dust Sampler	MSA	3-l/min gives 3.5 μm cut size.
Denuder	Annular Denuder Indoor Sampling System	URG	2, 4-l/min flow, 2.5 μm cut size. Includes pump.
Photometer	Handheld Aerosol Monitor (HAM)	PPM	Sensitivity to 1 μg/m^3. Real time.
Photometer	PCD-1	MDA	Sensitivity to 0.001 mg/m^3. Real time and averaging.
Photometer	Real Time Aerosol Monitor (RAM), Miniature Real Time Aerosol Monitor (MINIRAM)	MIE	2-l/min. Optional inlets for respirable dust. Instantaneous and average readings. Pump included.
Condensation Particle Counter	PortaCount	TSI	Detects particles 0.01 to 1.0 μm. Concentration range: 10^{-1}–10^5 particles/cm^3. Includes pump.
Piezobalance	Piezobalance Mass Monitor	TSI	1-l/min flow. 3.5 μm cut size. 0.01–10 μm particles collected. Near real time.
β-attenuation	Beta Gauge Automated Particle Sampler	WED	18.9 l/min. 2.5 μm inlet. Pump included.
Oscillating Element Microbalance	TEOM® Series 1200 Ambient Particulate Monitor	RP	0.5–5.0 l/min. Compatible with 2.5, 10 μm impactor heads. Includes pump and flow control. Near real time. Post collection of filters possible.

* Names of abbreviated sources are spelled out in Table 30-4

impaction, cyclone separation, and vertical elutriation. Inlets with specific cutpoints, such as that at 2.5 μm, have also been developed for indoor sampling (Marple and McCormack 1983; Marple et al. 1987, 1989; Hering 1989; Buckley et al. 1991) that use impaction on oil-soaked or greased, sintered metal surfaces. Cascade impactors can also be used to develop the mass size distribution, from which the contribution of specific size fractions can be estimated (Hering 1989; Martonen et al. 1984). Multichannel optical particle counters can similarly be used to provide the particle count distributions, from which the estimated mass distributions can be computed (Kamens et al. 1988).

TABLE 30-4 List of Commercial Sources for Microenvironmental (MEM) and Personal (PEM) Impactors

Code	Address	Telephone
ADI	Air Diagnostics and Engineering, Inc., R.R. 1, Box 445, Naples, ME 04055	(207) 583-4834
AND	Andersen Instruments, Inc., 4801 Fulton Ind. Blvd., Atlanta, GA 30336	(404) 691-1910 (800) 241-6898
BGI	BGI Incorporated, 58 Guinan St., Waltham, MA 02154	(617) 891-9380
ESI	Esoteric Systems Inc., P.O. Box 394, Newbury Park, CA 91320	(805) 497-2462
MDA	MDA Scientific, Inc., 405 Barclay Boulevard, Lincolnshire, IL 60069	(312) 634-2800 (800) 323-2000
MIE	Monitoring Instruments for Environment, Inc., 213 Burlington Road, Beford MA 01730	(617) 275-5444
MSA	Mine Safety Appliances Co., RDIC Industrial Park, 121 Gamma Drive, Pittsburg, PA 15238-2919	(800) 672-2222
MSP	MSP Corporation, 1313 Fifth St., SE, Suite 204, Minneapolis, MN 55414	(612) 379-3963
PPM	ppm, Inc., 11428 Kingston Pike, Knoxville, TN 37922	(615) 966-8796
RP	Rupprecht & Patashnick Co., Inc, 8 Corporate Circle, Albany, NY 12203	(518) 452-0065
TSI	TSI Inc., 500 Cardigan Road, P.O. Box 64394, St. Paul MN 55164	(612) 490-2888 (800) 876-9874
URG	URG, E. 118 Main Street, P.O. Box 368, Carrboro, NC 27510	(919) 942-2753
WED	Wedding & Associates, Inc., 209 Christman Dr., Fort Collins, CO 80524	(303) 221-0678 (800) 367-7610

Sampling residential indoor aerosols and measuring human exposure for risk assessment pose special problems because of the relatively low concentrations of the pollutants usually present ($<30\ \mu g/m^3$) and the necessity for lightweight PEMs and MEMs. Certain trade-offs must be made between the requirements necessary to conduct physical analytical sampling and the disruption (to the environment or life-styles of the subjects) caused by making the measurements. One must also consider the design of available samplers, the physical and chemical parameters that can be successfully measured, and the protocols required to limit the invasiveness of the procedures.

The quantity of sample necessary for analysis depends on the species of interest and the analysis method (chemical, microscopic, gravimetric). For example, using gravimetric weighing with microbalance samples of $<100\ \mu g$ results in poor precision because of the 3–10 μg composite error, even under the best of conditions, in weighing and handling samples.

A number of continuous particle counters are commercially available that include optical particle counters, photometers, aerodynamic analyzers, electrical mobility analyzers, and oscillating-element microbalances. Several portable photometers are available that read the mass concentration directly and are small enough for personal sampling (see Tables 30-3 and 30-4), (Baron 1988; Nagda, Rector, and Koontz 1987; Swift 1989). The TSI Portacount is a personal condensation nuclei counter. Many laboratory or ambient (outdoor) instruments can be used in microenvironments successfully, but may pose a space problem or demand excessive attention. The TEOM® (Rupprecht and Patashnick) continuous (direct-reading) monitor will collect a physical sample for subsequent chemical analysis and will directly determine the gravimetric mass. Beta attenuation instrument measurements are proportional to mass and also allow subsequent chemical analyses.

Recently, several new samplers have been developed that combine inertial impactors with annular denuders to separate particles and gases. The phase distribution of species that are present both as particles and gases may be measured with these devices (Koutrakis et al. 1989).

Elemental and Chemical Analysis Methods

The sample sizes for indoor air samples are often limited, as are financial resources. Therefore, a decision tree may be needed to determine those constituents that can be successfully analyzed. Different techniques provide different levels of sensitivity for various elements. Selecting the best method depends on the expected sample concentrations of the various elements of interest. Chemical analyses can be performed on aerosol sample extracts by gas chromatography (GC) or high-performance liquid chromatography (HPLC) and by any number of detection methods. Mass spectroscopy (MS) in combination with the chromatographic method (GC/MS) can resolve complex mixtures of semivolatile organic carbons (SVOCs) with high sensitivity, but at substantial costs (Rodes 1986). SVOCs have saturation vapor pressures in the range 10^{-6}–10^{-1} Pa at 25°C (1 torr = 133.322 Pa) and are present in the gas and particle phase (solid or liquid) under typical indoor conditions. PAHs, organochlorine pesticides, organophosphorus pesticides, polychlorinated biphenyls (PCBs), and chlorinated dibenzodioxins and furans have been detected in appreciable concentrations in home environments (Lewis et al. 1986, 1988; Immerman and Schaum 1990; Wilson et al. 1989; Lewtas et al. 1987*a*, *b*; Roberts et al. 1991; Sheldon et al. 1990).

Sampler design considerations for SVOCs include residence time, face velocity, volumetric flow rate required (often quite large), and choice of proper collection substrates—such as a low-background (quartz) filter substrate, followed by a trap. The filter and vapor trap capture both the particle and gas fractions. SVOCs may be collected by adsorption on sorbents such as polymeric resins (XAD-2, Tenax, or Porapaks) or polyurethane foam (PUF). PUF or XAD-2 collect polynuclear aromatics (PNAs) in the gas phase (Chuang, Hannan, and Wilson 1987; Wilson et al. 1985). Detectability is in the parts-per-billion (ppb) or parts-per-trillion (ppt) range. Particulate-phase SVOCs may be collected on a prefilter. Analytical procedures usually involve separate extraction of the filter and sorbent, with subsequent analysis by HPLC or GC/MS.

Although GC/MS, in combination with a variety of detectors, can detect a wide range of chemicals, analysis costs per sample limit the number of samples that can be analyzed. Costs can exceed \$1000 per sample when GC/MS is used with extensive precleaning.

Chromatographic techniques cannot be used to distinguish the phase distribution of the material in the test atmosphere because phase changes may occur during sampling. Because of problems with adsorption or desorption off the filter media, it may not be possible to directly extrapolate the measured concentrations to those present in the sampled environment. Sampling may disturb the phase distribution existing in air. Theoretical and empirical studies have produced models that can be used to estimate the vapor–aerosol partitioning (Zhang and McMurry 1987; Yamasaki, Kuwata, and Miyamoto 1982; Pankow 1988; Bidleman 1988; McDow and Huntzicker 1990), but serious concerns still arise in extracting from these vapor–aerosol distributions.

Annular aerosol denuders have been developed by Harvard University (Koutrakis et al. 1989), EPA (Vossler et al. 1988), and the University of Kansas (Randtke, Lane, and Baxter 1990) from a concept by Possanzini, Ribo, and Liberti (1983). The denuder can be used either as an MEM or a PEM. A preseparator, such as a cyclone or inertial impactor, is used to remove particles larger than 2.5 μm. The denuder removes gases of interest, and then an afterfilter removes fine particles. The large particles, adsorbed gases, and fine particles can then be analyzed separately. Larger versions are available for ambient outdoor studies. These may have serial denuders to measure a number of different species simultaneously (e.g., sulfur dioxide, nitric acid, nitrous acid, ammonia, sulfates, nitrates, acidity of fine particles, SVOCs, nicotine) (Vossler et al. 1988; Randtke, Lane, and Baxter 1990).

Particle Size Measurement Methods

The measurement of respirable particles that would be deposited in the alveolar region of the lungs can be approximated by using an impactor with a 2.5 μm cutpoint. This size fraction is the fine-particle size, often measured in outdoor settings. If a researcher is investigating reported cases of hypersensitivity pneumonitis in a building (and does not know the specific size fraction of particles responsible for a health response or illness), he or she might need to collect several size fractions for both viable and nonviable analyses to represent different areas of deposition and removal in the respiratory tract. Additionally, size separation for viable organisms can be used to help reveal the presence of organisms that have otherwise been obscured by competition, or to separate organisms by size class for different methods of analysis. Similarly, nonviable pollutants can be obscured by overlayering. Different types of inlets may be required to sample various environments, e.g., a microenvironment that differs from another, the indoors vs. the outdoors, and the environment vs. a person's body. For example, outdoor ambient samplers may be required to meet regulated performance standards, but are unacceptably large and noisy for indoor usage.

Limited data are available on the size distribution of residential indoor aerosols or on the distribution of particles to which people are commonly exposed (Kamens et al. 1988; Owen et al. 1990). Generally, one would expect a bimodal (fine, coarse) size distribution indoors, similar to that found outdoors. Indoor fine particles (<2.5 μm) typically originate from condensation and nucleation processes—smoking, heating, cooking, and infiltration of ambient particles. Coarse particles (>2.5 μm) can be inhalable and are generated from a wide variety of sources in the indoor environment. Mechanical (as opposed to chemical) generation and resuspension of dust are constantly occurring because of all human activities. Penetration of outdoor coarse particles to the indoor environment rarely occurs, because these particles are easily lost to surfaces. Aeroallergens and bioaerosols may be found in fine and coarse size ranges. The Marple Personal Sampler (Series 290) (see Table 30-3) is a personal cascade impactor which could be used to measure aerosol size distributions in settings where the aerosol concentration is sufficient to allow for gravimetric analysis of each of the stages.

When inlets or sampling devices are selected, the performance characteristics of the devices should be obtained. Information on characterization testing as a function of particle size may, however, not be available for some instruments. Testing may be necessary to measure the sampling efficiencies for the inlets in still air, at typical indoor air velocities, and at various wind speeds. The reported cut size of an inlet may have been calculated or estimated simply from the performance of an internal separator or impactor. The determination of the overall sampling efficiency may not have been considered. This efficiency would be the ratio of aerosols collected to those in the undisturbed sampled environment for individual particle sizes. In evaluating the performance of personal sampler inlets, it is also important to determine boundary and bluff body effects (Wiener and VanOsdell 1990; Vincent and Gibson 1981; Vincent, Emmett, and Mark 1985; Humphries and Vincent 1978). Particle bounce and internal-wall losses can introduce errors. The dispersion and uniformity characteristics of particles on the collection substrate are important for subsequent analyses, such as those done by X-ray fluorescence (XRF). The analytic field using XRF is only several square millimeters. If the sample is unevenly distributed on the substrate, errors in analysis will be introduced by measuring a nonrepresentative field.

For gravimetric methods, the characteristics of the filter media or impaction substrate become important. Glass fiber filters are acceptable if acid gases are not present. Acid gases react with glass filter surfaces and produce artifacts that can contribute substantially to the mass collected by personal samplers. Quartz filters minimize these artifacts, but are brittle and easily fray at the edges. Teflon filters of 2 μm pore size are generally acceptable for lower flow rates and can be used for subsequent XRF analyses. They are, however, subject to clogging when the sampler is collecting high loadings of certain particles, especially those from cigarette smoke.

Substrate handling and storage may cause a loss of large particles from filter surfaces. The sampling media should remain in an airtight enclosure during shipping and handling to prevent nonsampled exposures. If the presence of SVOCs is suspected, it is necessary to enhance their stability on the substrate through cold ($<0°C$), dark storage because of chemical and photochemical reactivity and artifact production.

Flow measurement and control are key factors in integrated sampling accuracy and precision. Changes in flow rate will alter the effective cut diameter, as well as change the amount of mass collected and the perceived concentration. Numerous microenvironmental and personal air sampling systems are available (Rubow and Furtado 1989; *Industrial Hygiene News* Staff 1991).

Direct-reading instruments, such as handheld optical particle counters, often have poor sensitivities at the low particulate concentrations measured in nonoccupational environments.

Transport, Ventilation, Mixing, and Air Exchange

Ventilation parameters are important for understanding (1) the mixing and dilution of aerosol sources and (2) the transportation from sources to receptors. The ventilation rate is measured in air changes per unit time and is governed by the HVAC system's design, the use of windows and doors, and the tightness of the building. Buildings with multiple mixing zones transport indoor pollutants throughout the building via HVAC or natural ventilation pathways. Aerosols can be removed by forced exhaust or natural sink effects. Make-up air introduces outside aerosols to the indoor environment and is an obvious source of ambient aerosols.

Air movement patterns within the zones are due to thermal effects (buoyancy, convective flow), air circulation (HVAC and natural), turbulence, and boundary effects. Building envelope air tightness will influence the air infiltration rate and the dilution rate. The humidification level may be a ventilation factor that is significant for the growth of hygroscopic particles and bioaerosols.

The methods and instrumentation used to measure building tightness include tracer gases such as sulfur hexafluoride (SF_6) or perfluorocarbons, blower doors (whole-house pressurization or component pressurization), and thermography (EPA 1990). The measurement of the efficiency of mechanical ventilation equipment and the characterization of fluid flow can be made with hot-wire/film anemometry, tracers, neutral-density bubbles, and smoke pencil testing (flow visualization). Acoustic anemometers have the necessary sensitivity for measurements of indoor velocity and turbulence intensity, without the bias of hot-wire anemometry (it self-heats, creating convective air movement).

Emissions from Indoor Sources

Only limited data are available for particulate emissions from indoor sources, except for ETS. There are few standardized test protocols for determining indoor source emission factors. No predictive models have been validated to apportion sources indoors. A few studies have begun to characterize unvented combustion and particles generated by home humidifiers, regarding the amount of adsorbed SVOCs; statistics on age, condition, and operational parameters of combustors;

leakage rates of particles from vented combustion sources; and studies of sink rates of particles on indoor materials (Highsmith, Rodes, and Hardy 1988; Highsmith, Zweidinger, and Merrill 1988; Girman et al. 1982; Leaderer, Boone, and Hammond 1990; Sexton, Spengler, and Treitman 1984*a*; Traynor et al. 1983, 1990; Traynor, Apte, and Carruthers 1987; Traynor and Nitschke 1984; Tu and Hinchliffe 1983).

Personal vs. Microenvironmental Sampling

Using fixed-location MEMs in indoor microenvironments to assess the 24 h exposure of residents to particles or SVOCs requires a knowledge of the expected spatial and temporal variability if meaningful extrapolations are to be made. Rodes, Kamens, and Wiener (1991) suggested that strong spatial gradients can cause substantial biases between MEM and PEM measurements. They also reported that there are lognormal relationships between microenvironmental and personal exposures when results are plotted as a frequency distribution (Fig. 30-3). The slopes of these relationships are hypothesized to be affected by a combination of source/receptor proximity and the spatial distribution of the sources. MEMs tend to underrepresent exposure most severely when the aerosol being measured is being generated by personal activities and when the MEM is being used as an area monitor.

Sampler Obtrusiveness and Participant Burden

Samplers can be obtrusive to individuals whose environments are being studied. The degree of obtrusiveness permissible and the annoyance caused (participant burden) are important factors in avoiding biasing normal activity patterns and in obtaining the cooperation of participants. Obtrusiveness depends on a number of factors, including the size (volume or square footage) of the microenvironment; sampler configuration (packaging, size, and weight); how the sampler is to be employed (fixed-location or mobile); the environment in which it will be used (i.e., children or pets present); the noise produced by the system relative to the background noise level (for homes or offices, the total instrument noise should be less than 50 dB to avoid nuisance; for bedrooms, the instruments must be quieter); conspicuousness (prefer dull color so as not to be interest-provoking); safety of operation (no sharp edges and tamper-resistant to prevent injury, sample bias, or destruction, and to reduce liability); power requirements (battery operation is the least obtrusive); and operational independence (does not require technical maintenance during sampling period). For PEMs, it is important that the sampling package be comfortable to carry for extended periods and not obstruct routine activities. The proper use of equipment depends on the participant's behavior. Minimizing the requirements for the participant's time and minimizing the interruptions in his or her daily activities (by judicious scheduling of field visits) further reduces participant burden and its resulting bias. It is imperative that the researcher balance the desire to constantly monitor the participant's activities and the need to not significantly alter the participant's behavior.

Other Sampling Considerations

In addition to meeting the study objectives, the sampling integration period must be selected based on other specific factors, including the battery life (if used), the expected temporal patterns of the aerosol source emissions and human activity patterns, and the time required to acquire enough material for analysis without overloading the filters. The levels of particulate matter and SVOCs in indoor environments can vary substantially because of strong local sources, such as ETS. If the analytical sensitivity is sufficient, it is usually desirable to stratify sampling by daytime and nighttime periods. This permits a better identification of the sources associated with occupant activity patterns and outdoor diurnal chemistry.

The accuracy and precision of the measurements determine the level of the researchers'

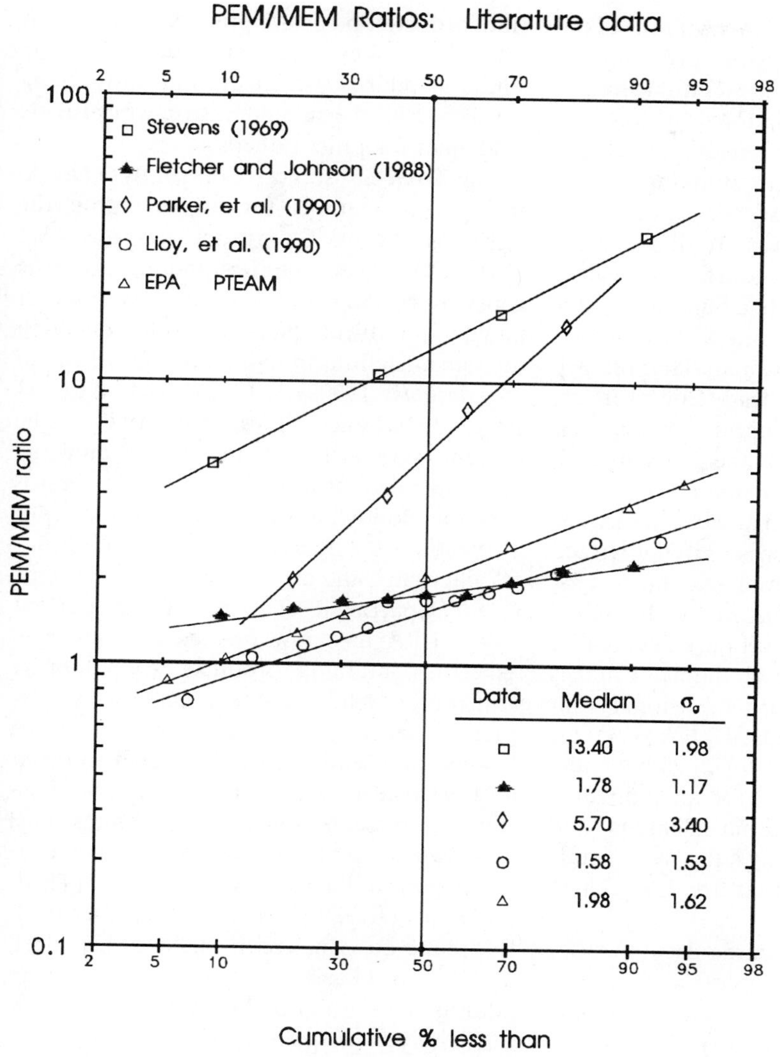

FIGURE 30-3. Personal Exposure Monitor to Microenvironmental Monitor (PEM/MEM) Ratios: Literature Data. (Source: Rodes, Kamens, and Wiener 1991.)

confidence in the data collected. Collocational placement of instruments (for replicate samples) are used to estimate precision. Coefficients of variation (CVs) for ambient aerosol monitors meeting outdoor equivalency requirements (*Code of Federal Regulations* 1990*c*) range from 3% to 10%, and those for indoor MEMs range from 3% to 15%. PEMs have poorer precision, primarily resulting from smaller sample collection and poorer flow controls, with CVs ranging from 5% to 25%. The accuracy of aerosol measurements can only be estimated by calibrating the instruments under controlled test atmosphere conditions, such as in wind tunnels. Accuracy is inferred from a comparison with isokinetic nozzle measurements.

REGULATORY ASPECTS

EPA, through the Radon Gas and Indoor Air Quality Research Act (RGIAQRA) of 1986 (*U.S. Code* 1986), has been given the authority to study indoor nonindustrial air quality and personal exposure. The EPA program is intended to complement existing programs at the Occupational Safety and Health Administration (OSHA) and the National Institute for Occupational Safety and Health (NIOSH), that have been engaged in studying, evaluating, and regulating workplace exposures to pollutants. EPA's objectives are to study the effects of nonindustrial exposures—such as in residences, offices, and public buildings—and to protect human health and welfare (minor health problems, esthetics, material damage). Except for asbestos and lead, federal, state, and local indoor air programs have been designed primarily to study the problems and to propose mitigation guidelines.

There are no mandatory federal IAQ or exposure standards for aerosols in the United States. The current outdoor standard is for inhalable aerosols in ambient air. Inhalable dust is defined to include all particles smaller than 10 μm in aerodynamic diameter and is referred to as PM_{10}. EPA's National Ambient Air Quality Standard (NAAQS) for PM_{10} is set at 50 μg/m^3 for a 24 h annual average and at 150 μg/m^3 for an individual 24 h daily limit. California (Jenkins and Phillips 1988) is one among a number of state and local governments that are establishing indoor air programs.

Aerosol exposure standards developed for occupational settings include the threshold limit values [time-weighted average (TLV-TWA), TLV-ceiling (TLV-C), TLV-short-term exposure limit (TLV-STEL)], which are informational standards recommended by the ACGIH. Permissible exposure limits (PELs) for various aerosols are established by OSHA and are mandatory. The National Academy of Science recommends an emergency exposure limit (EEL). Under the Clean Air Act, a National Emission Standard for Hazardous Air Pollutants (NESHAP) for asbestos requires control measures for asbestos in buildings undergoing demolition and renovation, but sets no standard other than requiring no visible emissions from such actions (*Code of Federal Regulations* 1990*a*). The American Society for Testing and Materials (ASTM) has a subcommittee responsible for providing standard methods and practices for indoor aerosol sampling (Subcommittee D22.05.03). These standards, which are currently being developed by the subcommittee, will provide guidelines for practitioners.

The Toxic Substances Control Act (TSCA) includes regulations for implementing the Asbestos Hazard Emergency Response Act (AHERA), which requires indoor airborne asbestos concentrations to be less than or equal to ambient concentrations, following abatement actions in schools (*Code of Federal Regulations* 1990*b*). OSHA and ACGIH have PELs and TLVs, respectively. The current standards for the PELs and the recommended proposed TLVs for asbestos are the following: amosite—0.2 fiber/cm^3, chrysotile—0.2 fiber/cm^3, crocidolite—0.2 fiber/cm^3, and other forms—0.2 fiber/cm^3 (U.S. Department of Labor 1988; ACGIH 1991). EPA has now banned a number of asbestos-containing products (e.g., flooring and roofing felt, asbestos concrete (a/c) corrugated sheet), with other products to be banned in a sequenced phase-down to occur in 1993 (sheet gaskets, automatic transmission components, drum brake linings, etc.) and 1996 (a/c pipe, commercial pipe, corrugated paper, millboard, a/c shingle, etc.) (*Federal Register* 1989). [As of 18 November 1991, this phase-down has been challenged in court, and the court has said that EPA must provide additional information before additional products may be banned (Hilleary 1991).]

STUDY CATEGORIES

Defined-Focus Studies

Defined-focus aerosol studies include those designed to understand better and evaluate the physical parameters governing specific aspects of the generation, distribution, and sampling of, and exposure to, indoor aerosols. The studies can be performed under controlled conditions in the laboratory or in field environments. Laboratory studies may include special studies to address aerosol source characteristics, sampler performance factors, aerosol distribution, and transport (Anderson et al. 1989). Quality assurance studies may be required for sampler inlet validation and method standardization. Precision can be obtained from sampler intercomparison and collocational testing of inlets and sampling systems (Anderson et al. 1989; Lioy et al. 1988). An estimation of sampling accuracy can only be made under controlled wind tunnel conditions. Studies of specific indoor aerosol properties (e.g., surface area, density, viability, carcinogenicity determinations) are best performed in the laboratory. The physical differences between the measurements from PEM and MEM sampling can be studied using aerosols in a test chamber.

Field studies are used to assess aerosol concentrations and exposures in actual indoor, microenvironmental, and macroenvironmental conditions and to determine source contributions (source/receptor analyses). Although the respiratory deposition of specific aerosols can be measured in the laboratory, the actual human exposure to aerosols cannot be studied directly. PEMs and MEMs are used as surrogate samplers to approximate the level of particulate matter to which subjects are exposed. Each microenvironment in which the subject might be significantly exposed must be considered. Most early residential IAQ studies used only fixed-location microenvironmental measurements. This was partly due to the additional costs and the burden imposed on participants when they used personal monitors. Also, the devices available were often not sensitive enough for residential concentration levels. Current studies use combinations of microenvironmental and personal sampling. Ambient samplers are typically placed outside and used to measure or represent large areas (i.e., counties or cities). Additional ambient samplers may be placed near the indoor environments to assess the contributions from outdoor aerosols.

Exposure studies also require the use of questionnaires, often referred to as survey instruments. Survey materials are used to obtain information from the study participants, including information on activity patterns, source activities, ventilation system settings, and health responses. These questionnaires and activity logs must be pretested using focus groups in laboratory situations and using pilot field tests, with participants from the population intended for study. In contrast with most laboratory or field studies, the success of human exposure assessment studies depends upon the degree of cooperation and the performance of the subjects in the study. In standard industrial hygiene sampling, the hygienist closely monitors the participant. Typically, the participant in an occupational setting is engaged in repetitive, well-defined tasks. When studying exposure in residential settings, the investigator must be as unobtrusive as possible, to minimize biases in participant activity patterns.

Fixed-location MEMs have a number of advantages, such as the following: they are less expensive to use; they allow the collection of larger samples, thus providing better precision in gravimetric measurements and subsequent analyses; they require less interaction with study participants; they permit the use of house current as a power source and are not restricted to batteries; and they tend to provide a larger proportion of valid samples. On the other hand, MEMs may be large and noisy and unacceptable to the occupants of the microenvironments. More importantly, the MEMs may also fail to provide representative measurements due to poor mixing or stratification, or to an improper sampling location.

PEMs provide the best estimates of the aerosol concentrations from integrated breathing zones. Unfortunately, they often have a number of drawbacks, including poor precision, poor sensitivity, anisokinetic sampling biases resulting from flow around the human body, and excessive size, weight, and noise.

Population Exposure Assessment Studies

An exposure survey study may be used to assess the distribution of exposure of a population. Inferences to a larger population (or frame) can be made by using statistical probability sampling methods to draw or select the study population. The survey and sampling instruments, methodologies, and protocols used in defined-focus/objective studies can be adapted for this type of study. The ingredients of exposure assessment include taking a representative probability sample, measuring pollutant concentrations, measuring body burden (not possible for generic aerosols without a regular and identifiable chemical constituency), and recording daily personal activities. The objectives of population exposure studies are to produce representative frequency distributions of the exposure of human beings to aerosols, to determine how exposures compare with existing regulatory standards or mitigation guidelines, to establish indoor/outdoor/personal relationships for aerosol exposure, and to determine the significant sources of aerosol exposure (Ott 1985, 1990; Wallace et al. 1986; Wallace 1987).

Each microenvironment that is encountered in daily exposures must be defined. Microenvironments typically consist of the home (with one or more cells), the workplace, the transportation vehicle, and the ambient environment near the home. Relationships between personal and microenvironmental sampling, spatial and temporal variance in aerosol concentrations, exposure and activity, and the frequency distribution of exposure must be determined for the test population (Ju and Spengler 1981).

Once exposure information has been collected, it can be used for risk assessment. Risk assessment requires determining the number of persons exposed, the sources and transport factors relating to the pollutant from its source to the receptor (person or sampler), the exposure-related significance of sampling data, and the health effects of exposure. Risk assessment also requires an estimate of the population at risk. The risk assessment can then be used to help define the relationship of exposure to outdoor standards.

Population Statistics

Survey design requires the use of population information or demographics. Participants in the study are selected from a known population, or frame, through a random draw. The frame must have a complete demographic evaluation. Random selection from the frame provides a probability sample. By this method, the participant data can later be used to make statistically valid inferences to the frame (U.S. EPA 1983). Particular factors of interest in the population, (e.g., smoking habits, vocation) can be used to stratify the sample. Weighting factors can then be applied so that specific subpopulations of interest can be represented in sufficient number to provide statistically significant approximations of exposure.

Many factors are critical in limiting the biases introduced through the participant selection process and through interviewing. The respondent rate must be kept high so that the test population is randomly represented. Low respondent rates suggest that possible biases may have been introduced to the study. Survey materials—such as diaries, logs, and questionnaires—must be free of loaded questions and response effects. Information collected from the subjects must be properly coded so that the data may be analyzed successfully. Questions requiring open-ended responses are typically the most difficult to code, and multiple-choice questions are among the easiest. Coding is essential to discern the relationship of exposure to activity. The questions asked must also be clearly relatable to physical exposure information (Converse and Presser 1986).

Sick Building Syndrome

The sick building syndrome (SBS) has been a key factor in promoting studies of IAQ and human exposure. *Sick building syndrome* is specifically defined by the World Health Organization (WHO 1983) as symptoms from which building occupants suffer during the time they spend in the building, which diminish when they leave the building. (Some symptoms may have a delayed onset and may occur after the occupant leaves the building; however, the symptoms diminish after he or she spends time away from the building.) These conditions cannot be readily traced to specific pollutants within the building. Typical health responses include the following: irritation of the eye, nose, and throat; dry mucous membranes and skin; erythema; mental fatigue; headache; airway infections; cough; hoarseness; wheezing; unspecific hypersensitivity reactions; nausea; and dizziness (WHO 1983; Molhave 1987).

Tight building syndrome (TBS) has similar symptoms to SBS and is due to poor ventilation practices, causing thermal discomfort or extremes in humidity, often resulting from energy conservation measures. The NIOSH Hazard Evaluation Program concluded that about one-third to one-half of sick buildings have this problem. At least 5% of the SBS and illnesses are due to biological aerosol contamination problems (Wallingford and Carpenter 1986).

Building-related illness (BRI) is defined as a discrete, identifiable disease or an illness that can be traced to a specific pollutant or source within a building. Aerosols (especially bioaerosols) are undoubtedly a leading cause of SBS, in combination with ventilation deficiencies (Brief and Bernath 1988; Burge et al. 1989; LaForce 1984).

The initial preparation for a sick building study is important. Discussions prior to sampling must be conducted with the occupants regarding the complaints and symptoms reported. An assessment must also be made of (1) the mitigation steps already taken to improve the IAQ and (2) the HVAC operation and maintenance schedules. A thorough inspection of the building must be conducted inside and outside, including the HVAC system. The following determinations must be made: the type of system; the configuration, location, and effectiveness of air handlers, chillers, cooling tower, filtration or particle-removal devices, air supply and air return ducts, and air circulation path; the location of outside air intake vents (check for cross-contaminants) and thermostats; and the presence of moisture, standing water, or microbial contaminants in the HVAC system.

The building must be surveyed for the sources and strengths of the suspected contaminants (e.g., tobacco smoke, combustion products), buildup of pollutants from poor air ventilation/high occupant density (use carbon dioxide monitoring as a surrogate), insulation materials (fiberglass, urea–formaldehyde foam, asbestos), age and cleanliness of carpets and furnishings, radon decay products from soil or concrete blocks, construction materials or activities, aerosol cleaning products, pesticides, moisture (condensation on windows, walls), biologicals (fungi or mold on walls or in closets or enclosed spaces), and the presence of unvented combustion heaters (kerosene heaters, unvented gas ranges). It is also desirable to determine if sociological or psychological factors have been introduced that would make personnel more receptive or responsive to sick building problems (Salisbury 1986).

SPECIFIC AEROSOLS

Environmental Tobacco Smoke

ETS is one of the most significant indoor aerosols. Passive smoke emission is the single largest nonoccupational source of exposure to indoor aerosols and is the greatest cause of air pollution health effects if active smokers are included in the exposed population. ETS is dominated by particles of size $<2.5\ \mu m$; in areas where smokers are present, 0.014–1.6 mg of respirable suspended particulate (RSP) per day of ETS is contributed passively to the exposure of nonsmokers. Hydrophobic components of the ETS aerosol often reach the alveolar portions of the lung, while hydrophilic components are either absorbed by the upper respiratory tract or removed by clearance mechanisms (U.S. EPA 1987, 5.2).

ETS exposure studies have been performed by a number of investigators (Vaughan and Hammond 1990; National Research Council 1986*a*, *b*; Surgeon General of the United States 1986). Repace and Lowrey (1983) estimated a mean exposure of 1.43 mg/day, with a range of 0–14 mg/day for a series of public buildings not randomly selected. The Harvard Six-Cities Study (Spengler et al.

1981; Samfield 1985) found the levels presented in Table 30-5. Radon daughters can attach to ETS, which produces a synergistic effect (Bergman, Edling, and Axelson 1986). ETS has the greatest effect on PAH levels and on mutagenicity for indoor samples (Wilson et al. 1985; Lewtas et al. 1987*a*, *b*).

Several methods are available to apportion ETS from the bulk RSP of indoor aerosol samples. The tests rely on marker species and metabolites to determine the degree of exposure to ETS. Nicotine is the most commonly used marker for ETS and is found almost exclusively in the vapor phase in air samples of ETS (Eatough et al. 1986; Eudy et al. 1986; Hammond et al. 1987). Nicotine has a short half-life (30 min), but is stable once frozen. A number of specific tests for nicotine in indoor air are available (Caka et al. 1990). Cotinine, a metabolite of nicotine with a long half-life (20 h), can be measured in urine or saliva as a surrogate for aerosol exposure. Cotinine analyses are expensive, but specific, because cotinine is endogenously produced. More than 3800 compounds have been identified from ETS, with many distributed between the vapor and aerosol phases. Some are secondary reaction products not found in side-stream or mainstream smoke.

Asbestos

Asbestos is defined as a group of impure hydrated silicate minerals that occur in serpentine and amphibole fibrous forms. Chrysotile is the only asbestos of serpentine form and accounts for greater than 90% of the asbestos material used in the United States. There are five amphiboles: amosite, crocidolite, and the asbestiform varieties of tremolite, anthophyllite, and actinolite. Chrysotile is a magnesium silicate that has fibers which are strong and flexible and can be spun. Amphibole asbestos compounds include various silicates of magnesium, iron, calcium, and sodium. The fibers of these compounds are brittle, but more resistant to heat. Asbestos is ubiquitous both because it occurs in nature and has been widely used. It is used for the following: fireproof fabrics, brake linings, gaskets, roofing compositions, electrical and heat insulations, paint fillers, chemical filters, and as a reinforcing agent in rubber and plastics. Asbestos is also present in many building materials (e.g., floor tiles, asbestos cements, roofing felts, and shingles).

TABLE 30-5 Comparison of Indoor and Outdoor Levels of Respirable Suspended Particulate from the Harvard Six Cities Study (Adapted from Spengler et al. 1981; Samfield 1985)

Location	Respirable suspended particulate (RSP) ($d_a < 2.5$ μm) mean concentration μg/m^3	I/O ratio*
Outdoor	21.1	
Indoor no smoking	23.4	1.16
Indoor 1 smoker	36.5	1.73
Indoor 2 smokers	70.4	3.34

* I/O ratio = ratio of indoor to outdoor RSP concentrations

Asbestos inside buildings is of two primary types: (1) cementitious, or plaster-like, sprayed as a slurry onto steel work or building surfaces and (2) loosely bonded fibrous mat, applied by blowing a dry mixture of fibers and binders through a water spray onto a target surface.

Sources of exposure include fallout from insulation due to physical disruption, adhesive degradation, vibration, humidity variation, air movement from heating and ventilating equipment, air turbulence and vibration caused by human activity, and static pressure differences between ducts and ambient outdoor or indoor air. An additional source of entrainment of fibers is weathered asbestos floor tile (U.S. EPA 1985*a*; Sebastien, Bignon, and Martin 1982; Litzistorf et al. 1985).

Air sampling data for exposure assessment are considered limited because they often represent a brief time period. The conditions at the time of sampling may not be representative of the variety of actions that take place in the indoor environment. Clearance air sampling for schools following an abatement (required by AHERA) is conducted under "aggressive" sampling conditions designed to create a worst-case atmosphere. These conditions are not practical or safe in ordinary building environments because they may create exposure hazards to building occupants (U.S. EPA 1985*a*, *b*; Yamate, Agarwal, and Gibbons 1984).

Phase contrast microscopy (PCM), scanning electron microscopy (SEM), and transmission electron microscopy (TEM) are commonly used for analysis. A phase contrast illumination microscope, is used at 400–450× magnification to detect fibers. PCM is more commonly used for monitoring asbestos in a workplace where asbestos materials are known to be in use. All fibrous forms with a length-to-width ratio of 3:1 or greater are assumed to be asbestos since the procedure cannot distinguish asbestos from nonasbestos materials. The PCM technique can detect only fibers with diameters greater than about 0.25 μm and is not suitable for many indoor monitoring efforts. SEM suffers from the same size-related limitation as PCM, but can be used with energy-dispersive spectroscopy (EDS) to distinguish mineral from nonmineral fibers. TEM is the method of choice for indoor asbestos monitoring because it can detect fibers with diameters as small as 0.025 μm. TEM can also be used with EDS to determine fiber composition, and crystal structure can be obtained by selected-area electron diffraction (SAED) analysis.

For TEM and SEM, polycarbonate filters with 0.4 μm pores or cellulose ester with 0.45–0.8 μm pore size are used. PCM fiber counting is performed on a cellulose ester membrane filter. Sampling may be performed at 2–15 lpm for 8 h. An appropriate solvent is used to dissolve the filter, and a count is made of fibers with aspect ratios of 3:1 that are 5 μm in length for the PCM method. For the TEM method, aspect ratios of 3:1 or 5:1 are used, and all fiber lengths or those >0.5 μm are counted.

Pesticides

Household dust can be a medium for the exposure to pesticides and other adsorbed SVOCs. Starr et al. (1974) concluded that this type of dust is a major reservoir for pesticides in the indoor environment. Roberts et al. (1991) sampled floor dust with a high-volume surface sampler, a vacuum cleaning device suspended on a rigid frame. High levels of pesticides were found associated with the indoor dust samples. Roberts, Ruby, and Warren (1987) showed that vacuum cleaner dust from rugs is mutagenic in the Ames *Salmonella* mutation assay and *E. coli* DNA (deoxyribonucleic acid) repair assay. Soil and indoor dust ingestion by small children has been estimated to range from 0.1 to 0.5 g/day (Roberts and Ruby 1989). The NOPES project measured indoor exposures to pesticides (combined particles and gases). The sampler used was a 4 lpm battery-powered pump with a PUF sorbent cartridge. Analyses were made for 58 pesticides, as well as PCBs, using GC/MS. The results showed a detectable limit of 0.01 μg/m^3 for some pesticides in air. The study found that levels indoors were greater than those outdoors for many pesticides. Peak values were 1.7–15 μg/m^3 indoors compared to 0.001–0.41 μg/m^3 outdoors (Immerman and Schaum 1990).

Bioaerosols and Viable Sampling

Indoor bioaerosols have received comparatively little attention in proportion to their probable role in SBS (Burge and Hoyer 1990). There are no standards for ambient or indoor levels, and typical levels of exposure have not been widely reported. Only a limited number of standardized protocols for sampling and analysis are available. The successful application of current bioaerosol analytical methods requires an extensive knowledge of microbiology. No general bioaerosol contamination survey method is currently available. In addition to the chapter on bioaerosols in this book, the ACGIH has recently published a guidance document for bioaerosol study (Burge et al. 1989).

Types of Bioaerosols and their Health Effects

Common types of indoor bioaerosols include molds (spores and toxins); bacteria; viruses; protozoa; algae; body parts and excreta of insects, acarids, and arachnids; dander and excreta from animals; and pollens from higher plants. Bioaerosols can range from being pathogenic and toxigenic substances to being mild allergens and inert particles. The fungi identified as the etiological agents for BRI and hypersensitivity reactions are *Aspergillus*, *Histoplasma*, *Penicillium*, and *Cladosporium* (Burge et al. 1989). The bacteria identified include *Legionella pneumophila*, *Bacillus*, *Thermoactinomyces*, and *Coxiella burnetii* (Burge et al. 1989; Gregory 1973). Some BRIs of interest include Legionnaires' disease, Pontiac fever, common cold, influenza, tuberculosis, and several fungal infections. Airborne bacteria are typically found in the 0.5–10 μm size range. Viruses are usually <0.1 μm in diameter. Spores and pollens tend to be in the 10–30 μm range (Hinds 1982). The inhalation of endotoxins produced by gram-negative bacteria has been associated with respiratory distress, headache, fever, and diarrhea (Olenchock, Lenhart, and Mull 1982), and mycotoxins produce an array of adverse health effects. The dust mite (*Dermatophagoides*) is one example of an organism that elicits a significant allergenic response. Mite fecal pellets disintegrate to form particles in the respirable size range (0.8–1.4 μm), which can be highly allergenic in sensitized individuals. Allergenic effects are very common and range in the degree of seriousness from rhinitis (hay fever) to hypersensitivity pneumonitis. Other allergenic conditions due to bioaerosols include conjunctivitis and bronchial asthma.

Monitoring of Biological Contaminants

Bioaerosol sampling can be divided into viable (cultures can be grown) and nonviable (organisms may be dead) methods. In the United States, the most common method is gravity collection on culture plates or sticky slides (U.S. EPA 1987, 2.146). Culture methods depend on the viability of collected particles, as well as on their ability to grow under the given culture conditions. Particles (e.g., actinomycetes, bacteria, viruses, many small-spored fungi) are usually cultured and then identified morphologically, biochemically, and immunologically. The choice of the culture media is critical. Cultural methods always underestimate the actual spore counts, and the underestimated count increases logarithmically with particle levels (Burge et al. 1977; U.S. EPA 1987, 2.146). In nonviable methods, particles can be visually counted and biochemically or immunologically analyzed. Morphological identification may also be made by optical or electron microscopy. Nonviable methods are best when the concentration of biological particles is of interest. For example, the total protein content of a dust sample is indicative of the concentration of biological material present in the sample (Buchan et al. 1973). Nonviable methods are also the only choice for measuring biological products (toxins, antigens, endotoxins, mycotoxins). Collectors include impactors, impingers, sieve samplers, settling plates, and slides. The collection media can be cultured and/or analyzed immunologically, biochemically, or morphologically.

The health effects due to bioaerosol contamination are numerous. Pathogenic effects are well documented (Burge et al. 1989; Gregory 1973). Some diseases of interest include Legionnaires' disease, Pontiac fever, common cold, influenza, tuberculosis, and some fungal infections. Mycotoxins have been found associated with toxigenic effects.

DESIGN CONSIDERATIONS FOR AEROSOL EXPOSURE STUDIES

Definition of Purpose

The most critical element in the planning of an aerosol exposure study is a thorough analysis of the purpose and objective. What is the specific contaminant to be measured? What population is being studied? How many types and numbers of samples will be needed to provide statistically significant correlations? When studying aerosols, questions regarding whether to include or exclude smokers are especially important. Including smoking participants confounds the relationship between personal and microenvironmental data. The most significant personal exposure to aerosol particles of the smoker is undoubtedly the mainstream tobacco smoke. On the other hand, excluding smokers jeopardizes the representativeness of the study population because of possible cluster effects. Smokers are not uniformly distributed in the population, and other demographic biases are incorporated into the study by the exclusion of smokers.

If the objective is to examine the exposure to combustion aerosols, using $PM_{2.5}$ inlets would gather samples that would be more representative of this exposure than using PM_{10} or total suspended particulate (TSP). If the purpose of the exposure study is for later risk analyses, what risks are important? Pesticides, PAHs, and other possible carcinogens may represent a fraction of the particulate matter sampled. Bioaerosols may not be carcinogenic or mutagenic, but clearly can be pathogenic and allergenic. Are especially susceptible populations, such as asthmatics, of interest? For asthmatics, the mite antigen would be an important type of particulate matter to assay.

Survey Instruments

Surveys and questionnaires provide a relatively inexpensive way to correlate activities with physical measurements of exposures. They are also relatively noninvasive. In addition, they allow a qualitative assessment of the exposure level. However, they also tend to be inaccurate regarding quantitative amounts of exposure and tend to be subject to reporting and recall bias. To be of analytic value, each question in a survey instrument requires careful consideration. The questions must be tested in focus groups by a survey statistician for appropriateness, ability to discriminate sources, the level of burden they pose to the participants, and their overall usefulness and integration in the study. Often, extensive survey instruments are developed, but the analytical requirements are left unmet and the data are, therefore, uninterpretable.

Physical Sampling Instrumentation

The assessment of human exposure to aerosols requires characterizing the levels of aerosol pollutants in particular microenvironments, determining the proportion of time spent and the activities conducted in those microenvironments, and calculating the degree of personal exposure to the aerosol over the time period of interest. To address these problems, a variety of instruments is required, including personal and microenvironment sampling systems.

Microenvironmental sampler location(s) must be defined by the study objectives and by the expected temporal and spatial variations in aerosol concentrations. Air movement patterns may help determine the placement of samplers within microenvironments. Additional factors, such as access to electrical power and the number of available sampling instruments, may limit the number of microenvironments studied.

The number of samples, types of samples, an analytical decision tree, and the number of replicates required to reach the data quality objectives mandated by the study goals must be understood before the sampling instruments are selected. The expected variance in the measurements will help determine the number of samples required for a specific level of confidence. This may require pretesting and evaluating the study instruments and sites to make reasonable estimates of the expected variation in aerosol concentrations (if historical or background data are unavailable).

The sampling instrumentation must be designed or selected with the study in mind. The need for a low level of unobtrusiveness must be addressed for portable, self-contained systems that do not require house current, and for lightweight, quiet, personal sampling systems.

Logistics

Implementation

Good field implementation is the result of successful logistics, i.e., planning, training, and field testing. Pretraining of field technicians and participants before the actual study begins is extremely important. In addition to the defined-focus/objective studies previously mentioned, small pilot field studies can be performed specifically to test instruments prior to their use in larger exposure programs. Field testing of equipment, methodologies, and protocols is necessary to address the hardiness of the sampling systems under operating conditions similar to those of the population exposure studies. An implementation plan should be elaborated that includes scheduling, training, contingency planning, and providing backups for equipment and for data collection efforts.

Workplace Sampling Considerations

An important element of exposure assessment is evaluating the contribution of workplace exposures to total exposure. [The methodologies for workplace sampling are discussed in detail in Lippmann (1989).] However, during population-based exposure studies, problems not normally accounted for in classical industrial hygiene studies are experienced. First, the participants may have been selected by a criterion other than their place of work (i.e., random sample, smoking habits). Second, the employer may not have agreed to allow sampling in the workplace, even if it is confined to a personally worn sampling system. The participant's occupational duties or setting may place unforeseen biases on the exposure study results. For example, occupations that may have security concerns (e.g., many in the aerospace industry) might limit the participant's access to the workplace facilities to be sampled if no explicit agreement is reached before the study begins. Individuals in the study frame may choose not to participate in the study if they feel their employer would object to their bringing sampling equipment into the workplace. Even if it is possible to get prior authorization from the employer, field logistics become difficult if technicians must accompany the participant to the worksite. Under these conditions, it might be necessary to define a separate residential study that excludes the occupational component.

MODELING

Exposure Modeling

Personal exposure monitoring studies to estimate population exposure levels can be expensive and, in some cases, impractical. An alternative approach combines limited-scale measurement studies and existing data with an appropriate exposure model. The most widely used exposure model (e.g., Letz, Ryan, and Spengler 1984; Spengler et al. 1985) is the simple summation model of the measurements of interest (e.g., mass concentration for a specific aerosol size fraction) in each microenvironment, weighted by the time spent in each microenvironment. These relationships are expressed as follows:

$$E = \sum (f_i c_i) \qquad (30\text{-}1)$$

where

E = mean exposure
f_i = fraction of time spent in microenvironment i
c_i = mean concentration (of any parameter) in microenvironment i

A study participant would be required to keep a time budget diary that provided information on the fraction of time spent in each compartment. The mean concentration measurement would be obtained either from a PEM carried by the participant while he or she was in the microenvironment or, less desirably, from a fixed-location monitor positioned to gather readings considered representative of the microenvironment. Because aerosols larger than about 2 μm tend to be lost more rapidly than gases, the size of the compartments assumed to be reasonably uniform is typically smaller than that for gases. More pronounced spatial stratifications also require careful selection of a representative aerosol sampling location.

As noted by Letz, Ryan, and Spengler (1984), the variance about the mean exposure can be estimated from the variances of the mean concentrations and exposure times. If these individual parameter variances are small relative to their means and the parameters are uncorrelated, Gauss's law of error propagation (see Bevington 1969) can be applied. The mean exposures and variance estimates representing specific microenvironments can be catalogued and used subsequently to predict the distributions of exposures for other combinations of activity patterns.

Compartmental Modeling

The average value of an aerosol parameter (most often mass concentration) for a microenvironment can be estimated using compartmental models. These models are based on differential mass balances and on the assumption that the contaminant accumulation in a room or compartment is equal to the sum of material entering, or being formed, minus that which is being lost. Analytical solutions are possible for the least complicated differential equations, although more complex situations typically use numerical methods to estimate the time-dependent or equilibrium concentration. Compartmental models compute an integrated average value that may not exist at any single point in the real microenvironment.

Several compartmental models have been proposed for aerosols in indoor environments, including those of Offermann et al. (1989), Raunemaa et al. (1989), and Nazaroff and Cass (1989). These models account for a variety of situations, including polydisperse size distributions, transport between multiple microenvironments, coagulation, gas-phase reaction kinetics, and wall losses. The Offermann et al. (1985) model was used to describe the relationship of the aerosol removal performance of air cleaners, as a function of particle diameter and the mean room concentration. Raunemaa et al. (1989) noted that a re-emission term for indoor aerosols released after loss to surfaces should be included in compartmental models. They also noted that the size distributions of these resuspended particles differ from those during deposition. Nazaroff, Salmon, and Cass (1990) applied the aerosol transport and fate model of Nazaroff and Cass (1989) to an estimate of aerosol soiling in museums from surface deposition.

In addition to estimating contaminant measurement parameters, a compartmental model input requires estimating the convective and diffusive flows between compartments. Ventilation models can be used to predict these flows if the flow model has been sufficiently validated for accuracy. Some of the most comprehensive indoor flow models have been developed for the National Institute of Standards and Technology (Axley 1990).

Dispersion Modeling

The prediction of concentrations in two and three dimensions for gas-phase contaminants has been applied extensively to outdoor situations. Some success in estimating outdoor aerosol mass concentrations downwind of point sources has been noted by Petersen and Lavdas (1986), who added a particle-settling term to a long-term transport model. Applications of a different type of dispersion model for indoor environments, using an advective–dispersive form of the Navier–Stokes equations, have recently been developed for contaminants by Yamamoto, Donovan, and Ensor (1989). A laminar flow version incorporating vorticity and potential flow terms is capable of predicting concentration profiles in rooms with complex flow paths, including recirculation zones. Indoor dispersion models that account for the transport and loss of aerosols are not yet available. A modification to the model of Yamamoto, Donovan, and Ensor (1989) has been proposed by Rodes, Kamens, and Wiener (1991) to consider aerosol-settling losses in short-range transport situations.

References

American Congress of Governmental and Industrial Hygienists (ACGIH). 1991. *Threshold Limit Values and Biological Exposure Indices for1991–1992.* Cincinnati: ACGIH.

Anderson, R., K. Kamens, C. Rodes, and R. Wiener.

1989. A collocation study of PM_{10} and $PM_{2.5}$ inertial impactors for indoor air exposure assessment. In *Proceedings of the 1989 EPA/AWMA (American Waste Management Association) Symposium on Measurement of Toxic and Related Air Pollutants*, eds. R. K. M. Jayanty and S. Hochheiser, Pub. No. VIP-13, pp. 464–69. Pittsburgh: AWMA.

Axley, J. 1990. Element assembly techniques and indoor air quality analysis. In *Proc. 5th International Conference on Indoor Air Quality and Climate, Vol. 4*, ed. D. S. Walkinshaw, pp. 115–20. Ottawa: Canada Mortgage and Housing Corporation.

Baron, P. A. 1988. Modern real-time aerosol samplers. *Appl. Ind. Hyg.* 3(3):97–103.

Bergman, H., C. Edling, and O. Axelson. 1986. Indoor radon daughter concentrations and passive smoking. *Environ. Int.* 12:17–19.

Bevington, P. R. 1969. *Data Reduction and Error Analysis for the Physical Sciences*. New York: McGraw-Hill.

Bidleman, T. F. 1988. Atmospheric processes: Wet and dry deposition of organic compounds are controlled by their vapor-particle partitioning. *Atmos. Environ.* 22(4):361–67.

Binder, R. E., C. A. Mitchell, H. R. Hosein, and A. Bouhuys. 1976. Importance of the indoor environment in air pollution exposure. *Arch. Environ. Health* 31:277–79.

Bond, A. E., R. G. Lewis, and F. W. Immerman. 1990. A study of residential exposure to pesticides in two urban areas of the United States. In *Proc. 5th International Conference on Indoor Air Quality and Climate, Vol. 2*, ed. D. S. Walkinshaw, pp. 683–88. Ottawa: Canada Mortgage and Housing Corporation.

Brief, R. S. and T. Bernath. 1988. Indoor pollution: Guidelines for prevention and control of microbiological respiratory hazards associated with air conditioning systems. *Appl. Ind. Hyg.* 3(1):5–10.

Bright, D. S. and R. A. Fletcher. 1983. New portable ambient aerosol sampler. *Am. Ind. Hyg. Assoc. J.* 44(7):528–36.

Buchan, R. N. et al. 1973. Atmospheric dispersion of particulate pollutants emitted from an activated sludge unit. *J. Environ. Health* 35(4):342.

Buckley, T. J., J. M. Waldman, N. C. G. Freeman, V. A. Marple, W. A. Turner, and P. J. Lioy. 1991. Calibration, intersampler comparison, and field application of a new PM-10 personal air sampling impactor. *Aerosol Sci. Technol.* 14:380–87.

Burge, H. A., Boise, J. R., Rutherford, J. A., and Solomon, W. R. 1977. Comparative recoveries of airborne fungus spores by viable and nonviable modes of volumetric collection. *Mycopathologia* 61:27–33.

Burge, H. A., J. C. Feeley, Sr., K. Kreiss, D. Milton, P. R. Morey, J. A. Otten, K. Peterson, and J. J. Tulis. 1989. *Guidelines for the Assessment of Bioaerosols in the Indoor Environment*. Cincinnati: American Conference of Governmental Industrial Hygienists.

Burge, H. A. and M. E. Hoyer. 1990. Indoor air quality. *Appl. Occup. Environ. Hyg.* 5(2):84–93.

Caka, F. M., D. J. Eatough, E. A. Lewis, H. Tang, S. K. Hammond, B. P. Leaderer, P. Koutrakis, J. D. Spengler, A. Fasano, J. McCarthy, M. W. Ogden, and J. Lewtas. 1990. An intercomparison of sampling techniques for nicotine in indoor environments. *Environ. Sci. Technol.* 24(8):1196–203.

Chuang, J. C., S. W. Hannan, and N. K. Wilson. 1987. Field comparison of polyurethane foam and XAD-2 resin for air sampling of polynuclear aromatic hydrocarbons. *Environ. Sci. Technol.* 21:798–804.

Code of Federal Regulations. July 1990*a*. EPA National Emissions Standards for Hazardous Air Pollutants Asbestos Regulations. 40 C.F.R. Part 61, Subpart M, Appendix C.

Code of Federal Regulations. July 1990*b*. EPA Interim Transmission Electron Microscopy Analytical Methods. 40 C.F.R. Part 763, Subpart E, Appendix A.

Code of Federal Regulations. July 1990*c*. EPA Ambient Air Monitoring Reference and Equivalent Methods. 40 C.F.R. Part 53.43(c).

Colome, S. D., J. D. Spengler, and S. McCarthy. 1982. Comparisons of elements and inorganic compounds inside and outside of residences. *Environ. Int.* 8:197–212.

Converse, J. M. and S. Presser. 1986. Survey Questions, Handcrafting the Standardized Questionnaire. Beverly Hills, CA: Sage Publications.

Dockery, D. W. and J. D. Spengler. 1981*a*. Personal exposure to respirable particulates and sulfates. *J. APCA* 31:153–59.

Dockery, D. W. and J. D. Spengler. 1981*b*. Indoor–outdoor relationships of respirable sulfates and particles. *Atmos. Environ.* 15:335–43.

Eatough, D. J., C. Benner, R. Mooney, D. Bartholomew, D. S. Steiner, L. D. Hansen, J. D. Lamb, E. A. Lewis, and N. L. Eatough. 1986. Gas and particle phase nicotine in environmental tobacco smoke. Paper no. 86-68.5 presented at the 79th Annual Meeting of the Air Pollution Control Association, 22–27 June 1986, Minneapolis, MN.

Eudy, L. W., F. A. Thome, D. L. Heavner, C. R. Green, and B. J. Ingebrethsen. 1986. Studies on the vapor–particulate phase distribution of environmental nicotine by selective trapping and detection methods. Paper no. 86-38.7 presented at the 79th Annual Meeting of the Air Pollution Control Association, 22–27 June 1986, Minneapolis, MN.

Federal Register. 1989. 54 Fed. Reg. 29,460.

Fletcher, R. A. 1984. A review of personal/portable airborne particulate monitors. *J. APCA* 34(10):1014–16.

Fletcher, R. A. and D. S. Bright. 1983. *NBS Portable Ambient Particulate Sampler*, Pub. No. NBSIR 82-2561. Gaithersburg, MD: National Bureau of Standards.

Fletcher, B. and A. E. Johnson. 1988. Comparison of personal and area concentration measurements, and the use of a manikin in sampling. In *Ventilation '88*, ed. J. H. Vincent, pp. 161–65. New York: Pergamon.

Girman, J. R., M. G. Apte, G. W. Traynor, J. R. Allen, and C. D. Hollowell. 1982. Pollutant emission rates from indoor combustion appliances and sidestream cigarette smoke. *Environ. Int.* 8:213–21.

Gregory, P. H. 1973. *The Microbiology of the Atmosphere*, 2nd edn. Aylesbury, Leonard Hill Books.

Hammond, S. K., B. P. Leaderer, A. C. Roche, and M. Schenker. 1987. Collection and analysis of nicotine as a marker for environmental tobacco smoke. *Atmos. Environ.* 21:457–62.

Hering, S. V. 1989. Inertial and gravitational collectors. In *Air Sampling Instruments for Evaluation of Atmospheric Contaminants*, ed. S. V. Hering, pp. 337–85. Cincinnati: American Conference of Governmental Industrial Hygienists.

Highsmith, V. R., C. E. Rodes, and R. J. Hardy. 1988. Indoor particle concentrations associated with use of tap water in portable humidifiers. *Environ. Sci. Technol.* 22:1109–12.

Highsmith, V. R., R. B. Zweidinger, and R. G. Merrill. 1988. Characterization of indoor and outdoor air associated with residences using woodstoves: A pilot study. *Environ. Int.* 14:213–19.

Hilleary, L. 1991. Federal appeals court overturns EPA's "ban and phase-out" proposal: Ruling could have a major impact on TSCA (asbestos). In *Indoor Air Review*, November 1991, pp. 1, 35. Rockville, MD: IAQ Publications.

Hinds, W. C. 1982. *Aerosol Technology: Properties, Behavior, and Measurement of Aerosol Particles.* New York: Wiley Interscience.

Humphries, W. and J. H. Vincent. 1976. An experimental investigation of the detention of airborne smoke in the wake bubble behind a disk. *J. Fluid Mech.* 73(3):453–64.

Humphries, W. and J. H. Vincent. 1978. The transport of airborne dusts in the near wake of bluff bodies. *Chem. Eng. Sci.* 33:1141–46.

Immerman, F. W. and J. L. Schaum. 1990. *Nonoccupational Pesticide Exposure Study (NOPES)*, Pub. No. EPA/600/3-90/003. Research Triangle Park, NC: U.S. Environmental Protection Agency.

Ind. Hyg. News Staff. 1991. Air sampling—Selection chart. *Ind. Hyg. News* 14(1):39.

Jenkins, P. L. and T. J. Phillips. 1988. The California Air Resources Board indoor air quality and personal exposure program. Presented at the 81st Annual Meeting of the Air Pollution Control Association, 20–24 June 1988, Dallas, TX.

Jenkins, P. L., T. J. Phillips, and E. J. Mulberg. 1990. Activity patterns of Californians: Use of and proximity to indoor pollutant sources. In *Proc. 5th International Conference on Indoor Air Quality and Climate, Vol. 2*, ed. D. S. Walkinshaw, pp. 465–70. Ottawa: Canada Mortgage and Housing Corporation.

Ju, C. and J. D. Spengler. 1981. Room-to-room variations in concentration of respirable particles in residences. *Environ. Sci. Technol.* 15(5):592–96.

Kamens, R., R. Wiener, C. Lee, and D. Leith. 1988. The characterization of aerosols in residential environments. In *Proceedings of the 1988 EPA/APCA (Air Pollution Control Association) Symposium on Measurement of Toxic and Related Air Pollutants*, eds. R. K. M. Jayanty and S. Hochheiser, pp. 89–97. Pittsburgh: APCA.

Koutrakis, P., A. M. Fasano, J. L. Slater, J. D. Spengler, J. F. McCarthy, and B. P. Leaderer. 1989. Design of a personal annular denuder sampler to measure atmospheric aerosols and gases. *Atmos. Environ.* 23(12):2767–73.

LaForce, F. M. 1984. Airborne infections and modern building technology. In *Indoor Air: Recent Advances in Health Sciences and Technology, Vol. 1*, eds. World Health Organization et al., pp. 109–19. Stockholm: Swedish Council for Building Research.

Leaderer, B. P., P. M. Boone, and S. K. Hammond. 1990. Total particle, sulfate, and acidic aerosol emissions from kerosene space heaters. *Environ. Sci. Technol.* 24:908–12.

Letz, R., P. B. Ryan, and J. D. Spengler. 1984. Estimated distributions of personal exposure to respirable particles. *Environ. Monit. Assess.* 4:351–59.

Lewis, R. G., A. E. Bond, T. R. Fitz-Simons, D. E. Johnson, and J. P. Hsu. 1986. Monitoring for non-occupational exposure to pesticides in indoor and personal respiratory air. Presented at the 79th Annual Meeting of the Air Pollution Control Association, 22–27 June 1986, Minneapolis, MN.

Lewis, R. G., A. E. Bond, D. E. Johnson, and J. P. Hsu. 1988. Measurement of atmospheric concentrations of common household pesticides: A pilot study. *Environ. Monit. Assess.* 10:59–73.

Lewtas, J., S. Goto, K. Williams, J. Chapell, and N. K. Wilson. 1987*a*. Mutagenicity of indoor air in a residential field study. *Environ. Mutagen.* 9(Suppl. 8):58–59.

Lewtas, J., S. Goto, K. Williams, J. C. Chuang, B. A. Petersen, and N. K. Wilson. 1987*b*. The mutagenicity of indoor air particles in a residential pilot field study: Application and evaluation of new methodologies. *Atmos. Environ.* 21:443–49.

Lioy, P. J. 1990. Assessing total human exposure to contaminants. *Environ. Sci. Technol.* 24(7):938–45.

Lioy, P. J., T. Wainman, W. Turner, and V. A. Marple. 1988. An intercomparison of the indoor air sampling impactor and the dichotomous sampler for a 10 μm cut-size. *J. APCA* 38:668–70.

Lioy, P. J., J. M. Waldman, T. Buckley, J. Butler, and C. Pietarinen. 1990. The personal, indoor and outdoor concentrations of PM-10 measured in an industrial community during the winter. *Atmos. Environ.* 24B:57–66.

Lippmann, M. 1983. Size-selective health hazard sampling. In *Air Sampling Instruments for Evaluation of Atmospheric Contaminants*, 6th edn., eds. P. J. Lioy and M. J. Lioy, pp. H1–22. Cincinnati: American Conference of Governmental Industrial Hygienists.

Lippmann, M. 1989. Size-selective health hazard sampling. In *Air Sampling Instruments for Evaluation of Atmospheric Contaminants*, 7th edn., pp. 163–98. Cincinnati: American Conference of Governmental Industrial Hygienists.

Litzistorf, G., M. P. Guillemin, P. Buffat, and F. Iselin. 1985. Influence of human activity on the airborne fiber level in paraoccupational environments. *J. APCA* 35(8):836–37.

Marple, V., B. Liu, S. Behm, B. Olson, and R. Wiener. 1988. A new personal impactor sampler inlet. Presented at the American Industrial Hygiene Conference, 15–20 May 1988, San Francisco, CA.

Marple, V., B. Liu, S. Behm, B. Olson, and R. Wiener. 1989. A personal environmental monitor (PEM). Presented at the American Association for Aerosol Research Annual Meeting, 10–13 October 1989, Reno, NV.

Marple, V. A. and J. E. McCormack. 1983. Personal sampling impactor with respirable aerosol penetration characteristics. *AIHA J.* 44(12):916–22.

Marple, V. A., K. L. Rubow, W. Turner, and J. D. Spengler. 1987. Low flow rate sharp cut impactors for indoor air sampling: Design and calibration. *J. APCA* 37:1303–7.

Martonen, T., M. Clark, D. Nelson, D. Willard, and E. Rossignal. 1984. Evaluation of a mini-cascade impactor for sampling exposure chamber atmospheres. *Fund. Appl. Toxicol.* 2:149–52.

McDade, J. E., C. C. Shepard, D. W. Fraser, R. T. Tsai, M. A. Redus, W. R. Dowdle, and the Laboratory Investigation Team. 1977. Legionnaires' disease: Isolation of a bacterium and demonstration of its role in other respiratory disease. *N. Engl. J. Med.* 297:1197–203.

McDow, S. R. and J. J. Huntzicker. 1990. Vapor adsorption artifact in the sampling of organic aerosol: Face velocity effects. *Atmos. Environ.* 24A:2563–71.

McFarland, A. R. and C. A. Ortiz. 1983. Evaluation of prototype-10 inlets with cyclonic fractionators. APCA *Proc. Ann. Meet.* 76:1–13.

McKenzie, R. L., D. S. Bright, R. A. Fletcher, and J. A. Hodgeson. 1982. Development of a personal exposure monitor for two sizes of inhalable particulates. *Environ. Int.* 8:229–33.

Molhave, L. 1987. The sick buildings—A subpopulation among the problem buildings. In *Indoor Air: '87 Proc. 4th International Conference on Indoor Air Quality and Climate, Vol. 2*, eds. B. S. Seifert et al., pp. 469–73. Berlin: Oraniendruck, GmbH Publishers.

Nagda, N. L., H. E. Rector, and M. D. Koontz. 1987. *Guidelines for Monitoring Indoor Air Quality.* Washington, DC: Hemisphere.

National Research Council, Comm. on Airliner Cabin Air Quality. 1986*a*. *Airliner Cabin: Environmental Air Quality and Safety.* Washington, DC: National Academy Press.

National Research Council, Comm. on Passive Smoking On-Board. 1986*b*. *Environmental Tobacco Smoke: Measuring and Assessing Health Effects.* Washington, DC: National Academy Press.

Nazaroff, W. W. and G. R. Cass. 1989. Mathematical modeling of indoor aerosol dynamics. *Environ. Sci. Technol.* 23(2):157–66.

Nazaroff, W. W., L. G. Salmon, and G. R. Cass. 1990. Concentration and fate of airborne particles in museums. *Environ. Sci. Technol.* 24(1):66–77.

Offermann, F. J., R. G. Sextro, W. J. Fisk, D. T. Grimsrud, W. W. Nazaroff, A. V. Nero, K. L. Revzan, and J. Yater. 1985. Control of respirable particles in indoor air with portable air cleaners. *Atmos. Environ.* 19:1761–71.

Offermann, F. J., R. G. Sextro, W. J. Fisk, D. T. Grimsrud, T. Raunemaa, M. Kulmala, H. Saari, M. Olin, and M. H. Kulmala. 1989. Indoor air aerosol model: Transport indoors and deposition of fine and coarse particles. *Aerosol Sci. Technol.* 11:11–25.

Olenchock, S. A., S. W. Lenhart, and J. C. Mull. 1982. Occupational exposure to airborne endotoxins during poultry processing. *J. Toxicol. Environ. Health* 9:339–49.

Ott, W. R. 1985. Total human exposure. *Environ. Sci. Technol.* 19(10):880–86.

Ott, W. R. 1990. Total human exposure: Basic concepts, EPA field studies, and future research needs. *J. Air Waste Manag. Assoc.* 40:966–75.

Owen, M. K., D. S. Ensor, L. S. Hovis, W. G. Tucker, and L. E. Sparks. 1990. Particle size distributions for an office aerosol. *Aerosol Sci. Technol.* 13:486–92.

Ozkaynak, H., J. D. Spengler, C. A. Clayton, E. Pellizzari, and R. W. Wiener. 1990. Personal exposure to particulate matter: Findings from the particle total exposure assessment methodology (PTEAM) prepilot study. In *Proc. 5th International Conference on Indoor Air Quality and Climate, Vol. 2*, ed. D. S. Walkinshaw, pp. 571–76. Ottawa: Canada Mortgage and Housing Corporation.

Pankow, J. F. 1988. The calculated effects of non-exchangeable material on the gas–particle distributions of organic compounds. *Atmos. Environ.* 22:1405–9.

Parker, R. C., R. K. Bull, D. C. Stevens, and M. Marshall. 1990. Studies of aerosol distributions in a small laboratory containing a heated phantom. *Annals Occup. Hyg.* 34:34–44.

Pellizzari, E. D., T. D. Hartwell, C. Leininger, H. Zelon, S. Williams, J. Breen, and L. Wallace. 1983. Human exposure to vapor-phase halogenated hydrocarbons: Fixed-site vs. personal exposure. In *Proceedings from the National Symposium on Recent Advances in Pollutant Monitoring of Ambient Air and Stationary Sources*, Pub. No. EPA/600/9-83/007, pp. 264–88. Research Triangle Park, NC: U.S. Environmental Protection Agency.

Petersen, W. B. and L. G. Lavdas. 1986. *INPUFF 2.0—A Multiple Source Gaussian Puff Dispersion Algorithm*, Pub. No. EPA/600/8-86/024. Research Triangle Park, NC: U.S. Environmental Protection Agency.

Possanzini, M., A. Rebo, and A. Liberti. 1983. New design of a high-performance denuder for the sampling of atmospheric pollutants. *Atmos. Environ.* 17(12):2605–10.

Randtke, S. J., D. D. Lane, and T. E. Baxter. 1990. *Development of a Sampling Procedure for Large Nitrogen- and Sulfur-Bearing Aerosols*, Pub. No. NTIS PB90-235 789/AS. Gaithersburg, MD: National Technical Information Service.

Raunemaa, T., M. Kulmala, H. Saari, M. Olin, and M. H.

Kulmala. 1989. Indoor air aerosol model: Transport indoors and deposition of fine and coarse particles. *Aerosol Sci. Technol.* 22:11–25.

Repace, J. L. 1987. Indoor concentration of environmental tobacco smoke: Field surveys. In *Environ. Carcinogens: Methods of Analysis and Exp. Meas., Vol. 9—Passive Smoking*, eds. I. K. O'Neill, K. Brunneman, B. Dodet, and D. Hoffman, Sci. Pub. No. 81, pp. 141–62. Lyon: World Health Organization, International Agency for Research on Cancer.

Repace, J. L. and A. H. Lowrey. 1980. Indoor air pollution, tobacco smoke and public health. *Science* 208:464–72.

Repace, J. L. and A. H. Lowrey. 1983. Modeling exposure of nonsmokers to ambient tobacco smoke. Paper no. 83-64.2 presented at the 76th Annual Meeting of the Air Pollution Control Association, 19–24 June 1983, Pittsburgh, PA.

Roberts, J. W., W. T. Budd, M. G. Ruby, A. E. Bond, R. G. Lewis, R. W. Wiener, and D. E. Camann. 1991. Development and field testing of a high volume sampler for pesticides and toxics in dust. *J. Exposure Anal. Environ. Epidemiol.* 1:143–55.

Roberts, J. W. and M. G. Ruby. 1989. *Development of a High Volume Surface Sampler for Pesticides in Floor Dust*, Pub. No. EPA/600/S4-88/036. Research Triangle Park, NC: U.S. Environmental Protection Agency.

Roberts, J. W., M. G. Ruby, and G. R. Warren. 1987. Mutagenic activity of house dust. In *Proceedings of the 1987 Symposium on Application of Short-Term Bioassays in the Analysis of Complex Environmental Mixtures*, eds. S. S. Sandhu, D. M. DeMarini, M. T. Mass, M. M. Moore, and J. L. Munford, pp. 355–67. New York: Plenum.

Rodes, C. E. 1986. Sampling and analysis for non-viable indoor particles and semi-volatiles. Presented at the Indoor Air Quality Symposium, 23–25 September 1986, Georgia Institute of Technology, Atlanta, GA.

Rodes, C. E., R. M. Kamens, and R. W. Wiener. 1991. The significance and characteristics of the personal activity cloud on exposure assessment measurements for indoor contaminants. *Indoor Air* 2:123–145.

Rubow, K. L. and V. C. Furtado. 1989. Air movers and samplers. In *Air Sampling Instruments for Evaluation of Atmospheric Contaminants*, ed. S. V. Hering, pp. 241–74. Cincinnati: American Conference of Government Industrial Hygienists.

Rubow, K. L., V. A. Marple, J. Olin, and M. A. McCawley. 1985. *A Personal Impactor: Design, Evaluation, and Calibration*, Pub. No. 469. Minneapolis: University of Minnesota, Particle Technol. Lab.

Ryan, B. 1986. Combustion products: Sources, sampling and analysis. Presented at the Indoor Air Quality Symposium, 23–25 September 1986, Georgia Institute of Technology, Atlanta, GA.

Ryan, P. B. and J. D. Spengler. 1988. Sequential box models for indoor air quality: Application to airliner cabin air quality. *Atmos. Environ.* 22(6):1031–38.

Salisbury, S. 1986. Indoor air quality (IAQ) investigation procedures. Presented at the Indoor Air Quality Symposium, 23–25 September 1986, Georgia Institute of Technology, Atlanta, GA.

Samfield, M. 1985. Importance of cigarette sidestream smoke—Further aspects. *Tabak J. Int.* 6:448.

Sebastien, P., J. Bignon, and M. Martin. 1982. Indoor airborne asbestos pollution: From the ceiling and the floor. *Science* 216:1410–13.

Sexton, K., J. D. Spengler, and R. D. Treitman. 1984*a*. Effects of residential wood combustion on indoor air quality: A case study in Waterbury, VT. *Atmos. Environ.* 18(7):1371–83.

Sexton, K., J. D. Spengler, and R. D. Treitman. 1984b. Personal exposure to respirable particles: A case study in Waterbury, VT. *Atmos. Environ.* 18(7):1385–98.

Sheldon, L. S., D. Whitaker, J. Sickles, E. Pellizzari, D. Westerdahl, and R. W. Wiener. 1990. Chemical characterization of indoor and outdoor air samples—PTEAM prepilot study. In *Proc. 5th International Conference on Indoor Air Quality and Climate, Vol. 2*, ed. D. S. Walkinshaw, pp. 765–70. Ottawa: Canada Mortgage and Housing Corporation.

Spengler, J. D., D. W. Dockery, W. A. Turner, J. M. Wolfson, and B. G. Ferris, Jr. 1981. Long-term measurements of respirable sulfates and particulates inside and outside homes. *Atmos. Environ.* 15:23–30.

Spengler, J., H. Ozkaynak, J. Ludwig, G. Allen, E. Pellizzari, and R. Wiener. 1989. Personal exposures to particulate matter: Instrumentation and methodologies (P-TEAM). *Proceedings of the 1989 EPA/-AWMA (American Waste Management Association) Symposium on Measurement of Toxic and Related Air Pollutants*, eds. R. K. M. Jayanty and S. Hochheiser, Pub. No. VIP-13, pp. 449–63. Pittsburgh: AWMA.

Spengler, J. D., R. D. Treitman, T. D. Tosteson, D. T. Mage, and M. L. Soczek. 1985. Personal exposures to respirable particulates and implications for air pollution epidemiology. *Environ. Sci. Technol.* 19(8):700–7.

Starr, H. G., Jr., F. D. Aldrich, W. D. MacDougall, and L. M. Mounce. 1974. Contribution of household dust to the human exposure of pesticides. *Pestic. Monit. J.* 8:209–12.

Stevens, D. C. 1969. The particle size and mean concentration of radioactive aerosols measured by personal and static air samples. *Ann. Occup. Hyg.* 12:33–40.

Surgeon General of the United States. 1986. *The Health Consequences of Involuntary Smoking: A Report of the Surgeon General.* Rockville, MD: U.S. Department of Health and Human Services, Public Health Service.

Swift, D. L. 1989. Direct-reading instruments for analyzing airborne particles. In *Air Sampling Instruments for Evaluation of Atmospheric Contaminants*, ed. S. V. Hering, pp. 477–506. Cincinnati: American Conference of Government Industrial Hygienists.

Szalai, A. 1972. *The Use of Time: Daily Activities of Urban and Suburban Populations in 12 Countries.* The Hague and Paris: Mouton.

Traynor, G. W., M. G. Apte, and A. R. Carruthers. 1987. Indoor air pollution due to emissions from woodburning stoves. *Environ. Sci. Technol.* 21(7):691–97.

Traynor, G. W., M. G. Apte, H. A. Sokol, J. C. Chuang,

W. G. Tucker, and J. L. Mumford. 1990. Selected organic pollutant emissions from unvented kerosene space heaters. *Environ. Sci. Technol.* 24:1265–70.

Traynor, G. W., J. R. Girman, M. G. Allen, A. R. Apte, J. Carruthers, J. F. Dillworth, and V.M. Martin. 1983. *Indoor Air Pollution from Kerosene-Fired Space Heaters*, Report No. LBL-15612. Berkeley, CA: Lawrence Berkeley Lab.

Traynor, G. W., and I. A. Nitschke. 1984. Field survey of indoor air pollution in residences with suspected combustion-related sources (chemical characterization and personal exposure). In *Proc. 3rd International Indoor Air Quality and Climate Conference, Vol. 14*, eds. World Health Organization et al., pp. 343–48. Stockholm: WHO.

Tu, K. W. and L. E. Hinchliffe. 1983. A study of particulate emissions from portable space heaters. *AIHA J.* 44:857–62.

Turner, W. A., J. D. Spengler, D. W. Dockery, and S. D. Colome. 1979. Design and performance of a reliable personal monitoring system for respirable particulates. *J. APCA* 29(7):747–74.

U.S. Code. 1986. Radon Gas and Indoor Air Quality Research Act. Superfund Amendments and Reauthorization Act (SARA), Public Law 99-499. 42 U.S.C. § 7401 et seq. [Radon Gas and Indoor Air Quality Research Act, 42 U.S.C. § 7401 note (1988)].

U.S. Department of Labor. 1988. Occupational Safety and Health (OSHA) Regulations. General Industry Asbestos Standard (29 C.F.R. § 1910.1001) and Construction Industry Asbestos Standard (29 C.F.R. § 1926.58). June 1986; Amended, September 1988.

U.S. Environmental Protection Agency (EPA). 1983. *Survey Management Handbook: Volume I. Guidelines for Planning and Managing a Statistical Survey*, Pub. No. EPA/230/12-84/00. Washington, DC: EPA.

U.S. Environmental Protection Agency (EPA). 1985*a*. *Guidance for Controlling Asbestos-Containing Materials in Buildings* (Purple Book), Pub. No. EPA/560/5-85/024. Washington, DC: EPA.

U.S. Environmental Protection Agency (EPA). 1985*b*. *Measuring Airborne Asbestos Following an Abatement Action* (Silver Book), Pub. No. EPA/600/4-85/0490. Research Triangle Park, NC: EPA.

U.S. Environmental Protection Agency (EPA). 1987. *EPA Indoor Air Quality Implementation Plan. Appendix A: Preliminary Indoor Air Pollution Information Assessment*, Pub. No. EPA/600/8-87/014. Research Triangle Park, NC: EPA.

U.S. Environmental Protection Agency (EPA). 1989. *Report to Congress on Indoor Air Quality, Volume II: Assessment and Control of Indoor Air Pollution*, Pub. No. EPA/400/1-89/000c. Research Triangle Park, NC: EPA.

U.S. Environmental Protection Agency (EPA). 1990. *Compendium of Methods for the Determination of Air Pollutants in Indoor Air*, Pub. No. EPA/600/S4-90/010. Research Triangle Park, NC: EPA.

Van Houdt, J. J. and J. S. M. Boleij. 1984. Mutagenic activity of indoor airborne particles compared to outdoors. In *Indoor Air: Proc. 3rd International Conference on Indoor Air Quality and Climate, Vol. 2*, eds. B. Berglund, T. Lindvall, and J. Sundell, pp. 169–76. Stockholm: Swedish Council for Building Research.

Vaughan, W. M. and S. K. Hammond. 1990. Impact of "designated smoking area" policy on nicotine vapor and particle concentrations in a modern office building. *J. Air Waste Manag. Assoc.* 40:1012–17.

Vincent, J. H., P. C. Emmett, and D. Mark. 1985. The effect of turbulence on the entry of airborne particles into a blunt dust sampler. *Aerosol Sci. Technol.* 4:17–29.

Vincent, J. H. and H. Gibson. 1981. Sampling errors in blunt dust samplers arising from external wall loss effects. *Atmos. Environ.* 15(5):703–12.

Vossler, T. L., R. K. Stevens, R. J. Paur, R. E. Baumgardner, and J. P. Bell. 1988. Evaluation of improved inlets and annular denuder systems to measure inorganic air pollutants. *Atmos. Environ.* 22(8):1729–36.

Vu Duc, T. and C. M. P. Favez. 1981. Characteristics of motor exhausts in an underground car park: Mass size distribution and concentration levels of particles. *J. Environ. Sci. Health* A16:647–60.

Wallace, L. A. 1987. *The Total Exposure Assessment Methodology (TEAM) Study: Summary and Analysis, Vol. 1*, Pub. No. EPA/600/6-87/002a. Washington, DC: U.S. Environmental Protection Agency.

Wallace, L. A., E. D. Pellizzari, T. D. Hartwell, R. Whitmore, C. M. Sparacino, and H. S. Zelon. 1986. Total Exposure Assessment Methodology (TEAM) study: Personal exposures, indoor–outdoor relationships, and breath levels of volatile organic compounds in New Jersey. *Environ. Int.* 12:369–87.

Wallingford, K. and J. Carpenter. 1986. NIOSH field experience overview: Investigating sources of IAQ problems in office buildings. In *IAQ '86: Managing Indoor Air for Health and Energy Conservation*, pp. 448–53. Atlanta: American Society for Heating, Refrigeration, and Air Conditioning Engineering.

Wedding J. B., M. A. Weigand, and Y. J. Kim. 1985. Evaluation of the Sierra–Andersen 10-μm inlet for the high-volume sampler. *Atmos. Environ.* 19(3):539–42.

Wiener, R. 1989. Particle total exposure methodology — An overview of planning and accomplishments. *Proceedings of the 1989 EPA/AWMA (American Waste Management Association) Symposium on Measurement of Toxic and Related Air Pollutants*, eds. R. K. M. Jayanty and S. Hochheiser, pp. 442–48. Pittsburgh: AWMA.

Wiener, R., J. Spengler, and E. Pellizzari. 1990. Review of the particle TEAM 9-home study. *Proceedings of the 1990 EPA/APCA (Air Pollution Control Association) Symposium on Measurement of Toxic and Related Air Pollutants*, eds. R. K. M. Jayanty and B. W. Gay, Pub. No. VIP-17, pp. 452–60. Pittsburgh: APCA.

Wiener, R. W. and D. VanOsdell. 1990. Validation of the personal cloud phenomenon. Presented at the 21st Annual Meeting of the Fine Particle Society, 21–25 August 1990. San Diego, CA.

Wilson, N. K., J. C. Chuang, M. R. Kuhlman, and G. A.

Mack. 1989. Measurement of PAH and their derivatives in indoor air. Presented at the 12th International Symposium on Polynuclear Aromatic Hydrocarbons, 19–21 September 1989, Gaithersburg, MD.

Wilson, N. K., R. G. Lewis, C. C. Chuang, B. A. Peterson, and G. A. Mack. 1985. Analytical and sampling methodology for characterization of polynuclear aromatic compounds in indoor air. In *Proc. 78th Annual Meeting of the Air Pollution Control Association*, Paper No. 85-30A.2. Pittsburgh, PA: Air Pollution Control Association.

Wolff, R. K., J. D. Sun, E. B. Barr,.S. J. Rotherberg, and H. C. Yeh. 1989. Lung retention and binding of [^{14}C]-1-nitropyrene when inhaled by F344 rats as a pure aerosol or adsorbed to carbon black particles. *J. Toxicol. Environ. Health* 26:309–25.

World Health Organization (WHO). 1983. *Indoor Air Pollutants: Exposure and Health Effects Assessment.* Euro Reports and Studies Working Group Report No. 78. Nordlingen, Copenhagen: WHO.

Yamamoto, T., R. P. Donovan, and D. S. Ensor. 1989. Optimization of clean room airflows. *J. Environ. Sci.* 31(6):24–27.

Yamasaki, H., K. Kuwata, and H. Miyamoto. 1982. Effects of ambient temperature on aspects of airborne polycyclic aromatic hydrocarbons. *Environ. Sci. Technol.* 16:189–94.

Yamate, G., S. G. Agarwal, and R. D. Gibbons. 1984. *Methodology for the Measurement of Airborne Asbestos by Electron Microscopy.* Draft report, EPA/ORD (Office of Research and Development) contract 68-02-3266. Washington, DC: U.S. Environmental Protection Agency.

Zhang, X. Q. and P. H. McMurry. 1987. Theoretical analysis of evaporative losses from impactor and filter deposits. *Atmos. Environ.* 21:1779–89.

31

Measurement of Aerosols and Clouds from Aircraft

Charles A. Brock and James Charles Wilson

Department of Engineering, University of Denver
Denver, CO, U.S.A.

and

W. Russell Seebaugh

Denver Research Institute, University of Denver
Denver, CO, U.S.A.

INTRODUCTION

The meteorological, chemical, and radiative roles played by aerosol and cloud particles in the atmosphere are important research topics. Links between the microphysical, optical, and chemical properties of aerosols and clouds, and the effects these systems have on larger-scale atmospheric processes are being studied. For example, aerosols and clouds are important in determining the earth's radiative budget; they serve as nuclei for cloud formation and precipitation; and they contribute to acid precipitation and ozone depletion.

In many studies, airplanes have been used as research platforms. In this chapter, cloud and aerosol research topics commonly pursued from aircraft are noted. Instruments used in this research and problems associated with their use on aircraft are reviewed and difficulties associated with airborne sampling of particles are discussed.

CURRENT RESEARCH INVOLVING AIRBORNE MEASUREMENTS OF AEROSOLS AND CLOUDS

Tropospheric Aerosol and Cloud Studies

The long-range transport of chemical species, the effects of such transport on cloud chemistry and deposition, and the fluxes of various species from continents and oceans are currently studied. Such studies are important in understanding the fate and impact of pollutants and the biogeochemical cycles of chemical species. On a smaller scale, the impact of plumes from point or regional sources, the vertical redistribution of pollutants, and the effects of aerosols on clouds have also been studied from aircraft. Airborne measurements of aerosols have also been utilized in comparisons with remote-sensing instruments on satellites.

Airborne measurement techniques are also being applied to cloud particle studies. Clouds are perhaps the least understood of major atmospheric phenomena, and are only crudely parameterized in the numerical simulations used as predictors of climate. Cloud systems are complex microphysical, chemical, radiative, and dynamical systems. In many cases, observations from airplanes are required to obtain measurements of the needed accuracy and resolution.

Airborne cloud investigations include studies of ice particle formation mechanisms in convective clouds, studies of various microphysical processes in precipitation formation, and studies of charge separation mechanisms which lead to electrification and lightning.

The role of aerosol particles in cloud formation is also being studied, particularly in remote, marine stratus clouds, where the abundance of cloud condensation nuclei (CCN) may be important in controlling the radiative, chemical, and microphysical properties of clouds. Additionally, the formation of acid precipitation and the scavenging, removal, and modification of aerosol particles by clouds are research topics of current interest.

Stratospheric Aerosol and Cloud Studies

The optical, physical, and chemical properties of the sulfate particles that dominate the stratospheric aerosol, and the formation of these particles from gaseous precursors such as OCS and SO_2 are often studied. Stratospheric sulfate particles may serve as nuclei for clouds composed of hydrates of nitric acid and water ice. The resulting polar stratospheric clouds (PSCs) play a role in ozone depletion and are studied. Injections and production of large quantities of sulfate particles after volcanic eruptions have been observed and have enabled a better understanding of the rates of horizontal and vertical mixing in the stratosphere. Particle emissions from current and proposed aircraft are being investigated, as are the emissions from the space shuttle and other orbital rocket launchers.

AIRBORNE AEROSOL AND CLOUD MEASUREMENT TECHNIQUES

Tables 31-1 and 31-2 list commercial instruments and suppliers of instruments specifically designed for aircraft use.

Aerosol Measurements Made Outside the Aircraft

Bringing particles into the aircraft for measurement frequently causes significant heating and often alters their size and composition. Therefore, it is often desirable to measure the particles in the airstream to the extent possible. Particle Measuring Systems' (1855 57th St., Boulder, CO, 80301) forward-scattering spectrometer probe (FSSP), Fig. 31-1, measures the optical diameter of single aerosol particles with diameters larger than about 0.4 μm in the airstream in real time. The instrument case is constructed so that the measuring volume is outside the aircraft and is surrounded with a cylindrical housing aligned with the airflow. The objective of this design is to perturb the airstream and particles as little as possible. Particles pass through a He–Ne laser beam and the forward-scattered light is measured to determine particle size. Concentration of particles is determined from the air speed and the size of the viewing volume.

Complications in the use of the FSSP arise from the nonmonotonic nature of the response function, the circular cross section of the laser beam, the variability of the energy density across the diameter of the laser beam, and the sensitivity of the instrument to the sphericity, homogeneity, and refractive index of the particles being measured. Discussions of aerosol and cloud measurements with the FSSP may be found in Knollenberg (1981), Pinnick, Garvey, and Duncan (1981), Baumgardner (1983), Cerni (1983), Dye and Baumgardner (1984), Baumgardner, Strapp,

TABLE 31-1 Commercial Instruments Described in this Chapter and Designed Specifically for Measurements from Aircraft

Measurement	Name	Source Code
Aerosol and Cloud Droplet Size	FSSP	PMS
Aerosol and Cloud Droplet Size	1-D OAP	PMS
Particle Image	2-D OAP	PMS
Cloud Water Content	J-W Probe	CTI, PMS
Supercooled Cloud Water Content	Model 871	RMI
Airflow Angle	Model 858	RMI
Ice Particle Concentration	Model IPC-IV	SPC
Ice Particle Concentration	Model 120	MEE

TABLE 31-2 Commercial Sources for Instruments in Table 31-1

Source Code	Address
PMS	Particle Measuring Systems, Inc., 1855 S. 57th Court, Boulder, CO 80301
CTI	Cloud Technology, Inc. 606 Wellsbury Ct Palo Alto, CA 94306.
MEE	Mee Industries Inc., 1629 S. Del Mar, San Gabriel, CA 91776
RMI	Rosemount Inc. Aerospace Division, 14300 Judicial Road, Burnsville, MN 55337-4898
SPC	SPEC, Inc., 5401 Western, Boulder, CO 80301

and Dye (1985), Brenguier and Amodei (1989), and Brenguier (1989).

Wire impactors are used to collect particles outside the boundary layer of the aircraft (Dunn and Renken 1987; Farlow et al. 1979). These impactors usually consist of single wires with diameters of a few tens to a few hundreds of micrometers. The wires are attached to a mount at the ends and deployed so that the middle of the wire is perpendicular to the airstream and exposed directly to it. After exposure, the wire is retracted and stored in a sealed container. The collected particles are analyzed with an electron microscope in the laboratory. Chemically reactive coatings on the wire (Pueschel et al. 1989) may be used to investigate the composition of the aerosol particles. Uncertainties associated with wire impactors include the impaction efficiency as a function of particle size (Farlow et al. 1979) and the sticking efficiency of the particles on the wire after the impaction (Dunn and Renken 1987).

Other simple impactor geometries have also been used. Bailey et al. (1984) and Ishizaka, Hobbs, and Radke (1989) have exposed grease-coated glass slides directly to the flow outside of the aircraft in order to collect coarse particles in the troposphere. Difficulties with the use of such impactors include uncertain collection efficiencies due to non-isoaxial sampling and flow distortion, the shattering of large particles upon impact, and the possibility of particle bounce-off.

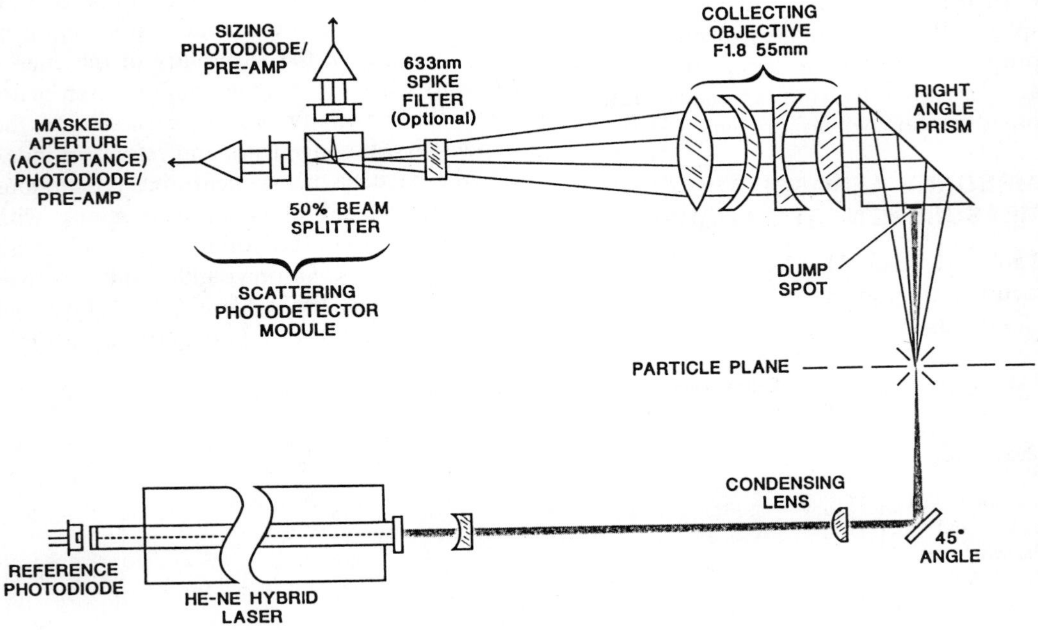

FIGURE 31-1. Optical Diagram of the Forward-Scattering Spectrometer Probe (FSSP). The Particle Plane is in the Free Stream. (Courtesy of Particle Measuring Systems Inc., Boulder, CO.)

Techniques Requiring Sampling and Transport of Aerosols to the Inside of the Aircraft

Many instruments normally used on the ground have been used at tropospheric pressures aboard aircraft. These instruments include Aitken nuclei counters (ANCs), condensation nuclei counters (CNCs), optical particle counters (OPCs), differential mobility analyzers (DMAs), diffusion batteries (DBs), cloud condensation nuclei counters, ice nuclei counters, aerodynamic particle sizers (APSs), electrostatic precipitators, nephelometers, optical extinction cells, aethelometers, impactors, quartz crystal microbalances, and filtration devices (Torgeson and Stern 1966; Radke and Hobbs 1969; Reagan et al. 1977; Farlow et al. 1979; Wilson, Hyun, and Blackshear 1983; Woods and Chuan 1983; Heintzenberg and Ogren 1985; Dreiling and Jaenicke 1988; Hudson 1989; Leaitch, Hoff, and MacPherson 1989; Brock, Radke, and Hobbs 1990; Noone and Hansson 1990).

When aerosol measuring instruments designed for use at one atmosphere pressure are operated at reduced pressures, the assumptions and approximations used in the design of the instruments may no longer hold. For example, Heintzenberg and Ogren (1985) recommend that the TSI Model 3020 CNC (TSI Inc., PO Box 64204, St. Paul, MN 55164) not be used at pressures less than ~ 250 mb. Noone and Hansson (1990) have calibrated the TSI Model 3760 CNC at various pressures and find that it also functions at pressures down to 250 mb. At higher altitudes and lower pressures, other designs are required (Wilson, Hyun, and Blackshear 1983). With some instruments such as diffusion batteries and impactors it is possible to predict the effects of pressure on instrument response. However, in the case of CNCs and instruments dependent on charging, it is not possible with current knowledge to predict the response of instruments as a function of pressure over a wide range of pressures. Thus, if the operating pressure of the instruments changes significantly, it is necessary to calibrate the instruments at the actual operating pressures.

Instruments used on aircraft are often exposed to varying temperatures and pressures. Controlling sample flow may require special effort. Instruments may encounter significant vibration and electronic noise, and in many cases, they may have to meet stringent airworthiness requirements. Arcing of high-voltage equipment may become a problem at low atmospheric pressures. Investigators must often take special care to control electronic noise, to monitor sample flow, to ensure that adequate heating or cooling is available, and to make measurements of the temperatures and pressures experienced in flight. Thus, they can calibrate and test instrument performance under the conditions experienced in the air. These steps are often critical for establishing confidence in the measurements.

Cloud Measurements Made Outside the Aircraft

The bulk liquid water content (LWC) of clouds, and the size distribution, and shape and phase of the cloud particles are often measured. A fairly comprehensive review of modern airborne cloud probing techniques may be found in Knollenberg (1981).

One common method of measuring the LWC of clouds is by sensing the amount of heat necessary to vaporize droplets which have been collected on a probe in the airstream. The Johnson–Williams (J–W) probe manufactured by Cloud Technology, Inc. (606 Wellsbury Ct., Palo Alto, CA 94306) (Strapp and Schemenauer 1982) and the King probe made by the Commonwealth Scientific and Industrial Research Organization of Australia (King, Parkin, and Handsworth 1978; King, Maher, and Hepburn 1981; King, Dye, and Strapp 1985) are accurate for cloud droplets with $d_p < 30$ μm, and underestimate the LWC for larger particles (Knollenberg 1981). The mass of water within the cloud and precipitation particles may also be measured by exposing a chemically treated electrolytic paper tape to impinging droplets and rolling the paper between two electrodes. The electrical resistance is related to the mass of liquid water absorbed by the paper. This method is most efficient for larger cloud droplets and raindrops, and is relatively insensitive at liquid water contents less than 0.5 g/m^3 (Knollenberg 1981).

The LWC of supercooled cloud droplets is commonly measured with Rosemount ice detectors (Model 871, Rosemount Inc. Aerospace Division, 14300 Judicial Road, Burnsville, MN 55337). This device is composed of an ultrasonically vibrating probe that extends into the flow around an aircraft. Supercooled droplets freeze upon contact with the sensor and change the frequency of oscillation. The LWC of supercooled cloud droplets is determined from the rate of accretion of ice.

The size distribution of liquid cloud particles is often measured in the free stream with optical instruments. The PMS FSSP (Particle Measuring Systems Inc., 1855 57 th St., Boulder, CO, 80301) has already been discussed above. Another instrument type includes the PMS one-dimensional optical array probes (1-D OAP). In these instruments, a linear array of small photodiodes, oriented perpendicularly to the airflow, is illuminated by a He–Ne laser. The viewing volumes of these instruments are exterior to the aircraft. Particles passing between the laser and the diodes occult individual diodes. The number of shadowed diodes is determined by the width of the particle. In some cases, particles may not be appropriately registered due to the limitation of the speed of the electronics. Ice crystals, particularly those with dendritic habits, can result in partial occultation of diodes, producing relatively large errors in the particle size reported by the OAP probes.

The cloud gun is used to determine the size distribution of small cloud droplets (Squires and Gillespie 1952; Clague 1965; Baumgardner 1983). A soot-covered slide is briefly exposed as the aircraft passes through a cloud or precipitation by shooting it across a sampling aperture located outside the boundary layer. Imprints left by the impact of droplets on the soot-covered slide are photographed and analyzed. Baumgardner (1983) estimates an uncertainty of 17% for particle number and 32% for total LWC using this device. Other techniques that are most effective for rain and snow particles include aluminum foil impactors (Schecter and Russ 1970), cloud particle cameras (Cannon 1974), and a device to artificially deposit and then measure electric charges on droplets (Keily and Millen 1960).

PMS 2-D OAP probes (Particle Measuring Systems Inc., 1855 57 th St., Boulder, CO, 80301), Fig. 31-2, have been used extensively to determine the size, shape, and crystalline habit of cloud particles. These instruments are similar to the PMS 1-D OAP probes, except that the output of the individual diodes of the array is recorded as a function of time as individual particles pass. Images of individual particles can then be constructed. The smallest particle that can be detected by one of these probes is $\sim 50\ \mu m$.

For highly detailed studies of cloud microphysics, it is often desirable to collect frozen particles or high-resolution replicas of the particles on impaction surfaces. The most common technique uses Formvar, a plastic material dissolved in solvent (Schaefer 1956). Formvar-coated film is exposed to the airstream. Ice crystals and liquid droplets impact and are collected, the Formvar solution flows over the particles, the solvent rapidly evaporates, and the particle usually evaporates or sublimates, leaving both a replica of the particle in plastic (MacCready and Todd 1964) and any nonvolatile chemical components contained in the original particle. Difficulties associated with Formvar replication include particle agglomeration and growth after impaction, maintaining the proper amount of solvent on the film, and the hazardous nature of common Formvar solvents.

It is often important to discriminate between liquid cloud droplets and frozen, although perhaps still nearly spherical, cloud particles at high-spatial and temporal resolution. The University of Washington (UW) ice particle counter (Turner and Radke 1973) uses an orthogonal pair of polarizing lenses to detect the presence of birefringent crystals.

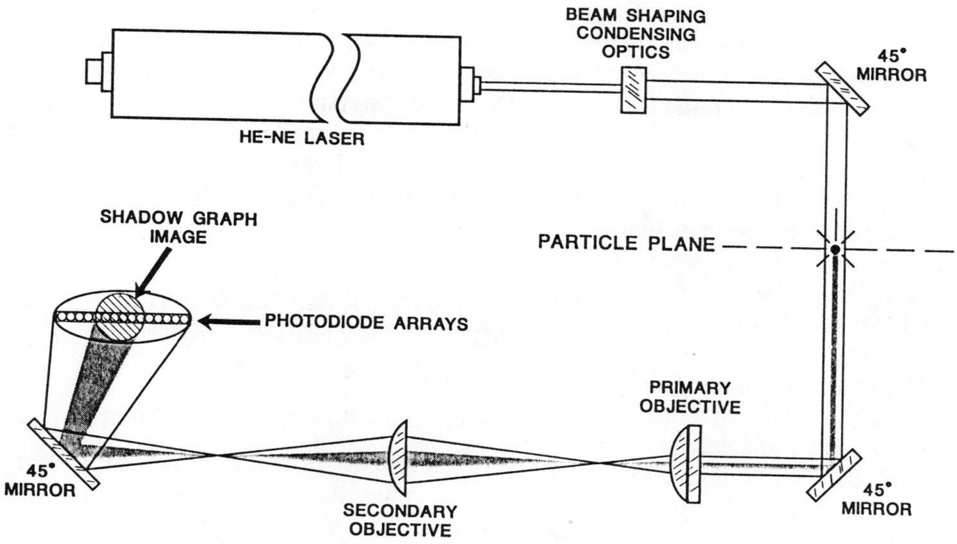

FIGURE 31-2. Optical Diagram of 2-D Optical Array Probe (OAP). The Particle Plane is in the Free Stream. (Courtesy of Particle Measuring Systems Inc., Boulder, CO.)

One lens, placed between the light source and the sensing volume, permits only plane polarized light to illuminate the particles. The second polarizer is orthogonal to the first and is in front of the photodetector. The polarization of any light reaching the photodetector has been rotated by scattering from birefringent crystals. Thus, in theory, the device only counts ice crystals. (In practice, multiple scattering of light on droplets within the viewing volume may lead to false readings of ice particle concentrations.) Two similar instruments—the Mee Industries Model 120 ice crystal counter (Mee Industries Inc., 1629 S. Del Mar, San Gabriel, CA 91776) and the SPEC Model IPC-IV ice particle counter (SPEC, Inc., 5401 Western, Boulder, CO 80301)—are commercially available.

A version of the PMS OAP-2-D cloud particle imaging probe also uses birefringent depolarization of light by ice crystals to discriminate between the liquid and solid phases (Knollenberg 1981). In the PMS instrument, the incident polarized light is provided by a polarized laser source, and a separate photodetector preceded by a Thompson polarizing prism is used to sense the depolarized light scattered by ice crystals.

Another device measures the electric charge generated upon the impact of ice particles with a wire sensor to infer the presence of ice in clouds (McTaggart-Cowan, Lala, and Vonnegut 1970; Jones, Grotbeck, and Vonnegut 1989). While there is some uncertainty regarding the mechanism of the charge generation, the technique appears to be capable of discriminating liquid water droplets from ice crystals, and gives at least a qualitative indication of the presence of ice in clouds. The Cannon Cloud Particle Camera (Cannon 1974) is also effective at discriminating between liquid and frozen cloud particles.

Cloud Measurements Requiring Collection of Samples

For studies of cloud chemistry, it is often necessary to collect cloud water in flight and return it to the laboratory for detailed analysis. A commonly used device for bulk cloud water sampling is the Mohnen slotted rod impactor which consists of cylindrical slotted Delrin or Teflon rods (Huebert and Baumgardner 1985; Huebert, Vanbramer, and Tschudy 1988). The rods are slightly tilted into the free stream during flight with the slots facing forward. Cloud droplets that strike the front face of the impactor are blown down the slots into a receptacle within the aircraft. Huebert et al. (1988) found collection efficiencies of 91% when total cloud LWC values were $> 0.1\ g/m^3$.

For the collection of supercooled cloud droplets, the Mohnen collector is ineffective, as most of the droplets will freeze upon contact with a surface. In such circumstances, simple cylindrical rods, usually made of Teflon, are exposed to the airstream where they accumulate ice through riming. The rods are then retracted and the ice is scraped into a receptacle for chemical analysis.

The counterflow virtual impactor (CVI) has been developed to sample cloud particles while rejecting the interstitial aerosol (Ogren, Heintzenberg, and Charlson 1985; Noone et al. 1988; Twohy et al. 1990). Air containing both cloud droplets and interstitial aerosol particles is impinged upon the tip of a circular sampling inlet (Fig. 31-3). A counterflow of a carrier gas is constantly emitted from the tip of the inlet against the onrushing airflow. Particles with large stopping distances (i.e., large Stokes numbers) penetrate the region of counterflowing air and are drawn into a sample airstream; smaller particles are excluded. The particulate residue from the evaporated cloud drops is then analyzed chemically, optically, or microscopically. Additionally, the gases evaporated from the droplets can be analyzed if sensitive enough techniques are available. By varying the exit flow from the CVI, and, hence, the location of the plane of stagnation within the inlet tip, the diameter of the smallest droplet sampled can be controlled. Thus, in principle, size-dependent samples of cloud droplet residues can be collected.

EFFECTS OF AIRFLOW ON ACCURATE AEROSOL AND CLOUD MEASUREMENTS

Effects of Airflow Around the Aircraft

There are two principal issues to be considered: flow distortion caused by the aerodynamic effects of the aircraft and sensor mount, and misalignment of the instrument or probe with the oncoming airstream.

Several experimental and theoretical studies have examined the effects of the airflow distortion on airborne measurements (Beard 1983; Baumgardner 1984; King 1984, 1986*a*, *b*; Drummond 1984; Drummond and MacPherson 1984, 1985; MacPherson and Baumgardner 1988; Norment 1988). Convergence and divergence of streamlines around the fuselage, wings, and instrument pods can lead to areas of enhanced particle concentrations as well as zones where concentrations are greatly reduced. These effects result from the fact that particles have significant inertia

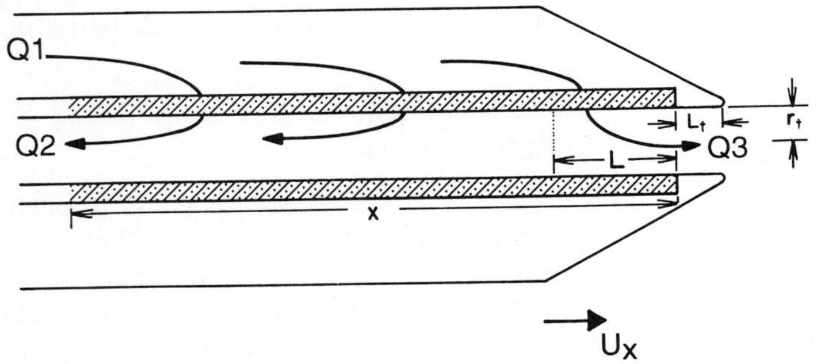

FIGURE 31-3. Counterflow Virtual Impactor (CVI). The Probe Moves Through the Air at a Velocity U_x. Flow $Q1$ is Filtered Air and Passes Through the Porous Section of Length x. Flow $Q3$ Goes Out the End. Particles with Sufficient Inertia to Traverse Distance $L_t + L$ are Entrained in Flow $Q2$ and Carried to Instruments. (Courtesy of C. Twohy, National Center for Atmospheric Research, Boulder, CO.)

and may not be able to follow the flow streamlines. The effects are more significant for particles with large Stokes numbers. Thus, one cannot assume that samples drawn from beyond the aircraft's boundary layer will contain a representative particle sample. Rather, if the streamlines are bent upstream of the sample point due to the aircraft or probe, the sampled concentration may differ from that in the free stream. The effects vary substantially from aircraft to aircraft depending on fuselage shape and diameter, wing thickness and aspect ratio, placement of the engines and propellers, and the location of appendages and of the sample point.

It is also possible to use flow distortion to achieve a desired result. For example, Fahey et al. (1989) used an aerodynamic shape to remove large particles from the air reaching the sample point.

Distortion of the airflow field can lead to rotational motion, reorienting or deforming the particles being measured. Rotation of ice crystals has been noted especially for 2-D imaging probes mounted near the wing tips of aircraft, where the developing wingtip vortices impart rotational and along-wing components to the flow. Figure 31-4 shows laser shadow images of ice particles from a PMS OAP-2-D probe mounted near the tip of the King Air aircraft operated by the National Center for Atmospheric Research. The ice particles have apparently been rotated $\sim 60°$ by the incipient wing tip vortex.

Many instruments and probes are designed to point into the airflow. Proper installation of such a sensor requires that the orientation of the streamlines at the sample point be known. Numerical modeling, wind tunnel testing, and in-flight measurements can be used to examine the airflow field around specific instrument locations on research aircraft. Care should be taken to understand the variations that can occur. The angle of attack of the aircraft may vary substantially, depending upon the altitude, the rate of descent or climb, aircraft weight, true air speed, and the presence of turbulence, while aircraft yaw may vary with wind speed, turn coordination, etc. It is possible to measure the wind vector at a particular location on the aircraft with an instrumented probe such as the Rosemount 858 Flow Angle Sensor (Rosemount Inc., Burnsville, MN). The wind vector should be characterized for all anticipated sampling conditions since angle of attack may vary considerably.

Sampling locations that minimize airflow distortion include: below the center of each wing, on the lower aircraft fuselage, on the upper fuselage ahead of the trailing edge of the wings but well behind the cockpit, and extending into the airstream ahead of the aircraft nose. These locations minimize the influences of wing tip vortices and the distorted flow created by the nose and windshield areas on the fuselage. Less preferable locations include: the wing tip, the near-wing tip, forward, dorsal fuselage mounts, and regions of propwash and engine exhaust. The area aft of the wing along the side of and over the top of the fuselage (a region which may be swept by horseshoe vortices produced at the juncture of the wing root and fuselage) should also be avoided, as should points on the airframe where electrical discharges are likely.

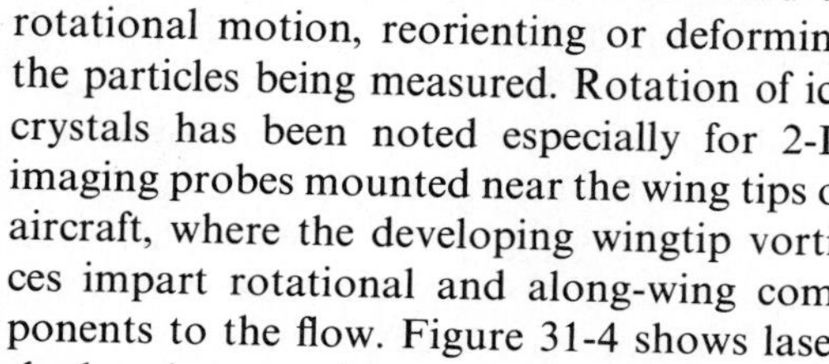

FIGURE 31-4. 2-D Optical Array Probe (OAP) Images of Ice Particles which have Been Rotated in a Flow Near the Wing Tip. (From MacPherson and Baumgardner 1988. Reproduced by permission of the American Meteorological Society.)

Effects of Sensors on Airflow

The sensor housing may also distort the airflow. The airflow effects of the PMS FSSP and OAP cloud and precipitation probes have been studied in some detail (MacPherson 1985; King 1986*b*; MacPherson and Baumgardner 1988; Norment 1988). As an example, the FSSP sensing volume is housed within a cylindrical tube. Based on detailed numerical simulation of the airflow around this tube, its support arms, and the FSSP nacelle itself (Norment 1988), the instrument was found to undercount particles with $d_p > 20\ \mu m$ by $\sim 10\%$. Norment also found that approximately half of the flux and speed distortions at the point of measurement were caused by the presence of the sensor housing. Additionally, Norment pointed out that distortions induced by the aircraft and those caused by the sensor interact synergistically, and, thus, should not be evaluated independently.

AEROSOL INLETS ON AIRCRAFT

In many instances, particles must be brought from the outside of the aircraft to the instruments located within the fuselage or instrument pods. The strictest requirements for such sampling are that the number, surface, mass, and volume mixing ratios be the same at the instrument in the aircraft as they are in the ambient air, and that the mass, volume, shape, phase, and composition of individual particles be unchanged. (Mixing ratios are defined as the measured quantity—such as aerosol number, surface, or volume—per unit *mass* of air sampled. Since aircraft sampling often involves compressible flow or changes in temperature and pressure, it is preferable to express measurements in terms of mixing ratios, rather than volume concentrations, as mixing ratios do not change if temperature and pressure change.)

In order to meet the above standard for sampling, it is necessary firstly that the sampling be representative and secondly that individual particles not be modified. Representative sampling occurs when particles are able to follow the mass flow streamlines from the ambient environment to the instrument. Particle modification occurs when changes in pressure or temperature cause particles to change phase or cause chemical species to evaporate from or condense on particles. There are no proven techniques for achieving representative sampling of unmodified particles with aircraft. In this section, the problems faced by those wishing to achieve this are briefly discussed.

Representative Sampling

Diffusing Inlets

The air velocity acceptable at the instrument or collection device in the aircraft is often ten or a hundred times less than the air speed of the aircraft. Diffusers are often employed near to or directly behind the sample intake to slow the flow (Fig. 31-5). Effects which can influence the performance of these commonly used inlets are discussed in this section.

The ability of the aircraft to alter the streamlines and to reduce or enhance concentrations along streamlines has been noted above. Sample inlets must be carefully located if the inlet is to draw samples from unperturbed air.

The velocity of air relative to the aircraft is often between 50 and 225 m/s. Due to the

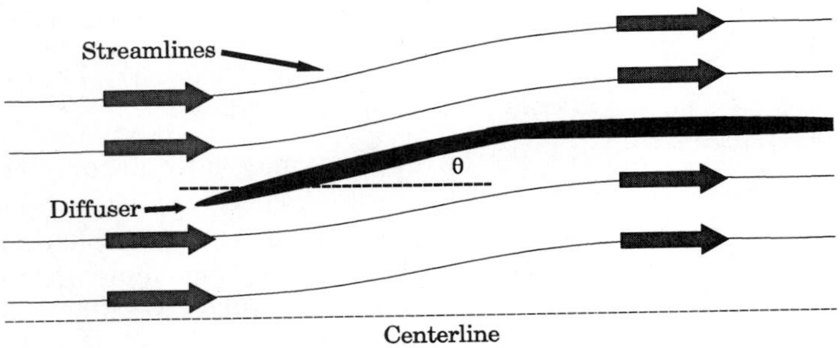

FIGURE 31-5. Diffusing Inlet. The Expansion Angle, θ, is shown. The Included Angle Equals 2θ.

large Stokes numbers which result from these velocities, many experimenters strive for isokinetic and isoaxial sampling at the inlet (see Chapter 7). Achieving isoaxial sampling is complicated by the effects, discussed above, of the flow around the aircraft.

Once inside the inlet, particles can be lost due to inertial deposition in both turbulent and laminar flows, and through diffusional losses. Although these effects are described in Chapter 7 for flow just inside tubular inlets and in long tubes and bends, aerosol loss mechanisms in diffusers have not been quantified. There is evidence that such losses can be significant. Huebert, Lee, and Warren (1990) report losses of 50%–90% of total mass inside the tip of sharp-edged diffusers. A surprising aspect of these results was that chemical species normally associated with particles having diameters less than 1.0 μm were heavily deposited on the inside of the diffuser. It had been previously assumed that particles of this size were sampled and transported to the instruments with high efficiency.

Specific mechanisms of particle deposition in conical, diffusing inlets have not been evaluated and quantitative studies of flows in conical diffusers with large area ratios (defined below) are lacking. However, qualitative discussions of important flow regimes and their likely impact on particle loss are possible. Flow regimes in diffusers are reviewed by Ward-Smith (1980) and are briefly discussed here.

In attached flow, the local flow velocity at all points is in the same direction as the mean flow; that is, there are no regions of reversed, or recirculating, flow. Designers of aerosol inlets often seek attached flow and assume that their design produces attached flow as a consequence of choosing an included angle (total angle) of 7°, which is an expansion angle (half-angle) of 3.5°. However, the occurrence of separation in a diffuser depends upon more than the expansion angle. The "area ratio"—the ratio of the cross-sectional area of the inlet after expansion to that prior to expansion—is also important. Using the data provided by Ward-Smith (1980), we would predict that separation occurs in turbulent flow in 2-D diffusers with 7° included angles and area ratios larger than about 2.5 at typical airplane velocities.

Figure 31-6 shows separated flow in a conical diffuser. In such a flow, velocities can be very erratic and gross fluctuations of the entire pattern are expected. Note that this regime both creates laminar velocity components perpendicular to the wall that may enhance inertial deposition and provides more random fluctuations which can enhance turbulent inertial deposition of particles. Although experimental comparisons of particle losses in attached and separated diffuser flow have not been published, it is likely that the attached flow would result in smaller particle losses.

Other factors influence whether or not flow separates in a diffusing inlet. The shape of the leading edge of the inlet is very important when there is any chance of misalignment of the inlet with the airstream. Many aerosol

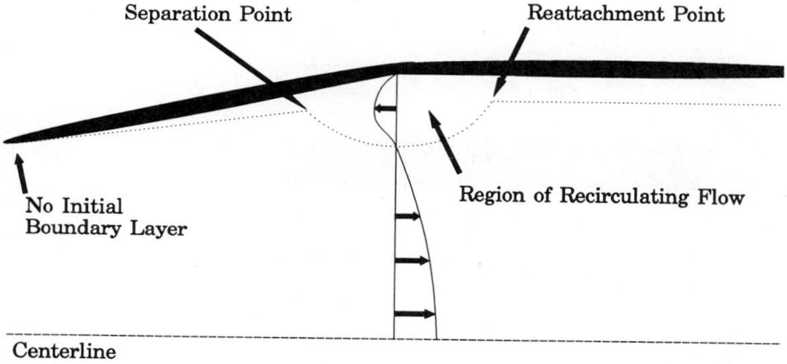

FIGURE 31-6. Separated Flow in a Diffusing Inlet.

scientists prefer sharp edges on inlets, since sharp edges avoid large stagnation regions which are associated with bent streamlines and nonrepresentative sampling. However, sharp edges on diffusing inlets can exacerbate conditions leading to separation in diffusers. Pena, Norman, and Thomson (1977) describe a rounded leading edge which appears to prevent separation at angles up to 8° between the airstream and the axis of the inlet.

Rader and Marple (1988) have studied the effects of blunt-tipped probes, and show that bluntness may improve the aerosol sampling efficiency of sub-isokinetic samplers, but degrades the performance of isokinetic samplers. Blunt-tipped probes are discussed in Chapter 7.

The optimum shape of the leading edge of a diffusing inlet has not been found, but is likely to be a compromise between the rounded edge which helps prevent separation in non-isoaxial flows and sharp edges which result in aspiration coefficients of unity at isokinetic conditions.

Turbulent flows within inlets may, from an aerosol sampling point of view, be desirable in some cases because laminar boundary layers are more susceptible to separation than are turbulent boundary layers. Although the initial boundary layer inside the diffuser is always laminar, the boundary layer further downstream may, depending on the airspeed, geometry, and altitude, become laminar, transitional, or turbulent. Although turbulent boundary layers favor attached flow, they may also enhance particle loss through turbulent deposition. Again, the data required to evaluate this issue are lacking.

Until data describing aerosol deposition in diffusers are available, it is likely that attached, laminar flow will be sought in inlet diffusers. The inlet designer should be cautioned, however, that simply specifying an included angle of 7° is not sufficient to guarantee attached, laminar flow.

Shrouded Inlets

Shrouded probes, Fig. 31-7, have been used on aircraft to reduce the effects of angle-of-attack variations and to reduce the velocity of the airstream (Torgeson and Stern 1966). McFarland et al. (1989) tested a shrouded probe in a wind tunnel at velocities up to 15 m/s and found that it performed better than a simple, isokinetic probe. Quantitative studies at aircraft wind speeds are needed to evaluate this approach as well.

High-Speed Inlets

Use of inlets in transonic and supersonic flows are reported by Martone, Daley, and Boubel (1980), Martone (1978), and Ivie Forney, and Roach (1990). Many of the same issues faced in subsonic diffusing inlets remain to be solved in the transonic and sonic regimes.

Modification of Particles

In cases where compressible effects are noticeable, the particles will be heated as they are decelerated. Even in incompressible sampling,

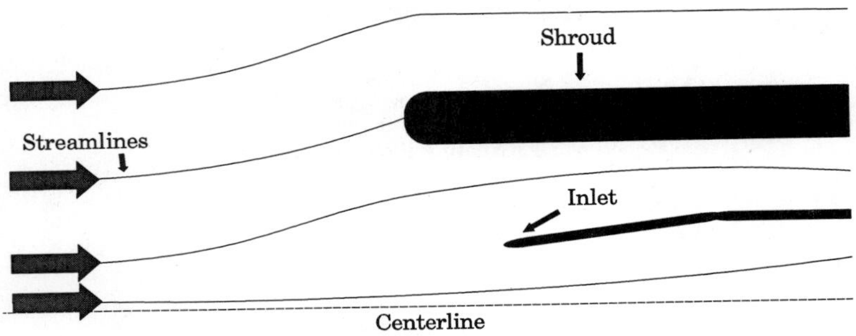

FIGURE 31-7. Flow Around a Shrouded Inlet

temperature differences between the inside and outside of the aircraft may be significant. Since ambient aerosols often contain components which are in equilibrium with a vapor at ambient conditions, changing the temperature during sampling can also change the composition, phase, size, shape, or optical properties.

Such sampling effects are exemplified in Fig. 31-8 for the case of stratospheric particles. For this case, particle sampling occurred at Mach 0.7; thus, the particles were heated by compressive heating from 197 K to approximately 217 K. The aerosol was transported at that temperature for 0.2 s and was then heated to 273 K for 0.01 s before reaching the beam of an optical particle counter. The secondary heating was caused by convective and conductive heat transfer from the relatively warm (273 K) instrument sampling chamber to the air sample. For this calculation, the particles were assumed to consist of a supercooled solution of sulfuric acid and water in equilibrium with ambient water vapor. The horizontal axis in Fig. 31-8 shows the diameter of the particles when they reach the laser, while the vertical axis shows the ratio of the diameter at ambient conditions to that at the laser as calculated for the described temperature history. This calculation shows that dramatic changes in composition, size, and in many cases, refractive index, can occur even during very rapid sampling and analysis of volatile particles.

CONCLUSIONS

Many aerosol instruments and techniques developed and used in the laboratory have been adapted for use in aircraft. Such adaptations are successful when the experimenters carefully evaluate the effects of the aircraft environment on instrument response and performance. Changes in pressure and temperature are often experienced in aircraft

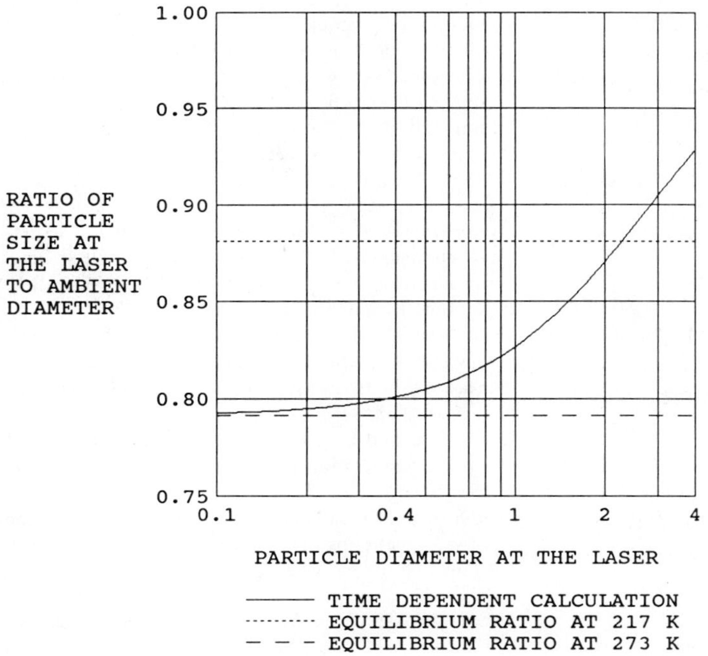

FIGURE 31-8. Changes in Sulfuric Acid and Water Particles During Sampling and Transport at Mach 0.7. The Time-Dependent Calculation Accounts for the Residence Times at Each Condition. The Equilibrium Calculations Assume that the Particles Adjust Completely to the Indicated Condition.

applications, and may affect the response of the instrument. When successful theories are available, as in the cases of diffusion batteries or incompressible flow impactors, these effects may be predicted with adequate accuracy. In other cases, the instrument response at the relevant temperatures and pressures should be determined in the laboratory.

Instrument performance may be degraded by vibration, electronic noise, and altered heat and mass transfer efficiencies experienced in flight. Experimenters should acquire sufficient housekeeping data to determine the actual conditions experienced in flight so that they can determine if the instrument is operating as expected.

The airflow around the aircraft introduces complications in sampling. Merely being outside the boundary layer of the aircraft does not ensure that the air sampled is representative of air away from the plane, since bending of streamlines by the aircraft can alter particle concentrations. Locations for sampling aerosol and cloud particles must be chosen with these effects in mind. Even instruments specifically designed for use on aircraft may exhibit unanticipated aerodynamic effects.

A number of problems associated with aircraft inlets for representative internal sampling of aerosol particles remain to be solved. At this point there is no widely accepted recipe for constructing aircraft inlets which provide representative aerosol samples. More work needs to be done in this area.

The sampling of aerosol particles for measurement in the airplane or for later analysis often subjects the particles to changing temperatures or relative humidities. Volatile components may condense or evaporate, even in the short times between sampling and measurement. These effects also need to be accounted for through both modeling and measurement.

References

Bailey, I. H., L. F. Radke, J. H. Lyons, and P. V. Hobbs. 1984. Airborne observations of arctic aerosols. II: Giant particles. *Geophys. Res. Letts.* 11:397–400.

Baumgardner, D. 1983. An analysis and comparison of five water droplet measuring instruments. *J. Clim. Appl. Meteorol.* 22:891–10.

Baumgardner, D. 1984. The effects of airflow distortion on aircraft measurement: A workshop summary. *Bull. Am. Meteor. Soc.* 65:1212–13.

Baumgardner, D., W. Strapp, and J. E. Dye. 1985. Evaluation of the forward scattering spectrometer probe. Part II: Corrections of coincidence and dead-time losses. *J. Atmos. Oceanic Technol.* 2:626–32.

Beard, K. V. 1983. Reorientation of hydrometeors in aircraft accelerated flow. *J. Clim. Appl. Meteor.* 22:1961–63.

Brenguier, J. L. 1989. Coincidence and dead-time corrections for particle counters. Part II: High concentration measurements with and FSSP. *J. Atmos. Oceanic Technol.* 6:585–98.

Brenguier, J. L. and L. Amodei. 1989. Coincidence and dead-time corrections for particle counters. Part I: A general mathematical formalism. *J. Atmos. Oceanic Technol.* 6:575–84.

Brock, C. A., L. F. Radke, and P. V. Hobbs. 1990. Sulfur in particles in arctic hazes derived from airborne in situ and lidar measurements. *J. Geophys. Res.* 95:22, 369–22, 387.

Cannon, T. W. 1974. A camera for photography of atomospheric (sic) particles from aircraft. *Rev. Sci. Instrum.* 45:1448–55.

Cerni, T. A. 1983. Determination of the size and concentration of cloud drops with an FSSP. *J. Appl. Meteor.* 22:1346–55.

Clague, L. F. 1965. An improved device for obtaining cloud droplet samples. *J. Appl. Meteorol.* 4:549–51.

Dreiling, V. and R. Jaenicke. 1988. Aircraft measurement with condensation nuclei counter and optical particle counter. *J. Aerosol Sci.* 19:1045–50.

Drummond, A. M. 1984. Aircraft flow effects on cloud droplet images and concentrations. Aeronautical Note NAE-AN-21 (NRC No. 23508), National Research Council Canada, p. 30.

Drummond, A. M. and J. E. MacPherson. 1984. Theoretical and measured airflow about the Twin Otter wing. Aeronautical Note NAE-AN-19 (NRC No. 33184), National Research Council Canada, p. 33.

Drummond, A. M. and J. E. MacPherson. 1985. Aircraft flow effects on cloud drop images and concentrations measured by the NAE Twin Otter. *J. Atmos. Oceanic Technol.* 2:633–43.

Dunn, P. F. and K. J. Renken. 1987. Impaction of solid aerosol particles on fine wires. *Aerosol. Sci. Technol.* 7:91–107.

Dye, J. E. and D. Baumgardner. 1984. Evaluation of the forward scattering spectrometer probe. Part I: Electronic and optical studies. *J. Atmos. Oceanic Technol.* 1:329–44.

Fahey, D. W., K. K. Kelly, G. V. Ferry, L. R. Poole, J. C. Wilson, D. M. Murphy, M. Loewenstein, and K. R. Chan. 1989. *In situ* measurements of total reactive nitrogen, total water and aerosol in a polar stratospheric cloud in the Antarctic. *J. Geophys. Res.* 94:11, 299–11, 315.

Farlow, N. H., G. V. Ferry, H. Y. Lem, and D. M. Hayes. 1979. Latitudinal variations of stratospheric aerosols. *J. Geophys. Res.* 84:733–43.

Heintzenberg, J. and J. A. Ogren. 1985. On the operation of the TSI-3020 condensation nuclei counter at altitudes up to 10 km. *Atmos. Environ.* 19:1385–87.

Hering, S. V. 1987. Calibration of the QCM impactor for stratospheric sampling. *Aerosol Sci. Technol.* 7:257–74.

Hudson, J. G. 1989. An instantaneous CCN spectrometer. *J. Atmos. Oceanic Technol.* 6:1055–65.

Huebert, B. J. and D. Baumgardner. 1985. A preliminary evaluation of the Mohnen slotted-rod water collector. *Atmos. Environ.* 19:843–46.

Huebert, B. J., S. Vanbramer, and K. L. Tschudy. 1988. Liquid cloudwater collection using modified Mohnen slotted rods. *J. Atmos. Chem.* 6:251–63.

Huebert, B. J., G. L. Lee, and W. L. Warren. 1990. Airborne aerosol inlet passing efficiency measurement. *J. Geophys. Res.* 95:16, 369–16, 381.

Ishizaka, Y. A., P. V. Hobbs, and L. F. Radke. 1989. Arctic hazes in summer over Greenland and the North American Arctic: II. Nature and concentrations of accumulation-mode and giant particles. *J. Atmos. Chem.* 9:149–59.

Ivie, J. J., L. J. Forney, and R. L. Roach. 1990. Supersonic particle probes: Measurement of internal wall losses. *Aerosol Sci. Technol.* 13:368–85.

Jones, J. J., C. Grotbeck, and B. Vonnegut. 1989. Airplane instrument to detect ice particles. *J. Atmos. Oceanic Technol.* 6:545–51.

Keily, D. P. and S. G. Millen. 1960. An airborne cloud-drop-size distribution meter. *J. Meteor.* 17:349–56.

King, W. D. 1984. Airflow and particle trajectories around aircraft fuselages. I: Theory. *J. Atmos. Oceanic Technol.* 1:5–13.

King, W. D. 1986*a*. Airflow and particle trajectories around aircraft fuselages. IV: Orientation of ice crystals. *J. Atmos. Oceanic Technol.* 3:439.

King, W. D. 1986*b*. Airflow around PMS canisters. *J. Atmos. Oceanic Technol.* 3:197–98.

King, W. D., D. A. Parkin, and R. J. Handsworth. 1978. A hot-wire liquid water device having fully calculable response characteristics. *J. Appl. Meteorol.* 17:1809–13.

King, W. D., C. T. Maher, and G. A. Hepburn. 1981. Further performance tests on the CSIRO liquid water probe. *J. Appl. Meteorol.* 20:195–202.

King, W. D., J. E. Dye, J. W. Strapp, D. Baumgardner, and D. Huffman. 1985. Icing wind tunnel tests on the CSIRO liquid water probe. *J. Atmos. Oceanic Technol.* 2:340–52.

Knollenberg, R. G. 1981. Techniques for probing cloud microstructure. In *Clouds, Their Formation, Optical Properties, and Effects*, eds. P. V. Hobbs and A. Deepak, pp. 15–91. New York: Academic Press.

Leaitch, W. R., R. M. Hoff, and J. I. MacPherson. 1989. Airborne and lidar measurements of aerosol and cloud particles in the troposphere over Alert, Canada in April 1986. *J. Atmos. Chem.* 9:187–211.

MacCready, P. B. Jr. and C. J. Todd. 1964. Continuous particle sampler. *J. Appl. Meteorol.* 3:450–60.

MacPherson, J. I. 1985. Wind tunnel calibration of a PMS canister instrumented for airflow measurement. Aeronautical Note NAE-AN-32 (NRC No. 24922), National Research Council Canada.

MacPherson, J. E. and D. Baumgardner. 1988. Airflow about King Air wingtip-mounted cloud particle measurement probes. *J. Atmos. Oceanic Technol.* 5:259–73.

Martone, J. A. 1978. Subisokinetic sampling errors for aircraft turbine engine smoke probes. *J. Air Pollut. Control Assoc.* 28:607–09.

Martone, J. A., P. S. Daley, and R. W. Boubel. 1980. Sampling submicrometer particles suspended in near sonic and supersonic free jets. *J. Air Pollut. Control Assoc.* 30:898–903.

McFarland, A. R., C. A. Ortiz, M. E. Moore, R. E. DeOtte, Jr., and S. Somasundaram. 1989. A shrouded aerosol sampling probe. *Environ. Sci. Technol.* 23:1487–92.

McTaggart-Cowan, J. D., G. G. Lala, and B. Vonnegut. 1970. The design, construction, and use of an ice crystal counter for ice crystal cloud studies by aircraft. *J. Appl. Meteorol.* 9:294–99.

Noone, K. J. and H.-C. Hansson. 1990. Calibration of the TSI 3760 condensation nucleus counter for nonstandard operating conditions. *Aerosol Sci. Technol.* 13:478–85.

Noone, K. J., R. J. Charlson, D. S. Covert, J. A. Ogren, and J. Heintzenberg. 1988. Design and calibration of a virtual impactor for sampling of atmospheric fog and cloud droplets. *Aerosol. Sci. Technol.* 8:235–44.

Norment, H. G. 1988. Three-dimensional trajectory analysis of two drop sizing instruments: PMS OAP and PMS FSSP. *J. Atmos. Oceanic Technol.* 5:743–56.

Ogren, J. A., J. Heintzenberg, and R. J. Charlson. 1985. *In situ* sampling of clouds with a droplet to aerosol converter. *Geophys. Res. Letts.* 12:121–24.

Pena, J. A., J. M. Norman, and D. W. Thomson. 1977. Isokinetic sampler for continuous airborne aerosol measurements. *J. Air Pollut. Control Assoc.* 27:337–41.

Pinnick, R. G., D. M. Garvey, and L. D. Duncan. 1981. Calibration of Knollenberg FSSP light-scattering counters for measurements of cloud droplets. *J. Appl. Meteorol.* 20:1049–57.

Pueschel, R. F., K. G. Snetsinger, J. K. Goodman, O. B. Toon, G. V. Ferry, V. R. Oberbeck, J. M. Livingston, S. Verma, W. Fong, W. L. Starr, and K. R. Chan. 1989. Condensed nitrate, sulfate, and chloride in Antarctic stratospheric aerosols. *J. Geophys. Res.* 94:11, 271–11, 284.

Rader, D. J. and V. A. Marple. 1988. A study of the effects of anisokinetic sampling. *Aerosol. Sci. Technol.* 8:293–99.

Radke, L. F. and P. V. Hobbs. 1969. An automatic cloud condensation nuclei counter. *J. Appl. Meteorol.* 8:105–09.

Reagan, J. A., J. D. Spinhirne, D. M. Byrne, D. W. Thomson, R. G. de Pena, and Y. Mamane. 1977.

Atmospheric particulate properties inferred from lidar and solar radiometer observations compared with simultaneous *in situ* aircraft measurements: A case study. *J. Appl. Meteorol.* 16:911–28.

Schaefer, V. J. 1956. Preparation of snow crystal replicas—VI. *Weatherwise.* 9:132–35.

Schecter, R. M. and R. G. Russ. 1970. The relationship between imprint size and drop diameter for an airborne drop sampler. *J. Appl. Meteorol.* 9:123–26.

Squires, P. and C. A. Gillespie. 1952. A cloud droplet sampler for use on aircraft. *Quart. J. Roy. Meteor. Soc.* 78:387–92.

Strapp, J. W. and R. S. Schemenauer. 1982. Calibrations of Johnson–Williams liquid water content meters in a high-speed icing tunnel. *J. Appl. Meteorol.* 21:98–108.

Torgeson, W. L. and S. C. Stern. 1966. An aircraft impactor for determining the size distributions of tropospheric aerosols. *J. Appl. Meteorol.* 5:205–10.

Turner, F. M. and L. F. Radke. 1973. The design and evaluation of an airborne optical ice particle counter. *J. Appl. Meteorol.* 12:1309–18.

Twohy, C. H., B. J. Huebert, P. A. Durkee, and R. J. Charlson. 1990. Airborne measurements of droplet chemistry in stratiform clouds. Preprints, American Meteorological Society Conference on Cloud Physics, 21–27 June, pp. 527–30. San Francisco, CA.

Ward-Smith, A. J. 1980. *Internal Fluid Flow: The Fluid Dynamics of Flow in Pipes and Ducts.* pp. 307–55. Oxford: Clarendon Press.

Wilson, J. C., J. H. Hyun and E. D. Blackshear. 1983. The function and response of an improved stratospheric condensation nucleus counter. *J. Geophys. Res.* 88:6781–85.

Woods, D. C. and R. L. Chuan. 1983. Size-specific composition of aerosols in the El Chichon volcanic cloud. *Geophys. Res. Lett.* 10(11):1041–44.

32

Measurement of High-Concentration and High-Temperature Aerosols

Pratim Biswas

Department of Civil and Environmental Engineering
University of Cincinnati
Cincinnati, OH 45221-0071, U.S.A.

INTRODUCTION

High-concentration aerosols are encountered in many industrial systems utilizing aerosol reactors. Such reactors are used in industry to make a wide variety of particulate commodities such as carbon black, pigments, and materials for high-technology applications such as optical waveguides and powders for advanced ceramics (Stamatakis et al. 1991). A similar situation is encountered in many other systems where a large quantity of the so-called "undesirable" aerosols are produced. Municipal waste incinerators, hazardous-waste incinerators, coke ovens, smelters, nuclear reactor accidents, and utility boilers are some examples. In spite of the inherent difference in the use of particles, a number of common features stand out. The aerosols are produced in high number concentrations, and in many cases are the result of a high-temperature process. In both cases, the particles need to be collected or deposited onto some substrate: in the first case, the particles might be used to fabricate a part for some application; in the latter case, the particles could be potentially very toxic and emission into the ambient atmosphere needs to be prevented. For efficient design of control devices, it is, therefore, essential to know the size distribution of the aerosol. Frequently, in the industrial aerosol processes, the focus is also on producing monodisperse particles or particles in very narrow size ranges (see Chapter 33). Measurements may, therefore, need to be carried out, to control the process parameters. Also, to control and understand particle formation and growth rates, there is a need to measure the evolution of the aerosol size distribution.

There are certain characteristics that must be considered in the measurement of the high-temperature, high-concentration aerosols. A number of dynamic physicochemical processes are in progress in the aerosol system. If a sampling system is used, it must be designed to quench all aerosol dynamics and chemistry to obtain a representative measurement. If the real-time measuring instruments described earlier in the book are to be used, the number concentration and temperature must be reduced to within the operational limits of the instrument. To alleviate some of these problems, *in situ* techniques can be used. The measurement of high-temperature and high-concentration aerosols using these two methods are described. Systems consisting of dilution probes and secondary diluters are described. Impactors used in high-temperature applications are also discussed. This is followed by a discussion of *in situ* techniques for measurement of high-temperature aerosols in combustion environments.

DILUTION SYSTEMS

The hot gases with the particles must be diluted and cooled in the sampling system. The aerosol dynamics must be rapidly quenched and so must the chemistry. High dilution ratios are often required to reduce the number concentration to the operational range of the aerosol instruments. Such high dilution ratios often require that multistage dilution systems be used.

Dilution systems used by various researchers are summarized in Table 32-1.

Primary Dilution Systems

A simple design of a primary dilution probe used by Linak and Peterson (1984) is shown in Fig. 32-1. Dry, filtered compressed air is used to entrain a constant volume sample by aspiration. Constant-pressure dilution air is supplied through an outer annulus, and as it enters the inner tube, a sample stream is drawn through the orifice due to the lower-pressure region that is created. This sample then mixes with the filtered and compressed dilution air. The dilution ratio is determined based on the capillary dimension, clearance between outer and inner annuli, and diluent gas pressure. The dilution ratio is not controlled directly (by flow control) and there may be associated fluctuations in the sample

TABLE 32-1 Summary of Dilution Systems

	Dilution Ratio				
Researchers	Primary	Secondary	Overall	Dilution Gas Used	Sampled Source
Newton et al. (1980)	5	n.a.	5	Air	Fluidized Bed Combustor
Pedersen et al. (1980)	n.a.	20	20	Air	Organic Aerosols, Automobiles
Ulrich and Riehl (1982)	n.r.	n.a.	n.r.	Nitrogen	Silica, Flame Reactor
Houck et al. (1982); Sousa et al. (1987)	n.a.	30	30	Air	Organic Aerosols, Stack
Linak and Peterson (1984)	21.2	39.5	837.4	Air	Pulverized Coal Combustion
Du and Kittleson (1984)	10	33	330	Nitrogen	Organic Aerosols, Diesel Engines
Bonfanti and Cioni (1986)	6	n.a.	6	Nitrogen, Air	Organic Aerosols, Stack
Wu and Flagan (1987)	n.r.	2000	2000	Nitrogen, Air	Silicon, Tubular Reactor
Biswas et al. (1989)	31	101	3131	Argon, Air	Silica, Tubular Reactor
Hildemann et al. (1989)	n.a.	25 to 100	25 to 100	Air	Organic aerosols, Stack
Sethi and Biswas (1990)	19.5	25	487.5	Argon, Air	Lead, Flame Incinerator
Pratsinis et al. (1990)	5	200	1000	Argon, Air	Titania, Tubular Reactor

n.a.: not applicable
n.r.: not reported

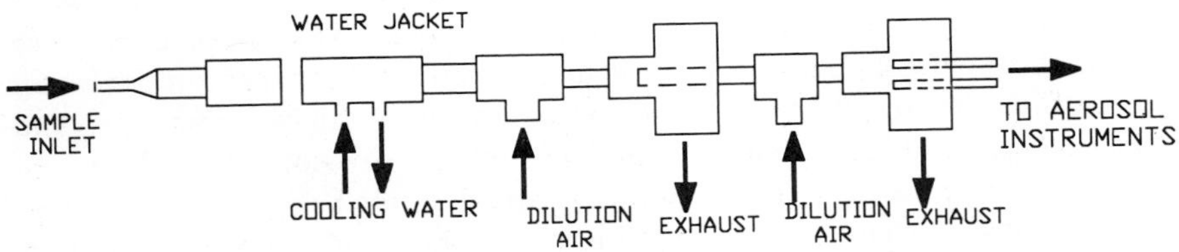

FIGURE 32-1. A Two-Stage Dilution System Consisting of a Particle Probe as the Primary Dilution System. (Adapted from Linak and Peterson 1984.)

flow. Isokinetic conditions may also be difficult to maintain in such systems.

An aspirated isokinetic probe consisting of a water jacketed air delivery system has been used by Peterson and coworkers (Scotto et al. 1988; Gallagher et al. 1990) for sampling aerosols for studying alkali metal partitioning in ash from pulverized coal combustion. Sampling conditions were isokinetic, thus allowing sampling of larger particles. The aerosol sample was rapidly quenched by a free turbulent jet which provided a high mixing rate.

The problem of not having direct control of the dilution ratio is alleviated by using a direct-flow-controlled primary sampling probe. Newton et al. (1980) describe a radially injected dilution probe for sampling in fluidized bed coal combustion systems. The inlet surfaces of the probe are chamfered to reduce edge effects. Dilution air enters through a porous stainless steel cylinder perpendicular to the direction of the sample flow. The dilution air flow rate is independently controlled and is particle-free. The system tends to compress the aerosol sample, allowing cooling and mixing to occur away from the walls, thus reducing losses. A sample flow of 25 lpm was diluted with a total air flow of 100 lpm, leading to a dilution ratio of 5. While thermophoretic losses are shown to be minimized, alteration of the aerosol by condensation and coagulation is not necessarily eliminated.

A design of a flow-controlled primary dilution probe is shown in Fig. 32-2 (Biswas, Li, and Pratsinis 1989). The probe is made of quartz and is constructed using two concentric tubes. Small 1/6″ (4 mm) diameter holes are drilled on the wall of the inside tube. The

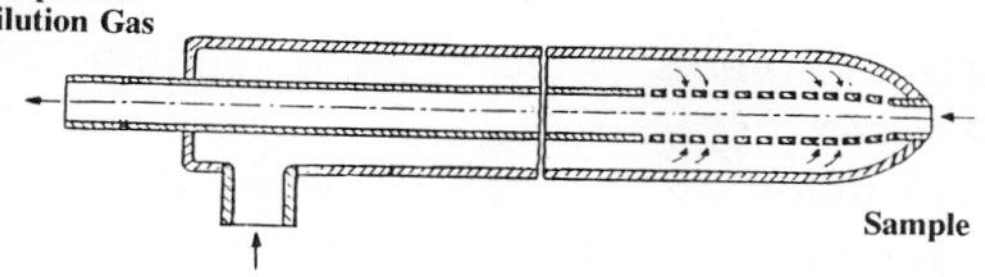

FIGURE 32-2. Schematic Diagram of a Primary Dilution Probe. The Dilution Gas Enters Through the Perforations in the Central Tube and Mixes with the Sample.

dilution gas can thus enter the inside tube and mix and dilute the sample aerosol. The probe is easily movable to sample at different locations in the system. In systems where chemical reactions are rapid an inert gas such as argon is used for dilution as it participates in no chemical reactions.

A more rugged version of a probe designed by Sethi and Biswas (1990) for measurement in flame environments is shown in Fig. 32-3. The following criteria are adopted in this modified design: Cross section of the probe tip should be small so as not to disturb the flow in the system at the sampling location; ease of cleaning and fabrication; and material should withstand flame environments. A stainless steel probe with a small outer diameter (1/8″, 3.2 mm) is used for the sampling end of the probe and is housed in a slightly larger diameter tube. The outer tube also serves as the outer jacket for the dilution gases. The sample is drawn out of the probe from a tube housed inside the outer tube. Thermophoretic and diffusional particle losses in the probe are negligible. The residence time in the sampling tube is short. Sampling losses due to anisokinetic sampling are estimated using the criteria for probe alignment and sampling-to-mainstream velocity ratio. The probe is aligned along the flame axis; hence, there are no losses due to misalignment. The Stokes number of particles sampled in the flame is very much smaller than 0.01; thus, the losses due to velocity differences are unimportant (Hinds 1982).

Ulrich and Riehl (1982) used a nitrogen-quenched, sonic expansion probe for collection of flame-generated silica particles for surface area analysis. The probe is constructed using two concentric tubes with nitrogen gas flowing in the region between the inner and outer tubes. A vacuum is applied at the inner tube which pulls some sample from the mainstream. The nitrogen gas not only cools and dilutes the sample but also prevents clogging of the probe tip.

In the above designs, a probe type configuration is used to draw a small sample from the mainstream flow and dilute it by mixing with some particle-free gas. This is done so

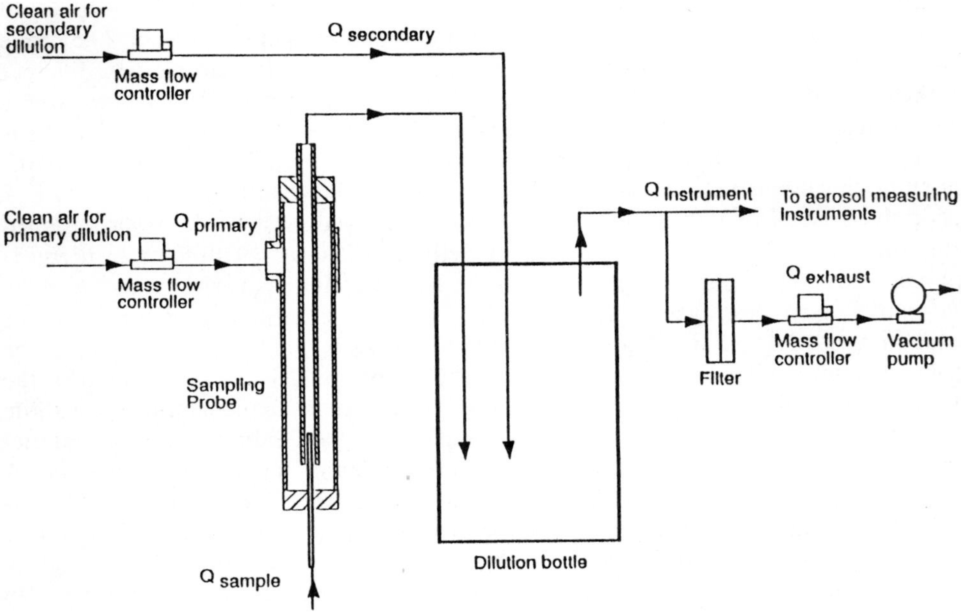

FIGURE 32-3. Schematic of a Dilution Probe for Flame Measurements. Also Shown is the Secondary Dilution System.

that higher dilution ratios can be attained. Designs where the entire reactor flow is diluted have been used by Alam and Flagan (1986), and then modified and used by Wu and Flagan (1987). A schematic diagram is shown in Fig. 32-4. As the aerosol comes out of the reactor, a coaxial flow is introduced from the sides of a sintered tube. This prevents the aerosol from coming into contact with the cooler walls, and depositing due to thermophoresis. Downstream, a large flow of diluent gas is mixed with the aerosol. The flow becomes turbulent and good mixing is obtained with the gases, causing both dilution and cooling simultaneously. A similar setup used by Pratsinis et al. (1990), is shown in Fig. 32-5. The reactor effluent is rapidly mixed and cooled with argon in a perforated quartz dilutor. The perforated quartz dilutor is used in place of the sintered tube used by Alam and Flagan (1986) for ease of fabrication and cleaning.

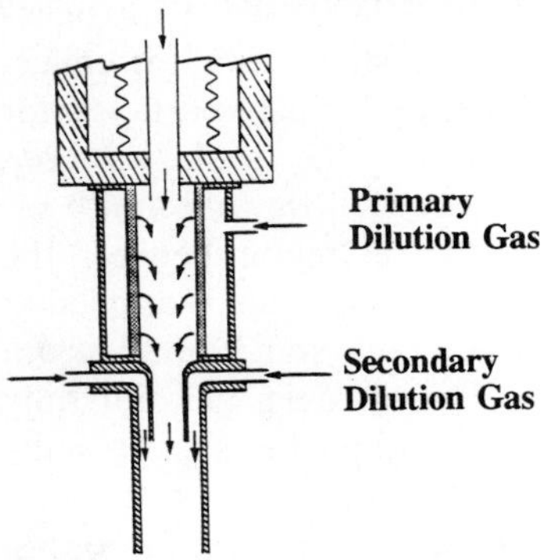

FIGURE 32-4. Dilution System (Primary and Secondary) Wherein the Entire Reactor Flow is Diluted. (Adapted from Alam 1984.)

Requirements of the Primary Dilution Probes

The following conditions must be met in the design of primary dilution probes:

1. Sampling under as close to isokinetic conditions as possible (see Chapter 6), matching the sample velocity to the free-stream velocity.

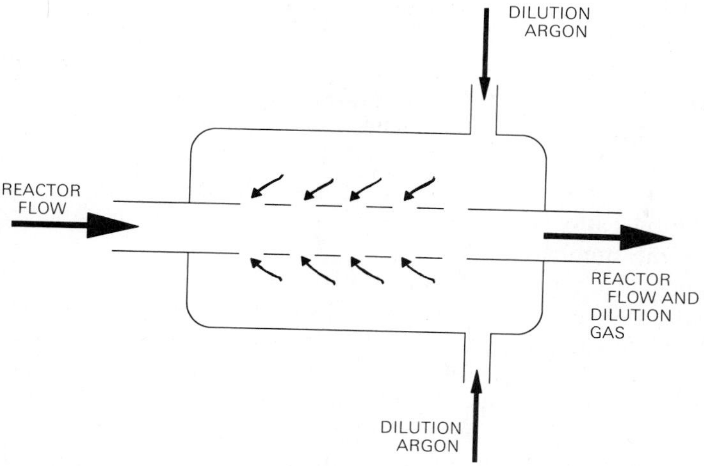

FIGURE 32-5. Details of a Primary Dilution System Wherein the Entire Reactor Flow is Diluted with a Volume of Inert Diluent Gas to Quench Chemical Reactions. (Adapted from Pratsinis et al. 1990.)

2. Choosing the dilution ratio to ensure minimal deposition losses due to thermophoresis and diffusion, and minimal alteration of the size distribution by various aerosol dynamic phenomena such as coagulation, nucleation, and condensation.

Secondary Dilution

The primary dilution probes should quickly cool and dilute a sample in order to quench aerosol growth and chemical reaction. As described earlier, in case of high main stream aerosol concentrations (10^{10} to 10^{15} particles/cm^3), a secondary dilution is often necessary as real-time aerosol instruments are limited to measuring a maximum of about 10^7 particles/cm^3. Secondary dilution can be done in configurations similar to the primary dilution system. The flow from the primary dilution system mixes with a large volume of particle-free air, and a segment of this flow is pulled into the aerosol measuring instrument (Linak and Peterson, 1984) (Fig. 32-1). As the residence times are relatively short, proper mixing has to be ensured to obtain a representative sample. In the dilution system developed by Alam and Flagan (1986), a coaxial flow is used to obtain secondary dilution (see Fig. 32-4). In both these systems, as the entire flow from the primary dilution system is being diluted, a relatively high flow rate of diluent gas must be chosen, depending upon the number concentration of the aerosol in the mainstream (see Example 32-1).

Example 32-1

(a) For the dilution system shown in Fig. 32-6, if the sample flow rate is $Q_{\text{sample, in}}$ and the primary dilution gas flow rate is $Q_{\text{primary, in}}$, compute the dilution ratio in the sampling probe. If the flow rate of clean air (particle free), mixed with a flow of $Q_{\text{sample, out}}$ coming from the probe, is $Q_{\text{secondary, in}}$, compute the overall dilution ratio.

The following are a set of flow rates used in the experiments described by Biswas et al. (1989)

$Q_{\text{primary, in}} = Q_{\text{primary, out}} = 3$ lpm;
$Q_{\text{secondary, in}} = 10$ lpm, $Q_{\text{secondary, out}} = 9.8$ lpm;
$Q_{\text{sample, in}} = Q_{\text{sample, out}} = 0.1$ lpm; and
$Q_{\text{instrument}} = 0.3$ lpm.

Answer. Primary Dilution: Assuming that the number concentration of the aerosol in the sample is c_n, and that in the diluted sample is c_{n1}, a mass balance yields

$$c_n Q_{sample,in} + 0 Q_{primary,in} = c_{n1}(Q_{sample,in} + Q_{primary,in})$$

where it is assumed that the diluent gas is particle-free (filtered). The dilution ratio is defined as the ratio by which the number concentration must be multiplied to obtain the number concentration of the aerosol in the system. Therefore, the dilution ratio in the sampling probe, DR_1, is

$$DR_1 = (Q_{sample,in} + Q_{primary,in})/Q_{sample,in}$$

Secondary Dilution: In a similar manner, the dilution ratio for the secondary system, DR_2, is

$$DR_2 = (Q_{secondary,in} + Q_{sample,out})/Q_{sample,out}$$

The overall dilution ratio, DR is thus given by

$$DR = DR_1 \times DR_2 = (Q_{sample,in} + Q_{primary,in}) \times (Q_{secondary,in} + Q_{sample,out})/ (Q_{sample,in} Q_{sample,out})$$

The number concentration of the aerosol in the system is then determined by multiplying the measured size distribution of the diluted aerosol by DR.

For the given flow rates, using the formulae above, one computes $DR = 3131$.

(b) If the number concentration in a system is 10^{10} particles/cm^3, what flow rates are to be used in the dilution system illustrated in Fig. 32-6 for measurement with commonly used real time aerosol instruments?

Answer. As most instruments have a measurement upper limit of around 10^7 particles/cm^3, and using a margin of safety, we choose an overall dilution ratio, $DR = 10^4$. Hence, for a number concentration in the system of 10^{10} particles/cm^3, the instrument would see a concentration of $10^{10}/10^4 = 10^6$ particles/cm^3. Let $Q_{primary,in} = Q_{primary,out} = 3$ lpm, $Q_{sample,in} = 0.1$ lpm as in Example 32-1a. Therefore, $DR_1 = 3.1/0.1 = 31$. Hence, the required $DR_2 = DR/DR_1 = 10^4/31 = 322.6$. If $Q_{sample,out} = 0.1$ lpm, then we obtain $Q_{secondary,in} = DR_2 Q_{sec} - Q_{sec} = 32.26 - 0.1 = 32.16$ lpm.

An alternative secondary dilution system developed by Biswas (1985) and used for particle sampling in high-temperature systems (Biswas, Li, and Pratsinis 1989; Sethi and Biswas 1990) is shown in Fig. 32-6. The dilution takes place in a 50 l bottle, with a small flow from the dilution system being drawn in and mixed with a large volume of particle-free air. Very high dilution ratios can be obtained as precise flow control is possible by using mass flow controllers. To minimize uncertainties in the dilution ratio, the dilution flow is often cooled, filtered, and recirculated (Biswas 1985, 125; Biswas and Flagan 1988; Wu and Flagan 1987). A schematic diagram of the setup indicating the various flows is shown in Fig. 32-6. The residence time is sufficiently long to allow good mixing. The expression developed by Crump, Flagan, and Seinfeld (1983) can be used to calculate the losses in the dilution vessel

$$\frac{c_n}{c_{n0}} = \exp(-\beta t) \tag{32-1}$$

where c_n is the number concentration of the aerosol after being resident in the dilution vessel for a time t, c_{n0} is the number concentration at time $t = 0$, and β is the wall loss coefficient.

Example 32-2

Consider the secondary dilution bottle system to be spherical with a volume of 50 l. Assuming a total flow rate of 10 lpm, estimate the fraction of particle loss in the bottle for 0.024, 0.042, 0.34, 0.51, and 0.79 μm diameter particles.

Answer. A simplistic calculation will be carried out, using Eq. 32-1 and the measured

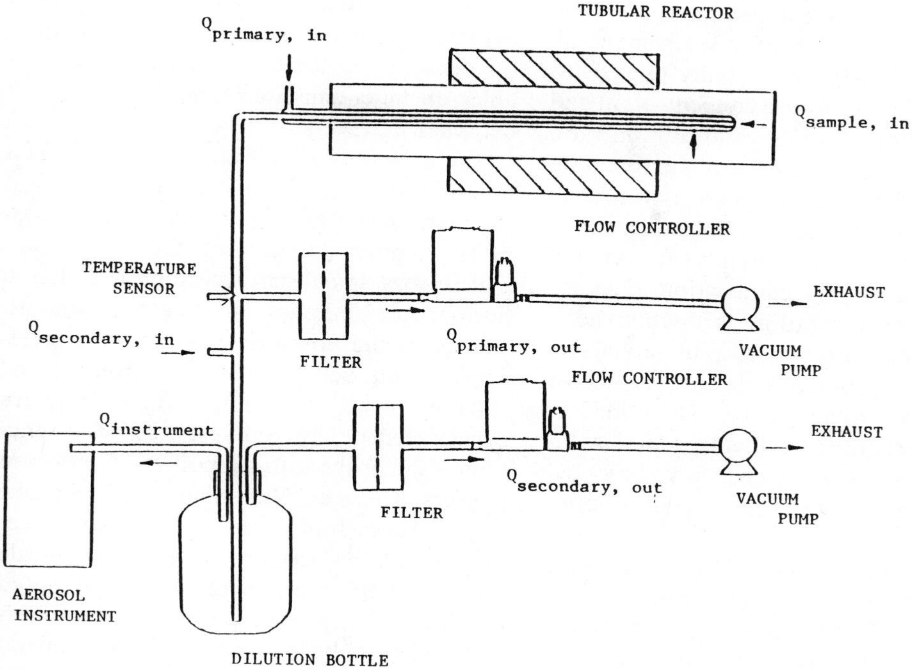

FIGURE 32-6. Entire Dilution System Used by Biswas et al. (1989), Showing Primary Dilution Probe and Secondary Dilution System. Also Indicated are the Various Flow Rates Through the Different Sections.

TABLE 32-2 Fraction of Particle Loss in the Dilution Bottle

Particle Diameter (μm)	β (sec^{-1}) Loss Coefficent (Crump et al. 1983)	Fraction of Particle Loss
0.024	1.8×10^{-4}	0.05
0.042	8.4×10^{-5}	0.02
0.34	1.5×10^{-5}	0.01
0.51	4.8×10^{-5}	0.02
0.79	1.26×10^{-4}	0.04

values of β reported by Crump et al. (1983). The residence time in the dilution bottle is $50/10 = 5$ min $= 300$ sec. The results, computed assuming that β remains constant for this time period for particles in a certain size range, are listed in Table 32-2.

Dilution Sampling of Organic Aerosols

Due to the toxic nature of organic compounds, it is of interest to determine the contribution of emission sources to ambient organic aerosol concentrations (Daisey, Cheney, and Lioy 1986). The ratio of the vapor phase to particulate organics is a strong function of the temperature. Direct filtration of hot gases may not lead to the capture of the organic aerosols that exist primarily as vapors at these temperatures, leading to underestimation of source organic aerosol contributions. Alternatively, the use of cooled impingers or cryogenic traps result in an overestimation of organic aerosol concentrations due to condensation of organics that normally would not condense. Thus, the sampling of organic aerosols from combustion sources require that dilution systems be used. Hildemann, Cass, and Markowski (1989) have reviewed the various systems used for organic aerosol sampling.

Large, nonportable dilution tunnels have been used for automotive emissions measurements. These systems were designed to simulate the dilution and cooling that take place in the atmosphere when exhaust gases leave the exhaust pipe of a vehicle (Pedersen et al. 1980). The exhaust was partially cooled (to 160°C to prevent moisture condensation) and mixed with dilution air with a dilution ratio of 20 to cool further the sample to about 35°C. Du and Kittleson (1984) used a sampling system capable of removing, quenching, and diluting the entire contents of a cylinder during actual engine combustion. The extracted gas was made to go through an adiabatic expansion into the sampling apparatus whereby it was diluted with nitrogen with a dilution ratio of about 10. These diluted combustion chamber contents were further diluted in a polyethylene sampling bag with nitrogen to give an overall dilution ratio of about 330.

Portable dilution systems for stack sampling of organic aerosols have been described by Houck, Cooper, and Larson (1982), Huynh et al. (1984), Bonfanti and Cioni (1986), Sousa et al. (1987), and Hildemann, Cass, and Markowski (1989) with dilution ratios in the range of 6–100. The important design issues for such systems have been summarized by Hildemann, Cass, and Markowski (1989):

1. Simulation of atmospheric dilution as closely as possible to collect a sample that includes all organics that would condense into the aerosol phase in the atmosphere. This is achieved by a high degree of dilution and cooling to ambient temperatures. The system should allow a long residence time for mixing.
2. Sampler surfaces should be inert and withstand rigorous cleaning and heat treatment.
3. The sampler should be configured and operated to minimize particle and vapor losses.

EPA METHOD 5 SAMPLING TRAIN

The EPA Method 5 Sampling Train is used to determine total particulate emissions on a mass basis from stationary sources such as utility plants, coal-fired boilers, and incinerators (U.S. EPA 1990). It is widely used for stationary-source sampling for air pollution control equipment performance evaluation, for determining the effect of operating variables on mass emissions, and for regulatory compliance tests.

The sampling train (Fig. 32-7) consists of a nozzle, probe, filter housing, impingers, pitot tube, and metering system. The nozzle is designed to operate under isokinetic conditions, and its size is chosen accordingly. It has a button hook design to create a smooth change in the flow direction by 90 degrees. Typical nozzle sizes range from 1/16″ (1.6 mm) to 1/2″ (12.7 mm) diameter, and are constructed of 316 stainless steel. The function of the probe is to support the nozzle and connect it to the filter housing. The probe length varies from 2 feet (0.61 m) to 12 feet (3.67 m), and is typically made of borosilicate glass. When long lengths are required (> 15 feet, 4.6 m), they are made of stainless steel. The probe and the filter assembly is maintained at 120°C to prevent moisture condensation. Glass fiber filters are used as the sampling media.

The recommended point of sample extraction should be ten duct diameters away from the nearest upstream disturbance (bends, dampers, duct outlets) and two duct diameters away from the nearest downstream disturbance. The number of sampling or traverse points ranges from 8 to 24, depending upon the stack diameter and the downstream distance from the nearest flow disturbance. The sampling time should be long enough to collect 0.6 m^3 at a rate no greater than 0.021 $m^3\,min^{-1}$. Leak tests also need to be performed to ensure that the correct volume of gas is sampled. The particulate sample is recovered from all components including the nozzle, probe, cyclone, filter holder, filter, rubber gasket, and impingers by washing with water or acetone. This extract is then dried and the amount of particulate matter weighed to determine the mass concentration.

HIGH-TEMPERATURE IMPACTORS

Inertial impactors have been discussed in Chapter 11. To prevent bounce and reen-

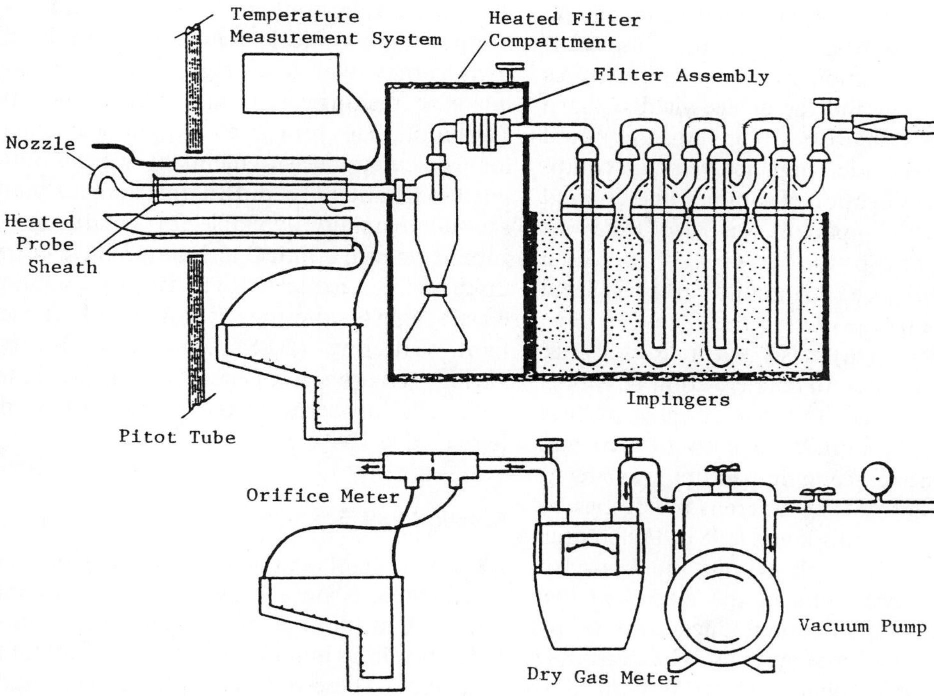

FIGURE 32-7. The EPA Method 5 Sampling Train Showing the Various Components. (Adapted from. U.S. EPA 1990.)

trainment of particles, the collection surfaces are often coated with a thin layer of grease or oil. These coatings are not practical for use in sampling of high-temperature gas streams. Other substrates such as ceramic fibers have been used at temperatures as high as 500°C with no apparent shift in the 50% cutoff size; however, the size cuts were considerably less sharp when compared to greased plate impaction at room temperature (Parker et al. 1981). Newton et al. (1982) used ungreased stainless steel substrates for high-temperature and high-pressure measurements; however, the impaction velocities were limited to prevent particle bounce. Pilat, Ensor, and Bosch (1970) designed a source stack cascade impactor for measuring particle size distributions in stacks and ducts of emission sources. By operating the impactor inside the duct, true isokinetic sampling could be achieved with minimum wall losses and condensation problems. However, the problem of particle bounce would remain in these impactors at temperatures greater than 150°C (about 300°F) as the collection surfaces cannot be coated with a sticky medium. Pilat and Steig (1983) used a version of this impactor with filter media as the collection surface for sampling in a pressurized fluidized-bed coal combustor. Although filter media reduce particle bounce in comparison with uncoated collection surfaces, impactor collection efficiency characteristics were found to be significantly different when compared to greased collection surfaces (Rao and Whitby 1977).

Virtual impactors are a good alternative, as the particles are collected off-line and greased plates do not need to be used (Schott and Ranz 1976; Marple and Chien 1980). A disadvantage of the virtual impactor is that secondary flows need to be controlled precisely at every stage and a multistage device becomes cumbersome and difficult to operate.

A particle trap impactor was developed by Biswas and Flagan (1988) for use in high-temperature systems. The particles were collected in a cavity instead of having a secondary flow as in a virtual impactor. Extensive calibration at room temperatures was carried out and the device yielded sharp efficiency characteristics. The impactor could also be used under high jet velocity conditions for classification of submicrometer-sized aerosols. The impactor was also tested at temperatures up to 500°C and indicated sharp cutoff characteristics (Biswas and Flagan 1988).

Distortions may also occur in the size measurements due to condensation of vapors and other species. The distortion of particle size distributions in both conventional and low-pressure cascade impactors was determined for aerosols at different humidities by Biswas, Jones, and Flagan (1987). Differential impaction of aerosols within the impactor may also cause shifts in the measured distributions as high number concentrations are encountered in these systems. The allowable upper limit in number concentration for sampling of such aerosols has been determined by Biswas (1988).

IN SITU MEASUREMENTS

In addition to the probe techniques described earlier, a variety of diagnostic techniques have been used to measure particle sizes in flames. Thermophoretic sampling techniques have been used by Megaridis and Dobbins (1989) to deposit particles on a cold probe and then examine them by electron microscopy. Many optical techniques have been used as they offer advantages of *in situ*, nonintrusive measurements and greater spatial resolution. The principles of light scattering for particulate measurements have been outlined in Chapters 15 and 16. Instruments that are commercially available for *in situ* measurements using optical techniques have been discussed in Chapter 16. The two techniques, elastic light scattering (ELS) and dynamic light scattering (DLS), used recently by various researchers for measurement in high-concentration aerosol systems are discussed here (Table 32-3).

Elastic Light Scattering

An elastic light scattering process is one in which there is no energy exchange between the incident photons of light and target particles, resulting in no shift in the incident light frequency. These techniques have been used by many researchers in various ways such as combination of scattering and extinction, angular dissymmetry, polarization ratio measurements, and two-color techniques. Light of different wavelengths (two colors) provide a small range of measurements and are costly to implement (Zachariah et al. 1989). They are not discussed in this chapter.

Combination of Scattering and Extinction

Santoro, Semerjian, and Dobbins (1983), Santoro et al. (1987) and Presser et al. (1990)

TABLE 32-3 Light Scattering Techniques for Particle Measurement

Elastic Light Scattering	
Scattering/Extinction	Mean particle size and number concentration for absorbing aerosols. No size limitation.
Angular Dissymmetry Polarization Ratio Two Color	Mean size, number concentration and spread in distribution. Measureable particle size greater than 0.1 λ. Good for nonabsorbing aerosols with known refractive index.
Nonelastic Llight Scattering	
Dynamic Light Scattering	No size limitation, sensitivity decreases with particle size. Determines diffusion coefficient, hence requires system temperature to compute particle size. Refractive index of particles not needed.

have summarized the measurement procedure for light absorbing aerosols using a combination of scattering and extinction measurements. The technique was demonstrated by measurement of soot particles formed by combustion of a gaseous fuel in a diffusion flame. The scattered light detection system allowed an absolute determination of the differential scattering cross section per unit volume. Light scattering and extinction measurements were obtained simultaneously at different radial and axial positions. The scattering cross section applicable for the vertically polarized light, $Q_{1,1}$, and the extinction coefficient, k_{ext}, are determined from the measurements.

Using the Rayleigh theory (Chapter 15), expressions for $Q_{1,1}$ and k_{ext} are obtained, and the ratio of the two is proportional to an average particle diameter, $d_{\text{p}6,3}$, defined by

$$d_{\text{p}6,3} = \frac{\int c_{\text{n}} f(d_{\text{p}}) d_{\text{p}}^6 \, \mathrm{d}d_{\text{p}}}{\int c_{\text{n}} f(d_{\text{p}}) d_{\text{p}}^3 \, \mathrm{d}d_{\text{p}}} = \frac{4E(m)Q_{1,1}}{\pi^2 F(m) k_{\text{ext}}} \tag{32-2}$$

where $f(d_{\text{p}})$ is the size distribution function, m is the complex refractive index of the particles, $E(m) = -\text{Im}[(m^2-1)/(m^2+2)]$, and $F(m) = |(m^2-1)/(m^2+2)|$. If a certain shape of the size distribution function is assumed, $d_{\text{p}6,3}$ can be related to the volume mean diameter, $d_{\text{p}3,0}$. Knowing $d_{\text{p}3,0}$, the particle number concentration, c_{n}, can be determined (Santoro, Semerfian, and Dobbins 1983):

$$c_{\text{n}} = \frac{\lambda k_{\text{ext}}}{\pi^2 E(m) d_{\text{p}3,0}^3} \tag{32-3}$$

Santoro, Semerjian, and Dobbins (1983) suggested that due to high number concentrations encountered in flames, the size distribution can be assumed to be self-preserving. They assumed $(d_{\text{p}6,3}/d_{\text{p}3,0})^3 = 2$ for the self preserving size distribution, and using the experimentally measured k_{ext}, determined the number concentration using Eqs. 32-2 and 32-3.

Angular Dissymmetry Measurements

The scattering to extinction ratio technique described above is not applicable to the study of nonabsorbing aerosols such as silica. The dissymetry technique utilizes the angular dependence of the scattered light intensity for particles of known refractive indices in the Mie regime. The intensity of the scattered light from the particles is a function of the measured angle, the refractive index of the particle, and particle size in the Mie scattering regime. The well-established relation between the particle size and the scattering intensity is described by van de Hulst (1981).

A description of the angular dissymetry technique to determine the mean particle size and number concentration is provided by Zachariah et al. (1989). The vertically polarized scattering intensity, $I_{1,1}$, can be measured and is given by

$$I_{1,1}(\alpha, \theta_i) = C_{1,1}(\alpha. m, \theta_i) c_{\text{n}} S(\theta) \tag{32-4}$$

where $\alpha = \pi d_{\text{p}}/\lambda$, λ is the wavelength of the incident light beam, $C_{1,1}$ is the Mie scattering cross section for spherical particles and the subscript (1, 1) denotes state of vertical polarization, c_{n} is the number concentration, and S is the system parameter. Zachariah et al. (1989) assumed the aerosol to be described by a mean size (monodisperse), and determined the dissymetry ratio from their measurements:

$$\frac{I_{1,1}(\alpha, \theta_1)}{I_{1,1}(\alpha, \theta_2)} = \frac{C_{1,1}(\alpha, m, \theta_1)}{C_{1,1}(\alpha, m, \theta_2)} \tag{32-5}$$

Using the Mie scattering cross section (Dave 1968), the above equation can be solved to determine α and, therefore, the mean particle size. Using the value of the mean particle size thus determined, Eq. 32-4 can be used to determine the number concentration, c_{n}, with $S(\theta)$ determined from calibration experiments. Thus, intensity measurements at two angles could provide information on the mean particle size and number concentration.

The angular dissymetry technique has been extended by Chang and Biswas (1990) to determine the spread of particle size distributions. The dissymetry ratio for a size distribution of particles is given by

$$\frac{I_{1,1}(\alpha,\theta_1)}{I_{1,1}(\alpha,\theta_2)} = \frac{\int C_{1,1}(\alpha,m,\theta_1)f(d_p)\,dd_p}{\int C_{1,1}(\alpha,m,\theta_2)f(d_p)\,dd_p} \tag{32-6}$$

As the above equation has as an unknown $f(d_p)$, the size distribution function, it cannot be solved directly as in the monodisperse case described above. Chang and Biswas (1990) represented a flame-generated silica aerosol with a lognormal distribution; and Eq. 32-6 then has two unknowns: the geometric mean size and the geometric standard deviation. Thus, two dissymmetry ratios were determined by scattered-intensity measurements at three angles, and then used to determine the geometric mean diameter and geometric standard deviation. Chang and Biswas (1991) further developed a statistical regression procedure for determining the best-fit parameters of the size distribution function using measurements at a number of angles. They also applied this technique for the measurement of multicomponent aerosols where the refractive index is unknown, and is used as a fitting parameter.

Polarization Ratio Measurements

The intensity of the parallel to perpendicular component of the scattered light for unpolarized incident light is termed the polarization ratio. Kerker and La Mer (1950) suggested that the polarization ratio at any angle could be used to determine the particle size. This technique has been used by several researchers, and has been reviewed by Bonczyk (1979). The polarization ratio $(I_{1,1}/I_{1,2})$ is a function of the refractive index, the size distribution of the scatterer, and the angle of measurement (Bonczyk and Sangiovanni 1983):

$$\frac{I_{1,1}}{I_{1,2}} = \frac{\int C_{1,1}(\alpha,m,\theta)f(d_p)\,dd_p}{\int C_{1,2}(\alpha,m,\theta)f(d_p)\,dd_p} \tag{32-7}$$

where $I_{1,1}$ is the vertically polarized scattered light intensity, $I_{1,2}$ is the horizontally polarized scattered light intensity (see Chapter 15), C is the Mie scattering cross section with subscripts 1,1 and 1,2 denoting states of vertical and horizontal polarization, respectively. If the aerosol is assumed to be monodisperse, the mean particle size can be determined from Eq. 32-7 using the expressions for the scattering cross section (Dave 1968) and the measured polarization ratio.

The advantage of the technique is that the ratio is independent of geometrical and system parameters. The disadvantage is that polarization ratio measurements may not be very reliable for high concentration aerosols due to the depolarization effects resulting from agglomerated nonspherical scatterers (Maron and Elder 1963).

Elastic Light Scattering Measurement System

This section is concluded by describing the components of an optical system used to carry out light scattering and transmission measurements. A schematic diagram of such a system is shown in Fig. 32-8. It consists typically of a laser light source with a mechanically chopped beam divided into two parts by a beam splitter. One beam is directed through the region of the flame for the light scattering and extinction measurements. The other beam is directed to a photodiode connected to a lock-in amplifier to obtain a reference signal and, thus, distinguish the scattered-light intensity signal from the unchopped background noise. If the polarization ratio is to be measured, the light source is not polarized. The scattered signal from the probe volume is measured using a photomultiplier tube (PMT) connected to a lock-in amplifier for signal conditioning. Two slits on the PMT allow a very small scattering volume to be defined, thus enhancing spatial resolution. The laser line filter centered at the light source wavelength with a $\Delta\lambda$ of around 1 nm is also mounted on the PMT to reject the extraneous signals from flame emissions. A polarizer is used before the PMT to ensure that the scattered light corresponds to the incident beam if the source is also polarized. The PMT is typically installed on a rotator so

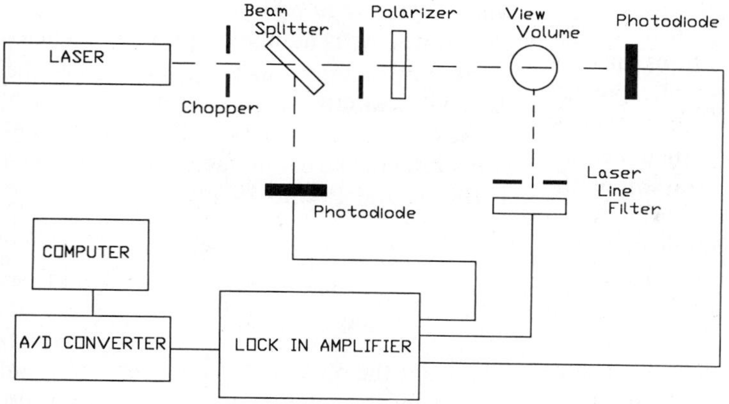

FIGURE 32-8. An Optical System for Light Scattering and Extinction Measurements. The Detection System is Rotated for Measurement at Different Angles.

that it can measure the scattered light intensities at various angles. The optical system or the burner can be installed on a moveable system which can be traversed for spatial measurements. The output signal from the lock-in amplifier is digitized with an A/D converter, and sent to a microcomputer for data storage and further processing.

Dynamic Light Scattering (DLS)

DLS, also known as photon correlation spectroscopy, has been used in the measurement of soot formation (Flower 1983*a*). Monochromatic light incident on particles in a defined probe volume results in a fluctuating scattering intensity due to the Brownian motion of the particles. These fluctuations are correlated to the mean speed or the diffusion coefficient of the particles, both of which are related to the particle diameter. A discussion of DLS is provided in Chapter 15. The procedure to obtain the particle size is discussed herewith (Flower and Hurd 1987; Zachariah et al. 1989).

Flower (1983*a, b*) used DLS techniques for measurement of soot particles in flames. The advantage of the DLS technique is that the measurement is independent of the particle refractive index. However, the technique provides a value for the particle diffusion coefficient which is used to compute the particle size. If the mean motion of the particles is small, the power spectra, $P(\omega)$ of the scattered light exhibits a Lorentzian function (Cummins and Swinney 1970)

$$P(\omega) = \frac{I^2 q^2 D/\pi}{(q^2 D)^2 + \omega^2} \tag{32-8}$$

where I is the intensity of scattered light, ω is the angular frequency, q is the magnitude of the scattering wave vector $[= 4\pi \sin(\theta/2)/\lambda_0]$, λ_0 is the wavelength of the incident light, D is the diffusion coefficient, and θ is the scattering angle measured in the forward direction. When the scattered light is focused on the photomultiplier tube, self-beating of the scattered light results in photocurrent fluctuations. From the above equation, the frequency at half-width is given by

$$\Delta f = \frac{\Delta\omega}{2\pi} = \frac{q^2 D}{2\pi} \tag{32-9}$$

The frequency at half-width, Δf can be determined from the spectral plot described above, and the diffusion coefficient can be computed using Eq. 32-9. Knowing the relationship of D to particle size ($D = kTC_c/3\pi\eta d_p$), the particle size d_p can be determined. However, an accurate estimate of the temperature and the properties of the medium are necessary.

Nonspherical Scatterers

Due to high number concentrations encountered in combustion environments, particle agglomerates may be formed. Light scattering has been used for examining such aerosol agglomerates. However, the relation of the scattered intensity to particle size from even a small aggregate is rather complicated. The fractal dimension can characterize statistically similar but geometrically irregular objects. Light scattering has been used to determine the fractal dimension of clusters in liquid media (Schaefer and Martin 1984) and in gas media (Martin, Schaefer, and Hurd 1986).

Mountain and Mullholland (1988) have established relationships of light scattering from simulated smoke agglomerates. The measured intensity and its angular dependence was related to the fractal dimension, the radius of gyration of the agglomerate, and the total number of primary particles in the agglomerate. They also reported the resulting error in the volume mean diameter between an effective sphere model and their proposed agglomerate model. The error increased with the number of primary particles in the agglomerate, with a 26% difference being obtained for a cluster with 119 primary particles, and a 8% difference for a cluster with 32 primary particles.

Hurd and Flower (1988) have studied flame-generated silica particle aggregates by *in situ* dynamic and static light scattering and reported fractal dimensions. A morphological description of flame-generated fractal particles has been provided by Megaridis and Dobbins (1990), and a description of the absorption and scattering of light by polydisperse aggregates has been discussed by Dobbins and Megaridis (1991).

Assuming Rayleigh–Debye scattering (Kerker 1969), the differential scattering coefficient, $Q_{1,1}$, for an aggregate of particles can be expressed as

$$Q_{1,1}(\theta) = c_N C_{1,1,\text{Rayleigh}} \cdot \sum_{j=1}^{N_p} e^{i\delta_j} = c_N C_{1,1,\text{Rayleigh}} \cdot N_p^2 \cdot S(q) \quad (32\text{-}10)$$

where c_N is the number concentration of agglomerates, N_p is number of primary particles in an agglomerate, and $C_{1,1,\text{Rayleigh}}$ is the Rayleigh scattering cross section of a primary particle in an agglomerate. For a single agglomerate, the structure factor $S(q)$ is defined as (Berne and Pecora 1976)

$$S(q) = \frac{1}{N_p^2} \sum_{j=1}^{N_p} e^{i\delta_j} \quad (32\text{-}11)$$

where δ is the phase lag of the scattered beam caused by the interference between two distinct primary particles in an agglomerate. Expanding the structure factor for small values of q leads to the relation (Berne and Pecora 1976)

$$S(q) = 1 - \frac{R_G^2 q^2}{3} + \ldots \quad (32\text{-}12)$$

where q is $4\pi/\lambda_0 \sin\theta/2$ (λ_0 is wavelength of beam and θ is scattering angle); R_G, the radius of gyration, is defined as

$$R_G = \sqrt{\frac{1}{N_p} \sum_{j=1}^{N_p} r_j^2} \quad (32\text{-}13)$$

where r_j is the distance from the center of mass of the agglomerate to the jth primary particle. For small values of q (so that $R_G q \ll 1$), Eq. 32-11 can be written as

$$S(q) \approx 1 - \frac{R_G^2 q^2}{3} \quad (32\text{-}14)$$

For larger values of q ($R_G q > 1$), the behavior of $S(q)$ becomes (Mountain and Mulholland, 1988)

$$S(q) \approx q^{-D_f} \quad (32\text{-}15)$$

D_f is the fractal dimension, and defines the mass distribution in an agglomerate by the relation

$$N_p = k_f \left(\frac{R_G}{d_p}\right)^{D_f} \quad (32\text{-}16)$$

where k_f is a constant which has a value of 5.8 as proposed by Mountain and Mulholland (1988) and d_p is the size of primary particle. For a fixed position, the intensity $Q_{1,1}(\theta)$ can be measured at different θ (or q). As $c_N C_{1,1,\text{Rayleigh}} N_p^2$ is a constant at a fixed location, the intensity measurement is proportional to $S(q)$. Using Eq. 32-15, a plot of $S(q)$ with respect to q on a logarithmic scale yields a slope which is equal to $-D_f$, the fractal dimension. At small q, $S(q)$ has a parabolic dependence on q (Eq. 32-14), and fitting the data yields a value of R_G, the radius of gyration. If the primary particle size is known (determined from microscopy), Eq. 32-16 can be used to determine N_p, the number of primary particles in an aggregate. These parameters can then be substituted in Eq. 32-10 to determine c_N, the number concentration of agglomerate particles.

References

Alam, M. K. 1984. Nucleation and condensational growth of aerosols: Application to silicon production. Ph.D. thesis, California Institute of Technology, Pasadena, CA.

Alam, M. K. and R. C. Flagan. 1986. Controlled nucleation aerosol reactors. Production of bulk silicon. *Aerosol Sci. Technol.* 5:237–48.

Berne, B. R. and R. Pecora. 1976. *Dynamics of Light Scattering*. New York: Wiley.

Biswas, P. 1985. Impactors for aerosol measurements: Developments and sampling biases. Ph.D. Thesis, California Institute of Technology, Pasadena, CA.

Biswas, P., C. L. Jones, and R. C. Flagan. 1987. Distortion of size distributions by condensation and evaporation in aerosol instruments. *Aerosol Sci. Technol.* 7:231–46.

Biswas, P. and R. C. Flagan. 1988. The particle trap impactor. *J. Aerosol Sci.* 19:113–21.

Biswas, P. 1988. Differential impaction of aerosols. *J. Aerosol Sci.* 19:603–10.

Biswas, P., X. Li, and S. E. Pratsinis. 1989. Optical waveguide preform fabrication: Silica formation and growth in a high temperature aerosol reactor. *J. Appl. Phys.* 65(6):2445–50.

Bonczyk, P. A. 1979. Measurement of particulate size by *in situ* laser-optical methods: A critical evaluation applied to fuel-pyrolyzed carbon. *Combust. Flame* 35:191–206.

Bonczyk, P. A. and J. J. Sangiovanni. 1984. Optical and probe measurements of soot in a burning fuel droplet stream. *Combust. Sci. Technol.* 36:135–47.

Bonfanti, L. and M. Cioni. 1986. Sampling of polynuclear aromatic hydrocarbons at stack with a dilution sampler. In *Aerosols: Formation and Reactivity*. pp. 952–55. 2nd International Aerosol Conference, Berlin.

Chang H. and P. Biswas. 1990. *In situ* light scattering dissymmetry measurements of the evolution of the aerosol size distribution in flames. Paper read at 1990 Annual Meeting of the American Association of Aerosol Research, Paper P1c.3. Also, *J. Colloid Interface Sci.*, vol. 152, 1992.

Chang, H. and P. Biswas. 1991. Multiangle light scattering dissymmetry measurements to determine evolution of silica particle size distributions in flames. Paper read at 1991 Annual Meeting of the American Association of Aerosol Research, Paper 9D.5. Also submitted to *Aerosol Sci. Technol.*

Crump, J., R. C. Flagan, and J. H. Seinfeld. 1983. Particle wall loss rates in vessels. *Aerosol Sci. Technol.* 2:303–9.

Cummins, H. Z. and H. L. Swinney. 1970. Light beating spectroscopy. *Prog. Optics* 8:135–38.

Daisey, J. M., J. L. Cheney, and P. J. Lioy. 1986. Profiles of organic particulate emissions from air pollution sources: Status and needs for receptor source apportionment models. *J. Air Polln. Control Assoc.* 36:17–33.

Dave, J. V. 1968. Subroutine for computing the parameters of the electromagnetic radiation scattered by a sphere. IBM Scientific Center Report No. 320-3237.

Dobbins, R. A. and C. M. Megaridis. 1991. Absorption and scattering of light by polydisperse aggregates. *Applied Optics*, to appear.

Du, C. J. and D. B. Kittleson. 1984. In-cylinder measurements of diesel particle size distributions. *Aerosols*, 744–48.

Flower, W. L. 1983*a*. Optical measurements of soot formation in premixed flames. *Combust. Sci. Technol.* 33:17–33.

Flower, W. L. 1983*b*. Measurements of the diffusion coefficient for soot particles in flames. *Phys. Rev. Lett.* 51(25):2287–90.

Flower, W. L. and A. J. Hurd. 1987. *In situ* measurement of flame formed silica particles using dynamic light scattering. *Appl. Opt.* 26:2236–39.

Gallagher, N. B., L. E. Bool, J. O. L. Wendt, and T. W. Peterson. 1990. Alkali Metal Partitioning in Ash from Pulverized Coal Combustion. *Combust. Sci. Technol.* 74:211–21.

Hildemann, L. M., G. R. Cass, and G. R. Markowski. 1989. A dilution stack sampler for collection of organic aerosol emissions: Design, characterization and field tests. *Aerosol Sci. Technol.* 10:193–204.

Hinds, W. C. 1982. *Aerosol Technology*. New York: Wiley.

Houck, J. E., J. A. Cooper, and E. R. Larson. 1982. Dilution sampling for chemical receptor source fingerprinting. Paper read at 75th Annual Meeting of the Air Pollution Control Association, New Orleans, LA, 82-61M.2.

Hurd, A. J. and W. L. Flower. 1988. *In situ* growth and structure of fractal silica aggregates in a flame. *J. Colloid Interface Sci.* 122:178–92.

Huynh, C. K., T. Vu Duc, C. Schwab, and H. Rollier. 1984. In-stack dilution technique for the sampling of polycyclic organic compounds. Application to effluents of a domestic waste incineration plant. *Atmos. Environ.* 18:255–59.

Kerker, M. 1969. *The Scattering of Light and Other Electromagnetic Radiation.* New York: Academic Press.

Kerker, M. and V. K. La Mer. 1950. Particle size distribution in sulfur hydrosols. *J. Am. Chem. Soc.* 72:3516–25.

Linak, W. P. and T. W. Peterson. 1984. Effect of coal type and residence time on the submicron aerosol distribution from pulverized coal combustion. *Aerosol Sci. Technol.*. 3:77–96.

Maron, S. H. and M. E. Elder. 1963. Determination of latex particle size by light scattering. *J. Colloid Sci.* 18:107–18.

Marple, V. A. and C. M. Chien. 1980. Virtual impactors: A theoretical study. *Environ. Sci. Technol.* 14:796–805.

Martin, J. E., D. W. Schaefer, and A. J. Hurd. 1986. Fractal geometry of vapor-phase aggregates. *Phys. Rev. A* 33(5):3540–43.

Megaridis, C. M. and R. A. Dobbins. 1989. Comparison of soot growth and oxidation in smoking and non-smoking ethylene diffusion flames. *Combust. Sci. Technol.* 66:1–16.

Megaridis, C. M. and R. A. Dobbins. 1990. Morphological description of flame generated materials. *Combust. Sci. Technol.* 71:95–109.

Mountain, R. D. and G. W. Mullholland. 1988. Light scattering from simulated smoke agglomerates. *Langmuir* 4:1321–26.

Newton, G. J., R. L. Carpenter, H. C. Yeh, and E. R. Peele. 1980. Respirable Aerosols from Fluidized Bed Coal Combustion. 1. Sampling Methodology for an 18-inch Experimental Fluidized Bed Coal Combustor. *Environ. Sci. Technol.* 14:849–53.

Newton, G. J., R. L. Carpenter, Y. S. Cheng, E. B. Barr, and H. C. Yeh. 1982. High temperature-high pressure cascade impactor design, performance and data analysis methods. *J. Colloid Interface Sci.* 87:279–90.

Parker, R., S. Calvert, D. Drehmel, and J. Abbott. 1981. Inertial impaction of fine particles at high temperature and high pressure. *J. Aerosol Sci.* 12:297–306.

Pedersen, P. S., J. Ingwersen, T. Nielsen, and E. Larsen. 1980. Effects of fuel, lubricant, and engine operating parameters on the emission of polycyclic aromatic hydrocarbons. *Environ. Sci. Technol.* 14:71–79.

Pilat, M. J., D. S. Ensor, and J. C. Bosch. 1970. Source test cascade impactor. *Atmos. Environ.* 4:671–79.

Pilat, M. J. and T. W. Steig. 1983. Size distribution of particulate emissions from a pressurized fluidized bed coal combustion facility. *Atmos. Environ.* 17:2429–33.

Pratsinis, S. E., H. Bai, P. Biswas, M. Frenklach, and S. V. R. Mastrangelo. 1990. Kinetics of titanium (IV) chloride oxidation. *J. Am. Ceramic Soc.* 73:2158–62.

Presser, C., A. K. Gupta, H. G. Semerjian, and R. J. Santoro. 1990. Application of laser diagnostic techniques for the examination of liquid fuel spray structure. *Chem. Engng Commun.* 90:75–102.

Rao, A. K. and K. T. Whitby. 1977. Nonideal collection characteristics of single stage and cascade impactors. *Am. Ind. Hyg. Assoc. J.* 38:174–79.

Santoro, R. J., H. G. Semerjian, and R. A. Dobbins. 1983. Soot particle measurements in diffusion flames. *Combust. Flame* 51:203–18.

Santoro, R. J., T. T. Yeh, J. J. Horvath, and H. G. Semerjian. 1987. The transport and growth of soot particles in laminar diffusion flames. *Combust. Sci. Technol.* 53:89–115.

Schaefer, D. W. and J. E. Martin. 1984. Fractal geometry of colloidal aggregates. *Phys. Rev. Lett.* 52(26):2371–74.

Schott, J. H. and W. E. Ranz. 1976. Jet-cone impactors as aerosol separators. *Environ. Sci. Technol.* 10:1250–56.

Scotto, M. V., E. A. Bassham, J. O. L. Wendt, and T. W. Peterson. 1988. Quench induced nucleation of ash constituents during combustion of pulverized coal in a laboratory furnace. In *Proceedings of 22nd International Symposium on Combustion*, pp. 239–47. The Combustion Institute.

Sethi V. and P. Biswas. 1990. Fundamental studies on particulate emissions form hazardous waste incinerators. In *Proceedings of the 16th Annual RREL Hazardous Waste Research Symposium.* EPA/600/9-90 073:59–67. Also, *Aerosol Sci. Technol.* 17:119–133, 1992.

Sousa, J. A., J. E. Houck, J. A. Cooper, and J. M. Daisey. 1987. The mutagenic activity of particulate organic matter collected with a dilution sampler at coal fired power plants. *J. Air Pollut. Control Assoc.* 37:1439–44.

Stamatakis, P., C. A. Natalie, B. R. Palmer, and W. A. Yuill. 1991. Research needs in aerosol processing. *Aerosol Sci. Technol.* 14:316–21.

Ulrich, G. D. and J. W. Riehl. 1982. Aggregation and growth of submicron oxide particles in flames. *J. Colloid Interface Sci.* 87:257–65.

USEPA. 1990. Code of Federal Register, CFR 42.

Van de Hulst, H. C. 1981. *Light Scattering by Small Particles.* New York: Dover.

Wu, J. J. and R. C. Flagan. 1987. Onset of runaway nucleation in aerosol reactors. *J. Ap. Phys.* 61:1365–71.

Zachariah, M. R., D. Chin, H. G. Semerjian, and J. L. Katz. 1989. Dynamic light scattering and angular dissymmetry for the in situ measurement of silicon dioxide particle synthesis in flames. *Appl. Opt.* 28(3):530–36.

33

Manufacturing of Materials by Aerosol Processes

Sotiris E. Pratsinis

Department of Chemical Engineering
University of Cincinnati
Cincinnati, OH 45221-0171, U.S.A.

and

Toivo T. Kodas

Department of Chemical and Nuclear Engineering
University of New Mexico
Albuquerque, NM, U.S.A.

AEROSOL PROCESSES

Numerous industrial systems exist in which aerosols are used to make materials such as optical fibers for telecommunications, carbon blacks, pigmentary titania, and fumed silica (Ulrich 1984). Furthermore, as the unique properties of matter in a finely divided state are being recognized, aerosol processes are receiving increased attention for the manufacture of diamonds, metallic fuels, polymers, and structural, optical, and electronic ceramics. In general, aerosol processes are attractive for their simplicity since they do not involve large volumes of liquid byproducts and can be used for the production of high-purity materials (Pratsinis and Mastrangelo 1989).

Most aerosol measurement techniques have been developed for aerosols in environmental systems. Industrial aerosols, however, are quite different from their environmental counterparts. Typically, high concentrations (10^8–10^{14} particles/cm^3) of irregular, fine particles suspended in a variety of gases are encountered in industrial processes. These aerosols are made at high temperatures (1000°C) and massive production rates (10 ton/h) in closed systems and experience a spectrum of physicochemical phenomena within a few seconds. Product yields vary widely. The yield, for example, for pigmentary titania is usually over 90% (Xiong and Pratsinis 1991), while that of optical fibers is under 50% (Bohrer et al. 1985).

Despite the economic significance of industrial aerosol processes and the importance of aerosol characterization in operation and control of these processes, very little has been done for the development of real-time instruments for aerosol characterization in manufacturing processes. There are three reasons for this oddity. First, industrial aerosols have notoriously high concentrations; furthermore, particle formation, growth, and transport take place simultaneously over very short time scales. Removing an industrial aerosol from its environment for sampling and analysis may drastically change its characteristics. For example, an aerosol with 10^{10} particles/cm^3 of average diameter

0.1 μm reduces its concentration by coagulation by 90% within 1 s (Friedlander 1977). Second, powders made by aerosol processes tend to be nonspherical aggregate structures making difficult their rapid characterization by conventional aerosol instruments. Third, because of these difficulties the development of industrial aerosol processes has historically been Edisonian rather than guided by the principles of aerosol science. In fact, the development of some of these processes was well ahead of aerosol science itself!

Aerosol routes in material synthesis can be classified as gas-to-particle and droplet-to-particle conversion, depending, upon the starting material (Friedlander 1982; Kodas 1989; Pratsinis 1990). In gas-to-particle conversion, precursor gases or vapors react forming product particles that grow further by coagulation and/or surface reactions. Powders made by gas-to-particle conversion have relatively narrow size distributions and consist of nonporous, spherical, primary particles. Production of chemical commodities such as carbon black, silica, and titania involves gas-to-particle conversion at high temperatures in the so-called flame reactors (Ulrich 1984).

In the droplet-to-particle route, solution or slurry droplets are suspended in gases by liquid atomization. These droplets are converted to powders by direct pyrolysis or by *in situ* reactions with another gas. The product powder distribution is determined primarily by the droplet distribution and in some cases by particle breakup during pyrolysis or drying. Powders made by this route are rarely agglomerated but can be porous, depending upon the precursor solute concentration and the drying rate (Zhang, Messing, and Borden 1990; Charlesworth and Marshall 1969; Leong 1981, 1987). Spray drying (Lukasiewicz 1989) and spray pyrolysis (Kodas 1989) are typical industrial processes employing droplet-to-powder conversion.

Solids can also be used as starting materials in aerosol processes involving either complete or partial vaporization or melting of the solid and subsequent formation of the product (Bolsaitis, McCarthy, and Mohiuddin 1987; Weimer et al. 1991). It should be noted, however, that employing solids as starting material creates several operational difficulties that have not been resolved on an industrial scale.

In this chapter, processes and materials employed in industrial scale manufacture of aerosols are reviewed and the most common techniques for particle characterization are outlined.

Gas-to-Powder Conversion

The gas-to-powder conversion route refers to "building" particles from individual molecules in the gas phase (Fig. 33-1). The particle formation process is driven by the generation of molecules by chemical reaction from precursor gases or by the rapid cooling of a superheated vapor. High temperatures are usually required to accomplish the reaction or to bring the vapor to the superheated state.

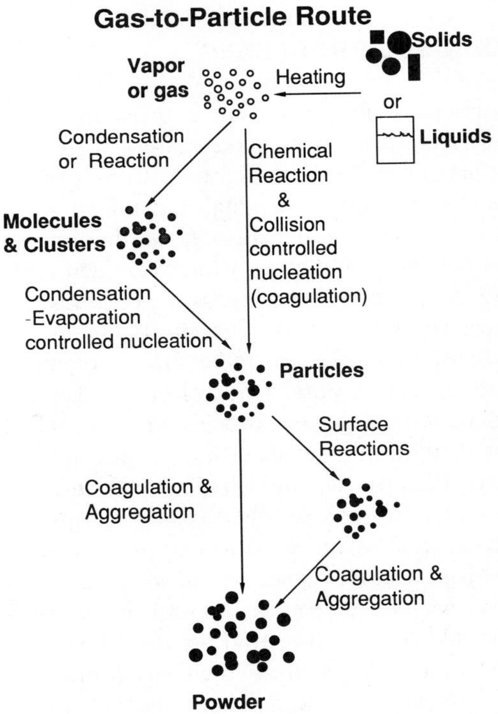

FIGURE 33-1. Physicochemical Processes Occurring During Powder Production from Gases (Gas-to-Particle Conversion). (Adapted from Pratsinis, 1990.)

Depending upon the thermodynamics of the process (Ulrich 1971; Rao and McMurry 1989; Xiong and Pratsinis 1991), the product molecules can form particles either by uninhibited collisions (collision-controlled nucleation) or by balanced condensation and evaporation to and from molecular clusters (condensation–evaporation-controlled nucleation). The newly formed particles grow further by collision with product molecules (condensation) and/or particles (coagulation). When the rate of particle collision is higher than that of particle coalescence (fusion), aggregates of spherical (primary) particles are formed. These are termed hard or soft aggregates (or agglomerates), depending upon how easy it is to break the bonds connecting the primary particles. Rapid cooling favors the formation of aggregates (Schaefer and Hurd 1990; Dobbins and Megaridis 1987; Hurd and Flower 1988; Hurd, Johnston, and Smith 1991; Ulrich and Riehl 1982).

Various energy sources are used to create the high temperatures required during gas-to-particle conversion. The name of these sources is frequently used to signify a specific aerosol reactor in which gas-to-particle conversion takes place. Thus, flame reactors use combustion of hydrocarbons in premixed or diffusion flames. Plasma reactors utilize the high energy of an ionized gas (plasma). Laser reactors employ the high energy and precision of a laser beam. Furnace reactors electrically heat the ceramic tubes through which precursor gases flow.

The gas-to-particle conversion route is used industrially for the production of pigmentary titania, fumed silica, and carbon blacks, and on a laboratory scale for the production of metals, semiconductors, and oxide and nonoxide ceramics. As a result, this route has been studied the most and the mathematical models describing particle formation in various gas-to-particle conversion processes have been developed. The effects of various process variables on the average powder properties (polydispersity and average diameter) are easily described by moment models (Dobbins and Mulholland 1984; Pratsinis 1988). The evolution of the detailed powder size distribution in gas-to-particle processes has been also modeled (Wu and Flagan 1988; Landgrebe and Pratsinis 1990). More recently, these models have been used to create design diagrams facilitating the understanding of the effect of process variables on powder size characteristics (Pratsinis, Landgrebe, and Mastrangelo 1989; Landgrebe, Pratsinis, and Mastrangelo 1990).

The gas-to-particle conversion route is most useful for the generation of single-component powders. It is not as convenient, however, for multicomponent materials, where differences in the vapor pressures and, subsequently, in the nucleation and growth rates of the various species may lead to nonuniform product composition from particle to particle and even within particles. The primary advantages of gas-to-particle routes are: small particle size, narrow particle size distribution, solid particles, and high purity. The disadvantages include the difficulties in producing multicomponent materials, and the problems in handling hazardous gases.

Flame Reactors

These reactors involve particle formation from gaseous precursors in diffusion or premixed flames (Ulrich 1984). Aromatic hydrocarbons are used as precursors of carbon blacks in fuel-rich flames (Medalia and Rivin 1976; Ivie and Forney 1988). A variety of metal oxides, most notably silica and titania, has been produced by oxidation of the respective metal halide vapors in hydrocarbon-supported flames (Ulrich and Riehl 1982; Kriechbaum and Kleinschmidt 1989; Formenti et al. 1972; Hurd and Flower 1988; Sokolowski et al. 1977; George, Murley, and Place 1973). Particle formation in these flames follows the earlier described gas-to-particle conversion route. Additives are introduced in the process to control the phase, shape, and size distribution of the product particles (Mezey 1966; Dannenberg 1971; Haynes, Jander, and Wagner 1979). Flame reactors provide an attractive method for particle generation since they are simple in construction and operation relative to other systems.

Particle formation in flame reactors closely follows the standard description of the gas-to-particle route. Flame-generated aggregates contain from a few up to many thousand primary particles. The typical size for the primary particles ranges from 1 to 500 nm. Maximum flame temperatures are usually of the order of 2500 K. In contrast to other gas-to-particle conversion systems, temperatures are lower in flame reactors than in plasmas but much higher than in furnace, evaporation/condensation, and laser reactors. The residence time and temperature determine whether the product powder is amorphous or crystalline and determines the degree of agglomeration and coalescence.

Advantages of flame reactors are: scale up has been demonstrated for carbon blacks, silica, titania, and other oxides; simplicity; ability to use volatile and nonvolatile precursors; high product purity; large range of diameters for particles in aggregates. The disadvantages of flame reactors are that hard agglomerates are formed under most conditions and that broad particle size distributions (coagulation controlled) are obtained.

Furnace, Laser, and Plasma Reactors

Furnace reactors involve driving chemical reactions of powder precursor vapors in a tubular, furnace-heated reactor (Masdiyasni, Lynch, and Smith 1965; Kanapilly, Raabe, and Newton 1970; Suyama and Kato 1976; Prochaska and Greskovich 1978; Alam and Flagan 1986; Wu and Flagan 1987; Okuyama et al. 1986; Biswas, Li, and Pratsinis 1989). The reactants are introduced into the reactor either as a gas or as an aerosol of a solution. In the case of solution droplets, the solution and precursors in the droplet evaporate in the reactor before gas phase reaction occurs. Thus, the aerosol facilitates introduction of low-volatility precursors into the reactor (Kagawa et al. 1983). Temperatures are usually less than 1700°C (a practical limit for most furnaces) and the reactors are operated at atmospheric pressure.

Particle formation in furnace reactors is reasonably well understood through detailed models for specific chemical systems, such as silicon particle formation by silane decomposition (Wu et al. 1988), silica by $SiCl_4$ oxidation (Kim and Pratsinis 1988, 1989), titania by oxidation of $TiCl_4$ or titanium isopropoxide (Landgrebe, Pratsinis, and Mastrangelo 1990; Okuyama et al. 1986, 1989). Furnace reactors provide excellent control of temperature and flow velocity; so they are frequently used for the measurement of global chemical reaction rates of particle-forming systems (SiO_2: Powers 1978; French, Pace, and Foertmeyer 1978; TiO_2: Pratsinis et al. 1990).

Lasers can be used to drive chemical reactions that result in particle formation by photothermal and photochemical processes (Cannon et al. 1982; Casey and Haggerty 1987). In the photothermal process, the laser energy is absorbed by the carrier gas and raises its temperature, thus driving the chemical reactions. In the photochemical process, the reactant molecules absorb the laser light and then dissociate to form condensable species. Thus, temperatures can be near ambient in contrast to the photothermal case, where temperatures are near 1000°C or more. The photochemical process has not been used extensively for particle formation since expensive reactants and lasers are required. In a typical photothermal process, a CO_2 laser is used to heat gas molecules by light absorption. The laser acts as a heat source since molecules are in thermal equilibrium with the gas. The absence of walls leads to high-purity product. The advantage of using a laser is that its narrow spectral width allows a coupling between the light source and the molecular precursor. The coupling requires near coincidence between laser light wavelength and wavelength of absorption. Efficiencies for the use of the laser power are about 15%. Models for particle formation in laser reactors have been developed, which indicate that particle coagulation determines the product powder size distribution (Flint and Haggerty 1990).

Plasma reactors use a plasma to provide the energy required to drive reactions that result in particle formation (Phillips and Vogt 1987; Pickles and McLean 1983; Gani and McPherson 1980; Girshick and Chiu 1990). A plasma is a system with a high-energy content in which a significant fraction of the species are ionized and are conductors of electricity. Two types of plasma systems are of interest, high-temperature equilibrium thermal plasmas and low-temperature non-equilibrium plasmas. In the thermal plasmas, the temperatures of the electrons and ions are equal. In low-temperature plasmas such as glow discharges, the temperature of the electrons is much greater than that of the ions. The most common type of systems for powder generation utilize thermal plasmas. Applications of thermal plasma processing of materials include plasma synthesis, spraying, and consolidation of materials. Glow discharges are not used extensively for powder generation or processing (Singh and Doherty 1990).

Numerous variations of plasma reactors exist depending upon how the reactants are introduced into the system. Temperatures (5000–25,000°C) are higher than in all other aerosol reactors and complete destruction of reactants is common. This allows the use of molecular and solid feed streams; so, in principle, any material can be processed. Common processes for all these methods are the formation of product species which nucleate to form particles during cooling while exiting the plasma, following the standard formation pathways of gas-to-particle conversion.

Two broad classes of thermal plasma reactors are used: dc arc jet and high-radiofrequency (rf) induction systems. In the case of the dc arc jet, current is supplied to the ionized gas (plasma) by physical contact with a metallic electrode surface. This system is relatively simple and inexpensive. However, the electrodes are consumed and end up in the product resulting in contamination. In the case of a high-radiofrequency induction plasma (rf plasma reactor), there is no contact between the plasma and its power source. The induction coil lies outside the reactor walls like the electrical thermal elements in furnace reactors. Energy transfer takes place through the electromagnetic field of the induction coil. An advantage of the rf plasma reactors is the fact that the electrodes are not consumed; so there is no contamination of the product. Variables that can be controlled are plasma composition and frequency. The most common case is an argon plasma operated at 200 kHz–20 MHz, with typical temperatures about 15,000°C. As a result, plasma reactors can handle high melting point materials and solid powder feeds.

Evaporation/Condensation with or without Chemical Reaction

Evaporation followed by condensation of a species is a convenient method for the formation of nanometer-sized particles. Metals and nanocrystalline ceramics with improved mechanical properties can be produced by these methods (Siegel 1990; Ramsey and Avery 1974; Granqvist and Buhrman 1976). A metal is vaporized into a host gas, where the vapor is cooled, resulting in particle formation by nucleation. Cooling of the vapor can be accomplished by contact with a cooler inert gas or by expansion through a nozzle. Expansion of a supercritical fluid through a nozzle can also be used for particle formation (Peterson, Matson, and Smith 1986; Matson, Peterson, and Smith 1987).

The process of evaporation/condensation is often carried out near atmospheric pressure in inert gases. A typical system involves a vacuum chamber which can be evacuated to pressures of 10^{-6}–10^{-10} mmHg in order to reduce contamination levels. Often a boat is used to hold a metal heated to vaporize it into the gas phase. A natural plume is established by free or forced convection from the metal-containing boat. This boat can be heated continuously or in pulses. Temperatures are usually lower than those in plasma and flame reactors.

In the evaporation/condensation/reaction process, reactions take place at the newly formed particle surface, altering the chemical and/or phase composition of the particles. Usually, a metal halide or an organometallic compound is vaporized into a carrier gas, which is then cooled, resulting in droplet formation by nucleation (Fig. 33-1) (Matijevic

1986; Ingebrethsen, Matijevic, and Partch, 1983; Visca and Matijevic 1979; Ingebrethsen and Matijevic 1980; Matijevic 1987; Kodas, Pratsinis, and Sood 1987). The droplets are then mixed with a gaseous reactant such as water which results in chemical reaction and precipitation of solid particles inside the droplets. The particles are then collected and heated to form the desired product.

Evaporation/condensation processes are reasonably well understood especially when precursors with known physicochemical properties are employed and well-defined flow patterns exist such as in tube flows. Usually, the chemical reaction step does not involve a drastic change in the droplet/particle size; so modeling is confined to droplet formation by condensation. Models for these processes in laminar flows have been developed by Davis and Liao (1975), Pesthy, Flagan, and Seinfeld (1983), and Phanse and Pratsinis (1989) and in turbulent flows by Kodas, Pratsinis, and Sood (1987). The process temperatures are usually less than 500°C and the systems are typically operated at atmospheric pressure.

Droplet-to-Particle Route

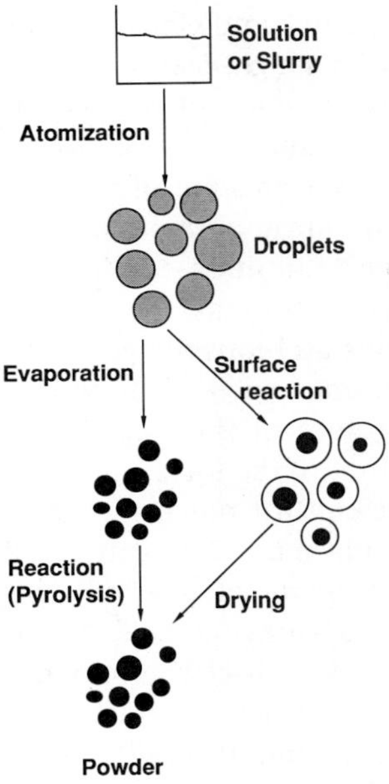

FIGURE 33-2. Physicochemical Processes Occurring During Powder Production from Liquid Droplets (Droplet-to-Particle Conversion). (Adapted from Pratsinis, 1990.)

Droplet-to-Particle Conversion

The most common processes for the generation of powders with liquid phase precursors are spray pyrolysis, spray drying, and freeze drying (Fig. 33-2). The common step in all these processes is the formation of droplets containing molecular or particulate precursors using some type of droplet generator.

In spray drying processes a variety of physical and chemical processes can occur while the particles are suspended in the gas phase. These processes include: droplet generation, evaporation of solvent from the droplets, initial crystallization of the solute in the droplet, and further evaporation from the droplet consisting of solution/solid along with further solute crystallization to form a dried particle.

In the case of spray pyrolysis, these steps are followed by reaction of the precursors in the dried particle to form the product powder and gaseous products, intraparticle transport processes resulting in changes of the particle morphology, evaporation of volatile metals, metal oxides, and other species, followed by condensation on reactor walls and on particles, and by new particle formation and coagulation.

During freeze drying, once the droplets are formed and collected in the cold liquid (hexane or nitrogen), the characteristics of the product powder are determined by the nature of the cold liquid, solvent, precursors, drying rate, and subsequent processing conditions.

In addition to the processes discussed above, coagulation, diffusion, sedimentation, impaction, and thermophoresis in the process vessels and manifold can modify the particle size distribution and influence the yields. Further operations may be required to form the final powder product when droplet-to-solid processes are involved. These processes are typically operated at atmospheric pressure and involve particles in the 0.1–100 μm range.

In general, the advantages of all these processes are: the ability to process organic and inorganic materials; the ability to form a variety of multicomponent materials; simplicity; cost effectiveness (only a few unit operations); many choices for inexpensive liquid phase precursors; doping is possible; scale-up has been already demonstrated to ton quantities; uniform chemical composition; relatively safe process since volatile precursors not required. The disadvantages for these processes are: porous or hollow particles can be formed at certain conditions and the spread of particle sizes is limited by the spread of the generated droplets.

Droplet Generation

The droplet generation process is critical to all droplet-to-particle processes since it largely controls the particle size distribution of the product. A variety of generators are available (Lefebvre 1989; Kerker 1975): pressure, rotary, air-assist, air-blast, ultrasonic, electrostatic, and vibrating-capillary atomizers. Other atomizers such as sonic, windmill, flashing liquid jets, and effervescent atomization are also used.

Each of these types of droplet generators produces different particle size distributions, and has different advantages and disadvantages. Typical average droplet diameters are 1–100 μm. Pressure atomizers work by discharging a liquid through a small aperture under high pressure into a slow-moving gas stream. Rotary atomizers operate by centrifugal atomization. A liquid is directed onto a rotating element that converts the liquid stream to droplets. Air-assist atomizers involve exposing the liquid to a stream of air or steam flowing at high velocity. Air-blast atomizers are similar to air-assist nozzles, with the exception that the former use much higher quantities of air or steam and at lower velocities. Ultrasonic atomizers use a transducer or horn which vibrates at ultrasonic frequencies to produce the short wavelength required for atomization. Flow rates are lower than can be obtained for air-blast, air-assist, pressure, and rotary atomizers. Electrostatic atomizers expose a liquid jet or film to a strong electrical pressure. The expansion in the droplet is opposed by surface tension. Although this method is capable of producing submicrometer droplets, it does not allow high liquid flow rates and is commonly used on a laboratory scale. Vibrating-capillary atomizers produce droplets down to 30 μm. Again, only low liquid flow rates are possible and the latter atomizers are mostly used on a laboratory scale.

Measurements of droplet size distributions have been routinely made by optical methods. The prediction of droplet size distributions is accomplished through empirical correlations, accounting for liquid properties, atomizer design, and fluid flows (Lefebvre 1989). The most relevant properties of a liquid are surface tension, viscosity, and density.

Spray Pyrolysis

Spray pyrolysis involves passing suspended precursor droplets through a high-temperature region (furnace or a flame). There, rapid evaporation of the volatile components of the droplet takes place first. Further residence in this region allows chemical reactions to occur in the solid phase that form the product powder. Depending upon the process conditions and the material properties, solid or porous powders are formed and collected on a filter. This process has been called evaporative decomposition, spray roasting, spray calcining, Aman process, atomizing burner technique, decomposition of misted solutions, aerosol or mist decomposition method, and others (Pebler and Charles 1989; Chadda et al. 1991; Wenckus and Leavitt 1957; Vollath 1990; Epstein 1976; Imai and Takami 1985; Kodas et al. 1988, 1989; Kodas, Engler, and Lee 1989; Biswas et al. 1989, 1990; Zhang, Messing, and Borden 1990; Gardner, Sporson, and Messing 1984).

Spray pyrolysis allows the generation of particles with a uniform chemical composition from particle to particle since the precursors are present in the correct stoichiometry in each droplet. Volatile precursors are not required; thus, a wide variety of reactants can be used while avoiding the need for carbon-containing precursors. This allows the use of inexpensive reactants such as metal nitrates, chlorides, fluorides, etc. The generation of multicomponent ceramic powders can be carried out on an industrial scale using only a few unit operations; the ability to scale up these systems for the production of ton quantities of conventional ceramic oxide powders has already been demonstrated (Ruthner 1979). The process is relatively safe since volatile precursors are not required. Unagglomerated particles are produced since particle growth rarely involves collisions between particles. High purities are also obtained since milling is not required.

It is possible to vary the porosity of the powders by changing the precursor solute concentration (Zhang, Messing, and Borden 1990), by controlling the temperature profile during pyrolysis (Ortega et al. 1991), and by introducing seed particles (nuclei) in the droplets.

The success of the particle-to-particle conversion route relies heavily on selecting the appropriate liquid precursors and solvents. A challenge is the development of the chemistry of these precursors! Multicomponent and composite powders can also be made by using mixtures of the corresponding precursors. Solid metal or oxide powders can be directly used as precursors for ceramic powders, as in carbothermal reduction of B_2O_3 for synthesis of B_4C (Weimer et al. 1991), and in nitridation of suspended Al powders for the synthesis of AlN.

Particle characterization techniques play a crucial role in understanding the effect of the various process variables on product particle characteristics. Key challenges include determining the morphology of the particles, their chemical composition, their phase composition, and their behavior in liquids during ceramic powder processing operations. Examples are presented here to demonstrate how the various characterization techniques can be used to examine these problems.

Spray Drying

Spray drying is similar to spray pyrolysis with the only difference that it involves heating the droplets to lower temperatures; so solid phase reactions do not occur (Masters 1972; Johnson 1981). An additional difference is that much larger droplets are used in spray drying. This is a consequence of the goal of spray drying, which is usually to form powder consisting of relatively large free-flowing granules (100 μm). Charlesworth and Marshall (1969) have discussed some of the possible particle morphologies and how they are produced in spray drying.

The primary problems involved in the characterization of spray-dried particles include determining the chemical homogeneity and particle morphology as in spray pyrolysis.

Freeze Drying

This process is similar to spray drying as far as droplet generation, but the droplets are frozen and then dried once in the solid state (Johnson, Gusman, and Rowcliffe 1987). The solution containing dissolved compounds is sprayed into liquid nitrogen or some other cold liquid, where the droplets freeze. The liquid is removed by sublimation and the dry product is converted to the desired product by further heating.

MATERIALS

Aerosol processes are used industrially to produce a variety of products. Each of the gas-to-particle conversion processes discussed in this chapter has its own advantages and disadvantages. Table 33-1 presents a comparison of these processes.

The materials that are made in the largest quantities include carbon black and ceramic powders. One of the most prominent examples of ceramic powder generation is TiO_2 for use as a pigment. Carbon blacks are the oldest manufactured aerosols. Today, they are made primarily in flame reactors. An aromatic ("feedstock") residual oil is sprayed into a hot gas–air flame in a furnace where carbon black is formed as soot. The flue gas is rapidly quenched and the carbon black is collected by electrostatic precipitators, cyclones, and baghouse filters (Medalia and Rivin 1976). The product carbon black is

TABLE 33-1 Comparison of Aerosol Processes for Powder Production

	Flame	Evap/ Cond/ Reaction	Laser	Plasma	Hot Wall	Spray Pyrolysis
Max Size, μm	1	0.1–10	1	1	10	0.10–100
Spread	broad	narrow	narrow	broad	narrow	broad
Morphology	solid agglomerates	solid	solid	agglomerates solid	spherical solid	spherical solid, porous hollow
Max T, K	2500	< 2000	2000	25,000	2000	1600
Material	oxides	metals and oxides	non-oxides oxides	non-oxides oxides	non-oxides oxides semiconductors	non-oxides oxides
Complexity	low	low	medium	high	low	low

made of aggregates, each containing several elementary spherical particles. Depending upon the commercial application, different grades of carbon black are manufactured by empirically controlling the primary particle size and the extent of aggregation. Fine particles, for example, can be made at short residence times and high temperatures by adjusting the air flow rate and the length-to-diameter ratio of the furnace (Mezey 1966). The aggregation of particles is usually controlled by the addition of traces of potassium salts into the flames (Dannenberg 1971). Particle formation proceeds by hydrocarbon oxidation in the diffusion flame, which is followed by soot particle formation by cyclization reactions at the fuel-rich side of the flame (Santoro and Miller 1987). The soot particles then grow by coagulation and surface reactions (material addition from the gas phase). Measurement problems that arise in this case include determination of particle size and morphology.

The other major application of aerosol reactors is the generation of pigmentary titania and fumed-silica powders in flame reactors. These reactors are similar to those used in the manufacture of carbon blacks. Precursor vapor ($TiCl_4$ or $SiCl_4$) with small amounts of hydrcarbon fuel are injected with nearly stoichiometric amounts of oxygen into the reactor. Rapid, exothermic oxidation of the vapor takes place forming oxide powders and chlorine gas (Clark 1977). These powders are loose aggregates of submicrometer-sized particles. The crystalline structure of the particles can be controlled by adding dopant vapors in the feed stream and by tailoring the temperature history of the reactor (Mezey 1966). Measurement problems of interest include determining the size distributions *in situ* and of collected particles, particle morphology, phase and chemical composition, and pigmentary properties.

A number of other examples of ceramic powder generation by aerosol routes exist. Spray pyrolysis is used on an industrial scale to produce simple and complex metal oxides (Ruthner 1979). These ceramic powders are used in a wide variety of applications for the manufacture of substrates for catalysts and integrated circuits, structural ceramics, and ceramics with specific optical, magnetic, and electrical properties. In most of these cases, manufacture of submicrometer-sized powders with precisely controlled physical and chemical properties is of utmost importance in making ceramic parts with the minimal number of imperfections (microcracks and flaws) and with optimal electrical, magnetic, and optical properties. Currently, ceramic powders are made primarily in spray pyrolysis and flame reactors. Other types of reactors such as furnace, plasma, and laser reactors are used on a much smaller scale. In all these cases, measurement problems include determination of particle size distributions and morphology, chemical and phase content, and microstructural information. Examples of different types of materials and the determination of their physical and chemical properties are presented in later sections.

Generation of powders using flame processes relies on introducing species into the flame that react to form gaseous species with low vapor pressures (Table 33-2). The reactants that have been used most often are metal chlorides because of their low cost and high volatility. Some processes have relied on using precursors in the aqueous phase. The solution is formed into droplets that are sent into the flame where evaporation takes place to provide gaseous reactants (Zachariah and Huzarewicz 1991).

Aerosol routes also allow direct production of thick and thin films of a variety of materials via aerosol deposition onto surfaces. A variety of aerosol processes exist for film formation. Films can be formed by either deposition of droplets or by deposition of solid particles onto surfaces. Examples of deposition of droplets onto surfaces include spray pyrolysis (reviews by Tomar and Garcia 1981; Mooney and Redding 1982). This route begins with a solution of molecular precursors which is converted into droplets. These droplets are directed onto a heated surface where the solvent evaporates and the precursors decompose, resulting in film formation. A related process involves delivery of volatile species into a reactor using an aerosol (Koukitu et al. 1989; Viguie and Spitz 1975; Siefert 1984; Tang et al. 1990; DeSisto et al. 1991). An example of solid particle deposition for film formation is the cluster deposition process (Takagi et al. 1980; Andres et al. 1989). Many other examples also exist including the formation of thick films of ceramics (Kodas, Engler, and Lee 1989; Baker, Hurng, and Steifink 1989; Komiyama et al. 1987; Adachi et al. 1988; Koguchi et al. 1990). However, the most well known and successful example of a solid particle deposition process is optical fiber generation, in which the unique capacity of aerosol processes for manufacture of high-purity particles has been exploited (MacChesney, O'Connor, and Presby 1974).

Optical fibers are made by fabrication of a preform glass rod, rod sintering, and fiber drawing and coating (Nagel, MacChesney, and Walker 1982). The key process with respect to the light transmission and mechanical properties of the fiber is the fabrication of the preform rod. The manufacturing goal of the latter process is to make a preform with a prescribed radial refractive index profile at the maximum yield. Preforms are made by silica and dopant particle deposition onto thin substrate rods (external processes) or into hollow, substrate glass tubes (internal processes).

In the external processes that were invented by Corning Glass Co., the particles are generated by a flame reactor and deposit primarily by thermophoresis onto the substrate. The flame-generated particles are too large for rapid Brownian diffusion and too small for inertial impaction to be important for particle deposition.

In the internal processes that were invented at AT&T, halide vapors and oxygen flow through a rotating, hollow, quartz tube, which is externally heated by a slowly, axially

TABLE 33-2 Powders Made by Flame Reactors

Product	Reactant	Author
Al_2O_3	Al acetylacetonate	Sokolowski et al. (1977)
SiO_2	$SiCl_4$	Ulrich (1984)
TiO_2	$TiCl_4$	George et al. (1973)
Carbon black	Alkanes	Dannenberg (1971)
Al_2O_3, Cr_2O_3, Fe_2O_3, GeO_2, SiO_2, SnO_2, TiO_2, V_2O_5, ZrO_2	Chlorides	Formenti et al. (1972)

traversing torch (or plasma). Inside the tube, the halide gases are oxidized, forming particles that either deposit on the tube wall by thermophoresis or exit the tube with the process gases. The preform has been made when the quartz tube is almost filled with glassy layers of particles that were fused by the torch. Currently, preforms are made on an industrial scale at rather low yields [50% for silica and much lower for costly dopants such as germania (Bohrer et al. 1985)].

A major challenge in the manufacture of optical fibers is to understand the relationship between deposit properties, glass particle size, and composition and process conditions. This is important since it has been observed that during sintering of multicomponent particulate deposits, small particles tend to be depleted of dopants much faster than large particles and bubbles can be present in deposits. This relationship is also important for improvement of the current low process yields. As the fiber optics market becomes more competitive, the low yields become a critical issue from both economic and environmental viewpoints (toxic halide vapors can be released when they are not converted to particles). Thus, there is a need for *in situ* measurements of particle size and composition.

MEASUREMENT TECHNIQUES

Physical and chemical characterization of materials produced by aerosol processes involves many techniques, summarized in Tables 33-3 and 33-4. Niessner and Fresenius (1990) have reviewed *in situ* aerosol measurement techniques, most of which are included in Chapters 12 and 13 of this book.

Physical Properties

Size

A wide variety of methods exist for the measurement of particle size distributions in the gas and liquid phases. Measurements of particle size distributions in the gas phase are discussed in Chapters 15, 16, and 18 of this book. Recent advances in particle size measurements for aerosol particles that have been collected and placed in liquids have been summarized by Miller and Lines (1988). Table 33-3 lists the techniques for measurement of particle size. A critical problem that arises when using several techniques for measurement of size is that each technique relies on different physical characteristics of the particles such as optical properties, electrical properties, and sedimentation velocity. As a result, particle sizes determined from different techniques can vary significantly. Examples of this will be given in later sections.

Light extinction by aerosols is a powerful method for the determination of particle size distributions. The method can be applied *in situ* in the gas phase and can also be used for particles that have been collected and suspended in liquids. Details of light absorption and scattering processes (static and dynamic) are discussed in Chapters 15 and 16 of this book.

The most basic particle size analysis involves a microscope. In recent years scanning and transmission electron microscopes (SEM and TEM) are used for direct size characterization of fine particles. These methods involve the examination of particle images followed by collection of the images by a computer and presentation to an image analyzer for automatic characterization. In addition to particle size distributions, particle shape is also best determined by SEM and TEM examination of particles.

The most common image analysis systems consist of a microscope, a video camera, and an image analyzing computer. The images of the particles can be collected with optical, scanning electron and transmission electron microscopes, depending upon the size range of interest. For example, TEM allows sizing of particles from the nanometer to the micrometer range. Only small amounts of materials are needed. For example, a filter can be inserted into a flow path to collect particles. A good dispersion of particles is required so only single particles are present on the filter surface. The filter surface can then be examined by a microscope. Figure 33-3 shows a

TABLE 33-3. Summary of Characterization Techniques for Collected Particles

Physical	Chemical
Size	***Phase Composition***
– Sedimentation velocity	– XRD
– Image analysis and microscopy	– Analytical TEM
– Light scattering	
– Sieving	
– Chromatography and field flow fractionation	
– Electronic methods	
Shape	***Elemental Composition***
– SEM	**Bulk**
– TEM	– X-ray fluorescence
– Optical microscopy	– Emission and absorption spectroscopies
– STM/AFM	– FTIR
	– PIXE
	– NRA
	– GC/MS
	– NMR
	Surface Sensitive
	– AES
	– EDS
	– XPS
	– Mass spectrometry
Microstructure/Porosity	***Thermochemical and Thermophysical***
– Mercury porosimetry	– DTA/DSC
– Surface area	– TGA
– He density	– TGA/MS
– Pellet Green density	
Microstructure/Grain size	***Assorted***
– TEM	– SAXS
– SEM	– SANS
– STM/AFM	– ESR
	– EPR
	– Raman spectroscopy
	– Magnetic and electrical properties
	– Optical properties (pigments)

TEM photograph of TiO_2 particles produced by a furnace aerosol reactor. A comparison of the results from SEM/image analysis and a differential-mobility particle size analyzer is shown in Fig. 33-4.

Sedimentation velocity measurements of particle size rely on the Stokes law (Reed 1988). Particles are collected and then placed in a fluid. The settling velocity of the particles is detected by measuring light or X-ray transmission. A calculation of particle size distributions assumes that the particles are solid and spherical since the Stokes law is used to relate velocity to diameter. In practice, particles greater than 0.1 μm can be sized by this method, while particles below 0.1 μm are influenced too strongly by Brownian motion to allow measurements, unless centrifuges are used. Particles larger than 50 μm do not follow the Stokes law, defining an upper limit for this technique. Figure 33-5 shows a particle size distribution obtained by this technique

TABLE 33-4 Particle Size Analysis Techniques

Method	Size	Sample Size
Sedimentation		
Velocity		
(Gravity)	0.2–100 μm	< 1 g
(Centrifuge)	0.02–100 μm	< 1 g
SEM	0.01–50 μm	mg level
TEM	0.001–1 μm	
Optical microscope	0.2–400 μm	
Sieving	30–5000 μm	5–20 g
Electronic Methods	0.3–400 μm	< 1 g
Chromatography and Field Flow Fractionation	1 nm–200 μm (depending on technique)	< 1 g

for $Ba_{0.86}Ca_{0.14}TiO_3$ particles produced by spray pyrolysis using a mixture of metal acetates, lactates, and nitrates (Ortega et al. 1991).

Sieving is a simple and easy way to fractionate particles according to size (Reed 1988). This method is limited to particles that are much larger than 1 μm, usually greater than about 50 μm. Agglomeration becomes a problem below this size. Smaller particle sizes can be sized using liquids, but these methods are often difficult to carry out reproducibly.

Chromatography and Field Flow Fractionation

These are separation methods whereby a liquid suspension of particles is classified into size fractions that must later be quantified in order to produce a particle size distribution (Miller and Lines 1988). The chromatographic techniques involve a separation of particles according to size within a flow channel, as in conventional chromatography. In field flow fractionation, the particle retention is induced by an electrical, centrifugal, or thermal gradient.

Hydrodynamic chromatography (HDC) involves passing particles through a packed bed of nonporous column material. The rate of transport of particles depends on the size of the particles, the size of the spheres in the bed, and the flow rate and ionic strength of the eluting solution. This technique is useful for particles from 20 nm to 2 μm. Size exclusion chromatography utilizes a porous bed. The flow through the porous material imposes a separation effect by steric exclusion. The upper limit of this technique is roughly 500 nm with a lower limit of essentially molecular size. Capillary particle chromatography is an extension of HDC. This method used a long capillary tube instead of a packed bed to

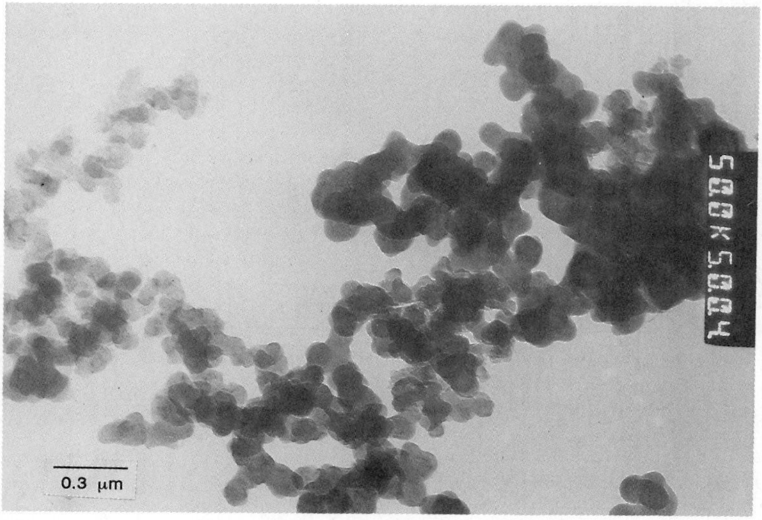

FIGURE 33-3. TEM of Titania Particles Produced at 1300 K by $TiCl_4$ Oxidation in a Furnace Aerosol Reactor. (Courtesy of M. K. Akhtar.)

provide the separation. Particles from 0.2 to 200 μm can be separated.

Field flow fractionation methods can be used for particles from 10 nm to 1 μm. Fractionation in the centrifugal sedimentation version of this process occurs within closely spaced parallel plates. The particles are caused to roll or tumble along the outer wall to produce the segregation.

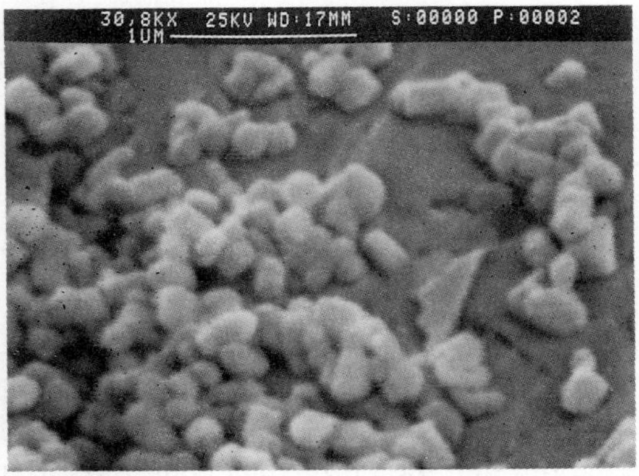

(a)

These methods have not been used extensively for the characterization of particles produced by aerosol processes, primarily because they require a resuspension of collected particles in a liquid. Resuspension of submicrometer particles into a liquid is not simple and can lead to significant errors in measurements of size distributions.

Electrical Sensing Methods (Coulter Counter)

Particles to be sized are suspended at low concentration in an electrolyte solution, which is drawn through a small aperture in an insulating wall, across which a current also flows (Miller and Lines 1988). Each particle that passes through the sensing zone displaces a certain amount of electrolyte solution, which results in a momentary change in the electrical impedance across the aperture. The volume of solution displaced by the particle is determined by the amplitude of the electrical pulse produced as the particle

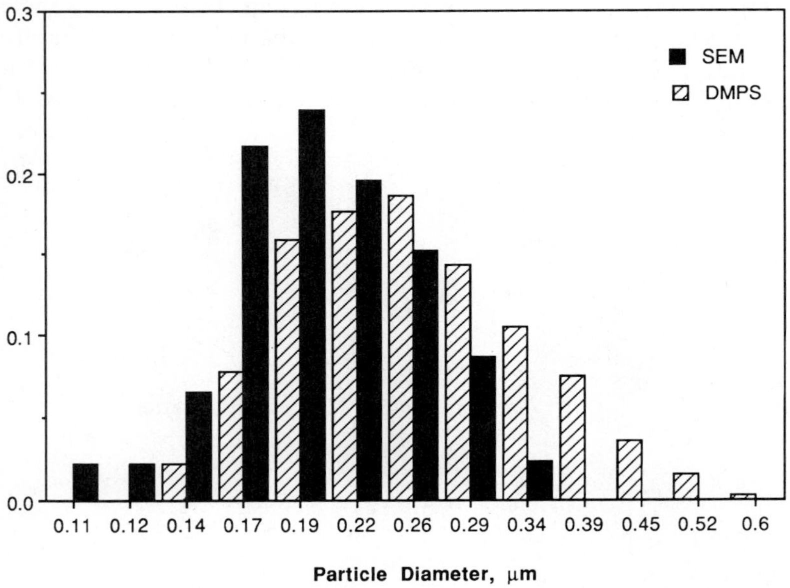

(b)

FIGURE 33-4. (a) Scanning Electron Micrograph (SEM) of Titania Particles Produced at 1723 K by $TiCl_4$ Oxidation in a Furnace Flow Reactor. (b) A Comparison of the Size Distributions Obtained by Image Analysis of the SEM Micrograph and those Obtained by the Differential-Mobility Particle Sizer. (Courtesy of M. K. Akhtar.)

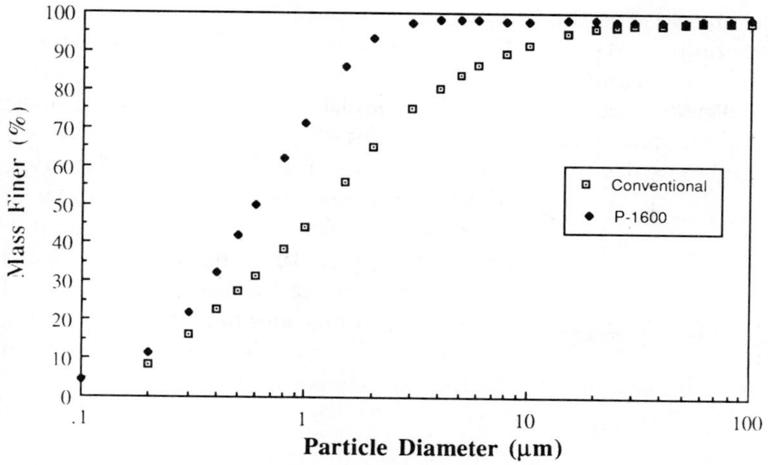

FIGURE 33-5. Particle Size Distribution Obtained by Sedimentation Velocity for $Ba_{0.86}Ca_{0.14}TiO_3$ Particles Produced by Spray Pyrolysis (P-1600) and by Conventional Milling of Ceramic Oxide Mixtures. (Adapted from Ortega et al. 1991.)

passes the sensing zone. Extreme dilution is required to avoid coincidence errors. Particles 0.1 to roughly 400 μm can be sized using different size apertures.

Microstructure

The description of the microstructure of particles involves characterization of porosity, surface area, density, and grain size. The microstructure of particles plays a critical role in determining the rates of densification of ceramic particles. Similarly, particles used for catalytic applications require a knowledge of the total surface areas.

Mercury Porosimetry

This technique utilizes mercury intrusion to obtain pore size distributions in powders and solids (Reed 1988). A sample is placed into the instrument, where mercury under a controlled pressure is forced into the pores of the material. As the pressure is increased, more mercury enters the pores. The accessible pore size corresponding to a given pressure is obtained from the Washburn equation, which relates the applied pressure to the intruded radius of a cylindrical pore. Pore size distributions from 2 nm to 200 μm can be easily obtained.

Surface Area

Surface areas of powders are usually measured using nitrogen or helium adsorption followed by analysis of the data using the Brunauer–Emmet–Teller isotherm (Reed 1988). The estimates of surface areas provide information about the porosity of particles and the primary particle size in particles that consist of agglomerates of smaller particles. The primary particle size can be related to surface area assuming that the particles are smooth, monodisperse spheres. Specific surface areas can range from a few m^2/g to hundreds of m^2/g in the case of highly porous materials or very fine particles.

Density

In many cases, it is useful to know the density of a material in particles. Helium density measurements provide the density of a material, assuming that no closed porosity is present (Reed 1988). A sample is weighed and placed in a vessel into which helium is introduced. The volume of the sample is obtained from the volume of gas displaced by the powder. The density is obtained from the volume and weight of the sample. For this reason, hollow particles with no porosity can lead to erroneously low densities. In contrast, if the density is known, He density measurements become a method for the determination of the amount of closed porosity in a sample.

Pellet Green Density

An indirect but practical measure of the quality of a powder is the green density that can be obtained by dry pressing or other powder processing techniques. Typically, about 100 mg of powder is needed. Hollow particles usually lead to low green densities, 20%–40%. In contrast, solid particles can provide green densities of 50%–60%. It must be noted that this is a crude measure since the type of processing can result in large variations in green density (Fig. 33-6). A related measurement is the tap density. This technique, which involves measuring the apparent density of powders, also provides a way of quantifying the packing characteristics of powders from an applications point of view.

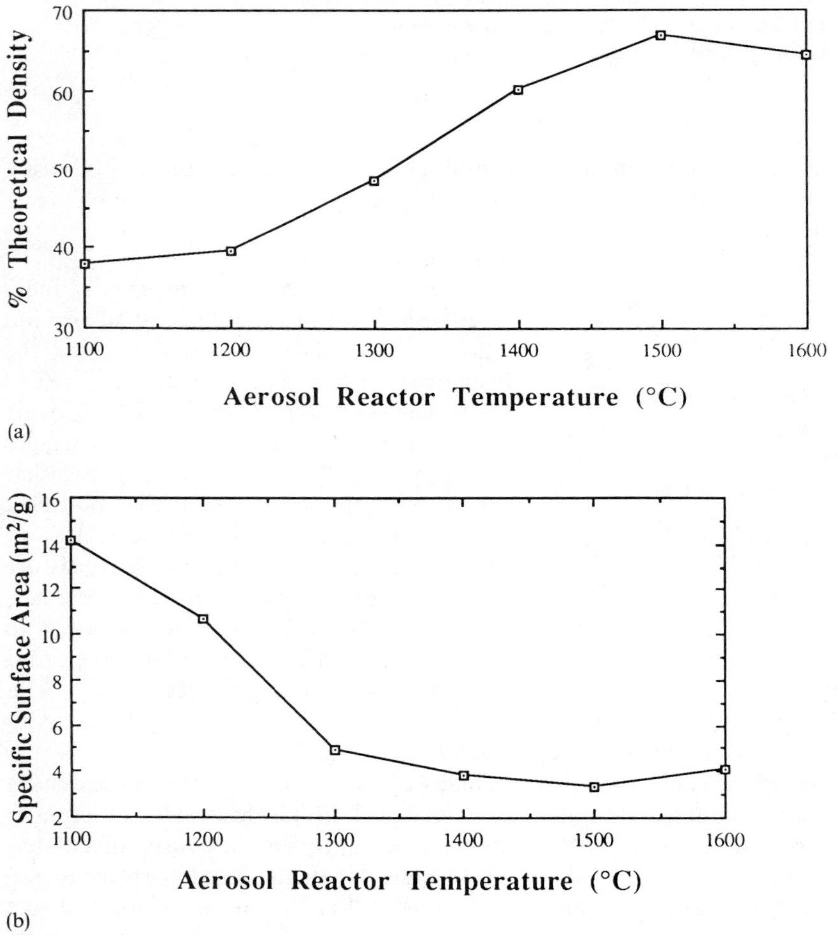

FIGURE 33-6. Green Density (a) and Surface Area (b) of $Ba_{0.86}Ca_{0.14}TiO_3$ Particles Produced by Spray Pyrolysis as a Function of Production Temperature. Particles Produced at Lower Temperatures are Hollow and Lead to Low Green Densities. (Adapted from Ortega et al. 1991.)

Transmission Electron Microscopy

Processing of ceramic or metallic powders involves particle sintering. Sintering rates and the grain size distribution of densified ceramics depend on the initial grain size distribution within each particle and the size distribution of the particles. For example, Edelson and Glaser (1988) have shown that sintering of spherical articles consisting of agglomerates of smaller particles can result in broad grain size distributions. Broad grain size distributions are usually undesirable (Barringer et al. 1984). Thus, it is useful to know the size of grains in particles. In addition, particles with smaller grain sizes have larger grain surface areas which provides a large driving force for increase of grain size (Kingery, Bowen, and Uhlmann 1976). Thus, particles consisting of nanocrystalline grains are usually more active than single-crystal particles. TEM provides a powerful method for the determination of grain sizes. Crystallites with dimensions ranging from nanometer to micrometer can be observed. Figure 33-7 shows how TEM can be used to distinguish if particles are nanocrystalline or single crystals (Feldman and Mayer 1986). The separate crystallites in the particles are visible within each particle.

Phase Composition

X-Ray Diffraction (XRD)

This is a standard tool for identification of crystalline phases in powder samples (Feldman and Mayer 1986). The crystalline phases in the sample diffract X-rays according to Bragg's equation which relates lattice spacing to the wavelength of the X-rays used as a probe. The amount of sample needed in practice is roughly 100 mg or more. Crystalline phases present at levels of about 1% or greater can be detected. The unit cell dimensions (a, b, c) of the crystal can also be determined and provide more information for complex structure materials (Biswas et al. 1990). Amorphous materials are not observed. In routine work, XRD relies on the availability of standards which allow the identification of peaks in the diffraction pattern. Figure 33-8 shows phases of aerosolized made titania and aluminum nitride identified by XRD (Akhtar, Xiong, and Pratsinis 1991).

TEM/Electron Diffraction

Microdiffraction with TEM allows a determination of the crystalline structure of phases present in particles (Feldman and Mayer 1986). Since single particles can be examined, only very small amounts of a sample are needed. The electron diffraction pattern allows a determination of whether the particles are polycrystalline (ring patterns), amorphous, or a single crystal (spot patterns). Figure 33-9 shows how different phases in a submicrometer-sized particle can be identified by electron diffraction by TEM (Carim, Doherty, and Kodas 1989).

Chemical Composition

A variety of techniques exist for determining the chemical composition of aerosol particles. A more extensive discussion of X-ray fluorescence, proton-induced X-ray emission, nuclear reaction analysis, and emission and absorption spectroscopies is presented in Chapter 13. This chapter outlines these methods and provides extensive reference materials in the context of material synthesis.

Methods for Bulk Materials

X-ray Fluorescence (XRF)

This is a method for the determination of the chemical composition of liquid and solid samples (Feldman and Mayer 1986). The sample is irradiated with X-rays, which results in fluorescence. The wavelength of the emitted light is characteristic of the elements in the sample. This technique can be used for examination of multicomponent solid samples where it is desirable to determine the ratio of the elements present in the sample. Elements with $Z > 11$ are detectable down to 10 ppm. Only 10–100 nanogram levels of particles are required on a filter in order to determine chemical composition.

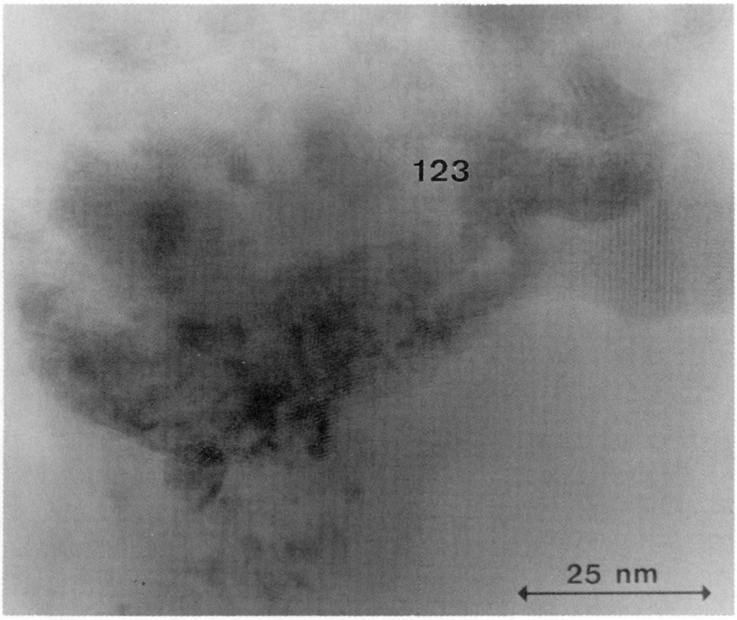

(a)

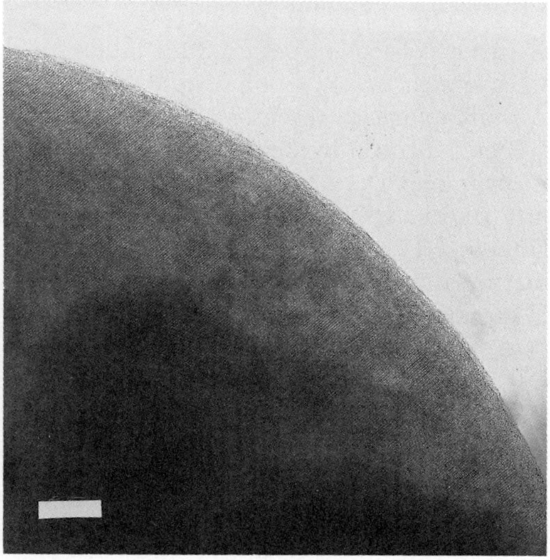

(b)

FIGURE 33-7. (a) $YBa_2Cu_3O_{7-x}$ Particles with Nanometer Grains. (b) Single-Crystal $YBa_2Cu_3O_{7-x}$ particles (bar = 10 nm). Particles were produced by spray pyrolysis. (Adapted from Carim et al. 1989.)

Fourier Transform Infrared Spectroscopy (FTIR)

This technique measures the amount of infrared radiation absorbed by a sample as a function of frequency (Reed 1988). It provides information about the different types of bonds present in a sample. The amount of sample required for obtaining a spectrum depends on the nature of the sample, but 100 mg amounts are often adequate.

Proton-Induced X-ray Emission (PIXE) Spectroscopy

This technique involves the use of ions to induce the emission of X-rays from a sample (Feldman and Mayer 1986). Elements from Na to U in the periodic table can be detected. PIXE allows the use of 10 μm beams for high-sensitivity detection of trace elements. PIXE also allows the identification of a number of elements simultaneously in complex samples.

Nuclear Reaction Analysis (NRA)

This technique (also known as instrumental neutron activation analysis) involves probing a sample using an ion beam and examining the light ion reaction products (Feldman and Mayer 1986). The probe beam induces radioactivity in the sample, which is examined

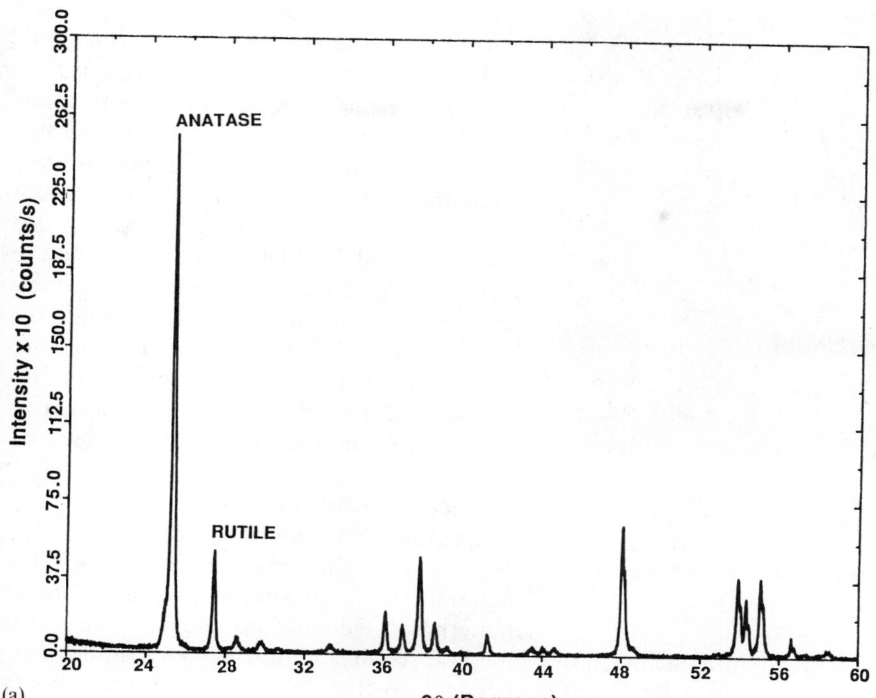

(a)

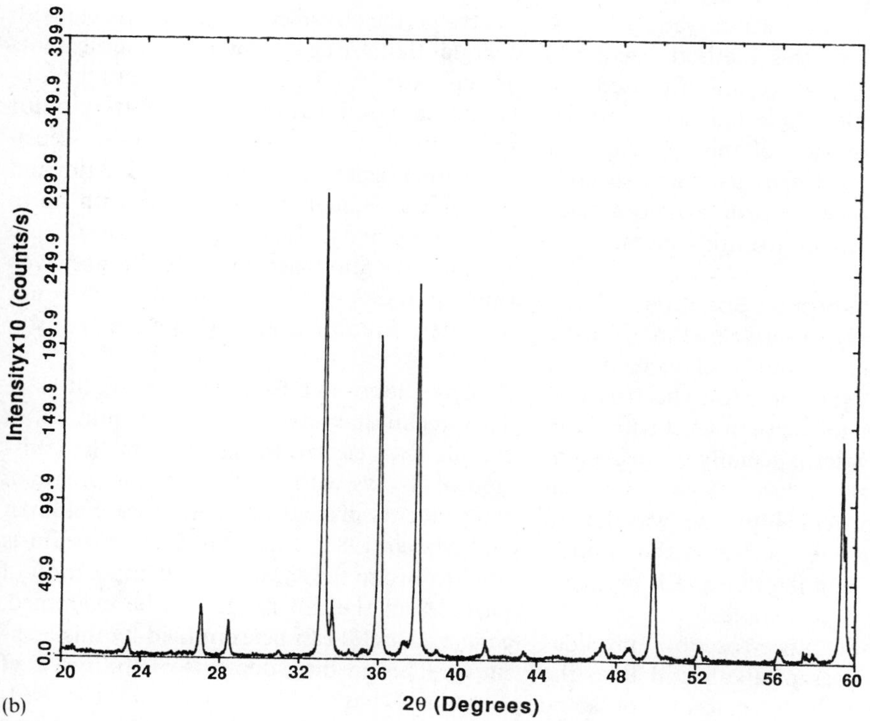

(b)

FIGURE 33-8. (a) X-ray Diffraction Patterns (XRD) for (a) Titania and (b) Aluminum Nitride Powder Produced in a Furnace Aerosol Reactor. (courtesy of M. K. Akhtar.)

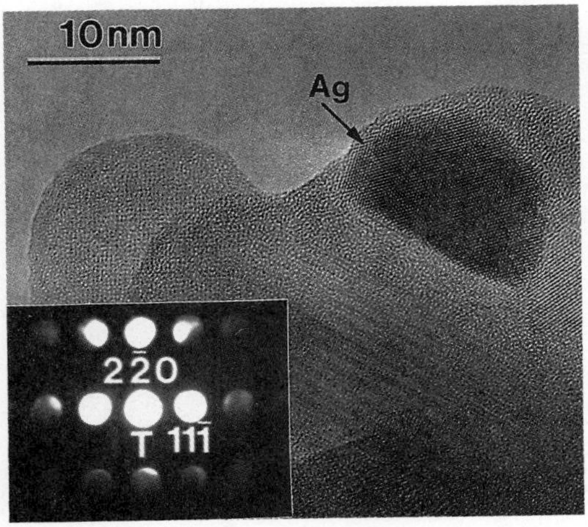

FIGURE 33-9. Microdiffraction Pattern for Crystallite in $YBa_2Cu_3O_{7-x}$ Particle. Spot Diffraction Pattern Demonstrates that Particle is a Single Crystal. (Courtesy of Larry Allard and Abhaya Datye.)

to determine the nature of the elements in the sample. Light elements such as H, Be, B, C, O, F can be detected. The detection limit for thermal neutron activation analysis is 10^{-8}–10^{-10} g. Since this method has poor spatial resolution, it is commonly used for samples of particles collected on filters of impactors. A drawback of this technique is that it requires a nuclear reactor and specialized expertise. However, several laboratories provide this service on a routine basis.

Emission and Absorption Spectroscopies

These are extremely sensitive methods for the detection of trace amounts of elements in samples (Reed 1988). Emission spectroscopy provides analyses to the ppm level with 5 mg samples. The powder is usually excited in an electric arc or by a laser flash. Chemical information is provided by the wavelength and intensities of the spectra of the emitted light. This technique is often used for quick qualitative surveys of samples.

Flame emission spectroscopy provides quantitative analyses of alkali and B to the ppm level, with ppb detectability for some elements. This technique is commonly used for liquid samples which can be sprayed into flames. Inductively coupled plasma emission spectroscopy also utilizes liquid samples and is a reliable, fast, and multielement method.

Atomic absorption spectroscopy can analyze 30–40 elements present at concentrations of $< 0.1\%$. This technique is the industry standard for quantitative analysis to ppm levels for solution samples, although it analyzes one element at a time whereas ICP can analyze about 40. The sample in solution form is sprayed into a flame where the radiation from a lamp is passed through the flame. Dissociated atoms in the flame absorb the radiation and reduce the transmitted intensity. The identity of the species present is determined from the absorption wavelengths.

Surface-Sensitive Methods

Auger Electron Spectroscopy (AES)

This is a surface-sensitive method for the determination of the chemical composition of materials (Feldman and Mayer 1986). This method requires an ultrahigh vacuum chamber and involves directing an electron beam at a sample, which results in the emission of electrons (the so-called Auger electrons) with energies that are characteristic of the elements in the sample. Depth profiling can be obtained in combination with sputtering for solid samples. This method is surface-sensitive with a penetration depth of 0.5–2 nm and can detect elements from Li to U with up to 0.3% sensitivity. Instruments with 0.1 μm spatial resolution are available. Since even single particles can be examined, the amount of material required is very small.

Energy-Dispersive Spectroscopy (EDS)

This technique involves the probing of a sample with electrons, resulting in the emission of X-rays with energies characteristic of the elements present in the sample (Feldman and Mayer 1986). Since an electron beam is used to probe the sample, the composition of particles in the nm range can be examined. Single particles can be examined by this technique. The probing depth is of the order of 1 μm. EDS has 0.1% detectability for $Z > 11$. Thus, elements with atomic weights greater than sodium can be detected in most instruments (Fig. 33-10). Resolution down to 1 μm

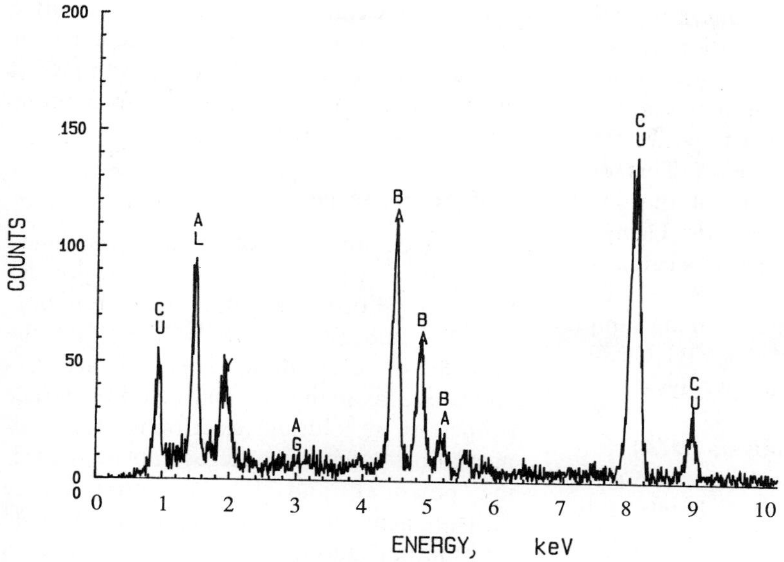

FIGURE 33-10. Energy Dispersive Spectrometer Results for a Ag–$YBa_2Cu_3O_{7-x}$ Particle. The Locations of the Peaks and their Intensities Determine the Element and the Relative Amount of the Element Present in the Probed Portion of the Sample. (Adapted from Carim et al. 1989.)

is available with SEM and down to 50 nm is available with TEM instruments. X-ray maps of a particle can be used to examine the chemical homogeneity of multicomponent particles. These maps show regions of high or low concentrations of a given element. The technique can also be used to examine a large number of particles and to look for variations in stoichiometry from particle to particle. Quantitative measurements usually require flat surfaces, but semiquantitative information can be obtained for particles.

X-ray Photoelectron Spectroscopy (XPS)

This technique (also known as electron spectroscopy for chemical analysis) can detect elements from Li to U (Feldman and Mayer 1986). This method irradiates the particles with X-rays, resulting in the emission of photoelectrons with energies characteristic of the elements in the sample and their chemical bonding. The effective probing depth is 3 nm and the method has 0.1 monolayer sensitivity. Spatial resolution of 10 μm can be obtained in state-of-the-art instruments. This limits the technique to relatively large particles. Since monolayer amounts can be detected, the amount of sample needed is small. The primary advantage of XPS over AES and EDS is that the chemical bonding of the elements can be determined.

Mass Spectrometric Methods

Mass spectrometers (MS) are extremely sensitive instruments for the detection of elements from H to U and can detect ppm levels (Feldman and Mayer 1988). The various mass spectrometric methods differ by the method used to introduce the sample into the MS. Particles on surfaces can be vaporized using pulsed lasers or can be introduced directly from the gas phase into ionization regions. These methods are usually not used on a routine basis because of their complexity.

Thermophysical and Thermochemical Analysis

Thermogravimetric Analysis (TGA)

This technique measures the change in the weight of a sample as it is heated at a known rate (Hench and Gould 1971). The primary use of TGA for materials produced by aerosol processes is for the determination of the extent of reaction and for the detection of species such as water vapor. For a determination of the identity of the evolved species, TGA/mass spectrometry can be used. This variation determines the identity of the species released at a certain temperature. Using rate of weight loss, kinetic parameters can also be determined (Biswas et al. 1989). TGA and DTS (below) typically require 50 mg samples. Figure 33-11 shows TGA for $YBa_2Cu_3O_{7-x}$ powder produced by spray pyrolysis.

Differential Thermal Analysis (DTA)

This technique measures the amount of heat released or absorbed by a sample as it is heated at a known rate (Hench and Gould 1971). When the enthalpy change is determined, the method is called differential scanning calorimetry (DSC). The presence of exothermic or endothermic processes at certain temperatures provides information about the nature of phase changes and chemical reactions occurring in the material as it is heated. DTA can often be used as a sensitive method for establishing the presence or absence of secondary phases in samples if these phases undergo phase transformations at known temperatures.

Surface Properties

The surface properties of powders suspended in liquids can play a critical role in powder processing. For this reason, it is often important to understand the behavior of the particles in solution. A variety of measurements are available to characterize surface properties in solution: electrophoresis, sedimentation potential, electroosmosis, streaming potential, and zeta potential. These measurements are related and can provide information about the nature of ionized functional groups on the surfaces of the particles.

Assorted Methods

A variety of other methods exist for the characterization of particles produced by

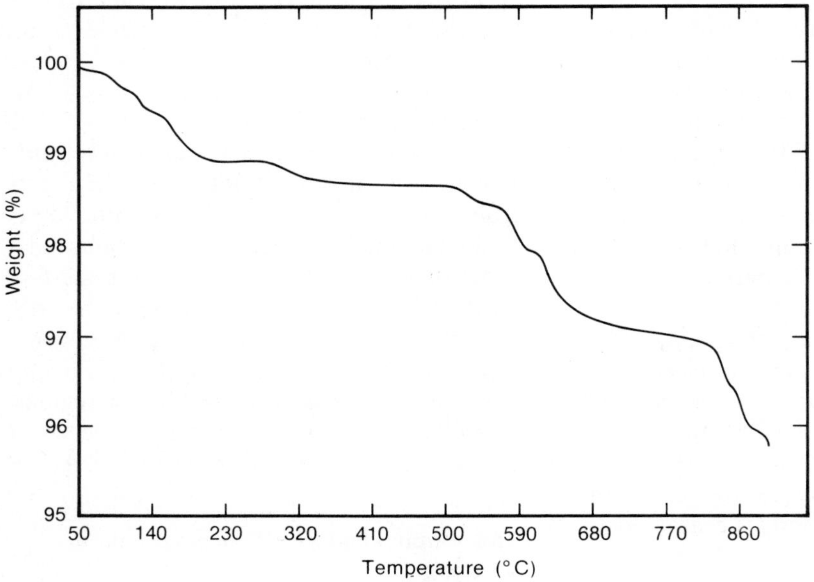

FIGURE 33-11. Weight Loss as a Function of Temperature for $YBa_2Cu_3O_{7-x}$ Powder. Weight Loss for Material with No Chemical Impurities would be 1.5%. Thus, Significant Amounts of Impurities are Present in the Sample. TGA/MS Identified these Impurities as Being Water (low *T*), Nitrates (mid *T*), and Carbonates (high *T*). (Adapted from Chadda et al. 1991.)

aerosol processes. These methods include Raman spectroscopy, electron spin resonance, small-angle X-ray scattering, small-angle neutron scattering, and other processes. These are relatively specialized techniques that are expensive and are not usually used on a routine basis.

A class of techniques that is being used on an increasing basis is scanning tunneling microscopy (STM) and atomic force microscopy (AFM). These techniques provide information about structure down to the atomic level (Neddermeyer 1990) for both conductors and insulators. This resolution is important in the case of nanometer-sized clusters on surfaces which may be difficult to image by techniques such as TEM.

References

Adachi, A., K. Okuyama, Y. Kousaka, and H. Tanaka. 1988. Preparation of gas sensor films by deposition of ultrafine tin oxide particles. *J. Aerosol. Sci.* 19:253–63.

Akhtar, M. K., Y. Xiong, and S. E. Pratsinis. 1991. Vapor synthesis of titania powder by oxidation of titanium tetrachloride. *A.I.Ch.E. J.* 37:1561–70.

Alam, M. K. and R. C. Flagan. 1986. Controlled nucleation aerosol reactors: Production of bulk silicon. *Aerosol Sci. Technol.* 5:237–48.

Andres, R. P., R. S. Averback, W. L. Brown, L. E. Brus, W. A. Goddard III, A. Kaldor, S. G. Louie, M. Moscovits, P. S. Peercy, S. J. Riley, R. W. Siegel, F. Spaepen, and Y. Wang. 1989. Research opportunities on clusters and cluster assembled materials. *J. Mater. Res.* 4:704–36.

Baker, R., W. Hurng, and H. Steinfink. 1988. Oriented high T_c superconductive layers on silver by devitrification of glasses formed in the Bi–Sr–Ca–Cu–O system. *Appl. Phys. Lett.* 54:371–73.

Barringer, E., N. Jubb, B. Gegley, R. L. Pober, and H. K. Bowen. 1984. In *Ultrastructure Processing of Ceramics, Glasses and Composites*, eds. L. L. Hench and D. R. Ulrich. New York: Wiley.

Biswas, P., X. Li, and S. E. Pratsinis. 1989. Optical waveguide preform fabrication: Silica formation and growth in a high-temperature aerosol reactor. *J. Appl. Phys.* 65:2445–50.

Biswas, P., D. Zhou, I. Zitkovsky, C. Blue, and P. Boolchand. 1989. Superconducting powders generated by an aerosol process. *Mat. Lett.* 8:233–37.

Biswas, P., D. Zhou, J. Grothaus, P. Boolchand, and D. McDaniel. 1990. Oxygen evolution from $YBa_2Cu_3O_{7-\delta}$ superconducting powders generated by aerosol routes. *Mat. Res. Symp. Proc.* 169:197–200.

Bohrer, M. P., J. A. Amelse, P. L. Narasimham, B. K. Tariyal, J. M. Turnipseed, R. F. Gill, W. J. Moebuis, and J. L. Bodeker. 1985. A process for recovering germanium from effluents of optical fiber manufacturing. *J. Lightwave Technol.* LT-3:699–705.

Bolsaitis, P. P., J. F. McCarthy, and G. Mohiuddin. 1987. Formation of metal oxide aerosols for conditions of high supersaturation. *Aerosol Sci. Technol.* 6:225–46

Cannon, W. R., S. C. Danforth, J. H. Flint, J. S. Haggerty, and R. A. Marra. 1982. Sinterable ceramic powders from laser-driven reactions. *J. Am. Ceram. Soc.* 65:324–35.

Carim, A., P. Doherty, and T. T. Kodas. 1989. Nanocrystalline $Ba_2YCu_3O_7$/Ag composite particles produced by aerosol decomposition. *Mat. Lett.* 8:335–39.

Casey J. and J. Haggerty. 1987. Laser-induced vapor phase synthesis of boron and titanium diboride powders. *J. Mater. Sci.* 22:737–44.

Chadda, S., T. T. Kodas, T. L. Ward, A. Carim, D. Kroeger, and K. C. Ott. 1991. Synthesis of $YBa_2Cu_3O_{7-x}$ and $YBa_2Cu_4O_8$ by aerosol decomposition. *J. Aerosol Sci.* 22:601–16.

Charlesworth, D. and W. Marshall. 1969. Evaporation from droplets containing dissolved solids. *A.I.Ch.E. J.* 6:9–23.

Clark, H. B. 1977. In *Treatise on Coatings: Pigments, 3*, eds. P. R. Myers and J. S. Long, New York: Marcel Dekker.

Dannenberg, E. M. 1971. Progress in carbon black technology. *J. IRI* 190–95.

Davis, E. J. and S. C. Liao. 1975. The growth kinetics and polydispersity of condensational aerosols. *J. Colloid Interface Sci.* 50:488–502.

DeSisto, W. J., R. L. Henry, M. Osofsky, and J. V. Marzik. 1991. $YBa_2Cu_3O_{7-\delta}$ thin films deposited by an ultrasonic nebulization and pyrolysis method. *Thin Solid Films*, 206:128–31.

Dobbins, R. A. and C. M. Megaridis. 1987. Morphology of flame-generated soot as determined by thermophoretic sampling. *Langmuir* 3:254–59.

Dobbins, R. A. and G. W. Mulholland. 1984. Interpretation of optical measurements of flame generated particles. *Combust. Sci. Technol.* 40:175–91.

Eastman, J. and R. Siegel. 1989. Nanophase synthesis assembled materials from atomic clusters. *Res. Dev.* Jan. 56–60.

Edelson L. and A. Glaser. 1988. Role of particle substructure in the sintering of monosized titania. *J. Am. Ceram. Soc.* 71:225–35.

Epstein, J. 1976. Utilization of the dead sea minerals. *Hydrometallurgy* 2:1–10.

Feldman, L. and J. Mayer. 1986. *Fundamentals of Surface and Thin Film Analysis.* North-Holland, N.Y.

Flint, J. H. and J. S. Haggerty. 1990. A model for growth of silicon particles from laser-heated gases. *Aerosol Sci. Technol.* 13:72–84.

Formenti, M., F. Juillet, P. Mereaudeau, S. Techner, and P. Vergnon. 1972. In *Aerosols and Atmospheric Chemistry*, ed. G. Hidy, New York: Academic Press.

French, W. G., L. J. Pace, and V. A. Foertmeyer. 1978.

Chemical kinetics of the reactions of $SiCl_4$, $SiBr_4$, $GeCl_4$, $POCl_3$ and BCl_3 with oxygen. *J. Phys. Chem.* 82:2191–94.

Friedlander, S. K. 1977. *Smoke, Dust, and Haze.* New York: Wiley.

Friedlander, S. K. 1982. The behavior of constant rate aerosol reactors. *Aerosol Sci. Technol.* 1:3–13.

Friedlander, S. K. 1983. Dynamics of aerosol formation by chemical reaction. *Ann. New York Academy Sci.* 404:354–64.

Gani, M. and R. McPherson. 1980. The structure of plasma-prepared Al_2O_3 and TiO_2 powders. *J. Mater. Sci.* 15:1915–25.

Gardner, T., D. Sproson, and G. Messing. 1984. Precursor effects on development of particle morphology during evaporative decomposition of solutions. *Mat. Res. Soc. Symp. Proc.* 32:227–32.

George, A. P., R. D. Murley, and E. R. Place. 1973. Formation of TiO_2 aerosol from the combustion supported reaction of $TiCl_4$ and O_2. *Faraday Symp. Chem. Soc.* 7:63–77.

Girshick, S. L. and C.-P. Chiu. 1990. Numerical study of MgO powder synthesis by thermal plasma. *J. Aerosol Sci.* 21:641–50.

Girshick, S. L. and C.-P. Chiu. 1989. Homogeneous nucleation of particles from the vapor phase in thermal plasma synthesis. *Plasma Chem. Plasma Process* 9:355–69.

Granqvist, C. and R. Buhrman. 1976. Ultrafine metal particles. *J. Appl. Phys.* 47:2200–19.

Haynes, B. S., H. Jander, and H. G. G. Wagner. 1979. In *Seventeenth Symposium (International) on Combustion*, pp. 1356–74. Pittsburgh, PA: The Combustion Institute.

Hench, L. L. and R. W. Gould. 1971. *Characterization of Ceramics.* New York: Marcel Dekker.

Hidy, G. M. and J. R. Brock. 1970. *The Dynamics of Aerocolloidal Systems.* New York: Pergamon.

Hinds, W. C. 1982. *Aerosol Technology.* New York: Wiley.

Hollabaugh, C. M., D. E. Hull, L. R. Newkirk, and J. Petrovic, 1983. RF plasma system for the production of ultrafine SiC powder. *J. Mater. Sci.* 18:3190–94.

Hurd, A. J. and W. L. Flower. 1988. *In-situ* growth and structure of fractal SiO_2 aggregates. *J. Colloid Interface Sci.* 122:178–92.

Hurd, A. J., G. P. Johnston, and D. M. Smith. 1991. In Characterization of Porous Solids II, eds. F. Rodriguez-Reinoso, J. Rouquerol, K. S. W. Sing, and K. K. Unger, p. 267. New York: Elsevier.

Imai, H. and K. Takami. 1985. Preparation of fine particles of carnegieite by a mist decomposition method. *J. Mater. Sci.* 20: 1823–27.

Ingebrethsen, B. J. and E. Matijevic. 1980. Preparation of uniform colloidal dispersions by chemical reactions in aerosols—2. Spherical particles of aluminum hydrous oxide. *J. Aerosol Sci.* 11:271–80.

Ingebrethsen, B. J., E. Matijevic, and R. Partch. 1983. Preparation of uniform colloidal dispersions by chemical reactions in aerosols—3. Mixed titania alumina colloidal spheres. *J. Colloid Interface Sci.* 95:228–39

Ivie, J. J. and L. J. Forney. 1988. A numerical model of the synthesis of carbon black by benzene pyrolysis. *A.I.Ch.E. J.* 34:1813–20.

Johnson, D. 1981. Non-conventional powder preparation techniques. *Ceramic Bulletin* 60:21–29.

Johnson, S. M., M. I. Gusman, and D. J. Rowcliffe. 1987. Freeze drying. *Adv. Ceram. Mater.* 2:237–41.

Kagawa, M., F. Honda, H. Onodera, and T. Nagae. 1983. The formation of ultrafine Al_2O_3, ZrO_2, and Fe_2O_3 by the spray ICP technique. *Mat. Res. Bull.* 18:1081–87.

Kanapilly, G., O. Raabe, and G. Newton. 1970. A new method for the generation of aerosols of insoluble particles. *Aerosol Sci.* 1:313–23.

Kerker, M. 1975. Laboratory generation of aerosols. *Adv. Colloid Interfacial Sci.* 5:105–72.

Kim, K.-S. and S. E. Pratsinis. 1988. Manufacture of optical waveguide preforms by modified chemical vapor deposition. *A.I.Ch.E. J.* 34:912–21.

Kim, K.-S. and S. E. Pratsinis. 1989. Modeling and analysis of modified chemical vapor deposition of optical fiber preforms. *Chem. Eng. Sci.* 44:2475–82.

Kingery, W. D., H. K. Bowen, and D. R. Uhlmann. 1976. *Introduction to Ceramics.* New York: Wiley.

Kodas, T. T. 1989. Generation of complex metal oxides by aerosol processes: Superconducting ceramic particles and films. *Angewandte Chemie: International Edition in English* 28:794–807.

Kodas, T. T., A. Datye, V. Lee, and E. Engler. 1989. Single-crystal $YBa_2Cu_3O_7$ particle formation by aerosol decomposition. *J. Appl. Phys.* 65:2149–51.

Kodas, T. T., E. Engler, and V. Lee. 1989. Generation of thick $YBa_2Cu_3O_7$ films by aerosol deposition. *Appl. Phys. Lett.* 54:1923–25.

Kodas, T. T., E. M. Engler, V. Lee, R. Jacowitz, T. H. Baum, K. Roche, S. S. P. Parkin, W. S. Young, S. Hughes, J. Kleder, and W. Auser. 1988. Aerosol flow reactor production of fine $YBa_2Cu_3O_7$ powder: Fabrication of superconducting ceramics. *Appl. Phys. Lett.* 52:1622–24.

Kodas, T. T., S. E. Pratsinis, and A. Sood. 1987. Submicron alumina powder production by a turbulent flow aerosol process. *Powder Technol.* 50:47–53.

Kodas, T. T. and A. Sood. 1990. In *Handbook of Science and Technology of Alumina Chemicals*, ed. L. Hart, p. 375. American Ceramics Society.

Koguchi, M., Y. Matsuda, E. Kinoshita, and K. Hirabayashi. 1990. Preparation of $YBa_2Cu_3O_{7-x}$ films by flame pyrolysis. *J. Appl. Phys. Japan* 29:L33–35.

Komiyama, A., T. Osawa, H. Kazi, and T. Konno. 1987. Rapid growth of AlN films by particle precipitation aided CVD. In *High Tech Ceramics*, ed. P. Vincenzini, pp. 667–76. Amsterdam: Elsevier.

Koukitu, A., Y. Hasegawa, H. Seki, H. Komijama, I. Tanaka, and Y. Kamioka. 1989. Preparation of YBaCuO superconducting thin films by the mist microwave plasma decomposition method. *J. Appl. Phys. Japan* 28:L1212–13.

Kriechbaum, G. and P. Kleinschmidt. 1989. Superfine oxide powders: Flame hydrolysis and hydrothermal synthesis. *Advanced Materials, Ang. Chemie* 10:330–37.

Landgrebe, J. D. and S. E. Pratsinis. 1990. A discrete-sectional model for powder production by gas-phase chemical reaction and aerosol coagulation in the free molecular regime. *J. Colloid Interface Sci.* 139:63–86.

Landgrebe, J. D., S. E. Pratsinis, and S. V. R. Mastrangelo. 1990. Nomographs for vapor synthesis of ceramic powders. *Chem. Eng. Sci.* 45:2931–41.

Lefebvre, A. H. 1989. *Atomization and Sprays.* New York: Hemisphere.

Leong, K. 1981. Morphology of aerosol particles generated from the evaporation of solution drops. *J. Aerosol Sci.* 12:417–35.

Leong, K. 1987. Morphology control of particle generated from the evaporation of solution drops. *J. Aerosol Sci.* 18:525–52.

Lukasiewicz, S. 1989. Spray drying of ceramic powders. *J. Am. Ceram. Soc.* 72:617–24.

MacChesney, J. B., P. B. O'Connor, and H. M. Presby. 1974. A new technique for the preparation of low-loss graded-index optical fibers. *Proc. IEEE* 62:1280–81.

Masters, K. 1972. *Spray Drying.* New York: Wiley.

Matijevic, E. 1986. Monodispersed colloids: Art and science. *Langmuir* 2:12–20.

Matijevic, E. 1987. In *High Tech Ceramics*, ed. P. Vincenzini, pp. 441–58. Netherlands: Elsevier.

Matson, D. W., R. Peterson, and R. Smith. 1987. Production of powders and films by the rapid expansion of supercritical solutions. *J. Mater. Sci.* 22:1919–28.

Masdiyasni, K., C. Lynch, and J. Smith. 1965. Preparation of ultra-high-purity submicron refractory oxides. *J. Am. Ceram. Soc.* 48:372–75.

Mayo, M., R. Siegel, A. Narayanasamy, Nix. 1990. Mechanical properties of nanophase TiO_2 as determined by nanoindentation. *J. Mater. Res.* 5:1073–81.

Medalia, A. I. and D. Rivin. 1976. In *Characterization of Powder Surfaces*, eds. G. D. Parfitt and K. S. W. Sing. New York: Academic Press.

Mezey, E. J. 1966. In *Vapor Deposition*, eds. C. F. Powell, J. H. Oxley, and J. M. Blocher, Jr. New York: Wiley.

Miller, B. and R. Lines. 1988. Recent advances in particle size measurements: A critical review. *CRC Crit. Rev. Analyt. Chem.* 20:75–116.

Mooney, J. and S. Redding. 1982. Spray pyrolysis. *Ann. Rev. Mat. Sci.* 12:81–90.

Nagel, S. R., J. B. MacChesney, K. L. Walker. 1982. An overview of the modified chemical vapor deposition. *IEEE J. Quantum Electron.* QE-18:459–76.

Neddermeyer, H. 1990. STM studies of nucleation and the initial stages of film growth. *Crit. Rev. Solid State Mater. Sci.* 16:309–35.

Niessner, R. and Fresenius. 1990. Chemical characterization of aerosols. *J. Anal. Chem.* 337:565–76.

Okuyama, K., J.-T. Jeung, Y. Kousaka, H. V. Nguyen, J. J. Wu, and R. C. Flagan. 1989. Experimental control of ultrafine TiO_2 particle generation from thermal decomposition of titanium tetraisopropoxide vapor. *Chem. Eng. Sci.* 44:1369–75.

Okuyama, K., Y. Kousaka, N. Tohge, S. Yamamoto, J. J. Wu, R. C. Flagan, and J. H. Seinfeld. 1986. Production of ultrafine metal oxide aerosol particles by thermal decomposition of metal alkoxide vapors. *A.I.Ch.E. J.* 32:2010–19.

Ono, T., M. Kagawa, and Y. Syono. 1985. Ultrafine particles of the ZrO_2–SiO_2 system prepared by the spray ICP technique. *J. Mater. Sci.* 20:2483–87.

Ortega, J., T. T. Kodas, S. Chadda, D. M. Smith, M. Ciftcioglu, and J. Brennan. 1991. Generation of dense barium calcium titanate particles by aerosol decomposition, *Chem. Mater.* 3:746–51.

Parker, J. and R. Siegel. 1990. Raman microprobe study of nanophase TiO_2 and oxidation induced spectral changes. *J. Mater. Res.* 5:1246–52.

Pebler A. and R. Charles. 1989. Synthesis of superconducting oxides by aerosol pyrolysis of metal EDTA solutions. *Mat. Res. Bull.* 24:1069–76.

Pesthy, A. J., R. C. Flagan, and J. H. Seinfeld. 1983. Aerosol formation and growth in laminar flows. *J. Colloid Interface Sci.* 91:525–45.

Petersen, R., D. Matson, and R. Smith. 1986. Rapid precipitation of low vapor pressure solids from supercritical fluid solutions: The formation of thin films and powders. *J. Am. Chem. Soc.* 108:2100–02.

Phanse, G. M. and S. E. Pratsinis. 1989. Theory for production of aerosols in laminar flow condensers. *Aerosol Sci. Technol.* 11:100–19.

Phillips, D. and G. Vogt. 1987. Plasma synthesis of ceramic powders. *MRS Bulletin* 54–58.

Pickles, C. and A. McLean. 1983. Production of fused refractory oxide spheres and ultrafine oxide particles in an extended arc. *Am. Ceram. Soc. Bull.* 62:1004–09.

Powers, D. R. 1978. Kinetics of $SiCl_4$ oxidation. *J. Am. Ceram. Soc.* 61:295–97.

Pratsinis, S. E. 1988. Simultaneous nucleation, condensation and coagulation in aerosol reactors. *J. Colloid Interface Sci.* 124:416–27.

Pratsinis, S. E. 1990. In *Ceramic Powder Science—III*, eds. G. L. Messing, H. Hausner, and S.-C. Hirano. Columbus, OH: American Ceramic Society.

Pratsinis, S. E., H. Bai, P. Biswas, M. Frenklach, and S. V. R. Mastrangelo. 1990. Kinetics of $TiCl_4$ oxidation. *J. Am. Ceram. Soc.* 73:2158–62.

Pratsinis, S. E., J. D. Landgrebe, and S. V. R. Mastrangelo. 1989. Design correlations for gas phase manufacture of ceramic powders. *J. Aerosol Sci.* 20:1457–60.

Pratsinis, S. E. and S. V. R. Mastrangelo. 1989. Material synthesis in aerosol reactors. *Chem. Eng. Prog.* 85(5):62–66.

Prochaska, S. and C. Greskovich. 1978. Synthesis and characterization of silicon nitride powder. *Ceram. Bulletin* 57:579–86.

Ramsey, D. and R. Avery. 1974. Ultrafine oxide particles prepared by electron beam evaporation. *J. Mater. Sci.* 9:1681–88.

Rao, N. R. and P. H. McMurry. 1989. Nucleation and growth of aerosol in chemically reacting systems: A theoretical study of the near collision-controlled regime. *Aerosol Sci. Technol.* 11:120–32.

Reed, J. S. 1988. *Introduction to the Principles of Ceramic Processing.* New York: Wiley.

Ruthner, M. I. 1979. Preparation and sintering characteristics of MgO, $MgO–Cr_2O_3$ and $MgO–Al_2O_3$. *Sci. Sintering* 11:203–08.

Ruthner, M. I. 1983. Industrial production of multicomponent ceramic powders by means of the spray roasting method. In *Ceramic Powders,* ed. P. Vincenzini, pp. 515–31. Amsterdam: Elsevier.

Santoro, R. J. and J. H. Miller. 1987. Soot particle formation in diffusion flames. *Langmuir* 3:244–54.

Schaefer, D. and A. J. Hurd. 1990. Growth and structure of combustion aerosols. *Aerosol Sci. Technol.* 12:876–90.

Seinfeld, J. H. 1986. *Atmospheric Chemistry and Physics of Air Pollution.* New York: Wiley.

Seitz, K., E. Ivers-Tiffee, H. Thomann, and A. Weiss, 1987. Influence of zinc acetate and nitrate salts on the characteristics of undoped ZnO powders. In *High Tech Ceramics.* ed. P. Vincenzini, pp. 1753–62. Amsterdam: Elsevier.

Siegel, R. 1990. Nanophase materials assembled from atomic clusters. *MRS Bull.* Oct. 60–67.

Siegel, R., R. Ramasamy, H. Hahn, Z. Li, L. Ting, and R. J. Gronsky. 1988. Synthesis, characterization and properties of nanophase TiO_2. *Mater. Res.* 3:1367–72.

Siefert, W. 1984. Properties of thin In_2O_3 and SnO_2 films by corona spray pyrolysis and a discussion of the spray pyrolysis process. *Thin Solid Films* 121:275–82.

Singh, R. and R. Doherty. 1990. Synthesis of TiN powders under glow discharge plasma. *Mat. Lett.* 9:87–89.

Sokolowski, M., A. Sokolowska, A. Michalski, and B. Gokieli. 1977. The in-flame reaction method for Al_2O_3 aerosol formation. *J. Aerosol Sci.* 8:219–30.

Suyama, Y. and A. Kato. 1976. TiO_2 produced by vapor-phase oxygenolysis of $TiCl_4$. *J. Am. Ceram. Soc.* 59:146–49.

Takagi, T., K. Matsubara, and H. Takaoka. 1980. Optical and thermal properties of BeO thin films prepared by reactive ionized cluster beam technique. *J. Appl. Phys.* 51:5419–29.

Takaoka, H., J. Ishikawa, and T. Takagi. 1988. Low temperature growth of AlN and Al_2O_3 films by the simultaneous use of a microwave source and an ionized cluster beam system. *Thin Solid Films* 157:143–58.

Tomar M. and F. Garcia. 1981. Spray pyrolysis in solar cells and gas sensors. *Prog. Cryst. Growth Charact.* 4:221–47.

Ulrich, G. D. 1971. Theory of particle formation and growth in oxide synthesis flames. *Combustion Sci. Technol.* 4:47–57.

Ulrich, G. D. 1984. Flame synthesis of fine particles. *Chem. Eng. News* Aug. 6:22–29.

Ulrich, G. D. and J. W. Riehl. 1982. Aggregation and growth of submicron oxide particles in flames. *J. Colloid Interface Sci.* 87:257–65.

Viguie, J. C. and J. Spitz. 1975. Chemical vapor deposition at low temperature. *J. Electrochem. Soc.* 122:585–88.

Visca, M. and E. Matijevic. 1979. Preparation of uniform colloidal dispersions by chemical reactions in aerosols. i. Spherical particles of titanium dioxide. *J. Colloid Interface Sci.* 68:308–19.

Vollath, D. 1990. Pyrolytic preparation of ceramic powders by a spray calcination technique. *J. Mater. Sci.* 25:2227–32.

Weimer, A. W., W. G. Moore, R. P. Roach, C. N. Haney, and W. Rafaniello. 1991. Rapid carbothermal reduction of boron oxide in a graphite reactor. *A.I.Ch.E. J.* 37:759–68.

Wenckus, J. and W. Leavitt. 1957. Preparation of ferrites by the atomizing burner technique. In *Magnetism and Magnetic Materials Conference,* 1956, pp. 526–30. Boston, MA: Amer. Inst. Elect. Engs.

Wu, J. J. and R. C. Flagan. 1988. A discrete-sectional solution to the aerosol dynamic equation. *J. Colloid Interface Sci.* 123:339–52.

Wu, J. J., H. V. Nguyen, R. C. Flagan, K. Okuyama, Y. Kousaka. 1988. Evaluation and control of particle properties in aerosol reactors. *A.I.Ch.E. J.* 34:1249–56.

Wu, J. J. and R. C. Flagan. 1987. Onset of runaway nucleation in aerosol reactors. *J. Appl. Phys.* 61:1365–71.

Xiong, Y. and S. E. Pratsinis. 1991. Gas phase production of particles in reactive turbulent flows. *J. Aerosol Sci.* 22:637–56.

Zachariah, M. R. and S. Huzarewicz. 1991. Aerosol processing of YBaCuO superconductors in a flame reactor. *J. Mater. Res.* 6:264–69.

Zhang, S. C., G. L. Messing, and M. Borden. 1990. Synthesis of solid spherical zirconia particles by spray pyrolysis. *J. Am. Ceram. Soc.* 73:61–67.

34

Clean-Room Measurements

Heinz Fissan, Wolfgang Schmitz, and
Andreas Trampe

Department of Electrical Engineering
Division of Process and Aerosol Measurement Technology
University of Duisburg
Germany

INTRODUCTION

Clean-room technology is an important field of application for aerosol measurement. This technology becomes more and more important in many fields such as the pharmaceutical industry or hospitals and especially the microelectronic industry. Clean-room technology is characterized by the goal of an aerosol-free environment. The loss of quality or the damage that occurs when particles deposit on a product have necessitated new technology. Using filtering and air flow techniques, the concentration of particles in a typical clean room remains by a factor of 10^8 lower than the typical indoor room concentration of 10^4–10^5 particles/cm^3.

Besides aerosol reduction, there are other important areas of contamination control in the clean-room environment. While this chapter will deal primarily with airborne particles, e.g., certification of ambient air, bulk or process gases, other sources and aspects of contamination must also be taken into account. Note that ambient air in the present context refers to air within the clean room but outside the equipment. These other contamination sources include hydrosols (e.g., control of deionized water and other process chemicals), biological contaminants (e.g., pyrogenic and endotoxin producing bacteria), ionic and radioactive components, condensation or diffusive products, and chemically reactive particles. These aspects will not be dealt with to any extent in this chapter.

The chapter consists of three parts. The first one deals with the different measurement tasks in clean-rooms. The second introduces the available techniques and devices for particle concentration measurements and the last discusses possible problems related to the statistics of extremely low concentration measurements as well as the relationship of surface deposition measurements to airborne measurements.

MEASUREMENT TASKS

The most important and routine measurements will be discussed in the light of various standards and guidelines. A list of these from several countries is given in Table 34-1. The object of the measurement is the clean-room, a limited area in which the level and sources of aerosol must be known and controlled. The necessary level of cleanliness in this area depends on the manufacturing process.

TABLE 34-1 National Standards and Guidelines Regulating Clean Rooms

Code	Country	Organisation
Fed. Std.	USA	General Service Administration, Federal Supply Service, Engineering Division, 819 Taylor Street, Fort Worth, TX 76102
ASTM	USA	American Society for Testing and Materials, 1916 Race Street, Philadelphia, PA 19103
IES RP CC	USA	Institute of Environmental Sciences, 940 East Northwest Highway, Mount Prospect IL 60056
BS	UK	British Standards Institution, 2 Park Street London W1A 2BS
JIS	Japan	Japanese Standard Association, 1-24 Akasaka 4, Minato, Tokyo 107
JACA	Japan	Japanese Air Cleaning Association, Tomoe-Ya Bldg. 2-14, 1-Chome, Uchi-Kanda, Chiyodaku, Tokyo 107
AFNOR	France	L'association francaise de normalisation afnor-tout europe, Paris
DIN	Germany	Deutsches Institut für Normung e.V., Burggrafenstrasse 4-10, 1000 Berlin 30
VDI	Germany	Verein Deutscher Ingenieure, Graf-Recke Strasse 84, 4000 Düsseldorf

Clean-Room Measurements

Particle concentration measurements are necessary for the assessment of contamination risks. The quality of a clean room is established by a classification system. The clean-room class is important for the planning, construction, and operation in the clean room. During operation, these classification guidelines supply limits that can be applied to the measurements. Clean-room classifications differ in various countries. Table 34-2 shows a comparison of clean-room classes defined by various standards. The left columns show the maximum concentration for four different particle sizes, with the corresponding clean-room classes indicated to the right. The large range of concentrations allowed by the various classes, depending upon the varying purposes of clean environments, is evident in this table. It should be noted that the class limits can vary slightly in different standards, due to different units. For instance, the U.S. Federal Standard 209D and the German VDI 2083 differ by 20% in concentration.

In the following discussion, we shall refer to the different standards where necessary, since the measurement task depends strongly on the corresponding standard. Nevertheless, when performing measurements, the reader

TABLE 34-2 Comparison of National Clean-Room Standards

Maximum Permissible Particle Concentration, $1/m^3$				Code Names				
				USA	Germany	Japan	United Kingdom	France
0.1 μm	0.5 μm	5.0 μm 209D	10.0 μm 2083	Fed. Std. B9920	VDI BS 5295	JIS X44101	BS	AFNOR
10	0.35	0	0	–	–	1	–	–
100	3.5	0	0	(0.1)	0	2	–	–
10^3	35	0	0	1	1	3	C	–
10^4	350	0	0	10	2	4	D	–
10^5	$3.5 \cdot 10^3$	30	0	100	3	5	E + F	4,000
10^6	$3.5 \cdot 10^4$	300	0	1,000	4	6	G + H	40,000
10^7	$3.5 \cdot 10^5$	$3 \cdot 10^3$	450	10,000	5	7	J	400,000
10^8	$3.5 \cdot 10^6$	$3 \cdot 10^4$	$4.5 \cdot 10^3$	100,000	6	8	K	4,000,000
–	–	$3 \cdot 10^5$	$4.5 \cdot 10^4$	–	7	–	L	–
–	–	–	$4.5 \cdot 10^5$	–	–	–	M	–

should consult the latest issues of the relevant standards. Although the principles in each standard are the same, the details may differ slightly. Figure 34-1 shows the clean-room limits as defined by the U.S. Federal Standard 209D and the German VDI 2083.

In the standards, maximum concentrations for ambient air within the clean room are defined as the number of particles larger than a given size per unit volume of air. These concentrations are plotted against particle diameter d_p on a log–log scale, resulting in a straight line. The German VDI 2083 is based on the Junge (1963) distribution. Each class definition uses a specific particle diameter: the U.S. Federal Standard 209D uses 0.5 μm, the German VDI 2083 uses 1.0 μm, and the Japanese JIS uses 0.1 μm. The different classes are scaled by shifting the distribution lines by a factor of 10. It should be noted that the allowed concentrations are not given for every particle size, but some standards allow interpolation if the measured particles have diameters between the defined class sizes.

The standards mentioned here do not describe a complete procedure for concentration measurement because of the difficulty of providing an entire recipe for every situation. Independent institutes, such as the Institute of Environmental Sciences (IES) in U.S.A. or the Institute for Environmental Technology and Analysis (Institut für Umwelttechnologie und Umweltanalytik, IUTA) in Germany (Bliersheimer 60, 4100 Duisburg 14), can give advice on special measurement problems. However, the operation conditions for initial determination of clean-room classification are similar for all the standards. For

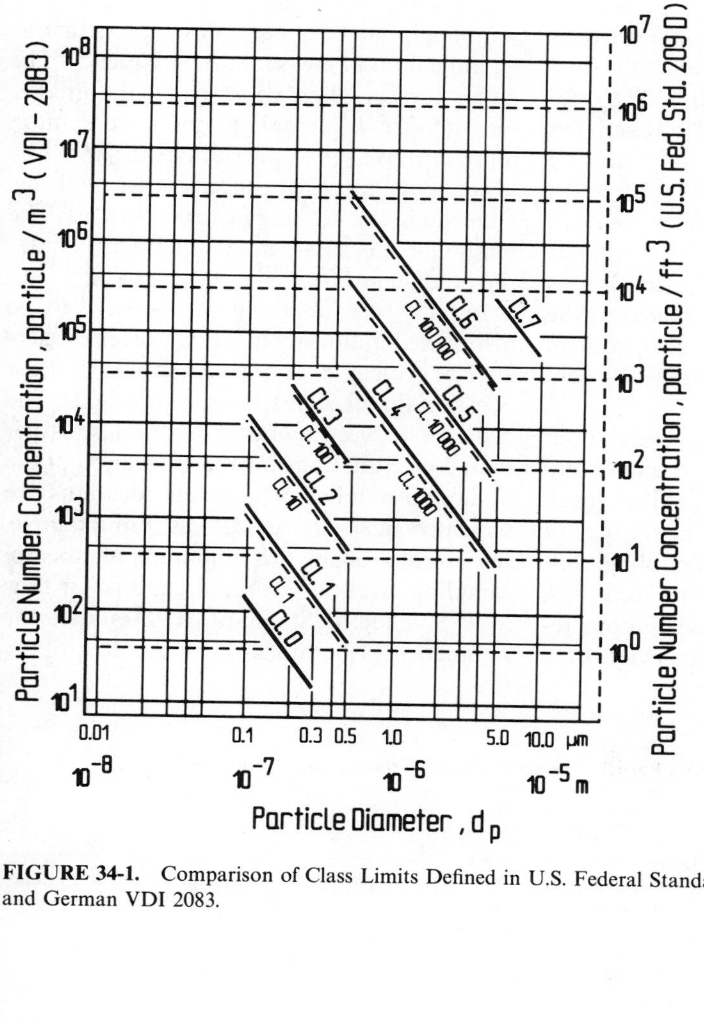

FIGURE 34-1. Comparison of Class Limits Defined in U.S. Federal Standard 209D and German VDI 2083.

instance, the U.S. Federal Standard 209D establishes the environmental test conditions as follows:

As built. A clean-room that is complete and ready for operation, with all services connected and functional, but without the production equipment or personnel within the facility.

At rest (in pharmaceutical industries: unmanned conditions). A clean-room that is complete and has the production equipment installed and operating, but without personnel within the facility.

In operation. A clean-room in normal operation, with all services functioning and with production equipment and personnel present and performing their normal work functions in the facility.

The principal recommendation regarding the frequency of particle measurement is simply to measure as often as possible. The only obstacle here is the expense. However, the optimum technical approach may not be necessary. The guidelines do not prescribe obligatory time intervals for measurement. Typically, the first measurements will be performed under one of the three conditions described above. Table 34-3 indicates the recommended time intervals for control measurements under the various clean-room classes.

The special case of continuous clean-room monitoring will be discussed in the next section. For the particle measurement itself, each of the following must be determined: type of device, sampling procedure, sampling volume, and the number of sampling locations. The guidelines give few recommendations regarding the choice of measuring device. Low detection limits and flow rates are implicit in the measurement conditions. Additional information on this topic is presented below in the section on measurement instruments.

During sampling, special attention should be paid to minimizing sampling losses. Therefore, short sampling tubes made of conductive material (e.g., electropolished stainless steel) should be used. Tubes of polymeric material should be avoided. If necessary, internally conductive tubing can be used. In most cases, isokinetic sampling is not possible because of the turbulence that occurs even in the so-called laminar-flow clean rooms. With the usual size range of particles measured in clean areas, isokinetic sampling may not be necessary, except for special situations such as source detection. Sampling problems and techniques are discussed in detail in Chapter 6.

The sampling volume and the number of sampling points can be dealt with together as an indication of measurement frequency. The larger the sampling volume and the greater the number of sampling points, the more representative the measurement and the lower the statistical uncertainties will be. Here again the limiting factor is the cost. The minimum measurement requirements under the guidelines often take several days.

Initially, the sampling points have to be distributed uniformly in a clean room, preferably at critical locations and should be kept constant for all successive measurements. The distance between points should not be larger than 2 m. Equations for the estimation of the minimum number of sampling locations are indicated in several guidelines. For example, the calculation of the minimum number of sampling locations for unidirectional air flow in U.S. Federal Standard 209D states: the number must be the lesser of the area of the

TABLE 34-3 Time Intervals for Regular Particle Measurements in Various Clean Room Classes

German VDI 2083 U.S. Fed. Std. 209D	6 100,000	5 10,000	4 1,000	3 100	2 10	1 1
Frequency	Half-Yearly	Monthly	14 Days	Weekly	Daily	Continuous

entrance plane divided by 25 ft^2 or the area of the entrance plane divided by the square root of the class designation. For non-unidirectional air flow, this standard demands the latter requirement.

Consideration should be given to providing additional sampling locations according to special geometries within the area and also to critical areas. The height of the sampling points should be at the working or product level. Table 34-4 has recommendations on the minimum sampling volume in liters according to the U.S. Federal Standard 209D.

The procedure and documentation of the measurements are facilitated by prescribed forms printed in several of the guidelines. The analysis of particle count data is performed by deriving a concentration. This concentration must remain below the limiting curve of particle number versus size (e.g., Fig. 34-1) for the applicable class. The statistical calculations are discussed below. Additional information about the corresponding statistics is given by Bzik (1986), Cooper (1988), and Cooper et al. (1990).

A problem occurs when the area over which the measurement is made cannot be defined precisely, e.g., for a piece of assembly equipment. In this case, the contamination risk can be masked by choice of too large an area, too high a flow rate, or too large a sample volume. In such situations, it is better to work directly with the particle number.

TABLE 34-4 Minimum Volume per Sample in Liters per Minute for the Air Cleanliness Class and Measured Particle Size (U.S. Fed. Std. 209D)

	Measured Particle Size				
Class	0.1 μm	0.2 μm	0.3 μm	0.5 μm	5.0 μm
1	17	85	198	566	N.A.
10	2.83	8.5	19.8	56.6	N.A.
100	N.A.	2.83	2.83	5.6	N.A.
1000	N.A.	N.A.	N.A.	2.83	85
10000	N.A.	N.A.	N.A.	2.83	8.5
100000	N.A.	N.A.	N.A.	2.83	8.5

N.A.—not applicable
2.83 l/min = 0.1 ft^3/min

Clean-Room Monitoring

This section discusses only clean-room monitoring tasks and the problems and approaches to their solution along with the realized monitoring systems. Definitive solutions to problems in monitoring procedures and data evaluation have not been developed.

Tasks

Particles should be monitored not only in the ambient air, but also within the equipment. The optimal monitoring system provides continuous and complete control of a process. To accomplish this, a sensor system is needed that can give an alarm quickly enough to feed back to the control system and prevent deposition of particles on the product. In addition to aerosol sensors, sensors for the media supplies and airflow systems need to be integrated into such a control system. This type of system would be very complex and difficult to implement. We are aware of only one such system in existence and so we will focus on just particle measurement in the following discussion. Currently, air quality is controlled mainly in the ambient air and at critical locations. The aim of clean-room monitoring is to gain information about the process, to determine the interdependence of different parameters, to elucidate the causes of low product quality, and to determine the effects of any actions on the product yield. From this information, one can estimate the level of aerosol that can be tolerated in a clean room without a decrease in product quality.

Monitoring Systems

The main components of monitoring systems currently in use are particle counters and environmental sensors interconnected via computers. Either optical particle counters (OPC) or condensation nucleus counters (CNC) can be used. A wide range of sensor capability is available, with some manufacturers producing instruments with no or few control capabilities, allowing lower cost devices that can be used in larger numbers. Table 34-5 lists several monitors and their principal features. A list of commercial sources is given in Table 34-6.

TABLE 34-5 Particle Monitors for Clean Room Measurements

Model	Company	Clean Room Suitability	False Count Rate, h^{-1}	Number of Channels	Flow Rate, l/min	Lower Detection Limit for X % efficiency
LPS A 101	PMS	class 10	ns	2–4	2.83	0.1 μm–50%
LPS A 1001	PMS	ns	ns	2–4	0.283	0.1 μm–50%
LPS A 201	PMS	class 10	ns	2–4	28.3	0.2 μm–50%
LPS A 310	PMS	class 10	ns	2–4	28.3	0.3 μm–50%
LPS A 510	PMS	class 100	ns	2–4	28.3	0.5 μm–50%
LPS C	PMS	ns	ns	ns	2.83	0.2 μm–50%
3751	TSI	class 1	< 6.0	2	28.3	0.5 μm–50%
3752	TSI	class 1	< 0.6	2	2.83	0.2 μm–50%
3753	TSI	class 1	< 6.0	2	2.83	0.3 μm–50%
3755	TSI	class 1	< 6.0	2	2.83	0.5 μm–50%
3756	TSI	class 1	< 6.0	2	2.83	0.5 μm–50%
3760 CNC	TSI	class 1	< 0.6	1	1.41	0.014 μm–50%
3761 CNC	TSI	class 1	< 1.2	1	2.83	0.02 μm–50%
PLD3	AAA	ns	ns	4	2.83	0.3 μm–ns
1100	HIA	ns	ns	6	28.3	0.5 μm–ns
1100A	HIA	class 10	ns	6	28.3	0.3 μm–ns
1200	HIA	ns	ns	6	2.83	0.3 μm–ns
CI3100OPT	CLI	ns	ns	2	2.83	0.3 μm–ns
202A	MET	ns	ns	2	2.83	0.5 μm–ns
202B	MET	ns	ns	2	2.83	0.3 μm–ns
202C	MET	ns	ns	2	2.83	0.25 μm–ns
202W	MET	ns	ns	2	28.3	0.2 μm–ns

ns—not specified
2.83 l/min = 0.1 ft^3/min

TABLE 34-6 List of Commercial Sources for Particle Counting

Code	Address	Products
AAA	Analytik Applikation Apparatebau GmbH, Krautstrasse 11, 7037 Magstadt, Germany, Tel: (0)7159/4888 FAX: (0)7159/44898	Stand alone counters for aerosols and hydrosols, monitors
AER	Aerometrics Inc., P.O. Box 308, Mountain View, CA 94042 USA, Tel: (415) 965 8887	Laser Doppler velocimetry
ACT	ACTOR Instrumentation Division, 2350 Charleston Road, Mountain View, CA 94043 USA, Tel: (415) 968 6080 FAX: (415) 961 5493	Portable particle counters, condensation nucleus counters
CLI	Climet Instruments Company, 1320 Colton Avennue, Redlands, CA 92373 USA, Tel: (714) 793 2788 FAX: (714) 793 1738	Stand alone counters for aerosols, monitors
DAI	DAN Industry Co. Ltd., Japan, Tel: 03 488 1111 FAX: 03 488 1118	Stand alone counters for aerosols
DAN	Dantec Electronic, Mileparken 22, 2740 Skovlunde Denmark, Tel: 45 28 422 11 FAX: 45 28 46 346	Laser Doppler velocimetry
DEH	Haan and Wittmer GmbH, Birkenstrasse 31, 7259 Friolzheim Germany, Tel: (0)7044/4064 FAX: (0)7044/4040	Stand alone counters for aerosols and hydrosols

TABLE 34-6 *(Continued)*

Code	Address	Products
HIA	HIAC ROYCO Division of Pacific Scientific, 2431 Linden Lane, Silver Spring MD 20910 USA, Tel: (301) 495 7000 FAX: (301) 495 0478	Stand alone counters for aerosols and hydrosols, monitors
HYT	High Yield Technology Division of HIAC ROYCO, 800 Maude Avenue, Mountain View, CA 94043 USA, Tel: (415) 960 3100	Particle flux counters
INS	INSITEC, 2100 Omega Road, Suite D, San Ramon, CA 94583 USA, Tel: (415) 837 1330 FAX: (415) 837 1361	Particle flux counters
KAN	Kanomax, Nihon Kagaku Kogyo Co. Ltd, Shimizu Suita 2-1, 565 Osaka Japan, Tel: (0) 6876-0693	Stand alone counters for aerosols in combination with condensation nucleus counters
LOT	LOT GmbH, Im Tiefen See 58, 6100 Darmstadt Germany, Tel: (0)6151/88060 FAX: (0)6151/84173	Stand alone (off line) counters for aerosols and hydrosols
MAL	Malvern Instruments LDT., Spring Lane South, Worcestershire WR14 1AQ, GB, Tel: (0)684 892456 FAX: (0)684 892789	Stand alone counters for aerosols
MET	MET ONE, 481 California Avenue, Grants Pass, OR 97526 USA, Tel: (503) 479 1248 FAX: (503) 479 3057	Stand alone counters for aerosols and hydrosols, monitors, condensation nucleus counters
MIE	MIE Monitoring Instruments for the Enviroment, 213 Burlington Road, Bedford, MA 1730 USA, Tel: (617) 275 5444 FAX: (617) 275 5747	Stand alone counters for aerosols
PAL	Palas GmbH, Partikel- und Lasermesstechnik, Haid und Neu Strasse 7, 7500 Karlsruhe Germany, Tel: (0)721/693433	Stand alone counters for aerosols, condensation nucleus counters
PMS	PMS Particle Measuring Systems Inc., 1855 South 57th Court, Boulder, CO 80301 USA, Tel: (303) 443 7100 FAX: (303) 449 6870	Stand alone counters for aerosols and hydrosols, monitors
POL	Polytec GmbH, P.O. Box 1140, 7517 Waldbronn Germany, Tel: (0)7243/604-0 FAX: (0)7243/69944	Stand alone counters for aerosols and hydrosols, laser Doppler velocimetry
PPM	Process Particulate Monitors Inc., 11428 Kingston Pike, Knoxville, TN 37922 USA, Tel: (615) 966 8796 FAX: (615) 675 4795	Stand alone counters for aerosols
RIO	RION Co., Ltd., Ikeda Bldg., 7-7 Yoyogi 2-chome, 151 Tokyo Japan, Tel: 33 79 23 52 FAX: 33 70 48 28	Stand alone counters for aerosols
SIB	SIBATA Scientific Technology Ltd., Ikenohata Taito-ku 3-1-25, 110 Tokyo Japan, Tel: 03 822 2111 FAX: 03 824 3045	Stand alone counters for aerosols
SIE	Siemens AG Energie- und Automatisierungstechnik, Balanstrasse 73, 8000 München 80 Germany, Tel: (0)89 4144-0 FAX: (0)89 4144-8002	Facility monitoring system
TSI	TSI Inc., 500 Cardigan Road, St.Paul, MN 55164 USA, Tel: (612) 490 2833 FAX: (612) 490 3825	Monitors for aerosols, condensation nucleus counters

In this table, the clean-room suitability indicates which standard the instrument can be used for. The false count rate limits the ability to determine accurately low concentrations and, thus, limits the class that can be detected by that instrument. The last column indicates the smallest particle size that can be detected at the indicated detection efficiency. This information is based on manufacturer's literature and has been taken from an IUTA database described below.

Another type of system, the so-called scanner, consists of a single counter connected to a multipoint sampling system. Such a system has the disadvantage of potential sampling errors caused by lengthy tubing. In most systems currently in use, a series of monitors are linked in a network to a central computer. The configuration of the network varies, e.g., using a ring configuration or a multiplexer. The capability of ring systems is limited by data flow considerations to about ten sensors while multiplexers allow the interconnection of up to 64 sensors. Some systems, e.g., the system shown in Fig. 34-2, have even higher capability. The system shown allows the interconnection of higher numbers of sensors by the use of multiplexers as well as expander boxes.

Such high numbers of sensors can cause reading intervals for each sensor of more than a minute, limiting the data available and causing problems in statistical analysis. With both ring systems and multiplexers, each manufacturer has a different data format so that systems are not directly compatible and can be connected only with considerable effort. Previous attempts to establish a standard for such networks have failed. A possible solution might be an existing standard network (e.g., Ethernet, IEEE 802.3).

In Table 34-7 some commercial sources of monitoring systems are given. Some manufacturers deliver only software and hardware of the network. Other manufacturers also produce the necessary sensors.

In addition to the problems described above, additional factors that remain include

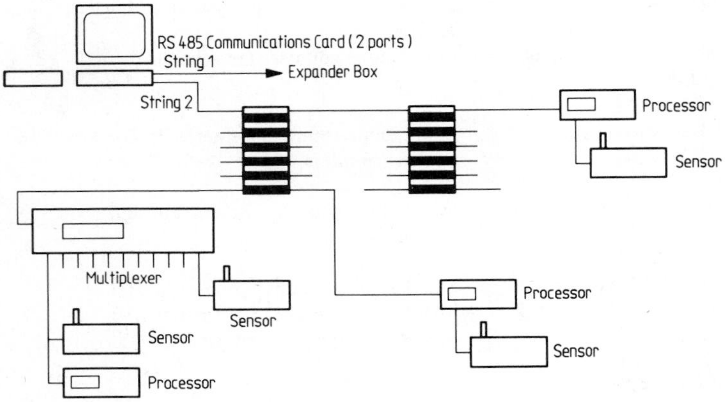

FIGURE 34-2. Multidrop Communications Clean Room Monitoring System.

TABLE 34-7 List of Commercial Sources for Facility Monitoring Systems (Software and Hardware)

Manufacturer	Monitoring Data						
Code	Particles in Air	Particles in Liquid	Temperature	Relative Humidity	Air Velocity	Pressure	Computer
ACT	Y	Y	X	X	X	X	PC-XT/DOS
AAA	Y	Y	X	X	X	X	PC-AT/DOS
CLI	Y	Y	Y	Y	Y	Y	PC-XT/DOS
HIA	Y	Y	Y	Y	Y	Y	PC-AT/DOS
MET	Y	Y	Y	Y	Y	Y	PC-AT/DOS
PMS	Y	Y	Y	Y	Y	Y	PC-AT/DOS
SIE	X	X	X	X	X	X	SICOMP M70/UNIX
TSI	Y	Y	Y	Y	Y	Y	PC-AT/DOS

X: only software available
Y: software and hardware (sensors) available

the sensor location in the clean room, the handling of large quantities of data, and the interpretation and representation of this data.

Choice of Sensor Location

Unfortunately, the optimum recipe for sensor location within the clean area does not exist (Cooper et al. 1990). However, some general rules and guidelines can be formulated. Some of the sensors should be used to control the clean room as a whole. From these, data can be obtained regarding the particle concentration in the make-up air, in the return air downstream of the prefilters, and from undisturbed parts of the clean room. Additional locations might be helpful depending upon the construction of the air flow system. After a sufficient period of time, generally more than a year (to take into account outside environmental variations), some areas may be understood well enough to allow removal of some of the sensors for use elsewhere. Most of the sensors should be placed at critical locations within the clean room. For media control (e.g., of supplied air or compressed gases), the point of use is the best sensor location.

The choice of additional sampling locations should be performed according to the following scheme. First, the air flow patterns must be analyzed to understand the air velocity and flow direction and, thus, to determine the area upstream of the product where any particle is critical. Possible sources of cross contamination should be determined for normal operating conditions. In this case, the sampling point can be located within the air stream lines towards the product surface. This procedure applies only to situations with constant operational conditions. Further critical areas are located where abrasion is known to occur. For example, measuring close to the bearings of transport devices or robots can provide information on contamination risks as well as maintenance requirements. In any case, optimal placement of sampling points can only be determined empirically.

Data Interpretation and Representation

A medium-sized monitoring system can deliver several megabytes of data per week. Current computer technology allows relatively inexpensive solutions to storage and handling of such amounts of data. However, the problem of how to interpret and represent this amount of data remains. For example, which group of data should be correlated to some other group, how are the representative means or other parameters determined, what does this resultant information mean, and how should the information be plotted or displayed?

An example of a frequently observed correlation in clean rooms is the interdependency of humidity and particle number concentration. One possible explanation for this is the growth of particles smaller than the lower detection limit of the measuring instrument due to condensation. Another explanation involves the influence of humidity on electrostatic effects and, therefore, on particle transport mechanisms and on particle-to-surface interactions. If a dependence of particle concentration on humidity is found, the particle concentration may be predicted by simply measuring the humidity. Additional interdependencies become evident via correlation analysis of each clean room. Knowledge about the interactions allows the prediction of contamination risk as well as prevention of that risk by taking appropriate measures.

Influences on Clean-Room Personnel

The psychological effect of sensor placement should not be neglected, especially when sensors provide a direct indication of excess concentrations visually or by audible alarm. In some cases, the particle concentration is not affected by the actions of the clean-room personnel. For instance, sampling might be performed within the assembly equipment where many particles are produced by abrasion. Then, from the operators point of view, the only chance to achieve tolerable levels might be to manipulate the sensor, e.g., to switch it off or change the alarm limits. Such situations can be quite complex and are indicated here to stress that the human aspect of monitoring systems can be quite important (Hauptmann and Hohmann 1991).

Special Tasks

Tests of Clean-Room Suitability

Any equipment or object, such as machinery, furniture or even measuring devices, brought into the clean room can be a source of particles. Such particle emission can take place constantly or under the influence of some specific action, i.e., movement, variation of temperature, pressure changes, humidity, air flow patterns, electrostatic disturbances, etc. The clean-room suitability of each item or its components should be tested by its producer and/or user.

Although the following discussion relates primarily to particle emissions, the reader must be aware that other sources of contamination can also occur and that interactions with other devices in the vicinity can affect the quality of production processes. Some of these interactions can be due to vibration, temperature effects, electrostatic or electromagnetic effects, or radiation.

The particle contamination risk depends on the position of a particle source; e.g., close to the product surface and upstream of the particle source is more critical than downstream of the product and further away.

Several procedures have been proposed to certify equipment and standards describing such certification measurements exist or will be published in the near future. The aim is to state the clean-room class that a particular piece of equipment is suitable for. According to proposed procedures (von Kahlden, Geissinger, and Degenhart 1989) two certification procedures are possible. In both cases the equipment is placed in a clean environment. Smaller devices or components can be tested in a certified test clean-room in which no other activities are taking place or in a specially designed test chamber supplied with clean air. Testing larger devices can take place only in a certified clean-room.

The first certification procedure can be employed only for smaller devices or components. Here a high flow rate particle counter is placed downstream of the test device and the total flow passing the piece is sampled with a funnel.

The second procedure involves a process of scanning the area that can be affected by the device with a high flow rate particle counter. The relevant area is determined by an analysis of the flow pattern around the device, defining an area of disturbed flow where particle influence can occur. With this scanning method, the maximum measured concentration will depend on the distance of the sampling point from the source as well as on the flow rate of the counter. If the particle concentration is very high, coincidence effects can reduce the apparent concentration. Compared to ambient-air sampling in a clean-room, measurements near a source may be prone to sampling errors. For instance, the size distribution close to the source may be relatively broad because deposition processes, such as diffusion for small particles and settling for large particles, have less time to take place. The temperature of the sampled air also may be different than that of the ambient air, causing thermophoretic losses.

For information on the statistical calculations needed to certify a test piece for a clean-room class on the basis of concentration data, the reader is referred to corresponding standards, such as the upcoming German VDI 2083 Part 8. There seems to be no universal recipe for this analysis because of the large range of concentrations that can appear over time in a set of sampling locations.

Particle Counting in Gases at Nonambient Pressures

Several types of clean-room monitoring may make it necessary to count particles at elevated or reduced pressures. High-pressure gases from supply systems or gas cylinders may need to be analyzed. Low-pressure devices or even vacuum systems, e.g., in the microelectronic industry, can also release particles.

Clean gases at higher pressures can be controlled either using high-pressure particle counters or by expanding the gases and measuring at ambient pressure. Several manufacturers provide high-pressure counters such as HIAC/ROYCO (Model 5400, 210 MPa or Model 2400, 70 MPa) and PMS

(Model HPGP, 1 MPa). Special pressure reducers for clean-room applications also exist (TSI Model 375547, HIAC/ROYCO Model 174). It must be kept in mind that expansion valves and nozzles can also be a source of particle emission. In addition, the gas expansion can also trigger particle generation by a condensation process. Any such device should be evaluated before use.

Filter Testing

In order to assess the performance of filtration systems and their components, the very low downstream concentration must be measured. Although some information is presented here, the reader should consult Chapter 10 for more detailed procedures on filter testing. Filter penetration is particle-size-dependent. If monodisperse particles are not used for evaluating a filter, an optical sizing spectrometer or sizing counter can be used. Several of the manufacturers listed in Table 34-6 provide instruments with sufficient resolution and lower particle size limit to perform such filter testing. Optical counters detecting the smallest particle sizes require high sensitivity and are, therefore, prone to higher false count rates. Thus, for high-efficiency filters, a test using monodisperse challenge particles and a CNC detector is recommended.

Source Detection

Possible sources of particles have been mentioned above several times. However, the main particle sources in clean-rooms are usually the clean-room personnel. Besides their clothing, their behavior can result in contamination. Therefore, education and motivation of the personnel are very important. Due to the availability of high-efficiency filters, particle transport into the clean room via various media is generally of minor importance, though a variety of defects can cause impurities in delivered gases, emissions from tubing and valves, etc. In addition, chemical impurities can lead to particle production by condensation or chemical reaction. These particles will tend to be submicrometer in size. Particles produced by shedding or abrasion tend to be much larger. Scanning measurements can be useful for searching for leaks in filter ceilings, defects in machinery, etc. Since such sources can produce particles with a typical size distribution, a sizing particle counter can "fingerprint" a detected distribution (Biswas, Tian, and Pratsinis 1989).

Useful information can also be gained by chemical analysis of sampled particles. In some cases, statistical analysis allows concentration patterns to be related to the contamination source. Although high flow rate counters can produce faster particle measurements, a low flow rate counter may give better spatial resolution. Detected particle sources can be eliminated by either repairing the problem or isolating or encapsulating the source.

AVAILABLE MEASURING TECHNIQUES

This section introduces the measurement techniques available for the tasks mentioned above. New detection principles under development will also be mentioned. Two types of measurements can be distinguished: the measurement of airborne particles and the measurement of deposited particles. Although the latter is not an aerosol measurement, it is commonly used for air certification. Such measurements also have the advantage of allowing subsequent analysis of the particles. However, since almost all surface measurements are off-line, they describe only an earlier situation.

Measurement of Airborne Particles

The optical particle counter (OPC), described in Chapter 15, allows the continuous measurement of airborne particles. The following sections will describe the principal features relevant to clean-room measurements. The counters can be divided into three groups: OPCs, CNCs, and particle flux meters. All the devices use the interaction of particles and light. CNCs can be distinguished from OPCs by the conditioning of the particles that takes place prior to detection. The main feature of the particle flux meter is that this device does not require a special sampling system.

Optical Particle Counters

Light scattering particle counters are typically used rather than extinction-type counters since the latter generally detect only particles larger than 1 μm diameter. By measuring the scattered light, information on particle number and size can be collected. The amount of scattered light reaching a detector from a particle depends on the particle shape and refractive index. The measured size is an "optical diameter", which further depends on the shape and refractive index of the particles used to calibrate the instrument. It should be noted that the difference between optical, geometric, and mobility size can frequently be greater than 30%. Table 34-8 lists the main features of particle counters related to clean-room applications and can be used to select a suitable counter.

Clean-room suitability in Table 34-8 indicates the contamination risk due to the counter itself. Not all counters are constructed so that no particle production occurs or that emitted particles are swept away by an additional air stream. Exactly how this suitability can be measured is not yet defined in standards or guidelines. The false count rate indicates the level of counts that can be detected while sampling particle-free gas and limits the lowest concentration that can be measured. Specifications provided by the manufacturer can often be exceeded dramatically due to a "memory effect" that depends on the usage history of the counter. For instance, a high-concentration peak can cause continued particle detection even after the air concentration has returned to normal. A ratio of noise count rate to particle count rate of 1:10 or less is desirable. The upper concentration limit of the counter can be important even if the normal measured concentration is low because of peak concentrations that can occur near a source.

The reader should keep in mind that the lower particle size detection limit is given for a certain counting efficiency, typically 50%. Particles in the size range of the lower detection limit are counted with a certain probability, increasing with particle size. This results in an error in the defined concentration, especially for very small particles. In general, only calibrated instruments should be used. The calibration procedure should be repeated at fixed intervals. The calibration should include flow rate, size resolution, counting efficiency versus particle size, and false count rate. For some high-pressure instruments, alternating external pressures can result in unsteady internal flows as well as changing temperatures within the instrument. This can lead to an error in the defined concentration or particle size.

Condensation Nucleus Counters

If very small particles (< 0.1 μm) must be detected, CNCs can be used. The detection principle is similar to that of the scattered-light OPC. However, small particles are enlarged by vapor condensation such that all particles grow to approximately the same size. This allows efficient detection of a wide

TABLE 34-8 Main Features of Particle Counters for Clean Room Measurements

Feature	Typical	Optimum
Clean room suitability	?	?
False count rate	0.5–1 h^{-1}	→ 0
Upper concentration limit	5 cm^{-3}–10 cm^{-3}	20 cm^{-3}
Lower detection limit (LDL)	0.3 μm–0.2 μm	< 0.1 μm
Flow rate	2.8 or 28.3 l/min	> 28.3 l/min
Counting efficiency	50% at LDL	100% at LDL

2.83 l/min = 0.1 ft^3/min

range of particle sizes (Table 34-8) though all information about the original particle size is lost.

Two types of CNCs are available: a continuous-flow CNC and a cycled-flow CNC. The latter CNC type, which samples the air, conditions it and then measures it, is not usually appropriate for clean-room applications. The main features of OPCs (Table 34-8) can also be applied to CNCs. Table 34-9 gives an overview of several CNCs currently available for clean-room applications. In addition to these devices, special instruments have been developed for specific tasks, such as rapid testing of filter elements (Holländer, Dunkhorst, and Lödding 1990).

Particle Flux Meters

A variety of particle flux meters are currently undergoing development. These devices will play an important role in the field of *in situ* measurement in clean rooms in the near future. Figure 34-3 provides a schematic of three types of particle flux meters (Fissan and Trampe 1989). In the first type (A) a laser beam is expanded by a system of cylindrical lenses to form a rectangular beam. The second type (B) produces a light sheet by multiple reflection of a laser beam between two mirrors. The last type (C) is a scanner, where a laser beam is deflected by a rotating polygon mirror.

The main differences between OPCs and particle flux meters are that the latter do not take samples, have the capability of high effective flow rates, and can be used to control

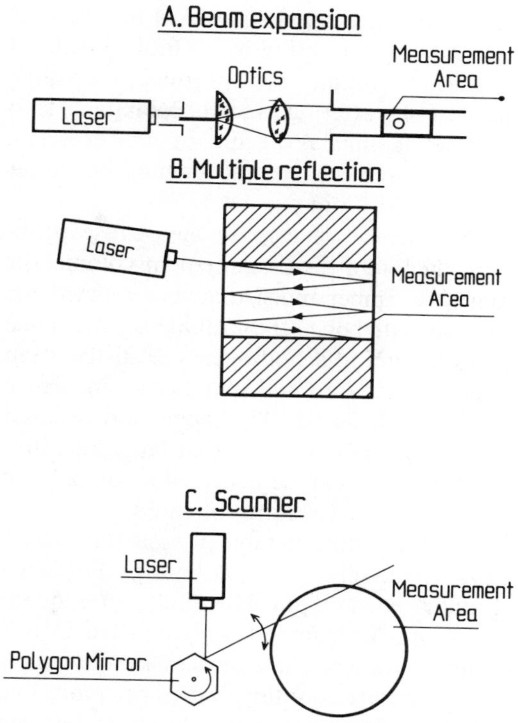

FIGURE 34-3. Working Principle of a Particle Flux Meter: (A) Expansion; (B) Multiple reflection; (C) Scanner.

a large measurement area. With a particle flux meter, the aerosol is not drawn through the instrument but rather the detection volume is superimposed on the aerosol.

One of the problems in the construction of these devices is the opposing requirements of a low detection limit and a large measurement area. A high sensitivity can be realized

TABLE 34-9 Clean Room CNCs

Company	Model	Flow Rate, l/min	Lower Detection Limit, μm	Maximum Concentration, 1/cm³	False Count rate, h⁻¹
TSI	3760	1.41	0.014	$1.6 \cdot 10^3$	< 0.6
TSI	3761	2.83	0.02	$1.0 \cdot 10^3$	< 1.2
MET	1100	1.41	0.01	353	ns
KAN	3851	2.83	0.01	353	ns
KAN	3861	5.0	0.05	317	ns
ACT	5000	2.83	0.05	353	ns

ns—not specified
2.83 l/min = 0.1 ft³/min

only if an extremely low level of background light occurs. Scattering by air molecules from a 1 mm^3 volume can produce the same amount of scattering as an 0.22 μm particle. If particles as small as 0.1 μm are to be detected, the measurement volume cannot be larger than 6×10^{-3} mm^3.

Coincidence effects can also cause problems: the larger the measurement volume, the larger the probability that more than one particle will be in the volume at the same time. Assuming a homogeneous distribution of particles in time and space, the coincidence can be calculated (Umhauer and Raasch 1984). Figure 34-4 shows that large measurement volumes can be achieved if coincidence levels of up to 10% are tolerated.

Currently, the available particle flux meters have relatively low counting efficiency (Caldow et al. 1988). Hopefully, subsequent generation instruments will improve in this regard. However, these devices are not meant to be accurate counters, but to act more like indicators of the changing levels of particle concentration.

Measurement of Deposited Particles

In contrast to the on-line measurement techniques described above, the following offer the advantage of a particle at rest. An analysis of the deposited particles allows a determination of the particle shape and composition, which provide information about the source. For a more detailed discussion of particle analysis, the reader is referred to Chapter 13. The present discussion will deal only with counting the deposited particles. Electron microscopy typically is not used for this purpose because of the long counting times involved when large areas must be viewed. The typical field area that can be viewed with transmission electron microscope at a useful magnification is 1 μm^2; thus, for a 1 cm^2 total viewing area 10^8 fields must be observed.

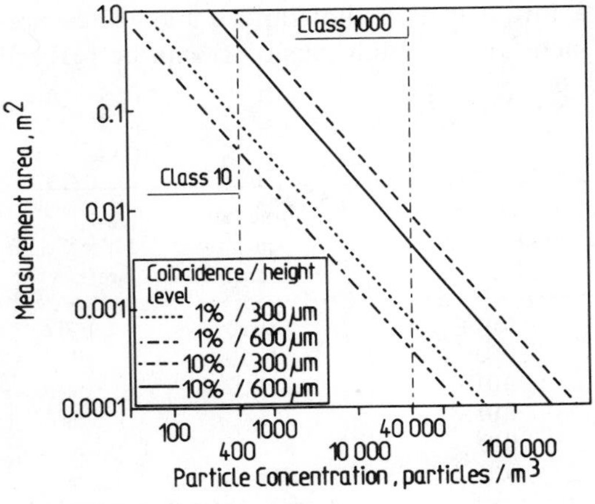

FIGURE 34-4. Measurement Area for Particle Flux Meter versus Particle Concentration.

The use of the so-called witness wafers is common in clean-room technology, especially in the microelectronic industry. It could also be applied to many other industries, using another typical product surface besides the semiconductor wafer. The procedure is to place this surface at critical locations in the working area or to move it through the process like the product itself. After a certain period of time, the deposited particles are counted.

The calculation of the corresponding airborne concentration assumes that each particle that was within the air above the surface deposited with an average deposition velocity (averaged over all particle sizes) (Welker 1988). The result, a surface contamination rate with units of particles/(cm^2 h), may be incorrect due to the various assumptions. Several parameters affect the deposition velocity, including the particle size and charge, surface temperature, and surface charge. This effect will be discussed in the last section of this chapter.

Optical Microscopes

Figure 34-5 shows the setup for using an optical microscope for surface particle counts. Transmitted light and reflected light illumination can be used. The setup must allow surface motion along all the three axes. The resolution of the positioning device must be at least one-tenth of the minimum image field to avoid false counts at field borders and poor repeatability. The resolution in the *z*-direction should be 20–40 nm to allow focusing on every plane.

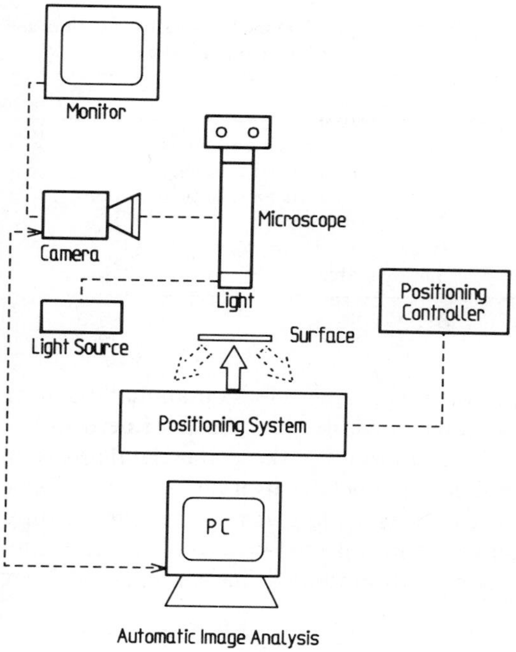

FIGURE 34-5. Set Up to Use an Optical Microscope for Particle Counts on Surfaces.

Optical microscopes can be used for counting particles as small as 100 nm (Schmidt, Schmidt, and Fissan 1991*a*), which is much lower than the limits specified in several standards. The detection limit can be lowered even further by using a particle growing process similar to that in the CNC. Particle size on the surface can be amplified by coating the surface with a reactive substance (Schmidt, Schmidt, and Fissan 1990; 1991*b*). Once this is done, particle sizing is no longer possible. Particles are often out of focus, but are still visible as indicated in Fig. 34-6. Here a photograph of a fluorescent latex particle (d_p = 80 nm) taken under an optical microscope is shown.

Digital filtering of the image can be used to reduce the halo around unfocused particles. The size distribution on the surface can be calculated following the filtering procedure (Fu 1980; Aggarwal, Duda, and Rosenfeld 1977; Huang 1979).

Surface Scanners

The use of a surface scanner is a common technique in the microelectronic industry (Lilienfeld 1986). A laser beam is used to scan the product surface and the light scattered from a particle is detected as in an OPC. Currently, there is no theory that gives a complete description of the scattering process. Therefore, each instrument must be calibrated individually. As with OPCs, the signal-to-noise level determines the lower detection limit, though in this case the noise is primarily due to scattering from the substrate surface. When substrate structure is minimal,

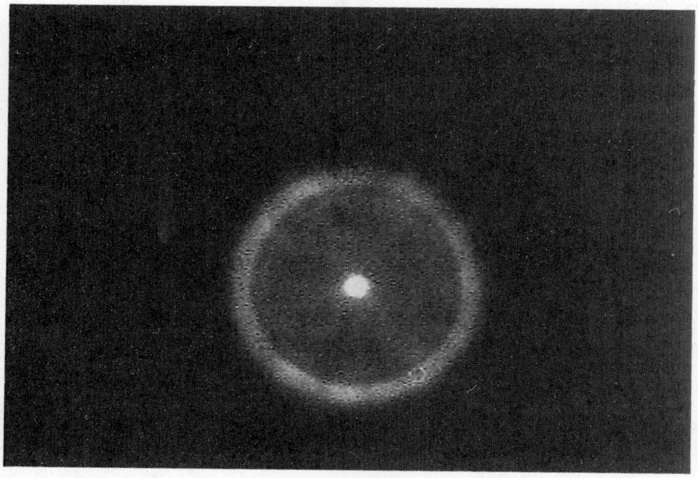

FIGURE 34-6. Fluorescent Latex Particle d_p = 80 nm, Photograph Taken Under an Optical Microscope.

the lower detection limit can be as low as 0.1 μm. The counting efficiency tends to be poor for surface scanners, often less than 10%. Finally, the slope of the counting efficiency versus particle diameter curve is low. As with OPCs, surface scanners count particles in the size range near the lower detection limit with relatively poor efficiency. For particles below 1 μm this must be taken into account.

TABLE 34-10 Content of the Data Base for Aerosol and Hydrosol Instrumentation

Device group
Aerosol single particle counters
Hydrosol single particle counters
Software products for facility monitoring
Aerosol generators
Hydrosol generators
Surface scanners

Data Base for Aerosol and Hydrosol Instrumentation

The database for aerosol and hydrosol instrumentation (DAI) has been developed to facilitate the choice of the best monitoring instrumentation (Trampe and Fissan 1990). Table 34-10 indicates the current contents of the database. A variety of instruments useful for assessing particle contamination have been included, not just those for aerosols.

An important task of the database is to present an objective overview of the available instruments. Data given by the manufacturer have been collected without further evaluation and presented in comparable form,

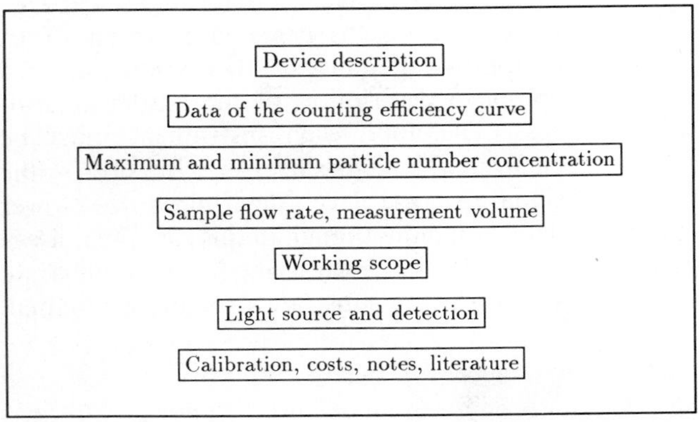

FIGURE 34-7. Structure of a Data Set for Single-Particle Counters.

TABLE 34-11 Attributes for Device Description

Device description		
Principle	:	Light scattering
Object of measurement	:	Particle size, particle number per volume
Concept	:	Stand alone
Display/control desk	:	Keyboard 24 keys, LCD Display 8 lines 40 characters/line
Sample	:	Internal pump
Sampling system	:	Inlet at the front, tube stainless steel 9 mm
Clean room suitability	:	Yes, to class 10
Gases	:	Air and inert gases
Sensor dimensions	:	W: 300 mm, H: 400 mm, L: 445 mm
Weight	:	17.5 kg
Data interface	:	RS 232C
Explosion-proof	:	No

using consistent terms and units. Besides presenting the information in a more comparable format, information gaps are also indicated. This means that data not present in the manufacturer's materials can be included when it becomes available. A further task of the database is to support market analyses and trend studies. The manufacturers receive feedback through user enquiries. Journal articles that describe application of the instruments have also been included in the database, providing the reader with experiences of other users without a comprehensive literature search.

A total of 81 attributes have been stored in the database for each optical counter. Figure 34-7 indicates the structure of the data blocks. The data blocks for other types of instrumentation are similar. The following example expands the information present in the first data block. Table 34-11 lists the attributes of a fictitious instrument.

PROBLEMS

Statistics of Low-Concentration Measurements

A particle measurement in a clean room typically is limited in volume and time. Assuming that the measurement is performed correctly, i.e., there are no sampling errors, the concentration of interest is inferred from the measurement. Particles emitted into a gas become randomly distributed due to effects like diffusion or mixing. Therefore, the number of particles counted in the analyzed volume can vary statistically. This means that the concentration of interest, e.g., in a clean area or downstream of a filter, can exceed an established limit even though the measured concentration is below the limit. In many cases the uncertainty of the concentration measurement is derived from Poisson statistics. The necessary preconditions include:

Random distribution of particles. There must be sufficient time for distribution processes such as diffusion or mixing to occur.

Low probability of a particle count. The concentration must be low.

If, under these preconditions, a large number of concentration measurements in volumes of equal size were made, the particle number of one volume would be independent of the number from the previously measured one. For an arithmetic mean of particles per volume μ and a large number of measurements, the relative frequency of counting a certain number of particles n is equal to the probability $P(n)$ as given by the Poisson distribution:

$$P(n) = \frac{\mu^n \cdot e^{-\mu}}{n!} \tag{34-1}$$

For events distributed according to Poisson, the statistical uncertainty depends only on the number of counted particles. This can be demonstrated by having a look at the upper, and lower confidence limits, of a certain number of counted particles (Johnson and Leone 1977). If 20 particles are counted in a certain volume of a gas with randomly distributed particles, the average number of particles in this volume could be 30.8 (95% upper confidence limit, C_u) or it could be 12.2 (95% lower confidence limit, C_l). In other words, the actual average particle number can be 54% larger (in the following defined as a relative upper confidence limit c_u) or 39% lower than the measured one (in the following defined as a relative lower confidence limit c_l). In the case of only a single-particle count, the upper and lower limits are 3.7 and zero, or 270% higher (c_u) and 100% lower (c_l), respectively. It becomes evident that for the determination of a concentration within certain limits that a sufficient number of particles must be counted. That particle number can be expected to occur within a certain measurement time, depending upon the instrument used and the clean area. Assuming a random distribution of particles in clean rooms of different classes according to the U.S. Federal Standard 209D and assuming certain OPC parameters (2.8 l/min flow rate, 0.5 μm lower detection limit), the necessary particle number can be expected in a certain sampling time. The relative particle count confidence limits c_u and c_l calculated as described above for several clean-room classes are given in Fig. 34-8 as a function of the sampling time.

In the U.S. Federal Standard 209D and other standards, the volume to be measured is dependent on the class. For instance, Table 34-4 requires that at least 20 particles be measured, resulting in the confidence limits mentioned above.

For larger numbers of detected particles, the Poisson distribution approaches the Gaussian distribution. In this case, confidence limits can be derived using the standard deviation. For Poisson-distributed events, the standard deviation is equal to the square root of the particle number. If, for example, 50 particles are counted, the upper and lower 95% confidence limits are 37 and 65.9, respectively. According to the Gaussian distribution, these limits are 35.1 and 63.9 ($50 \pm 1.96 \cdot (50)^{0.5}$). For lower particle numbers the latter approach is not correct and the calculated limits diverge even more.

Determination of Distribution

Unfortunately for many concentration measurements, one of the preconditions for Poisson statistics, namely the random distribution of particles, is not met. In a clean room in operation, some particle counts will appear higher due to a process in the vicinity. Whether a Poisson distribution is present can be checked by comparison of the frequency distribution in a large number of samples with the distribution calculated by Eq. 34-1 (Schmitz and Fissan 1989).

The three graphs in Fig. 34-9 show the frequency distribution of particle numbers from identical sampling volumes taken in a clean workbench.

The data shown in Fig. 34-9A were obtained in an empty clean bench and agree with a Poisson distribution. The subsequent distributions were obtained after a particle source was introduced. The data shown in Fig. 34-9B were obtained at a larger horizontal (h) and vertical distance (a) than the data shown in Fig. 34-9B. Both of the experimentally determined distributions do not agree with the Poisson distribution. The discrepancy increases with decreasing distance from the source. The random distribution of particles passing the workbench filters is modified by the nonrandomly distributed particles from the source. Nevertheless, an estimate of the mean of the random distribution μ_0 can be derived from the frequency of zero counts. According to Eq. 34-1, the mean

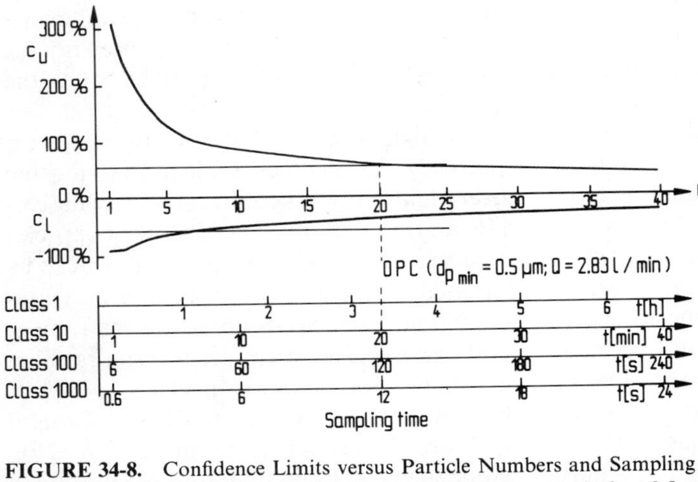

FIGURE 34-8. Confidence Limits versus Particle Numbers and Sampling Time of a Certain Optical Particle Counter (OPC) (Lower Detection Limit $d_p = 0.5$ µm, Flow Rate $Q = 2.83$ l/min, $c_u = C_u - n/n$, $c_l = (C_l - n)/n$ Where C_u = Upper Confidence Limit and C_l = Lower Confidence Limit, n = Particle Number).

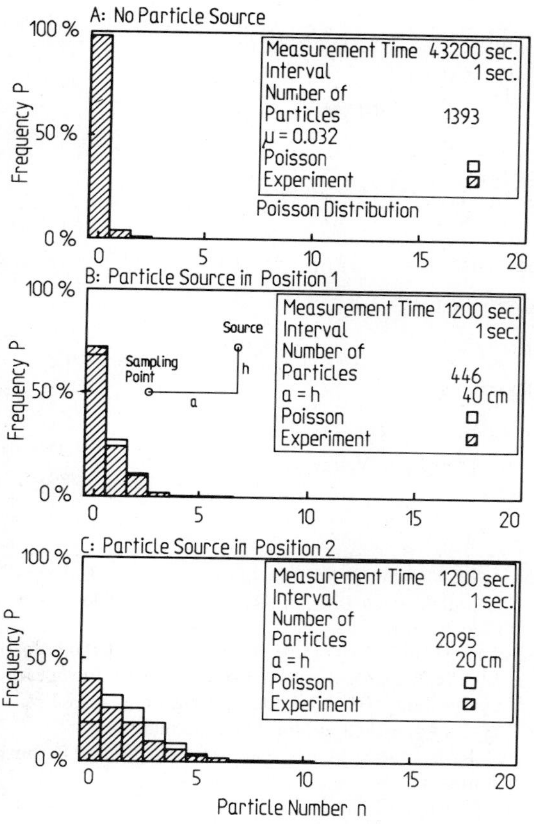

FIGURE 34-9. Frequency *P* of Particle Number *n* in Time Intervals in a Clean Workbench with Particle Source: (A) No Particle Source; (B) Particle Source in Position 1, Horizontal and Vertical Distance 40 cm; (C) Particle Source in Position 2, Horizontal and Vertical Distance 20 cm.

of a Poisson distribution μ_0 with the frequency of zero counts $P(0)$ is given by

$$\mu_0 = -\ln P(0) \tag{34-2}$$

In a clean room, this simple analysis can be employed to monitor the quality of the ambient air, even if no particle counter is available close to the filter exhaust in the ceiling.

The causes of nonrandomly distributed counts include:

Particles from sources. The number of particles can vary considerably due to lack of sufficient sampling time or to turbulence.

Noise pulses. Noise pulses can reach the counter from other equipment via the power supply; these are often systematic with time. If the sum of counts from different counters are analyzed statistically, noise pulses often appear as higher count numbers for two instruments recording at the same time.

In most cases, it cannot be assumed that a Poisson distribution occurs at a certain location or that the concentration is uniform throughout an entire clean room. Then the described statistics can only be applied to obtain an approximate estimate of the true concentration. The calculations in the U.S. Federal Standard 209D demand a minimum volume measured at a single location as well as at a number of locations, depending upon the area of the clean area. Thus, a class limit is met if the calculated upper confidence limit as well as the measured concentration are within the class limit. This confidence limit (based on the Student's *t*-distribution) assumes a higher concentration calculated on the basis of a limited number of sampling locations. The confidence limits can be tightened by increasing the number of sampling locations.

An attempt to describe an entire clean area over an extended period of time is unlikely to be successful. Even if a distribution can be found that describes the measurement data, the applicability of the calculated parameters and confidence limits remains dubious.

Correlation between Clean-Room Concentration and Particle Deposition

As mentioned above, there are at least several parameters that influence the deposition velocity V_d of particles onto a surface. If the contamination risk in a clean area is to be assessed, this must be taken into account. The airborne concentration derived from a test surface can deviate from the real data as can the calculated deposition onto the product surface. Parameters of influence include particles size and charge as well as surface temperature and charge. For example, Fig. 34-10 shows the influence of the surface temperature on the deposition velocity versus particle diameter curve. Particles in the range

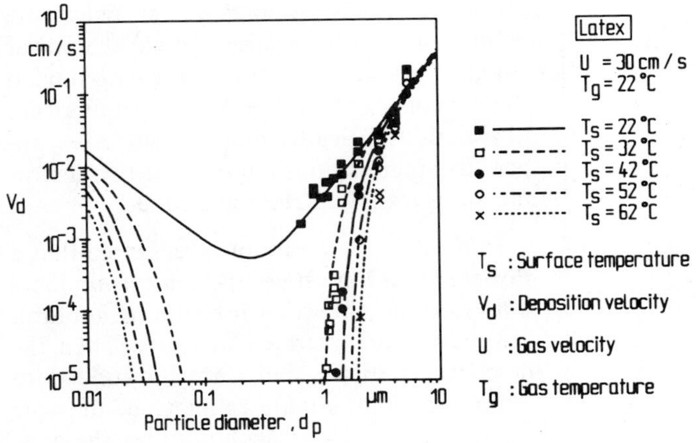

FIGURE 34-10. The Influence of Surface Temperature on the Deposition Velocity V_d (U = Gas Veolcity, T_g = Gas Temperature).

0.1–1.0 µm can have very low deposition velocities and will not deposit, but will rather follow the streamlines of the airflow. Additional effects, such as electric field or temperature gradients can shift the deposition velocity (Ye et al. 1991). For detailed calculations, the limiting parameters of temperature, electrical charge, etc., of each particle size must be measured.

References

Aggarwal, J. K., R. O. Duda, and A. Rosenfeld. 1977. *Computer Methods in Image Analysis.* New York: IEEE-Press.

Biswas, P., Y. Tian, and S. E. Pratsinis. 1989. Receptor modeling of microcontamination in clean rooms. *J. Aerosol Science* 20(8):1361–64.

Bzik, T. J. 1986. Statistical management and analysis of particle count data in ultraclean environments part II. *Microcontamination* 6(4):34–41.

Caldow, R., D. Y. H. Pui, W. W. Szymanski, and B. Y. H. Liu. 1988. Performance of the high yield technology PM-100 *in-situ* particle flux monitor. Presented at 7th Annual Meeting of the American Association of Aerosol Research, AAAR, 10–13 October 1988 in Chapel Hill, NC.

Cooper, D. W. 1988. Statistical analysis relating to recent federal standard 209 cleanrooms revision. *J. Environ. Sci.* 31(2):32–36.

Cooper, D. W., S. J. Grotzinger, C. R. Nackman, and V. Srinirason. 1990. Selecting nearly optimal sampling locations throughout an area. Application to clean rooms and Federal Standard 209. Presented at 36th Annual Technical Meeting of the Institute of Environmental Sciences, IES, 30th April–4th May 1990 in New Orleans, LA.

Fissan, H. and A. Trampe. 1989. Optical methods for particle flux measurements. Presented at 4th European Symposium Particle Characterization PARTEC, 19–21 April 1989 in Nürnberg, Germany.

Fu, K. S. 1980. *Digital Pattern Recognition.* Berlin: Springer.

Hauptmann, G. and R. Hohmann. 1991. *Handbook of Clean Room Technology and Human-Ressources.* Landsberg: Ecomed.

Holländer, W., W. Dunkhorst, and H. Lödding. 1990. A condensation nucleus counter array for clean room applications. *Swiss Contamination Control* 3(4a):21–22.

Huang, T. S. 1979. Picture Processing and Digital Filtering. Berlin: Springer.

Johnson, N. L. and Leone, F. C. 1977. *Statistics and Experimental Design in Engineering and Physical Science.* New York, London, Sydney: Wiley.

Junge, C. E. 1963. *Air Chemistry and Radioactivity.* New York: Academic Press.

von Kahlden, T., J. Geißinger, and E. Degenhart. 1989. Standardisierung von Untersuchungen auf Reinraumtauglichkeit, Richtlinienentwurf. *Reinraumtechnik* 3(5):42–45.

Lilienfeld, P. 1986. Optical detection of particle contamination on surfaces: A review. *Aerosol Sci. Technol.* 5(2):145–65.

Schmidt, F., K. G. Schmidt, and H. Fissan. 1990. Methods for detection of submicron particles by using lightmicroscope. *J. Aerosol Sci.* 21(S1):535–38.

Schmidt, F., K. G. Schmidt, and H. Fissan. 1991*a*. Submikrone Partikel auf Oberflächen mit dem Lichtmikroskop nachweisen. *Reinraumtechnik* 5(1):38–39.

Schmidt, F., K. G. Schmidt, and H. Fissan. 1991*b*. Nachweis der Partikelabscheidung mit Hilfe chemischer Reaktionen Chemie. *Chem. Ing. -Technik*

63(3):276–77.

Schmitz, W. and H. Fissan. 1989. The lower detection limit of particle number concentrations in pure gases. Presented at 4th European Symposium Particle Characterization PARTEC, 19–21 April 1989 in Nürnberg, Germany.

Trampe, A. and H. Fissan. 1990. Testing and calibration of optical particle counters. Swiss Contamination Control 3(4a):28–31.

Umhauer, H. and J. Raasch. 1984. *Der Koinzidenzfehler bei der Streulicht-Partikelgrößen-Zählanalyse, Fortschrittberichte der VDI Zeitschriften Reihe 3,Verfahrenstechnik*. Düsseldorf: VDI-Verlag.

Welker, R. W. 1988. Equivalence between surface contamination rates and class 100 conditions. Presented at 34th Annual Technical Meeting of the Institute of Environmental Sciences, IES, 3–5 May 1988 in King of Prussia, Pennsylvania.

Ye, Y., D.Y.H. Pui, B.Y.H. Liu, S. Opiolka, S. Blumhorst, and H. Fissan. 1991. Thermophoretic effect of particle deposition on a free-standing semiconductor wafer in a clean room. *J. Aerosol Sci.* 22(1):63–72.

35

Radioactive Aerosols

Mark D. Hoover and George J. Newton

Inhalation Toxicology Research Institute
Lovelace Biomedical and Environmental Research Institute[1]
P.O. Box 5890
Albuquerque, NM, U.S.A.

INTRODUCTION

Measurement of radioactive aerosols involves most of the standard tools of aerosol science and technology, as well as a number of specialized techniques that take advantage of the unique physical properties of radioactive materials. Instruments and techniques for characterizing radioactive aerosols have been described extensively in numerous manuscripts, books, and standards, including those by the International Commission on Radiological Protection (ICRP 1965, 1968, 1982, 1985), the American National Standards Institute (ANSI 1969), the International Commission on Radiation Units and Measurements (ICRU 1971), Raabe (1972), the International Atomic Energy Agency (IAEA 1978), the National Council on Radiation Protection (NCRP 1978*a*, *b*), the American Conference of Governmental Industrial Hygienists Air Sampling Instruments books (see, for example, Cohen 1989), the U.S. Department of Energy Operational Health Physics Training Manual (U.S. DOE 1988), and numerous nuclear radiation and health physics text books (see, for example, Price 1965; Eicholz and Posten 1979; Shapiro 1981; Cember 1983; Turner 1986; Eisenbud 1987; Turner et al. 1988). The requirements for a proper use of these instruments have been further described in documents such as the American National Standard for Radiation Protection Instrumentation Test and Calibration (ANSI 1978), the standard for Performance Specifications for Reactor Emergency Radiological Monitoring Instrumentation (ANSI 1979), the standard for Specification and Performance of On-Site Instrumentation for Continuously Monitoring Radioactivity in Effluents (ANSI 1980*a*), the standard for Performance Criteria for Instrumentation Used for Inplant Plutonium Monitoring (ANSI 1980*b*), and the Performance Specifications for Health Physics Instrumentation (ANSI 1989*a*, *b*, *c*). Published proceedings of many scientific meetings have been devoted to describing the measurement of radioactive aerosols (see, for example, the Proceedings of the Committee on the

1. The preparation of this manuscript was supported by the Office of Health Physics and Industrial Hygiene Programs and by the Office of Health and Environmental Research of the U.S. Department of Energy under Contract No. DE-AC04-76EV01013.

Safety of Nuclear Installations, Meeting on Nuclear Aerosols in Reactor Safety, U.S. Nuclear Regulatory Commission (1980). A detailed review and application of this information, as well as the development of new techniques and applications, continue to occupy the careers of many aerosol scientists and health protection professionals.

This chapter provides an overview of the principles, techniques, and applications of measuring radioactive aerosols. The information needed for a basic understanding is presented, and several newer measuring techniques are described.

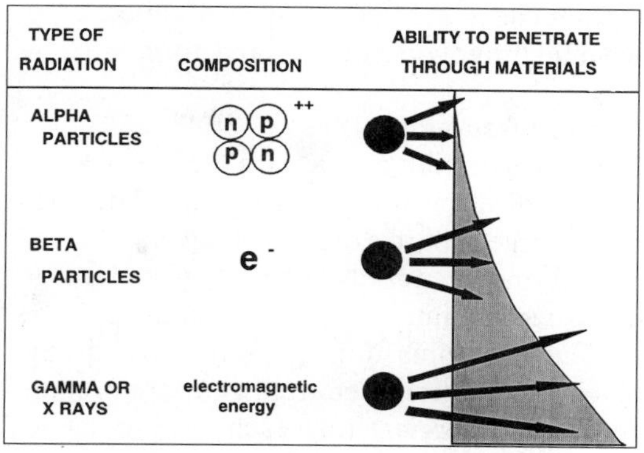

FIGURE 35-1. Characteristics of the Three Major Forms of Radiation of Concern for Measurements of Radioactive Aerosols: Alpha Particles, Beta Particles, and Gamma Radiation.

RADIATION AND RADIOACTIVE DECAY

Aerosols of radioactive materials have all the physical and chemical forms of nonradioactive aerosols. They can range from ultrafine metal fumes to large liquid droplets. Their physical form has the usual influence on their aerodynamic behavior and on the choice of measurement technique. Their chemical form has a similarly important influence on their biological or environmental behavior and on the technique that will be used to confirm their chemical form after collection. It is their radioactive properties that can make them both easier to detect, yet, in many cases, more hazardous to handle.

Types of Radiation

Three types of radiation are of concern in studies of airborne radioactive materials: alpha, beta, and gamma radiation (see Fig. 35-1). Neutrons and positrons are also of concern in some special circumstances. Details of the origin and characteristics of all types of radiation as well as the radioactive decay schemes for radionuclides are presented in references such as the Table of Isotopes (Lederer, Hollander, and Perlman 1978) and the Health Physics and Radiological Health Handbook (Nucleon Lectern Associates 1984).

Alpha Radiation

Alpha radiation is the least penetrating, but the most highly ionizing type of radiation. Alpha particles consist of two neutrons and two protons, carry two positive charges, are identical to a helium nucleus, and are created spontaneously during the radioactive decay of high atomic number elements such as radium, uranium, or plutonium. Alpha-emitting radionuclides can be identified by the characteristic energy of their emissions. They are of special concern when they are inhaled and deposited in the respiratory tract. If inhaled in large quantities, damage to cells in the lung or in other organs to which the material is translocated can cause acute health effects such as fibrosis and long-term health effects such as cancer (see, for example, Hobbs and McClellan 1986).

Beta Radiation

Beta radiation is more penetrating than alpha radiation and consists of negatively charged particles that are identical to electrons. Beta radiation can penetrate the skin and, like alpha radiation, is of concern when large amounts of beta-emitting radionuclides are deposited in the body. The energy spectrum of beta emissions covers a broad range up to a characteristic maximum. Iodine-131, cesium-137, and strontium-90 are beta-emitting radionuclides that are of special concern for accidents involving nuclear reactors.

Gamma Radiation

Gamma radiation and X-rays are penetrating quanta of electromagnetic energy. They are not charged particles, but they can cause ionization in materials and biological damage in tissues. Gamma radiation originates in the nucleus of atoms during radioactive decay. Gamma radiation is emitted at discrete, characteristic energies for each radionuclide. Many alpha- and beta-emitting radionuclides also emit gamma radiation. X-rays originate in disruptions of electrons from their orbits around the nucleus. X-rays can also arise as a secondary form of radiation called bremsstrahlung (braking radiation). Bremsstrahlung is created when the path of a negatively charged beta particle is altered by the coulombic attraction to positively charged nuclei of the absorbing material. This alteration of direction causes a radial acceleration of the beta particle, which is accompanied, in accordance with classical theory, by a loss of electromagnetic energy at a rate proportional to the square of the acceleration (see, for example, Cember 1983, 106).

Half-Life for Radioactive Decay

The rate at which a radioactive material decays is described by its half-life: the time for half of the atoms present to undergo a spontaneous nuclear transformation. Half-life is an important consideration in determining how to collect, handle, and quantify samples of a radioactive material. For example, samples of short-lived materials may need to be analyzed immediately after collection, or even in real time. Conversely, samples of long-lived radionuclides may remain radioactive for years, centuries, or even millennia. The experimental determination of half-life is also one way of helping to identify a given radionuclide.

The equation describing radioactive decay is

$$A(t) = A_0 e^{-0.693t/t_{1/2}} \tag{35-1}$$

where

$t_{1/2}$ is the decay half-life for the radionuclide (in time units)
t is the elapsed time (in the same time units as $t_{1/2}$)
e is the base of the Naperian logarithm system (2.718)
0.693 is the natural logarithm of 2
A_0 is the activity of the sample at time $t = 0$ (in disintegrations per second) and
$A(t)$ is the activity of the sample after an elapsed time of t

Another way of writing the relationship involves the decay constant, λ, where

$$\lambda = 0.693/t_{1/2} \quad \text{and} \quad A(t) = A_0 e^{-\lambda t} \tag{35-2}$$

In terms of the number of elapsed half-lives, n, the amount of activity remaining can be determined as

$$A = A_0(\tfrac{1}{2})^n \tag{35-3}$$

Thus, after two half-lives, only 1/4 (25%) of the original activity is present, and after seven half-lives, only 1/128 (0.8%) of the original activity remains.

Specific Activity

The specific activity of a radioactive material is the rate of decay per unit mass. Materials with a short half-life have a high specific activity. The historical unit for specific activity, the Curie (2.22×10^{12} disintegrations per minute), was defined in terms of the specific activity of radium (1 Ci/g). The system international (SI) unit for radioactive decay is the Becquerel, which is 1 disintegration per second. Thus, 1 nanocurie is 37 Bq. Specific activity is of biological concern for radioactive materials in the body because it determines the rate at which energy will be deposited and damage will occur in tissues. Specific activity is of practical concern for aerosol measurement because it determines the amount of radioactivity that will be present in a sample of a given mass, and the mass or number of particles that will be associated with a sample of a given activity.

The dependence of airborne mass concentration, c_m (g/m^3), on activity concentration, c_a (Bq/m^3), involves the specific activity of the material, A (Bq/g), in the following way:

$$c_m = c_a/A \qquad (35\text{-}4)$$

The number of particles per cubic meter, c_n (particles/m^3), required to provide a given activity air concentration, c_a, involves a similar relationship of specific activity, A, particle density, ρ (g/cm^3), and the volume of the airborne particles. For a simple example in which particles are assumed to be monodisperse in size and spherical in shape (volume $= \pi d_p^3/6$), this relationship is

$$c_n = c_a/(A\rho(\pi d_p^3/6)) \qquad (35\text{-}5)$$

Inhalation Exposure Limits

The specific activity and biological behavior of materials are considered in determining acceptable Annual Limits on Intake (ALIs) for workers exposed to radioactive aerosols (see ICRP 1979 and its addendums). A related concept is the Derived Air Concentration (DAC). A DAC is the calculated air concentration to which a worker could be exposed 8 h per day, 5 days per week, for 50 weeks (an entire work year of 2000 h), without exceeding the ALI for the radionuclide. The worker is assumed to breathe at the rate of 20 l/min as given for the ICRP Reference Man (ICRP 1975). Thus, the DAC is the ALI divided by the volume of air inhaled during the working year (2400 m^3).

Particle number concentrations for several materials at their DAC are illustrated in Table 35-1 as a function of particle size. These materials include three that are radioactive ($^{238}PuO_2$, $^{239}PuO_2$, and enriched uranium), and one that is nonradioactive (beryllium, a toxic metal of concern in many nuclear facilities). Insoluble ^{238}Pu has a specific activity of 6.44×10^{11} Bq/g and a DAC of 0.3 Bq/m^3. Insoluble ^{239}Pu has a specific activity of 2.26×10^9 Bq/g and a DAC of 0.2 Bq/m^3. For enriched uranium, the specific activity is 2.35×10^6 Bq/g (dominated by the contribution from ^{234}U, which is present at 1% by mass), and the DAC for the insoluble material is 0.6 Bq/m^3. The occupational exposure limit for the nonradioactive beryllium is 2 μg/m^3. This example demonstrates that the number of particles of concern for different radionuclides covers a broad range.

RADIATION DETECTION

Radioactive materials can be detected in a number of ways. Most of them depend on the ionization process that occurs when gamma rays or charged particles pass through a gas,

TABLE 35-1 Number of Particles per Cubic Meter as a Function of Monodisperse Particle Size for Selected Toxic Materials at Their Derived Air Concentration (DAC)

Particle Diameter (μm)	$^{238}PuO_2$	$^{239}PuO_2$	Enriched Uranium	Beryllium Metal
10	0.0001	0.02	54	2065
5	0.0008	0.15	433	16,518
3	0.004	0.7	2007	76,471
1	0.1	19	54,180	2,064,715
0.5	0.8	150	433,443	16,517,716

a. Insoluble ^{238}Pu has a specific activity of $6.44 \cdot 10^{11}$ Bq/g and a DAC of 0.3 Bq/m^3
Insoluble ^{239}Pu has a specific activity of $2.26 \cdot 10^9$ Bq/g and a DAC of 0.2 Bq/m^3
For enriched uranium, the specific activity is $2.35 \cdot 10^6$ Bq/g (dominated by the contribution from ^{234}U, which is present at 1% by mass), and the DAC is 0.6 Bq/m^3.
The occupational exposure limit for beryllium is 2 μg/m^3.

liquid, or solid and form ion pairs by disrupting atoms and electrons in the material. Radiation detection methods can be used as direct tools for quantifying radioactive aerosols and as support tools for the safe handling of radioactive materials. The measurement of alpha-emitting radionuclides generally requires direct, or nearly direct, contact between the alpha-emitting material and the detection device. Proximity requirements are less severe for the more penetrating beta and gamma forms of radiation. Investigators must have a thorough understanding of the decay schemes of the radionuclides of interest. In many cases the radioactive emissions of progeny radionuclides are of greater biological concern than the emissions of the parent radionuclide. For example, although strontium-90 has a relatively long half-life (28.8 yr) and emits beta radiation at a relatively low energy (0.54 MeV), its progeny radionuclide yttrium-90 has a short half-life (64.2 h) and is a high-energy beta emitter (2.28 MeV). An overview of the measuring processes for all three types of radiation is given below. Additional detail can be found in resources such as Evans (1955), the chapter on health physics instrumentation in Cember (1983), or the section on radiation detection principles in the Operational Health Physics Training Manual (U.S. DOE 1988).

The Cloud Chamber

Condensation tracks are visible in a cloud chamber as charged particles pass through a supercooled vapor (carbon dioxide from dry ice, for example). Photographs of the particle tracks observed in such devices readily reveal the manner in which charged particles travel through a medium and deposit their energy (see, for example, Rutherford, Chadwick, and Ellis 1930; Rasetti 1947). A popular, simple experiment involves placing a camping lantern mantle (Welsbach mantle) in a cloud chamber and observing the tracks from the naturally occurring uranium and thorium progeny in the phosphors of the mantle. The concern for inhalation hazards from aerosols of these materials has led to the warnings that new lantern mantles should be lighted in well-ventilated areas. Cloud chambers provide a useful insight into the physical processes that make radiation detection devices work. Improved versions of the cloud chamber are still used in high-energy physics experiments.

Scintillation Counting

Phosphors are materials that absorb energy during the ionization process and reemit a fraction of it as light flashes (scintillations). Light emission occurs when electrons are elevated to a higher-energy state and then make the transition back to the ground state. Scintillation bursts can be detected with a photodiode or photomultiplier device and counted to determine the number of charged particles or gamma rays that passed through the material. The intensity and duration of the scintillation can also be analyzed to determine the energy of the radiation being detected.

A number of different phosphors are suitable for radiation detection. For gamma or low-level beta detection, the most notable are solid phosphors, such as thallium-doped sodium iodide (NaI(Tl)), cesium iodide (CsI(Tl)), or potassium iodide (KI(Tl)). The thallium activator serves as an impurity in the crystal structure and converts the energy absorbed in the crystal to light. Zinc sulfide (usually doped with silver), ZnS(Ag), is an excellent detector for alpha radiation. Scintillations from the interactions of alpha particles with ZnS are bright enough to be viewed with the human eye in a darkened room. ZnS must be used as a thin layer (usually a fine crystalline coating on a clear plastic film) because it has a low transmission factor for visible light. Plastic or liquid scintillators such as trans-stilbene are also available. Samples containing radioactive materials can be dissolved in a scintillation "cocktail" involving a solvent such as toluene to improve the contact between the radionuclides being detected and the phosphors. There are a wide range of commercially available cocktail media, many with optimized features for dissolving the test material or functioning at special conditions of pH or salt content.

Ionization Chamber Devices

Gas proportional counters and other ionization chamber devices can detect the number and, in some cases, the energy of radioactive emissions. When gamma rays or charged particles pass through the chamber, ion pairs are formed in the gas. These bursts of ion pairs are drawn by an electric field to a charged sensor, and the number and energy of the emissions can be recorded. Counting efficiency, fidelity of energy determination, and counting rate limits are available for standard ionization devices.

Solid-State Detection

More recently, solid-state detectors have been developed that use special layers of semiconducting materials to provide a site for ion-pair production. These high-efficiency devices provide excellent energy resolution and can be used with a multichannel analyzer to identify the energy spectrum of the detected radiation. In detectors such as germanium–lithium (GeLi) or high-purity germanium (HPGe), the sensitive region extends far enough into the surface that they are highly effective for gamma rays. Diffused-junction or surface-barrier detectors have a thin, sensitive layer that is especially useful for alpha and beta radiation detection. When these detectors are used for alpha spectroscopy, the energy resolution is highest when the alpha-emitting material is placed directly on the detector surface, or when the air in the gap between the alpha source and the detector is evacuated. New developments in surface coatings for solid-state detectors are making them highly resistant to damage from liquids, acids, bases, or abrasion.

Other Detection Systems

Radiation-induced ionization can cause chemical changes in materials like photographic film. The film can be developed to reveal the location of the damage tracks. This process is known as autoradiography; the radioactive material provides its own exposure of the film. Solid-state track recorders can also be made from materials such as polycarbonate or cellulose nitrate. Following exposure to radiation, the surface of the material can be chemically etched to reveal the location and length of the damage tracks. Other detection systems use materials that respond to ionizing radiation by undergoing optical density changes, radiophotoluminescence, thermoluminescence, or conductivity changes (see, for example, U.S. DOE 1988, Chapter 13).

Calibration Considerations

Calibration is a critical part of the detection process. Background samples are analyzed, followed by counting of calibration standards of known activity. The normal calibration process involves the use of sealed sources or electroplated sources with radioactivity levels that are traceable to the National Institute of Standards and Technology. Quantities of radioactivity are generally chosen to correspond to those expected in the samples to be analyzed. This allows appropriate corrections to be made for nonlinear phenomena such as instrument dead time (the inability of a system to detect a new event until it has finished reporting or recovering from a previous event). At high count rates (several million counts per minute or more, depending upon the instrument), effects such as dead time can be substantial, but at very low count rates such errors are negligible. When instruments are to be used at very low count rates, higher activity standards are used to ensure that statistically valid counting efficiencies are determined for the instruments. Note that detection uncertainty is generally calculated from Poisson statistics as being proportional to the square root of the number of events detected. Thus, the relative uncertainty is high for low-activity sources or for short counting times. The influence of background interference must also be taken into account. For errors of less than 1%, net counts of 10,000 or more are generally required (Price 1965).

SOURCES OF RADIOACTIVE AEROSOLS

Ambient Air and Soil

Radioactive aerosols occur naturally in the environment from the interaction of cosmic rays with the atmosphere, producing radionuclides such as carbon-14 and beryllium-7; from the dispersion of soil particles containing potassium-40 or uranium and thorium and their radioactive decay products; and from the emanation from soil and building materials of radioactive noble gases [radon-222 from the uranium-238 decay series and radon-220 (also known as thoron) from the thorium-232 decay series]. Radioactive aerosols also arise from the use of phosphate fertilizers mined from deposits containing naturally occurring thorium and uranium. Chapter 36 contains additional information on radon and radon progeny.

Tobacco Smoke and Fossil Fuel Effluents

Radioactive aerosols are also released by cigarette smoking and combustion of fossil fuels. Tobacco contains minute amounts of uranium and thorium decay products in its plant matrix and in residual soil and phosphate fertilizer contamination on leaf surfaces. Lead-210, a beta-emitting radionuclide, is taken up naturally from the soil by tobacco plants because lead behaves like calcium. Polonium-210, the alpha-emitting progeny of ^{210}Pb, is responsible for most of the radioactivity inhaled as a part of cigarette smoke. It has been postulated that ^{210}Po in cigarette smoke contributes to lung cancer in smokers (see, for example, Martell 1975).

Coal, oil, and natural gas also contain radioactivity, that is released during combustion. Based on the amounts of radioactive aerosol routinely released per unit of electrical energy produced, coal combustion is believed to provide a higher cancer risk for people than nuclear power (see, for example, Cuddihy et al. 1977).

Nuclear Power Plants

The use of nuclear fission power for energy production involves the exposure of uranium miners to radon gas and radon progeny (see, for example, Cothern and Smith 1987). Lung cancer has been observed in uranium miners and other hard-rock miners, especially those who smoked and those who were exposed to radon progeny at high levels in poorly ventilated mines. The use of nuclear fission power also involves the exposure of uranium milling workers to small amounts of aerosol during the extraction of uranium from ore (see, for example, Eidson 1984), exposures of workers to aerosols during fabrication of nuclear fuel (see, for example, Hoover et al. 1983), and exposures to workers and the general public from a routine operation of nuclear electric generation facilities. Mechanisms and probabilities for reactor accidents have been studied in a number of programs (see, for example, U.S. NRC 1975; Gieseke et al. 1977). Overall, the use of nuclear power for electrical energy production has contributed very little to the total health risks for people (see, for example, the exposure comparisons and discussion in Shapiro 1981, Part VI). It now appears that even the accidents at the Three Mile Island and Chernobyl nuclear plants produced more economic and societal disruption from emergency response, risk assessment, and cleanup activities than actual health threats from the release of radioactive aerosols.

The Accident at Three Mile Island

The accident at the Three Mile Island Unit Two nuclear generating station in Middletown, Pennsylvania, began on 28 March 1979, as the result of a complex sequence of mechanical circumstances and operator responses. Ironically, damage to the nuclear core of the reactor occurred when cooling to the core was inadvertently curtailed as operators attempted to prevent the core from overheating. Much of the early concern centered around determining whether the reactor core had suffered a "meltdown" and whether there had been significant releases of radioactive particles and gases to the environment. Aerosol sampling and monitoring devices were not in place to measure such releases. Radiation monitoring around the plant involved thermoluminescent dosimeters, which are integrating beta–gamma detectors. The controversy over the type and amounts of radiation releases continued for many years. Eventually, there was consensus that the major releases from the plant involved the noble gases xenon-133 and krypton-85, which were quickly dispersed in the environment. Radioiodine releases were much smaller than noble gas releases. Most of the iodine was prevented from being released because it combined chemically with the cesium that was present as a fission product in the nuclear fuel. A large portion of the reactor core had, in fact, undergone a meltdown, but releases of radioactive particles from the accident were negligible. The accident demonstrated that improved operator training and instrumentation were required, that there are still things to be learned about the physical and chemical processes of reactor accidents, and that adequate aerosol sampling and monitoring instrumentation is needed in and around nuclear facilities to provide technically defensible release information in the event of an accident. Although there is no evidence of harm to the public around the Three Mile Island reactor, epidemiological and low-dose health effects studies continue (see, for example, TMI Public Health Fund 1990).

The Accident at Chernobyl

Unlike the accident at Three Mile Island, the 26 April 1986 accident at the Chernobyl nuclear generating station in the Soviet Union did involve substantial releases of radioactive aerosols. Because of the graphite moderator construction of the reactor, and because of the limited-containment design of the facility, an explosion and fire in the reactor core dispersed radioactive aerosols into the environment. Low concentrations of radioactive aerosols were detected at sampling stations worldwide. Wind and rain patterns caused fallout concentrations to vary widely. Higher concentrations found in some locations were due to rain washout of the aerosol. Areas surrounding the reactor site were evacuated because of heavy contamination of the air, water, and soil.

The accident occurred when the operators were preparing to conduct a nonroutine series of low-power tests. Some of the controversy surrounding the causes and health consequences of the accident were recently summarized (Nuclear Engineering International 1991). The official Soviet assessment was that the operators had violated operating procedures and disabled safety systems. The operators contend that procedures were not violated, but that the reactor did not meet Soviet design rules and had a fundamental problem in its control system. They claim that the accident occurred because the safety shutdown system actually caused an increase in power, instead of the decrease it was intended to provide. According to the IAEA International Safety Advisory Group: "The accident took place because of deficiencies in operating practices, arising for whatever reason, in combination with specific poor reactor design features." Although environmental releases of radioactive aerosols have been clearly documented, the IAEA advisory group notes that radionuclide concentrations in air, water, and most foods outside the immediate accident area are below guideline levels and often below detection. Immediate casualties from the accident involved workers and persons involved in firefighting and other emergency response activities, and did not involve members of the general population. The IAEA's International Chernobyl Project has concluded that there are, as yet, "no health disorders that could be attributed directly to radiation exposure" among people living on land contaminated by the Chernobyl accident. Nevertheless, they point to a total lack of public confidence in the way the response to the accident was handled. Studies of environmental contamination and possible health effects continue.

Nuclear Weapons Production and Testing

National programs to produce, test, and deploy nuclear weapons have routinely and accidently released radioactive aerosols into the workplace and the environment. Eisenbud (1987) summarizes these releases, including those from a chemical explosion in a plutonium processing pilot plant at the Oak Ridge National Laboratory in Tennessee; from a major fire at the Rocky Flats weapons manufacturing plant in Colorado; and from atmospheric testing in the South Pacific. Atmospheric fallouts of radioactive aerosols from the tests were measured and documented in a series of reports by the U.S. Atomic Energy Commission's Health and Safety Laboratory, which is now the U.S. Department of Energy's Environmental Measurements Laboratory (see, for example, HASL 1977). Human radiation exposures related to the nuclear weapons industries have been reviewed by Cuddihy and Newton (1985).

Releases of plutonium aerosols from the Rocky Flats Plant were unique because they involved fires in the high-efficiency particulate air (HEPA) filters designed to prevent aerosol releases, and dispersion of plutonium particles during the "cleanup" of contaminated machining lubrication oils that had leaked onto the ground from corroded storage drums at an outdoor storage area. In one case the protection system was the source of the release, and in the second case the releases only occurred when vegetation was scraped off and burned, and the ground was disturbed in the cleanup effort. The extent of the releases during the remedial action effort should provide a warning to use adequate containment and handling procedures for any future cleanup activities. Fortunately, a fairly extensive network of air sampling devices around the facility documented the time and magnitude of these releases. These measurements, in combination with data from soil sampling around the Rocky Flats Plant, provided evidence that off-site contamination was not of concern for health consequences. In fact, contamination levels were found to be so low that an offsite area has since been converted into a reservoir for a nearby municipality. Analyses of radioactivity in soil samples around the plant have included comparisons of isotopic ratios of the radionuclides found in soils to help identify the source of the plutonium contamination. In general, soil contamination levels around the Rocky Flats Plant are not significantly different from the levels of plutonium in soil that had been caused by the worldwide fallout of plutonium from atmospheric testing of nuclear weapons.

Accidents Involving Nuclear Weapons

Most nuclear weapons consist of a subcritical assembly of fissile material and an array of conventional high explosives inside a protective housing. The fissile material is typically kilogram quantities of plutonium-239 or uranium-235. Uranium-233 can also be used if it is available in sufficient mass. Accidental nuclear detonation of a weapon is unlikely, but accidental detonation of the high explosive has occurred during accidents such as plane crashes or fires, and detonation of the high explosives has been known to disperse the fissile material. Because of the pyrophoric nature of metallic plutonium or uranium, a very fine oxide aerosol (less than 10 μm aerodynamic diameter) can be formed. Such accidents involving nuclear weapons are called "Broken Arrows". A number of Broken Arrows have occurred (see U.S. DOD 1981), with two notable incidents receiving worldwide attention.

The first widely publicized Broken Arrow occurred over Palomares, Spain, on 17 January 1966. A KC-135 tanker plane collided in mid-air with a B-52 bomber carrying nuclear weapons in a strike configuration during an air refueling operation. One nuclear weapon fell into the Mediterranean Ocean and sank in 2600 feet of water about 5 miles off the Spanish coast. It was recovered after several weeks of search at a cost of over $100 million. A second nuclear weapon landed intact in sand in a dry stream bed. Part of the conventional high explosives in two additional weapons detonated upon impact with the ground, thereby aerosolizing plutonium and contaminating a large area of land in the vicinity of the village of Palomares. The subsequent cleanup costs exceeded $80 million.

A second incident occurred on 21 January 1968, when a B-52 bomber carrying nuclear weapons in a strike configuration developed an on-board fire and crashed on the off-shore sea ice near the United States Air Force Base at Thule, Greenland. The chemical high explosives in the weapons were detonated during the crash and subsequent fire, aerosolizing plutonium and contaminating a large area of sea ice. Costs of the cleanup approached $100 million.

Following these accidents and other near-misses, a major program was undertaken to develop and deploy weapons with "insensitive" high explosives that would not be susceptible to inadvertent detonation by impact or fire. This development has lessened the risks of aerosol releases from accidents involving modern nuclear weapons.

Use of Nuclear Power in Space

Radioisotope thermoelectric generators (RTGs) have been used to provide electrical power for earth-orbiting satellites and for lunar and deep-space missions. RTGs produce electricity by the thermoelectric conversion of the decay heat from high-specific-activity alpha-emitting radionuclides such as plutonium-238 or polonium-210. Small radioisotope heat sources are also used at many locations within space vehicles to provide thermal energy to maintain a proper operating temperature for critical instruments and equipment. Nuclear reactors have been used to a more limited extent to provide electrical power on earth-orbiting satellites. Although the successes of space nuclear power are many, a few notable failures have released radioactive materials to the environment.

A plutonium RTG on board a U.S. navigational satellite was dispersed in the atmosphere over the Indian Ocean on 21 April 1964, when the space vehicle failed to reach orbit. At that time the reentry philosophy was "dilute and disperse". Subsequently, the U.S. developed a contain-and-recover philosophy and the later failure of a meteorological satellite on 18 May 1968 resulted in recovery of its intact RTGs from the ocean.

The most notable reactor systems to fail and burn up in the atmosphere have been part of the Soviet radar ocean reconnaissance satellite (RORSAT) program. The Soviet satellite Cosmos 954 reentered the earth's atmosphere on 24 January 1978, and spread debris over sparsely populated areas of the Great Slave Lake region of northern Canada. On 7 February 1983, another Soviet RORSAT (Cosmos 1402) entered the dense layers of the earth's atmosphere over the southern part of the Atlantic Ocean and was burned up and dispersed into the environment. A third Soviet satellite, Cosmos 1900, appeared destined to the same ignominious fate in the fall of 1988, but built-in safety systems successfully boosted the malfunctioning satellite to a high-enough orbit (500 yr orbital decay time to allow for radioactive decay of the fission products in the reactor core) to prevent the early reentry. Fortunately, the releases from Cosmos 954 and Cosmos 1402 were diluted and dispersed over large areas and have not posed significant health risks to the people.

Use of Radionuclides in Industrial or Medical Processes

Some industrial processes use radioactive materials for special applications. These include the manufacture and use of thickness gauges that use attenuation of beta or gamma emissions to determine or monitor the density and thickness of soils or materials and the manufacturing of smoke alarms that use attenuation of alpha emissions from small americium-241 sources to detect the presence of smoke particles in the air. Aerosol releases during manufacturing, handling, and disposal of these devices have occasionally occurred due to mismanagement. In 1957, iridium-192 dust was released from a work enclosure during the opening of a container of iridium pellets in a commercial laboratory engaged in the preparation of encapsulated sources for gamma cameras (Eisenbud 1987). Employees dressed in street clothing inadvertently spread the dust beyond the laboratory to homes and automobiles. Other notable industrial exposures of workers to radionuclides occurred prior to World War II during the production of luminous dial devices involving radium. These exposures were mainly due to an ingestion of radium by workers who used their tongues to "tip" their brushes to provide a fine point for applying the radium paint. Exposures may also have involved an airborne component from radium paint dusts.

There are many medical applications of radioactive materials, including radioactive particles and gases, that are conducted routinely without unwanted aerosol releases. For example, radioactive tracers such as technetium-99m and the radioactive isotope of xenon gas (^{131}Xe) are widely used with spatial imaging devices such as a gamma camera in human medicine to diagnose lung disease. Radioactive xenon gas can be inhaled to reveal areas of reduced ventilation in the lung, and microaggregated albumin particles radiolabeled with technetium can be injected into the blood to become trapped in the capillary bed of the lung and reveal areas of reduced blood flow. In addition, radioactive materials are used as tracers to determine the deposition, clearance, metabolism, and toxicity of other materials. Examples include the studies of toxic metals such as beryllium (labeled with beryllium-7, a gamma emitter with a 54 day half-life; see Hoover et al. 1988*a*; Finch et al. 1988), nickel-63 (Benson et al. 1989), and selenium-75 (Medinsky, Cuddihy, and McClellan 1981). Other examples include toxic organic chemicals such as benzene or benzo[a]pyrene that can be synthesized with a carbon-14 or tritium radiolabel (Bond et al. 1987). These are normally handled without incident, but unwanted exposures of workers and other people are always a possibility.

SAFE HANDLING OF RADIOACTIVE AEROSOLS

The handling techniques for radioactive aerosols are usually very stringent. Personal protection and contamination control are major concerns. Contamination of people, equipment, other samples, the workplace, and the environment must be avoided.

Facility Design and Licensing

Radioactive materials are generally handled in special facilities with access control, HEPA filtration, and ventilation systems that provide successively more negative air pressures from the outside environment, to general office areas, to laboratory or work areas, and finally to any special enclosures in which actual handling occurs. Figure 35-2 illustrates a laboratory design with graded pressure zones and access control.

The operation of facilities that handle radioactive materials may require federal, state, and local approvals. Laboratories involved in measuring radioactive aerosols must meet the same standards for worker protection, posting of warning signs, effluent monitoring, and waste disposal as the production or reactor facilities to which they provide measuring services.

Use of Special Enclosures

Special laminar flow hoods, glovebox enclosures, and handling rooms are generally needed for work involving radioactive materials. Flow hoods for handling radioactive

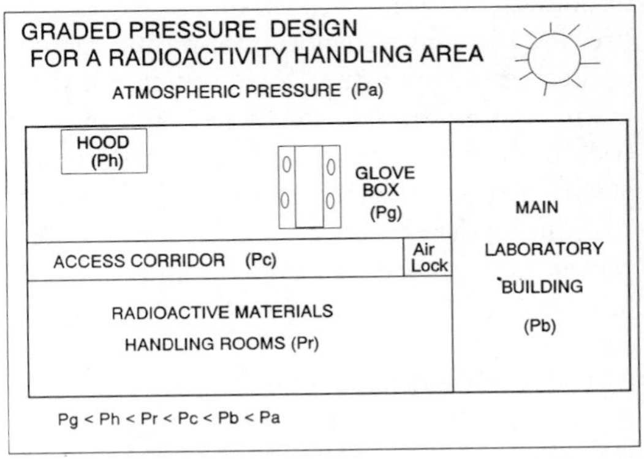

FIGURE 35-2. Illustration of the Successively Decreasing Pressure Conditions in a Facility Designed for Handling Radioactive Aerosols.

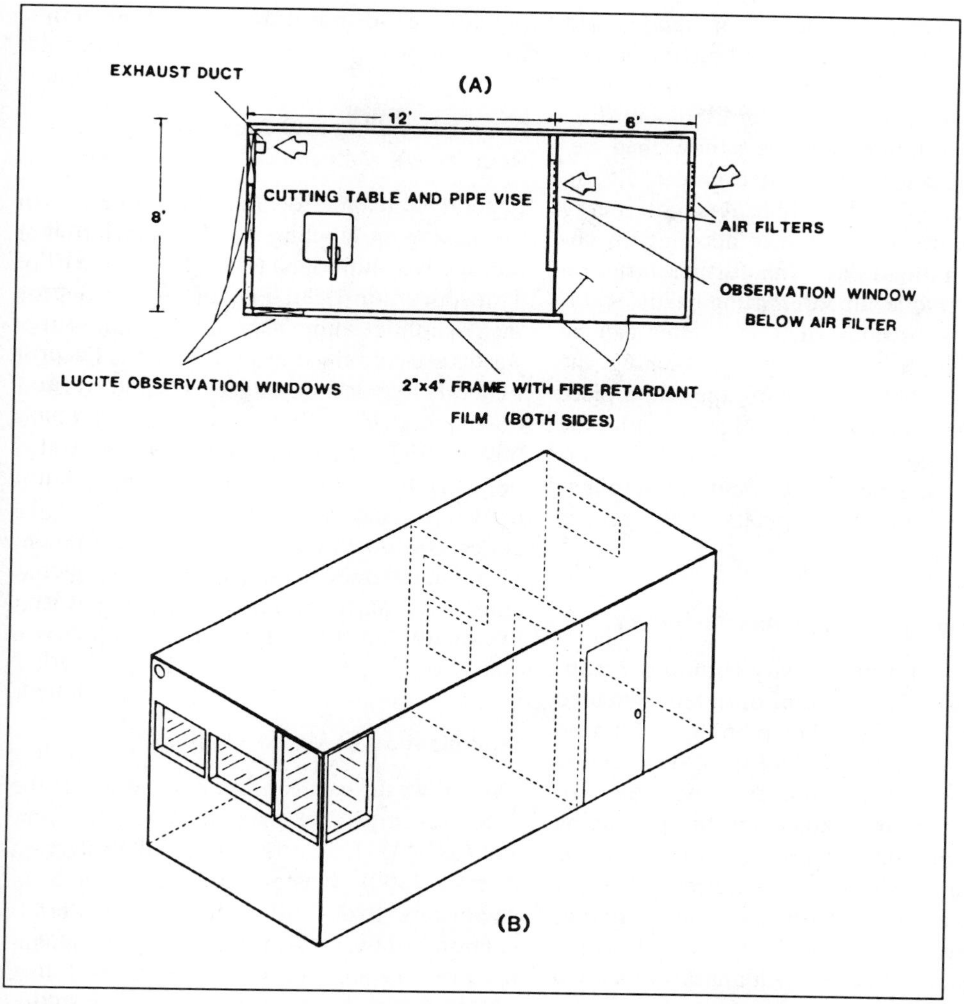

FIGURE 35-3. Sketches of the Temporary Containment Structure ("Tent") Built to Enable Controlled Cutting of Radioactively Contaminated Pipe. Sampling Probes were Placed at Breathing Zone Levels for a Worker at the Cutting Table.

materials draw room air into the handling area and exhaust it through HEPA filtration. Temporary, special-use enclosures may be required for nonroutine activities, such as maintenance or decommissioning, that are being performed (see, for example, Newton et al. 1987; Fig. 35-3). Such systems may have to be built when the existing radioactivity handling areas are being fitted with special features for aerosol sampling or when special sampling efforts are underway. Portable aerosol sampling systems may include local ventilation and filtration around the sampling instruments (see, for example, Hoover et al. 1983).

Contamination Control for Radioactive Aerosol Samples

Samples of radioactive aerosols collected for radiation detection can be stabilized for sample preservation and contamination control by several techniques. For example, filters containing gamma-emitting samples can be stored in sealed containers or sandwiched between tape without degrading the radioactivity counting results. For low-energy beta-emitting materials or for alpha-emitting materials, samples may be sandwiched between tape and a thin (for example, 1.5 μm thick) piece of Mylar® film. A slight energy correction must be made to account for energy loss in the Mylar®. In a further variation of this technique, alpha-emitting particles collected on substrates such as a filter can be prepared for counting by sandwiching the substrate between a piece of tape and a piece of ZnS(Ag)-coated clear plastic with the scintillator side facing the particles. This stabilizes the sample and prevents contamination of the counting equipment or laboratory.

Requirements for Radiation Shielding

Workers and radioactivity-counting equipment must also be appropriately shielded from radiation. As shown in the radiation penetration illustration in Fig. 35-1, shielding requirements for alpha-emitting radionuclides are minimal; normally, the packaging used for contamination control is adequate. Conversely, shielding requirements for gamma-emitting radionuclides may involve several centimeters of lead, steel, or other high atomic number (high Z) material. Shielding for beta radiation involves two considerations: attenuation of the beta particles themselves, and attenuation and minimization of secondary X-rays from bremsstrahlung. Because bremsstrahlung production increases with the atomic number of the absorber, low-Z materials (less than atomic number 13, aluminum) are normally used to shield beta-emitting radionuclides. Thus, a typical storage unit for aerosol samples of beta-emitting radionuclides involves an inner layer of several millimeters or centimeters of methylmethacrylate acrylic plastic or graphite for beta attenuation, supplemented with an outer layer of high-Z material for the absorption of any bremsstrahlung photons. Failure to provide adequate shielding for radioactive samples can cause unnecessary radiation exposures of people and may interfere with proper radioactivity counting.

Proper Handling and Disposal of Radioactive Materials

Federal, state, and local laws must be met in the packaging, labeling, and transportation of radioactive materials (U.S. Department of Transportation 1990). Federal regulations for the handling and disposal of radioactive wastes are contained in Title 10 of the Code of Federal Regulations Part 20, and related documents (U.S. Nuclear Regulatory Commission 1991). State and local laws must also be consulted. Even when sampling radioactive particles from airstreams that have concentrations below legal limits for environmental release, the process of concentrating the particles may create samples that eventually must be treated as radioactive waste.

Minimization of Mixed Wastes

"Mixed waste" is a concept defined under the U.S. Resource Conservation and Recovery Act (RCRA) (U.S. Environmental Protection Agency 1990). Wastes that are mixtures of chemically toxic and radioactive materials cannot, by law, be disposed of in either chemical disposal sites or radioactive disposal sites. The lead used for shielding samples of radioactive aerosols and many of the chemicals, such as toluene, used for scintillation cocktails to count radioactive samples are listed as hazardous wastes under RCRA. The production of mixed wastes should be minimized, because they must be handled at special processing facilities, usually at a great cost. The mismanagement of mixed wastes can result in fines and/or imprisonment.

OBJECTIVES FOR MEASURING RADIOACTIVE AEROSOLS

As is true for all air sampling and air monitoring, the selection of methods for measuring radioactive aerosols should be determined by the underlying objectives of the sampling

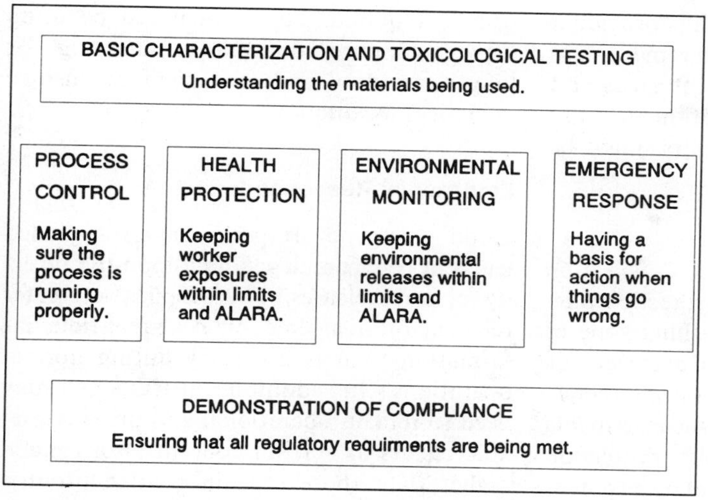

FIGURE 35-4. Illustration of Six Major Objectives for Sampling Radioactive Aerosols. The Objectives are not Mutually Exclusive, and Other Special Objectives may Also Exist.

effort. "What information is really needed?" This concept seems trivial, but it must be emphasized so that needed data will not be missed, and time and fiscal resources will not be wasted. As shown in Fig. 35-4, sampling objectives can be grouped into six areas: basic characterization and toxicological testing, process control, health protection, environmental monitoring, emergency response, and demonstration of compliance. These objectives are not mutually exclusive, and other special objectives may also exist.

Basic Characterization and Toxicological Testing

Basic characterization and toxicological testing are ideally done before initiation of any process involving radioactive materials. They involve both collecting and characterizing relevant samples of selected materials, and creating laboratory surrogate aerosols that have well-controlled and well-characterized properties. The full spectrum of aerosol sampling tools is usually applied to measure parameters such as particle size, concentration, morphology, chemical composition, and solubility. This approach provides a technical basis for selecting appropriate requirements for designing and managing the entire process in order to keep the exposure of people within limits and as low as reasonably achievable (ALARA). It also provides a technical basis for balancing the relative risks of internal radiation exposures from inhalation or ingestion and external radiation exposures from working or living in a radiation area. Because both internal and external radiation doses pose risks to people, choices must be made. Historically, the choice has been to accept larger-than-necessary external exposures in an effort to avoid inhalation exposures. This is not reasonable in light of the underlying risks, and an equal treatment of both types of exposures is recommended by bodies such as the International Commission on Radiological Protection (ICRP) (see, for example, ICRP 1973).

Process Control

Routine sampling or monitoring for aerosols in processes such as nuclear fuel fabrication, reactor operations, or radioactive-waste disposal assures that the processes are working properly, or provides an early warning that conditions are changing. Sampling refers to methods that involve off-line analysis, and monitoring refers to methods that provide a real-time response. Such measurements are usually made within containment areas and generally focus on concentration or size measurements. Other parameters may be measured periodically.

Health Protection

The measurements for health protection purposes can be made in the breathing zone of workers, using fixed or personal samplers, or in the general work area. Basic characterization and process control measurements provide the guidance for health protection measurements; processes involving potentially small source terms of low-toxicity materials do not require as extensive a measurement program as would be appropriate for potentially large source terms of high-toxicity materials. In all cases, the objective is to keep worker exposures within limits and ALARA. Concerns for particle solubility are generally high because of the influence it has on the biological behavior of inhaled material, and because a knowledge of solubility is needed to correctly apply biokinetic models and interpret bioassay information from urine, fecal, or blood samples obtained from workers. Other measurement techniques, such as radiation monitoring for hand and foot contamination or for contamination of workplace surfaces, are also part of a total health protection program. They often signal increased airborne concentrations before worker exposures become excessive.

Environmental Monitoring

Because the Clean Air Act sets limits on allowable releases of radioactive materials to the environment (U.S. Environmental Protection Agency 1991), effluent monitoring or stack sampling is required. This must be done routinely to ensure that environmental releases are within limits and ALARA. Measurements are often made both on-site and off-site. Because concentration limits for environmental releases are generally lower than for the workplace, greater sensitivity is usually required. This is achieved by using higher sampling flow rates, sampling for longer periods of time, or using more sensitive analytical techniques.

Emergency Response

Good emergency response involves graded levels of reaction, depending upon the severity of any releases. This implies a greater reliance on real-time or near-real-time information than is necessary during normal operations. In addition, instruments may need to remain operational and provide useable results at aerosol concentrations much higher than those seen in routine process, health protection, or environmental monitoring. It is sometimes necessary to deploy portable or mobile instrumentation. Instruments may also need more frequent cleaning or replacement.

Demonstration of Compliance

Throughout all the measuring regimes, it is necessary to demonstrate that the instruments are working properly and to document actual releases of airborne radionuclides in comparison to statutory release limits. In a sense, this effort is a subset of all other measuring objectives. Once the operator is convinced of what has occurred, the regulators and other interested parties must be convinced.

APPLICATION OF STANDARD MEASURING TECHNIQUES

Nearly all standard aerosol sampling techniques can be applied to the measurement of radioactive aerosols. Some limitations are related to the amount of material that may be significant or available. For example, the mass amounts of concern for radionuclides may be below the limits of detection of piezoelectric mass monitoring systems or optical monitoring devices. On the other hand, radioactivity levels associated with particles collected for electron microscopy may require the use of designated and restricted equipment. Considerations such as these are discussed below for a number of standard measuring techniques.

Optical Particle Counting

Optical particle counters have not been widely used for radioactive aerosols, but there are circumstances under which they might provide useful information to help protect the workers from exposures to radioactive aerosols, especially to warn them of unusual particle releases. Optical particle counters have two main advantages: they sample airstreams continuously, and they provide real-time information. A number of commercially available instruments are being used to detect nonradioactive aerosols in clean rooms, to monitor work area dust levels in industries such as mining and textiles, and to provide quality control monitoring for processes such as paint pigment preparation that fabricate or use fine particles. Optical monitoring has also been used in systems for generating and characterizing inhalation exposure atmospheres of radioactive aerosols (see, for example, Hoover et al. 1988*a*). Issues related to their application for radioactive aerosols include: (1) the level of detection compared to allowed air concentrations for radioactive aerosols; (2) the level of detection compared to background levels of nonradioactive aerosols in the work place; (3) aerosol characterization requirements to determine relationships between radioactive and nonradioactive aerosols in the work place; (4) calibration requirements to quantify instrument response to specific aerosols; and (5) health protection management strategies for using optical monitoring information in a total program for workplace control and worker protection.

Figure 35-5 illustrates a graphical scheme recently developed (Hoover and Newton 1991) to evaluate whether a given optical particle counter can meet useful requirements for a level of detection. The abscissa of the two-dimensional field describes the level of an

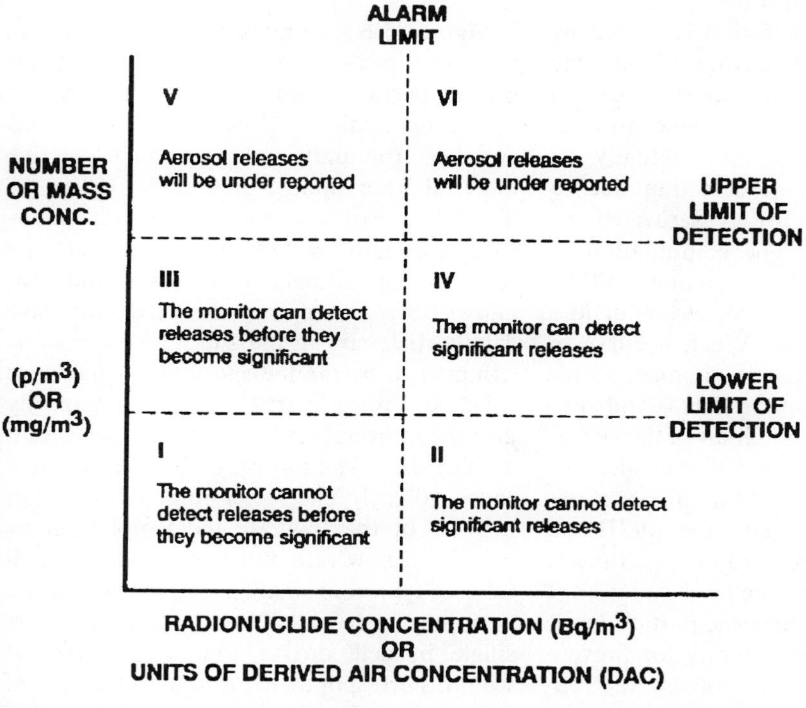

FIGURE 35-5. Format of a Graphical Scheme for Evaluating Whether an Optical Monitor Could be Useful for Detecting Radioactive or Other Toxic Aerosols.

airborne radionuclide concentration in relevant units such as Bq/m^3. This permits the placement of a vertical line for any concentration level of interest, such as 1 DAC for a specific radionuclide. For convenience, this level-of-interest line will be referred to as the "alarm limit". This line divides the field in half. The ordinate of the two-dimensional field describes the particle number concentration (particles/m^3) or mass concentration (mg/m^3) that corresponds to a given radionuclide concentration. The ordinate uses the units in which the optical counter measures or reports information and it permits placement of two horizontal lines corresponding to the lower limit of detection and the upper limit of detection for the instrument. The lower limit of detection depends intrinsically on the sensitivity of the instrument (a function of flow rate, sensing volume dimensions, and internal signal-to-noise ratio) and it also depends in practice on the concentration of background dust (external signal-to-noise ratio). Except in very clean environments, the lower limit of detection is due to the background aerosol concentration.

The upper limit of detection is caused by saturation of the counter (more than one particle in the sensing volume at once). In very dusty environments, the concentration of background aerosols may actually be *above* the upper limit of detection due to coincidence (the monitor is saturated by background aerosols). The combination of the horizontal limit of detection lines and the vertical concentration of interest line divides the field into six sectors. When a curve of mass versus activity or particle number versus activity (assuming a particle size) is added to this plot, it reveals the usefulness of the instrument for detecting a given radionuclide.

The ideal situation would involve a monitor whose effectiveness spans Sectors III and IV. This would involve a robust instrument that tracks routine airborne levels, allows the detection of increased airborne particle levels (thus, providing an opportunity for preventive action or progressive responses), and provides a reliable alarm when airborne concentrations exceed an action level.

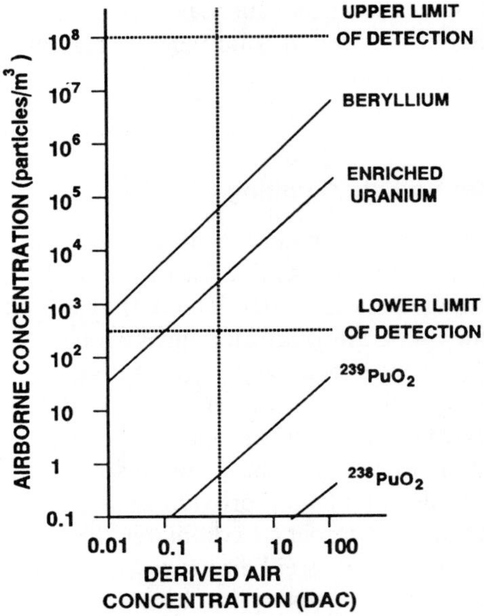

FIGURE 35-6. Illustration of the Evaluation Scheme for Usefulness of Optical Particle Monitors. Particles are Assumed to be Monodisperse with Diameters of 3 μm. Sampling Flow Rate is 3 l/min (Nominal 0.1 cfm).

Figure 35-6 illustrates a practical application of the evaluation scheme for applying an optical particle monitor to the three radioactive materials ($^{238}PuO_2$, $^{239}PuO_2$, and enriched uranium), and one nonradioactive material (beryllium) presented earlier in Table 35-1. For convenience, the abscissa units of Fig. 35-6 are DAC. This equalizes the scale for all radioactive materials and also allows treatment of nonradioactive materials. The particle size in the Fig. 35-6 example is assumed to be monodisperse 3 μm diameter spheres. Optical particle counters typically give the information as "number greater than a given size". The saturation limit is assumed to be 10^8/m^3. The lower limit of detection depends on the flow rate and sampling interval. For a flow rate of 0.03 m^3/min (nominal 0.1 cfm, typical of small, portable units) and a sampling interval of 1 min, detection of a single particle during the sample interval would correspond to a concentration of 347 particles/m^3. Figure 35-6 indicates that optical particle counting will not be effective for ^{238}Pu and ^{239}Pu because number concentrations of concern are below the normal limits of detection. However, optical particle counting may be useful for enriched uranium and beryllium metal (assuming that background levels of other dusts are not excessive).

Parameters that still need to be evaluated include the unique response of light scattering devices to each material. This is of special concern because the particles are unlikely to be spherical. Correlations of optical diameter or real diameter with aerodynamic diameter are also important because movement of the aerosol through the workplace and inhalation of the aerosol by workers will depend on aerodynamic, not optical, diameter. Strategies also need to be developed on precisely how optical monitors would be integrated into a total monitoring program (placement, worker response to alarm, record keeping, and other considerations). The use of optical monitors as early warning devices to detect radioactive releases appears promising, especially for special applications, such as during maintenance operations to detect unusual leaks and to guide in the placement of local ventilation. An early warning could speed up the reactions to protect the workers and to regain control of the workplace.

Particle Collection for Microscopy

The collection of radioactive particles for morphological examination by transmission or scanning electron microscopy can easily be done using standard instruments such as the point-to-plane electrostatic precipitator (Morrow and Mercer 1964). A small hood is normally sufficient for sample handling. The air flow rate through the hood opening should be maintained at the lowest effective level (approximately 75 linear feet per minute) to avoid problems in handling the small, fragile electron microscope grids. Standard formvar®-coated copper grids can be used for transmission electron microscopy. Degradation of the formvar® by radiation damage is usually only a problem for high-specific-activity radionuclides such as plutonium-238 (half-life = 87.7 yr).

Care should always be taken to use contamination control features, such as liquid nitrogen cold fingers, to minimize the spread of contamination within the microscope. High-specific-activity alpha-emitting radionuclides such as plutonium-238 are prone to migration from the collection grid by radioactive decay-induced recoil and spallation. The need for respiratory protection and radiation monitoring should be considered before servicing any microscope used for radioactive materials.

Filtration

Filtration is the most widely used method for collecting samples of radioactive aerosols The methods and equipment range from high-volume samplers (sampling rates up to about 60 m^3/h) for environmental or short-term workplace sampling to low-volume, miniature lapel samplers (1 l/min or less) for collecting aerosols in the breathing zone of individual workers (U.S. DOE 1988). Low-pressure-drop cellulose filters are commonly used, and samples can be easily reduced to ash or dissolved for analysis by analytical chemistry or radiochemistry.

The concern for the penetration of particles into the filter matrix is a function of the type of filter, the type of radiation, and the radiation-counting method being used. Membrane filters with their superior front-surface collecting characteristics are preferred over fiber-type filters when alpha particle spectroscopy is applied. Shielding by the filter media is seldom a concern for the detection of gamma radiation. Although energy degradation concerns are greatest for alpha-emitting radionuclides, we have found that even glass microfiber filters such as the Gelman A/E glass (Gelman Sciences, Inc., Ann Arbor, MI) collect particles near enough to the filter surface that radioactivity counting results from the ZnS method are as accurate as radiochemical results.

Long-term storage of filter samples for archival purposes is not always feasible for high-specific-activity alpha-emitting radionuclides such as plutonium-238. Radiation damage to the filter, packaging tape, or plastic container may allow release of radioactivity. Care should be taken when handling samples that have aged.

Inertial Sampling

Inertial sampling using cascade impactors, spiral-duct centrifuges, and cyclones has been the major approach for characterizing the aerodynamic particle size distribution of radioactive aerosols. A number of specialized versions of these instruments have been developed specifically for use with radioactive aerosols (for example, Mercer, Tillery, and Newton 1970; Kotrappa and Light 1972). Special requirements for instruments used in handling radionuclides generally include their compactness and ease to assemble and disassemble while wearing protective gloves in a confined space such as a glovebox enclosure. They also need to be easy to clean. Low sample collection rates are usually adequate for the collection of small sample masses. Analytical methods for collected samples are straightforward by radioactive counting. The spiral-duct centrifuge has been widely used to estimate the density or shape factor of individual particles. This works equally well for radioactive and nonradioactive particles. Because the aerodynamic diameter associated with each particle deposition location is known, electron microscopy can be used to determine the physical size and shape of particles found at these locations, and the density or shape factor can be calculated (see, for example, Stoeber and Flachsbart 1969).

Real-time inertial techniques, such as time-of-flight measurements of particles accelerated through a nozzle, are also useful for radioactive aerosols. This assumes a willingness to purchase dedicated instruments for use with radioactive aerosols, because decontamination of equipment for return to unrestricted use is not always easy.

Measurement of Electrical Properties

The measurement of aerosol electrostatic charge distribution can be done using a standard aerosol charge spectrometer (see, for example, Yeh et al. 1976). The theory of Yeh et al. (1976, 1978) predicts that self-charging due to alpha or beta emission in radioactive aerosols will occur in addition to friction charging due to comminution. Even when aerosols are created by highly charging processes like grinding, radioactive decay processes such as alpha or beta decay may quickly result in a charge distribution that is near the Boltzmann equilibrium. The measurements of particle charge distribution on a plutonium–uranium aerosol obtained by Yeh et al. (1978) with and without the use of a krypton-85 discharge unit were identical. At high alpha radioactivity concentrations (> 25 nCi/l), it is likely that sufficient ion pairs are present to reduce the charge on the aerosols to near Boltzmann equilibrium.

At lower radioactivity concentrations, this equilibrium condition may not be reached. Raabe et al. (1978) reported the anomalous results of two cascade impactor samples taken without the use of a ^{85}Kr discharge unit in the blending step of a mixed plutonium–uranium oxide fuel preparation process. The alpha radioactivity concentration at the time those samples were taken was only 1–2 nCi/l. Because the activity median aerodynamic diameter of these samples was larger than observed in samples taken with a discharger, it is likely that anomalous deposition on the upper stages of the impactors occurred as a result of electrostatic charge effects. It is, therefore, recommended, without advance information on the radioactivity concentration being sampled, to include an in-line ^{85}Kr discharge unit as a standard procedure, even though it may not be needed to reduce the charge distribution to the Boltzmann equilibrium. The fabrication and use of ^{85}Kr discharge units are described by Teague, Yeh, and Newton (1978).

Volumetric Grab Samples, Impingers, Cold Traps, and Adsorbers

Sampling techniques such as evacuated volumes, liquid bath impingers, cold traps, and activated charcoal adsorbers work equally well for capturing radioactive and nonradioactive vapors and particles. Standard radioactivity-counting techniques can be applied, depending upon sample geometry. The Lucas cell is an interesting variation of a grab sampling approach (see Fig. 36-4). Radioactive particles or gases are drawn into a chamber whose interior walls are coated with a layer of crystalline ZnS(Ag) or some other scintillator. Nearly 100% of the alpha rays reaching the scintillator will result in a flash of light. Any flashes of light occurring within the chamber are observed by a detector. This method can be applied to radon gas, radon progeny, or other alpha-emitting radionuclides that can be drawn into the chamber. The half-life of the radionuclides being sampled influences the delay time or cleaning requirements before the cells can be reused.

Analytical Chemical Techniques

Traditional analytical chemistry methods such as infrared spectrometry, flame or furnace atomic absorption spectrometry, energy-dispersive X-ray analysis, electron or neutron diffraction, and inductively coupled plasma (with either emission spectroscopy or mass spectroscopy) have sensitivities that are compatible with the small sample sizes usually associated with radioactive aerosols. Dedicated equipment is usually required for handling radionuclides, and appropriate controls must be applied. Techniques requiring tens or hundreds of milligrams, such as X-ray diffraction, have a much more limited application for radioactive aerosols.

SPECIAL TECHNIQUES FOR RADIOACTIVE AEROSOLS

Detection of Individual Particles by Autoradiography

The interaction of ionizing radiation with photographic film or nuclear track detector foils such as CR-39 can be used to detect the presence and activity level of individual particles. The observation of nuclear tracks on the film or foil by scanning electron microscopy or light microscopy can be used to determine the position of individual radioactive particles. Applications include studies of radioactive particles from industrial sources. In one example, Voigts et al. (1986) used this technique to identify single aerosol particles as the alpha sources from industrial plume samples with particle number concentrations of 2000 particles/mm^2. Tracks emanating from the radioactive particles are recorded in the film or foil and reveal the location and activity level of the individual particles. Cohen, Eisenbud, and Harley (1980) used cellulose nitrate track etch film to measure the alpha radioactivity on human autopsy specimens of the bronchial epithelium. In combination with microscopy, autoradiography can be used to determine how cells and organelles are irradiated by particles that are inhaled and deposited in the body. Figure 35-7 shows an example of autoradiography used with histological slides of lung tissue to determine the microdosimetry of an inhaled uranium–plutonium oxide aerosol.

Measurement of Particle Solubility and Biological Behavior

Particle solubility and biokinetic studies can be done on all classes of aerosol particles, but the detection of dissolved material is especially straightforward for radioactive aerosols (see, for example, Kanapilly et al. 1973). Samples can be sandwiched between filters and subjected to continuous solvent flow (dynamic systems) or placed sequentially in fresh containers of solvent (static systems). Particles can also be placed in a tube with the solvent and periodically centrifuged to concentrate the particles at the bottom of the tube and to allow sampling of dissolved material from the supernatant. Radioactivity counting provides the very low limits of detection needed for an accurate determination of particle dissolution, especially for highly insoluble materials, and radioactive aerosols have played a unique role in many unusual discoveries about aerosol behavior. The high degree of measurement sensitivity

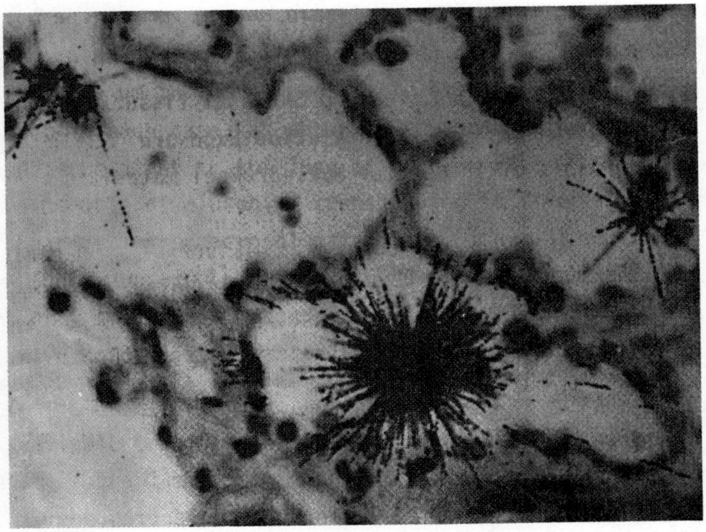

FIGURE 35-7. Autoradiograph of Alpha-Emitting Particles of Varied Specific Activity in the Lungs of a Rat Following Inhalation Exposure to Mixed Uranium–Plutonium Oxide Aerosols. (Adapted from Mewhinney 1978, 80.)

was a major factor in the studies of Mewhinney, Eidson, and Wong (1987*a*), that showed a rapid initial release of material whenever particles were reintroduced into a solvent. That work provided an insight into the possible environmental effects of wet and dry weathering on particles released to the biosphere.

Density Measurement by Isopycnic Gradient Ultracentrifugation

Isopycnic density gradient ultracentrifugation has been shown to be a useful technique for measuring the density of small quantities (0.1–5 mg) of a variety of particles (Allen and Raabe 1985; Finch et al. 1989). Normal density measurement techniques, such as air or gas pycnometry or liquid displacement, are not suitable for the small sample volumes normally associated with radioactive aerosols. Thallium formate has been the usual heavy-metal solution for this technique, but sodium metatungstate has been shown to be an economical, nontoxic alternative (Hoover, Finch, and Castorina 1991). Figure 35-8 illustrates the technique. Particles are added to a centrifuge tube containing the heavy liquid. The tube is subjected to centrifugation to form a gradient of density from top to bottom. The density near the top of the tube normally approaches that of water and the higher density near the bottom of the tube can be more than 3.0 g/cm^3. Particles move to the location within the tube where their density equals that of the surrounding liquid. Successive samples of known volume are then removed, weighed to confirm the density of the liquid in the sample, and then analyzed by radioactivity counting or other suitable methods to determine the fraction of the particle material in the sample. Refractive index measurements of the liquid samples can also be used to determine the density of the liquid samples, but this requires a dedicated refractometer and is usually more time-consuming than simple weighing. The density gradient ultracentrifugation technique provides information about the density distribution of particles within an aerosol sample.

Surface Area Measurement by Krypton-85 Adsorption

Particle specific surface area (m^2/g) influences the rate of surface phenomena such as dis-

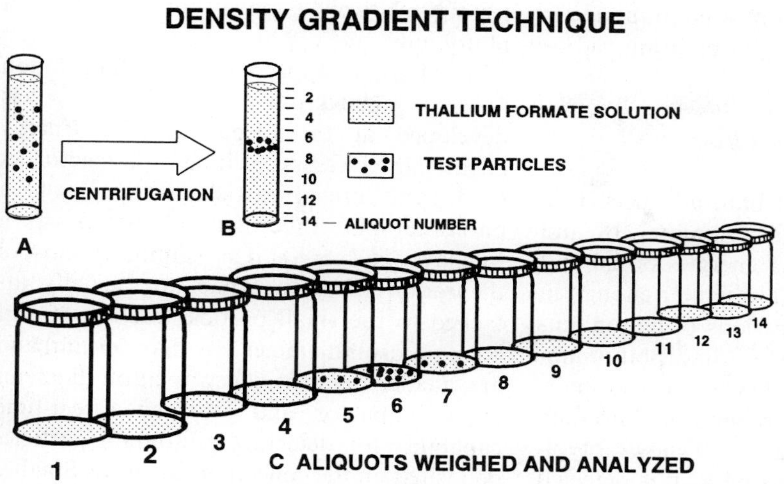

FIGURE 35-8. Illustration of the Density Gradient Ultracentrifugation Technique for Determining the Density of Individual Particles. (Adapted from Finch et al. 1989.)

solution. When adequate sample masses are available (10–50 mg, or more), measurement of specific surface area is reliable and straightforward. The most widely used approach involves the Brunauer, Emmett, and Teller method of calculating nitrogen adsorption onto the surface of a sample of known mass. A number of commercial instruments are available. Rothenberg et al. (1982, 1987) have focused attention on the special problems of surface area measurement when sample size is less than 10 mg, which is a typical restriction when characterizing radioactive particles. They have described and evaluated a method for adsorbing ^{85}Kr gas onto the sample surface (Rothenberg et al. 1987). The radioactive decay of ^{85}Kr emits a 0.514 MeV gamma ray that can be readily detected by a standard scintillation method such as NaI(Tl). They note that a 1-cm^2 monolayer of ^{85}Kr gas having a specific activity of 10 Ci/g of gas will give approximately 10,000 disintegrations per min, that can easily be measured with an uncertainty of less than 1%. The major statistical uncertainty is associated with blank correction for the sample holder. The ^{85}Kr-adsorption technique has been successfully applied to the characterization of small samples of mixed uranium and plutonium dioxide particles (Mewhinney et al. 1987*b*), and recommend the technique for samples as small as 1 mg, with specific surface areas greater than 1 m^2/g. However, the disadvantage of the method rests in the lack of a commercially available instrument and the significant effort required for a user to set up the method.

Real-Time Monitoring for Airborne Radionuclides

A large number of instruments for real-time monitoring of airborne radioactive aerosols are commercially available. The standard configuration involves a radiation detector in close proximity to a filter that is collecting aerosols. Radioactivity concentration information can be accessed at the instrument itself or instruments can be networked to central monitoring stations. Radiation shielding and background correction for external radiation sources are required for beta and gamma radiation detection systems and a range of background correction, calibration, and geometry considerations come into play for alpha radiation detection systems. Some of the important considerations for design, calibration, and operation of continuous air monitors (CAMs) for alpha emitting radionuclides are given below.

Mitigation of Interference from Radon Progeny

The alpha emissions of naturally occurring radon progeny such as polonium-218 and bismuth-212 (with alpha energies of 6.0 and 6.08 MeV, respectively) are similar enough in energy to the alpha emissions of plutonium-239 (alpha energy 5.2 MeV) and plutonium-238 (alpha energy 5.5 MeV), to cause interference or false positive reports of plutonium air concentrations. Figure 35-9 shows a typical radon progeny spectrum as it is detected by an alpha spectrometer. Early alpha CAM designs did not include spectrometry, but used a single-channel analyzer to detect radioactivity in a plutonium energy region of interest (ROI), and a second analyzer to both detect radon progeny activity in a second energy region and to allow a simple background correction for the counts seen in the plutonium ROI. The correction method was crude and the limit of detection high, but it provided a useful, real-time means for detecting relatively large airborne releases of plutonium.

An alternate approach to handling radon progeny background interference was developed at the Department of Energy Savannah River Site. It used an impactor jet to deposit aerosol particles directly onto a photomultiplier tube that was coated with a thin layer of ZnS(Ag). This eliminated most of the radon progeny, which are usually attached to the small particle fraction of ambient aerosol (diameter less than 0.3 μm) and are smaller than the effective cutoff diameter of the impactor, and provided a real-time capability for detecting plutonium particles deposited on the collection substrate. Smaller plutonium particles were also undetected, but most releases were considered to be dominated by larger particles. Radioactivity counting efficiency for the collected plutonium was approximately 50% (nearly 100% of the emissions occurred in the direction of the ZnS layer) because the light emissions from the ZnS(Ag) are emitted isotropically, and all emissions can be seen with equal likelihood by a photomultiplier tube or photodiode detector. A recent variation of the Savannah

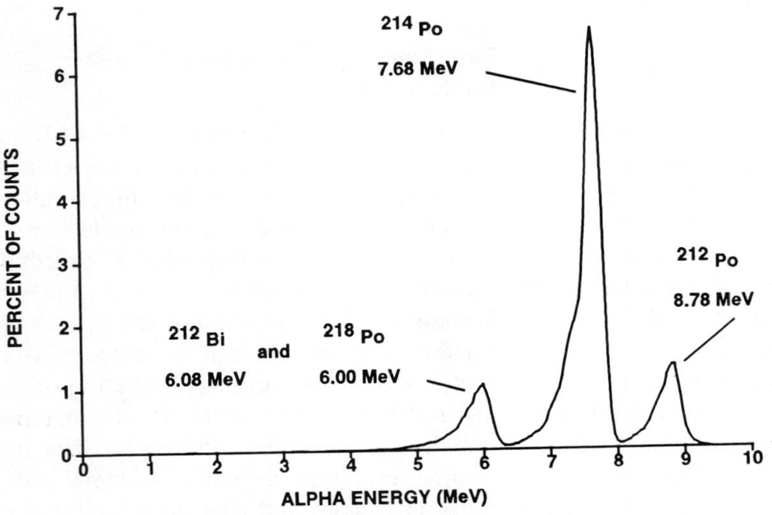

FIGURE 35-9. Typical Radon Progeny Spectrum Collected in an Alpha Continuous Air Monitor. The Energy Region for ^{239}Pu (5.2 MeV) and ^{238}Pu (5.5 MeV) is in the Tail of the ^{218}Po Peak.

River approach uses an impactor jet to deposit the aerosol directly onto the surface of a solid-state detector (Kurz Model 8300, Kurz Instruments, Inc., Monterey, CA). Detection efficiency is excellent, and direct detection of the alpha particles causes good separation of the plutonium and radon progeny peaks. The particle bounce off the detector surface degrades the collection efficiency of large particles (we found that more than 90% of the 10 μm aerodynamic diameter particles are lost to bounce), but this might be solved by use of a virtual impactor, or particle trap approach [see Biswas and Flagan (1988) for a description of the particle trap].

Recent advances in microcomputers have enabled a major development effort to include alpha spectroscopy in solid-state detector CAMs for real-time monitoring of alpha-emitting radionuclides. Again, a solid-state detector is placed just above the surface of a filter onto which alpha-emitting aerosols are drawn, but the detector output goes to an embedded spectrometer in the CAM, rather than to single-channel analyzers. In a batch operation, the air gap between the filter and the detector can be evacuated. This prevents attenuation of the alpha particle energies as they pass through the air gap and minimizes any spectral overlap of the plutonium and radon progeny alpha energy peaks. For real-time sampling of the aerosol, however, the region between the filter and the detector cannot be evacuated, and attenuation of the alpha particle energies occurs in the air gap.

The algorithm for subtraction of radon progeny interference in the plutonium energy region was developed by Unruh (1986) at Los Alamos National Laboratory. Four ROIs are established: ROI-1 on the lower-energy tail of the ^{218}Po peak, where plutonium is expected; ROI-2 on the upper-energy tail of the ^{218}Po peak; ROI-3 on the lower-energy tail of the ^{214}Po peak; and ROI-4 on the upper-energy tail of the ^{214}Po peak. Basically, the ^{218}Po and ^{214}Po peaks are assumed to be congruent in shape, and the radon progeny activity in the plutonium ROI (ROI-1) can be estimated by taking a ratio of the radon progeny counts in the other three regions of interest:

$$(\text{ROI-1}) = k \cdot \text{ROI-2} \cdot \text{ROI-3}/\text{ROI-4} \quad (35\text{-}6)$$

This approach is used in the Eberline Alpha 6 (Eberline Instruments, Santa Fe, NM) and in the Victoreen Model 758 (Victoreen Inc., Cleveland, OH). The Radeco Model 452 (SAIC/Radeco, San Diego, CA) uses a similar detector and filter arrangement with a triple-window background subtraction algorithm (counts in regions above and below the plutonium region are used to make the background subtraction). The capabilities of all of these instruments, along with several British, German, and French devices, are being evaluated at a number of laboratories, including our own. The concern relates to particle collection efficiency and the stability of the radon correction algorithms over a range of radon progeny concentrations. Although very promising, many consider these devices to be prototypic in nature.

Filter Requirements for Continuous Alpha Air Monitoring

An active area of recent work has related to the selection of appropriate filter media for alpha CAMs. The quality of the alpha energy spectrum reported by a solid-state detector system is strongly dependent on the type of filter substrate used for particle collection. Membrane filters have long been accepted as having both the excellent particle collection efficiency and the excellent surface collection characteristics needed for alpha spectroscopy. Lindeken et al. (1964) demonstrated that, compared to submicrometer-pore diameter filters, large-pore membrane filters show no serious sacrifice in collection efficiency until the pore diameters exceed 5 μm. Thus, larger-pore filters, with their lower pressure drop, are preferred.

The filter that has been widely recommended by CAM manufacturers is the Millipore 5 μm pore size membrane filter (mixed cellulose ester, type SMWP, Millipore Corporation, Bedford, MA). However, we found the fragile nature of that filter to be unacceptable. Particle collection efficiency and gravimetric confirmation of particle mass collection were unreliable because of filter breakage under field conditions.

We sought a more rugged membrane filter that would have a reasonably low pressure drop and excellent surface collection efficiency. We evaluated the Versapor-3000 (an acrylic copolymer on a nonwoven nylon fiber support, Gelman Sciences, Ann Arbor, MI) and found its performance to be very similar to that of the Millipore SMWP. The Durapore 5 μm pore size polyvinylidene fluoride membrane filter from Millipore is acceptable. We found the performance of the Fluoropore FSLW filter from Millipore to be superior to all the others. It is a polypropylene-backed, polytetrafluoroethylene (PTFE) filter that is extremely rugged and provides excellent spectral separation of the collected radionuclides. The quality of radon progeny spectra collected with the Fluoropore and Millipore membrane filters is compared in Fig. 35-10a. Note the much lower tail of the ^{218}Po peak in the plutonium region of interest (around

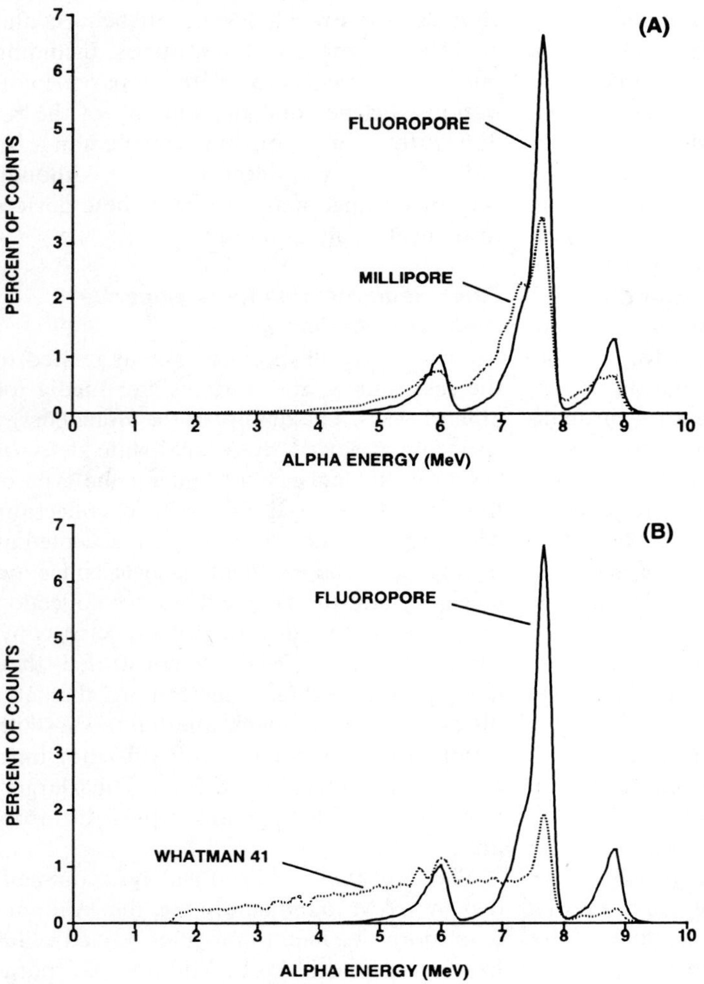

FIGURE 35-10. Illustration of the Influence of Filter Type on the Quality of the Radon Progeny Energy Spectrum in an Alpha Continuous Air Monitor. The Fluoropore Filter Provides Superior Resolution.

5.2 MeV). As a warning, Fig. 35-10b shows the extremely poor spectral quality obtained when a conventional fiber filter such as the Whatman 41 (Whatman Lab Sales, Hillsboro, OR) is used in an alpha CAM. Thus, the use of fiber-type filters must be avoided.

A useful new technology involves mounting each filter in a thin cardboard carrier. The card protects the filter during handling and can be labeled to readily identify the sample. Bar code labeling and computer data logging can be used for both the filter cards and the sampling instruments. This improves chain-of-custody control and facilitates long-term archiving of samples. If samples are to be destructively analyzed instead of archived, they can be easily removed from the cards. The choice of filter for use in the cards depends on the quality of the energy spectrum that is required and on compatability requirements of the filter for the analytical technique to be used.

Mitigation of Interference from Airborne Dust

The accumulation of ambient dust on the collection filter of an alpha CAM leads to attenuation of alpha energy, just as the air gap above the filter degrades the alpha energy. Such a burial of plutonium leads to an underreporting of air concentrations, ranging from 10% to 100% when airborne dust concentrations are greater than 1 mg/m^3 (Hoover et al. 1988*b*, 1990). Alpha particles from plutonium that is buried by 2 mg/cm^2 of salt on a filter are prevented from reaching the detector. This does not prevent the CAM from responding to large puff releases of radioactivity, but it does raise the limit of detection for slow, continuous releases. Dust concerns are primarily associated with decommissioning activities where metal piping and structures are being cut (see Newton et al. 1987), with environmental restoration activities where soil is being disturbed, and with special activities such as in the underground salt mine environment of the Department of Energy Waste Isolation Pilot Plant near Carlsbad, New Mexico.

Efficiency Considerations for Filter/Detector Geometry

Concerns for dust loading make it necessary to consider some tradeoffs on filter and detector size. Many CAMs use a 2.5 cm diameter detector and a 2.5 cm diameter filter. This arrangement has an overall detection efficiency of approximately 20%. Retaining the same detector size but increasing the diameter of the filter collection surface to 4.3 cm, cuts the detection efficiency by half (down to 10%), but increases the collection surface area by nearly a factor of 4. At first, this appears to be a reasonable tradeoff.

However, a closer examination of detection efficiency as a function of filter diameter reveals that material collected at the filter edges contributes very little to overall efficiency. This is because of solid-angle considerations that reduce both the efficiency at which alphas are intercepted by the detector and the energy at which they are detected. With a larger filter, alphas from the filter edge lose a significant amount of energy as they travel to the detector. Alphas traveling from one edge of the filter to the opposite edge of the detector have the longest path through air, and thus suffer the greatest energy loss, often enough to remove them from the plutonium energy region. Plutonium collected at the center of the filter is detected in the plutonium ROI at an efficiency of 30%, with little energy degradation. At a diameter of 2.5 cm, detection efficiency has dropped to 15% but only marginal energy degradation has occurred. At a diameter of 4.3 cm, only 0.04% of the emitted alpha particles reach the detector and have energies in the plutonium ROI. Thus, the larger filter has marginal utility, especially in the presence of dust, where additional energy degradation will degrade any alphas from the filter edge to energies below the plutonium region.

Remote Detection of Radioactive Particles

A new technique has been developed at the Los Alamos National Laboratory for remotely detecting alpha radioactivity contamination on internal surfaces of equipment or in areas that cannot be directly surveyed by radiation instrumentation (MacArthur and Allander 1991). It uses the fact that ion pairs persist in the air after they are formed by the movement of positively charged alpha particles through air. This persistence was a surprise to many who believed that the recombination of ion pairs was nearly instantaneous. By drawing clean air across a contaminated surface, ion pairs can be carried to an electrometer and detected. This new technique should receive wide use in facilities having airborne alpha-emitting aerosols, and makes it possible to identify contamination with a greater degree of certainty than was previously possible.

PRACTICAL OPTIONS FOR DATA TRANSMISSION AND NETWORKING

For many years, radiation monitoring instruments were considered stand-alone devices that provided information locally. A new emphasis on demonstrating "control of the work place" is leading many health protection professionals to consider networking and central monitoring capabilities as essential. Applications in nuclear power plants have long demonstrated the advantages of networking, but these applications have largely involved fixed physical situations with hardwired cables installed during original construction. The development of new networking capabilities is especially difficult when cables must be installed in radioactive areas, or when work tasks change and new monitoring locations are needed. The cost of installing the cables for data transmission can be prohibitive.

Some of the most notable breakthroughs in recent years are in new radiation monitoring devices that provide continuous dose rate information by radio signal to a remotely located individual. To accomplish this, the most beneficial devices use spread spectrum radio transmission. This is basically the same high-technology approach used for cellular telephones and remote microphones. For the first time, it is possible to cut radiation doses into half by removing the health physics technician from direct proximity to the work site. Someone totally outside a contaminated area can now monitor air concentrations and worker whole-body exposures and provide instructions and guidance for minimizing the dose. This is an area of rapid progress.

ADEQUACY OF THE EXISTING AEROSOL SCIENCE DATA BASE

An ever-increasing emphasis is being placed on having a "technically defensible basis" for measuring radioactive aerosols in the workplace and environment. New terms such as "Conduct of Operations" and "Total Quality Management" have been coined to provide new paradigms for management and worker attitudes and responsibilities, record keeping and quality assurance requirements, and demonstration of compliance. Some of the notable deficiencies in the area of radioactive aerosol measurement are quality and handling of radioactive calibration sources, selection of appropriate instrumentation for specific tasks, placement of instrumentation in the workplace or environment, and criteria for collection of an adequate sample. Improvements in the limit of detection for workplace and environmental releases are also needed. This might come through wider application and improvements in virtual impaction techniques to concentrate airborne particles prior to collection.

The pressure for improved quality and capabilities can be expected to be the greatest in four classes of nuclear applications: environmental restoration of radioactively contaminated facilities such as are located throughout the Department of Energy weapons production complex; design, construction, and operation of facilities for safe disposal of low-level and high-level radioactive waste; monitoring and handling of radioactive materials in medical settings; and continued operation or design of new nuclear facilities for electricity generation.

Opportunities are available for improving the quality of regulations and standards and for preparing guidance documents for state-of-the-art sampling and monitoring. Revision of the American National Standard for Guide to Sampling Airborne Radioactive Materials in Nuclear Facilities (ANSI 1969) is underway and it is likely that new recommendations will minimize reliance on simplistic concepts such as isokinetic sampling (considered by many to be broadly misapplied in turbulent sampling conditions) and will emphasize "qualification" of the sample. Continued work will be needed to understand the basic behavior of radioactive aerosols and to synthesize total measurement approaches that are technically defensible and cost-effective.

CONCLUSIONS

The measurement of radioactive aerosols is a challenging and specialized subset of aerosol science. The basic physics of radiation and radioactivity provides some unique technical advantages and disadvantages for accomplishing the task. Overall, the advantages outweigh the disadvantages. At the same time, a range of political, regulatory, and emotional obstacles arise from both the justifiable and the unreasonable health and environmental concerns that surround radioactivity and the use of radioactive materials. Because the health, economic, psychological, and political costs to society can be high, the tools of aerosol science must be used in wise and technically defensible ways when dealing with radioactivity.

References

Allen, M. D. and O. G. Raabe. 1985. Slip correction measurements of spherical solid particles in an improved Millikan apparatus. *Aerosol Sci. Technol.* 4:269–86.

ANSI. 1969. *American National Standard Guide to Sampling Airborne Radioactive Materials in Nuclear Facilities*, ANSI N13.1-1969. New York: American National Standards Institute.

ANSI. 1978. *American National Standard for Radiation Protection Instrumentation Test and Calibration*, ANSI N323-1978. New York: American National Standards Institute.

ANSI. 1979. *American National Standard for Performance Specifications for Reactor Emergency Radiological Monitoring Instrumentation*, ANSI N320-1979. New York: American National Standards Institute.

ANSI. 1980*a*. *American National Standard for Specification and Performance of On-Site Instrumentation for Continuously Monitoring Radioactivity in Effluents*, ANSI N42.18-1980. New York: American National Standards Institute.

ANSI. 1980*b*. *American National Standard for Performance Criteria for Instrumentation Used for Inplant Plutonium Monitoring*, ANSI N317-1980. New York: American National Standards Institute.

ANSI. 1989*a*. *American National Standard on Performance Specifications for Health Physics Instrumentation—Portable Instrumentation for Use in Normal Environmental Conditions*, ANSI N42.17A-1989. New York: American National Standards Institute.

ANSI. 1989*b*. *American National Standard on Performance Specifications for Health Physics Instrumentation—Occupational Airborne Radioactivity Monitoring Instrumentation*, ANSI N42.17B-1989. New York: American National Standards Institute.

ANSI. 1989*c*. *American National Standard on Performance Specifications for Health Physics Instrumentation—Portable Instrumentation for Use in Extreme Environmental Conditions*, ANSI N42.17C-1989. New York: American National Standards Institute.

Benson, J. M., D. G. Burt, Y. S. Cheng, F. F. Hahn, P. J. Haley, R. F. Henderson, C. H. Hobbs, J. A. Pickrell, and J. K. Dunnick. 1989. Biochemical responses of rat and mouse lung to inhaled nickel compounds. *Toxicology* 57:255–66.

Biswas, P. and R. C. Flagan. 1988. The particle trap impactor. *J. Aerosol Sci.* 19:113–21.

Bond, J. A., J. M. Benson, J. D. Sun, and M. D. Hylarides. 1987. Use of radiolabeling in environmental and toxicological research. In *Hazard Assessment of Chemicals*, Vol. 5, pp. 29–59. New York: Hemisphere.

Cember, H. 1983. *Introduction to Health Physics*. New York: Pergamon.

Cohen, B. S., M. Eisenbud, and N. H. Harley. 1980. Measurement of the α-radioactivity on the mucosal surface of the human bronchial tree. *Health Phys.* 39:619–32.

Cohen, B. 1989. Sampling airborne radioactivity. In *Air Sampling Instruments for Evaluation of Atmospheric Contaminants*. 7th edn. Cincinnati: American Conference of Governmental Industrial Hygienists.

Cothern, C. R. and J. E. Smith, Jr. 1987. *Environmental Radon*. New York: Plenum.

Cuddihy, R. G., R. O. McClellan, M. D. Hoover, V. L. Dugan, L. D. Chapman, and J. R. Wayland. 1977. Radiation risks from plutonium recycle. *Environ. Sci. Technol.* 11:1160–65.

Cuddihy, R. G. and G. J. Newton. 1985. *Human Health Exposures Related to Nuclear Weapons Industries*. LMF 112, Lovelace Biomedical and Environmental Research Institute. Springfield, VA: National Technical Information Services.

Eicholz, C. G. and J. S. Poston. 1979. *Principles of Nuclear Radiation Detection*. Ann Arbor, MI: Ann Arbor Science.

Eidson, A. F. 1984. *Biological Characterization of*

Radiation Exposure and Dose Estimates for Inhaled Uranium Milling Effluents. U.S. Nuclear Regulatory Commission Report, NUREG/CR-2539, LMF-108. Springfield, VA: Natl. Technical Information Center.

Eisenbud, M. 1987. *Environmental Radioactivity from Natural, Industrial, and Military Sources.* San Diego: Academic Press.

Evans, R. D. 1955. *The Atomic Nucleus.* New York: McGraw-Hill.

Finch, G. L., J. A. Mewhinney, A. F. Eidson, M. D. Hoover, and S. J. Rothenberg. 1988. *In vitro* dissolution characteristics of beryllium oxide and beryllium metal aerosols. *J. Aerosol Sci.* 19:333–42.

Finch, G. L., M. D. Hoover, J. A. Mewhinney, and A. F. Eidson. 1989. Respirable particle density measurements using isopycnic density gradient ultracentrifugation. *J. Aerosol Sci.* 20:29–36.

Gieseke, J. A., H. Jordan, K. W. Lee, B. Vaishnavi, and L. D. Reed. 1977. *Aerosol Measurements and Modelling for Fast Reactor Safety: Annual Report for FY 1977.* BMI-NUREG-1989, 23 December 1977. Columbus, OH: Battelle Columbus Laboratory.

Health and Safety Laboratory. 1977. *Final Tabulation of Monthly ^{90}Sr Fallout Data: 1954–1976.* HASL-329, 1 October 1977. New York, NY: Health and Safety Laboratory.

Hobbs, C. H. and R. O. McClellan. 1986. Toxic effects of radiation and radioactive materials. In *Casarette and Doull's Toxicology: The Basic Science of Poisons.* 3rd edn., eds. C. D. Klaassen, M. O. Amdur, and J. Doull. New York: MacMillan.

Hoover, M. D., G. J. Newton, H. C. Yeh, and A. F. Eidson. 1983. Characterization of aerosols from industrial fabrication of mixed-oxide nuclear reactor fuels. In *Aerosols in the Mining and Industrial Work Environments.* eds. V. A. Marple and B. Y. H. Liu. Ann Arbor: Ann Arbor Science.

Hoover, M. D., A. F. Eidson, J. A. Mewhinney, G. L. Finch, B. J. Greenspan, and C. A. Cornell. 1988*a*. Generation and characterization of respirable beryllium oxide aerosols for toxicity studies. *Aerosol Sci. Technol.* 9:83–92.

Hoover, M. D., G. J. Newton, H. C. Yeh, F. A. Seiler, and B. B. Boecker. 1988*b*. *Evaluation of the Eberline Alpha-6 Continuous Air Monitor for Use in the Waste Isolation Pilot Plant: Phase I Report,* 21 December 1988. Albuquerque, NM: Inhalation Toxicology Research Institute.

Hoover, M. D., G. J. Newton, H. C. Yeh, F. A. Seiler, and B. B. Boecker. 1990. *Evaluation of the Eberline Alpha-6 Continuous Air Monitor for Use in the Waste Isolation Pilot Plant: Report for Phase II* 31 January 1990. Albuquerque, NM: Inhalation Toxicology Research Institute.

Hoover, M. D., G. L. Finch, and B. T. Castorina. 1991. Sodium metatungstate as a medium for measuring particle density using isopycnic density gradient ultracentrifugation. *J. Aerosol Sci.* 22: 215–21.

Hoover, M. D. and G. J. Newton. 1991. *Preliminary Evaluation of Optical Monitoring for Real-Time Detection of Radioactive Aerosol Releases.* Albuquerque, NM: Inhalation Toxicology Research Institute.

IAEA. 1978. *Particle Size Analysis in Estimating the Significance of Airborne Contamination.* Technical Report Series No. 179. Vienna: International Atomic Energy Agency.

ICRP. 1965. *Principles of Environmental Monitoring Related to the Handling of Radioactive Materials.* International Commission on Radiological Protection Publication 7. Oxford: Pergamon.

ICRP. 1968 and 1982. *General Principles of Monitoring for Radiation Protection of Workers.* International Commission on Radiological Protection Publications 12 and 35. Oxford: Pergamon.

ICRP. 1973. *Implications of Commission Recommendations That Doses Be Kept as Low as Reasonably Achievable.* International Commission on Radiological Protection Publication 22. Oxford: Pergamon.

ICRP. 1975. *Reference Man: Anatomical, Physiological and Metabolic Characteristics.* International Commission on Radiological Protection Publication 23. Oxford: Pergamon.

ICRP. 1979. *Limits on Intakes of Radionuclides by Workers.* International Commission on Radiological Protection Publication 30 and addendums. Oxford: Pergamon.

ICRP. 1985. *Principles of Monitoring for the Radiation Protection of the Population.* International Commission on Radiological Protection Publication 43. Oxford: Pergamon.

ICRU. 1971. *Radiation Protection Instrumentation and its Application.* ICRU Report 20. Washington, DC: International Commission on Radiation Units and Measurement.

Kanapilly, G. M., O. G. Raabe, C. H. T. Goh, and R. A. Chimenti. 1973. Measurement of the *in vitro* dissolution of aerosol particles for comparison to *in vivo* dissolution in the respiratory tract after inhalation. *Health Phys.* 24:497–507.

Kotrappa, P. and M. E. Light. 1972. Design and performance of the Lovelace aerosol particle separator. *Rev. Sci. Instrum.* 43:1106–12.

Lederer, C. M., J. M. Hollander, and I. Perlman (eds.). 1978. *Table of Isotopes.* 7th edn., New York: Wiley.

Lindekin, C. L., F. K. Petrock, W. A. Phillips, and R. D. Taylor. 1964. Surface collection efficiency of large-pore membrane filters. *Health Phys.* 10:495–99.

Martell, E. A. 1975. Tobacco radioactivity and cancer in smokers. *American Scientist* 63: 404–12.

MacArthur, D. W. and K. S. Allander. 1991. *Long-Range Alpha Detectors.* LA-12073-MS. Los Alamos, NM; Los Alamos National Laboratory.

Medinsky, M. A., R. G. Cuddihy, and R. O. McClellan. 1981. Systemic absorption of selenious acid and elemental selenium aerosols in rats. *J. Toxic Environ. Health* 8:917–28.

Mercer, T. T., M. I. Tillery, and G. J. Newton. 1970. A multi-stage low flow rate cascade impactor. *J. Aerosol Sci.* 1:9–15.

Mewhinney, J. A. 1978. *Radiation Exposure and Risk Estimates for Inhaled Airborne Radioactive Pollutants including Hot Particles*, Annual Progress Report 1 July 1976–30 June 1977. NUREG/CR-0010, Albuquerque, NM: Inhalation Toxicology Research Institute.

Mewhinney, J. A., A. F. Eidson, and V. A. Wong. 1987*a*. Effect of wet and dry cycles on dissolution of relatively insoluble particles containing Pu. *Health Phys.* 53:337–84.

Mewhinney, J. A., S. J. Rothenberg, A. F. Eidson, G. J. Newton, and R. Scripsick. 1987*b*. Specific surface area determination of U and Pu particles. *J. Colloid Interface Sci.* 116:555–62.

Morrow, P. E. and T. T. Mercer. 1964. A point-to-plane electrostatic precipitator for particle size sampling. *Am. Ind. Hyg. Assoc. J.* 25:8–14.

NCRP. 1978*a*. *Instrumentation and Monitoring Methods for Radiation Protection.* NCRP Report No. 57. Bethesda, MD: National Council on Radiation Protection.

NCRP. 1978*b*. *A Handbook of Radiation Protection Measurements Procedures.* NCRP Report No. 58. Bethesda, MD: National Council on Radiation Protection.

Newton, G. J., M. D. Hoover, E. B. Barr, B. A. Wong, and P. D. Ritter. 1987. Collection and characterization of aerosols from metal cutting techniques typically used in decommissioning nuclear facilities. *Am. Ind. Hyg. Assoc. J.* 48:922–32.

Nuclear Engineering International. 1991. World news: Chernobyl health consequences. *Nuclear Engr. Int.* 36(444):2–4.

Nucleon Lectern Associates. 1984. *The Health Physics and Radiological Health Handbook.* Olney, MD: Nucleon Lectern Associates.

Price, W. S. 1965. *Nuclear Radiation Detection.* New York: McGraw-Hill.

Raabe, O. G. 1972. Instruments and methods for characterizing radioactive aerosols. *IEEE Transactions on Nuclear Science.* NS-19(1):64–75.

Raabe, O. G., G. J. Newton, C. J. Wilkenson, and S. V. Teague. 1978. Plutonium aerosol characterization inside safety enclosures at a demonstration mixed-oxide fuel fabrication facility. *Health Phys.* 35:649–61.

Rasetti, F. 1947. *Elements of Nuclear Physics.* New York: Prentice-Hall.

Rothenberg, S. J., P. B. Denee, Y. S. Cheng, R. L. Hanson, H. C. Yeh, and A. F. Eidson. 1982. Methods for the measurement of surface areas of aerosols by adsorption. *Adv. Colloid Interface Sci.* 15:223–49.

Rothenberg, S. J., D. K. Flynn, A. F. Eidson, J. A. Mewhinney, and G. J. Newton. 1987. Determination of specific surface area by krypton adsorption, comparison of three different methods of determining surface area, and evaluation of different specific surface area standards. *J. Colloid Interface Sci.* 116:541–54.

Rutherford, E., J. Chadwick, and C. D. Ellis. 1930. *Radiation from Radioactive Substances.* New York: Macmillan.

Shapiro, J. 1981. *Radiation Protection, A Guide for Scientists and Physicians.* Cambridge: Harvard University Press.

Stoeber, W. and H. Flachsbart. 1969. Size-separating precipitation of aerosols in a spinning spiral duct. *Environ. Sci. Technol.* 3:1280–96.

Teague, S. V., H. C. Yeh, and G. J. Newton. 1978. Fabrication and use of krypton-85 aerosol discharge devices. *Health Phys.* 35:392–95.

TMI Public Health Fund. 1990. *1989–1990 Annual Report.* Philadelphia: The Three Mile Island Public Health Fund.

Turner, J. E. 1986. *Atoms, Radiation, and Radiation Protection.* New York: Pergamon.

Turner, J. E., J. S. Bogard, J. B. Hunt, and T. A. Rhea. 1988. *Problems and Solutions in Radiation Protection.* New York: Pergamon.

U.S. Department of Defense. 1981. *Narrative Summaries of Accidents Involving Nuclear Weapons 1950–1980.* Washington, DC: U.S. Department of Defense.

U.S. Department of Energy. 1988. *Operational Health Physics Training.* ANL-88-26. Prepared by Argonne National Laboratory, Argonne, IL, for the U.S. Department of Energy Assistant Secretary for Environment, Safety, and Health. Oak Ridge, TN: National Technical Information Service.

U.S. Department of Transportation. 1990. *Regulations for Transportation of Hazardous and Radioactive Materials.* Title 49, Code of Federal Regulations, Parts 171-177, 1 October 1990. Washington, DC: U.S. Government Printing Office.

U.S. Environmental Protection Agency. 1990. *Hazardous Waste Management Regulations.* Title 40, Code of Federal Regulations, Parts 260 through 268, 1 July 1990. Washington, DC: U.S. Government Printing Office.

U.S. Environmental Protection Agency. 1991. *National Emission Standards for Hazardous Air Pollutants.* Title 40, Code of Federal Regulations, Part 61, 1 July 1991. Washingtons DC: U.S. Government Printing Office.

U.S. Nuclear Regulatory Commission. 1975. *Reactor Safety Study: An Assessment of Accident Risks in U.S. Commercial Nuclear Power Plants.* WASH-1400 (NUREG-75/014), October 1975. Washington, DC: U.S. Nuclear Regulatory Commission.

U.S. Nuclear Regulatory Commission. 1980. *Proceedings of the CSNI Specialists Meeting on Nuclear Aerosols in Reactor Safety.* NUREG/CR-1724. Washington, DC: U.S. Nuclear Regulatory Commission.

U.S. Nuclear Regulatory Commission. 1991. *Standards for Protection Against Radiation.* Title 10, Code of Federal Regulations, Part 20, 21 May 1991. Washington, DC: U.S. Government Printing Office.

Unruh, W. P. 1986. *Development of a Prototype Plutonium CAM at Los Alamos*. LA-UR-90-2281, 15 December 1986. Los Alamos, NM; Los Alamos National Laboratory.

Voigts, Chr., G. Siegmon, M. Berndt, and W. Enge. 1986. Single alpha-emitting aerosol particles. In *AEROSOLS, Formation and Reactivity, 2nd Int. Aerosol Conf.* Berlin, p. 1153. Oxford: Pergamon.

Yeh, H. C., G. J. Newton, O. G. Raabe, and D. R. Boor. 1976. Self-charging of ^{198}Au-labeled monodisperse gold aerosols studied with a miniature electrical mobility spectrometer. *J. Aerosol Sci.* 7:245–53.

Yeh, H. C., G. J. Newton, and S. V. Teague. 1978. Charge distribution on plutonium-containing aerosols produced in mixed-oxide reactor fuel fabrication and the laboratory. *Health Phys.* 35:500–03.

36

Radon and Its Short-Lived Decay Product Aerosols

Beverly S. Cohen

New York University Medical Center
Institute of Environmental Medicine
Long Meadow Road
Tuxedo, NY, U.S.A.

INTRODUCTION

Radon, a radioactive noble gas, with its short-lived progeny, or "daughters", are known to cause cancer in uranium and other underground miners. Radon is present at elevated concentrations in many homes and workplaces and concentration measurements are required for evaluation and control. Recent measurements have demonstrated radon levels in some homes high enough to deliver a radiation dose to the bronchial epithelium equal to that known to produce cancer in underground miners. Exposure in most homes is substantially lower, but experience with ionizing radiation demonstrates that any dose, however small, confers some risk. The large number of people potentially exposed to elevated radon concentrations has led to a recognition of the need for measurement of indoor as well as outdoor concentrations.

Measurement of radon and its daughters presents unique problems because there occurs a serial transformation of the samples. The airborne concentrations that are biologically significant are extremely low and difficult to measure. In addition, radon is a gas and the progeny are particles, so that sampling methods differ. This chapter presents the background information needed to understand the reason for radon sampling, briefly defines the relationships between radon, its progeny and radiation dose because these interactions shape the sampling strategy, and describes methods for sampling and measuring both radon and the short-lived decay products.

RADON IN THE ENVIRONMENT

Radon is formed in the decay chains that begin with the naturally occurring primordial radionuclides Uranium-238, Thorium-232, and Uranium-235 present in the earth's crust. When radon is formed, the gas may diffuse from the rocks and soils to enter the atmosphere. The amount of radon formed depends only on the concentration of the parent nuclide. The fraction that emanates to the atmosphere depends on the rock or soil matrix, water content, atmospheric pressure, and other geological and climatic factors. Also important is the half-life of the radon, i.e., the time needed for half of a given quantity to decay. The half-life is different for each radon isotope (Table 36-1). The concentration of a radioactive element may be designated by mass, i.e., mg/kg or ppm, or by activity, i.e., Bq/kg or pCi/g.

TABLE 36-1 Parent Nuclide and Radon Isotope of the Natural Radioactive Series*

Series	Long-lived Parent		Crustal Abundance			Radon Isotope			
	Isotope	Half-life	ppm	Bq/kg	pCi/g	Isotope	Common Name	Half-life	α-Particle energy
Uranium	^{238}U	4.5×10^9 y[a]	2.7	33	0.89	^{222}Rn	Radon	3.82 d	5.49 MeV
Thorium	^{232}Th	14.1×10^9 y	8.5	34	0.92	^{220}Rn	Thoron	55.6 s	6.29 MeV
Actinium	^{235}U	0.7×10^9 y	0.02	1.5	0.04	^{219}Rn	Actinon	3.96 s	6.82 MeV

* Adapted from NCRP (1988)
a. y (year), d (day), s (second)

The unit of activity is the becquerel (Bq):

$$1 \text{ Bq} = 1 \text{ s}^{-1} \tag{36-1}$$

Thus, 1 Bq represents one transformation or disintegration per second. The conventional unit of activity is the curie (Ci):

$$\begin{aligned} 1 \text{ Ci} &= 3.7 \times 10^{10} \text{ s}^{-1} \\ &= 3.7 \times 10^{10} \text{ Bq} \end{aligned} \tag{36-2}$$

Further conversions are shown in Table 36-2.

TABLE 36-2 Units and Conversion Factors for Radon and Radon Progeny

ACTIVITY

SI Unit	Bequerel	$1 \text{ Bq} = 1 \text{ s}^{-1}$
Conventional Unit	Curie	$1 \text{ Ci} = 3.7 \times 10^{10} \text{ s}^{-1}$
Useful Conversions:		
	$1 \text{ Ci} = 3.7 \times 10^{10}$ Bq	1 Bq = 27 pCi
	$1 \text{ pCi} = 3.7 \times 10^{-2}$ Bq	$100 \text{ Bq m}^{-3} = 2.7 \text{ pCi l}^{-1}$
	$1 \text{ pCi l}^{-1} = 37 \text{ Bq m}^{-3}$	

SPECIAL UNITS

SI Units

^{222}Rn
- WL[a] $= 2.8 \times 10^{-5}$ (A) $+ 1.4 \times 10^{-4}$ (B) $+ 1.0 \times 10^{-4}$ (C)
- EEC[b] in Bq m^{-3} = 0.105 (A) + 0.516 (B) + 0.379 (C)
  - where A, B, C = concentrations of ^{218}Po, ^{214}Pb, and ^{214}Bi in Bq m^{-3}
- EEC in Bq m^{-3} = (3,700) (WL)

^{220}Rn
- WL $= 3.3 \times 10^{-3}$ (B) $+ 3.1 \times 10^{-4}$ (C)
- EEC in Bq m^{-3} = 0.91 (B) + 0.09 (C)
  - where B, C = concentrations of ^{212}Pb and ^{212}Bi in Bq m^{-3}. The contribution of ^{216}Po to WL and EEC is negligible
- EEC in Bq m^{-3} = 275 (WL)

Either
- J m^{-3} $= 5.6 \times 10^{-9}$ (EEC in Bq m^{-3})
- J h m^{-3} = (J m^{-3}) (hours exposed)

Conventional Units

^{222}Rn
- WL $= 1.05 \times 10^{-3}$ (A) $+ 5.16 \times 10^{-3}$ (B) $+ 3.79 \times 10^{-3}$ (C)
- EEC in pCi l^{-1} = 0.105 (A) + 0.516 (B) + 0.379 (C)
  - where A, B, C = concentrations of ^{218}Po, ^{214}Pb, and ^{214}Bi in pCi l^{-1}
- EEC in pCi l^{-1} = (100) (WL)

a. WL = working level
b. EEC = Equilibrium Equivalent Concentration

Both ^{238}U and ^{232}Th are present in all rocks and soils at concentrations of roughly 1–5, and 2–12 ppm, respectively, or about 7–60 Bq/kg of soil for both nuclides. The ^{232}Th concentration of some rocks is as high as 20 ppm (NCRP 1988). The crustal abundance (by mass) of ^{232}Th averages about three times that of ^{238}U but the activity concentrations are about equal because of the much longer half-life of ^{232}Th (Table 36-1). These nuclides and their progeny will be in radioactive equilibrium unless separated by a physical process, such as diffusion of the gas radon away from its site of origin.

The most significant isotope with respect to exposure of the population to background radiation is ^{222}Rn, of the ^{238}U decay series. Its half-life of 3.82 days allows time for substantial diffusion to soil gas and to the atmosphere. ^{220}Rn (thoron), with a half-life of 55 s, contributes less significantly to the airborne inventory of radon. ^{219}Rn (actinon) is the least abundant isotope both because of its short half-life (3.96 s) and because the content of ^{235}U in crustal matter is less than 1% that of ^{238}U. The following discussion will center on sampling and evaluation of airborne ^{222}Rn because (1) the atmospheric concentration far exceeds that of ^{220}Rn, and (2) the radiation dose that results from exposure to ^{220}Rn is only about 1/3 as great as that from an equal concentration of ^{222}Rn. For special circumstances (e.g., geological formations with highly elevated thorium concentrations or manufacturing processes utilizing thorium), thoron exposure may be significant and must be recognized if radon sampling is required. A few methods are available for sampling thoron progeny (NCRP 1988) but the need is rare and will not be considered further in this chapter. Detailed information on concentrations of the radon isotopes, sources, emanation rates, atmospheric mixing and transport, the radiation dose resulting from exposure and the measurement of radon and radon daughters in air can be found in a series of reports issued by the National Council on Radiation Protection and Measurements (NCRP 1984*a*, 1984*b*, 1987, 1988, 1989). These authoritative reports have been freely consulted in the preparation of this chapter. Primary sources for material referenced to NCRP documents can be found in those reports.

Sources

Direct emanation to the atmosphere from soils and rocks is the major source of atmospheric radon (Table 36-3). Ground water is the second most significant source. Radium, the immediate predecessor of radon is dissolved by ground water and carried into streams and rivers; the decay of the radium provides a reservoir of potential atmospheric radon. Lesser amounts are contributed from the ocean and other radium-containing crustal materials. The major source of indoor radon is the infiltration of soil gas into a structure drawn by pressure differentials between the indoor air and the soil gas. Small pressure differentials caused by temperature differences between indoor and outdoor air, or surface winds, draw soil gas from large areas into a dwelling (NCRP 1989).

Airborne Concentrations

Typical atmospheric concentrations of radon in the United States range from about 4 to 15 $Bq\,m^{-3}$ (0.1–0.4 $pCi\,l^{-1}$) (NCRP 1988) with an average of about 7 $Bq\,m^{-3}$ (NCRP 1989). Concentrations are much lower over the oceans. Indoor concentrations are reported to be lognormally distributed with a geometric mean of about 40 $Bq\,m^{-3}$ (1.0 $pCi\,l^{-1}$) and a geometric standard deviation (GSD) of 2.8 (Nero et al. 1986). This implies that about 2.3% of homes exceed

TABLE 36-3 Major Sources of Worldwide Atmospheric Radon-222*

Source	Ci/y
Emanation from soil	2×10^9
Ground water (potential)	5×10^8
Emanation from oceans	3×10^7
Phosphate residues	3×10^6
Uranium tailings piles	2×10^6

* Adapted from NCRP 1984*b*

300 Bq m^{-3} (8 pCi l^{-1}), the level at which NCRP (1984*a*) suggests remedial action be considered. These concentrations represent a few thousand atoms of radon and only a few atoms of the short-lived progeny per liter so that identification by chemical methods is not possible. The radiometric properties of the nuclides are used for detection; these radiometric properties also confer the potential for induction of biological effects.

RADIOMETRIC PROPERTIES OF RADON AND DAUGHTERS

Radon, an alpha particle emitter, transforms to a polonium isotope, that further decays through a series of short-lived isotopes of bismuth, lead, and polonium (also thallium for ^{220}Rn and ^{219}Rn). (Figs. 36-1, 36-2, 36-3). The short-lived progeny of ^{222}Rn include the series from ^{218}Po to ^{214}Po (RaA to RaC′). This portion of the decay chain has an overall half-life of about 30 min. Various alpha, beta, and gamma rays emitted from these nuclides are detectable (Table 36-4) and specific radiations may be used to quantitate the air concentration of individual progeny. The next nuclide in the series, ^{210}Pb, has a half-life of 21 years, is largely removed from the atmosphere before disintegrating, and ultimately decays to stable lead.

AEROSOL PROPERTIES OF RADON AND DAUGHTERS

Unattached Fraction

When ^{222}Rn decays, the atom of freshly formed ^{218}Po may quickly attach to a particle of the ambient aerosol. Those atoms that remain unattached will rapidly grow by reacting with trace gases, or, if charged, they will attract ambient water molecules to form clusters. These small, highly diffusive particles, or

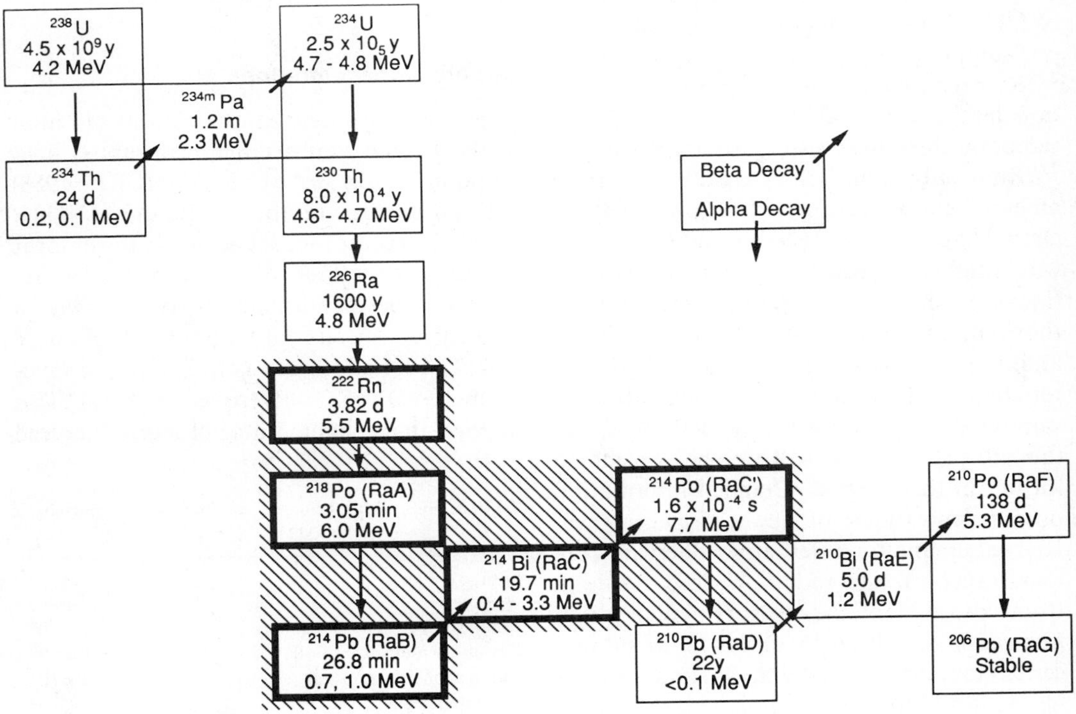

FIGURE 36-1. Decay Scheme of the Uranium-238 Series. The Hatchmarked Area Highlights Radon and Its Short-Lived Progeny.

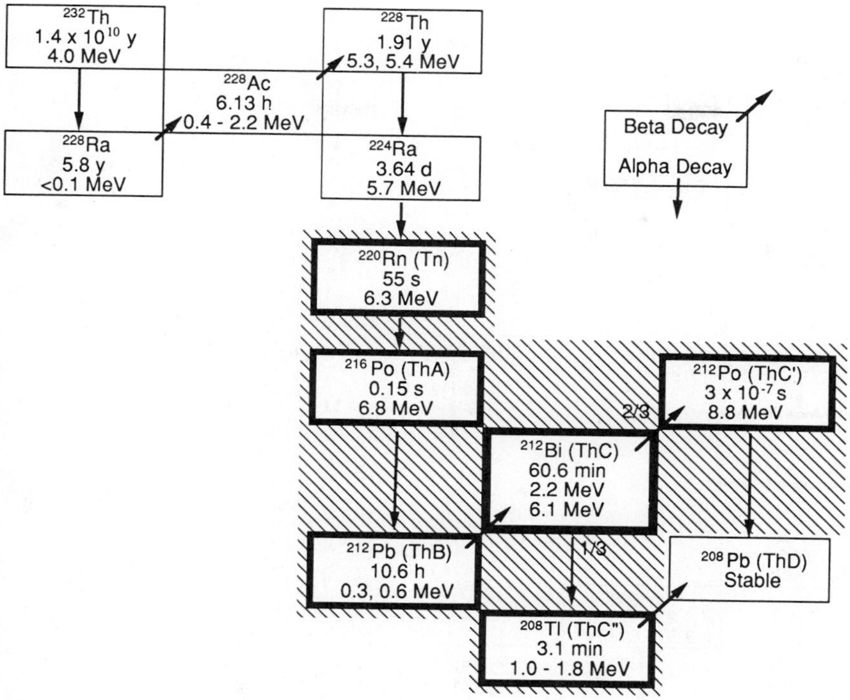

FIGURE 36-2. Decay Scheme of the Thorium-232 Series. The Hatchmarked Area Highlights Thoron and Its Progeny.

the resulting clusters, are known as the "unattached fraction". A review of the diffusion coefficients measured for this size fraction will be found in NCRP 1988 and Hopke 1989. They range from 0.0025 to 0.08 $cm^2\ s^{-1}$. This corresponds to a size range of about 0.5–5.0 nm. Measurements in homes have documented a progeny particle mode in the range of 10–40 nm and it has been suggested that these larger diameters represent radon progeny attached to a particle size mode present in the environment that was frequently unrecognized. These highly mobile particles diffuse rapidly and deposit on larger atmospheric particles or plate out to the walls and other surfaces, but since new RaA is constantly formed, there is always a small percentage of unattached ^{218}Po. Similarly, a few ^{214}Pb or ^{214}Bi (RaB or RaC,C′) remain unattached, but the fraction is even lower than that of the RaA.

It is estimated that, after formation, about 90% of the RaA atoms have a single positive charge at the end of the recoil path, the rest being neutral (Hopke 1989; NCRP 1988). Although the ^{218}Po is rapidly neutralized there remains some charged fraction. NCRP (1988) reports that a positive charge is regained during growth or attachment to ambient aerosol.

Attached Fraction

Most of the radon progeny attach to ambient aerosol particles, the magnitude of the fraction depending on the number of airborne particles available. Their subsequent physical behavior tracks that of the stable airborne particles except for radioactive decay. The latter is the primary mode of removal from the atmosphere because the overall radiological half-life, 30 min, is faster than removal by deposition or plateout.

The important measurable quantities for the attached fraction are the concentration and particle size. The size distribution of the

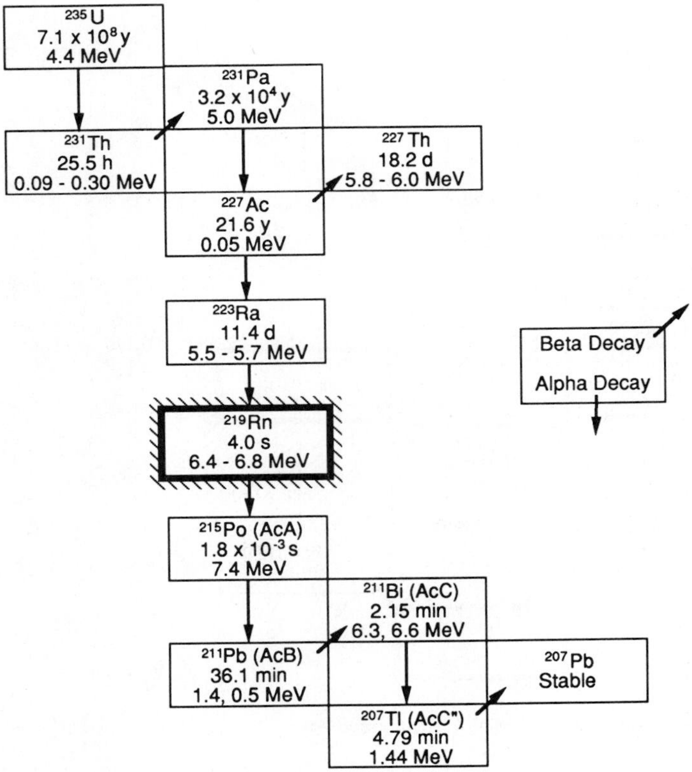

FIGURE 36-3. Decay Scheme of the Uranium-235 Series.

aerosol determines its physical behavior in the environment including particle transport and deposition. Particle size is the most important factor in determining whether the particles are inhaled and where they will deposit in the respiratory tract. For these reasons, particle size is a critical determinant of the radiation dose to the respiratory airways. The size distribution, and, thus, the radiation dose differs for mine, outdoor, and indoor air (Table 36-5). For radioactive aerosols the diameter of interest is the activity median diameter (AMD), that is the geometric mean diameter determined by measuring the radioactivity of the size fractionated sample, assuming a lognormal distribution. The AMD reported for radon progeny varies with ambient conditions, as seen from the representative ranges shown in Table 36-5, but it is always less than 1 μm. AMD is one of the most significant variables when comparing the dose per unit exposure for underground miners with that of the general population.

HUMAN EXPOSURE PARAMETERS

When atmospheric radon decays, both attached and unattached short-lived progeny are inhaled and deposit on the bronchial epithelium, where they decay before they can be removed by natural clearance processes. The dose delivered by the alpha particles emitted after deposition is the significant quantity in determining the carcinogenic potential of a particular radon atmosphere. For ^{222}Rn the alpha emitters are ^{218}Po (RaA) and ^{214}Po (RaC′) (Table 36-4). The ^{214}Pb (RaB) and ^{214}Bi (RaC), both beta particle emitters, are not significant dosimetrically but their behavior determines the ultimate concentration of ^{214}Po.

TABLE 36-4 Uranium Series from Rn-222 to Pb-210

Nuclide	Historical Name	Half-life*	Major Radiation Energies (MeV) and Intensities$^{(s, y)}$ α	β	γ
$^{222}_{86}Rn$	Emanation Radon (RN)	3.823 d	5.49 (99.92%)	—	—
$^{218}_{84}Po$	Radium A	3.05 m	6.00 (99.98%)	—	—
99.98% → $^{214}_{82}Pb$; 0.02% → $^{216}_{85}At$					
$^{214}_{82}Pb$	Radium B	26.8 m	—	0.178 (2.4%) 0.665 (46.1%) 0.722 (40.8%) 1.02 (9.5%)	0.295 (18.4%) 0.352 (35.4%) 0.768 (1.04%)
$^{216}_{85}At$	Astatine	~2 s	6.65 (6.4%) 6.99 (90%) 6.76 (3.6%)	—	—
$^{214}_{83}Bi$	Radium C	19.9 m	—	1.06 (5.56%) 1.15 (4.25%) 1.41 (8.15%) 1.50 (16.9%) 1.54 (17.5%) 1.89 (7.56%) 3.27 (19.8%)	0.609 (44.8%) 0.768 (4.76%) 1.12 (14.8%) 1.24 (5.83%) 1.76 (15.3%) 2.20 (4.98%)
99.98% → $^{214}_{84}Po$; 0.02% → $^{210}_{81}Tl$					
$^{214}_{84}Po$	Radium C′	164 μs	7.69 (100%)	—	—
$^{210}_{81}Tl$	Radium C″	1.30 m	—	1.86 (24%) 2.02 (10%) 2.41 (10%) 4.20 (30%) 4.38 (20%)	0.298 (79.1%) 0.800 (99.0%) 1.07 (12%) 1.21 (17%) 1.36 (21%)
$^{210}_{82}Pb$	Radium D	22.3 y	—	0.0165 (87%) 0.063 (18%)	0.046 (4.18%)

* Common time units: y (years); d (days); m (minutes); s (seconds)

Reprinted with permission. From Cohen, B. Sampling airborne radioactivity. In *Air Sampling Instruments for Evaluation of Atmospheric Contaminants*, 7th Edition, Ed. Suzanne V. Hering, pp. 73–109. Cincinnati, OH: American Conference of Governmental Industrial Hygienists, Inc.

TABLE 36.5 Characteristics of Radon Progeny Aerosols[a]

Atmosphere	Number Concentration No. cm^{-3}	AMD[b] (nm)	Equilibrium Factor
Outdoor	10^3–10^5	30–500	0.7[c], 0.8[d]
Indoor	10^3–10^5	5–150	0.4[c, d]
Mines	10^4–10^5	90–300	0.3[d]

a. Adapted from NCRP 1988
b. AMD = Activity Median Diameter
c. NCRP 1987
d. UNSCEAR 1988

Units of Exposure

The historical unit of exposure is the working level (WL). This is equivalent to an exposure atmosphere containing 100 pCi l^{-1} of ^{222}Rn in equilibrium with all of the short-lived progeny. The alpha particle energy that would eventually be emitted into this air volume is 1.3×10^5 Mev l^{-1} (Table 36-6). One WL is any combination of short-lived daughters in one liter of air that will result in the emission of 1.3×10^5 MeV of potential alpha energy. An exposure atmosphere with this quantity of

TABLE 36-6 ^{222}**Rn Short-Lived Decay Products**

Isotope	Alpha Ray Energy (MeV)	Half-life	Number of Atoms/100pCi	Ultimate Alpha Ray Energy (MeV)	Total Alpha Ray Energy (MeV)
^{222}Rn	5.49	3.82d	1.76×10^6	—	—
^{218}Po	6.00	3.05m	977	6.00 + 7.68	1.34×10^4
^{214}Pb	0	26.8m	8580	7.68	6.50×10^4
^{214}Bi	0	19.7m	6310	7.68	4.85×10^4
^{214}Po	7.68	2.7×10^{-6} m	8.5×10^{-4}	7.68	0.007
					12.7×10^4

1. The total amount of alpha ray energy that will ultimately be emitted by the short lived decay products of the 100pCi of ^{222}Rn = 1.3×10^5 MeV

potential alpha particle energy is called a WL regardless of whether equilibrium exists. Note that WL is a measure of the air concentration of short-lived decay products, not of the radon gas.

The unit of cumulative exposure is the working level month (WLM) and is equal to exposure in WL multiplied by exposure duration in multiples of the occupational month (170 h). To calculate the cumulative exposure of a member of the population in terms of WLM, the hours must be adjusted since exposure times exceed 170 h/month. Then

$$\text{WLM} = \text{WL}[(\text{hours exposed})/170] \quad (36\text{-}3)$$

Atmospheric clearance processes, such as deposition to surfaces, usually result in disequilibrium between the concentration of ^{222}Rn and its offspring. The airborne activity concentration of ^{218}Po is usually less than that of radon and so on for each successive decay product, except that the extremely short half-life of ^{214}Po (RaC′) assures that it remains in equilibrium with ^{214}Bi (RaC). The resulting airborne radioactivity concentration may be described as an equilibrium equivalent concentration (EEC). The EEC is the concentration of radon that in equilibrium with each of the daughters would have the same potential alpha energy per unit volume as the actual mixture. By calculating the fractional alpha energy shown in the last column of Table 36-6, it can be seen that

$$\text{EEC} = 0.105\,[A] + 0.516[B] + 0.379[C] \quad (36\text{-}4)$$

where, [A], [B], and [C] are the concentrations of ^{218}Po, ^{214}Pb, and ^{214}Bi, respectively in either Bq m^{-3} or pCi l^{-1}. EEC and WL are related as follows:

$$\text{WL} = \text{EEC}(\text{Bq m}^{-3})/3700 \quad (36\text{-}5)$$

$$\text{WL} = \text{EEC}\,(\text{pCi l}^{-1})/100 \quad (36\text{-}6)$$

The state of equilibrium is also described by the equilibrium factor, F. F is the ratio of the potential alpha energy concentration that exists in the mixture to that which would exist if all the decay products were in equilibrium with the radon present. Then F is the ratio of EEC to the radon concentration:

$$F = \text{EEC}/[\text{Rn}] \quad (36\text{-}7)$$

where [Rn] is the concentration of radon gas. If F is about 0.4, a level commonly found in homes in the U.S. (Table 36-5, NCRP 1987), in an atmosphere with 100 pCi l^{-1} Rn, the EEC is 40 pCi l^{-1} and the exposure is 0.4 WL; in an atmosphere with 4 pCi l^{-1} of Rn, the EEC is 1.6 for an exposure of about 0.02 WL.

AIR SAMPLING FOR RADON AND ITS SHORT-LIVED DECAY PRODUCTS

Sampling Strategy

When air sampling is planned, the reason for making a measurement will determine the type of measurement that is needed. Additional considerations are the estimated amount of the activity present and whether suitable equipment is available. For any measurement, it is necessary to have adequate sensitivity and adequately calibrated instruments.

The purpose of the measurement, in order of complexity, may range from simply determining the indoor or outdoor air concentration, through quantitation of sources, to a health effects evaluation. If a determination is to be made as to whether there is a potential for people to be significantly exposed, then decision-making criteria will be required. Various national and international regulatory bodies have recommended limits for air concentrations of radon and radon daughters in buildings. These range from 74 to 400 Bq m^{-3} (0.02 to 0.11 WL). *The Indoor Radon Abatement Act* of 1988, which is an amendment to the *Toxic Substances Control Act*, established a long-term national goal of reducing radon in U.S. buildings to ambient outside levels. There is some controversy about the appropriate limit for initiating remedial action, but there is a general agreement that remedial action be taken if radon concentrations are greater than about 800 Bq m^{-3}, or 0.1 WL (20 pCi l^{-1} Rn, EEC = 10 pCi l^{-1}).

Activity concentrations can be determined by measuring radon, the daughters, or both. Measurement of Rn is easier and will give an upper limit to the amount of potential alpha energy (WL) in the atmosphere. In addition, if an average equilibrium factor for homes of about 0.4 is assumed, the radiation dose which will result from inhalation exposure can be estimated. Determining the sources of radon is more complex. This may require a variety of measures to assess the emanation rate from surfaces, soil, building materials, water, etc. To do a complete health effects evaluation requires the measurement of both the concentration and size distribution of the radon progeny.

Sampling Methods

Sampling for either radon or progeny may be accomplished by grab sampling, continuous sampling, or collecting an integrated sample. Grab samples are collected by filling a small container and transfering the sample to the laboratory for analysis. These are either instant or very short period samples. Either radon or daughter samplers may be used. Grab samples are useful for screening on a small scale. Samples can be collected over several seasons, and seasonal average concentrations can be assessed from a few measurements taken in each season.

Continuous sampling with an instrument that produces sequential on-site readout is more descriptive because there are spatial and temporal variations in radon and progeny concentrations. It is necessary for any in-depth assessment of exposure, but requires skilled personnel, a high level of quality control, and relatively complex instrumentation. Continuous sampling is essential, however, for remedial work so that sources and ventilation effects may be correlated with the observed concentrations.

Integrated sampling will result in a single average concentration value over the duration of the sampling. Average sampling times may be days, weeks, or months. Three important advantages for radon sampling are that (1) integrated samples enable the measurement of concentrations that are too low for grab sampling, (2) many long-duration samples can be collected at the same time, and (3) the samples can be analyzed at a convenient time and place. Both passive and active integrated samplers for radon are available. These samplers are generally less expensive than continuous samplers.

Most measurements are undertaken to measure the air concentration of radon. Separation of radon from its daughters is feasible because radon is a gas and the progeny are particles. To sample radon only, the ambient daughters are removed by filtration. The progeny that are subsequently formed by the decay of the isolated radon are either counted or quantitatively removed from the sample volume. Counting the progeny increases the sensitivity of the sampling system because it provides three alpha particles for each ^{222}Rn decay (see Fig. 36-1 or Table 36-4). It is only necessary to wait for radioactive equilibrium to be established. This will occur in about six half-lives, or approximately 3 h for the ^{222}Rn short-lived decay products.

Grab samplers and integrating samplers have been developed for both radon and the short-lived decay products. Continuous monitors are available for radon, but only semi-continuous monitors are available for the short-lived progeny. The operation of these samplers is based on a few collection and detection principles, but there are multiple variations of many of the basic types of monitors. This chapter presents at least one example of each type of monitoring system (grab, continuous, and integrating) and some of the more common variations. Details will be given where the information is instructive of the sampling principles. Standard methods for indoor air measurements are presented; however, since new monitors are developed frequently, it is advisable to review the recent literature before selecting a sampling method for any extensive radon detection program.

Detection Principles

Most detection systems rely on counting of the alpha particles emitted from radon, RaA, and RaC′. The methods are limited by the range of the alpha particles, which can penetrate only a few cm in air and less than 0.1 mm in a unit-density material. However, this limited range permits the development of very low background detection systems for alpha particle counting. When counting the gamma rays emitted from RaB and RaC, it is more difficult to reduce the always-present background radiation and, thus, more difficult to detect low levels of activity. To calculate WL, it is necessary to know the concentrations of RaA, RaB, RaC, and RaC′. Either alpha, beta, or gamma spectrometers may be used to identify the radiation emitted at a specific energy, for the identification of individual decay products. The different half-lives of the radon progeny provide another means of identification, and several sampling systems utilize the temporal pattern of the decay of total alpha activity to separate out RaA, RaB, RaC, and RaC′.

Lower Limits of Detection

An important measure of system performance is the smallest quantity of radon or progeny that can be detected. This will generally depend on both the volume sampled and the duration of the sampling. For example, if radon gas is concentrated from an air volume, increasing the air volume will allow adequate detection of a lower initial concentration. Similarly, increasing the exposure time of an integrating sampler will generally permit quantitation of lower concentrations. The determinants of the lowest detectable amount are the sensitivity with which the instrument responds to radiation (i.e., counts/minute/unit activity) and the background count rate. A good measure is the lower limit of detection (LLD) as defined by Pasternack and Harley (1971). For a 5% risk of concluding falsely that activity is present and for a 95% predetermined degree of confidence for detecting its presence, a simplified expression is,

$$\mathrm{LLD} = 3.29\, \gamma \sigma_n \tag{36-8}$$

where γ is a calibration constant to convert counts into activity and σ_n is the standard deviation of the net count rate.

Grab Sampling for Radon

A common and popular device for grab sampling of radon is the scintillation cell or flask (Fig. 36-4). It consists of a 100–1000 ml glass, plastic, or metal chamber with a flat transparent bottom and either one or two valves. The flask is either evacuated at the laboratory and filled by opening the valve in the field, or an air sample may be drawn through the cell. The inside surfaces, except for the base, are

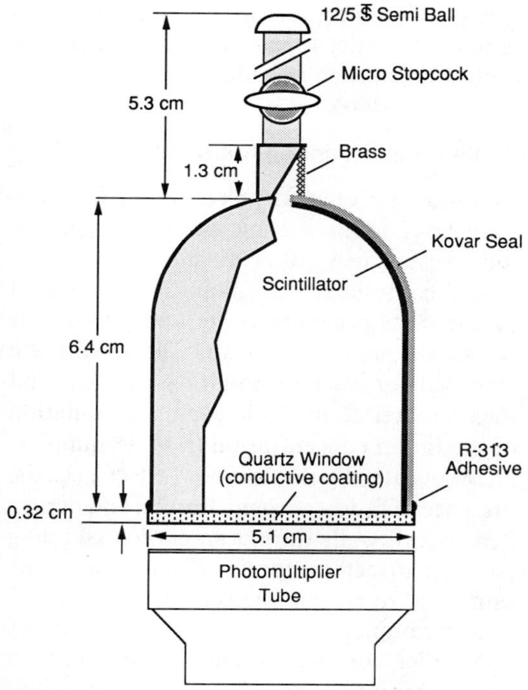

FIGURE 36-4. Radon Grab Sampler. Alpha Particles Emitted by ^{222}Rn, Po^{218} and Po^{214} in the Sampled Air Volume Interact with Phosphor Coated on the Inner Walls to Produce Scintillations. The Light Pulses are Transmitted through the Optically Clear Base to a Photomultiplier Tube and Converted to Electrical Pulses. The Signal is then Amplified and Counted. (Adapted from Lucas 1977)

coated with ZnS phosphor. The phosphor emits a flash of light when struck by an alpha particle. The air sample is held for six half-lives to allow the decay of any progeny originally present and for buildup of equilibrium with the trapped radon. Counting is done by viewing the scintillations with a photomultiplier tube through the optically transparent base. Detection limits are about 10 pCi l^{-1}, but can be as low as 0.1 pCi l^{-1} with longer counting periods. The system is simple and reliable as long as the containers are thoroughly leak-tested. The drawbacks are that the flasks are fragile and that time must be allowed for decay and to establish radioactive equilibrium.

A second grab-sampling method for radon is the "two-filter method" (Fig. 36-5). Air is drawn through a tube with filters at both inlet and outlet. The first filter removes all airborne decay products. As the radon traverses the tube, some of it decays and the resulting progeny are collected on the second filter. Knowing the time for flow through the tube, the tube volume, and the deposition loss to the walls as a function of the time of formation of the RaA, assay of the second filter permits a calculation of the radon concentration. The sensitivity depends on the tube dimensions, but can be as low as 0.1 pCi l^{-1}.

Grab Sampling for Radon Daughters

Determination of EEC or WL is made by collecting the short-lived progeny from a known air volume onto a filter and measuring the amount of each decay product collected on the filter. The collected progeny continue to decay during sampling; this complicates the analysis. In addition, the amount collected is often close to the LLD of the detection system; so counting errors are relatively large. A method still in use for mine atmospheres was developed by Kusnetz (1956). He used the simple approach of counting the filter sample at 60 min after sampling. The basis of the

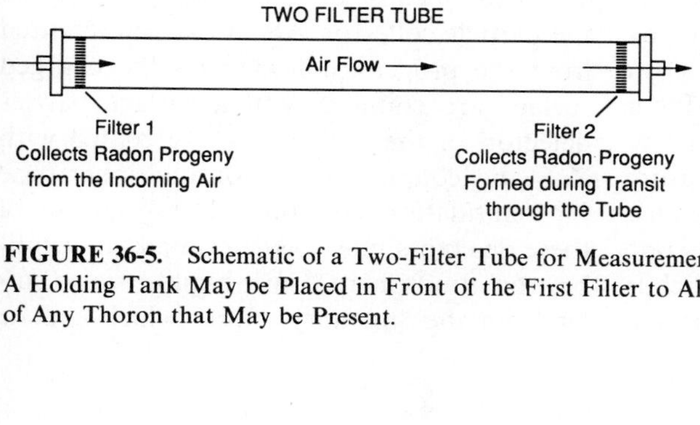

FIGURE 36-5. Schematic of a Two-Filter Tube for Measurement of Radon in Air. A Holding Tank May be Placed in Front of the First Filter to Allow Time for Decay of Any Thoron that May be Present.

method is that if an atmosphere of 100 pCi l^{-1} of radon is in equilibrium with the short-lived decay products (1 WL), an air filter will collect 100 pCi l^{-1} of each daughter. At 60 min after collection there will be 112.6 disintegrations per minute (dpm) on the filter, all due to RaC′. The number of multiples of 112.6 dpm l^{-1} is a measure of the number of WL. The complications are that the equilibrium ratios in the air vary, the samples may not be counted at precisely 60 min, and the buildup and decay occur on the filter during sampling. Kusnetz showed that for a sample collected for 5 or 10 min, the error due to buildup and decay on the filter will be $< 12\%$ under normal mine conditions, and provided correction factors for counting times from 40–90 min. The method has been modified and improved by several investigators (Harley and Pasternack 1969; Rolle 1972) and can measure down to 0.0005 WL with a reproducibility of 35%. At 0.04 WL the reproducibility is about 4%. If a substantial amount of thoron is present the filter may be counted again after about 4 h have elapsed (after decay of the short-lived ^{222}Rn daughters) to obtain a correction factor. The later count will result from ThC, and ThC′ that are formed after decay to ThB (10.6 h half-life).

Another grab-sampling method, known as the Thomas method (Thomas 1972), is a modification of an older method (Tsivolglou, Ayers, and Holaday 1953). A filter sample is collected for exactly 5 min, and the total number of alpha counts is determined for the periods 2–5, 6–20, and 21–30 min. These counts are entered into three simultaneous equations, together with the air sampling rate in l min^{-1} and the counting efficiency of the system, from which the concentration in air of RaA, RaB, and RaC may be individually determined. This method does not make any assumptions about equilibrium, but it is slow. Scott (1981) has adapted the method for a reduced counting time.

Alpha spectrometry can directly measure the concentration of each of the progeny on a filter sample. It requires the use of solid-state detectors and a spectrometry system which is more expensive than a simple alpha counting system. Counting at two separate times is required in order to calculate the initial concentration of each nuclide.

Continuous Sampling for Radon

Continuous radon samplers count continuously, but totals are recorded for specified time periods (e.g., 30 or 40 min counts).

The continuous scintillation cell is based on the same principle as the Lucas flask. Air passes through the cell and the counts are recorded for 30 min counting periods, and then another count cycle begins. Calculation of the radon concentration in these samplers is complicated by the "memory effect", i.e., the presence of radon progeny formed and deposited on cell walls during an earlier sampling period. Corrections for this effect are made by sampling from a test radon atmosphere with a clean sampler (Thomas and Countess 1979).

Another continuous sampler for radon is an automated version of the two-filter method. Air flows continuously through a two-filter tube and an alpha particle detector is placed facing the second filter. Alternatively, the second filter may be replaced by a moving tape so that the sample is collected for a specified time period, after which the tape moves the collected sample under an alpha particle detector and is replaced by a clean filter.

A useful class of radon samplers operate by allowing radon to diffuse into a detection volume from which the progeny are excluded. The atmospheric particles are filtered by a diffusion barrier of foam or other material. The progeny that subsequently form in the volume are collected by a charged surface. The shapes and sizes of these monitors vary, as do the means for maintaining charge on the particle collector. Alpha particles emitted from the progeny collected on the charged surface are counted with a surface barrier detector, or the surface may be coated with ZnS and coupled to a photomultiplier tube for scintillation counting. The response of these detectors must be determined in a calibration atmosphere. These monitors also suffer from the "memory effect" which can be troublesome in a changing radon atmosphere. The same principle is applied in several integrating radon monitors that use film or thermoluminescent dosimeters as alpha detectors. This popular method is sensitive to relative humidity which alters the collection efficiency of the progeny by the charged surface, thus changing the calibration of the system. The details of individual monitors based on this method can be found in NCRP (1988).

An alternate approach used by Chittaporn, Eisenbud, and Harley (1981) is to collect the progeny on a surface that keeps the alpha particles out of the range of the detector. The radon progeny are collected by an electret that can maintain a field of 100 V cm^{-1} in the sensitive volume. The sensitive volume, coated with scintillation phosphor, detects only the alpha particles from the decay of the parent radon. Removing the daughters from the counting volume avoids any changes in efficiency due to humidity. The monitor can detect 0.03 pCi l^{-1} for a 1 h counting time.

To avoid counting thoron in continuous radon monitors, a decay chamber may be placed in front of the detection volume. The volume needed will depend on the sampling flow rate and must be large enough to allow the decay of the 55 s half-life thoron. Passive radon samplers, i.e., those that depend on diffusion of the gas into the sensitive volume can reduce thoron detection by use of a diffusion barrier thick enough to delay the thoron beyond its average life.

Continuous Samplers for Radon Daughters

Radon daughters are sampled by collecting them onto a filter and then counting the alpha particles emitted as they decay. The samplers can only be semicontinuous, but sampling can be done for 2–5 min and counting for 2–3 min, so that a measurement can be made every 10 min. Membrane filters with pore sizes < 1 μm are usually used because they are highly efficient for all particle sizes, and because the counting efficiency of the samples thus collected is not too variable. Some other filters exhibit a variability in counting efficiency caused by self-absorption of alpha particles in the filter sample. As described for the grab samplers, the filter may be counted in place, or it may move to a counting position and be replaced by the next filter collector. Counting the filter in place avoids mechanical problems that are troublesome in harsh environments. Such instruments may incorporate computer capability to directly calculate the concentrations of each of the progeny and the WL from the measured RaA and RaC′ alpha particles or decay of the beta count rate. Several commercial continuous "working level monitors" are available.

Integrated Sampling for Radon

Integrating samplers for radon may be either passive or active. Passive samplers that require no power supply are extremely useful field instruments. A significant advantage is that they are quiet and, thus, acceptable to occupants when measurements are needed in homes. There are currently four types of passive radon monitors. As with other radon samplers, the progeny particles are excluded by diffusion barriers. The two most common are activated-charcoal-containing collectors and monitors containing etched-track detectors. Activated-charcoal-containing monitors are perforated cannisters 2.5–5 cm (1–2 in) high and 5–10 cm in diameter. The container is sealed at the laboratory, then opened to the test atmosphere for a fixed time period, usually 4–7 days for a cannister with > 50 g of charcoal, less for smaller quantities. Atmospheric radon diffuses onto the charcoal where it is adsorbed; the container is resealed and returned to the laboratory for counting. Counting is usually done by gamma-ray spectrometry of the entire cannister. Charcoal adsorbs water vapor as well as radon, so that for open-faced carbon containers an experimentally derived correction factor is needed when the humidity is high. Each sampler must be weighed before and after exposure, any difference is attributed to adsorbed water vapor, and an experimentally derived correction factor is applied. These detectors can be reused after regeneration by heating. The heating drives off the water vapor before the sampler is resealed. George and Weber (1990) report that if a sintered metal filter is used as a diffusion barrier over the inlet, no correction is needed for adsorbed water vapor under most conditions. Activated charcoal adsorbant is satisfactory for most sampling environments, but if the concentration of radon varies dramatically, sample loss may occur by desorption when air concentration declines. The sintered filter cannister is less sensitive to changing Rn concentrations than other open-faced or diffusion barrier cannisters (George and Weber 1990).

Etched-track detectors are plastic films, usually cellulose nitrate (Kodak LR115, Eastman Kodak, Rochester, NY) or allyl diglycol carbonate (CR-39, Tech/OPs Landauer, Inc., Glenwood, IL), that develop tracks when traversed by an alpha particle. Chemical etching enlarges the tracks; so they can be optically counted. These detectors do not respond to beta or gamma radiation. The detectors are placed in a container into which the radon diffuses. The alpha particles from the subsequent progeny expose the detector. The results are in terms of tracks/cm^{-2}/day^{-1}. Calibration must be done in a test atmosphere, because the geometry of the detector and container will determine the detection efficiency. They are usually left in place for several months or a year, thus giving a seasonal or annual average exposure.

Another passive monitor is the passive environmental radon monitor (PERM). This unit developed by Breslin and George (1978) is similar to the continuous passive diffusion radon monitors described above. Radon diffuses through a silica gel and filter barrier, the progeny are collected by electrostatic attraction to a rod on which a voltage (900 V) is maintained by batteries, and the radiation dose is monitored by a thermoluminescent dosimeter (TLD). The PERM sampler is usually deployed for a few weeks, then the TLD is collected and read out on a TLD analyzer.

A new, small passive monitor, the environmental gamma and radon detector (EGARD) (Maiello and Harley 1987) also utilizes TLDs to record exposure. The detector consists of a small aluminum container providing a 1 l volume into which the radon diffuses. There are three sets of TLDs. One set is covered with aluminized mylar to prevent alpha particles from reaching the TLDs, and records background gamma radiation. The second set is placed on an electret to which the progeny are attracted and records both background gamma and the alpha radiation. The third set is for quality control. This small detector is easy to mail and has an LLD of 100 pCi l^{-1}-days.

Active sampling into gas sampling bags or sampling tanks may be used for radon as for other gases. The sample is transported to the laboratory, where the gas is transfered to a detection system.

Integrated Sampling for Radon Daughters

There are no passive samplers for radon daughters. Since radon progeny aerosols must be sampled onto filters, power is required. Once collected, the resulting radiation may be measured by an integrating detector such as a film, a TLD or an etched-track detector. One such monitor, the RPISU (for radon progeny integrating sampling unit), collects daughters on a filter near a TLD (Schiager 1974).

Other Monitors

Several active and passive samplers have been developed that are small enough, sensitive enough, and reliable enough to be useful as personal monitors, while others are under development (NCRP 1988). The operating principle of these monitors is basically the same as of those described above.

A personal radon monitor that can also be used in homes has recently been developed (Harley, Chittaporn, and Scarpitta 1991). The monitor is a 7.5 cm diameter, 3.0 cm high cylinder made of lightweight conducting plastic. Radon diffuses into the cylinder through a conducting foam barrier. CR-39 plastic is the radon detector. A TLD is mounted beneath the CR-39 to detect gamma ray exposure. Both are covered with thin aluminized mylar to nullify charge effects. Personal exposure measurements and simultaneous ^{222}Rn exposure measurements on different levels in a home were made by Harley et al. (1991) using the newly developed personal radon monitor. They report that personal exposure correlated well with exposure on the first floor of the homes tested. The reported ratio of personal to first floor exposure was 0.71.

A system for measuring the fraction of airborne radon decay products that will deposit in the nasal cavity or on the bronchial tree has been suggested (Hopke, Ramamurti, and Knutson 1990). Wire screen collectors are used preceding a filter. The wire screen sizes and flow rates are selected based on lung deposition models to mimic respiratory deposition. A parallel filter sampler measures the total airborne activity. The activity deposited is then obtained from the difference between the amount of alpha activity measured on the two filters.

Special techniques have been developed for measuring the flux of radon from soils and building materials. It is difficult to measure radon flux without perturbing the system under study. Various methods are described in NCRP (1988). Detection of the radon and progeny for this special class of samplers is also based on the principles described above.

CALIBRATION

All sampling and counting systems must be calibrated to assure accurate measurement. Flow calibration is required for active sampling systems (see Lippmann 1989). Calibration of counting systems may be accomplished with standard reference sources traceable to the National Institute of Standards and Technology (NIST). Determination of the counting efficiency of any detector, i.e., the counts per minute per unit activity of radon or daughters, is an essential part of any quality control program. Two reports present detailed information on calibration procedures (NCRP 1988; Beckman 1972). The counting efficiency of radon samplers may be determined by emanating a known amount of radon into the sampler from a standard solution of ^{226}Ra obtainable from NIST. A few radon chambers exist in which carefully controlled atmospheres are maintained. These atmospheres are standardized by using samplers that have been standardized with reference sources. Among these are a chamber at the Environmental Measurements Laboratory of the U.S. Department of Energy (New York, NY), two chambers maintained by the U.S. Environmental Protection Agency (EPA) (Montgomery, AL and Las Vegas, NV) and one maintained by the U.S. Bureau of Mines (Denver, CO). Arrangements can usually be made to calibrate the sampling equipment at one of these chambers. For radon progeny there is no standard atmosphere available. Thus, alpha-emitting standard sources are used for comparison with laboratory counters of known efficiency; alternatively, beta or gamma standards of appropriate energy are used. If counters and detectors are properly calibrated, it is then assumed that the progeny count rates are correct. In addition to counting efficiency it may be necessary to determine the collection efficiency of a sampling system, e.g., by passive monitors containing etched-track detectors. This can only be done in a chamber with a known radon atmosphere. If any changes are made in the detection system, the calibration may no longer be valid and should be reevaluated.

PROTOCOLS FOR INDOOR MEASUREMENT

The U.S. Environmental Protection Agency (U.S. EPA 1986) has prepared a set of protocols to be used for obtaining standardized indoor measurements. The protocols require standardized house conditions in order to maintain a stable environment. Sampling is done in a closed house with minimum ventilation. The purpose is to permit an intercomparison of the measurements. The values of airborne radon and decay product concentrations obtained using these protocols do not necessarily reflect the concentrations to which occupants would be exposed. The EPA gives specifications for measurements of radon with continuous radon monitors, charcoal cannisters, alpha track detectors, and grab samplers. For radon daughters, procedures have been specified for the continuous working-level monitors, the RPISU, and for grab samplers. For each of these, requirements are listed for the location of the measurement device and the house conditioning; a set of minimum requirements for quality control are established for each procedure.

SUMMARY

The study of radon and its progeny is challenging. There are few atoms of either radon or its daughters in air, even at concentrations that are considered high in terms of the potential for causing harm to lung tissue. The size distribution of the progeny aerosol particles is of special significance because it is an important determinant of the efficiency with which they will deposit in the lung. This, in turn, determines the radiation dose that results from a particular exposure, and it is the radiation dose that confers the health risk. A variety of sensitive methods for radon (and progeny) sampling have been developed, many of which have been noted in this chapter. These methods, and those for sample analyses, must take into account the formation and decay of the short-lived progeny. The time frame (minutes to a few hours) for the buildup and decay of these nuclides complicates the measurements. It is not possible to simply allow decay until only one nuclide is present, or to assay samples quickly before the buildup of interferences. New sampling instruments are rapidly being developed and marketed because of the current intense interest in determining the exposure of the general population. Thus, the reader who plans to sample for these airborne materials is advised to investigate the current literature, as well as sources that compile and update a list of available instruments (e.g., Hering 1989).

References

Beckman, R. T. 1972. Calibration procedures for radon and radon-daughter measurement equipment. Informational Report 1005, United State Department of the Interior, Denver Technical Support Center, Denver, CO.

Breslin, A. J. and A. C. George. 1978. An improved time integrating radon monitor Presented at the NEA Specialist Meeting on Personal Dosimetry and Area Monitoring Suitable for Radon and Daughter Products, 20–22 November 1978, Paris.

Chittaporn, P., M. Eisenbud, and N. H. Harley. 1981. A continuous monitor for the measurement of environmental radon. *Health Phys.* 41:405–10.

Cohen, B. S. 1989. Sampling airborne radioactivity. In *Air Sampling Instruments for Evaluation of Atmospheric Contaminants*, ed. S. V. Hering, 7th edn., pp. 221–40. Cincinnati, OH: American Conference of Governmental Industrial Hygienists.

George, A. C. 1986. Instruments and methods for measuring indoor radon progeny concentrations. In *Radon, Proceedings of an APCA International Specialty Conference.* Pittsburgh, PA: Air Pollution Control Assocation.

George, A. C. and T. Weber. 1990. An improved passive activated C collector for measuring environmental ^{222}Rn in indoor air. *Health Phys.* 58:583–89.

Harley, N. H. and B. S. Pasternack. 1969. The rapid estimation of radon daughter working levels when daughter equilibrium is unknown. *Health Phys.* 17:109–14.

Harley, N. H., P. Chittaporn, and S. C. Scarpitta. 1991. The influence of time-activity patterns and lifestyle on human exposure to radon in air. Part I: Development of a personal radon monitor. Report to New Jersey Department of Environmental Protection, Trenton, NJ.

Harley, N. H., P. Chittaporn, M. H. Roman, and J. Sylvester. 1991. Personal and home ^{222}Rn and gamma-ray exposure measured in 52 dwellings. *Health Phys.* (in press).

Hering, S. V. (ed.) 1989. *Air Sampling Instruments for Evaluation of Atmospheric Contaminants*, 7th edn. Cincinnati, OH: American Conference of Governmental Industrial Hygienists.

Hopke, P. K. 1989. Initial behavior of ^{218}Po in indoor air. *Environmental International* 15:299–308.

Hopke, P. K., M. Ramamurthi, and E. V. Knutson. 1990. A measurement system for Rn decay product lung deposition based on respiratory models. *Health Phys.* 58:291–95.

Kusnetz, H. L. 1956. Radon daughters in mine atmospheres—A field method for determining concentrations. *Ind. Hyg. Quart.* 3:85–88.

Lippmann, M. 1989. Calibration of air sampling instruments. In *Air Sampling Instruments for Evaluation of Atmospheric Contaminants*, ed. S. V. Hering, 7th edn, pp. 73–109. Cincinnati, OH: American Conference of Governmental Industrial Hygienists.

Litt, B. R., J. M. Waldman, N. H. Harley, and P. Chittaporn. 1991. Validation of a personal radon monitor for use in residential exposure studies. *Health Phys.* (in press).

Lucas, Henry F. 1977. Alpha scintillation counting. In *Atomic Industrial Forum Workshop on Methods for Measuring Radiation In and Around Uranium Mills*, ed. E. D. Harwood. Vol. 3, No. 9. Washington, DC: Atomic Industrial Forum.

Maiello, M. L. and N. H. Harley. 1987. EGARD: An environmental gamma-ray and radon detector. *Health Phys.* 53:301–5.

NCRP (National Council on Radiation Protection and Measurements). 1984*a*. Exposures from the uranium series with emphasis on radon and its daughters. NCRP Report No. 77. National Council on Radiation Protection and Measurements, Bethesda, MD.

NCRP (National Council on Radiation Protection and Measurements). 1984*b*. Evaluation of occupational and environmental exposures to radon and radon daughters in the United States. NCRP Report No. 78. National Council on Radiation Protection and Measurements, Bethesda, MD.

NCRP (National Council on Radiation Protection and Measurements). 1987. Exposure of the population in the United States and Canada from Natural background radiation. NCRP Report No. 94. National Council on Radiation Protection and Measurements, Bethesda, MD.

NCRP (National Council on Radiation Protection and Measurements). 1988. Measurement of radon and radon daughters in air. NCRP Report No. 97. National Council on Radiation Protection and Measurements, Bethesda, MD.

NCRP (National Council on Radiation Protection and Measurements). 1989. Control of radon in houses. NCRP Report No. 103. National Council on Radiation Protection and Measurements, Bethesda, MD.

Nero, A. V., M. B. Schwehr, W. W. Nazaroff, and K. L. Revzan. 1986. Distribution of airborne radon-222 concentrations in U.S. homes. *Science* 234:992–96.

Pasternack, B. S. and N. H. Harley. 1971. Detection limits for radionuclides in the analyses of multi-component gamma-ray spectrometer data. *Nucl. Instrum. Methods* 91:533–40.

Ramamurthi, M. and P. K. Hopki. 1989. On improving the validity of wire screen "unattached" fraction Rn daughter measurements. *Health Phys.* 56:189–94.

Rolle, R. 1972. Rapid working level monitoring. *Health Phys.* 22:233–38.

Schiager, K. J. 1974. Integrating radon progeny air sampler. *Am. Ind. Hyg. Assoc. J.* 35:165.

Scott, A. G. 1981. A field method for measurement of radon daughters in air. *Health Phys.* 41:403–5.

Thomas, J. W. 1972. Measurement of radon daughters in air. *Health Phys.* 23:783–89.

Thomas, J. W. and R. J. Countess. 1979. Continuous radon monitor. *Health Phys.* 36:734–38.

Thomas, J. W. and P. C. LeClare. 1970. A study of the two filter method for radon 222. *Health Phys.* 8:113–22.

Tsivoglou, E. E., H. E. Ayers, and D. A. Holaday. 1953. Occurrence of non-equilibrium atmospheric mixtures of radon and its daughters. *Nucleonics* 11:40–5.

UNSCEAR. (United Nations Scientific Committee on the Effects of Atomic Radiation). 1988. Sources, effects and risks of ionizing radiation. Report to the General Assembly, United Nations.

USEPA. (United States Environmental Protection Agency). 1986. Interim indoor radon and radon decay product measurement Protocols. EPA 520/1-86-04, Office of Radiation Programs, Washington, DC.

37

Aerosol Measurement in the Health Care Field

David L. Swift

Division of Environmental Health Engineering
Johns Hopkins University
615 N. Wolfe Street
Baltimore, Maryland, MD, U.S.A.

INTRODUCTION

There are a number of instances in which aerosols occur in the health care field and where aerosol measurement is needed. These instances may conveniently be grouped into three categories: (1) the use of aerosols in therapy; (2) the use of aerosols in diagnostic procedures; and (3) the inadvertent exposure to aerosols and its control in various areas of health care.

Aerosols encountered in the health care field share many common characteristics with aerosols in other applications. For examples, they have a life history consisting of generation, maturation with its physicochemical changes, and ultimate demise due to deposition or some other physical process. It is not the intent of this chapter to reiterate the techniques of measurement which have been extensively described for other applications, but to emphasize the unique aspects of aerosols in health care and to describe how aerosol measurement techniques have been devised, adopted, or adapted to these special circumstances. Some special aerosol measurement problems in health care applications requiring further future development will also be discussed.

Near neighbors to aerosol measurement techniques in the health care industry are some aerosol measurement applications in industrial hygiene, indoor and outdoor air pollution, and radioactive aerosols. All these applications involve potential human exposure to aerosols, whether for good or ill, and it is not uncommon to see aerosol techniques in one field being adapted for another. Of these fields, the study of aerosol use in the health care field as an organized discipline is the most recent, even though applications of aerosols in health care stretch back at least as far as the pioneering work of Lord Lister, who employed phenol aerosols in his original disinfection studies.

Among aerosol measurement applications, aerosols in the health care field have certain unique features which bear on the appropriate measurement methods. One important feature is the presence in many (but not all) such aerosols of substances which are extremely biologically potent, including macromolecules of various configurations and weights. Some of these substances have the capacity to initiate marked biological changes at very small concentration, and are able to be detected and quantified with extremely sensitive and specific tests. Such macromolecules and other biologically active substances are often physicochemically fragile and their existence in the aerosol state may alter their biological activity.

Other aerosol substances in the health care field carry living organisms such as fungi, bacteria, or viruses. Such aerosols can be detected and quantified with sensitive and selective methods of bacteriology and virology. They are similarly often physiologically fragile in certain environments, and these features are the concern of a significant historical interest in airborne infection. Although the introduction of vaccines and other preventive agents has markedly reduced the concern about many classical diseases spread by aerosols, the concern with newly discovered pathogens has renewed the interest in airborne infection, and the continuing inability to prevent infection from the common cold or influenza provides impetus to understand the airborne route. Despite a lack of any positive evidence to date, there is continuing interest in the possibility that HIV infection may have an airborne component, and many diseases, such as tuberculosis and measles, are known to be transmitted in air.

MEASUREMENTS OF INHALED THERAPEUTIC AEROSOLS

Types of Therapeutic Aerosols

A therapeutic aerosol for inhalation is intended to deliver a substance to the respiratory tract that will bring about some desired physiological change (Newman 1984). The respiratory tract may be the locus of activity or it may be the portal of entry for systemic distribution of a therapeutic drug. Several different types of therapeutic agents may be delivered to the respiratory tract. These include bioactive molecules which act locally or systemically at the cellular level, substances which physically alter the properties of respiratory secretions, and substances which treat infection. They range in chemical complexity from macromolecular solutions to water or saline aerosols.

For systemic drugs, it is acceptable to deliver the substance to any site of the respiratory tract that will permit the desired absorption characteristics (usually rapid absorption). Therapeutic agent aerosols which act locally usually require a more specific site of delivery, depending upon the desired site of action. For example, aerosol bronchodilator drugs which are acting locally must be delivered to the bronchial airways in sufficient quantity to achieve the desired result. If the aerosol is entirely deposited in the airways above the trachea or predominantly delivered to the alveolar spaces, the local bronchodilation may not occur.

The considerable body of knowledge which now exists concerning the deposition of "ideal" aerosols in the respiratory tract (Heyder and Rudolph 1984) has been drawn upon to predict and design aerosol delivery systems for therapeutic substances. However, as will be detailed below, there are a number of "non idealities" concerning many therapeutic aerosol systems which hinder reliable prediction, even though some general statements about deposition can be made. Measurement methods for therapeutic aerosols must take these nonidealities into account in linking aerosol properties to deposition sites.

Generation Methods of Therapeutic Aerosols

A number of different techniques for aerosol generation are employed for therapeutic aerosols, each of which bears on the appropriate measurement methods for these aerosols. The initial physical state of the substance to be aerosolized is either a bulk dry powder, a solution, an emulsion, or a solid-particle suspension. Depending upon the method of generation, the liquid vehicle may be either water or a liquid propellant.

Several methods exist for aerosolizing bulk powders of therapeutic agents. A device known as a Spinhaler® was originally described by Bell, Hartley, and Cox (1971). Bulk powder in a gelatin capsule is mounted in a rotor holder inside the inhalation tube. Two holes for powder dissemination are punched just prior to use. The rapid inspiratory flow of the patient causes the rotor to spin and vibrate on the undersize shaft, causing the powder to be discharged out of the capsule into the airstream. In its original form, the bulk powder was a mixture of the micronized drug and powdered lactose, the latter intended to enhance aerosolization by preventing excessive cohesion of the drug particles. Currently, the micronized drug is loaded into the capsule without lactose or other diluent.

The Rotohaler® is a similar device for aerosolizing a dry powdered drug, in which the powder is discharged into a small cylindrical chamber prior to inspiration and is aerosolized during rapid inspiration (causing high air velocity and turbulence in the chamber) and transported through a mouth tube to the respiratory tract (Kjellman and Wirenstrand 1981).

Both these devices required a loading step prior to inspiration. In the Turbohaler®, a disc containing many powder charges (up to several hundred) can be rotated for each application to locate a fresh powder charge in the path of inspired air. Entry of the inspired air through the device suspends the powder charge. The turbulent flow through flow channel to the mouth tube is intended to disperse and mix the powder so that primarily individual powder particles enter the respiratory tract (Jaegfeldt et al. 1987).

Many therapeutic aerosols are generated by nebulizers (Mercer 1981). Jet nebulizers employ high-velocity gas (usually compressed air) to shear bulk liquid into fine particles. The principle was originally applied to generate liquid aerosols at least 150 years ago and all such devices are linearly descended from the original. It is well known that for a given liquid, the median particle size produced decreases with increasing air velocity, reaching a practical limit at sonic velocity. For a given air velocity, the median-size particle decreases with decreasing liquid viscosity and surface tension. Modern nebulizers achieve smaller median particle sizes by inertial removal of larger particles, thus "recycling" a large fraction (usually $\sim 99\%$) of the liquid mass atomized and, therefore, reducing the airborne droplet mass concentration. All aerosols produced by jet nebulizers are polydisperse such that the majority of the aerosol mass (even after inertial removal) is usually found in a relatively small number of large particles whose size is much greater than the count median diameter.

The other major type of nebulizer employed for therapeutic aerosols is the ultrasonic nebulizer, in which ultrasonic energy originating from an electric transducer is focussed within a bulk liquid, producing airborne particles whose median size is related to the liquid properties and the ultrasonic frequency. The amount of aerosol produced is determined by the ultrasonic energy level and the aerosol mass concentration is determined by the generation rate and air flow dilution rate which can be independently set. This is the primary practical difference between jet and ultrasonic nebulizers, in that the aerosol mass concentration of the jet is limited by the jet flow for atomization.

A widely used method of generation for therapeutic aerosols is the metered dose inhaler (MDI), in which the drug is aerosolized by the release of a mixture of drug, propellant and (in some instances) another substance to maintain stable dispersion of the drug in the liquid propellant within the canister prior to release. This is the well-known "aerosol can" concept (used to dispense numerous commercial products) miniaturized and fitted with a dosing valve which dispenses a given liquid volume on each depression of the valve cap. In most instances, the drug is maintained as a suspension of micronized powder within the propellant. Some drugs are liquid emulsions or solutes. The propellant normally makes up most of the material within the canister; a surfactant to prevent particle agglomeration is sometimes a minor constituent.

The MDI enjoys wide popularity as a therapeutic drug generator because it is self-contained (requires no air pressure, battery, or other energy source), pocket-sized, generates a relatively constant mass of drug and is relatively easy to use. The phase-out of widely used chlorofluoromethane propellants for environmental reasons has clouded the future use of MDIs. Alternative propellants are now in development that should have similar generative properties, provided they do not demonstrate undesirable physicochemical interactions with therapeutic agents.

Some aerosols for localized delivery in the nasal or oral airways are generated by liquid spray devices. These generation devices are known technically as hydraulic atomizers as opposed to the jet nebulizer; in the former a liquid is mechanically forced through a small orifice and breaks up to form large droplets (50–200 μm) while, as described above, jet nebulizers use air streams to break up liquids, producing much smaller particles. Liquid spray devices differ from MDIs in that no explosive evaporation of the droplets takes place to disperse the particles and reduce particle size.

Delivery Systems for Therapeutic Aerosols

Because therapeutic aerosols are meant to be breathed, a system is needed to couple the aerosol generator to the breathing pattern of the user (Swift 1989). Account must be taken both of the generative process of the aerosol and the breathing pattern of the user. A mismatch between these may result in non-optimal conditions under which either aerosol is generated but not breathed or breathing takes place without sufficient aerosol being provided. Since the object of aerosol therapy is to deliver an optimum amount of agent to a specified respiratory tract location, non optimal conditions can result in the potentially undesirable conditions of underdose or overdose.

In the case of dry-powder aerosol generators, the delivery system is simply a sutiably sized tube to be inserted into the oral or nasal passage. Ideal coupling between breathing and aerosol generation is ensured by making generation of the aerosolized powder breath-actuated. The only condition of failure occurs when the user cannot achieve a high enough inspiratory flow rate to aerosolize the dry powder. Exhaling through the system before inhalation could humidify the powder and cause particle adhesion, reducing the ability to aerosolize individual particles; therefore, such devices usually include some means to reduce or obviate such conditions.

Jet and ultrasonic nebulizers normally run at a constant flow rate, thus the necessity to provide a path for exhaled air and aerosol generated during the exhalation phase. Because inspiratory flow rate varies, it is also necessary to provide for an extra "makeup" air when the inspiratory rate exceeds the generation flow rate (and vice versa). In some jet nebulizer systems, an air pressure bypass is provided to permit generation only during inspiration; the user simply covers the bypass by finger pressure during the inspiratory phase to produce aerosol. This does eliminate aerosol generation during exhalation, but does not provide for flow matching during inspiration. Since ultrasonic generation does not require a constant flow for aerosol generation, the flow through the generator can be inspiratory-flow-driven, resulting in some temporal variation in aerosol mass concentration.

The temporal generation pattern of MDIs is quite different from the above types; when the actuator is pressed, an aerosol is formed during a very short period of time ($\sim$ 0.1 s). This aerosol generation phase is accompanied by a very high air flow rate resulting from the rapid evaporation of the propellant and the induction of air. It is impossible to match the temporal flow pattern of the MDI to the inspiratory flow of the user. Thus, if the canister is fired directly into the oral cavity through a short mouth tube, the particles, which consisit of the evaporating propellant and agent, have a very high velocity and a strong tendency to preferentially impact on the posterior pharyngeal wall. In studies of deposition of MDI aerosols in the human respiratory tract, it has been shown (Newman et al. 1982) that only 6–14% of the aerosol reaches the thoracic region of the respiratory tract distal to the trachea.

Efforts to reduce high pharyngeal deposition and improve the matching between aerosol generation and inspiratory breathing have centered on "spacers". These are conduits of various geometries and volumes that are placed between the MDI and the mouth into which the MDI is discharged. The aerosol is then inspired from the spacer (Moren 1985). The spacer is intended to allow the aerosol particles to both slow down and evaporate before they are inhaled, resulting in greater deep-lung delivery. It is observed that spacers do not result in greater thoracic deposition, but do reduce pharyngeal deposition; larger particles impact or settle in the spacer, and the remaining aerosol is inhaled with less resulting oral-cavity deposition. This delivery system allows generation and breathing to be temporally separated.

A special issue associated with delivery systems for therapeutic aerosols is product utilization. Many currently used drug substances are relatively expensive, and there is considerable motivation to reduce the cost by utilizing as much of the drug as possible. The ultimate measurement of utlilization is how much of the provided therapeutic substance ends up at the desired site in the respiratory tract during the breathing periods. Often, the utilization efficiency is a very small percent, even in the case of MDIs, in which practically all of the agent in the container is available for aerosolization.

Other therapeutic aerosol systems have lower utilization for a variety of reasons. Optimal design of generation–delivery systems should strive for high utilization as well as proper particle size (Swift 1989).

Measurement Methods for Therapeutic Aerosols Prior to Inhalation

The properties of therapeutic aerosols necessary to describe their behavior are similar to properties of other aerosols, i.e., size distribution (number, surface, and volume), number and mass concentration, electric charge, hygroscopicity, and the distribution of the active ingredient among the particles. Because inhaled particle transport and deposition depend on particle aerodynamic behavior in a flowing gas, particles in the inertial regime are properly characterized by their aerodynamic equivalent diameter. The conditions of generation, inhalation, and deposition in the respiratory tract are such that for isometric particles with $d_a \geq 0.5$ μm this is the appropriate size descriptor.

There are three major problems with therapeutic aerosols that complicate the measurement process. The first is that, in many instances, the distance between the generator and inhalation site of the aerosol is very short, and there is not an opportunity to place aerosol measuring devices in this location. It is conceivable to sample (remove) the aerosol being generated before it is inhaled, but the influence of sampling (for subsequent measurement) on inhalation of the aerosol is not known in general.

A second related issue is the matching of flows between generation, inhalation, and measurement involving a sampling flow. As discussed above, generation of therapeutic aerosols may be either at a constant flow rate (e.g., nebulizers) or under pulse flow conditions (e.g., MDIs). In the first case, the matching between the generation and sampling can be achieved, but the cyclical breathing pattern is not matched to either generation or sampling. In the case of pulse generation, neither sampling nor breathing can be properly flow-matched to the pattern of generation, and this leads to potential errors in aerosol characterization.

In the case of dry-powder aerosol generation, the generation is forced to match the inspiratory flow, but if sampling is envisioned for aerosol characterization, the sampling flow cannot readily be matched to the inspiratory flow. Thus, under any circumstance, it is not feasible to match generation, sampling, and breathing flows simultaneously.

The third problem of therapeutic aerosols that make aerosol characterization difficult is the unstable nature of the liquid aerosols produced by nebulizers and MDIs. Aqueous aerosols produced by jet or ultrasonic nebulizers are subject to rapid changes in particle diameter and other properties, due to evaporation or growth of aqueous particles as they seek equilibrium at the changing conditions of temperature and humidity within the delivery system. This is particularly true when the delivery systems encounter both inspiratory and expiratory flows; the expiratory flow consists primarily of air at near to the body core temperature and fully saturated, while aerosol generated by nebulizers is often at 5–7°C below the ambient temperature (Mercer 1981).

In the case of MDIs the aerosol emerging from the canister consists of propellant particles containing the active ingredient and other substances (e.g., surfactant). These particles are travelling at high velocity and are rapidly evaporating, with attendant cooling. The particle size distribution which might be measured at the front end of the distribution tube is likely not the same as the distribution at the distal end of the oral passage.

Because of these experimental difficulties in the measurement of aerosols entering the respiratory tract during actual breathing, the usual practice has been to connect the delivery system output directly to an appropriate instrument for aerosol measurement. The most common instrument used for nebulizer-generated aerosols is the cascade impactor (Chapter 11). This size separation device is operated at a constant flow rate, which can be set according to one of the several mutually exclusive criteria. The first is that of exactly matching the output flow rate of the delivery system so that all the aerosol generated enters the impactor, but this requires that the impactor be capable of such matching, a situation not possible with all commercially available impactors. Next, the flow rate into the impactor can be less than the output flow of the nebulizer, with the excess allowed to escape. Third, if the impactor flow rate exceeds the output from the delivery system, additional air must be drawn in, simulating the possible real situation during the breathing cycle when inspiratory flow exceeds nebulizer output flow.

As alluded to above, no one of these criteria is totally satisfactory when the inspiratory breathing is cyclical and the nebulizer output is constant, especially if the size and concentration of the aerosol depend strongly on the makeup air temperature and humidity. No impactor presently exists which can be operated predictably under cyclical flow conditions to match the breathing profiles.

A further difficulty with using cascade impactors to measure aerosols from nebulizers is the apparent evaporation of the particles taking place as they traverse the stages of the impactor. It is observed that when aqueous aerosols containing a solute such as 0.9% NaCl (isotonic with respect to extracellular fluids) are passed through a cascade impactor, the upper-stage deposition is that of liquid particles while the final stage(s) collect solid salt particles. The evaporation which takes place may be primarily due to the pressure drop through the impactor in conjunction with the increase in the vapor pressure of submicrometer-sized particles (Kelvin effect).

The situation when dry particle aerosols are generated by breath actuation presents many of the same difficulties when the cascade impactor is employed to measure the particle size distribution during actual or stimulated inspiration. A sample can be drawn from the delivery system to an impactor during this period of generation and delivery, but it would be normally done at constant flow, whereas the flow, particle concentration, size distribution, and static pressure are likely to vary markedly during the inspiratory period. The equal weighting of all time periods during inspiratory flow by the cascade impactor will not likely reflect the highly time-dependent character of the aerosol.

Aerosols produced from MDIs have similar problems when sampled by instruments such as cascade impactors during their passage from the canister to the respiratory tract. In this case, the very rapid evaporation of the propellant and the pulsatile velocity of the generated particles, as described above, are not consistent with conditions ideal for cascade impactor sampling.

In order to avoid the practical and theoretical difficulties and difficulties of sampling all these therapeutic aerosols in a "true" fashion outside the respiratory tract, many investigators have elected to deliver the aerosol into a large-volume chamber from which the residual aerosol, after a short period of evaporation or mixing, is sampled into a cascade impactor for size and mass analysis (Mercer, Tillery, and Chow 1968). If the mixed aerosol is sampled, integration over the time of generation has already been performed, and the aerosol sampled represents the mixed average throughout the entire period of generation. While this method offers a means of comparison for various methods and conditions of aerosol generation and delivery, the aerosol which is "eventually" sampled is the "fossil remains" of the aerosol which is delivered to the respiratory tract in the actual therapeutic application. It remains, therefore, for the investigator to calculate or estimate backward in time and space to the aerosol condition at the point of delivery in order to determine what aerosol actually entered the respiratory tract at a particular time during the inspiratory maneuver.

The use of size-selective instruments such as cascade impactors faces similar interpretive difficulties when other occupational environmental aerosols are sampled due to their labile nature and spatial and temporal variability. However, these difficulties seem to be more severe in the case of therapeutic aerosols because less time is available between their generation and inhalation compared to other situations.

Measurement of Therapeutic Aerosol Delivery within the Respiratory Tract

Because of the difficulties of therapeutic aerosol measurement outside the respiratory tract and because the objective of aerosol therapy is to deliver a specified quantity of an agent to some respiratory-tract region, it is often desirable to measure aerosol deposition within the respiratory tract. This has led to the development of an expanding technology of aerosol deposition measurement.

In almost all cases, this is accomplished by means of gamma-emitting radioisotopes, which are either physically or chemically attached to the aerosol particles inhaled. The isotopes emit gamma rays which are detected outside the body by a radiation detector (Taplin and Chopra 1978).

The type of technology used depends on the level of spatial resolution required to describe the location of the aerosol particle. In the simplest example, one might wish simply to estimate the quantity of aerosol deposited in the entire respiratory tract. Even then, it is usually necessary to separate the respiratory tract into the thoracic and extrathoracic compartments, the division line usually taken to be at the mid trachea. In this division, the extrathoracic airways include the upper trachea, larynx, pharynx, oral, and nasal passages. These are alternatively known as the head airways or (taking the division slightly more proximal) the nasal–oral–pharyngeal–laryngeal (NOPL) region. The simplest measurement of deposition in this extrathoracic region is obtained with a single scintillation detector, with some collimation to exclude thoracic deposition oriented either laterally or anterior posterior. Similarly, an estimate of the thoracic deposition can be made with a single scintillation detector aimed at the lungs and situated far enough from the chest to accept gamma rays from the lung periphery. Such measurements provide an estimate of the total amount of deposition in one or the other major region, without any further spatial information.

With this degree of spatial information, the amount of activity required to exceed background radiation is not great; subjects can inhale aerosol containing a radioisotope and receive a dose within the annual limit considered safe for the general population. In cases where the isotope used is ^{99m}Tc, several inhalations can be performed by the same subject, with a total dose still well below the annual exposure limit of 0.5 rem.

If a greater degree of spatial resolution is desired, the deposited aerosol can be detected by a gamma scintillation camera. Because of the collimation required to attain good spatial resolution ($\sim$ 0.5 cm) the amount of radioactivity required for a detailed image of the thoracic or extrathoracic region is much greater than with an uncollimated detector, and the dose received for a single study limits the number of studies that can be performed with the same subject to keep within the annual dose requirement. Even so, the information from such a study can only be grouped into about three concentric regions for each lung or three regions in the head airways. A planar (two-dimensional) image of activity in the lung does not allow a good degree of separation between the peripheral (alveolar) region and the medium and small bronchial airways since these regions overlap in such a planar view.

An even more detailed spatial look at the respiratory-tract deposition of aerosol can be obtained using the technique of SPECT (single photon emission computed tomography) (Phipps et al. 1989). This technique allows the activity in the lung or extrathoracic region to be assigned to three-dimensional volume elements, "voxels", the equivalent of two-dimensional "pixels", so that activity distributions in various planes can be visualized and quantified. This added spatial information is obtained at the cost of additional radioactivity, a measurement time for a single image of $\sim$ 20 min (compared to $\sim$ 5 min for a gamma-camera scintigram), and additional computer capacity to process all of the information from the multiposition scan about the region. The scan time of 20 min may be unacceptably long because rapid upper bronchial or extrathoracic airway clearance is taking place. For example, aerosol deposition onto the mucociliary surface of the nasal passage is normally cleared to the pharynx and swallowed within about 15 min. The SPECT procedure is suitable for aerosol deposition and clearance measurement only when the temporal scales of clearance are much greater than the scanning time.

In some cases of therapeutic aerosol delivery to the respiratory tract the spatial distribution is not necessary, and the uptake of the agent can be monitored by serial measurements of blood or urine concentration of the agent or its metabolites (Walker et al. 1972). Likewise, in the extrathoracic region, the aerosol deposition can sometimes be measured by washing out the deposited material for quantitative analysis of an appropriate tracer. For example, it is well known that MDIs deliver large fractions of drug to the posterior pharyngeal wall, which can readily be removed by oral rinse and gargle for analysis.

Nonideal Behavior of Therapeutic Aerosols

The quantitative measurement of aerosol deposition of therapeutic agents is important because in many instances it is difficult to predict how therapeutic aerosol will behave based on well-established models of respiratory deposition. This is because such models are based on a number of simplifying assumptions defining an "ideal" aerosol. This means that the particles are of spherical shape, are solid, nonhygroscopic, nonevaporating, noninteracting, and are present in the air at moderate concentration, and are travelling through the airways essentially at the velocity of the inspired or expired air. In most therapeutic aerosols, one or more of these conditions does not hold. (Newman 1984).

For example, the aerosol from an MDI begins as a very high velocity jet of large, rapidly evaporating propellant particles at high concentrations. After losing most of the propellant, the particles are often hygroscopic and, thus, subject to growth by accretion of water at the temperature and humidity of the central airways. The jet begins in the center of the inhaled air and may not be spatially uniform during part of its path. Because of its short duration, the inspired dose is an aerosol bolus, not a continuous volume of aerosol, subject to dispersion within the respiratory tract. Other therapeutic aerosols consisting of liquid droplets or dispersed solid particles are likewise "nonideal" to a degree as a result of their physicochemical properties and/or their mode of generation and exposure.

An additional factor that makes the prediction of therapeutic aerosol deposition difficult is the nonideal nature of the respiratory tract in many people using such aerosols to treat respiratory diseases. Even in "normal" individuals, aerosol deposition exhibits significant biological variability, but in many individuals using therapeutic aerosols, lung morphology and/or breathing pattern is markedly abnormal. For example, individuals who have cystic fibrosis inhale bland saline aerosols to "liquify" the secretions. However, observations suggest that the preferential sites of deposition of these aerosols are in the "healthy" regions of the lung while the diseased regions receive little aerosol because they are poorly ventilated (Alderson et al. 1974). Much work remains to be done to clarify this situation.

AEROSOL MEASUREMENT OF DIAGNOSTIC AEROSOLS FOR INHALATION

Types of Diagnostic Aerosols

Diagnostic aerosols differ from therapeutic aerosols in that the intention for their use is to learn something about the state of the respiratory tract rather than the delivery of an agent to produce a specific local or systemic effect (Wagner 1976). It is usually important to characterize properly the properties of such a diagnostic aerosol, since the information to be determined may be critically dependent on the quantity and site of deposition of the aerosol.

There are two widely used diagnostic procedures involving aerosols: the measurements of pulmonary ventilation and those of airway reactivity. The remainder of diagnostic methods employing aerosols are rather more specialized and, thus, limited to research laboratory settings where special equipment and expertise is available.

Ventilation measurements using aerosols may have several purposes, including the diagnosis of pulmonary embolism (PE) and the identification of regions of either acute or chronic airway constriction. Ventilation is defined as the actual quantity of air in various compartments of the lungs compared to what would be expected if the lung were operating normally with respect to the inspired air. The measurement of ventilation for diagnosis of PE is carried out in tandem with a lung perfusion measurement. Lung perfusion is likewise defined as the quantity of blood delivered to various vascular compartments of the lung compared to the expected quantity. A specific region of the lung which has both a ventilation and a perfusion defect, as evidenced by a significantly reduced amount of radioactivity of both aerosol and vascular-injected labelled microspheres, is considered a positive indication for PE. The use of aerosols for ventilation measurements is not universal; many clinicians avoid using aerosols (despite their superiority in several respects) and prefer to measure ventilation with ^{133}Xe or ^{82}Kr gas. While these gas methods do not require aerosol generation equipment, the radiation dose required to obtain several views of the thoracic region for thorough analysis is greater than that for aerosols. In the case of aerosol use, one administration of the radiolabelled aerosol is adequate for all views which can be taken sequentially. It is common to use a jet nebulizer (below) for such aerosols.

The other common diagnostic aerosol procedure is the measurment of airway hyperreactivity, often called bronchoprovocation. In this method aerosol containing a specific activating substance, usually methacholine or histamine, is administered to the subject in progressively concentrated solutions. Measurement of pulmonary mechanical behavior is made after each administration. When the mechanical behaviour has changed by a significant amount, the test is stopped, noting the total dose required to achieve the change. It is found that allergic and asthmatic individuals have a markedly greater sensitivity to these agents compared to individuals who have no such clinical history, and the test is a convenient means to quantify the degree of sensitivity. Other environmental agents may modulate the response of sensitive individuals. This provides a convenient means for investigating the interaction between common environmental airborne agents (indoor or outdoor air pollutants) and specific reactivity producing agents or the conditions to which individuals may be exposed.

Aerosols are also employed in diagnostic procedures for measuring the permeability of the alveolar membrane, the degree of constriction of medium and small bronchi, the mucociliary system, and the mixing of tidal and residual air in the lungs (Agnew 1984). All these methods are rather more specialized, requiring a greater degree of understanding of the aerosol characertistics and the means for delivery. Aerosols used for alveolar permeability must not only have the proper biological properties to differentiate high from low permeability, but must be of such size and nature to achieve predominantly alveolar deposition. High degrees of bronchoconstriction measured by aerosol penetrance, the ratio of the quantity of aerosol in the lung periphery to the quantity in the central hilar zone, are difficult to make without well-defined aerosols and techniques such as gamma-camera scintigraphy or SPECT (see above). Measurements of lung mucociliary clearance are usually made with radioisotope-labelled particles which are not readily dissolved in the lung fluid or otherwise suffer transport away from the airway epithelium by means other than mucus transport. The particles must deposit predominantly on ciliated airway surfaces. Similarly, measurements of mucociliary clearance in the nasal airway requires particles to deposit or be transported to ciliated epithelium. Aerosol mixing experiments require that the aerosols suffer minimal respiratory deposition, a condition that can be realized with a highly monodisperse aerosol of a specific size range.

Generation of Diagnostic Aerosols

For the commonly used aerosol diagnostic methods, it is usual for the aerosol to be generated by a jet or ultrasonic nebulizer, with the liquid containing the radioisotope in solution or colloidal suspension (Newman 1984). Although this may not be the ideal method, the equipment and expertise required to generate stable monodisperse designed aerosols for specific procedures is not usually available. It is likely that the aerosol generated for bronchoprovocation tests has a large mass fraction which deposits in the alveolar region; even so, the test is reasonably reproducible for a single subject. Bronchoconstriction aerosol measurements are also made with polydisperse radiolabelled aerosols generated from jet or ultrasonic nebulizers. If the mucociliary clearance of large bronchi is to be measured, large-diameter particles ($d_a > 5$ μm) must predominate in the aerosol which must be breathed at high inspiratory rate to achieve significant deposition at bifurcations of large airways. Under such conditions, a significant fraction of the aerosol is deposited in the posterior oral cavity, reducing effective delivery. Nebulization of suspended large particles (3–10 μm diameter) can be employed for such procedures. For mixing studies, it is necessary to measure aerosol concentration profiles during inspiration and expiration. This is normally accomplished by photometric light scattering measurement of a monodisperse aerosol generated by an evaporation condensation (Sinclair–LaMer) generator. This method of generation requires expertise in determining the degree of aerosol monodispersity.

Delivery Systems for Diagnostic Aerosols

Most delivery systems for diagnostic aerosols are similar to certain delivery systems for therapeutic aerosols. The primary difference in most cases is that the material limitations on substances used for diagnosis are usually less stringent. The exception is that the utilization of radioisotopic aerosols used in diagnosis is very low. Thus, most of the isotope remains behind after the procedure. This is a situation which may be costly or involve high activity, which must be shielded during handling and use.

In some diagnostic procedures it is important to measure the breathing characteristics during delivery to estimate aerosol deposition mass. Since diagnosis is usually done in a clinical laboratory, the space and portability features of delivery systems which are important for some therapeutic aerosols are not a factor (Heyder 1988). The MDI is rarely, if ever, used in diagnosis employing aerosols. In bronchoprovocation aerosol tests the usual delivery system is a jet nebulizer operating at a constant flow rate, but allowing the subject to inhale at whatever rate is convenient. This is similar to the delivery systems for therapeutic aerosols such as pentamidine for the treatment of pneumocystic carinii pneumonia (PCP) secondary to acquired immunodeficiency syndrome (AIDS).

Control of breathing volume or rate is sometimes required in diagnostic procedures in order to assure aerosol delivery to a specific depth or region. Aerosol properties, especially of aqueous or hygroscopic aerosols, may change during breathing or be different, depending upon the amount of dilution room air added to the output of the aerosol generator during inhalation. As with therapeutic aerosols, it is often difficult to measure directly the aerosol properties entering the respiratory tract because the distance from the generator to mouth is very short. Likewise, directing the aerosol output into an aerosol size classifier (e.g., cascade impactor) may not correctly simulate its behavior in the airways.

The ideal situation for a diagnostic aerosol application is to produce an aerosol of known properties in a reproducible and reliable fashion so that the measured property reflects changes in the desired respiratory feature, not the variability in the aerosol being used. The delivery systems design should reflect this ideal to the extent possible.

Specific Measurement Techniques for Diagnostic Aerosols

Because many diagnostic aerosol systems are rather similar to therapeutic aerosol systems, the measurement methods used, such as gamma-camera scintillation detectors, are often similar. In some diagnostic procedures, the time scale is unique to the measurement. For example, in the measurements of mucociliary clearance, periods from 6–24 h may be required to construct a clearance curve of thoracically deposited aerosol activity (Pavia 1984). In some diagnostic aerosol applications, such as mixing of tidal and reserve air, the aerosol must be monodisperse and have a diameter that has minimum deposition during normal breathing, approximately 0.5 μm.

Measurement of aerosol concentration instantaneously throughout the inhalation–exhalation cycle is necessary to analyze aerosol mixing. This can be accomplished either by measuring the aerosol concentration with a photometer or with an optical particle counter. Placing the photometer just outside the mouthpiece leading to the aerosol storage chamber on one side and the respiratory tract on the other permits the measurement to be made without any losses other than those occurring within the respiratory airways.

Such measurements of aerosol mixing can be made in two ways: (1) by providing aerosol throughout the inspiratory phase and exhaling all the way to residual volume to determine how much aerosol transports to the reserve volume; (2) by introducing the aerosol as a limited-volume bolus somewhere within the tidal inspiration. The methods are physically and theoretically equivalent, but in practice the bolus technique appears to give better results. For example, when the bolus is placed such that it readily penetrates to the pulmonary region, a readily observable difference between normal nonsmokers and smokers in terms of bolus dispersion can be demonstrated.

Determination of lung deposition patterns during normal breathing of radiolabelled aerosols is a useful technique for measuring the degree of central airway constriction (Agnew 1984). The measure of penetrance, determined either by planar or SPECT scintigraphy, is found to be a sensitive indicator of constriction and has the added potential advantage of locating the sites of constriction by deposition at hot spots (Phipps et al. 1989). This gives aerosol techniques advantages over the conventional pulmonary functional tests such as specific airway resistance or the ratio of one-second forced expiratory volume (FEV_1) to vital capacity (VC), in that these conventional test are global (i.e., have no spatial information about the constrictive site).

AEROSOL MEASUREMENT OF NONINHALED THERAPEUTIC AEROSOLS

Some therapeutic agents appear in aerosol form although they are not intended to be inhaled. Usually, these aerosols are intended to be transported to some surface where their therapeutic action occurs. This was the case with Lister's aerosol generation of phenol, which presumably had its antiseptic action by depositing on biological and other surfaces to render harmful pathogens impotent. It is not imagined that inhalation of the phenol had any therapeutic benefit; in fact, it was ultimately abandoned because of the irritant effect of the aerosol on the eyes and mucous membranes.

Such agents which may be used at present to deliver therapeutic agent by aerosol to skin and other body surfaces (such as during surgery) must have as desirable characteristics a high probability of being directly deposited on the intended surface, with a minimum of "overspray" which might lead to other undesirable effects or (at minimum) less than optimal utilization of the agent.

Agents could be in the form of dry dusts, solutions or suspensions, but in order to achieve even coverage of a surface, the use of propellants as "drivers" and dispersers of the agent from pressure-packaged canisters ("aerosol" cans) seems the most ideal approach. As has been detailed above, the aerosol emerging from such a device is composed primarily of large propellant drops containing the active substance and travelling at high velocity ($\sim 40\,\mathrm{m\,s^{-1}}$). According to the theory of inertial deposition, these are the ideal conditions for a high percentage of the droplets to deposit on a surface perpendicular to the spray axis and near enough to prevent total evaporation and deceleration by viscous drag.

The major disadvantage of this method of administration of aerosols to biological surfaces is the cooling effect on the surface as a result of the propellant evaporation. However, this is not ordinarily a significant problem if a distance is maintained to allow some evaporation to take place in the gas phase before deposition. In fact, some applications of propellant-driven droplets are actually intended to retard inflammation and pain by "freezing" the tissues, similar to the effect of cold water or other cooling upon burns.

Little information exists in the scientific literature on the optimal conditions for therapeutic aerosol delivery to surfaces, but manufacturers of these and other consumer products applied to surfaces by propellant atomization (e.g., waxes, cleansers, polishes, and hair sprays) have determined by experiment the conditions maximizing even-spray application and minimizing overspray. Some of these products are irritant to some individuals when inhaled. Notable among these products are nonstick spray containing lecithin and certain hair sprays.

Therapeutic skin sprays contain agents such as bactericides, steroids for control of inflammation and itch, antibiotics and antifungal agents. In their normal use, the lifetime of the aerosol between generation and surface deposition is short enough to render standard methods of aerosol sampling and measurement ineffective. Aerosol overspray of such applications can be characterized in a standard form using area cascade impactors, filter cassettes or personal aerosol samplers with cyclones to determine the respirable fraction. However, the irritant effect of such aerosols is likely to be observed for any region of the respiratory tract.

In summary, no aerosol measurement techniques unique to the noninhalation aerosol area have been developed in spite of the importance of proper minimization of overspray for many of the agents used.

INADVERTENT EXPOSURE TO AEROSOLS IN HEALTH CARE

There are a number of potential and actual aerosol exposure situations in the health care field which are unintended and can result in a risk of ill effect. In such cases, aerosol measurement provides the estimates of potential exposure and expected benefit of exposure control methodology. Some of these instances involve chemical therapeutic agents which have potent biological effects while other instances involve living organisms or their bioactive remains which may result in infections, transmission of disease, or other ill effects.

The anecdotal observations of airborne (aerosol) transmission of disease has an illustrious history dating back at least to the Greeks, despite their lack of knowledge of the nature of the agents (usually attributing such transmission to "bad airs"). Modern biochemistry, physiology, bacteriology, virology, and parasitology have provided the tools to understand the nature of the airborne transmission of many diseases ranging from the common cold to tuberculosis. The understanding of immunologic defense mechanisms has led to the development of vaccines and other strategies to prevent or attenuate the harmful spread of these diseases.

Despite these advances, the appearance of "new" diseases such as AIDS, the proliferation of potent therapeutic agents such as antitumorals, and the increased processing of human and other biological tissues and fluids in diagnostic and research laboratories has made an understanding of airborne exposure and transmission as important now as in the days when tuberculosis, influenza, and pneumonia were responsible for pandemics.

Human–Human Aerosol Transmission of Disease

It is a well-established observation that aerosols are produced from the respiratory tract by a number of means including sneezing, coughing, conversation, and singing (Wells 1955). Particles thus emitted have a wide size range, the larger of which ($d_a > 10\ \mu m$) settle rapidly to horizontal surfaces. Smaller particles emitted as such or resulting from evaporation remain airborne for longer time periods. These "droplet nuclei," which may contain any of the numerous biological organisms, are the means by which disease is transmitted over distance by air between humans. Depending upon their environmental "hardiness", such droplets may even be transmitted through ventilation systems and infect individuals quite remote from the source individual (Riley 1974).

Two approaches toward aerosol characterization have been used for such agents: (1) the number concentration of organisms collected as a suitable medium and determined by colony (or equivalent) counts; and (2) the empirical number of "infectivity units," each of which contains the minimum number of organisms needed to produce an active disease case. In some cases this may be only a single viable organism, while in other cases many organisms are needed. Control of transmission by air dilution, killing of organisms, or physical removal (filtration) can be measured by the reduction of organisms or disease units by standard measurement methods. Various schemes for killing organisms have been proposed. For example, reduction of viable aerosol particles by ultraviolet light has been quantified with liquid impingers or agar-plate cascade impactors (see Chapter 21).

With extensive use of vaccines and effective chemotherapy for tuberculosis, it was thought, not long ago, that airborne transmission of disease in developed countries was limited to the common cold, influenza, and other treatable illnesses. However, the reoccurrence of tuberculosis as a side effect of AIDS and other immune deficiency conditions has revived the concern, particularly among health care personnel, that more serious effects of airborne transmission may still occur. Although there is no evidence that AIDS itself is transmitted by the airborne route, there are other opportunistic pathogens which deserve attention with respect to airborne spread. Thus, viability techniques to assess the magnitude of the problem and the effect of control technology employing aerosol measurement need to be developed (Riley 1972).

Inadvertent Exposure to Therapeutic-Agent Aerosols

There is a growing concern that the production, handling, and administration of therapeutic agents may pose health hazards for health care workers and others who routinely or infrequently are exposed by several routes, including the formation of aerosols. Measurement techniques for bioaerosols which result during bioprocessing are discussed in Chapter 21. However, it is important to emphasize that the rapid development of new therapeutic agents (many of which result from developments in genetic engineering and molecular biology) should be accompanied by an examination of the consequences resulting from exposure of pharmaceutical workers and health care providers to these agents. These include not only the agents for treating human disease, but an increasingly expanding array of agents used in animal husbandry and veterinary medical practice. Means must be developed to assure that exposure, including that by airborne particles, is controlled to minimize or eliminate ill effects.

The detection and quantification of airborne concentration of such substances can be carried out with standard aerosol collection techniques as long as the collection substrates do not interfere with the sensitive and selective bioassay methods available for many of these substances. One of the advantages of environmental sampling for such agents is that their biological potency permits highly sensitive and selective detection tests to be used; these detection capabilities are already in place in many laboratories where the agents are employed.

Inadvertent Exposure to Bioaerosols from Tissue, Cell, and Fluid Tests

A concern similar to that referred to above is the increasing number of routine tests being carried out with tissues, cells, and body fluids (including blood) of living organisms (mainly human) and the possibility that many routes of infection may contribute to disease transmission to health care workers (Zimmerman et al. 1981). Although high on the list of transmission routes is the introduction of substances directly into the blood via inadvertent needle sticks, etc., the possibility of aerosol formation and transmission in such settings has received considerable attention. Numerous processes such as centrifugation, stirring, pipetting, etc., can lead to aerosol formation as well as poor housekeeping procedures such as spilling liquids which may dry and become aerosolized. Even procedures which are intended to minimize other transmission routes, such as hypodermic needle clipping, can produce aerosols directly or result in fluid spills on surfaces (Binley et al. 1984).

These possibilities have achieved heightened public interest with the advent of AIDS. Even without the fear of AIDS transmission, the possible transmission of much more robust organisms such as hepatitis virus ought to be a reason enough to understand and develop control measures to eliminate such pathways. The health literature contains several occupational epidemiology studies demonstrating that transmission takes place in such settings, but little environmental sampling including the airborne route has been done. As with therapeutic agents, good microbiological techniques for the detection of most organisms are available and can be combined with appropriate aerosol sampling technology when indicated.

Inadvertent Exposure to Bioaerosols in Dental Procedure

It has been known for some time that many dental procedures such as tooth drilling, abrasive cleaning, and certain oral surgical procedures can produce airborne particles containing irritant dust and pathogenic flora from the mouth (Macdonald 1987). The newer technologies being introduced into dentistry, such as laser cutting, water abrasive cleaning, and dry powder abrasive cutting, are all sources of aerosols which could pose hazards for the dental patient, the dentist and staff, and other patients being treated in the same area (Pagniano et al. 1986). Greater drilling efficiency with very high speed drills produces finer-size dust and the escape of this dust from the mouth due to its high velocity leads to concern for aerosol exposure. Aerosol samples taken in dental suites have demonstrated the existence of viable and pathogenic airborne organisms. This has led to the introduction of exhaust ventilation and other methods to eliminate or minimize exposure. A pratical problem in developing such exhaust systems is the difficulty of getting good capture efficiency without interfering with access to the work area.

With new techniques and materials, it is important to evaluate the degree of aerosol exposure (as well as other routes) in dental practice and minimize the risk of infection. The amount of inhaled aerosol producing an effect may be very small for some highly infectious and pathogenic organisms.

Inadvertent Exposure to Aerosols in Surgery

Advances in surgical techniques, as in dentistry, are sometimes accompanied by new modes of transmission, affecting not only the patient but also health care workers. This has been demonstrated in the case of the increasing use of lasers and electrocautery in many surgical procedures. Both these techniques, as well as bone and tissue cutting procedures, are capable of producing airborne particles which can transmit infections and pathogenic organisms both from one region of the patient to another and from the patient to other individuals (Hoye, Riggle, and Ketcham 1968).

This is a reversal of the traditional concept of infection transmission, where concern was (and still is) centered on preventing infection agents arising from surgeons or other operating-room personnel from gaining access to the patient via an open surgical wound. This was attempted (not always successfully) by using clean clothing, skin cleansing and disinfection, surgical masks, and (more recently) filtered laminar flow of air over the site of surgery as well as the application of disinfecting materials directly upon the wound. This becomes a difficult task, especially in large-area surgical procedures such as hip replacement or open-heart surgery.

Aerosol characterization of smokes and other aerosols produced during surgery is very limited and mostly anecdotal at present, but it has been demonstrated that viable organism can be aerosolized by these procedures (Merritt and Myers 1991). Since both lasers and electrocautery produce high temperatures at the site of cutting, it might be questioned how aerosol generated by such processes can indeed remain viable. It appears that boiling of tissues adjacent to the cutting site results in the emission of cells and tissue fragments into the air. No studies linking particle size to the nature of the viable organisms have been performed although such investigations do not appear to be beyond the present capability of aerosol sampling and analysis by one of the several biological tests for viable species.

Inadvertent Exposure to Aerosol in Aerosol Therapy

There are several therapeutic aerosol procedures currently used in which aerosol production, either from the therapeutic agent source or from the patient, poses a concern for health workers (Arnold and Buchan, 1991). One of these which has achieved some attention is the increasing use of aerosol pentamidine in the treatment of AIDS-related pneumocystis carinii pneumonia (PCP). This is presently carried out in hospitals where several patients are normally placed in a room, each with an aerosol delivery system containing an aqueous solution of pentamidine. Although it is the intent of the systems to deliver all the pentamidine to the patient's alveolar compartment, it is often the case that pentamidine aerosol escapes the delivery system and is transported throughout the room. Health care workers in the room and nearby are exposed to this aerosol, which is designed to be accessible to the deep lung, and there are anecdotal reports of cough and irritation among health care workers (Boulard 1991). The possible long-term effects of this exposure are unknown.

Aerosol measurements of the pentamidine have been performed in a few instances employing fairly crude means for area aerosol sampling, and these confirm that pentamidine aerosol of inhalable size is present. Several methods to reduce exposure have been proposed, but none has been evaluated; these include increased room ventilation, isolation of the patients in booths containing a separate air supply and filtered exhaust, and respiratory protection devices for the health care personnel. The disadvantages of all these proposed control methods have been recognized, but ideal solutions which do not entail high cost or questionable benefits have not been proposed.

Another widely used aerosol treatment method is the administration of saline aerosol to induce sputum production for the collection of bronchial cells and fluid and cytological examination. It is well known that the coughing which accompanies the aerosol administration and aids in bringing the bronchial secretions upward also produces aerosols which may contain pathogenic organisms. As above, little quantitative aerosol characterization has been performed to test the control methods to reduce exposure.

It is likely that in the future many more drug substances will be administered to patients in similar settings and it is important that aerosol measurement technology be employed to evaluate both the effectiveness of the treatment and the possibility that health care workers may be inadvertantly exposed at harmful levels.

FUTURE AEROSOL MEASUREMENT NEEDS IN HEALTH CARE

It is clear that advances in many of the health care areas alluded to above carry with them both the possibility of increased effectiveness of treatment and diagnosis employing aerosolized substances and the possibility of undesirable exposure of health care workers to biologically potent aerosol agents. Aerosol measurement technology has an important part to play in these developments both to guide in the most effective and efficient use of therapeutic and diagnostic aerosols and in the evaluation of the degree of exposure and the effectiveness of exposure control strategies.

There is a long history of combining effective aerosol measurement and biological assay techniques to characterize aerosols of importance to health. Some of this effort has been directed toward health care worker protection, but much more has been directed toward workers in the "dusty trades". The design of methods to deliver therapeutic and diagnostic aerosols has borrowed heavily from the knowledge base developed by industrial hygienists, health phyicists, and air pollution researchers.

These interactions ought to be fostered by continued professional contacts through conferences, journals, and collaborative research projects, and ought to be supported both by public and private funding. The results of these efforts will be health practice beneficial to health care consumers and health care providers, both in cost and life quality. These developments give further strength to the widening role of aerosol science and technology, including aerosol measurement, as a scientific discipline and practice which touches many important fields of endeavor.

References

Agnew, J. E. 1984. Aerosol contributions to the investigations of lung structure and ventilation function. In *Aerosols and the Lung*, eds. S. W. Clarke, and D. Pavia. London: Butterworths.

Alderson, P. O., R. H. Secher-Walker, D. B. Strominger, J. Markham, and R. L. Hill. 1974. Pulmonary deposition of aerosols in childern with cystic fibrosis. *Pediatrics*, 84;479–89.

Arnold, S. D. and R. M. Buchan. 1991. Exposure to ribavirin aerosol. *Appl. Occp. Environ. Hyg.* 6:271–79.

Bell, J. H., P. S. Hartley, and J. S. G. Cox. 1971. Dry power aerosols I: A new powder inhalation device. *J. Pharm. Sci.* 60:1159–64.

Binley, R. J., D. O. Fleming, D. L. Swift, and B. S. Tepper. 1984. Release of residual material during needle cutting. *Am. J. Infection Control* 12:202–88.

Boulard, M. 1991. Evaluation of health care worker's exposure to aerosolized pentamidine in the treatment of HIV-positive individuals. Paper at 1991 Am. Ind. Hyg. Conf., Salt Lake City, VT, 18–24 May.

Heyder, J. and G. Rundolph. 1984. Mathematical models of particle deposition in the human respiratory tract. *J. Aerosol Sci.* 15:697–707.

Heyder, J. 1988. Assessment of airway geometry with inert aerosols. *J. Aerosol Med.* 1:167–71.

Hoye, R. C., G. C. Riggle, and A. S. Ketcham. 1968. Laser

destruction of experimental tumors: State of the art and protection of personnel. *Am. Ind. Hyg. Assc. J.* 29:173–80.

Jaegfeldt, A., J. Anderson, E. Trofast, and K. Welterlin. 1987. A new concept in inhalation therapy. In *Proc. Int. Workshop on a New Inhaler*, eds. S. Newman, F. Moren, and G. Kropmton. London: Medicom.

Kjellman, N. and B. Wirenstrand. 1981. Letter to the editor. *Allergy* 36:437–38.

Macdonald, G. 1987. Hazards in the dental workplace. *Dental Hyg.* 61:212–18.

Mercer, T. T. 1981. Production of therapeutic aerosols: Principles and Techniques. *Chest* 80: Supplement 1, 813–18.

Mercer, T. T., M. I. Tillery, and H. Y. Chow. 1968. Operating characteristics of some compressed air nebulizers. *Am. Ind. Hyg. Assoc. J.* 29:66–78.

Merritt, W. H. and W. R. Myers. 1991. Real time aerosol monitoring during laser surger. Paper at 1991 Am. Ind. Hyg. Conf., Salt Lake City, VT, 18–24 May.

Moren, F. 1985. Aerosol dosage forms and formulations. In *Aerosols in Medicine*, eds. F. Moren, M. T. Newhouse, and M. B. Dolovich. Amsterdam: Elsevier.

Newman, S. 1984. Therapeutic Aerosols. In *Aerosols and the Lung*, eds. S. W. Clarke and D. Pavia. London: Butterworths.

Pagniano, R. P., R. C. Scheid, S. Rosen, and F. M. Beck. 1986. Reducing airborne microbes in the pre-clinical dental laboratory. *J. Dental Education* 50:234–35.

Pavia, D. 1984. Lung mucocilliary clearance. In *Aerosols and the Lung*, eds. S. W. Clarke. and D. Pavia. London: Butterworths.

Phipps, P., I. Gonda, D Bailey, P. Borham, G. Bautovich, and S. Anderson. 1989. Comparison of planar and tomographic gamma scintigraphy to measure the penetration index of inhaled aerosols. *Am. Rev. Resp. Dis.* 139:1516–23.

Riley, R. L. 1972. The ecology of indoor atmospheres: Airborne infections in hospitals. *J. Chron. Dis.* 25:421–30.

Riley, R. L. 1974. Airborne infection *Am. J. Med.* 57:466–75.

Swift, D. L. 1989. Design of aerosol delivery systems to optimize regional deposition and agent utilization. *J. Aerosol Med.* 2:211–20.

Taplin, G. V. and S. K. Chopra. 1978. Inhalation lung imaging with radioactive aerosols and gases. *Prog. Nuclear Med.* 5:119–20.

Wagner, H. N. 1976. The use of radioisotopes techniques for the evaluation of patients with pulmonary disease. *Am. Rev. Resp. Dis.* 113:203–18.

Walter, S. R., M. Evans, M. E. Richards, and J. W. Paterson. 1972. The fate of (^{14}C) disodium cromoglycate in man. *J. Pharm. Pharmacol.* 24:525–31.

Wells, W. F. 1955. Airborne contagion and air hygiene. Cambridge: Harvard University Press.

Zimmerman, P. F., R. K. Larsen, E. W. Barkley, and J. F. Gallelli. 1981. Recommendations for the safe handling of injectable antineoplastic drug products. *Am. J. Hosp. Pharm.* 38:1693–95.

38

Inhalation Toxicology: Sampling Techniques Related to Control of Exposure Atmospheres

Owen R. Moss

Chemical Industry Institute of Toxicology
Research Triangle Park, NC, U.S.A.

INTRODUCTION

In contrast to environmental and industrial monitoring, a critical consideration for aerosol, gas, and vapor measurement in support of inhalation toxicology research is that the measurement process must not significantly change the concentration or total airflow through the system. Control of exposure conditions must be made to ensure an accurate estimation of the inhaled dose and, subsequently, the best possible correlation between the concentration of the material presented and the biological response at the molecular, cellular, and tissue level.

In this chapter, the goal of sampling without changing the chamber concentration and the total airflow in the system is discussed, with the understanding that the preceding chapters in this book have provided the technical basis for accurate and effective sampling. The basic structure and properties of the exposure systems used in inhalation toxicology are described. Generic sampling techniques are then presented for providing flexibility in the use of a variety of sampling instrumentation. Finally, the sampling techniques are incorporated into basic sampling strategies that meet the goal of no change in concentration and a minimum change in flow during the measurement of exposure concentration of aerosols and/or gases. For a broader background in the measurement and control of exposure atmospheres related to inhalation toxicology, the reader is referred to the latest editions of texts by Phalen (1984), McClellan and Henderson (1989), the American Conference of Governmental Industrial Hygienists (Hering 1989), and the materials and methods sections in journals such as Inhalation Toxicology, Journal of Aerosol Science, Aerosol Science and Technology, Journal of Aerosols in Medicine, Toxicology and Applied Pharmacology, and Fundamental and Applied Toxicology.

BASIC EXPOSURE ATMOSPHERE GENERATION AND CONTROL SYSTEMS

The types of exposure systems used in inhalation toxicology are whole-body exposure chambers, nose-only and head-only exposure systems, or single- or multiple-animal closed-loop metabolism chambers. The construction and operation of these devices have been described elsewhere (Phalen 1984; Hinner, Burkart, and Punte 1968; Moss 1989;

McClellan and Henderson 1989). In this section the basic properties of exposure systems are discussed, including the properties of well-mixed chambers, laminar-flow chambers, nose-only exposure systems, and closed-loop exposure systems, the latter mainly being used to determine the metabolism rates of inhaled material. This section is concluded with a discussion of several properties of the exposure systems; the concept of air change per hour, the concept of half-time in a well-mixed system, and the concept of half-time in a laminar flow or plug flow system.

Whole-Body Exposure Chamber

Whole-body exposure chambers are either well-mixed systems or laminar-flow or plug flow systems (Fig. 38-1). In a well-mixed system, it is assumed that any incremental volume of air entering the chamber will be instantaneously and completely mixed with all other volume elements present (in the chemical engineering literature, this is known as a continuous stirred-tank reactor). With this assumption, the exponential buildup and decay curves described by Silver (1946) can be calculated. The equations are based on the basic exponential relations used to predict the reaction rates in continuously stirred-tank reactors and the heat transfer and mixing rates in heating and air-conditioning systems. Airflow through a well-mixed or completely mixed whole-body exposure chambers is produced in either a "push," a "pull," or a "push–pull" mode. Energy (to create compression) is used to push exposure air into the inlet of the chamber, and/or energy (to create suction) is used to pull exposure air from the outlet of the same chamber. In the "push–pull" mode of operation two driving systems, one on the inlet side and the other on the exhaust side, are adjusted to maintain a proper atmospheric pressure inside the chamber. In a "pull" system, energy is used to pull exposure atmosphere from the chamber while the total input into the chamber is partially supplied from a pressure-driven exposure atmosphere generation system and partially from passive suction of air from the room or supply duct into the chamber. Although seldom used, a "push-only" system is operated by pushing exposure atmosphere into the chamber and letting the increased pressure drive air out of the chamber, through filtration systems and into an exhaust area. The latter approach is not commonly used because it would result in the chamber being operated above atmosphere pressures, increasing the possibility for contamination of the surrounding room and personnel. The "pull" system naturally operates at lower than ambient pressures and the "push–pull" system is generally adjusted to ensure that the chamber is at pressures lower than ambient. Thus, if there are leaks in this system, room air will be sucked into the chamber preventing the exposure atmosphere from escaping.

Laminar-Flow (or Plug Flow) Exposure Chamber

In a laminar-flow or plug flow exposure chamber, an incremental element of exposure air volume, upon entering the chamber, maintains its uniqueness throughout its passage through the chamber to the exhaust side. It is assumed that there is no mixing, horizontally or vertically, for any element of volume and that the flow is simple plug flow either from top to bottom or from side to side. Such systems normally operate on the push–pull principle (exposure air is pushed into the horizontal or vertical laminar-flow portion of the exposure chamber and then pulled from the chamber by equipment on the exhaust side). The pressure of the chamber and the total flow through the chamber are controlled. A laminar-flow whole-body exposure chamber can also be operated using a pull system (whereby the exposure atmosphere is pulled from the chamber by equipment on the exhaust side and the exposure air is introduced into the laminar-flow system, in part by a pressure-driven aerosol generation system and in part by passive suction of the material from the surrounding room or open supplied air duct (Fig. 38-1)).

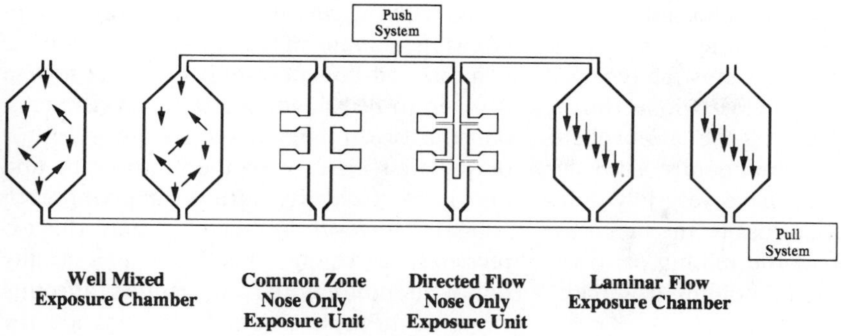

FIGURE 38-1. Configurations of Four Basic Exposure Systems

Nose-Only and Head-Only Exposure System

Nose-only and head-only exposure systems have important applications in inhalation toxicology, especially when only the head or nose of the animal is to be exposed to eliminate surface contamination or when very little research chemical is available for study. These systems have been described previously by Phalen (1984), McClellan and Henderson (1989), Cannon, Blanton, and McDonald (1983), and Hemmenway and Jakab (1990). Nose-only exposure systems, by their very design, are push–pull systems for they all include some point where compressed-air flow is balanced or exceeded by suction. There are two basic types of nose-only exposure systems: the ones with a common distribution zone and the ones that incorporate a directed-flow system (Fig. 38-1). Nose-only exposure systems with common distribution zones are designed and operated essentially around a small cavity that is several times larger than the head of the animal. The nose or the head of the animal is placed through the wall of this cavity, which may have the shape of a rectangular parallelepiped or a cylinder and air is directed through the cavity. The exhaled breath from the animal is captured in this region and is carried past other animals prior to passing into the exhaust side of the nose-only exposure system. Usually, rebreathing of air from other animals is avoided by staggering the position of the animals in the common distribution zone. The assumption is made that there is very little mixing of the exposure atmosphere with the cleaner air exhaled by the animal and, thus, very little dilution to the subsequent levels of animals.

Directed-flow nose-only exposure systems were designed to eliminate the potential dilution of exposure air by the exhaled air of other animals and to reduce the consumption of material used in the exposure. Such systems have a separate, small-volume exposure air distribution line that feeds short delivery tubes positioned and sized so that the flow through them is directed at the breathing zone of the animal. The nose of the animal is positioned in a short side-port of an exhaust tube or duct. The exhaled breath of the animal is directed into the exhaust tube or duct and carried away. Since the delivery and the exhaust systems are completely separate in such exposure units, each animal is exposed to a unique, well-controlled exposure atmosphere, undiluted by its own breath or the breath of other animals present in the system. Such systems as those described by Cannon, Blanton, and McDonald (1983), Baumgartner and Coggins (1980), and Hemenway and Jakab (1990) operate on the premise that the directed exposure air has, at least, the minimum velocity and mass flow to dominate the air space in front of the nose of the animal at the beginning of each breath.

Closed-Loop Exposure Systems

In single- or multiple-animal exposure systems used in metabolism studies, the animal is placed in a closed vessel which is attached to a closed-loop air circulation system. All air withdrawn from the system must be replaced in order to preserve the closed loop during sampling and analysis of the atmosphere. Such systems are predominantly used for gases and aerosols that have very low diffusion and sedimentation loss to the fur of the animal, to the sides of the tubing, or to the walls of the small animal-holding chamber.

PROPERTIES OF EXPOSURE SYSTEMS

An exposure system has two basic measurable properties: volume (liters) and flow (liters/minute). In discussing the operation of an exposure system, the flow through the system is normally described in terms of air changes per hour. The flow, Q (l/min), is equal to n, the number of air changes per minute, times the volume, v, of the system (Fig. 38-2). The number of air changes per hour is equal to $(60\ \text{min/h}) \times n$, where $n = Q/v$. The air change per hour is a very useful term for comparing the operation of one exposure system to another.

The definition of volume, v, is the key in this normalization. For example, the volume of a single-animal closed-loop exposure system is the entire volume of the animal containment vessel and all tubing and lines associated with the closed recirculating air system minus the volume of the animals. The volume of a directed-flow nose-only exposure system is taken to be the volume of the system contained between the point where the generator output first mixes with the dilution air and the tip of the delivery tube supplying each animal. However, in such systems the expression "air changes per hour" is normally not used because the key factors affecting exposure in such nose-only systems are the animal's minute ventilation and the peak flow rate that occurs during each breath. Directed-flow nose-only exposure systems must be set so that the flow through the delivery tube is greater than the maximum inspiratory flow rate of the animal. This is necessary in order to prevent dilution of the delivered exposure air by the exhaled air of the animal.

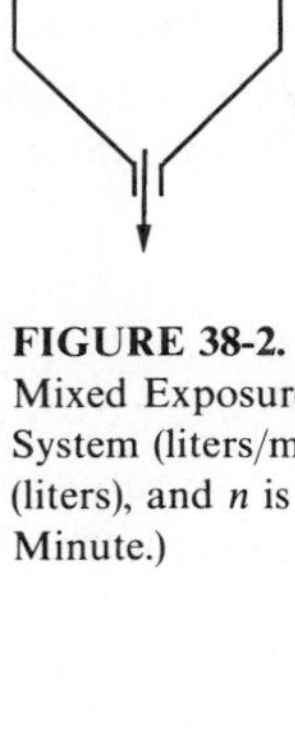

Air change per minute

$$n = \frac{Q}{v}$$

Chamber concentration buildup

$$C = C_{(\text{inlet})}(1 - e^{-n\,t})$$

Chamber concentration decrease for $C_{(\text{inlet})} = 0$

$$C = C_{(\text{chamber at } t=0)}e^{-n\,t}$$

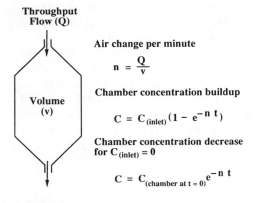

FIGURE 38-2. Airflow and Concentration in a Well-Mixed Exposure Chamber. (Q is the Flow Through the System (liters/min), v is the Relevant Chamber Volume (liters), and n is the Number of Air Changes per Minute.)

In common distribution zone nose-only exposure systems, the volume of the system is the volume of the common distribution zone in which the nose of the animal or the head of the animal is placed. In a laminar-flow whole-body exposure system, the volume of the system is taken to be the volume in which the animals are housed. And finally, in a well-mixed whole-body exposure chamber, the chamber volume is defined as that volume above and below the location of the animals where there is contribution (mixing) to the inhaled exposure atmosphere. This definition is essential in order to meet the basic assumption of complete and thorough mixing made in the operation of these chambers. This assumption also leads to subtle differences in the definitions of volumes for whole-body exposure chambers.

In exposures where air is drawn from and mixed in both the top and bottom of the chamber, the entire volume of the chamber and its transition pieces must be taken to be the volume of the chamber. In exposure systems where there is no back-flux of exposure air, once the housing level of the animal is passed, the exposure system volume is the volume of the top transition piece plus the volume of the region housing the animals.

Although the expression "air changes per hour" is used to describe the total flow through the system, it does not truly indicate either the half-time or the rate at which the concentration builds up and decays in the chamber. The term half-time comes from the equations used to describe concentration change in a well-mixed exposure system. In such systems, the rate of change in the concentration is directly proportional to its current value. This relation is a direct consequence of the assumption that upon entering the chamber volume any incremental element of exposure atmosphere is instantaneously and thoroughly mixed with the entire remaining volume of the chamber. Under such ideal conditions, concentration buildup is equal to the concentration of the airflow entering the chamber times the quantity $1 - e^{-nt}$ (where n is the air change rate). When the incoming concentration is set to zero, the concentration in such a chamber at any subsequent time is equal to C_0e^{-nt}, where C_0 is the concentration in the chamber at the start of the time period. In both buildup and decay of concentration, the time it takes to move from any starting concentration to 50% closer to the equilibrium concentration (i.e., the target concentration for buildup and zero for decay) is given as the half-time, $t_{1/2}$, and is equal to $(\ln 2)/n$ (i.e., $0.693/n$). For example, if it takes three minutes for one chamber volume of air to be pulled into the chamber inlet (i.e., $n = Q/v = (1/3)\ \text{min}^{-1}$ and $t_{1/2} = 0.693 \times 3 = 2.1$ min). After 2.1 min from any starting concentration, the chamber concentration comes 50% closer to the equilibrium concentration. Two half-times or 4.2 min from the start, the concentration would be 75% closer to the equilibrium concentration. The relation can be generalized. The time it takes to change concentration from the start to some percentage, P, of the difference between the starting concentration and the equilibrium concentration can be calculated as $t_{(P/100)} = -(1/n)*\ln(1 - P/100)$. For example, $t_{(1/2)} = 0.693/n$; and similarly $t_{0.75} = 1.386/n$; $t_{0.90} = 2.3/n$; $t_{0.95} = 3/n$; and $t_{0.99} = 4.61/n$.

The term half-time does not accurately apply to a true laminar-flow or plug flow system. In a perfect laminar-flow or plug flow system, the concentration changes instantaneously from its current value to the new equilibrium value as the "plug" moves past the levels of the animals and the sampling probe. In such an ideal system, the characteristic time of the system is taken to be the time for an increment of air to reach the animals after it leaves the point where the generator air is first diluted. In such cases this is n, the air change rate for the system. However, there is usually some level of mixing in such systems, and there is a buildup or decay of concentration near the breathing zone of the animal which can be discussed on the basis of a half-time, as described above for the well-mixed system, where the volume of the system is taken to be the volume around the housing of the animals.

BASIC SAMPLING TECHNIQUES AND STRATEGIES

The sampling and analyses of aerosol and gas concentrations in support of inhalation toxicology must be carried out so that the measurement process does not significantly change concentration or total airflow through the system. The latter is important in dealing with well-mixed systems because changing the total airflow can significantly change the half-time of the system and concentration level. Nearly all exposure systems are constant-flow systems except for the case of exposure of a single animal by head-only or intratracheal intubation through a non-rebreathing valve. In such cases the flow in the system fluctuates with the breathing cycle of the animal.

Sampling System for Constant-flow Exposure Systems

In addition to constant exposure system concentration and flow, a basic operating goal for a generic multiuse sampling system is to be able to stop sampling while leaving the sampling device turned on and the sampling system connected to the exposure chamber. This is especially important when very short, accurately timed samples are to be taken.

Sampling of exposure air is initiated or stopped by directing clean air away from or into the sampling device. In the commonly used "Direct Sample System," Fig. 38-3, this is not possible. A stable vacuum source or vacuum pump internal to the sampling device is used to pull a sample at flow rate Q_s from the exposure chamber. Exhaust from the sampling device is passed through a cleanup system and then through a flow control valve to a stable vacuum source and then into the facility exhaust system. An optional source of clean compressed air at flow rate Q_d may be used to continuously dilute the sample prior to entry of the combined flow ($Q_{sd} = Q_s + Q_d$) into the sampling device. Such direct sampling systems are turned on or off by closing the air and vacuum valves, shutting off the sampling device, or unplugging the sampling line from the exposure system. The operation of such direct sampling systems may change the concentration and airflow rate in the exposure system. However, in most cases Q_s is less than 2% of the total airflow through the chamber and the effect on concentrations can be ignored.

The goal of stabilizing the sampling device by leaving it on at all times can be met by incorporating an additional clean air line of flow Q_s in the direct sample system (Figs. 38-3 and 38-4). Sampling of exposure air in such a system is turned off by diverting the flow of clean air into the sampling device. The total clean-air flow rate matches the air flow rate through the sampling device and shuts off flow from the sampling inlet; eliminating the need for a valve in the sampling line. Clean air is sampled at a rate Q_s directly into the sampling device after passing a Y connector. The vacuum source and control valve are not needed if the pump and control device are contained in the sampling unit. Clean air is provided to one or two rotameters at the front of the system as shown. Sampling is initiated when the flow Q_s of clean replacement air through the first rotameter (just below the three-way valve C in Fig. 38-4a) is diverted to an exhaust system. The total flow Q_{sd} being pumped by the sampling device is maintained by removing an equal flow rate, Q_s, from the exposure chamber (Fig. 38-4a and 38.4b).

The flows in Fig. 38-4a are "balanced" when the flow through the first rotameter is diverted to the sampling device and set equal to the sampling flow, Q_s, and $Q_s + Q_d = Q_{sd}$. One way to set this rotameter is to place a magnehelic or soap bubble volume meter at the sample inlet. The flow is balanced when there is no pressure difference on the magnehelic or when the soap bubble in the soap bubble volume meter does not move.

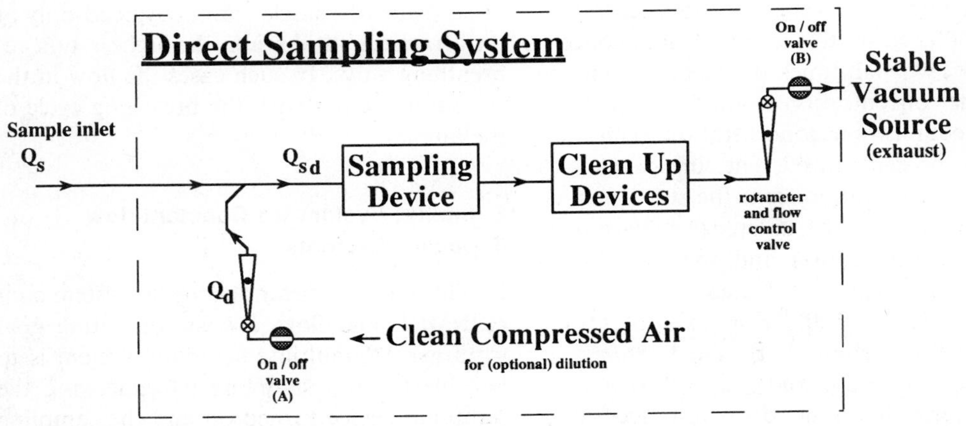

FIGURE 38-3. Direct Sample System for Measuring Concentrations from Inhalation Exposure Systems. (Q_{sd} is the Total Flow Through the Sampling Device, Q_d is the Flow of Dilution Air (If Required), and Q_s is the Sample Flow.)

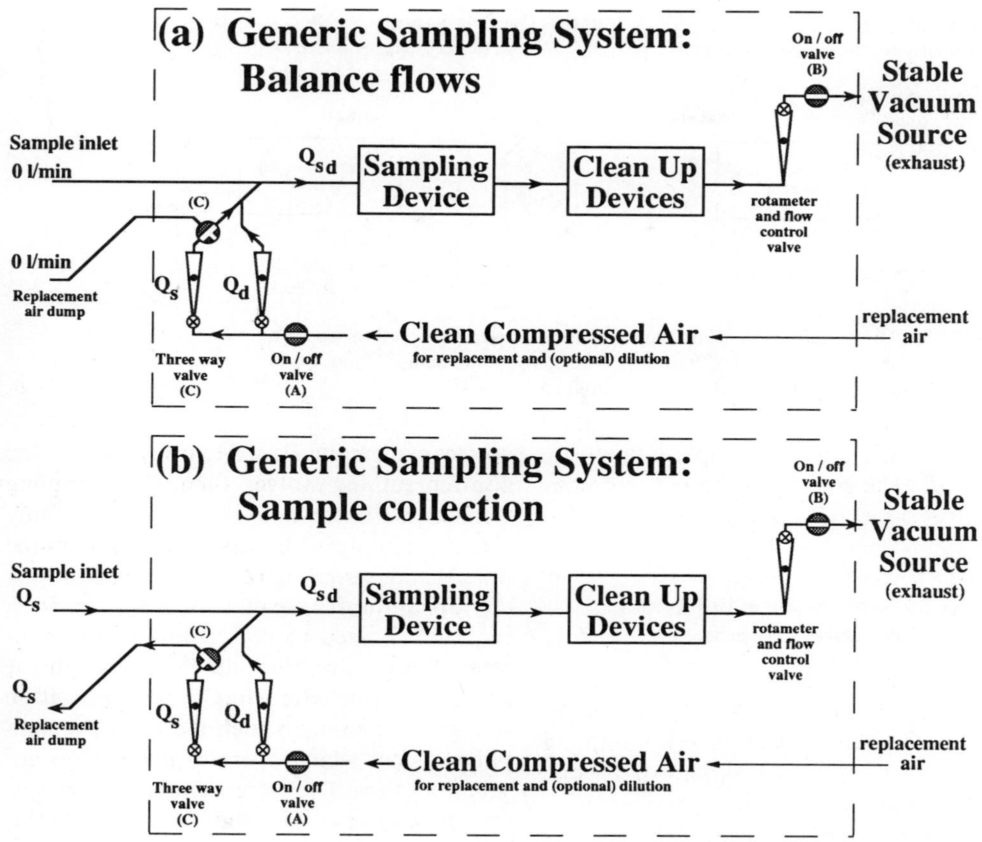

FIGURE 38-4. Generic Sampling System: (a) Diversion of Q_s to the Sample Device with Valve (C) Allows Continuous Operation of the Sampling Device Without Pulling a Sample from the Exposure System; (b) Diversion of Q_s to the Sample Exhaust with Valve (C) Allows Chamber Air to Enter the Sampling Device with the Same Flow Rate Q_s. Valves (A) and (B) Together can Remain On or Off and will not Affect the Chamber Concentration. (Q_{sd} is the Total Flow Through the Sampling Device, Q_d is the Flow of Dilution Air (If Required), and Q_s is the Sample Flow.)

Sampling System for Constant Flow: Push–Pull Exposure Systems

A direct sampling system (Fig. 38-3) should not be used for those cases where the sample flow rate, Q_s, is greater than 2–5% of the total flow through the chamber. The generic sampling system, shown in Figs. 38-4 and 38-5, can be hooked to an exposure chamber, with the sample being pulled from the breathing zone of the animals and system flow balance maintained by attaching the "clean replacement air" exhaust line to the chamber exhaust (Fig. 38-5). Sampling commences by turning the three-way valve so that "clean replacement air," Q_s, flows from the rotameter to the chamber exhaust. With such a generic system, both the total flow and the concentration within the exposure chamber remain constant. The concentration in the exhaust line, of course, decreases downstream of the entry of clean air at flow rate Q_s. On the other hand, if the direct sample system shown in Fig. 38-3 is attached directly to a "pull" only exposure system (Figs. 38-1 and 38-5) then during the sampling cycle, the flow through the system is increased by an amount Q_s. The new total flow, Q_{t2}, through the chamber, is equal to the

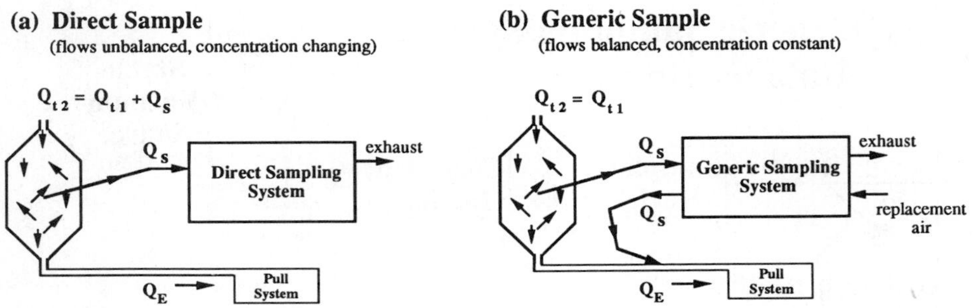

FIGURE 38-5. Direct and Generic Sampling System Attached to a Well-mixed Exposure Chambers Operated in the "Pull-only" Mode. (Q_s is the Sample Flow, Q_t is the Total Flow Before (Q_{t1}) and During (Q_{t2}) Sampling.)

original flow, Q_{t1} plus the sampling flow: $Q_{t2} = Q_{t1} + Q_s$. The concentration will decrease by a factor $1 - Q_s/(Q_{t1} + Q_s)$.

Sampling System for Directed-Flow Nose-Only Exposure Systems

In sampling directed-flow nose-only exposure systems, special care must be taken to maintain the flow balance. The generic sampling system must be attached as shown in Fig. 38-6. The clean replacement air exhaust is so attached that it vents into the exhaust duct near the sampling site. The inlet sampling point must be so placed that it is in the same relative position as would be the nose of the animal.

Sampling System for Pulsed-Flow Exposure Systems

Pulsed flows are produced when an exposure system is attached to a single-direction non-rebreathing valve being used by a subject or an animal. If sampling can be accomplished upstream from the T used to divert flow to the non-rebreathing valve, then the sampling techniques for whole-body or nose-only chambers apply as discussed above. If, on the other hand, sampling of concentration in the delivery air must take place downstream from the T or Y used to divert flow to the non-rebreathing valve, then the flow through the system will fluctuate while the concentration remains constant. In such a case, for a sampling system which operates under constant sampling flow, Q_s, the generic sampling system shown in Fig. 38-7 should be used. In this system, the fluctuating total flow, Q_t, is the result of the difference in the constant flow from the generator, Q_g, minus the fluctuating inhalation flow, Q_i: $Q_t = Q_g - Q_i$. The total flow, Q_t, is divided between the sampling device and the exhaust system. The sample flow, Q_s, is kept less than Q_t, even though Q_t fluctuates because of Q_i. The loss of flow caused by the inspiration of the animal, Q_i, is passively replaced by suction of clean air from the room through a low-pressure cleanup device, $Q_m = -Q_i$. In this way, the flows in the system can be fully controlled for accurate sampling. The amount of exposure atmosphere sampled at flow Q_s is a true measure of the concentration inhaled through the non-rebreathing valve.

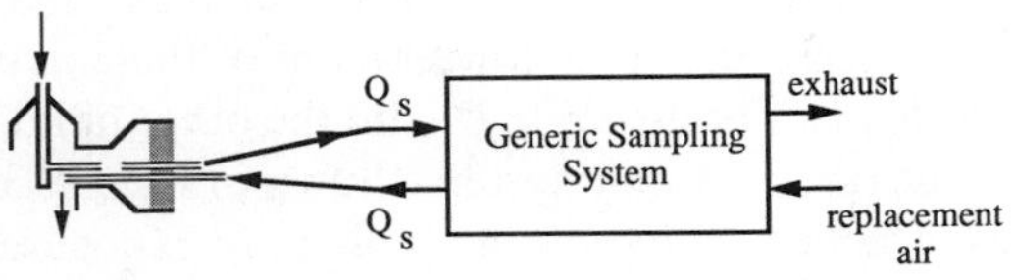

FIGURE 38-6. Generic Sampling System Attached to a Directed-Flow Nose-only Exposure System. (Q_s is Both the Sample Flow and the Replacement Air Exhaust Flow.)

In order to measure the total amount of material exhaled from the exhaust side of a non-rebreathing valve, a different sampling train must be used. Constant total flow, Q_t, in this sampling system is maintained by using a flow of clean compressed air, Q_a, that is either

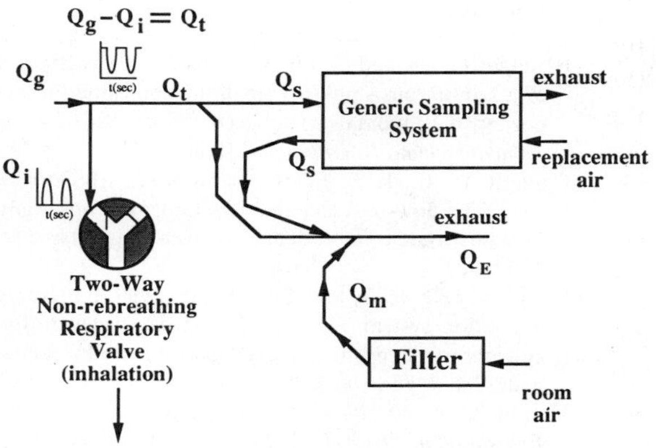

FIGURE 38-7. Sampling for Concentration of a Fluctuating Air Stream Where the Total Flow in the Air Stream is Always Greater than the Sampling Flow. (Q_g is the Total Flow Entering the System from the Generator; Q_i is the Pulsating Suction from the Non-Rebreathing Valve; Q_s is the Constant Sample Flow; Q_m is the Flow of Makeup Air from the Room; and Q_E is the Exhaust Flow.)

vented into the room (or hood) or sucked into the sampling system. In the schematic diagram shown in Fig. 38-8, the flow of exposure air, Q_i, is from the exhaust side of the non-rebreathing valve. The flow of Q_a through the rotameter is set to be equal to or greater than the peak flow rate coming from Q_i, the non-rebreathing valve ($Q_a \geq Q_i$). As the animal exhales, an additional portion of the clean-airflow, Q_a, is diverted through the filter, flow Q_f, and into the room or into the exhaust system ($Q_i \leq Q_f$). As the animal stops exhaling and Q_i decreases to zero, the total flow through the system is made up by Q_a. This

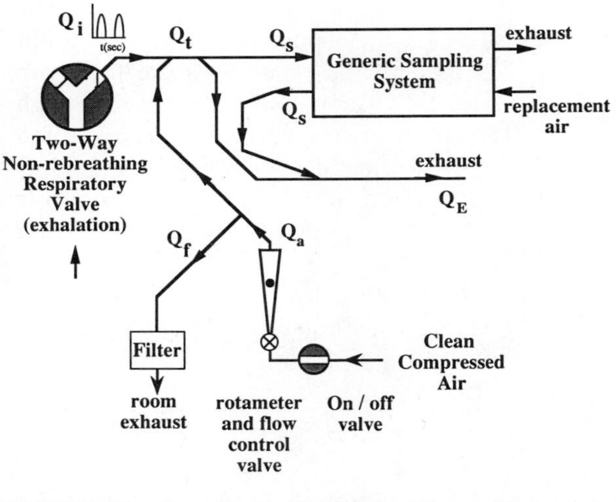

FIGURE 38-8. Sampling for Total Output from a Fluctuating Air Stream Where the Total Flow Periodically Drops to Zero. (Q_i is the Flow Entering the System from the Non-Rebreathing Valve: Q_a is the Flow of Clean Air into the System; Q_s is the Constant Sample Flow; Q_f is the Flow of Clean Air out to the Room; and Q_E is the Exhaust Flow.)

dynamic air flow system allows a constant flow, Q_t, to be established at the middle of the system dividing the sampling side from the non-breathing valve. The flow through the generic sampling system, Q_s, is less than or equal to Q_t.

SUMMARY

Even though the sampling device itself may be very accurate, the measurement may be inaccurate because the process of sampling changes the equilibrium of the exposure system. The basic sampling systems shown meet the general requirement of obtaining a concentration measurement from a variety of exposure systems without any significant change in concentration or total airflow. For large flow systems, where the sampling flow, Q_s, is small compared to the total flow through the system, the use of the direct sample system will not change the exposure concentration. However, for lower flow systems, or for systems requiring special airflow balancing, such as directed flow, nose-only, and closed-loop exposure systems, care must be taken to allow sampling to proceed in an accurate and efficient manner.

References

Baumgartner, H. and C. R. E. Coggins. 1980. Description of a continuous-smoking inhalation machine for exposing small animals to tobacco smoke. *Beiträge zur Tabakforschung International*, 10(3):69–74.

Cannon, W. C., E. F. Blanton, and K. E. McDonald. 1983. The flow-past chamber: An improved nose-only exposure system for rodents. *Am. Ind. Hyg. Assoc. J.* 44(12):923–8.

Hemenway, D. R. R. and G. J. Jakab. 1990. Nose-only inhalation system using the fluidized-bed generation system for coexposures to carbon black and formaldehyde. *Inhal. Toxicol.* 2:69–89.

Hering, S. V. ed. 1989. *Air Sampling Instruments for Evaluation of Atmospheric Contaminants*, 7th edn, Cincinnati, OH: American Conference of Governmental Industrial Hygienists, Inc.

Hinner, R. G., J. K. Burkart, and E. L. Punte. 1968. Animal inhalation exposure chambers. *Arch. Environ. Health* 16:194–206.

McClellan, R. O. and R. F. Henderson, eds. 1989. *Concepts in Inhalation Toxicology*. NY: Hemisphere Publishing Corporation.

Moss, O. R. 1989. Sampling in calibration and exposure chambers. *In Air Sampling Instruments for Evaluation of Atmospheric Contaminants*, 7th edn, ed. Susan V. Hering, pp. 157–62. Cincinnati, OH: American Conference of Governmental Industrial Hygienists, Inc.

Phalen, Robert F. 1984. *Inhalation Studies: Foundations and Techniques*. Boca Raton: CRC Press.

Silver, S. D. 1946. Constant flow gassing chambers: Principles influencing design and operation. *J. Lab. Clin. Med.* 31:1153–61.

APPENDIX A. GLOSSARY OF TERMS

Term	*Explanation*
Absorption	A process whereby gas or vapor molecules are transferred to the liquid phase
Accumulation mode	A mode in the atmospheric particle size distribution, formed primarily by coagulation of smaller particles
Accuracy	An indication of the correctness of a measurement
Actinomycetes	A group of bacteria with a morphology resembling fungi
Active sampling	Aerosol measurement using an air-moving device to draw the air into the collector or detector; as opposed to passive sampling
Activity	A measurable property of a particle population, e.g., number, surface, mass, especially radioactivity
Activity coefficient	A measure of the deviation from ideal solution behavior
Activity concentration	Amount of radioactive material in a unit of air
Activity median diameter	Particle diameter with 50% of airborne activity above and 50% below this size
A/D convertor	An electronic device used to change an analog signal to a digital one
Adsorption	Transfer of gas or vapor molecules from the surrounding gas to a solid surface
Adsorption isotherm	A function relating the volume of vapor adsorbed on a surface to the pressure of the vapor in the gas phase at a fixed temperature
Aerodynamic (equivalent) diameter	Diameter of a unit-density sphere having the same gravitational-settling velocity as the particle in question
Aerodynamic particle sizer	A particle spectrometer that uses an acceleration system to differentiate particles by aerodynamic diameter and a laser velocimeter to detect particles
Aerosol	An assembly of liquid or solid particles suspended in a gaseous medium long enough to be observed and measured; generally, about 0.001–100 μm in size
Aethalometer	An instrument used to measure the optical absorption of collected aerosol samples
Air monitoring	Sampling and analysis of air to determine the quantity of pollutants present
Air pollution	Condition of materials present in the air at levels detrimental to the health and/or welfare of human beings
Air quality standards	Level beyond which air pollutants can cause damage to humans, animals, plants, or materials
Agglomerate	A group of particles held together by van der Waals forces or surface tension
Aggregate	A heterogeneous particle in which the various components are not easily broken apart
Airy disk	A disk-like image of a small point produced by an optical system with diffraction-limited resolution
Aitken nuclei	Atmospheric particles in the approximate size range of 0.01–0.1 μm
Algae	A microscopic plant
Aliasing	Observation of a low-frequency signal because of inadequate sampling or measurement rate; see Nyquist frequency

Alveolar	Part of the respiratory system in which gas exchange occurs; alveoli are small sacs at the end of the bronchioles
Ambient air	Surrounding air
Anisoaxial sampling	Sampling condition in which the air flowing into an inlet has a different direction from the ambient air flow
Area sample	A sample taken in a fixed location assumed to be representative of the area being investigated
Aspiration efficiency	Fraction of particles entering an inlet from the ambient environment
Atomizer	A device used to produce droplets by mechanical disruption of a bulk liquid
Autocorrelation	Relationship of the measured values to previously measured ones
Bacteria	Single-celled microorganisms; some genera produce endospores
Beta gauge	A method of mass measurement which relies on the attenuation of a beam of beta particles
Beta particle	A energetic electron emitted in certain nuclear decay processes
BET method	Brunauer–Emmett–Teller method; procedure using the adsorption isotherm of a material to measure its surface area
Bias	Consistent difference between a measurement and a true or accepted value
Bimodal size distribution	Particle size distribution with two distinct maxima
Bioaerosol	A suspension of particles of biological origin; viable or dead cells; spores or pollen grains; fragments, products, or residues of organisms
Bipolar ion field	Region in which ions of both polarities exist
Boltzmann charge distribution	Residual or minimum charge distribution on particles after exposure to a bipolar ion field
Boundary layer	Region of flow near the bounding surface, where the flow is dominated by friction forces resulting in reduced flow velocity relative to the free stream
Breathing zone sample	A sample taken as close as possible to the point at which the subject inhales air; represents a subject's inhaled air
Brownian motion	Random motion of particles due to collisions with gas molecules
Bubble meter	A tube with a defined volume into which bubbles are injected to measure flow rate
Bulk analysis	Analysis of a sample in its entirety versus analysis of individual particles
Capillary pore filter	A filter consisting of a solid membrane with an array of cylindrical holes of uniform size penetrating the membrane
Carcinogen	An agent that causes cancer
Cascade impactor	A device that uses a series of impaction stages with decreasing particle cut size so that particles can be separated into relatively narrow intervals of aerodynamic diameter; used for measuring the aerodynamic size distribution of an aerosol
Centrifuge	A device in which particles are removed by centrifugal forces from an aerosol flowing in a helical path; the device is usually characterized by high-resolution particle size separation
Closed-face sampler	A filter cassette sampler with the inlet smaller than the filter, as used in in-line liquid filtration

Colony-forming units	Number of colony-forming units (e.g., bacteria, fungi) in a unit volume of air
Cloud	An assembly of particles with an aerosol density that is more than about 1% higher than the density of the gas alone
Coagulation	An aerosol growth process resulting from the collision of aerosol particles with each other
Coarse particle mode	Largest particle mode (> 2 μm) in atmospheric particle size distributions, consisting primarily of particles generated by mechanical processes
Coincidence	Simultaneous presence of two or more particles in the sensing volume of a particle counter
Comminution	Breakup of particles by mechanical action
Condensation	A process with more vapor molecules arriving at a particle's surface than leaving the surface, resulting in a net growth of the particle
Condensation nuclei counter	A device in which submicrometer-sized particles are grown by vapor supersaturation to a larger size and are detected by light scattering
Confidence limits	Values defining a range around a sample statistic
Continuum flow	Flow governed by the macroscopic properties of the gas or fluid such as viscosity and density
Corona	Region of intense ionization, often surrounding an electrode at high voltage
Coulter counter	An instrument that measures individual particle volume in a liquid by measuring the change in resistivity of the liquid as it passes through an orifice
Cowl	A cylindrical tube used in front of a filter cassette to prevent direct impaction or contamination of samples; used primarily for asbestos fiber sampling
Critical orifice	An orifice through which there is a constant airflow when a sufficient pressure drop across the orifice causes sonic flow
Cunningham slip correction factor	Same as slip correction factor
Cyclone	A device in which particles are removed by centrifugal forces in a cyclonic path
Cutoff particle diameter	Diameter of a particle which has 50% probability of being removed by the device or stage and 50% probability of passing through; also called 50% cut point, d_{50}, or the effective cutoff diameter
Dichotomous impactor	A virtual impactor with two emerging aerosol flows
Differential-mobility classifier	A device that removes all particles except those within a narrow range of electrical mobility
Diffraction	Change in direction and amplitude of radiation after passing near an object or through an orifice
Diffusion	Net movement of particles or gas from a higher to a lower concentration
Diffusion battery	An aerosol spectrometer used for submicrometer-sized aerosols, in which size is measured by the diffusive loss of particles in an arrangement of ducts (e.g., tubes, filters, screens)
Diffusion charging	A process by which airborne particles acquire charge from ions undergoing Brownian motion
Diffusion denuder	A device that passes particles (low diffusivity) and removes gases (high diffusivity)
Diffusion (equivalent) diameter	Diameter of a unit-density sphere with the same rate of diffusion as the particle in question

Diffusiophoresis	Particle motion under the influence of a concentration gradient
Dilution ratio	Factor by which measured concentration is multiplied to obtain mainstream concentration
Dilution system	A system wherein aerosol is mixed with particle-free dilution gas in a known volumetric ratio to reduce concentration
Disinfection	Destruction of the majority of microorganisms, not necessarily of all the spores
Dispersion	A system consisting of particles suspended in a fluid
Drag force	Resistance experienced by a particle when moving in a fluid
Drag coefficient	A coefficient that relates the particle's drag force to the velocity pressure
Dust	Solid particles formed by erosion or other mechanical breakage of a parent material; generally consists of particles of irregular shape and larger than about 0.5 μm
Dust generator	A device used to disperse dry particles in the air in a controlled fashion
Dynamic shape factor	Ratio of the drag force on a particle to that on a sphere of equivalent diameter
Effective density	Density of a particle with voids, in contrast to a compact particle's bulk material density
Elastic light scattering	A process in which there is no energy exchange between incident photons of light and target particles
Electrical aerosol analyzer	An aerosol size spectrometer in which the particles are separated by removing those with an electrical mobility greater than a selected value
Electrical aerosol classifier	An aerosol size spectrometer in which the particles are separated by selecting those within a narrow range of electrical mobilities
Electrical mobility	An aerosol parameter that indicates a particle's ability to move in an externally applied flow field
Electrical mobility (equivalent) diameter	Diameter of a unit-density spherical particle moving at the same velocity in an electric field as the particle in question
Electrodynamic balance	A device which uses superimposed ac and dc fields to levitate particles
Electrophoresis	Charged particle motion induced by an electric field
Electrostatic balance	A device which uses a dc field to levitate particles, e.g., the Millikan condenser
Electrostatic precipitator	A device in which airborne particles are charged in a unipolar ion field and desposited with a high-voltage electric field
Elutriator	A device used to separate particles by aerodynamic diameter by allowing them to settle in a moving air stream
Emission	Material being discharged into the outdoor atmosphere
Endotoxin	Toxic cell wall component of gram-negative bacteria
Envelope (equivalent) diameter	Diameter of a sphere composed of the particle bulk material and included voids that has the same mass as the particle in question
Epiphaniometer	An instrument that measures the surface area of aerosol particles
Evaporation	Process with more vapor molecules leaving a particle's surface than arriving at the surface, resulting in shrinkage of the particle
Extinction coefficient	A measured parameter given by the amount of light scattered and

	absorbed by a particle divided by the amount incident upon it
Extrathoracic	Region of the respiratory system above the larynx, containing the nose and the mouth
Equivalent diameter	Diameter of a sphere having the same value of a specific physical property (activity) as the particle in question
Fabric filter	A filter consisting of a woven or felted fabric
Feret's diameter	Particle dimension determined by the projection of the particle's silhouette onto a selected axis
Fibrous filter	A filter consisting of a mat of individual fibers
Field charging	A process by which particles are charged by ions moving in a strong electric field
Filter	A porous membrane or mat of fibers used to collect particles from the air
Fine particle	Particles less than about 2 μm in size, consisting of particles in the nuclei and accumulation modes; term used in describing atmospheric aerosols
Flocculate	A group of particles very loosely held together, often by electrostatic forces; flocculates can easily be broken apart by shear forces within the air
Fluidized-bed generator	A device using air pressure to fluidize a powder to release dust particles
Fly ash	Particles of ash entrained in flue gas produced by fossil fuel combustion
Fog	A liquid particle aerosol, typically formed by condensation of supersaturated vapors
Fractal dimension	A measure of the complexity of a particle's shape
Free molecular flow	Flow governed by discrete impacts of gas molecules
Fume	Small particles that are usually the result of condensed vapor (often from combustion), with subsequent agglomeration
Fungi	Multicellular organisms that produce spores
Gaussian curve	A profile of distribution or a curve similar to that observed for the normal distribution
Geiger–Müller tube	A radiation sensing instrument which relies on an avalanche process initiated by the production of electron–ion pairs
Geometric	Refers to a size parameter on a logarithmic size scale, where a given ratio of two sizes appears as the same linear distance
Geometric standard deviation	A measure of dispersion in a lognormal distribution (always ≥ 1)
Graticule	A transparent disk with a calibrated scale placed in the focal plane of an optical system, e.g., a microscope, used for the measurement of particles or other objects
Gravitational deposition parameter	Ratio of particle settling distance during transport in the sampling inlet region to the diameter of the inlet
Gravitational settling velocity	Particle motion in a gravitational field after an equilibrium between gravity and aerodynamic drag forces has been reached
Half-life	Time interval required to reduce the rate of emission of a radioisotope by a factor of two
Hatch–Choate equations	Expressions that, given a characteristic diameter and geometric standard deviation of a distribution, allow the calculation of any other characteristic diameter of the distribution

Heterogeneous	Consisting of individual components that may differ from each other in size, shape, and chemical composition
Heterogeneous nucleation	Formation of droplets on condensation nuclei (existing submicrometer particles)
Homogeneous nucleation	Formation of droplets in the absence of condensation nuclei; also called self-nucleation
Horizontal elutriator	A horizontal channel through which aerosol flows and particles above a given size or size range are removed by gravitational settling
Hot-wire anemometer	A device used to measure air velocity by measuring the change in resistance of a heated wire
House-dust mites	Common insects living in mattresses and carpets; excreta are common allergens
Hydraulic diameter	Hypothetical diameter of an object equal to four times the object's cross-sectional area divided by the perimeter of that area
Hydrosol	Suspension of particles in a liquid
Hygroscopicity	Property of a chemical that indicates its tendency to absorb water from the air
Hyphae	A chain of fungal cells
Ideal fluid	A hypothetical fluid having no viscosity
Impactor	A device in which aerosol particles with sufficiently high inertia in a deflected air stream are impacted onto a surface
Impinger	A device in which particles are removed by impacting the aerosol particles into a liquid
Inhalable	Fraction of an aerosol that can enter the human respiratory system
Inlet efficiency	Fraction of ambient particles that is delivered to the aerosol transport section of a sampling system by the inlet; it is the product of the aspiration and transmission efficiencies
Inspirable	Same as inhalable; inhalable is the currently preferred term
Interception	Collision with and deposition of a particle on an object when the particle passes within one particle radius of the object
Ionization chamber	A radiation sensing instrument which relies on the detection of free electron–ion pairs
Isoaxial sampling	Sampling condition in which the air flowing into an inlet has the same direction as the ambient air flow
Isokinetic sampling	Sampling condition in which the air flowing into an inlet has the same velocity and direction as the ambient air flow
Jet nebulizer	A nebulizer employing air pressure to aerosolize a bulk liquid
Kelvin effect	Increase in partial vapor pressure for a particle's curved surface required to maintain mass equilibrium relative to the vapor pressure above a flat liquid surface
Knudsen number	Ratio of gas molecular mean free path to the physical dimension of the particle; indicator of free molecular flow versus continuum gas flow
Kuwabara flow	Solution of the two-dimensional viscous flow field for a system of cylinders perpendicular to the flow, taking into account the interference effects of neighboring fibers (has also been applied to

	spheres); used to model flow in fibrous filters
Laminar flow	Gas flow with a smooth, nonturbulent pattern of streamlines, with no streamline looping back on itself; usually occurs at very low Reynolds numbers
Light scattering	Change in the direction of light radiation due to reflection, diffraction, and refraction from a particle
Lognormal size distribution	Particle size distribution characterized by a bell-shaped or Gaussian distribution shape when plotted on a logarithmic-size scale
Lung model	Representation of the respiratory system used to make quantitative estimates of particle deposition
Mach number	Ratio of gas to acoustic velocity; indicator of compressibility
Manometer	A device used to measure pressure differences
Martin's diameter	Length of the horizontal line bisecting particle cross section into equal areas
Mass (equivalent) diameter	Diameter of a sphere composed of the particle bulk material with no voids that has the same mass as the particle in question
Mass median size	Size with an equal mass of particles above and below this value (see also median size)
Mean free path	Mean distance a molecule in a gas travels before colliding with another molecule
Mean size	Average of all sizes, i.e., the sum of all sizes divided by the number of particles
Mechanical mobility	Aerosol parameter that indicates a particle's ability to move in a suspending medium; see mobility
Median size	Size with an equal number of particles above and below this value (see also mass median size)
Membrane filter	A filter that is formed as a gel from a colloidal suspension; characterized by tortuous air passages
Microparticles	Particles with sizes of the order of micrometers
Micronize	A process by which coarse powders are mechanically reduced to a particle size suitable for redispersion as an aerosol from a solvent or propellant
Mildew	A visible fungal growth on surfaces
Mist	A liquid particle aerosol, typically formed by physical shearing of liquids, such as in nebulization, spraying, or bubbling
Mobility	Ratio of particle velocity to the force producing that velocity
Mobility (equivalent) diameter	Diameter of a spherical particle with the same dynamic mobility as the particle in question
Mode	Value occurring most often in a distribution of values; peak value of a distribution
Mold	A visible fungal growth on surfaces
Monodisperse	Composed of particles with a single size or a small range of sizes
Mycelium	A mass of fungal hyphae
Mycotoxin	A toxic chemical produced by fungi
Nanophase particles	Particles with sizes of the order of nanometers
Nasopharyngeal compartment	A portion of the respiratory tract between the epiglottis and the anterior nares
Nebulizer	A device in which droplet aerosols are produced by dispersion of a bulk liquid
Nephelometer	An instrument that measures the amount of light scattered from a particle cloud; also called a photometer

Neutralizing	Reduction in electronic charge on particles by exposure of the aerosol to ion clouds (often produced by radioactive sources)
Normal size distribution	Particle size distribution characterized by a bell-shaped or Gaussian distribution shape when plotted on a linear size scale
Nucleation	Process of initial formation of particles from a vapor
Nuclei mode	Smallest mode in atmospheric particle size distributions, formed by condensation of atmospheric gases or emissions from hot processes, typically containing particles < 0.1 μm in size
Nyquist frequency	Highest frequency component in a signal; twice this frequency is the minimum sampling rate that can be used without biasing the measured values
Opacity	Degree to which an aerosol obscures an observer's view
Open-face sampler	A filter cassette sampler with the inlet approximately the same size as the filter
Optical (equivalent) diameter	Diameter of a calibration particle that scatters as much light in a specific instrument as the particle being measured
Optical (single) particle counter	An aerosol size spectrometer that differentiates particles by the amount of light scattered by each particle
Orifice meter	A device used to measure flow rate in a duct by measuring the pressure drop across a calibrated constriction
Owl	A device used to measure diameter of monodisperse aerosol particles by illumination with white light and detection of higher-order Tyndall spectra
Packing density	Ratio of fiber or membrane volume of a filter to its total volume; also solidity
Particle	A small discrete object, often having a density approaching the intrinsic density of the bulk material; it may be chemically homogeneous or contain a variety of chemical species; it may consist of solid or liquid materials or both
Particle bounce	Rebound of particles that fail to adhere after impacting on a collecting surface
Particle size distribution	A relationship expressing the quantity of a particle property (activity) associated with particles in a given size range
Particulate	A particle; this term is also used as an adjective indicating that the material in question has particle-like characteristics
Partial pressure	Pressure which a vapor would exert if it were the only component present in a volume of gas
Passive sampling	Aerosol measurement using natural convection or diffusion to draw the air into the measurement device; as opposed to active sampling
Pathogen	A microorganism that causes disease
Peclet number	Ratio of a particle's convective to diffusive transport
Personal sampler	A device attached to a person in order to sample air in the person's immediate vicinity
Phantom particles	Particles that appear in a measured distribution that are due to coincidence or other nonideal aspects of

the measurement process and not due to real particles

Photometer	An instrument that measures the amount of light scattered from a particle cloud; also called a nephelometer
Photophoresis	Particle motion under the influence of asymmetric light absorption within a particle
Pitot tube	A device used to measure velocity pressure in a flow stream
Plume	Flow of visible effluent from an outlet, e.g., a stack or vent
Point-to-plane precipitator	An electrostatic precipitator using a corona from a single point to deposit particles on a flat grounded plane
Poiseuille flow	Laminar flow with a parabolic velocity profile occurring in a circular duct; the gas velocity in the center of the tube equals twice the average velocity in the tube
Poisson distribution	Mathematical function relating the number of particles in a given volume element to the average concentration of randomly distributed particles in the entire volume
Polydisperse	Composed of particles with a range of sizes
Porosity	(1 − packing density)
Precision	An indication of the degree of variation in the results of repeated measurements of a variable
Preclassifier	A device that removes particles ahead of an aerosol sensor, often in a manner similar to the particle removal occurring ahead of the respiratory region of interest; also called a preseparator or a precutter
Primary particle	A particle introduced into the air in solid or liquid form
Projected-area (equivalent) diameter	Diameter of a circle that has the same area as the projected area of a particle seen under a microscope
Pulmonary compartment	Portion of the respiratory tract in which gas exchange occurs (includes alveoli and respiratory bronchioles)
Radiometric force	Force produced by light pressure
Rayleigh scattering	Scattering of radiation occurring when the size of the scattering object is much smaller than the radiation wavelength
Re-entrainment	Return of particles to an air stream after deposition on a collecting surface
Refractive index	Ratio of the speed of light in a vacuum to that in a material in question
Refraction	Change in speed and direction of radiation passing from one medium into another
Relative settling velocity	Ratio of the terminal gravitational-settling velocity to sampling air velocity in an inlet
Relaxation time	Time for a particle to reach $1/e$ of its final velocity from an initial velocity or from rest when subjected to an external force; an indicator of a particle's ability to adjust to changes in flow velocity
Respirable fraction	Fraction of aerosol that can reach the gas exchange region of the human respiratory system
Reticle	A transparent disk with lines or other marks placed in the focal plane of optical systems for calibration or alignment
Reynolds number	Flow similitude parameter, expressed as the ratio of the inertial force of the gas to the friction force of the gas moving over the surface of an object; flow

	Reynolds number the gas flow in a tube and particle Reynolds number describes the gas flow around a particle
Rotameter	A device used to measure the flow rate, as indicated by the height of a float centered in a vertical tapered tube
Sampling ratio	Ratio of the ambient air velocity to the air velocity in an inlet
Sampling probe	A device to withdraw aerosol from a system
Saturation ratio	Ratio of the partial pressure of a vapor to its saturation vapor pressure
Saturation vapor pressure	Partial pressure of a liquid's vapor required to maintain the vapor in equilibrium with the condensed liquid or solid; also referred to as vapor pressure
Sauter mean diameter	Diameter of a droplet whose surface to volume ratio is equal to the mean of all the surface-to-volume ratios of the droplets in a spray distribution; also referred to as the surface area mean diameter or the mean volume–surface diameter
Schmidt number	Ratio of the Peclet number to the Reynolds number, or the ratio of kinematic viscosity to diffusion coefficient
Scintillation spectrometer	A radiation sensing instrument which relies on the excitation of optical emission followed by detection with a photomultiplier tube
Secondary particle	A particle formed in the air, usually by gas-to-particle conversion; also sometimes used to describe agglomerated or redispersed particles
Sedimentation	Movement of particles by the influence of gravity
Semiconductor detector	A radiation sensing instrument which relies on the generation of free carriers in a semiconductor material
Shape factor	A factor that relates the drag force on a particle to that on an equivalent sphere
Sherwood number	Dimensionless mass transfer coefficient that relates the particle's diffusive deposition velocity to the particle's diffusion coefficient
Sinclair–LaMer generator	A device that produces monodisperse aerosols by condensation of vapor onto nuclei
Slip correction factor	A factor which allows slip flow behavior to be calculated using continuum gas flow equations
Slip flow regime	Transition between free molecular flow and continuum gas flow
Smog	An aerosol consisting of solid and liquid particles, created, at least in part, by the action of sunlight on vapors; the term smog is a combination of the words smoke and fog and often refers to the entire range of such pollutants, including the gaseous constituents
Smoke	A solid or liquid aerosol, the result of incomplete combustion or condensation of supersaturated vapor; most smoke particles are submicrometer in size
Snell's law	Fundamental principle in optics that the sines of the angles of incidence and refraction are in a constant ratio to one another
Solidity	See packing density
Soot	A conglomeration of particles formed by incomplete combustion of carbonaceous material

Source apportionment	Analysis of an aerosol sample so that fractions of the aerosol can be assigned specific sources
Source sampling	Collection of materials emitted from an air pollutant generating source
Specific surface	Particle surface area per unit mass or volume of particles
Spinning disk atomizer	A device that produces monodisperse droplets from the breakup of a thin film of liquid ejected from the surface of a spinning disk
Spirometer	A device used to measure the gas volume (or flow rate with a timer) using an expandable can sealed with a liquid
Spores	Dormant cells of microorganisms
Standard	Maximum allowable level of an air contaminant established by law
Stephan flow	A special case of diffusiophoresis, with particle motion towards or away from evaporating or condensing surfaces; also written as the Stefan flow
Sterilization	Complete destruction of microorganisms and their spores
Stokes diameter	Diameter of a spherical particle with the same density and settling velocity as the particle in question
Stokes' law flow	Flow around a body under the influence of viscous, but not inertial forces
Stokes' number	Ratio of a particle's stopping distance to a characteristic dimension; generally used as an indicator of similitude in particle behavior in a given aerosol flow configuration
Stokes regime	Condition for which Stokes' law applies
Stopping distance	Product of relaxation time and the initial particle velocity; an indicator of a particle's ability to adjust to directional changes in aerosol flow
Subisokinetic sampling	Sampling condition in which the air flowing into an inlet has a lower velocity than the ambient air flow
Superisokinetic sampling	Sampling condition in which the air flowing into an inlet has a higher velocity than the ambient air flow
Surface barrier detector	A type of semiconductor detector used primarily for charged particle emissions
Terminal settling velocity	Equilibrium velocity of a particle, approached when falling under the opposing influences of gravity and fluid drag
Thermal precipitator	A device that deposits particles using a temperature gradient
Thermophoresis	Particles motion in a temperature gradient, i.e., from a hotter to a colder region
Thoracic	Region of the respiratory tract from the larynx down
Tidal volume	Volume of gases inhaled or exhaled during each breath
Total lung capacity	Volume of air contained in the lung at maximum inspiration
Tracheobronchial compartment	Region of the respiratory tract from the larynx to the terminal bronchioles
Transmission efficiency	Fraction of aspirated particles that is transmitted through an inlet to the rest of the sampling system
Turbulent flow	Chaotic flow with streamlines looping back on themselves; less "well-behaved" than laminar flow
Ultrasonic nebulizer	A nebulizer employing focused sound waves to aerosolize a liquid into droplets

Ultra-Stokesian	Condition in which flow relative to an object is high enough to be outside the Stokes regime
Unipolar ion field	Region containing ions of only one polarity
Vapor pressure	Partial pressure of a liquid's vapor required to maintain the vapor in equilibrium with the condensed liquid or solid; also referred to as saturation vapor pressure
Variability	Measure of spread of repeated measurements of a parameter
Variance	Square of the standard deviation; a measure of variability
Vena contracta	Flow contraction with flow separation from the wall, usually occurring after constriction of a flow channel or just downstream of the entry point of an inlet
Venturi meter	A device used to measure flow rate in a duct by measuring the pressure drop across a calibrated streamlined constriction
Vertical elutriator	A vertical channel that gravitationally retains or removes particles above a given size or size range and emits the remaining airborne particles
Virtual impactor	A device in which particles are removed by impacting them through a virtual surface into a stagnant volume, or a volume with a slowly moving airflow, so that large particles remain in this volume while smaller particles are deflected with the bulk of the original air flow; the dichotomous impactor is a frequently used virtual impactor
Virus	A microorganism that needs a complete cell to reproduce
Vital capacity	Maximum volume of gas that can be exhaled from the lung after maximum inhalation
Wall loss	Deposition of particles in a sampler on surfaces other than those designed for particle collection
Weighting	Application of a factor to one particle activity to obtain another activity, e.g., count, surface, volume, or mass
Yeast	A unicellular fungus

APPENDIX B. CONVERSION FACTORS

Length

1 micrometer (μm) $= 10^{-6}$ m $= 10^{-4}$ cm
$= 10^{-3}$ mm $= 10^{3}$ nm $= 10^{4}$ Å
$= 3.937 \times 10^{-5}$ in
$= 3.281 \times 10^{-6}$ ft

1 nanometer (nm) $= 10^{-3}$ μm $= 10^{-9}$ m

1 Ångstrom (Å) $= 10^{-4}$ μm $= 10^{-10}$ m

1 inch (in) = 2.540 cm

1 foot (ft) = 12 in = 0.3048 m

Volume

1 μm^3 $= 10^{-15}$ l $= 10^{-18}$ m^3 $= 6.102 \times 10^{-14}$ in^3
$= 3.531 \times 10^{17}$ ft^3

1 liter (l) $= 10^{15}$ μm^3 $= 10^{-3}$ m^3 = 61.02 in^3
$= 3.531 \times 10^{-3}$ ft^3

1 m^3 $= 10^{18}$ μm^3 $= 10^{3}$ l $= 6.102 \times 10^{4}$ in^3
= 35.31 ft^3

1 in^3 $= 5.787 \times 10^{4}$ ft^3 $= 1.639 \times 10^{3}$ μm^3
$= 1.639 \times 10^{-2}$ l $= 1.639 \times 10^{-5}$ m^3

1 ft^3 $= 1.728 \times 10^{3}$ in^3 $= 2.832 \times 10^{16}$ μm^3
= 28.32 l $= 2.832 \times 10^{-2}$ m^3

Force

1 dyne $= 10^{-5}$ N $= 2.248 \times 10^{-6}$ lb
$= 1.021 \times 10^{-3}$ g

1 Newton (N) $= 10^{5}$ dyn = 0.2248 lb = 102 g

1 pound (lb) $= 4.448 \times 10^{5}$ dyn = 4.448 N
= 453.6 g

1 gram (g) force = 980.7 dyn $= 9.807 \times 10^{-3}$ N
$= 2.205 \times 10^{-3}$ lb

1 grain (gr) = 63.55 dyn
1 poundal = 1.383×10^4 dyn

Temperature

degree Celsius (°C) = T K − 273.16
= 5/9(T °F − 32)
degree Kelvin (K) = T °C + 273.16
= 5/9 (T °F + 459.69)
degree Fahrenheit (°F) = 1.8 T °C + 32
= 1.8 T K − 459.69
degrees Rankine (R) = T °F + 459.69
where T is temperature in the indicated units

Pressure

1 atmosphere (atm) = 1.013×10^6 dyn/cm²
= 1.013×10^5 N/m²
= 14.70 lb/in²
= 760 mmHg = 406.8 inH_2O
= 1.013×10^5 Pa = 101.3 kPa
1 dyne/cm² = 9.869×10^{-7} atm = 0.1 N/m²
= 1.450×10^{-5} lb/in²
= 7.501×10^{-4} mmHg
= 4.015×10^{-4} inH_2O

1 inch of water (inH_2O) (at 4°C)
= 2.458×10^{-3} atm = 2491 dyn/cm²
= 249.1 N/m² = 3.613×10^{-3} lb/in²
= 1.868 mmHg
1 mm of mercury (mmHg) (at 0°C)
= 1.316×10^{-3} atm = 1.333×10^3 dyn/cm²
= 1.333×10^2 N/m² = 0.535 in H_2O
= 1.934×10^{-2} lb/in²
1 Pascal (Pa) = 10 dyn cm² = 1 N/m²
1 torr = 1 mmHg

Viscosity

1 poise (P) = 1 g/cm s = 1 dyn s/cm² = 0.1 Pa s

Electrical Units

1 ampere (amp) = 2.998×10^9 statamp
1 statampere (statamp) = 3.336×10^{-10} amp
1 volt (V) = 3.336×10^{-3} statV
1 statvolt (statV) = 299.8 V
1 farad (F) = 10^6 μF = 8.987 statF
1 statfarad (statF) = 1.113×10^{-12} F
1 ohm = 1.113×10^{-12} statohm
1 statohm = 8.987×10^{11} ohm

APPENDIX C. COMMONLY ENCOUNTERED CONSTANTS

Boltzmann's constant	k	1.381×10^{-16} dyn cm/K
Avogadro's number	N_a	6.022×10^{23} molecules/mol
Gas constant	R	8.314×10^7 dyn cm/mol K, 82.06 cm³ atm/mol K
Stefan–Boltzmann constant	s	5.670×10^{-5} dyn/cm s K⁴
Elementary charge	e	1.602×10^{-19} C, 4.803×10^{-10} statC
Permittivity of free space	ε_0	1 electrostatic unit, 8.854×10^{-12} F/m
Speed of light in vacuum	c	2.998×10^{10} cm/s
Gravitational acceleration	g	980.7 cm/s²

APPENDIX D. COMMON PROPERTY VALUES OF AIR AND WATER

Air at 20°C and 1 atm (NTP)

Density (ρ_g)	1.205×10^{-3} g/cm³ = 1.205 g/l = 0.075 lb/ft³
Viscosity (ν)	1.832×10^{-4} P = 1.832×10^{-5} Pa s
Mean free path (λ)	0.0665 μm
Average molecular weight ($\bar{M}$)	28.96 g/mol
Specific heat ratio (γ)	1.40
Diffusion coefficient (D)	0.19 cm²/s

For some properties of other gases, see Table 3-1.

Composition of Dry Air by Volume

Gas	Content (volume %)	Molecular Weight (g/mol)
N_2	78.08	28.01
O_2	20.95	32.00
A	0.934	39.95
CO_2	0.033[a]	44.01
Other	< 0.003	

a. CO_2 concentration may vary from place to place

Water at 20°C

Viscosity	0.01002 dyn s/cm^2
Surface tension	72.75 dyn/cm
Vapor pressure	17.54 mmHg = 2.338 kPa

Water Vapor at 20°C

Diffusion coefficient	
Density	0.75×10^{-3} g/cm^3

APPENDIX E. DIMENSIONLESS NUMBERS

Knudsen (Kn) $= \frac{\lambda}{d_p}$ — Mean free path of molecules/particle diameter

Mach (Ma) $= \frac{U}{U_{sonic}}$ — Flow velocity/ sonic velocity

Peclet (Pe) $= \frac{LU}{D}$ — Bulk mass transfer/ diffusive mass transfer

Prandtl (Pr) $= \frac{\nu}{\alpha}$ — Momentum diffusivity/ thermal diffusivity

Reynolds (Re) $= \frac{d_p U}{\nu}$ — Inertia force/ viscous force

Schmidt (Sc) $= \frac{\nu}{D}$ — Momentum diffusivity/ mass diffusivity

Stokes (Stk) $= \frac{\tau U}{L}$ — Stopping distance/ characteristic flow dimension

APPENDIX F. FREQUENTLY USED AEROSOL PROPERTIES AT 20°C AND 1 atm

Unit density spheres at NTP (20°C, 1 atm)

Particle Diameter[2] (μm)	Slip Correction Factor	Settling Velocity (cm/s)	Diffusion Coefficient (cm^2/s)	Mobility [cm/s dyn)]	rms Brownian Displacement (cm)[1]
0.001	226.20	6.82E − 07	5.37E − 02	1.33E + 12	1.036571
0.002	113.33	1.37E − 06	1.35E − 02	3.33E + 11	0.518800
0.005	45.605	3.44E − 06	2.17E − 03	5.36E + 10	0.208147
0.01	23.039	6.95E − 06	5.47E − 03	1.35E + 10	0.104612
0.02	11.770	1.42E − 05	1.40E − 04	3.46E + 09	0.052873
0.05	5.0506	3.81E − 05	2.40E − 05	5.93E + 08	0.021905
0.1	2.8658	8.65E − 05	6.81E − 06	1.68E + 08	0.011667
0.2	1.8405	2.22E − 04	2.19E − 06	5.41E + 07	0.006612
0.5	1.3067	9.86E − 04	6.21E − 07	1.54E + 07	0.003523
1	1.1516	3.48E − 03	2.74E − 07	6.76E + 06	0.002339
2	1.0758	0.0130	1.28E − 07	3.16E + 06	0.001598
5	1.0303	0.0777	4.89E − 08	1.21E + 06	0.000989
10	1.0152	0.306	2.41E − 08	5.96E + 05	0.000694
20	1.0076	1.21	1.20E − 08	2.96E + 05	0.000489
50	1.0030	7.57	4.76E − 09	1.18E + 05	0.000309
100	1.0015	30.2	2.38E − 09	5.88E + 04	0.000218

1. Effective diameter of air molecule
2. Displacement in 10 s

APPENDIX G. GEOMETRICAL PROPERTIES OF PARTICLES

Circle

Circumference $= \pi d$

Area $= \pi r^2 = \dfrac{\pi d^2}{4}$

Ellipse

Circumference (approx.) $= 2\pi \sqrt{\dfrac{a^2 + b^2}{2}}$

Area $= \pi ab$

where a and b are the major and minor semi-axes, respectively.

Sphere

Surface area $= 4\pi r^2 = \pi d^2$

Volume $= \dfrac{4\pi r^3}{3} = \dfrac{\pi d^3}{6}$

Ellipsoids: Prolate

Surface area $= 2\pi b^2 + \dfrac{2\pi ab}{\varepsilon} \sin^{-1} \varepsilon$

where $\varepsilon = \dfrac{\sqrt{a^2 - b^2}}{a}$

Volume $= \dfrac{4\pi ab^2}{3}$

Ellipsoids: Oblate

Surface area $= 2\pi a^2 + \dfrac{\pi b^2}{\varepsilon} \ln\left(\dfrac{1 + \varepsilon}{1 - \varepsilon}\right)$

Volume $= \dfrac{4\pi a^2 b}{3}$

Right cylinder

Surface area $= 2\pi rL + 2\pi r^2 = \pi dL + \dfrac{\pi d^2}{2}$

where L is the length.

Volume $= \pi r^2 L = \dfrac{\pi d^2 L}{4}$

APPENDIX H. BULK DENSITY OF COMMON AEROSOL MATERIALS

Material	Density (g/cm^3)
Solids	
Aluminum	2.70
Corundum (Al_2O_3)	4.0
Ammonium sulfate	1.77
Asbestos	2.4–3.3
Calcite ($CaCO_3$)	2.7–2.9
Coal	1.2–1.8
Coal fly ash	ca. 2.0
Glass	2.4–2.8
Granite	2.4–2.7
Iron	7.86
Iron oxide	5.2–5.7
Limestone	2.1–2.9
Lead	11.3
Lead oxide	8.0–9.5
Methylene blue	1.26
Mineral wool	ca. 2.7
Plant particles	1.1–1.5
Paraffin	0.9
Pollen	ca. 1.4
Polystyrene	1.05
Polyvinyl toluene	1.03
Portland cement	3.2
Potassium biphthalate	1.64
Rock wool	ca. 2.5
Quartz	2.64–2.66

Materials	Density (g/cm^3)
Sodium chloride	2.17
Sulfur	2.07
Starch	1.5
Talc	2.6–2.8
Titanium dioxide	4.26
Uranine dye	1.53
Wood	ca. 1.5
Zinc oxide	5.61
Liquids	
Isopropyl alcohol	0.7855
Dibutyl phthalate	1.043
Dioctyl phthalate (DOP)	0.981
Dioctyl sebecate	0.915
Hydrochloric acid	1.19
Mercury	13.6
Oils	0.88–0.94
Oleic acid	0.894
Polyethylene glycol	1.13
Sulfuric acid	1.84

Additional references for particle densities and other properties:

McCrone, W. C. and J. G. Delly. 1973. *The Particle Atlas: Vol. IV*. Ann Arbor: Ann Arbor Science Publishers.

Weast, R. C. 1991. *Handbook of Chemistry and Physics*. Cleveland: CRC Press.

Index

Note that many terms are listed primarily in terms of their acronyms. This has been done to include definitions and references for acronyms in the Index rather than in a separate list. See also Appendix A for the glossary of terms.
* Indicates commercial designations of products. All trademark rights held by respective trademark holders.